Lecture Notes in Computer Science 5690

Commenced Publication in 1973
Founding and Former Series Editors:
Gerhard Goos, Juris Hartmanis, and Jan ˈ

Sourav S. Bhowmick   Josef Küng
Roland Wagner (Eds.)

Database and Expert Systems Applications

20th International Conference, DEXA 2009
Linz, Austria, August 31 – September 4, 2009
Proceedings

 Springer

Volume Editors

Sourav S. Bhowmick
Nanyang Technological University
50 Nanyang Avenue, Singapore 639798
E-mail: assourav@ntu.edu.sg

Josef Küng
Roland Wagner
University of Linz
Altenbergerstraße 69, 4040 Linz, Austria
E-mail: {jkueng, rrwagner}@faw.at

Library of Congress Control Number: 2009932140

CR Subject Classification (1998): J.2, I.2.5, H.2, J.1, H.4, H.3

LNCS Sublibrary: SL 3 – Information Systems and Application, incl. Internet/Web
and HCI

ISSN   0302-9743
ISBN-10   3-642-03572-8 Springer Berlin Heidelberg New York
ISBN-13   978-3-642-03572-2 Springer Berlin Heidelberg New York

springer.com

© Springer-Verlag Berlin Heidelberg 2009
Printed in Germany

Typesetting: Camera-ready by author, data conversion by Scientific Publishing Services, Chennai, India
Printed on acid-free paper   SPIN: 12734429   06/3180   5 4 3 2 1 0

Preface

The annual international conference on Database and Expert Systems Applications (DEXA) is now well established as a reference scientific event. The reader will find in this volume a collection of scientific papers that represent the state of the art of research in the domain of data, information and knowledge management, intelligent systems, and their applications.

The 20th edition of the series of DEXA conferences was held at the Johannes Kepler University of Linz, from August 31 to September 4, 2009.

Several collocated conferences and workshops covered specialized and complementary topics to the main conference topic. Seven conferences——the 11th International Conference on Data Warehousing and Knowledge Discovery (DaWaK), the 10th International Conference on Electronic Commerce and Web Technologies (EC-Web), the 8th International Conference on Electronic Government (EGOV), the 6th International Conference on Trust, Privacy, and Security in Digital Business (Trust-Bus), the 4th International Conference on Industrial Applications of Holonic and Multi-Agent Systems (HoloMAS), the First International Conference on eParticipation (ePart), and the Second International Conference on Data Management in Grid and P2P Systems (GLOBE)——and 14 workshops are collocated with DEXA.

These events formed a unique international forum with a balanced depth and breadth of topics. Its much appreciated conviviality fostered unmatched opportunities to meet, share the latest scientific results and discuss the latest technological advances in the area of information technologies with both young scientists and engineers and senior world-renown experts.

This volume contains the papers selected for presentation at the DEXA conference. Each submitted paper was reviewed by at least three reviewers, members of the Program Committee or external reviewers appointed by members of the Program Committee. Based on the reviews, the Program Committee accepted two categories of papers: 35 regular papers and 35 short papers of the 202 originally submitted papers. Regular papers were given maximum 15 pages in the proceedings to report their results as well as a 25-minute presentation slot in the conference. Short papers were give an 8-page limit and a 15-minute presentation slot.

The excellence brought to you in these proceedings would not have been possible without the efforts of numerous individuals and the support of several organizations.

First and foremost, we thank the authors for their hard work and for the quality of their submissions. We also thank the members of the Program Committee, the reviewers, and many others who assisted in the organization for their contribution to the success and high standard of DEXA 2009 and of these proceedings.

Finally we thank the DEXA Association, the Austrian Computer Society, the Research Institute for Applied Knowledge Processing (FAW), and the Johannes Kepler University of Linz for making DEXA 2009 happen.

June 2009

Sourav S. Bhowmick
Josef Kung

Organization

Honorary Chairperson

Makoto Takizawa   Seikei University, Japan

General Chairperson

Roland R. Wagner   FAW, University of Linz, Austria

Conference Program Chairpersons

Sourav S Bhowmick   Nanyang Technological University, Singapore
Josef Küng   University of Linz, Austria

Publication Chairperson

Vladimir Marik   Czech Technical University, Czech Republic

Workshop Chairpersons

A Min Tjoa   Technical University of Vienna, Austria
Roland R. Wagner   FAW, University of Linz, Austria

Program Committee

Talel Abdessalem   TELECOM ParisTech, France
Ajith Abraham   Norwegian University of Science and Technology, Norway
Witold Abramowicz   The Poznan University of Economics, Poland
Osman Abul   TOBB University, Turkey
Rafael Accorsi   University of Freiburg, Germany
Hamideh Afsarmanesh   University of Amsterdam, The Netherlands
Patrick Albert   ILOG, France
Riccardo Albertoni   CNR-IMATI-GE, Italy
Paolo Alencar   University of Waterloo, Canada
Rainer Alt   University of Leipzig, Germany
Toshiyuki Amagasa   University of Tsukuba, Japan

Badrish Chandramouli	Microsoft Research , USA
Chin-Chen Chang	Feng Chia University, Taiwan
Amitabh Chaudhary	University of Notre Dame, USA
Cindy Chen	University of Massachusetts Lowel, USA
Jinjun Chen	Swinburne University of Technology, Australia
Lei Chen	Hong Kong University of Science and Technology, Hong Kong
Phoebe Chen	Deakin University, Australia
Shu-Ching Chen	Florida International University, USA
Hao Cheng	University of Central Florida, USA
James Cheng	The Chinese University of Hong Kong, Hong Kong
Jingde Cheng	Saitama University, Japan
Reynold Cheng	The University of Hong Kong, China
Max Chevalier	IRIT - SIG, Université de Toulouse, France
Byron Choi	Hong Kong Baptist University, Hong Kong
Henning Christiansen	Roskilde University, Denmark
Soon Ae Chun	City University of New York, USA
Christophe Claramunt	Naval Academy Research Institute, France
Eliseo Clementini	University of L'Aquila, Italy
Sara Cohen	Hebrew University of Jerusalem, Israel
Martine Collard	University of Nice, France
Gao Cong	Microsoft Research Asia, China
Emilio Corchado	University of Burgos, Spain
Oscar Corcho	Universidad Politécnica de Madrid, Spain
Bin Cui	Peking University, China
Carlo A. Curino	Politecnico di Milano, Italy
Emiran Curtmola	University of California, San Diego, USA
Alfredo Cuzzocrea	University of Calabria, Italy
Deborah Dahl	Conversational Technologies, worldwide
Ernesto Damiani	University of Milan, Italy
Violeta Damjanovic	Salzburg Research Forschungsgesellschaft m.b.H., Austria
Jérôme Darmont	Université Lumière Lyon 2, France
Valeria De Antonellis	Università di Brescia, Italy
Andre de Carvalho	University of Sao Paulo, Brazil
Vincenzo De Florio	University of Antwerp, Belgium
Guy De Tré	Ghent University, Belgium
Olga De Troyer	Vrije Universiteit Brussel, Belgium
Roberto De Virgilio	Università Roma Tre, Italy
Paul de Vrieze	SAP Research, Switzerland
John Debenham	University of Technology, Sydney, Australia
Hendrik Decker	Universidad Politécnica de Valencia, Spain
Hepu Deng	RMIT University, Australia
Zhi-Hong Deng	Peking University, China
Vincenzo Deufemia	Universit`a degli Studi di Salerno, Italy
Alin Deutsch	Univesity of California at San Diego, USA
Beniamino Di Martino	Seconda Universita' di Napoli, Italy

Elisabetta Di Nitto   Politecnico di Milano, Italy
Claudia Diamantini   Università Politecnica delle Marche, Italy
Juliette Dibie-Barthélemy   AgroParisTech, France
Ying Ding   Indiana University, USA
Zhiming Ding   Chinse Academy of Sciences, China
Gillian Dobbie   University of Auckland, New Zealand
Peter Dolog   Aalborg University, Denmark
Dejing Dou   University of Oregon, USA
Marek J. Druzdzel   University of Pittsburgh, Bialystok Technical
                                        University, USA, Poland
Cedric du Mouza   CNAM, France
Arjan Durresi   Indiana University-Purdue University Indianapolis,
                                        USA
Curtis Dyreson   Utah State University, USA
Silke Eckstein   Technical University of Braunschweig, Germany
Johann Eder   University of Vienna, Austria
Suzanne M. Embury   The University of Manchester, UK
Christian Engelmann   Oak Ridge National Laboratory, USA
Jianping Fan   University of North Carolina at Charlotte, USA
Cécile Favre   University of Lyon, France
Bettina Fazzinga   University of Calabria, Italy
Leonidas Fegaras   The University of Texas at Arlington, USA
Yaokai Feng   Kyushu University, Japan
Stefano Ferilli   University of Bari, Italy
Eduardo Fernandez   Florida Atlantic University, USA
Filomena Ferrucci   Università di Salerno, Italy
Flavius Frasincar   Erasmus University Rotterdam, The Netherlands
Ada Fu   Chinese University of Hong Kong, China
Mariagrazia Fugini   Politecnico di Milano, Italy
Hiroaki Fukuda   Keio University, Japan
Benjamin Fung   Concordia University, Canada
Gabriel Fung   The University of Queensland, Australia
Steven Furnell   University of Plymouth, UK
Renata de Matos Galante   UFRGS - Federal University of Rio Grande do Sul,
                                        Brazil
Fabien Gandon   INRIA, France
Aryya Gangopadhyay   University of Maryland Baltimore County, USA
Sumit Ganguly   Indian Institute of Technology, Kanpur, India
Maria Ganzha   Polish Academy of Sciences, Poland
Bin Gao   Microsoft Research Asia, China
Yunjun Gao   Singapore Management University, Singapore
Dragan Gasevic   Athabasca University, Canada
Mário J. Gaspar da Silva   University of Lisbon, Portugal
Elli Georgiadou   Middlesex University, UK
Manolis Gergatsoulis   Ionian University, Greece
Shahram Ghandeharizadeh   University of Southern California, USA
Anastasios Gounaris   Aristotle University of Thessaloniki, Greece

Naga Govindaraju	Microsoft Corporation, USA
Bernard Grabot	LGP-ENIT, France
Fabio Grandi	University of Bologna, Italy
Carmine Gravino	University of Salerno, Italy
Nathan Griffiths	University of Warwick, UK
Sven Groppe	Lübeck University, Germany
Crina Grosan	Babes-Bolyai University Cluj-Napoca, Romania
William Grosky	University of Michigan, USA
Le Gruenwald	University of Oklahoma, USA
Volker Gruhn	Leipzig University, Germany
Stephane Grumbach	INRIA, France
Jerzy Grzymala-Busse	University of Kansas, USA
Francesco Guerra	Università degli Studi Di Modena e Reggio Emilia, Italy
Giovanna Guerrini	University of Genova, Italy
Levent Gurgen	National Institute of Informaatics (NII), Japan
Adolfo Guzman-Arenas	Tribunal Electoral del Poder Judicial de la Federacion, Mexico
Antonella Guzzo	University of Calabria, Italy
Saman Kumara Halgamuge	University of Melbourne, Australia
Abdelkader Hameurlain	Paul Sabatier University, Toulouse, France
Ibrahim Hamidah	Universiti Putra Malaysia, Malaysia
Hyoil Han	Drexel University, USA
Sung-Kook Han	Won Kwang University, Korea
Wook-Shin Han	Kyungpook National University, Korea
Takahiro Hara	Osaka University, Japan
Theo Härder	TU Kaiserslautern, Germany
Aboul Ella Hassanien	Cairo University, Egypt
Igor T. Hawryszkiewycz	University of Technology, Sydney, Australia
Saven He	Microsoft Research at Asia, China
Francisco Herrera	University of Granada, Spain
Rattikorn Hewett	Texas Tech University, USA
Stijn Heymans	Vienna University of Technology, Austria
Birgit Hofreiter	University of Vienna, Austria
Steven Hoi	Nanyang Technological University, Singapore
Vagelis Hristidis	Florida International University, USA
Estevam Rafael Hruschka Jr	Carnegie Mellon University, USA
Wynne Hsu	National University of Singapore, Singapore
Yu Hua	Huazhong University of Science and Technology, China
Jimmy Huang	York University, Canada
Xiaoyu Huang	South China University, China
Yan Huang	University of North Texas, USA
Ela Hunt	University of Strathclyde, UK
San-Yih Hwang	National Sun Yat-Sen University, Taiwan
Ionut Emil Iacob	Georgia Southern University, USA
Renato Iannella	National ICT Australia (NICTA), Australia

Sergio Ilarri	University of Zaragoza, Spain
Abdessamad Imine	University of Nancy, France
Yoshiharu Ishikawa	Nagoya University, Japan
Mizuho Iwaihara	Kyoto University, Japan
Anne James	Coventry University, UK
Vandana Janeja	University of Maryland Baltimore County, USA
Adam Jatowt	Kyoto University, Japan
Wie Jie	University of Manchester, UK
Peiquan Jin	University of Science and Technology, China
Jan Jurjens	Open University and Microsoft Research, UK
Janusz Kacprzyk	Polish Academy of Sciences, Poland
Urszula Kaczmar	Wroclaw University of Technology, Poland
Ejub Kajan	High School of Applied Studies, Serbia
Panagiotis Kalnis	National University of Singapore, Singapore
Vana Kalogeraki	University of CA, Riverside, USA
Rushed Kanawati	University of Paris North, France
Ken Kaneiwa	National Institute of Information and Communications Technology(NICT) Japan
Anne Kao	Boeing Phantom Works, USA
Dimitris Karagiannis	University of Vienna, Austria
Kamal Karlapalem	Indian Institute of Information Techology, India
George Karypis	University of Minnesota, USA
Stefan Katzenbeisser	Technical University of Darmstadt, Germany
Yiping Ke	Chinese University of Hong Kong, Hong Kong
Anastasios Kementsietsidis	IBM T.J. Watson Research Center, USA
Etienne Kerre	University of Ghent, Belgium
Myoung Ho Kim	KAIST, Korea
Sang-Wook Kim	Hanyang University, Korea
Markus Kirchberg	Institute for Infocomm Research, A*STAR, Singapore
Hiroyuki Kitagawa	University of Tsukuba, Japan
Carsten Kleiner	University of Applied Sciences&Arts Hannover, Germany
Christian König	Microsoft Research, USA
Ibrahim Korpeoglu	Bilkent University, Turkey
Harald Kosch	University of Passau, Germany
Michal Krátký	VSB-Technical University of Ostrava, Czech Republic
Petr Kroha	Technische Universität Chemnitz-Zwickau, Germany
Arun Kumar	IBM India Research Lab., India
Ashish Kundu	Purdue University, USA
Josef Küng	University of Linz, Austria
Axel Küpper	Ludwig-Maximilians-Universität München, Germany
Lotfi Lakhal	University of Marseille, France
Eric Lam	City University of Hong Kong, Hong Kong

Brahim Medjahed	University of Michigan - Dearborn, USA
Carlo Meghini	ISTI-CNR, Italy
Xiaofeng Meng	Renmin University, China
Rosa Meo	University of Turin, Italy
Paolo Merialdo	Universita' degli Studi Roma Tre, Italy
Elisabeth Metais	CNAM, France
Farid Meziane	Salford University, UK
Sanjay Misra	Atilim University, Turkey
Jose Mocito	INESC-ID/FCUL, Portugal
Mohamed Mokbel	University of Minnesota, USA
Lars Mönch	FernUniversität in Hagen, Germany
Anirban Mondal	University of Tokyo, Japan
Hyun Jin Moon	UCLA Computer Science, USA
Yang-Sae Moon	Kangwon National University, Korea
Reagan Moore	San Diego Supercomputer Center, USA
Mirella M. Moro	Universidade Federal de Minas Gerais, Brazil
Franck Morvan	IRIT, Paul Sabatier University, Toulouse, France
Tadeusz Morzy	Poznan University of Technology, Poland
Kyriakos Mouratidis	Singapore Management University, Singapore
Yi Mu	University of Wollongong, Australia
Tetsuya Murai	Hokkaido University, Japan
Mirco Musolesi	University of Cambridge, UK
Tadashi Nakano	University of California, Irvine, USA
Ullas Nambiar	IBM India Research Lab, India
Ismael Navas-Delgado	University of Málaga, Spain
Rimma V. Nehme	Purdue University, USA
Wolfgang Nejdl	University of Hannover, Germany
Wilfred Ng	University of Science and Technology, Hong Kong
Daniela Nicklas	University of Stuttgart, Germany
Christophe Nicolle	University of Burgundy, France
Barry Norton	Open University, UK
Chris Nugent	University of Ulster, UK
Selim Nurcan	University Paris 1 Pantheon Sorbonne, France
Byung-Won On	Pennsylvania State University, USA
Jose Antonio Onieva González	University of Malaga, Spain
Joann Ordille	Avaya Labs Research, USA
Mehmet Orgun	Macquarie University, Australia
Luís Fernando Orleans	Federal University of Rio de Janeiro, Brazil
Mourad Oussalah	University of Nantes, France
Gultekin Ozsoyoglu	University Case Western Research, USA
Claus Pahl	Dublin City University, Ireland
George Pallis	University of Cyprus, Cyprus
Christos Papatheodorou	Ionian University, Corfu, Greece
Paolo Papotti	Università Roma Tre, Italy
Marcin Paprzycki	Polish Academy of Sciences, Warsaw Management Academy, Poland
Vamsi Paruchuri	University of Central Arkansas, USA

Paolo Santi	Istituto di Informatica e Telematica, Italy
Ismael Sanz	Universitat Jaume I, Spain
Marýa Luýsa Sapino	Università degli Studi di Torino, Italy
N.L. Sarda	I.I.T. Bombay, India
Sumit Sarkar	University of Texas at Dallas, USA
Marinette Savonnet	University of Burgundy, France
Raimondo Schettini	Università degli Studi di Milano-Bicocca, Italy
Ingo Schmitt	University of Magdeburg, Germany
Harald Schöning	Software AG, Germany
Erich Schweighofer	University of Vienna, Austria
Florence Sedes	IRIT Toulouse, France
Valeria Seidita	University of Palermo, Italy
Nazha Selmaoui	University of New Caledonia, France
Luciano Serafini	FBK-irst, Italy
Heng Tao Shen	The University of Queensland, Australia
Lei Shu	National University of Ireland, Ireland
Patrick Siarry	Université Paris 12 (LiSSi), France
Gheorghe Cosmin Silaghi	Babes-Bolyai University of Cluj-Napoca, Romania
Hala Skaff-Molli	Université Henri Poincaré, France
Giovanni Soda	University of Florence, Italy
Leonid Sokolinsky	South Ural State University, Russia
MoonBae Song	Sungkyunkwan University, Korea
Adrian Spalka	CompuGROUP Holding AG, Germany
Bala Srinivasan	Monash University, Australia
Umberto Straccia	Italian National Research Council, Italy
Darijus Strasunskas	Norwegian University of Science and Technology (NTNU), Norway
Martin J. Strauss	Michigan University, USA
Lena Stromback	Linköpings Universitet, Sweden
Heiner Stuckenschmidt	Mannheim University, Germany
Aixin Sun	Nanyang Technological University, Singapore
Raj Sunderraman	Georgia State University, USA
Ashish Sureka	Infosys Technologies Limited, India
Jun Suzuki	University of Massachusetts, Boston, USA
Makoto Takizawa	Tokyo Denki University, Japan
Wei Tan	University of Chicago and Argonne National Laboratory, USA
Katsumi Tanaka	Kyoto University, Japan
Jie Tang	Tsinghua University, China
David Taniar	Monash University, Australia
Cui Tao	Brigham Young University, USA
Maguelonne Teisseire	LIRMM, University of Montpellier 2, France
Sergio Tessaris	Free University of Bozen-Bolzano, Italy
Olivier Teste	IRIT, University of Toulouse, France
Stephanie Teufel	University of Fribourg, Switzerland
Jukka Teuhola	University of Turku, Finland
Taro Tezuka	Ritsumeikan University, Japan

Clement Yu	University of Illinios at Chicago, USA
Ting Yu	North Carolina State University, USA
Xiaohui Yu	York University, Canada
Zhiwen Yu	Northwestern Polytechnical University, China
Gian Piero Zarri	University Paris IV, Sorbonne, France
Xiao-Jun Zeng	University of Manchester, UK
Zhigang Zeng	Huazhong University of Science and Technology, China
Xuan F. Zha	Extension Systems International (ESI), National Inst. of Standards and Tech. (NIST) , USA
Ji Zhang	CSIRO ICT Centre, Australia
Xiuzhen (Jenny) Zhang	RMIT University Australia, Australia
Yanchang Zhao	University of Technology, Sydney, Australia
Yu Zheng	Microsoft Research Asia, China
Xiao Ming Zhou	Sybase Engineering, worldwide
Xiaofang Zhou	University of Queensland, Australia
Qiang Zhu	The University of Michigan, USA
Yi Zhuang	Zhejiang Gongshang University, China
Ester Zumpano	University of Calabria, Italy

External Reviewers

Sven Hartmann
Giorgos Stoilos
Simon S. Msanjila
Ekaterina E. Ermilova
Chun Ruan
Lina Tutkute
Gang Qian
Adegoke Ojewole
Navin Viswanath
Ana Marilza Pernas
Sérgio Mergen
Devis Bianchini
Michele Melchiori
Camelia Constantin
Horst Pichler
Julius Köpke
Christian Meilicke
Sven Casteleyn
Carlos Buil
Andres García
Boris Villazón-Terrazas
Bei Pan
Zonghui Lian
Dino Ienco

Kaushik Chakrabarti
Luciana Cavalcante de Menezes
Sang Se Lee
Jongwoo Lim
Dongwoo Won
Matthiew Damigos
Arnab Bhattacharyya
Muhammad Aamir Cheema
Wenwu Qu
Chaoming Li
Shariq Bashir
Giorgio Terracina
Shahin Shayandeh
Yang Liu
Zi Huang
Gabriel Fung
Ke Deng
Jun Miyazaki
Ermenlinda Oro
Shiping Chen
Zaki Malik
Wanita Sherchan
Yousuke Watanabe
Wee Hyong Tok

Table of Contents

Temporal, Spatial, and High Dimensional Databases (Short Papers)

Invited Talk

Web, Semantics and Ontologies II

Database and Information System Architecture, Performance and Security (Short Papers)

Web, Semantics and Ontologies III

XML and Databases III (Short Papers)

Query Processing and Optimization I

Semantic Web and Ontologies IV (Short Papers)

Invited Talk

Query Processing and Optimization II

Query Processing and Optimization III

Data and Information Integration and Quality

Data Mining and Knowledge Extraction (Short Papers)

Data and Information Streams

Data Mining Algorithms

Data and Information Modelling

Information Retrieval and Database Systems
(Short Papers)

Database and Information System Architecture and Performance

Query Processing and Optimization IV (Short Papers)

Management of Information Supporting Collaborative Networks

Hamideh Afsarmanesh[1] and Luis M. Camarinha-Matos[2]

[1] Informatics Institute, University of Amsterdam, Science Park 107,
1098 XG, Amsterdam, The Netherlands
h.afsarmanesh@uva.nl
[2] Faculty of Sciences and Technology, New University of Lisbon,
Quinta da Torre, 2829-516, Monte Caparica, Portugal
cam@uninova.pt

Abstract. Dynamic creation of opportunity-based goal-oriented Collaborative Networks (CNs), among organizations or individuals, requires the availability of a variety of up-to-date information. In order to effectively address the complexity, dynamism, and scalability of actors, domains, and operations in opportunity-based CNs, pre-establishment of properly administrated strategic CNs is required. Namely, to effectively support creation/operation of opportunity-based VOs (Virtual Organizations) operating in certain domain, the pre-establishment of a VBE (Virtual organizations Breeding Environment) for that domain plays a crucial role and increases their chances of success. Administration of strategic CN environments however is challenging and requires an advanced set of inter-related functionalities, developed on top of strong management of their information. With the emphasis on information management aspects, a number of generic challenges for the CNs and especially for the administration of VBEs are introduced in the paper.

Keywords: Information management for Collaborative Networks (CNs), virtual organizations breeding environments (VBEs).

1 Introduction

Collaborative networks as collections of geographically dispersed autonomous actors, which collaborate through computer networks, has led both organizations and individuals to effectively achieving common goals that go far beyond the ability of each single actor, and providing cost effective solutions, and value creating functionalities, services, and products. The paradigm of "Collaborative Networks (CN)" represents a wide variety of networks where each one has distinctive characteristics and features, as presented in the taxonomy of existing CNs [1].

There is a wide diversity in **structural forms, duration, behavioral patterns,** and **interaction forms,** manifested by different CNs. The structural diversities range from the *process-oriented chain structures* as observed in supply chains, to those *centralized around dominant entities,* and the *project-oriented federated networks* [2; 3]. The duration and behavioral patterns of different CNs may range from the *short life cycle of the goal-oriented dynamic VOs* to the *long term strategic alliances,* such as the

S.S. Bhowmick, J. Küng, and R. Wagner (Eds.): DEXA 2009, LNCS 5690, pp. 1–6, 2009.

VBEs that aim at sufficiently supporting their actors with the configuration and establishment of new opportunity-driven VOs [4]. Similarly, interaction forms within different CNs vary in intensity, from merely networking to cooperation as well as collaboration, which also represent different levels of their collaboration maturity [4].

1.1   Challenges in Managing the Information in CNs

Even if all information within CNs, related to its actors, domain of activity, and operation, were semantically and syntactically homogeneous, still a main generic challenge remains related to *assuring the availability of strategic information about the actors within the CN*. Such information is required in CNs for its proper coordination and decision making. This can be handled through the enforcement of a push/pull mechanism and establishment of proper mapping strategies and components, between the information managed at different sites belonging to actors in the network and all those systems (or sub-systems) that support different functionalities of the CN and its activities during its life cycle. Therefore, it is necessary that from autonomous actors involved in the CNs, various types of distributed information are collected. Such information shall then be processed and organized, to become accessible within the network, both for navigation by different CN stakeholders as well as for processing by different software systems running at the CN. However, although the information about actors evolves in time - which is typical of dynamic environments such as CNs - and therefore needs to be kept up to date, there is no need for *continuous flow* of all the information from each legacy system to the CN. This would generate a major overload on the information management systems at the CN. Rather, for effective CN's operation and management, only at some intervals, partial information needs to be pulled/pushed from/to legacy systems to the CN. Similarly, needs to access information also vary depending on the purpose for which it is requested. These variations in turn pose a second generic information management challenge that is related to the *classification, assessment, and provision of the required information based on intended use cases in CNs*. These generic information challenges in CNs shall be addressed ad supported for all key required CN functionalities, e.g. for its common ontology engineering, competency management, and trust management, among others.

A third generic information management challenge is related to *modeling and organizing the variety and complexity of the information that needs to be processed* by different functionalities, which support the management and operation of the CNs. While some of these functionalities deal with the information that is known and stored within different sites of network's actors (e.g. data required for actors' competency management), the information required for some other functionalities of CNs may be unknown, incomplete, or imprecise, for which soft computing approaches, such as causal analysis and reasoning or similar techniques introduced in computational intelligence, shall be applied to generate the needed information, mostly qualitatively.

There is however a number of other generic challenges related to the management of the CN information, and it can be expected that more challenges will be identified in time as the need for other functional components unfolds in the research on supporting the management and operation of the CNs. Among other identified generic challenges, we can mention: ensuring the consistency among the locally managed semantically and syntactically heterogeneous information at each organization's legacy systems and the information managed by the management system of the CNs as well as their availability for access by authorized CN stakeholders (e.g. individuals or organizations) when necessary. At the heart of this challenge lies the establishment of needed interoperation infrastructure, as well as a federated information management system to support inter-linking of autonomous information management systems at different sites of the actors in CNs. Furthermore, challenges related to update mechanisms among autonomous nodes are relevant. Nevertheless, these generic challenges are in fact common to many other application environments and are not specific to the CN's information management area.

2 Establishing Collaborative Networks – Base Requirements

A generic set of requirements, including: (1) definition of common goal and vision, (2) performing a set of initiating actions, and (3) establishing common collaboration space, represent the base pre-conditions for the setting up of the CNs. Furthermore, the CN environment needs to properly operate, for which its daily supervision as well as distribution of tasks among its actors shall be supported that represent other challenges: (1) performing coordination, support, and management of activities, and (2) achieving agreements and contracts.

Following five sub-sections briefly summarize these main basic requirements, as identified and addressed within the CN area of research for the next generation of CNs, and emphasize their information management challenges in italic:

i. Defining a common goal and vision. Collaboration requires the pre-existence of a motivating common goal and vision to represent the joint/common purpose for establishment of the collaboration [4]. *Establishing a well-conceived vision for CNs needs involvement of all its actors, and in turn requires the availability of up-to-date information regarding many aspects of the network. Development of ontology for CNs, to support the management of required information for visioning as well as to assure its effective accessibility to all actors within the network is challenging.*

ii. Performing a set of initiating actions. There are a number of initiating actions that need to be taken as a pre-condition to establishing CNs. These actions, typically taken by the founder of the CN include [5; 6]: defining the scope of the collaboration and its desired outcomes; defining roles, responsibilities; setting the plan of actions; task scheduling and milestones; and defining policies, e.g. for handling conflicts. *Typically most information related to the initiating actions is strategic and considered proprietary, to be accessed only by the CN's administration. The classification of information in CNs to ensure its confidentiality level and privacy, while guaranteeing enough access to the level required by each CN stakeholder is challenging.*

iii. Substantiating a common collaboration space. Establishing CNs require the pre-establishment of their common collaboration space that refers to all needed elements, principles, infrastructure, etc. that together provide the needed environment for CN actors to be able to cooperate/collaborate. These include achieving:

- *Common concepts and terminology* (e.g. *meta-data defined for databases or an ontology, specifying the collaboration environment and purpose) [7].*
- *Common communication infrastructure and protocols* for interaction and *data/information sharing and exchange (e.g. the internet, GRID, open or*

commercial tools and protocols for communication and information exchange, document management systems for information sharing, etc.) [8]

– **Common working and sharing principles, value system, and policies** (e.g. procedures for cooperation/collaboration and sharing of different resources, assessment of collaboration preparedness, measurement of the alignment between value systems, etc.) [9]. *The CN related policies are typically modeled and stored by its administration and are available to its stakeholders.*

– **Common set of base trustworthiness criteria** (e.g. identification, modeling and specification of periodic required measurements of partners' performance to identify their trust level [10]. *It is necessary to model, specify, store, and manage entities and concepts related to trust establishment and their measurements related to different actors.*

– **Harmonization/adaptation of heterogeneities among stakeholders due to external factors** such as those related to actors from different regions involved in virtual collaboration networks, e.g. differences in time, language, laws/regulations, and socio-cultural aspects [9]. *Some of these heterogeneities affect the sharing and exchange of information among the actors in the network, for which proper mappings and/or adaptors shall be developed and applied.*

Certain other specific characteristics of CNs also require to be supported by their common collaboration space. For example some CNs may require simultaneous or synchronous collaboration, while others depend on asynchronous collaboration [11].

*iv. **Substantiating coordination, support, and management of activities.*** A well defined approach is needed for coordination of CN activities, and consequently establishment of mechanisms, tools, and systems are required for common coordination, support, and management of activities in the CN. As a starting point, considering the wide variety of terms and concepts introduced and applied in the CN environments, e.g. within the VBEs, specification and management of the **ontology** for these environments is crucial [7]. As another example, in almost all CNs, the involved actors need to know about each others' capabilities, capacities, resources, etc. that is referred to as the **competency** of the involved actors in [12]. In VBEs for instance, such competency information constitutes the base for partner search by the broker/planner, who needs to match partners' competencies against detailed characterization of an emerged opportunity, in order to select the best-fit partners. Similarly, as an antecedent to any collaboration, some level of **trust** must pre-exist among the involved actors in the CN and needs to be gradually strengthened depending on the purpose of the cooperation/collaboration. Therefore, as a part of the CN management system, rational measurement of the performance and achievements of CN actors can be applied to determine their trustworthiness, from different perspectives [10]. *Considering these and other advanced functionalities [13], as represented in Figure 1, which are needed for effective management of the CNs [14], classification, storage, and manipulation of their related information e.g. for competencies of actors and their trust-related criteria need to be effectively supported and is challenging.*

*v. **Achieving agreements and contracts among actors.*** Successful operation of the CN requires reaching common agreements/contracts among its actors [6]. For instance, during the CN's operation stage, clear agreements should be reached among the actors on distribution of tasks and responsibilities, extent of commitments, sharing of resources, and the distribution of both the rewards and the losses and liabilities [15]. *Due to their relevance and importance for the successful operation of the CNs, detailed information about all agreements and contracts established with its actors are stored and preserved by CN administration. Furthermore, some CNs with advanced management systems, model and store these agreements and contracts within a system so that they can be semi-automatically enforced, for example such a system can issue automatic warnings when a CN actor has not fulfilled or has violated some timely terms of its agreement/contract. Organizing, processing, and interfacing the variety of information to different stakeholders, required to support both reaching agreement as well as enforcing them, is quite challenging.*

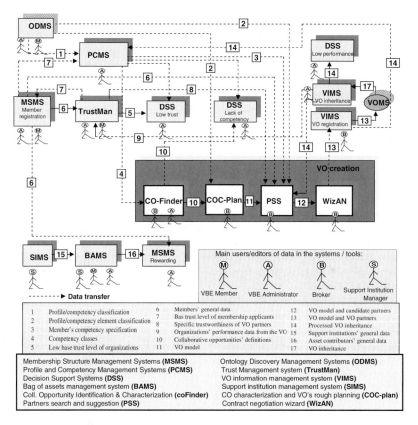

Fig. 1. VBE management system and its constituent subsystems

3 Conclusion

A main challenging criterion for the success of collaborative networks is the effective management of the wide variety of information that needs to be handled inside the CNs. The paper briefly addresses the main challenges of the next generation of CNs, while addressing their requirements for management of information. Furthermore, the paper focuses down on the strategic alliances and specifically on the management of the VBEs, in order to introduce some of the complexity of their needed functionality. As illustrated by these examples, collaborative networks raise complex challenges, requiring effective modeling and management of large amounts of heterogeneous information from autonomous sources, where some information is incomplete and/or imprecise, therefore, requiring a combination of different approaches such as distributed/federated database approaches, ontology engineering, as well as applying approaches introduced by computational intelligence and qualitative modeling.

References

1. Camarinha-Matos, L.M., Afsarmanesh, H.: Collaborative Networks: Reference Modeling. Springer, New York (2008)
2. Afsarmanesh, H., Camarinha-Matos, L.M.: The ARCON modeling framework. In: Collaborative networks reference modeling, pp. 67–82. Springer, New York (2008)
3. Katzy, B., Zang, C., Loh, H.: Reference models for virtual organizations. In: Virtual organizations – Systems and practices, pp. 45–58. Springer, Heidelberg (2005)
4. Camarinha-Matos, L.M., Afsarmanesh, H.: A comprehensive modeling framework for collaborative networked organizations. The Journal of Intelligent Manufacturing 18(5), 527–615 (2007)
5. Afsarmanesh, H., Camarinha-Matos, L.M.: On the classification and management of virtual organization breeding environments. The International Journal of Information Technology and Management – IJITM 8(3), 234–259 (2009)
6. Giesen, G.: Creating collaboration: A process that works! Greg Giesen & Associates (2002)
7. Afsarmanesh, H., Ermilova, E.: Management of Ontology in VO Breeding Environments Domain. To appear in Int. Journal of Services and Operations Management Inderscience (2009)
8. Rabelo, R.: Advanced collaborative business ICT infrastructure. In: Methods and Tools for collaborative networked organizations, pp. 337–370. Springer, New York (2008)
9. Romero, D., Galeano, N., Molina, A.: VO breeding Environments Value Systems, Business Models and Governance Rules. In: Methods and Tools for collaborative networked organizations, pp. 69–90. Springer, New York (2008)
10. Msanjila, S.S., Afsarmanesh, H.: Trust Analysis and Assessment in Virtual Organizations Breeding Environments. The International Journal of Production Research 46(5), 1253–1295 (2008)
11. Winkler, R.: Keywords and Definitions Around "Collaboration". SAP Design Guild, 5th edn. (2002)
12. Ermilova, E., Afsarmanesh, H.: Competency modeling targeted on promotion of organizations towards VO involvement. In: Proceedings of PRO-VE 2008 – 9th IFIP Working Conference on Virtual Enterprises, Poznan, Poland, pp. 3–14. Springer, Boston (2008)
13. Afsarmanesh, H., Camarinha-Matos, L.M., Msanjila, S.S.: On Management of 2nd Generation Virtual Organizations Breeding Environments. Journal of Annual Reviews in Control (in press, 2009)
14. Camarinha-Matos, L.M., Afsarmanesh, H., Ollus, M. (eds.): Methods and Tools for Collaborative Networked Organizations Collaborative. Springer, New York (2008)
15. Oliveira, A.I., Camarinha-Matos, L.M.: Agreement negotiation wizard. In: Methods and Tools for collaborative networked organizations, pp. 191–218. Springer, New York (2008)

A Low-Storage-Consumption
XML Labeling Method for
Efficient Structural Information Extraction

Wenxin Liang[1,*], Akihiro Takahashi[2,**], and Haruo Yokota[2,3]

[1] School of Software, Dalian University of Technology
[2] Department of Computer Science, Tokyo Institute of Technology
[3] Global Scientific Information and Computing Center, Tokyo Institute of Technology
wxliang@dlut.edu.cn, akihiro@de.cs.titech.ac.jp, yokota@cs.titech.ac.jp

Abstract. Recently, labeling methods to extract and reconstruct the structural information of XML data, which are important for many applications such as XPath query and keyword search, are becoming more attractive. To achieve efficient structural information extraction, in this paper we propose C-DO-VLEI code, a novel update-friendly bit-vector encoding scheme, based on register-length bit operations combining with the properties of Dewey Order numbers, which cannot be implemented in other relevant existing schemes such as ORDPATH. Meanwhile, the proposed method also achieves lower storage consumption because it does not require either prefix schema or any reserved codes for node insertion. We performed experiments to evaluate and compare the performance and storage consumption of the proposed method with those of the ORDPATH method. Experimental results show that the execution times for extracting depth information and parent node labels using the C-DO-VLEI code are about 25% and 15% less, respectively, and the average label size using the C-DO-VLEI code is about 24% smaller, comparing with ORDPATH.

1   Introduction

The Extensible Markup Language (XML) has rapidly become the *de facto* standard for representing and exchanging data on the Internet, because it is portable for representing different types of data from multiple sources. An XML document can be modeled as a nested tree structure in which each element is treated as a node with a start tag and end tag. An example XML document and its corresponding XML tree are shown in Figure 1 and Figure 2, respectively.

To handle the hierarchical tree model of XML data, XML labeling methods are required to extract and reconstruct the structural information of the XML document, which are commonly used for many applications such as XPath

---

* This work was partially done when the author was with Japan Science and Technology Agency (JST) and Tokyo Institute of Technology.
** The author is currently with NTT Data Corporation.

S.S. Bhowmick, J. Küng, and R. Wagner (Eds.): DEXA 2009, LNCS 5690, pp. 7–22, 2009.

```
<BOOK ISBN = "1-55860-438-3">
  <SECTION>
    <TITLE>Bad Bugs</TITLE>
    Nobody loves bad bugs
    <FIGURE CAPTION="Sample bug" />
  </SECTION>
  <SECTION>
    <TITLE>Tree Frogs</TITLE>
    All right-thinking people
    <BOLD>love</BOLD>free frogs.
  </SECTION>
</BOOK>
```

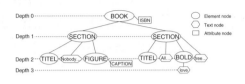

Fig. 1. Example XML document   **Fig. 2.** Example XML tree

query [20] and keyword search [16]. XML labeling methods assign a label to each node of the XML tree, and each labeled node is stored as a tuple containing the tag information for element nodes or the Parsible Character Data values of text nodes together with the label. Structural information, such as containment relationship, order of sibling and depth of nodes in the XML tree, is possible to be obtained from the regularity of the labels.

A number of XML labeling methods based on numbering schemes such as the preorder–postorder [10] and Dewey Order (DO) [20] have been proposed. The preorder–postorder method assigns preorder and postorder numbers to all nodes to maintain the containment and order information between nodes. The DO method uses a delimiter to separate the label of a parent node from the code that expresses the sibling order among its children. The parent–child and ancestor–descendant relationships, and the relative sibling order, can be deduced by comparing the labels at corresponding positions determined by the delimiters. The DO method is effective for obtaining the labels of parent and ancestor nodes and for determining common ancestors of multiple nodes by using the delimiters. These properties of the DO method are useful for achieving efficient XPath query [20] and keyword search [16]. For example, the parent, ancestor and depth information can be directly determined by analyzing the delimiters, which is valuable to enable efficient XPath axis determination in XPath queries and efficient LCA or SLCA extraction in keyword search. However, using simple numerical values for the sibling order in the preorder–postorder and DO methods has the problem of expensive update costs. If a new node is inserted in an intermediate position of the XML tree, many node labels have to be renumbered to maintain the proper sibling order.

To address the update problem, several update-friendly XML labeling methods such as DO Variable-Length-Endless-Insertable (DO-VLEI) code [13] and ORDPATH [17] have been proposed. The DO-VLEI labeling method inherits features of the DO method, but reduces the update cost for insertion operations. It uses the VLEI code [13] for expressing the sibling order by a unique magnitude relationship, which enables the unlimited insertion of new nodes with no relabeling of other nodes being required. ORDPATH, which is implemented in Microsoft® SQL Server^{TM} 2005, is also an update-friendly labeling method based on the DO number. [18] reported that ORDPATH achieves both better structural information extraction and storage consumption than the existing prefix-based labeling methods such as LSDX [8] and persistent labeling schemes [9, 12]. DO-VLEI achieves efficient structural information extraction as ORDPATH does. However, the size of DO-VLEI codes becomes larger when numbering large-scale XML documents.

Other two update-friendly labeling schemes, namely QED [14] and CDBS [15], are more efficient than ORDPATH in respect to data updates. However, QED works worse in some cases and CDBS results in both worse structural information extraction and storage consumption than ORDPATH. Therefore, it is essential to have both efficient structural information extraction and low storage consumption and update-friendly methods for large-scale XML data. The OR-DPATH method uses a Compressed binary representation (C-ORDPATH) and a prefix schema to reduce the size of node labels and to achieve effective structural information extraction. However, as criticized in MonetDB/XQuery [5, 6], C-ORDPATH has problems of expensive manipulation and storage costs. On the structural information extraction point of view, the features of C-ORDPATH cause the problem of high manipulation cost because decoding of C-ORDPATH requires to traverse from the head through the whole code and refer to the prefix schema. With respect to the storage consumption, the label in the C-ORDPATH cannot be optimized during the initial labeling, because negative integers and even numbers are reserved for node insertions. In addition, the labels available for assignment to the XML nodes are limited by the prefix schema. Therefore, the prefix schema used for C-ORDPATH must be changed when handling large XML documents, which causes further expensive manipulation and storage costs.

To achieve both efficient structural information extraction and low storage-consumption for large XML documents with high update frequency, in this paper we propose the Compressed-bit-string DO-VLEI (C-DO-VLEI) code, an update-friendly XML labeling method. To achieve efficient structural information extraction, we propose a novel and efficient method based on register-length bit operations combining with the properties of DO numbers, which cannot be implemented in other bit-vector encoding schemes including ORDPATH. Meanwhile, C-DO-VLEI also achieves lower storage consumption by using a compressed binary representation using DO numbering schemes without any prefix schema. We performed experiments to evaluate and compare the storage consumption and performance of the proposed method with those of the C-ORDPATH method. Experimental results show that the proposed method outperforms the ORD-PATH method with respect to both structural information extraction and storage consumption, particularly for documents of large size and depth.

The main contributions of this paper can be listed as follows.

1. We propose a novel update-friendly bit-vector encoding scheme, C-DO-VLEI code, for efficiently extracting the structural information between nodes from the C-DO-VLEI code. The proposed method enables efficient structural information extraction based on register-length bit operations combining with the properties of DO numbers instead of bit-by-bit comparisons utilized in the C-ORDPATH method.

2. The C-DO-VLEI code is also able to shorten the label length and thereby reduce the storage consumption required for the labeling of large-scale XML

documents, because it does not require either prefix schema or any reserved codes for node insertion.

3. We performed experiments to compare the performance and storage consumption of the proposed method with those of C-ORDPATH, as verified in [18], outperforms other existing methods in both storage consumption and performance. Experimental results indicate that: 1) the execution times for extracting depth information and parent node labels using the C-DO-VLEI code are about 25% and 15% less, respectively, than for C-ORDPATH, which enables the proposed method to achieve high performance in such applications as XPath query and keyword search; 2) the label size using the C-DO-VLEI code is about 24% smaller than that using C-ORDPATH, which means that the proposed method also outperforms the ORDPATH method with respect to storage consumption.

The remainder of the paper is organized as follows. In Section 2, we briefly introduce related work. Section 3 proposes the C-DO-VLEI code. Section 4 describes the method for extracting structural information from the C-DO-VLEI code. Section 5 describes the experiments and the evaluation of the proposed methods. Finally, Section 6 concludes the paper.

2   Related Work

One traditional labeling method is the preorder–postorder method [10], which assigns preorder and postorder numbers to all nodes. Another traditional method is the DO method [20], which uses a delimiter "." to separate a parent label from its child's label. However, when simple consecutive integers are used for node labeling in these methods, many nodes must be relabeled to satisfy sibling order following the insertion of a new node into an intermediate position.

To tackle this problem, several methods have been proposed. The Quartering-Regions Scheme [4] uses floating-point numbers with the preorder–postorder method. Range labeling methods [7] prepare appropriate intervals between preorder–postorder labels in advance. However, these methods still require relabeling of many nodes when the prepared intervals for node insertions are used up. MonetDB is an open-source database system for high-performance applications in data mining, OLAP, GIS, XML query, text and multimedia retrieval [1]. MonetDB/XQuery [5, 6] encodes XML documents based on the preorder–postorder method but achieves reasonable update costs by using a technique based on logical pages. However, MonetDB/XQuery still cannot completely avoid renumbering the *size* and *level* fields of some nodes when updating the document. Therefore, it is predicted that the update costs will degrade the performance of MonetDB/XQuery, especially when processing large-scale XML documents with high update frequency.

Labeling methods that are capable of unrestrictedly inserting nodes without relabeling any nodes, have been proposed [13,14,15,17,21]. In [21], the label of a newly inserted node is made from the product of a prime number and the label of its parent node. The label of the ancestor node can be extracted by factorizing the label of the child node. However, the factorization is still quite expensive. In [18], it was reported that the ORDPATH method [17] achieves both better structural information extraction and storage consumption than the existing prefix-based labeling methods such as LSDX [8] and persistent labeling schemes [9,12]. Other two update-friendly labeling schemes, namely QED [14] and CDBS [15], are more efficient than the ORDPATH method in respect to data updates. However, QED has worse performance in some cases and CDBS results in both worse structural information extraction and storage consumption than the ORDPATH method. The DO-VLEI code [13] can achieve efficient structural information extraction as the ORDPATH method does. However, the size of DO-VLEI codes becomes larger when labeling large-scale XML documents. The DO-VLEI code [13], which is the basis of our proposed method, and the ORDPATH method, which is the baseline of comparison in our experiments, will be described in detail in the following two subsections.

2.1   DO-VLEI Code

We first introduce the definition of the VLEI (Variable Length Endless Insertable) code as follows.

Definition 1 (VLEI Code). *A bit string $v = 1 \cdot \{0|1\}^*$ is a VLEI code, if the following condition is satisfied.*

$$v \cdot 0 \cdot \{0|1\}^* < v < v \cdot 1 \cdot \{0|1\}^*$$

Assume 1 as the reference point, the VLEI code will be smaller than 1 when 0 is added behind it, and on the contrary, it will be larger than 1 when 1 is added behind it; that is $10<1<11$. In the same way, for any bit sequence, it will be smaller than the original bit sequence if 0 is added behind it, and will be larger if 1 is added behind it. Besides, the bit sequence, behind which 0 or 1 is added, is assumed to succeed to the same containment relationship as that of the original bit sequence. For example, if 0 and 1 are added behind two VLEI codes 10 and 11, respectively, it will be true that $100 < 10 < 101$, $110 < 11 < 111$. Since we have already known that $10 < 1 < 11$, so we can infer that $100 < 1$, $101 < 1$, and $110 > 1$, $111 > 1$, namely $100 < 10 < 101 < 1 < 110 < 11 < 111$. A new VLEI code can be unrestrictedly generated from two arbitrary adjacent VLEI codes by an effective algorithm [13]. The VLEI code can be used for both the preorder–postorder and the DO methods. Here, we focus on the combination of the DO method and the VLEI code. It is named DO-VLEI code and defined as follows.

Definition 2 (DO-VLEI Code)
1. The DO-VLEI code of the root node $C_{root} = 1$.
2. The DO-VLEI code of a non-root node $C = C_{parent} \cdot C_{child}$, where C_{parent} denotes the DO-VLEI code of its parent and C_{child} denotes the VLEI code satisfying the appropriate sibling order.

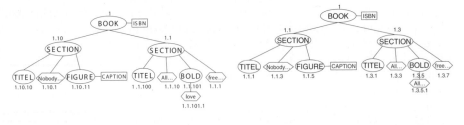

Fig. 3. Labeling by DO-VLEI code   **Fig. 4.** Labeling by ORDPATH

The DO-VLEI code of the nth ancestor node can be obtained by extracting the character string between the leftmost character and the nth delimiter. An example XML document tree labeled by DO-VLEI code is illustrated in Figure 3.

2.2 ORDPATH

ORDPATH [17] is an update-friendly XML labeling method implemented in Microsoft® SQL Server^{TM} 2005, in which arbitrary node insertions require no relabeling of the existing nodes. The ORDPATH labels are also based on the DO method, in that a child label is made from a sibling code, a delimiter, and the label of its parent node. Figure 4 shows an XML tree labeled by ORDPATH. In ORDPATH, only positive, odd numbers are assigned for the initial labeling; negative integers and even numbers are reserved for insertions. The end of a sibling code must be an odd number. For example, when the ORDPATH label of a node is newly inserted between nodes labeled "1.3.1" and "1.3.3", the label "1.3.2.1", which is made by adding "1" to "1.3.2", is assigned to the node. The sibling code of "1.3.2.1" is "2.1".

ORDPATH is implemented by a compressed binary representation using bit strings $\{0,1\}$, and is called C-ORDPATH. A pair of bit strings L_i and O_i are used to represent an integer by using a prefix schema (see reference [17] for details). An ORDPATH label is a string of delimited integers, and i is the order of the integer within the string. L_i is a bit string specifying the number of bits of O_i. L_i is specified by analysis of the tree using the prefix schema. O_i is treated as a binary number within a range set by L_i. For example, ORDPATH "1.5" is represented in C-ORDPATH as (0111001). (0111001) is divided into (01,110-01) by sequential analysis using the prefix schema in [17]. (01) and (110-01) are converted into "1" and "5" based on the O_i value range.C-ORDPATH represents delimiters by delimiting the bit string according to the prefix schema and compresses the label size. An XML tree labeled by the C-ORDPATH is shown in Figure 5.

3 C-DO-VLEI Code

In this section, we propose a compressed binary representation of the DO-VLEI code, called C-DO-VLEI code, which aims to reduce the label size. In this paper,

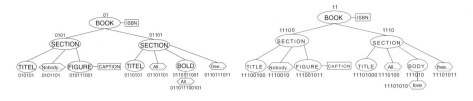

Fig. 5. Labeling by C-ORDPATH   **Fig. 6.** Labeling by C-DO-VLEI code

"" denotes a character string, and () denotes a bit string. For example, "01" represents a concatenation of character *0* and *1*, and (01) represents a bit sequence of bit *0* and *1*.

3.1   Components of the DO-VLEI Code

A DO-VLEI label is a variable-length character string constructed from the three characters ".", "1", and "0", which satisfies the following three conditions.

(1) Consecutive "." characters do not appear.
(2) "." does not exist at the tail end of a label.
(3) The VLEI code starts with "1".

According to the above conditions, the "1" that is the beginning of a VLEI code always appears after ".". Therefore, "." and "1" can be combined as ".1", meaning that a DO-VLEI code is composed of the three elements ".1", "1", and "0".

3.2   Definition of C-DO-VLEI Code

The C-DO-VLEI code is defined as follows.

Definition 3 (C-DO-VLEI Code). *The three elements of a DO-VLEI code: ".1", "1", and "0", are represented by the bit strings: (10), (11), and (0), respectively. This compressed binary representation of DO-VLEI codes is called C-DO-VLEI code.*

Note that the shortest bit string (0) is assigned to the "0" that can be manually adjusted to appear more frequently than the other two components, and that the set of bit strings: (0), (10), and (11) are prefix codes. A prefix code is a set of words such that no word of the set is a prefix of another word in the set. By sequentially analyzing the bit string from the prefix code, C-DO-VLEI codes can be uniquely decoded into the original DO-VLEI codes[1]. Figure 6 shows an XML tree labeled by C-DO-VLEI codes.

---

[1] Due to the space limitation of this paper, we do not discuss the end-of-label detection methods. Please refer to Reference [19] for details.

Table 1. XPath axes

Axis	Description
ancestor	all the ancestors of the context node (parent, grandparent, etc.)
preceding	all the nodes that precede the context node in the document except any ancestor nodes
descendant	all the descendants of the context node (children, grandchildren, etc.)
following	all the nodes that appear after the context node except any descendant nodes
parent	the single node that is the parent of the context node
preceding-sibling	all the nodes that have the same parent as the context node and appear before the context node
child	the children of the context node
following-sibling	all the nodes that have the same parent as the context node and appear after the context node

4   Structural Information Extraction

In this section, we propose an efficient method, based on register-length bit operations combining with the properties of Dewey Order numbers, for extracting structural information between nodes from the C-DO-VLEI code. The extracted structural information, such as depth, parent information, are necessary and valuable for determining XPath axes in XPath query and for detecting LCAs or SLCAs in XML keyword search. However, due to the space limitation, here we take XPath query as an example to show how the extracted structural information works to determine the XPath axes. The most important and frequently used XPath axes are listed in Table 1, in which parent, preceding-sibling, child and following-sibling are subsets of ancestor, preceding, descendant, and following, respectively.

Next, we explain how to obtain this information from the C-DO-VLEI codes by using the properties of DO and simple code comparisons.

4.1   Using Properties of DO

Since C-DO-VLEI codes use the DO method, information about 1) depth of node, 2) parent's label and 3) ancestors' labels can be obtained by the following three operations that use the properties of DO, respectively.

DO-1: Count the number of delimiters.
DO-2: Extract the prefix code before the rightmost delimiter.
DO-3: Repeat *DO-2* until the last delimiter is reached.

It is important to find the locations and count the number of the delimiters in C-DO-VLEI to implement *DO-1*–*DO-3*. The most straightforward method is to decode the label from the beginning to the end to detect the bit string (10) that is assigned for the delimiter. However, this is expensive when handling long codes. To reduce the processing cost, we propose an efficient method for detecting the delimiters by using register-length bit operations instead of bit-by-bit comparison.

Since the delimiter in C-DO-VLEI codes is assigned to bit string (10), any place where the bit changes from (1) to (0) is a delimiter candidate. However, when "10" appears in the DO-VLEI code, the corresponding C-DO-VLEI will be (110). That is, the last two bits are also (10). In order to distinguish these two patterns of (10), we focus on the number of consecutive (1s) before the (0). In the fragment of C-DO-VLEI code shown in Figure 7, wherever an odd number of consecutive (1s) appears before (0), the last two bits (10) represent a delimiter,

Fig. 7. Example fragment of a C-DO-VLEI code

Table 2. Bit operation notation

notation	description
$x \ll y$	left shift
$x \overset{u}{\gg} y$	right shift (logical shift)
$\bar{x}$	NOT
$x \& y$	AND
$x \mid y$	OR

while (10) ending at an even number of consecutive (1s) represents the DO-VLEI code "10". Therefore, it is possible to detect the delimiters by counting the number of consecutive (1s). Next, we describe algorithms for operations **DO-1**–**DO-3** based on delimiter detection. Notations for the bit operations used in the proposed algorithms are shown in Table 2.

Algorithm for DO-1. The operation **DO-1**, namely the operation for detecting the node depth, can be implemented by the following steps.

D-1: Generate a bit string, *endPointOfOne*, in which the initial values of all the bits are (0), and only the last bit at the place where an odd number of consecutive (1s) appears is set to (1). For the C-DO-VLEI code v shown in Figure 8, the bit string *endPointOfOne* can be generated by $v \& \overline{v \ll 1}$.

D-2: Generate a bit string *endPointOfOne'*, in which the initial values of all the bits are (0), and only the bit at the place where (0) changes to (1) in the C-DO-VLEI code is set to (1). For the C-DO-VLEI code v shown in Figure 8, the bit string *endPointOfOne'* can be generated by $endPointOfOne \overset{u}{\gg} 1$.

D-3: Assuming that each bit of the code is assigned an integer ID, as shown in Figure 8, then the bit (1) in *endPointOfOne* is a delimiter if the condition $ID(i) - ID(j) = oddnumber$ is satisfied, where $ID(i)$ and $ID(j)$ denote the ID of (1) in *endPointOfOne* and its nearest leftmost (1) in *endPointOfOne'*, respectively. For example, the $5th$ bit (ID=4) in *endPointOfOne* is a delimiter because the ID of its nearest leftmost (1) in *endPointOfOne'* is 1. However, the $9th$ bit (ID=8) in *endPointOfOne* is not a delimiter because the ID of its nearest leftmost (1) *endPointOfOne'* is 6. Algorithm 1 shows the details of delimiter detection. Algorithm 1 outputs a bit string *pointOfDelimiter*, in which the initial values of all the bits are (0), and only the bit indicating the place of delimiters is set to (1).

D-4: After the delimiters are found by Algorithm 1, the node depth can be determined by counting the number of (1s) in *pointOfDelimiter*. Algorithm 2 shows the details for counting the number of (1s) in a bit string based on the divide-and-conquer algorithm [11].

Algorithm 1. Delimiter detection

Input: C-DO-VLEI code v
Output: pointOfDelimiter
1: endPointOfOne $\leftarrow v \& \bar{v} \ll 1$
2: delimiterEven $\leftarrow \bar{v} \& (v +$
   $(\text{endPointOfOne} \& 0x5555)) \& 0xAAAA$
3: $v' \leftarrow v \overset{u}{\gg} 1$
4: endPointOfOne $\leftarrow$ endPointOfOne $\overset{u}{\gg} 1$
5: delimiterOdd $\leftarrow \bar{v'} \& (v' + (\text{endPointOfOne} \leftarrow 0x5555)) \& 0xAAAA$
6: pointOfDelimiter $\leftarrow$ delimiterEven|delimiterOdd

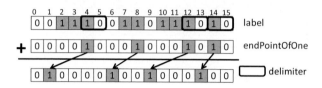

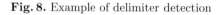

Fig. 8. Example of delimiter detection

Algorithm for _DO-2_ and _DO-3_. _DO-2_ finds the last delimiter from _pointOfDelimiter_ generated by Algorithm 1, which is implemented by the following steps.

_P-1_: Generate _pointOfDelimiter_.

_P-2_: Generate a _mask_ by
$(\text{pointOfDelimiter} \& - \text{pointOfDelimiter}) - 1|label$, as shown in Figure 9.

_P-3_: The length of the parent label L_p can be obtained by counting the number of consecutive (1s) from the head of the _mask_. Algorithm 3 shows the details of how to count the number of (1s) from the head of a C-DO-VLEI code. Finally, the parent label can be generated by outputting the L_p bits from the head of the original label.

In addition, **_DO-3_**, namely obtaining the ancestors' labels, can be implemented by repeating the above operations.

4.2   Using Code Comparison

Theorem 1. _Let the bit strings_ v_0, v_d, _and_ v_1 _represent the three elements of the C-DO-VLEI code "0", ".1" and "1", respectively. According to Definition 3,_

Algorithm 2. Depth detection

Input: v (16-bit unsigned integer)
Output: number of (1) in v (depth)
1: $v \leftarrow v - ((v \overset{u}{\gg} 1) \& 0x5555)$
2: $v \leftarrow (v \& 0x3333) + ((v \overset{u}{\gg} 2) \& 0x3333)$
3: $v \leftarrow ((v + (v \overset{u}{\gg} 4)) \& 0x0F0F)$
4: $v \leftarrow v + (v \overset{u}{\gg} 8)$
5: $v \leftarrow v + (v \overset{u}{\gg} 16)$
6: depth $\leftarrow v \& 0x001F$
7: **return** depth

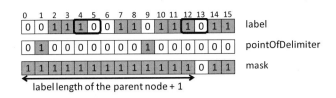

Fig. 9. Example of mask for extracting parent label

Algorithm 3. Count number of consecutive (1s) from the head of code

---

Input: v (16-bit unsigned integer)
Output: length of parent node, L_p
1: $v \leftarrow \bar{v}$
2: $n \leftarrow 16$
3: $c \leftarrow 8$
4: **repeat**
5:     $x \leftarrow v \overset{u}{\gg} c$
6:     **if** $x \neq 0$ **then**
7:         $n \leftarrow n - c$
8:         $v \leftarrow x$
9:     **end if**
10:     $c \leftarrow c \overset{u}{\gg} 1$
11: **until** $c \neq 0$
12: **return** $L_p = n - x$

$v_0 = (0)$, $v_d = (10)$ and $v_1 = (11)$. *Using Algorithm 4, comparison of the C-DO-VLEI codes for two nodes yields the proper document order if v_0, v_d and v_1 satisfy the following equation*[2].

$$1 \cdot v_0 < 1 < 1 \cdot v_d < 1 \cdot v_1 \qquad (1)$$

Moreover, according to Definition 3 and the conditions described in Section 3.1, the code b for a descendant node of the node with code a is $b = a \cdot 10 \cdot \{11|0\}^*$. Therefore, according to Definition 1, $a \cdot 10 \cdot \{0\}^* \leq b < a \cdot 11 \cdot \{0\}^*$. That is, the codes of a and b satisfy the following equation, where $\prec_{CDV}$ denotes the comparison based on Algorithm 4.

$$a \prec_{CDV} b \prec_{CDV} a \cdot 1 \qquad (2)$$

Similarly, the ancestor, preceding, and following nodes can be also obtained by the code comparison using Algorithm 4. Therefore, as shown in Figure 10, using the code a of a context node and $a \cdot 1$ can divide the nodes into three ranges **I**, **II** and **III** for ancestor and preceding nodes, descendant nodes, and following nodes, respectively.

4.3   XPath Axis Determination Using Structural Information

Combining the code comparison method described in Section 4.2 and the operations introduced in Section 4.1, all the XPath axes listed in Table 1 can

---

[2] Here we do not give the proofs of this theorem and the other relevant ones due to the space limitation.

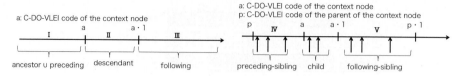

Fig. 10. Code range for ancestor, preceding, descendant and following

Fig. 11. Code range for parent, preceding-
sibling, child and following-sibling

Algorithm 4. Code comparison

Input: two C-DO-VLEI codes, v,w
    (length(v)≤length(w))
Output: $v' \geq w$ is true or false
1: $l' \leftarrow length(w) - length(v)$
2: $v' \leftarrow v \cdot 1 \cdot \{0\}^{l'-1}$
3: **if** $v' \geq w$ **then**
4:      **return true**
5: **else**
6:      **return false**
7: **end if**

Table 3. XPath axis extraction method

Axis	Extraction method
ancestor	*DO-3*
preceding	(node-set extracted from range **I**) - ancestor
descendant	node-set extracted from range **II**
following	node-set extracted from range **III**
parent	*DO-2*
preceding-sibling	nodes ∈ range **IV** ∩ nodes with the same depth as the context node (*DO-1*)
child	nodes ∈ range **II** ∩ nodes whose depth is one more than the context node (*DO-1*)
following-sibling	nodes ∈ range **V** ∩ nodes with the same depth as the context node (*DO-1*)

be determined. First, the descendant and following nodes can be obtained by
extracting nodes from ranges **II** and **III**, respectively. Ancestor nodes can be
obtained by using *DO-3*, and then preceding nodes can be obtained by excluding ancestor nodes from range **I**. The parent node can be directly determined
by *DO-2*. Using the codes of the context node and its parent, the ranges **IV**
and **V** for preceding-sibling and following-sibling nodes can be determined by
using Algorithm 4, as shown in Figure 11. As the depths of preceding-sibling and
following-sibling nodes are the same as the context node, and the child nodes
are deeper by one than the context node, the preceding-sibling, following-sibling,
and child nodes can be extracted from the ranges **IV**, **V** and **II** by using *DO-
1*. The extraction methods for XPath axes listed in Table 1 are summarized in
Table 3.

5   Experimental Evaluation

We performed experiments within the environment described in Table 4 to evaluate the storage consumption and performance of *DO-1–DO-3*, comparing the

Table 4. Experimental Environment

CPU:	Dual-Core Intel Xeon 5110 (1.60GHz)
Memory:	DDR2 ECC FB-DIMM 9GB(2GB×4, 512MB×2)
Storage:	D-RAID RAID 0+1
HDD:	SEAGATE ST3750640AS (750GB, 7200rpm, 3.5inch)
OS:	Linux 2.6.18 CentOS 5(Final)
Compiler:	gcc (Red Hat 4.1.1-52)
RDB:	PostgreSQL 8.2.4

Table 5. XML documents

XML document	size (byte)	elements	max. depth	avg. depth	max. fanout	avg. fanout
SwissProt.xml	114820211	2977031	5	3.556711	50000	2.034511
dblp.xml	133862772	3332130	6	2.902279	328858	1.943555
ebay.xml	35525	156	5	3.75641	380	5.391305
item0.xml(xmlgen SF=1)	118552713	1666315	12	5.547796	25500	2.10692
item1.xml(xmlgen SF=0.1)	11875066	167865	12	5.548244	2550	2.103751
item2.xml(xmlgen SF=0.01)	1182547	17132	12	5.506012	255	2.102092
item3.xml(xmlgen SF=0.001)	118274	1729	12	5.717756	25	2.04965
lineitem.xml	32295475	1022976	3	2.941175	60175	1.941175
nasa.xml	25050288	476646	8	5.583141	2435	2.000728
orders.xml	5378845	150001	3	2.899987	15000	1.899987
part.xml	618181	20001	3	2.899905	2000	1.899905
partsupp.xml	2241868	48001	3	2.833295	8000	1.833295
psd7003.xml	716853012	21305818	7	5.15147	262526	1.818513
reed.xml	283619	10546	4	3.199791	703	1.809579
treebank_e.xml	86082517	2437666	36	7.872788	56384	1.571223
uwm.xml	2337522	66729	5	3.952435	2112	1.952276
wsu.xml	1647864	74557	4	3.157866	3924	2.077534

C-DO-VLEI code with C-ORDPATH. The XML documents used for the experiments were generated by xmlgen [3], using scale factors (SF) from 0.001 to 1, and were sourced from the *XML Data Repository* [2]. Table 5 shows the details of these XML documents. For the evaluation of performance, the labels of all the nodes for each document were stored in a *document* table, and the labels of the nodes in all documents of the same depth were stored in a *depth* table.

We compared the C-DO-VLEI code with C-ORDPATH to evaluate the performance of **DO-1–DO-3**[3]. Figure 12 shows the execution time ratio for extracting depth information, parent labels, and ancestor labels from the depth tables, using the C-DO-VLEI code and C-ORDPATH[4]. From this figure, we can see that the greater the node depth, the better is the performance by the C-DO-VLEI code in extracting node depth and parent labels. This is because delimiter detection in ORDPATH requires traversal from the head through the whole code and refers to the prefix schema for determining each delimiter. On the other hand, there is no significant difference in extracting the ancestor labels, because, even for the C-DO-VLEI code, output of all the ancestors requires handling of

---

[3] There is no significant performance difference between C-DO-VLEI and C-ORDPATH in detecting other XPath axes, because they can determine such axes as descendant or following by simple code comparison. Therefore, only the experimental evaluation of **DO-1–DO-3** is discussed in this paper.

[4] Note that the label size using C-ORDPATH is set to 1 (same in Figure 13 and 14).

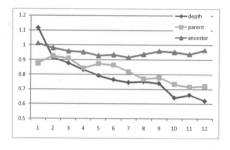
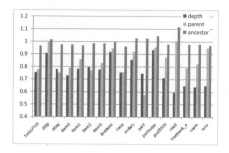

Fig. 12. Execution time ratio of C-DO-VLEI to C-ORDPATH (horizontal axis: depth of node)

Fig. 13. Execution time ratio of C-DO-VLEI to C-ORDPATH (horizontal axis: XML document)

all delimiters one by one. Figure 13 shows the execution time ratio using the *document* tables. From these results, we deduce that the execution times for extracting depth information and parent labels using the C-DO-VLEI code are about 25% and 15% less than execution times using C-ORDPATH, respectively, which indicates that the proposed method can achieve high performance in such applications as XPath query and keyword search.

We then measured and compared the total label size of each XML document labeled by the C-DO-VLEI code and C-ORDPATH. Figure 14 shows the label size ratio of the C-DO-VLEI code to C-ORDPATH for each XML document, from which we can learn that the average label size using the C-DO-VLEI code is about 24% on the average smaller than that using the C-ORDPATH, which means that the proposed method also outperforms the ORDPATH method with respect to storage consumption.

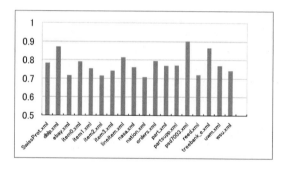

Fig. 14. Label size ratio of C-DO-VLEI to C-ORDPATH (horizontal axis: XML documents)

6   Conclusions

Recently, labeling methods to determine and reconstruct the structural information of XML data are becoming more attractive. The recent increase in large-scale XML data storage is making the update problem in XML labeling methods more critical. Therefore, update-friendly XML labeling methods becomes more paramount and require not only low storage consumption by the node labels but high-performance XML data query handling.

To achieve both efficient structural information extraction and low storage-consumption for large XML documents with high update frequency, in this paper we have proposed the C-DO-VLEI code, an update-friendly XML labeling method. To achieve efficient structural information extraction, we have proposed a novel and efficient method based on register-length bit operations combining with the properties of DO numbers, which are not able to be implemented in other bit-vector encoding schemes including ORDPATH. Meanwhile, the C-DO-VLEI code also achieves lower storage consumption by using a compressed binary representation using DO numbering schemes without any prefix schema.

We have performed experiments to evaluate and compare the storage consumption and performance of the proposed method with those of the C-ORDPATH method. Experimental results indicate that: 1) the execution times for extracting depth information and parent node labels using the C-DO-VLEI code are about 25% and 15% less, respectively, than for C-ORDPATH, which enables the proposed method to achieve high performance in such applications as XPath query and keyword search; 2) the label size using the C-DO-VLEI code is about 24% smaller than that using C-ORDPATH, which means that the proposed method also outperforms the ORDPATH method with respect to storage consumption.

Acknowledgments

This work was partially supported by CREST of JST (Japan Science and Technology Agency), by the Grant-in-Aid for Scientific Research of MEXT Japan (#19024028), and by the start-up funding (#1600-893313) for newly appointed academic staff of Dalian University of Technology, China.

References

1. MonetDB, http://monetdb.cwi.nl/
2. XML Data Repository, http://www.cs.washington.edu/research/xmldatasets/
3. Xmlgen, http://monetdb.cwi.nl/xml/downloads.html
4. Amagasa, T., Yoshikawa, M., Uemura, S.: QRS: A Robust Numbering Scheme for XML Documents. In: Proc. of ICDE, pp. 705–707 (2003)
5. Boncz, P., Flokstra, J., Grust, T., van Keulen, M., Manegold, S., Mullender, S., Rittinger, J., Teubner, J.: MonetDB/XQuery—consistent and efficient updates on the pre/Post plane. In: Ioannidis, Y., Scholl, M.H., Schmidt, J.W., Matthes, F., Hatzopoulos, M., Böhm, K., Kemper, A., Grust, T., Böhm, C. (eds.) EDBT 2006. LNCS, vol. 3896, pp. 1190–1193. Springer, Heidelberg (2006)
6. Boncz, P.A., Grust, T., van Keulen, M., Manegold, S., Rittinger, J., Teubner, J.: MonetDB/XQuery: A Fast XQuery Processor Powered by a Relational Engine. In: Proc. of SIGMOD Conference, pp. 479–490 (2006)

7. Cohen, E., Kaplan, H., Milo, T.: Labeling Dynamic XML Trees. In: Proc. of PODS, pp. 271–281 (2002)
8. Duong, M., Zhang, Y.: LSDX: A New Labeling Scheme for Dynamically Updating XML Data. In: Proc. of ADC, pp. 185–193 (2005)
9. Gabillon, A., Fansi, M.: A persistent labelling scheme for XML and tree databases. In: Proc. of SITIS, pp. 110–115 (2005)
10. Gerdemann, D.: Parsing As Tree Traversal. In: Proc. of COLING, pp. 396–400 (1994)
11. Steele Jr., G.L.: Hacker's Delight. Addison-Wesley Professional, Reading (2003)
12. Khaing, A., Thein, N.L.: A Persistent Labeling Scheme for Dynamic Ordered XML Trees. In: Proc. of Web Intelligence, pp. 498–501 (2006)
13. Kobayashi, K., Liang, W., Kobayashi, D., Watanabe, A., Yokota, H.: VLEI code: An Efficient Labeling Method for Handling XML Documents in an RDB. In: Proc. of ICDE, Tokyo, Japan, pp. 386–387 (2005) (poster)
14. Li, C., Ling, T.W.: Qed: a novel quaternary encoding to completely avoid re-labeling in xml updates. In: Proc. of CIKM, pp. 501–508 (2005)
15. Li, C., Ling, T.W., Hu, M.: Efficient Processing of Updates in Dynamic XML Data. In: Proc. of ICDE, p. 13 (2006)
16. Liang, W., Miki, T., Yokota, H.: Superimposed code-based indexing method for extracting mCTs from XML documents. In: Bhowmick, S.S., Küng, J., Wagner, R. (eds.) DEXA 2008. LNCS, vol. 5181, pp. 508–522. Springer, Heidelberg (2008)
17. O'Neil, P.E., O'Neil, E.J., Pal, S., Cseri, I., Schaller, G., Westbury, N.: ORDPATHs: Insert-Friendly XML Node Labels. In: Proc. of ACM SIGMOD Conference, pp. 903–908 (2004)
18. Sans, V., Laurent, D.: Prefix Based Numbering Schemes for XML: Techniques, Applications and Performances. In: Proc. of VLDB, pp. 1564–1573 (2008)
19. Takahashi, A., Liang, W., Yokota, H.: Storage Consumption of Variable-length XML Labels Uninfluenced by Insertions. In: Proc. of ADSS, pp. 571–573 (2007)
20. Tatarinov, I., Viglas, S., Beyer, K.S., Shanmugasundaram, J., Shekita, E.J., Zhang, C.: Storing and Querying Ordered XML Using a Relational Database System. In: Proc. of ACM SIGMOD Conference, pp. 204–215 (2002)
21. Wu, X., Lee, M.-L., Hsu, W.: A Prime Number Labeling Scheme for Dynamic Ordered XML Trees. In: Proc. of ICDE, pp. 66–78 (2004)

Inclusion Dependencies in XML: Extending Relational Semantics

Michael Karlinger[1], Millist Vincent[2], and Michael Schrefl[1]

[1] Johannes Kepler University, Linz, Austria
[2] University of South Australia, Adelaide, Australia

Abstract. In this article we define a new type of integrity constraint in XML, called an XML inclusion constraint (XIND), and show that it extends the semantics of a relational inclusion dependency. This property is important in areas such as XML publishing and 'data-centric' XML, and is one that is not possessed by other proposals for XML inclusion constraints. We also investigate the implication and consistency problems for XINDs in complete XML documents, a class of XML documents that generalizes the notion of a complete relation, and present an axiom system that we show to be sound and complete.

1 Introduction

Integrity constraints are one of the oldest and most important topics in database research, and they find application in a variety of areas such as database design, data translation, query optimization and data storage [3]. With the adoption of the eXtensible Markup Language (XML) [4] as the industry standard for data interchange over the internet, and its increasing usage as a format for the permanent storage of data in database systems [5], the study of integrity constraints in XML has increased in importance in recent years.

In this article we investigate the topic of *inclusion constraints* in XML. We use the syntactic framework of the `keyref` mechanism in XML Schema [4], where both the LHS and RHS of the constraint include a *selector*, which is used to select elements in the XML document, followed by a sequence of *fields*, which are used to specify the descendant nodes that are required to match in the document. This general idea of requiring selected elements in a document to have matching descendant nodes is also found in other approaches towards inclusion constraints in XML [6,7,8].

While the syntactic framework of the `keyref` mechanism in XML Schema is an expressive one, both it and other proposals for XML inclusion constraints [6,7,8,9,15] have some important limitations from the perspective of semantics. In particular, these proposals for XML inclusion constraints do not always allow one to extend the semantics of a relational inclusion dependency (IND), which we now illustrate by the example of an XML publishing scenario.

Fig. 1 shows two relations `teaches` and `offer`. Relation `teaches` stores the details of courses taught by lecturers in a department, where `lec` is the name of

S.S. Bhowmick, J. Küng, and R. Wagner (Eds.): DEXA 2009, LNCS 5690, pp. 23–37, 2009.

the lecturer, cno is the identifier of the course they are teaching, day is the day of the week that the course is being taught and sem is the semester in which the course is taught. Relation offer stores all the courses offered by the university, where cno, day and sem have the same meaning as in teaches. We note that the key for offer is {cno, day, sem} and the key for teaches is {lec, cno, day, sem}, thus more than one lecturer can teach a course. The database also satisfies the IND teaches[cno, day, sem] $\subseteq$ offer[cno, day, sem], which specifies that a lecturer can only teach courses that are offered by the university.

Suppose we now map the relational data to XML by first mapping the two relations to separate documents with root nodes offer and teaches and then combining these documents to a single document with root node uni, as shown in Fig. 1. In particular, the tuples in relation offer were directly mapped to elements with tag course. Concerning the relation teaches, a nesting on {lec, day, sem} preceded the direct mapping of the (nested) tuples, within which tags course and info were introduced.

```
Flat Relations    Nested Relation              XML Document
                                    <uni>
teaches                               <teaches>
 lec cno day sem  cno {lec day sem}     <course cno="C1">          3
 L1  C1  TUE 09S   C1   L1 TUE 09S        <info lec="L1" day="TUE" sem="09S"/>
 L1  C1  MON 08W        L1 MON 08W        <info lec="L1" day="MON" sem="08W"/>
                                        </course>        1          2
offer                                 </teaches>
 cno day sem                          <offer>
 C1  TUE 09S                            <course cno="C1" day="TUE" sem="09S"/>
 C1  MON 08W                            <course cno="C1" day="MON" sem="08W"/>
                                      </offer>
                                    </uni>
```

Fig. 1. Example Relations and XML Document

Now the XML document satisfies an inclusion constraint because of the original IND, but one cannot express this inclusion constraint by

$$\chi = ((\texttt{uni.teaches.course}, [\texttt{cno}, \texttt{info.day}, \texttt{info.sem}]) \subseteq$$
$$(\texttt{uni.offer.course}, [\texttt{cno}, \texttt{day}, \texttt{sem}])),$$

where uni.teaches.course and uni.offer.course are the LHS and RHS selector, and [cno, info.day, info.sem] and [cno, day, sem] are the LHS and RHS fields, and applying the semantics given in [4,6,7,8,9,15]. In particular, the keyref mechanism [4] requires that there exists for each selector node at most one descendant node per field. This however does not hold in our example, since for the LHS selector uni.teaches.course, there exist two descendant info.sem nodes, which are marked with 2 and 3 in Fig. 1. The approaches in [6,8] require the field nodes to be attributes of the selector nodes. In our example, LHS fields info.day and info.sem are no attributes of the LHS selector uni.teaches.course and therefore the approaches in [6,8] cannot express the constraint χ. Finally, in the proposals in [7,9,15] it is required that every possible combination of nodes from [cno, info.day, info.sem] within a uni.teaches.course node must have matching nodes in [cno, day, sem] within a uni.offer.course node. So, since one combination is {C1, TUE, 08W}, this requires a uni.offer.course node with child nodes {C1, TUE, 08W}, which clearly does not hold.

In this paper, we propose different semantics so that the constraint χ holds in our example. The key idea is that we do not allow arbitrary combinations of nodes from the LHS fields, we only allow nodes that are closely related by what we will define later as the *closest* property. So, for example in Fig. 1, the day and sem nodes marked with $\boxed{1}$ and $\boxed{2}$ satisfy the *closest* property, but not nodes $\boxed{1}$ and $\boxed{3}$. The motivation for this restriction is that in the relational model data values that appear in the same tuple are more closely related than those that belong to different tuples, and our *closest* notion extends this idea to XML, and hence allows relational semantics to be extended.

Having an XML inclusion constraint that extends the semantics of an IND is important in several areas. Firstly, in the area of XML publishing [14], where a source relational database has to be mapped to a single predefined XML schema, knowing how relational integrity constraints map to XML integrity constraints allows the XML document to preserve the original semantics. This argument also applies to 'data-centric' XML [5], where XML databases (not necessarily with predefined schemas) are generated from relational databases.

The first contribution of this article is to define an XML inclusion constraint (called *XIND*) that extends the semantics of an IND. While the constraint is defined for any XML document (tree), we show that in the special case where the XML tree is generated by first mapping complete relations to nested relations by arbitrary sequences of nest operations, and then directly to an XML tree, the relations satisfy the IND if and only if the tree satisfies the corresponding XIND.

The second contribution of this article is to address the *implication problem*, i.e. the question of whether a new XIND holds given a set of existing XINDs, and the *consistency problem*, i.e. the question of whether there exists at least one non-empty XML tree that satisfies a given set of XINDs. We consider in our analysis a class of XINDs which we call *core* XINDs. This class excludes from all XINDs the very limited set of XINDs that interact with structural constraints of an XML document and impose as a consequence a constraint comparable to a *fixed value* constraint in a DTD [4]. We do not address this exceptional case, since we believe that it is not the intent of an XIND to impose a fixed value constraint. Also, we focus in our analysis on a class of XML trees introduced in previous work by one of the authors [10], called *complete XML trees*, which is intuitively the class of XML trees that contain 'no missing data'. Within this context, we present an axiom system for XIND implication and show that the system is sound and complete. While our axiom system contains rules that parallel those of INDs [3], it also contains additional rules that have no parallel in the IND system, reflecting the fact, as we soon discuss, that complete XML trees are more general than complete relations. Our proof techniques are based on a chase style algorithm, and we show that this algorithm can be used to solve the consistency problem and to provide a decision procedure for XIND implication.

The motivation for complete XML trees is to extend the notion of a complete relation to XML. A complete XML tree is however a more general notion than a complete relation since it includes trees that cannot be mapped to complete relations, such as those that contain duplicate nodes or subtrees, and trees that contain element leaf nodes rather than only text or attribute leaf nodes. Our motivation for considering the implication and consistency problems related to XINDs in complete XML trees is that while XML explicitly caters for irregularly structured data, it is also widely used in more traditional business applications involving regularly structured data [5], often referred to as 'data-centric' XML, and complete XML trees are a natural subclass in such applications.

The rest of the paper is organized as follows. Preliminary definitions are given in Sect. 2 and our definition of an XIND in Sect. 3. Section 4 shows that an XIND extends the semantics of an IND, and the implication and consistency problems are addressed in Sect. 5. Finally, Sect. 6 discusses related work.

2   XML Trees, Paths and Reachable Nodes

In this section we present some preliminary definitions. First, following the model adopted by XPath and DOM [4], we model an XML document as a tree as follows. We assume countably infinite, disjoint sets $\mathbf{E}$ and $\mathbf{A}$ of element and attribute labels respectively, and the symbol $\mathcal{S}$ indicating text. Thereby, the set of labels that can occur in the XML tree, $\mathbf{L}$, is defined by $\mathbf{L} = \mathbf{E} \cup \mathbf{A} \cup \{\mathcal{S}\}$.

Definition 1. *An XML tree $\mathbb{T}$ is defined by $\mathbb{T} = (\mathbf{V}, E, \text{lab}, \text{val}, v_\rho)$, where*

- $\mathbf{V}$ *is a finite, non-empty set of nodes;*
- *the total function* lab $: \mathbf{V} \to \mathbf{L}$ *assigns a label to every node in $\mathbf{V}$. A node v is called an* element node *if* $\text{lab}(v) \in \mathbf{E}$, *an* attribute node *if* $\text{lab}(v) \in \mathbf{A}$, *and a* text node *if* $\text{lab}(v) = \mathcal{S}$;
- $v_\rho \in \mathbf{V}$ *is a distinguished element node, called the* root node, *and* $\text{lab}(v_\rho) = \rho$;
- *the parent-child relation $E \subset \mathbf{V} \times \mathbf{V}$ defines the directed edges connecting the nodes in $\mathbf{V}$ and is required to form a tree structure rooted at node v_ρ. Thereby, for every edge $(v, \bar{v}) \in E$,*
  - *v is an* element node *and is said to be the* parent *of $\bar{v}$. Conversely, $\bar{v}$ is said to be a* child *of v;*
  - *if $\bar{v}$ is an attribute node, then there does not exist a node $\tilde{v} \in \mathbf{V}$ and an edge $(v, \tilde{v}) \in E$ such that $\text{lab}(\tilde{v}) = \text{lab}(\bar{v})$ and $\tilde{v} \neq \bar{v}$;*
- *the partial function* val $: \mathbf{V} \to \text{string}$ *assigns a string value to every attribute and text node in $\mathbf{V}$.*

In addition to the parent of a node v in a tree $\mathbb{T}$, we define its ancestor nodes, denoted by ancestor(v), to be the transitive closure of parents of v.

An example of an XML tree is presented in Fig. 2, where $\mathbf{E} = \{\rho, \text{offer}, \text{teaches}, \text{course}, \text{info}\}$ and $\mathbf{A} = \{\text{cno}, \text{day}, \text{lec}, \text{sem}\}$.

The notion of a path, which we now present together with some frequently required operators on paths, is central to all work on XML integrity constraints.

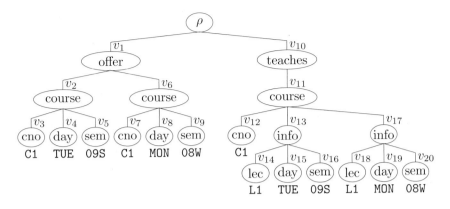

Fig. 2. Example XML Tree

Definition 2. *A path $P = l_1.\cdots.l_n$ is a non-empty sequence of labels (possibly with duplicates) from* **L**. *Given paths $P = l_1.\cdots.l_n$ and $\bar{P} = \bar{l}_1.\cdots.\bar{l}_m$ we define*

- *P to be a legal path, if $l_1 = \rho$ and $l_i \in$ **E** for all $i \in [1, n-1]$[1].*
- *P to be a prefix of $\bar{P}$, denoted by $P \subseteq \bar{P}$, if $n \le m$ and $l_i = \bar{l}_i$ for all $i \in [1, n]$.*
- *P to be a strict prefix of $\bar{P}$, denoted by $P \subset \bar{P}$, if $P \subseteq \bar{P}$ and $n < m$.*
- *length$(P) = n$ to denote the length of P.*
- *the concatenation of P and $\bar{P}$, denoted by $P.\bar{P}$, to be $l_1.\cdots.l_n.\bar{l}_1.\cdots.\bar{l}_m$.*
- *the intersection of P and $\bar{P}$ if both are legal paths, denoted by $P \cap \bar{P}$, to be the longest path that is a prefix of both P and $\bar{P}$.*

For example, referring to Fig. 2, offer.course and ρ.cno.course are paths but not legal ones, whereas ρ.offer.course is a legal path. Also the path ρ.offer is a strict prefix of ρ.offer.course, and if $P = \rho$.offer.course.cno and $\bar{P} = \rho$.offer.course.sem, then $P \cap \bar{P} = \rho$.offer.course.

We now define a path instance, which is essentially a downward sequence of nodes in an XML tree.

Definition 3. *A path instance $p = v_1.\cdots.v_n$ in a tree $\mathbb{T} = (\mathbf{V}, E, \mathrm{lab}, \mathrm{val}, v_\rho)$ is a non-empty sequence of nodes in* **V** *such that $v_1 = v_\rho$ and for all $i \in [2, n]$, $v_{i-1} = \mathrm{parent}(v_i)$. The path instance p is said to be defined over a path $P = l_1.\cdots.l_n$, if $\mathrm{lab}(v_i) = l_i$ for all $i \in [1, n]$.*

For example, referring to Fig. 2, $v_\rho.v_1.v_2$ is a path instance, and this path instance is defined over path ρ.offer.course.

The next definition specifies the set of nodes reachable in a tree $\mathbb{T}$ from the root node by following a path P.

Definition 4. *Given a tree $\mathbb{T} = (\mathbf{V}, E, \mathrm{lab}, \mathrm{val}, v_\rho)$ and a legal path P, the function $\mathrm{N}(P, \mathbb{T})$ returns the set of nodes defined by $\{v \in \mathbf{V} \mid v$ is the final node in path instance p and p is defined over $P\}$.*

---

[1] $[1, n]$ denotes the set $\{1, \ldots, n\}$.

For instance, if $\mathbb{T}$ is the tree in Fig. 2 and $P = \rho.\texttt{offer.course.day}$, then $\mathrm{N}(P, \mathbb{T}) = \{v_4, v_8\}$. We note that it follows from our tree model that for every node v in a tree $\mathbb{T}$ there is exactly one path instance p such that v is the final node in p and therefore $\mathrm{N}(P, \mathbb{T}) \cap \mathrm{N}(\bar{P}, \mathbb{T}) = \emptyset$ if $P \neq \bar{P}$. We therefore say that P is *the* path such that $v \in \mathrm{N}(P, \mathbb{T})$.

3   Defining XML Inclusion Dependencies

In this section we present the syntax and semantics of our definition of an XIND, starting with the syntax.

Definition 5. *An XML Inclusion Dependency is a statement of the form*

$$((P, [P_1, \ldots, P_n]) \subseteq (P', [P'_1, \ldots, P'_n]))$$

where P and P' are paths called LHS and RHS selector, and $P_1, \ldots, P_n$ and $P'_1, \ldots, P'_n$ are non-empty sequences of paths, called LHS and RHS fields, such that $\forall i \in [1, n]$, $P.P_i$ and $P'.P'_i$ are legal paths ending in an attribute or text label.

We now compare this definition to the `keyref` mechanism in XML, which is the basis for the syntax of an XIND. (i) We only consider simple paths in the selectors and fields, whereas the `keyref` mechanism allows for a restricted form of XPath expressions. (ii) In contrast to an XIND, the `keyref` mechanism also allows for relative constraints, whereby the inclusion constraint is only evaluated in part of the XML tree. (iii) The restrictions on fields means that we only consider inclusion between text/attribute nodes, whereas the `keyref` mechanism also allows for inclusion between element nodes.

We should mention that the restrictions discussed in (i) - (iii) are not intrinsic to our approach, and our definition of an XIND can easily be extended to handle these extension. Our reason for not considering these extensions here is so that we can concentrate on the main contribution of our paper, which is to apply different semantics to an XIND so as to extend relational semantics.

To define the semantics of an XIND, we first make a preliminary definition (first presented in [10]) that is central to our approach. The intuition for it, which will be made more precise in the next section, is as follows. In defining relational integrity constraints such as FDs or INDs, it is implicit that the relevant data values from either the LHS or RHS of the constraint belong to the same tuple. The *closest* definition extends this property of two data values belonging to the same tuple to XML, that is if two nodes in the XML tree satisfy the *closest* property, then 'they belong to the same tuple'.

Definition 6. *Given nodes v_1 and v_2 in an XML tree $\mathbb{T}$, the boolean function* $\mathrm{closest}(v_1, v_2)$ *is defined to return true, iff there exists a node v_2^1 such that $v_2^1 \in$ aancestor(v_1) and $v_2^1 \in$ aancestor(v_2) and $v_2^1 \in \mathrm{N}(P_1 \cap P_2, \mathbb{T})$, where P_1 and P_2 are the paths such that $v_1 \in \mathrm{N}(P_1, \mathbb{T})$ and $v_2 \in \mathrm{N}(P_2, \mathbb{T})$, and the aancestor function is defined by aancestor$(v) = \mathrm{ancestor}(v) \cup \{v\}$.*

For instance, in Fig. 2 closest(v_3, v_4) is true. This is because $P = \rho$.offer.course.cno and $\bar{P} = \rho$.offer.course.day are the paths such that $v_3 \in N(P, \mathbb{T})$ and $v_4 \in N(\bar{P}, \mathbb{T})$, and v_3 and v_4 have the common ancestor node $v_2 \in N(\rho$.offer.course$, \mathbb{T})$, where ρ.offer.course $= P \cap \bar{P}$. However, closest(v_3, v_8) is false since v_3 and v_8 have no common ancestor node in $N(\rho$.offer.course$, \mathbb{T})$, and closest(v_3, v_7) is false because v_3 and v_7 have no common ancestor node in $N(\rho$.offer.course.cno$, \mathbb{T})$.

This leads to the definition of the semantics of an XIND.

Definition 7. *An XML tree $\mathbb{T}$ satisfies an XIND $\sigma = ((P, [P_1, \ldots, P_n]) \subseteq (P', [P'_1, \ldots, P'_n]))$, denoted by $\mathbb{T} \models \sigma$, iff whenever there exists an LHS selector node v and corresponding field nodes $v_1, \ldots, v_n$ such that:*

i) $v \in N(P, \mathbb{T})$,
ii) for all $i \in [1, n]$, $v_i \in N(P.P_i, \mathbb{T})$ and $v \in$ ancestor(v_i),
iii) for all $i, j \in [1, n]$, closest(v_i, v_j) = true,

then there exists an RHS selector node v' and field nodes $v'_1, \ldots, v'_n$ such that

i') $v' \in N(P', \mathbb{T})$,
ii') for all $i \in [1, n]$, $v'_i \in N(P'.P'_i, \mathbb{T})$ and $v' \in$ ancestor(v'_i),
iii') for all $i, j \in [1, n]$, closest(v'_i, v'_j) = true,
iv') for all $i \in [1, n]$, val(v_i) = val(v'_i).

For instance, the XML tree in Fig. 2 satisfies the XIND $\chi = ((\rho$.teaches.course, [cno, info.day, info.sem]) $\subseteq (\rho$.offer.course, [cno, day, sem])). This is because the only sequences of LHS field nodes that pairwise satisfy the *closest* property are v_{12}, v_{15}, v_{16} and v_{12}, v_{19}, v_{20}, and v_{12}, v_{15}, v_{16} is value equal to the sequence of RHS field nodes v_3, v_4, v_5 and v_{12}, v_{19}, v_{20} is value equal to v_7, v_8, v_9. The essential difference between an XIND and other proposals is that we require the sequence of field nodes (both LHS and RHS) generated by the cross product to also satisfy the *closest* property, whereas other proposals do not contain this additional restriction [9,7,15]. As a consequence, the constraint χ is violated in the XML tree in Fig. 2 according to the other proposals.

We also make the point that we do not address the situation where there may be no node for a LHS field. We only require the inclusion between LHS and RHS field nodes, when there is a node for every LHS field. This is consistent with other work in the area, like for example the proposal for XML keys in [11].

4 Extending Relational Semantics

In this section we justify our claim that an XIND extends the semantics of an IND by showing that in the case where the XML tree is generated from a complete database by a very general class of mappings, then the database satisfies the IND if and only if the XML tree satisfies the corresponding XIND. To show this, we first define a general class of mappings from complete relational databases to XML trees. The presentation of the mapping procedure presented here will be abbreviated because of space requirements, and we refer the reader to [10] for a more detailed presentation if needed.

The first step in the mapping procedure maps each initial flat relation to a nested relation by a sequence of nest operations. To be more precise, we recall that the nest operator $\nu_{\mathbf{Y}}(\mathbb{R}^*)$ on a nested (or flat) relation $\mathbb{R}^*$, where $\mathbf{Y}$ is a subset of the schema $\mathbf{R}$ of $\mathbb{R}^*$, combines the tuples in $\mathbb{R}^*$ which are equal on $\mathbb{R}^*[\mathbf{R} - \mathbf{Y}]$ into single tuples [2]. So if the initial flat relation is denoted by $\mathbb{R}$, we perform an arbitrary sequence of nest operations $\nu_{\mathbf{Y}_1}, \ldots, \nu_{\mathbf{Y}_n}$ on $\mathbb{R}$ and so the final nested relation $\mathbb{R}^*$, is defined by $\mathbb{R}^* = \nu_{\mathbf{Y}_n}(\cdots \nu_{\mathbf{Y}_1}(\mathbb{R}))$.

For instance, in the introductory example the flat relation teaches is converted to a nested relation $\mathbb{R}^*$ by $\mathbb{R}^* = \nu_{\texttt{lec,day,sem}}(\texttt{teaches})$.

The next step in the mapping procedure is to map the nested relation to an XML tree by converting each sub-tuple in the nested relation to a subtree in the XML tree, using a new element node for the root of the subtree, as illustrated in the introductory example. While we don't claim that our method is the only way to map a relation to an XML tree, it does have two desirable features. First, it allows the initial flat relation to be nested arbitrarily, which is a desirable feature in data-centric applications of XML [5]. Second, it has been shown that the mapping procedure is invertible [10], and so no information is lost by the transformation.

In the context of mapping multiple relations to XML, we extend the method just outlined as follows. We first map each relation to an XML tree as just discussed. We then replace the label in the root node by a label containing the name of the relation (which we assume to be unique), and construct a new XML tree with a new root node and with the XML trees just generated being principal subtrees. This procedure was used in the introductory example.

This leads to the following important result which justifies our claim that an XIND extends the semantics of an IND.

Theorem 1. *Let complete flat relations $\mathbb{R}_1$ and $\mathbb{R}_2$ be mapped to an XML tree $\mathbb{T}$ by the method just outlined. Then $\mathbb{R}_1$ and $\mathbb{R}_2$ satisfy the IND $\mathbf{R}_1[A_1, \ldots, A_n] \subseteq \mathbf{R}_2[B_1, \ldots, B_n]$, where $\mathbf{R}_1$ and $\mathbf{R}_2$ are the schemas of $\mathbb{R}_1$ and $\mathbb{R}_2$, iff $\mathbb{T}$ satisfies the XIND $((\rho.\mathbf{R}_1, [P_{A_1}, \ldots, P_{A_n}]) \subseteq (\rho.\mathbf{R}_2, [P_{B_1}, \ldots, P_{B_n}]))$, where $\rho.\mathbf{R}_1.P_{A_1}, \ldots, \rho.\mathbf{R}_1.P_{A_n}, \rho.\mathbf{R}_2.P_{B_1}, \ldots, \rho.\mathbf{R}_2.P_{B_n}$ represent the paths over which the path instances in $\mathbb{T}$ that end in leaf nodes are defined.*

For instance, we deduce from the IND teaches[cno, day, sem] $\subseteq$ offer[cno.day.sem] the XIND $((\rho.\texttt{teaches}, [\texttt{course.cno, course.info.day, course.info.sem}]) \subseteq (\rho.\texttt{offer}, [\texttt{course.cno, course.day, course.sem}]))$ and, from the inference rules to be given in the next section, this XIND is equivalent to the XIND given in the introductory example, namely $((\rho.\texttt{teaches.course}, [\texttt{cno, info.day, info.sem}]) \subseteq (\rho.\texttt{offer.course}, [\texttt{cno, day, sem}]))$.

The proof of this theorem is based on a result established in [10], which shows that if a relation is mapped to an XML tree by the procedure just outlined, then a set of data values appear in the same tuple of the relation if and only if the nodes representing these values in the tree pairwise satisfy the *closest* property.[2]

---

[2] We omit detailed proofs throughout this paper because of space requirements and refer the reader to the technical report [1].

5   Reasoning about XML Inclusion Dependencies

We focus in our reasoning on *core* XINDs in *complete* XML trees, and we first define these key concepts in Sect. 5.1. We then present in Sect. 5.2 a chase algorithm and use this algorithm in Sect. 5.3 to solve the implication and consistency problems related to core XINDs in complete XML trees.

5.1   The Framework: Core XINDs in Complete XML Trees

From a general point of view, before requiring the data in an XML document to be complete, one first has to specify the structure of the information that the document is expected to contain. We use a set of legal paths $\mathbf{P}$ to specify the structure of the expected information in an XML document, and now define what we mean by an XML tree conforming to $\mathbf{P}$.

Definition 8. *A tree* $\mathbb{T}$ *is defined to* conform *to a set of legal paths* $\mathbf{P}$, *if for every node* v *in* $\mathbb{T}$, *if* P *is the path such that* $v \in \mathrm{N}(P, \mathbb{T})$, *then* $P \in \mathbf{P}$.

For example, if we denote the subtree rooted at node v_1 in Fig. 2 by $\mathbb{T}_1$, then $\mathbb{T}_1$ conforms to the set of paths $\mathbf{P}_1 = \{\texttt{offer},\texttt{offer.course}, \texttt{offer.course.cno},\texttt{offer.course.day},\texttt{offer.course.sem}\}$.

We now introduce the concept of a complete XML tree, which extends the notion of a complete relation to XML. To understand the intuition, consider again the subtree $\mathbb{T}_1$ in Fig. 2 and the set of paths $\bar{\mathbf{P}}_1 = \mathbf{P}_1 \cup \{\texttt{offer.course.max}\}$. Then $\mathbb{T}_1$ also conforms to $\bar{\mathbf{P}}_1$, but we do not consider it to be complete w.r.t. $\bar{\mathbf{P}}_1$ since the existence of the path $\texttt{offer.course.max}$ means that we expect every $\texttt{course}$ in $\mathbb{T}_1$ to have a $\texttt{max}$ number of students, which is not satisfied by nodes v_2 and v_6 in Fig. 2. We now make this idea more precise.

Definition 9. *If* $\mathbb{T}$ *is a tree that conforms to a set of paths* $\mathbf{P}$, *then* $\mathbb{T}$ *is defined to be* complete *w.r.t.* $\mathbf{P}$, *if whenever* P *and* $\bar{P}$ *are paths in* $\mathbf{P}$ *such that* $P \subset \bar{P}$, *and there exists node* $v \in \mathrm{N}(P, \mathbb{T})$, *then there also exists node* $\bar{v} \in \mathrm{N}(\bar{P}, \mathbb{T})$ *such that* $v \in \mathrm{ancestor}(\bar{v})$.

For instance, as just noted, $\mathbb{T}_1$ is not complete w.r.t. $\bar{\mathbf{P}}_1$ but it is complete w.r.t. $\mathbf{P}_1$. This example also illustrates an important point. Unlike the relational case, the completeness of a tree is only defined w.r.t. a specific set of paths and so, as we have just seen, a tree may conform to two different sets of paths, but may be complete w.r.t. one set but not the other. We also note that if a tree $\mathbb{T}$ is complete w.r.t. a set of paths $\mathbf{P}$, then $\mathbf{P}$ is what we call *downward-closed*. That is, if P and $\tilde{P}$ are paths and $P \in \mathbf{P}$, then $\tilde{P} \in \mathbf{P}$ if $\tilde{P} \subset P$. For example the sets $\mathbf{P}_1$ and $\bar{\mathbf{P}}_1$ are downward-closed.

We now turn to the class of XINDs that we consider in our reasoning. It is natural to expect that if an XIND σ is intended to apply to an XML tree $\mathbb{T}$, then the constraint imposed by σ should belong to the information represented by $\mathbb{T}$. We incorporate this idea by requiring that the paths in an XIND are taken from the set of paths to which the targeted tree conforms, which we now define.

Definition 10. *An XIND* $\sigma = ((P, [P_1, \ldots, P_n]) \subseteq (P'([P'_1, \ldots, P'_n]))$ *is defined to conform to a set of paths* **P**, *if for all* $i \in [1, n]$, $P.P_i \in \mathbf{P}$ *and* $P'.P'_i \in \mathbf{P}$.

We also place another restriction on an XIND, motivated by our belief that an XIND σ should not enforce, as a hidden side effect, that each node in a set of nodes in a tree $\mathbb{T}$ must have the same value. Suppose then that for the RHS selector of σ, $P' = \rho$, and for some $i \in [1, n]$, the RHS field P'_i is an attribute label. Then since there is only one root node, and in turn at most one attribute node in $N(P'.P'_i, \mathbb{T})$, the semantics of σ means, that every node in $N(P.P_i, \mathbb{T}) \cup N(P'.P'_i, \mathbb{T})$ must have the same value. We believe that this not the intent of an XIND, and that such a constraint should be specified instead explicitly in a DTD or XSD. Since the study of the interaction between structural constraints and integrity constraints is known to be a complex one [6], and outside the scope of this paper, we exclude such an XIND and this leads to the following definition.

Definition 11. *An XIND* $((P, [P_1, \ldots, P_n]) \subseteq (P', [P'_1, \ldots, P'_n]))$ *is a core XIND, if in case that* $P' = \rho$, *then there does not exist a RHS field* P'_i *such that* $length(P'_i) = 1$ *and* P'_i *ends in an attribute label.*

5.2   The Chase for Core XINDs in Complete Trees

The *chase* is a recursive algorithm that takes as input (i) a set of paths **P**, (ii) a tree $\mathbb{T}$ that is complete w.r.t **P**, and (iii) a set of XINDs Σ that conforms to **P**, and adds new nodes to $\mathbb{T}$ such that $\mathbb{T} \vDash \Sigma$. From a bird-eyes view, the chase halts if the input tree $\mathbb{T}_s$ for a (recursive) step s satisfies Σ, and otherwise it

1. chooses an XIND $\sigma_s = ((P, [P_1, \ldots, P_n]) \subseteq (P', [P'_1, \ldots P'_n,]))$ from Σ, such that $\mathbb{T}_s \nvDash \sigma_s$ because of a sequence of nodes $[v_0, v_1, \ldots, v_n]$, where v_0 is a LHS selector node for σ_s and $v_1, \ldots, v_n$ are corresponding field nodes, and
2. creates new nodes in $\mathbb{T}_s$, such that the resulting tree $\mathbb{T}_{s+1}$ contains a RHS selector node v'_0 and field nodes $v'_1, \ldots, v'_n$ that remove the violation.

We now illustrate a step s in the chase by the example depicted in Fig. 3. Here the XIND σ is violated in tree $\mathbb{T}_s$ by the sequences of nodes $[v_1, v_2]$ and $[v_3, v_4]$. Given that the sequence $[v_1, v_2]$ is chosen, the chase creates nodes v_8, v_9 in tree $\mathbb{T}_{s+1}$, which remove the violation, and then adds node v_{10} as a child of v_8, in order that $\mathbb{T}_{s+1}$ is complete w.r.t. the set of paths **P**.

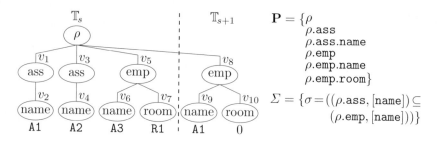

Fig. 3. Example Chase Step

in: A downward-closed sequence of legal paths $\mathbf{P} = [R_1, \ldots, R_m]$ ordered by length
      A tree $\mathbb{T} = (\mathbf{V}, E, \mathrm{lab}, \mathrm{val}, v_\rho)$ that is complete w.r.t. $\mathbf{P}$
      A sequence of core XINDs Σ that conforms to $\mathbf{P}$
out: Tree $\bar{\mathbb{T}}$ that is complete w.r.t. $\mathbf{P}$ and satisfies Σ

1: **if** $\mathbb{T} \vDash \Sigma$ **then return** $\mathbb{T}$; **end if**
2: let $\sigma = ((P, [P_1, \ldots, P_n]) \subseteq (P', [P'_1, \ldots, P'_n]))$ be the first XIND in Σ such that $\mathbb{T} \nvDash \sigma$;
3: let $\mathbf{Y}$ be the set of all sequences of nodes that violate σ in $\mathbb{T}$;
4: **for** $i := 0$ **to** n **do**   ▷ choose violation
5:     **repeat**
6:         choose sequences $[v_0, v_1, \ldots, v_n]$ and $[\hat{v}_0, \hat{v}_1, \ldots, \hat{v}_n]$ from $\mathbf{Y}$;
7:         **if** $v_i \prec \hat{v}_i$ **then** remove $[\hat{v}_0, \hat{v}_1, \ldots, \hat{v}_n]$ from $\mathbf{Y}$; **end if**
8:     **until** no more change to $\mathbf{Y}$ is possible
9: **end for**
10: let $[v_0, v_1, \ldots, v_n]$ be the remaining sequence of nodes in $\mathbf{Y}$;
11: let $\mathbf{X}$ be a set that exclusively contains the root node v_ρ;
12: **for** $i := 1$ **to** m **do**   ▷ remove violation
13:     **if** there exists path $P'_x \in [P'_1, \ldots, P'_n]$ such that $R_i \cap P'.P'_x \neq \rho$ **then**
14:         create a new node v and add v to both $\mathbf{V}$ and $\mathbf{X}$;
15:         let $l_1. \cdots .l_k$ be the sequence of labels in R_i;
16:         set $\mathrm{lab}(v) = l_k$;
17:         let $\hat{v}$ be the node in $\mathrm{N}(l_1. \cdots .l_{k-1}, \mathbb{T}) \cap \mathbf{X}$;
18:         add $(\hat{v}, v)$ to E such that v is the last child of $\hat{v}$;
19:         **if** there exists path $P'_y \in [P'_1, \ldots, P'_n]$ such that $R_i = P'.P'_y$ **then**
20:             set $\mathrm{val}(v) = \mathrm{val}(v_y)$;   ▷ $v_y \in [v_1, \ldots, v_n]$
21:         **else if** l_k is an attribute or text label **then**
22:             set $\mathrm{val}(v) = $ "0";
23:         **end if**
24:     **end if**
25: **end for**
26: **return** $\mathrm{Chase}(\mathbf{P}, \mathbb{T}, \Sigma)$;

Fig. 4. Algorithm $\mathrm{Chase}(\mathbf{P}, \mathbb{T}, \Sigma)$

The procedure just outlined is non-deterministic since both the choice of σ_s and the choice of a violating sequence of nodes $[v_0, v_1, \ldots, v_n]$ in a step s is random. These choices however essentially determine the characteristics of the trees generated by the steps of the chase, and thus our proof techniques, which are based on certain characteristics of the generated trees. We therefore designed a deterministic chase algorithm, depicted in Fig. 4, that results in a *unique* tree. The essential prerequisite is the following, simplified version of document-order.

Definition 12. *In a tree $\mathbb{T}$, node v is defined to precede node $\bar{v}$ w.r.t. document-order, denoted by $v \prec \bar{v}$, if v is visited before $\bar{v}$ in a pre-order traversal of $\mathbb{T}$.*

We now illustrate how uniqueness of the tree $\mathbb{T}_{s+1}$ resulting from a step s of the chase is achieved. First, the choice of σ_s at Line 2 in Fig. 4 is deterministic, given that σ_s is the first XIND in Σ that is violated in $\mathbb{T}_s$, and that Σ is expected to be a sequence, rather than a set, of XINDs. Second, in case that there is more than one sequence of violating nodes, then the one that is, roughly speaking, in the top-left of tree $\mathbb{T}_s$ is chosen at Lines 3 - 9. Referring to the example in Fig. 3, the chase deliberately chooses the sequence of violating nodes $[v_1, v_2]$, since $v_1 \prec v_3$. Third, the procedure for removing the violation in a step s is deterministic. In particular, the chase loops for this purpose over the paths in $\mathbf{P}$ and creates path instances accordingly (cf. Lines 12 - 25), such that the resulting tree $\mathbb{T}_{s+1}$ is complete w.r.t. $\mathbf{P}$ and contains a sequence of nodes $[v'_0, v'_1, \ldots, v'_n]$ that removes the violation. The desired uniqueness is basically achieved, since (i) paths $\mathbf{P}$ are expected to be a sequence, rather than a set of paths, and therefore the succession in which the paths in $\mathbf{P}$ are iterated in the loop at Line 12 is deterministic, and (ii) a new node is always added to the parent as the last child w.r.t. document-order (cf. Line 18).

We then have the following result on the procedure of the chase.

Lemma 1. *An application of* Chase$(\mathbf{P}, \mathbb{T}, \Sigma)$ *terminates and returns a unique tree* $\tilde{\mathbb{T}}$ *which contains* $\mathbb{T}$ *as a sub-tree, is complete w.r.t.* $\mathbf{P}$ *and satisfies* Σ.

Thereby, the argument for the termination of the chase bases on the following observation on the set $\mathbf{U}$ of distinct values of attribute and text nodes in a tree $\mathbb{T}$. A tree $\mathbb{T}$ satisfies an XIND $\sigma = (P, (P_1, \ldots, P_n) \subseteq (P', (P'_1, \ldots, P'_n))$, if $\mathbb{T}$ contains for every sequence of values $u_1, \ldots, u_n \in \mathbf{U} \times \cdots \times \mathbf{U}$, a sequence of nodes $[\tilde{v}_1 \in N(P'.P'_1, \mathbb{T}), \ldots, \tilde{v}_n \in N(P'.P'_n, \mathbb{T})]$ that pairwise satisfy the *closest* property, such that for all $i \in [1, n]$, val$(\tilde{v}_i) = u_i$. Since the number of values in the initial tree is finite and the chase introduces at most one new value (cf. Line 22), the number of distinct sequences of values in $\mathbf{U} \times \cdots \times \mathbf{U}$ is finite. In turn, the chase terminates, since it adds in every step, w.r.t. an XIND in Σ, a sequence of pairwise *closest* RHS field nodes $v'_1, \ldots, v'_n$, and val$(v'_1), \ldots,$ val$(v'_n) \in \mathbf{U} \times \cdots \times \mathbf{U}$.

5.3 Consistency and Implication of Core XINDs in Complete Trees

We formulate the consistency problem in our framework as the question of whether there exists a tree $\mathbb{T}$, for any given combination of a downward-closed set of paths $\mathbf{P}$ and a set of core XINDs Σ that conforms to $\mathbf{P}$, such that $\mathbb{T}$ is complete w.r.t. $\mathbf{P}$ and satisfies Σ. We have the following result.

Theorem 2. *The class of core XINDs in complete XML trees is consistent.*

The correctness of Theorem 2 follows from the fact that there always exists a tree $\tilde{\mathbb{T}}$ that is complete w.r.t. a given set of paths $\mathbf{P}$, and the result in Lemma 1 that the tree $\tilde{\mathbb{T}}$ returned by Chase$(\mathbf{P}, \tilde{\mathbb{T}}, \Sigma)$, is complete w.r.t. $\mathbf{P}$ and satisfies the given set of core XINDs Σ.

We now turn to the implication of core XINDs. We use $\Sigma \models \sigma$ to denote that Σ implies σ, i.e. that given a set of paths $\mathbf{P}$ to which $\Sigma \cup \{\sigma\}$ conforms to, there does not exist a tree $\mathbb{T}$ that is complete w.r.t. $\mathbf{P}$ such that $\mathbb{T} \models \Sigma$ but $\mathbb{T} \nvDash \sigma$.

In order to discuss the implication problem, which we formulate as the question whether $\Sigma \models \sigma$ (and also $\Sigma \nvDash \sigma$) is decidable, we start by giving in Table 1

Table 1. Inference Rules for Core XINDs

R1 **Reflexivity**
$\{\} \vdash ((P, [P_1, \ldots, P_n]) \subseteq (P, [P_1, \ldots, P_n]))$

R2 **Permutated Projection**
$((P, [P_1, \ldots, P_n]) \subseteq (P', [P'_1, \ldots, P'_n])) \vdash$
$((P, [P_{\pi(1)}, \ldots, P_{\pi(m)}]) \subseteq (P', [P'_{\pi(1)}, \ldots, P'_{\pi(m)}]))$, if $\{\pi(1), \ldots, \pi(m)\} \subseteq \{1, \ldots, n\}$

R3 **Transitivity**
$((P, [P_1, \ldots, P_n]) \subseteq (\bar{P}, [\bar{P}_1, \ldots, \bar{P}_n])) \wedge ((\bar{P}, [\bar{P}_1, \ldots, \bar{P}_n]) \subseteq (P', [P'_1, \ldots, P'_n])) \vdash$
$((P, [P_1, \ldots, P_n]) \subseteq (P', [P'_1, \ldots, P'_n]))$

R4 **Downshift**
$((P, [P_1, \ldots, P_n]) \subseteq (P'.R, [P'_1, \ldots, P'_n])) \vdash ((P, [P_1, \ldots, P_n]) \subseteq (P', [R.P'_1, \ldots, R.P'_n]))$

R5 **Upshift**
$((P, [P_1, \ldots, P_n]) \subseteq (P', [R.P'_1, \ldots, R.P'_n])) \vdash ((P, [P_1, \ldots P_n]) \subseteq (P'.R, [P'_1, \ldots, P'_n]))$

R6 **Union**
$((P, [P_1, \ldots, P_m]) \subseteq (\rho, [P'_1, \ldots, P'_m])) \wedge ((P, [P_{m+1}, \ldots, P_n]) \subseteq (\rho, [P'_{m+1}, \ldots, P'_n])) \vdash$
$((P, [P_1, \ldots, P_n]) \subseteq (\rho, [P'_1, \ldots, P'_n]))$, if $\rho.P'_i \cap \rho.P'_j = \rho$ for all $i, j \in [1, m] \times [m+1, n]$

a set of inference rules, where symbol $\vdash$ denotes that the XINDs in the premise derive the XIND in the conclusion.

Rules R1 - R3 correspond to the well known inference rules for INDs [3], which is to be expected given Theorem 1 and the fact that XML trees generated from a complete relational database from the mapping described in Sect. 4 are a subclass of complete XML trees. The remaining rules have no parallels in the inference rules for INDs, and we now discuss them.

Rule R4 allows one to shift a path from the end of the RHS selector in an XIND down to the start of the RHS fields. For example, by applying R4 to the XIND $((\rho.\text{teaches.course}, [\text{cno}]) \subseteq (\rho.\text{offer.course}, [\text{cno}]))$, we derive the XIND $((\rho.\text{teaches.course}, [\text{cno}]) \subseteq (\rho.\text{offer}, [\text{course.cno}]))$, whereby the last label in the RHS selector $\rho.\text{offer.course}$ has been shifted down to the start of the RHS fields. Rule R5 is the reverse of R4, whereby a path from the start of the RHS fields is shifted up to the end of the RHS selector.

Rule R6 is a rule that, roughly speaking, allows one to union the LHS fields and the RHS fields of two XINDs, provided that the RHS fields intersect only at the root. For example, given the XINDs $((\rho.\text{teaches.course},$ $[\text{cno}]) \subseteq (\rho, [\text{offer.course.cno}]))$ and $((\rho.\text{teaches.course}, [\text{info.lec}]) \subseteq$ $(\rho, [\text{department.lec}]))$, then we can derive $((\rho.\text{teaches.course}, [\text{cno}, \text{info.lec}])$ $\subseteq$ $(\rho, [\text{offer.course.cno}, \text{department.lec}]))$, since $\rho.\text{offer.course.cno}$ $\cap$ $\rho.\text{department.lec}$ $=$ ρ. However, the XINDs $((\rho.\text{teaches.course},$ $[\text{cno}]) \subseteq (\rho, [\text{offer.course.cno}]))$ and $((\rho.\text{teaches.course}, [\text{info.sem}]) \subseteq$ $(\rho, [\text{offer.course.sem}]))$ do not imply $((\rho.\text{teaches.course}, [\text{cno}, \text{info.sem}])$ $\subseteq (\rho, [\text{offer.course.cno}, \text{offer.course.sem}]))$ since $\rho.\text{offer.course.cno}$ $\cap$ $\rho.\text{offer.course.sem} \neq \rho$.

We have the following result on the soundness $(\Sigma \vdash \sigma \Rightarrow \Sigma \models \sigma)$ and completeness $(\Sigma \models \sigma \Rightarrow \Sigma \vdash \sigma)$ of our inference rules.

Theorem 3. *The set of inference rules R1 - R6 is sound and complete for the implication of core XINDs in complete XML trees.*

Thereby, rule R1 is trivially sound and we can show soundness of rules R2 - R6 by the contradiction that if $\Sigma \vdash \sigma$ and $\mathbb{T}$ is a tree such that $\mathbb{T} \nvDash \sigma$, then $\mathbb{T} \nvDash \Sigma$. The key idea for the proof of completeness is that we first construct a special initial tree $\mathbb{T}_\sigma{}^3$, which essentially has LHS field nodes with distinct values w.r.t. the XIND σ and is empty elsewhere. We then show by induction that the only XINDs satisfied by any intermediate tree during the chase are those derivable from Σ using rules R1 - R6. That is, if $\bar{\mathbb{T}}_\sigma \vDash \sigma$, where $\bar{\mathbb{T}}_\sigma$ is the final tree returned by the application Chase$(\mathbf{P}, \mathbb{T}_\sigma, \Sigma)$, then $\Sigma \vdash \sigma$ and thus $\Sigma \vDash \sigma \Rightarrow \Sigma \vdash \sigma$, since $\bar{\mathbb{T}}_\sigma \vDash \sigma$ if $\Sigma \vDash \sigma$ from Lemma 1.

Given that $\Sigma \vdash \sigma$ if $\bar{\mathbb{T}}_\sigma \vDash \sigma$ it follows that $\bar{\mathbb{T}}_\sigma \vDash \sigma \Rightarrow \Sigma \vDash \sigma$ since $\Sigma \vdash \sigma \Rightarrow \Sigma \vDash \sigma$. Also, $\Sigma \vDash \sigma \Rightarrow \bar{\mathbb{T}}_\sigma \vDash \sigma$, since if to the contrary $\bar{\mathbb{T}}_\sigma \nvDash \sigma$ then $\Sigma \nvDash \sigma$ from Lemma 1. Combining this with the result in Lemma 1 that the chase terminates yields that the chase is a *decision procedure* for the implication of core XINDs in complete XML trees, i.e. $\Sigma \vDash \sigma$ iff $\bar{\mathbb{T}}_\sigma \vDash \sigma$, and we therefore finally have the following result on the implication problem.

Theorem 4. *The implication problem for the class of core XINDs in complete XML trees is decidable.*

6   Discussion and Related Work

In recent years, several types of XML Integrity Constraints (XICs) such as functional dependencies or keys for XML have been investigated. Because of space requirements, we restrict our attention in this section to inclusion type constraints and refer the reader to [13] for a survey of other types of XICs.

An early type of XICs are path constraints [12]. A path inclusion constraint (PIC) essentially requires that whenever a node is reachable over one path, it must also be reachable over another path. In contrast, an XIND asserts that given a set of nodes, there also exist other nodes with corresponding values. Because of this basic difference, one cannot directly compare a PIC and an XIND.

Closer to XINDs are the XML Foreign Keys defined in [6,7]. Translated to the selector/field framework, these XICs constrain the fields to point to attribute or text nodes that are children of the selector nodes and so cannot express for example the constraint $\tau = ((\rho.\texttt{teaches}, [\texttt{course.cno}]) \subseteq (\rho.\texttt{offer}, [\texttt{course.cno}]))$.

The keyref mechanism of XSD [4] is limited with respect to the possible number of matching nodes per field. For instance, referring to τ and Fig. 2, the semantics of the keyref mechanism requires that any offer node has at most one descendant course.cno node, which is clearly not satisfied in Fig.2.

The XICs in [7,9] overcome these limitations. However, again translated to the selector/field framework, these XICs regard every sequence of field nodes as relevant as long as they are descendants of one selector node. As a consequence,

---

[3] A detailed construction procedure is presented in [1].

these XICs do not always preserve the semantics of an IND, which we have illustrated in detail in the introductory example.

The limitation of not always preserving the semantics of an IND also applies to the XML inclusion constraint in [15] developed by a subset of the authors, which in fact motivated the present work of defining an XIND. Compared to the XIND defined in this paper, the approach in [15] is less expressive and does not use the selector/field framework. Further, the semantics used in [15], although partly based on the *closest* concept, is nevertheless different from the semantics used in this paper, and as a result an XIND preserves the semantics of an IND.

In further research, we will relax some of the restrictions on the syntax of an XIND and address the implication and consistency problems related to XINDs that allow for path expressions rather than simple paths in the selectors and fields, and a relative constraint that is only evaluated in parts of the XML tree.

References

1. Karlinger, M., Vincent, M., Schrefl, M.: Inclusion Dependencies in XML. Technical Report 09.01, Dept. of Business Informatics - DKE, JKU Linz (2009)
2. Atzeni, P., DeAntonellis, V.: Relational Database Theory. Benjamin Cummings (1993)
3. Abiteboul, S., Hull, R., Vianu, V.: Foundations of Databases. Addison Wesley, Reading (1995)
4. Möller, A., Schwartzbach, M.: An Introduction to XML and Web Technologies. Addison-Wesley, Reading (2006)
5. Vakali, A., Catania, B., Maddalena, A.: XML Data Stores: Emerging Practices. IEEE Internet Computing 9(2), 62–69 (2005)
6. Arenas, M., Fan, W., Libkin, L.: On the Complexity of Verifying Consistency of XML Specifications. SIAM J. Comput. 38(3), 841–880 (2008)
7. Fan, W., Kuper, G.M., Siméon, J.: A Unified Constraint Model for XML. Computer Networks 39(5), 489–505 (2002)
8. Fan, W., Siméon, J.: Integrity constraints for XML. J. Comput. Syst. Sci. 66(1), 254–291 (2003)
9. Deutsch, A., Tannen, V.: XML Queries and Constraints, Containment and Reformulation. Theor. Comput. Sci. 336(1), 57–87 (2005)
10. Vincent, M.W., Liu, J., Mohania, M.: On the Equivalence between FDs in XML and FDs in Relations. Acta Informatica 44(3-4), 207–247 (2007)
11. Buneman, P., Davidson, S.B., Fan, W., Hara, C.S., Tan, W.C.: Keys for XML. Computer Networks 39(5), 473–487 (2002)
12. Abiteboul, S., Vianu, V.: Regular Path Queries with Constraints. J. Comput. Syst. Sci. 58(3), 428–452 (1999)
13. Fan, W.: XML Constraints: Specification, Analysis, and Applications. In: DEXA Workshops, pp. 805–809. IEEE Computer Society, Los Alamitos (2005)
14. Fan, W.: XML publishing: Bridging theory and practice. In: Arenas, M., Schwartzbach, M.I. (eds.) DBPL 2007. LNCS, vol. 4797, pp. 1–16. Springer, Heidelberg (2007)
15. Vincent, M.W., Schrefl, M., Liu, J., Liu, C., Dogen, S.: Generalized inclusion dependencies in XML. In: Yu, J.X., Lin, X., Lu, H., Zhang, Y. (eds.) APWeb 2004. LNCS, vol. 3007, pp. 224–233. Springer, Heidelberg (2004)

The Real Performance Drivers behind XML Lock Protocols

Sebastian Bächle and Theo Härder

University of Kaiserslautern, Germany
{baechle,haerder}@cs.uni-kl.de

Abstract. Fine-grained lock protocols should allow for highly concurrent transaction processing on XML document trees, which is addressed by the taDOM lock protocol family enabling specific lock modes and lock granules adjusted to the various XML processing models. We have already proved its operational flexibility and performance superiority when compared to competitor protocols. Here, we outline our experiences gained during the implementation and optimization of these protocols. We figure out their performance drivers to maximize throughput while keeping the response times at an acceptable level and perfectly exploiting the advantages of our tailor-made lock protocols for XML trees. Because we have implemented all options and alternatives in our prototype system XTC, benchmark runs for all "drivers" allow for comparisons in identical environments and illustrate the benefit of all implementation decisions. Finally, they reveal that careful lock protocol optimization pays off.

1 Motivation

Native XML database systems (XDBMSs) promise tailored processing of XML documents, but most of the systems published in the DB literature are designed for efficient document retrieval only [17]. However, XML's standardization and, in particular, its flexibility (e.g., data mapping, cardinality variations, optional or non-existing structures, etc.) are driving factors to attract demanding write/read applications, to enable heterogeneous data stores and to facilitate data integration. Because business models in practically every industry use large and evolving sets of sparsely populated attributes, XML is more and more adopted by those companies which have even now launched consortia to develop XML schemas adjusted to their particular data modeling needs.[1] For these reasons, XML databases currently get more and more momentum if data flexibility in various forms is a key requirement of the application and they are therefore frequently used in collaborative or even competitive environments [11]. As a consequence, the original "retrieval-only" focus – probably caused by the first proposals of XQuery respectively XPath where the update part was left out – is not enough anymore. Hence, update facilities are increasingly needed in

---

[1] World-leading financial companies defined more than a dozen XML vocabularies to standardize data processing and to leverage cooperation and data exchange [20].

S.S. Bhowmick, J. Küng, and R. Wagner (Eds.): DEXA 2009, LNCS 5690, pp. 38–52, 2009.

XDBMSs, i.e., fine-grained, concurrent, and transaction-safe document modifications have to be efficiently supported. For example, workloads for *financial application logging* include 10M to 20M inserts in a 24-hour day, with about 500 peak inserts/sec. Because at least a hundred users need to concurrently read the data for troubleshooting and auditing tasks, concurrency control is challenged to provide short-enough response times for interactive operations [11].

Currently, all vendors of XML(-enabled) DBMSs support updates only at document granularity and, thus, cannot manage highly dynamic XML documents, let alone achieve such performance goals. Hence, new concurrency control protocols together with efficient implementations are needed to meet these emerging challenges. To guarantee broad acceptance, we strive for a *general solution* that is even applicable for a spectrum of XML language models (e.g., XPath, XQuery, SAX, or DOM) in a multi-lingual XDBMS environment. Although predicate locking for declarative XML queries would be powerful and elegant, its implementation rapidly leads to severe drawbacks such as undecidability and application of unnecessarily large lock granules for simplified predicates – a lesson learned from the (much simpler) relational world. Beyond, tree locks or key-range locks [5,12] are not sufficient for fine-grained locking of concurrently evaluated stream-, navigation- and path-based queries. Thus, we necessarily have to map XQuery operations to a navigational access model to accomplish fine-granular locking supporting other XML languages like SAX and DOM [3], too, because their operations directly correspond to a navigational access model.

To approach our goal, we have developed a family consisting of four DOM-based lock protocols called the taDOM group by adjusting the idea of multi-granularity locking (MGL) to the specific needs of XML trees. Their empirical analysis was accomplished by implementing and evaluating them in XTC (XML Transaction Coordinator), our prototype XDBMS [9]. Its development over the last four years accumulated substantial experience concerning DBMS performance in general and efficient lock management in particular.

1.1 The taDOM Protocol Family

Here, we assume familiarity of the reader with the idea of multi-granularity locking (MGL) – also denoted as hierarchical locking [6] – which applies to hierarchies of objects like tables and tuples and is used "everywhere" in the relational world. The allow for fine-grained access by setting R (read) or X (exclusive) locks on objects at the lower levels in the hierarchy and coarse grained access by setting the locks at higher levels in the hierarchy, implicitly locking the whole subtree of objects at smaller granules. To avoid lock conflicts when objects at different levels are locked, so-called intention locks with modes IR (intention read) or IX (intention exclusive) have to be acquired along the path from the root to the object to be isolated and vice versa when the locks are released [6].

Although an MGL protocol can also be applied to XML document trees, it is in most cases too strict, because both R and X mode on a node, would always lock the whole subtree below, too. While this is the desired semantics for part-of object hierarchies as in relational databases, these restrictions do not apply to

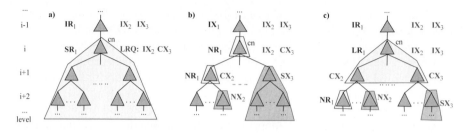

Fig. 1. Example of the taDOM3+ protocol

XML where transactions must not necessarily be guaranteed to have no writers in the subtree of their current node. Hence, MGL does not provide the degrees of concurrency that could be achieved on XML documents.

For ease of comprehension, we will give a brief introduction into the essentials of our taDOM lock protocols (beyond the MGL modes just sketched), which refine the MGL ideas and provide tailored lock modes for high concurrency in XML trees [9]. To develop true DOM-based XML lock protocols, we introduced a far richer set of locking concepts, beyond simple intention locks and, in our terms, subtree locks. We differentiate read and write operations thereby renaming the well-known (IR, R) and (IX, X) lock modes with (IR, SR) and (IX, SX) modes, respectively. We introduced new lock modes for *single nodes* called NR (node read) and NX (node exclusive), and for *all siblings under a parent* called LR (level read). As in the MGL scheme, the U mode (SU in our protocol) plays a special role, because it permits lock conversion. The novelty of the NR and LR modes is that they allow, in contrast to MGL, to read-lock only a node or all nodes at a level (under the same parent), but not the corresponding subtrees.

To enable transactions to traverse paths in a tree having (levels of) nodes already read-locked by other transactions and to modify subtrees of such nodes, a new intention mode CX (child exclusive) had to be defined for a context (parent) node. It indicates the existence of an SX or NX lock on some direct child nodes and prohibits inconsistent locking states by preventing LR and SR locks. It does not prohibit other CX locks on a context node c, because separate child nodes of c may be exclusively locked by other transactions (compatibility is then decided on the child nodes themselves). Altogether these new lock modes enable serializable transaction schedules with read operations on inner tree nodes, while concurrent updates may occur in their subtrees.[2] An important and unique feature (not applicable in MGL or other protocols) is the optional variation of the *lock depth* which can be dynamically controlled by a parameter. Lock depth n determines that, while navigating through the document, individual locks are acquired for existing nodes up to level n. If necessary, all nodes below level n are locked by a subtree lock (SR, SX) at level n.

Continuous improvement of these basic concepts led to a whole family of lock protocols, the taDOM family, and finally resulted in a highly optimized protocol

---

[2] Although edge locks [9] are an integral part of taDOM, too, they do not contribute specific implementation problems and are, therefore, not considered here.

called taDOM3+ (tailor-made for the operations of the DOM3 standard [3]), which consists of 20 different lock modes and "squeezes transaction parallelism" on XML document trees to the extent possible. Correctness and, especially, serializability of the taDOM protocol family was shown in [9,18].

Let us highlight by three scenarios taDOM's flexibility and tailor-made adaptations to XML documents as compared to competitor approaches. Assume transaction *T1* – after having set appropriate intention locks on the path from the root – wants to read-lock context node *cn*. Independently of whether or not *T1* needs subtree access, MGL only offers a subtree lock on *cn*, which forces concurrent writers (*T2* and *T3* in Fig. 1a) to wait for lock release in a lock request queue (LRQ). In the same situation, node locks (NR and NX) would allow greatly enhance permeability in *cn*'s subtree (Fig. 1b). As the only lock granule, however, node locks would result in excessive lock management cost and catastrophic performance behavior, especially for subtree deletion [8]. A frequent XML read scenario is scanning of *cn* and all its children, which taDOM enables by a single lock with special mode (LR). As sketched in Fig. 1c, LR supports write access to deeper levels in the tree. The combined use of node, level, and subtree locks gives taDOM its unique capability to tailor and minimize lock granules. Above these granule choices, additional flexibility comes from lock-depth variations on demand – a powerful option only provided by taDOM.

1.2   Related Work and Our Own Contribution

To the best of our knowledge, we are not aware of contributions in the open literature dealing with implementation of an XML lock manager. So far, most publications just sketch ideas of specific problem aspects and are less compelling and of limited expressiveness, because they are not implemented and, hence, cannot provide empirical performance results [4,14,15]. Four Natix lock protocols [10] focus on DOM operations, provide node locks only, and do not enable direct jumps to inner document nodes and effective escalation mechanisms for large documents. Together with four MGL implementations supporting node and subtree locks, their lock protocol performance was empirically compared against our taDOM protocols [8]. The taDOM family exhibited for a given benchmark throughput gains of 400% and 200% compared to the Natix resp. MGL protocols which clearly confirmed that availability of node, level, and subtree locks together with lock modes tailored to the DOM operations pays off.

While these publications only address ideas and concepts, no contribution is known how to effectively and efficiently implement lock protocols on XML trees. Therefore, we start in Sect. 2 with implementation and processing cost of a lock manager. In Sect. 3, we emphasize the need and advantage of prefix-based node labeling for efficient lock management. Sect. 4 outlines how we coped with runtime shortcomings of protocol execution, before effectiveness and success of tailored optimizations are demonstrated by a variety of experimental results in Sect. 5. Finally, Sect. 6 summarizes our conclusions.

2   Lock Manager Implementation

Without having a reference solution, the XTC project had to develop such a component from scratch where the generic guidelines given in [6] were used. Because we need to synchronize objects of varying types occurring at diverse system layers (e.g., pinning pages by the buffer manager and locking XML-related objects such as nodes and indexes), which exhibit incomparable lock compatibilities, very short to very long lock durations, as well as differing access frequencies, we decided to provide specialized lock tables for them (and not a common one). Where appropriate, we implemented lock tables using suitable lock identification (see node labeling scheme, Sect. 3) and dynamic handling of lock request blocks and queues. Lock request scheduling is centralized by the lock manager. The actions for granting a lock are outlined below. Otherwise, the requesting transaction is suspended until the request can be granted or a time-out occurs. Detection and resolution of deadlocks is enabled by a global wait-for graph for which the transaction manager initiates the so-called transaction patrol thread in uniform intervals to search for cycles and, in case of a deadlock, to abort the involved transaction owning the fewest locks.

2.1   Lock Services

Lock management internals are encapsulated in so-called lock services, which provide a tailored interface to the various system components, e.g., for DB buffer management, node locks, index locks, etc. [9]. Each lock service has its own lock table with two pre-allocated buffers for lock header entries and

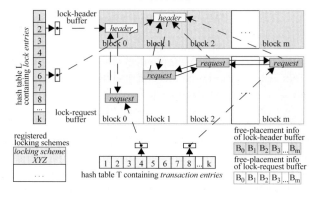

Fig. 2. Data structures of a lock table

lock request entries, each consisting of a configurable number (m) of blocks, as depicted in Fig. 2. This use of separate buffers serves for storage saving (differing entry sizes are used) and improved speed when searching for free buffer locations and is supported by tables containing the related free-placement information. To avoid frequent blocking situations when lock table operations (look-up, insertion of entries) or house-keeping operations are performed, use of a single monitor is not adequate. Instead, latches are used on individual hash-table entries (in hash tables T (for transactions) and L (for locks)) to protect against access by concurrent threads thereby guaranteeing the maximum parallelism possible. For each locked object, a lock header is created, which contains name and current mode of the lock together with a pointer to the lock queue where all lock requests for the object are attached to. Such a lock request carries among administration information the requested/granted lock mode together with the transaction ID. To speed-up lock release, the lock request entries are doubly chained and contain a separate pointer to the lock header, as shown in Fig. 2. Further, a transaction entry contains the anchor of a chain threading all lock request entries, which minimizes lock release effort at transaction commit.

To understand the general principles, it is sufficient to focus on the management of node locks. A lock request of transaction $T1$ for a node with label $ID1$ proceeds as follows. A hash function delivers $hT(T1)$ in hash table T. If no entry is present for $T1$, a new transaction entry is created. Then, $hL(ID1)$ selects (possibly via a synonym chain) a lock entry for node $ID1$ in hash table L. If a lock entry is not found, a lock header is created for it and, in turn, a new lock request entry; furthermore, various pointer chains are maintained for both entries. The lock manager enables protocol adaptation to different kinds of workloads by providing a number of registered lock schemes [9]. For checking lock compatibility or lock conversion, a pre-specified lock scheme is used.

2.2 Cost of Lock Management

Lock management for XML trees is hardly explored so far. It considerably differs from the relational multi-granularity locking, the depth of the trees may be much larger, but more important is the fact that operations may refer to tree nodes whose labels – used for lock identification – are not delivered by the lock request. Many XML operations address nodes somewhere in subtrees of a document and these often require direct jumps "out of the blue" to a particular inner tree node. Efficient processing of all kinds of language models [3,19] implies such label-guided jumps, because scan-based search should be avoided for direct node access and navigational node-oriented evaluation (e. g., getElementById() or getNextSibling()) as well as for set-oriented evaluation of declarative requests (e.g., via indexes). Because each operation on a context node requires the appropriate isolation of its path to the root, not only the node itself has to be locked in a sufficient mode, but also the corresponding intention locks on all ancestor nodes have to be acquired. Therefore, the lock manager often has to procure the labels for nodes and their contexts (e.g., ancestor paths) requested. No matter what labeling scheme is used, document access cannot always be avoided (e.g., getNextSibling()). If label detection or identification, however, mostly need document access (to disk), dramatic overhead burdens concurrency control. Therefore, node labeling may critically influence lock management cost.

In a first experiment, we addressed the question how many lock requests are needed for frequent use cases and what is the fraction of costs that can be attributed to the labeling scheme. For this purpose, we used the xmlgen tool of the XMark benchmark project [16] to

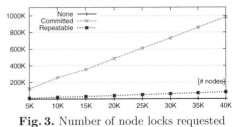

Fig. 3. Number of node locks requested

generate a variety of XML documents consisting of 5,000 up to 40,000 individual XML nodes. These nodes are stored in a B*-tree – a set of doubly chained pages as document container (the leaves) and a document index (the inner pages) – and reconstructed by consecutive traversal in depth-first order (i.e., document order corresponds to physical order) within a transaction in single-user mode.

To explore the performance impact of fine-grained lock management, we have repeated this experiment under various isolation levels [6]. Furthermore, we have reconstructed the document twice to amplify the differing behavior between isolation levels *committed* and *repeatable read* (in this setting, *repeatable* is equivalent to *serializable*). Because of the node-at-a-time locking, such a traversal is very inefficient, indeed, but it drastically reveals the lock management overhead for single node accesses. Depending on the position of the node to be locked, *committed* may cause much more locking overhead, because each individual node access acquires short read locks on all nodes along its ancestor path and their immediate release after the node is delivered to the client. In contrast, isolation level *repeatable read* sets long locks until transaction commit and, hence, does not need to repetitively lock and unlock ancestor nodes. In fact, they are already locked due to the depth-first traversal. Fig. 3 summarizes the number of individual node locks requested for the various isolation levels.

In our initial XTC version, we had implemented SEQIDs (including a level indicator) where node IDs were sequentially assigned as integer values. SEQIDs allow for stable node addressing, because newly inserted nodes obtain a unique, ascending integer ID. Node insertions and deletions preserve the document order and the relationships to already existing nodes. Hence, the relationship to a parent, sibling, or child can be determined based on their physical node position in the document container, i.e., within a data page or neighbored pages. While access to the first child is cheap, location of parent or sibling may be quite expensive depending on the size of the current node's subtree. Because intention locking requires the identification of all nodes in the ancestor path, this crude labeling scheme frequently forces the lock manager to locate a parent in the stored document. Although we optimized SEQID-based access to node relatives by so-called on-demand indexing, the required lock requests (Fig. 3) were directly translated into pure lock management overhead as plotted in Fig. 4a. Hence, the unexpectedly bad and even "catastrophic" traversal times caused a rethinking and redesign of node labeling in XTC (see Sect. 3).

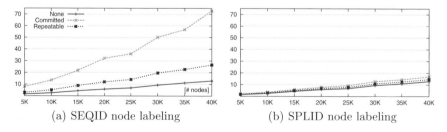

(a) SEQID node labeling   (b) SPLID node labeling

Fig. 4. Documents traversal times (sec.)

2.3   Lower Isolation Levels Are Not Always Superior

As compared to *repeatable*, isolation level *committed* provides for higher degrees of concurrency with (potentially) lesser degrees of consistency concerning read/write operations on shared documents. Hence, if chosen for a transaction program, the programmer must be carefully consider potential side-effects, because he accepts responsibility (because of early releases and later re-acquisitions of the same locks) to achieve full consistency. As shown in Fig. 4, *committed* may cause higher lock management overhead at the system side. Nevertheless, the programmer expects higher transaction throughput – as always obtained for isolation level *committed* in relational systems – compensating for his extra care.

In a dedicated experiment, we went into this matter whether or not the potentially high locking overhead for isolation level *committed* can be compensated by reduced blocking in multi-user mode. For this scenario, we set up a benchmark with three client applications on separate machines and an XDBMS instance on a forth machine. The clients are executing for a fixed time interval a constant load of over 60 transactions on the server. The workload – repeatedly executed for the chosen isolation levels and the different lock depths – consisted of about 16 short transaction types with an equal share of reader and writer transactions, which processed common access patterns like node-to-node navigation, child and descendant axes evaluation, node value modifications, and fragment deletions.

Fig. 5a shows the results of this benchmark run. Isolation level *none* means that node and edge locks are not acquired at all for individual operations. Of course, processing transactions without isolation is inapplicable in real systems, because the atomicity property of transactions (in particular the transaction rollback) cannot be guaranteed. Here, we use this mode only to derive the upper bound for transaction throughput in a given scenario. Isolation level *repeatable* acquires read and write locks according to the lock protocol and lock depth used, whereas *committed* requires write locks but only short read locks. Note, *committed* leads to a fewer number of successful transactions than the stronger isolation level *repeatable* – and obtains with growing lock depth (and, hence, reduced conflict probability) an increasing difference. Although less consistency guarantees are given in mode *committed* to the user, the costs of separate acquisitions and immediate releases of entire lock paths for each operation reduced the transaction throughput. Running short transactions, this overhead may not be amortized by higher concurrency.

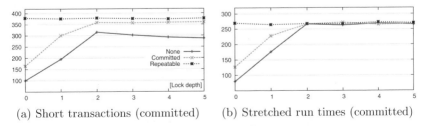

(a) Short transactions (committed)   (b) Stretched run times (committed)

Fig. 5. Transaction throughput controlled by different isolation levels

After we had artificially increased the run times of the same transactions by programmed delays (in this way simulating human interaction), of course, the overall transaction throughput decreases, but meets the expectations of traditional transaction processing (see Fig. 5b): A lower degree of isolation leads to a higher transaction throughput. If few access conflicts are occurring (lock depths 2 and higher), lock management costs hardly influence the "stretched" transaction durations and transaction throughput is decoupled from the chosen isolation level. Hence, higher throughput on XML trees is not given for granted using isolation level *committed*. Surprisingly, *committed* seems to be inappropriate for large lock depths and short transactions.

3   Prefix-Based Node Labeling Is Indispensable

Range-based and prefix-based node labeling [2] are considered the prime competitive methods for implementation in XDBMSs. A comparison and evaluation of those schemes in [7] recommends prefix-based node labeling based on the Dewey Decimal Classification. Each label represents the path from the document's root to the related node and the local order w.r.t. the parent node. Some schemes such as OrdPaths [13], DeweyIDs, or DLNs [7] provide immutable labels by supporting an overflow technique for dynamically inserted nodes. Here, we use the generic name *stable path labeling identifiers* (SPLIDs) for them.

Whenever a node, e.g., with SPLID 1.19.7.5, has to be locked, all its ancestor node labels are needed for placing intention locks on the entire path up to the root. Hence, they can be automatically provided: 1.19.7, 1.19, and 1 for the example. Because such lock requests occur very frequently, the use of SPLIDs is *the key argument* for locking support.[3] Referring to our lock table (see Fig. 2), intention locks for the ancestors of 1.19.7.5 can be checked or newly created using hL(1.19.7), hL(1.19) and hL(1) without document access. Because of the frequency of this operation, we provide a function which acquires a lock and all necessary intention locks at a time. A second *important property* for stable lock management is the immutability of SPLIDs, i.e., they allow the assignment of new IDs without the need to reorganize the IDs of nodes present.

We have repeated document traversal using SPLID-based lock management (see Fig. 4b). Because the difference between *none* and *committed/repeatable* is caused by locking overhead, we see drastic performance gains compared to SEQIDs. While those are responsible for an up to ~600% increase of the reconstruction times in our experiment, SPLIDs keep worst-case locking costs in the range of ~10 – ~20%. SEQIDs have fixed length, whereas SPLIDs require handling of variable-length entries. Coping with variable-length fields adds some complexity to SPLID and B*-tree management. Furthermore, prefix-compression of SPLIDs is a must [7]. Nevertheless, reconstruction time remained stable when SPLIDs were used – even when locking was turned off (case *none*).

---

[3] Range-based schemes [21] cause higher locking overhead than SPLIDs. They enable the parent label computation, but not those of further ancestors. An optimization would include a parent label index to compute that of the grandparent and so on.

Comparison of document reconstruction in Fig. 4a and b reveals for identical XML operations that the mere use of SPLIDs (instead of SEQIDs) improved the response times by a factor of up to 5 and more. This observation may convince the reader that node labeling is of utmost importance for XML processing. It is not only essential for internal navigation and set-based query processing, but, obviously, also most important for lock manager flexibility and performance.

4   Further Performance Drivers

Every improvement of the lock protocol shifts the issue of multi-user synchronization a bit more from the level of logical XML trees down to the underlying storage structures, which is a B*-tree in our case. Hence, an efficient and scalable B*-tree implementation in an *adjusted infrastructure* is mandatory.

D1: B*-tree Locking. Our initial implementation revealed several concurrency weaknesses we had to remove. First, tree traversal locked all visited index pages to rely on a stable ancestor path in case of leaf page split or deletion. Thus, update operations lead to high contention. Further, the implemented page access protocol provoked deadlocks under some circumstances. Although page locking itself was done by applying normal locks of our generic lock manager, where deadlocks could be easily detected and resolved, they had a heavy effect on the overall system performance. Thus, we re-implemented our B*-tree to follow the ARIES protocol [12] for index structures, which is completely deadlock-free and can therefore use cheap latches (semaphores) instead of more expensive locks. Further, contention during tree traversals is reduced by *latch coupling*, where at most a parent page and one of its child pages are latched at the same time.

D2: Storage Manager. Navigational performance is a crucial aspect of an XML engine. A B*-tree-based storage layout, however, suffers from indirect addressing of document nodes, because every navigation operation requires a full root-to-leaf traversal, which increases both computational overhead and page-level contention in the B*-tree. Fortunately, navigation operations have high locality in B*-tree leaves, i.e., a navigation step from a context node to a related node mostly succeeds in locating the record in the same leaf page. We exploit this property, by remembering the respective leaf page and its version number for nodes accessed as a hint for future operations. Each time when re-accessing the B*-tree for a navigation operation, we use this information to first locate the leaf page of the context node. Then, we quickly inspect the page to check if we can directly perform the navigation in it, i.e., if the record we are looking for is definitely bound to it. Only if this check fails, we have to perform a full root-to-leaf traversal of the index to find the correct leaf. Note, such an additional page access is also cheap in most cases, because the leaf page is likely to be found in the buffer due to locality of previous references.

D3: Buffer Manager. As shown in [7], prefix-compression of SPLIDs is very effective to save storage space when representing XML documents in B*-trees. As with all compression techniques, however, the reduced disk I/O must be paid with higher costs for encoding and decoding of compressed records. With page-wide prefix compression as in our case, only the first record in a page is guaranteed to be fully stored. The reconstruction of any subsequent entry potentially requires to decode all of its predecessors in the same page. Accordingly, many entries will have to be decoded over and over again, when a buffered page is frequently accessed. To avoid this unnecessary decoding overhead and to speed up record search in a page, we enabled buffer pages to carry a cache for already decoded entries. Using page latches, the page-local cache may be accessed by all transactions and does not need further considerations in multi-user environments. Although such a cache increases the actual memory footprint of a buffered disk page, it pays off when a page is accessed more than once – the usual case, e.g., during navigation. Further, it is a non-critical auxiliary structure that can be easily shrinked or dropped to reclaim main memory space.

A second group of optimizations was concerned with XML lock protocols for which empirical experiments identified *lock depth* as the most performance-critical parameter (see Sect. 1.1). Choosing lock depth 0 corresponds to document-only locks. In the average, growing lock depth refines lock granules, but enlarges administration overhead, because the number of locks to be managed increases. But, conflicting operations often occur at levels closer to the document root (at lower levels) such that fine-grained locks at levels deeper in the tree do not always pay off. A general reduction of the lock depth, however, would jeopardize the benefits of our tailored lock protocols.

D4: Dynamic Lock Depth Adjustment. Obviously, optimal lock depth depends on document properties, workload characteristics, and other runtime parameters like multiprogramming level, etc., and has to be steadily controlled and adjusted at runtime. Therefore, we leveraged *lock escalation/deescalation* as the most effective solution: The fine-grained resolution of a lock protocol is – preferably in a step-wise manner – reduced by acquiring coarser lock granules (and could be reversed by setting finer locks, if the conflict situation changes). Applied to our case, we have to dynamically reduce lock depth and lock subtrees closer to the document root using single subtree locks instead of separate node locks for each descendant visited. Hence, transactions initially use fine lock granules down to high lock depths to augment permeability in hot-spot regions, but lock depth is dynamically reduced when low-traffic regions are encountered to save system resources. Using empirically proven heuristics for conflict potential in subtrees, the simple formula $threshold = k * 2^{-level}$ delivered escalation thresholds, which takes into account that typically fanout and conflicts decrease with deeper levels. Parameter k is adjusted to current workload needs.

D5: Avoidance of Conversion Deadlocks. Typically, deadlocks occurred when two transactions tried to concurrently append new fragments under a node already read-locked by both of them. Conversion to an exclusive lock involved both transactions in a deadlock. Update locks are designed to avoid such conversion deadlocks [6]. Tailored to relational systems, they allow for a direct upgrade to exclusive lock mode when the transaction decides to modify the current record, or for a downgrade to a shared lock when the cursor is moved to the next record without any changes. Transactions in XDBMS do not follow such easy access patterns. Instead, they often perform arbitrary navigation steps in the document tree, e.g., to check the content child elements, before modifying a previously visited node. Hence, we carefully enriched our access plans with hints when to use update locks for node or subtree access.

5 Effects of Various Optimizations

We checked the effectivity of the infrastructure adjustments (D1, D2, and D3), before we focused on further XML protocol optimizations (D4 and D5). Based on our re-implemented B*-tree version with the ARIES protocol (D1), we verified the performance gain of navigation optimizations for D2 and D3. We stored an 8MB XMark document in a B*-tree with 8K pages and measured the average execution time of the dominating operations *FirstChild* and *NextSibling* during a depth-first document traversal. We separated the execution times for each document level, because locality is potentially higher at deeper levels.

The results in Fig. 6a and b confirm that the optimizations of D2 and D3 help to accelerate navigation operations. With speed-ups of roughly 70% for all non-root nodes, the benefit of both is nearly the same for the *FirstChild* operation. The fact that the use of cached page entries results even in a slightly higher performance boost than the drastic reduction of B*-tree traversals through the use of page hints, underlines the severeness of repeated record decoding. The actual depth of the operation does not play a role here. In contrast, the page-hint optimization shows a stronger correlation to the depth of a context node. As expected, page hints are less valuable for the *NextSibling* operation at lower levels, because the probability that two siblings reside in the same leaf page is lower. For depths higher than 2, however, this effect completely disappears. For the whole traversal, the hit ratio of the page hints was 97.88%. With documents of other size and/or structure, we achieved comparable or even higher hit ratios. Even for a 100MB XMark document, e.g., we still obtained a global hit ratio of 93.34%. With the combination of both optimizations, we accomplished a performance gain in the order of a magnitude for both operation types.

To examine and stress-test the locking facilities with the lock protocol optimizations of D4 and D5, situations with a high blocking potential had to be provoked. We created mixes of read/write transactions, which access and modify

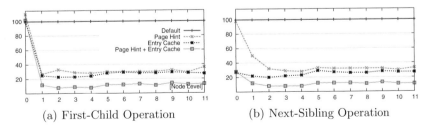

(a) First-Child Operation   (b) Next-Sibling Operation

Fig. 6. Relative execution times of navigation operations (%)

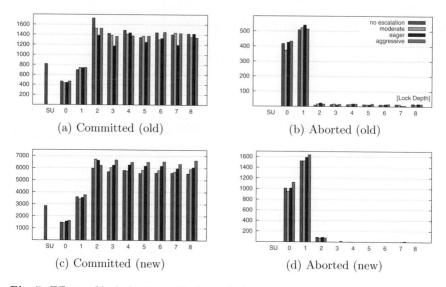

(a) Committed (old)

(b) Aborted (old)

(c) Committed (new)

(d) Aborted (new)

Fig. 7. Effects of lock depth and lock escalation on transaction throughput (tpm)

a generated XMark document at varying levels and in different granules [1]. We again chose an initial document size of only 8 MB and used a buffer size large enough for the document and auxiliary data structures. Further details of the specific workloads are not important here, because we only aim at a differential performance diagnosis under identical runtime conditions. To get insight in the behavior of the lock-depth optimization D4, we measured the throughput of transactions per minute (tpm) and ran the experiments for three escalation thresholds (moderate, eager, aggressive) in single user mode (SU) and in multi-user mode with various initial lock depths (0–8).

To draw the complete picture and to reveal the dependencies to our other optimizations, we repeated the measurements with two XTC versions: XTC based on the old B*-tree implementation and XTC using the new B*-tree implementation together with the optimizations D2 and D3. To identify the performance gain caused by D1–D3, we focused on transaction throughput, i.e., commit and abort rates, and kept all other system parameters unchanged. Fig. 7 compares the experiment results. In single-user mode, the new version improves throughput by a factor of 3.5, which again highlights the effects of D2 and D3. The absence of deadlocks and the improved concurrency of the latch-coupling protocol in the B*-tree (D1) becomes visible in the multi-user measurements, where throughput speed-up even reaches a factor of 4 (Fig. 7a and c) and the abort rates almost disappear for lock depths > 2 (Fig. 7b and d).

Deadlocks induced by the old B*-tree protocol were also responsible for the fuzzy results of the dynamic lock depth adjustment (D4). With a deadlock-free B*-tree, throughput directly correlates with lock overhead saved and proves the benefit of escalation heuristics (Fig. 7c and d).

In a second experiment, we modified the weights of the transactions in the previously used mix to examine the robustness of the approach against shifts in

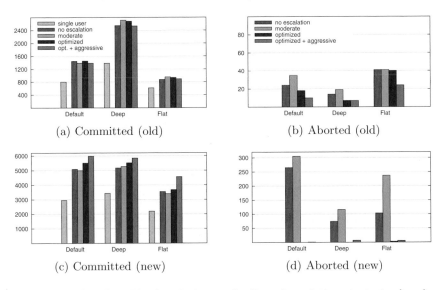

Fig. 8. Results of workload variations and adjusted escalation strategies (tpm)

the workload characteristics. In the default workload, the document was modified both in lower and deeper levels. In contrast, the focus in the two other workloads is on nodes at deeper (deep) and lower (flat) levels, respectively. Additionally, we provided optimized, update-aware variants of the transaction types to examine the effect of careful use of update locks. For simplicity, we ran the multi-user measurements only with initial lock depth 8.

The results in Fig. 8 generally confirm the observations of the previous experiment. But, throughput comparison between old and new B*-tree variants attest the new one a clearly better applicability for varying workloads. The value of update locks is observable both in throughput and in abort rates. The optimized workloads are almost completely free of node-level deadlocks, which directly pays off in higher throughput. Here, the performance penalty of page-level deadlocks in the old B*-tree becomes particularly obvious. Further, the results show that our lock protocol optimizations by the performance drivers D4 and D5 complement each other, similar to the infrastructure optimizations D1–D3.

6   Conclusions

In this paper, we outlined the implementation of XML locking, thereby showing that the taDOM family is perfectly eligible for fine-grained transaction isolation on XML document trees. We disclosed lock management overhead and emphasized the performance-critical role of node labeling, in particular, for acquiring intention locks on ancestor paths. In the course of lock protocol optimization, we have revealed the real performance drivers: adjusted measures in the system infrastructure and flexible options of the lock protocols to respond to the workload characteristics present. All performance improvements were substantiated by numerous measurements in a real XDBMS and under identical runtime conditions which enabled performance comparisons of utmost accuracy – not reachable by comparing different systems or running simulations.

References

1. Bächle, S., Härder, T.: Implementing and Optimizing Fine-Granular Lock Management for XML Document Trees. In: Proc. DASFAA Conf., Brisbane (2009)
2. Christophides, W., Plexousakis, D., Scholl, M., Tourtounis, S.: On Labeling Schemes for the Semantic Web. In: Proc. 12th Int. WWW Conf., pp. 544–555 (2003)
3. Document Object Model (DOM) Level 2 / Level 3 Core Specification. W3C Recommendation (2004), http://www.w3.org/TR/DOM-Level-3-Core
4. Grabs, T., Böhm, K., Schek, H.-J.: XMLTM: Efficient transaction management for XML documents. In: Proc. CIKM Conf., pp. 142–152 (2002)
5. Graefe, G.: Hierarchical locking in B-tree indexes. In: Proc. German Database Conference (BTW 2007), LNI P-65, p. 18–42 (2007)
6. Gray, J., Reuter, A.: Transaction Processing: Concepts and Techniques. Morgan Kaufmann, San Francisco (1993)
7. Härder, T., Haustein, M.P., Mathis, C., Wagner, M.: Node Labeling Schemes for Dynamic XML Documents Reconsidered. Data & Knowl. Eng. 60(1), 126–149 (2007)
8. Haustein, M.P., Härder, T., Luttenberger, K.: Contest of XML Lock Protocols. In: Proc. VLDB Conference, Seoul, pp. 1069–1080 (2006)
9. Haustein, M.P., Härder, T.: Optimizing lock protocols for native XML processing. Data & Knowl. Eng. 65(1), 147–173 (2008)
10. Helmer, S., Kanne, C.-C., Moerkotte, G.: Evaluating Lock-Based Protocols for Cooperation on XML Documents. SIGMOD Record 33(1), 58–63 (2004)
11. Loeser, H., Nicola, M., Fitzgerald, J.: Index Challenges in Native XML Database systems. In: Proc. German Database Conf. (BTW), LNI (2009)
12. Mohan, C.: ARIES/KVL: A key-value locking method for concurrency control of multiaction transactions operating on B-tree indexes. In: Proc. VLDB Conf., pp. 392–405 (1990)
13. O'Neil, P., O'Neil, E., Pal, S., Cseri, I., Schaller, G., Westbury, N.: OrdPaths: Insert-Friendly XML Node Labels. In: Proc. SIGMOD Conf., pp. 903–908 (2004)
14. Pleshachkov, P., Chardin, P., Kuznetsov, S.O.: XDGL: XPath-based concurrency control protocol for XML data. In: Jackson, M., Nelson, D., Stirk, S. (eds.) BNCOD 2005. LNCS, vol. 3567, pp. 145–154. Springer, Heidelberg (2005)
15. Sardar, Z., Kemme, B.: Don't be a Pessimist: Use Snapshot based Concurrency Control for XML. In: Proc. Int. Conf. on Data Engineering, p. 130 (2006)
16. Schmidt, A., Waas, F., Kersten, M.L., Carey, M.J., Manolescu, I., Busse, R.: XMark: A Benchmark for XML Data Management. In: Proc. VLDB Conf., pp. 974–985 (2002)
17. Schöning, H.: Tamino – A DBMS designed for XML. In: Proc. Int. Conf. on Data Engineering, pp. 149–154 (2001)
18. Siirtola, A., Valenta, M.: Verifying parameterized taDOM+ lock managers. In: Geffert, V., Karhumäki, J., Bertoni, A., Preneel, B., Návrat, P., Bieliková, M. (eds.) SOFSEM 2008. LNCS, vol. 4910, pp. 460–472. Springer, Heidelberg (2008)
19. XQuery Update Facility, http://www.w3.org/TR/xqupdate
20. XML on Wall Street, Financial XML Projects, http://lighthouse-partners.com/xml
21. Yu, J.X., Luo, D., Meng, X., Lu, H.: Dynamically Updating XML Data: Numbering Scheme Revisited. World Wide Web 8(1), 5–26 (2005)

XTaGe: A Flexible Generation System for Complex XML Collections

María Pérez, Ismael Sanz, and Rafael Berlanga

Universitat Jaume I, Spain
{mcatalan,isanz,berlanga}uji.es

Abstract. We introduce XTaGe (XML Tester and Generator), a system for the synthesis of XML collections meant for testing and micro-benchmarking applications. In contrast with existing approaches, XTaGe focuses on complex collections, by providing a highly extensible framework to introduce controlled variability in XML structures. In this paper we present the theoretical foundation, internal architecture and main features of our generator; we describe its implementation, which includes a GUI to facilitate the specification of collections; we discuss how XTaGe's features compare with those in other XML generation systems; finally, we illustrate its usage by presenting a use case in the Bioinformatics domain.

1 Introduction

Testing is an essential step in the develoment of XML-oriented applications and in most practical settings, this requires the creation of synthetic data.

Existing XML generators focus on either the creation of collections of a given size (for stress testing and workload characterization purposes) or with a fixed schema and little variation. These systems do not suit the requirements of an emerging class of important applications in fields such as Bioinformatics and GIS, which have to deal with large collections that present complex structural features, and specialized content such as protein sequences or vectorial map data.

In this context, the main drawback of existing systems in our application context is the lack of extensibility, since all systems are limited by the support of a limited number of predefined generation primitives. Another limitation is the uneven support for the introduction of *controlled* variability in generated structures, useful for example for micro-benchmarking purposes. Finally, the specification of collections is generally done through the manual creation of a text-based specification file, which can be tedious and error-prone.

In this paper we introduce XTaGe (XML Tester and Generator), which focuses on the creation of collections with complex structural constraints and domain-specific characteristics. XTaGe contributes (i) a flexible component-based framework to create highly tailored generators, (ii) a ready-made set of components that model common patterns that arise in complex collections, (iii) easy adaptability to new use cases using a high-level language (XQuery itself) The resulting system

S.S. Bhowmick, J. Küng, and R. Wagner (Eds.): DEXA 2009, LNCS 5690, pp. 53–67, 2009.
© Springer-Verlag Berlin Heidelberg 2009

makes it possible to generate test collections whose characteristics would be very difficult, or impossible, to replicate using the existing generic XML generators.

Related Work. As indicated above, current approaches for generating synthetic XML data can be classified as either *schema-unaware* or *template-based*. The former are based on the specification of a few global structural characteristics, such as maximum depth, fan-out and amount of textual data. They are commonly used in benchmarking applications. Examples include Niagdatagen [2], *xmlgen* (developed for the XMark [16]) and *genxml* (used in the X007 Benchmark [5]).

In contrast, template-based generators use as input an annotated schema that precisely describe the desired structure of the output documents. The best-known example is ToXgene [4], which defines an extension of XML Schema, the *Template Specification Language* (TSL) to describe the generated XML document content. It has some support for generating variability through the use of probability distributions to create heterogeneous structures. Many benchmarks applications, such as [17] and [14], generate their testing collections using ToXgene. Other examples of template-based generators are: VeXGene [10], MeMBeR [3] and [6], which bases the XML data generation on a DTD and examples of XML instances and support the specifications of constraints of the generated collection, such required XPath expressions.

As a special case, other approaches attempt to create new collections by transforming existing ones, such as [7], which can adapt existing documents for experiments meant to evaluate semantic query optimization methods; they provide a set of four transformations to adapt existing XML documents. Another relevant system is [15], which can modify the content of XML documents by creating duplicates or by removing content of the documents in order to create "dirty" documents suitable for testing data cleaning algorithms.

Outline of the Paper. The remainder of this paper is structured as follows. First, the foundations of XTaGe are presented in Section 2. Then, the XTaGe component-based framework is described in detail in Section 3. In Section 4 we present the prototype and a use case that shows an application of the generation model. Finally, Section 5 presenta a short discussion of XTaGe's features and introduces directions of future work.

2   Foundations of XTaGe

One of the main goals of XTaGe is to provide precise control of the generated data when creating heterogeneous collections and, as a consequence, we will use template-based techniques as a basis for our approach. In this section we provide a formal basis for the definition of XML generators, which will allow us to define a flexible mechanism for the creation and adaptation of XML generators for complex domains.

We adopt the *XML Store* [9] as a suitable abstraction of XML collections, which is commonly used in the context of XML update languages. Following [8], we will use the following notations: the set $\mathcal{A}$ denotes the set of all atomic values, $\mathcal{V}$ is the set of all nodes, $\mathcal{S} \subseteq \mathcal{A}$ is the set of all strings, and $\mathcal{N} \subseteq \mathcal{S}$ is the set of strings that may be used as tag names. The set $\mathcal{V}$ is partitioned into the sets of document nodes ($\mathcal{V}^d$), element nodes ($\mathcal{V}^e$), attribute nodes ($\mathcal{V}^a$), and text nodes ($\mathcal{V}^t$).

Definition 1 (XML Store). *An* XML store *is a 6-tuple* $St = (V, E, <, \nu, \sigma, \delta)$ *where:*

- *V is a finite subset of $\mathcal{V}$; we write V^d for $V \cap \mathcal{V}^d$ (resp. V^e for $V \cap \mathcal{V}^e$, V^a for $V \cap \mathcal{V}^a$, V^t for $V \cap \mathcal{V}^t$);*
- *(V, E) is an acyclic directed graph (with nodes V and directed edges E) where each node has an in-degree of at most one, and hence it is composed of trees; if $(m, n) \in E$ then we say that n is a child of m; we denote by E^* the reflexive transitive closure of E;*
- *$<$ is a strict partial order on V that compares exactly the different children of a common node. Hence for two distinct nodes n_1 and n_2 it holds that $((n_1 < n_2) \vee (n_2 < n_1)) \Leftrightarrow \exists m \in V((m, n_1) \in E \wedge (m, n_2) \in E))$*
- *$\nu : V^e \cup V^a \to \mathcal{N}$ labels the element and attribute nodes with their node name*
- *$\sigma : V^a \cup V^t \to \mathcal{S}$ labels the attribute and text nodes with their string value*
- *$\delta : \mathcal{S} \to \mathcal{V}^d$ a partial function that associates a document node with an URI or a file name. It is called the* document function. *This function represents all the URIs of the Web and all the names of the files, together with the documents they contain. We suppose that all these documents are in the store.*

The following properties must hold for an XML store: each document node of V^d is the root of a tree and has only one child element; attribute nodes of V^a and text nodes of V^t do not have any children; in the $<$-order attribute children precede the element and text children, i.e. if $n_1 < n_2$ and $n \in V^a$ then $n_1 \in V^a$; there are no adjacent text children, i.e. if $n_1, n_2 \in V^t$ and $n_1 < n_2$ then there is an $n_3 \in V^e$ with $n_1 < n_3 < n_2$; for all text nodes n_t of V^t holds $\sigma(n_t) \neq ""$; all the attribute children of a common node have a different name, i.e. if $(m, n_1), (m, n_2) \in E$ and $n_1, n_2 \in V^a$ then $\nu(n_1) \neq \nu(n)_2$.

Given an XML Store St we will use following auxiliary notations and functions:

- $V_{St}, E_{St}, \nu_{St}, \sigma_{St}$ and δ_{St} return the corresponding components of St. We also define V_{St}^d, V_{St}^e and V_{St}^a.
- $genDocNode()$, $genElement()$, $genAttribute()$ return members from $\mathcal{V}^d$, $\mathcal{V}^e$ and $\mathcal{V}^a$ which do not exist in V_{St}^d, V_{St}^e and V_{St}^a respectively. These functions are abstractions of the creation of new element document and attribute nodes. Note that text nodes will be generated by appropriately specific functions.
- $root(St)$ is the root node of the store.
- $descendants_{St}(n)$ is the set of all nodes in St which are descendants of n.

For simplicity, and without loss of generality, in the remainder of this paper we will ignore the partial order $<$, and we will not indicate the name of the XML Store when it is obvious from context.

2.1   Creating XML Documents from Scratch

We will model XML generators as functions that create XML Stores. Like in other schema-based systems, the generation will be based on the specification of a *base model*, whose expressivity must be the same of XML Schema languages ([13]).

Definition 2 (base model, generator, interpretation). *An* XTaGe *base model, M, is a tree that represents a* generating functional expression *(or generator, for short) f whose interpretation $Gen_M(f)$ is an XML store.*

A base model is, therefore, conceptually similar to the expression trees that appear when parsing programming languages. For example, consider the following operation:

Example 1. Given a generator model tree M, The *element(name)* component is a generating functional expression whose interpretation $Gen_M(element(name))$ generates a XML Store $(V, E, <, \nu, \sigma, \delta)$ such that:

- $newNode = genElement()$, that is, a new node
- $V = \{newNode\} \cup V_{Gen(c_i)}$ for each $c_i \in children_T(c)$
- $E = \{(newNode, root(Gen(c_i)))\} \cup (\bigcup E_{Gen(c_i)})$ for each $c_i \in children_T(c)$
- $\nu = \{(newNode, name)\} \cup (\bigcup \{(a, b) : (a, b) \in \nu_{Gen(c_i)}\})$ for each $c_i \in children_T(c)$
- $\delta = \bigcup \{(a, b) : (a, b) \in \delta_{Gen(c_i)}\}$ for each $c_i \in children_T(c)$
- $\sigma = \bigcup \{(a, b) : (a, b) \in \sigma_{Gen(c_i)}\}$ for each $c_i \in children_T(c)$

where $children_M(c)$ represents the children of node c in the generation model tree M. We treat the *attribute* nodes similarly.[1]

Figure 1 represents a simple generation tree that uses the *element* generator.

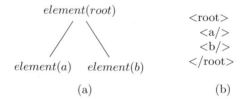

(a)   (b)

Fig. 1. A simple tree generation using the *element* generator component and the corresponding XML tree

In order to provide a functionality similar to basic XML Schema, we introduce the following generators that account for the possible content models:

- *sequence(name, attr, n, minOccurs, maxOccurs)*: A functional component that is a generalization of the previously introduced *element*, including support for attributes (*attr*) a number of repetitions (*n*) and cardinality constraints (*minOccurs* and *maxOccurs*).

---

[1] The main difference with the treatment of *element* nodes is that the *attribute* nodes cannot be nested.

- *choice(name, attr, n, minOccurs, maxOccurs)*: A functional equivalent of the XML Schema *choice* content model, represented by a bar (|) in DTDs.

The features of the generation model presented so far support the creation of XML documents based on a fixed XML-like schema. We now introduce two features which are specifically designed to introduce controlled variability in collections: *distributions* and *probability-labeled arcs*.

Value Distributions. In XTaGe, *every* value is extracted from a probability distribution, including constant numbers and strings (which are considered to be extracted from a suitably defined constant distribution). This includes:

- Parameters of functional components, for example the number of repetitions n in sequences or choices.
- Synthetic content in attributes or text nodes.

This allows us to easily express empirical properties of the generated data, such as that the number of children of a given node is normally distributed, or Zipf-like distribution for words in textual content.

Probability-Labeled Tree Arcs. Another mechanism to introduce controlled variability in generated XML is introducing the notion of probabilistic labeling. Each arc (u, v) in a generator tree M is labeled with a probability value $p((u, v)) \in [0, 1]$. The meaning of a probability-labeled arc is that the child functional expression v will be ignored with probability $1 - p((u, v))$.

Example 2. Figure 2 shows a simple generator tree that includes probability-labeled arcs, and some of the XML trees that could be generated by it.

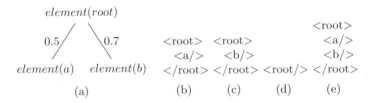

Fig. 2. A generator tree with probabilistic arcs and some possible generated trees

2.2 Transforming XML Documents

The second mechanism by which a new XML collection may be generated is by the controlled transformation of an existing one. Analogously to tree generation model trees, we introduce a *transforming functional expressions* (or *transformations*, in short). Moreover, these transformations provide the foundations required to model global contraints over XML documents. Since there are many transformation languages available for XML (XSLT, XDuce), we will focus in the introduction of controlled variability into XML collections through the definition of transforms.

Definition 3 (locator). *A locator is a function that takes an XML Store St and returns an XML Store St' such that:*

- $V'_{St} \subseteq V_{St}$
- $E'_{St} \subseteq E_{St}$
- *St' is well formed according to Definition 1*

Example 3. Given a locator and an XML Store St, the *delChild* transformation removes a random child of the root node of a tree with probability p, and returns a new XML Store St'. We can define it as follows:

- Choose a node $n \in children(root(l))$, where l is the subtree induced by the locator, with probability p.
- $V_{St'} = V_l \backslash (\{n\} \cup descendants_l(n))$
- $E_{St'} = E_l \backslash \{(u, v) \text{ such that } u, v \in \{n\} \cup descendants(n)\}$
- $\nu_{St'}$, $\delta_{St'}$ and $\sigma_{St'}$ are suitably modified.

Probability-labeled tree arcs are also used in transformation trees to determine if a transformation will be applied or not in the XML documents. It is important also to remark that transformations are applied on the original XML Store, and not on a "different" XML store induced by the locator, which must then be grafted on the original tree. This property allows us to define meta-transformation operations, which can be used to combine different transformations into complex operations. The main meta-transformation is the macro-transformation:

Definition 4 (macro-operation). *Given an XML Store St and list of (locator, transformation) pairs, the macro-operation $macro_{St}[(l_1, t_1), \dots, (l_n, t_n)]$ is defined as the sequential application of all pairs to St.*

3   Component-Based Framework

The concepts outlined in the previous section have been realized in XTaGe by means of a lightweight component-oriented software model, outlined in Figure 3. The use of components as an abstraction of the generation and transformation functional expressions has a number of benefits directly related to the goals of XTaGe.

First of all, it makes metadata about components and their relationships explicit. This facilitates greatly the construction of support tools such as GUIs, and it also allows the simplified creation of new components without requiring the modification of the XTaGe code. This can be accomplished by a simple two-stage process: (*i*) create the function in a high-level language (XQuery in our case, as we will explain presently) and (*ii*) register it by filling in the appropriate metadata. For the most complex cases (usually involving calling external libraries), new components can be coded by implementing the appropriate interface. Libraries of related components (e.g. for testing of biomedical data sets) may be put together and maintained independently.

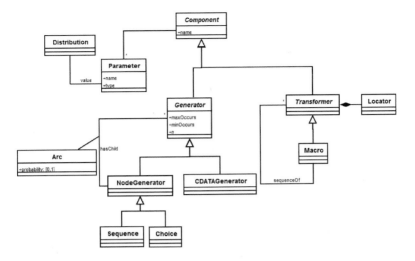

Fig. 3. Simplified UML diagram of the XTaGe component architecture

3.1 Generators in XTaGe

In addition to the basic XML Schema-related components described in Section 2, a number of components useful for the generation of collections with controlled variability are pre-defined:

XOR. This generator chooses one node between all its descendants according to their *xor probability* values. The descendants of a XOR constructor have an additional probability parameter, *xor probability*, which determines the likelihood of a node of being chosen by its parent.

Combi. This generator creates a new node whose tag is a combination of the tags of its descendants.

DminDmax. The functionality of this generator is creating a new node located n levels below its ancestor. The value of n depends on the values of the attributes *dmin* and *dmax* of the constructor. The value of n is a random number between *dmin* and *dmax*.

IfAncestor. This generator determines if a node appears in the new document depending on the tag of its ancestor.

An XML node generated by one of these components is assigned a unique id value, which can be specified by the user, automatically by the system, or by a user-defined function. Attributes (and IDREFs, which are treated as a special case) can be defined by the user in the generator model tree.

As mentioned above, the preferred way to create new components is by means of the creation of an XQuery functions. XQuery was chosen because it is inherently well-suite to define operations on XML trees. The function must conform to the following signature:

declare function *component−name*($comp **as** node()) **as** node()∗;

where $comp represents the component metainformation (serialized as XML), including its parameters.

To support the creation of components, a library of XQuery functions has been defined that permits access to the defined distributions and the structure of the model tree.

Example 4. The following XQuery function implements the component *Dmin Dmax*, using an auxiliary function to handle recursion:

declare function dmindmax($comp **as** node())) **as** node()∗ {
   let $dist := $comp/param/dist
   return dmindmax_aux($comp, $dist) }

declare function dmindmax_aux($comp **as** node()∗, $dist)) **as** node()∗ {
   if ($dist = 1)
   then xtg:create_node($comp)
   else(**let** $actual_dist := xs:integer($dist) − 1
          let $random_tag := xtg::randomTag()
          return element{$random_tag} {dmindmax_aux($comp, $actual_dist)})
}

where library function `xtg:create_node($comp)` creates an element based on the parameters of the component and `xtg:randomTag()` returns random strings.

Note the use of XPath to extract the value of the parameters from the component metadata object. Besides the `$comp` parameter, the component accepts `$dist`, that is a number obtained from a user-defined distribution that must lie between the values of the parameters *dmin* and *dmax* of the component.

3.2   Transformations in XTaGe

In order to apply controlled transformations, XTaGe includes a few pre-defined XML transformation components:

Add. This transformation component takes as input two XML trees, t_1 and t_2. The component adds t_2 to t_1 as a descendant of the node or nodes of t_1 determined by the component *locator*.

Delete. This component removes the node determined by the component *locator* of the XML document tree. The component has a parameter called *recursive*, whose value determines if the operation is executed recursively or not. If its value is 1, the node and all its descendants are removed; if its value is 0, only the node is removed and its descendants occupy its place. In case the nodes have references, XTaGe allows the user to specify whether the references must be automatically re-calculated.

Change Order. This component changes the order of a node and one of its siblings by changing their positions. The user has to determine the ancestor of the node that is going to be changed and, optionally, the node that is going to change its position. If the user does not specify this node, the component chooses a descendant of the *ancestor node* randomly. The user can also determine the new position of the node; in case the user does not specify it, the component determines randomly the new position.

Change Level. This component changes the position of a node and one of its descendants, chosen by the user or randomly. The descendant will be now the ancestor of its siblings and its ancestor will become one of its descendants.

XTaGe also allows the definition of new transformations using XQuery. The functions must conform to the following signature:

declare function *component−name*($context **as** node(), $locator **as** node()*,
$comp **as** node()) **as** node()*;

where $context represents the current context node in the source XML document, that defines where the transformation will be applied; $locator is the set of nodes induced by the locator; and $comp contains metainformation about the component.

Example 5. The XQuery function that implements the functionality of the component *Change_order* may be implemented as follows:

```
declare function change_order($context, $loc, $config) as node()* {
    let $newpos:= xtg:newPos($config)
    let $child:= xtg:child($context, $config)
    let $sibling:= xtg:sibling($context, $config)
    return element{fn:local−name($context)}
             {for $att in $context/@*
                 return attribute{fn:local−name($att)}{$att},
              for $c at $pos in $context/*
                 return if ($pos = $newpos)
                         then xtg:traverse($child, $context, $config)
                         else ( if($c is $child)
                                 then xtg:traverse($sibling, $context, $config)
                                 else xtg:traverse($c,$context, $config))}
}
```

where:

newPos($config) returns the new position of the node, if this value is not set in the parameter *NewPos* of the component Change_order, the function returns a random value in the range of [1, number of descendants of the *context* node].

child($context, $config) returns the element that is going to be changed. This node is retrieved by the *locator* component.

sibling($context, $config) returns the element that now occupies the position *newpos*.

traverse($context, $loc, $config) is a function that traverses the XML document, element by element.

This set of transformations does not allow the user to model global constraints but it is possible to create and add new components that support non-local contraints models.

4   Prototype and Use Cases

The next section presents a use case in the Bioinformatics domain in which XTaGe is applied. It addresses the problem of evaluating techniques based on the modification of specific characteristics of a XML collection.

4.1   Generating Controlled Testing Collections

We consider a problem of exploratory search of XML collections in the Bioinformatics domain. This domain is characterized by the existence of a great number of complex, large and heterogeneous XML data sources, which poses serious issues in data integration applications [11]. Usually, a first step in the design of these applications is the characterization of a sample of these collections, which requires the use of approximate querying techniques due to the lack of a schema and the presence of complex, domain-specific data such as protein sequences, which inherently require approximate matching algorithms.

Testing of these systems is difficult, since (i) they may not correspond well with the expected work load (ii) they might not exercise all possible structural variations that might appear in the production system or (iii) they may contain errors which need to be corrected [12].

In such a context, the features of XTaGe for the introduction of controlled variability in collections may facilitate the design process greatly. Consider the case of an application trying to integrate information coming from the complex BioPAX collection [1], which is derived from OWL specifications and exhibits an essentially free-form schema when translated into XML.

To account for the possible variations, XTaGe can be used to automatically generate test cases that can be used to check for unexpected structural variations. For example, Figure 4 shows an example of a user-defined schema that generates XML structures with BioPAX-derived information. Figure 5 shows different XML structures generated by this schema. Note how the structure of the documents varies due to the probabilities and the patterns of the components.

This is also useful to help assess the performance of the approximate techniques being used for data exploration, in particular in the presence of characteristics of interest. This calculation requires the generation of different versions of a same collection, each one exhibiting a different characteristic. To achieve this goal, XTaGe can be used to generate such a set of XML collections.

Figure 6 shows the steps to generate the new versions of the XML collection. The approach is based in a multi-step process. First, a "background collection" is determined; this can be synthetic, or a sample of existing. Next, to facilitate comparisons, a XML structure suitable for transformation is determined. A number of transformations are written, in order to exercise the different structural

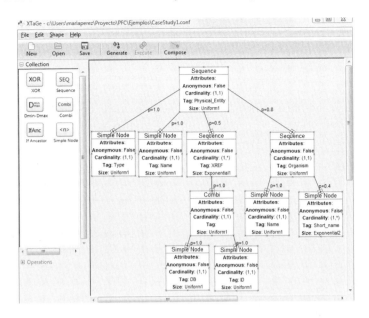

Fig. 4. An example of a user-defined schema. Note that the GUI shows the list of available components (left pane) which can be dragged and dropped to create the schema.

characteristics that should be tested (e.g. presence/absence of nodes or subtrees, changes in ordering, and so on). Finally, new versions of the background collections are created and subjected to testing.

Next we explain with further details the two main steps required to obtain the different versions of the background collection and we clarify them with an example.

Creating the New XML Structure. In this first phase the user has to define the schema of the XML fragment that, in the following phase, is going to be modified and finally, added to the background collection. The result of executing this schema is an XML document, which we call *synthetic XML document*, whose structure and tags are well-known by the user.

In our example, we have designed a XML document that contains information about publications, its schema is shown in Figure 7. Figure 8b shows the synthetic XML document generated by it.

Creating Synthetic Collections. In this phase, the user has to define the schemas of the transformations that are going to be executed in the synthetic XML document in order to exercise the different structural characteristics that should be tested. The goal is to create a set of n synthetic collections where each one differs from the others in a known characteristic. The steps to do that are the following:

1. Create n transformation schemas, one per each characteristic to be analyzed. Each schema is composed by a set of transformations whose execution modifies a specific characteristic of the synthetic XML document.
2. Execute each transformation schema on the synthetic XML document. The result is a set of n modified versions of this synthetic XML document.
3. Add each modified version of the synthetic XML document to the background collection. The result of this last step is a set of n versions of the background collection ready to testing experiments.

①	②	③
`<Physical_Entity>`	`<Physical_Entity>`	`<Physical_Entity>`
`<Type />`	`<Type />`	`<Type />`
`<Name />`	`<Name />`	`<Name />`
`<XREF>`	`<Organism>`	`<XREF>`
`<DB_ID>`	`<Name />`	`<DB_ID>`
`<DB />`	`</Organism>`	`<DB />`
`<ID />`	`</Physical_Entity>`	`<ID />`
`</DB_ID>`		`</DB_ID>`
`</XREF>`		`</XREF>`
`<Organism>`		`</Physical_Entity>`
`<Name />`		
`</Organism>`		
`</Physical_Entity>`		

Fig. 5. Structures generated by the schema shown in Figure 4

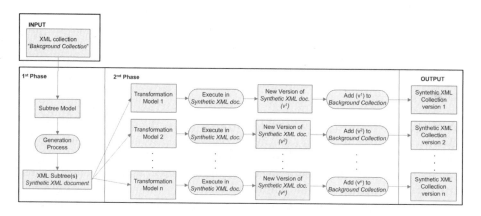

Fig. 6. Steps to generate new versions of an existing XML collection

Figure 8a shows the transformation schema we have created to modify the synthetic XML document. Then, Figure 8b shows the result of executing this transformation schema on the synthetic XML document generated by the generator schema shown in Figure 7. The execution of this schema consists on:

1. Change Level: The execution of this component will change the positions of the elements "Publication" and "ID", being now the element "ID" the ancestor and "Publication" the descendant.

2. Change order: The element "Author" will occupy the second position in its siblings set wherever it appears, due to the ancestor's path specified in the *location* parameter, ".$// * /[Author]$".

Later, this modified XML structure will be added to the background collection by using an *Add* transformation component.

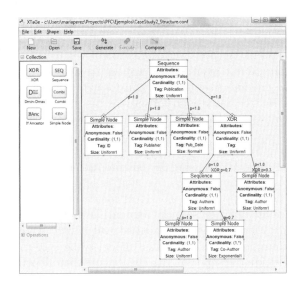

Fig. 7. A schema that generates XML documents with information about publications

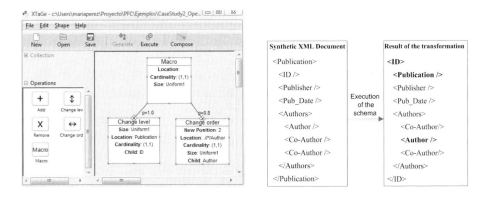

Fig. 8. (a) A transformation schema and (b) a resulting document

5   Discussion and Conclusions

We have presented how the XTaGe XML generator can be used to overcome the limitations of existing system when dealing with complex collections.

The first issue that XTaGe addresses is the lack of adaptability of the current generators to new domains or new use cases. Most of them cannot be adapted to new domains because they have been designed for specific purposes in a specific domain. XTaGe provides a flexible component-based framework that makes possible to adapt it to new specifications. The user can add new components that implement new functionalities in order to fulfill the new requirements.

In addition, XTaGe also supports creating different versions of an existing XML collection by applying a set of user-defined transformations as [7] does. However, [7] supports only a limited set of 4 functions that have been implemented for their specific purpose, the evaluation of semantic query optimization techniques, and they don't mention any way to expand this set with new functions. XTaGe also provides a set of basic transformation components that encapsulate typical XML tree-based transformations but, in contrast to [7], XTaGe allows the user to expand this set with other components according to the new requirements.

Future directions for research include extending and generalizing the features of XTaGe. We are currently focusing on the automatic generation of preliminary generation models by examining existing samples of collections. Another important issue is the lack of support for non-local constraint specification (as in [6]), although XTaGe's architecture sets the foundations to model these constraints, specific components have to be implemented and added. Another relevant direction is the extension of the component model to be able to better organize collections of pre-defined components and managing their dependencies. Finally, we aim to design specific generators, such as OWL instances generator and GIS data generator, based on XTaGe. The implementation of these specific generators will be based on an MDA architecture, where the specific generator models will be transformed into the XTaGe's model.

In conclusion, XTaGe builds upon the main features of existing schema-aware generators, and extends them in order to provide support for complex collections. The resulting system has excellent support for the creation of *controlled variability*, which is useful in testing complex and highly specific features of XML collections in particular domains. The component-based architecture is the basis for a GUI, which facilitates the specification of new collections.

Acknowledgements

This work has been partially supported by the Ministry of Science and Innovation (TIN2008-01825).

References

1. Biopax, http://www.biopax.org/
2. Aboulnaga, A., Naughton, J.F., Zhang, C.: Generating Synthetic Complex-structured XML Data. In: WebDB 2001 (2001)

3. Afanasiev, L., Manolescu, I., Michiels, P.: MemBeR XML Generator, http://ilps.science.uva.nl/Resources/MemBeR/member-generator.html
4. Barbosa, D., Mendelzon, A.O.: Declarative generation of synthetic XML data. Software: Practice and Experience 36, 1051–1079 (2006)
5. Bressan, S., Lee, M.L., Li, Y.G., Lacroix, Z., Nambiar, U.: The XOO7 Benchmark. In: Efficiency and Effectiveness of XML Tools, and Techniques (EEXTT 2002), pp. 146–147. Springer, London (2002)
6. Cohen, S.: Generating XML Structure Using Examples and Constraints. In: VLDB (2008)
7. Geng, K., Dobbie, G.: An XML Document Generator for Semantic Query Optimization Experimentation. In: iiWAS 2006, pp. 367–376 (2006)
8. Hidders, J., Marrara, S., Paredaens, J., Vercammen, R.: On the expressibility of functions in XQuery fragments. Information Systems 33, 435–455 (2008)
9. Hidders, J., Michiels, P., Paredaens, J., Vercammen, R.: LiXQuery: A formal foundation for XQuery research. SIGMOD Record 34(4), 21–26 (2005)
10. Jeong, H.J., Lee, S.H.: A Versatile XML Data Generator. International Journal of Software Effectiveness and Efficiency 1, 21–24 (2006)
11. Mesiti, M., Jiménez-Ruiz, E., Sanz, I., Berlanga, R., Valentini, G., Perlasca, P., Manset, D.: Data integration issues and opportunities in biological XML data management. In: Open and Novel Issues in XML Database Applications: Future Directions and Advanced Technologies. IGI Global (2009)
12. Mlynkova, I., Toman, K., Pokorny, J.: Statistical Analysis of Real XML Data Collections. In: COMAD 2006 (2006)
13. Murata, M., Lee, D., Mani, M., Kawaguchi, K.: Taxonomy of XML schema languages using formal language theory. ACM Trans. Internet Techn. 5(4), 660–704 (2005)
14. Nicola, M., Kogan, I., Schiefer, B.: An XML Transaction Processing Benchmark. In: SIGMOD 2007 (2007)
15. Puhlmann, S., Naumann, F., Weis, M.: The Dirty XML Generator
16. Schmidt, A., Waas, F., Kersten, M., Carey, M.J., Manolesc, I., Busse, R.: XMark: A Benchmark for XML Data Management. In: VLDB, pp. 974–985 (2002)
17. Yao, B.B., Tamer Özsu, M., Keenleyside, J.: XBench - A Family of Benchmarks for XML DBMSs (2003)

Utilizing XML Clustering for Efficient XML Data Management on P2P Networks

Panagiotis Antonellis, Christos Makris, and Nikos Tsirakis

Computer Engineering and Informatics Department,
University of Patras, Rio 26500, Greece
{adonel,makri,tsirakis}@ceid.upatras.gr

Abstract. Peer-to-Peer (P2P) data integration combines the P2P infrastructure with traditional scheme-based data integration techniques. Some of the primary problems in this research area are the techniques to be used for querying, indexing and distributing documents among peers in a network especially when document files are in XML format. In order to handle this problem we describe an XML P2P system that efficiently distributes a set of clustered XML documents in a P2P network in order to speed-up user queries. The novelty of the proposed system lies in the efficient distribution of the XML documents and the construction of an appropriate virtual index on top of the network peers.

Keywords: XML, P2P, XML clustering, XML queries, XML management.

1 Introduction

1.1 Background and Related Work

Since the emergence of file sharing applications, the P2P model has been increasingly popular along with the deployment in distributed directory service, storage and grid computing [18], [22]. The model refers to communications between similar processes running in different computers, or communication between devices that are equivalent with regard to how they exchange information and control communications. Among the main qualities that distinguish P2P networks, we recall dynamicity of data sources, robustness, scalability, reliability, no central administration, and no control over data placement.

In the context of P2P computing many methods have been proposed for data management. As far as concerning the content-based full-text search which is a challenging problem in Peer-to-Peer (P2P) systems, Tang et al. [31] developed a P2P information retrieval system called pSearch, in which document semantics are computed by latent semantic indexing in a vector space. The pSearch system can achieve performance comparable to centralized information retrieval systems by searching only a small number of nodes. Aberer et al. [1] proposed PGrid that builds a trie and clusters semantically similar data, thereby providing in-network indexing. Their algorithm is an efficient, completely decentralized approach which supports the fast, parallel construction of structured overlay networks. BATON [16], a balanced tree overlay structure which supports both exact queries and range queries efficiently.

S.S. Bhowmick, J. Küng, and R. Wagner (Eds.): DEXA 2009, LNCS 5690, pp. 68–82, 2009.

Each node of the tree is stored on exactly one peer, and each node has links to its parent, children, adjacent nodes, and selected neighbors at the same level. P-Ring [9] proposes a P2P range index for efficiently supporting equality and range queries. Viglas in [32] addresses the issue of building scalable distributed structures over peer-to-peer overlay networks. In this concept he proposes schemes to maintain these structures such as B+-trees and heap files in a DHT. P2P indexes have been proposed for multi-dimensional data in [22], [12]. In these approaches, the entire multi-dimensional space is partitioned and merged as peers join and leave the P2P system. Sartiani et al. [27] proposed a p2p XML data management system called XPeer. XPeer is currently being implemented on top of an existing persistent XML query engine. In this system more powerful peers take up extended responsibilities for a group of peers. Peers export a summary of their XML data in the form of a tree-shaped DataGuide [14]. XP2P [4] is a P2P framework for answering XPath queries which also builds on a DHT framework and allows peers to store whole or fragments of XML documents locally, whose path expressions are encoded by Rabin's finger-printing method [26] and stored in the DHT. Garces et al. [13] describe techniques for indexing data stored in peer-to-peer DHT networks. Their system built a hierarchy of indexes using a DHT containing query-to-query mappings, such that a user can look up more specific queries for a given broader query, thereby refining his or her inter-ests. Skobeltysn et al. [29] proposed a solution for the efficient support of structured queries, more specifically, XPath queries, in large-scale structured P2P systems based on the approach of the P-Grid structured overlay network. Abiteboul et al. [2] present KADOP which is a distributed infrastructure for warehousing XML resources in a P2P framework. This system allows a user to publish XML resources, search for them and declaratively built thematic portals. Some challenges arise in specific tasks, such as the need of an efficient query language that can handle the nature of these XML documents which may be incomplete, of different schemes and being distrib-uted in a network. XQuery [7] is designed to provide a flexible and standardized way of searching through (semi-structured) data that is either physically stored as XML or virtualized as XML. Also the XKeyword [15] provides efficient keyword proximity queries on large XML graph databases and XSearch [8] is a semantic search engine for XML based also on keyword search. Structure based queries work effectively when data have the same structure but require from the user to know each time the scheme of the data in order to perform right queries. When data have different schemes then keyword queries are more suitable.

Another challenge is techniques for indexing XML documents for this type of ap-plications. There are three basic types of P2P indexes. First the no index type [19] just floods data in the network for routing information but this results to a network's over-load. This means that the peer where the query is formed contact with neighbor peers until there is a result. The second type is a centralized index [23] where the information for the data that peers handle, is being stored in a single peer. For example, if a peer enters the network, it has to send its data information to the central peer of the network in order to be aware for future queries. This type has the drawback of central index server bottleneck and can only be solved in a degree with the creation of copies of this single peer. Finally a distributed index can provide better results and it depends on the nature of the network. In structured networks each peer stores index information with a hash function, about the data that handles. Recent research such as [28], [31] extend structured P2P systems by exploiting the content of documents for determining the keys. In [11] the authors propose a distributed catalog framework based on Chord [30]. XP2P [4] also extends Chord for XML data while RDFPeers [5] are based on Multi-Attribute Addressable Network [6] which extends again Chord to answer multi-attribute and range queries. A DHT-based approach is presented in [33] while a non-DHT P2P architecture is presented in [24]. In unstructured networks there are used routing indexes such as those presented in [10]. These routing indexes of a peer store information about the data that neighbor peers have. In [21] the authors present two architectures for routing XML documents, while in [20] Koloniari and Pitoura present content-based routing of path queries using Bloom filters for indexing. Finally, the authors at [25] present a new system, called psiX that runs on top of an existing distributed hashing framework. PsiX supports efficient location of relevant XML documents into the network according to user-submitter XPATH queries by creating algebraic signatures of both XML documents and user queries.

1.2 Paper Motivation and Contribution

Most of the previous work on indexing and querying XML data over a P2P network is based on path decomposition. Complex user queries are decomposed in separate paths and those paths are looked-up over the network. The results of each look-up are then merged in order to identify the matching XML documents. However, this approach may lead in a vast increase of the number of hops required to identify relevant XML documents in the network. In addition, all of the previous works suppose random distribution of the XML documents among the network peers. Although this approach imposes no restrictions of what XML documents can be stored in every peer, it results on more complex indexing and querying algorithms.

Based on these notions, in our work we introduce an innovative scheme for storing and querying XML data over a P2P network, called *PeerXML*. The proposed scheme differentiates from the previous works in the sense that instead of assuming random distribution of XML documents along the network peers, it introduces specific rules of what XML documents can be stored in every network peer. More specifically, the proposed scheme is based on clustering of the stored XML documents and then efficiently distributing them along the network's peers. Each peer can store only XML documents belonging to the same cluster, thus ensuring a more homogeneous distribution of the XML documents along the P2P network. In addition, peers who are physically close to each other, store XML documents of the same cluster in order to reduce costly hops between distant peers.

Rather than storing a centralized index of the XML documents, *peerXML* builds a hierarchical index of the stored XML documents inspired by the VBI-tree which utilizes multi-level Bloom filters for reducing the size of the index. An indexed approach has been aloes used by psiX [25]. However, psiX requires an existing DHT network to work, while our index has no such restrictions. Finally, our querying algorithm allows a query to be processed holistically - even in the presence of '//', thus eliminating additional hops when searching for matching XML documents.

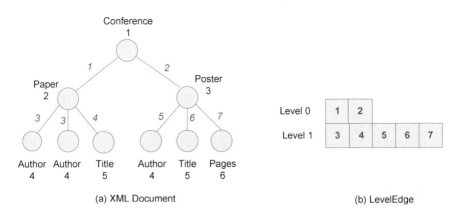

Fig. 1. Example of a LevelEdge representation

The contribution of this work can be summarized as follows:

- Clustering of the XML documents using the XEdge clustering algorithm.
- Distribution of the XML documents in the network's peers based on the belonging cluster.
- Multi-level indexing of the network's peers inspired by the VBI-tree [17].
- Utilization of multi-level Bloom filters for quickly testing if a query may match with a set of XML documents.
- Efficient routing and processing of the incoming user queries.
- Efficient handling of deletion/insertion/alteration of stored XML documents.

To our knowledge, this is the first work that utilizes XML clustering for efficiently distributing XML documents into the P2P network peers.

2   LevelEdge Representation and XEdge Clustering Algorithm

Our system utilizes the LevelEdge representation and XEdge clustering algorithm introduced in [3] in order to efficiently cluster the stored XML documents, before distributing them to the underlying P2P network. Below, we briefly describe LevelEdge and XEdge along with their use in our system.

The LevelEdge representation groups the distinct edges for each level in the XML document. It is organized as a vector of levels, where each level contains a list of distinct edges. Each distinct edge is uniquely defined by its two distinct point-nodes. The distinct edges are first encoded as integers and those integers are used in order to construct the LevelEdge representation of an XML document. Figure 1(b) presents the LevelEdge representation of the XML document in Figure 1(a). The integer numbers in the side of each edge in the XML document are the encodings of the corresponding edges. For example, all the Paper-Author edges are encoded as 3, while the Poster-Author edge is encoded as 5.

Although the LevelEdge cannot be used for fully reconstructing the original XML document, it is compact enough and it can be used for quickly answering if a query doesn't exist in a given XML document. If the query's structure doesn't match with the LevelEdge summarized information, it is certain that the query is not contained in the underlying XML document. However, depending on the structure of the query and the underlying document, the LevelEdge may provide a false positive answer. Although this case may lead to an overhead in our system, the system's accuracy is not affected because every positive answer leads to a full query checking against the corresponding XML document, thus a false positive answer will be rejected after fully checking the query against the underlying XML document.

Our system utilizes the structure of the LevelEdge representation to construct multi-level Bloom filters for quickly checking if a query is contained in the stored XML documents of a peer, as we describe in a later section.

The XEdge clustering algorithm is a modified version of k-Means where each XML document is represented by its LevelEdge and which utilizes the previously described distance metric in order to calculate the distance between two LevelEdge representations. In addition, for every cluster we define its cluster representative. A cluster representative is a LevelEdge representation that summarizes all the LevelEdge representations of the XML documents belonging to the corresponding cluster. More precisely, each level of the cluster representative contains all the distinct edges in that level of all the cluster's LevelEdge representations.

XEdge consists of the initialization phase and the main phase. In the initialization phase, k clusters are formed and the initial centroid for each cluster is calculated. During the main phase, every LevelEdge representation is checked again each cluster and is assigned to the closest cluster. The distance between a LevelEdge representation and a cluster is defined as the distance between and the cluster's representative. After assigning all the LevelEdge representations, the cluster representatives are recalculated. The main phase is repeated until no cluster representative is changed.

Our system utilizes XEdge for clustering the XML documents before distributing them to the underlying P2P network and the cluster representatives for quickly checking if a user query may have an answer in the XML documents belonging to a specific cluster. Thus, we check the query against only documents that belong to matched clusters, as described later.

3   Multi-level Bloom Filters

In order for quickly checking if a twig query is contained in a set of XML documents, we propose an extension of Bloom filters based on the LevelEdge representation. The LevelEdge Bloom filter (LBF) for a LevelEdge representation L with n levels is a set of n Bloom filters $\{LBF_0, LBF_1, \ldots LBF_{n-1}\}$. In each LBF_i Bloom filter we insert all the edges appearing in the i-th level of L. If the original LevelEdge representation summarizes the structure of only one XML document, then the corresponding LBF can be utilized for checking if a twig query exists in the underlying XML document. On the other hand, if the original LevelEdge representation summarizes a set of XML documents, then the LBF can be utilized for checking the existence of a twig query in any document of the underlying set.

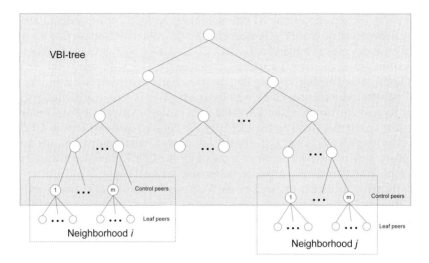

Fig. 2. Utilized indexing scheme

The filtering of a twig query q through an LBF LF is the process of checking whether there is possibility that q is contained in any of the documents that are summarized by LF. If the filtering process returns a positive result, then we fully process the corresponding query against every document summarized by the underlying LevelEdge representation of LF. Our model is orthogonal to any method of twig query processing against a set of XML documents.

In order to filter a twig query against a LBF we first query the edges at the first level of the query with every level of the LBF. If we find some level(s) of the LBF that contain all the edges of the first level of the query we continue the filtering process, otherwise we abort it. For every such matched level of LBF we apply the following steps: We proceed to the next level of the query and check the current edges with the next level of the LBF. If the edges are contained in the LBF we repeat until we reach the end of the query or until the edges are not contained in the LBF. For edges with the ancestor/descendant axis '//', the query is split at the //, and the sub-queries are processed at all the appropriate levels. All matches are stored and compared to determine whether there is a match for the whole query. If the query was matched against the LBF, then there is big possibility that the query is contained in some of the underlying XML documents; otherwise we are certain that the query is not contained in any of the underlying XML documents.

4   Document Distribution and Index Construction

The clustering of the initial set of XML documents is performed once in a central server, before initializing our system. The utilized clustering algorithm is the XEdge algorithm described in Section 2. The number of formed clusters may vary and depends mainly on how many different "topics" of XML documents are included in our collection. From now on we make the assumption that the clustering algorithm has formed k distinct clusters.

For every formed cluster, the XEdge algorithm creates a corresponding LevelEdge representation, called cluster representative as described in Section 2. Our system creates for every cluster representative an LBF as described in Section 3. Those LBFs are used for quickly testing if a query is contained in the XML documents of a specific cluster, thus avoiding routing the query in peers that we are certain they don't contain XML documents matching with the query. This is a main advantage of utilizing the formed clusters for minimizing the hops during query routing and processing. From now on, we will refer to the LBFs for the cluster representatives as $CLBF_1$ ... $CLBF_k$, where $CLBF_i$ is the LBF for the i-th cluster representative. As we describe later, every peer of the network has a copy of all the CLBFs in order to speed up the query routing and processing phase and minimize the total hops.

Our indexing scheme is inspired by the VBI-tree framework [17] which introduces a P2P framework for multidimensional indexing schemes. Below we describe in details the proposed scheme:

The underlying P2P network is divided into k *neighborhoods*, with each neighborhood storing and managing the XML documents of a single cluster. Each neighborhood consists of physical neighbor peers, in order to minimize the total number of required hops during query routing and processing. Thus, neighbor peers store similar XML documents, while distant peers store different XML documents. Every neighborhood is organized in a two-level hierarchy to help optimize the query routing as well as the insertion/deletion/update of the stored XML documents. The low-level peers of a neighborhood N_i are called *leaf peers* and are used for actually storing the XML documents of the i-th cluster. Each leaf peer stores d_i / n_i documents, where d_i is the total number of XML documents belonging to the i-th cluster and n_i is the total number of leaf peers in N_i. The top-level peers of N_i are called *control peers* and are used for query routing through the current neighborhood as well as through different neighborhoods in the network. Every control peer is responsible for a subset of the leaf peers in N_i, called its leaf *subset peers*. The total number of control peers is much smaller than the total number of leaf peers, but not too small otherwise they will become the bottleneck of the query processing procedure described later. The control peers of each neighborhood know all their leaf subset peers and can redirect any query to all of them. On the other hand, the leaf peers know only their control peer as well all their sibling leaf peers, which are the peers belonging to the same control peer.

All the control peers of our network are organized in a multi-level indexing scheme inspired by VBI-tree. Thus, every control peer is assigned a pair of VBI-Tree nodes: a routing node and a data node, in which the data node is the left adjacent node of the routing node (in the in-order traversal of the tree). We utilize this scheme for efficient and balanced query routing through the different control peers in our network. Each routing node of the indexing scheme maintains links to its parent, its children, its adjacent nodes and its sideways routing tables as in VBI-tree. In addition every routing node stores a LevelEdge structure that summarizes the XML documents of all its children data nodes (e.g. control peers) and a corresponding LBF. This LBF is used for checking the cover area of each routing node in the VBI-tree. Thus the root node stores an LBF that covers all the XML documents stored in the network.

A query is said to match with the cover area of a routing node if and only if it matches with the corresponding node's LBF. The routing algorithm described in the original VBI-tree utilizes this cover-area check in order to identify in which nodes the query should be forwarded.

In order for the query processing to be efficient, every control and leaf peer initializes and utilizes some extra structures, described below:

Each leaf peer creates an LBF for all its XML documents, called *Local LBF*, which is used for quickly determining if a query is likely to match with any stored XML document in that peer. On the other hand, each control point stores the cluster representative LBF for the neighborhood's cluster along with an LBF for all the XML documents that are stored in its leaf subset peers.

5 Query Routing and Processing

The query processing algorithm of our system utilizes the indexing structure to efficiently forward the user queries to the appropriate control peers of the neighborhoods that are possible to match the query. The set of XML documents is considered as the total search space and each data node (control peer) of the indexing scheme contains a region of the search space that corresponds to the XML documents belonging to its leaf subset peers. A query intersects with a region of the search space if and only if it matches with any of the XML documents belonging to that region. The indexing scheme in order to apply the VBI-tree range query algorithm for forwarding the query to the appropriate control peers should also be able to check if two regions (sets of XML documents) intersect with each other. This can be easily done by checking the LevelEdge structures of the corresponding routing nodes. If the two LevelEdges contain at least one common edge in any of their levels, then the corresponding regions (sets of XML documents) intersect with each other.

When a query is submitted to a peer p_j of the network, the peer p_j is automatically responsible for processing and answering the submitted query. p_j checks at first its LBF to see if the query is likely to be contained in its XML documents. If so, it performs a full query processing against all its stored XML documents using an XML search algorithm and stores the results in its cache. Then, it forwards the query to its parent control peer for further routing.

When a query reaches a control peer c_j, it first checks its LBF to see if the query is likely to be contained in the XML documents stored in its leaf subset peers. If so, it forwards it to all its leaf subset peers. Next, it checks the cluster representative LBF to see if the query is likely to be contained in the XML documents of the neighborhood's cluster. If so, it forwards the query to the rest control peers of the neighborhood. Those peers will check their LBFs and if the query matches, they will forward the query to their leaf subset peers. Finally, the control peer uses the VBI-tree to forward the query to any other control peer which its LBF contain the query. The query routing is done as proposed in the original VBI-tree framework [17], with the difference that the routing nodes check their LBFs to decide where to forward the query.

Algorithm 1. QueryProcessing(node n, query q)

---

if (n is leaf_peer) **then**
    if (*q matches LBF of n*)
        perform full match against the documents of n
    end if
    QueryProcessing(parent(n), q) /*Forward the query to the parent of n*/
else /*n is a control peer*/
    if (*q matches LBF of n*)
        /*Forward the query to the leaf subset peers of the control peer*/
        for (*each leaf l_n in leaf subset peers of n*) **do**
            QueryProcessing(l_n, q)
        end
    end if

    if (*q matches CLBF of n*)
        /*Forward the query to the all control peers of the neighborhood*/
        for (*each control peer c_p of the neighborhood(n)*) **do**
            QueryProcessing(c_p, q)
        end
    end if

    /*Now use the VBI-tree routing algorithm to forward the query into*/
    /*appropriate peers*/
    VBI_query_process(n, q)
end if

Every peer that matches any of its XML documents with the query propagates the results back to the original peer, because this peer is responsible for gathering the total results. The peer can utilize any ranking or top-k algorithm before displaying the results to the end-user. From the previously described query routing process, it is clear that the query is propagated only to the appropriate cluster neighborhood leaf peers, thus reducing the total hops and checks.

6   Updates Handling

The proposed P2P scheme has the advantage of efficiently handling inser-tion/deletion/alteration of the underlying XML documents. Any of those updates is handled locally in a single neighborhood and includes updates only in the control peers of the corresponding neighborhood and in a single leaf peer. Thus, the overhead is always constant and relatively small as it only affects a very small number of peers in the total P2P network. Below we describe the handling of each supported operation:

Insertion/Alteration. When a new XML document is inserted or altered in a leaf peer, the peer itself updates its LBF in order to include the newly added/altered XML docu-ment. In addition, it forwards the XML document to its parent control peer, which in turn updates its LBF as well as the CLBF of the corresponding cluster. At next, it forwards the altered CLBF to the rest of control peers in the current neighborhood. Finally, the VBI-tree nodes are updated accordingly in order to reflect the changes in the data nodes (control peers). It is important to note that every altered VBI-tree node also updates its stored LBF and LevelStructure in order to reflect the changes. Thus, the change is propagated only on nodes belonging to the path from the root to the affected leaf peer.

Deletion. When an XML document is deleted from a leaf peer, the peer itself updates its LBF in order to exclude the removed XML document. In addition, it forwards the XML document to its parent control peer, which in turn updates its LBF as well as the CLBF of the corresponding cluster. At next, it forwards the updated CLBF to the rest of control peers in the current neighborhood. Finally, the VBI-tree nodes are updated accordingly in order to reflect the changes in the data nodes (control peers). Again the change is propagated only on nodes belonging to the path from the root to the affected leaf peer.

The insertion/removal of a node from the network is handled accordingly to the original VBI tree and is beyond the scope of our work.

7   Experimental Study

We have built a prototype P2P emulator to evaluate the performance of our proposed indexing system over large-scale networks. The prototype was implemented in Java 6 and the experiments were performed in a machine with 2.2 Core2Duo processor and 2 GB of RAM.

In order to evaluate the performance of the proposed system, we performed two different experiments and we counted the average number of hops required for each query in order to reach the appropriate peers in the network. In each experiment we utilized the Niagara XML dataset along with additionally created synthetic XML documents to form a dataset of about 2000 XML documents. Those documents were first clustered and then distributed in the P2P network as described. In addition we created a varied set of user queries, with each query matching either with some XML documents of a specific cluster or with none cluster.

For each query, which was propagated in a random leaf peer, we counted the number of hops required in order to reach the appropriate leaf peers which contain the XML documents that match with it.

7.1   Varying Number of Peers

In this experiment, we wanted to study the relationship between the number of peers in the network and the number of hops required for each query to be processed. Thus, we created 8 clusters of totally 2000 XML documents which were distributed in 100, 200, 500 and 1000 number of peers in the network. For each case we counted the average number of hops for each query in the query set. The experimental results are shown in Table 1 and Figure 3.

Table 1. Results of first experiment

#Peers	#Total hops	#Neighbor hops	#Routing hops	Percentage (#Total hops / #peers)
100	11	5	6	11%
200	19	15	4	9.5%
500	57	51	6	11.4%
1000	108	97	11	10.8%

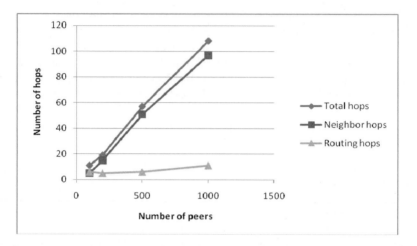

Fig. 3. Number of hops in relation to the number of peers

As we can observe, the average number of hops required for each query to be processed is increasingly analogously to the number of peers in the P2P network. This was an expected result because as the number of peers increases, the number of peers that contain XML documents which match with the query increases, so the query needs to be propagated to more nodes. In addition, due to the fact that the XML documents belonging to each utilized cluster were very similar to each other, each query related to a specific cluster was propagated to all the leaf peers of the corresponding network neighborhood. This leads to the identified relationship between the number of hops and the number of peers in the P2P network.

Table 2. Results of second experiment

#Clusters	#Total hops	#Neighbor hops	#Routing hops
4	145	141	4
8	108	97	11
16	107	91	16

However, from Table 1 it is clear that the average number of peers engaged in the processing of a single query is about 10% of the total peers in all cases. This means that no matter how large is the P2P network, each single query will reach only 10% of

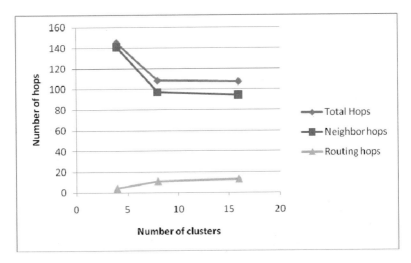

Fig. 4. Number of hops in relation to the number of clusters

the peers, thus the traffic is reasonably small in all cases. However, the most important result of that experiment is that most of the required hops per query (about 90% in most cases) are between leaf peers in the same neighborhood of the P2P network and only about 10% are hops between routing peers. For example, in the case of 1000 peers, only 11 hops are between routing nodes, while the rest 97 are hops between leaf peers in the same neighborhood. This means that each query requires very little hops in order to reach the matched appropriate network neighborhood which is related to it. After reaching it, it is being propagated to all the neighborhood's peers as described in Section 5.

Based on the assumption that peers belonging to the same network's neighborhood are physically close to each other, the cost of the neighbor hops is much less than the cost of the routing hops. Thus the proposed scheme achieves to process user queries with a very small number of routing hops (about 1% of the total number of peers in the P2P network).

7.2   Varying Number of Clusters

In this experiment, we wanted to study the relationship between the number of formed clusters (neighborhoods) in the network and the number of hops required for each query to be processed. Thus, we emulated a P2P network of 1000 peers and formed 4, 8 and 16 clusters of totally 2000 XML documents. For each case we counted the average number of hops for each query in the query set. The experimental results are shown in Table 2 and Figure 4.

As we can observe, the average number of hops required for each query to be processed is at first decreasing as the number of clusters increases and then converges to a constant value. This was an expected result because as the number of clusters increases, the number of control peers and the size of the VBI tree index increases but the size of each neighborhood is decreasing, so the query requires more routing hops but much less neighbor hops. Thus, the total number of hops is decreasing between 4 and 8 clusters but remains about the same between 8 and 16 clusters. The later result comes from the fact that in the case of 16 clusters some clusters were very similar to each other, so some queries matched with more than one cluster. This prevented the number of hops of decreasing as expected. If the clusters were totally distinct to each other, then the number of hops would decrease again between 8 and 16 clusters. In addition, this experiment confirms the observation that the number of routing hops is about 10% of the total number of hops required per query. The rest of them are hops between leaf peers of the same network's neighborhood.

Both experiments showed that each query requires a very small number of hops (~1% of peers) between routing nodes of different neighborhoods. The rest of the required hops are between nodes in the same neighborhood and are necessary, because a query that matches with a cluster's signature (LBF) may match any document belonging to that cluster. Thus, the query should be forwarded to every node in the corresponding network neighborhood in order to acquire all the possible matches. However, between nodes in the same neighborhood that are physically located close to each other, the cost of those hops is small relatively with the cost of hops between nodes in different neighborhoods. Thus, the main contribution of the proposed scheme is that it eliminates the costly hops between different neighborhoods, thus reducing the total processing time of each query.

8   Conclusions and Future Work

In this work we have presented a novel scheme for storing and querying XML data over a P2P network. The proposed scheme is based on clustering of the stored XML documents for efficiently distributing them along the network's peers. The proposed scheme utilizes the XEdge clustering algorithm for clustering the XML documents and then distributes the documents of the same cluster into peers belonging to the same network neighborhood. This distribution is based on the assumption that peers belonging to the same network neighborhood are physically close to each other, thus we are able to eliminate messages between peers belonging to different network neighborhoods. This is achieved by implementing an efficient index structure on top of the network's neighborhoods, inspired by the VBI-tree index and by efficiently propagated only to the neighborhoods that match with. The experimental results showed that the total number of hops required per query is about 10% of the total number of peers, but only 10% of them are hops between peers in different neighborhoods.

As future work, we intend to improve our P2P network simulator in order to perform more detailed experiments and integrate into it other approaches too; moreover we aim to extend our routing process in order to efficiently take into consideration not only the query structure but the query value predicates as well.

Acknowledgements

Panagiotis Antonellis' work was supported in part by the Hellenic State Scholarships Foundation (IKY).

References

1. Aberer, K., Datta, A., Hauswirth, M., Schmidt, R.: Indexing Dataoriented Overlay Networks. In: Proc. of the 31st VLDB Conference, Trondheim, Norway, pp. 685–696 (2005)

2. Abiteboul, S., Manolescu, I., Preda, N.: Constructing and Querying Peer-to-Peer Warehouses of XML Resources. ICDE: 1122-1123 (2005)

3. Antonellis, P., Makris, C., Tsirakis, N.: XEdge: Clustering Homogeneous and Heterogeneous XML Documents Using Edge Summaries. In: 23rd Annual ACM Symposium on Applied Computing, Fortalezza, Brazil (2008)

4. Bonifati, A., Matrangolo, U., Cuzzocrea, A., Jain, M.: XPath Lookup Queries in P2P Networks. In: The 6th annual ACM Intl. Workshop on Web Information and Data Management (WIDM 2004), Washington, DC, November 2004, pp. 48–55 (2004)

5. Cai, M., Frank, M.: RDFPeers: A Scalable Distributed Repository based on a Structured Peer-to-Peer Network. In: WWW (2004)

6. Cai, M., Frank, M., Chen, J., Szekely, P.: MAAN: A Multi-Attribute Addressable Network for Grid Information Services. J. Grid Comput. 2(1), 3–14 (2004)

7. Chamberlin, D.: XQuery: An XML query language. IBM System Journal 41 (2003)

8. Cohen, S., Mamou, J., Kanza, Y., Sagiv, Y.: XSEarch: A semantic Search Engine for XML. In: VLDB (2003)

9. Crainiceanu, A., Linga, P., Machanavajjhala, A., Gehrke, J., Shanmugasundaram, J.: P-Ring: An Efficient and Robust P2P Range Index Structure. In: Proc. of the 2007 ACM-SIGMOD Conference, Beijing, China, pp. 223–234 (2007)

10. Crespo, A., Garcia-Molina, H.: Routing Indices for Peer-to-Peer Systems. In: ICDCS (2002)

11. Galanis, L., Wang, Y., Jeffrey, S., DeWitt, D.: Locating Data Sources in Large Distributed Systems. In: VLDB (2003)

12. Ganesan, P., Yang, B., Garcia-Molina, H.: One Torus to Rule them All: Multi-dimensional Queries in P2P Systems. In: Seventh Intl. Workshop on the Web and Databases, Paris, France (June 2004)

13. Garces-Erice, L., Felber, P.A., Biersack, E.W., Urvoy-Keller, G., Ross, K.W.: Data Indexing in Peer-to-peer DHT Networks. In: Proc. of the 24th IEEE Intl. Conference on Distributed Computing Systems, Tokyo, March 2004, pp. 200–208 (2004)

14. Goldman, R., Widom, J.: DataGuides: Enabling Query Formulation and Optimization in Semistructured Databases. In: Proc. of the 23rd VLDB Conference, Athens, Greece, August 1997, pp. 436–445 (1997)

15. Hristidis, V., Papakonstantinou, Y., Balmin, A.: Keyword Proximity Search on XML Graphs. In: ICDE (2003)

16. Jagadish, H., Ooi, B.C., Vu, Q.H.: BATON: A Balanced Tree Structure for Peer-to-Peer Networks. In: Proc. of the 31st VLDB Conference, Trondheim, Norway (2005)

17. Jagadish, H.V., Ooi, B.C., Vu, Q.H., Zhang, R., Zhou, A.: VBI-Tree: a Peer-to-Peer Framework for Supporting Multi-Dimensional Indexing Schemes. In: ICDE (2006)

18. Jiang, H., Jin, S.: Exploiting Dynamic Querying like Flooding Techniques for Unstructured Peer-to-peer Networks. In: Proceedings of IEEE ICNP (2005)

19. Knowbuddy's Gnutella faq (2009), http://www.rixsoft.com/Knowbuddy/gnutellafaq.html (Accessed January 10, 2009)

20. Koloniari, G., Pitoura, E.: Content-based routing of path queries in peer-to-peer systems. In: Bertino, E., Christodoulakis, S., Plexousakis, D., Christophides, V., Koubarakis, M., Böhm, K., Ferrari, E. (eds.) EDBT 2004. LNCS, vol. 2992, pp. 29–47. Springer, Heidelberg (2004)

21. Koudas, N., Rabinovich, M., Srivastava, D., Yu, T.: Routing XML Queries. In: ICDE (2004)
22. Liu, B., Lee, W.C., Lee, D.L.: Supporting Complex Multi-Dimensional Queries in P2P Systems. In: Proc. of the 25th IEEE Intl. Conference on Distributed Computing Systems, Columbus, OH, June 2005, pp. 155–164 (2005)
23. Napster (2009), http://www.napster.com (Accessed January 10, 2009)
24. Nejdl, W., Wolpers, M., Siberski, W., Schmitz, C., Schlosser, M., Brunkhorst, I., Loser, A.: Super-Peer-Based Routing and Clustering Strategies for RDF-Based Peer-to-Peer Networks. In: WWW (2003)
25. Rao, P.R., Moon, B.: Locating XML Documents in a Peer-to-Peer Network using Distributed Hash Tables. IEEE Transactions on Knowledge and Data Engineering (January 08, 2009)
26. Rabin, M.O.: Fingerprinting by Random Polynomials. Harvard University, Cambridge, MA 02138, Tech. Rep. TR 15-81 (1981)
27. Sartiani, C., Manghi, P., Ghelli, G., Conforti, G.: XPeer: A Self-Organizing XML P2P Database System. In: Intl. Workshop on Peer-to-Peer Computing and Databases, Greece (2004)
28. Schmidt, C., Parashar, M.: Flexible Information Discovery in Decentralized Distributed Systems. In: HPDC (2003)
29. Skobeltsyn, G., Hauswirth, M., Aberer, K.: Efficient Processing of XPath Queries with Structured Overlay Networks. In: The 4th Intl. Conference on Ontologies, DataBases, and Applications of Semantics, Aiga Napa, Cyprus (October 2005)
30. Stoica, I., Morris, R., Karger, D., Kaashoek, M.F., Balakrishnan, H.: Chord: A Scalable Peer-to-Peer Lookup Service for Internet Applications. In: SIGCOMM (2001)
31. Tang, C., Xu, Z., Dwarkadas, S.: Peer-to-Peer Information Retrieval Using Self-Organizing Semantic Overlay Networks. In: Proc. of the 2003 ACM-SIGCOMM Conference, Germany, August 2003, pp. 175–186 (2003)
32. Viglas, S.: Distributed File Structures in a Peer-to-Peer Environment. In: Proc. of the 23th IEEE Intl. Conference on Data Engineering, Cancun, Mexico, pp. 406–415 (2007)
33. Wang, Q., Oszu, M.: A Data Locating Mechanism for Distributed XML Data over P2P Networks. Technical report, CS-2004-45, University of Waterloo, School of Computer Science, Waterloo, Canada (2004)

On the Termination Problem for Declarative XML Message Processing

Tadeusz Litak and Sven Helmer[*]

School of Computer Science and Information Systems
Birkbeck, University of London
Malet Street, Bloomsbury, London WC1E 7HX, UK
{tadeusz,sven}@dcs.bbk.ac.uk

Abstract. We define a formal syntax and semantics for the Rule Definition Language (RDL) of DemaqLite, which is a fragment of the declarative XML message processing system Demaq. Based on this definition, we prove that the termination problem for any practically useful sublanguage of DemaqLiteRDL is undecidable, as any such language can emulate a Single Register Machine—a Turing-complete model of computation proposed by Shepherdson and Sturgis.

1 Introduction

An important way to model active systems, i.e. systems that not only respond to queries but trigger actions themselves (such as reordering stock when supply levels fall below a certain threshold), is based on event processing. Event-based systems employ event-condition-action (ECA) rules to model the application logic, which means that actions are only triggered when certain events occur and certain conditions are met. The concept of rule-based processing has also been picked up by distributed systems (such as Web Services and their infrastructure) to orchestrate their activities. Message-oriented middleware, for example, provides communication based on message queues monitoring and reacting to incoming messages.

We focus on Demaq (DEclarative Messaging And Queuing), an XML-based message queue management system proposed by Böhm et al. [5] that uses a declarative rule language to specify the application logic. While a prototype of Demaq has been implemented, no formal semantics of the rule language has been provided yet and all available specifications of the Demaq Rule Definition Language (RDL) and the rule execution semantics are of an informal character. There is also no analysis of the termination of rules in Demaq. The concept of termination, i.e. determining whether rule execution will cease at some point given a certain input, is very important for systems employing active rules. Oversights in formulating rules can lead to unwanted side effects such as a subset of rules triggering each other endlessly. The ability to analyze and handle termination issues is an important prerequisite for a systematic study of rule rewriting and

[*] The authors gratefully acknowledge the support of an EPSRC grant EP/F002262/1.

S.S. Bhowmick, J. Küng, and R. Wagner (Eds.): DEXA 2009, LNCS 5690, pp. 83–97, 2009.

optimization in active systems of this kind—see [1] for an analysis of termination in active relational databases and Section 5 below for more references. Our work takes place in this context and we make the following contributions:

- We give a clear-cut formal description of the syntax and semantics for a core fragment of Demaq RDL, which we call DemaqLiteRDL.
- Working with this formal specification, we show that for any practically useful fragment of DemaqLiteRDL the termination problem is undecidable. Our undecidability proof relies on a less-known, but useful and elegant Turing-complete model of computation proposed by Shepherdson and Sturgis [16].

Our goal was to obtain an undecidability result which relies as much as possible on the active aspects of the language and as a little as possible on the power of the underlying XML query and transformation language. In other words, the design of DemaqLiteRDL is fairly minimalist. For example, the query equivalence problem for the fragment of XPath used by DemaqLiteRDL is decidable—and would remain decidable even if more powerful constructs were allowed. See [18] for a particularly powerful yet decidable superset and [17] for an overview.

The remainder of this paper is organized as follows. In Section 2 we give a brief introduction to Demaq by presenting its basic message queue management functionality. This is followed by a formal description of the syntax and semantics of DemaqLiteRDL in Section 3. We prove that the termination problem is undecidable for DemaqLiteRDL in Section 4. Section 5 covers the related work and Section 6 concludes the paper.

2   A Brief Introduction to Demaq

Demaq, which stands for "DEclarative Messaging And Queuing", is a system for managing XML messages in the context of a native XML database system (the one specifically intended by the authors of Demaq was Natix [7]). The message queue management is integrated into a database system in order to reuse some of the functionality needed for reliable message queue management, such as transaction management and recovery mechanisms. An application is made up of sets of rules, which are defined on XML message queues and govern the flow of messages through the system. The rule language itself is based on XQuery to allow for native processing of XML messages, avoiding issues concerning impedance mismatch. It was also designed as a declarative language to make it easier for developers to write applications: there is no need to connect many different components such as message-oriented middleware, local application logic, and relational database systems. In the following, we give a brief overview of the main features of Demaq; see [5] for a more detailed description.

2.1   Queues

Demaq uses queues both in order to communicate with remote systems (*incoming gateway queues* and *outgoing gateway queues*) and to model the actual business logic of applications via rule sets (*basic queues*). Queues are created using the Demaq Queue Definition Language (QDL). Here are three simple examples to illustrate Demaq queues:

```
create queue customerData kind basic;
create queue incomingMsg kind incoming;
create queue outgoingMsg kind outgoing;
```

In full-blown Demaq, queues can have a number of properties like *modes* or *interfaces*, which help to improve reliability of the system or to describe the way gateway queues interact with remote systems.

2.2 Rules

The rule-based language of Demaq is an extension to XQuery, adding message-processing functionality to the latter. The most important features of this extension cover the assignment of rules to queues, the querying of the content of messages and queues, and enqueuing (new) messages into queues. Here is an example of a rule that registers information about a new customer arriving at the gateway queue incomingMsg:

```
create rule registerNewCust for incomingMsg
   if (//registerNewCustomer) then (
        enqueue message . into customerData;
        enqueue message <result> new customer inserted </result> into outgoingMsg;)
   else ();
```

The rule registerNewCust is attached to the queue incomingMsg and first checks whether a received message contains an element called registerNewCustomer or not. In case it does, this incoming message is then enqueued into the basic queue customerData and a response is enqueued into the gateway queue outgoingMsg. In case it does not contain such an element, this rule does not perform any action (indicated by the empty else-branch).

2.3 Additional Features Not Covered by DemaqLite

There are some additional features of rules and messages, which we will only cover very briefly here and leave out from the syntax and semantics of DemaqLiteRDL. They are not essential for understanding the basics of rule execution semantics and the discussion of the termination problem. Compared to other event-condition-action languages, Demaq is much simpler, basically containing only rules for insertions—but not for deletion or update operations.

In addition to the content, a message can be annotated with *properties*, which consist of key/value pairs. These are kept separate from the actual payload and, once assigned, are fixed for the rest of the lifetime of a message. Properties help in identifying certain sets of messages when accessing them some time after they have been inserted.

Another concept, called a *slicing*, is used to simplify the access to specific subsets of messages by grouping logically related messages together. A slicing

Table 1. Syntax of DemaqLiteRDL (full)

ProgramEx	::=	RuleCreateEx+
RuleCreateEx	::=	"on-enqueue-at" QueueName ConditionalEx EnqueueEx+
ConditionalEx	::=	"if (" BoolCondEx ")"
EnqueueEx	::=	"enqueue" MessageCreateEx "into" QueueName";"
MessageCreateEx	::=	MessageCreateFun | SMessageEx
MessageCreateFun	::=	"message { " NodeCreateEx " }"
NodeCreateEx	::=	"element" StringEx "{ " (ContentCreateEx ("," ContentCreateEx)*)? " }"
ContentCreateEx	::=	NodeCreateEx | AbsPathEx
BoolCondEx	::=	"fn:true()" | "fn:false()" | AbsPathEx |
		| BoolCondEx "and" BoolCondEx | "not ("BoolCondEx")"
AbsPathEx	::=	AnyMessageEx "/" RelPathEx
AnyMessageEx	::=	MMessagesEx | SMessageEx
MMessageEx	::=	QueueFun
SMessageEx	::=	MMessagesEx "[" PositionTest "]" | "qs:message()"
QueueFun	::=	"qs:queue ("QueueName ")"
RelPathEx	::=	AxisStep ("/" AxisStep)*
AxisStep	::=	ForwardStep FilterEx*
ForwardStep	::=	ForwardAxis "::" NameTest
ForwardAxis	::=	"child" | "descendant" | "following-sibling" | "following"
NameTest	::=	NodeName | WildCardOp
WildCardOp	::=	"*"
FilterEx	::=	"["BoolFilterEx"]"
BoolFilterEx	::=	"fn:true()" | "fn:false()" | AbsPathEx
		| RelPathEx | LocFilterEx | "not ("BoolFilterEx")"
LocFilterEx	::=	PositionTest | AxisTest
PositionTest	::=	" fn:position()=" ("-")? **NUM**
AxisTest	::=	ForwardAxis | ReverseAxis
ReverseAxis	::=	"parent" | "ancestor" | "preceding-sibling" | "preceding"
StringEx	==	**STRING**
NodeName	==	**NAME**
QueueName	==	**NAME**

can range over messages from different queues (e.g. a customer together with all their invoices). It is similar to parameterized views for relational databases, which define a family of views [19].

3 DemaqLiteRDL and Its Formal Semantics

3.1 Syntax of DemaqLiteRDL vs. Demaq RDL

In the previous section we have already hinted that a number of features of Demaq are going to be left out from the small fragment under consideration here: we wanted the undecidability result to depend solely on the active aspects of the rule language and to keep the formal semantics as simple as the undecidability proof would allow. In particular, we are not interested in the Queue Definition Language (QDL) mentioned in Section 2.1 above and all features presented in Section 2.3 are left out. Besides, we use a very restricted fragment of XPath and XQuery. For example, we do not allow the use of variables—hence there is no counterpart of FLWOR expressions. Finally, some minor modifications are introduced to the Demaq RDL syntax in order to make it analogous to standard ECA (event-condition-action) active database languages. The full syntax of the DemaqLite Rule Definition Language (DemaqLiteRDL) is presented in Table 1. To give some intuition of this language, here is a reformulation of the `RegisterNewCust` rule presented in the previous section:

```
on-enqueue-at incomingMsg
    if (qs:message()/descendant::registerNewCustomer )
        enqueue qs:message() into customerData;
        enqueue message {
            element result {
                element new_customer_inserted {  }
        }  } into outgoingMsg;
```

In addition to the differences mentioned above, the example makes it clear that we do not give names to rules and do not consider text nodes—hence the contents of the `result` node in the original example had to be replaced with an element node. We are going to see more examples of DemaqLiteRDL rules in the proof of Theorem 2 below.

3.2 Formal Semantics

The semantic types we use in evaluating DemaqLiteRDL are presented in Table 2. The semantics itself is provided by the function EV-RUL presented in Table 3. It utilizes auxiliary functions EV-AT, EV-CON, EX-ENQ and EX-NOD—these are to be discussed in more detail below. This is an example of what is known as (typed) *operational semantics*, just like the formal semantics of XQuery—see [9] or [14] for more on operational semantics in general and [12], [20] on the formal semantics of XQuery. The notation differs somewhat to the one most commonly used (i.e., styled after logical inference rules), however we found the present notational convention most readable and precise.

Table 2. Semantic Types

Bool	::= "fn:true()" | "fn:false()"
ElementSgl	::= "element" NodeName "{" ElementSeq "}"
ElementSeq	::= ElementSgl (","ElementSgl)* | "fs:empty()"
MessageSgl	::= "message {" ElementSgl "}" | "fs:empty()"
MessageSeq	::= MessageSgl (","MessageSgl)* | "fs:empty()"
Queue	::= "fs:queue" QueueName "{" MessageSeq? "}" | "fs:empty()"
EnqStat	::= MessageSgl fs:into QueueName | "fs:empty()"
Schedule	::= "fs:schedule {" (EnqStat (","EnqStat)*)? "}"
Database	::= "fs:database {" (Queue (","Queue)*)? "}"
OutcomeSgl	::= Database "+" Schedule
OutcomeInDet	::= "fs:outcomes {" OutcomeSgl* "}"

Just like in the case of formal semantics of XQuery [20], our semantics introduces a number of auxiliary abstract entities, which themselves are not part of the DemaqLiteRDL language. In such cases, we will follow the W3C convention of using an italicized prefix fs:. Items preceded by this prefix are used only for specification purposes; more specifically, to define entities which have no explicit DemaqLiteRDL constructors (see Table 2). These are, namely:

- queues (*fs:*queue).
- their collections—which we call *databases* (*fs:*database).
- *schedules* (*fs:*schedule)—lists of pending primitive update events, i.e., messages to be enqueued together with target queue ID (preceded by the keyword *fs:*into).
- *possible outcomes* (*fs:*outcomes)—possible results yielded by a given set of rules P triggered by a given primitive enqueue event, depending on the reordering of rules inside P. The need for *fs:*outcomes arises because of the assumed indeterminism: there are no relative priorities between rules in a given set.
- Finally, *fs:*empty() is used universally to denote both null items inside queues, databases, and outcomes and the SKIP command inside schedules. We need it for technical reasons: mostly to define evaluation results for trivial or redundant (but well-formed) inputs.

While *fs:*queue { ... } and *fs:*schedule { ... } are ordinary lists (i.e., both the order and the number of occurrences of items does matter), *fs:*outcomes { ... } is to be interpreted as *a set constructor*: the order and number of occurrences of items inside this constructor is irrelevant. *fs:*database { ... } is to be interpreted as a *partial mapping constructor* between queue identifiers and queue contents.

Thus, our interpretation of entities prefixed with *fs:* entails that some strings of distinct syntactic shape may be in fact equivalent in the formal semantics. To be precise, we write $P \equiv R$ and say that the two expressions are *semantically equivalent* if one can be obtained from the other by the following standard transformations:

- Adding or removing an arbitrary number of *fs:*empty() items inside all four constructors in the *fs:* namespace
- Reordering, adding or removing duplicates of items inside the *fs:*database { ... } and *fs:*outcomes { ... } constructors

Let $NORM$ denote the *normalization function* removing duplicates and occurrences of *fs:*empty() wherever appropriate so that $\mathrm{NORM}(S) \equiv S$. Also, for all four constructors, the $\in$ notation denotes being equivalent to an item listed in the constructor. E.g.,

$(S,$ *fs:*schedule { *fs:*empty() }$) \in$ *fs:*outcomes { $(S,$ *fs:*schedule { *fs:*empty(), *fs:*empty() }$)$ }

and

*fs:*outcomes { $(S,$ *fs:*schedule { *fs:*empty() }$)$ }$=$ NORM({ $(S,$ *fs:*schedule { *fs:*empty(), *fs:*empty() }$), (S,$ *fs:*schedule { *fs:*empty(), *fs:*empty(), *fs:*empty() }$)$ })

Now we are in the position to define the semantics of program execution. Evaluation of a set of rules against database D annotated with schedule S (i.e., an object of semantic type OutcomeSgl) is performed by the function EV-RUL presented in Table 3. It executes in D the first of the operations scheduled in S (this is performed by the auxiliary function EX-ENQ) and returns the resulting database annotated with an updated schedule. By the latter we mean the removal of the recently performed action from the head of the schedule and adding at the tail of schedule all the actions initiated by the rules fired by this update. However, as there are no relative priorities between those rules (recall the indeterministic nature of Demaq), we have to consider all possible orders in which the corresponding sequences of actions can be combined. This is why EV-RUL returns results of type OutcomeInDet, that is, collections of objects of type OutcomeSgl.

The auxiliary function EV-AT evaluates which rules were fired by a given primitive update event and what actions were induced by these rules. In other words, it translates from the DemaqLiteRDL language to the semantic metalanguage specified in Table 2 given an object of the type Database and an object of the type EnqStat. To do so, it has to evaluate both the Demaq-specific syntactic constructs and the fragment of XPath allowed in DemaqLite. Table 3 focuses on the former task, delegating the job of evaluating XPath expressions to yet two other auxiliary functions EV-CON and EX-NOD described briefly in Table 4. They require an additional argument called *a context vector*: a vector representing the position of *context nodes* in $\{D, E\}$ (see [20] and other W3C documents for the notion of a context node and its semantic role). As each context node is represented as a vector of integers describing its position at subsequent nesting levels of "{ }" in constructors, the context vector is a vector of vectors. The only reason why we replace a single context node with a context vector is the desire to avoid a FLWOR-like construct for iterating through a given sequence in our variable-free language. Other than that, our semantics for navigational XPath expressions is perfectly standard and we do not give the details.

With all those formal prerequisites, we are ready to define the basic notions of the semantics of DemaqLiteRDL program execution: *input, execution path* and the exact meaning of *termination*.

Definition 1

- *An* input *consists of a database D and a primitive enqueue event E, i.e., it is a pair of semantic types Database and EnqStat, respectively (see Table 2).*
- *An* execution path *for a given collection of rules P (of syntactic type ProgEx in Table 1) and a given input (D, E) is a (possibly finite) sequence of elements $D_i + S_i$ s.t. $D_0 = D$, $S_0 = fs\text{:}schedule\{ E \}$ and for every i, $D_{i+1} + S_{i+1} \in EV\text{-}RUL(D_i + S_i, P)$. If for some k, $S_k \equiv fs\text{:}schedule\{ fs\text{:}empty() \}$ then D_k is called a terminal state for $D + fs\text{:}schedule\{ E \}$ and the branch itself is said to* terminate.

When we restrict our attention to normalized schedules, we can define termination in an alternative way: an execution path reaches a terminal state iff it *stabilizes* or *reaches a fixed point*. This is shown as follows:

Lemma 1. *For an arbitrary argument $D+S$ of type OutcomeSgl s.t. $D=NORM$ (D) and an arbitrary collection of rules P,*

$$D + S \in EV\text{-}RUL(D + S, P) \text{ iff } S \equiv fs\text{:}schedule\{ fs\text{:}empty() \}.$$

Moreover, if $D' \not\equiv D$, then $D' + S' \notin EV\text{-}RUL(D + \{ fs\text{:}empty() \}, P)$ for any S'.

Table 3. Formal Semantics: Executions of Programs (Nondeterministic Rule Sets)

EV-RUL: (OutcomeSgl,ProgEx) →OutcomeInDet

Fix $D := fs$:database$\{ Q_0, \ldots, Q_{n-1} \}$, $S := D + fs$:schedule$\{ E_0, \ldots, E_{k-1} \}$

EV-RUL$(S, R_0; \ldots; R_{l-1}) = $NORM$(fs$:outcomes$\{ P \})$ where $P :=$ the list of all
EX-ENQ$(D,E_0) + fs$:schedule$\{ E_1, \ldots, E_{k-1},$ EV-AT$(D,E_0,R_{f(0)}), \ldots,$ EV-AT$(D,E_0,R_{f(l-1)}) \}$ for each possible permutation f of $\{0, \ldots, l-1\}$

EX-ENQ: (Database,EnqStat) $\mapsto$ Database (auxiliary mapping)
Keep the same D as above
EX-ENQ$(D,M'\ fs$:into $Q_j) = D$ with fs:queue $Q_j \{ M \}$ replaced with fs:queue $Q_j \{ \overline{M}, M' \}$ and   EX-ENQ$(D, fs$:empty$()) = D$

AbsExpr ::= NodeCreateEx | AbsPathEx | MessageCreateFun | ConditionalEx | EnqueueEx | RuleCreateEx
AbsOutType ::= ElementSgl | ElementSeq | MessageSgl | Bool | EnqStat | Schedule
EV-AT: (Database, EnqStat, AbsExpr) $\mapsto$ AbsOutType
Fix $D := fs$:database $\{ Q_1, \ldots, Q_n \}$ and E of type EnqStat

NodeCreateEx	evaluates to type ElementSgl	
EV-AT$(D, E,$ element $N \{ S \})$	$= \Big\{$ element $N \{ \}$	if EV-AT$(D,E,S) \equiv fs$:empty$()$
	element $\{$ EV-AT$(D,E,S) \}$	else
AbsPathEx	evaluates to type ElementSeq	
EV-AT$(D, E, A/R)$	$=$ EX-NOD(EV-CON$(\{ D, E \}, A/R,((0))))$	
MessageCreateFun	evaluates to type MessageSgl	
EV-AT$(D, E,$ message $\{ S \})$	$= \Big\{$ message $\{$ element $El \}$	if EV-AT$(D, E, S) =$ element El
	fs:empty$()$	else
	note: we are thus demanding every message has exactly one element child (*root element*)	
ConditionalEx	evaluates to type Bool	
EV-AT$(D, E,$ if $(A/R))$	$= \Big\{$ fn:false$()$	if EV-AT$(D, E,A/R) \equiv fs$:empty$()$
	fn:true$()$	else
EV-AT$(D, E,$if (fn:true$()$))$	$=$ fn:true$()$	
EV-AT$(D, E,$if (fn:false$()$))$	$=$ fn:false$()$	
clauses for booleans	standard	
EnqueueEx	evaluates to type EnqStat	
EV-AT$(D, E,$enqueue $S\ fs$:into $\underline{Q_i})$	$= \Big\{$ message $Me\ fs$:into Q_i if EV-AT$(D,E,S) =$ message Me and $Q_i = Q_j$ and "fs:queue Q_i" occurs in D	
	fs:empty$()$	else
RuleCreateEx	evaluates to type (EnqStat ("," EnqStat)?)*	
EV-AT$(D, E,$on-enqueue-at$\underline{Q_k}$ if (B)	$= \Big\{$ EV-AT$(D,E,S_0), \ldots,$ EV-AT(D,E,S_{m-1}) if EV-AT$(D,E,B) =$ fn:true$()$,
$S_0; \ldots, S_{m-1};)$		$E = M\ fs$:into Q_k for some M and $\underline{Q_k}$ occurs in D
	fs:empty$()$	else

See Table 4 for auxiliary functions EV-CON and EX-NOD.

Table 4. Formal Semantics Continued: Auxiliary Functions EV-CON and EX-NOD

ContextExpr   ::= RelPathEx | AbsPathEx | AnyMessageEx
ContextResultType ::= ElementSgl | ElementSeq | MessageSgl
EV-CON: (Database, EnqStat, ContextExpr, $((\mathbb{N}^*)^* \mid fs\text{:empty}()))) \mapsto ((\mathbb{N}^*)^* \mid fs\text{:empty}())$
EX-NOD: (Database, EnqStat, $((\mathbb{N}^*)^* \mid fs\text{:empty}()))) \mapsto$ ContextResultType

Fix $D := fs\text{:database}\{ Q_0, \ldots, Q_{n-1} \}$ and E of type EnqStat
EV-CON($\{ D,E \}$, S,$fs\text{:empty}()) = fs\text{:empty}()$
An auxiliary function LOCAT($\{ D,E \}$, T) returns the position of the item denoted by T inside the nested list NORM($\{ D,E \}$) as a $\mathbb{N}^*$-vector assuming T occurs exactly once and $fs\text{:empty}()$ otherwise
Another auxiliary function ELEMS-AT($\{ D,E \}$, $\overline{n}$) returns the length of the sequence pointed by the vector $\overline{n}$ inside the nested list NORM($\{ D,E \}$) if the node pointed to exists and $fs\text{:empty}()$ otherwise
As we keep $\{ D,E \}$ fixed in this table, we drop it in our notation

$$\text{EV-CON}(\text{qs:message}(),\overline{v}) = \begin{cases} (1) & \text{if } E \not\equiv fs\text{:empty}() \\ fs\text{:empty}() & \text{else} \end{cases}$$

$$\text{EV-CON}(\text{qs:queue}(Q_i[\text{fn:position}()=m]),\overline{v}) = \begin{cases} ((\text{LOCAT}(\text{qs:queue}(Q_i),((0))),m-1)) & \text{if} \\ \qquad \text{LOCAT}(\text{qs:queue}(Q_i),((0))) \neq fs\text{:empty}() \\ \qquad \text{and } m < \text{ELEMS-AT}(\text{LOCAT}(\text{qs:queue}(Q_i))) \\ fs\text{:empty}() & \text{else} \end{cases}$$

$$\text{EV-CON}(\text{qs:queue}(Q_i[\text{fn:position}()=-m]),\overline{v}) = \begin{cases} ((\text{LOCAT}(\text{qs:queue}(Q_i),((0))),k-m+1)) & \text{if} \\ \qquad \text{LOCAT}(\text{qs:queue}(Q_i),((0))) \neq fs\text{:empty}() \text{and } m < k \\ \qquad \text{where } k = \text{ELEMS-AT}(\text{LOCAT}(\text{qs:queue}(Q_i))) \\ fs\text{:empty}() & \text{else} \end{cases}$$

We give just one example for the evaluation of XPath navigational expressions:
$\text{EV-CON}(\text{child::}N[B],((v_0^0, \ldots, v_{j_0-1}^0), \ldots, (v_0^{k-1}, \ldots, v_{j_{k-1}-1}^{k-1}))) =$
$((v_0^0, \ldots, v_{j_0-1}^0, v_{g_0^0}^0), \ldots, (v_0^0, \ldots, v_{j_0-1}^0, v_{g_f^0(0)}^0), \ldots$
$\ldots (v_0^{k-1}, \ldots, v_{j_{k-1}-1}^{k-1}, v_{g_0^{k-1}}^{k-1}), \ldots, (v_0^{k-1}, \ldots, v_{j_{k-1}-1}^{k-1}, v_{g_f^{k-1}(k-1)}^{k-1})),$

where for every $i \in \{0, \ldots, k-1\}$, $g_0^i, \ldots, g_{f(i)-1}^i$ is the list of those indices on the list pointed to by LOCAT$(v_0^i, \ldots, v_{j_i-1}^i)$ (i.e., all children of the corresponding node) which are labeled by N and satisfy the condition B, while $f(i) \leq \text{ELEMS-AT}(v_0^i, \ldots, v_{j_i-1}^i)$ stores the number of such children items. If no item of those pointed to by $(v_0^0, \ldots, v_{j_0-1}^0), \ldots, (v_0^{k-1}, \ldots, v_{j_{k-1}-1}^{k-1})$ has such a child, then the result is $fs\text{:empty}()$.
Finally, the function EX-NOD($\overline{v}$) extracts the items pointed to by vectors in $\overline{v}$.

Proof (Sketch). The "only if" direction relies on the fact that insertions are the only primitive events in our framework. If the head of the normalized schedule NORM(S) is a non-trivial enqueue statement, i.e., is of the form M $fs\text{:into}\, Q_i$ for some Q_i occurring in D, then the definition of EX-ENQ forces that for every $D' + S' \in \text{EV-RUL}(D + S, P)$, $D' \not\equiv D$. If the head of the schedule NORM(S) evaluates to $fs\text{:empty}()$, it is removed from S' via the normalization procedure if S' contains any other scheduled event; thus S' is strictly shorter than S. Note that the head of the normalized schedule NORM(S) cannot be equal to $fs\text{:empty}()$. The "if" direction of the equivalence is a useful exercise for the reader to understand the working of the formal semantics. □

It is an interesting question whether termination depends on the order of execution of the rules. However, for the purposes of the present paper it is actually irrelevant, as the set of rules used in the undecidability proof is going to be *deterministic* in the sense defined below and thus the issue of non-equivalent possible execution sequences cannot arise at all.

Definition 2

- *EV-RUL is said to be* deterministic *for a given triple* $(D + S, P)$ *if the the value of EV-RUL for this triple is unique up to equivalence.*
- *A collection of rules P is* deterministic *if EV-RUL$(D + S, P)$ is deterministic for every $D + S$.*

Thus, Lemma 1 entails that EV-RUL(D + *fs:*`schedule`{ *fs:*`empty()` },P) is deterministic for any D and P. However, this is a trivial reason for determinism: if no events are scheduled, no rules are triggered. More importantly, when no more than one rule is triggered or when at most one rule can *fire* even if more than one rule is triggered, then clearly no ambiguities concerning scheduling can arise. As we are going to encounter the last situation in what follows, we record it as a separate

Lemma 2 (Determinism). *If a collection of rules is of the form*

$$P = \text{on-enqueue-at } \underline{Q_0} \text{ if } (B_0)\dots; \text{ on-enqueue-at } \underline{Q_{n-1}} \text{ if } (B_{n-1})\dots;$$

and for any (D, E), there is at most one $i < n$ s.t. EV-AT$(D, E, \text{if}(B_i) \dots) = $ `fn : true()`, *then P is deterministic.*

Proof. Straightforward application of semantic rules in Table 3—see the proof of Lemma 1 above for a similar reasoning. □

4   The Undecidability of the Termination Problem

As our undecidability proof relies on a model of computation described by Shepherdson and Sturgis [16, Section 5], we first give a brief introduction to their *Single Register Machine*. This is followed by the actual proof.

4.1   Shepherdson-Sturgis Single Register Machine (SSSRM)

We fix an alphabet of symbols $\mathbf{A} = \{a_0, a_1, \dots, a_m\}$ for $m \geq 1$, symbols a_0 and a_m having a special role. We follow the convention of representing a natural number n by a sequence of $n + 1$ a_0's. Thus, $\underline{0} := a_0$, $\underline{1} := a_0 a_0 \dots$ and so on. The second special symbol a_m is used only as a separator and can be written as ",". Those are the only symbols whose presence is crucial; other ones are only for convenience and from a theoretical point of view can be eliminated. [16] shows how to generalize the notion of a partial recursive function in case of $m > 1$, i.e., for alphabets containing more than one non-separator symbol. The set of words over alphabet $\mathbf{A}$ is, as usual, denoted as $\mathbf{A}^*$.

A *Shepherdson-Sturgis Single Register Machine (SSSRM) program over alphabet* $\mathbf{A}$ is defined as a mapping $S : \{0, \dots, n-1\} \mapsto I_m$ for some $m \in \mathbb{N}_{\geq 2}$, where I_m is the set of instructions in the following language (its meaning is going to be explained below):

$$\langle \textbf{push } a_i \rangle \quad | \quad \langle \textbf{pop} \rangle \quad | \quad \langle \textbf{if\_first } = a_i \textbf{ goto\_line } j \rangle,$$

where $i < m$, $j \in \mathbb{N}$ (in fact, given a specific upper bound n on the number of lines, we can restrict attention to $j \leq n$). The value of $S(j)$ will be called *the j-th line of S*. The set of all SSSRM-programs over alphabet $\mathbf{A}$ is denoted as PROGRAMS($\mathbf{A}$); we are going to drop $\mathbf{A}$ from the notation wherever it does not lead to ambiguities. For a given $S \in$ PROGRAMS of the form $S : \{0, \ldots, n-1\} \mapsto I_m$, we define maxline($S$) = n (that is, the number of lines occurring in S) and maxletter(S) = m (that is, the upper bound of the index of letters occurring in the program code).

A pair (A, j) where $a \in \mathbf{A}^*$ and $j \in \mathbb{N}$ is going to be called *a (normal) SSSRM-state*. The set of all normal states is denoted as NORMSTATES. In addition, we have a special *halting constant* Halt. The set of *halting states* HALTSTATES is defined as $\{(A, \mathsf{Halt}) \mid A \in \mathbf{A}^*\}$. We denote ALLSTATES = NORMSTATES $\cup$ HALTSTATES. The semantics of SSSRM-programs is specified by the *atomic action mapping* $\Rightarrow$ from PROGRAMS $\times$ NORMSTATES to ALLSTATES defined as follows:

$$S, (a^0 \ldots a^{k-1}, j) \Rightarrow \begin{cases} \text{if } j \geq \mathsf{maxline}(S) & : & (a^0 \ldots a^{k-1}, \mathsf{Halt}) \\ \text{else if } S(j) = \langle \mathbf{push}\, a_i \rangle & : & (a^0 \ldots a^{k-1} a_i, j+1) \\ \text{else if } S(j) = \langle \mathbf{pop} \rangle & : & (a^1 \ldots a^{k-1}, j+1) \\ \text{else if } S(j) = \langle\, \mathbf{if\_first} =_{a_i} \mathbf{goto\_line}\, j' \rangle \text{ and } a^0 = a_i & : & (a^0 \ldots a^{k-1}, j') \\ \text{else if } S(j) = \langle\, \mathbf{if\_first} =_{a_i} \mathbf{goto\_line}\, j' \rangle \text{ and } a^0 \neq a_i & : & (a^0 \ldots a^{k-1}, j+1) \end{cases}$$

Note that we write $S, (A, j) \Rightarrow (A', j')$ rather than $\Rightarrow (S, (A, j)) = (A', j')$. Intuitively, the first coordinate (A, j) denotes the input word and the second coordinate—the number of the next line the program would execute. If there exists a sequence $(A_0, j_0), \ldots, (A_k, j_k)$ s.t. $j_0 = 0$, $j_k =$ Halt and for every $l < k$, $S, (A_l, j_l) \Rightarrow (A_{l+1}, j_{l+1})$, then we say *S halts on input A_0 with A_k as the result*.

As it turns out, this simple model of computation can compute all the functions computable by Turing machines. Formally, for $S : \{0, \ldots, n-1\} \mapsto I_m$, let us define a function $f_S : \mathbf{A}^* \mapsto \mathbf{A}^*$ as follows:

– if S halts on input A with B as a result, $f_S(A) = B$;
– otherwise, $f_S(A)$ is undefined.

Shepherdson and Sturgis [16, Theorem 8.1] show that all partial recursive functions over $\{a_0, \ldots, a_{m-1}\}$ can be represented as f_S for some S over $\{a_0, \ldots, a_{m-1}, \text{","}\}$. In the simplest case of $\mathbf{A} = \{a_0, \text{","}\}$, it means that SSSRM-programs can compute all partial recursive functions over $\mathbb{N}$ (recall how we represent the natural numbers!) and hence their halting problem is as undecidable as the halting problem for Turing machines. To put it formally,

Theorem 1. *For any $m \geq 1$, there is no algorithm to decide whether a given $S : \{0, \ldots, n-1\} \mapsto I_m$ will halt for a given $A \in \{a_0, \ldots, a_m\}^*$.*

In fact, just like for all models of computation equivalent to Turing machines, one can prove a stronger result. It is easily seen that one can define a recursive encoding of all SSSRM-programs over a given alphabet. Thus, there exists an SSSRM-program equivalent to the universal Turing machine: a program which given the code number of another SSSRM-program S and a word A performs exactly the same action as S would on A. The halting problem would be undecidable then for this particular SSSRM-program: there is no algorithm to decide whether it eventually halts on a given input. However, for our purposes, even the weaker version of the result presented above would do.

4.2   Undecidability Proof

The undecidability of the halting problem for SSSRM-programs will be used by us to show the undecidability of the termination problem for active rules in DemaqLiteRDL. We represent an SSSRM-state $(a^0 \ldots a^{k-1}, j)$ $(j \leq n+1)$ as an XML tree whose root element node (recall it is the single child of the virtual message root node) is labeled with j and the immediate children of the root are labeled with subsequent letters of $\bar{A}$ as shown below:

```
message { element j { element a⁰ { }, element a¹ { }, ..., element aᵏ⁻¹ { } } }
```

Theorem 2. *The question whether a given set of DemaqLiteRDL rules will terminate on a given XML tree is undecidable, even under the following restrictions:*

- *only deterministic sets of rules are allowed—conditions are required to be mutually disjoint, i.e., the assumptions of Lemma 2 have to be satisfied*
- *queries are not allowed to scan the content of messages enqueued in previous stages—the use of* qs:queue *is not allowed*
- *no more than one queue name is allowed*
- *the body of every rule contains only one enqueue statement*
- *the use of backward axis is allowed only in axis tests and no other axis or location tests are allowed. That is, neither the use of forward axes nor of* fn:position() *is allowed in boolean filter expressions.*

Proof. Assume the SSSRM-program to be encoded has n lines. We fix a single queue—let it be called "Default". Line numbers correspond to root labels of messages, as described above. For each $j < n$, the translation of j-th line looks as follows:

- if $S(j) = \langle \mathbf{push}\, a_i \rangle$ for some $i \leq m$:

  ```
  on-enqueue-at Default if (qs:message() /child::j )
     enqueue message { element j+1
     { qs:message() /child::*/child::*, element aᵢ { } } }
     into Default;
  ```

- if $S(j) = \langle \mathbf{pop} \rangle$:

  ```
  on-enqueue-at Default if (qs:message() /child::j )
     enqueue message { element j+1 { qs:message() /child::*/child::*[preceding-sibling] } }
     into Default;
  ```

- If $S(j) = \langle\, \mathbf{if\_first} = a_i\, \mathbf{goto\_line}\, j' \rangle$ for some $i \leq m$, $j' \in \mathbb{N}$:

  ```
  on-enqueue-at Default if (qs:message() /child::j and
  qs:message() /child::*/child::aᵢ[not (preceding-sibling)] )
     enqueue message { element j' { qs:message() /child::*/child::*}}
     into Default;
  ```

```
on-enqueue-at Default if (qs:message()/child::j and
not (qs:message()/child::*/child::aᵢ[not (preceding-sibling))])
   enqueue message { element j+1 { qs:message()/child::*/child::*}}
   into Default;
```

We need to verify that the behavior of this set of rules on an input encoding an SSSRM-state will mimic the behavior of the corresponding program. Our set of rules satisfies the assumptions of Lemma 2: on a given input, the condition of at most one of them can evaluate to `fn:true()` (recall that messages cannot have more than one root element). Thus, the output is going to be deterministic (equivalent to an element of the form *fs:*outcomes{ OutcomeSgl }) and as each of those rules involves a single `enqueue` operator, the schedule part of the single outcome will consist of a single enqueue statement—either of the form "M' *fs:*into Default" or "*fs:*empty()". It is straightforward, if tedious, to verify that for an input of the form "M *fs:*into Default" where M is a message encoding an SSSRM-state (A, j) as described above, the schedule part of the single outcome is going to be of the former shape where M' is a message encoding the SSSRM-state (A', j') s.t. $S, (A, j) \Rightarrow (A', j')$. ☐

It is possible to prove an analogous theorem for other fragments of DemaqLi-teRDL: the most straightforward is to replace [preceding-sibling] (and its negation) in the formulation of rules above with [fn:position()=1] (and its negation, accordingly). Other encodings are also possible, e.g., replacing axes other than child with the attribute axis and more involved string operations— namely, the operations of taking substrings and concatenation. Finally, allowing advanced use of the counting predicate fn:count would enable us to replace the use of the SSSRM machine with the well-known 2CM-machine model of computation along the lines sketched in [8] and [17].

5   Related Work

Looking at a fragment or a core of a language is a common approach when dealing with such issues as the decidability of a certain problem. For example, there are numerous papers on the decidability of XPath query containment covering various subsets of the language ([8,13,17,21] just to name a few, for an overview see [15]). That is one reason why we decided to investigate the active, rule-based part of Demaq rather than reiterating results already established for XPath and XQuery.

Attempts to isolate a core sublanguage have also been made for XQuery and XQuery Update by Hidders et al. [10,11]. The focus, however, was on suitability for educational and research purposes rather than on combination of decidability and practical usefulness. We believe that DemaqLiteRDL is well-suited for similar goals (e.g., research on query optimization or expressive power of sub-languages). Nevertheless, there is rather little overlap with the present paper.

Recently, there has been work on combining ECA rules with XML. Bailey et al. [4] look at data consistency when updating XML documents while Bonifati et al. examine ECA rules in an e-commerce context [6]. However, these publications focus on updates within XML documents (this also applies to [10] above), not on message queue management. More closely related to our work is the research on rule execution semantics in active database systems by Bailey et al. [2]. The main differences to our work are the relative simplicity of Demaq language (we only consider insertions into message queues) and the indeterminism of the order in which Demaq rules are processed. It is worth mentioning, however, that our approach to semantics of indeterministic ECA rules was inspired by the concluding remarks of that paper [2, Section 8.3].

6   Conclusion and Outlook

We have shown that the termination problem for DemaqLiteRDL, a very restrictive subset of Demaq, is undecidable. From a practical point of view, DemaqLiteRDL is already too restricted to be of much use (e.g. it does not allow the use of variables or FLWOR expressions), hence reducing the functionality even further in order to arrive at a language for which the termination problem is decidable as it was done in, e.g., [1] is not going to improve the situation.

Consequently, we want to turn our attention to rule analysis and optimization via abstract interpretation, similar to what has been done in the context of active (relational and functional) databases—see, e.g., [3]. On the one hand our job is made simpler by the fact that we only have to consider insertion operations, but on the other hand we have to be able to cope with a highly indeterministic rule evaluation.

References

1. Bailey, J., Dong, G., Ramamohanarao, K.: Decidability and undecidability results for the termination problem of active database rules. In: 17th ACM SIGMOD-SIGACT-SIGART PODS Symposium, Seattle, Washington, pp. 264–273 (1998)
2. Bailey, J., Poulovassilis, A.: Abstract interpretation for termination analysis in functional active databases. J. Intell. Inf. Syst. 12(2-3), 243–273 (1999)
3. Bailey, J., Poulovassilis, A., Courtenage, S.: Optimising active database rules by partial evaluation and abstract interpretation. In: Ghelli, G., Grahne, G. (eds.) DBPL 2001. LNCS, vol. 2397, pp. 300–317. Springer, Heidelberg (2002)
4. Bailey, J., Poulovassilis, A., Wood, P.: An event-condition-action language for XML. In: 11th International World Wide Web Conference, pp. 486–495 (2002)
5. Böhm, A., Kanne, C.C., Moerkotte, G.: Demaq: A foundation for declarative XML message processing. In: 3rd Biennial Conference on Innovative Data Systems Research (CIDR), Asilomar, California, pp. 33–43 (2006)
6. Bonifati, A., Ceri, S., Paraboschi, S.: Active rules for XML: A new paradigm for e-services. VLDB Journal 10(1), 39–47 (2001)
7. Fiebig, T., Helmer, S., Kanne, C.C., Moerkotte, G., Neumann, J., Schiele, R., Westmann, T.: Anatomy of a native XML base management system. VLDB Journal 11(4), 292–314 (2002)

8. Geerts, F., Fan, W.: Satisfiability of XPath queries with sibling axes. In: Bierman, G., Koch, C. (eds.) DBPL 2005. LNCS, vol. 3774, pp. 122–137. Springer, Heidelberg (2005)
9. Hennessy, M.: The Semantics of Programming Languages: an Elementary Introduction using Structural Operational Semantics. John Wiley and Sons, New York (1990)
10. Hidders, J., Paredaens, J., Vercammen, R.: On the expressive power of XQuery-based update languages. In: Amer-Yahia, S., Bellahsène, Z., Hunt, E., Unland, R., Yu, J.X. (eds.) XSym 2006. LNCS, vol. 4156, pp. 92–106. Springer, Heidelberg (2006)
11. Hidders, J., Paredaens, J., Vercammen, R., Demeyer, S.: A light but formal introduction to XQuery. In: Bellahsène, Z., Milo, T., Rys, M., Suciu, D., Unland, R. (eds.) XSym 2004. LNCS, vol. 3186, pp. 5–20. Springer, Heidelberg (2004)
12. Katz, H., Chamberlin, D., Kay, M., Wadler, P., Draper, D.: XQuery from the Experts: A Guide to the W3C XML Query Language. Addison-Wesley Longman Publishing Co. Inc., Boston (2003)
13. Miklau, G., Suciu, D.: Containment and equivalence for an XPath fragment. In: 21st Symposium on Principles of Database Systems (PODS), pp. 65–76 (2002)
14. Plotkin, G.D.: A Structural Approach to Operational Semantics. Technical Report DAIMI FN-19, University of Aarhus (1981)
15. Schwentick, T.: XPath query containment. SIGMOD Record 33(1), 101–109 (2004)
16. Shepherdson, J.C., Sturgis, H.E.: Computability of recursive functions. Journal of ACM 10(2), 217–255 (1963)
17. ten Cate, B., Marx, M.: Navigational XPath: calculus and algebra. SIGMOD Record 36(2), 19–26 (2007)
18. ten Cate, B., Marx, M.: Axiomatizing the logical core of XPath 2.0. Theory of Computing Systems (to appear) Open access: http://www.springerlink.com/content/m62011j670270282/fulltext.pdf
19. Toyama, M.: Parameterized view definition and recursive relations. In: 2nd Int. Conf. on Data Engineering (ICDE), Los Angeles, California, pp. 707–712 (1986)
20. W3C. XQuery 1.0 and XPath 2.0 formal semantics. W3C recommendation. http://www.w3.org/TR/xquery-semantics/
21. Wood, P.T.: Containment for XPath fragments under DTD constraints. In: Calvanese, D., Lenzerini, M., Motwani, R. (eds.) ICDT 2003. LNCS, vol. 2572, pp. 297–311. Springer, Heidelberg (2002)

Consistency Checking for Workflows with an Ontology-Based Data Perspective

Gabriele Weiler[1,2], Arnd Poetzsch-Heffter[2], and Stephan Kiefer[1]

[1] Fraunhofer Institute for Biomedical Engineering, St. Ingbert, Germany
{gabriele.weiler,stephan.kiefer}@ibmt.fraunhofer.de
[2] Software Technology Group, University of Kaiserslautern, Germany
poetzsch@informatik.uni-kl.de

Abstract. Static analysis techniques for consistency checking of workflows allow to avoid runtime errors. This is in particular crucial for long running workflows where errors, detected late, can cause high costs. In many classes of workflows, the data perspective is rather simple, and the control flow perspective is the focus of consistency checking. In our setting, however, workflows are used to collect and integrate complex data based on a given domain ontology. In such scenarios, the data perspective becomes central and data consistency checking crucial.

In this paper, we motivate and sketch a simple workflow language with an ontology-based data perspective (SWOD), explain its semantics, classify possible inconsistencies, and present an algorithm for detecting such inconsistencies utilizing semantic web reasoning. We discuss soundness and completeness of the technique.

1 Introduction

Workflows describe the processing of data according to a well-defined control flow. Workflows often interact with humans and integrate them into the processing. We consider workflows in which the processed data has semantic metadata in terms of a given domain ontology, like e.g. for medical trials. Such semantic metadata is not only crucial to allow data integration, but can also help to improve data quality. In our work we show that it can be utilized to detect data inconsistencies in the workflow definition at design time of the workflow. Such *static* checking prevents executing faulty workflows where errors might occur only months after the workflow was started, causing potentially huge costs. In this paper we focus on the detection of data inconsistencies. Integration with existing control flow checking techniques, such as deadlock detection (e.g. [1], [2]) is considered as future work. We divide data inconsistencies into two categories:

1. *Data-Dependent Control Flow Inconsistencies* causing e.g. undesired abortion of workflow executions or unreachable tasks due to unsatisfiable conditions.
2. *Semantic Data Inconsistencies* causing data collected during workflow execution to be inconsistent with the knowledge of the underlying domain that we assume to be given by a domain ontology[1].

---

[1] Similarly, one could check consistency with complex XML-schema information.

S.S. Bhowmick, J. Küng, and R. Wagner (Eds.): DEXA 2009, LNCS 5690, pp. 98–113, 2009.

To guarantee reliable workflow executions, it is important to detect and eliminate the first kind of inconsistencies. Avoiding the second kind of inconsistencies guarantees that the collected data is consistent, a prerequisite to get reliable results from data analysis and to enable integration of data collected in different information or workflow systems.

Few existing algorithms are able to detect data inconsistencies in workflows. Sun et al. [3] developed a framework for detecting data flow anomalies, which is e.g. capable of detecting missing data in conditions, but is not able to check the satisfiability of conditions in a workflow. Eshuis [4] describes a framework for verification of workflows based on model checking. His framework is able to check satisfiability of conditions, but these checks consider only boolean expressions and their dependencies. Our central contribution is an algorithm to check both categories of data inconsistencies, in particular to verify consistency with domain ontologies. The main aspects of this contribution are:

- **Workflow Language.** The considered class of workflows consists of human processable tasks, comprising forms that users have to fill in at execution time. We present a language for this class of workflows with a formally defined data perspective based on an existing domain ontology. We provide a well defined semantics for the language.
- **Categories of Inconsistencies.** We provide a precise definition and categorization of potential data inconsistencies that can cause problems during workflow execution.
- **Consistency Checking Algorithm.** We describe a static consistency checking technique that utilizes description logic and its reasoning services to detect the defined inconsistencies. We discuss soundness and completeness of the algorithm.

Our approach has various strengths. Semantic annotations in terms of the domain ontology can be assigned automatically to data collected in workflows encoded in our workflow language. That allows to query the data in terms of the ontology, enabling meaningful interpretation of the data and inference of new knowledge from the data. Technologies developed for the semantic web are reused, e.g. the Web Ontology Language (OWL) [5] and description logic reasoners. Consistency checking avoids that contradictory data is collected that can not be utilized for information integration and increases reliability of the workflow execution avoiding e.g that workflow execution aborts unexpectedly.

Our workflow language shall not be seen as a substitute to existing powerful workflow languages, but as an example, how these languages can be augmented with ontology-based data perspectives and profit from techniques described in this work. The interested reader is refered to the Technical Report accompanying this paper [6], where detailed formal descriptions and examples can be found.

The paper is structured as follows: In Sec. 2, we describe our motivating application scenario the ontology-based trial management system ObTiMA. In Sec. 3, we specify the workflow language. In Sec. 4, data inconsistencies and in Sec. 5 an algorithm to detect these inconsistencies are described. We conclude with discussion and related work.

2 ObTiMA – An Ontology-Based Trial Management System

The motivating application for this work is ObTiMA, an ontology-based trial management system [7], which has been developed for the European project ACGT (Advancing Clinico Genomic Trials on Cancer) [8]. ACGT aims to provide a biomedical grid for semantic integration of heterogeneous biomedical databases using a shared ontology for cancer trials, the ACGT Master Ontology [9].

ObTiMA allows trial chairmen to set up patient data management systems with comprehensive metadata in terms of the ACGT Master Ontology to facilitate integration with data collected in other biomedical data sources. Forms to collect patient data during the trial can be designed and questions on the forms can be created from the ontology in a user friendly way. Furthermore, treatment plans can be designed in ObTiMA, which are workflows to guide doctors through the treatment of a patient. A form with ontology annotation can be assigned to each task of these treatment plans to document the patient's treatment. The algorithm described in this work will enable ObTiMA to detect data inconsistencies in the treatment plans automatically.

3 Workflow Language

In this section we describe a simple workflow language with an ontology-based data perspective (SWOD). A workflow description in SWOD consists of a workflow template describing the control flow and the forms, and a workflow annotation containing the semantic description of the data.

3.1 Workflow Template

A workflow template describes tasks and their transitions. Tasks describe a piece of work which has to be executed by a human user. Each task contains one form that has to be filled in to document the piece of work. A form contains one or more questions also called items. SWOD supports acyclic workflows with the basic control flow patterns sequence, XOR-/AND-split and XOR-/AND-join (for description of control flow patterns s. [10]). Each outgoing transition of an XOR-split has an associated condition described in terms of an ontology (s. Sec. 3.2). For reasons of simplicity we describe in the following SWOD, inconsistencies and the algorithm for workflows without concurrency. An extension for concurrent workflows can be found in [6]. To illustrate our language and our algorithm, we use a simple example treatment plan (for outline s. Fig. 1).

3.2 Workflow Annotation

A workflow annotation describes the semantic annotation of the data in terms of an existing domain ontology. It assigns a description from the ontology to each item (item annotation) and to each condition (condition annotation). With

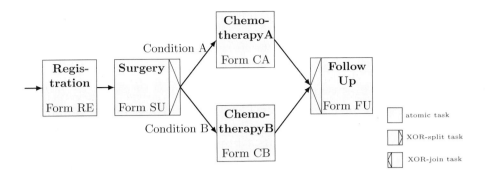

Fig. 1. Example workflow: "example treatment plan"

this information, the data, stored in form based data sources, can be queried in terms of the ontology.

Each workflow has a so-called *focal point*, which denotes the subject of a workflow execution (e.g. for a treatment plan it is a patient). Each annotation of an item or a condition refers to that focal point.

The workflow annotation of the example treatment plan is shown in Lst. 1. It describes the item annotations of items "Age of mother", "Type of tumor" "Name of mother" and "Does patient have metastasis?" and the condition annotations of Condition A and B. The underlying domain ontology is a simple example ontology, which is partly shown in the next paragraph. In the following, classes of the domain ontology are prefixed with 'd:'.

```
     WFAnnotation ExampleTreatmentPlan
       focal PPatient;
       FormAnnotation FormRE
       ItemAnnotation(IAge OntoPath(d:HumanBeing(PPatient) hasMother(PPatient,
  5      PMother) d:HumanBeing(PMother) hasAge(PMother, PAge) integer(PAge))
         Value(Max(130)))
       ItemAnnotation(ITum OntoPath(d:HumanBeing(PPatient) hasTumor(PPatient,
         PTumor) d:Tumor(PTumor)) Specify(Case(breasttumor d:Breasttumor)
         Case(nephroblastoma d:Nephroblastoma)
 10      Case(other NOT(d:Breasttumor OR d:Nephroblastoma)))
       FormAnnotation FormSU
       ItemAnnotation(IName OntoPath(d:HumanBeing(PPatient) hasMother(PPatient,
         PMother) d:HumanBeing(PMother) hasName(PMother, MName) string(MName)
         Value())
 15    ItemAnnotation(IMet OntoPath((d:HumanBeing(PPatient) hasMetastasis(PPatient,
         PMet) d:Metastasis(PMet) Exist())
       ConditionAnnotation(ConditionA
       d:Metastasis(?met)∧ hasMetastasis(PPatient, ?met))
       ConditionAnnotation(ConditionB
 20      (=0 hasMetastasis.d:Metastasis)(PPatient))
```

Listing 1. Extract from workflow annotation of example treatment plan

Domain Ontology. We have chosen description logics (DL) as language for the domain ontology, because this family of knowledge representation formalisms provides the base of most modern ontology languages, as e.g. OWL 2, the semantic web ontology language. A basic understanding of DL is required in the following, and we refer to [11] for a detailed description.

The basic elements in DL are *individuals*, *atomic concepts* representing sets of individuals, and *roles* representing binary relationships between individuals. In DL, an *ontology* $\mathcal{O}$ introduces the terminology of an application domain. An extract from the domain ontology for our example treatment plan is shown below, which describes that a human being has at most one mother and his age may be at most 150. A breast tumor is a tumor and it is located in the breast:

HumanBeing $\sqsubseteq$ (($\leq$ 1 hasMother.HumanBeing) $\sqcap$ ($\forall$ hasAge.integer[$\leq$ 150]))
Breasttumor $\sqsubseteq$ (Tumor $\sqcap$ ($\forall$ locatedIn.Breast))

An ontology introduces the terminology as a set of axioms of the form C $\sqsubseteq$ D (i.e. D subsumes C) and C $\equiv$ D (i.e C $\sqsubseteq$ D and D $\sqsubseteq$ C), where C and D are general concepts. DL languages can be distinguished by the constructors they provide for defining general concepts. We currently consider the DL language ALCQ(D) for the domain ontologies. This language provides amongst others the constructors number restriction (e.g. $\leq$ 1 hasMother.HumanBeing), value restriction (e.g. $\forall$ locatedIn.Brain) and data type restriction (e.g. $\forall$ hasAge.integer[$\leq$ 150]). The signature of an ontology Sig($\mathcal{O}$) is the set of concepts, roles and individuals that occur in $\mathcal{O}$.

In DL an *ABox* $\mathcal{A}$ describes assertions about individuals in terms of an ontology. An ABox contains concept assertions C(a) (i.e. individual a is an instance of concept C), and role assertions R(a,b) (i.e. individual b is a filler of the role R for a).

DL-reasoners exist, which can e.g. check if one concept subsumes another, if an ontology is consistent, if an ABox is consistent wrt. an ontology, i.e. they are not contradictory, or if an ontology implies an axiom β (written $\mathcal{O} \models \beta$).

In most modern ontology languages, concepts are called classes and roles are called properties. We use this notation in the following sections.

Item Annotation. An item annotation is the ontology description of an item. It consists of an ontology path and an item constructor.

Ontology Path. An ontology path is the basic ontology description of an item. It describes the individuals and constants, for which information is queried in the item, and their relations to the focal point of the workflow. In a simplified notation it can e.g. be "Patient hasTumor Tumor" or "Patient hasTumor Tumor hasWeight Weight". Formally an ontology path consists of variable assertions, which represent the individuals and constants, and relation assertions. A *variable assertion* consists of a variable name and a type, which can be a class or a primitive data type from the domain ontology. Variables of the former kind are called *object variables*, variables of the later kind *data type variables*. E.g. in Lst. 1, line 5 the object variable PMother of type d:HumanBeing and the data type variable PAge of type integer are described. Variables with distinct names denote distinct individuals or constants. Relations between variables can be expressed with *relation assertions* (e.g. in Lst. 1, line 5 PMother and PAge are related with hasAge). A relation between an object variable and a data type (resp. an object) variable has to be a data type (resp. an object) property from the domain ontology. For each workflow description one focal variable is declared, e.g. PPatient (Lst. 1, line 2). Each ontology path of an item starts with this variable.

Item Constructors. Different kinds of item descriptions can be assembled from an ontology path depending on how the value of the item is considered in the semantics (s. Sec. 3.3). Thus, we defined different item constructors. The last variable in the ontology path is the associated variable of the item constructor.

1. *Value-Items* query values of data type properties from the ontology, e.g. "age of mother" (Lst. 1, line 4-6) or "weight of patient". Numerical value-items can have range restrictions, denoted with Min and Max. The associated variable has to be a data type variable. E.g. the associated variable of the item "age of mother" is "PAge" and the maximum value is 130.
2. *Exist-Items* query if an individual for the associated variable exists, e.g. "Does patient have metastasis?" (Lst. 1, line 15-16). The associated variable has to be an object variable.
3. *Specify-Items* are multiple choice items, which restrict the associated variable for different answers with different classes from the domain ontology, e.g. "Type of tumor?" with answer possibilities: "breasttumor", "nephroblastoma", "other" (Lst. 1, line 7-10). The associated variable has to be an object variable. The answer possibilities need to have associated class descriptions from the ontology which have to be subsumed by the ontology classes which are defined as type of the associated variable.

```
itemAnnotation ::= ItemAnnotation(itemID ontoPath itemConstructor);
ontoPath ::= OntoPath(focPointAssert ontoPathPart);
ontoPathPart ::= (relAssert varAssert)*;
itemConstructor ::= existItem|specifyItem|valueItem;
valueItem ::= Value(Min(minv)? Max(maxv)?);
specifyItem ::= Specify(cases);
cases ::= case | cases case;
case ::= Case(acode ontdescr);
existItem ::= Exist();
focPointAssert ::= varAssert;
relAssert ::= rel(srcvar, tarvar);
varAssert ::= type(var);
```

Listing 2. Grammar for an item annotation.

The grammar for an item annotation is depicted in Lst. 2. We have chosen the notation according to this grammar here, because it is well suited to describe the algorithms, although for storage of SWOD workflow descriptions a less redundant notation can be used (e.g. declaring variable and relation assertions for each form).

Condition Annotation. A condition annotation is the ontology description of a condition. It is formalized similar to bodies of SWRL-rules (Semantic Web Rule Language [12]), but in a simplified form. Conditions refer to the focal variable of the workflow also called focal variable of the condition. Furthermore, the conditions refer to condition variables, which are prefixed with "?" (e.g. ?v). Conditions consist of a conjunction of atoms of the form class(x), dataType(x), objProp(x, y), dataProp(x, y) or cmpOp(x, y), where class is a class description, dataType a data type, objProp is an object property, dataProp a data type property from the domain ontology, cmpOp is a comparison operator like $\leq$ and x and y are either the focal variable of the condition, condition variables or data values. Each condition variable has to be related by property atoms to the focal variable. E.g. "Patient has a tumor with weight greater than 4" is formalized as follows:

d:Tumor(?tum) $\wedge$ hasTumor(PPatient, ?tum) $\wedge$ hasDiameter(?tum, ?dia) $\wedge$ float(?dia) $\wedge$ >(?dia, 4)

3.3   Semantics of Workflow Description

We describe the semantics of a workflow description by defining workflow executions. A workflow execution starts with the start task of the workflow description. Then tasks are executed in the order they are related with transitions. For an XOR-split the task succeeding the condition, which is satisfied for the current execution, is executed next. This must be exactly one in a consistent workflow description. During execution of a task a user has to fill in a value into each item of the associated form of the task.

Since we do not consider concurrency, at each point of execution the executed part of the workflow can be described by a sequence of already filled items (described by their item annotation and the filled in value). We call such a sequence executed workflow data path (EWDP).

From such an EWDP an ABox $\mathcal{A}_p$ can be calculated, representing the state, i.e. the data which has been collected until this point in the workflow execution. This ABox is the base to query the collected data in terms of the ontology.

To derive $\mathcal{A}_p$ from an EWDP starting with an empty ABox, we defined a calculus, called "ABoxRules". The calculus extends $\mathcal{A}_p$ for each filled item in EWDP as follows (illustrated with item "age of mother" (Lst. 1, line 4-6) and value 61):

- *Ontology Path.* For each object variable in the ontology path an individual is created in $\mathcal{A}_p$, which is represented with the same name as the variable. Each variable assertion type(var) is added as a concept assertion to $\mathcal{A}_p$ (e.g. d:HumanBeing(PPatient) and d:HumanBeing (PMother)). Each object relation assertion rel(srcvar, tarvar) is added as a role assertion to $\mathcal{A}_p$ (e.g. hasMother(PPatient, PMother)). The individual created for the focal variable of the workflow is called focal individual.
- *Item constructor.* For an item into which value v is filled in and for which the last part of the associated ontology path is rel(srcvar, tarvar) type(tarvar), $\mathcal{A}_p$ is extended according to the item constructor as follows:

For a value-item, the according individual in $\mathcal{A}_p$ is related with the appropriate data type relation to the value of the item, i.e. rel(srcvar, v) is added to $\mathcal{A}_p$. For the example item hasAge(PMother, 61) is added to $\mathcal{A}_p$.

For a specify-item, the according individual is restricted with the class description c, which is associated to v, i.e. c(tarvar) is added to $\mathcal{A}_p$.

For an exist-item, an individual for the associated variable is only created if value is yes. Else a concept assertion $(=0 \; \text{rel.type})(\text{srcvar})$ is added to $\mathcal{A}_p$.

Whether a condition is satisfied for an execution (i.e. the task after it is selected for further execution), can be determined with the help of the ABox $\mathcal{A}_p$ calculated from the EWDP ending with the last item before the condition. A condition is satisfied for the execution if it has a valid binding to $\mathcal{A}_p$. Here, we describe valid bindings informally and refer to [6] for a formal description. A valid binding between a condition and $\mathcal{A}_p$ exists if each of the variables from the condition can be bound to an individual of $\mathcal{A}_p$, where the focal variable of the condition is bound to the focal individual, and certain binding- and condition-specific constraints hold on $\mathcal{A}_p$. E.g. Condition A (s. Lst. 1, line 18) has a valid binding to any ABox $\mathcal{A}_p$, with individuals p and m, for which holds that p is the focal individual and assertions d:Metastasis(m) and hasMetastasis(p,m) can be inferred from $\mathcal{A}_p$. Then a valid binding exists, where p is bound to PPatient and m is bound to ?met.

4   Data Inconsistencies

In this section, we describe the kinds of data inconsistencies, which can occur in a workflow description, and define them based on the semantics proposed in Sec. 3.3.

4.1   Semantic Data Inconsistencies

If at any point of a workflow execution the calculated ABox $\mathcal{A}_p$ is inconsistent wrt. the domain ontology, a Semantic Data Inconsistency occurs. In such a situation the workflow description contradicts the domain ontology causing collected data to be erroneous. We distinguish Semantic Data Inconsistencies by the restrictions, which are violated through them in the domain ontology. In the following, we list some Semantic Data Inconsistencies and describe them with an example.

- *Violation of disjoint class restrictions.* E.g. a variable on a form is declared to be of type d:Breasttumor, a variable with the same name on another form of type d:Nephroblastoma. Both forms can occur in the same workflow execution. The classes d:Nephroblastoma and d:Breasttumor are declared to be disjoint in the domain ontology.
- *Violation of number restrictions.* E.g. in the domain ontology is defined that a human being can have at most one mother. Two variables for the patient's mother with different names are defined on the forms.

– *Violation of data type restrictions.* E.g. in the domain ontology is defined that the height of a patient must not be greater than 250 cm. A value between 0 and 300 cm may be filled into the item, which queries the height of a patient.

4.2 Data-Dependent Control Flow Inconsistencies

Unsatisfiable conditions A condition is unsatisfiable if it is not satisfied for any workflow execution (has no valid binding to the ABox $\mathcal{A}_p$ of the appropriate EWDP). That means that for each workflow execution it is either not satisfied or, due to missing data, it can not be determined if it is satisfied or not. Unsatisfiable conditions can lead to unreachable tasks after the condition.

Example: A condition is fulfilled if the weight of the patient's tumor is greater than 2000 g. For the item querying the weight of the tumor, a maximum input value of 1500 g is defined.

XOR-stall. If for a workflow execution none or more then one of the conditions at an XOR-split are satisfied, the task to be executed next can not be determined unambiguously. We call such a situation XOR-stall. In such a case workflow execution aborts.

Example: Item "type of tumor" (Lst. 1, line 7-10) is declared on a form. An XOR-split, following the form, has two outgoing transitions with conditions "Patient has breasttumor" and "Patient has nephroblastoma". If answer possibility "other" is selected for the item, an XOR-Stall occurs.

5   Consistency Checking Algorithm

In the following, we describe an algorithm, which is able to detect the described inconsistencies. We describe in detail how the algorithm detects Semantic Data Inconsistencies and in principle how Data-Dependent Control Flow Inconsistencies can be detected. The most obvious algorithm for this problem is a simulation of all possible workflow executions. We have described such an algorithm, called CCABoxes-algorithm, in [6]. This algorithm calculates for each possible workflow path the set of possible ABoxes, called $\mathcal{X}_p$, and checks the ABoxes for inconsistencies according to their definitions described in Sec. 4. This algorithm is by definition sound and complete, but it does not always terminate. This is due to the fact that primitive data types (e.g. integer) have infinite data ranges, and e.g. into a value-item with data type integer an infinite number of different values can be filled in. Since in the CCABoxes-algorithm an ABox is created for each possible execution, this results in the creation of an infinite number of ABoxes. Therefore, we need to replace the calculated set of ABoxes with a finite abstraction, which preserves the information needed to detect the described inconsistencies, to gain a terminating algorithm.

We use an ontology as abstraction, which describes all ABoxes sufficiently to detect the inconsistencies. The resulting "CCOnto-algorithm" calculates a so-called path ontology $\mathcal{O}_p$ for each workflow path. The idea of our abstraction is to represent individuals by newly created classes in $\mathcal{O}_p$. Each individual created

---

Algorithm 1. CCOnto-algorithm

---

input: SWOD WF description swodWf, domain ontology $\mathcal{O}_D$
Set $\mathcal{W} \leftarrow$ determineWFDPs(swodWf);
foreach *WFDP* $\in \mathcal{W}$ **do**
$\quad \mathcal{O}_{pi} \leftarrow \mathcal{O}_D$;
$\quad \mathcal{O}_p \leftarrow$ ontoRules($\mathcal{O}_{pi}$, WFDP);
$\quad$ **if** $\mathcal{O}_p \equiv ERROR(kind, id)$ **then**
$\quad\quad$ **if** *kind* $\equiv$ *"XOR-Stall" OR "Semantic Data Inconsistency"* **then**
$\quad\quad\quad$ ABORT with ERROR(kind, id)
$\quad\quad$ **if** *kind* $\equiv$ *"Condition unsatisfiable for WFDP"* **then**
$\quad\quad\quad$ delete WFDP from $\mathcal{W}$

foreach *condition con in swodWf* **do**
$\quad$ **if** *con not in any of WFDP* $\in W$ **then**
$\quad\quad$ ABORT with ERROR("Condition unsatisfiable", id);

---

during any workflow execution is represented by at least one class, called its corresponding class. The information about the relations of the individual and the classes it belongs to are preserved in $\mathcal{O}_p$ by appropriate restrictions on its corresponding class. That results in the fact that Semantic Data Inconsistencies can be detected, because in any ABox of $\mathcal{X}_p$ a Semantic Data Inconsistency occurs if and only if $\mathcal{O}_p$ is inconsistent or a violation of a data value restriction occurs. A class in $\mathcal{O}_p$ can have an infinite number of corresponding individuals from different ABoxes $\mathcal{A}_p$.

Furthermore, to detect Data-Dependent Control Flow Inconsistencies, a so-called condition ontology is created in the CCOnto-algorithm as abstraction for a condition. In such an ontology a focal condition class is created to represent the condition and this class is restricted according to the axioms in the condition. The focal condition class is constructed such that a focal leaf class in $\mathcal{O}_p$ is a subclass of the focal condition class if and only if it has a corresponding focal individual in an ABox which has a valid binding to the condition.

5.1   Outline of Algorithm

In the following, we describe the CCOnto-algorithm (see Alg. 1) in more detail. The input of the algorithm is the SWOD workflow description swodWf and the domain ontology $\mathcal{O}_D$. If the algorithm does not abort with an error, the workflow description is consistent.

Determine WFDPs. To calculate $\mathcal{O}_p$ for a workflow path we have to consider the items in the flow as well as each condition, because the ABoxes $\mathcal{A}_p$ for which the condition is not satisfied are not any more possible after its application. We describe this information by a workflow data path (WFDP), which is a sequence of item and condition annotations, according to the following grammar, where conditionAnnot is a condition annotation:

wfdp::= itemAnnotation wfdp | conditionAnnot wfdp | itemAnnotation | conditionAnnot;

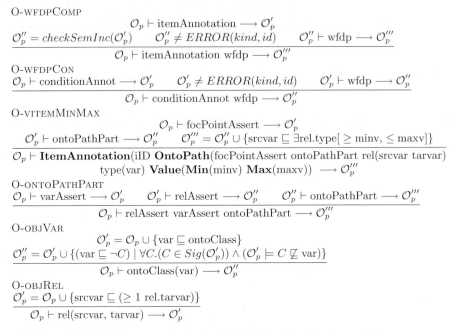

O-wfdpComp
$$\frac{\mathcal{O}_p \vdash \text{itemAnnotation} \longrightarrow \mathcal{O}'_p \qquad \mathcal{O}''_p = checkSemInc(\mathcal{O}'_p) \quad \mathcal{O}''_p \neq ERROR(kind, id) \qquad \mathcal{O}''_p \vdash \text{wfdp} \longrightarrow \mathcal{O}'''_p}{\mathcal{O}_p \vdash \text{itemAnnotation wfdp} \longrightarrow \mathcal{O}'''_p}$$

O-wfdpCon
$$\frac{\mathcal{O}_p \vdash \text{conditionAnnot} \longrightarrow \mathcal{O}'_p \qquad \mathcal{O}'_p \neq ERROR(kind, id) \qquad \mathcal{O}'_p \vdash \text{wfdp} \longrightarrow \mathcal{O}''_p}{\mathcal{O}_p \vdash \text{conditionAnnot wfdp} \longrightarrow \mathcal{O}''_p}$$

O-vitemMinMax
$$\frac{\mathcal{O}_p \vdash \text{focPointAssert} \longrightarrow \mathcal{O}'_p \qquad \mathcal{O}'_p \vdash \text{ontoPathPart} \longrightarrow \mathcal{O}''_p \qquad \mathcal{O}'''_p = \mathcal{O}''_p \cup \{\text{srcvar} \sqsubseteq \exists \text{rel.type}[\geq minv, \leq maxv]\}}{\mathcal{O}_p \vdash \textbf{ItemAnnotation}(\text{iID } \textbf{OntoPath}(\text{focPointAssert ontoPathPart rel(srcvar tarvar) type(var) } \textbf{Value}(\textbf{Min}(minv) \textbf{Max}(maxv)) \longrightarrow \mathcal{O}'''_p}$$

O-ontoPathPart
$$\frac{\mathcal{O}_p \vdash \text{varAssert} \longrightarrow \mathcal{O}'_p \qquad \mathcal{O}'_p \vdash \text{relAssert} \longrightarrow \mathcal{O}''_p \qquad \mathcal{O}''_p \vdash \text{ontoPathPart} \longrightarrow \mathcal{O}'''_p}{\mathcal{O}_p \vdash \text{relAssert varAssert ontoPathPart} \longrightarrow \mathcal{O}'''_p}$$

O-objVar
$$\frac{\mathcal{O}'_p = \mathcal{O}_p \cup \{\text{var} \sqsubseteq \text{ontoClass}\} \qquad \mathcal{O}''_p = \mathcal{O}'_p \cup \{(\text{var} \sqsubseteq \neg C) \mid \forall C.(C \in Sig(\mathcal{O}'_p)) \wedge (\mathcal{O}'_p \models C \not\sqsubseteq \text{var})\}}{\mathcal{O}_p \vdash \text{ontoClass(var)} \longrightarrow \mathcal{O}''_p}$$

O-objRel
$$\frac{\mathcal{O}'_p = \mathcal{O}_p \cup \{\text{srcvar} \sqsubseteq (\geq 1 \text{ rel.tarvar})\}}{\mathcal{O}_p \vdash \text{rel(srcvar, tarvar)} \longrightarrow \mathcal{O}'_p}$$

Fig. 2. Excerpt of calculus OntoRules, for all rules see [6]

The first step of the CCOnto-algorithm is to determine all possible workflow paths from swodWF and store the corresponding WFDPs in a set $\mathcal{W}$. E.g. one of the two WFDPs for the example consists of the item annotations IAge, ITum, IName, IMet and the condition annotation ConditionA.

Create $\mathcal{O}_p$s for each WFDP in $\mathcal{W}$. The path ontology $\mathcal{O}_p$ is created by the function ontoRules($\mathcal{O}_{Pi}$, WFDP), where $\mathcal{O}_{Pi}$ is the initial path ontology comprising the axioms from the domain ontology (s. Sec. 5.2). The result of the function is either the consistent path ontology or an error term describing an inconsistency which occurred during creation of $\mathcal{O}_p$. If the result is an error term ERROR(kind id) with kind "XOR-Stall" or "Semantic Data Inconsistency", the algorithm aborts with an error message. If kind is "Condition unsatisfiable for WFDP", the processed WFDP is deleted from $\mathcal{W}$ since the corresponding workflow path can never be taken during workflow execution.

After $\mathcal{O}_p$ is created for each WFDP, it is checked if a condition is unsatisfiable for the workflow by checking if any of the conditions does not appear in any of the WFDPs left in $\mathcal{W}$, which represent the possible flows through the workflow. In that case the algorithm aborts with an appropriate error.

5.2   Creation of Path Ontology

Function ontoRules($\mathcal{O}_{Pi}$, WFDP) creates $\mathcal{O}_p$ as the longest possible derivation with the rules of the calculus OntoRules (part of calculus is shown in Fig. 2).

In principle $\mathcal{O}_p$ is created as follows from WFDP and the initial path ontology utilizing the calculus. WFDP is split into its item and condition annotations (described in rules O-WFDPCOMP and O-WFDPCON), which are processed in the order they appear in the WFDP:

- *Item Annotation.* For each item annotation the following steps are processed (illustrated with item "age of mother" (Lst. 1, line 4-6)):

  • *Object variable assertion.* As described in rule O-OBJVAR, for each object variable in the ontology path a class is created, which is represented with the same name as the object variable, called corresponding class of the variable. The class is declared to be a subclass of the type of the variable and to be disjoint to classes created from variables with distinct names. The class created for the focal variable is called top focal path class and its subclasses are called focal path classes.
  For the example item {PPatient $\sqsubseteq$ d:HumanBeing $\sqcap \neg$ PMother, PMother $\sqsubseteq$ d:HumanBeing} is added to $\mathcal{O}_p$. PPatient is the top focal path class.

  • *Object relation assertion.* As described in rule O-OBJREL, for each object relation objProp(X, Y) in the ontology path the axiom (X $\sqsubseteq$ ($\geq$1 objProp.Y)) is added to $\mathcal{O}_p$ (e.g. PPatient $\sqsubseteq$ ($\geq$1 hasMother.PMother)}).

  • *Item constructor.* We describe in detail how value-items and in principle how exist- and specify-items are processed.
  For a value-item with minimum value mi, maximum value ma and for which the last part of the associated ontology path is rel(X, Y) type(Y), the axiom (X $\sqsubseteq \exists$ rel.type[$\geq$ mi, $\leq$ ma]) is added to $\mathcal{O}_p$, as described in rule O-VITEMMINMAX. For the example item, PMother $\sqsubseteq \exists$ hasAge.integer[$\leq$ 130] is added to $\mathcal{O}_p$. This shows that using an ontology allows the CCOnto-algorithm to work with data ranges instead of single data values, which enables the algorithm to terminate always.
  For exist-, and specify items, classes for each answer possibility are created, which represent the individuals in the ABoxes, for the workflow executions, in which the according answer possibilities are filled out. In a first step classes for the associated variables of the item are created and in a second step, which is called expansion step, classes for each other variable in the associated ontology path is created. The created classes are subclasses of the corresponding class of the variable they are created for and of classes, which are specific for the answer possibility they are created for.

  • *Semantic Data Inconsistency.* A Semantic Data Inconsistency occurs after applying an item annotation if $\mathcal{O}_p$ is inconsistent or a violation of a data value restriction occurs. In this case $\mathcal{O}_p$ is set to an appropriate error term. A violation of a data value restriction occurs if dataRange $\not\sqsubseteq$ dataRangeC holds, for any axiom (X $\sqsubseteq \exists$ datarel.dataRange) $\in \mathcal{O}_p$, for which a data range dataRangeC exists and ($\mathcal{O}_p \models$ X $\sqsubseteq \forall$ datarel.dataRangeC) holds.

- *Condition Annotation.* When a condition is applied, the classes are deleted from $\mathcal{O}_p$, which are corresponding to individuals in ABoxes $\mathcal{A}_p$ for which the condition is not satisfied. Therefore, the condition ontology is created and all focal path classes are deleted from $\mathcal{O}_p$, except for these which are subsumed by the focal condition class or are a superclass of such a subsumed class. Then recursively each class Y is deleted from $\mathcal{O}_p$ for which holds ($\nexists$C $\in$ Sig($\mathcal{O}_p$).$\mathcal{O}_p \models$ C $\sqsubseteq$ (≥ 1 objProp.Y)), for any object property objProp.

- *Condition unsatisfiable.* If $\mathcal{O}_p$ has no more classes after applying a condition, the condition is unsatisfiable for this WFDP and $\mathcal{O}_p$ is set to an appropriate error term.
- *XOR-Stall.* Before the first condition of an XOR-split is applied, $\mathcal{O}_p$ is checked for an XOR-stall. Therefore, it is checked if any of the focal path classes, which is a leaf class, is a subclass of exactly one of the focal condition classes of the conditions at the XOR-split. If any focal path class is a subclass of none condition class, then it exists a possible ABox for which none of the conditions is satisfied. If any focal path class is a subclass of more then one of the condition classes, then it exists a possible ABox for which more then one of the conditions is satisfied. In both cases $\mathcal{O}_p$ is set to an appropriate error term.

Example Inconsistency. The example treatment plan is consistent. In the following, we illustrate how a Semantic Data Inconsistency can be detected. Therefore, the workflow annotation is changed as follows. In line 5 of Lst. 1, variable PMother is replaced by PMother1. During execution of the changed workflow in any ABox two different individuals for mother are created. Since it is declared in the domain ontology that a human being can have at most one mother, each possible ABox $\mathcal{A}_p$ is not consistent wrt. the domain ontology. That can be detected with the CCOnto-algorithm as follows. For the original workflow description in $\mathcal{O}_p$ only one class is created for mother. For the changed workflow description in $\mathcal{O}_p$ two disjoint classes for mother are created:

PPatient $\sqsubseteq$ (≥ 1 hasMother.PMother) $\sqcap$ (≥ 1 hasMother.PMother1) $\sqcap$ d:HumanBeing...
PMother1 $\sqsubseteq \neg$ PMother

Since in the domain ontology it is declared that a human being can have at most one mother, class PPatient is unsatisfiable and thus $\mathcal{O}_p$ is inconsistent. The algorithm aborts with error "Semantic Data Inconsistency".

5.3   Soundness and Completeness

In this section we sketch the proof that the CCOnto-algorithm is complete, i.e. all inconsistencies listed in Sec. 4 are detected, and sound, i.e. only the described inconsistencies are detected. From the definition of inconsistencies follows soundness and completeness for the CCABoxes-algorithm. Completeness and soundness of the CCOnto-algorithm is proved by showing that it detects the same inconsistencies as the CCABoxes-algorithm, as outlined in the following.

The steps of the two algorithms are the same, with the difference, that CCOnto utilizes $\mathcal{O}_p$ instead of a set of ABoxes, $\mathcal{X}_p$, to detect inconsistencies. $\mathcal{O}_p$ derived from WFDP is called an abstraction of $\mathcal{X}_p$ derived from the same WFDP. We can show that following criteria hold between $\mathcal{X}_p$ and its abstraction $\mathcal{O}_p$: Each individual in any ABox of $\mathcal{X}_p$ has a corresponding class in $\mathcal{O}_p$ and each leaf class in $\mathcal{O}_p$ has a corresponding individual. With the help of these criteria we are able to prove for each type of the inconsistencies that exactly those inconsistencies are detected with the CCOnto as with the CCABoxes-algorithm (for full proof and definition of corresponding individuals and classes s. [6]).

6   Discussion

In this paper, we described a workflow language with an ontology-based data perspective, called SWOD, and defined its semantics. We classified possible data inconsistencies, which can occur in SWOD-workflow descriptions and described an algorithm, called CCOnto-algorithm, to detect them during design time.

Implementation. A prototypical implementation of the CCOnto-algorithm is currently developed in Java using the OWL API [13] and the DL-reasoner Pellet [14]. We plan to integrate it into ObTiMA, to support users in defining consistent treatment plans.

Related Work. Consistency checking algorithms exist to check structural consistency (e.g. [1], [2]), but few integrate data (e.g. [3], [15] or [4]). We are not aware of an algorithm, which is able to detect inconsistencies in complex data perspectives based on semantic annotations. Nonetheless, semantic annotation and its applications are an active research field crucial for the realization of the semantic web. Existing work on semantic annotations for web services is especially interesting, since web services can be composed to workflows, also called composite services. Various languages for semantic annotations of Web Services have been defined (e.g [16], [17]), which amongst others describe data, like input and output values, with metadata from ontologies. E.g. SAWSDL (Semantic Annotations for WSDL and XML), recommended by W3C, defines the syntax for semantic annotations, but does not define a formal semantics. The main aims of SAWSDL is automatic service discovery and composition, but enriched with appropriate semantics it can be used for data consistency checking of complex workflows composed from web services. Therefore, in the future we aim to investigate how to integrate semantic web services in our work. However, currently we are interested in allowing collection of data and focus on the semantic annotation for forms. Few related work on generating forms from an ontology exist. These approaches mostly base the forms on the structure of the ontology and do not allow for flexible item creation (e.g. Protege form generation [18]). More flexible approaches allow to define arbitrary items but do not allow to define relational metadata from the ontology and only link items to classes (e.g. caCore FormBuilder [19]).

Future Work. We plan to extend the data perspective e.g. by allowing to create more complex items from the ontology and to define constants or constraints between items. We aim for a more expressive control flow perspective comprising e.g. cyclic workflows with respect to research on workflow-patterns [10]. We plan to consider time in the data and control flow perspectives.

Conclusion. Ontology-based data perspectives in data intensive workflows are well suited to provide the basis for static data analysis. In this work we defined a semantics for ontology-based workflows based on ABoxes, which provides the basis for algorithms to find data inconsistencies. We have shown how to use ontologies to finitely abstract infinitely many ABoxes, representing infinite data ranges in such an algorithm. Integrated with existing algorithms for checking structural consistency (e.g. [1], [2]) the technique described in this paper can have the capability to guarantee soundness of complex workflows.

References

1. Verbeek, H., van der Aalst, W., ter Hofstede, A.: Verifying Workflows with Cancellation Regions and OR-joins: An Approach Based on Relaxed Soundness and Invariants. Computer Journal 50(3), 294–314 (2007)
2. Qian, Y., Xu, Y., Wang, Z., Pu, G., Zhu, H., Cai, C.: Tool Support for BPEL Verification in ActiveBPEL Engine. In: Proceedings of the 2007 Australian Software Engineering Conference, pp. 90–100 (2007)
3. Sun, X., Zhao, J., et al.: Formulating the Data-Flow Perspective for Business Process Management. Information Systems Research 17(4), 374–391 (2006)
4. Eshuis, R.: Semantics and Verification of UML Activity Diagrams for Workow Modelling. PhD thesis, University of Twente (2002),
   http://www.ctit.utwente.nl/library/phd/eshuis.pdf
5. W3C OWL Working Group: OWL 2 Web Ontology Language. W3C Working Draft (2009), http://www.w3.org/TR/owl2-overview/
6. Weiler, G.: Consistency Checking for Workflows with an Ontology-Based Data Perspective (unpublished) (2009),
   http://softech.informatik.uni-kl.de/twiki/bin/view/Homepage/
   PublikationsDetail?id=133
7. Weiler, G., Brochhausen, M., Graf, N., Schera, F., Hoppe, A., Kiefer, S.: Ontology Based Data Management Systems for Post-Genomic Clinical Trials within a European Grid Infrastructure for Cancer Research. In: Proc. of the 29 Annual International Conference of the IEEE EMBS, August 2007, pp. 6434–6437 (2007)
8. Tsiknakis, M., Brochhausen, M., et al.: A Semantic Grid Infrastructure Enabling Integrated Access and Analysis of Multilevel Biomedical Data in Support of Postgenomic Clinical Trials on Cancer. IEEE Transactions on Information Technology in Biomedicine 12(2), 205–217 (2008)
9. Brochhausen, M., Weiler, G., et al.: The ACGT Master Ontology on Cancer - A new Terminology Source for Oncological Practice. In: Proc. of 21st IEEE International Symposium on Computer-Based Medical Systems, pp. 324–329 (2008)
10. Russel, N., ter Hofstede, A., van der Aalst, W., Mulyar, N.: Workflow Control-Flow Patterns: A Revised View. Technical report, BPMcenter.org (2006)
11. Baader, F., Calvanese, D., et al. (eds.): The Description Logic Handbook, Theory Implementation and Applications. Cambridge University Press, Cambridge (2003)
12. Horrocks, I., Patel-Schneider, P.F., Boley, H., Tabet, S., Grosof, B., Dean, M.: SWRL: A Semantic Web Rule Language. W3C Member Submission (2004)
13. Horridge, M., Bechhofer, S., Noppens, O.: Igniting the OWL 1.1 Touch Paper: The OWL API. In: OWLED 2007, 3rd OWL Experiences and Directions Workshop, Innsbruck, Austria (June 2007)

14. Sirin, B., Parsia, B., Grau, B., Kalyanpur, A., Katz, Y.: Pellet: A practical OWL-DL reasoner. Journal of Web Semantics 5(2) (2007)
15. Sundari, M., Sen, A., Bagchi, A.: Detecting Data flow Errors in Workflows: A Systematic Graph Traversal approach. In: Proc. 17th Workshop on Information Technology & Systems (2007)
16. Farrell, J., Lausen, H.: Semantic Annotations for WSDL and XML Schema. W3C Recommendation (2008)
17. Martin, D., Burstein, M., et al.: OWL-S: Semantic Markup for Web Services. W3C Member Submission (2004)
18. Rubin, D., Knublauch, H., et al.: Protege-owl: Creating ontology-driven reasoning applications with the web ontology language. In: AMIA Annu. Symp. Proc. (2005)
19. Whitley, S., Reeves, D.: Formbuilder: A tool for promoting data sharing and reuse within the cancer community. Oncological Nursing Forum 34 (2007)

Conceptual and Spatial Footprints
for Complex Systems Analysis:
Application to the Semantic Web

Bénédicte Le Grand[1], Michel Soto[2], and Marie-Aude Aufaure[3]

[1] Université Pierre et Marie Curie– Paris 6 CNRS
Benedicte.Le-Grand@lip6.fr
[2] Université Paris Descartes
Michel.Soto@lip6.fr
[3] MAS Laboratory, Ecole Centrale Paris
Marie-Aude.Aufaure@ecp.fr

Abstract. This paper advocates the use of Formal Concept Analysis and Galois lattices for complex systems analysis. This method provides an overview of a system by indicating its main areas of interest as well as its level of specificity/generality. Moreover, it proposes possible entry points for navigation by identifying the most significant elements of the system. Automatic filtering of outliers is also provided.

This methodology is generic and may be used for any type of complex systems. In this paper, it is applied to the *Topic Map* formalism which can be used in the context of the Semantic Web to describe any kind of data as well as ontologies. The proposed Conceptual and Spatial Footprints allow the comparison of Topic Maps both in terms of content and structure. Significant concepts and relationships can be identified, as well as outliers; this method can be used to compare the underlying ontologies or datasets as illustrated in an experiment.

Keywords: Topic Maps, Semantic Web, Ontologies, Navigation, Visualization, Clustering, Formal Concept Analysis, Galois lattices, Conceptual footprint.

1 Introduction

Semantic Web [1] standards have been developed to enhance information retrieval on the Web by adding explicit semantics to its content. Various formalisms exist, with different levels of complexity and expressiveness, from simple annotation syntaxes to sophisticated reasoning capabilities. In this paper the "lower" layers of the Semantic Web stack are considered, i.e. semantic annotation syntaxes, such as RDF [2] or Topic Maps [3] [4]. These formalisms allow describing data's content and relationships, thus providing it with a semantic structure. However, these syntaxes can also be used at a more abstract level to formalize ontologies themselves. Topic Maps and RDF can inter-operate at a fundamental level [5] and each language may be used to model the other. The analysis method proposed in this paper is illustrated on systems represented as Topic Maps but it may thus also be used with RDF.

S.S. Bhowmick, J. Küng, and R. Wagner (Eds.): DEXA 2009, LNCS 5690, pp. 114–127, 2009.
© Springer-Verlag Berlin Heidelberg 2009

Although semantics and relationships are explicit in Topic Maps (or ontologies), some problems remain in terms of information retrieval. A large number of ontologies have indeed been developed for various purposes and their comparison and/or integration is a challenge. In order to compare ontologies, one should first have a clear idea of each ontology's content and structure. At a more general level, the conceptual analysis method proposed in this paper provides an overview of the complex network under study, through the identification of its most significant elements, as well as a characterization in terms of homogeneity / heterogeneity. Navigation facilities also provide intuitive and interactive means to have a better knowledge of the system. An automatic outliers filtering function is also proposed.

This paper proposes a methodology and tools based on Galois lattices [6] [7] [8] for the analysis of Topic Maps or ontologies at a semantic, conceptual and structural level. More precisely, the goal of this methodology is therefore to help users' information retrieval and navigation in large Topic Maps and ontologies by:

- giving information about their specificity / generality,
- identifying their most significant elements,
- simplifying their structure by eliminating outliers,
- providing an overview and showing the relative impact of the various elements of the semantic structure.

This analysis is particularly useful to compare or integrate several Topic Maps or ontologies.

This paper is organized as follows. Section 2 introduces the context of this work, i.e. the notion of complex system in general and, in particular, the Topic Maps which are a possible formalism for the Semantic Web. Section 3 presents the proposed conceptual analysis method, which relies on Galois lattices. Experimental results are analyzed in Section 4 and Section 5 finally concludes and discusses the perspectives and extensions of this work.

2 Context

2.1 Complex Systems

The etymology of the word "complex" has for origin the Latin word "complexus" which means "weaving". A complex system is composed of elements connected by numerous and diverse links. The number and the diversity of these links make the system difficult to understand. Such a system is made of heterogeneous elements associated (weaved) in an inseparable way and between which numerous actions exist, interactions, feedback, determinations and sometimes fates [9]. In order to understand a complex system neither a global analysis nor the analysis of the elementary parts should be privileged: both have to be performed [10]. From a methodological point of view, it ensues that the analysis of a complex system has to be conducted by exploring frames of complexity which weave the system by making as much round trips as necessary between the global and the elementary parts [11]. In the domain of visual data mining Shneiderman [12] has formulated this methodology as: "*The Visual Information Seeking Mantra is: Overview first, zoom and filter, then details-on-demand*". This requires tools such as the ones presented in section 3 et 4.

The methodology proposed in this paper and its associated tools are illustrated on the Semantic Web which can be considered as a complex system.

2.2 Topic Maps

Many candidate techniques have been proposed to add semantic structures to the Web, such as semantic networks [13], conceptual graphs [14], the W3C Resource Description Framework (RDF) [2] and Topic Maps [3]. Semantic networks are basically directed graphs (networks) consisting of vertices linked by edges which express semantic relationships between vertices.

The conceptual graphs theory developed by Sowa [15] is a language for knowledge representation based on linguistics, psychology and philosophy.

RDF data consists of nodes and attached attribute/value pairs. Nodes can be any Web resource (pages, servers, basically anything identified by a URI), or other instances of metadata. Attributes are named properties of the nodes, and their values are either atomic (character strings, numbers, etc.), metadata instances or other resource. This mechanism allows building labelled directed graphs.

Topic Maps, as defined in ISO/IEC 13250 [3], are used to organise information in a way which can be optimised for navigation. Topic Maps were designed to solve the problem of large quantities of unorganised information, as information is useless if it cannot be found or if it is related to nothing else. The Topic Map formalism comes from the paper publishing world, where several mechanisms exist to organise and index the information contained within a book or a document. Topic Maps constitute a kind of semantic network above the data themselves and can thus be thought of as the online equivalent of printed indexes; they are indeed much more than this: Topic Maps are also a powerful way to manage link information, much as glossaries, cross-references, thesauri and catalogues do in the paper world. The Topic Map formalism can also be used to represent ontologies themselves as explained in the following.

The original Topic Map language, based on SGML, is often considered as too complex and a specification aiming at applying the Topic Map paradigm to the Web has been written: XTM (XML Topic Maps) [4], which allows structuring data on the Web so as to make Web mining more efficient.

As stated in the introduction, all the formalisms described in this section have the same goals and many of them are compatible. RDF and Topic Maps both add semantics to existing data without modifying them. They are two compatible formalisms. Moore, in [5], stated that RDF could be used to model Topic Map and vice versa. There are slight differences, e.g. the notion of scope -contest- exists in Topic Maps and not in RDF. RDF is more synthetic and more suited to queries whereas Topic Maps more suited for navigation. In the following, this paper will focus on XML Topic Maps, called *Topic Maps* in the remainder of the article for simplicity.

A Topic Map defines concepts (*topics*), for example ontological concepts if the Topic Map describes an ontology, and original data elements are *occurrences* of these topics. It is also possible to specify relationships (*associations*) among topics, i.e. among abstract concepts themselves. Ontologies can be formalized as Topic Maps because associations are not limited to the "is-a" relationship. Thanks to the links provided by associations, users may navigate at the (abstract) topic level and "go down" to original data once they have found an interesting concept. Finding a topic of interest is easier than navigating in the original data layer, thanks to the semantic structure provided by the Topic Map.

An example Topic Map will be used to illustrate the notions presented in this paper. This Topic Map, called *Music*, is dedicated to music, *The Clash* band in particular[1]. This is a very small Topic Map which contains 46 topics, 6 associations and 4 occurrences. Figure 1 illustrates the resource and the semantic levels of a Topic Map; original data elements belong to the resource level and are related to the semantic level through occurrences relationships (which may be of different types). The semantic level consists of topics (which may be of different types) and the associations between topics (which may also be of different types). In a Topic Map, almost everything is a topic: topics of course, but also topic types, association types and occurrences types.

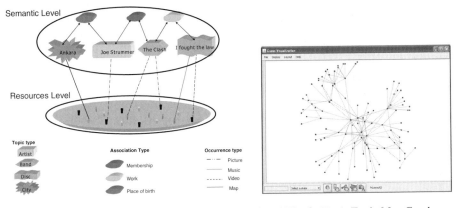

Fig. 1. Topic Map Structure   **Fig. 2.** *Music* Topic Map Graph

One of Topic Maps' strengths is that they do not modify data itself; this semantic annotation and structure may thus be performed by other persons than the owners of original data. Moreover, the semantic layer is independent of the underlying data, and can thus be used on different datasets or ontologies. Conversely, a given dataset/ontology may be described by different Topic Maps.

Topic Maps significantly increase the relevance of query results as the semantics avoid ambiguities. Moreover, advanced queries, based on the various types of relationships, may be performed, for example: *"Which members of The Clash band were not born in Ankara?"*. *Tolog* [16] is an example of Topic Maps query language.

The Figure 2 represents the graph of concepts (linked by associations) contained in the *Music* Topic Map. If the number of topics, associations and occurrences increases, this graph may rapidly become cluttered and therefore difficult to interpret.

The approach followed in this article relies on the exploitation of all existing relationships: explicit ones defined in a Topic Map –as associations or occurrences- and implicit ones, inferred from data elements' similarity. The method developed to identify these implicit relationships is based on a specific clustering performed by Galois

---

[1] This Topic Map was written by Kal Ahmed, http://www.techquila.com/

lattices (see Section 3). The clustering provided by a Galois lattice has several advantages. First, it is exhaustive as it computes all possible clusters. Moreover, these clusters may be overlapping, which is another advantage of this algorithm. Finally, the semantics of each cluster is explicit, as it consists of the common properties of objects in the corresponding concept.

3 Conceptual Approach for Topic Maps Analysis

This section presents the conceptual method proposed in this paper for Topic Maps analysis. Early ideas and preliminary results have been presented in [17]; since then, substantial refinement and optimizations have been presented, as presented in this paper. Section 3.1 introduces Galois lattices and section 3.2 describes the objects and properties' generation process. The conceptual statistics computed from Galois lattices are then presented in section 3.3, in particular the *Relatedness* and *Closeness* parameters which constitute the basis of the proposed *conceptual footprints* and *conceptual distributions*. Finally, a hierarchical and multiple-scale visualization is proposed in Section 3.4.

3.1 Formal Concept Analysis and Galois Lattices

FCA is a mathematical approach to data analysis which provides information with structure. FCA may be used for conceptual clustering as shown in [8] and [18]. The notion of Galois lattice to describe a relationship between two sets is the basis of several conceptual classification methods. This notion was introduced by [7] and [6]. Galois lattices group objects into classes which materialize concepts of the domain under study. Individual objects are discriminated according to the properties they have in common.

Consider two finite sets D (a set of *objects*) and M (the set of these objects' *properties*), and a binary relation $R \subseteq DxM$ between these two sets. Let o be an object of D and p a property of M. We have oRp if the object o has the property p.

According to Wille's terminology [19]:

$$Fc = (D, M, R)$$

is a formal context which corresponds to a unique Galois lattice, representing natural groupings of D and M elements.

Let P(D) be the powerset of D and P(M) the powerset of M. Each element of the lattice is a couple, also called *concept*, noted (O, A). A concept is composed of two sets $O \in P(D)$ and $A \in P(M)$ which satisfy the two following properties:

$$A = f(O), \text{ where } f(O) = \{a \in M \mid \text{for all } o \in O, oRa\}$$
$$O = f'(A), \text{ where } f'(A) = \{o \in D \mid \text{for all } a \in A, oRa\}$$

O is called the *extent* of the concept and A its *intent*. The extent represents a subset of objects and the intent contains these objects' common properties.

An example of Galois lattice generated from a binary relation is illustrated on Figure 3. The concept *({1, 2}; {a, c})* contains the objects *1* and *2* in its extent, and the common properties of these two objects are *a* and *c*. The concept may be generalized in two ways, either with *({1, 2, 3}; {a})* or *({1, 2, 4}; {c})*, depending on which common property is chosen. It may also be specialized in two different manners, either with *({1}; {a, c, f, h})* or *({2}; {a, c, g, i})*, depending on which object is chosen.

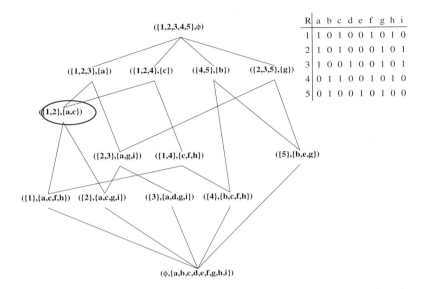

R	a	b	c	d	e	f	g	h	i
1	1	0	1	0	0	1	0	1	0
2	1	0	1	0	0	0	1	0	1
3	1	0	0	1	0	0	1	0	1
4	0	1	1	0	0	1	0	1	0
5	0	1	0	0	1	0	1	0	0

Fig. 3. Example of Galois lattice

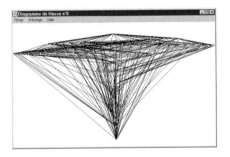

Fig. 4. Galois lattice of the *Music* Topic Map

A partial order on concepts is defined as follows:

$$\text{Let } C1=(O1, A1) \text{ and } C2=(O2, A2), C_1 < C_2 \Leftrightarrow A_1 \subseteq A_2 \Leftrightarrow O_2 \subseteq O_1.$$

Galois lattices are very well fitted to showing the various types of relations –explicit and implicit- among topics of a Topic Map. However, the number of concepts in a lattice grows exponentially with the number of objects and properties, which makes the interpretation with traditional Hasse diagrams impossible. The Figure 4 shows the Galois lattice generated for the *Music* Topic Map. Although it "only" contains 124 concepts, its Hasse diagram is very cluttered. The authors of [20] have defined interest measures to reduce the size of large concept lattices and apply their method to healthcare social communities. Our approach exploits the whole lattice, in order to follow an exhaustive approach, leading to the definition of a *Conceptual Distribution* and a *Conceptual Footprint*, as presented in the following Section.

3.2   Objects and Properties Generation

As explained in Section 2, Topic Map consists of *topics* related to other topics through *associations* and to raw data through *occurrences*. Topics my also be *instances* of other topics (i.e. they are subclasses of these topics). In the context of Galois lattice construction algorithm, objects are topics and their properties are their XML attributes, superclasses and associations. In this case, the Galois lattice then consists of concepts comprising sets of topics (*objects*) described by their common attributes, superclasses and associations (common *properties*).

Let a specific topic: *t-disc-the-clash*, which is the topic related to discs made by The Clash band. This topic is an instance of the *tt-disc* topic and it has an occurrence –*clash.gif-*, which is the picture on a CD jacket. Moreover, it is associated to the topic *t-the-clash* –which is itself an instance of the topic *tt-band,* through the association with id *N304* (which is an instance of the *at-recorded* type of association).

The *t-disc-the-clash* object's properties are its occurrences, the topic(s) it is an instance of and the identifiers of the associations it is involved in, i.e. *tt-disc, clash.gif* and *N304*. The topics with which *t-disc-the-clash* is associated (e.g. *t-the-clash*) do not directly appear as its properties, but the relationship with them will appear in Galois lattices as they have the *N304* association as a common property.

In practice, the Topic Map needs to be parsed twice to generate objects and properties: the first step identifies all objects (i.e. all identifiers of topic elements in the Topic Map) and their intrinsic properties (i.e. the values of the attributes of the topic elements, basically their ids, names and superclasses, as well as the values of the attributes of their children elements as the XML file structure is hierarchical). The second step completes the list of objects properties with the identifiers of all associations they are involved in.

3.3   Conceptual Distribution and Conceptual Footprint

Consider an object o (i.e. a *topic*). In the methodology proposed in this paper, it is characterized by two parameters called *Relatedness* and *Closeness* computed from the Galois lattice.

Relatedness indicates the proportion of objects with which o has some properties in common; such objects are called o's *related* objects. However, the "resemblance" between o and its related objects might be very little -or on the contrary very high.

Closeness precisely provides this information, by indicating the proportion of common properties between o and its related objects. The computation of these two parameters –which constitute the object o's *Conceptual Distribution*-, is explained in the following.

Relatedness. Let $C(o)$ be the set of the lattice G's concepts containing o in their extent. Let $C'(o)$ be the subset of concepts from $C(o)$ which have a least one other object than o in their extent and at least one property in their intent. The value of the object o's Relatedness (*Relatedness(o)*) is the average number of objects with which o is clustered into C'(o) concepts, divided by the total number of objects in the system (i.e. in the Topic Map). The Relatedness value indicates if o is connected to many other objects.

$$relatedness(o) = \frac{\frac{1}{Card(C'(o))} \sum_{i=1}^{Card(C')} Card(Extent(C'_i(o)))}{Card(Extent(Inf(G)))}$$

Closeness. Let S be the set of objects which are grouped with o in one –or more– concepts of the lattice (i.e. the set of o's *related* objects)[2]; these objects have at least one of o's properties (by construction). The value of the object o's Closeness (*Closeness(o)*) is the average number of properties o shares with other objects, divided by the total number of properties in the system. This parameter indicates whether o's resemblance to other objects from S is little or high (in terms of common properties).

$$closeness(o) = \frac{\frac{1}{Card(C'(o))} \sum_{i=1}^{Card(C')} Card(Intent(C'_i(o)))}{Card(Intent(Sup(G)))}$$

Conceptual Distribution. The couple (*Relatedness(o)*, *Closeness(o)*) constitutes the object o's *conceptual distribution* (illustrated of Figure 6).

Conceptual Footprint. Each object of a given system is therefore characterized by the Relatedness and Closeness parameters (i.e. its conceptual distribution). The average value for all objects of the dataset is the conceptual footprint of this system (illustrated on Figure 5).

Let *syst* be the complex system under study –in this case a Topic Map- containing N objects.

$$conc\_footprint(syst) = (relatedness(syst), closeness(syst))$$

$$\text{where } relatedness(syst) = \frac{\sum_{i=1}^{N} relatedness(o_i)}{N}$$

$$\text{and } closeness(syst) = \frac{\sum_{i=1}^{N} closeness(o_i)}{N}$$

3.4 Automatic Objects Filtering

In a Topic Map, some objects may share "many" common properties with "many" other objects. These objects are called *regular* objects and they are semantically more significant than others. The meaning of the words "many" (properties) and "many" (objects) depends on the values of relatedness and closeness of the Topic Map's conceptual footprint.

---

[2] C'(o) is a subset of S, consisting of elements from S which have at least one property in their intent.

On the other hand, a Topic Map may contain topics which are not semantically significant or not much related to others. These topics are called *marginal* topics and may be considered as outliers. They may be eliminated from the Topic Map in order to simplify it. The filtering process was automated by eliminating from the original dataset all marginal objects, i.e. objects whose Relatedness and Closeness values are below the system's Relatedness and Closeness values minus α*standard deviation for each parameters, where α may vary.

An object *o* is marginal if it verifies the following conditions:

$$relatedness\ (o)\ \leq relatedness\ (syst) - \alpha * std.dev\ (relatedness\ (syst)\ )$$

and

$$closeness\ (o)\ \leq\ closeness\ (syst) - \alpha * std.dev\ (closeness\ (syst)\ ),$$

where $std.dev(relatedness(syst)) = \dfrac{\sum\limits_{i=1}^{N}\left|relatedness(o_i) - relatedness(syst)\right|}{N}$

and

$$std.dev\ (closeness\ (syst\ )) = \frac{\sum\limits_{i=1}^{N}\left|closeness\ (o_i\ ) - closeness\ (syst\ )\right|}{N}$$

The Relatedness and Closeness parameters are computed again for all objects of the new dataset and marginal elements are eliminated, and so on. The filtering algorithm converges when the dataset no longer contains marginal elements, i.e. when the Relatedness and Closeness values of all remaining objects are sufficiently homogeneous (or when there is only one topic left if the network is highly heterogeneous).

3.5   Galois Lattice's Concepts Selection

The number of objects in a Topic Map after the convergence of the filtering algorithm may still be high and consequently the number of concepts of the corresponding Galois lattice too. It would thus be interesting to select the most significant concepts, organized in different levels of details –or *scales*- in order to make navigation and visualization easier.

The selection of concepts proposed here consists in extracting a tree from the original lattice. This tree provides a hierarchical representation of the complex system with different levels of detail corresponding to the depth of the tree. The result is therefore a hierarchical clustering where clusters may be overlapping (as they are selected concepts from the lattice). The root of the tree contains all objects; the next level groups some objects together (with possible overlaps), the next level is a finer grouping of objects, etc.

The construction of the tree starts from the finest level of detail: the leaf clusters of the tree are the most specific concepts of the lattice –i.e. the parent concepts of the upper bound.

For each leaf, one unique parent concept is selected which is a generalisation of the leaf concept. This selection is done according to a hierarchy of criteria in case a concept has several parent concepts in the lattice. The first criterion consists in selecting the parent with the lower distance to the lower bound of the lattice. If several parents meet this criterion, another one aims at minimizing the total number of concepts in the extracted tree. Another selection criterion relies on possible weights assigned to properties. If several candidates still remain, one of them is selected randomly.

A unique parent is then selected for each selected concept, and so on until the lower bound of the lattice is reached. At the end of this process, a tree is created. Each level of the tree contains clusters which correspond to a specific level of detail. The tree extracted from the *Music* Topic Map's Galois lattice is shown on Figure 9.

The methodology and features presented above are illustrated in an experiment in the following section.

4   Results

Four Topic Maps – of different sizes and subjects- were analyzed; these datasets' features are summarized in Table 1.

Table 1. Description of Studied Topic Maps

Topic Map	Subject	Nb of objects	Nb of properties	Nb of associations	Nb of occurrences	Nb of concepts in lattice
Music	Music	46	112	6	4	124
Discovery	Cooking	48	79	9	18	172
SHCC	Theatre	76	125	25	3	185
ICC	XTM	25	84	6	28	72

4.1   Conceptual Footprints and Distributions

Figure 5 represents the conceptual footprints of the four studied Topic Maps. These conceptual footprints consist in the average *Relatedness* and *Closeness* values for each system. Figure 5 shows that *ICC* has the highest average Relatedness value, then *Music, Discovery* and finally *SHCC*. This means that the topics from *ICC* dataset have common properties with a higher proportion of topics in their Topic Map than the topics of the three other samples.

The Closeness values show whether the resemblance between topics is strong or not: Relatedness indeed indicates that there is at least one common property; Closeness specifies if there are few or many common properties (it thus represents the *strength* of the relationship). Results presented on Figure 5 show that even in samples with a quite high average Relatedness value, relationships amongst topics are not very strong as the average Closeness values are very low.

In order to analyze more precisely these Topic Maps (as conceptual Footprints represent average values), it is interesting to study the conceptual distributions of individual topics. On Figure 6 each bubble represents a set of objects of the Topic Map. The (x, y) coordinates of a bubble's center correspond respectively to the Relatedness and Closeness values of the corresponding object(s). If several objects have the same Relatedness and Closeness values then they are represented by a unique bubble which size is proportional to the number of objects.

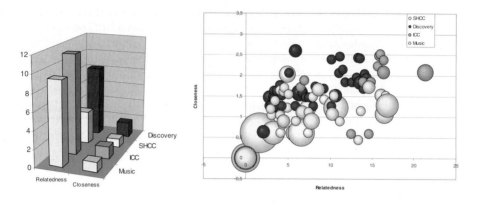

Fig. 5. Conceptual Footprints   **Fig. 6.** Topics Conceptual Distribution

Figure 6 illustrates various types of Conceptual Distributions (homogeneous / heterogeneous). *ICC*'s Relatedness and Closeness values are slightly more homogeneous (i.e. the bubbles are located in a smaller zone of the graphic) than those of the other Topic Maps for which several groups may be identified. If the Topic Map formalism is used to describe ontologies, the comparison of these ontologies' individual conceptual distributions indicates which ones are the more specific / general.

4.2 Automatic Topic Maps Filtering

Music, SHCC and *Discovery's* conceptual distributions contain topics with very low values of both Relatedness and Closeness parameters; these correspond to *marginal* objects which have very few common properties with very few other topics of their respective Topic Map (these objects correspond to the bottom-left corner bubbles in Figure 6).

The Figure 7 shows the various steps of the automatic filtering algorithm. At Step 1 –i.e. before any filtering- the *Music* Topic Map contains 100% of its objects. At Step 2, 89% of original objects are present, then 85%, 83% and finally 78% after which the algorithm converges.

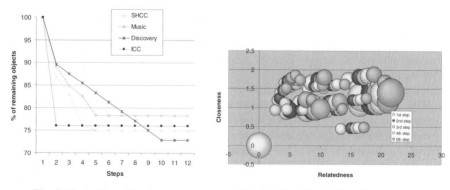

Fig. 7. Topic Maps Filtering   **Fig. 8.** *Music* Topic Map filtering process

The convergence is faster for the *ICC* Topic Map, which is not surprising as this was the most homogeneous dataset (see Figure 6). After the end of the filtering process, more than 70% of original objects remain in the topic maps; this means that there was little "noise" (or outliers) in these datasets.

The Figure 8 shows the conceptual distributions of the *Music* Topic Map at the various steps of the filtering process. The Topic Map obviously contains marginal elements contained in the bubble at the bottom-left corner; this bubble disappears at the second filtering step. The evolution of conceptual distributions between steps 2 and 5 is minor: the values of Relatedness and Closeness of the remaining objects slightly increase after each step.

4.3   Galois Lattice's Concepts Selection

Figure 9 shows the result of the tree extraction algorithm applied to the *Music* Topic Map. This interactive visualization (in SVG format [21]) indicates the most relevant topics and associations of this Topic Map and thus provides an intuitive overview of its most significant content.

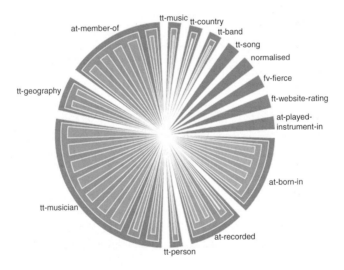

Fig. 9. Clusters Visualization

The Figure 9 shows the tree of concepts extracted from the lattice, displayed as a disc (one disc for each level of detail, where the more general discs have the larger diameters). The labels displayed on the Figure 9 correspond to first level of detail of the tree –i.e. the most general layer. The size of each portion of the disc is proportional to the number of objects in the cluster. This Figure thus shows that most topics are about musicians (*tt-musician* label) and the bands they belong to (*at-member-of*), as well as the records they made (*at-recorded*). It also indicates that a significant proportion of topics deal with their place of birth (*at-born-in* and *tt-geography*). This Topic Map is therefore dedicated to music but not restricted to the musical content only. This hierarchical visualization is very useful to provide an intuitive overview of the complex system's content and structure.

5 Conclusion and Perspectives

The goal of this paper was to propose a generic method relying on Formal Concept Analysis and Galois lattices for complex systems analysis. Although this method may be applied to any type of complex system, this article has focused on its application to the Semantic Web, in particular to complex systems (e.g. ontologies) formalized with the Topic Map paradigm.

Conceptual parameters - called *Relatedness* and *Closeness*- were designed to provide an overview of the datasets, help identify significant elements and conversely eliminate outliers automatically. The proposed method has been experimented on four Topic Maps in order to illustrate its operation and results. Ontologies could be compared in terms of content and structure, as well as regarding their level of specificity/generality.

A conceptual algorithm was finally defined to select the most significant concepts of a Galois lattice by extracting a tree of concepts from the original lattice. An interactive hierarchical and multiple-scale visualization was proposed to help users identify the Topic Map's content and structure. The labels of the clusters at the most general level of detail provide an overview of the Topic Map by indicating the most significant concepts.

One perspective of this work is to integrate an additional dimension to the proposed conceptual parameters in order to reflect the generalization/specialization relationship between the concepts of a lattice. Relatedness and Closeness are currently based on Galois concepts' extents and intents; they should be enriched with information provided by links among concepts so as to fully exploit the information provided by the lattice.

The complexity of Galois algorithm limits the size of the analyzed complex systems. We are currently studying how this conceptual approach may be combined to more scalable clustering algorithms in order to solve this problem. A distributed computation of our conceptual parameters is also investigated. Another possible approach consists in building small lattices and connect them through semantic bridges [22].

Finally, this method will also be used to analyze datasets with neither structure nor semantics, as Galois lattices precisely provide data with a structure and help identify implicit relationships among objects. Such a method could automate (part of) the semantic annotation process –and therefore of Topic Map and ontologies generation.

References

1. Berners-Lee, T.: A roadmap to the Semantic Web (September 1998),
   `http://www.w3.org/DesignIssues/Semantic.html`
2. World Wide Web Consortium. Resource Description Framework (RDF) Model and Syntax Specification, W3C Recommendation, February 22 (1999)
3. International Organization for Standardization, ISO/IEC 13250, Information Technology-SGML Applications-Topic Maps, ISO, Geneva (1998)

4. TopicMaps.Org XTM Authoring Group, XTM: XML Topic Maps (XTM) 1.0: TopicMaps.Org Specification, March 3 (2001)

5. Moore, G.: RDF and Topic Maps, An Exercise in Convergence. In: XML Europe 2001, Berlin, Germany, May 21-25 (2001)

6. Barbut, M., Monjardet, B.: Ordre et classification, Algèbre et combinatoire, Tome 2, Hachette (1970)

7. Birkhoff, G.: Lattice Theory, First Edition, Amer. Math. Soc. Pub. 25, Providence, R. I (1940)

8. Carpineto, C., Romano, G.: Galois: An order-theoretic approach to conceptual clustering. In: Proc. Of the 10th Conference on Machine Learning, Amherst, MA, pp. 33–40. Morgan Kaufmann, San Francisco (1993)

9. Morin, E., Motta, R., Ciurna, E.-R.: Eduquer pour l'aire planétaire – la pensée complexe comme méthode d'apprentissage dans l'erreur et l'incertitude humaine, Balland Ed (2003)

10. Morin, E.: Sciences avec conscience, Points Sciences, Seuil Ed (1990)

11. De Freitas, L., Morin, E., Nicolescu, B.: Charte de la transdisciplinarité, 1er Congrès Mondial de la Trandisciplinarité, Convento da Arrábida, Centre International de Recherches et Études Transdisciplinaires, Portugal (1994)

12. Shneiderman, B.: The eyes have it: a task by data type taxonomy for information visualizations. In: Proceedings of 1996 IEEE Visual Languages, Boulder, CO, pp. 336–343 (1996)

13. Woods, W.A.: What's in a link: foundations for semantic networks. In: Bobrow, D.G., Collins, A.M. (eds.) Representation and Understanding: Studies in Cognitive Science, pp. 35–82. Academic Press, New York (1975)

14. Chein, M., Mugnier, M.-L.: Conceptual Graphs: Fundamental Notions. Revue d'intelligence artificielle 6(4), 365–406 (1992)

15. Sowa, J.F.: Conceptual Information Processing in Mind and Machine. Addison-Wesley, Reading (1984)

16. Garshol, L.M.: "tolog" – A Topic Map Query Language. In: XML Europe 2001, Berlin, Germany, 21-25 May (2001)

17. Le Grand, B., Soto, M.: XML Topic Maps and Semantic Web Mining. In: Proceedings of Semantic Web Mining Workshop, jointly with ECML/PKDD 2001 conference, Freiburg, Germany (September 2001)

18. Wille, R.: Line diagrams of hierarchical concept systems. Int. Classif. 11, 77–86 (1984)

19. Wille, R.: Concept lattices and conceptual knowledge systems. Computers & Mathematics Applications 23(6-9), 493–515 (1992)

20. Jay, N., Kohler, F., Napoli, A.: Analysis of Social Communities with Iceberg and Stability-Based Concept Lattices. In: Medina, R., Obiedkov, S. (eds.) ICFCA 2008. LNCS, vol. 4933, pp. 258–272. Springer, Heidelberg (2008)

21. World Wide Web Consortium, Scalable Vector Graphics (SVG) 1.0 Specification, W3C Candidate Recommendation, November 2 (2000)

22. Polaillon, G., Aufaure, M.-A., Le Grand, B., Soto, M.: FCA for contextual semantic navigation and information retrieval in heterogeneous information systems. In: Proceedings of the workshop on Advances in Conceptual Knowledge Engineering, in conjunction with DEXA 2007, Regensburg, Germany (2007)

Automatic Extraction of Ontologies Wrapping Relational Data Sources

Lina Lubyte and Sergio Tessaris

KRDB Research Centre for Knowledge and Data – Free University of Bozen-Bolzano

Abstract. Describing relational data sources (i.e. databases) by means of ontologies constitutes the foundation of most of the semantic based approaches to data access and integration. In spite of the importance of the task this is mostly carried out manually and, to the best of our knowledge, not much research has been devoted to its automatisation. In this paper we introduce an automatic procedure for building ontologies starting from the integrity constraints present in the relational sources.

Our work builds upon the wide literature on database schema reverse engineering; however, we adapt these techniques to the specific purpose of reusing the extracted schemata (or ontologies) in the context of semantic data access. In particular, we ensure that the underlying data sources can be queried through the ontologies and the extracted ontologies can be used for semantic integration using recently developed techniques in this area.

In order to represent the extracted ontology we adopt a variant of the *DLR-Lite* description logic because of its ability to express the mostly used modelling constraints, and its nice computational properties. The connection with the relational data sources is captured by means of sound views. Moreover, the adoption of this formal language enables us to prove that the extracted ontologies preserve the semantics of the integrity constraints in the relational sources. Therefore, there is no data loss, and the extracted ontology constitutes a faithful wrapper of the relational sources.

1 Introduction

Recent research on ontology languages tailored to data access demonstrates that ontologies (or conceptual models) can be very effective in overcoming several limitations of traditional database systems. In particular, the capability of handling incomplete information has been proved crucial in several important applications of databases, including federated databases [1], data warehousing [2], information integration through mediated schemas [3], and the Semantic Web [4] (for a survey see [5]).

In order to take advantage of semantics charged techniques to access or reuse legacy data, one of the pre-requisites is the definition of wrappers providing formal and machine readable descriptions of the semantics of the underlying data sources. These wrappers are often ontologies which make explicit the assumptions (or constraints) over the stored data. The definition of these wrappers, in

S.S. Bhowmick, J. Küng, and R. Wagner (Eds.): DEXA 2009, LNCS 5690, pp. 128–142, 2009.

spite of the fact that this is a crucial and error prone process, is usually performed manually with little or no automatic support. Moreover, it requires at the same time a deep understanding of the adopted ontology language and good knowledge of the data source being wrapped.

In this paper we propose a technique that enables the automatic extraction of an ontology from a relational data source; in addition, our algorithm provides mappings which connect the terms from the ontology to the actual data. These mappings are defined as sound views over the logical schema of the relational data source; i.e., similar to the global-as-view (GAV) approach in the information integration literature [3]. We show that the extracted ontologies capture all the constraints of the underlying data sources, and the availability of mappings enables the use of these ontologies to query and integrate the wrapped data.

The adopted ontology language – a variant of the *DLR-Lite* family of languages (see [6]) – is expressive enough to capture commonly used features from Entity-Relationship (ER) [7] and UML class diagrams[1], and at the same time is compatible with the Semantic Web ontology language OWL[2].

Our ontology extraction technique relies on the availability of the logical schema of the relational data sources as well as constraints (e.g. foreign keys, uniqueness, etc.) providing the actual semantics of the data. In most of the cases this information can be automatically extracted from any DBMS; but we are aware of the fact that often these constraints are not stored in the actual DBMS but enforced by the programs accessing and updating the data or by ad hoc triggers and stored procedures. To account for these cases we provide the possibility to manually annotate the logical schema in order to specify the constraints. Our experience in several projects showed us that while it is relatively easy for data analysts to provide the constraints of a specific database application, the process of writing an ontology describing the same data can be daunting. Moreover, most of the standard database constraints can be discovered by analysing the actual data.

The contributions of this paper are the adaptation of a Description Logic [8] based ontology language compatible with OWL[3] including the definition of mappings over relational data (Section 2); the definition of an algorithm to extract an ontology given a relational data source (Section 3); and the formal proof that the extracted ontology fully capture the meaning of the data source using the general concept of *information capacities* (Section 3.1).

2   Formal Framework

In this section we define a formal framework for describing relational sources and their wrapping ontologies. For the input relational source, we adopt a standard relational model with integrity constraints. In order to represent the extracted ontology, we use a variant of *DLR-Lite* [6] description logic detailed below.

---

[1] http://www.uml.org
[2] See http://www.w3.org/TR/owl-ref/
[3] Demonstrated by the availability of a Protégé plugin in Section 5.

2.1   Relational Model, Constraints and Queries

We assume the reader is familiar with the basic notions of relational databases [9]. A *relational schema* $\mathcal{R}$ consists of an alphabet of *relation* symbols, each one with a fixed set of attributes (assumed to be pairwise distinct) with associated datatypes. The number of attributes denotes the *arity* of a relation. We assume that the *database domain* is a fixed denumerable set of elements Δ representing real world objects, and that every element in Δ is denoted uniquely by a constant symbol, called its *standard name* [10]. Moreover, we consider Δ to be partitioned into the datatypes D_i and to contain a special constant null, called the null value[4]. Then, a *database instance* (or simply a *database*) $\mathcal{D}$ over a relational schema $\mathcal{R}$ is an (interpretation) function that maps each relation R in $\mathcal{R}$ into a set $R^{\mathcal{D}}$ of total functions from the set of attributes of R to Δ^5.

The ontology extraction procedure takes as input a relational source. We abstract from any specific database implementation by considering an abstract *relational source* $\mathcal{DB}$, which is a pair (Ψ, Σ), where Ψ is a relational schema as defined above and Σ is a set of *integrity constraints*, i.e., assertions on the relations that express conditions that are intended to be satisfied by database instances. A database $\mathcal{D}$ over Ψ is said to *satisfy* a set of integrity constraints Σ expressed over Ψ if every constraint in Σ is satisfied by $\mathcal{D}$. Given a relation r in Ψ and s attribute of r, let A denote the sequence of attributes of r and $r[s]$ the *projection* of r on attribute s [9]. The database integrity constraints that we consider in our framework are the following (for more details see [11]):

- *nulls-not-allowed constraints*, written $nonnull(r, A)$, satisfied in a database when null is not contained in any attribute in A of r;
- *unique constraints*, written $unique(r, A)$, satisfied in a database when the sequence of attributes A is unique in a relation r. If in addition we have $nonnull(r, A)$, then these correspond to *key constraints*, denoted $key(r, A)$;
- *inclusion dependencies*, written $r_1[s_1] \subseteq r_2[s_2]$[6], satisfied in a database when projections over s_1, s_2 of relations r_1 and r_2, respectively, are included one in the other. If in addition we have $key(r_2, s_2)$, we call them *foreign key constraints*;
- *exclusion dependencies*, written $(r_1[s_1] \cap r_2[s_2]) = \emptyset$, satisfied in a database when the intersection of the projections over s_1, s_2 of relations r_1, r_2 is empty set;
- *covering constraints*, written $(r_1[s_1] \cup \ldots \cup r_m[s_m]) \subseteq r_0[s_0]$, satisfied in a database when the projection of the relation r_0 over s_0 is included in the union of the projections of the respective relations in the set.

---

[4] We consider a null value to be different from any other constant and from a null value in any other tuple. Assuming this semantics is not crucial though, different ones can be accommodated.

[5] I.e., each total function represents a single tuple in $R^{\mathcal{D}}$. We assume set semantics.

[6] For simplicity, we restrict inclusion, exclusion and covering constraints to projections over single attribute; see last paragraph of Section 2.2.

$$R[s] \sqsubseteq R'[s'] \quad \pi_s R^{\mathcal{D}} \subseteq \pi_{s'} R'^{\mathcal{D}} \qquad \qquad \text{Inclusion}$$

$$R[s] \text{ disj } R'[s'] \quad \pi_s R^{\mathcal{D}} \cap \pi_{s'} R'^{\mathcal{D}} = \emptyset \qquad \qquad \text{Disjointness}$$

$$\text{key}(R[s_1, \ldots, s_k]) \quad \text{for all } \phi_1, \phi_2 \in R^{\mathcal{D}} \text{ with } \phi_1 \neq \phi_2, \text{ we have} \qquad \qquad \text{Key}$$
$$\phi_1(s_i) \neq \phi_2(s_i) \text{ for some } s_i, 1 \leq i \leq k$$

$$R_1[s_1], \ldots, R_k[s_k] \text{ cover } R[s] \quad \pi_s R^{\mathcal{D}} \subseteq \bigcup_{i=1 \ldots k} \pi_{s_i} R_i'^{\mathcal{D}} \qquad \qquad \text{Covering}$$

Fig. 1. Syntax and semantics of *DLR-DB* axioms

2.2   Ontology Language

In this section we present the ontology language we shall deal in the rest of the paper, and we give its semantics in terms of relational models. The ontology language adopted can be seen as an alternative to the use of standard modelling paradigms of ER or UML class diagrams and enables to represent their commonly used modelling constructs (see [12]). The advantage over these formalisms lies on the fact that our adopted ontology language, besides enabling the use of automatic reasoning to support the designer, also represents models that preserve the relational ones (see Section 3.1).

We call a *DLR-DB* system $\mathcal{S}$ a tuple $\langle \mathcal{R}, \mathcal{K} \rangle$, where $\mathcal{R}$ is a *relational schema* as described in Section 2.1 and $\mathcal{K}$ is a set of assertions involving names in $\mathcal{R}$. The *DLR-DB* ontology language, used to express the constraints in $\mathcal{K}$, is based on the idea of modelling the domain by means of *axioms* involving the projection of the relation over the attribute. We call $\mathcal{K}$ an *ontology*.

An *atomic formula* is a projection of a relation R over one of its attributes, denoted by $R[s]$. The attributes involved in the projections correspond to key attributes of the respective relations. This reflects the fact that in conceptual models non key attributes are not considered relevant to identify an element of an entity or a relationship (see Example 1). Two attributes are *compatible*, if their datatypes are equal. Then we say that two atomic formulae $R[s]$ and $R'[s']$ are *compatible* iff the two corresponding attributes s and s' are compatible.

Given the atomic formulae $R[s]$, $R'[s']$, $R_i[s_i]$, an *axiom* is an assertion of the form specified in Figure 1[7], where all the atomic formulae involved in the same axiom must be compatible. In the same figure, we give the semantics of a *DLR-DB* system $\langle \mathcal{R}, \mathcal{K} \rangle$, which is provided in terms of relational models for $\mathcal{R}$, where $\mathcal{K}$ plays the role of constraining the set of "admissible" models. A database $\mathcal{D}$ is said to be a *model* for $\mathcal{K}$ if it satisfies all its axioms. The above conditions are well defined because we assumed the compatibility of the atomic formulae involved in the axioms.

Example 1. To provide the intuition on the use of the *DLR-DB* formalism we show a simple example exhibiting some of the modelling constructs defined above. Consider the ER diagram shown in Figure 2, and assume, for the sake of exposition, that we have the underlying relational source containing a relation for

---

[7] In relational algebra, $\pi_s R^{\mathcal{D}}$ denotes the projection of $R^{\mathcal{D}}$ over attribute s [9].

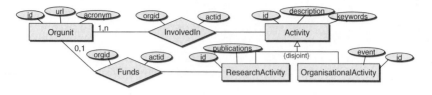

Fig. 2. ER diagram for Example 1

(1) InvolvedIn[orgid] ⊑ Orgunit[id]   (7) key(Activity[id])
(2) InvolvedIn[actid] ⊑ Activity[id]   (8) key(ResearchActivity[id])
(3) Funds[orgid] ⊑ Orgunit[id]   (9) key(OrganisationalActivity[id])
(4) Funds[actid] ⊑ ResearchActivity[id]   (10) key(Funds[orgid])
(5) Orgunit[id] ⊑ InvolvedIn[orgid]   (11) ResearchActivity[id] ⊑ Activity[id]
(6) key(Orgunit[id])   (12) OrganisationalActivity[id] ⊑ Activity[id]
(13) ResearchActivity[id] disj OrganisationalActivity[id]

Fig. 3. *DLR-DB* axioms corresponding to the ER diagram in Figure 2

each entity and relationship in the diagram. That is, we have relations Orgunit of arity 3, InvolvedIn of arity 2, etc. Then, to model the constraints reflected in the given ER diagram, we will define the axioms shown in Figure 3 that constrain the relational schema. In particular, axioms (1)–(4) represent role typing constraints, stating that the projection of InvolvedIn and Funds relations on the orgid (resp. actid) attribute is of type Orgunit (resp. Activity and ResearchActivity). Axiom (5) instead states mandatory participation, meaning that instances of Orgunit projected over the id attribute participate to the relation InvolvedIn as value for its projection over the orgid attribute. The key axioms in (6)–(10) express that, for instance, an object can appear in the orgid attribute of Funds relation only once. Axioms (11) and (12) correspond to is-a relationships among the respective relations, while axiom (13) states disjointness among the objects of the corresponding projections of relations.

The adopted ontology language is close to the *DLR* family of DLs [13]. This means that the same reasoning mechanism used for *DLR* can be employed for *DLR-DB*. The ability of employing correct and complete automated reasoning enables us to provide well-founded tools to support the maintenance and evolution of the extracted ontology (see [14]). More importantly, by taking away the covering axioms and allowing only unary key constraints, this language corresponds to *DLR-Lite$_{\mathcal{F}}$* [6]. This implies that we can use the same efficient query answering technique (LOGSPACE in the size of the data) to evaluate conjunctive queries mediated by the ontology.

For the sake of simplicity, in this paper we restricted the atomic formulae to projections over a single attribute; however, our original definition of a *DLR-DB* system captures the notion of composite keys in relational databases. Assume for example that in our ontology we need to represent the inclusion axiom corresponding to the foreign key constraint spanning over several attributes. To account for these cases, we associate to each relation in the relational schema a set of *components* (each one with a sequence of attributes) which partitions the set of attributes of the relation. Then, the axioms involve the projections of relations over their components instead of single attributes (see [11] for details). Importantly, we can still show that the reasoning mechanism used for $\mathcal{DLR}$ and *DLR-Lite$_{\mathcal{F}}$* can be employed for *DLR-DB* extended with components.

3   Ontology Extraction

The principles of our ontology extraction process are based on ideas used in database reverse engineering (DBRE) literature.[8] Roughly speaking, the essence of our extraction technique is to *reverse* the standard database modelling process [15], namely, that of translating ER model to the relational one. The benefit of such approach is that it can be shown that our algorithm, though heuristic in general, is able to reconstruct the original ER diagram. In this way, we can formally prove that our extraction procedure preserves semantics of constraints in the relational database (see Section 3.1).

The ontology extraction algorithm consists of two steps: *(i)* a classification scheme for relations is derived by analysing the constraints in the relational source, *(ii)* the actual ontology is generated, together with a set of sound views (i.e., GAV mappings [3]) that connect the extracted ontology with the source schema. Specifically, given a relational source $\mathcal{DB} = (\Psi, \Sigma)$ as input, the algorithm generates the *DLR-DB* system $\mathcal{S} = \langle \mathcal{R}, \mathcal{K} \rangle$ with an ontology $\mathcal{K}$ and a set of views in $\mathcal{R}$, defined over the source schema Ψ of $\mathcal{DB}$. So, $\mathcal{R}$ can be seen as a new schema containing set of view definitions, and axioms of the extracted ontology $\mathcal{K}$ are over names in $\mathcal{R}$. In such setting, every ontology term has an associated view over the data sources (see Section 4 for an example).

Given a relation r in the source schema Ψ, in the following we will denote by A sequence of all attributes of r, K the set of key attributes of r such that $key(r, K)$, and **FK** the set of all foreign keys of r such that $r\,[FK_i] \subseteq r'\,[K']$, for each foreign key $FK_i \in$ **FK** that references key K' of relation r', where $1 \leq i \leq n$ and n – number of foreign keys of r. Then, each relation r in Ψ is classified as one of the following:

- *base relation*, if K and **FK** do not share attributes;
- *specific relation*, if K is among **FK** (i.e., key is also foreign key) and if *one* of the following holds
  (a) $|\textbf{FK}| = 1$, i.e., r has single foreign key, or
  (b) $r'\,[FK'_i] \subseteq r\,[K]$, i.e., r is referred to by the foreign key of other relation;
- *relationship relation*, if K is entirely composed of foreign keys and $|\textbf{FK}| > 1$;
- *ambiguous relation*, if it does not satisfy any of the above conditions.

The intuition for the above classification scheme comes directly from the process of translating the ER model to relational model. In particular, a base relation

---

[8] We discuss how our proposed framework relates to DBRE approaches in Section 6.

Table 1. Summary of the extraction procedure

Relation type	Views in $\mathcal{R}$	Corresponding axiom in $\mathcal{K}$
base relation r $key(r,K)$	$R = \pi_{A-\mathbf{FK}}(r)$	$key(R[K])$
specific relation r $key(r,K)$ $r[K] \subseteq r'[K']$	$R = \pi_{A-\mathbf{FK}}(r)$	$key(R[K])$ $R[K] \sqsubseteq R'[K']$
$r_1[K_1] \cap r_2[K_2] = \emptyset$ $r'[K'] \subseteq r_1[K_1] \cup \ldots \cup r_m[K_m]$		$R_1[K_1]$ disj $R_2[K_2]$ $R_1[K_1], \ldots, R_m[K_m]$ cover $R'[K']$
base or specific relation r $r[FK_i] \subseteq r'[K'],\ FK_i \neq K$ $nonnull(r, FK_i)$ $unique(r, FK_i)$ $r'[K'] \subseteq r[FK_i]$	$R'' = \pi_{K,K'}(r' \bowtie r)$	$key(R''[K])$ $R''[K] \sqsubseteq R[K];\ R''[K'] \sqsubseteq R'[K']$ $R[K] \sqsubseteq R''[K]$ $key(R''[K'])$ $R'[K'] \sqsubseteq R''[K']$
relationship relation r $r[FK_i] \subseteq r'[K']$	$R = \pi_A(r)$	$R[FK_i] \sqsubseteq R'[K']$
$r'[K'] \subseteq r[FK_i]$		$R'[K'] \sqsubseteq R[FK_i]$
ambiguous relation r	repeat steps as for relationship relations	

r results from mapping an entity to a relation, and its foreign key FK_i – from "embedding" a one-to-one or one-to-many (i.e., functional) relationship between entities corresponding to (base or specific) relations r and r'. A specific relation instead follows from translating a sub-entity. The condition *(b)* in discovering a specific relation is needed due to the fact that such relation (i.e., with key being a foreign key) may also represent a one-to-one relationship between two entities mapped to a single relation. A relationship relation results from mapping a many-to-many relationship between entities corresponding to base or specific relations. Finally, note that in order not to mislay any relation during the extraction process, we put all "non-standard" relations to the ambiguous relations category.

Once relations in the source $\mathcal{DB}$ are classified, the actual algorithm derives the ontology and views (i.e. mappings) by means of *DLR-DB* system. Table 1 shows the corresponding axioms and view definitions that are generated for each category of relations and integrity constraints imposed on them. Note that the table also reflects the order in which the distinct relations are processed, that is, we start by creating axioms and mappings for base relations, then specific ones, followed by the corresponding structures for the foreign keys of base and specific relations, etc. Furthermore, observe that for ambiguous relations our automated algorithm uses heuristics which "prefers" to recover elements corresponding (in ER terms) to many-to-many, possibly n-ary relationships. However, we also provide the possibility for a user to manually define their intended meaning (see Section 5 for discussion).

It is easy to see that the whole two-step process is linear in the number of relations in the source schema.

3.1   Correctness and Completeness of the Technique

Our proposed ontology extraction technique can be seen as a *schema transformation* as defined in [16]. An important consideration in such a process (i.e., transforming one data model into another) is the potential for loss of information. We evaluate the correctness of our schema extraction procedure using the relative *information capacities* of the source and target schemata. In this section we briefly outline the main principles of this analysis; for full details and the actual proofs the reader is referred to [11].

In the following we denote by S and T source and target schemata corresponding to the input relational source $\mathcal{DB}$ and relational schema $\mathcal{R}$ of the extracted *DLR-DB* system. Let $\mathcal{D}_S$ and $\mathcal{D}_T$ be consistent instances of schemata S and T, respectively. An *equivalence preserving mapping* between the instances of S and T is a bijection $\mu : \mathcal{D}_S \rightarrow \mathcal{D}_T$. Then S and T are said to be *equivalent* via μ, denoted $S \equiv T$. Given schemas S and T, a *(schema) transformation* is a total function $\mathcal{M} : S \rightarrow T$. $\mathcal{M}$ is an *equivalence preserving transformation* if it induces an equivalence preserving mapping. To this end, we con show the following:

Theorem 1. *The ontology extraction procedure is an equivalence preserving schema transformation.*

The actual proof of the above theorem can be found in [11]. Roughly speaking, we devise a bijective transformation for the respective models and we show that the constraints of the original schema and extracted ontology are satisfied by the models generated by this transformation.

The fact that our extraction procedure is equivalence preserving not only shows that there is no information loss, so the extracted schema can be used to access the data, but that we can evaluate queries expressed using the extracted ontology by simply expanding the generated views. This is no longer true in the case that the ontology is going to be modified; in this case, more sophisticated query answering techniques must be adopted in order to guarantee completeness (e.g. query rewriting [6]).

4   Ontology Extraction by Example

Consider the relational schema with constraints detailed below (keys are underlined).

Orgunit(<u>id</u>, url, acronym, actid)	Activity(<u>id</u>, description, keywords)
ResearchActivity(<u>id</u>, publications)	OrganisationalActivity(<u>id</u>, events)
InvolvedIn(<u>orgid</u>, <u>actid</u>)	
(1) InvolvedIn [orgid] $\subseteq$ Orgunit [id]	(4) Orgunit [id] $\subseteq$ InvolvedIn [orgid]
(2) InvolvedIn [actid] $\subseteq$ Activity [id]	(5) *unique*(Orgunit, actid)
(3) Orgunit [actid] $\subseteq$ ResearchActivity [id]	(6) ResearchActivity [id] $\cap$
	OrganisationalActivity [id] $= \emptyset$

At the initial step of the extraction process, relations *Orgunit* and *Activity* are classified as base relations, *ResearchActivity* and *OrganisationalActivity* as specific relations, while relation *InvolvedIn* satisfies the condition required for relationship relations.

The ontology generated from this relational source is the one given in Example 1 of Section 2. The extracted schema with view definitions is given below (we denote with serif and *slanted* font, respectively, relation names in the extracted and source schema).

$$\text{Orgunit} = \pi_{id,url,acronym}(\textit{Orgunit})$$
$$\text{Activity} = \pi_{id,description,keywords}(\textit{Activity})$$
$$\text{ResearchActivity} = \pi_{id,publications}(\textit{ResearchActivity})$$
$$\text{OrganisationalActivity} = \pi_{id,events}(\textit{OrganisationalActivity})$$
$$\text{Funds} = \pi_{id,id}(\textit{Orgunit} \bowtie \textit{Activity})$$
$$\text{InvolvedIn} = \pi_{orgid,actid}(\textit{InvolvedIn})$$

Now, consider the extracted ontology provided in Example 1 and the set of views above denoting mappings between the ontology and the actual sources. Suppose we want to know the the pairs of organisational units and research activities that those organisational units fund. The corresponding conjunctive query we would formulate is

$$q(x, y) \leftarrow \text{Orgunit}(x, w, z), \text{Funds}(x, y), \text{ResearchActivity}(y).$$

To answer this query, it is enough to substitute each atom in the body of q with its corresponding query in the view definition, and to evaluate it over the actual data sources (we recall the reader that a conjunctive query can be translated into an equivalent SQL select-project-join (SPJ) query using standard translation [9]).

5   Implementation and Case Study

In order to evaluate the applicability of our approach, it is crucial to test it with real-world schemas. To automate the ontology extraction process for wrapping the underlying data sources, we have implemented the ontology extraction algorithm in a prototype system. The implementation allows to display the extracted ontology with annotated views using the ICOM Ontology Design tool [17]. However, in order to fully leverage the available and well-established techniques for using such wrappers for data access, we have also implemented an automatic ontology extraction support plug-in on top of the OBDA plug-in[9] for Protégé[10]. The OBDA plug-in provides facilities to design Ontology Based Data Access (OBDA) system components (i.e., data sources and mappings). It supports the definition of relational data sources and GAV like mappings to link the concepts in the $DL\text{-}Lite_A$ ontology [18] to the data in the defined sources. It also provides support for conjunctive query answering (by using SPARQL syntax), a service commonly offered by OBDA centric reasoners. A notable aspect of $DL\text{-}Lite_A$ description logic used as ontology language in OBDA plug-in is that it admits

---

[9] http://obda.inf.unibz.it/protege-plugin
[10] http://protege.stanford.edu

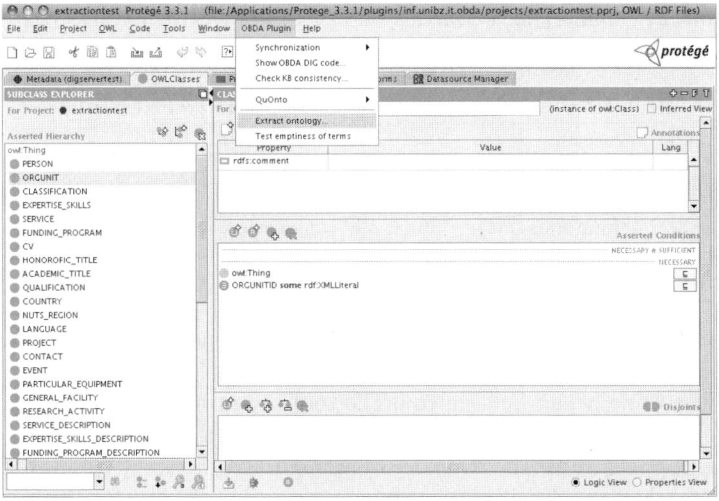

Fig. 4. Ontology extraction plug-in for Protégé

query answering (with incomplete information) that is LOGSPACE in the size of the data at the sources. Even more importantly, it allows to reformulate query answering in terms of the evaluation of suitable SQL queries issued over the sources. Our ontology extraction framework fits thus very well into such OBDA setting (see [19] for interesting scenarios).

The goal of our plug-in is to provide a framework for the automated support for deriving the wrapping ontology form existing data sources together with an automatic generation of mappings. The ontology engineer can then explore the obtained ontology, possibly refine it, and formulate conjunctive queries over the this ontology using the OBDA plug-in. Figure 4 shows the screenshot of the ontology (displayed with OWL Protégé plug-in) automatically extracted from the source schema, where the data sources are specified by using the *Datasource Manager* tab of the OBDA plug-in. The generated mappings (i.e., views associated to ontology terms) are also manifested in the latter tab. The user at this point can pose queries over the resulting ontology, and the answers are returned from the underlying data source by taking into account the mappings (see OBDA plug-in website and [20] for details). It is worth noting however that since our ontology language supports n-ary roles, while $DL\text{-}Lite_{\mathcal{A}}$ (and OWL) does not, the extracted ontology must be *reified* [21].

We next report the results of a case study with CERIF database schema. CERIF (Common European Research Information Format)[11] is the standard EU recommendation used to harmonise databases on research projects. The schema is strongly structured containing 123 relations and is fairly rich in terms of integrity constraints explicitly declared through the DDL code. Table 2 shows the outcome of analysing the constraints over CERIF schema that result in

---

[11] http://cordis.europa.eu/cerif/src/toolkit.htm

Table 2. Summary for extracting ontology from CERIF database schema

(a) Classified relations		(b) Extracted axioms	
Relation type	**#Classified**	**Axiom**	**#Extracted**
Base	52	Inclusion	68
Relationship	25	Key	63
Specific	0	Disjointness	0
Ambiguous	46	Covering	0

classified relations (a) and the corresponding axioms derived for each class of relations (b). Note that those numbers do not include axioms for ambiguous relations.

From the table it can be seen that the tool produced correct[12] ontology constructs for 77 relations (out of 123). We were particularly interested in the usefulness of our approach for the category of ambiguous relations. We have identified that all 46 relations classified as ambiguous can be divided into two types: *(i)* those having their set of foreign keys properly included in their keys, and *(ii)* those having keys properly included in the set of their foreign keys, where the number of foreign keys is at least 3 and keys span at least 2 foreign keys. For instance,

(i) PERSON_RESEARCH_INTEREST(LANGUAGE,TRANS_TYPE,PER_ID,KEYWORDS),
(ii) PROJ_PERSON(PROJ_ID,PER_ID,PROJ_PER_ROLE,PROJ_PER_START,PROJ_PER_END)

where keys are underlined and PER_ID in *(i)* is the single foreign key, while in *(ii)* all PROJ_ID,PER_ID and PROJ_PER_ROLE are foreign keys. By carefully analysing the schema, we have derived that the intended meaning behind both types of the above relations are, in ER terms, relationships between entities corresponding to the referenced relations. In particular, LANGUAGE,TRANS_TYPE can be treated as a "hidden" foreign key referencing a "hidden" relation.

We are currently extending the original extraction algorithm in order to capture the above cases. The idea is to use an *iterative* approach for the extraction process. That is, first derive the initial classification as described in Section 3 with the corresponding axioms, then analyse ambiguous relations again by taking into account the above cases and manifest to the user possible suggestions.

6   Related Work

As we have mentioned, we build our method on top of the existing results in the area of *database reverse engineering* (DBRE) [22]. DBRE is defined as a process of recovering a conceptual model that represents the meaning of the logical schema by examining an existing database system to identify the database contents and their interrelationships. Approaches to recovering a conceptual

---

[12] With "correct" we mean the intended meaning when following the principled methodologies of relational database design from ER diagrams.

schema from a relational database have appeared in the literature over the years [23,24,25,26]. Four main sources (and their combination) have been explored for finding evidence to construct a conceptual schema from a logical database: structures and integrity constraints of the database schema [23,24], application programs that access the database [25], data instances stored in the database [26], and users and designers [27]. Moreover, because reverse engineering of relational databases is a complex task, all existing approaches are conditioned by a set of restrictive assumptions, namely, relational schemas are supposed to be normalised (3NF, BCNF), and the constraints on the schema are available (e.g., keys and foreign keys, inclusion and exclusion dependencies).

Even though there is a close connection between this area and the framework that we propose, there are however important differences. First, DBRE approaches usually produce just a pictorial representation of a conceptual model, without formal mappings that link the obtained schema to the database, and are thus used for "documenting" the database. Our approach, instead, is tailored for the direct use of the extracted ontology – that of accessing the data. In this setting, views generated during the extraction process that connect the derived ontology with the data sources play a crucial role. Second, most of DBRE methods are informal and do not specify quality of the outcome. On the contrary, we provide formal results showing that the extracted ontology represents all information sources and does not represent any extra information not present in the sources (see Section 3.1).

There are several works coming closer to ours that arose in the context of the Semantic Web and Information Integration, and that bring together relational databases and ontologies. Astrova [28] uses DBRE techniques to build ontologies with the purpose of *migrating* relational database content into the ontology (i.e., data-intensive web sites to the Semantic Web). Thus, while our framework uses the extracted ontology for accessing the data stored in external sources by means of sound views, here data is to be stored in the reverse engineered ontology. The work by Volz et al. [29][13] provides a framework for creating metadata by generating Web pages from an available database which finally leads to the *deep annotation* of the database. Then, such annotations can be used for two purposes: for querying the database through an ontology, and for migrating database content to ontology-based instance data. The database structure is first manually described on a Web page and then one of the means to create mappings between the ontology and the database is a semi-automatic process that integrates DBRE techniques. The advantage of our approach over this work, as well as the aforementioned one, is that our framework enables the representation of a *formal* ontology wrapping relational sources which allows for *automated reasoning* to support the designer. The proposals on data source ontology wrapping as SWARD [31] and Virtuoso[14] systems support the automated generation of RDF views over relational data sources, enabling to access the underlying data using RDF query languages. The advantage of our technique over all the above

---

[13] Later version of this work has also appeared as part of WonderWeb project (see [30]).
[14] http://virtuoso.openlinksw.com/wiki/main/

mentioned works is that we ensure *faithfulness* of the obtained model, meaning that it fully captures the meaning of the data source being wrapped.

D2R MAP [32] and R2O [33] provide means to declaratively state, respectively, ontology-to-database and database-to-ontology mappings. The mapping language of D2R MAP is based on RDF[15], while R2O uses XML. Both languages allow to define expressive and explicit correspondences between components of the two models. A similar work in [34] presents an approach to map data stored in relational databases into the Semantic Web using RDF query language (e.g., SPARQL [35]). A different approach of [36] describes a (semi-)automatic mapping discovery between database relations and ontologies. In [18] a set of pre-existing sources is linked to the ontology (expressed in description logic) by defining expressive mappings. These works however require an existing target ontology the relations are mapped onto.

7   Conclusions and Current Work

We have described an heuristic procedure for extracting an ontology from a relational database schema. The mappings between the extracted ontology and the underlying database are defined by associating views over the original data to each term in the ontology. To represent the extracted ontology, instead of a graphical notation, we employ an ontology language thus retaining its precise semantics. Our extraction procedure integrates and enhances standard database reverse engineering techniques by relying on constraints defined over the database schema, (i.e., key and foreign key structure, restrictions on attributes and dependencies between relations), as well as standard methodologies for database design. We ensure that the underlying data sources can be queried through the extracted ontology by expanding the defined views.

We are aware of the fact that many databases have not been designed following the disciplined methodologies, so their schemas exhibit "idiosyncrasies", as coined by Blaha et al [37]. To this purpose we are starting to experiment with other real database schemas to identify other design patterns and enrich with them our extraction algorithm. We also realise that often databases do not include all the relevant constraints on the data that were planned at design phase. Indeed, in most of the cases these constraints are enforced by the code accessing and updating the data. To the same extent, we are designing a tool to provide a support to explore the logical schema of a database and to facilitate the annotation of the schema by means of standard database constraints.

References

1. Sheth, A.P., Larson, J.A.: Federated database systems for managing distributed, heterogeneous and autonomous databases. ACM Computing Surveys 22(3), 183–236 (1990)

---

[15] http://www.w3.org/RDF/

2. Calvanese, D., Giacomo, G.D., Lenzerini, M., Nardi, D., Rosati, R.: Data integration in data warehousing 10(3), 237–271 (2001)
3. Lenzerini, M.: Data integration: A theoretical perspective. In: Proc. of PODS 2002, pp. 233–346 (2002)
4. Heflin, J., Hendler, J.: A portrait of the semantic web in action. IEEE Intelligent Systems 16(2), 54–59 (2001)
5. Wache, H., Vogele, T., Visser, U., Stuckenschmidt, H., Schuster, G., Neumann, H., Hubner, S.: Ontology-based integration of information - a survey of existing approaches. In: Proc. of IJCAI 2001 Workshop: Ontologies and Information Sharing, pp. 108–117 (2001)
6. Calvanese, D., Giacomo, G.D., Lembo, D., Lenzerini, M., Rosati, R.: Tractable reasoning and efficient query answering in description logics: The dl-lite family. J. of Automated Reasoning 39(3), 385–429 (2007)
7. Chen, P.: The entity-relationship model: Toward a unified view of data. ACM Transactions on Database Systems (TODS) 1(1), 9–36 (1976)
8. Baader, F., Calvanese, D., McGuinness, D., Nardi, D., Patel-Schneider, P.F. (eds.): The Description Logic Handbook: Theory, Implementation and Applications. Cambridge University Press, Cambridge (2003)
9. Abiteboul, S., Hull, R., Vianu, V.: Foundations of Databases. Addison-Wesley, Reading (1995)
10. Levesque, H.J., Lakemeyer, G.: The Logic of Knowledge Bases. MIT Press, Cambridge (2001)
11. Lubyte, L., Tessaris, S.: Extracting ontologies from relational databases. Technical report, KRDB group – Free University of Bozen-Bolzano (2007), http://www.inf.unibz.it/krdb/pub/TR/KRDB07-4.pdf
12. Berardi, D., Calvanese, D., De Giacomo, G.: Reasoning on uml class diagrams. Artificial Intelligence 168(1), 70–118 (2005)
13. Calvanese, D., De Giacomo, G., Lenzerini, M.: Identification constraints and functional dependencies in description logics. In: Proc. of the 17th Int. Joint Conf. on Artificial Intelligence (IJCAI 2001), pp. 155–160 (2001)
14. Lembo, D., Lutz, C., Suntisrivaraporn, B.: Tasks for ontology design and maintenance. Deliverable D05, TONES EU-IST STREP FP6-7603 (2006)
15. Elmasri, R., Navathe, S.B.: Fundamentals of Database Systems, 4th edn. Addison Wesley Publ. Co., Reading (2004)
16. Miller, R.J., Ioannidis, Y.E., Ramakrishnan, R.: The use of information capacity in schema integration and translation. In: Proc. of VLDB 1993, pp. 120–133. Morgan Kaufmann Publishers Inc., San Francisco (1993)
17. Fillottrani, P.R., Franconi, E., Tessaris, S.: The new icom ontology editor. In: Proc. of the 19th Int. Workshop on Description Logics, DL 2006 (2006)
18. Poggi, A., Lembo, D., Calvanese, D., Giacomo, G.D., Lenzerini, M., Rosati, R.: Linking data to ontologies. J. on Data Semantics X, 133–173 (2008)
19. Rodriguez-Muro, M., Lubyte, L., Calvanese, D.: Realizing ontology based data access: A plug-in for protégé. In: Proc. of the Workshop on Information Integration Methods, Architectures, and Systems (IIMAS 2008), pp. 286–289 (2008)
20. Calvanese, D., Giacomo, G.D., Horridge, M., et al.: Software tools for ontology interoperation. Deliverable D25, TONES EU-IST STREP FP6-7603 (2008)
21. Noy, N., Rector, A.: Defining n-ary relations on the semantic web. Technical report, W3C Recommendation (2006), http://www.w3.org/TR/swbp-n-aryRelations/
22. Hainaut, J.L.: Database reverse engineering: models, techniques and strategies. In: Proc. of the 10th Conference on ER Approach (1998)

23. Markowitz, V.M., Makowsky, J.A.: Identifying extended entity-relationship object structures in relational schemas. IEEE Transactions on Software Engineering 16(8), 777–790 (1990)
24. Chiang, R.H.L., Barron, T.M., Storey, V.C.: Reverse engineering of relational databases: extraction of an eer model from a relational database. Data and Knowledge Engineering 12(2), 107–142 (1994)
25. Andersson, M.: Extracting an entity-relationship schema from a relational database through reverse engineering. In: Loucopoulos, P. (ed.) ER 1994. LNCS, vol. 881, pp. 403–419. Springer, Heidelberg (1994)
26. Alhajj, R.: Extracting an extended entity-relationship model from a legacy relational database. Information Systems 26(6), 597–618 (2003)
27. Johannesson, P.: A method for transforming relational schemas into conceptual schemas. In: Proc. of the Int. Conf. on Data Engineering (ICDE 1994), pp. 190–201 (1994)
28. Astrova, I.: Reverse engineering of relational databases to ontologies. In: Bussler, C.J., Davies, J., Fensel, D., Studer, R. (eds.) ESWS 2004. LNCS, vol. 3053, pp. 327–341. Springer, Heidelberg (2004)
29. Volz, R., Handschuh, S., Staab, S., Stojanovic, L., Stojanovic, N.: Unveiling the hidden bride: deep annotation for mapping and migrating legacy data to the semantic web. Web Semantics 2(1), 187–206 (2004)
30. Volz, R., Handschuh, S., Staab, S., Studer, R.: Ontolift demonstrator. Deliverable Del 12, WonderWeb IST-2001-33052 (2004)
31. Petrini, J., Risch, T.: Processing queries over RDF views of wrapped relational databases. In: Proc. of the 1st Int. Workshop on Wrapper Techniques for Legacy Systems, WRAP 2004 (2004)
32. Bizer, C.: D2R MAP - a database to RDF mapping language. In: Int. World Wide Web Conference, WWW 2003 (2003)
33. Barrasa, J., Corcho, O., Gomez-Perez, A.: An extensible and semantically based database-to-ontology mapping language. In: Bussler, C.J., Tannen, V., Fundulaki, I. (eds.) SWDB 2004. LNCS, vol. 3372. Springer, Heidelberg (2005)
34. de Laborda, C.P., Conrad, S.: Database to semantic web mapping using RDF query languages. In: Embley, D.W., Olivé, A., Ram, S. (eds.) ER 2006. LNCS, vol. 4215, pp. 241–254. Springer, Heidelberg (2006)
35. Prud'hommeaux, E., Seaborne, A.: SPARQL query language for RDF. Technical report, W3C Recommendation (2008), http://www.w3.org/TR/rdf-sparql-query/
36. An, Y., Borgida, A., Mylopoulos, J.: Inferring complex semantic mappings between relational tables and ontologies from simple correspondences. In: Int. Conf. on Ontologies, Databases and Applications of Semantics (ODBASE 2005), pp. 1152–1169 (2005)
37. Blaha, M.R., Premerlani, W.J.: Observed idiosyncracies of relational database designs. In: Proc. of the Working Conf. on Reverse Engineering (1995)

A Query Cache Tool for Optimizing Repeatable and Parallel OLAP Queries

Ricardo Jorge Santos[1] and Jorge Bernardino[1,2]

[1] CISUC – Centre of Informatics and Systems of the University of Coimbra, Portugal
[2] ISEC – Superior Institute of Engineering of Coimbra, Portugal
lionsoftware.ricardo@gmail.com, jorge@isec.pt

Abstract. On-line analytical processing against data warehouse databases is a common form of getting decision making information for almost every business field. Decision support information oftenly concerns periodic values based on regular attributes, such as sales amounts, percentages, most transactioned items, etc. This means that many similar OLAP instructions are periodically repeated, and simultaneously, between the several decision makers. Our Query Cache Tool takes advantage of previously executed queries, storing their results and the current state of the data which was accessed. Future queries only need to execute against the new data, inserted since the queries were last executed, and join these results with the previous ones. This makes query execution much faster, because we only need to process the most recent data. Our tool also minimizes the execution time and resource consumption for similar queries simultaneously executed by different users, putting the most recent ones on hold until the first finish and returns the results for all of them. The stored query results are held until they are considered outdated, then automatically erased. We present an experimental evaluation of our tool using a data warehouse based on a real-world business dataset and use a set of typical decision support queries to discuss the results, showing a very high gain in query execution time.

1 Introduction

Over the last decades, data warehouses (DW) have become excellent decision-support resources for almost every business field. Decision making information is mainly obtained using tools performing On-Line Analytical Processing (OLAP) against DW databases. These databases usually store the whole business history, having a frequent huge number of rows, and grow to gigabytes or terabytes of storage size, making query performance one of the most important issues in data warehousing. The author in [14] refers that standard decision making OLAP queries which are executed periodically at regular intervals are, by far, the most usual form of obtaining decision making information. This implies that this kind of information is usually based on the same regular SQL instructions. This makes it relevant and important to optimize the performance of predefined decision support queries, which would be executed repeatedly at any time, by a significant number of OLAP users. Most research proposals for optimizing parallel and repeatable query execution focus on issues such as data and hardware balancing, to take advantage of multi-threading and multi-core processors

S.S. Bhowmick, J. Küng, and R. Wagner (Eds.): DEXA 2009, LNCS 5690, pp. 143–152, 2009.

[6, 9]. The proposed solutions are somewhat complex and expensive. In this paper, we propose a solution at the data and SQL level, which is farther more simple, understandable and inexpensive.

Our proposal consists on a method for speeding up the execution of two types of queries: periodically repeatable queries, which keep their original OLAP instruction; and two or more similar query instructions which are executed simultaneously. This is done by storing the latest results of the frequently used OLAP queries. Therefore, only the most recent factual data is used for processing incremental results, which will be joined with the previous results in order to supply the OLAP queries' response. Our proposal also avoids spending time and resources of the DBMS in processing simultaneously similar OLAP instructions. This is done by looking into the query cache, for every OLAP query to be executed, to see if there is any similar query being executed at the same time. If there is, the latest user is put on hold and will receive the results as soon as it finishes processing for the first user who started the execution. As it can be seen in the results provided in the experimental evaluation, this method provides very high gains in query response time and resource consumption for repeatable and parallel querying, for several number of simultaneous users.

The remainder of this paper is organized as follows. Section 2 presents our proposal, describing how the query caching method works and is used in the Query Cache Tool. In Section 3 we present an experimental evaluation and discuss its results. Section 4 presents related work on parallel query execution, query caching and other research related with the solutions used in our proposal. Finally, section 5 presents conclusions and future work.

2   The Query Cache Tool

Traditionally, it has been well accepted that DW databases are updated periodically – typically in a daily, weekly or even monthly basis [19]. In our experience, the daily updates seem to be the most used approach. These updates consist on integrating new data into the DW databases and rebuilding all associated optimization data structures, such as indexes, materialized views, etc. While these update procedures are executed, the databases are offline, i.e., unavailable to end users such as decision makers and OLAP tools. Between these updates, i.e., while the databases are available, the existing data is static in its contents and structures.

Now suppose that several decision makers need to execute the same queries among each other along the day, for instance, consulting how much was the total sales amount of the day before the current. During that same day, the existing data in the DW databases does not change. This brings up a very relevant question: Why should we request the execution of similar queries more than once, between DW updates, if the data is always the same? The results are also always the same! Therefore, if we store the results for the most recently executed queries, which decision makes will probably need to consult repeatedly, we already have fast direct access to the results and do not need to process those queries once more. Furthermore, the new data which is integrated in the databases is always incremental, i.e., it adds new records and never changes previously stored data [10]. Therefore, if a repeatable query is executed before a data update, and a user requests its execution afterwards, in order to obtain its results we should only query the most recent added factual data and join the results with the previously stored ones from query's prior execution. We can also avoid over-time and resource consumption. Comparing real-time simultaneous query execution between the DW users, we can see if there are any similar queries requested to execute at the same time. Therefore, if we consider a set of users trying to execute the same queries, and put the latest users on hold until the first conclude query process-ing, then returning results for all of them, we efficiently avoid overuse of resource consumption and processing time, minimizing query response time.

The Query Cache Tool (QCT) deals with all of the mentioned issues, looking to optimize all repeatable and parallel querying. In the following subsections, we shall explain what data structure is used for managing the query execution history and how the query cache algorithm works.

2.1 The Query Cache Tool Data Schema

To store all the information for the QCT, we use the data schema shown in Figure 1.

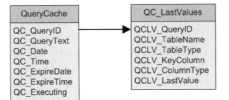

Fig. 1. Query Cache Tool data schema

Table `QueryCache` is the master table for the QCT, storting one row for each query executed by the QCT. Column `QC_QueryID` is a unique identifier for each SQL query instruction and `QC_QueryText` stores a copy of the instruction. Columns `QC_Date` and `QC_Time` store the date and time when the respective query was first executed. Columns `QC_ExpireDate` and `QC_ExpireTime` allow defining when will the respective query's result become overdue or irrelevant. When this happens, the QCT will automatically delete all references and results to it, using what we call the `QueryCacheCleanSweep` procedure, which we will explain further on. Column `QC_Executing` is a logical flag attribute which indicates if the respective query is currently being executed or not. Table `QC_LastValues` is a detail table which will store the last values of the data in each dimension and fact tables which are needed for processing each query. Column `QCLV_QueryID` references the query identifier `QC_QueryID` for the query in master table `QueryCache`. Column `QCLV_TableName` indicates the name of a table which is needed for query identified by `QCLV_QueryID`, and column `QCLV_TableType` indicates if that table is dimensional (D) or factual (F). Columns `QCLV_KeyColumn` and `QCLV_ColumnType` respectively indicate the name and type of a key column existing in table `QCLV_TableName`, while column `QCLV_LastValue` stores the greatest recorded value for that `QCLV_KeyColumn` in table `QCLV_TableName`, acting as a row stamp for distinguishing new data since the query was last executed. For the QCT, each query execution generates a table named QCacheResponse<u>X</u>, that stores the corresponding result, where <u>x</u> is the value of the query's identifier QC_QueryID in the QueryCache table. For instance, if it receives a query to execute to which it associates QC_QueryID = 1, the corresponding results of its execution is stored in an isolated table QCacheResponse1, in the QCT database.

We shall now explain how our QCT algorithm uses this data schema in order to optimize repeatable and parallel OLAP query execution.

2.2 The Query Cache Tool Algorithm

As we mentioned before, the QCT assumes that if no new data has been added to the DW database, the results for any query <u>x</u> which has already been executed is stored in one of the formerly saved QCacheResponse<u>X</u> tables. Therefore, there is no need to execute these queries again, just to supply the results by returning the rows in the correspondent QCacheResponse<u>X</u> table which relates to the desired query, saving time and resource consumption. This makes supplying results for repeated queries an extremely fast task for the QCT. Suppose a certain user A, which starts execution of an OLAP query X. If another user B, has previously started executing an OLAP query Y, similar to query X, which is currently being processed, the QCT does not execute query X. Instead, it discards the execution of query X and puts user A on hold while query Y finishes being processed, and then returns the same results to both users. This allows minimizing time and resource consumption for simultaneous similar query execution, speeding up response time for this type of parallel querying. The algorithm also needs to insure the creation and storage of the results from the first execution of each different query, along with the latest values of each dimensional and factual table needed in processing those results, for identifying in the future if the data in the needed tables has changed or not. It also needs to define the validity of each query results, for automatically disposing those which become overdue.

The QCT algorithm for OLAP query execution is showed on the next page. Due to space constraints, the algorithm is presented in a summarized manner, for its complete code is too long to include in this paper. We have highlighted the instructions which distinguish its major sections. The first highlighted IF instruction verifies if the submitted query *QueryN* has already been executed earlier, meaning that it has already been stored in QueryCache and its results are stored in a corresponding QCacheResponse table. If *QueryN* exists in QueryCache, the second IF instruction checks if it is currently being executed on behalf of other user, and if this is true, waits until the execution finishes. Otherwise, it processes the query against the data which has been added to the database since it was last executed and joins those results to the previously stored ones, saving new results in the corresponding QCacheResponse table. The actual last recorded values of each key column for each table in the query are recorded in QC_LastValues with QC_QueryID of the each current query, for future comparison in data content updates. If data has not changed since the query's last execution, no processing is needed, since the results are already stored in the corresponding QCacheResponse table. If first highlighted IF instruction is FALSE, this means it is the first execution of *QueryN*. Consequently, a new QC_QueryID value is given to the query, recorded in a new row in QueryCache for identification, along with the query's features (complete SQL instruction, current execution date and time, expiring date and time, and `QC_Executing` flag attribute as TRUE). The actual last recorded values of each key column for each table in the query are recorded in `QC_LastValues` with `QC_QueryID` of the current query for future comparison in data content updates. The results of the query's execution are then stored in the corresponding `QCacheResponse` table. The results to all users which submitted the query are given by querying the `QCacheResponse` table, independently if it is a first time execution, a waiting process or an incremental join to previously stored results.

```
PROCEDURE ExecuteQuery(QueryN: SQL Query Instruction)
BEGIN
    IF THERE IS A ROW IN QueryCache WHERE QC_QueryText = QueryN THEN
        QID = QC_QueryID FOR QueryN
        IF QueryN IS ALREADY BEING PROCESSED (QC_Executing = TRUE) THEN
            WAIT
                DELAY Y SECONDS
                VERIFY QC_Executing VALUE FOR QueryN
            UNTIL QC_Executing FOR QueryN IS EQUAL TO FALSE
        ELSE
            SAVE QC_Executing = TRUE IN QueryCache FOR QC_QueryID = QID
            ReQuery = FALSE
            FOR EACH TABLE NEEDED IN QueryN
                LOOKUP LAST RECORDED VALUES IN EACH KEY COLUMN
                IF VALUES ARE DIFFERENT FROM
                    RECORDED VALUES IN QC_LastValues FOR QueryN THEN
                    LOOKUP LAST RECORDED VALUES IN EACH KEY COLUMN
                    SAVE THOSE LAST RECORDED VALUES IN QC_LastValues
                    ReQuery = TRUE
                END IF
            NEXT
            IF ReQuery = TRUE THEN
                FOR EACH FactTable IN QueryN
                    BUILD TmpFactTable WITH ALL THE NEW ROWS INSERTED
                        SINCE LAST EXECUTION OF QueryN
                NEXT
                EXECUTE QueryN AGAINST TmpFactTables
                JOIN RESULTS WITH PREVIOUSLY STORED QCacheResponseX
                    WHERE X = QC_QueryID FOR QueryN
                RECREATE QCacheResponseX WITH NEW RESULTS
            END IF
            SAVE QC_Executing=FALSE IN QueryCache FOR QC_QueryID=QID
        END IF
    ELSE
        DETERMINE A NEW QC_QueryID FOR QueryN
        INSERT A NEW ROW IN QueryCache FOR QueryN WITH QC_Executing = TRUE
        FOR EACH TABLE NEEDED IN QueryN
            LOOKUP LAST RECORDED VALUES IN EACH KEY COLUMN
            SAVE THOSE LAST RECORDED VALUES IN QC_LastValues
        NEXT
        EXECUTE QueryN AND SAVE RESULTS IN QCacheResponseX
            WHERE X = QC_QueryID FOR QueryN
        SAVE QC_Executing = FALSE IN QueryCache FOR QC_QueryID = QID
    END IF
    RETURN RESULTS BY SELECTING ALL ROWS FROM QCacheResponseX
        WHERE X = QC_QueryID FOR QueryN
END
```

Joining the results from a previous execution of a query with new processed results requires taking several issues under consideration. Queries containing the SUM and COUNT aggregation functions do not need to be changed. The first stored results just need to be added to the new ones. The final results of the queries with aggregation functions is computed in a similar way as in DW stripping, presented in [3, 4]. The average function AVG is calculated dividing a SUM by COUNT, and if there is a need for obtaining STDDEV and VARIANCE, they are determined by usage of COUNT, VARIANCE, SUM and COUNT functions, as shown in the previous mentioned papers.

As time goes by, the number of QCacheResponse tables and storage space they take up need to be dealt with. This is done by looking for query results which have been considered overdue or obsolete, checking the values of the QC_ExpireDate and QC_ExpireTime columns. To perform this, the QCT executes a procedure which we have called the QueryCacheCleanSweep. This procedure seeks for all the rows in QueryCache referring queries that are not currently being executed (QC_Executing = FALSE) and where the current server date/time considers overtime (already past the values of the QC_ExpireDate and QC_ExpireTime columns). The procedure is automatically executed every X seconds, where X should be defined by the DBA after consulting with decision makers as to which is the minimum period of interestingness. The simplified algorithm for the QueryCacheCleanSweep is shown below.

```
PROCEDURE QueryCacheCleanSweep
BEGIN
    FOR EACH Row IN QueryCache WHERE QC_Executing = FALSE
        QID = VALUE OF KEY COLUMN QC_QueryID IN CURRENT QueryCache ROW
        IF (CurrentDate() > QC_ExpireDate) OR
           (CurrentDate() = QC_ExpireDate AND
            CurrentTime() >= QC_ExpireTime) THEN
            DELETE ALL ROWS IN QC_LastValues WHERE QCLV_QueryID = QID
            DELETE CURRENT ROW IN QueryCache
            DROP TABLE QCacheResponseX WHERE X = QID
        END IF
    NEXT
END
```

3   Experimental Evaluation

We have implemented a real-world sales DW, based on a star schema with one fact table (Sales) and four dimension tables (Time, Customers, Products, and Promotions). The dimension features of the database, corresponding to one year of commercial data, are shown in Table 1. To build the DW, we used Oracle 10g DBMS on a 2.8 GHz Pentium IV CPU, with 1 GByte RAM and a 180 GByte 7200 rpm hard disk.

Table 1. Dimensional features of the Commercial Sales Business Enterprise Data Warehouse

	Times	Customers	Products	Promotions	Sales
Number of Rows	8 760	250 000	50 000	89 812	31 536 000
Storage Size	0,12 MB	90 MB	7 MB	10 MB	1 927 MB

To obtain results for aiding business decision making, a set of 12 OLAP queries was selected. These queries represent a sample of decision making information which is typically needed in the business, such as customer product and promotion sales daily, monthly, quartery and anually values. The set of these 12 queries represents the workload used in each experimental scenario. We have tested the QCT for each day of December, 2008, considering 4 execution possibilities for each query:

a) **Standard Execution:** traditional execution of the query workload in the Oracle SQL*Plus interface in a standard manner;

b) **QCT First Execution:** execution of the query workload by QCT, assuming that the query cache database is empty, i.e., each query is executed for the first time by QCT;

c) **QCT Incremental Execution:** execution of the workload by the QCT, where a first execution has been previously made and their results are already stored in the query cache database, and after inserting an entire day of new data in the DW fact table (which stands for an average of 86746 new rows in Sales), to join these new results with the previously stored ones;

d) **QCT Sequential Execution:** execution of the workload by the QCT a second time after it has already been executed and their results are already stored in the query cache database, with no change in the DW tables' contents.

Assuming that the DW is updated in a daily fashion, first we shall present the results concerning the usage of the QCT during one day, against traditional query workload execution. Tables 2, 3, and 4 present the results for comparing standard query workload execution on 31-12-2008, against each of the three presented execution possibilities using the QCT on the same day.

Table 2. Standard workload execution time vs. QCT workload first exec. time on a day

	Standard Exec. Time	QTC First Exec. Time	Time Difference	Times Faster/Slower
1 User	764 s	810 s	+ 46 s	1.06 times slower
2 Users	1336 s	832 s	- 504 s	1.61 times faster
4 Users	2050 s	1212 s	- 838 s	1.69 times faster
8 Users	4206 s	1868 s	- 2338 s	2.25 times faster
16 Users	7807 s	3797 s	- 4010 s	2.06 times faster

Table 3. Standard workload exec. time vs. QCT workload incremental exec. time on a day

	Standard Exec. Time	QTC Incremental Exec. Time	Time Difference	Times Faster/Slower
1 User	764 s	49 s	- 715 s	15.6 times faster
2 Users	1336 s	89 s	- 1247 s	15.0 times faster
4 Users	2050 s	164 s	- 1886 s	12.5 times faster
8 Users	4206 s	304 s	- 3902 s	13.8 times faster
16 Users	7807 s	583 s	- 7224 s	13.4 times faster

Table 4. Standard workload exec. time vs. QCT workload sequential exec. time on a day

	Standard Exec. Time	QTC Sequential Exec. Time	Time Difference	Times Faster/Slower
1 User	764 s	13 s	- 751 s	58.8 times faster
2 Users	1336 s	26 s	- 1310 s	51.4 times faster
4 Users	2050 s	48 s	- 2002 s	42.7 times faster
8 Users	4206 s	87 s	- 4119 s	48.3 times faster
16 Users	7807 s	160 s	- 7647 s	48.8 times faster

As it can be seen from the results, the QCT is much faster than the standard query execution for all cases, except for the first execution with only 1 user querying, showed in *QCT First Time Execution*. This happens because the QCT has to execute the first workload with 1 user in a standard manner and still has to create and store the initial results. However, for more than 1 user, the QCT takes advantage of checking if there are any similar queries executing simultaneously, dismissing parallel querying for those queries, contrarily to the standard execution, which reexecutes all of the queries. This means that the more the users, the better QCT outperforms the standard execution. This can be confirmed by observing Figure 2. By analyzing the previous tables and the figure, we can see that if a query which has been stored in the query cache database is repeated, showed by the *QCT Sequential Execution*, the QCT can supply the results around 50 times faster, for it only needs to access the previously stored results in order to process the query. It is also much faster to join new calculated results from new added data with previously stored ones to supply query results, showed by the *QCT Incremental Execution*, than reexecuting the queries against the whole amount of data.

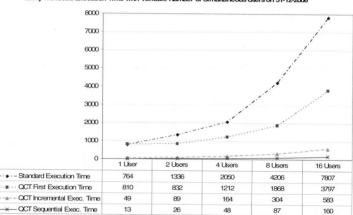

Query Workload Execution Time with Variable Number of Simultaneous Users on 31-12-2008

	1 User	2 Users	4 Users	8 Users	16 Users
– · ♦ · – Standard Execution Time	764	1336	2050	4206	7807
· · · ■ · · · QCT First Execution Time	810	832	1212	1868	3797
– ▲ – QCT Incremental Exec. Time	49	89	164	304	583
—×— QCT Sequential Exec. Time	13	26	48	87	160

Fig. 2. Query workload execution of standard execution vs. QCT execution for 31-12-2008

4   Related Work

Most of the research work done in this area is focused on optimizing data search methods and physical data distribution. Our method aims for the OLAP SQL instruction level. Many DBMS vendors claim to support parallel data warehousing to various degrees, e.g. Oracle10g R2 [13], IBM/Informix Red Brick [15], and Microsoft SQL Server [9]. Most of these products, however, do not use dimensionality of data that exists in a DW and it remains unclear to what extent multidimensional fragmentation is exploited to reduce query work. None of the aforementioned vendors provide sufficient information or even tool support on how to determine an adequate data allocation for star schemas. The effective use of parallel processing in this environment can be achieved only if we are able to find innovative techniques for parallel data placement using the underlying properties of data in the warehouse.

The most common choice consists of systems that offer massive parallel processing capabilities [1, 18], as Massive Parallel Processing (MPP) systems or Symmetric MultiProcessing (SMP) systems. Due to the high price of this type of systems, less expensive alternatives have already been proposed and implemented [5, 7, 12]. One of those alternatives is the DW Stripping (DWS) technique [3, 4]. A large amount of research has been performed for processing and optimizing queries over distributed data (see, e.g. [2, 11, 16, 17]). However, this research has focused mainly on distributed join processing rather than distributed computation. Only recently [6] we have a new architecture and optimizations for parallel SQL execution in the Oracle 10g database and a practical solution for parallelizing query optimization in the multi-core processor architecture, including a parallel join enumeration algorithm and several alternative ways to allocate work to threads to balance their load. This solution has been prototyped in PostgreSQL [9]. The approach we explore in this paper marries the concepts of distributed processing and parallel OLAP queries to provide a fast and reliable relational DW.

5   Conclusions and Future Work

We have explained how our query cache tool works, optimizing the execution of repeatable OLAP queries and simultaneous similar query executions. We have also shown that our query cache tool is efficient, significantly reducing query execution time and processing resources. The presented results in the experimental evaluation show that the query cache method is much better than the standard query workload execution, for this type of queries. As future work, we intend to enhance the method for including features which can deal with queries which represent incremental column results that can be added to the results of other previously processed queries. We also mean to work on similar query recognition, for identifying similar OLAP instructions which are not written exactly the same way, but aim for similar results.

References

[1] Agosta, L.: Data Warehousing Lessons Learned: SMP or MPP for Data Warehousing. DM Review Magazine (2002)

[2] Akinde, M.O., Bhlen, M.H., Johnson, T., Lakshmanan, L.V.S., Srivastava, D.: Efficient OLAP query processing in distributed data warehouses. Information Systems 28, 111–135 (2003)

[3] Bernardino, J., Madeira, H.: Experimental Evaluation of a New Distributed Partitioning Technique for Data Warehouses. In: Int. Symposium on Database Engineering and Applications, IDEAS 2001 (2001)

[4] Bernardino, J., Furtado, P., Madeira, H.: Approximate Query Answering Using Data Warehouse Striping. Journal of Intelligent Information Systems – Integrating Artificial Intelligence and Database Technologies 19(2), 145–167 (2002)

[5] Critical Software SA, DWS, http://www.criticalsoftware.com

[6] Cruanes, T., Dageville, B., Ghosh, B.: Parallel SQL Execution in Oracle 10g. In: ACM SIG International Conference on Management of Data, SIGMOD (2004)

[7] DATAllegro, DATAllegro v3™, http://www.datallegro.com

[8] Galindo-Legaria, C.A., Grabs, T., Gukal, S., Herbert, S., Surna, A., Wang, S., Yu, W., Zabback, P., Zhang, S.: Optimizing Star Join Queries for Data Warehousing in Microsoft SQL Server. In: Int. Conf. on Data Engineering (ICDE 2008), pp. 1190–1199 (2008)

[9] Han, W.S., Kwak, W., Lee, J., Lohman, G.M., Markl, V.: Parallelizing Query Optimization. In: International Conference on Very Large Data Bases, VLDB (2008)

[10] Kimball, R., Ross, M.: The Data Warehouse Toolkit, 2nd edn. John Wiley & Sons, Chichester (2002)

[11] Kossman, D.: The state of the art in distributed query processing. ACM Computing Surveys 32(4), 422–469 (2000)

[12] Netezza, The Netezza Performance Server® DW Appliance,
     http://www.netezza.com
[13] Oracle Data Warehousing Guide 10g R2,
     http://downloadwest.oracle.com/docs/cd/B1930601/server.102/
     b14223.pdf
[14] Pedersen, T.B.: How is BI used in industry?: Report from a knowledge exchange net-
     work. In: Kambayashi, Y., Mohania, M., Wöß, W. (eds.) DaWaK 2004. LNCS, vol. 3181,
     pp. 179–188. Springer, Heidelberg (2004)
[15] RedBrick White Paper,
     ftp://ftp.software.ibm.com/
     -software/data/informix/pubs/whitepapers/
     redbrickwp040904.pdf
[16] Schewe, K.D., Zhao, J.: Balancing redundancy and query costs in distributed data ware-
     houses – an approach based on abstract state machines. In: Hartmann, S., Stumptner, M.
     (eds.) 2nd Asia-Pacific Conference on Conceptual Modelling (ER), Austral. CRPIT,
     vol. 43, pp. 97–105. Computer Society (2005)
[17] Stanoi, I., Agrawal, D.P., El Abbadi, A.: Modeling and Maintaining Multi-view Data
     Warehouses. In: Akoka, J., Bouzeghoub, M., Comyn-Wattiau, I., Métais, E. (eds.) ER
     1999. LNCS, vol. 1728, pp. 161–176. Springer, Heidelberg (1999)
[18] Sun Microsystems, Data Warehousing Performance with SMP and MPP Architectures,
     White Paper (1998)
[19] Zurek, T., Kreplin, K.: SAP Business Information Warehouse – From Data Warehousing
     to an E-Business Platform. In: International Conference on Data Engineering, ICDE
     (2001)

Efficient Map Portrayal
Using a General-Purpose Query Language
(A Case Study)

Peter Baumann, Constantin Jucovschi, and Sorin Stancu-Mara

Jacobs University Bremen
{p.baumann,c.jucovschi,s.stancumara}@jacobs-university.de

Abstract. Fast image generation from vector or raster data for map
navigation by Web clients is an important geo Web application today.
Raster data obviously account for the larger volume of the underlying
data sets served through WMS and other such interfaces. Dedicated
server implementations prevail because an often heard argument is that
general-purpose server software, such as database systems, cannot be
efficient enough for such high-volume application scenarios.

In this paper we refute that. We investigate just-in-time compilation
of query fragments in two variants, for CPU and GPU, as implemented in
the general purpose raster DBMS rasdaman. Results suggest that array
databases are suitable for realtime geo raster services.

1 Introduction

GoogleMaps, Virtual Earth, and several similar services have become a commodity for everybody, thereby stretching use of geo data from a mere expert focus to the general Internet user. Actually, web-based map navigation currently constitutes one of the most widely used Web GIS (Geographic Information System) functionality and accounts for a large part of the geo service traffic generated.

For portrayal in a map, one or more layers are superimposed. We disregard vectorial data and concentrate on raster data where four types of layers are common:

- Greyscale or color raster imagery. These can be displayed more or less directly.
- Elevation and bathymetry data. They lack an immediate visual semantics and, hence, have to be classified, i.e., height/depth values are replaced by some color indicator.
- Thematic raster maps. In these boolean images a pixel value of *false* indicates transparency, while a value of *true* indicates that an object covers this location. During portrayal, *true* values get replaced by some color value.

Obviously, fast portrayal of raster maps represents an extra challenge due to the base data set's size which typically ranges into multi-Terabyte volumes, thereby exceeding vector data volumes by several orders of magnitudes. Elevation and

S.S. Bhowmick, J. Küng, and R. Wagner (Eds.): DEXA 2009, LNCS 5690, pp. 153–163, 2009.

bathymetry data, in turn, constitute the computationally most expensive part due to the classification task. Hence, in this contribution we will focus on speed-ups particularly for these processing intensive map data type.

We inspect such services from a database perspective to see how query languages can act as a tool interface for data access. Array databases provide a suitable abstraction for raster services, including declarative, optimizable query models [2][14][15][29]; our analysis is based on the rasdaman array algebra [5] and DBMS [3] which is in operational use as a map server since many years and on multi-Terabyte image objects. Concretely, we evaluate a recently implemented optimization technique, just-in-time (JIT) compilation. This method starts out by pre-clustering of operations in the query tree, and then generates either CPU or GPU (Graphics Processing Unit) code.

The remainder of this contribution starts, in Section 2, with a brief array model introduction as far as needed here. Section 3 discusses the use case. Section 4 presents the JIT optimization techniques. The state of the art is reviewed in Section 5, and Section 6 concludes the paper.

2   The rasdaman Array DBMS

In this Section we give a brief overview of the rasdaman data model and query processing, as far as required for our discussion. The rasdaman language [20] and algebra [5], various optimization techniques [10][13][11][32], architecture, further applications [3][21][23], and its impact on standardization [4] are presented elsewhere.

2.1   Array Model and Query Language

Arrays are modeled as functions mapping from an d-dimensional domain to some value set. Anticipating its embedding into the relational model, the rasdaman conceptual model consists of tables ("collections") with two columns, one holding the array and the other one a unique object identifier (OID) allowing for foreign key relationships. The query language [20], rasql, allows to compose expressions on arrays embedded into the select/from/where style of SQL. For easier comprehension we only use rasql syntax here; see [2][5] for more background.

Three core operations cover the complete range of expressive power [5]: MARRAY constructs an array; CONDENSE represents a summation functional; SORT, finally, sorts slices within an array. All other operations can be reduced to these.

Of particular interest for the web map query patterns investigated here are *induced operations*. They apply some unary or binary operation which is defined on the cell (pixel) type of the array to all cells simultaneously. For example, $a + b$ performs a pixelwise addition of arrays a and b which obviously must have matching domains. The induced comparison $a > 0$ returns a boolean array with *true* values for all cells where a has positive values.

Lateron we will make use of induced multiplication with composite pixels to lift binary or greyscale images to color images. To this end, we allow nested records of values. The expression $a * \{1, 0, 0\}$ takes an array a with an atomic, integer cell type and extends it to a 3-component RGB array with first component holding the a values while all others are zero, i.e., black. The result can be interpreted as a colored in red.

Specifically for the support of geo applications the induced *overlay* operator has been added to rasdaman. It overlays two same-size arrays and takes the first operand's cell value whenever this is not null, otherwise takes the second operand's cell value.

Further operations and their efficient implementation, such as scaling [9] are discussed elsewhere. In rasql, such operations are embedded in SQL-style queries, such as

```
select png( (AirborneImage.red + AirborneImage.green
+ AirborneImage.blue) / 3
from   AirborneImage
```

AirborneImage is assumed to be a collection of color-valued arrays. Each array is inspected in turn. The *select* clause first applies the induced operations which effectively convert the color image into a grayscale image, and then encode the result image into PNG format for shipping to the client.

2.2 Array Query Processing

The rasdaman engine follows the classical query evaluation scheme: incoming queries are parsed, analyzed syntactically and semantically, heuristic and cost-based optimizations are applied [22][10], and the resulting query program is executed against the data [30]. For storage, arrays are partitioned into sub-arrays called *tiles* which form the unit of disk access and array processing [6]. The query tree generated consists of operator nodes representing, among others, the above introduced operations. The derived operators do not just constitute syntactic sugar, but are internally mapped to specifically optimized implementations - hence, they are not transformed into MARRAY nodes. During evaluation of the query, nodes recursively fetch and process data based on the *open-next-close* (ONC) protocol whenever possible.

3   Web Mapping as Database Queries

A first simple request retrieves a color image from *AirborneImage*. By convention, each collection holds exactly one array tuple. We assume that the translation from geo to pixel coordinates has been performed already; let the source bounding box be given by coordinates $(x_0, y_0$ and (x_1, y_1) and the result image size by 300×300.

```
select png(scale(AirborneImage[x_0:x_1,y_0:y_1], [0:299,0:299]))
from   AirborneImage
```

Elevation data add significant complexity to such queries. For portrayal, their floating point pixel values, which indicate height or depth in some unit of measure, need to be mapped to color codes. The corresponding lookup tables typically contain several dozens of entries. Assuming a suitable collection *Elevation* and n threshold height values $h1, ..., hn$ and grey levels $\{i, i, i\}$ for $i \in \{0..255\}$ we obtain

```
select png(                        (Elevation <= h1) * {1,1,1}
           + (h1 < Elevation and Elevation <= h2) * {2,2,2}
           + (h2 < Elevation and Elevation <= h3) * {3,3,3}
           ...
           + (hn < Elevation)                     * {n,n,n}
           )
from    Elevation
```

Obviously, query complexity easily encompasses hundreds of operations in presence of overlays and elevation layers.

4 Dynamic Compilation of WMS Queries

The above query patterns convey a regular, albeit expensive schema in terms of the number of operations. In particular induced operations constitute the CPU intensive part. Our approach to optimizing such computationally complex queries first clusters suitable query fragments and subsequently compile them to run on either CPU or GPU.

4.1 Operator Node Conflation

In a query tree representing some algebraic expression, normally each node is responsible for exactly one operation. For physical optimization, however, it is common to introduce special node types which combine, e.g., a table scan with attribute filtering. We extend this principle by introducing a *conflation* node type which represents not just some fixed operation combination, but an expression subtree [12] (there called *group iterator*).

Node conflation determines maximal subtrees in the query tree which are suitable for grouping and substitutes them with the according conflation node. The algorithm for replacing query fragments by conflation nodes walks the query tree in a bottom-up fashion as follows. Assume the algorithm is inspecting node N. Then,

– if N is a leaf node (i.e., either an array reader or a constant) then it is transformed into a new conflation node C containing this one node operation.
– if non-leaf N is a singleton operation, like scaling, then its operation is incorporated into a new conflation node which replaces N. The conflation node is locked against any further conflation.

– if N is a non-leaf node supporting conflation then N together with its sub-tree is replaced by a single conflation node. Assume N has as children the non-locked conflation nodes $C_1, ..., C_m$ and further nodes $N_1, ..., N_n$ (which, by construction, can be either locked conflation nodes or node types not supported by conflation) for some $m, n > 0$, then the new node C has N's root operation with C_i as operation subexpressions arguments and N_j as tile input streams for $i \leq m, j \leq n$.

Nodes supporting conflation are those which receive array-valued input, i.e.: unary and binary induced operations, aggregates, and scaling. Induced operations, like comparison and arithmetics, take arrays and deliver arrays, hence can appear anywhere inside a conflation node; they are simply conflated into one node when encountered. Aggregates (which are not of relevance in WMS-type queries) deliver scalars and, consequently, can appear only at the root of the tree fragment collected, not as a conflationr-internal or leaf node. An aggregate is added to the conflation node on hand, which subsequently is locked. All other query tree nodes remain unchanged, except for scaling nodes which cannot be merged into a loop and anyway are expensive enough to individually benefit from just-in-time compilation; hence, these are boxed into separate single-operation conflation nodes to flag them for compilation.

This merging per se does not lead to an improved performance, as it again would require interpretation of its contents. A benefit, however, is that iteration over some input array (i.e., a tile) now has only one cell iteration variable instead of the previous situation where every node required its own iteration.

4.2   Dynamic CPU Compilation

In the first variant, from each conflation node a piece of C source code is generated, compiled into a shared library, linked into the server, and executed on the conflation node's parameters.

The code generation algorithm [12] is implemented as a recursive function which first generates C code for its children and then concatenates their codes in a way that the overall result is provided as the result of the current operation. A challenge is presented by the complex type system of rasdaman which actually supports any valid C/C++ struct as array cell type. To make the C program generated agnostic against complex types all complex structures are linearized into separate operation units.

Elevation data classification may serve as an example. Consider the following query fragment:

```
(T > -15 and  T<0) * {10,40,100}
```

The code generated essentially looks like this:

```
void process(int units, void *data, void *result) {
    int iter; void* dataIter = data; void* resIter = result;
    for (iter=0; iter<units; ++iter, dataIter+=8, resIter +=12) {
```

```
    float var0 = *(float*)dataIter;
    bool c = (var0>-15) && (var0<0);
    *((int*)resIter)   = (c?10c:0c);
    *((int*)resIter+4) = (c?40c:0c);
    *((int*)resIter+8) = (c?100c:0c);
  }
}
```

Register allocation, peephole optimization, instrumentation for processor-specific speedup like pipelining, etc. are exploited through the compiler's builtin optimization.

Preliminary performance results were obtained on an Intel Pentium QuadCore CPU 6600 running at 2.4 GHz. The compiler used was GNU gcc 4.3 under a vanilla Debian Etch Linux with kernel 2.6.25. The effect of JIT optimization has been evaluated both standalone and embedded into the rasdaman system. In the standalone scenario, data were already in memory; in the integrated scenario, times measured include tile loading from the 7200 rpm disk. Each scenario was tested in two modes, COLD and HOT, resp. In COLD mode the C program had to be generated and compiled prior to query evaluation. In HOT mode the shared library was available and can be loaded and executed right away. For comparison additionally a handcrafted, manually tuned C code version ("TAILORED") was implemented.

The dataset used consists of one 512×512 array of double-precision floats stored in the database as one tile upon which the elevation classification query was applied. Measurements were made for queries containing $n = 2^0, .., 2^7$ classification fragments as presented above, each one applying five operations per pixel.

Figure 1 shows left the results for the standalone version of the algorithm and right the results of the integrated test.

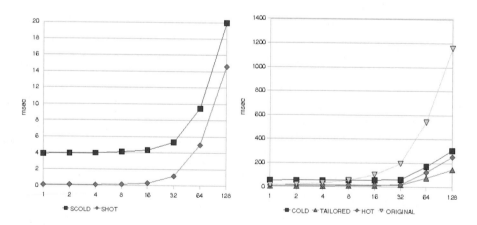

Fig. 1. Standalone (left) and integrated (right) performance results for n classification query fragments

Repeated occurrence of the query fragment pattern which the conflation node represents is beneficial in several respects: When the same fragment arrives next, time the code still is loaded and can be executed immediately; in this case, runtime is close to precompiled code. In case the server instance has been restarted since, the dynamic library remains cached in the file system, this still means a significant boost compared to interpretation. Further, if parallel server instances run on a server then all of them will benefit from the first compilation of the library; this is particularly beneficial in a parallel architecture like rasdaman has.

4.3   Dynamic GPU Compilation

GPU tasking with array query fragments [26][25] takes conflation nodes, generates GPU code for each node, ships code and tiles to the GPU, and fetches back results for regular further processing in the query tree. We inspect each step in turn.

For each conflation node an OpenGL Shading Language (GLSL) [24] source code unit is generated. On principle, the GLSL code generated looks as follows where "..." is the placeholder for the expression to be evaluated on each pixel (which is rather similar to conventional CPU code):

```
uniform sampler2D data;
void main() {
    float f = texture2D(data,vec2(gl_FragCoord.x
    /512.0,gl_FragCoord.y/512.0)).r;
    f = ...;
    gl_FragColor = vec4( f, 1, 1, 1 );
}
```

In the GPU programming model, input and output data form *textures*; the program to be executed in the GPU is coded as a *shader*. The above code first declares `data` to be a 2-D texture variable. Inside the `main()` routine first a 2-D texture is generated and bound to the data, with the extent given by the system-provided global variable `gl_FragCoord`. The next expression resembles the operation to be executed on each pixel, in a syntax very much like C code. Then, a 4-channel red/green/blue/alpha pixel is generated as prescribed by the programming model. After rendering, the result is copied back into main memory, ready for passing it on to the conflation node's parent.

Compilation of the GLSL code into GPU machine code is done by the graphics card driver. The OpenGL driver manages all concurrency and load balancing, so no preparation is required in the source code. As OpenGL lacks any intuition about further usage of the textures (i.e., image data) and shader programs, a special-purpose cache manager performs this task.

Tasking a GPU with array processing is beneficial from several perspectives. First, GPUs are optimized for image processing operations like the ones on hand. Further, GPUs provide a large number – typically hundreds – of cores which come with builtin scheduling which makes them a very cost-efficient method of increasing hardware parallelism. Finally, the CPU can perform other tasks in parallel.

For performance evaluation, an Intel Core2 2.5 GHz on an nForce 790i SLi motherboard was used with 1 GB RAM with 1.333 GHz front side bus running a vanilla Debian Etch 2.6.25. Graphics cards under test were Nvidia 8800GT, 8800GT SLI, 8800GTX, and 9800GX2 (a high-end card at the time of this writing). Disk access was excluded from measurements. GPU code compilation was found to be almost always below 1ms.

Two classes of tests were performed, one varying the number of instructions, the other one varying the number of textures used. For each test, with results averaged over 100 runs, the following scenarios were investigated:

- Compile query, ship query, and input texture to GPU; process texture and ship back result (single query, cold shader cache).
- Process a texture already in place, with a shader program readily compiled and loaded (single query, hot shader cache). This together with the previous measurement allowed to separate net processing time from time spent in logistics.
- Process one texture 100 times, with both texture and shader already in place. This allowed to observe the parallelization effect.

Figure 3 gives the corresponding benchmark results for the existing, interpreted query processing algorithm running on the CPU. It clearly shows that time consumed depends directly on the number of operations.

Of particular interest in presence of GPU parallelism is how the number

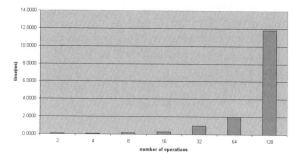

Fig. 2. CPU query processing time without JIT

of operations within a query fragment impacts response time. Figure 3; shows results of the classification query used earlier; "number of operations" refers to

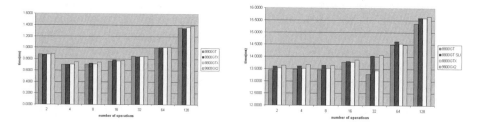

Fig. 3. Single query in a cold pipeline (left) and 100 queries in a hot pipeline (right)

the number of multiplications per pixel. Interestingly, almost no difference between low-end and high-end GPUs is obvserved; further research is required to explain this.

5   State of the Art

For heavy-traffic services usually response time is optimized through large-scale hardware with pre-materialized map imagery (which is what GoogleMaps and similar services do) and/or specialized data architectures, such as tailored image formats [19][16][17].

A large part of research in array database query processing focuses on optimization by finding a semantically equivalent sequence of operations which would determine the same result faster. In most cases optimizations are limited to reordering, restructuring, and sometimes joining query tree nodes [8][15]. There are not many papers which focus on the optimal implementation of the actuall execution part of the query [31]. This is unfortunate as some papers, when comparing their results with tailored solution written in C or C++ find themselves 5-181 times slower [15] even though the strategies of computing are the same.

Similar attempts to optimize the evaluation parts of a query were made by introducing embedded (or stored) procedures [18][1][28]. Indeed, many complex applications reported increased performance as well as better modularity and easier management. The obvious disadvantage, however, is that manual work and depp expertise is required to identify optimization candidates and implement them subsequently.

In the domain of supercomputing, array processing has a long tradition. Loop fusion is one of the techniques successfully applied there, see, e.g., [7]. Further, optimization of main memory array operations has been investigated, e.g., in the context of APL [33]. Our approach is to adopt such techniques to cluster array iterations into maximal query fragments for subsequent native code generation.

6   Conclusion and Outlook

Map portrayal requires flexible, scalable services on large objects – hence, a classical database tasks. In this contribution we have shown first results from applying non-standard optimization techniques to Web mapping requests implemented through array database queries. The benchmark results are encouraging as they seem to indicate that the high number of operations can be optimized effectively by CPU and GPU just-in-time code generation.

Scaling currently is not yet performed on the GPU, although it clearly is a candidate. Among further aspects to be researched is a comprehensive evaluation of CPU and GPU JIT, including best practices which can serve as a decision basis for an intelligent optimizer.

Still, the optimization potential of array databases by far is not exhausted. Among further optimizations not addressed here are n-D image pyramids and cluster-based parallelization. While pyramids are common practice since long, we are working on transferring results from OLAP preaggregation, thereby extending the pyramid concept to the multi-dimensional case and to flexible, demand driven maintenance of pre-computed data [10]. Cluster-based parallelization transparently distributes incoming queries over a network of servers.

Finally, now that rasdaman is IO-bound, further disk optimization gets on the agenda again. Among the work in progress are tile sequencing strategies [27].

References

1. Acheson, A., et al.: Hosting the.net runtime in microsoft sql server. In: Proc. ACM SIGMOD, pp. 860–865. ACM, New York (2004)
2. Baumann, P.: On the management of multi-dimensional discrete data. VLDB Journal Special Issue on Spatial Database Systems 4(3), 401–444 (1994)
3. Baumann, P.: Large-scale raster services: A case for databases (invited keynote). In: Roddick, J., Benjamins, V.R., Si-said Cherfi, S., Chiang, R., Claramunt, C., Elmasri, R.A., Grandi, F., Han, H., Hepp, M., Lytras, M.D., Mišić, V.B., Poels, G., Song, I.-Y., Trujillo, J., Vangenot, C. (eds.) ER Workshops 2006. LNCS, vol. 4231, pp. 75–84. Springer, Heidelberg (2006)
4. Baumann, P.: The ogc web coverage processing service (wcps) standard. Geoinformatica (2009) (accepted for publication)
5. Baumann, P.: A database array algebra for spatio-temporal data and beyond. In: Tsur, S. (ed.) NGITS 1999. LNCS, vol. 1649, pp. 76–93. Springer, Heidelberg (1999)
6. Furtado, P., Baumann, P.: Storage of multidimensional arrays based on arbitrary tiling. In: Proc. ICDE, pp. 328–336 (1999)
7. Gao, G.R., Olsen, R., Sarkar, V., Thekkathdw, R.: Collective loop fusion for array contraction. In: Banerjee, U., Gelernter, D., Nicolau, A., Padua, D.A. (eds.) LCPC 1992. LNCS, vol. 757, pp. 281–295. Springer, Heidelberg (1993)
8. Graefe, G.: Query evaluation techniques for large databases. ACM Comput. Surv. 25(2), 73–169 (1993)
9. Gutierrez, A.G.: The Application of OLAP Pre-Aggregation Techniques to Speed Up Query Processing in Raster-Image Databases. Phd thesis (2009)
10. Gutierrez, A.G., Baumann, P.: Computing aggregate queries in raster image databases using pre-aggregated data. In: Proc. ICCSA (2008)
11. Hahn, K., Reiner, B., Höfling, G., Baumann, P.: Parallel query support for multidimensional data: Inter-object parallelism. In: Hameurlain, A., Cicchetti, R., Traunmüller, R. (eds.) DEXA 2002. LNCS, vol. 2453, p. 820. Springer, Heidelberg (2002)
12. Jucovschi, C.: Precompiling Queries in a Raster Database System. Bachelor thesis, Jacobs University Bremen (2008)
13. Jucovschi, C., Baumann, P., Stancu-Mara, S.: Speeding up array query processing by just-in-time compilation. In: Proc. IEEE SSTDM, pp. 408–413 (2008)
14. Libkin, L., Machlin, R., Wong, L.: A query language for multidimensional arrays: design, implementation and optimization techniques. In: ACM SIGMOD, pp. 228–239 (1996)
15. Marathe, A.P., Salem, K.: Query processing techniques for arrays. VLDB Journal 11(1), 68–91 (2002)
16. n. n. Ecw – ermapper compress wavelets (.ecw), gdal.org/frmt_ecw.html (accessed June 13, 2009)

17. n. n. Jpeg2000, `www.jpeg.org/jpeg2000/` (accessed June 13, 2009)
18. Neugebauer, L.: Optimization and evaluation of database queries including embedded interpolation procedures. SIGMOD Rec. 20(2), 118–127 (1991)
19. n.n. Mrsid – multi-resolution seamless image database, `en.wikipedia.org/wiki/MrSID#External_links` (accessed June 13, 2009)
20. n.n. rasdaman query language guide, 7.0 ed. rasdaman GmbH (2008)
21. Pisarev, A., Poustelnikova, E., Samsonova, M., Baumann, P.: Mooshka: a system for the management of multidimensional gene expression data in situ. Information Systems 28, 269–285 (2003)
22. Ritsch, R.: Optimization and Evaluation of Array Queries in Database Management Systems. Phd thesis (1999)
23. Roland, P., Svensson, G., Lindeberg, T., Risch, T., Baumann, P., Dehmel, A., Frederiksson, J., Halldorson, H., Forsberg, L., Young, J., Zilles, K.: A database generator for human brain imaging. Trends in Neurosciences 24(10), 562–564 (2001)
24. Rost, R.J.: OpenGL shading language. Addison-Wesley, Reading (2006)
25. Stancu-Mara, S.: Method for server-side data processing using graphic processing units (2007)
26. Stancu-Mara, S.: Using Graphic Cards for Accelerating rater Database Query Processing. Bachelor thesis, Jacobs University Bremen (2008)
27. Stancu-Mara, S.: Optimization Support for Linear Indexed Queries in Raster Databases. Master thesis, Jacobs University Bremen (2009)
28. Trissl, S., Leser, U.: Fast and practical indexing and querying of very large graphs. In: Proc. ACM SIGMOD, pp. 845–856. ACM, New York (2007)
29. van Ballegooij, A.R.: RAM: A multidimensional array DBMS. In: Lindner, W., Mesiti, M., Türker, C., Tzitzikas, Y., Vakali, A.I. (eds.) EDBT 2004. LNCS, vol. 3268, pp. 154–165. Springer, Heidelberg (2004)
30. Widmann, N.: Efficient Operation Execution on Multidimensional Array Data. Phd thesis (2000)
31. Widmann, N., Baumann, P.: Efficient execution of operations in a DBMS for multidimensional arrays. In: Proc. SSDBM, pp. 155–165 (1998)
32. Widmann, N., Baumann, P.: Performance evaluation of multidimensional array storage techniques in databases. In: Proc. IDEAS (1999)
33. Wiedmann, C.: A performance comparison between an apl interpreter and compiler. In: Proc. APL, pp. 211–217. ACM, New York (1983)

On Low Distortion Embeddings of Statistical Distance Measures into Low Dimensional Spaces

Arnab Bhattacharya, Purushottam Kar, and Manjish Pal

Department of Computer Science and Engineering,
Indian Institute of Technology Kanpur, India
{arnabb,purushot,manjish}@cse.iitk.ac.in

Abstract. In this paper, we investigate various statistical distance measures from the point of view of discovering low distortion embeddings into low dimensional spaces. More specifically, we consider the Mahalanobis distance measure, the Bhattacharyya class of divergences and the Kullback-Leibler divergence. We present a dimensionality reduction method based on the Johnson-Lindenstrauss Lemma for the Mahalanobis measure that achieves arbitrarily low distortion. By using the Johnson-Lindenstrauss Lemma again, we further demonstrate that the Bhattacharyya distance admits dimensionality reduction with arbitrarily low additive error. We also examine the question of embeddability into metric spaces for these distance measures due to the availability of efficient indexing schemes on metric spaces. We provide explicit constructions of point sets under the Bhattacharyya and the Kullback-Leibler divergences whose embeddings into any metric space incur arbitrarily large distortions. To the best of our knowledge, this is the first investigation into these distance measures from the point of view of dimensionality reduction and embeddability into metric spaces.

1 Introduction

The problem of embedding distance measures into normed spaces arises in applications dealing with large amounts of high dimensional data where performing point, range or nearest-neighbor (NN) queries in the ambient space entails enormous computational costs (curse of dimensionality [1]). The problem of indexing and searching is magnified if the distance measures being imposed on the data objects do not form a metric. Various approaches have been proposed to solve this problem including easily estimable upper/lower-bounds on the distance measures [2] and finding embeddings that allow specific proximity queries to be efficiently carried out [3]. These methods have been found to be crucial for database retrieval algorithms in obtaining speedups over naïve search techniques.

An interesting, and often more difficult, situation arises in the case of statistical distance measures which are widely used in database and pattern recognition applications. It has been found that in many scenarios, especially in similarity based search in image retrieval [4], statistical distance measures like the Mahalanobis and Bhattacharyya measures give better performance than the standard l_2 distance. The Mahalanobis distance has found also application in face

S.S. Bhowmick, J. Küng, and R. Wagner (Eds.): DEXA 2009, LNCS 5690, pp. 164–172, 2009.
© Springer-Verlag Berlin Heidelberg 2009

recognition tasks [5]. The Bhattacharyya class of distance measures which include the Bhattacharyya distance and the Hellinger distance are used in diverse database scenarios such as nearest-neighbor classification [6] and detecting voice over IP floods [7]. Another important statistical similarity measure is the Kullback-Leibler divergence which has been shown to be well suited for use in real-time image segmentation algorithms [8]. This measure is also interesting from a theoretical perspective as well because of its information-theoretic roots. These distance measures have received a lot of attention recently and have been examined from several perspectives including clustering [9] and sketching [10].

We examine these distance measures from another interesting perspective – that of low-distortion embeddings into metric spaces and dimensionality reduction. The lack of inherent "geometric" properties make them harder candidates for such embeddings.

Our Contributions: In this paper, we examine three statistical distance measures with the goal of obtaining low distortion, low dimensional embeddings for them. First, we consider the Bhattacharyya distance and develop a technique to prove that there cannot exist low-distortion embeddings for the Bhattacharyya distance into a metric space in Section 3. We also provide a satisfactory positive result by providing an embedding into the l_2^2 space. In Section 4 we develop another technique that, along with the previous technique, allows us to prove lower bounds on the distortion of any embedding of the Kullback-Leibler divergence into a metric space. Finally, in Section 5, we investigate the Mahalanobis distance and develop a dimensionality reduction scheme for the more general family of Quadratic Form Distances. Due to lack of space we do not provide complete proofs of the theorems stated. The proofs appear in the complete version of the paper [11].

2   Preliminaries

We begin by defining the concept of distortion for embeddings of metric spaces.

Definition 1 (D-embedding and Distortion). *Given two metric spaces* (X, ρ) *and* (Y, σ), *a mapping* $f : X \longrightarrow Y$ *is called a D-embedding where* $D \geq 1$, *if there exists a number* $r > 0$ *such that for all* $x, y \in X$, *we have*

$$r \cdot \rho(x, y) \leq \sigma\left(f(x), f(y)\right) \leq D \cdot r \cdot \rho(x, y).$$

The infimum of all numbers D *such that* f *is a D-embedding is called the* distortion *of* f.

It is easy to see that this notion of distortion can be naturally extended to non-metric spaces as well. A classic result widely used in the field of metric embeddings is the Johnson-Lindenstrauss Lemma which makes it possible for large point sets in high dimensional Euclidean spaces to be embedded into low dimensional Euclidean spaces with arbitrarily small distortion [12]. This result was made more accessible for use in databases by a result of Achlioptas [13] which showed that one can use a projection matrix with each entry chosen independently from the distribution $U\{-1, +1\}$. This is most suited to a database application where the random projection can now be applied using simple SQL queries. We now state the main result of Achlioptas which assures that our algorithmic results are readily applicable to database situations as well.

Lemma 1 ([13]). *Let $R = (r_{ij})$ be a random $d \times k$ matrix, such that each entry r_{ij} is chosen independently according to $U\{+1, -1\}$. For any fixed unit vector $u \in \mathbb{R}^d$, and any $\epsilon > 0$, let $u' = \sqrt{\frac{d}{k}}\left(R^T u\right)$. Then, $E\left[\|u'\|^2\right] = 1 = \|u\|^2$ and*

$$\Pr\left[(1 - \epsilon)\|u\|^2 < \|u'\|^2 < (1 + \epsilon)\|u\|^2\right] \geq 1 - e^{\frac{-k}{2}\left(\frac{\epsilon^2}{2} - \frac{\epsilon^3}{3}\right)}.$$

Corollary 1. *Let u, v be unit vectors in $\mathbb{R}^d$. Then, for any $\epsilon > 0$, a random projection of these vectors to yield the vectors u' and v' respectively satisfies*

$$\Pr\left[u \cdot v - \epsilon \leq u' \cdot v' \leq u \cdot v + \epsilon\right] \geq 1 - 4e^{\frac{-k}{2}\left(\frac{\epsilon^2}{2} - \frac{\epsilon^3}{3}\right)}.$$

Proof. Apply Lemma 1 to the vectors u, v and $u - v$. The result follows from using simple facts concerning inner products. □

We shall refer to the process of mapping high dimensional point sets to low dimensional ones via random projections as *JL-type embeddings*. In the discussion below, we assume the histograms to be normalized, i.e., they correspond to probability distributions.

Definition 2 (Representative vector). *Given a d-dimensional histogram $P = (p_1, \ldots p_d)$ the representative vector of P is the unit vector $\sqrt{P} = (\sqrt{p_1}, \ldots, \sqrt{p_d})$.*

Definition 3 (α-constrained histogram). *A histogram $P = (p_1, p_2, \ldots p_d)$ is said to be α-constrained if $p_i \geq \frac{\alpha}{d}$ for all $i \in \{1, 2, \ldots, d\}$.*

The above definition ensures that the α-constrained histograms have a level of "smoothness" to them. It can be easily seen that the inner product between the representative vectors of two α-constrained histograms P and Q is at least α. For convenience, we will denote $\frac{\alpha}{d}$ by β. A β-constrained distribution will imply an α-constrained distribution with $\alpha = \beta \cdot d$. We next examine three statistical distance measures starting with the Bhattacharyya class of distance measures.

3   The Bhattacharyya Class of Distance Measures

In the field of pattern classification, more specifically Bayesian decision theory, the Bhattacharyya bound is an upper-bound on the expected error rate of a Bayesian decision process [14]. For two histograms $P = (p_1, p_2, \ldots, p_d)$ and $Q = (q_1, q_2, \ldots q_d)$, the *Bhattacharyya coefficient* is defined as $BC(P, Q) = \sum_{i=1}^{n} \sqrt{p_i q_i}$. Using this coefficient, two distance measures can be defined as follows. The *Bhattacharyya distance* is defined as $BD(P, Q) = -\ln BC(P, Q)$. This measure does not form a metric. Another distance measure in this class, namely the *Hellinger distance* is defined as $H(P, Q) = 1 - BC(P, Q) = \frac{1}{2}\left(\|\sqrt{P} - \sqrt{Q}\|\right)^2$. The fact that $H(P, Q)$ is the Euclidean distance between representative vectors allows us to state the following theorem upon application of Lemma 1.

Theorem 1. *The Hellinger distance admits a low-distortion dimensionality reduction.*

3.1 Dimensionality Reduction for the Bhattacharyya Distance

We now consider the possibility of extending this idea to the Bhattacharyya distance. The following theorem shows that indeed such an embedding incurs only a small additive error.

Theorem 2. *The Bhattacharyya distance measure for α-constrained histogram admits a JL-type embedding with arbitrarily low additive error.*

Proof. By Corollary 1, we have the following with high probability: $\left\langle \sqrt{P}, \sqrt{Q} \right\rangle - \epsilon' \leq \left\langle \sqrt{P'}, \sqrt{Q'} \right\rangle \leq \left\langle \sqrt{P}, \sqrt{Q} \right\rangle + \epsilon'$. Taking $-\ln()$ throughout and using the definition of the Bhattacharyya distance, we have $BD(P', Q') \geq BD(P, Q) - \ln\left(1 + \frac{\epsilon'}{\langle \sqrt{P}, \sqrt{Q} \rangle}\right)$ and $BD(P', Q') \leq BD(P, Q) + \ln\left(\frac{1}{1 - \frac{\epsilon'}{\langle \sqrt{P}, \sqrt{Q} \rangle}}\right)$. Since the distributions are α-constrained, we have $\left\langle \sqrt{P}, \sqrt{Q} \right\rangle \geq \alpha$. Hence, $BD(P', Q') \geq BD(P, Q) - \ln\left(1 + \frac{\epsilon'}{\alpha}\right)$ and $BD(P', Q') \leq BD(P, Q) + \ln\left(\frac{1}{1 - \frac{\epsilon'}{\alpha}}\right)$. For any x, $e^x \geq 1 + x$. Hence, $\ln\left(1 + \frac{\epsilon'}{\alpha}\right) \leq \frac{\epsilon'}{\alpha}$. Also, the function $f(x) = 2x - \ln\left(\frac{1}{1-x}\right)$ is positive for all $x \leq \frac{1}{2}$. Hence, for $\frac{\epsilon'}{\alpha} \leq \frac{1}{2}$ (which is true since $\epsilon' = \frac{\epsilon \cdot \alpha}{2}$ and $\epsilon \leq 1$), we have $\ln\left(\frac{1}{1 - \frac{\epsilon'}{\alpha}}\right) \leq \frac{2\epsilon'}{\alpha}$. Hence, $BD(P, Q) - \frac{\epsilon'}{\alpha} \leq BD(P', Q') \leq BD(P, Q) + \frac{2\epsilon'}{\alpha}$ which gives us the desired result since $\epsilon' = \frac{\epsilon \cdot \alpha}{2}$. □

We now explore whether the Bhattacharyya distance, being a non-metric, also admits low distortion embeddings into metric spaces due to the availability of efficient indexing schemes in metric spaces [1]. We next develop a proof technique that shows that the distortion incurred by any embedding of point sets under the Bhattacharyya distance into a metric space can be made arbitrarily large by including appropriately chosen histograms.

3.2 The Relaxed Triangle Inequality Technique

We first define the notion λ-*relaxed triangle inequality* for a distance measure which parallels the definition of a *relaxed metric* as defined in [9].

Definition 4 (λ-Relaxed Triangle Inequality). *A distance measure $d : X \times X \longrightarrow \mathbb{R}^+ \cup \{0\}$ defined on a set X is said to satisfy the λ-relaxed triangle inequality if for all triplets $p, q, r \in X, d(p, r) + d(r, q) \geq \lambda \cdot d(p, q)$ for some constant $\lambda \leq 1$.*

Lemma 2. *Any embedding of a distance function d violating the λ-relaxed triangle inequality into a metric space incurs a distortion of at least $\frac{1}{\lambda}$.*

Proof. Let X contain points p, q, s that violate the inequality. Let f be a D-distortion embedding of (X, d) into a metric space (Y, ρ). Hence $\rho(f(p), f(s)) + \rho(f(s), f(q)) \geq \rho(f(p), f(q))$ since (Y, ρ) is a metric space. However the distortion bounds tell us that for all points $x, y \in X$, we have $r \cdot d(x, y) \leq \rho(f(x), f(y)) \leq D \cdot r \cdot d(x, y)$. This yields $D > \frac{1}{\lambda}$.   □

3.3   Lower Bound on Distortion for Embeddings into Metric Spaces

We now appeal to the relaxed triangle inequality argument by constructing point sets under the Bhattacharyya distance that fail to satisfy the relaxed triangle inequality and then applying Lemma 2 to get a lower bound on the distortion. Our result is characterized by the following theorem.

Theorem 3. *There exist d-dimensional β-constrained distributions such that any embedding of these distributions under the Bhattacharyya distance measure into a metric space must incur a distortion of $\Omega\left(\frac{\ln \frac{1}{d\beta}}{\ln d}\right)$ when $\beta > \frac{4}{d^2}$ and $\Omega\left(\frac{\ln \frac{1}{\beta}}{\ln d}\right)$ when $\beta \leq \frac{4}{d^2}$.*

Proof (Sketch). Consider the β-constrained distributions : $P = \left(\frac{1}{d}, \frac{1}{d}, \ldots, \frac{1}{d}\right)$, $Q = (1 - (d-1)\beta, \beta, \ldots, \beta)$ and $R = (\beta, 1 - (d-1)\beta, \ldots, \beta)$. An application of Lemma 2 along with some manipulations gives us the desired result. In the following section, we demonstrate that this bound is tight upto a $O(d \ln d)$ factor. The complete proof can be found in [11].   □

3.4   A Metric Embedding for the Bhattacharyya Distance

In this section, we first show that the Bhattacharyya distance is very closely related to the Hellinger distance measure. Since the Hellinger distance forms a metric in the positive orthant, this allows us to get an upper bound on the distortion which, for a fixed dimension, approaches the lower bound.

Theorem 4. *For any two d-dimensional β-constrained distributions P and Q with $\beta < \frac{1}{2d}$, we have $H(P, Q) \leq BD(P, Q) \leq \frac{d}{1-2\beta d} \ln \frac{1}{(d-1)\beta} H(P, Q)$.*

Proof. For two distributions P, Q, recall $BD(P, Q) = -\ln\left(\sum_{i=1}^{d} \sqrt{p_i}\sqrt{q_i}\right) = -\ln(1 - H(P, Q)) = \sum_{k=1}^{\infty} \frac{H(P,Q)^k}{k}$. To arrive at the lower bound, we truncate the infinite series at the first term. For the upper bound, we use the fact that the function $f(x) = -\ln(1 - x)$ is convex. The maximum Hellinger distance between any two β-constrained distributions is $2(\sqrt{1 - (d-1)\beta} - \beta)^2$. Let, $a = (\sqrt{1 - (d-1)\beta} - \beta)^2$. Due to convexity of f, the line mx lies above the curve $-\ln(1 - x)$ where $m = \frac{f(a)}{a}$. Therefore, we have $BD(P, Q) = -\ln(1 - H(P, Q)) \leq \frac{1}{a} \ln\left(\frac{1}{1-a}\right) H(P, Q)$. Also, $1 - a = (d-1)\beta + 2\beta\sqrt{1 - (d-1)\beta} - \beta^2 \geq (d-1)\beta$ since $2\sqrt{1 - (d-1)\beta} - \beta \geq 0$. Thus we get $BD(P, Q) \leq \frac{d}{1-2\beta d} \ln\left(\frac{1}{(d-1)\beta}\right) H(P, Q)$.

This implies that the identity embedding of a point set under the Bhattacharyya distance into one under the Hellinger distance incurs a distortion of $\frac{d}{1-2\beta d} \ln \frac{1}{(d-1)\beta}$. □

This gives the distortion of the identity embedding into the Hellinger distance. For constant d and sufficiently small β, the lower bound presented in Section 3.3 is essentially $\Omega\left(\ln \frac{1}{\beta}\right)$, whereas the embedding presented in this section has a distortion of $O\left(\ln \frac{1}{\beta}\right)$ which implies that the lower bound is tight. In general it can be seen using Theorems 3 and 4 that for sufficiently small β the lower bound presented is tight upto a factor of $O(d \ln d)$. Further, the result presented in Theorem 1 can be used to perform dimensionality reduction as well.

4   The Kullback-Leibler Divergence

Given two histograms $P = \{p_1, p_2, \ldots, p_d\}$ and $Q = \{q_2, q_2 \ldots q_d\}$, the Kullback-Leibler divergence $KL(P, Q) = \sum_{i=1}^{d} p_i \ln \frac{p_i}{q_i}$. The Kullback-Leibler divergence is non-symmetric and unbounded. In order to avoid these singularities, we assume that the histograms are β-constrained.

Lemma 3. *Given two β-constrained histograms P, Q, $0 \le KL(P, Q) \le \ln \frac{1}{\beta}$.*

Proof. The lower bound follows directly from Jensen's inequality [14]. Since $\frac{p_i}{q_i} \le \frac{1}{\beta}$, we have $KL(P, Q) = \sum_{i=1}^{d} p_i \ln \frac{p_i}{q_i} \le \sum_{i=1}^{d} p_i \ln \frac{1}{\beta} = \ln \frac{1}{\beta}$. □

In the following, we develop a technique to prove lower bounds akin to those in Section 3.2. Together with the relaxed triangle inequality technique, we use it to prove the non-existence of low distortion embeddings into metric spaces for the Kullback-Leibler divergence.

4.1   The Asymmetry Technique

We present a general result that can be used to prove lower bounds on the embedding distortion of a non-symmetric distance measure into a metric space. Consider the following definition.

Definition 5 (γ-Relaxed Symmetry). *A set X equipped with a distance function $d : X \times X \longrightarrow \mathbb{R}^+ \cup \{0\}$, is said to satisfy γ-relaxed symmetry if there exists $\gamma \ge 0$ such that for all point pairs $p, q \in X, |d(p, q) - d(q, p)| \le \gamma$.*

Lemma 4. *Any embedding of a bounded distance function d (i.e., $d(x, y) \le M$ for all $x, y \in X$) violating the γ-relaxed symmetry into a metric space incurs a distortion of at least $1 + \frac{\gamma}{M}$.*

Proof. Let X contain points p, q that violate the γ-relaxed symmetry. Without loss of generality, assume that $d(p, q) > d(q, p) + \gamma$. Let f be a D-distortion embedding of (X, d) into a metric space (Y, ρ). Hence $\rho(f(p), f(q)) = \rho(f(q), f(p))$ since (Y, ρ) is a metric space. However the distortion bounds tell us that for all points $x, y \in X$, we have $r \cdot d(x, y) \le \rho(f(x), f(y)) \le D \cdot r \cdot d(x, y)$. This yields $D > 1 + \frac{\gamma}{M}$. □

4.2   Lower Bounds on Distortion for Embeddings into Metric Spaces

We now apply the above lemma to show that one cannot obtain an almost iso-
metric embedding of the Kullback-Leibler divergence into any metric space. We
show the existence of two histograms P and Q such that $|KL(P,Q) - KL(Q,P)|$
is large. The result is formally stated in the following theorem.

Theorem 5. *For sufficiently large d and small β, there exists a set S of d-
dimensional β-constrained histograms and a constant $c > 0$ such that any embedding of S into a metric space incurs a distortion of at least $1 + c$.*

Proof (Sketch). Consider the distributions $P = \{\frac{1}{d}, \frac{1}{d}, \dots, \frac{1}{d}\}$ and $Q = \{1 - (d - 1)\beta, \beta, \dots, \beta\}$. For large d, an application of Lemma 4 gives us a lower bound of
$1 + \Omega(1)$ for both large β, – say $\beta = \frac{1}{\Theta(d)}$ as well as small β, – say $\beta = o\left(\frac{1}{d^4}\right)$.
The complete proof can be found in [11].   □

It turns out that one cannot get significant improvement on the above bound
by choosing different points. However, an application of the relaxed triangle
inequality technique shows that the situation is much worse as is demonstrated
below.

Theorem 6. *For sufficiently large d, there exist d-dimensional β-constrained
distributions such that embedding these under the Kullback-Leibler divergence
into a metric space must incur a distortion of $\Omega\left(\dfrac{\ln \frac{1}{d\beta}}{\ln\left(d \ln \frac{1}{\beta}\right)}\right)$.*   □

Proof. We construct three β-constrained distributions that fail to satisfy the
relaxed triangle inequality under the Kullback-Leibler divergence. Consider the
following distributions. The parameters ϵ and c will be fixed later. Let $P = \left(\frac{1}{d}, \frac{1}{d}, \dots, \frac{1}{d}\right)$, $Q = (1 - (d-1)\epsilon, \epsilon, \dots, \epsilon)$ and $R = (1 - (d-1)e^{-c}, e^{-c}, \dots, e^{-c})$
where $\frac{1}{d} \geq \epsilon > e^{-c} \geq \beta$. We have,

$$KL(P,Q) = \left(1 - \frac{1}{d}\right)\ln\frac{1}{d\epsilon} + \frac{1}{d}\ln\frac{1}{d(1-(d-1)\epsilon)} \leq \ln\frac{1}{d\epsilon}$$

$$KL(Q,R) = (1 - (d-1)\epsilon)\ln\frac{1-(d-1)\epsilon}{1-(d-1)e^{-c}} + (d-1)\epsilon\ln(\epsilon e^{c})$$
$$\leq (d-1)\epsilon\ln\epsilon + (d-1)c\epsilon$$

$$KL(P,R) = \left(1 - \frac{1}{d}\right)\ln\frac{1}{de^{-c}} + \frac{1}{d}\ln\frac{1}{d(1-(d-1)e^{-c})}$$
$$\geq \frac{1}{2}(c - \ln d) + \frac{1}{d}\ln\frac{1}{d} = \Omega(c - \ln d) - O(1)$$

Using the above inequalities, $\lambda = \frac{KL(P,Q)+KL(Q,R)}{KL(P,R)} = O\left(\frac{\ln\frac{1}{d\epsilon}+(d-1)\epsilon\ln\epsilon+(d-1)c\epsilon}{c-\ln d}\right)$.
Hence, any point set containing these three points violates the λ-relaxed triangle
inequality. Now, using Lemma 2, the distortion for the Kullback-Leibler diver-
gence is $D > \frac{1}{\lambda} = \Omega\left(\frac{c-\ln d}{\ln\frac{1}{d\epsilon}+d\epsilon\ln\epsilon+dc\epsilon}\right)$. Since $\epsilon\ln\epsilon < 0$, hence $D = \Omega\left(\frac{c-\ln d}{\ln\frac{1}{d\epsilon}+dc\epsilon}\right)$.
Consider the function $f(c,\epsilon) = \frac{c-\ln d}{\ln\frac{1}{d\epsilon}+dc\epsilon}$. It turns out that $\frac{\partial f}{\partial c} > 0$ for all values
of c. Hence, the maxima is achieved at the maximum value of c which is $\ln\frac{1}{\beta}$.
Furthermore we find that $\frac{\partial f}{\partial \epsilon} = 0$ at $\epsilon = \frac{1}{dc} = \frac{1}{d\ln\frac{1}{\beta}}$. It can be confirmed that
this extrema is actually a maxima. For a fixed value of d we can choose β small

enough to make sure that the value of ϵ is at least β. For these values of c and ϵ we get the lower bound as $D = \Omega\left(\frac{\ln\frac{1}{d\beta}}{\ln\left(d\ln\frac{1}{\beta}\right)+1}\right)$. Thus, the result follows.   □

Interpreting the Lower Bounds: The above bounds indicate that near the uniform distribution, asymmetry makes the Kullback-Leibler divergence hard to approximate by a metric but as we move away from the uniform distribution the hardness is because of the violation of the relaxed triangle inequality. More formally, it can be seen that for point sets which are β-constrained for large β (say $\beta = \Omega\left(\frac{1}{d}\right)$), the lower bound using the asymmetry argument gives a $1 + \Theta(1)$ bound whereas the triangle inequality argument gives a $o(1)$ bound. For smaller β (say $\beta = o\left(\frac{1}{d^4}\right)$) we get a better lower bound using the relaxed triangle inequality argument. This lower bound behaves asymptotically as $\frac{\ln\frac{1}{\beta}}{\ln\ln\frac{1}{\beta}}$ which can be made arbitrarily large as against the constant bound given by the asymmetry technique.

4.3   An Embedding for the Kullback-Leibler Divergence

In this section, we examine the properties of the identity embedding of point sets under the Kullback-Leibler divergence into the l_2^2 distance measure. Our result is characterized by the following theorem.

Theorem 7. *For any two d-dimensional β-constrained distributions P and Q,* $\frac{l_2^2(P,Q)}{2} \leq KL(P,Q) \leq \left(\frac{1}{2\beta} + \frac{1}{3\beta^5}\right) l_2^2(P,Q).$

Proof. Present in the complete version of the paper [11].   □

Although this embedding does not give us a provably low distortion, due to the embedding being into l_2^2 space, it allows for low distortion dimensionality reduction via JL-type embeddings.

5   The Class of Quadratic Form Distance Measures

Given a $d \times d$ positive definite matrix A, the *Quadratic Form Distance measures (QFDs)* define a distance measure over $\mathbb{R}^d$. If $x, y \in \mathbb{R}^d$, then $Q_A(x,y)$ is defined to be $Q_A(x,y) = \sqrt{(x-y)^T A(x-y)}$. The family of quadratic form distances corresponding to positive definite A form a metric. We now show that every metric QFD can be embedded into a low dimensional space with low distortion.

Theorem 8. *The family of quadratic form distance measures admit a low-distortion JL-type embedding.*

Proof. Since every positive definite matrix A can be subjected to a Cholesky Decomposition of the form $A = L^T L$, $Q_A(x,y)$ is essentially the Euclidean distance between the points Lx and Ly. Hence the result follows from Lemma 1.   □

Since the Mahalanobis distance is special QFD measure where the A is the covariance matrix of some multivariate distribution, it too, admits a low-distortion JL-type embedding.

6   Conclusions

We examined various statistical distance measures from the point of view of dimensionality reduction and embeddability into metric spaces. In particular we presented dimensionality reduction schemes for the Bhattacharyya distance, the Hellinger distance and the Mahalanobis distance. We developed two novel techniques that were used to prove lower bounds on the distortion of embeddings of non-metrics into metric spaces and applied these to the Bhattacharyya distance and the Kullback-Leibler divergence. For the Bhattacharyya distance, we demonstrated that the lower bound presented is almost tight. We performed a similar exercise for the Kullback-Leibler divergence where the embedding provided is of practical significance since it allows for dimensionality reduction.

References

1. Samet, H.: Foundations of Multidimensional and Metric Data Structures. Morgan Kaufmann Publishers, San Francisco (2005)
2. Ljosa, V., Bhattacharya, A., Singh, A.K.: Indexing spatially sensitive distance measures using multi-resolution lower bounds. In: Ioannidis, Y., Scholl, M.H., Schmidt, J.W., Matthes, F., Hatzopoulos, M., Böhm, K., Kemper, A., Grust, T., Böhm, C. (eds.) EDBT 2006. LNCS, vol. 3896, pp. 865–883. Springer, Heidelberg (2006)
3. Indyk, P., Thaper, N.: Fast Image Retrieval via embeddings. In: 3rd International Workshop on Statistical and Computational Theories of Vision (2003)
4. Rahman, M.M., Bhattacharya, P., Desai, B.C.: Similarity searching in image retrieval with statistical distance measures and supervised learning. In: Pattern Recognition and Data Mining, pp. 315–324 (2005)
5. Fraser, A., Hengartner, N., Vixie, K., Wohlberg, B.: Incorporating invariants in Mahalanobis distance based classifiers: Application to Face Recognition. In: International Joint Conference on Neural Networks (2003)
6. Lee, C.H., Shin, D.G.: Using Hellinger Distance in a nearest neighbour classifier for relational databases. Knowledge-Based Systems 12, 363–370 (1999)
7. Sengar, H., Wang, H., Wijesekera, D., Jajodia, S.: Detecting VoIP Floods using the Hellinger Distance. IEEE Transactions on Parallel and Distributed Systems 19(6), 794–805 (2008)
8. Mathiassen, J.R., Skavhaug, A., Bø, K.: Texture Similarity Measure Using Kullback-Leibler Divergence between Gamma Distributions. In: Proceedings of the 7th European Conference on Computer Vision-Part III, pp. 133–147 (2002)
9. Chaudhuri, K., McGregor, A.: Finding Metric Structure in Information-Theoretic Clustering. In: Proceedings of COLT 2008, pp. 391–402 (2008)
10. Guha, S., Indyk, P., McGregor, A.: Sketching Information Divergences. In: Bshouty, N.H., Gentile, C. (eds.) COLT. LNCS, vol. 4539, pp. 424–438. Springer, Heidelberg (2007)
11. Bhattacharya, A., Kar, P., Pal, M.: On Low Distortion Embeddings of Statistical Distance Measures into Low Dimensional Spaces (Manuscript) (May 2009), http://www.cse.iitk.ac.in/users/purushot/low-dist.pdf
12. Johnson, W.B., Lindenstrauss, J.: Extensions of Lipschitz maps into a Hilbert Space. Contemporary Mathematics 26, 189–206 (1984)
13. Achlioptas, D.: Database-friendly Random Projections. In: ACM Symposium on Principles of Database Systems (2001)
14. Duda, R.O., Hart, P.E., Stork, D.G.: Pattern Classification, 2nd edn. Wiley Publications, Chichester (2000)

Real-Time Traffic Flow Statistical Analysis Based on Network-Constrained Moving Object Trajectories

Zhiming Ding[1] and Guangyan Huang[2]

[1] Institute of Software, Chinese Academy of Sciences, Beijing 100190, P.R. China
[2] School of Engineering & Science, Victoria University, Melbourne, Australia
zhiming@iscas.ac.cn, abysshuang@gmail.com

Abstract. In this paper, we propose a novel traffic flow analysis method, Network-constrained Moving Objects Database based Traffic Flow Statistical Analysis (NMOD-TFSA) model. By sampling and analyzing the spatial-temporal trajectories of network constrained moving objects, NMOD-TFSA can get the real-time traffic conditions of the transportation network. The experimental results show that, compared with the floating-car methods which are widely used in current traffic flow analyzing systems, NMOD-TFSA provides an improved performance in terms of communication costs and statistical accuracy.

Keywords: Database, Spatiotemporal, Moving Object, Statistics.

1 Introduction

With the recent advances in mobile computing and in Intelligent Transportation Systems (ITS), the statistical analysis of traffic flow has become a key research issue. To improve traffic conditions and to manage transportation systems more effectively, various techniques have been adopted to collect traffic data, such as stationary sensor/camera based methods (monitoring from traffic sensors or optical devices), air/space borne methods (monitoring from airplanes or satellites), and floating-car based methods (monitoring from floating/probe cars).

However, the above mentioned methods have a lot of limitations. For example, stationary sensor/camera based methods can only measure traffic data at fixed positions. To get the traffic information of the whole transportation network, a large number of detectors are needed so that the system can be very expensive. Air/space borne methods can monitor traffic conditions over large areas, but the data are available only when the air/space borne detectors are flying over the monitored areas.

To solve the above problems, increasing research interests are focused on the Floating-Car Method (or FCM for short) in recent years, with a lot of feasible solutions achieved [1-6]. In the FCM system, certain kinds of vehicles are equipped with GPS and wireless communication interfaces, and periodically report to the central server their locations, velocities, and directions (these data are called Floating-Car Data, or simply FCD). In every certain time interval the server launches a statistical process to match the FCD with the traffic network so that traffic flow parameters (for instance, average speed, travel time, and traffic jam of each route) of the network can

S.S. Bhowmick, J. Küng, and R. Wagner (Eds.): DEXA 2009, LNCS 5690, pp. 173–183, 2009.

be computed and refreshed. Since floating cars collect traffic data when they move, FCM is a "moving sensor" based method.

Even though FCM is becoming increasingly popular because of its flexibility, it still has some obvious limitations. (1)The data sampling method in FCM is fixed-time-based (sampling FCD in fixed time interval, say in every 2 minutes) or fixed-distance-based (sampling FCD in fixed distance, say in every 500 meters). As a result, the dilemma between communication cost and accuracy can not be easily solved, as stated in [7, 8]. (2)The precision of traffic flow analysis in current floating car systems is not satisfactory. In FCM, the geographical path of a moving object between two consecutive FCD sampling points is approximated by the shortest path, which can cause errors in traffic statistical analysis. Actually, multiple paths can coexist between two points over the network and the driver often take a path which he/she is familiar with, instead of taking shortest path always. (3)In FCM, the traffic flow parameters are refreshed in certain time interval (say in 5 minutes) instead of in real-time so that considerable time delays can exist.

To solve the above problems, we propose a new "moving-sensor" based traffic flow analysis method, *Network-constrained Moving Objects Database based Traffic Flow Statistical Analysis* (NMOD-TFSA) model, in this paper. Through a network-based location update mechanism, NMOD-TFSA can track the precise network-matched trajectories of moving objects so that the precision can be greatly improved while the communication cost can still be kept to the minimum.

The remaining part of this paper is organized as follows. Section 2 defines the data model, Section 3 describes the NMOD-TFSA traffic flow statistical analysis methods and the statistical analysis data structure CTSAG, Section 4 discusses performance evaluation results, and Section 5 finally concludes the paper.

2   Modeling Traffic-Parameterized Road Networks and Moving Object Trajectories

In this section, we propose a two-layered, route-ARS based traffic-parameterized road network framework, in which the network is modelled as a set of routes plus a set of junctions. Each route is again composed of a set of directed Atomic Route Sections (ARS) which are equivalent to directed edges of the traffic network. With every ARS or junction a set of traffic parameters is associated so that the current traffic condition of the read network can be expressed. Figure 1 illustrates the two-layered road network framework.

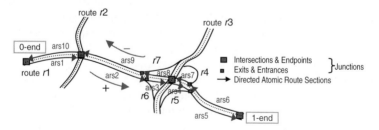

Fig. 1. Two-layered route-ARS based transportation network

In Figure 1, route $r1$ has two traffic flow directions, "+" and "–" ("+" traffic flow is from 0-end to 1-end, and "–" is from 1-end to 0-end, see Definition 2). Therefore, the directed atomic route sections are also in two directions. ARSs of the two route sides can be asymmetrical (for instance, ars2 and ars9, ars3 and ars8).

Each ARS or junction of the traffic network has a set of traffic parameters associated to express its current traffic conditions (such as number of moving objects, traffic-jam status, and average travel time), and these parameters are refreshed in real-time through location updates.

Definition 1 (Traffic-parameterized Road Network). A traffic-parameterized road network N is defined as a pair:

$N = (Routes, Juncts)$, where $Routes$ is a set of traffic-parameterized routes (see Definition 2), and $Juncts$ is a set of traffic-parameterized junctions (see Definition 4).

Definition 2 (Traffic-parameterized Route). A traffic-parameterized route of network N, denoted by r is defined as follows:

$$r = (rid,\ geo,\ len,\ ((jid_j,\ pos_j))\ _{j=1}^{m},\ ARS),\ \text{where}\ rid \in \text{int is the identifier of}\ r;$$

$geo \in$ polyline is the geometry of r (the beginning point and the end point of geo are called "0-end" and "1-end" respectively); $len \in$ real is the length of r, $(jid_j,\ pos_j)$ $(1 \le j \le m)$ describes the jth junction inside the route where $jid_j \in$ int is the identifier of the junction and $pos_j \in [0, 1]$ is the relative position of the junction's center (suppose that the total length of each route is 1, then any position inside the route can be presented by a real number $pos \in [0, 1]$, which is called "relative position" inside the route); and ARS is the set of directed atomic route sections (each directed atomic route section has a set of traffic parameters associated, see Definition 3) of the route.

Definition 3 (Directed Atomic Route Section). A directed Atomic Route Section (ARS), denoted by ars, is a directed edge which connects two nearby junctions of the network and does not contain any other junctions in between along the same traffic flow direction, which is defined as:

$$ars = (aid,\ (jid_s,\ pos_s),\ (jid_e,\ pos_e),\ Paraa),\ \text{where}\ aid \in \text{int is the identifier of}\ ars;$$ $(jid_s,\ pos_s)$ and $(jid_e,\ pos_e)$ describe the starting point and the end point of ars respectively (jid_s and jid_e are junction identifiers, and pos_s and pos_e are the starting position and end position of ars which are measured just outside the junction area (see Definition 4) borders); $Paraa = \{\eta_{mo},\ \tau,\ \beta\}$ is a set of basic traffic flow parameters describing the current traffic condition of ars ($\eta_{mo} \in$ int is the number of moving objects currently inside ars; $\tau \in$ real is the average travel time of moving objects passing through ars; $\beta \in \{0, 1\}$ is the traffic jam status of ars, which can be either 1 or 0, indicating "blocked" or "unblocked" respectively).

Through these basic traffic parameters and some additional parameters of the transportation network (such as the portion of moving objects among all vehicles, the capacity of each route section, and so forth), the system can derive more complicated traffic information of ars, such as flux, vehicle density, and so forth.

Definition 4 (Traffic-parameterized Junction). A traffic-parameterized junction of traffic network N, denoted by j, can correspond to an intersection, an exit/entrance, or a route endpoint of real-world traffic networks, which is defined as follows:

$j = (jid, loc, ((rid_i, pos_i))_{i=1}^{n}, \gamma, matrix, Paraj)$, where $jid \in$ int is the identifier of j; $loc \in$ point is the location of j; (rid_i, pos_i) $(1 \le i \le n)$ describes the ith route connected by j, where $rid_i \in$ int is the route identifier, $pos_i \in [0,1]$ is the relative position of j's center inside the route; γ is the radius of the junction area, which describes the size of the junction area. *matrix* is the traffic-parameterized connectivity matrix of j. It contains possible matches of traffic flows in the routes connected by the junction, and the element associated with each match takes the form (φ, τ, β), where φ can assume either 0 or 1, indicating whether moving objects can transfer from the "in" traffic flow to the "out" traffic flow through this junction [7, 9], τ describes the average travel time through the junction area along the corresponding traffic flow, and β describes the traffic jam status of the corresponding traffic flow inside the junction area. $paraj = \{\eta_{mo}\}$ where η_{mo} is the number of moving objects currently inside the junction.

Based on the above route-ARS based traffic-parameterized road network framework, we can then define the trajectories of network-constrained moving objects. A trajectory is defined as a sequence of trajectory segments, with each segment describing a continuous movement of the moving object.

Definition 5 (Network Position). A position inside the network N, denoted by *npos*, is defined as a pair:

$npos = (rid, pos)$, where $rid \in$ int is a route identifier, and $pos \in [0, 1]$ is a relative position inside the route. Since the geometry of each route is kept in the database as a poyline (see Definition 2), *npos* can be transformed to the (x, y) form easily.

Definition 6 (Network Route Section). A network route section *nrs* is a part of a route, and is defined as the form:

$nrs = (rid, S)$, where $rid \in$ int is a route identifier isomorphic to integer, and $S \subseteq [0, 1]$ is a interval over $[0, 1]$ specifying a section of the route.

Definition 7 (Motion Vector). Motion vectors are snapshots of moving object's movements and are generated through location updates. A motion vector, *mv*, is defined as the following form:

$mv = (t, (rid, pos), \vec{v}, actv)$, where $t \in$ real is a time instant, (rid, pos) is a network position describing the location of the moving object at time t, and $\vec{v} \in$ real is the speed measure of the moving object at time t. $\vec{v}$ contains both speed and direction information. Its absolute value is equal to the speed of the moving object at time t, while its sign (either positive or negative) indicates the traffic flow direction the moving object belongs to at time t. If the moving object is moving from 0-end towards 1-end, then the sign is positive. Otherwise, the sign is negative. $actv \in$ bool is a flag indicating whether the motion vector is the active motion vector [8].

Definition 8 (Trajectory Segment of Moving Object). A trajectory segment of moving object *mo*, denoted as *trseg*, is a sequence of motion vectors sent by *mo* through location updates during its journey. *trseg* describes a continuous online movement of *mo* (by "online" we mean that the moving object is continuously tracked by the server) and is defined as follows:

$$trseg = (mv_j)_{j=1}^m = ((t_j, (rid_j, pos_j), \vec{v}_j, actv_j))_{j=1}^m, \text{ where for } \forall i \in \{1, \dots n\text{-}1\}: t_i < t_{i+1}.$$

Definition 9 (Trajectory of Moving Object). The trajectory of a moving object *mo*, denoted as *traj*, is a sequence of trajectory segments generated when *mo* is running in the traffic network, which is defined as:

$$traj = (trseg_i)_{i=1}^n = (((t_{ij}, (rid_{ij}, pos_{ij}), \vec{v}_{ij}, actv_{ij}))_{j=1}^{mi})_{i=1}^n$$

In the trajectory, only the last motion vector can be active [8]. If the last motion vector is active, then we know that the moving object is currently online, and the last motion vector (called "active motion vector" in this case) contains the key information for computing the current or near future locations of the moving object and for triggering the next location update. If the last motion vector is inactive, then the moving object is currently offline (for instance stopped at night).

A trajectory can describe the movement of a moving object for a long period of time (for instance 3 months), with each trajectory segment describing a continuous online movement. As explained in [7, 8], the trajectory is generated through location updates. In network-constrained moving objects databases, we have defined three kinds of location updates, that is, ID-Triggered Location Update (IDTLU), Distance-Threshold Triggered Location Update (DTTLU), and Speed-Threshold Triggered Location Update (STTLU) [7, 8]. DTTLU and STTLU are triggerd when the moving object exceeds the distance threshold Th_D and the speed threshold Th_S respectively, and IDTLU is triggered when the moving object transfers from one route to another via a junction. These three kinds of location updates work together to finish the trajectory data sampling process of the moving object.

3   Real-Time Statistical Analysis of Traffic Parameters in NMOD-TFSA

As stated earlier, each ARS or junction of the transportation network in NMOD-TFSA has a set of traffic parameters associated to describe its current traffic condition. These basic parameters are refreshed whenever a location update related to the corresponding ARS/junction occurs.

Suppose that the functions route(*rid*), ars(*rid*, *aid*), and junct(*jid*) return the route, ARS, and junction corresponding to the identifiers respectively.

3.1   Trajectory Transformation Functions

Assume that *traj* is a trajectory, and its last motion vector is $mv_n = (t_n, (rid_n, pos_n), \vec{v}_n, actv_n)$. The function appcurr(*traj*) appends the current motion vector mv_{now} to the end of *traj*. If the last motion vector of *traj* is active (that is $actv_n$ = true), then appcurr(*traj*) first computes the location of the moving object at the current time instant t_{now}, denoted as pos_{now}, and then generates a new motion vector $mv_{now} = (t_{now}, (rid_n, pos_{now}), \perp, false)$ and appends it to *traj*. If the last motion vector of *traj* is inactive (that is $actv_n$ = false), the function will do nothing.

Function truncate_t(*traj*, *I*) returns part of *traj* (the result is still a trajectory) which is corresponding to the given time interval $I=[t_1, t_2]$ temporally. Function truncate_g(*traj*, *ars*) returns part of *traj* which is corresponding to the given atomic route section *ars* geographically. Function truncate_v(*traj*, v_{slow}) returns part of *traj* during which the speed of the moving object is slower than v_{slow}. Necessary interpolation may be required to get the end points of the resulted trajectory.

Function project_t(*traj*) projects *traj* on the time axle and returns a set of time intervals. Function project_g(*traj*) projects *traj* on the geographical plane and returns a set of network route sections (see Definition 6).

3.2 Traffic Parameter Refreshing Algorithms for ARSs and Junctions

Let's first consider how to compute traffic parameters for ARSs. When the traffic parameter refreshing process for a certain directed atomic route section *ars* is triggered, the system will check all trajectories of the moving objects that have stayed in or passed through *ars* in the last Δt time (Δt is a time period of 5-10 minutes, called "statistics window"). From each trajectory, the system can derive the travel time and the current position of the corresponding moving object. Therefore, *ars*'s traffic parameters η_{mo} and τ can be computed accordingly.

To get *ars*'s traffic jam status β, the system first computes the jammed area of *ars*:

$$\alpha_{jam} = \bigcap_{i=1}^{n} (\text{project\_g}(\text{truncate\_v}(\text{truncate\_g}(traj_i, ars), v_{slow})))$$

where $\cap$ is the spatial intersection operator between network route sections.

From the formula we can see that α_{jam} is a section of *ars* through which all moving objects move with speed slower than v_{slow} in the last Δt time. Therefore, if α_{jam} is not NULL, then *ars* is blocked. Otherwise, no blockage exists in *ars*. That is:

$$\beta = \begin{cases} true; & (\text{if } \alpha_{jam} \neq \varnothing) \\ false; & (\text{if } \alpha_{jam} = \varnothing) \end{cases}$$

The traffic parameter refreshing algorithm for ARSs is given in Algorithm 1.

Algorithm 1. Traffic Parameter Refreshing Algorithm for ARSs

INPUT: *rid*, *aid* ;

1. $TrajSet \leftarrow \text{GetRTraj}(rid, \Delta t)$;
2. $ars = \text{ars}(rid, aid)$;
3. $\eta_{mo} = \tau_{sum} = \tau_{num} = mo_{num} = 0$;
4. FOR $traj \in TrajSet$ DO
5. mo_{num}++;
6. $traj = \text{addcurr}(traj)$;
7.   IF inside(GetLastMV(*traj*).pos, *ars*) THEN η_{mo}++; ENDIF;
8. $slowseg = \text{trancate\_v}(\text{trancate\_g}(traj\_ars, ars), v_{slow})$;
9.   IF $mo_{num} = 1$ THEN $\alpha_{jam} = slowseg$; ELSE $\alpha_{jam} = \alpha_{jam} \cap slowseg$; ENDIF;
10. $t_{in} = \text{at}(traj, ars.pos_s)$; $t_{out} = \text{at}(traj, ars.pos_e)$;
11.   IF defined(t_{in}) AND defined(t_{out}) THEN $\tau_{mo} = t_{out} - t_{in}$; $\tau_{sum} = \tau_{sum} + \tau_{mo}$; τ_{num}++; ENDIF;
12. ENDFOR;
13. $\tau = \tau_{sum} / \tau_{num}$;
14. IF $\alpha_{jam} \neq \varnothing$ THEN $\beta = true$ ELSE $\beta = false$; ENDIF;
15. Refresh(*ars*, η_{mo}, τ, β).

The algorithm first retrieves all the trajectories geographically passing through route(rid) in the last Δt time by calling the GetRTraj(rid, Δt) function (line 1). Through the statistical data structure (see Subsection 3.3), this function can be supported efficiently. For each trajectory in GetRTraj(rid, Δt), the algorithm checks whether its latest position in route(rid), GetLastMV($traj$).pos, is inside ars, and adjusts η_{mo} accordingly (line 7). Then the algorithm computes the slow speed segments for every trajectory (line 8) and gets the union of them into α_{jam} (line 9). The jam status β of ars can then be determined from the final result of α_{jam} (line 14). Lines 10-11 compute the travel time of each moving object through ars by computing two time instants t_{in} and t_{out} (the entering time and exiting time of the moving object on ars), so that ars's average travel time τ can be derived (line 13). When the statistics is finished, the parameters of ars are refreshed with the new values (line 15).

The traffic parameter refreshing method for junctions is similar to that of ARSs. For example, η_{mo} can be computed by counting the moving objects whose current position is inside the junction area. The difference is that τ and β need to be computed for each traffic flow inside the junction.

Suppose that $\xi_{\mu v}$ is a traffic flow of the junction $junct$ (the in-flow and out-flow are μ and v respectively). When computing the average travel time of the junction along $\xi_{\mu v}$, denoted as $\tau_{\mu v}$, we only need to consider the moving objects running along $\xi_{\mu v}$. The traffic jam status of $\xi_{\mu v}$, denoted as $\beta_{\mu v}$, can be derived from $\tau_{\mu v}$. If $\tau_{\mu v}$ is longer than a predefined threshold ψ, then $\beta_{\mu v}$=true, and otherwise $\beta_{\mu v}$=false. The traffic parameter refreshing algorithm for junctions is given in Algorithm 2.

Algorithm 2. Traffic Parameter Refreshing Algorithm for Junctions

INPUT: jid

1. $TrajSet \leftarrow$ GetJTraj(jid, Δt);
2. $junct =$ junct(jid); $\eta_{mo}= 0$;
3. FOR $\mu \in junct.inflows$ AND $v \in junct.outflows$ DO
4. $\tau_{sum}(\mu, v) = \tau_{num}(\mu, v)=0$;
5. ENDFOR;
6. FOR $traj \in TrajSet$ DO
7. $traj =$ addcurr ($traj$);
8.   IF inside(GetLastMV($traj$).pos, $junct$) THEN η_{mo}++; ENDIF;
9.   GetInOutFlows($traj$, $junct$, μ, v);
10. $t_{in} =$ at($traj$, $junct$, μ); $t_{out} =$ at($traj$, $junct$, v);
11.   IF defined(t_{in}) AND defined(t_{out}) THEN $\tau_{mo} = t_{out} - t_{in}$; $\tau_{sum}(\mu, v) = \tau_{sum}(\mu, v) + \tau_{mo}$; $\tau_{num}(\mu, v)$++; ENDIF;
12.   ENDFOR
13.   FOR $\mu \in junct.inflows$ AND $v \in junct.outflows$ DO
14. $matrix(\mu, v).\tau = \tau_{sum}(\mu, v) / \tau_{num}(\mu, v)$;
15.   IF $matrix(\mu, v).\tau > \psi$ THEN $matrix(\mu, v).\beta =$ true; ELSE $matrix(\mu, v).\beta =$ false; ENDIF;
16.   ENDFOR;
17.   Refresh($junct$, η_{mo}, $matrix$).

Algorithm 2 first retrieves all the trajectories geographically passing through junct(jid) in the last Δt time by calling the GetJTraj(jid, Δt) function (line 1). For each trajectory in GetJTraj(jid, Δt), the algorithm first checks whether its latest position, GetLastMV($traj$).pos is inside $junct$, so that η_{mo} can be computed accordingly (line 8).

After that, the in and out traffic flows will be determined (line 9), so that the trajectory will only contribute to the statistical computation of the traffic flow it belongs to (lines 10-11, 13-16). The parameters are kept in *matrix* which is used to refresh the junction parameters when the statistics is finished (line 17).

3.3   NMOD-TFSA Statistical Data Structure and Refreshing Method

To speed up the statistical analysis, we propose a statistical data structure, called the Current Traffic-status Statistical Analysis Graph (CTSAG), in this subsection. Figure 2 illustrates the structure of CTSAG.

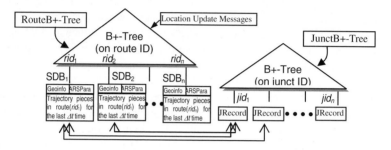

Fig. 2. Structure of CTSAG

As shown in Figure 2, CTSAG includes two B+-Trees, RouteB+-Tree and JunctB+-Tree, which are interconnected with each other at the bottom.

RouteB+-Tree organizes the route records on the *rid* attribute into a B+-Tree structure. The leaf nodes contain records of the form (*rid, SDBPointer*), where *rid* is the identifier of the route, and *SDBPointer* is a pointer to SDB(*rid*), the statistical data block (SDB) of route(*rid*). Each SDB takes the form (*geo, len*, (*aid_i*, (*jid_{si}, pos_{si}*), (*jid_{ei}, pos_{ei}*), *Paraa_i*) $_{i=1}^{n}$, *datasource*), where (*geo, len*, (*aid_i*, (*jid_{si}, pos_{si}*), (*jid_{ei}, pos_{ei}*), *Paraa_i*) $_{i=1}^{n}$) is the route record (with ARS information included, see Definitions 2 and 3), and *datasource* is a set of trajectory pieces acting as the data source for the statistical computation. Each trajectory piece in SDB(*rid*) is still in a trajectory form (see Definition 9), but it only contains the motion vectors corresponding to route(*rid*). For the sake of efficiency, only the recent Δt time trajectory data corresponding to route(*rid*) are kept in *datasource*. Each SDB has a set of pointers ((*jpointer_j, pos_j*)) $_{j=1}^{m}$ leading to the records of junctions within the route.

JunctB+-Tree organizes the junction records on the *jid* attribute into a B+-Tree structure. The leaf nodes contain records of the form (*jid, JRecordPointer*), where *jid* is the junction identifier, and *JRecordPointer* is a pointer to the junction record of the form (*jid, loc, γ, matrix, Paraj*) (see Definition 4). Each junction has a set of pointers ((*SDBPointer_i, pos_i*)) $_{i=1}^{n}$ leading to the SDBs of the routes connected by the junction.

When a location update occurs with a moving object *mo*, the system will first save the newly generated motion vector(s) to the corresponding SDB(s) and then refresh the traffic parameters of the related ARSs and junctions by calling algorithms 1 and 2 respectively.Suppose that the last motion vector of *mo* is $mv_n = (t_n, (rid_n, pos_n), \vec{v}_n, actv_n)$, and the new location update occurs at position (rid_u, pos_u). We notate the geographical path that *mo* has covered from (rid_n, pos_n) to (rid_u, pos_u) as $path_{nu}$.

If *mo* triggers a DTTLU or an STTLU, then a new motion vector $mv_u = (t_u, (rid_u, pos_u), \vec{v}_u, actv_u)$ will be generated (with $rid_u = rid_n$). In this case, the system will first save mv_u to *mo*'s trajectory piece in SDB(rid_n), and meanwhile, discard motion vectors in SDB(rid_n) which are older than Δt. After that, the traffic parameters of all ARSs and junctions that intersect $path_{nu}$ will be refreshed.

If *mo* transfers from route(rid_n) to route(rid_u) via junct(jid_{nu}) and triggers an IDTLU, then three motion vectors $mv_{u1} = (t_{u1}, (rid_{u1}, pos_{u1}), \vec{v}_{u1}, actv_{u1})$, $mv_{u1} = (t_{u1}, (rid_{u1}, pos_{u1}), \vec{v}_{u1}, actv_{u1})$, and $mv_{u1} = (t_{u1}, (rid_{u1}, pos_{u1}), \vec{v}_{u1}, actv_{u1})$, will be generated (with $rid_{u1} = rid_n$, $rid_{u2} = rid_{u3} = rid_u$). In this case, the system will save mv_{u1} to *mo*'s trajectory piece in SDB(rid_n) and save mv_{u2} and mv_{u3} to *mo*'s trajectory piece in SDB(rid_u), and refresh parameters for all ARSs and junctions that intersect $path_{nu}$.

Since all the trajectory pieces associated with route(rid) are kept together in SDB(rid), the system can support the GetRTraj(rid, Δt) and GetJTraj(jid, Δt) functions (see Algorithms 1 and 2) through CTSAG efficiently so that the performance of statistical analysis can be improved.

4 Performance Evaluation

The above stated NMOD-TFSA model has been implemented as a prototype system in C++, running in a Pentium IV processor (512M RAM, 1.6G HZ) under Linux. To evaluate the performance, we have conducted a series of experiments based on the prototype system and some additionally implemented modules.

The traffic network data sets for the experiments are real GIS data of Beijing. To generate network-constrained moving object trajectories, we have implemented a network constrained moving objects generator, NMO-Generator, which can simulate the movements of network-constrained moving objects. In the experiments, each simulation run involves 3000 moving objects running for 16 hours.

Figure 3 shows the average number of samplings per moving object/floating car in collecting traffic data, which can reflect the data sampling efficiency. From the figure we can see that, compared with FCM (sampling FCD in fixed distance), NMOD-TFSA can considerably reduce the communication cost in the data sampling process. The reason is that in NMOD-TFSA, the location update mechanism is motion vector based, which samples data according to moving parameters. If the moving object moves roughly according to the parameters, then nothing happens even though it runs for a long distance. Only when the motion vector becomes out-of-date a new location update is initiated. In this way, the frequency for data sampling is reduced.

Figure 4 shows the experiment result in terms of statistical precision λ. λ is defined based on travel time precision (with other traffic parameters we get similar results):

$$\lambda = \left(1 - \sum_{i=1}^{u} \left(\frac{|ars_{i}.\tau_{mf} - ars_{i}.\tau_{g}|}{\max(ars_{i}.\tau_{mf}, ars_{i}.\tau_{g})}\right) \middle/ u \right) \times 100\%$$

where μ is the number of ARSs that the moving object has passed through, $ars_{i}.\tau_{mf}$ is the computed average travel time of ars_i through NMOD-TFSA or through FCM,

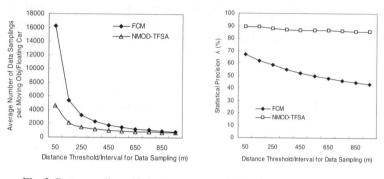

Fig. 3. Data sampling efficiency   **Fig. 4.** Statistical precision

$ars_i.\tau_g$ is the actual average travel time of ars_i computed from the original trajectory data generated by NMO-Generator. λ can reflect in what extent the statistical results conform to the reality.

From Figure 4 we can see that, NMOD-TFSA has better performance in terms of statistical precision compared with FCM. The reason is that NMOD-TFSA is based on moving object trajectories, which can better describe the movement of moving objects, while FCM can incur errors because of the shortest path approximation.

5 Conclusion

With the recent advancement in mobile computing, sensor networks, and intelligent transportation systems, the network dynamic traffic flow statistical analysis has become a hot research issue. However, current traffic flow analysis methods have a lot of limitations such as high communication costs, low statistical precision, and considerable time delay. To solve these problems, we propose an NMOD-TFSA model in this paper. The experimental results show that compared with floating car methods which are widely used in real-world applications, NMOD-TFSA provides better performance in terms of data sampling efficiency and statistical precision. In the future work, the traffic-aware continuous query based on NMOD-TFSA and the dynamic traffic data broadcasting mechanisms will be dealt with. Also, data warehousing and data mining techniques base on NMOD-TFSA will be studied.

Acknowledgments. The work was partially supported by NSFC under grand number 60573164, and by SRF for ROCS, SEM.

References

1. Cowan, K.W., Gates, G.: Floating Vehicle Data System – A Smart Move. In: Proc. of 9th World Congress on Intelligent Transport Systems, Chicago (2002)
2. Fouladvand, M.E., Darooneh, A.H.: Statistical Analysis of Floating-Car Data: An Empirical Study. The European Physical Journal B 47 (2005)

3. Torday, A.: Link Travel Time Estimation with Probe Vehicles in Signalized Networks. In: Proc. of Swiss Transport Research Conference, Ascona (2003)

4. Lahrmann, H.: Floating Car Data for Traffic Monitoring. In: Proc. of i2TERN conference, Aalborg, Denmark (2007)

5. Lo, C.-H., Chen, C.-W., Lin, T.-Y., Lin, C.-S., Peng, W.-C.: CarWeb: A Traffic Data Collection Platform. In: Proc. of MDM 2008, Beijing, China (2008)

6. Yoon, J., Noble, B., Liu, M.: Surface Street Traffic Estimation. In: Proc. of MobiSys 2007, San Juan, Puerto Rico (2007)

7. Ding, Z., Güting, R.H.: Managing Moving Objects on Dynamic Transportation Networks. In: Proc. of SSDBM 2004, Santorini, Greece (2004)

8. Ding, Z., Zhou, X.: Location update strategies for network-constrained moving objects. In: Haritsa, J.R., Kotagiri, R., Pudi, V. (eds.) DASFAA 2008. LNCS, vol. 4947, pp. 644–652. Springer, Heidelberg (2008)

9. Güting, R.H., Almeida, V.T., Ding, Z.: Modeling and Querying Moving Objects in Networks. VLDB Journal 15(2) (2006)

Data Management for Federated Biobanks*

Johann Eder[1], Claus Dabringer[1], Michaela Schicho[1], and Konrad Stark[2]

[1] Alps Adria University Klagenfurt, Department of Informatics Systems
{Johann.Eder,Claus.Dabringer,Michaela.Schicho}@uni-klu.ac.at
[2] University of Vienna, Department of Knowledge and Business Engineering
Konrad.Stark@univie.ac.at

Abstract. Biobanks store and manage collections of biological material (tissue, blood, cell cultures, etc.) and manage the medical and biological data associated with this material. Biobanks are invaluable resources for medical research. The diversity, heterogeneity and volatility of the domain make information systems for biobanks a challenging application domain. The European project BBMRI (Biobanking and Biomolecular Resources Research Infrastructure) has the mission to network European biobanks, to improve resources for biomedical research, an thus contribute to improve the prevention, diagnosis and treatment of diseases. We present the challenges and discuss some architectures for interconnecting European biobanks and harmonizing their data.

Keywords: Biobanks, heterogeneity, federation, CSCW.

1 Introduction

Biobanks are collections of biological material (tissue, blood, cell cultures, etc.) together with data describing this material and their donors and data derived from this material. Biobanks are of eminent importance for medical research - for discovering the processes in living cells, the causes and effects of diseases, the interaction between genetic inheritance and life style factors, or the development of therapies and drugs. Information systems are an integral part of any biobank and efficient and effective IT support is mandatory for the viability of biobanks.

For an example: A medical researcher wants to find out why a certain liver cancer generates a great number of metastasis in some patients and in others not. This knowledge would help to improve the prognosis, the therapy, the selection of therapies and drugs for a particular patient, and help to develop better drugs. For such a study the researcher needs besides biological material (cancer tissue) an enormous amount of data: clinical records of the patients donating the tissue, lab analysis, microscopic images of the diseased cells, information about the life style of patients, genotype information (e.g. genetic variations), phenotype information (e.g. gene expression profiles), etc. Gathering all these data in the

---

* The work reported here was partially supported by the European Commission 7th Framework program - project BBMRI and by the Austrian Ministry of Science and Research within the program Gen-Au - project GATIB.

S.S. Bhowmick, J. Küng, and R. Wagner (Eds.): DEXA 2009, LNCS 5690, pp. 184–195, 2009.

course of a single study would be highly inefficient and costly. A biobank is supposed to deliver the data needed for this type of research and share the data and material among researchers.

A major challenge for biobank information systems is to integrate various forms of data stemming from very different autonomous sources. So biobanks are foremost integration and interoperability projects. Another important issue is the dynamics of the field: new insight leads to more differentiated diagnosis, new analysis methods allow the assessment of additional measurements, or improve the accuracy of measurements. So an information system for biobanks will be continuously evolving. And last but not least, biobanks store very detailed personal information about donors. To protect the privacy and anonymity of the donors is mandatory and misuse of the stored information has to be precluded.

In recent years biobanks have been set up in various organizations, mainly hospitals and medical and pharmaceutical research centers. Since the availability of material and data is a scarce resource for medical research, the sharing of the available material and data within the research community increased. This leads to desire to organize the interoperation of biobanks in a better way.

The European project BBMRI (Biobanking and Biomolecular Resources Research Infrastructure) has the mission to network European biobanks to improve resources for biomedical research an thus contribute to improve the prevention, diagnosis and treatment of diseases. BBMRI is organized in the framework of European Strategy Forum on Research Infrastructures (ESFRI).

In this paper we give a broad overview of the requirements for IT systems for biobanks, present the architecture of information systems supporting biobanks, discuss possible integration strategies for connecting European biobanks and discuss the challenges for this integration. Furthermore, we show how such an infrastructure can be used and present a support system for medical research using data from biobanks. An extended version of this paper will appear in [11].

2   What Are Biobanks?

Biobanks are biorepositories in which biological material with associated data is collected, stored, processed and distributed. Human biological samples in combination with donor-related clinical data are essential resources for the identification and validation of biomarkers and the development of new therapeutic approaches, especially in the development of systems for biological approaches to study the disease mechanisms. Nowadays, they are prominently used to explore and understand the function and medical relevance of human genes, their interaction with environmental factors and the molecular causes of diseases [7]. Biological material (samples) can include any kind of tissue, fluid or other material that can be obtained from an individual. Usually, biospecimens in a biobank are blood and blood components (serum), solid tissues such as small biopsies and so on. The stored data from a donor, which come along with the collected sample includes:

- General information (e.g. race, gender, age, ...)
- Lifestyle and environmental information (e.g. smoker - non smoker, living in a big city with high environmental pollution or living in rural areas)
- History of present illnesses, treatments and responses (e.g prescribed drugs and the reactions of adverse)
- Longitudinal information (e.g. a sequence of blood tests after tissue collection in order to test the progress behavior of diseases)
- Clinical outcomes (e.g. success of the treatment: Is the donor still living?)
- Data from gene expression profiles, laboratory data,...

Donors of biological materials must be informed about purpose and intended use of their samples. Typically, the donor signs an informed consent [4] which allows the use of samples for research and obliges the biobank institution to guarantee privacy of the donor.

Biobanks contribute to avoid redundant analysis and achieve the most efficient and effective use of non-renewable biological material [8]. A common and synergetic usage of this resource will enable lots of research projects especially in case of rare diseases with very limited material available. The aim is to answer as many research questions as possible without access to the samples themselves. Therefore, already acquired data of samples are stored in databases and shared among interested researchers (in silico experiments [23]). So modern biobanks offer the possibility to decrease long-term costs of research and development as well as effective data acquisition and usage.

Biobanks may contain various types of collections of biological materials. Apart from the organizational challenges, an elaborated information system is required for capturing all relevant information of samples, managing borrow and return activities and supporting complex search enquiries. If a biobank is built on the basis of existing resources (material and data), a detailed evaluation is essential. The collection process, the inventory and documentation of samples has to be assessed, evaluated and optimized. The increasing number of biobanks all over the world has drawn the attention of international organizations, encouraging the standardization of processes, sample and data management of biobanks. The Organization for Economic Cooperation and Development (OECD) released the definition of Biological Resource Centers (BRC) which "must meet the high standards of quality and expertise demanded by the international community of scientists and industry for the delivery of biological information and materials" [3]. BRCs are certified institutions providing high quality biological material and information. The model of BRCs may assist the consolidation process of biobanks defining quality management and quality assurance measures [16].

3   Data Integration in Biobanks

Since biobanks may involve many different autonomous datasources and manages very different types data ranging from typical record keeping, over text and various forms of images to gene vectors and 3-D chemical structures, it is obvious that heterogeneity is ever-present. Biobanks may comprise interfaces to sample management systems, labor information systems, research information systems, etc. The origin of the heterogeneity lies in different data sources (clinical, laboratory systems, etc), hospitals, research institutes and also in the evolution of involved disciplines. Heterogeneity appearing in biobanks comes in various forms and thus can be divided into two different classes.

Heterogeneity between different data sources. This kind of heterogeneity is mostly caused by the independent development of the different datasources. Here we have to deal with several different types of mismatches which all lead to heterogeneity between the systems as shown in [14,21]. Typical mismatches can be found in the attribute namings, different attribute encodings, content and precision of attributes, attribute granularity, different modeling of schemata, multilingualism, quality of the data stored and in the aspect of semi-structured data (incompleteness, plain-text,...).

Heterogeneity within one data source. [17] showed, that the longer data will be kept in biobanks the greater its scientific value is. On the other hand keeping data in biobanks for a long time leads to heterogeneity because medical progress leads to changes in database structures and the modeled domain. Changes that typically arise in this context are changes in disease codes, progress in biomolecular methods results in higher accuracy of measurements, extension of relevant knowledge (e.g. GeneOntology [2] is changed daily), treatments and standard procedures change or the quality of sample conservation increases, etc. Furthermore, also the technical aspects within one biobank are volatile: data structures, semantics of data, parameters collected, etc.

Evolution. Biobanks need a mechanism to represent the changes mentioned above and to correctly deal for the best possible exploitation of the collection. Wherever possible biobanks should provide transformations to map data between different versions. Using ontologies to annotate content of biobanks can be quite useful. By providing mapping support between different ontologies the longevity problem can be addressed. Further on, versioning and transformation approaches can help to support the evolution of biobanks. Techniques from temporal databases and temporal data warehouses can be used for the representation of volatile data together with version mappings to transform all data to a selected version [9,10,12,13,24]. This knowledge can be directly applied to biobanks as well.

3.1   Example Architecture – MUG Biobank

In the context of the MUG (Medical University of Graz) biobank several types of information systems are accessed, as illustrated in figure 1. The different data sources are integrated in a database federation, whereas interface wrappers have been created for the relevant data. On the one hand, there are large clinical information systems which are used for routine diagnostic and therapeutical activities of medical doctors. Patient records from various medical institutes are stored in the OpenMedocs sytem, pathological data in the PACS system and laboratory data in the laboratory information system LIS. On the other hand research databases from several institutes (e.g. the Archimed system) containing data about medical studies are incorporated as well as the biological sample management system SampleDB and diverse robot systems. Further, survival data of patients is provided by the external institution Statistics Austria. Clinical and routine information systems (at the bottom of figure 1) are strictly separated from operational information systems of the biobank. That is, sensitive patient-related data is only accessible for medical staff and anonymized otherwise. The MUG Biobank operates an own documentation system in order to protocol and coordinate all cooperation projects. The CSCW system (at the top of figure 1) provides a scientific workbench for internal and external project partners, allowing to share data, documents, analysis results and services. A modified version of the CSCW workbench will be used as user interface for the European Biobank initiative BBMRI, described in section 4.

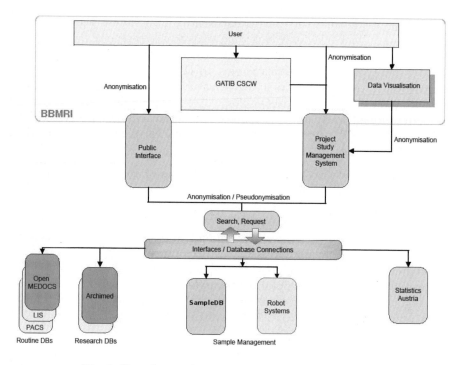

Fig. 1. Data Integration in context of the MUG Biobank

3.2 Related Work

UK-Biobank. The aim of UK Biobank is to store health information about 500.000 people from all around the UK who are aged between 40-69. UK Biobank has evolved over several years. Many subsystems, processes and even the system architecture have been developed from experience gathered during pilot operations. UK Biobank integrated many different subsystems to cooperate [6].

caBIG. The cancer Biomedical Informatics Grid (caBIG) has been initiated by the National Cancer Institute (NCI) as a national-scale effort in order to develop a federation of interoperable research information systems. The federated interoperability is reached by a service oriented middleware infrastructure, called caGrid. A key characteristic of the framework is its focus on metadata and model driven service development. This aspect of caGrid is particularly important for the support of syntactic and semantic interoperability across heterogeneous applications [18].

CRIP. The concept of CRIP (Central Research Infrastructure for molecular Pathology) enables biobanks to annotate projects with additional necessary data and to transfer them into valuable research resources. CRIP offers a virtual simultaneous access to tissue collections of participating pathology archives. Annotated valuable data comes from different heterogeneous data sources and is stored in a central CRIP database [19].

4   Biobanking and Biomolecular Resources Infrastructure

To benefit European health-care, medical research, and ultimately, the health of the citizens of the European Union the European Commission is funding a biobank integration project called BBMRI. The aim of BBMRI is to build a coordinated, large scale European infrastructure of biomedically relevant, quality-assessed mostly already collected samples as well as different types of biomolecular resources. In addition to biological materials and related data, BBMRI will facilitate access to detailed and internationally standardised data sets of sample donors (clinical data, lifestyle and environmental exposure) as well as data generated by analysis of samples using standardised analysis platforms [1].

Benefits. The benefits of BBMRI are versatile. Talking in *short-terms* BBMRI leads to an increased quality of research as well as to a reduction of costs. The *mid-term* impacts of BBMRI can be seen in an increased efficacy of drug discovery/development. *Long-term* benefits of BBMRI are improved health care possibilities in the area of personalized medicine/health care [5].

Data Harmonisation and IT-infrastructure. An important part of BBMRI is responsible for designing the IT-infrastructure and database harmonisation, which includes also solutions for data and process standardization. The harmonization of data deals with the identification of the scope of needed information and data structures. Further on, it analysis how available nomenclature and coding systems can be used for storing and retrieving (heterogenous) biobank information. Several controlled terminologies and coding systems may be used for organizing the information about biobanks [5,15]. Since not all medical information is fully available in the local databases of biobanks the retrieval of data involves big challenges. That implies the necessity of flexible data sharing and collaboration between centers.

4.1   Use Cases in a Federation of Biobanks

In section 3 we discussed issues within one biobank as integration project. Now we are concerned with a set of heterogenous biobanks as integration project. There exist several different proposals for the handling of enquiries within the BBMRI project. In the following we distinguish between five different kinds of use cases:

1. *Identification of biobanks.* Retrieves a list with contact data from participating biobanks which have desired material for a certain study. This use case only operates on the meta-databases which contains a small set of attributes. A good candidate, for example, is the attribute "*diagnose*" standardized as ICD-Code because it may be very useful to know which biobank(s) store information about specific diseases.
2. *Identification of cases.* Retrieves the pseudonym identifiers of cases[1] stored in local biobanks which correspond to a given set of parameters. Since the meta-database located on the hosts does not store any case or donor related information it is necessary to additionally query the local databases. Within this use case no donor related information is sent to the researcher.
3. *Retrieval of data.* Obtains available information (material, data, etc.) directly from a biobank for a given set of parameters. To realize this use case further legal and ethical questions have to be answered.
4. *Upload or linking of data.* Connecting samples with data generated from this sample internally and externally. Quality of data as well as data provenance are important issues within this use case.
5. *Statistical queries.* Performs analytical queries on a set of biobanks.

We assume that researchers use the contact information and pseudonym identifiers to retrieve data from a biobank. Within BBMRI the focus lies on the first two use cases mentioned before. The following query is an exemplary query for use case 2. *A researcher requires the ID of about 20 cases and their location (biobank) with the following characteristics:*

− *paraffin tissue*
− *with diagnose breast cancer*
− *staging T1 N2 M0*
− *from donors of age 40-50 years*
− *including follow-up data (e.g. therapy) from the oncology*

A special case within the *identification of cases* is determined by a slight variance in the result set. Depending on a certain policy the result set can also contain only a list of biobanks with their contact information as discussed in scenario 1.

---

[1] In our context a case is a set of jointly harvested samples of one donor.

4.2   Data Sharing and Collaboration between Different Biobanks

There exists several different approaches how to realize the IT-infrastructure within BBMRI. We are going to present the most important ones and discuss their pros and cons.

Peer to Peer. Within this approach the biobanks are connected via a peer to peer infrastructure. All biobanks must provide a query interface because queries are sent to all participating biobanks by the requestor. The exchange format can be defined or even undefined. An undefined exchange format leads to inter-operability problems - but defining an exchange format upfront could lead to a domination of the biobank with the smallest schema. Therefore defining the format is a critical success factor for the peer to peer architecture. One disadvantage of this approach is that small biobanks which are not permanently online can not take part in the federation. The major drawback of the peer to peer approach is that there exists data in biobanks that is not allowed to leave the biobank until it is aggregated and anonymized.

Centralized with Integration Hub. The centralized integration hub works as a mediator within the federation. This central hub distributes queries within the federation and it is also responsible for integrating results from the different biobanks. This architecture is also able to solve the disadvantage of the peer to peer approach with the use of a data warehouse. Here the data is stored in an aggregated and anonymized form and it is available for the federation. Further on small biobanks could export the needed data and thus do not need to be online all the time. Nevertheless the centralized integration hub represents a single point of failure.

Combined Approach. To overcome the before mentioned problems we designed an architecture for the collaboration between different biobanks as a hybrid of peer to peer and a hub and spoke structure. In our approach a BBMRI-Host (figure 2) represents a domain hub in the IT-infrastructure and uses a meta structure to provide data sharing. Several domain hubs are connected via a peer-to-peer structure and communicate with each other via standardized and shared Communication Adapters. Each participating European biobank is connected with its specific domain hub resp. BBMRI-Host via hub and spoke-structure.

Biobanks provide their obtainable attributes and contents as well as their contact data and biobank specific information via the *BBMRI Upload Service* of the associated BBMRI-Host. A Mediator coordinates the interoperability issues between BBMRI-Host and the associated biobanks. The information about uploaded data from each associated biobank is stored in the BBMRI Content-Meta-Structure. Permissions related to the uploaded data as well as contracts between a BBMRI-Host and a specific biobank are managed by the Disclosure Filter. A researcher can use the *BBMRI Query Service* for sending requests to the federated system.

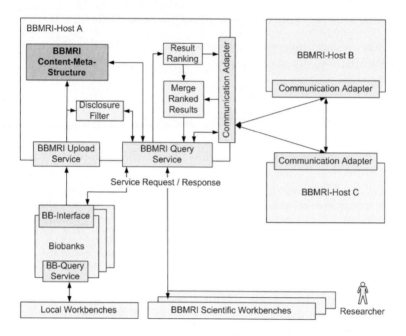

Fig. 2. Architecture of BBMRI IT-infrastructure

4.3   Lookup Data-Mart

Our approach for the BBMRI Content-Meta-Structure (see figure 2) was to accomplish a hybrid-solution of a federated system and an additional data warehouse as a kind of index to primarily reduce the query overhead. This decision led to the design of a class (named ContentInformation, figure 3) which contains attributes with different meanings, similar to online analytical processing (Olap), including:

- *Content-attributes.* Are a small set of attributes, which provide information about their content in the local database (cf. 4.1). The data-type must be an enumeration like ICD-Code, patient sex etc.
- *Number of Cases (NoC).* Is an order of magnitude for all available cases of a specific disease in combination with all content-attributes.
- *Existence-attributes* accept two different kinds of characteristics:
  1. **Value as quantity.** Represented by a numeric value greater than a defined k-value for an aggregated set of cases.
  2. **Value as availability.** Storage of values does not take place in an aggregated form like mentioned above, but as bitmap - with 0 not available and 1 available.

To avoid data overkill we designed a kind of lower-bound schema that contains attributes usually occurring in most or even all of the participating biobanks.

In comparison to an Olap data-cube our class *ContentInformation* (see figure 3) acts as the *fact-table* with the content-attributes as dimensions and

		Content-Attributes		Number of Cases	Existence-Attributes					
Id	Diagnose	PatSex	BMI-Cat	NoC	Cryo Tissue	Paraffin Tissue	StagingT	StagingN	StagingM	...
1	C50.8	female	overweight	5000	1000	2000	500	500	500	
2	C50.9	female	overweight	1000	700	300	1000	null	null	
...	...	...			...	...	...	...	...	...

Fig. 3. Example data of class *ContentInformation*

existence-attributes (including the Number of Cases) as measures. Depending on the type of the biobank (cancer, metabolic, ...) the datamodel is able to store different sets of attributes. With the proposed data model it is possible to support different kinds of content information depending on the needs of the biobanks. Besides once a new attribute is introduced, this does not lead to changes in the database schema. The datamodel enables a dynamic generation of the ContentInformation. I.e. Each biobank first declares the attributes they store in their local databases. Especially they declare which of them are content-attributes and which of them are existence-attributes. However this could affect requests on material, therefore one must be careful with the declaration.

- *Requests on existence of attributes.* For a request on the existence of several attributes it does not matter whether the requested attributes are declared as content-attribute or existence-attribute. The only fact to get a query-hit for that request is that the searched attributes are declared by a biobank.
- *Requests on content of attributes.* For a request on the content of several attributes, all requested attributes must be declared as content-attribute by a biobank in order to get a query-hit.

The semi-structured data within the federation of biobanks complicates query answering. It will often happen that there are only a few entries in the result because different data is available in different biobanks. For the researchers not only full matches but also partial or near matches could contain interesting material. To cope with that problem we suggest an approximate query answering with integrated result ranking. With the help of an intelligent result ranking we are able to include partial and near matches in the result set by assigning them lower ranks.

5   Working with Biobanks

Medical research is a collaborative process in an interdisciplinary environment that may be effectively supported by a CSCW system allowing flexible integration of data, analysis services and communication mechanisms. Persons with different expertise and access rights cooperate in mutually influencing contexts (e.g. clinical studies, research cooperations). Thus, appropriate virtual environments are needed to facilitate context-aware communication, deployment of biomedical tools as well as data and knowledge sharing. Leveraging on the service oriented CSCW middleware Wasabi (www.open-steam.org), we developed a CSCW system for medical research [20,22] with the following key features:

- $R(1)$ **User and Role Management.**
- $R(2)$ **Transparency of physical Storage.**
- $R(3)$ **Flexible Data Presentation.**
- $R(4)$ **Flexible Integration and Composition of Services.**
- $R(5)$ **Support of cooperative Functions.**
- $R(6)$ **Data-coupled Communication Mechanisms.**
- $R(7)$ **Knowledge Creation and Knowledge Processing.**

We tested this workbench supporting a workflow for gene expression analysis for a Breast Cancer project. A detailed breast cancer data set was annotated at the Pathology Graz. In this context much emphasis is put on detecting deviations in the behavior of gene groups. We support the entire analysis workflow by supplying an IT research platform allowing to select and group patients arbitrarily, pre-process and link the related gene expressions and finally perform state-of-the-art analysis algorithms. We developed an appropriate database structure with import/export methods allowing to manage arbitrary medical data sets and gene expressions. We also implemented web service interfaces to various gene expression analysis algorithms. The workflow consisted of the following steps: (1) Selecting cases with appropriate documentation for the study, (2) Normalization of Gene Expression Profiles, (3) gene annotation, (4) extend data with gene function groups from Gene Ontology, (5) link with annotated patient data, (6) group cases according to medical parameters and (7) analysis to identify significant genes.

We are able to show that a service-oriented CSCW system provides the functionality to build a workbench for medical research supporting the collaboration of researchers, allowing the definition of workflows and gathering all necessary data for maintaining provenance information. According to our medical collaborator the increase in performance using this system was dramatic and reduced essential steps of the study from weeks to days.

6   Conclusion

Biobanks are challenging application areas for advanced information technology. The foremost challenges for the information system support in a network of biobanks as envisioned in the BBMRI project are the following: We presented biobanks and discussed the requirements for biobank information systems. We have shown that many different research areas within the IS field contribute to this endeavor. We were just able to show some examples: advanced information modeling, (semantic) interoperability, federated databases, approximate query answering, result ranking, computer supported cooperative work (CSCW), and security and privacy. Some well known solutions from different application areas have to be revisited given the size, heterogeneity, diversity dynamics, and complexity of data to be organized in biobanks.

References

1. Biobanking and Ciomolecular Resources Research Infrastructure,
   http://www.bbmri.eu
2. Geneontology, http://www.geneontology.org

3. Oecd: Underpinning the future of life sciences and biotechnology (2001)
4. Nih guide: Informed Consent in Research Involving Human Participants (2006)
5. Bbmri: Construction of new Infrastructures - Preparatory Phase. In: INFRA–2007–2.2.1.16: European Bio-Banking and Biomolecular Resources (April 2007)
6. Uk biobank: Protocol for a large-Scale Prospective Epidemiological Resource. Protocol No: UKBB-PROT-09-06 (March 2007)
7. Asslaber, M., Abuja, P., Stark, K., et al.: The Genome Austria Tissue Bank (GATIB). In: Pathobiology 2007, vol. 74, pp. 251–258 (2007)
8. Asslaber, M., Zatloukal, K.: Biobanks: Transnational, European and Global Networks. Briefings in Functional Genomics & Proteomics 6(3), 193–201 (2007)
9. Chamoni, P., Stock, S.: Temporal structures in data warehousing. In: Mohania, M., Tjoa, A.M. (eds.) DaWaK 1999. LNCS, vol. 1676, pp. 353–358. Springer, Heidelberg (1999)
10. Eder, J., Koncilia, C.: Evolution of Dimension Data in Temporal Data Warehouses. In: Kambayashi, Y., Winiwarter, W., Arikawa, M. (eds.) DaWaK 2001. LNCS, vol. 2114, pp. 284–293. Springer, Heidelberg (2001)
11. Eder, J., Dabringer, C., Schicho, M., Stark, K.: Information Systems for Federated Biobanks. Transactions on Large Scale Data and Knowledge Centered Systems 1(1) (2009) (in print)
12. Eder, J., Koncilia, C., Morzy, T.: The COMET metamodel for temporal data warehouses. In: Pidduck, A.B., Mylopoulos, J., Woo, C.C., Ozsu, M.T. (eds.) CAiSE 2002. LNCS, vol. 2348, p. 83. Springer, Heidelberg (2002)
13. Goos, G., Hartmanis, J., Sripada, S., Leeuwen, J.V., Jajodia, S.: Temporal Databases: Research and Practice. Springer, New York (1998)
14. Litwin, W., Mark, L., Roussopoulos, N.: Interoperability of Multiple Autonomous Databases. ACM Comput. Surv. 22(3), 267–293 (1990)
15. Muilu, J., Peltonen, L., Litton, J.: The Federated Database - a basis for Biobank-Based Post-Genome Studies, Integrating Phenome and Genome Data from 600 000 Twin Pairs in Europe. European Journal of Human Genetics 15, 718–723 (2007)
16. Rebulla, P., Lecchi, L., Giovanelli, S., Butti, B., Salvaterra, E.: Biobanking in the Year 2007. Transfusion Medicine and Hemotherapy 34, 286–292 (2007)
17. Riegman, P., Morente, M., Betsou, F., de Blasio, P., Geary, P.: Biobanking for Better Healthcare. In: The Marble Arch International Working Group on Biobanking for Biomedical Research (2008)
18. Saltz, J., Oster, S., Hastings, S.: Design and Implementation of the Core Architecture of the Cancer Biomedical Informatics Grid. Bioinformatics (2006)
19. Schroeder, C.: Vernetzte Gewebesammlungen f. d. Forschung. Laborwelt, 5 (2007)
20. Schulte, J., Hampel, T., Stark, K., Eder, J., Schikuta, E.: Towards the Next Generation of Service-oriented Flexible Collaborative Systems – a basic framework applied to medical research. In: ICEIS 2008 - Proc. of the 10th Int. Conf. on Enterprise Information Systems (2008)
21. Sheth, A.P., Larson, J.A.: Federated Database Systems for Managing Distributed, Heterogeneous, and Autonomous Databases. ACM Comput. Surv. (1990)
22. Stark, K., Schulte, J., Hampel, T., Schikuta, E., Zatloukal, K., Eder, J.: GATiB-CSCW, medical research supported by a service-oriented collaborative system. In: Bellahsène, Z., Léonard, M. (eds.) CAiSE 2008. LNCS, vol. 5074, pp. 148–162. Springer, Heidelberg (2008)
23. Stevens, R., Zhao, J., Goble, C.: Using Provenance to Manage Knowledge of in Silico Experiments. Briefings in bioinformatics 8(3) (2007)
24. Yang, J.: Temporal Data Warehousing. Stanford University (2001)

Peer-to-Peer Semantic Wikis

Hala Skaf-Molli, Charbel Rahhal, and Pascal Molli

INRIA Nancy-Grand Est
Nancy University, France
{skaf,charbel.rahal,molli}@loria.fr

Abstract. Wikis have demonstrated how it is possible to convert a community of strangers into a community of collaborators. Semantic wikis have opened an interesting way to mix web 2.0 advantages with the semantic web approach. P2P wikis have illustrated how wikis can be deployed on P2P wikis and take advantages of its intrinsic qualities: fault-tolerance, scalability and infrastructure cost sharing. In this paper, we present the first P2P semantic wiki that combines advantages of semantic wikis and P2P wikis. Building a P2P semantic wiki is challenging. It requires building an optimistic replication algorithm that is compatible with P2P constraints, ensures an acceptable level of consistency and generic enough to handle semantic wiki pages. The contribution of this paper is the definition of a clear model for building P2P semantic wikis. We define the data model, operations on this model, intentions of these operations, algorithms to ensure consistency and finally we implement the SWOOKI prototype based on these algorithms.

1 Introduction

Wikis are the most popular tools of Web 2.0, they provide an easy to share and contribute to global knowledge. The encyclopedia Wikipedia is a famous example of a wiki system. In spite of their fast success, wiki systems have some drawbacks. They suffer from search and navigation [1], it is not easy to find information in wikis [2]. They have also scalability, availability and performance problems [3,4] and they do not support offline works and atomic changes [5]. To overcome these limitations, wiki systems have evolved in two different ways: semantic wikis and peer-to-peer wikis.

Semantic Wikis. Semantic wikis are a new generation of collaborative editing tools.They allow users to add semantic annotations in the wiki pages. Users collaborate not only for writing the wiki pages but also for writing semantic annotations. Usually, this is done by annotating the links between wikis pages. Links in semantic wikis are typed. For instance, a link between the wiki pages "France" and "Paris" may be annotated by "capital". Semantic wikis provide a better structuring of wikis by providing a means to navigate and search based on annotations. These annotations express relationships between wikis pages, they are usually written in a formal syntax so they are processed automatically by

S.S. Bhowmick, J. Küng, and R. Wagner (Eds.): DEXA 2009, LNCS 5690, pp. 196–213, 2009.

machines and they are exploited by semantic queries. Many semantic wikis are being developed such [1,6,2].

P2P wikis. Wikis on a peer-to-peer network attempt to reconcile the benefits of mass collaboration and the intrinsic qualities of peer-to-peer network such as scalability, fault-tolerance, better performance and resistance to censorship.There are currently many research proposals to build P2P wikis such [3,7,5,8]. The basic idea is to replicate wiki pages on the peers of a P2P network and the main problem is to ensure the consistency of copies.

In this paper, we propose to build the first peer-to-peer semantic wiki, called SWOOKI. SWOOKI combines advantages of P2P wikis and Semantic wikis. The main problem for building such a system is to maintain consistency of replicated semantic wiki pages. We propose an algorithm that ensures the CCI consistency model. This model [9] is a well established consistency model for group editors.

  – *Causality preservation:* operations ordered by a precedence relation will be executed in same order on every peer.
  – *Convergence:* When the system is idle, all copies are identical.
  – *Intention and Intention preservation:* The intention of an operation is the effects observed on the state when the operation was generated. The effects of executing an operation at all sites are the same as the intention of the operation. The effect of executing an operation does not change the effects of independent (not causally dependent) operations.

Building a P2P semantic wiki based on the CCI model is challenging. The fundamental problem is to provide an optimistic replication algorithm that (1) is compatible with P2P constraints, (2) ensures the CCI model and (3) is generic enough to manage semantic wiki pages. In this paper, we define formally the semantic wiki page data type, we specify its operations and we define the intentions of these operations. We extend the WOOT [10] algorithm to take into account semantic annotations and finally we build SWOOKI the first P2P semantic wiki based on this algorithm.

The paper is organized as follow. Section 2 presents use cases. Section 3 presents some related works. Section 4 details the general approach for building a P2P semantic wiki. It defines a data model and editing operations. Section 5 defines the causality and intentions of operations used to edit semantic data. Section 6 develops the integration algorithm. Section 7 gives an overview of the architecture. The last section concludes the paper and points to future works.

2   Use Cases for P2P Semantic Wikis

We have identified three interesting use cases for P2P semantic wikis. We aim to develop a peer to peer semantic wiki that supports these use cases.

Mass Collaboration. In this case, a P2P semantic wiki system is deployed as a Usenet network[11]. A thousands of semantic wiki servers can be deployed within organizations or universities. Any user can connect to any semantic wiki server. This allows : (1) to handle a large number of users by dividing the load on the whole network. Semantic queries can be performed locally on each semantic wiki server. (2) to tolerate many faults. A crash of one semantic wiki server does not stop the service. (3) to share the cost of the infrastructure. Wikis are set up and maintained by different organizations. Therefore, it is not necessary to collect funds just to maintain the infrastructure. For instance, Wikipedia foundation has to collect 150000 \$ every three months just to maintain the Wikipedia infrastructure. (4) to resist to censorship. An organization controls only one semantic wiki server and not all data.

Off-line work and transactional changes. Adding off-line capabilities to web applications is currently a major issue. For instance, the development of Google gears and Firefox3 off-line capabilities demonstrate the need of the off-line work. Wikis are web applications and the need for off-line wiki editing is real. Current technologies for adding off-line capabilities to web applications focus on Ajax applications. However, the off-line mode of these web applications does not provide all features available in the on-line mode. This can be an obstacle for a wiki system. For instance, the off-line mode of the wiki allows navigation but it does not allow editing. A P2P semantic wiki tolerates naturally off-line work by means of an integrated merge algorithm. The off-line mode enables also transactional changes.

Ad-hoc Collaborative Editing. This scenario is derived from the previous one. Imagine several off-line wiki users have a meeting. Unfortunately, there is no Internet connection available in the meeting room. Therefore, they decide to set up an ad-hoc network within the meeting room. A P2P semantic wiki is able to propagate changes within the ad-hoc network and allows collaborative editing just for these off-line users. Of course, when the meeting is finished and users return to their organizations, their semantic wiki systems will re-synchronize with the whole P2P network.

The above use cases illustrate the importance of optimistic replication algorithms.

3   Related Work

The fundamental problem for building a P2P semantic wiki is to provide an optimistic replication algorithm that (1) supports collaborative editing, (2) manages a semantic wiki data type, (3) ensures the CCI model and (4) is compatible with P2P constraints. Many researches have been done in P2P semantic web [12,13,14,15,16]. These works focus on sharing, querying and synchronizing RDF resources rather than collaborative editing of RDF resources. Sharing is different from collaboration. In sharing, some peers publish data while others can only read these data and concurrent updates are not managed. In P2P semantic wikis, some peers publish data, others can read and write these data and a synchronization algorithm integrates concurrent updates while maintaining consistency of these data. Data replication in collaborative P2P systems mainly relies on optimistic replication. Existing approach can not be apply to the P2P semantic wikis context. For instance, the *Bayou* system [17] is suitable for deploying a collaborative application on a decentralized network. Unfortunately, in order to ensure the convergence of copies, *Bayou* has to arrange eventually operations in the same order. To achieve this, it relies on a primary site that will enforce a global continuous order on a growing prefix of history. Using such a primary site is not compatible with P2P network constraints. Other systems such as Usenet [11] apply the Thomas write rule [18] to ensure eventual consistency. They ensure that, when the systems are idle *i.e* all operations have been sent and received by all sites, all copies are identical. Unfortunately, in case of two concurrent write operations the rule of "the last writer wins" is applied. This means that a modification of a user is lost. Collaborative editing cannot be easily achieved if the system can loose some changes just to ensure eventual consistency.

Many algorithms have been developed by the Operational Transformation community [9] such as SOCT2, GOTO, COT etc. They are designed to verify the CCI model. But only few of them support P2P constraints such as MOT2 [19]. However MOT2 algorithm suffers from a high communication complexity [20] and no transformation functions for a semantic wiki page are available.

Some existing P2P wikis such as *giki* or *git-wiki* use distributed version control systems (DVCS) to manage data. Wiki pages are stored as text files. DVCS manage them as code files. They ensure causal consistency. This implies that concurrent write operations can be seen in a different order on different machines. In this case, if two sites observe 2 write operations in different order then copies on both sites can diverge. DVCS systems are aware of this problem and delegate the problem to external merge algorithms for managing concurrent operations. However, as existing merge algorithms are not intrinsically deterministic, commutative and associative so convergence cannot be ensured in all cases. Wooki [3] is the only available P2P wiki that ensures the CCI consistency. Wooki relies on the WOOT [10] algorithm to ensure the CCI consistency for wiki pages. However, WOOT cannot be applied directly to a P2P semantic wiki. WOOT is designed to synchronize linear structures, it can not synchronize a mix of text and RDF graphs. We propose to extend the WOOT algorithm to handle collaborative writing on replicated RDF data model.

4   P2P Semantic Wiki Approach

A P2P semantic wiki is a P2P network of autonomous semantic wiki servers (called also peers or nodes) that can dynamically join and leave the network. Every peer hosts a copy of all semantic wiki pages and an RDF store for the semantic data. Every peer can autonomously offer all the services of a semantic wiki server. When a peer updates its local copy of data, it generates a corresponding operation. This operation is processed in four steps:

1. It is executed immediately against the local replica of the peer,
2. it is broadcasted through the P2P network to all other peers,
3. it is received by the other peers,

4. it is integrated to their local replica. If needed, the integration process merges this modification with concurrent ones, generated either locally or received from a remote server.

The system is correct if it ensures the CCI consistency model (see section 1).

4.1   Data Model

The data model is an extension of Wooki [3] data model to take in consideration semantic data. Every semantic wiki peer is assigned a global unique identifier named $NodeID$. These identifiers are totally ordered. As in any wiki system, the basic element is a semantic wiki page and every semantic wiki page is assigned a unique identifier $PageID$, which is the name of the page. The name is set at the creation of the page. If several servers create concurrently pages with the same name, their content will be directly merged by the synchronization algorithm. Notice that a URI can be used to unambiguously identify the concept described in the page. The URI must be global and location independent in order to ensure load balancing. For simplicity, in this paper, we use a string as page identifier.

Definition 1. *A semantic wiki page Page is an ordered sequence of lines $L_B L_1$, $L_2, \ldots L_n L_E$ where L_B and L_E are special lines. L_B indicates the beginning of the page and L_E indicates the ending of the page.*

Definition 2. *A semantic wiki line L is a four-tuple $< LineID$, content, degree, visibility $>$ where*

    $-$ $LineID$ is the line identifier, it is a pair of $(NodeID, logicalclock)$ where $NodeID$ is the identifier of the semantic wiki server and $logicalclock$ is a logical clock of that server. Every semantic wiki server maintains a logical clock, this clock is incremented when an operation is generated. Lines identifiers are totally ordered so if $LineID_1$ and $LineID_2$ are two different lines with the values $(NodeID_1, LineID_1)$ and $(NodeID_2, LineID_2)$ then $LineID_1 < LineID_2$ if and only if (1) $NodeID_1 < NodeID_2$ or (2) $NodeID_1 = NodeID_2$ and $LineID_1 < LineID_2$.

    $-$ $content$ is a string representing text and the semantic data embedded in the line.

    $-$ $degree$ is an integer used by the synchronization algorithm, the degree of a line is fixed when the line is generated, it represents a kind of loose hierarchical relation between lines. Lines with a lower degree are more likely generated earlier than lines with a higher degree. By definition the degree of L_E and L_B is zero.

    $-$ $visibility$ is a boolean representing if the line is visible or not. Lines are never really deleted they are just marked as invisible. For instance, suppose there are two lines in a semantic wiki page about "France" , "France" is the identifier of the page.

France is located **in** [locatedIn :: Europe]
The capital of France is [hasCapital :: Paris]

Suppose these two lines are generated on the server with $NodeID = 1$ in the above order and there are no invisible lines, so the semantic wiki page will be internally stored as.

L_B
((1,1), France is located **in** [locatedIn :: Europe], 1, true)
((1,2), The capital of France is [hasCapital :: Paris], 2, true)
L_E

Text and semantic data are stored in separate persistent storages. Text can be stored in files and semantic data can be stored in RDF repositories, as described in the next section.

Semantic data storage model. RDF is the standard data model for encoding semantic data. In P2P semantic wikis, every peer has a local RDF repository that contains a set of RDF statements extracted from its wikis pages. A statement is defined as a triple (Subject, Predicate, Object) where the subject is the name of the page and the predicates (or properties) and the objects are related to that concept. For instance, the local RDF repository of the above server contains: R = {("France", "locatedIn", "Europe"), ("France", "hasCapital", "Paris") }. As for the page identifier, a global URI can be assigned to predicates and objects of a concept, for simplicity, we use a string. We define two operations on the RDF repositories:

– insertRDF(R,t): adds a statement t to the local RDF repository R. – deleteRDF(R,t): deletes a statement t from the local RDF repository R.

These operations are not manipulated directly by the end user, they are called implicitly by the editing operations as shown later.

4.2   Editing Operations

A user of a P2P semantic wiki does not edit directly the data model. Instead, she uses traditional wiki editing operations, when she opens a semantic wiki page, she sees a view of the model. In this view, only visible lines are displayed. As in a traditional semantic wiki, she makes modifications i.e. adds new lines or deletes existing ones and she saves the page(s). To detect user operations, a diff algorithm is used to compute the difference between the initial requested page and the saved one. Then these operations are transformed into model editing operations. A *delete* of the line number n is transformed into a *delete* of the nth visible line and an *insert* at the position n is transformed into *insert* between the $(n-1)^{th}$ and the nth visible lines. These operations are integrated locally and then broadcasted to the other servers to be integrated. There are two editing operations for editing the wiki text: *insert* and *delete*. An update is considered as a delete of old value followed by an insert of a new value. There are no special operations for editing semantic data. Since semantic data are embedded in the text, the RDF repositories are updated as a side effect of text replication and synchronization. (1) *Insert(PageID, line, l_P , l_N)* where *PageID* is the identifier of the page of the inserted line. *line* is the line to be inserted. It is a tuple containing $<$ LineID, content, degree, visibility $>$. l_P is the identifier of the line that precedes the inserted line. l_N is the identifier of the line that follows the inserted line. During the insert operation, the semantic data embedded in the line are extracted, RDF statements are built with the page name as a subject and then they are added to the local RDF repository thanks to the function *insertRDF(R, t)*. (2) The *delete(PageID, LineID)* operation sets the visibility of the line identified by *LineID* of the page *PageID* to false. The line is not deleted physically, it is just marked as deleted. The identifiers of deleted lines must be kept as a tombstones. During the delete operation, the set of RDF statements contained in the deleted line is deleted from the local RDF repository thanks to the *deleteRDF(R, t)*.

5   Correction Model

This section defines causal relationships and intentions of the editing operations for our P2P semantic wiki data model.

5.1   Causality Preservation

The causality property ensures that operations ordered by a precedence relation will be executed in the same order on every server. In WOOT, the precedence relation relies on the *semantic causal dependency*. This dependency is explicitly declared as preconditions of the operations. Therefore, operations are executed on a state where they are legal *i.e.* preconditions are verified. We define causality for editing operations that manipulate text and RDF data model as:

Definition 3. *insert Preconditions* *Let Page be the page identified by PageID, let the operation op=Insert(PageID, newline, p , n), newline =< LineID, c, d, v> generated at a server NodeID, R is its local RDF repository. The line newline can be inserted in the page Page if its previous and next lines are already present in the data model of the page Page.*

$$\exists i \ \exists j \ LineID(Page[i]) = p \wedge LineID(Page[j]) = n$$

Definition 4. *Preconditions of delete operation* *Let Page be the page identified by PageID, let op = Delete(PageID, dl) generated at a server NodeID with local RDF repository R, the line identified by dl can be deleted (marked as invisible), if its dl exists in the page.*

$$\exists i \ LineID(Page[i]) = dl$$

When a server receives an operation, the operation is integrated immediately if its pre-conditions are evaluated to true else the operation is added to a waiting queue, it is integrated later when its pre-conditions become true.

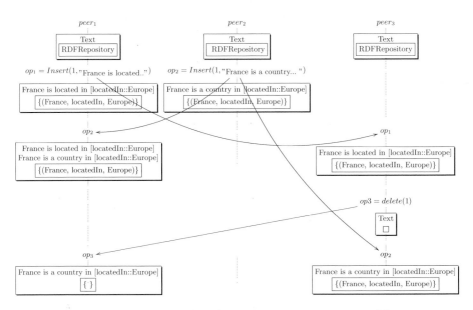

Fig. 1. Semantic inconsistency after integrating concurrent modifications

5.2   Intentions and Intentions Preservation

The intention of an operation is the visible effect observed when a change is generated at one peer, the intention preservation means that the intention of the operation will be observable on all peers, in spite of any sequence of concurrent operations. We can have a naive definition of intention for *insert* and *delete*:

– The intention of an insert operation *op= Insert(PageID, newline, p , n)* when generated at site $NodeID$, where *newline* =< nid, c, d, v> is defined as: (1) The content is inserted between the previous and the next lines and (2) the semantic data in the line content are added to the RDF repository of the server.
– The intention of a delete operation *op= delete(pid, l)* when generated at site S is defined as : (1) the line content of the operation is set to invisible and (2) the semantic data in the line content are deleted from the RDF repository of the server.

Unfortunately, it is not possible to preserve the previous intention definitions. We illustrate a scenario of violation of these intentions in figure 1. Assume that three P2P semantic wiki servers, $peer_1$, $peer_2$ and $peer_3$ share a semantic wiki page about "France". Every server has its copy of shared data and has its own persistence storage repository. At the beginning, the local text and the RDF repositories are empty. At $peer_1$, $user_1$ inserts the line "France is located [located In::Europe]" at the position 1 in her copy of the "France" page. Concurrently, at $peer_2$ $user_2$ inserts a new line "France is a country in [located In::Europe]" in her local copy of "France" page at the same position and finally at $peer_3$ $user_3$ deletes the line added by $user_1$. When op_2 is integrated at $peer_1$, the semantic annotation is present two times in the text and just one time in the RDF repository. In fact, the RDF repository cannot store twice the same triple. When op_3 is finally integrated on $peer_1$, it deletes the corresponding line and the semantic entry in the RDF repository. In this state, the text and the RDF repository are inconsistent. Concurrently, $peer_3$ has integrated the sequence $[op_1;op_3;op_2]$. This sequence leads to a state different than the state on $peer_1$. Copies are not identical, convergence is violated.

The above intentions cannot be preserved because the effect of executing op_3 changes the effect of op_2 which is independent, of op_3 i.e. op_3 deletes the statement inserted by op_2, but op_3 has not seen op_2 at generation time.

5.3   Model for Intention Preservation

It is not possible to preserve intentions if the RDF store is defined as a set of statements. However, if we transform the RDF store into multi-set of statements, it becomes possible to define intentions that can be preserved.

Definition 5. RDF repository *is the storage container for RDF statements, each container is a* multi-set *of RDF statements. Each RDF repository is defined as a pair* (T, m) *where* T *is a set of RDF statements and* m *is the multiplicity function* $m : T \rightarrow \mathcal{N}$ *where* $\mathcal{N} = 1, 2......$

For instance, the multi-set $R = \{$ ("France", "LocatedIn", "Europe"),("France", "LocatedIn", "Europe"),("France", "hasCapital", "Paris") $\}$ can presented by $R=\{$ ("France", "LocatedIn", "Europe" $)^2$, ("France", "hasCapital", "Paris" $)^1 \}$ where 2 is the number of occurrence of the first statement and 1 is this of the second one.

Definition 6. Intention of insert operation *Let* S *be a P2P semantic wiki server,* R *is its local RDF repository and Page is a semantic wiki page. The intention of an insert operation op=* Insert$(PageID, newline, p, n)$ *when generated at site* S, *where newline* $=< nid, c, d, v>$ *and* T *is the set (or multi-set) of RDF statements in the inserted line, is defined as: (1) The content is inserted between the previous and the next lines and (2) the semantic data in the line content are added to* R.

$$\exists i \wedge \exists i_P < i \ LineID(Page[i_P]) = p \tag{1}$$

$$\wedge \ \exists i \leq i_N \ LineID(Page[i_N]) = n \tag{2}$$

$$\wedge \ Page'[i] = newline \tag{3}$$

$$\wedge \ \forall j < i \ Page'[j] = Page[j] \tag{4}$$

$$\wedge \ \forall j \geq i \ Page'[j] = Page[j-1] \tag{5}$$

$$\wedge \ R' \leftarrow R \uplus T \tag{6}$$

Where $Page'$ and R' are the new values of the page and the RDF repository respectively after the application of the insert operation at the server S and $\uplus$ is the union operator of multi-sets. If a statement in T already exists in R so its multiplicity is incremented else it is added to R with multiplicity one.

Definition 7. *Intention of delete operation* *Let S be a P2P semantic wiki server, R is the local RDF repository and Page is a semantic wiki page. The intention of a delete operation op= delete(PageID, ld) where T is the set (or multi-set) of RDF statements in the deleted line, is defined as: (1) the line ld is set to invisible and (2) the number of occurrence of the semantic data embedded in ld is decreased by one, if this occurrence is equal to zero which means these semantic data are no more referenced in the page then they are physically deleted from the R.*

$$\exists i \quad \wedge \; PageID(Page'[i]) = ld \tag{7}$$

$$\wedge \; visibility(Page'[i]) \leftarrow false \tag{8}$$

$$\wedge \; R' \leftarrow R - T \tag{9}$$

Where $Page'$ and R' are the new values of the page and the RDF repository respectively after the application of the delete operation at the server S and $-$ is the difference of multi-sets. If statement(s) in T exists already in R so its multiplicity is decremented and deleted from the repository if it is equal to zero. Let us consider again the scenario of the figure 1. When op_2 is integrated on $peer_1$, the multiplicity of the statement $("France", "locatedIn", "Europe")$ is incremented to 2. When op_3 is integrated on $peer_1$, the multiplicity of the corresponding statement is decreased and the consistency between text and RDF repository is ensured. We can observe that $Peer_1$ and $Peer_3$ now converge and that intentions are preserved.

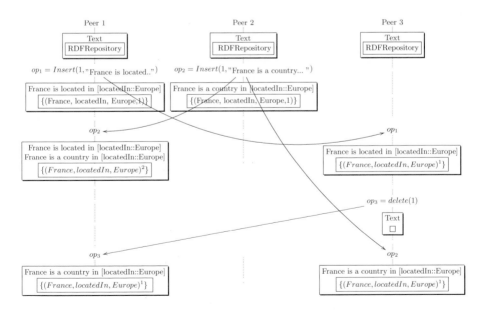

Fig. 2. Convergence after integrating concurrent modifications

6   Algorithms

As any wiki server, a P2P semantic server defines a *Save* operation which describes what happens when a semantic wiki page is saved. In addition, it defines *Receive* and *Integrate* operations. The first describes what happens upon receiving a remote operation and the second integrates the operation locally.

Save operation. During saving a wiki page, a *Diff* algorithm computes the difference between the saved and the previous version of the page and generates a patch. A *patch* is the set of delete and insert operations on the page ($Op = Insert(PageID, line, l_P, l_N)$ or $Op = Delete(PageID, LineID)$). These operations are integrated locally and then broadcasted to other sites in order to be executed as shown below.

```
Upon Save(page, oldPage) :
    let P ← Diff(page, oldPage)
  for each op ∈ P do
    Receive(op)
  endfor
Broadcast(P)
```

At this level of description, we just make the hypothesis that $Broadcast(P)$ will eventually deliver the patch P to all sites. More details are given in section 7.

Delivery Operation. When an operation is received (cf figure 3) its preconditions are checked (cf figure 4). If they are not satisfied, the operation is added to the waiting log of the server, else according to the type of the operations some steps are executed. The waiting log is visited after the integration and the operations that satisfy their preconditions are removed from the log and integrated. The function $ContainsL(PageID, id)$ tests the existence of the line in the page, it returns true if this is the case. The function $isVisible(LineID)$ tests the visibility of the line.

```
Upon Receive(op) :
  if isExecutable(op) then
    if type(op) = insert then
      IntegrateIns (op)
    if type(op) = delete then
      IntegrateDel(op)
  else
      waitingLog ← waitingLog ∪ {op}
  endif
```

```
isExecutable(op) :
  if type(op) = del then
    return
        containsL(PageID,LineID)
        and isVisible(LineID)
  else
        return ContainsL(PageID,l_P)
          and ContainsL(PageID, l_N)
  endif
```

Fig. 3. Receive operation   **Fig. 4.** isExecutable Operation

Integrate operation. The integration of an operation is processed in two steps (cf figure 5): (1) text integration and (2) RDF statements integration. To integrate a text *delete* operation (cf. figure 6), the visibility flag of the line is set to false whatever is its content. To integrate RDF statements (cf figure 7), a counter is used to implement a multi-set RDF repository. A counter is attached to every RDF triple, the value of the counter corresponds to the number of occurrence of the triple in the repository. During the delete operation, the counter of the deleted statements is decreased, if the counter is zero the statements are physically deleted from the repository.

```
IntegrateDel(LineID) :
    IntegrateDelT(LineID)
    IntegrateDelRDF(LineID)
```

Fig. 5. IntegrateDel operation

```
IntegrateDelT(LineID) :
    Page[LineID]. visibility ←false
```

Fig. 6. IntegrateDelT Operation

```
IntegrateDelRDF(LineID) :
let  S ← ExtractRDF(LineID)
if S ≠ ∅ then
for each triple ∈ S do
        triple .counter− −
        if  triple .counter = 0 then
            deleteRDF(R,triple)
        endif
endif
```

Fig. 7. IntegrateDelRDF operation

```
IntegrateIns(PageID, line, l_P , l_N ) :
    IntegratedInsT(PageID, line, l_P , l_N )
    IntegrateInsRDF(line)
```

Fig. 8. IntegrateIns Operation

To integrate an insert operation (cf figure 8) the *line* has to be placed among all the lines between l_P and l_N, some of these lines can be previously deleted or inserted concurrently and the inserted semantic data are integrated. To integrate a line in a wiki page, we use the integration algorithm defined in [3]. This algorithm (cf. figure 9) selects the sub-sequence S' of lines between the previous and the next lines, in case of an empty result, the line is inserted before the next line. Else, the sub-sequence S' is filtered by keeping only lines with the minimum degree of S'. The remaining lines are sorted according to the line identifiers order relation $<_{id}$ [10], therefore, *line* will be integrated in its place according $<_{id}$ among remaining lines, the procedure is called recursively to place *line* among lines with higher degree in S'. To integrate the semantic data (cf figure 10), the RDF statements of the inserted line are extracted and added to the local RDF repository. If the statements exist already in the repository, their counter is incremented, otherwise, they are inserted into the RDF repository with a counter value equals to one as shown below.

```
IntegrateInsT(PageID, line, l_P , l_N) :
let S' ←
subseq(Page[PageID]), l_P, l_N)
if S = ∅ then
    insert(PageID, line, l_N)
else
    let i ← 0
    let d_min ← min(degree(S'))
    let F ← filter(S', degree = d_min)
    while (i < |F| − 1) and (F[i] <_id line)
    do  i ← i +1
        IntegrateInsT(PageID,line,F[i−1],F[i])
endif
```

```
IntegrateInsRDF(line) :
let  S ← ExtractRDF(line)
if S ≠ ∅ then
    for each triple ∈ S do
        if  Contains(triple) then
            triple .counter++
        else
            insertRDF(R,triple)
        endif
endif
```

Fig. 9. Integrate insert text operation   **Fig. 10.** IntegrateInsRDFOperation

To summarize, causality as defined in section 5.1 is ensured by the *Receive* algorithm. Convergence for text is already ensured by the WOOT algorithm [10]. Convergence for semantic data is trivially ensured by the multi-set extension of the RDF repository. The intention preservation for a text is demonstrated in [10]. Here, we are concerned with the intention of semantic data as defined in 5.2. The intention of an *insert* operation is trivially preserved by the algorithm *IntegrateInsRDF*. Since a possible way to implement a *multi-set* is to associate a counter to every element. In the same way, the algorithm *IntegrateDelRDF* preserves the intention of the *delete* operation. The basic idea behind all these algorithms is to reach convergence and preserve intentions whatever is the order of reception of operations. This implies that these algorithms "force" commutativity of operations. If operations are commuting then all concurrent executions are equivalent to a serial one. In our system, users can start a transaction just by switching to the offline mode and end a transaction by switching to online mode. We chose this way to interact with users in order to keep the system simple. If a user produces a consistent change, as all operations of any transaction are commuting and ensure the same effects, then all concurrent execution of transactions generate a correct state.

7 Implementation and Discussion

We have implemented the first peer-to-peer semantic wiki called SWOOKI based based on algorithms presented in section 6. The SWOOKI prototype has been implemented in Java as servlets in a Tomcat Server and demonstrated in [21]. This prototype is available with a GPL license on sourceforge at http://sourceforge.net/projects/wooki and it is also available online at: http://wooki.loria.fr/wooki1

SWOOKI Architecture. A SWOOKI server is composed of the following components (cf. figure 11):

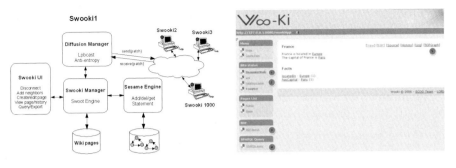

Fig. 11. SWOOKI Architecture   Fig. 12. User Interafce of SWOOKI

User Interface. It is basically a regular wiki editor (cf. figure 12). It allows users to edit a view of a page by getting the page from the SWOOKI manager. Users can disconnect their peer to work in an offline mode (feature1) and they can add new neighbors in their list to work with (feature2). In addition, the UI allows users to see the history of a page, to search for pages having some annotation (feature3), to execute semantic queries (feature4), and to export the semantic annotations of the wiki pages in an RDF format (feature5).

SWOOKI Manager. The SWOOKI manager implements the SWOOKI algorithm. Its main method is `Integrate(Patch)` that calls the `Receive()` algorithm for all operations contained in the patch.

Sesame Engine. We use Sesame 2.0 [22] as RDF repository. Sesame is controlled by the SWOOKI manager for storing and retrieving RDF triples. We used a facility of the Sesame interface to represent RDF triples as multi-set. This component allows also generating dynamic content for wiki pages using queries embedded in the wiki pages. It provides also a feature to export RDF graphs.

Diffusion Manager. In order to ensure the CCI model, we made the hypothesis that all operations eventually reach all sites of the unstructured P2P network. The diffusion manager is in charge to maintain the membership of the unstructured network and to implement a reliable broadcast. Membership and reliable broadcast of operations are ensured by an implementation of the Lpbcast algorithm [23]. This algorithm ensures that all connected sites receive messages and that there is no partition in the P2P network. Disconnected sites cannot be reached, so we added an anti-entropy mechanism based on [24]. The anti-entropy algorithm selects randomly a neighbor in the local table of neighbors and sends a digest of its own received messages. The receiver returns missing messages to the caller. Using the anti-entropy implies that each server keeps received messages in a log, as this log can grow infinitely, the log is purged as detailed in [3].

Discussion. Every SWOOKI server provides all the services of a semantic wiki server. We analyze our system with respect to the following criteria, more detailed analyzing results can be found in [23] and [20].

Availability, fault tolerance and load balancing. Data are available on every peer so they are accessible even when some of the peers are unavailable. The global naming of the wiki pages (concepts) and their associated properties and objects and the respect of the CCI model ensure to have the same data at any node. So if a server is unavailable or slow, it is possible to access to another server.

Performance. We analyze the performance with respect to messages necessary to execute query, propagate modification and synchronize data.

    — *Query execution:* Every server can execute every query locally without generating network traffic for resolving it.

    — *Messages delivery:* As our algorithm generates no traffic for ensuring CCI consistency, the traffic cost for our system is the traffic cost of Lpbcast [23] and the traffic cost of the anti-entropy classical algorithm. As logs of messages can be purged safely, the traffic cost, even for anti-entropy, is bounded. Currently, the volume of change generated by Wikipedia is less than 10Mb by month for all wikipedia in all languages [1]. In 2008, the total size of wikipedia in french was less than 2,5Gb.

    — *Data synchronization:* The complexity of the integration of n operations is $O(n * l^2)$ [20], where l is the number of lines that have been inserted in the wiki page. In fact, as deleted lines are just marked as deleted and there is no garbage collecting algorithm compatible with P2P constraints, the size of the wiki page is growing infinitely. However, traditional wikis such as Wikipedia keeps all changes in log history and never delete it. Consequently, in the context of a wiki, our solution seems acceptable.

Scalability. SWOOKI scales with respect to the number of peers. The number of peers is not a parameter of the complexity in time and space of our algorithm. However, It does not support solution for scalability with the size of data, the storage capacity is limited by the storage capacity of each node. For achieving this scalability, a solution based on partial replication is better. But in this case, offline editing and transactional changes are much more difficult to obtain.

Offline-work and transactional changes. Users can work disconnected if they lack internet connection or if they decide to disconnect directly from the user interface.While disconnected, a user can change many semantic wiki pages in order to produce a consistent change. By this way she generates a transaction. All changes performed in disconnected mode are kept in the diffusion manager component. As our optimistic replication algorithm forces all operations to commute (according to the CCI consistency) then, the concurrent execution of several transactions is always equivalent to a serial one. Thus, a consistent state is produced in all cases.

Cost sharing. The deployment of SWOOKI network is very similar to the deployment of the Usenet P2P network. A trusted peer of any organization can join the network, take a snapshot of replicated data and start answering wiki requests. The proposed architecture can be easily deployed on the Internet

---

[1] http://stats.wikimedia.org/EN/TablesDatabaseEdits.htm

across different organizations. In the contrast to the Wikipedia infrastructure that requires a central site with costly hardware and high bandwidth, the cost of the underlying infrastructure of our system can be shared by many different organizations.

8 Conclusion, Open Issues and Perspectives

Peer-to-peer semantic wikis combines both advantages of semantic wikis and P2P wikis. The fundamental problem is to develop an optimistic replication algorithm that ensures an adequate level of consistency, supports P2P constraints and manages semantic wiki page data type. In this paper, we proposed such an algorithm. By combining P2P wikis and semantic wikis, we are able to deliver a new work mode for people working on ontologies: transactional changes. This work mode is useful to help people to produce consistent changes in semantic wikis. Often, managing a semantic wiki requires to change a set of semantic wiki pages. These changes can take a long time and if intermediate state results are visible, it can be confusing for other users and it can corrupt the result of semantic requests. We believe that this working mode is crucial if we want to use semantic wikis for collaborative ontologies building. However, this approach has many open issues and perspectives:

– Security issues are an important aspect. Replication makes security management more difficult. Often in wikis, security is represented as page attributes. If wiki pages are replicated, it means that security policies are replicated. In this case, it is possible to produce concurrent changes on security policy itself. If we re-centralize security management, we loose the benefits of replication. This problem can be solved by applying the approach proposed in this paper. To manage security policy: define the security policy data type, its operations, the intentions of these operations and update the replication algorithm with these new operations.
– An alternative way for managing security issues is to deploy a P2P semantic wiki on the web of trust. Instead on relying on Lpbcast to build an unstructured P2P network, users can organize themselves in the topology of a network based on trusted relationships. Consequently, the resulting system is a P2P semantic wiki based on a social network that promotes privacy.
– We explored also the combination of an unconstrained semantic wiki with a P2P wiki system based on total replication. We motivated this choice by pointing out to the need of transactional changes. However, even if it is more difficult, it is possible to achieve the same objective with a P2P wiki based on partial replication and consequently take advantage of partial replication benefits such as reduced traffic, infinite storage and cheap join procedure.

References

1. Krötzsch, M., Vrandecic, D., Völkel, M., Haller, H., Studer, R.: Semantic wikipedia. Journal of Web Semantic 5(4), 251–261 (2007)
2. Buffa, M., Gandon, F.L., Ereteo, G., Sander, P., Faron, C.: Sweetwiki: A semantic wiki. Journal of Web Semantic 6(1), 84–97 (2008)

3. Weiss, S., Urso, P., Molli, P.: Wooki: a p2p wiki-based collaborative writing tool. In: Web Information Systems Engineering, Nancy, France. Springer, Heidelberg (2007)
4. Morris, J.: DistriWiki: a distributed peer-to-peer wiki network. In: Proceedings of the 2007 international symposium on Wikis, pp. 69–74 (2007)
5. Du, B., Brewer, E.A.: Dtwiki: a disconnection and intermittency tolerant wiki. In: 17th international conference on World Wide Web, pp. 945–952. ACM, New York (2008)
6. Schaffert, S.: Ikewiki: A semantic wiki for collaborative knowledge management. In: WETICE, pp. 388–396. IEEE Computer Society, Los Alamitos (2006)
7. Git: git based wiki (2008), http://atonie.org/2008/02/git-wiki
8. Patrick Mukherjee, C.L., Schurr, A.: Piki - a peer-to-peer based wiki engine. In: Eighth International Conference on Peer-to-Peer Computing, pp. 185–186. IEEE, Los Alamitos (2008)
9. Sun, C., Jia, X., Zhang, Y., Yang, Y., Chen, D.: Achieving Convergence, Causality Preservation, and Intention Preservation in Real-Time Cooperative Editing Systems. ACM Transactions on Computer-Human Interaction 5(1), 63–108 (1998)
10. Oster, G., Urso, P., Molli, P., Imine, A.: Data Consistency for P2P Collaborative Editing. In: Proceedings of the ACM Conference on Computer-Supported Cooperative Work - CSCW 2006, Banff, Alberta, Canada. ACM Press, New York (2006)
11. Spencer, H., Lawrence, D.: Managing Usenet. O'Reilly, Sebastopol (1988)
12. Nejdl, W., Wolf, B., Qu, C., Decker, S., Sintek, M., Naeve, A., Nilsson, M., Palmér, M., Risch, T.: Edutella: a p2p networking infrastructure based on rdf. In: 11th international conference on World Wide Web, pp. 604–615. ACM, New York (2002)
13. Morbidoni, C., Tummarello, G., Erling, O., Bachmann-Gmür, R.: Rdfsync: efficient remote synchronization of rdf models. In: 6th International Semantic Web Conference and 2nd Asian Semantic Web Conference. Springer, Heidelberg (2007)
14. Cai, M., Frank, M.: Rdfpeers: a scalable distributed rdf repository based on a structured peer-to-peer network. In: 13th international conference on World Wide Web, pp. 650–657. ACM, New York (2004)
15. Chirita, P.-A., Idreos, S., Koubarakis, M., Nejdl, W.: Publish/Subscribe for RDF-based P2P networks. In: Bussler, C.J., Davies, J., Fensel, D., Studer, R. (eds.) ESWS 2004. LNCS, vol. 3053, pp. 182–197. Springer, Heidelberg (2004)
16. Staab, S., Stuckenschmidt, H. (eds.): Semantic Web and Peer-to-peer. Springer, Heidelberg (2005)
17. Petersen, K., Spreitzer, M.J., Terry, D.B., Theimer, M.M., Demers, A.J.: Flexible update propagation for weakly consistent replication. In: Proceedings of the sixteenth ACM symposium on Operating systems principles, pp. 288–301. ACM Press, New York (1997)
18. Johnson, P., Thomas, R.: RFC677: The maintenance of duplicate databases (1976)
19. Cart, M., Ferrie, J.: Asynchronous reconciliation based on operational transformation for P2P collaborative environments. In: International Conference on Collaborative Computing: Networking, Applications and Worksharing, pp. 127–138. IEEE Computer Society, Los Alamitos (2008)
20. Ignat, C.L., Oster, G., Molli, P., et al.: A Comparison of Optimistic Approaches to Collaborative Editing of Wiki Pages. In: Proceedings of the International Conference on Collaborative Computing: Networking, Applications and Worksharing. IEEE Computer Society, Los Alamitos (2007)
21. Rahhal, C., Skaf-Molli, H., Molli, P.: Swooki: A peer-to-peer semantic wiki. In: The 3rd Semantic Wikis workshop, co-located with the 5th Annual European Semantic Web Conference (ESWC), Tenerife, Spain (2008)

22. Broekstra, J., Kampman, A., van Harmelen, F.: Sesame: A generic architecture for storing and querying rdf and rdf schema. In: First International Semantic Web Conference (2002)
23. Eugster, P.T., Guerraoui, R., Handurukande, S.B., Kouznetsov, P., Kermarrec, A.M.: Lightweight Probabilistic Broadcast. ACM Transactions on Computer Systems 21(4), 341–374 (2003)
24. Demers, A., Greene, D., Hauser, C., Irish, W., Larson, J., Shenker, S., Sturgis, H., Swinehart, D., Terry, D.: Epidemic Algorithms for Replicated Database Maintenance. In: Proceedings of the ACM Symposium on Principles of Distributed Computing, Vancouver, British Columbia, Canada, pp. 1–12. ACM Press, New York (1987)

VisiNav: Visual Web Data Search and Navigation

Andreas Harth[*,**]

National University of Ireland, Galway
Digital Enterprise Research Institute

Abstract. Semantic Web technologies facilitate data integration over a large number of sources with decentralised and loose coordination, ideally leading to interlinked datasets which describe objects, their attributes and links to other objects. Such information spaces are amenable to queries that go beyond traditional keyword search over documents. To this end, we present a formal query model comprising six atomic operations over object-structured datasets: keyword search, object navigation, facet selection, path traversal, projection, and sorting. Using these atomic operations, users can incrementally assemble complex queries that yield a set of objects or trees of objects as result. Results can then be either directly displayed or exported to application programs or online services. We report on user experiments carried out during the design phase of the system, and present performance results for a range of queries over 18.5m statements aggregated from 70k sources.

1 Introduction

Keyword search over hypertext documents is an established technology and is used by a large majority of web users [6]. Search engines are popular because i) users are accustomed to the concept of hypertext: documents and links, and ii) search engines employ a simple conceptual model: the engines return those documents that match the specified keywords. Search engines operate over millions of documents which have been collected automatically, however, the functionality is limited: the engine returns only links to web pages but not directly the actual answer or data items sought. Typical keyword phrases used for search are insufficient to specify a complex information need since they consist mostly of only a few words [6]; moreover, information expressed in documents in natural language is ambiguous and thus hard to process automatically. Data formats such as RDF[1] provide more structure, however, there is the open question of how end users should express complex queries over such datasets.

Natural language question answering interfaces are judged preferable to other interfaces by users [11], however, are not in common use today because the approach is fraught with usability issues: despite user training with regards to the

---

[*] Current affiliation: Institute AIFB, University of Karlsruhe (TH), Germany.
[**] This work has been supported by Science Foundation Ireland (SFI/08/CE/I1380).
[1] Resource Description Framework, http://www.w3.org/RDF/

S.S. Bhowmick, J. Küng, and R. Wagner (Eds.): DEXA 2009, LNCS 5690, pp. 214–228, 2009.

capabilities and limitations of a natural language system, users quickly develop negative expectations about the system due to the relatively high error rates in parsing and interpreting natural language [17]. Users are unable to understand the limitations of such systems, that is, to distinguish between conceptual coverage (i.e. does the dataset contain the answer?) and linguistic coverage (i.e. is the system capable of parsing the query?).

A promising approach is to use a menu-based dialogue system in which users incrementally construct the query [17] [19]. Offering only valid choices ensures that the user can only pose queries which can be satisfied by the available data, preventing empty result sets. Designing an interaction model and developing a useable system for interrogating collaboratively-edited datasets raises a number of challenges:

1. Intuitive Use: both occasional users and subject-matter experts should be able to interact with the data immediately. The user interface should be consistent and allow users to quickly derive results with a few clicks.
2. Universality: previous attempts at using structured information have been restricted to manually crafted domain-specific datasets since the data on the web lacked quantity (no general-domain information available) and quality (no shared identifiers, no inter-linkage).
3. Zero Configuration: data on the web comes in an abundance of formats and vocabularies. Consequently, manual intervention is a labour intensive task. In addition, web data is often chaotic and may contain duplicates, erroneous items, malformed syntax and incorrect formatting.
4. Scalability: since we target the web as a data source the system has to scale competently, which has implications on the architecture and implementation of the system.
5. User Satisfaction: the system should be visually appealing and users should be able to import the results of their information seeking task into application programs to get a sense of achievement immediately.

In this paper, we describe VisiNav, a fully implemented system[2] based on a visual query construction paradigm. The users of the system can construct complex queries from several atomic operations. Our system is unique in that it is the first system which offers these features in combination over datasets collected from a large number of web sources. To leverage existing familiarity of users with search engines, the first step in our interaction model is typically a keyword search to locate objects. In subsequent steps, users refine their query based on the navigation primitives; as such, the interaction model leads to an explorable system that can be learned through experimentation. Since the system calculates the possible next steps based on the current state, only legal choices are displayed and thus the user can only compose queries which the system can answer.

Our contributions are as follows:

- We define and formalise a set of atomic query operations on object-orientated data models which can be combined to form complex queries.

---

[2] http://visinav.deri.org/

- We introduce the notion of result trees which extend single-set results to multiple result sets containing result paths.
- We describe the architecture and implementation of a prototype system to investigate the practicality of the interaction model.
- We define the notion of topical subgraphs, subsets of the data which contain both the answer to the query and auxiliary information required to derive prospective choices and render the results.
- We propose a set of indices supporting the atomic operations and a query processing algorithm with top-k processing, and present a performance evaluation of the system on a web dataset with 18.5m statements.

We provide an overview of the user interface and preliminary definitions in Section 2, define and formalise the atomic operations and result trees in Section 3, present architecture and implementation in Section 4 and discuss experiments and evaluation in Section 5. Section 6 covers related work, and Section 7 concludes.

2   Overview and Preliminaries

In the following, we describe the characteristics of the target dataset collected from the web, present example queries, and introduce the conceptual model and the user interface.

2.1   Web Data

Common to data currently found on the web in structured formats (microformats, XML, RDF) is that data publishers take a loosely object-centred view. RDF in particular uses URIs[3] as global identifiers for objects, which, if multiple sources reuse identifiers, leads to an interconnected object space encoded in a graph-structured data format. Currently, reuse of identifiers is particularly common in social networking and social media data, expressed in FOAF[4] for people, SIOC[5] for online community sites, and DC[6] for documents. While a large number of current RDF files use a mix of these vocabularies, data publishers use a plethora of other vocabularies. Our dataset of 18.5m statements from 70k sources contains over 21k different vocabulary URIs.

Given the wide availability of information about people and communities, we use the social network scenario to study user interfaces on collaboratively-edited datasets. However, the interaction model and the implemented system are domain independent. We list a number of example queries – that can be answered with currently available web data – with increasing complexity in Table 1[7]. We

---

[3] Uniform Resource Identifiers, http://www.rfc-editor.org/rfc/rfc3305.txt
[4] Friend-of-a-Friend, http://foaf-project.org/
[5] Semantically Interlinked Online Communities, http://sioc-project.org/
[6] Dublin Core, http://dublincore.org/
[7] timbl:i expands to http://www.w3.org/People/Berners-Lee/card#i; we assume the standard namespace prefixes for foaf, sioc and dc.

Table 1. Example queries. Users typically start with a keyword query ("tim berners-lee") and subsequently select the URI identifying the intended object. Further choices are made from a menu of valid selections.

Query	Description
1	objects matching the keyword phrase "tim berners-lee"
2	information available about `timbl:i`
3	objects `foaf:made` by `timbl:i`
4	`sioc:Posts foaf:made` by `timbl:i`
5	objects that `timbl:i foaf:knows`
6	objects `foaf:made` by objects that `timbl:i foaf:knows`
7	query 6, results sorted by `dc:date`

Query — Home   Log in

      [Tim Berners-Lee]✗ [knows]✗ [made]✗

View — Detail List Table Graph Timeline   previous next Results 1 - 5 of 749

title
name
description
mbox_sha1sum   **Edd Dumbill's Weblog: Behind the Times**
label   http://times.usefulinc.com/
Datatype   subject   channel Document Document   Results
properties   date   Thoughts and comment from Edd Dumbill, technology writer and free
nick   software hacker.
title   Edd Dumbill Edd Dumbill
comment   2008-11-21T14:22:20Z
title

                          Ontology-Based Data Integration For Biomedical Research.
                          http://bibsonomy.org/uri/bibtexkey/conf/swb/KashyapCDSMLSHH07/d...
                          Document Document InCollection 6 more
Object   type   Matthias Samwald Raymond Hookway Iván Herman 7 more
properties   homepage   2007
                          Foaf:knows
                          subject
                          maker   **tt0325710-jhendler**
                          seeAlso   http://www.cs.umd.edu/~hendler/2003/foaf.rdf#tt0325710-jhendler
                          primaryTopic   Review
                          workplaceHomepa... This movie was rated 3 stars on a 4 star scale. That has been rounded to a
                          based_near   7.5 on a 10 point scale.
                          mbox   Jim Hendler
                          items   2006-05-18

                          Dave Beckett's RDF Resource Guide
                          http://planetrdf.com/guide/
                          channel Document Document
                          A comprehensive guide to resources about RDF.
                          Dave Beckett
                          2005-09-23

                          Do You See What I Mean?
                          http://bibsonomy.org/uri/bibtexkey/journals/cga/DukeBDH05/dblp
                          Document Article Document 4 more
                          Iván Herman Ken W. Brodlie Iván Herman 2 more
                          2005

Export
    JSViz Timeline RSS
                                          previous next Results 1 - 5 of 749

Fig. 1. User interface displaying "objects `foaf:made` by objects that `timbl:i foaf:knows`, sorted by `dc:date`" (Query 7). The interface consists of three main sections: i) the current result set in the main content area, ii) the current query in the top part and iii) the prospective choices on the left, divided into datatype properties and object properties to reflect the different operations possible on each.

use these queries instantiated with different names and object URIs to conduct performance tests in Section 5.

2.2   Conceptual Model

Our conceptual model for navigation assumes an object-oriented view, describing objects (U), their attributes and links to other objects. Attributes of objects are expressed using datatype properties (P_D), and links to other objects are specified using object properties (P_O)[8]. Please note that there is no clear distinction between instance-level objects and schema-level ones – classes and properties can be instances themselves.

Users are able to search over the object space yielding objects as a result set. Users can restrict the result set to objects matching specified facets - combinations of properties and objects or datatype values (L). In addition, query operations can be used to navigate in the result set along object properties, yielding another set of objects. The individual object sets form, in combination, a result tree, and previous result sets can be used later in the search process. The current result set can be modified by projecting out datatype properties and sorting the object result set according to datatype properties. Users can choose to display the result set in detail, list, and table view; optionally, a map, timeline, or graph view are available if the result objects contain suitable information for the view. Users are able to export the results to application programs or services. We discuss each operation in detail in Section 3.

The interface in Figure 1 shows results for the query "objects `foaf:made` by objects that `timbl:i foaf:knows`, sorted by `dc:date`" (Query 7). The results displayed have been aggregated and integrated from multiple web sources.

3   Search and Navigation Operations

In the following we introduce our query operations and a grammar describing how to compose complex queries from atomic operations, and present a formalisation of query results using trees.

3.1   Query Operations

– **Keyword Search:** A search session can start with the user specifying keywords to pinpoint objects of interest. The operation leads to an initial set of results based on a broad matching of string literals connected to objects. We perform matching on keywords without manually extending the query for synonyms or other natural language processing techniques. Rather, we leverage the noise in web data, ie. the fact that the same resource might be annotated using different spellings or different languages.

---

[8] As specified in OWL, Web Ontology Language, http://www.w3.org/2004/OWL/

- **Object Navigation:** The object navigation operation is similar to following a hypertext link in a web browser. The user either starts with a set of results or a single result, and clicks on a node to bring it into focus. The operation yields always a result set with a single element.
- **Facet Selection:** Another way of restricting the result set is via selecting facets. A facet is a combination of a property and a literal value or an object (distinguishing between datatype and object properties). Facets are calculated relative to the current result set. Based on derived facets, the user can reformulate the query and obtain increasingly specific result sets.
- **Path Traversal:** Rather than arriving at a single result by performing a object navigation operation, users are also able to navigate along an object property to establish a new set of results. Users can select an object property which allows them to perform a set-based focus change, ie. they follow a certain link, either from a single result or a set of results.
- **Projection:** For views which display individual values of datatype properties (such as the table view), our framework includes a projection operation to select only a number of datatype properties for display.
- **Sorting:** Users are often required to sort the result set according to specific datatype property values. In our model, users can select one or more sorting criteria which can be applied to the current result set.

Users start a query building process via specifying a keyword or a URI of the object to bring into focus. The Extended Backus-Naur Form grammar in Figure 2 describes how the individual operations can be combined (via interactions with the user interface) to form complex queries.

```
<query>            ::= <init> { <refine> | <modify> } ;
<init>             ::= keyword search | object navigation ;
<refine>           ::= <facet> | path traversal ;
<facet>            ::= datatype facet | object facet;
<modify>           ::= project | sort ;
keyword search     ::= specify keyword ;
object navigation  ::= specify object focus ;
datatype facet     ::= restrict result by P_D, L facet ;
object facet       ::= restrict result by P_O, U facet ;
path traversal     ::= traverse path P_O ;
project            ::= add P_D to projection criteria ;
sort               ::= sort results according to P_D ;
```

Fig. 2. EBNF grammar describing queries. Terminals describe end user actions.

3.2   Result Trees

Iterative application of the restriction and navigation operations leads to a set of focus nodes R. Using one result set is sufficient for keyword searches, object navigation (specifying a single-element result set), and faceted browsing (incrementally reducing the size of the result set). The path traversal (or set-based navigation) operation is different: often, users are interested in the objects on the navigation path that led them to the current result set. Thus, the system adds a new result set R_i whenever the user performes a path traversal operation. The result of multiple path traversal operations are result sets $R_0 \ldots R_n$ where n is the number of path traversal operations in a query. Users are able to select result sets $R_0 \ldots R_n$ for display; R_n is the result set displayed as default.

Consider, for example, the query "objects foaf:made by objects that timbl:i foaf:knows" (Query 6). That query yields three result sets R_0, R_1, R_2. We assume that the query was constructed in the following way: the user teleports to Tim Berners-Lee $R_0 =$ timbl:i, from there performs a path traversal along the foaf:knows property $R_1 =$ people that Tim knows, and from there again perform a path traversal along the foaf:made property yielding $R_2 =$ things made by people Tim knows. When inspecting the results, users might be interested not only in the things made by Tim's acquaintances, but also in retaining the connection between the things and the person who made them. To this end, we incorporate the notion of result trees, ie. objects connected to each other based on the path traversal steps performed. Figure 3 shows an example result tree.

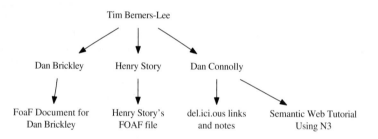

Fig. 3. Partial result tree for query "objects foaf:made by objects that timbl:i foaf:knows" (Query 6). Labels displayed instead of URIs for clarity.

4   Architecture and Implementation

To verify our ideas, we implemented a prototype system as a Java web application. We present first the architecture, describe our indexing and query processing component, explain how we generate a set of prospective choices for the current result set, and finally describe the rendering pipeline.

4.1   System Architecture

The architecture is based on the Model-View-Controller (MVC) paradigm. The Controller, implemented as servlet, receives queries from the users and retrieves the topical subgraph (Section 4.2) from the database – we use top-k processing over specialised index structures described in Section 4.3. The Model classes parse the statements comprising the topical subgraph for the query into Java objects and generate the set of prospective choices (Section 4.4). Finally, in the View, the results are rendered and returned to the web browser (Section 4.5). Optionally the user can request the results in a format suitable for import in external services or applications.

4.2   Topical Subgraphs

We use the notion of a "topical subgraph", which contains all information required to firstly display the results tree and secondly calculate possible next steps for navigation. Figure 4 depicts a topical subgraph. To retrieve the topical subgraph we first calculate the sets of focus nodes: the nodes directly matching the query criteria. Then, we expand the sets by following outgoing links up to a specified limit ϵ. The parameter ϵ denotes how much information in the neighbourhood of the focus nodes should be returned as input to the subsequent processing steps. In our current implementation we use $\epsilon = 2$, however, applications might require larger portions of the graph to operate. To be able to track the sources of a given piece of data, the topical subgraph refers to a directed labeled graph of data with context [4] derived for a particular query.

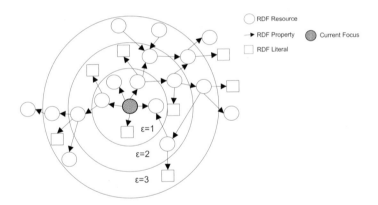

Fig. 4. Topical subgraph for one focus node with $\epsilon = 1, 2, 3$

Please observe that our notion of topical subgraph does not contain in-links to nodes. Given that the notion of directionality of links in semantic graphs is both somewhat arbitrary and difficult to communicate to end users, we just assume out-links from nodes. In case browsing is required both from and to a node, we assume that a property is either specified as symmetric or has its inverse property defined. Upon reasoning [8], a link is inferred in both directions.

4.3   Indexing and Query Processing

The query processing component performs top-k processing of the queries, based on a set of rankings for all identifiers in the system. Top-k processing – which is not sufficiently supported in current RDF query processors – is a crucial feature for the application since the intermediate result sizes become large, leading to performance degradation, and a simple cut-off of unranked identifiers leads to suboptimal results. Ranking is out of scope for this paper, however, one can assume a simple frequency-based ranking where the rank of an identifier depends on how often the identifier occurs.

We devise a set of index structures to match the navigation primitives offered to the user. Conceptually, our index structures match the <key, posting list> structure known from Information Retrieval systems. The current prototype utilises the following indices:

- Statement Index (<subject, poc list>): store a list of predicate/object/ context tuples (poc) per subject. This index is used for topical subgraph lookups where $\epsilon = 1$.
- Path Index (<subject, path list>): store the topical subgraph with $\epsilon = 2$ per subject. This index is used for topical subgraph lookups where $\epsilon = 2$.
- Text Index (<term, subject list>): store a list of subjects per term. This inverted index is used for keyword lookups, intersecting the postings list in case of multiple search terms.
- Facet Index (<po, subject list>): store a list of subjects per facet (predicate/object pair). The index is used for facet restrictions.
- Out-link Index (<sp, object list>): store a list of objects per subject/ predicate (sp) pair. The index is used for the path traversal operation.

The Statement Index in combination with the Path Index is used for topical subgraph lookup. In case there is no Path Index available, or subgraphs with $\epsilon > 2$ are requested, the query processor computes the joins between Statement Index and Path Index in a breadth-first manner.

The query processing is carried out as follows: execute each navigation operation using the respective index, sort the posting lists according to the global ranks, intersect the posting lists (starting with the smaller one), and look up the topical subgraph for the resulting focus nodes. During query processing, each navigation primitive is applied to an index, which returns a set of focus nodes, which are in turn again used as input for the next operation. Lastly, the topical subgraph for the final result set is retrieved, and returned as set of statements. In case of path traversals, the topical subgraphs for the multiple result sets are retrieved, and information to link together the objects on the results path is added. We use the sets of statements abstractions rather than storing objects directly to be able to optionally plug in an RDF store as the back-end, or perform lookups on live RDF sources.

The Model component converts the information in the topical subgraphs to Java objects that other components can conveniently process the data.

4.4   Computing Prospective Choices

The users should be able to refine their queries relative to the current result set (now encoded in Java objects). The system computes prospective choices (possible facets and path traversals) from the current result set. Similarly to result sets, we rank the properties and also rank the values and objects that are part of a facet based on their global rank.

4.5 Result Rendering Pipeline

Having processed and ranked the dataset, the View components prepare the display of information to the user. The system can present the results using different visualisation views, ie. detail view, list view, table view. The table view is similar to a spreadsheet program, where users are allowed to specify projections to show only selected properties of the returned objects. Depending on the types of objects returned, a map view (for geographic coordinates) or a timeline view (for objects with associated date) can be selected. In case users performed path traversal operations, they can optionally select a graph view which renders the result tree in a node-link diagram.

There are three ways of rendering results:

- Results display in the web browser: the web application renders the view in XML; the browser then applies XSL and CSS to finally render the view to the user.
- XML-based data export: the web application renders the view in XML and returns the file to the requester.
- Text-based data export: the web application renders the results in plain text returns the file to the requester.

In addition, the system offers to generate certain files in matching data formats for subsequent processing by the user via software programs. For example, geographic coordinates can be exported to KML[9] or objects with associated dates to iCalendar format (RFC 2445).

Displaying and exporting based on the result types requires export plug-ins to process and convert the objects to the target file format. This is the inverse to data integration systems where wrappers are used to convert the data to a common data format. With Semantic Web data, the objects are already described in the common data format RDF, so export plugins are becoming important. The system currently allows to export objects containing RDF literals of type `xsd:date` in RSS and Timeline[10] formats, `geo:lat` and `geo:long` in KML, and result trees in JSViz[11]. We provide rendering views of these formats in Timeline, Google Maps, and JSViz widgets directly in the user interface.

5 Experiments and Evaluation

We implemented a series of prototypes operating on a number of datasets to validate and refine our design ideas. Our methodology was iterative: once we

---

[9] Keyhole Markup Language, http://www.opengeospatial.org/standards/kml/
[10] http://simile.mit.edu/timeline/
[11] JavaScript graph visualisation, http://www.jsviz.org/

received feedback on a version of the implementation, we incorporated the user feedback into the next prototype. We tested the system on a number of datasets: the Mondial database[12] consisting of information about countries, an RDF version of CrunchBase[13] containing information about technology startups, and an RDF web crawl, seven hops from a seed URI[14], containing information mainly about people. We first provide anecdotal evidence of the utility of our system, and then present measurements evaluating the performance of query processing and view rendering.

5.1   Iterative Design and Continuous Feedback

We initiated the design process using the Mondial dataset and several queries (e.g. "islands in calabria", "gdp of countries bordering italy"). We asked in total ten participants to evaluate early versions of the interface design based on several user tasks and a questionnaire. One session took around 20 minutes; we asked users to interact with the system immediately without a training or introduction phase since visitors to the web site would not receive training either.

The setup was the participants' laptop together with a projector so that the evaluator could track the user actions. We utilised the "thinking aloud" method to gain insight into what the users would expect from the system. The results were mixed: some users picked up quickly the conceptual model behind the user interface and were able to complete all queries, while the majority were able to retrieve the right answers only for about half of the queries.

The suggestions and comments of the first round of evaluations were taken into account for subsequent versions of the interface, and a second round of evaluations were conducted on a new user interface using a different dataset. This time, we used CrunchBase as the dataset, and performed only a few tests to verify the changes made were actually benefitting the users (which the small study confirmed).

Finally, we performed a series of user tests with the current version of the user interface on the web dataset with five participants. All five participants were able to find the correct answers to queries over the web dataset ("find `foaf:Person` X", "find `foaf:Persons` that X `foaf:knows`", "find objects that X `foaf:made`"). While the initial user studies have proved very valuable during the design phase, we plan to conduct larger, more rigorous user studies.

5.2   Performance Evaluation

For the performance evaluation we use the queries from Table 1 as templates; we inserted into the query templates the names and URIs of six selected people for which a sizable amount of information is available. We measure separately the time elapsed in query processing and rendering the view on the server; data transmission time and rendering performance on the client are independent of

---

[12] http://www.dbis.informatik.uni-goettingen.de/Mondial/
[13] http://cb.semsol.org/
[14] http://www.w3.org/People/Berners-Lee/card

our method and thus not covered in the measurements. The measurements were carried out on a machine with a 2.2GHz AMD Opteron CPU, 4GB of main memory and a 160GB hard disk. The servlet container was Tomcat 5.5 in combination with a Sun Java 1.6.0. Figure 5 shows the average performance of query processing and view rendering.

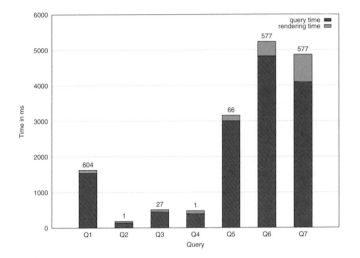

Fig. 5. Average query processing and view rendering performance for queries from Table 1. The numbers above the bars denote the average result sizes for the final layer in the result tree.

The results indicate that the majority of time spent is on query processing; we indeed removed a previous bottleneck due to the use of a method which renders the XML in-memory rather than stream-based. Queries with many path traversal steps (Q5 - Q7) are the most expensive ones due to the join method used. We expect major performance improvements by replacing the current hash join implementation with an index nested loops join algorithm.

6   Related Work

NLMenu [17] is an early system advocating the use of multi-step query construction based on menus. Faceted browsing [19], while less expressive in terms of the complexity of queries, has become popular and is used on e-commerce sites such as Ebay.com. Polaris [16] provides complex query and aggregation operations, however, operates over relational data and thus requires a priori knowledge about the schema used.

A number of systems exist that operate over graph-structured data, which range from quite basic browsing facilities (e.g. Disco[15] allows only object navigation) to systems allowing complex constructs such as negation [14] or nested facets [18]. Table 2 provides a feature-set comparison of related systems.

Table 2. Feature comparison of related systems

System	Keywords	Facets	Paths	Results	Ranking	Configuration	Data sources
Flamenco [19]	x	x	-	set	-	manual	one
Magnet [15]	x	x	-	set	-	schema	one
MuseumFinland [10]	x	x	-	set	-	rules	several
GRQL [1]	-	o	x	set	-	schema	one
/facet [7]	x	x	-	set	group-by	auto	one
BrowseRDF [14]	x	x	-	set	facets	auto	one
ESTER [2]	x	x	-	set	top-k	auto	one
SWSE [5]	x	o	x	set	top-k	auto	web
TcruziKB [13]	x	-	x	set	-	schema	several
Humboldt [12]	x	x	x	sets	-	auto	one
Parallax [9]	x	x	x	sets	-	auto	several
VisiNav	x	x	x	trees	top-k	auto	web

In general, system designs have to balance a trade-off between ease of use and query expressiveness. Our system uses the combined set of query primitives offered by a range of established browsing and navigation systems for graph-structured data, providing evidence that the selection of features in our system represent a consensus in the community. This suggests that a sizeable user community is able to understand the operations.

The systems most closely related to our system in terms of features are GRQL [1], Humboldt [12] and Parallax [9]. GRQL relies on schema information rather than automatically deriving the schema from the data itself, a feature required for web data which does not necessarily adhere to the vocabulary definitions. GRQL lacks keyword search, a useful feature when operating on arbitrary data, since keywords are independent of any schema. Rather than allowing arbitrary facets, GRQL allows to restrict based on the rdf:type predicate. GRQL is, to our knowledge, the earliest system that provides functionality to perform set-based navigation. Parallax [9] is a recent system which exhibits browsing features similar to ours. However, Parallax operates over the Freebase dataset which is manually curated; our system operates over RDF data collected from the web. Parallax lacks ranking, a crucial feature when operating on web data. Our system

---
[15] http://www4.wiwiss.fu-berlin.de/bizer/ng4j/disco/

prioritises facets, navigation axes and results based on global ranks. Although Parallax uses multiple result sets, the connections between the result sets are not propagated to the level of the user interface; our system maintains result paths in the results trees. Finally, we provide a set of export plug-ins which allows to directly load result sets into application programs and online services for display or further processing.

Regarding methodology, our system can be described in terms of the Semantic Hypermedia Design Method [3]. We describe our abstract interface – the information exchange between users and system – in terms of user operations, formalised in EBNF, and result trees. Our concrete interface – the look and feel – is implemented using a multi-layered rendering pipeline spanning server (RDF, queries and XML) and client (XSLT and CSS).

7 Conclusion

Established efforts such as the Linked Open Data[16] already provide large corpora of structured data in various domains, and more efforts are underway[17]. Projects such as FOAF and SIOC provide vocabularies and best practices, enabling both individuals and organisations to publish high-quality data on the web. More structured and interlinked data, in combination with a search and navigation system as presented in this paper, represents an opportunity to bring novel and powerful ways for interacting with data to the web. To this end, we have demonstrated VisiNav, a system based on a formal interaction model that empowers users to search, browse, and navigate a large, domain-independent dataset collectively created by a global user community. Future work includes further improvements of the usability of the system; in particular we would like to enhance the query response times and streamline the user experience based on insights obtained through additional user testing.

References

1. Athanasis, N., Christophides, V., Kotzinos, D.: Generating on the fly queries for the semantic web: The ics-forth graphical rql interface (grql). In: 3rd International Semantic Web Conference, November 2004, pp. 486–501 (2004)
2. Bast, H., Chitea, A., Suchanek, F., Weber, I.: Ester: efficient search on text, entities, and relations. In: 30th ACM SIGIR Conference on Research and Development in Information Retrieval, pp. 671–678 (2007)
3. de Moura, S.S., Schwabe, D.: Interface development for hypermedia applications in the semantic web. In: Joint Conference 10th Brazilian Symposium on Multimedia and the Web & 2nd Latin American Web Congress, pp. 106–113 (2004)
4. Harth, A., Decker, S.: Optimized index structures for querying RDF from the web. In: 3rd Latin American Web Congress, pp. 71–80 (2005)

---

[16] http://linkeddata.org
[17] e.g. http://openflydata.org/ and http://www.w3.org/2001/sw/hcls/

5. Harth, A., Hogan, A., Delbru, R., Umbrich, J., O'Riain, S., Decker, S.: SWSE: Answers before links! In. In: Semantic Web Challenge, 6th International Semantic Web Conference (2007)
6. Henzinger, M.: Search Technologies for the Internet. Science 317(5837), 468–471 (2007)
7. Hildebrand, M., van Ossenbruggen, J., Hardman, L.: Facet: A browser for heterogeneous semantic web repositories. In: 5th International Semantic Web Conference, November 2006, pp. 272–285 (2006)
8. Hogan, A., Harth, A., Polleres, A.: SAOR: Authoritative reasoning for the web. In: 3rd Asian Semantic Web Conference, pp. 76–90 (2008)
9. Huynh, D.F., Karger, D.: Parallax and companion: Set-based browsing for the data web. Technical report
10. Hyvnen, E., Mkel, E., Salminen, M., Valo, A., Viljanen, K., Saarela, S., Junnila, M., Kettula, S.: Museumfinland – finnish museums on the semantic web. Journal of Web Semantics 3(2), 25 (2005)
11. Kaufmann, E., Bernstein, A.: How useful are natural language interfaces to the semantic web for casual end-users? In: 6th International Semantic Web Conference, November 2007, pp. 281–294 (2007)
12. Kobilarov, G., Dickinson, I.: Humboldt: Exploring linked data. In: Linked Data on the Web Workshop (2008)
13. Mendes, P., McKnight, B., Sheth, A., Kissinger, J.: Tcruzikb: Enabling complex queries for genomic data exploration. In: IEEE International Conference on Semantic Computing, August 2008, pp. 432–439 (2008)
14. Oren, E., Delbru, R., Decker, S.: Extending faceted navigation for RDF data. In: 5th International Semantic Web Conference (November 2006)
15. Sinha, V., Karger, D.R.: Magnet: supporting navigation in semistructured data environments. In: ACM SIGMOD International Conference on Management of Data, pp. 97–106 (2005)
16. Stolte, C., Tang, D., Hanrahan, P.: Polaris: A system for query, analysis, and visualization of multidimensional relational databases. IEEE Transactions on Visualization and Computer Graphics 8(1), 52–65 (2002)
17. Thompson, C.W., Ross, K.M., Tennant, H.R., Saenz, R.M.: Building usable menu-based natural language interfaces to databases. In: 9th International Conference on Very Large Data Bases, pp. 43–55 (1983)
18. Tvarozek, M., Bielikova, M.: Adaptive faceted browser for navigation in open information spaces. In: 16th International Conference on World Wide Web, pp. 1311–1312 (2007)
19. Yee, K.-P., Swearingen, K., Li, K., Hearst, M.: Faceted metadata for image search and browsing. In: SIGCHI Conference on Human factors in Computing Systems, pp. 401–408 (2003)

Diagnosing and Measuring Incompatibilities between Pairs of Services*

Ali Aït-Bachir and Marie-Christine Fauvet

LIG, University of Grenoble, France
{Ali.Ait-Bachir,Marie-Christine.Fauvet}@imag.fr

Abstract. This paper presents a technique which detects all behavioural incompatibilities between two service interfaces (a client and a provider). This may happen because the provider has evolved and its interface has been modified. It may also happen because the client decided to change for another provider which addresses the same needs but offers a different interface. Unlike prior work, the proposed solution does not simply check whether two services are incompatible or not, it rather provides detailed diagnosis, including the incompatibilities and for each one the location in the service interfaces where these incompatibilities occur. A measure of similarity between interfaces which considers outputs from the detection algorithm is proposed too.

1   Introduction

A service *interface* is defined as the set of messages the service can receive and send, and the inter-dependencies between these messages. Service interfaces can be seen from at least three perspectives: structural, behavioural and non-functional. The structural interface of a service describes the types of messages that the service produces or consumes and the operations underpinning these message exchanges. In the case of web services, the structural interface of a service can be described for example in WSDL [20]. The behavioural interface refers to the order in which the service produces or consumes messages. This can be described for example using BPEL ([20]) business protocols, or more simply using state machines as discussed in this paper. Finally, the non-functional interface refers to reliability, security and other aspects that are not considered to be part of the functional requirements of a service. The work presented here focuses on behavioural interfaces and is complementary to other work which has studied the problem of structural interface incompatibility [17]. These incompatibilities lead to the situation where the interface provided by a service no longer matches the interfaces that its peers expect from it. This may result in stopping relationships between the service provider and her/his clients. Actually, each time an incompatibility occurs a new client application has to be implemented. Developing such pieces of software is a costly and tedious task.

---

* This work is partially funded by the Web Intelligence Project, Rhône-Alpes French Region.

S.S. Bhowmick, J. Küng, and R. Wagner (Eds.): DEXA 2009, LNCS 5690, pp. 229–243, 2009.
© Springer-Verlag Berlin Heidelberg 2009

Our approach aims at providing a tool which is capable of reporting incompatibilities between two service interfaces. Its main contributions are:

- An algorithm which detects *all* differences that cause two service interfaces not to be compatible from a behavioural viewpoint.
- A measure of similarity between behavioural interfaces of services which is based on the outputs of the detection algorithm. This measure evaluates the degree of similarity between two interfaces.
- A tool which implements the algorithm and the similarity measure and provides business process designers a visual diagnosis, resulting from the incompatibility detection process applied on two interfaces.

The paper is structured as follows. Section 2 frames the problem addressed and introduces a motivating example. In Section 3 we show how we model service interfaces according to their behavioural dimension. Section 4 presents the principle of the proposed approach while Section 5 details the detection algorithm and discusses implementation details and experiments. Section 6 compares the proposal with related ones, and Section 7 concludes and sketches further work.

2   Motivation

As a motivating example, we consider services that handle purchase orders processed either online or offline. In Figure 1 the behavioural interfaces are described using UML activity diagram notation that captures control-flow dependencies between message exchanges (i.e. activities for sending or receiving messages). The figure distinguishes between the *provided* interface that a service exposes, and its *required* interface as it is expected by its clients or peers. Specifically, Figure 1-a shows the provided interface P of a service S. S interacts with a client application C that requires an interface R. We consider the scenario where C wishes to interact with another service S' whose interface is P' while meeting the same needs then S (see Figure 1-b). In this setting, and considering client applications or peers of the service S, the questions that we address are: (i) do the differences between P and P' cause incompatibilities between S' and client(s) of S? and if so, (ii) which differences lead to these incompatibilities? Specifically, we consider three situations: (1) an operation[1] is defined in P while it is not in P', (2) conversely, an operation is defined in P' while it is not in P, (3) an operation is defined in P and changed with another one in P'. We argue that other changes can be described in terms of these ones.

In Figure 1, we observe that the flow which loops from *Receive OfflineOrder* back to itself in P does not appear in P'. In other words, customers of S' are not allowed to alter offline orders. This is a source of incompatibility since clients that rely on interface P may attempt to send messages to alter their offline order while the service S' does not expect a new order after the first one. On the

---

[1] We use the terms *operation* and *message* interchangeably, while noting that strictly speaking, messages are events that initiate or result from operations.

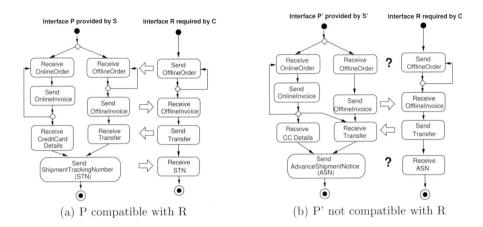

(a) P compatible with R   (b) P' not compatible with R

Fig. 1. Differences between two service interfaces

other hand, message *ShipmentTrackingNumber* (STN in short) has been replaced in P' by message *AdvanceShipmentNotice* (ASN in short). This difference will certainly cause an incompatibility *vis-a-vis* of S's clients and peers. Another difference is that paying by bank transfer is offered in service S' while it is not in service S. However, this difference does not lead to any incompatibilities since S's clients have not been designed to use this option. A difference between P' and P only leads to an incompatibility if it causes P' not to simulate P.

3   Modelling Behavioural Dimension of Service Interfaces

In our approach, the detection of incompatibilities relies on an abstract representation of service interfaces with an emphasis on behavioural aspects. Thus, we consider order dependencies between messages but we do not look into the schema of these messages. Accordingly, we model the behaviour of a web service interface using *Finite State Machines (FSM [5,16])*. Our choice of FSMs is motivated by the following reasons:

- It is arguably the simplest and most widely understood model of system behaviour and it has been used in several previous work in the area of behavioural service interface analysis [6,4,15].
- It is sufficiently powerful to capture most forms of behaviour encountered in service interfaces, including race conditions and interleaved parallelism.
- There exist transformations from other notations for service behaviour modelling to FSMs. In particular several transformations from BPEL to FSMs are implemented in existing tools such as WS-Engineer [9].

Following [5,14], we adopt a simple yet effective approach to model service interface behaviour using *Finite State Machines* (FSMs). In the FSMs we consider, transitions are labelled with messages (to be sent or received). When a message is sent or received, the corresponding transition is fired. Figure 2 depicts FSMs of provided interfaces P and P' of the running example presented in Section 2. The message m has prefix $>$ (respectively $<$) when it is sent (respectively received). Each conversation initiated by a client starts an execution of the corresponding FSM. The figure shows also all differences between P and P'. The latter will be discussed in the next section.

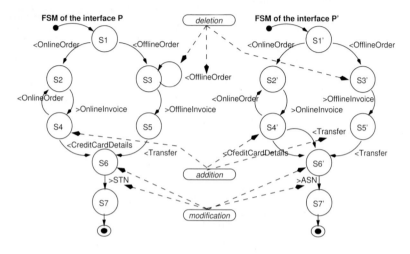

Fig. 2. FSMs modelling P and P'

Definitions and notations: An FSM is a tuple (S, L, T, s_0, F) where: S is a finite set of states, L a set of events (actions), T the transition function ($T : S \times L \longrightarrow S$). s_0 is the initial state such as $s_0 \in S$, and F the set of final states such as $F \subset S$. The transition T associates a source state $s_1 \in S$ and an event $l_1 \in L$ to a target state $s_2 \in S$.

To check whether or not differences between an interface P (of service S, seen as a reference) and another one P' (of service S') lead to incompatibilities, it is necessary to identify situations when P' does not simulate P. Actually, if P' simulates P then each interface R required by the clients of S, which are compatible with P remain compatible with P' (see [2] for a proof).

Assumptions: (1) Even thought web service communication is not always synchronous, we assume synchronous communication as it provides, to a certain extent, a suitable basis for analysing service behaviour. First of all, synchronous communication is more restrictive than asynchronous communication. Therefore, incompatibilities that arise within the asynchronous case arise in the synchronous case as well. Second, for a relatively large class of interfaces, it has been shown that adopting the synchronous communication model leads to the same analysis results than adopting the asynchronous model [10]. (2) We focus on interfaces that expose only externally visible behaviour. In particular, internal actions or timeouts do not appear in the service interface unless they are externalised as messages. (3) We assume messages with the same structure to be semantically equivalent.

4   Detection of Differences

To detect differences between P and P', their respective FSMs are traversed synchronously starting from their respective initial states s_0 and s_0'. The traversal seeks for two states s and s' (belonging respectively to P and P') which are such as the sub-automaton starting from s in P and the one starting from s' in P' are *incompatible* (details are given in Section 5.1). We first discuss and illustrate the conditions that need to be evaluated when P has an operation which does not exist in P' (for the sake of simplicity we call this situation, a deletion, see Section 4.1) and when an operation in P is replaced with another one in P' (this is called a modification, see Section 4.2). We do not detail here the situation when P' has an operation which does not exist in P as it is transposed from the addition mentioned above.

4.1   Deletion of an Operation

Figure 3 depicts two situations where an operation appears in P and not in P'. First in Figure 3-a, we observe that all operations enabled in state $S1'$ are also enabled in state $S1$. Moreover, there is an operation (namely $>R(m)$) enabled in state S that has no match in state $S1'$. Hence we conclude that, considering the pair of states $S1$ and $S1'$, $>R(m)$ is missing in P'. Once this difference has been detected, the pairs of states to be examined next in the process of comparing P and P' are $\langle S2, S2' \rangle$ and $\langle S3, S3' \rangle$: $S2$ in P and $S2'$ in P' are targets of transitions both labelled by the same operation: $>X(m)$. The same remark applies to $S3$ and $S3'$ with the operation $<Z(m)$.

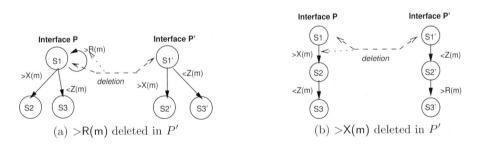

(a) $>R(m)$ deleted in P' (b) $>X(m)$ deleted in P'

Fig. 3. Diagnosis of deletions

In Figure 3-b we note that first, the operation $<Z(m)$ is enabled in $S1'$ and not in $S1$, and second the operation $>X(m)$ is enabled in $S1$ but not in $S1'$. There are two reasons for this mismatch: either operation $>X(m)$ has been modified and has become $<Z(m)$, or $>X(m)$ has been deleted. In this example, we can discard the former possibility because $<Z(m)$ appears downstream in the FSM of P' (it labels an outgoing transition of state $S2$). Hence, $<Z(m)$ can not be considered as a replacement for $>X(m)$. Thus, we conclude that $>X(m)$ has been deleted in P'. Once this difference has been detected, the pair of states to be examined next in the process of comparing P and P' is $\langle S2, S1'\rangle$.

Formally, when comparing two interface FSMs P and P', the fact an operation is defined in P and missing in P' is diagnosed in a pair of states $\langle s, s'\rangle$ (respectively belonging to P and P) if the following condition holds (each part of this condition is explained further down).

$$\|Label(s\bullet) - Label(s'\bullet)\| \geqslant 1 \wedge \|Label(s'\bullet) - Label(s\bullet)\| = 0 \qquad (1)$$

$$\vee \; \exists t \in s\bullet, \exists t' \in s'\bullet : Label(t) \notin Label(s'\bullet) \wedge ExtIn(t', (t\circ)\bullet) \qquad (2)$$

In the previous equations, the notations given below apply (examples refer to Figure 3):

- $s\bullet$ is the set of outgoing transitions of s
  (e.g. $S1\bullet = \{\langle S1, >X(m), S1\rangle, \langle S1, <Z(m), S3\rangle, \langle S1, >R(m), S2\rangle\}$
- $t\circ$ is the target state of the transition t. (e.g. $\langle S1, <Z(m), S2\rangle\circ = S2$).
- $Label(t)$ is the label of t. (e.g. $Label(\langle S1, <Z(m), S2\rangle) = <Z(m)$)
- $\| X \|$: cardinality of X.
- The $\circ$ operator (respectively $\bullet$) is generalised to a set of transitions (respectively states). For example, if $T = \bigcup_{i=1}^{n}\{t_i\}$ then $T\circ = \bigcup_{i=1}^{n}\{t_i\circ\}$; where $n = \| T \|$. Similarly, operator $Label$ is generalised to a set of transitions.

A deletion is detected in state pair (s, s') in two cases. The first one (line 1) is when every outgoing transition of s' can be matched to an outgoing transition of s, but on the other hand, there is an outgoing transition of s that can not be matched to a transition of s'. A second case is when there exists a pair of outgoing transitions t and t' (of states s and s' respectively) such that: (i) transition t can not be matched to any outgoing transition of s'; and (ii) the label of t' occurs somewhere in the FSM rooted at the target state of t (line 2).[2] This second condition is tested in order to determine whether the non-occurrence of t's label among the outgoing transitions of s' should indeed be interpreted as a deletion, as opposed to a modification or an addition. To check if a transition label occurs somewhere in the FSM rooted at the target of a given transition, we use the following recursive Boolean function: $ExtIn(t, T) \equiv T \neq \emptyset \wedge (Label(t) \in Label(T) \vee \bigcup_{i=1}^{\|T\|} ExtIn(t, (T_i\circ)\bullet))$. In other words, $ExtIn(t, T)$ (where t is a transition and T is a set of transitions) evaluates to true if either transition t's label appears among the labels of transitions in T ($Label(t) \in Label(T)$) or, there exists a transition taken in T which has a target state whose set of outgoing transitions (namely $T1$) is such that $ExtIn(t, T1)$ evaluates to true. The way it is defined, this recursive function does not converge

---

[2] By *FSM P rooted at s* we mean FSM P in which the initial state is set to be s. This means that we ignore any state or transition that is not reachable from s.

if the FSM has cycles, but it can be trivially extended to converge by adding an input parameter to store the set of visited states and to ensure that each state is only visited once.

4.2 Modification of an Operation

Figure 4 shows a situation where we can diagnose that operation $>X(m)$ has been replaced by operation $>Y(m)$ (i.e. a modification). The reason is that the operation $>X(m)$ is enabled in $S1$ but not in $S1'$, and conversely $>Y(m)$ is enabled in $S1'$ but not in $S1$. Moreover, the transition labelled $>X(m)$ does not match to any transitions t' in state $S1'$ such that operation $>X(m)$ occurs downstream along the branch starting with t', and symmetrically, $>Y(m)$ does not match any transitions t of state $S1$ such that $>Y(m)$ occurs downstream along the branch starting with t. Thus we can not diagnose that $>X(m)$ has been deleted, nor can we diagnose that $>Y(m)$ has been added.

In this case, the pairing of transition $>X(m)$ with transition $>Y(m)$ is arbitrary. If state $S1'$ had a second outgoing transition labelled $>Z(m)$, we would just as well diagnose that $>X(m)$ has been replaced by $>Z(m)$. Thus, when we diagnose that $>X(m)$ has been replaced by $>Y(m)$, all we capture is that $>X(m)$ has been replaced by another operation, possibly $>Y(m)$. The output produced by the proposed technique should be interpreted in light of this.

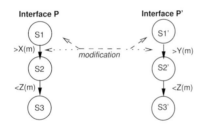

Fig. 4. Diagnosis of a modification/replacement

The state pair to be visited next in the synchronous traversal of P and P' is such that both transitions involved in the modification are traversed simultaneously. In this example, $\langle S2, S2' \rangle$ should be visited next.

Formally, a modification is diagnosed in state pair (s,s') if the following condition holds:

$$\exists t1 \in s\bullet, \exists t1' \in s'\bullet : Label(t1) \notin Label(s'\bullet) \wedge Label(t1') \notin Label(s\bullet)$$
$$\wedge \neg \exists t2 \in s\bullet : ExtIn(t1', (t2\circ)\bullet)) \wedge \neg \exists t2' \in s'\bullet : ExtIn(t1, (t2'\circ)\bullet))$$

5   Implementation Details and Experiments

The detection algorithm presented below (see Section 5.1) is implemented in a tool whose main feature is to detect differences between two behavioural interfaces that cause that the second interface does not simulate the behaviour of the first one[3][1].

5.1 Detection Algorithm

The algorithm implementing the detection illustrated in the previous section is detailed in Figure 5. Given two interface FSMs P and P', the algorithm traverses P and P' synchronously starting from their respective initial states s_0 and s'_0. At each step, the algorithm visits a state pair consisting of one state from each of the two FSMs. Given a state pair, the algorithm determines if an incompatibility exists and if so, it classifies it as an addition, deletion or modification. If an *addition* is detected (e.g. an operation is enabled from s'_0 in P' and not from s_0 in P), the algorithm progresses along the transition of the operation in the interface it has been added. Conversely, if the change is a *deletion* (e.g. an operation is enabled from s_0 in P and not from s'_0 in P'), the algorithm progresses along the transition of the deleted operation in. However, if a *modification* is detected, the algorithm progresses along both FSMs simultaneously. While traversing the two input FSMs, the algorithm accumulates a set of differences represented as tuples of the type Difference defined as below:

> type Difference: < State, Transition, State, Transition >
> { *Let* $\langle s, t, s', t' \rangle$ *be of type Difference: s and s' are states respectively belonging to FSMs P and P' to be compared.* $t = null \Longleftrightarrow t' \neq null \wedge t'$ *is enabled in P' while it is not in P (t' added in P'),* $t' = null \Longleftrightarrow t \neq null \wedge t$ *is enabled in P while it is not in P' (t is deleted),* $t \neq null \wedge t' \neq null \Longleftrightarrow t$ *in P is modified by t' in P'. }*

For instance, the detection algorithm applied on the motivating example (see Figure 2) returns the set of tuples $\{\langle$S2, <OfflineOrder, S2', null$\rangle$, $\langle$S4, null, S4', <Transfer$\rangle$ $\langle$S6, >STN, S6', >ASN$\rangle\}$ which summarises the differences found when comparing P' to P. It is worth noting that comparing P to P' returns $\{\langle$S2',<null, S2, OfflineOrder$\rangle$, $\langle$S4',<Transfer, S4, null$\rangle$ $\langle$S6',>ASN, S6, >STN$\rangle\}$.

The algorithm proceeds as a depth-first algorithm over state pairs of the compared FSMs. Two stacks are maintained: one with the visited state pairs and another with state pairs to be visited (see Figure 5, line 5). These state pairs are such that the first state belongs to the FSM of Pi while the second state belongs to the FSM of Pj. The first state pair to be visited is the one containing the initial states of Pi and Pj (line 6). Once a pair of states is visited it will not be visited again. To ensure this, the algorithm uses the variable visited to memorise the already visited pairs of states (line 10). Labels in common among those of outgoing transitions of si and labels of outgoing transitions of sj are considered as unchanged (no change to detect). Thus, a set of state pairs is built where states are target states of common labels (line 11). Also, the algorithm reports all differences between the outgoing transitions of si and the outgoing transitions of sj (line 12). The two set differences of transitions are put in two variables difPiPj (transitions whose labels belongs to $Label(si\bullet)$ but do not belongs to $Label(sj\bullet)$) an difPjPi (transitions whose labels belong to $Label(sj\bullet)$ but do not belong to

---

[3] See http://mrim.imag.fr/ali.ait-bachir/webServices/webServices.html

Detection (Pi: FSM, Pj: FSM): {Difference}
2 { Detection (Pi,Pj) is the set of differences between Pi and Pj. }
3 setRes: { Difference } { the result }
4 si, sj: State { auxiliary variables }
5 visited, toBeVisited: Stack of type <State, State>
      { pairs of states that have been visited / must be visited }
7 toBeVisited.push($< initState(Pi), initState(Pj) >$)
8 while notEmpty(toBeVisited)
9     $< si, sj > \leftarrow$ toBeVisited.pop()
10   visited.push($< si, sj >$) { $< si, sj >$ is now considered as visited }
11   combEqual $\leftarrow \{(ti, tj) \in si\bullet \times sj\bullet \mid Label(ti) = Label(tj)\}$
      { pairs of matching transitions }
12   difPiPj $\leftarrow \{ti \in si\bullet \mid Label(ti) \notin Label(sj\bullet)\}$
      difPjPi $\leftarrow \{tj \in sj\bullet \mid Label(tj) \notin Label(si\bullet)\}$
13   combPiPj $\leftarrow$ difPiPj $\times$ difPjPi
      { all pairs of si and sj uncorresponding outgoing transitions. }
14   If $\|$difPiPj$\| \geqslant 1$ and $\|$difPjPi$\| = 0$ then   { deletion }
15   For each t in difPiPj do setRes.add($< si, t, sj, null>$)
16   If($(to, sj) \notin$ visited) then toBeVisited.push((to, sj))
17   If $\|$difPjPi$\| \geqslant 1$ and $\|$difPiPj$\| = 0$ then   { addition }
18   For each t in difPjPi do
19   If (polarity(t) = 'send') then setRes.add($< si, null, sj, t>$)
            { otherwise this addition does not lead to incompatibility }
20   If ($(si, to) \notin$ visited) then toBeVisited.push((si, to))
21   For each (ti, tj) in combPiPj do
22   If ExtIn$(ti, (tjo)\bullet)$ then   { addition }
23   setRes.add($< si, null, sj, tj>$)
24   If ($(si, tjo) \notin$ visited) then toBeVisited.push((si, tjo))
25   If ExtIn$(tj, (tio)\bullet$) then   { deletion }
26   setRes.add($< si, ti, sj, null$,'deletion'$>$)
27   If ($(tio, sj) \notin$ visited) then toBeVisited.push((tio, sj))
28   If (($\neg\exists tj' \in sj\bullet : ExtIn(ti, (tj'o)\bullet))$
           $\wedge(\neg\exists ti' \in si\bullet : ExtIn(tj, (ti'o)\bullet)))$ then   { modif. }
29   setRes.add($< si, ti, sj, tj>$)
30   if($(tio, tjo) \notin$ visited) then toBeVisited.push((tio, tjo))
31   For each (ti, tj) in combEqual do
         If ($(tio, tjo) \notin$ visited) then toBeVisited.push((tio, tjo))
32 Return setRes

Fig. 5. Detection algorithm

$Label(si\bullet)$). Line 13 calculates all combinations of transitions whose labels are not in common among $Label(si\bullet)$ and $Label(sj\bullet)$.

Lines 14 to 16 are dedicated to detect a deletion when an outgoing transition of si does not match any transition in $sj\bullet$. The result is returned as set of tuples $< si, t, sj, null >$ where t is one of the outgoing transitions of si whose label does not appear in any of sj's outgoing transitions. As mentioned in Section4.1, when an operation is deleted in Pj FSM the algorithm progresses in Pi FSM, along the branch of the transition which does not exist in Pj, but remains in the same state in Pj FSM.

The detection of an addition is quite similar to the detection of a deletion (lines 17 to 20).

The variable combPiPj contains transition pairs such that the label of the first transition ti belongs to $si\bullet$ but does not belong to $Label(sj\bullet)$ while the label of the second transition tj belongs to $sj\bullet$ but not to $Label(si\bullet)$. For each transition pair satisfying this condition, the algorithm checks the conditions for diagnosing an *addition* (lines 22 to 24), a *deletion* (lines 25 to 27) or a *modification* (lines 28 to 30).

Finally, the algorithm also progresses along pairs of matching transitions, i.e. pairs of transitions with identical labels (line 31). In fact, if no incompatibilities are detected in the current state pair, the algorithm will only progress along pairs of transitions that match one another.

5.2   Complexity of the Detection Algorithm

Let P and P' be two interface FSMs given as input to the detection algorithm, P (respectively P') has n (resp. n') states and m (resp. m') transitions. Also, let w and w' be the number of distinct transition labels appearing in P and P' respectively. We observe that the algorithm performs a depth-first search over the space of state pairs $\langle s, s' \rangle$ such that s is a state of P and s' is a state of P'. The algorithm visits each state pair at most once, therefore one component of the complexity is $O(n * n')$. We then observe that for each visited state pair, the algorithm examines transitions pairs $\langle t, t' \rangle$ such that t is an outgoing transition of s and t' is an outgoing transition of s'. Also, when a transition t in one FSM can not be matched to a transition in the other FSM, we examine t individually. Overall each transition pair $\langle t, t' \rangle$ such that t is a transition of P and t' is a transition of P' is examined at most once. Additionally, each transition t in P and t' in P' is examined at most once individually. Thus another component of the complexity is $O(m * m' + m + m')$. Since the first term dominates the other two, this can be written as $O(m * m')$. Thus, the complexity of the traversal is $O(n * n' + m * m')$.

For each visited pair $\langle t, t' \rangle$ of transitions a condition is evaluated. This condition is based on the transition labels and, in some cases, it also involves a "look-ahead" operation. The purpose of this look-ahead is to find, for a given label, whether or not this label appears in the FSM rooted at either the target of t or the target of t'. This look-ahead can be avoided as follows. In a pre-processing stage, we traverse each of the two FSMs individually using a breadth-first search algorithm. During this traversal, we construct a look-up table that maps each state s to a list of pairs $\langle l, b \rangle$ where l is a transition label and b is a Boolean value indicating whether or not l is the label of a transition reachable from s. For each state s, we calculate the value of b for each label, based on the corresponding values of b for each direct successor of s. This step is linear on the number of labels appearing in the FSM. Thus, the complexity of this pre-processing is $O((n + m) * w)$ for P and $O((n' + m') * w')$ for P'. Since the number of distinct labels in an FSM is bounded by the number of transitions, the complexity of the pre-processing stage is bounded by $O(n * m + (m)^2 + n' * m' + (m')^2)$.

Adding up the complexity of the pre-processing and the detection algorithm, the overall complexity is $O(n * m + (m)^2 + n' * m' + (m')^2 + n * n' + m * m')$. Assuming the number of transitions in an FSM is greater than the number of states (which, modulo one transition, holds because the FSMs are connected graphs), the complexity is bounded by $O((m + m')^2)$. Thus the worst-case complexity is quadratic on the total number of transitions in both FSMs.

5.3   Measure of Similarity

This section presents a measure meant to give a quantitative evaluation of *how much* an interface is different from another one. This measure relies on a function $QS : VStates \rightarrow [0..1]$ where *VStates* is the set of state pairs visited by the detection algorithm ($VStates \subseteq S \times S'$, S being the set of states in P and S' the set of those in P'). Given a pair of states $\langle s, s' \rangle \in VState$, $QS(\langle s, s' \rangle)$ measures incompatibilities detected at $\langle s, s' \rangle$ relatively to the number of transitions in common between s and s'. The formulæ is (see explanations below):

$$QS(\langle s, s' \rangle) = \begin{cases} 1 & \text{if } s\bullet = \emptyset \\ \dfrac{\| LC \| + \sum_{d \in Diff(\langle s,s' \rangle)} Weight(d)}{\| LC \| + \| Diff(\langle s, s' \rangle) \|} & \text{otherwise} \end{cases}$$

$LC = Label(s\bullet) \cap Label(s'\bullet)$ is the set of labels in common in transitions whose sources are s and s'. $Diff(\langle s, s' \rangle)$ is the set of differences pinpointed from the state pair $\langle s, s' \rangle$. The function $Weight : \mathsf{Difference} \rightarrow [0..1[$ is such as $Weight(d)$ is the penalty associated with d. Penalties are arbitrary chosen and depend on whether the difference is an addition, a deletion or a modification.

When s does not have any outgoing transitions, $QS(\langle s, s' \rangle) = 1$. Otherwise, QS tends toward zero as the weight of incompatibilities, evaluated relatively to the global number of transitions in common, rooted at s and s'. For a fixed number of these transitions, more differences are found at $\langle s, s' \rangle$ higher is the dividend and closer to 0 is $QS(\langle s, s' \rangle)$. The divisor, which is meant to keep QS in $[0, 1]$, is never equal to 0: either s has no outgoing transition ($QS(\langle s, s' \rangle) = 1$), or s has at least one outgoing transition and it corresponds to a difference ($\| Diff(\langle s, s' \rangle) \| > 0$) or not ($\| LC \| \geq 1$).

For example, in Figure 2, assuming the penalty for the deletion is set to 0.5, thus: $QS(\langle$ S3, S3'$\rangle) = (1+0.5)/(1+1) = 0.75$ while $QS(\langle$ S1, S1'$\rangle) = (1+0)/(1+0) = 1$

Eventually, to quantitatively compare P and P', we propose to calculate the mean of values returned when applying QS on each pair of states visited by the algorithm. This is done by the function MQS. $MQS(P, P') = 1$ means that P' simulates P.

$$MQS(P, P') = \sum_{p \in VStates} QS(p) / \| VStates \|$$

In the running example, if the penalty values are set to 0.5 then the mean quantitative simulation is: $MQS(P, P') = 0.875$.

5.4   Experimental Results

For validation purposes, we built a test collection of 15 behavioural interfaces derived from the textual description of choreographies expressed in the standard xCBL[4]. The experiment consisted in comparing interfaces to each other.

Table 1 gives a fragment of the results obtained when comparing service interfaces. Each line reports the comparison between the interface seen as a reference and a particular interface given by its id number (see column *Interface*). In the column *MQS* is displayed the value returned when applying the function *MQS* (see above) to the list of differences built by the detection algorithm. The number of items in this list is given in column *Nb diff* while the column *States* (resp. *Transitions*) shows how many states (resp. transitions) where found in the interface to be compared. Each interface has between 3 and 16 transitions. The interface given as a reference has 11 states and 13 transitions.

Table 1. Fragment of experimental results

Interface	MQS	States	Transitions	Nb diff
♯12	1	11	13	0
♯14	0.977	11	13	1
♯13	0.875	10	13	3
♯1	0.43	4	3	11
♯3	0.37	6	6	16
♯5	0.30	8	11	21
♯11	0.233	10	14	19

The interface whose id is ♯11 has 10 states and 14 transitions. It has 19 differences with the interface given as the reference. The value returned by *MQS* is 0.233 which is lower then the one returned when comparing the interface whose id is ♯5. The interface ♯5 has a better score (0.30) then the one which id is ♯11, even thought ♯5 has less differences then ♯11. The interface ♯12 scores 1 and has no difference with the reference, thus it simulates the reference interface.

6   Related Work

The issues tackled in this paper have been partially addressed before, with various points of view. Web service interactions may fail because of interface incompatibilities according to their structural dimension. In this context, reconciling incompatible interactions leads towards transforming message types (using for instance Xpath, XQuery, XSLT). Issues that arise in this context are similar to those widely studied in the data integration area. A mediation-based approach is proposed in [3]. While this approach relies on a mediator (called *virtual supplier*) it focuses on structural dimension of interfaces only.

---
[4] XML Common Business Library (http://www.xcbl.org/).

In [14], authors introduced a technique to diagnosis message structure mismatches between service interfaces and to fix them with adaptors. An extension of this technique is applied to resolve mismatches between service protocols. The proposed iterative algorithm builds a mismatch tree to help developers to choose the suitable adapter each time and incompatibility is detected. However, this technique can only be applied to protocols which describe a sequence of operations. More complex flow controls such as iterative or conditional compositions are not taken into consideration. The solution proposed in this paper does not have this limitation. Another drawback of this approach is that adaptors have no control logic and can not resolve complicated protocol mismatches, such as extra condition, missing condition, or iteration structure, etc.

Compatibility test of interfaces has been widely studied in the context of Web service composition. Most of the approaches which focus on the behavioural dimension of interfaces rely on equivalence and similarity calculus to check, *at design time*, whether or not interfaces described for instance by automata are compatible (see for example [6,11]). The behavioural interface describes the structured activities of a business process. Checking interface compatibility is thus based on bi-similarity algorithms [13]. In [19], authors analyze the compatibility of two services by using the colored Petri net of service interfaces. The idea is to build the reachability graph of boths services and to verify if the graph is well-formed (*i.e.* services are compatible) or not (*i.e.* services are incompatible). These approaches do not deal with pinpointing exact locations of incompatibilities as our proposition does.

Recent research has addressed interface similarity measure issues. In [18], authors present a similarity measure for labelled directed graphs inspired by the simulation and bi-simulation relations on labelled transition systems. The presented algorithm returns a value of a simulation measure but does not give the location of the incompatibilities which have been detected. Its complexity is exponential or factorial to the number of states of the graphs to be compared. According to this theoretical result, our algorithm is more efficient. In [12], the author presents a similarity measure for labelled directed graphs inspired by the simulation and bi-simulation relations on labelled transition systems. The author applies this technique to detect and correct deadlocks. A similar algorithm with the same limitations and complexity has been used in service discovery as introduced in [8]. More specifically, some algorithms for detecting incompatibilities have been proposed, but they focus only on structural aspect of interfaces and do not address their behavioural dimension [7].

In [15], authors propose an operator *match* which is a similarity function comparing two interfaces for finding correspondences between models. This function is the same as the one introduced in [18] which consider the behavioural semantics. The similarity measure is a heuristic which returns a value which calculated according to changes involved by the addition and by the deletion of an operation. However, the result do not pinpoints the exact location of these changes.

In [21], the authors propose an approach to business process matchmaking based on automata extended with logical expressions associated to states. Their algorithm determines if the languages of two automata have a non-empty intersection. This technique for detecting process differences returns a Boolean output. It does not provide detailed diagnosis.

7   Conclusion and Further Study

In this paper we have presented both design and implementation of a tool intended to detect differences (*addition*, *deletion* or *modification* of an operation) that give rise to behavioural incompatibilities between two service interfaces. The main originality of the proposed solution is that the detection algorithm does not stop at the first incompatibility encountered but keeps searching further to identify all incompatibilities leading up to the final state of one of the interfaces to be compared. We have introduced a measure of similarity between interfaces. This measure is meant to be used to select, among a set of services, which one has the closest interface to a given service interface.

Ongoing work aims at extending the proposed solution toward two directions: (i) detecting complex types of incompatibilities (e.g. the order of two operations is swapped or an entire branch is deleted); and (ii) assisting business process designers in determining how to address an incompatibility. Also, communications are currently assumed to be synchronous. Future work will aim at extending the technique to address the asynchronous case. This extension can be achieved by maintaining a buffer of unconsumed messages during the traversal, as it is proposed in [14].

References

1. Aït-Bachir, A., Dumas, M., Fauvet, M.-C.: BESERIAL: Behavioural service interface analyser. In: Dumas, M., Reichert, M., Shan, M.-C. (eds.) BPM 2008. LNCS, vol. 5240, pp. 374–377. Springer, Heidelberg (2008)
2. Ait-Bachir, A., Dumas, M., Fauvet, M.-C.: Detection behavioural incompatibilities between pairs of services. In: Proc. of the 4th WESOA 2008 in conj. with the 6th ICSOC, Sydney, Australia (2008)
3. Altenhofen, M., Boerger, E., Lemcke, J.: An execution semantics for mediation patterns. In: Proc. of the BPM'2005 Workshops: Workshop on Choreography and Orchestration for Business Process Managament, France (2005)
4. Benatallah, B., Casati, F., Grigori, D., Nezhad, H.R.M., Toumani, F.: Developing adapters for web services integration. In: Pastor, Ó., Falcão e Cunha, J. (eds.) CAiSE 2005. LNCS, vol. 3520, pp. 415–429. Springer, Heidelberg (2005)
5. Beyer, D., Chakrabarti, A., Henzinger, T.A.: Web service interfaces. In: Proc. of the 14th WWW int. conf, Japan. ACM Press, New York (2005)
6. Bordeaux, L., Salaün, G., Berardi, D., Mecella, M.: When are two web services compatible? In: Shan, M.-C., Dayal, U., Hsu, M. (eds.) TES 2004. LNCS, vol. 3324, pp. 15–28. Springer, Heidelberg (2005)
7. Champin, P.-A., Solnon, C.: Measuring the similarity of labeled graphs. In: Ashley, K.D., Bridge, D.G. (eds.) ICCBR 2003. LNCS, vol. 2689. Springer, Heidelberg (2003)

8. Corrales, J.C., Grigori, D., Bouzeghoub, M.: BPEL processes matchmaking for service discovery. In: Meersman, R., Tari, Z. (eds.) OTM 2006. LNCS, vol. 4275, pp. 237–254. Springer, Heidelberg (2006)

9. Foster, H., Uchitel, S., Magee, J., Kramer, J.: WS-Engineer: A tool for model-based verification of web service compositions and choreography. In: Proc. of the IEEE Int. Conf. on Software Engineering, China (2006)

10. Fu, X., Bultan, T., Su, J.: Synchronizability of conversations among web services. IEEE Transactions on Software Engineering 31(12) (2005)

11. Haddad, S., Melliti, T., Moreaux, P., Rampacek, S.: Modelling web services interoperability. In: Proc. of the 6th Int. Conf. on Enterprise Information Systems, Portugal, vol. 4. ICEIS Press (2004)

12. Lohmann, N.: Correcting deadlocking service choreographies using a simulation-based graph edit distance. In: Dumas, M., Reichert, M., Shan, M.-C. (eds.) BPM 2008. LNCS, vol. 5240, pp. 132–147. Springer, Heidelberg (2008)

13. Martens, A., Moser, S., Gerhardt, A., Funk, K.: Analyzing compatibility of bpel processes. In: Proc. of the Advanced Int. Conf. on Telecommunications and International Conference on Internet and Web Applications and Services, French Caribbean. IEEE Computer Society Press, Los Alamitos (2006)

14. Motahari-Nezhad, H.-R., Benatallah, B., Martens, A., Curbera, F., Casati, F.: Semi-automated adaptation of service interactions. In: Proc. of the 16th Int. Conf. on World Wide Web, Canada. ACM Press, New York (2007)

15. Nejati, S., Sabetzadeh, M., Chechik, M., Easterbrook, S., Zave, P.: Matching and merging of statecharts specifications. In: Proc. of the 29th Int Conf on Software Engineering, USA. IEEE Computer Society Press, Los Alamitos (2007)

16. Pathak, J., Basu, S., Honavar, V.: Modeling web service composition using symbolic transition systems. In: Proc. of the 21st Conf. on Artificial Intelligence. Workshop on AI-driven Technologies for Service-Oriented Computing, USA (2006)

17. Ponnekanti, S.R., Fox, A.: Interoperability among independently evolving web services. In: Jacobsen, H.-A. (ed.) Middleware 2004. LNCS, vol. 3231, pp. 331–351. Springer, Heidelberg (2004)

18. Sokolsky, O., Kannan, S., Lee, I.: Simulation-based graph similarity. In: Hermanns, H., Palsberg, J. (eds.) TACAS 2006. LNCS, vol. 3920, pp. 426–440. Springer, Heidelberg (2006)

19. Tan, W., Fan, Y., Zhou, M.: A petri net-based method for compatibility analysis and composition of web services in business process execution language. IEEE Transactions on Automation Science and Engineering 6(1) (2009)

20. Weerawarana, S., Curbera, F., Leymann, F., Storey, T., Ferguson, D.: Web Services Platform Architecture. Prentice-Hall, Englewood Cliffs (2005)

21. Wombacher, A., Fankhauser, P., Mahleko, B., Neuhold, E.: Matchmaking for business processes based on choreographies. In: Proc. of the IEEE Int. Conf. on Multimedia and Expo, Taiwan. IEEE, Los Alamitos (2004)

Scaling-Up and Speeding-Up Video Analytics Inside Database Engine

Qiming Chen[1], Meichun Hsu[1], Rui Liu[2], and Weihong Wang[2]

[1] HP Labs, Palo Alto, California, USA
[2] HP Labs, Beijing, China
Hewlett Packard Co.
{qiming.chen,meichun.hsu,liurui,weihong.wang}@hp.com

Abstract. Most conventional video processing platforms treat database merely as a storage engine rather than a computation engine, which causes inefficient data access and massive amount of data movement. Motivated by providing a convergent platform, we push down video processing to the database engine using User Defined Functions (UDFs).

However, the existing UDF technology suffers from two major limitations. First, a UDF cannot take a set of tuples as input or as output, which restricts the modeling capability for complex applications, and the tuple-wise pipelined UDF execution often leads to inefficiency and rules out the potential for enabling data-parallel computation inside the function. Next, the UDFs coded in non-SQL language such as C, either involve hard-to-follow DBMS internal system calls for interacting with the query executor, or sacrifice performance by converting input objects to strings.

To solve the above problems, we realized the notion of Relation Valued Function (RVF) in an industry-scale database engine. With tuple-set input and output, an RVF can have enhanced modeling power, efficiency and in-function data-parallel computation potential. To have RVF execution interact with the query engine efficiently, we introduced the notion of RVF *invocation patterns* and based on that developed *RVF containers* for focused system support.

We have prototyped these mechanisms on the Postgres database engine, and tested their power with Support Vector Machine (SVM) classification and learning, the most widely used analytics model for video understanding. Our experience reveals the value of the proposed approach in multiple dimensions: modeling capability, efficiency, in-function data-parallelism with multi-core CPUs, as well as usability; all these are fundamental to converging data-intensive analytics and data management.

1 Introduction

Video has become an indispensable carrier of information for business perception, decision and actions that enterprises cannot afford to ignore. However, the existent video analysis applications generally fail to scale. A major reason is that almost all the video processing platforms treat database merely as a storage engine rather than a computation engine. As a result, the transfer of massive amount of video data between

S.S. Bhowmick, J. Küng, and R. Wagner (Eds.): DEXA 2009, LNCS 5690, pp. 244–254, 2009.

the storage platform and the computation platform causes serious problems in performance and scalability. With demand for near real-time responses to enable Operational BI (OpBI) in face of bigger and bigger data sets, more and more complex transformation and analysis, the separation of data-intensive transformation/analytics and data management is increasingly recognized as the performance bottleneck.

Motivated by providing a convergent platform, we push down video processing to the database engine for fast data access and reduced data transfer, and rely on User Defined Functions (UDFs) to perform video analysis and search operations which are beyond the relational database operations.

1.1 The Problems

For wrapping computation the current UDF technology has several limitations. One limitation lies in the lack of formal support of relational input and output. Existing SQL systems offer scalar, aggregate and table functions, where a scalar or aggregate function cannot return a set; a table function does return a set but its input is limited to a single-tuple argument. These types of UDFs are not relation-schema aware, unable to model complex applications, and cannot be composed with relational operators in a SQL query. Further, they are typically executed in the tuple-wise pipeline in query processing, which may incur performance penalty for certain applications, and prohibits data-parallel computation inside the function body. Although the notion of relational UDF has been studied by us [3] and others [9], it is not yet realized in any product due to the difficulty in interacting with the query executor.

Next, there exists a dilemma between UDF execution efficiency and ease of coding. Executed inside a DBMS core by the query executor, a UDF (in a non-SQL language such as C) must deal with the system internal data objects with system specific calls, which are hard to follow by users. In some systems such complexity is alleviated by a system utility that converts DBMS internal data into strings to pass to UDFs. This approach, not only is limited to simple UDFs, but also incurs significant overhead, particularly in per-tuple processing. As reported in [10], with SQL Server, no matter how simple a UDF is, it significantly underperforms system functions and expressions. To the other extreme, in some other database systems such as Postgres, UDFs are coded in exactly the same way as system functions. While such UDFs can run efficiently, the UDF developer must have the knowledge of DBMS internal data structures and system calls to deal with argument passing, memory management, etc, which actually keeps UDFs out of reach from most users.

1.2 The Proposed Solutions

To solve the above problems, our solutions start with *supporting Relation-Valued Functions (RVF) at SQL language level.* The output of RVFs can serve as relational data sources in the dataflow of query processing; their abilities to receive multiple relations as input, and to return a relation as output are required by modeling and scaling most video analysis and other analytics applications. We shall use Support Vector Machine (SVM), which is widely used as the classification method in video understanding, to illustrate the power and to quantify the performance gain of using RVFs. We will also use multi-core enabled SVM learning to show the potential of using RVFs for in-function parallel processing.

An RVF is invoked within a query. Like other relational operators in the query, the RVF may be called once for returning an entire tuple-set, or called multiple times, one for each returned tuple. During execution, an RVF interacts with the query executor in several aspects for resolving arguments, caching initial data for multiple calls, managing memory life span, etc. Without any constraint, in a function body, the code for system utilities and the code for user logic may be interleaving, thus it is hard to provide high-level APIs to free the UDF development from DBMS internal details in a deterministic way. We address this challenge in the following way.

- We explicitly define the specific mechanisms for applying RVFs to their input relations as *RVF invocation patterns*. Regulating RVF execution to well-defined patterns ensures well-understood behavior and system interface.
- We develop invocation pattern-oriented *RVF containers* for running RVFs with focused system support.

These solutions have been prototyped on the open-sourced Postgres database engine. We tested our approach with the Support Vector Machine (SVM) based classification and learning, which are widely used in video and image understanding. Our experience reveals the value of the proposed approach in multiple dimensions: modeling capability, efficiency, in-function data-parallelism with multi-core CPUs, as well as usability; all these are fundamental to converging data-intensive analytics and data management.

The rest of this paper is organized as follows: Section 2 discusses the limitation of the conventional UDF approach with in-database SVM classification as an example; Section 3 reviews the notion of RVF and introduces RVF invocation patterns; Section 4 describes RVF container as the query executor extension for supporting RVF invocations; Section 5 illustrates experimental results; Section 6 concludes the paper.

2 UDF Limitation in Supporting Video Analysis

In this section we use a widely adopted analytics application for video understanding, Support Vector Machine (SVM) based image classification, to explain how video analysis is pushed down to the database engine as a UDF; we reveal the limitation of the conventional UDF technology in supporting such applications.

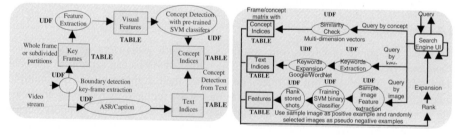

Fig. 1. Video "pattern recognition" process (left) and retrieval process (right)

Let us observe two general processes in video information management where SVM model plays a key role.

- The *pattern recognition process* (Fig. 1 left) that segments video frames, identifies key frames for each segment, extracts (up to 60) visual *features* from each key frame, and recognizes multiple high-level *concepts* from the key frames by applying their features against *a priori* knowledge or pre-trained model, which is typically the SVM model.

  The resulting features and concepts represent patterns of the images, which can be compared with existing patterns and serve as indices of the video frames.

- The *retrieval process* (Fig. 1 right) that returns the meaning of newly captured video streams or a ranked list of N segments or key-frames in answering a user's keyword-, concept- or image based query. In image-based searching, the sample image first goes through a classifier, i.e., SVM in our system, generating a concept vector, which is compared with existent concept vectors using multidimensional similarity (dot product). If the sample image gets annotated by the user in relevance feedback, a new SVM will be trained on the fly, using that image as a positive example, along with 300 randomly selected negative examples.

These processes are analogous to *load* (or *ETL*) and *query* processes in a database, although more complicated.

Below we will show how SVM classification can be expressed in SQL with UDFs. Then we will reveal the performance problem using conventional scalar UDFs, and indicate the need of RVFs. In a later section, we will also use SVM learning as an example to show the potential of using RVF for in-function parallel processing.

2.1 Support Vector Machines

SVM was first introduced in 1992 [1] and is now regarded as an important example of "kernel methods", one of the key area in machine learning. In video processing, SVM is used for mapping the extracted image features to the high-level concepts, the key step for image understanding.

For image understanding, the low-level features of images are binary-classified to high-level concepts based on SVM models. Multiple types of low-level visual features such as color map, color histogram, etc, can be extracted from a key-frame image. A feature is a *vector*, or *multidimensional point*. A SVM "model" contains vectors that can be stored as first class object in a database. Our experience showed

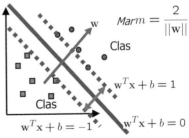

that such integration is a precondition to making video analysis models shared and reused in a rich way with near-real-time response.

For each feature type, a SVM model is pre-trained for labeling a feature instance of this type with a concept with a probability measure. Therefore, given K feature types and M concepts, the classification has two phases:

- In the classification phase, the K features of an image are used as input to the pre-trained SVM models to yield K•M scores, each represents the degree of nearness of a feature of the image to a particular concept;
- In the fusion phase, the nearness measures obtained from multiple features of the same image are aggregated, say, by average, to yield an aggregate score for each concept.

Let us express SVM classification in SQL using UDF. Suppose the input data are stored in tables *Features* and *Models* while the output data are kept in table *Labels*:

Features [featureID, imageID, featureType, feature]
Models [modelID, featureType, concept, model]
Labels [imageID, concept, nearness]

We create Postgres composite UDTs (User Defined Types) for vector and vector arrays as

```
CREATE TYPE FloatVectorType AS (        CREATE  TYPE SVMModelType AS (
    mask BIT VARYING(8),                    floatVectorArray FloatVectorType [],
    floatVector float4 []                   weight float4[],
);                                          vector_num int
                                        );
```

In the *Features* table, a feature is a FloatVectorType object, and a model in the *Models* table is a SVMModelType object. The *weight* array stores the weights **w** of support vectors. Our implementation is based on the well-known LIBSVM, which represents sparse vectors with (index:value) pairs to avoid storing too many 0's. For example, (1:32 2:44 4:69 6:89) is used to represent the vector (32, 44, 0, 69, 0, 89).

The SQL statement for SVM classification expresses two steps:

- For each feature of each image, its nearness score to each concept is computed;
- The resulting nearness measures are aggregated by an average function (or any other appropriate aggregate function) and grouped by image and concept.

2.2 Limitations of the Current UDF Technology

Using a conventional scalar UDF, *classify0,* which returns a nearness score, the above two steps of SVM classification can be expressed in SQL as follows:

[Query 0: Classify using conventional scalar UDF]

```
SELECT imageID, concept, AVG (nearness) FROM
    (SELECT imageID, featureID, concept, classify0 (f.featureType, m.concept, f.feature,
    m.model) AS
    nearness FROM Features f, Models m WHERE f.featrureType = m.featrureType)
GROUP BY imageID, concept;
```

Since the UDF used in this query is a scalar UDF, it is evaluated on a per-feature basis but unable to receive a set of models for the input feature. As a result, the set of relevant models are not cached but retrieved repeatedly for each feature. Such *relation fetch overhead* is caused by the lack of relation input argument for UDFs, and is proportional to the number of feature points, as for each feature, the set of models must be reloaded. Introducing block operations, such as hash join, provides a limited, but far from general, solution to this *kind* of problems.

In addition to the above problem, some applications cannot be modeled without the presence of whole relations (such as minimal spanning tree computation). Further, feeding a UDF only one tuple may not make full use of the power of multi-core or GPU for data-parallel computation inside the function. All these have motivated us to support a more general form of UDFs called RVFs.

3 RVF and Invocation Pattern

We have discussed the notion of RVF in [3]; it was also studied in the OODB context [10]. In fact, a SQL query can be viewed as a limited form of an RVF where the function body consists of standard relational operations. Here, for completeness, we provide a short description of this notion, while motivate and justify our choice of using RVF to support data-intensive analytics inside database engine.

3.1 RVFs as Relational Operators

The conventional scalar, aggregate and table UDFs are unable to express relational transformations and cannot be composed with other relational operators in a query, since neither their inputs nor outputs are relations. In query processing, they are typically processed with tuple-wise input which may incur modeling difficulty or execution inefficiency [3]. In order to overcome these limitations, we introduce RVFs at the *SQL language level*. An RVF is a kind of UDF which takes a list of relations as input and returns a relation as output. For instance, a simple RVF definition can be

```
DEFINE RVF f (R₁, R₂, k) RETURN R₃ {
    Relation R₁ (/*schema*/); Relation R₂ (/*schema*/);
    int k; Relation R₃ (/*schema*/);
    PROCEDURE fn(/*dll name*/);
    RETURN MODE SET_MODE; INVOCATION PATTERN BLOCK
}
```

where the relation schemas R_1, R_2 and R_3 denote the *"schema"* of f, in the sense that the actual relation instances or query results compliant to those schemas can be bound to f as actual parameters. In a SQL statement, a relational argument of an RVF can syntactically be expressed by a relation name, view name or query statement.

An RVF produces a relation as output (although it can have database update effects in the function body) just like a standard relational operator, thus can be naturally composed with other relational operators or sub-queries in a SQL query, such as

```
SELECT * FROM RVF₁(RVF₂(Q₁, Q₂), Q₃);
```

where Q_1, Q_2, Q_3 are views or queries.

Resuming our previous example, SVM classification can make use of a generic RVF **classify1** (*Features*, *Models*), as expressed in Query 1 in Fig 2. It has two input relations and returns a set of <imageID, featureID, concept, nearness> tuples.

[Query 1: RVF with relation input and output]

SELECT imageID,concept,AVG(nearness)
  FROM (SELECT imageID, featureID,
  concept, nearness FROM **classify1**(
   "SELECT * FROM Features",
   "SELECT concept, model, featureType
    FROM Models"))
GROUP BY imageID, concept;

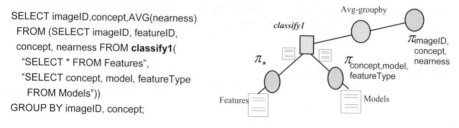

Fig. 2. SVM by RVF with entire relations as input

3.2 RVF Invocation Patterns

In a relational database engine, the argument of a relation operator may be fed in tuple by tuple (e.g. at the probe site of hash-join), or by a set of tuples (e.g. at the build-site of hash-join). If an operator has a tuple-wise input, it is called multiple times w.r.t. that input during execution. The query is thus evaluated tuple by tuple in a pipelined fashion, where a parent operator demands its child operator to return and supply the "next" tuple, and recursively the child operator demands its own child operator to return the "next" tuple, etc., in the top-down demand driven and bottom-up dataflow fashion. How to deal with input/output relation data constitutes the *invocation patterns*.

RVFs and relational operators can be composed in a query; the notion of invocation patterns can be applied to RVFs. An RVF pattern represents a specific mechanism for applying the RVF to its input/output relations. The simplest pattern, *PerTuple,* can be defined such that applying *PerTuple* to RVF *f* with a single input relation R means *f* is to be invoked for every tuple in R (pipelined). Under the *Block* pattern, as shown in Query 1 in Fig. 2, an RVF is called only once in processing a query, with all input relations re-trieved and cached up front. The block pattern underlies "in-RVF data parallel computa-tion"; however, when the input relation is sizable, this invocation mode is inappropriate as the system may run out of memory. In that case, a more complex pattern, *CartProd-Probe* (Cartesion product probe), can be used. Applying this pattern to RVF *f* with 2 in-put relations R_{left} and R_{right}, means that *f* is to be invoked for every combination of tuples in R_{left} and R_{right}, where R_{left} is invoked tuple by tuple (pipelined), and R_{right} is small enough that one can assume that a data structure representing all tuples in R_{right} can reside in memory. In Query 2 shown below (Fig 3), we have specified the invocation pattern of RVF *classify2* as *CartProdProbe*; for each given feature, it returns a set of <featureID, imageID, concept, nearness> tuples.

[Query 2: *CartProdProbe* pattern]

SELECT r.imageID, r.concept, AVG(r.nearness)
  FROM (Features f CROSS APPLY **classify2** (
   f.featureID, f.featureType, f.feature, "SELECT
   concept, model, featureType FROM Models")) r
GROUP BY r.imageID, r.concept;

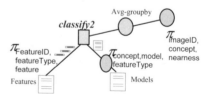

Fig. 3. Features table is fed into RVF tuple by tuple; Models table fed in as a whole

RVF patterns are a generalization of the limited forms of declarations existent to-day on some implementation of user-defined aggregate functions. We have defined richer patterns to provide benefits in optimized data flow. This is because explicitly declaring RVF "*invocation pattern*" can ensure that its interaction with the query executor is defined at a high level, therefore making it possible to provide focused system support and high-level APIs to shield the UDF developers from tedious DBMS system internal details.

4 RVF Container

In order for the RVF to be executable in the query processing environment, certain system support is needed. An *RVF container* is an extension of query executor for offering such support. RVF containers are invocation pattern-specific; each container provides specific facilities for building and running the contained RVFs based on a designated pattern, in argument evaluation, return value wrapping, memory context switching, data conversion, initial data preparation, cross-call data passing, and final cleanup.

When an RVF is defined and registered, its name, arguments (scalars, relations) and return mode: TUPLE_MODE or SET_MODE is stored in a system table.

When the RVF is invoked, several *handle* data structures are provided by sub-classing the corresponding ones in query executor, which can be outlined abstractly below.

– Handle of RVF Execution (*hFE*) keeps track of, at a minimum, the information about actual input/output relation arguments: schema, values (as C array), return mode, result set, etc.
– Handle of RVF Invocation Context (*hFIC*) is used to control the execution of the RVF across calls. hFIC has a pointer to the hFE, and at a minimum keeps track of the information about number of calls, end-of-data status, memory context (e.g. life span over one or multi-calls), etc. A pointer to user-provided context known as *scratchpad* for retaining certain application data between calls, is provided.

During function execution, the RVF container uses several system functions and macros to manipulate the hFE and hFIC structure for performing RVF execution. For instance, in the case of multi-calls, an RVF invocation includes the following steps.

– On the first call (only), initialize the hFIC to persist across calls; evaluate each relation argument expressed by a relation name or a query, by launching a query evaluation sub-process where the argument query is parsed, planned and executed; convert the complex DBMS internal tuple structures to an array of simple data structures to be passed into the "*user-function*"; initialize other arguments and possibly the scratchpad.
– On every function call, including the first, set the environment up for using the hFIC and clearing any previously returned data left over from the previous pass; get non-static input argument values; invoke **user-function** where the input and re-turned relations are array of structures defined in the corresponding header files; convert the data generated by user-function back to DBMS internal data structures, and store them in the result-set pointed to by hFE. If the return mode is

TUPLE_MODE, return the first tuple in the result-set to the caller; otherwise if the return mode is SET_MODE, return the entire result-set.
- At the end, perform clean up and terminate the RVF.

5 Experiments

Our video processing platform is a server cluster running multiple PostgreSQL engines in parallel, with a centralized planner for controlling the communication of these engines. The server is HP ProLiant DL360 G4 with 2 x 2.73 Ghz CPUs and 7.74 GB RAM, running Linux 2.6.18-92.1.13.el5 (x86_64). In the given SVM computation, the relation *Features* is hash partitioned by imageID over multiple nodes, and the relation *Models* is replicated on every node. With this arrangement, each node has sufficient information for its local SVM classification computation in the share-nothing paradigm. In SVM learning, however, the local results need to be aggregated. For the computations that require the communication between the root node planner and the regular nodes, we carefully design the algorithm based on sufficient statistics.

5.1 Performance Gain in SVM Classification by Using RVF

We support parallel classification. The dll code and registration information of an RVF are made available, and the RVF container capabilities are supported on each node. Since the SVM classification can be made self-contained in every node, in this experiment we measure the per-node performance. The performance comparison between using RVF (Query 1) and using scalar UDF (Query 0) for SVM classification, demonstrates scalability and better performance of the proposed RVF approach. Both queries calculate the nearness score between each image and each concept. We ran the two queries with different data load (number of feature vectors) and 39 concepts. The performance comparison is shown in Fig 4.

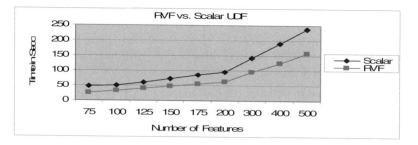

Fig. 4. Using RVF in SVM query over-performs that using conventional scalar UDF

Both queries scale linearly, while our RVF based SVM classification outperforms the query based on conventional UDFs by 35-40% The inability of conventional UDFs of receiving whole relations forces the engine to perform multi-scan joins and fetch models repeatedly wrt each feature. With RVF, in contrast, the above overhead is avoided and hence the performance gain.

5.2 Support In-RVF Data-Parallel SVM Learning

The use of RVF allows sets of tuples to be manipulated in the user function, which potentially supports *in-function data parallelism*. To illustrate this advantage, we implemented parallel *SVM learning* inside an RVF using multi-core CPUs. The learning procedure is outlined as follows.

- A training set of key-frame images are provided where each image is labeled by its nearness scores to a given concept. For example, an image may be labeled as 100% for "out-door", 70% for "sports", 50% for "boys", and 40% for "school". These data are stored in the table *TrainLabels [imageID, concept, nearness]*.
- The labels of an image apply to all the features of that image, stored in *TrainFeatures [imageID, featureType, feature]*.
- A join of these two relations on imageID forms the base data with schema [featureType, feature, concept, nearness] for SVM model learning.
- Models are stored in **Models** [*modelID, featureType, concept, model*]. A model is unique with respect to a featureType and a concept.

For a given *'feature_type'* and *'concept_name'* the SVM learning process is expressed as the following query:

```
INSERT INTO Models
SELECT modelID + 1, 'feature_type', 'concept_name', svm_learning (
    "SELECT feature, nearness FROM TrainFeatures f, TrainLables l WHERE
    l.imageID = f.imageID AND l.concept = 'concept_name' AND f.featureType = 'feature_type'")
FROM Models WHERE modelID = (SELECT max(modelID) from Models);
```

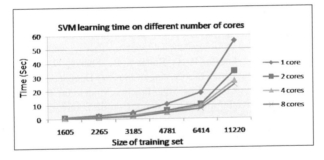

Fig. 5. SVM learning speed up in multi-core RVF

With data structure conversion handled by the RVF container, the developer only needs to care about the learning algorithm itself, *i.e.,* the user function *svm_learning()*. The returned support vectors are automatically mapped into VectorArrayType defined in section 2.1. Feeding sets of tuples into the RVF allows in-function data parallel computation for gaining high performance. In this experiment, the training procedure is parallelized on multi-core CPUs using the Intel Threading Building Blocks C++ template library. Each model was obtained in iterations similar to cascade SVM [9]. Fig. 5 shows the execution times of the above query on a dual quad-core CPU workstation, when the number of cores simultaneously evaluating *svm_learning()* increased from 1 to 8. Our experiments show that performance scales well from 1 core to 2 cores; however 4 cores and 8 cores show diminishing returns. This is consistent with the intrinsic parallelism available in the SVM learning algorithm employed in this experiment, and confirms our RVF's ability to utilize parallelism for in-database computation.

6 Conclusions

Embedding data-intensive analytics in the database layer for fast data access and reduced data transfer is an active research field [2-7]. This work aims to build a video analysis system inside a database engine for achieving a powerful combination. Since query processing engines are primarily used for relational query evaluation, the reach of more general applications relies on UDFs. In this research we tackled two major limitations found in the existing UDF technology: lack of set-oriented input or output which makes application modeling difficult, causes the inefficiency of execution, and mingling of system code and application logic. With RVFs, a language level extension, we gain improvement in application modeling and efficiency in the use of cache and computation parallelism. With RVF container and its associated APIs, analytics logic is well separated from system administration and programming efforts. Prototyped on the Postgres database engine, our experience reveals the benefits of the proposed approaches in enhancing UDF's modeling power for executing complex analytics applications, in significant performance gain, and in ease of user function development, which makes the convergence of video analytics and database engine a reality.

References

1. Boser, B.E., et al.: A Training Algorithm for Optimal Margin Classifiers. In: Proceedings of the Fifth Annual Workshop on Computational Learning Theory, vol. 5, pp. 144–152 (1992)
2. Chaiken, R., Jenkins, B., Larson, P.-Å., Ramsey, B., Shakib, D., Weaver, S., Zhou, J.: SCOPE: Easy and Efficient Parallel Processing of Massive Data Sets. In: VLDB 2008 (2008)
3. Chen, Q., Hsu, M.: Data-Continuous SQL Process Model. In: Proc. 16th International Conference on Cooperative Information Systems, CoopIS 2008 (2008)
4. Chen, Q., Hsu, M.: Inter-Enterprise Collaborative Business Process Management. In: Proc. of 17th Int'l Conf on Data Engineering (ICDE 2001), Germany (2001)
5. Dayal, U., Hsu, M., Ladin, R.: A Transaction Model for Long-Running Activities. In: VLDB 1991 (1991)
6. Dean, J.: Experiences with MapReduce, an abstraction for large-scale computation. In: Int. Conf. on Parallel Architecture and Compilation Techniques. ACM, New York (2006)
7. DeWitt, D.J., Paulson, E., Robinson, E., Naughton, J., Royalty, J., Shankar, S., Krioukov, A.: Clustera: An Integrated Computation And Data Management System. In: VLDB 2008 (2008)
8. Graf, H.P., Cosatto, E., Bottou, L., Durdanovic, I., Vapnik, V.: Parallel Support Vector Machines: The Cascade SVM. In: NIPS 2004 (2004)
9. Jaedicke, M., Mitschang, B.: User-Defined Table Operators: Enhancing Extensibility of ORDBMS. In: VLDB 1999 (1999)
10. Novick, A.: Drilling Down into Performance Problem. Transact-SQL User-Defined Functions, ch. 11, pp. 235–244. Wordware Publishing (2004) ISBN 1-55622

Experimental Evaluation of Processing Time for the Synchronization of XML-Based Business Objects

Michael Ameling[1], Bernhard Wolf[1], Thomas Springer[2], and Alexander Schill[2]

[1] SAP Research CEC Dresden
[2] Technische Universität Dresden
{michael.ameling,b.wolf}@sap.com,
{thomas.springer,alexander.schill}@tu-dresden.de

Abstract. Business objects (BOs) are data containers for complex data structures used in business applications such as Supply Chain Management and Customer Relationship Management. Due to the replication of application logic, multiple copies of BOs are created which have to be synchronized and updated. This is a complex and time consuming task because BOs rigorously vary in their structure according to the distribution, number and size of elements. Since BOs are internally represented as XML documents, the parsing of XML is one major cost factor which has to be considered for minimizing the processing time during synchronization. The prediction of the parsing time for BOs is an significant property for the selection of an efficient synchronization mechanism. In this paper, we present a method to evaluate the influence of the structure of BOs on their parsing time. The results of our experimental evaluation incorporating four different XML parsers examine the dependencies between the distribution of elements and the parsing time. Finally, a general cost model will be validated and simplified according to the results of the experimental setup.

1 Introduction

Business applications such as Supply Chain Management and Customer Relationship Management are placed on business application servers. The business application data is organized as business objects (BOs) such as a `Sales Order` which represent object data containers specified for business processes. Within multi-tier architectures BOs are stored at the data tier and are cached at the middle tier for efficient processing. Due to the replication of application logic in distributed environments, multiple copies of BOs (called replica) are created to achieve fast local access, scalability and availability. Thus, concurrent changes of BO data might occur. Once a BO is modified the changes have to be forwarded to all other replicas. Changes have to be identified, forwarded to and integrated in other replicas. However, the processing of BOs for change detection and integration is a complex and time consuming task because BOs not only differ in their content. They rigorously vary in their structure according to the distribution, number and size of elements caused by their definition and usage. Since BOs are internally represented as XML documents, the parsing of XML is one major cost factor during the synchronization [1]. Therefore, the prediction of the parsing time for BOs is one significant property for the selection of an efficient synchronization mechanism. In [2]

S.S. Bhowmick, J. Küng, and R. Wagner (Eds.): DEXA 2009, LNCS 5690, pp. 255–262, 2009.

we introduced a cost model for an efficient replication of BOs. The parameters for a system model were defined to express the processing time for all elements of BOs. Finally, simplifications based on an experimental evaluation for the system model were done for one parser implementation.

In this paper, we present a method to evaluate the influence of the structure of XML-based BOs on their processing time. We adopt our evaluation method for an experimental evaluation incorporating four XML parsers to examine the dependencies between the distribution of elements in XML-based BOs and their parsing time. Finally, the cost model provided in [2] can be validated and simplified according to the results of the experimental setup in this paper. In the following, we discuss the general structure of XML documents and particularly XML-based BOs, the relation of the structure, and the resulting parsing time (Section 2). Next we describe the setup and results of our experimental evaluation (Section 3). Moreover, the results of our measurements are discussed (Section 4). After that related work is discussed in Section 5. We end the paper with a summary and an outlook to future work.

2   Processing of Business Objects

2.1   Structure of Business Objects

A BO is a data container for complex data structures of the business world. A BO instance belongs always to one BO type. The BO types can be compared to a class in object oriented programming. The BO instances are the real BOs and are filled with content. XML-based BOs consist of *nodes, node values, attributes, attribute values* and *links*. The structure of a BO is defined as a tree-structure and describes the amount and size of elements as well as how elements are distributed within the tree structure. The distribution of elements is described by the location of elements at certain levels and positions.

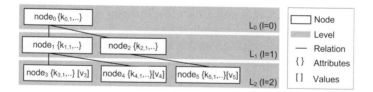

Fig. 1. BO Structure

In Figure 1 an abstract BO structure is depicted. A node, denoted as $node_n$, has the index n. The mandatory *root node* has the index $n = 0$. Each node can have none, one or multiple *subnodes* which are also called child nodes. In the abstract example, the nodes $node_1, node_2$ are subnodes of the root node $node_0$. Nodes that do not have a subnode are *leaf nodes*. Leaf nodes can have a *node value*. Since BOs can link to other BOs a leaf node can have a *link* as a node value. The node values v_n and links f_n have the index of the node. Moreover, nodes may have attributes. An *attribute* $(k_{n,\lambda})$

is indexed by the index of the node n and the position λ within the list of attributes of one particular node. Each attribute has an *attribute value*. The size of the attribute value is $w_{n,\lambda}$. The connection between a node (parent node) and subnode (child node) is defined as *relation*. A *level* describes the distance (number of relations) between a node and the root node. The variable l describes the level a particular element is located in the tree-structure of the BO. The root node is placed at level $l = 0$.

2.2   Processing of Business Objects

During the synchronization services which returns BO data in a XML document in a serialized form are used. Thus, the time for processing the BO is determined by the time for accessing the document elements, i.e. parsing of and navigation through the document, and the time for performing the operation. Since the goal of our work is to examine the influence of the BO structure on the processing time, we focus in the follOwing experiments on the time for accessing all elements. Thus, the term processing time refers to the access time for the elements of a BO. We assume that the processing time for each element type is determinable. It can be distinguished in: a - the processing time for a node, b - the processing time for an attribute, c - the processing time for an attribute value, d - the processing time for a (node) value, and e - the processing time for a link.

The influence of the distribution of elements is has to be defined in dependency of the position. The indexes l for the level of an element, ν for the position of an element within a list of elements at a certain level, n for the node the element is placed and λ for the position of an element at a node are used for the processing time of elements as follows: $a_{l,\nu}$, $b_{n,\lambda}$, $c_{n,\lambda}$, d_n and e_n. As proposed in [2] the dependency of processing time on the structure varies between parser implementations. However, for some parser implementations we assume that the system model can be simplified. For example, the processing time for a node $a_{l,\nu}$ at level l and position ν can be the same for all positions: $a_{l,\nu} = a_l$. The dependency of the processing time can be evaluated experimentally. A detailed description of a suitable experiment setup to determine the processing time for all element types on their position follows.

3   Experimental Evaluation

3.1   Description of Experiments

The influence of the distribution across levels and number of elements will be shown by moving or adding elements levelwise. Additionally, the size of node values and attribute values is modified. Within one experiment class only one type of element is adjusted. For each experiment a group of documents is created. Within one group of documents only one property of an element is adjusted. Elements are adjusted by adding elements, moving elements or changing the size of an element. The experiments are labeled in capital letters. Each of the following created XML documents for a BO consists of a header with a fixed size and a body. The XML header is equal for all XML documents and is not considered as BO data. The body of the document contains the BO with the BO name as root node. The processing of the header is defined as the *offset*.

Node Experiments. The following class of experiments deals with the modification of nodes. (Figure 2). The number of node and the position of nodes is changed separately. The node experiments show the influence of additional nodes and the position of nodes on the processing time. For each experiment a group of 1001 documents was created.

Experiment A_{1000} shows the influence of additional nodes at the same level on the processing time. In this experiment 1000 child nodes are added stepwise to the root node. Finally, 1000 nodes exist at the first level (L_1). Figure 2a depicts the increase of nodes. Finally, 1001 XML documents are created where the first document just contains the root node. The last document contains the root node having 1000 equal child nodes.

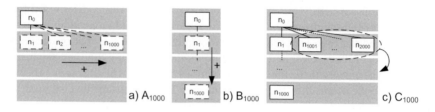

Fig. 2. Node Experiments

Experiment B_{1000} shows the influence of the distribution of nodes across levels on the processing time. The number of nodes is increased stepwise but each additional node is child node of the last added node (Figure 2b). This way, the number of level increases proportionally with each additional node. Finally, the last document within this group of documents has 1000 additional nodes distributed across 1000 levels ($L_1 - L_{1000}$; root at L_0). Experiment C_{1000} shows the influence of the position of nodes at a certain level. The number of nodes is constant for this group of documents. There always exist 1000 child nodes distributed over 1000 levels. Additionally there exist a group of 1000 nodes as child nodes of the root node at the first level. The group of 1000 nodes is moved stepwise one level down through the levels (Figure 2b). Finally there exist 1001 nodes at level 1000 (L_{1000}).

Attribute and Node Value Experiments. For attribute experiments and node value experiments we created similar documents where we stepwise added attributes or node values to one child node (D_{1000}, D_1) or to many child nodes a the same level (E_{1000}, K_{1000}) or to many child nodes on different levels (F_{1000}, M_{1000}). Experiment G_1 shows the influence of the attribute size on the processing time. The attribute value is increased stepwise for one child node by $1kB$ up starting with $0kB$ and ending with $1000kB$ attribute size.

3.2 Experimental Setup

The system used for measurements has a 1.6GHz CPU and 2GB RAM. We did the same experiments on an Itanium server. Since we did not focus on absolute values we got the same results. For processing of the XML documents the parser implementations of JDOM [3], DOM4J [4], SAX [5] and XPP [6] were used. The measured process time includes the loading of the document and the serializing (marshalling) of the whole XML document. Each experiment has been repeated 100 times. The results represent the minimum values of the processing times measured in nanoseconds. The average values are not used since system processes can increase the measured values significantly.

3.3 Measurements

Node Experiments. The measurements of experiment A_{1000} (Figure 3) depict a linear increase of the processing time for nodes added at the first level with all four parser implementations. The slopes of the lines of discrete points enable the determination of the processing time for one node. The intersection with the ordinate is indicates the processing time for the offset. SAX performs the best for nodes since the slope is the lowest. However, the time for processing the offset with SAX is higher than with XPP. The slope for JDOM and DOM4J is equal. The measurements of experiment B_{1000} (Figure 3) depict a linear increase of the processing time for added nodes at sublevels $(L_1 - L_{1000})$ with DOM4J, SAX and XPP. The processing time increases quadratically with JDOM. The intersections with the ordinate are the same as in experiment A_{1000} since the first documents of both groups are equal. The experiment shows that with JDOM the processing time for a node increases when the node is placed at a higher level. The level has no influence for DOM4J, SAX and XPP. The measurements of experiment C_{1000} (Figure 4) also reflect the results of B_{1000}.

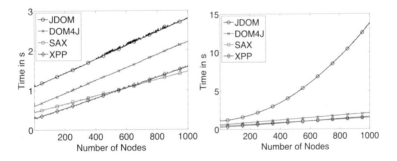

Fig. 3. Measurements Experiment A_{1000} and B_{1000}

Attribute and Node Value Experiments. The measurements of experiment D_{1000} (Figure 4) depict a linear increase of the processing time for additional attributes for SAX and DOM4J. The processing time of attributes increases quadratically with the position λ using JDOM or XPP. The experiment also shows that the position λ at a node has no influence with SAX and DOM4J. The measurements of experiment E_{1000} and F_{1000} depict a linear increase of the processing time for additional attributes at the first level (L_1) or rather across levels with all four implementations. The measurements of experiment G_1 and H_1 depict the linear increase of the processing time for increasing attribute and node vales with all four implementations. The stepwise extension of the Java heap size causes the jumps for the processing time with JDOM at certain points.

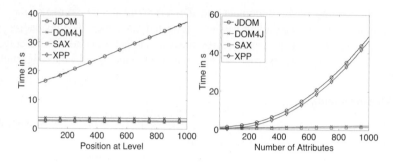

Fig. 4. Measurements Experiment C_{1000} and D_{1000}

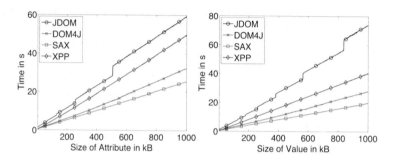

Fig. 5. Measurements Experiment G_1 and H_1

The measurements of experiments K_{1000} and M_{1000} showed for all implementations a linear increase of the processing time for adding node values at the same level or across levels, respectively.

4   Discussion

The results of the experimental evaluation allow simplifying the system model for the processing time of elements. Table 1 summarizes the simplification of the processing time parameters for elements which can be done based on the experimental evaluation for each of the four parser implementations. The experiments show the influence of the structure of BOs on the processing time of elements. In dependency of the used implementation the position of an element has influence on its processing time. Especially the level where a node is placed and the position where an attribute is placed within a node can have crucial impact. We showed an experimental evaluation how to find the dependencies for four parser implementations. The procedure can be also used with any other implementation e.g.,: Xerces [7] and Electric XML as well. Once the influence of the structure of BOs on the processing time of elements is identified the introduced experiment setup allows exactly determining the processing time for all type of elements. Since the processing time for elements depends on the running machine they have to be determined for each system.

Table 1. Implementation Dependent Simplified System Model

RESULT	JDOM	DOM4J	SAX	XPP	DESCRIPTION
$a_{l,\nu} = a_l$	X	X	X	X	The processing time for a node without any attributes is same for all nodes ν at level l
$a_l = a$	-	X	X	X	The processing time for a node without any attributes is same for all nodes at all levels l
$b_{l,\lambda} = b_\lambda$	X	X	X	X	The processing time for an attr. with fixed size at position λ is the same for nodes at all levels l
$b_{\nu,\lambda} = b_\lambda$	X	X	X	X	The processing time for an attr. with fixed size at position λ is as same as for nodes at all pos. ν
$b_\lambda = b$	-	X	X	-	The processing time for an attribute with fixed size is the same at all positions λ
$c_{n,\lambda} = c$	X	X	X	X	The processing time for the attribute value is the same for all nodes n at all levels l
$d_n = d$	X	X	X	X	The processing time for node values is the same at all nodes n

To determine the processing time for a node the created documents can be used. Firstly, the offset can be measured using the first document from the group in experiment A_{1000}. Secondly, in dependency of the implementation the parameter a or the parameter a_l with influence of the level position of the node has to be determined. The parameter a_l equals a linear function $f(l) = m*l+r$. The constant value r presents the processing time for the node at level 0. The slope m presents the increase of the node processing time for one level. The slope m can be determined with experiment C_{1000}. The slope of the curve of discrete points divided by the number of moved nodes (1000) is m. For parser implementations where the level has no influence on the processing time of nodes the value m is zero. The parameter r can be determined with experiment A_{1000}. The discrete point for one node minus the offset equals r. The processing time for the other elements can be determined equally. To achieve a proper accuracy several documents should be measured for parameter determination. Finally, the knowledge of the processing time for elements for a system allows predicting the time for processing a BO. The amount of elements and the positions have to be known which can be done by profiling BOs [1]. A full model of all parameters for a BO profile is described in [2].

5   Related Work

A cost model for replication BOs at the application layer was introduced in [2]. The adaptive synchronization approach is described in [1]. Other approaches implementing a replication at the middle-tier are [8] (Middle-R), [9] (Ganymed) and [10,11] (CORBA). However, they do not provide a cost model for the synchronization process of BOs and focus on strategies such as snapshot isolation [12]. An overview of middle-ware based data replication can be found in [13]. To improve the performance of data transfer through the network compression can be used to decrease the size of messages. However, XML compression such as binary XML [14] are not focused since it is more network layer related and they take place after the processing of BOs.

6   Conclusions and Outlook

A cost model can be used to determine the processing time of BOs. The structure of BOs has influence on the processing time of their elements. In this paper, we proposed an experimental setup to validate and determine the influence of the position of elements on their processing time. A comprehensive experimental evaluation was done for the elements of BOs with four parser implementations. We described how to determine dependencies on the structure. A procedure how determine the parameters for the processing time of elements was introduced. Furthermore, we described a procedure to predict the processing time of elements which is an important step for the cost estimation of synchronizing BOs. The experimental setup is generally applicable for different implementations. Furthermore, our solution can be even used to switch between parser implementations to reach more efficiency based on the used BOs. Another contribution we did not focused but which is suitable as well is the design of BOs which are less complex and easy to process.

References

1. Ameling, M., Wolf, B., Springer, T., Schill, A.: Adaptive synchronization of business objects in service oriented architectures. In: ICSOFT (2009)
2. Ameling, M., Wolf, B., Armendariz-Inigo, J.E., Schill, A.: A cost model for efficient business object replication. In: WAMIS 2009. IEEE, Los Alamitos (2009)
3. JDOM: JDOM project, http://jdom.org/
4. dom4j, http://dom4j.org/index.html
5. SAX: Simple API for XML, http://www.saxproject.org
6. XML Pull Parser: XPP,
   http://www.extreme.indiana.edu/xgws/xsoap/xpp/
7. Apache XML: Xerces Java Parser, http://xerces.apache.org/xerces-j/
8. Patiño-Martinez, M., Jiménez-Peris, R., Kemme, B., Alonso, G.: Middle-R: Consistent database replication at the middleware level. ACM Trans. Comput. Syst. (2005)
9. Plattner, C., Alonso, G.: Ganymed: Scalable replication for transactional web applications. Middleware, 155 – 174 (2004)
10. Killijian, M.O., Fabre, J.C.: Implementing a reflective fault-tolerant CORBA system. In: SRDS (2000)
11. Felber, P., Guerraoui, R., Schiper, A.: Replication of CORBA objects. In: Krakowiak, S., Shrivastava, S.K. (eds.) BROADCAST 1999. LNCS, vol. 1752, p. 254. Springer, Heidelberg (2000)
12. Daudjee, K., Salem, K.: Lazy database replication with snapshot isolation. In: VLDB 2006, VLDB Endowment, pp. 715–726 (2006)
13. Cecchet, E., Candea, G., Ailamaki, A.: Middleware-based database replication: the gaps between theory and practice. In: SIGMOD 2008, pp. 739–752. ACM, New York (2008)
14. Martin, B., Jano, B.: Wap binary xml content format (W3C),
   http://www.w3.org/TR/wbxml/

SimulPh.D.: A Physical Design Simulator Tool

Ladjel Bellatreche[1], Kamel Boukhalfa[1], and Zaia Alimazighi[2]

[1] LISI/ENSMA Poitiers University Futuroscope, France
{bellatreche,boukhalk}@ensma.fr
[2] USHTB Algiers, Algeria
alimazighi@wissal.dz

Abstract. The importance of physical design has been amplified as query optimizers became sophisticated to cope with complex decision support applications. During the physical design phase, the database designer (DBD) has to select optimization techniques to improve query performance and to well manage different resources assigned for his/her databases. The decision of using these optimization techniques is taken either before or after creating the database schema. Once this decision taken, DBD has to perform three main tasks: (i) choosing one or several optimization techniques, (ii) managing interdependencies among the chosen techniques and (iii) choosing a selection algorithm for each technique. Faced to these crucial choices, the development of simulators intended to improve the quality of the physical design represents challenging issue. In this paper, we propose a simulator, called SimulPh.D that assists DBD to perform different choices thanks to user friendly graphical interfaces.

1 Introduction

Designing advanced database applications is complex and time-consuming. The lifecycle of designing such applications requires four main phases: *conceptual, logical, physical* and *tuning*. During the conceptual phase, database designer (DBD) identifies properties used by the future application. Conceptual model is translated to logical one using rules. These two phases can be done without knowing the target DBMS. Both are usually performed by a designer tool. Physical design determines how *efficiently* a priori known queries are executed on a database thanks to optimization techniques. Tuning phase monitors and diagnoses the use of configuration (set of optimization techniques) generated by the physical phase. It is usually performed when the database is *under exploitation*. As for conceptual and logical design phases, some tasks of physical design may be done also without having a precise idea on the target DBMS. Conceptual, logical and physical phases are strongly related, since, the inputs of each phase are the outputs of the previous one. For instance, the physical design uses information coming from conceptual and logical models, such as *number of tables, number of attributes per table, length of each attribute*, etc. During the physical design phase, DBD has to choose optimization techniques among a large spectrum: *materialized views, advanced indexing schemes, data partitioning, data compression, parallel processing*, etc. The decision of selecting an optimization technique

S.S. Bhowmick, J. Küng, and R. Wagner (Eds.): DEXA 2009, LNCS 5690, pp. 263–270, 2009.

is done either before or after the creation of the database schema. An example of optimization techniques selected when creating the database schema is *horizontal partitioning* [2,10] (when it is applied on tables). Two main examples of optimization techniques selected when exploiting the database are: materialized views and indexing schemes. Algorithms used for selecting optimization techniques are usually driven by cost models estimating the execution cost of queries in the presence of these techniques. We distinguish two types of cost models: *academic cost models* and *industrial cost models*. The first type is mainly used by research community to validate their proposed algorithms [5], whereas the second one by commercial DBMS [2,13].

To assist administrators in their physical design tasks, several commercial advisor tools were proposed. We can cite *Microsoft AutoAdmin, Database Tuning Advisor* (which is part of Microsoft SQL Server 2005) [2,1], *DB2 Design Advisor* [13] and Oracle Advisor. These tools use cost models of their optimizers. They are *DBMS-dependent* and mainly concentrated in developing self-managing systems that can relegate many of the database designer's more mundane and time-consuming tasks [13]. Therefore, they cannot be applied easily during the physical phase design because they suppose the existence of the target DBMS. Some take into account interdependencies between materialized views and indexes such as in [13], but they ignore interdependencies between other optimization techniques such as bitmap join indexes and horizontal partitioning [6]. Selection algorithms used by these tools are *encapsulated* in the optimizer.

Since physical design can be done without having a precise idea on the target DBMS, the use simulation may contribute in getting efficient database applications. Simulation has proven to be highly effective tool for evaluating database designs. It has been used in 80's in order to facilitate the process of designing centralized and distributed databases [7,8] and recently for assisting students to understand the process of evaluating queries [3]. Brownsmith [7] proposed a database system simulator helping database designers and analysts in designing their logical and physical models using a language for data description and data access. Allenstein et al. [3] proposed a query simulation system for general computer science education community that offers means to understand the query execution process on Oracle. The idea behind this simulator motivates us to propose a physical design simulator, where students are replaced by designers. The main contributions of the use of simulators during the physical design are: (i) aiding in guarantying an efficient physical design, since DBD can test/evaluate several optimization scenarios and (ii) helping DBD in choosing the target DBMS based on the proposed recommendations (for instance, if the simulator recommends the use of referential horizontal partitioning which is only supported by Oracle, and if the DBD is convinced by this solution, he/she may adopt Oracle DBMS for his/her application). Note that simulators and commercial advisors may be conjointly used to ensure a high performance of database applications.

In this paper, we present a simulator, called, *SimulPh.D* offering DBDs possibility to choose their favourite optimization technique(s), to evaluate their benefit and to measure the used resources (e.g., storage). Once these choices done,

SimulPh.D proposes DBDs recommendations summarizing different information regarding optimization techniques. If DBD is satisfied with this recommendation, s/he by a simple click generates appropriate scripts that will execute on the target DBMS (if it is available).

The remainder of this paper is organized as follows. Section 2 presents functionalities that we consider important to develop a simulation tool. An overview of the *SimulPh.D* design and its architecture illustrated in Section 3, where each component is illustrated by screenshots. Section 4 validates our tool. Section 5 concludes the paper by suggesting some future work.

2   Requirements for Designing Simulator Tool

Any physical design simulator has to offer DBDs at least three following functionalities: (1) the *choice of optimization techniques*, (2) the *choice of their selection mode* and (3) the *choice of selection algorithms*.

1. Choice of Optimization Techniques. A simulator shall propose to DBS a large variety of optimization techniques supporting by main DBMS. Note that some of these techniques are similar like horizontal partitioning and bitmap join indexes [6]. This similarity complicates their selection processes [13].

2. Choice of the Selection Mode. Once optimization technique(s) chosen, two selection modes are possible: *sequential* and *combined*. In the sequential selection, each technique is selected in isolation. The main drawback of this approach is its ignorance of the interactions between different optimization techniques. In the combined selection mode, a *joint searching* is performed directly in the combined search space of optimization techniques. Choosing one of these modes for selecting optimization techniques adds another degree of difficulty to DBD.

3. The Choice of Selection Algorithms. Once the DBD chooses optimization techniques and their selection mode, he/she shall choose their selection algorithms. To select an optimization technique in the sequential mode, an important number of algorithms are available. In the combined mode, few algorithms exist and concern materialized views and indexes [11,12]. Note that each selection algorithm has its parameters, utilization context, advantages and limitations. For optimization techniques like horizontal partitioning and indexing, DBD shall also choose *table(s)* that will be partitioned/indexed and then *attributes*. Commercial DBMS advisors do not allow designers to choose their favourite algorithms.

3   SimulPh.D Overview

This section describes the different components of our tool. The main objectives of *SimulPh.D* that we fix are:

– displaying the current state of the database (the schema, attributes, size of each table, definition of each attribute, etc.) and the workload (description of queries, number of selection operations, selection predicates, etc.).

– offering two types of administration: *zero administration* and *personalized administration*. If DBD chooses zero administration, *SimulPh.D* selects different optimization techniques in a transparent manner without designer intervention (this mode is well adapted when DBD wants an auto-administration of his/her database). In the personalized administration, DBD chooses selection techniques, algorithms and sets different parameters that s/he considers important.

– supporting both sequential and combined selection modes.

– improving iteratively the selected optimization techniques based on the proposed recommendation based on feedback. *SimulPh.D* displays the quality of each optimization technique. This quality is based on a cost model estimating the number of inputs outputs required for executing each query [5]. Therefore, if some queries do not get benefit from the suggested optimization techniques, DBD can refine some parameters in order to satisfy them.

– generating scripts for each optimization technique. They can be directly executed on the database, in the case, where the DBD is satisfied with the suggested recommendations. This task is used when the target DBMS is available, otherwise, this generation will be postponed till the presence of DBMS.

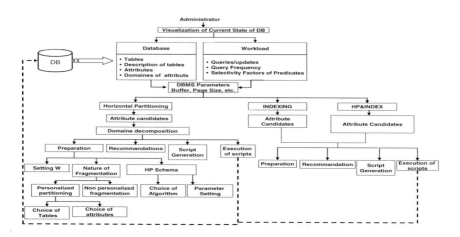

Fig. 1. Architecture Overview

3.1   Design Methodology of SimulPh.D

The main difficulty in designing interactive applications is the identification of different requirements of DBD and their representations. To perform this essential step (named the *task analysis*), we use one of the task model formalisms [4]. They first allow analyzing interactive applications focusing on their use and then, to express the activities/tasks that user wants be able to carry out. Among existing models, we choose K-MAD[1], since it was developed in partnership with the HCI team of our Laboratory and INRIA. It offers simulation tool such as

---

[1] http://kmade.sourceforge.net

CTTE. In addition, it allows the definition of logical conditions (pre, post, iteration) taken into account when simulating the use of *SimulPh.D*. The task modeling helps us to identify the different needs of the future use of *SimulPh.D*. In order to validate the task scheduling, we used the simulation tool supported by K-MAD which produces use scenarios. Based on these scenarios, we obtain the global architecture of the tool *SimulPh.D* (Figure 1).

3.2   Components of SimulPh.D

SimulPh.D supports three techniques: primary horizontal partitioning, derived horizontal partitioning and join indexes. It is mainly composed of three components, where each one corresponds to a non elementary task: (1) visualization of the current state of the database, (2) horizontal partitioning and (3) indexing.

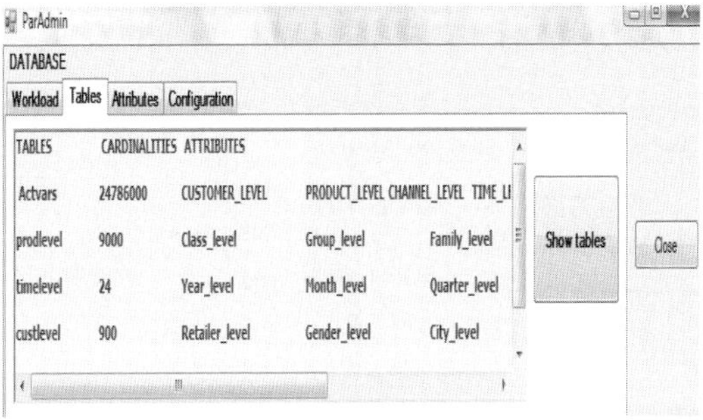

Fig. 2. Visualization of Current State of the Database

Visualization of Current State of Database. This task allows DBD to visualize the current state of his/her database that concerns three aspects: (i) *tables*: their descriptions, definition and domain of each attribute, (ii) *workload*: type of queries (search, update), their SQL descriptions, access frequency of each query, selectivity factors of selection and join predicates, and (iii) *resources* required by physical design phase: size of the buffer, page size, storage required for redundant techniques, etc. Information regarding tables is obtained by accessing meta-base of the database.

Data Partitioning Component. This component has six sub tasks are identified: (a) display non key attributes, (b) cut, (c) prepare data warehouse, (d) recommend, (e) generate script and (f) perform script[2]. The first subtask allows to display different non key attributes candidate for fragmenting dimension table(s) using the primary partitioning mode. The *cut subtask* partitions each non key attribute domain in sub domains [5]. The *prepare subtask* is a complex since it requires three phases: (1) determine W, (2) choose partitioning type and (3) select fragmentation algorithm. Determine W concerns mainly the derived

---

[2] K-MAD requires to use verb to naming tasks.

horizontal fragmentation of the fact table based on partitioning schemas of dimension tables. To realize this phase, DBD shall first choose which dimension table(s) to be partitioned and then on which attributes. DBD has the possibility to control the number of fragments of fact fragments (denoted by W) in order to avoid its explosion. The *"choose partitioning type"* phase offers to DBD two modes to perform her/his partitioning: (a) *personalized partitioning* and (b) *non-personalized partitioning*. In the *personalized mode*, the DBD is free to choose attributes candidate and the partitioning algorithm. Three selection algorithms are supported by our tool: *hill climbing*, *genetic* and *simulated annealing* algorithms [5]. DBD can also set different parameters of the selected algorithm. In this mode, the choice of the partitioning algorithm is mandatory. To ensure this operation, we used the formal pre-condition offered by K-MAD. Figure 3 gives a screen shot of this task. In *non-personalized mode*, *SimulPh.D* performs partitioning process using all attributes candidate and using a partitioning algorithm per default. The *prepare subtask* generates the partitioning schema.

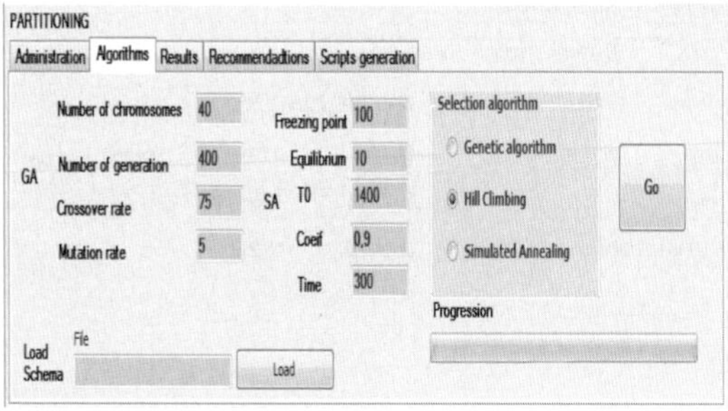

Fig. 3. Choosing Algorithms and their Setting

Once the "prepare subtask" finished, the DBD can visualize the recommendations proposed by *SimulPh.D* (*subtask recommends*). They concern the number of fragments of different tables (those partitioned by primary mode and those by derived mode), decomposition of each domain attribute, reduction obtained by horizontal partitioning (which is estimated using a cost model [6] computing the inputs/outputs required for executing a set of queries), queries getting benefit or not from horizontal partitioning, etc. If DBD is not satisfied with the suggested optimization techniques, s/he can use another algorithm and modify its parameters. This rollback is essential in the physical design phase. If s/he is satisfied, the partitioning process ends and *SimulPh.D* generates partitioning scripts (*subtask generate script*) that may be applied directly on the warehouse (*subtask perform script*). Figure 4 shows interface of personalized partitioning.

Indexing the Database. Indexing task allows the DBD to select bitmap join indexes. This selection can be done either sequentially or mixed with horizontal partitioning. In the combined mode, *SimulPh.D* proposes to DBD a list of indexable attributes candidate for performing this optimization and the storage

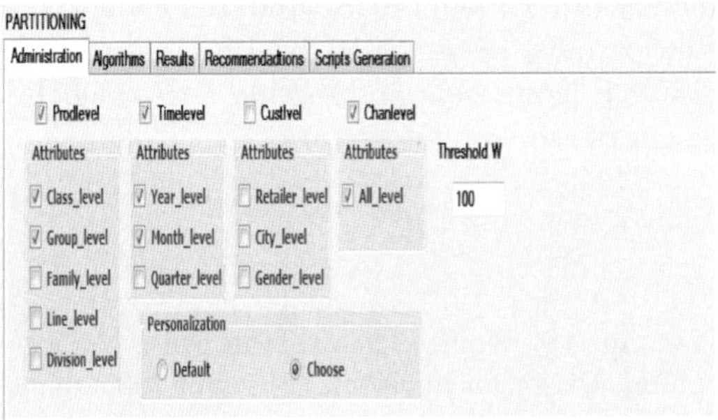

Fig. 4. Personalized Partitioning

capacity. This task is quite similar to the partitioning one. The only difference is the *cut subtask* which not required for selecting indexes. Two indexing modes are available: *non personalized indexing* and *personalized indexing*. Two types of selection algorithms are supported: a greedy (considered as per default algorithm for non personalized mode) and a data mining driven algorithm. The recommendations given by *SimulPh.D* concern indexing attributes, reduction obtained by the selected indexes, storage cost consumed by indexes, available storage, etc. In the combined mode, instead of proposing all indexable attributes candidate, only attributes which are not used by the partitioning are proposed to DBD.

4   Implementation and Validation Aspect of SimulPh.D

All components of our simulator have been implemented using Visual C++. Before the implementation phase, we have simulated our tool using K-MAD tool. This simulation allows us to fix different needs of using *SimulPh.D* and make ordering for different tasks and subtasks. We have established many use scenarios to validate our tool with Ph.D. students of our laboratory (LISI). These scenarios are based on data set of the ABP-1 benchmark and OLAP workload [9]. Each student using a user friendly interface proposed by *SimulPh.D* can visualize the tables and queries defined on the data set of this benchmark.

5   Conclusion and Future Work

In this paper, we have highlighted the difficulties that DBD might encounter during the phases of physical design and tuning. These difficulties are numerous, since they involve multiple levels of design. Given these difficulties, we have identified the need to develop a simulation tool to assist designers to meet the needs in terms of choices. We proposed a tool, called *SimulPh.D* supporting three optimization techniques: primary horizontal fragmentation, referential fragmentation and bitmap join indexes. To design the interfaces, we use a methodology borrowed from Ergonomics and Human-Computer Interaction (HCI) community, called K-MAD which is a task model. It is validated using the APB1 benchmark, where several scenarios have been tested by Ph.D. students of LISI lab. This preliminary work gives a new research direction regarding the use of simulators during physical design phase.

References

1. Agrawal, S., Chaudhuri, S., Kollár, L., Marathe, A.P., Narasayya, V.R., Syamala, M.: Database tuning advisor for microsoft sql server 2005: demo. In: SIGMOD, pp. 930–932 (2005)
2. Agrawal, S., Narasayya, V.R., Yang, B.: Integrating vertical and horizontal partitioning into automated physical database design. In: SIGMOD, June 2004, pp. 359–370 (2004)
3. Allenstein, B., Yost, A., Wagner, P., Morrison, J.: A query simulation system to illustrate database query execution. In: Proceedings of the 39th SIGCSE Technical Symposium on Computer Science Education (SIGCSE 2008), pp. 493–497 (2008)
4. Balbo, S., Ozkan, N., Paris, C.: Choosing the right task-modeling notation: A taxonomy. In: Diaper, D., Stanton, N. (eds.) The Handbook of Task Analysis for Human-Computer Interaction. Lawrence Erlbaum Associates (LEA), Mahwah (2004)
5. Bellatreche, L., Boukhalfa, K., Abdalla, H.I.: SAGA: A combination of genetic and simulated annealing algorithms for physical data warehouse design. In: Bell, D.A., Hong, J. (eds.) BNCOD 2006. LNCS, vol. 4042, pp. 212–219. Springer, Heidelberg (2006)
6. Bellatreche, L., Boukhalfa, K., Mohania, M.: Pruning search space of physical database design. In: Wagner, R., Revell, N., Pernul, G. (eds.) DEXA 2007. LNCS, vol. 4653, pp. 479–488. Springer, Heidelberg (2007)
7. Brownsmith, J.D.: The database system simulator (dbss): Data description and data access capabilities. In: Proceedings of the 15th annual symposium on Simulation, pp. 265–276 (1982)
8. Chaturvedi, A.R., Gupta, S., Bandyopadhyay, S.: Simds: A simulation environment for the design of distributed database systems. Database 29(3), 65–81 (1998)
9. OLAP Council. Apb-1 olap benchmark, release ii (1998), http://www.olapcouncil.org/research/bmarkly.htm
10. Eadon, G., Chong, E.I., Shankar, S., Raghavan, A., Srinivasan, J., Das, S.: Supporting table partitioning by reference in oracle. In: SIGMOD 2008, pp. 1111–1122 (2008)
11. Sanjay, A., Surajit, C., Narasayya, V.R.: Automated selection of materialized views and indexes in microsoft sql server. In: VLDB 2000, September 2000, pp. 496–505 (2000)
12. Talebi, Z.A., Chirkova, R., Fathi, Y., Stallmann, M.: Exact and inexact methods for selecting views and indexes for olap performance improvement. In: EDBT 2008, pp. 311–322 (March 2008)
13. Zilio, D.C., Rao, J., Lightstone, S., Lohman, G.M., Storm, A., Garcia-Arellano, C., Fadden, S.: Db2 design advisor: Integrated automatic physical database design. In: VLDB, August 2004, pp. 1087–1097 (2004)

Protecting Database Centric Web Services against SQL/XPath Injection Attacks

Nuno Laranjeiro, Marco Vieira, and Henrique Madeira

CISUC, Department of Informatics Engineering
University of Coimbra, Portugal
{cnl,mvieira,henrique}@dei.uc.pt

Abstract. Web services represent a powerful interface for back-end database systems and are increasingly being used in business critical applications. However, field studies show that a large number of web services are deployed with security flaws (e.g., having SQL Injection vulnerabilities). Although several techniques for the identification of security vulnerabilities have been proposed, developing non-vulnerable web services is still a difficult task. In fact, security-related concerns are hard to apply as they involve adding complexity to already complex code. This paper proposes an approach to secure web services against SQL and XPath Injection attacks, by transparently detecting and aborting service invocations that try to take advantage of potential vulnerabilities. Our mechanism was applied to secure several web services specified by the TPC-App benchmark, showing to be 100% effective in stopping attacks, non-intrusive and very easy to use.

Keywords: Web services, vulnerabilities, security attacks, SQL Injection, XPath Injection, code instrumentation.

1 Introduction

Web services are now widely used to support many businesses, linking suppliers and clients in sectors such as banking and financial services, transportation, or automotive manufacturing, among others. Web services are self-describing components that can be used by other software in a platform-independent manner, and are supported by standard protocols such as SOAP (Simple Object Access Protocol), WSDL (Web Services Description Language) and UDDI (Universal Description, Discovery, and Integration) [1]. In a service-based environment, providers offer a set of services that frequently access a back-end database and can be explored and used by service consumers. The web services technology provides a clear interface for consumers, and this is frequently used to enable the aggregation of services in compositions [2], where a security failure in a component may compromise the whole composition.

A recent McKinsey report indicates web services and SOA as one of the most important trends in modern software development [10]. However, the wide use and exposure of web services results in any existing security vulnerability being most probably uncovered and exploited by hackers. In fact, command injection attacks (e.g., SQL or XPath injection) are frequent types of attacks in the web environment

S.S. Bhowmick, J. Küng, and R. Wagner (Eds.): DEXA 2009, LNCS 5690, pp. 271–278, 2009.

[11]. These attacks take advantage of improperly coded applications to change queries sent to a database, enabling, for instance, access to critical data.

Vulnerabilities allowing SQL/XPath injection attacks are particularly relevant in web services [15], as their exposure is high and they frequently use a data persistence solution [14] based either in a traditional relational database or in a XML database. Currently major database vendors and several open-source efforts provide XML databases (e.g., Oracle XML DB, SQL Server 2008, Apache Xindice, etc) and, frequently, the access to this type of databases uses XPath expressions. While the goal of XPath Injection is to maliciously explore any existing vulnerabilities in XPath expressions used by an application (for instance to access an XML database), SQL Injection tries to change the SQL statements in a similar manner [11].

Although web services are increasingly being used in complex business-critical systems, current development support tools do not provide practical ways to protect applications against security attacks. In this paper, we present a phased approach that is able to: 1) **characterize** the web service in terms of security vulnerabilities; 2) **learn** the profile of regular client requests by transforming requests into invariant statements; 3) **protect** web service applications from SQL/XPath injection attacks by matching incoming requests with the valid set of codes previously learned. Regard that this work focuses on source code vulnerabilities and not any specific security mechanisms, such as authentication or data encryption.

A common way to remove SQL/XPath Injection vulnerabilities is to separate the query structure from the input data by using parameterized queries (e.g. prepared statements or parameterized Xpath expressions). In [12] an approach for replacing the SQL statements by secure prepared statements is described. Code inspection and static analysis were used to disclose code prone to SQL injection, which was then replaced by generated secure code. An approach for converting SQL statements into prepared statements is presented in [13]. However, the conversion algorithms are limited and need to be improved to reduce the large number of unhandled cases.

AMNESIA (Analysis and Monitoring for NEutralizing SQL-Injection Attacks) [4] is a tool that uses a model-based approach designed to detect SQL injection attacks, and combines static analysis and runtime monitoring. Static analysis is used to build a model of the legitimate queries that an application can generate. At runtime, when a query that violates the model is detected, it is classified as an attack and is prevented from accessing the database. Our approach learns the profile of legitimate queries at runtime, which may represent a richer, more realistic learning profile, overcoming the intrinsic limitations of static analysis (e.g., requiring access to source code).

The proposed approach is extremely effective, has a quite low overhead, and does not require any access to the source code of the application. Instead, we use bytecode instrumentation for transparently performing the necessary modifications to protect the target service. To show the effectiveness of our approach we have used two implementations of the web services specified by the TPC-App performance benchmark. A large number of security problems have been disclosed and fully corrected, showing that our approach is effective and a powerful tool for developers and system administrators.

The structure of the paper is as follows. Next section presents the technique for fixing security problems and Section 3 presents the experimental evaluation. Section 4 concludes the paper.

2 Security Improvement Approach

To perform SQL Injection the attacker exploits an unchecked input in order to modify the structure of a SQL command [11]. Usually, the attacker starts by adding an extra condition in the 'where' clause of a SQL command to gain a privileged access. Then the attacker executes a SQL command returning valuable information (typically using a union clause with the malicious select), disrupting the database by performing inserts, deletes or updates. Regarding XPath, the attack approach is basically the same and only the expression syntax differs. This way, our proposal for identifying potential SQL and XPath injection attacks is based on anomaly detection, which consists of searching for deviations from an historical (learned) profile of good commands, and includes three major phases:

1. **Service assessment** – Consists of using penetration testing, automated static code analysis, or human code inspection to disclose SQL/XPath Injection vulnerabilities and thus characterize the service in terms of these vulnerabilities.
2. **Statement learning** – The goal is to identify the valid set of valid SQL statements or XPath expressions. It is composed of two steps:
   2.1. Workload generation, execution, and measurement.
   2.2. Service instrumentation to learn valid SQL statements and XPath expressions used by the application.
3. **Service protection** – Consists of instrumenting the service to provide protection against SQL/XPath Injection attacks. Afterwards, the developer may revisit phase 1 to verify if the previously detected vulnerabilities were effectively protected.

2.1 Service Assessment

The goal of this first phase is to assess the security of the web service application in terms of SQL/XPath injection vulnerabilities. This initial characterization phase is optional, as the developer may simply wish to apply the security mechanism as a regular attack barrier, without searching the service for potential vulnerabilities.

Any of the following alternatives can be used for vulnerabilities detection: penetration testing (by using scanners or fuzzers) [11]; static code analysis [9] (a developer can easily use tools such as FindBugs [5]); or, in more difficult cases (or in cases where a high degree of confidence is needed), human code inspections by security assurance teams [3]. The outcome of this phase is essentially a set of SQL/XPath injection vulnerabilities in the service code. This information can be used later to verify the effectiveness of the proposed protection scheme by re-running this phase over the protected service.

2.2 Statement Learning

This phase includes 2 steps. The first step in this phase consists of **generating and executing a workload**, which is essentially inspecting the service description document, the WSDL file. This XML file is automatically processed to obtain the list of operations, parameters and associated data types and domains. In this context, we use a language, named 'Extended Domain Expression Language – EDEL', that enables web services to fully express their operations domains (including complex parameter domains dependencies) [8]. EDEL can be used to create workloads that respect the operations' domains, hence greatly increasing their coverage.

After having collected the necessary information, the workload generation is conducted in an automatable way, as proposed in [8]. In summary, the procedure consists of generating a set of XML objects, which are the service inputs (created in compliance with the WSDL file) and integrating them into unit tests in an automatable way. Our goal is to exercise as many source code points as possible (ideally, the complete set of data access SQL/XPath statements present in the code). The final step consists of executing the workload and using a test coverage analysis tool, such as Cobertura (http://cobertura.sourceforge.net/), to get a metric of the code coverage. If the developer is not satisfied with the coverage then more service calls are required.

The second step consist of **learning the SQL/XPath commands profile**. We start by exercising the web service by executing the generated workload. This enables us to automatically identify all the locations in the web service code where the SQL and XPath commands are executed. This is achieved by using AOP (Aspect Oriented Programming) [6] to intercept all the calls to a set of method signatures that correspond to well-known APIs for executing SQL commands (e.g., Java's JDBC API, the Spring Framework JDBC API, etc.) and evaluating XPath expressions (e.g., Java's JAXP API). Besides this set of well-known APIs, virtually any API can be easily added to the learning mechanism, as the only requirement is to know the full signature of the method to be intercepted.

At runtime, each data access call is intercepted and delivered to a dispatcher that determines if the application is in learning or protection mode. During learning, SQL and XPath commands are parsed in order to remove the data variant part (if any) and a hash code is generated to uniquely identify each command. In other words, the information used does not represent the exact command text, since commands may differ slightly in different executions, while keeping the same structure. For example, in the SQL command "SELECT * from EMP where job like 'CLERK' and SAL >1000", the job and the salary in the select criteria (job like ? and sal > ?) depend on the user's choices. This way, instead of considering the full command text, we just represent the invariant part of it. After removing the variant part of each command it is possible to calculate the command signature using a hash algorithm. We associate each hash with a code entry point (provided that the code being tested was compiled with code line information, which is generally the case).

2.3 Service Protection

Service protection at runtime (i.e., after deployment) consists in performing one security check per each data access command executed. All SQL and XPath commands are intercepted and hashed. The request flow is very similar to the learning phase, but obviously, the calculated hash codes are not added to the learned command set. Instead, they are compared to the hash values of the learned valid commands for the code point at which the command was submitted.

In practice, the matching process consists in looking up the current source code origin and getting the list of hash codes of the valid (learned) commands for that point. This list (generally quite small) is then searched for an element that exactly matches the hash of the command that is being executed. Execution is allowed to proceed if a match is found. Otherwise, a security exception (the unqualified name for this exception is *SecurityRuntimeException*) is thrown and, in this way, code execution is kept from proceeding, which prevents the potential attack. If the source code origin is not found in the lookup process, code execution is also kept from proceeding in a similar manner (in this case, a different exception is thrown –*CodePointNotTrainedRuntimeException*). This case strongly indicates that the learning phase is incomplete (test coverage was not good enough) and that an extended workload is probably required.

To verify if the security mechanism is working properly the web service should be re-assessed using a security analysis approach (similar to phase 1). The goal is to check if any of the initially identified vulnerabilities still exist and the expectation is that our mechanism stops any injection attempts by raising the appropriate security exception. If a security vulnerability is detected it means that the workload coverage was not good enough and that the learning phase is incomplete. In this case, the workload should be extended and the learning process repeated.

Finally, the developer may want to re-execute the original workload to verify if the service behavior remains correct. Problem indicators include responses outside the expected domains. For certain services, responses that are different from those obtained during the first workload execution are also problem indicators.

3   Experimental Evaluation

In this section we present the experimental evaluation performed over an initial Java prototype tool (available at [7]). To demonstrate our approach we have used the following subset of the web services specified by the standard TPC-App [14] performance benchmark: Change Payment Method, New Customer, New Product, and Product Detail. TPC-App is a performance benchmark for web services and application servers that is widely accepted as representative of real environments. Two versions of each service (versions A and B) were created by independent programmers, and the setup consisted of two nodes (client and server) that were deployed on two machines connected over an isolated Fast Ethernet network.

The **first phase** of the experimental evaluation consisted of performing a **services assessment** to try to identify potential vulnerabilities. Initially, we opted to use automated tools (vulnerability scanners and static code analyzers), however, due to the poor results obtained we decided to perform a code inspection by a team of security experts with different experience backgrounds. Table 1 summarizes the results. All detected vulnerabilities correspond entirely to SQL injection issues, as the TPC-App specification does not include any XPath usage. However, the assessment approach is essentially the same, as the main difference resides on the syntax of each language. As discussed below, FindBugs, the static analyzer used, was unable to provide individual results per service.

Table 1. Vulnerabilities detected by the different methods

Service	Scanner		FindBugs		Code Inspection	
	A	B	A	B	A	B
ChangePaymentMethod	0	0 (3 FP)[1]			2 (2 FP)	0
NewCustomer	1 + 1	0 (3 FP)	2	0	19 (1 FP)	0
NewProducts	0	0			1 (1 FP)	0
ProductDetail	0	0			0	0

We used a well-known commercial <u>vulnerability scanner</u> that was able to identify 2 critical vulnerabilities in version A. Both were manually checked and in fact corresponded to SQL Injection vulnerabilities (although one was originally identified by the scanner as a database error). The scanner also indicated 6 vulnerabilities in version B. An important aspect is that version B was using SQL prepared statements (with exception of one statement that, however, does not add any security concern as it is a static SQL command). As prepared statements are the most powerful way of preventing SQL Injection, we were expecting no issues in this version. Anyway, we decided to examine the scanner responses and the code of version B. We found that the reported errors indicated in all these cases a 'value to large for column' error message. However, even if a smaller attack expression had been used, it would still pose absolutely no threat as the prepared statement engine escapes offending characters like (').

As vulnerability scanners are known to present poor results in this kind of environments [15] we decided to use also a well known <u>static code analysis tool</u> (FindBugs) for disclosing SQL Injection vulnerabilities. As we can see in Table 1, FindBugs was able to mark 2 vulnerabilities for version A, and none for version B as expected. Considering version A, the developer created a set of methods for database access and FindBugs marked the last point of the source code where a non constant string was passed to an execute SQL method. We then analyzed the database access methods to try to distribute the vulnerabilities per service, which was not possible, as some services did use the database methods in a vulnerable way, while others did not.

To obtain more accurate results we asked a team of security experts to disclose SQL Injection Vulnerabilities in the source code by executing a thorough <u>code inspection</u> and <u>penetration tests.</u> The security analysis team was composed of 5 elements. Three of these elements are developers with more than 2 years of experience on developing database centric business critical web applications in Java. The remaining two are security researchers, one junior (one year of experience) and one senior (four years working on security related topics). Table 1 presents the summary of the vulnerabilities detected by the team (results represent the union of the vulnerabilities detected by each team member). One vulnerability was counted per each web service input parameter used in a given SQL statement in a vulnerable way. It is important to mention that we double-checked the vulnerabilities pointed out by each participant (under the form of an example service request) to discard false-positives.

As we can see, 3 of the services were vulnerable in version A, and one in particular had 19 security flaws. This large number is due to a large number of user input

---

[1] FP: False positives.

parameters, being used in more than one SQL statement throughout the code. As expected, Version B presented no security vulnerabilities.

For the **second phase (statement learning)** we analyzed the WSDL and XML schema (XSD) of each web service and, for each input and output parameter, we manually extended the service definitions to include domain restrictions while fully respecting the TPC-App specification. EDEL [8] was applied to express the final domains. The workload was defined based on a set of web service requests (a total of 5 requests for the 4 services). Before continuing we analyzed the coverage using Cobertura, and found out that, the coverage was in general above 80% (except in one case), a value accepted as representative by the developers.

The workload was then applied to exercise each TPC-App version in order to learn the expected SQL commands. After the learning process, we manually checked whether all possible SQL commands executed by the service application were correctly learned by our mechanism, and that was effectively the case. Note that, the learning process is quite important in our approach and is directly influenced by the coverage of the workload used. If there were commands not learned we would have to increase the size (and coverage) of the workload.

After this, we proceeded to the **third phase** by configuring our mechanism to enter the protective state and detect maliciously modified commands, thus **improving security**. The vulnerability scanner was then used to re-test all services for security vulnerabilities. The results were a total zero disclosed SQL/XPath injection vulnerabilities for all services. All new malicious requests were indeed stopped, preventing any further service execution and possible security consequences. Security tests over version B presented the same initial erroneous results discussed before, so for our purposes the total sum of security issues is zero.

Due to the instrumentation technique we were using, we did not re-run FindBugs, as static analysis is not able to detect that our protection mechanism blocks particular data access statement executions. So, we replayed all malicious requests crafted by our code inspection participants. All attempts to inject SQL code were again aborted.

To verify if the security improvement mechanisms changed the services' functionality we re-ran the workload for all three versions. The web services responses were analyzed for deviations from the valid output domains. No problem was identified, providing a strong indicator that we did not change the application's normal behavior.

Finally, we executed a test to assess the performance impact related to executing the security system. As we were expecting small values for the security improvement, we tested the worst case scenario found in the services and executed 100000 invocations using that worst-case scenario. Our mechanism took on average 0,052 ms ($\pm$ 0,029) to execute, less than 0,3% of the total time for the fastest executing service.

In summary, our learning mechanism was able to stop all security attacks with a negligible overhead. This is a very significant result, as besides effectively securing the target application, it implied absolutely no extra-effort from the developers that implemented the original services.

4 Conclusion

Previous works on web application security have shown that SQL/XPath Injection attacks are extremely relevant in web service applications. This paper presents an approach for improving web services security. The proposed approach consists of learning the profile of valid data access statements (SQL and XPath) and using this profile to later prevent the execution of malicious client requests. The approach was illustrated using two different TPC-App implementations. Various security issues were disclosed and corrected without additional development effort. In fact, while introducing an extremely low performance overhead, our approach proved to be 100% effective, as it was able to abort all attacks attempted in our experiments.

During the whole experimental process, no extra complexity was added to the source code. In fact, as source code is not needed, the mechanism can also be used to easily protect legacy services, which would otherwise require a difficult to implement and hard to maintain procedure. These facts make it an extremely useful tool for developers and service administrators.

References

1. Curbera, F., et al.: Unraveling the Web services web: an introduction to SOAP, WSDL, and UDDI. IEEE Internet Computing 6, 86–93 (2002)
2. Erl, T.: Service-Oriented Architecture: Concepts, Technology, and Design. Prentice Hall Professional Technical Reference (2005)
3. Fagan, M.: Design and code inspections to reduce errors in program development. Software pioneers: contributions to software engineering, pp. 575–607. Springer, Heidelberg (2002)
4. Halfond, W., Orso, A.: Preventing SQL injection attacks using AMNESIA. In: 28th international conference on Software engineering, pp. 795–798. ACM, Shanghai (2006)
5. Hovemeyer, D., Pugh, W.: Finding bugs is easy. ACM SIGPLAN Notices, 39 (2004)
6. Kiczales, G., et al.: Aspect-Oriented Programming. In: 11th European Conf. on Object-oriented Programming (1997)
7. Laranjeiro, N., Vieira, M., Madeira, H.: EDEL and Security Improvement for Web Services (2009),
   http://eden.dei.uc.pt/~cnl/papers/edel-security-tool.zip
8. Laranjeiro, N., Vieira, M., Madeira, H.: Improving Web Services Robustness. In: International Conference on Web Services (ICWS). IEEE Computer Society, Los Angeles (2009)
9. Livshits, V., Lam, M.: Finding security vulnerabilities in java applications with static analysis. In: Proceedings of the 14th conference on USENIX Security Symposium, vol. 14, p. 18. USENIX Association, Baltimore (2005)
10. McKinsey&Company: Enterprise Software Customer Survey (2008)
11. Stuttard, D., Pinto, M.: The Web Application Hacker's Handbook: Discovering and Exploiting Security Flaws. Wiley, Chichester (2007)
12. Thomas, S., Williams, L., Xie, T.: On automated prepared statement generation to remove SQL injection vulnerabilities. Information and Software Technology 51, 589–598 (2009)
13. Thomas, S., Williams, L.: Using Automated Fix Generation to Secure SQL Statements. In: Third International Workshop on Software Engineering for Secure Systems (2007)
14. Transaction Processing Performance Council: TPC BenchmarkTM App (Application Server) Standard Specification, Version 1.1 (2005),
    http://www.tpc.org/tpc_app/
15. Vieira, M., Antunes, N., Madeira, H.: Using Web Security Scanners to Detect Vulnerabilities in Web Services. In: Intl. Conf. on Dependable Systems and Networks, Estoril, Lisbon (2009)

Reasoning on Weighted Delegatable Authorizations

Chun Ruan[1] and Vijay Varadharajan[1,2]

[1] School of Computing and Mathematics,
University of Western Sydney, Penrith South DC, NSW 1797 Australia
{chun,vijay}@scm.uws.edu.au
[2] Department of Computing,
Macquarie University, North Ryde, NSW 2109 Australia
vijay@ics.mq.edu.au

Abstract. This paper studies logic based methods for representing and evaluating complex access control policies needed by modern database applications. In our framework, authorization and delegation rules are specified in a Weighted Delegatable Authorization Program (WDAP) which is an extended logic program. We show how extended logic programs can be used to specify complex security policies which support weighted administrative privilege delegation, weighted positive and negative authorizations, and weighted authorization propagations. We also propose a conflict resolution method that enables flexible delegation control by considering priorities of authorization grantors and weights of authorizations. A number of rules are provided to achieve delegation depth control, conflict resolution, and authorization and delegation propagations.

1 Introduction

Access control comprises all system mechanisms that are required to decide whether an access request issued by a particular user is allowed or not. It is needed in any secure database system that provides for controlled sharing of information among multiple users. Access control models, or authorization models, provide a formalism and framework for specifying, analyzing and evaluating security policies that determine how an access is granted and delegated among users.

Several issues need to be considered before developing an authorization model. The first issue is about the types of authorizations to be allowed in the model. Basically there are two types of authorizations, positive and negative. Positive authorization means permission, whereas negative authorization means prohibition. Many systems consider only positive authorizations whereas some consider only negative authorizations. A more comprehensive system needs to consider both positive and negative authorizations, and require a policy to resolve conflicts. The second issue is about the type of privilege administration paradigm to

S.S. Bhowmick, J. Küng, and R. Wagner (Eds.): DEXA 2009, LNCS 5690, pp. 279–286, 2009.

be adopted by the model, centralised or decentralised. Centralised administration allows only one central authorization unit to grant access to subjects, while decentralised administration allows many grantors to grant access to subjects, and may further allow grantors to delegate administrative privilege to subjects. Decentralised administration is usually more flexible and more suitable for modern database applications, but is also more difficult to manage. The third issue is whether implicit authorizations are to be supported. In a system that supports implicit authorizations, not every access has to be explicitly granted. Authorizations may be derived from inheritance relationships, which are very popular in an object oriented databases. Authorizations may also be derived from specified logic rules through reasoning techniques. Allowing implicit authorizations can usually greatly reduce the size of explicit authorization set.

This paper studies the problem of resolving conflicts properly in a decentralized authorization administration context that allows both positive and negative authorizations; especially when the administrative privilege can be delegated between subjects. There has not been much work in literature that addresses this problem. In this paper, we propose a logic program based model that supports both positive and negative authorizations, authorization delegation with delegation depth control, and authorization inheritance. A comprehensive conflict resolution method is provided to solve conflicts in authorization delegations, which will consider the priorities of grantors as well as those of authorizations. As you will see, most of the current conflict resolution methods are special cases of our conflict resolution method.

On the other hand, logic based authorization models have been studied by many researchers for the purpose of formalizing authorization specifications and evaluations [1,3,5]. The advantage of this methodology is to separate policies from implementation mechanisms and give policies precise semantics. We will develop our framework based on extended logic programs [2], which supports both negation as failure and classical negation. We extend our previous work in [4] by allowing the weights to be expressed in authorizations and delegations, by enforcing delegation depth control, which is an important issue in authorization delegations, and by providing a more comprehensive conflict resolution method etc. In our framework, authorization rules are specified in a weighted delegatable authorization program (WDAP) which is an extended logic program associated with different types of partial orderings on the domain for inheritance relationships. A number of domain-independent rules are provided to achieve delegation depth control, conflict resolution, and authorization and delegation propagations. The semantics of WDAP is defined based on the well-known answer set semantics. The framework provides users a useful way to express complex security policies in a database system.

The paper is organised as follows. Section 2 describes basic ideas about the model, and Section 3 presents the syntax of the weighted delegatable authorization program (WDAP). Section 4 defines the semantics of the program, while Section 5 concludes the paper.

2   Basic Ideas

To develop a formal semantics of a WDAP, five aspects will be taken into consideration in the process of evaluating a WDAP: delegation correctness, delegation propagation, authorization correctness, authorization propagation, and conflict resolution.

Administrative privilege delegation correctness

Definition 1. *We say that an authorization set is* delegation correct *if it satisfies the following two conditions: A subject s can delegate other subjects the privilege to grant an access right a over object o with depth d if and only if s is the owner of o or s has been delegated the privilege to grant a over o with delegation depth d + 1.*

Authorization correctness

Definition 2. *We say that an authorization set is* authorization correct *if it satisfies the following condition: subject s can grant other subjects an access right a of type + or − over object o if and only if s is the owner of o or s has been delegated the right to grant a over o.*

Authorization and delegation propagations

Rule based authorization specification allows implicit authorizations to be derive from the authorization set, and hence this can greatly reduce the size of explicit authorization set. In our model, we also support the implicit authorizations by permitting authorization inheritance. We consider the authorization propagations along hierarchies of subjects, objects and access rights represented by the corresponding partial orders.

Conflict resolution

 - Solving conflicts using weighted authorization path. We say that $(s_1, s_2, ..., s_{n-1}, s_n)$ is an authorization path from s_1 to s_n on an access right r over an object o if there exists delegations on r over o from s_i to s_{i+1}, $i = 1, ..., n-2$, and an authorization from s_{n-1} to s_n. The weighted length of the path is $w_1 + w_2 + ... + w_{n-1}$, where w_i is the weight of the grant from s_i to s_{i+1}. If two authorizations a_1 and a_2 are conflicting and their grantee is s, then we say a_1 overrides a_2 if the weighted length of the path from the root (owner) to s via a_1 is shorter than that of the path via a_2.
 - Conflicts that are unsolvable. If the two conflicting authorizations have the same shortest weighted path, we treat this conflict as *unsolvable*.

3   Syntax of Weighted Authorization Programs

Our language $\mathcal{L}$ is a multi-sorted first order language, with five disjoint *sorts* $\mathcal{S}, \mathcal{O}, \mathcal{A}, \mathcal{T}$, and $\mathcal{N}$ for subject, object, access right, authorization type and weight or depth respectively. Variables are denoted by strings starting with lower case letters, and constants by strings starting with upper case letters. In addition, three partial orders $<_S, <_O$ and $<_A$ are defined on sorts $\mathcal{S}, \mathcal{O}$ and $\mathcal{A}$ respectively, which represent the hierarchical structures of subjects, objects and access rights. There are two authorization types denoted by $-, +$, where $-$ means *negative*, $+$ means *positive*. A negative authorization specifies that the access must be forbidden, while a positive authorization specifies that the access must be granted. N is a set of non-negative numbers. In the following, we will normally use the following vocabulary:

1. *Sort subject*: with subject *constant* poset $(S, <_S)$: $S, S_1, S_2, ...$, and subject *variables* $s, s_1, s_2, ...$.
2. *Sort object*: with object *constant* poset $(O, <_O)$: $O, O_1, O_2, ...$, and object *variables* $o_1, o_2, o_3, ...$.
3. *Sort access right*: with access right *constant* poset $(A, <_A)$: $A, A_1, A_2, ...$, and access right *variables* $a, a_1, a_2, ...$.
4. *Sort authorization type*: with authorization type *constant* set $T = \{-, +\}$, and authorization type *variables* $t, t_1, t_2, ...$.
5. *Sort non-negative number*: with *constant* set $N = 0, 1, 2$, and *variables* $w, w_1, w_2, ..., d, d_1, d_2, ...$.
6. *Predicate Symbol* set P

P consists of a set of ordinary predicates defined by users, and three built-in predicate symbols, *grant*, *can-grant*, and *own*, for authorization, administrative privilege delegation, and object ownership respectively. *grant* is a 6-term predicate symbol with type $S \times O \times T \times A \times S \times N$. Intuitively, $grant(s, o, t, a, g, w)$ means that s is granted by g a t type of access right a on object o with weight w. w represents the degree of certainty for this authorization, with smaller weight denoting the higher certainty. *can-grant* is a 6-term predicate symbol with type $S \times O \times A \times S \times N \times N$. Intuitively, $can\text{-}grant(s, o, a, g, w, d)$ means that s is granted by g the right to further grant access right a on object o for the maximum depth d and the grant weight is w . w represents the priority of the delegatee, in terms of their further granting, given by the delegator, with smaller weight denoting the higher priority. d represents the allowed depth of the administrative privilege delegation. If $d = 0$, the delegatee cannot further grant administrative privilege to other subjects. If $d = 1$, the delegatee can further grant administrative privilege with maximum depth of 0. If $d = 2$, the delegatee can further grant administrative privilege with maximum depth of 1, and etc. *own* is a 2-term predicate symbol with type $S \times O$. Intuitively, $own(s, o)$ means s is the owner of o.

A *term* is either a variable or a constant. In general, we prohibit function symbols in our language for the sake of simplicity, but allow some simple built-in arithmetic functions to be used.

An *atom* is a construct of the form $p(t_1, ..., t_n)$, where p is a predicate of arity n in P and $t_1, ..., t_n$ are terms.

A *literal* is either an atom p or the negation of the atom $\neg p$, where the negation sign $\neg$ represents classical negation. Two literals are *complementary* if they are of the form p and $\neg p$, for some atom p. For simplicity, we forbidden the negation form of the authorization predicate *grant*. The restriction will not affect the expressive power of the language since the type argument in *grant* can denote the opposite meanings. Two authorizations $grant(s, o, +, r, g, w)$ and $grant(s', o', -, r', g', w')$ are *conflicting* if $s = s'$, $o = o'$, $r = r'$, and $g \neq g'$.

A *rule* r is a statement of the form:

$$b_0 \leftarrow b_1, ..., b_k, not\, b_{k+1}, ..., not\, b_m, m >= 0$$

where $b_0, b_1, ..., b_m$ are literals, and not is the negation as failure symbol. The b_0 is the *head* of r, while the conjunction of $b_1, ..., b_k, not\, b_{k+1}, ..., not\, b_m$ is the *body* of r. Obviously, the body of r could be empty. Correspondingly, when b_0 is an authorization literal, the rule is called *authorization rule*.

A *Weighted Authorization Program*, WDAP, consists of a finite set of rules.

A term, an atom, a literal, a rule or program is *ground* if no variable appears in it.

Example 1. let $S=\{Dean, Alice, Bob, Mary, HOS, A/Dean, Casual; Casual <_S Alice, HOS <_S Bob, A/Dean <_S Mary\}, O = \{Staff\}, A = \{select, update; update <_A select\}$. Then the following is an example WDAP.

(r_1). $can\text{-}grant(HOS, Staff, select, Dean, 2, 1) \leftarrow$
(r_2). $can\text{-}grant(A/Dean, Staff, select, Dean, 1, 2) \leftarrow$
(r_3). $grant(Casual, Staff, -, select, Dean, 3) \leftarrow$
(r_4). $grant(Alice, Staff, +, select, Bob, 0) \leftarrow$
(r_5). $grant(Casual, Staff, -, select, Mary, 0) \leftarrow$
(r_6). $own(Dean, Staff) \leftarrow$

4   Formal Semantics

In this section, we first define a set of general rules. We then present the formal semantics for WDAP, which is based on answer set semantics, and give our access control policy.

4.1   Domain-Independent Rules

In this section, we define a set of domain-independent rules to formally achieve delegation correctness, authorization correctness, authorization and delegation propagations, and conflict resolution discussed before.

Rules for administrative privilege delegation correctness

The following rules are used to guarantee that the administrative privileges are properly delegated in terms of the eligible delegators and the valid delegation depths. The first rule means the owner's any administrative privilege delegation is accepted (represented by predicate *can-grant1*). The second rule means if a subject has been delegated the right to grant for the maximum delegation depth d', then the subject's further administrative delegation with depth less than d' is accepted.

(w_1). $can\text{-}grant1(s, o, a, g, w, d) \leftarrow can\text{-}grant(s, o, a, g, w, d), own(g, o)$

(w_2). $can\text{-}grant1(s, o, a, g, w, d) \leftarrow can\text{-}grant(s, o, a, g, w, d),$
$$can\text{-}grant1(g, o, a, g', w', d'),$$
$$d' > d, g \neq s$$

Rules for administrative privilege propagation

The next three rules are about administrative privilege propagations. It means that a subject's administrative privilege on some object and access right would propagate automatically to its next lower level subjects represented by $<_S$ relation, next lower level objects represented by $<_O$ and next lower level access rights represented by $< A$.

(w_3). $can\text{-}grant1(s, o, a, g, w, d) \leftarrow can\text{-}grant1(s', o, a, g, w, d), s' <_S s$

(w_4). $can\text{-}grant1(s, o, a, g, w, d) \leftarrow can\text{-}grant1(s, o', a, g, w, d), o' <_O o$

(w_5). $can\text{-}grant1(s, o, a, g, w, d) \leftarrow can\text{-}grant1(s, o, a', g, w, d), a' <_A a$

Rules for authorization correctness

The following two rules mean any grant from the owner or a grantor holding the right to grant is accepted (represented by predicate $grant1$).

(w_6). $grant1(s, o, t, a, g, w) \leftarrow grant(s, o, t, a, g, w), own(g, o)$

(w_7). $grant1(s, o, t, a, g, w) \leftarrow grant(s, o, t, a, g, w),$
$$can\text{-}grant1(g, o, a, g', w', d'), g \neq s$$

Rules for authorization propagation

The following rules are used to achieve authorization propagation along subjects, objects and access rights inheritance hierarchies. The first rule means any authorization given to a subject would propagate to its subordinate subjects represented by the $<_S$ relation. The second rule means any authorization on an object would propagate to its sub-objects represented by the $<_O$ relation. The third and fourth rules mean any authorization on an access right would propagate to its next lower level or next higher level access right depending on its authorization type being positive or negative. Note that unlike other propagations that are downward along the hierarchies, when the grant type is $-$, the propagation is upward along the access right hierarchy.

(w_8). $grant1(s, o, t, a, g, w) \leftarrow grant1(s', o, t, a, g, w), s' <_S s$

(w_9). $grant1(s, o, t, a, g, w) \leftarrow grant1(s, o', t, a, g, w), o' <_O o$

(w_{10}). $grant1(s, o, +, a, g, w) \leftarrow grant1(s, o, +, a', g, w), a' <_A a$

(w_{11}). $grant1(s, o, -, a, g, w) \leftarrow grant1(s, o, -, a', g, w), a <_A a'$

Rules for conflict resolution

As mentioned before, the conflict resolution is based on the priority of the grantor and the weight of the authorization. The priority of a grantor is the weighted length of the shortest delegation path from the owner to it, since there may exist multiple paths to it. In the following rules, we use predicate *priority* to represent the priority of a grantor, which is the length of the shortest path to it. We use *priorities* to represent all the priorities that a grantor received from their delegators, which are lengths of all possible paths. Predicate *exist-higher-priorities* means that the corresponding *priorities* is not the highest one. It is introduced to avoid the existential quantifier to be used in (w_{15}), as in an extended logic program all the variables in clauses are considered to be universally quantified. For the two conflicting authorizations, we will compare the sum of the grantor's priority and weight of the authorization, and the one with smaller value will win. Predicate *overridden* is introduced to indicate that the corresponding authorization is overridden by some other authorizations. Predicate *hold* means the corresponding authorization holds as it is not overridden by any other authorizations.

(w_{12}). $priority(g, o, a, 0) \leftarrow own(g, o)$

(w_{13}). $priorities(g, o, a, x + w) \leftarrow can\text{-}grant1(g, o, a, g', w, d),$
$$priority(g', o, a, x)$$

(w_{14}). $exist\text{-}higher\text{-}priorities(g, o, a, x) \leftarrow priorities(g, o, a, x),$
$$priorities(g, o, a, y), y < x$$

(w_{15}). $priority(g, o, a, x) \leftarrow priorities(g, o, a, x),$
$$not \; exist\text{-}higher\text{-}priorities(g, o, a, x)$$

(w_{16}). $overridden(s, o, t, a, g, w) \leftarrow grant1(s, o, t, a, g, w),$
$$grant1(s, o, t', a, g', w'),$$
$$priority(g, o, a, x),$$
$$priority(g', o, a, y), y + w' < x + w$$

(w_{17}). $hold(s, o, t, a, g, w) \leftarrow grant1(s, o, t, a, g, w),$
$$not \; overridden(s, o, t, a, g, w)$$

Let W denote all the general rules, i.e. $W = \{w_1, ..., w_{17}\}$

4.2 Formal Semantics

Let Π be a WDAP, the *Base* B_Π of Π is the set of all possible ground literals constructed from the system reserved predicates and predicates appearing in the rules of Π, the constants occurring in $\mathcal{S}, \mathcal{O}, \mathcal{A}, \mathcal{T}, \mathcal{N}$. A *ground instance* of r is a rule obtained from r by replacing every variable W in r by $\delta(x)$, where $\delta(x)$ is a mapping from the variables to the constants in the same sorts. Let $G(\Pi)$ denote all ground instances of the rules occurring in Π. Two ground literals are *conflicting* on subject S, object O and access right A if they are of the form $hold(S, O, +, A, G, W)$ and $hold(S, O, -, A, G', W')$. A subset of the Base of B_Π is *consistent* if no pair of complementary or conflicting literals is in it. An *interpretation* I is any consistent subset of the Base of B_Π.

Definition 3. *Given a WDAP Π, an interpretation for Π is any interpretation of $\Pi \cup W$.*

Definition 4. *Let I be an interpretation for a WDAP $G(\Pi)$, the reduction of Π w.r.t I, denoted by Π^I, is defined as the set of rules obtained from $G(\Pi \cup W)$*

by deleting (1) each rule that has a formula not L in its body with $L \in I$, and (2) all formulas of the form not L in the bodies of the remaining rules.

Given a set R of ground rules, we denote by $pos(R)$ the positive version of R, obtained from R by considering each negative literal $\neg p(t_1, ..., t_n)$ as a positive one with predicate symbol $\neg p$.

Definition 5. *Let M be an interpretation for Π. We say that M is an answer set for Π if M is a minimal model of the positive version $pos(\Pi^M)$. If M is an answer set for Π, then its subset of all the literals with predicate name hold is called authorization answer set for Π, denoted by A.*

Example 2. Let us consider the WDAP given in Example 1. To evaluate the authorizations holding, it should be combined with the general rules in W. In the process of answer set generation, the authorization and delegation correctness is enforced, the authorization and delegation propagations are carried out along the defined inheritance hierarchies in the domain, and the conflicts are resolved based on our resolution policy. For instance, for Alice, three authorizations will be derived from Bob (directly), Mary and Dean (indirectly through the inheritance from the Casual)respectively, and the Mary's authorization will override.

5   Conclusions

There are several interesting issues that worth further exploring. First, we are considering to extend our current model by considering dynamic trust relationships in authorizations and delegations. Second, we plan to develop a dynamic prioritized conflict resolution method which can support different policies by using various semantic properties of subjects, objects and access rights. Finally, we also plan to implement a prototype of our framework using answer set logic programming techniques.

References

1. Bertino, E., Buccafurri, F., Ferrari, E., Rullo, P.: A logical framework for reasoning on data access control policies. In: Proceedings of the 12th IEEE Computer Society Foundations Workshop, pp. 175–189. IEEE Computer Society Press, Los Alamitos (1999)
2. Gelfond, M., Lifschitz, V.: Classical negation in logic programs and disjunctive databases. New Generation Computing 9, 365–385 (1991)
3. Jajodia, S., Samarati, P., Subrahmanian, V.S.: A logical language for expressing authorizations. In: Proceedings of the 1997 IEEE Symposium on Security and Privacy, pp. 31–42. IEEE Computer Society Press, Los Alamitos (1997)
4. Ruan, C., Varadharajan, V., Zhang, Y.: Logic-based reasoning on delegatable authorizations. In: Proceedings of the 13th International Symposium on Methodologies for Intelligent Systems (2002)
5. Woo, T., Lam, S.: Authorization in distributed systems: a formal approach. In: Proceedings of IEEE on Research in Security and Privacy, pp. 33–50 (1992)

Annotating Atomic Components of Papers in Digital Libraries: The Semantic and Social Web Heading towards a Living Document Supporting eSciences

Alexander García Castro[1], Leyla Jael García-Castro[2], Alberto Labarga[3], Olga Giraldo[4], César Montaña[5], Kieran O'Neil[6], and John A. Bateman[1]

[1] University of Bremen, Bibliothekstrasse 1,
28359 Bremen, Germany
cagarcia@uni-bremen.de
[2] Universität der Bundeswehr München, Werner-Heinsenberg-Weg 39,
85779 Neubiberg, Germany
leyla.garcia@ebusiness-unibw.org
[3] University of Granada,
Madrid, Spain
alberto.labarga@scientifik.info
[4] International Center for Tropical Agriculture,
Palmira, Valle, Colombia
oxgiraldo@cgiar.org
[5] IT&BI Consulting Services,
Bogota, Colombia
ca.montana70@egresados.uniandes.edu.co
[6] Terry Fox Laboratory, British Columbia Cancer Research Centre,
Britsh Columbia, Canada
koneill@bccrc.ca

Abstract. Rather than a document that is being constantly re-written as in the wiki approach, the Living Document (LD) is one that acts as a document router, operating by means of structured and organized social tagging and existing ontologies. It offers an environment where users can manage papers and related information, share their knowledge with their peers and discover hidden associations among the shared knowledge. The LD builds upon both the Semantic Web, which values the integration of well-structured data, and the Social Web, which aims to facilitate interaction amongst people by means of user-generated content. In this vein, the LD is similar to a social networking system, with users as central nodes in the network, with the difference that interaction is focused on papers rather than people. Papers, with their ability to represent research interests, expertise, affiliations, and links to web based tools and databanks, represent a central axis for interaction amongst users. To begin to show the potential of this vision, we have implemented a *novel web prototype* that enables researchers to accomplish three activities central to the Semantic Web vision: *organizing*, *sharing* and *discovering*. **Availability**: http://www.scientifik.info/

Keywords: Semantic Web, Social Web, Knowledge Integration, Knowledge Representation, Folksonomy, Ontology.

S.S. Bhowmick, J. Küng, and R. Wagner (Eds.): DEXA 2009, LNCS 5690, pp. 287–301, 2009.
© Springer-Verlag Berlin Heidelberg 2009

1   Introduction

eScience is an umbrella term describing converging sets of trends and technologies that have the potential to radically transform the conduct of science [1]. In particular, sharing and linking data in order to support the extraction of information and knowledge discovery is one central pillar in the realization of eSciences. Within Life Sciences scholarly production is predominantly stored in databases (DBs) and digital libraries. For instance genomic and proteomic databases store not only sequences but also other related information; furthermore, these resources are highly interrelated and it is sometimes possible to formulate and execute queries that are distributed across several databases; this process is transparent for the end user.

Although papers and DBs store complementary information, the relationships currently defined across the information contained within papers, existing DBs and online resources are negligible. Digital libraries within the biomedical domain store information related to methods, defined material, topics, statements of problems being addressed, hypotheses, results, etc. However, retrieving papers addressing the same topic and for which similar biomaterial has been used is not a trivial task. In order to improve search and retrieval, and also to enrich the available metadata, digital libraries should provide facilities by means of which links can be established between atomic components of papers (domain terminologies, concepts) and resources over the Web capable of processing and/or adding meaning to them. For instance, for papers containing data types such as proteins, genes and metabolic pathways, digital libraries should link them to the corresponding data entries in DB's, Knowledge Organization Systems (KOS), and/or existing online resources. Once a data type has been identified within a text, available infrastructure such as BioMoby [2] also makes it possible to determine what resources can consume it. Furthermore, existing infrastructure can also support the replication of experiments if the raw data generated and used by researchers becomes part of the paper; in this way could not only the review process be more open, but also collaboration could be boosted. Biomedical ontologies—such as those from the Open Biomedical Ontologies (OBO, http://www.obofoundry.org) initiative —could be used as an anchor point over which links are established and further expanded by collaborative tagging. Combining social tagging and ontology-based marking improves information retrieval and facilitates the enrichment of metadata [3-5]; it also would serve to encourage methods that shift the creation of metadata from the individual to a collective [6].

Tags as such allow users to organize resources into categories, *i.e.* groups of resources with the same tag by means of which the retrieval of the tagged information becomes easier [5]. For instance, Delicious (http://delicious.com), Bibsonomy (http://www.bibsonomy.org) and Connotea (http:// www.connotea.org) facilitate the tagging of online resources and bibliographic references. These online tools harness the collective knowledge that is modelled by the collective tagging. Collaboration is thus based on similarities in tags and tagged objects. Although tagging papers, as a whole is now possible, this is still insufficient. Marking and linking atomic components of papers, *idem* words, pieces of images or segments of video, is also necessary in order to enrich the metadata structure and to facilitate concept-based social interaction. Annotation of this kind is not yet possible, however, within a collaborative environment.

The research reported here investigates how to support the annotation of the atomic components of research papers in the life sciences by combining ontology-based and user-generated tags within a social network built upon these tagged concepts. To this end we propose the Living Document (LD), a document that lives on the web by interacting with other papers and resources related to the data types it hosts. In this manner researchers can tag individual components of their papers drawing on categories from ontologies such as the Gene Ontology [7] via automatic tagging systems, without being exposed to the complexity of the ontology; at the same time, they are also able to generate their own tags, or extend existing ones. The Living Document (LD) is therefore not one that is constantly being written as in the wiki approach, but one that acts as a document router by drawing on both structured and organized social tagging and existing ontologies. We argue that papers should be understood as containers of knowledge and that, as such, the hosted knowledge should be easily networked with related resources. As an example, the sentence *"IGFBP-2 expression is negatively regulated by PTEN and positively regulated by phosphatidilinositol 3-kinase (PI3K) and Akt activation [8]"* contains valid ontology terms that could be linked to existing databases and relevant online resources. By the same token, domain experts reading this paper could enhance the annotation by providing, for instance, more external resources relevant to the identified terms.

The LD supports this notion of rich semantic annotation by applying the social tagging paradigm to a document management environment encompassed in a document-information centric social network. The LD is conceived as a web service for storing, sharing, relating and discovering knowledge in documents, research papers in this case. The framework is heavily based on tags and a Semantic Web (SW) technology layer making use of these tags.

From the perspective of software functionality, there are two main types of tags: i) predefined tags, those coming from existing ontologies, and ii) user-generated tags, those created by the community, which together mediate the interaction and provide the semantic tissue between papers. The combination of these two types of tags also makes it possible to build upon the knowledge contained in existing ontologies by association with new user-defined tags. For instance, a paper in which draught tolerance is compared between *Oryza Sativa* (rice) and Barley in saline soils will have genes, metabolic pathways, geographical locations, atmospheric conditions attached to locations and specific periods of time, soil conditions, breeding conditions, etc., could be enriched with both, manual and automatic tags. Within the biomedical domain there are ontologies describing all these concepts; moreover, there are also databases with additional relevant information.

Due to the interoperability infrastructure bioinformaticians have built, biological DBs are highly interrelated; this makes it possible to execute crossed queries that retrieve integrated views: thus relating, for instance, molecular markers associated to those genes that were studied, or geographic locations to online resources such as Google Maps. The underlying interest of the authors is then how to move from the *collected* intelligence closer to the *collective* intelligence within life sciences. As Berners-Lee highlights in his definition of the SW, better enabling computers and people to work in cooperation requires an environment where information is given a well-defined meaning [9]. For this, we assume that structured and targeted collaboration is key to the realization of the SW vision. Moreover, life sciences pose an ideal scenario for making this real as there are i) communities of practice actively engaged in the development of their resources, ii) ontologies being used by databases for annotation purposes and developed by communities of practice, iii) highly interlinked and interrelated databases and iv) several analysis tools related to these databases. These features make it feasible to generate a concept-centric social network where papers more accurately define relevant interaction paths between social agents, and over which papers can be easily linked to external resources capable of consuming those data types forming part of the documents.

2   Related Work

Within the biomedical community the notion of community annotation has recently started to be adopted. For instance, WikiProteins [10] delivers an environment in which it is possible to address a biological problem: the annotation of proteins; this allows the wider research community to directly benefit from the generation and peer-review of knowledge at minimal cost. The more annotations the systems receives, the better the chances are for users to interact with other researchers who share similar interests/problems, for example working on the same motif, crystallographic method etc. [10]. WikiProteins allows the annotation of proteins as a whole; however, annotating valid biological parts of a protein is not possible—*e.g.* positional features. Within the context of a paper these valid biological parts could be assimilated to words/specialized-terminology/ontology-terms.

Another example that illustrates the usefulness of harnessing the collective intelligence is BIOWiki [11]; this collaborative ontology annotation and curation framework facilitates the engagement of the community with the sole purpose of improving an ontology. Similarly to both WikiProteins and BIOWiki, myExperiment [12] provides an environment in which the community interacts on the basis of a common problem; sharing and reusing workflows is this system's main goal. Moreover, within the publishing industry there has also been a series of efforts in promoting Social Networks. BioMedExperts (BME, http://www.biomedexperts.com), for example, is a professional network in which literature references are used to support interaction. Although this system does not support tagging by users, it does support automatic tagging based on a reference terminology—thus allowing the identification of researchers with similar interests. Nature Network works in a similar way; however it does not facilitate any controlled vocabulary for annotating the literature references.

Interestingly, the widely known PubMed system (http://www.ncbi.nlm.nih.gov/pubmed) does not offer any kind of tagging system; nor does it make use of existing ontologies to classify documents. Although highly interrelated to DBs and analysis resources provided by the National Center for Biotechnology Information (NCBI), PubMed does not provide a direct relationship between the data types available in the abstracts and those NCBI resources. How semantic descriptors of resources, user profiles and enriched metadata may improve the usability of digital libraries has been investigated within the JeromeDL (http://www.jeromedl.org) project; this system offers a richer retrieval system and applies Semantic Web principles to the management of digital libraries. The Annotea project (http://www.w3.org/2001/Annotea) enhances collaboration further via shared metadata based Web annotations, bookmarks, and their combinations. The LD reuses concepts from Annotea within a more focused scenario.

Most of the investigated sites offer tagging systems for bio-related documents; however none of their functionalities addresses the problem of tagging atomic components of some paper within a social concept-centric network. The pattern of relationships amongst the information contained in such papers, in particular how that information inherently gives rise to networks of concepts across papers, which can facilitate more accurate assessments of similarity as well as linking to external resources capable of consuming this information, is not represented. Interestingly, within the publishing industry in Life Sciences, publishers offer limited integration between the information contained in published papers and the Web [13]; no automatic or semi-automatic tagging systems are available, nor are there knowledge management facilities built over the papers and consistent with the data types sitting on them. The programmatic access, APIs that the majority of publishers offer, is restricted, as are the social network capabilities built over papers. All of these deficits represent clear opportunities for more advanced solutions, such as that we propose here.

3   The Living Document and Paper-of-a-Paper Ontology

One disappointing aspect of digital libraries is that papers, although in digital format, are neither interconnected nor interoperate with other valid resources capable of consuming the data types available in papers within life sciences. Most systems limit their functionalities to syntactic features such as: "which other papers were published by an author"; given this, users are able to "jump" to a list of papers; also, users can "click&jump" on bibliographic references, and then again "jump" to another paper. These are very limited functionalities for digital documents that live on the web. The LD goes beyond this not only by allowing researchers to define the network environment of a paper within its research context but also by predefining a network based on existing ontologies and resources.

Papers are conceptually related to each other and to external resources. Papers have readers, writers, images, tables, etc. and, since people have interests, papers reflect the interests of those who write them as well as of those who read them. In order to facilitate the representation of the syntactic structure of a paper and the conceptual networks that papers support, we have created the Paper-Of-A-Paper Ontology (POAP, http://poap-project.org). POAP represents the network of concepts and associated external resources derived from the tagging activity. POAP has been conceived to be interoperable with models such as the Meaning Of a Tag (MOAT) ontology [14] and the Social Semantic Cloud of Tags (SCOT) ontology [15] as they both offer representations for the tag as well as for the social tagging activity itself. This model was carefully designed to play a similar role to that played by FOAF in human-centric social networks. The intersection between POAP and FOAF facilitates accurate interaction based on documents, in our case, research papers.

POAP extends these models in order for tags to represent networks of concepts across papers and external resources. A simple view that illustrates some sections of POAP and how it is interoperable with other models is presented in Fig. 1. Ultimately POAP aims to support, in coordination with software such as BioMoby [2], the discovery of services capable of consuming those valid biological data types available in papers. For instance, for those valid biological data types relevant information may be retrieved from BioMoby.

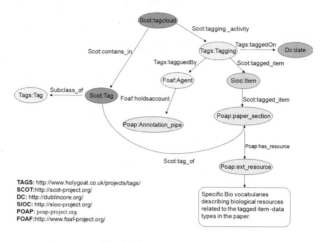

Fig. 1. Overview of POAP

POAP combines input from several existing ontologies in a way well suited to supporting the functionalities intended for the LD. For example, it has classes such as *defined_material*, which is being reused from the Ontology for Biomedical Investigations (OBI, http://purl.obofoundry.org/obo/obi). In this manner the biomaterial used can be explicitly declared as well as other relevant experimental information modeled by OBI. Citations are also being enriched by reusing the Citation Typing Ontology (CiTO) CiTO models citations is terms of the nature or type of the citation relationship (*e.g.* reuses, usesMethodIn). The nature or type or work, *e.g.* research, review, is also considered. Authors and their affiliations are modelled by using FOAF; in this manner, arrangements of authors, concepts and affiliations can be easily represented as FOAF networks. Expressions relating concepts pertaining ontology classes such as *defined_material, study design, author, tag, tagger, geographical_location, isCitedBy*, and *research* can be easily represented by FOAF, CiTO, EnvO (http://environmentontology.org/) and OBI.

3.1   POAP and the LD within the Publishing Workflow

A part of the publishing workflow is usually supported by software that assists users in the submission process. Authors usually submit manuscripts, figures, tables and, sometimes, raw data in the form of, for instance, spreadsheets, specialized generated proprietary files from machines processing samples, etc. Once the article has undergone the review process it is then published. Some publishers, such as PLOS, allow readers to comment published papers, engage in discussions with authors and others reading the same paper. Although authors generate metadata, the data is not improved during the review process. Moreover, during the process of writing, ontology based metadata could be generated as proposed by Fink *et al* (http://ucsdbiolit.codeplex.com)—see discussion section for more information. The available metadata is, in spite of the multiple possible stages at which it could be generated and enriched, scant. The LD and POAP aim to support the enrichment of structured metadata so that, for instance, authors can tag figures, tables, text and raw data. Fig. 2 illustrates how both, POAP and the LD support the overall process. FOAF networks are extracted from the information provided by authors (names, affiliations, emails, URLs, etc). Before submitting a paper, authors can add metadata; tools supporting such process are available, see for instance http://www. codeplex.com/ucsdbiolit. By the same token, authors can annotate figures and tables; furthermore, they can link text, tables and/or figures to the corresponding raw data. Once the paper has been published automatic annotation pipelines such as WhatIZit can easily extend the metadata; the community of readers can also add metadata. In this manner researchers can choose the type of metadata they are most interested in, for instance disease related metadata, metabolic pathways, combinations of controlled vocabularies, etc. Equally important, the metadata thus generated can be linked to external resources, in this manner data sets can be linked to algorithms capable of processing them, gene and protein names can be linked to diseases or specialized databases.

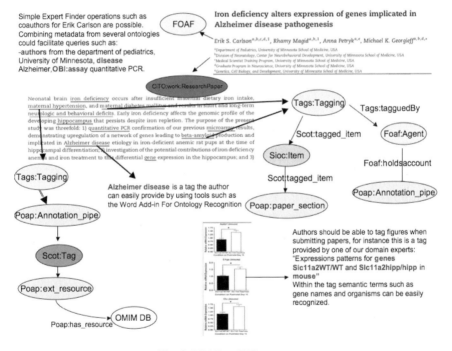

Fig. 2. POAP and LD support

3.2 Architecture and Implementation

In order to support the LD functionality, we have developed a prototype that makes use of annotation pipelines such as WhatIzIt (http://www.ebi.ac.uk/whatizit) [16, 17];

Reflect (http://reflect.ws) will soon also be incorporated. This makes it possible for users to generate their own tags over specific sections (words, sentences) of papers/documents. There are currently over 500 papers in the system and new documents can be easily loaded. Users can share and reuse tags for better-defining queries; it is also possible for them to specify external resources associated with tags and sections of the papers.

A general overview of the LD is presented in Fig. 3. Documents have sections (images, tables, words, phrases) and these can be tagged by users or by automatic pipelines such as WhatIzIt or Reflect. Other workflows can be easily added via the LD Application Programming Interface (API). Usually tags are consumed by an external resource such as a database or an analysis tool. For instance, by means of WhatIzIt users can tag the word "NADPH"; this tag has a default external resource, UniProt. Other users could have found it important to manually tag the term "Ca Channels" and link it to a different external resource.

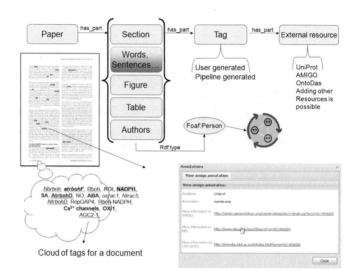

Fig. 3. General overview of the Living Document

For the purposes of our prototype we have developed infrastructure for the storage, indexing, annotation and retrieval of articles. Initially we are working with papers from the Elsevier digital collection, but adding new data sources, such as PubMed, DBLP (http://dblp.uni-trier.de) or any other XML based digital library, is also possible. The search and retrieval module has been developed on top of the Lucene (http://lucene.apache.org/) project framework; this is an open-source, high-performance, full-featured text search engine library written entirely in Java. The system allows the user to search globally across Elsevier journals or individually in selected resources, and filtering by authors or dates by using an advanced search. Fig. 4 illustrates an overview of the architecture we have implemented.

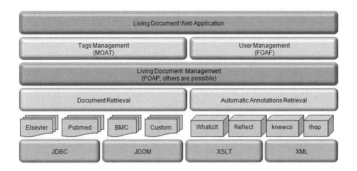

Fig. 4. General system architecture

In order to support other digital libraries and different annotation pipelines, the architecture supports the Service Provider Interface (SPI) paradigm for retrieval and annotation. The SPI is a software mechanism that supports replaceable components via a set of hooks. On top of these SPIs, we have build a semantic layer supporting MOAT, FOAF POAP and other controlled vocabularies. This makes it possible for new metadata to be managed so that more expressive queries can be supported.

As indicated above, we distinguish two types of tags, those generated by an automatic workflow, such as WhatIzIt, and those generated by human users. The automatic annotation infrastructure is built around the Monq software package (www.ebi.ac.uk/~kirsch/monq-doc/monq); this Java library enables the processing of text input streams based on regular expressions. The library binds regular expressions to actions that are automatically executed whenever a match occurs in the text stream being processed. A filter server is a computer program, also Java technology, which accepts TCP connections on a particular port from clients across the network. Each filter server specializes in recognizing the vocabulary of a particular terminology. Clients connect to a filter server and send a stream of text. The server runs its embedded DFA on the incoming text to recognize and tag the terminology with XML tags.

Multiple filter servers can be cascaded to form processing pipelines, whereby the output of one filter becomes the input of the next. In this case the XML tags added by each server carry the data needed to accomplish the tasks of complex distributed text mining algorithms. Currently, we support the annotation of Swissprot protein/gene names, drug names, organism names, disease names, chemical entities and Gene Ontology terms. The pipeline considers some disambiguation based on acronym resolution and term frequency. Protein/Gene names resembling acronyms, for instance NPY (neuropeptide Y), are analyzed in order to disambiguate whether the target name is really a Protein/Gene name. If this is unclear, then the pipeline will assume that names with a high frequency in the British National Corpus (http://www.natcorp. ox.ac.uk) are common enough to be considered relevant in the biomedical field. The result of this process is an extended XML document with the annotated information.

For instance, users can tag the paper using GO [7], as illustrated in Fig. 5; once the set of tags has been generated users are free to add or modify the set of predefined tags. The functionality embedded within the set of predefined tags also includes a set of predefined links built over each tag. For instance, the GO term "membranes" has a direct reference to its corresponding DB entry in AMIGO (http://amigo.geneontology.org), as

Fig. 5. Users are able to generate sets of automatically generated tags (with references to external resources) by using existing ontologies and/or annotation pipelines

illustrated in the middle of Fig. 5. By default these tags are automatically linked to a set of external resources such as ENTREZ (http://www.ncbi.nlm.nih.gov/Entrez), AMIGO, SRS (http://srs.ebi.ac.uk) and others. Both the ontologies being used to support the automatic tagging, and the corresponding external resources being used by default, can readily be increased.

Adding new external resources, such as URIs capable of either consuming tags as input data types or just providing additional information is currently achieved by typing the information into a text field. By adding new tags and external resources, the clouds of tags are constantly being updated for each paper. Furthermore, users can use the clouds of tags in order to find more information and build more expressive and accurate queries, as suggested in Fig. 6. In addition, users can identify tags that are associated with two or more documents. This tag association is a measure that indicates, based on the tags, and how closely related the documents are. More importantly, users can browse through the surrounding area of the tags so that they can contextually discover how valid the coincidence in tags may be. For instance, two documents may share the user-generated tag "<X>" but this coincidence alone may not be an indicator that both documents are actually related because one tag may have multiple meanings or be applied in different contexts.

Fig. 6. Using tags to refine queries

An interesting side effect of social tagging is the semi-automatic generation of social trust rankings. As users tag documents, these tags are easily identified by ownership; in this way any user can ask for someone else's tags without including in the cloud of tags those generated by anybody else. Experts know who the experts are in their field. A user can select only those tags generated for a given paper by user X, Y, and W; people tend to trust those tags generated by particular members of the community.

4 Discussion, Conclusions and Future Work

We have developed a web prototype that allows the ontology-based or user-generated tagging of atomic components within the structure of scientific papers in life sciences. The prototype also makes it possible for users to use tags in order to better filter search results; users can always refine their searches by adding terms from the cloud of tags. Finding related papers is supported not only by those available tags, but also by the prototype's use of eTBlast (http://invention.swmed.edu/etblast); thus allowing the user to input an entire paragraph and receive MEDLINE abstracts that are similar to it. The LD enables authors to easily add scientific hyperlinks to their documents and research papers as semantic annotations, drawn from ontologies or provided by their peers; this has the additional function of linking their papers to the Web in a meaningful way. The LD is complementary to the MS Word Add On (http://ucsdbiolit.codeplex.com) recently released (May 11/09); however it goes one step further as it involves a massive collaboration in the process of generating semantic annotations; furthermore, it is appropriate for existing digital libraries.

An informal controlled experiment was conducted; our collaborators in this evaluation were plant biologists, Researchers A and B, see Fig. 7. Both were requested to independently tag some of the papers we gave them. Once they had tagged the papers these were exchanged, allowing each researcher to see the other's tags. Interestingly, via the generated tags they discovered how two of these papers were related. More importantly, it was clear how tags made it easier for Researcher A to find information that could have taken longer for him to gather – external bibliographic references and links to DBs. Also, and equally important, was the use Researcher A made of some tags when further constructing queries. Researcher A made combined tags from both clouds, as our system uses query previews only combinations of tags with non-empty results sets were allowed.

We requested from another plant biologist (Researcher C) to tag between 3 to 5 papers. The main interest for this researcher was hormones involved in defensive responses to pathogens. Researcher C tagged without any controlled vocabulary. Initially, he tagged and then identified external resources for most of the generated tags. Some interesting outcomes arose from this exercise: i) synonymies not previously reported. These had to be found manually by establishing the relationship between the concept as reported in the paper and that stored in the DB. For instance, flg22 is a motif that belongs to the N-terminal conserved domain of *bacterial flagellin*; this was not explicit in the papers nor was it clears in any of the DBs –PFAM, UniProt, AMIGO, AraCyc. ii) Researcher C manually tagged genes, protein families and metabolic pathways. Once the tags were generated, it became clear for Researcher C how the clouds of tags could facilitate the construction of more targeted queries such as "genes involved in signaling pathways in response to different pathogens", "specific genes for each pathogen" and "genes involved in pathway cross talking".

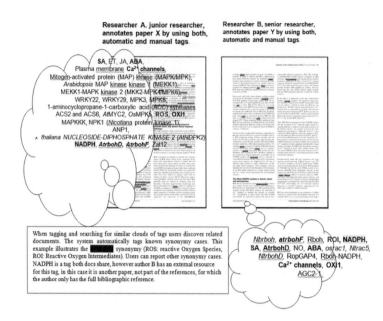

Fig. 7. An initial test for LD

Interestingly, for those papers we used in our tests the recommendations given by the original digital library were not always consistent with the papers discovered by the generated tags. The tag-based recommendation was more accurate. The axes, tag, tagged_object, and tagger, over which the tagging process was happening were consistent with those described by MOAT and SCOT; such simple models proved to be easily extendible so that semantic and syntactic components of a scientific paper could be represented by POAP and coherently orchestrated with MOAT, SCOT and other existing tag-related models as well as biomedical ontologies. Although POAP does not support the modeling of rhetorical structures within the paper and the publishing workflow, it has been conceived to interoperate with such ontologies.

Folksonomies have recently gained attention from the research community [18], partly because of their rapid and spontaneous growth and partly because of the need for structuring and classifying information. Although social tagging is widely used, as demonstrated in numerous applications, clouds of tags *per se* are not formal classification systems; rather, they are a complement for organization systems. For the task of finding information, taxonomies tend to be rigid and purely text-based search is not optimal. By contrast, tags introduce distributed human intelligence into the system [19]; also, tags are axes over which collaboration is supported.

When researchers collaborate, the structure of the collaboration is based upon similarities in their work; an aspect of this similarity may be defined by the literature that researchers are reading. It has been observed that if two researchers use similar literature they are working on similar or conceptually related problems. Bibliographic references, keywords and abstracts are therefore a valuable starting point for supporting information retrieval across large digital libraries and interaction based on similar interests. However, they do not offer the possibility for establishing networks of associated concepts across papers (NACAP), nor do they offer the facility for linking concepts from the paper to external resources (P2ext).

Both of these features, NACAP and P2ext, are central to the structure of real collaboration amongst researchers. Extending support for this interaction by using social tagging over entire documents augments the support for the collaboration based on similar research interests. Moreover, the metadata thus generated can help researchers to retrieve an article when descriptive elements are not known. The main function of these tags is not then to support a fixed taxonomical classification, but a dynamic one [9], allowing the relevance and content of articles to keep pace with the evolution of scientific discovery. This dynamic annotation is the serendipity that may allow researchers to find other researchers working on similar areas as they find papers via generated tags. Providing a platform where both NACAP and P2ext can be exploited, both at the time of authoring the paper and post-publication, will greatly enhance social networking and information retrieval in life sciences. This enhancement allows knowledge to be discovered more expediently and facilitates the formation of subnetworks of collaboration over specific knowledge units, defined by tags. Social tagging provides direct insight into the knowledge conveyed within the body of the scientific paper.

Ideally social networks and ontologies should help in offering an environment in which researchers can take advantage of collective knowledge. In principle, efforts such as Delicious and Connotea facilitate both, social interaction and harvesting the collective intelligence. Delicious offers a collective annotation facility for bookmarks in which the community interacts via the annotations they generate over their own bookmarks; the community as a whole benefits from everybody's knowledge as it is always possible to access everybody's annotations. Connotea is a more targeted environment in which users share their bibliographic references as well as those annotations that describe the shared set of references. The Connotea approach relies on the assumption that users with similar interests should have bibliographic references in common. The limitation of Connotea is that it assumes that an annotation, independently from the nature of the annotated object, always has a structure similar to that of a bibliographic reference. In order to offer insights into papers, systems similar to Connotea facilitate the manipulation of abstracts. However, abstracts *per se* do not provide a full summary of the work described in some document, nor do they offer any way to integrate the document into existing knowledge. Connotea and similar systems do not support the entire structure of collaboration that is usually found within and across communities of researchers.

An approach similar to the one presented in this paper is Concept Web Linker (http://conceptweblinker.wikiprofessional.org/, CWL); both approaches assume a document deeply interconnected with other documents and with the Web. However, unlike CWL, the LD approach embraces a generative technology so users can actively generate the tools they need for the kind of information they want to manage. Furthermore Concept Web Linker does not allow users to better-define queries based on available tags; thus limiting the usability of the cloud of tags. The generative capacity of a system should be understood as "its capacity to produce unanticipated change through unfiltered contributions from broad and varied audiences" [20]. Currently, the generative capacities in the Concept Web Linker are limited; for instance it is not possible to add annotation pipelines. More importantly, Concept Web Linker does not make use of social networking technology. Generative technology enables an open market for which specialized plug-ins can be developed; in this way better and more specialized mining tools will be built over digital libraries.

Loading XSLT files is supported by our prototype; this flexibility was deliberately designed into the LD as we are planning to load PubMed and DBLP for further testing. We are also carrying out more detailed evaluations not only concerning the tagging and its related operations but also regarding the added valued required for communities to adopt and actively participate in scientific oriented folksonomies. An important aspect we are currently improving is better scoping of POAP; we are narrowing POAP so it models only conceptual relationships across papers. The Structural Ontology for Document Annotation (SODA) should represent the other aspects. SODA incorporates rhetoric, tagging, and document structure, and inherits the interplay with biomedical ontologies; SODA is inspired by the Semantic Annotation of Latex (SALT, http://salt.semanticauthoring.org/ontologies.html) ontology. Our interest is the same, the intersection between the SW and the Social Web in life sciences.

An interesting aspect that arose from our work with biologists was the need for intelligent interfaces —those consistent with the information they are delivering; we are consequently redesigning our interfaces to cater to this requirement. Support for browsers other than IE as well as the release of the project to the source-forge community is also on our to-do list. And finally, we are conducting further research into the relationship between ontologies and tagsonomies.

Acknowledgments. Funding: Alexander Garcia acknowledges the financial support of the EU: Project OASIS Integrated Project, Grant Agreement # 215754.

References

1. Wright, M., Sumner, T., Moore, R., Koch, T.: Connecting digital libraries to eScience: the future of scientific scholarship. International Journal of Digital Libraries, 1–4 (2007)
2. Wilkinson, M.: BioMOBY: an open-source biological web services proposal. Briefings In Bioinformatics 3, 331–341 (2002)
3. Bindelli, S., Criscione, C., Curino, C.A., Drago, M.L., Eynard, D., Orsi, G.: Improving search and navigation by combining ontologies and social tags. In: Meersman, R., Tari, Z., Herrero, P. (eds.) OTM-WS 2008. LNCS, vol. 5333, pp. 76–85. Springer, Heidelberg (2008)
4. Chen, M., Liu, X., Qin, J.: Semantic Relation Extraction from Socially-Generated Tags: A methodology for Metadata Generation. In: Proc. Int'l Conf. on Dublin Core and Metadata Applications (2008)
5. Hunter, J., Khan, I., Gerber, A.: HarvANA - Harvesting Community Tags to Enrich Collection Metadata. In: Pittsburgh, P., USA (ed.) ACM IEEE Joint Conference on Digital Libraries, JCDL 2008, Pittsburgh, PA (2004)
6. Berners-Lee, T., Hendler, J., Lassila, O.: The Semantic Web. Scientific American (2001)
7. Ashburner, M., Ball, C., Blake, J., Botstein, D., Butler, H., Cherry, J., Davis, A., Dolinski, K., Dwight, S., Eppig, J., Harris, M., Hill, D., Issel, T.L., Kasarskis, A., Lewis, S., Matese, J., Richardson, J., Ringwald, M., Rubin, G., Sherlock, G.: Gene Ontology: tool for the unification of biology. The Gene Ontology Consortium. Nature Genetics 25, 25–29 (2000)

8. Mehrian-Shai, R., Chen, C.D., Shi, T., Horvath, S., Nelson, S.F., Reichardt, J.K.V., Saw-yers, C.L.: Insulin growth factor-binding protein 2 is a candidate biomarker for PTEN status and PI3K/Akt pathway activation in glioblastoma and prostate cancer. PNAS 104, 5563–5568 (2007)

9. Gendarmi, D., Abbattista, F., Lanubile, F.: Fostering knowledge evolution through com-munity-based participation. In: 16th International World Wide Web Conference, Banff, Alberta, Canada (2007)

10. Mons, B., Ashburner, M., Chichester, C., Van Milligen, E., Weeber, M., den Dunnen, J., Bairoch, A.: Calling on a million minds for community annotation in WikiProteins. Ge-nome Biology (2008)

11. Backhaus, M., Kelso, J.: BIOWiki - a collaborative annotation and ontology curation framework. In: 16th International World Wide Web Conference, Banff, Alberta, Canada (2007)

12. De Roure, D., Goble, C., Aleksejevs, S., Bechhofer, S., Bhagat, J., Cruickshank, D., Michaelides, D., Newman, D.: The myExperiment Open Repository for Scientific Work-flows. Open Repositories Atlanta, Georgia, US (2009)

13. Shotton, D., Portwin, K., Klyne, G., Miles, A.: Adventures in Semantic Publishing: Exem-plar Semantic Enhancements of a Research Article. PLOS Computational Biology 5 (2009)

14. Passant, A.: The Meaning Of A Tag (2008), http://moat-project.org/

15. Kim, L.-H., Breslin, J., Scerri, S., Deker, S., Kim, H., Yang, S.-K.: SCOT Ontology Speci-fication (2008), http://scot-project.org/scot/

16. Rebholz-Schuhmann, D., Arregui, M., Gaudan, M., Kirsch, H., Jimeno, A.: Text process-ing through Web Services: Calling Whatizit. Bioinformatics 24 (2007)

17. Labarga, A., Franck, V., Anderson, M., Lopez, R.: Web Services at the European Bioin-formatics Institute. Nucleic Acid Research 35 (2007)

18. Kim, L.-H., Scerri, S., Breslin, J., Decker, S., Kim, H.: The state of the Art in Tag Ontolo-gies: A Semantic Model for Tagguing and Folksonomies. In: International Conference on Dublin Core and Metadata Applications, Berlin, Germany (2008)

19. Gruber, T.: Ontology of Folksonomy: a Mash-up of apples and Oranges. International Journal of Semantic Web and Information Systems 3 (2005)

20. Zittrain, J.: The Future of the Internet and how to stop it. Yale Universoty Press, New Heaven (2008)

Web Navigation Sequences Automation in Modern Websites

Paula Montoto, Alberto Pan, Juan Raposo, Fernando Bellas, and Javier López

Department of Information and Communication Technologies, University of A Coruña
Campus de Elviña s/n 15071 A Coruña, Spain
{pmontoto,apan,jrs,fbellas,jmato}@udc.es

Abstract. Most today's web sources are designed to be used by humans, but they do not provide suitable interfaces for software programs. That is why a growing interest has arisen in so-called web automation applications that are widely used for different purposes such as B2B integration, automated testing of web applications or technology and business watch. Previous proposals assume models for generating and reproducing navigation sequences that are not able to correctly deal with new websites using technologies such as AJAX: on one hand existing systems only allow recording simple navigation actions and, on the other hand, they are unable to detect the end of the effects caused by an user action. In this paper, we propose a set of new techniques to record and execute web navigation sequences able to deal with all the complexity existing in AJAX-based web sites. We also present an exhaustive evaluation of the proposed techniques that shows very promising results.

Keywords: Web automation, web integration, web wrappers.

1 Introduction

Web automation applications are widely used for different purposes such as B2B integration, web mashups, automated testing of web applications or business watch. One crucial part in web automation applications is to allow easily generating and reproducing navigation sequences. We can distinguish two stages in this process:

- Generation phase. In this stage, the user specifies the navigation sequence to reproduce. The most common approach, cf. [1,9,11], is using the 'recorder' metaphor: the user performs one example of the navigation sequence using a modified web browser, and the tool generates a specification which can be run by the execution component. The generation environment also allows specifying the input parameters to the navigation sequence.
- Execution phase. In this stage, the sequence generated in the previous stage and the input parameters are provided as input to an automatic navigation component which is able to reproduce the sequence. The automatic navigation component can be developed by using the APIs of popular browsers (e.g. [9]). Other systems like [1] use simplified custom browsers specially built for the task.

S.S. Bhowmick, J. Küng, and R. Wagner (Eds.): DEXA 2009, LNCS 5690, pp. 302–316, 2009.

Most existing previous proposals for automatic web navigation systems (e.g. [1,9,11]) assume a navigation model which is now obsolete to a big extent: on one hand, the user actions that could be recorded were very restrictive (mainly clicking on elements and filling in form fields) and, on the other hand, it was assumed that almost every user action caused a request to the server for a new page.

Nevertheless, this is not enough for dealing with modern AJAX-based websites, which try to replicate the behavior of desktop applications. These sites can respond to a much wider set of user actions (mouse over, keyboard strokes, drag and drop…) and they can respond to those actions executing scripting code that manipulates the page at will (for instance, by creating new graphical interface elements on the fly). In addition, AJAX technology allows requesting information from the server and repainting only certain parts of the page in response.

In this paper, we propose a set of new techniques to build an automatic web navigation system able to deal with all this complexity. In the generation phase, we also use the 'recorder' metaphor, but substantially modified to support recording a wider range of events; we also present new methods for identifying the elements participating in a navigation sequence in a change-resilient manner.

In the execution phase, we use the APIs of commercial web browsers to implement the automatic web navigation components (the techniques proposed for the recording phase have been implemented using Microsoft Internet Explorer (MSIE) and the execution phase has been implemented using both MSIE and Firefox); we take this option because the approach of creating a custom browser supporting technologies such as scripting code and AJAX requests is effort-intensive and very vulnerable to small implementation differences that can make a web page to behave differently when accessed with the custom browser. In the execution phase, we also introduce a method to detect when the effects caused by a user action have finished. This is needed because one navigation step may require the effects of the previous ones to be completed before being executed.

2 Models

In this section we describe the models we use to characterize the components used for automated browsing. The main model we rely on is DOM Level 3 Events Model [3]. This model describes how browsers respond to user-performed actions on an HTML page currently loaded in the browser. Although the degree of implementation of this standard by real browsers is variable, the key assumptions our techniques rely on are verified in the most popular browsers (MSIE and Firefox). Therefore, section 2.1 summarizes the main characteristics of this standard that are relevant to our objectives. Secondly, section 2.2 states additional assumptions about the execution model employed by the browser in what regards to scripting code, including the kind of asynchronous calls required by AJAX requests. These assumptions are also verified by current major browsers.

2.1 DOM Level 3 Events Model

In the DOM Level 3 Events Model, a page is modelled as a tree. Each node in the tree can receive events produced (directly or indirectly) by the user actions. Event types exist for actions such as clicking on an element (*click*), moving the mouse cursor over it (*mouseover*) or specifying the value of a form field (*change*), to name a few. Each node can register a set of event listeners for each event type. A listener executes arbitrary code (typically written in a script language such as Javascript). Listeners have the entire page tree accessible and can perform actions such as modifying existing nodes, removing them, creating new ones or even launching new events.

The event processing lifecycle can be summarized as follows: The event is dispatched following a path from the root of the tree to the target node. It can be handled locally at the target node or at any target's ancestors in the tree. The event dispatching (also called event propagation) occurs in three phases and in the following order: capture (the event is dispatched to the target's ancestors from the root of the tree to the direct parent of the target node), target (the event is dispatched to the target node) and bubbling (the event is dispatched to the target's ancestors from the direct parent of the target node to the root of the tree). The listeners in a node can register to either the capture or the bubbling phase. In the target phase, the events registered for the capture phase are executed before the events executed for the bubbling phase. This lifecycle is a compromise between the approaches historically used in major browsers (Microsoft IE using bubbling and Netscape using capture).

The order of execution between the listeners associated to an event type in the same node is registration order. The event model is re-entrant, meaning that the execution of a listener can generate new events. Those new events will be processed synchronously; that is, if l_i, l_{i+1} are two listeners registered to a certain event type in a given node in consecutive order, then all events caused by l_i execution will be processed (and, therefore, their associated listeners executed) before l_{i+1} is executed.

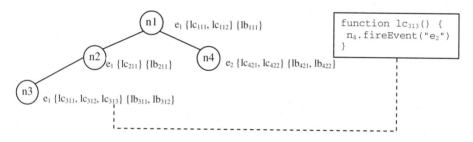

$$\{lc_{111}, lc_{112}, lc_{211}, lc_{311}, lc_{312}, lc_{313}, \{lc_{421}, lc_{422}, lb_{421}, lb_{422}\}, lb_{311}, lb_{312}\, lb_{211}, lb_{111}\}$$

Fig. 1. Listeners Execution Example

Example 1: Fig. 1 shows an excerpt of a DOM tree and the listeners registered to the event types e_1 and e_2. The listeners in each node for each event type are listed in registration order (the listeners registered for the capture phase appear as l_{cxyz} and the ones registered for the bubbling phase appear as l_{bxyz}). The figure also shows what listeners and in which order would be executed in the case of receiving the event-type e_1 over the node n_3, assuming that the listener on the capture phase l_{c313} causes the event-type e_2 to be executed over the node n_4.

DOM Level 3 Events Model provides an API for programmatically registering new listeners and generating new events. Nevertheless, it does not provide an introspection

API to obtain the listeners registered for an event type in a certain node. As we will see in section 3.1, this will have implications in the recording process in our system.

2.2 Asynchronous Functions and Scripts Execution Model

In this section we describe the model we use to represent how the browser executes the scripting code of the listeners associated to an event. This model is verified by the major commercial browsers.

The script engine used by the browser executes scripts sequentially in single-thread mode. The scripts are added to an execution queue in invocation order; the script engine works by sequentially executing the scripts in the order specified by the queue.

When an event is triggered, the browser obtains the listeners that will be triggered by the event and invokes its associated scripts, causing them to be added to the execution queue. Once all the scripts have been added, execution begins and the listeners are executed sequentially.

The complexity of this model is slightly increased because the code of a listener can execute asynchronous functions. An asynchronous function executes an action in a non-blocking form. The action will run on the background and a callback function provided as parameter in the asynchronous function invocation will be called when the action finishes.

The most popular type of asynchronous call is the so-called AJAX requests. An AJAX request is implemented by a script function (i.e. in Javascript, a commonly used one is *XMLHTTPRequest*) that launches an HTTP request in the background. When the server response is received, the callback function is invoked to process it.

Other popular asynchronous calls establish timers and the callback function is invoked when the timer expires. In this group, we find the Javascript functions *setTimeout(ms)* (executes the callback function after ms milliseconds) and *setInterval(ms)* (executes the callback function every ms milliseconds). Both have associated cancellation functions: *clearTimeout(id)* and *clearInterval(id)*.

It is important to notice that, from the described execution model, it is inferred the following property:

Property 1: The callback functions of the asynchronous calls launched by the listeners of an event will never be executed until all other scripts associated to that event have finished.

The explanation for this property is direct from the above points: all the listeners associated to an event are added to the execution queue first, and those listeners are the ones invoking the asynchronous functions; therefore, the callback functions will always be positioned after them in the execution queue even if the background action executed by the asynchronous call is instantaneous.

3 Description of the Solution

In this section we describe the proposed techniques for automated web navigation. First, we deal with the generation phase: section 3.1 describes the process used to record a navigation sequence in our approach. Section 3.2 deals with the problem of identifying the target DOM node of a user action: this problem consists in generating a path to the node that can be used later at the execution phase to locate it in the page and section 3.3 deals with the execution phase.

3.1 Recording User Events

The generation phase has the goal of recording a sequence of actions performed by the user to allow reproducing them later during the execution phase.

A user action (e.g. a click on a button) causes a list of events to be issued to the target node, triggering the invocation of the listeners registered for them in the node and its ancestors, according to the execution model described in the previous section. Notice that each user action usually generates several events. For instance, the events generated when the user clicks on a button include, among others, the mouseover event besides of the click event, since in order to click on an element it is previously required to place the mouse over it. Recording a user action consists in detecting which events are issued, and in locating the target node of those events.

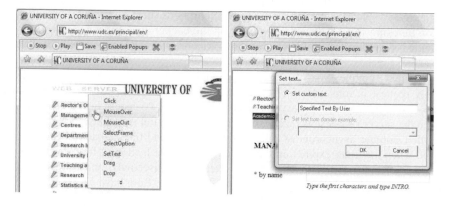

Fig. 2. Recording Method

In previous proposals, cf. [1,6,9], the user can record a navigation sequence by performing it in the browser in the same way as any other navigation. The method used to detect the user actions in these systems is typically as follows: the recording tool registers its own listeners for the most common events involved in navigations (mainly clicks and the events involved in filling in form fields) in anchors and form-related tags. This way, when a user action produces one of the monitored event-types e on one of the monitored nodes n, the listener for e in n is invoked, implicitly identifying the information to be recorded.

Nevertheless, the modern AJAX-based websites can respond to a much wider set of user actions (e.g. placing the mouse over an element, producing keyboard strokes, drag and drop...); in addition, virtually any HTML element, and not only traditional navigation-related elements, can respond to user actions: tables, images, texts, etc.

Extending the mentioned recording process to support AJAX-based sites would involve registering listeners for every event in every node of the DOM tree (or, alternatively, registering listeners for every event in the root node of the page, since the events execution model ensures that all events reach to the root). Registering listeners for every event has the important drawback that it would "flood" the system by recording unnecessary events (e.g. simply moving the mouse over the page would generate hundreds of *mouseover* and *mouseout* events); recall that, as mentioned in section 2, it is not possible to introspect what events a node has registered a listener for; therefore, it is not possible to use the approach of registering a listener for an event-type *e* only in the nodes that already have other listeners for *e*.

Therefore, we need a new method for recording user actions. Our proposal is letting the user explicitly specify each action by placing the mouse over the target element, clicking on the right mouse button, and choosing the desired action in the contextual menu (see Fig. 2). If the desired action involves providing input data into an input element or a selection list, then a pop-up window opens allowing the user to specify the desired value (see Fig. 2). Although in this method the navigation recording process is different from normal browsing, it is still fast and intuitive: the user simply changes the left mouse button for the right mouse button and fills in the value of certain form fields in a pop-up window instead of in the field itself.

This way, we do not need to add any new listener: we know the target element by capturing the coordinates where the mouse pointer is placed when the action is specified, and using browser APIs to know what node the coordinates correspond to. The events recorded are implicitly identified by the selected action.

Our prototype implementation includes actions such as *click, mouseover, mouseout, selectOption* (selecting values on a selection list), *setText* (providing input data into an element), *drag* and *drop*. Since each user action actually generates more than one event, each action has associated the list of events that it causes: for instance, the *click* action includes, among others, the events *mouseover, click* and *mouseout*; the *setText* action includes events such as *keydown* and *keyup* (issued every time a key is pressed) and *change* (issued when an element content changes).

This new method has a problem we need to deal with. By the mere process of explicitly specifying an action, the user may produce changes in the page before we want them to take place. For instance, suppose the user wishes to specify an action on a node that has a listener registered for the *mouseover* event; the listener opens a pop-up menu when the mouse is placed over the element. Since the process of specifying the action involves placing the mouse over the element; the page will change its state (i.e. the pop-up menu will open) before the user can specify the desired action. This is a problem because the process of generating a path to identify the target element at the execution phase (described in detail in section 3.2) cannot start until the action has been specified. But, since the DOM tree of the page has already changed, the process would be considering the DOM tree after the effects of the action have taken place (the element may even no longer exist because the listeners could remove it!).

We solve this problem by deactivating the reception of user events in the page during the recording process. This way, we can be sure that no event alters the state of the page before the action is specified. Once the user has specified an action, we use the browser APIs to generate on the target element the list of events associated to the action; this way, the effects of the specified action take place in the same way as if the user would have performed the action, and the recording process can continue.

Another important issue we need to deal with is ensuring that a user does not specify a new action until the effects caused by the previous one have completely finished. This is needed to ensure that the process for generating a path to identify at the execution phase the target element of the new action has into account all the changes in the DOM tree that the previous action provokes. Detecting the end of the effects of an action is far from a trivial problem; since it is one of the crucial issues at the execution phase, we will describe how to do it in section 3.3.

3.2 Identifying Elements

During the generation phase, the system records a list of user actions, each one performed on a certain node of the DOM tree of the page. Therefore, we need to generate an expression to uniquely identify the node involved in each action, so the user action can be automatically reproduced at the execution phase.

An important consideration is that the generated expression should be resilient to small changes in the page (such as the apparition in the page of new advertisement banners, new data records in dynamically generated sections or new options in a menu), so it is still valid at the execution stage.

To uniquely identify a node in the DOM tree we can use an XPath [15] expression. XPath expressions allow identifying a node in a DOM tree by considering information such as the text associated to the node, the value of its attributes and its ancestors. For our purposes, we need to ensure that the generated expression on one hand identifies a single node, and on the other hand it is not too specific to be affected by the formerly mentioned small changes. Therefore, our proposal tries to generate the less specific XPath expression possible that still uniquely identifies the target node. The algorithm we use for this purpose is presented in section 3.2.1.

In addition, the generated expressions should not be sensible to the use of session identifiers, to ensure that they will still work in any later session. Section 3.2.2 presents a mechanism to remove session identifiers from the generated expressions.

3.2.1 Algorithm for Identifying Target Elements

This section describes the algorithm for generating the expression to identify the target element of a user action.

As it has already been said, the algorithm tries to generate the less specific XPath expression possible that still uniquely identifies the target node. More precisely, the algorithm first tries to identify the element according to its associated text (if it is a leaf node) and the value of its attributes. If this is not enough to uniquely identify the node, its ancestors (and the value of their attributes) will be recursively used. The algorithm to generate the expression for a node n consists of the following steps:

0)   Initialize $X_{[n]}$ (the variable that will contain the generated expression) to the empty string. Initialize the variable n_i to the target node n.

1)   Let m be the number of attributes of n_i, T_{ni} be the tag associated to n_i and t_{ni} be its associated text. Try to find a minimum set of r attributes $\{a_{ni1},...,a_{nir}\}_{r<=m}$, of $_{ni}$ such that the following expression ('+' represents the concatenation of two strings):

"//" $+ T_{ni} [@a_{ni1}=v_{ni1}$ and... and $@a_{nir}=v_{nir}$ and $@text()=t_{ni}] + X_{[n]} + "/"$

uniquely identify n. (NOTE:The fragment and $text()=t_{ni}$ of the expression would only be added if n_i is a leaf node, since only leaf nodes have associated text).

2)   If the set is found then

     2.1)   return the expression from step 1.

else

2.2)   Let $\{a_{ni1},...,a_{nim}\}$ be the set of all attributes of n_i. Set $X_{\{n\}} = $ "/"$+T_{ni}$ $[@a_{ni1}=v_{ni1}$ and... and $@a_{nim}=v_{nim}$ and $@text()=t_{ni}] + X_{\{n\}}$; that is, we add conditions by all the attributes of n_i to the expression.

3)   If n_i is not the root of the DOM tree then

3.1)   Set $n_i=parent(n_i)$ and go to step 1

else

3.2)   Obtain the relative position j of n in the page with respect to all the nodes verifying the current expression $X_{\{n\}}$. Return "/"$+ X_{\{n\}}+ [j] + $"/".

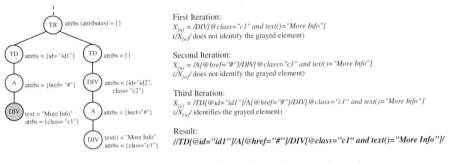

Fig. 3. Algorithm for Identifying Target Elements Example

Fig. 3 shows an example sub-tree and the $X_{\{n\}}$ value in each iteration of the algorithm to generate the XPath expression to identify the grayed *DIV* node.

Now, we provide further detail about some of the steps. The step 1 of the algorithm tries to identify the minimum set of attributes of the currently considered node n_i, that allow completing the identification of n. To do this, we add attributes one by one until either n is uniquely identified or all the attributes of n_i, have been added. To decide the order in which we add the attributes, we have defined an order for the attributes of each HTML tag based on its estimated selectivity (that is, how much they contribute to narrow the selection). For instance, we consider the *id* and *name* attributes highly selective for all HTML tags and the *href* attribute highly selective for the *A* tag, while we consider the class attribute as of low selectivity.

Step 3.2 considers the case when the algorithm reaches the root, and the generated expression still does not uniquely identify n. In that case, the algorithm adds to the XPath expression the relative position in the page of n with respect to the rest of elements identified by the expression.

3.2.2 Removing Session IDs

Many websites use session identifiers in URL attributes to track user sessions. In these sites, the values of attributes containing URLs may vary in each session. Since our method to identify target elements at the execution phase relies on attribute values, this causes a problem for our approach.

Our prototype implementation recognizes the main standard formats for including session identifiers in URLs. Unfortunately, many websites do not use any standard, but include the session identifier using arbitrary query parameters.

Therefore, we propose an algorithm to generalize the value of attributes containing URLs. The algorithm is based on two observations: 1) a query parameter acting as session identifier must take the same value in all the URLs of the page in which it appears; 2) if a query parameter takes the same value in all URLs with the same host and query parts, then it is irrelevant for the purpose of identifying an element in the DOM tree by the value of its attributes.

The basic idea of the algorithm derives directly from the above observations: find all the query parameters that take the same value in all the URLs in which they appear and ignore their values for identification purposes. Although some of the identified query parameters may not be session identifiers, according to observation 2 it is safe to ignore their values anyway.

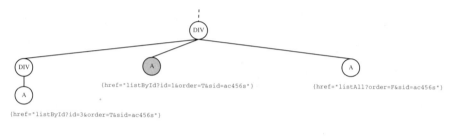

//A[matches(@href,"listById?id=1&order=[^&=]+&sid=[^&=]+")]/

Fig. 4. Removing Session IDs Example

Fig 4 shows a simple example of the algorithm where *n* is the grayed node in the figure. The query parameters named *order* and *sid* take the same value in all the URLs with the same path (in the example the page does not contain other URLs with the same path). Therefore, they are considered irrelevant for node identification purposes. (NOTE: *matches()* is XPath function for applying regular expressions).

3.3 Execution Phase

The generation phase generates a program capturing the navigation sequence recorded by the user. The execution phase runs the program in the automatic navigation component.

A first consideration is that we opt to use the APIs of commercial web browsers to implement the automatic web navigation components instead of building a simplified custom-browser. The main reason for taking this option is that web 2.0 sites make an intensive use of scripting languages and support a complex event model. Creating a custom browser supporting those technologies in the same way as commercial browsers is very effort-intensive and, in addition, is extremely vulnerable to small implementation differences that can make a web page to behave differently when accessed with the custom browser than when accessed with a "real" browser. Our techniques for the execution phase have been implemented in both MSIE and Firefox.

To reproduce an action in the navigation sequence, there are three steps involved:

1. Locating the target node in the DOM tree of the page.
2. Generating the recorded event (or list of events) on the identified node.

3. Wait for the effects of the events to finish. This is needed because the following action can need the effects of the previous ones to be completed (e.g. the action *n+1* can generate an event on a node created in the action *n*).

The implementation of 1) and 2) is quite straightforward using browser APIs and given the output of the recording process. Step 1) uses the XPath expression produced by the process described in section 3.2, and step 2) uses the events recorded in the process described in section 3.1.

In turn, step 3) is difficult because browser APIs do not provide any way of detecting when the effects on the page of issuing a particular event have finished. These effects can include dynamically creating or removing elements in the DOM tree, maybe also having into account the response to one or several AJAX requests to the server. Previous works have addressed this problem by establishing a timer after the execution of an event before continuing execution. This solution has the usual drawbacks associated to a fixed timeout in a network environment: if the specified timeout is short, then when the response to an asynchronous AJAX request is slower than usual (or even if the machine is very heavily loaded), the sequence may fail. If, in turn, we use a higher timeout valid even in those circumstances, then we are introducing an unnecessary delay when the server is responding normally.

The remaining of this section explains the method we propose to detect when the effects caused directly or indirectly by a certain event have finished. This way, the system waits the exact time required. The correctness of the method derives from the assumptions stated in section 2, which are verified by the major commercial browsers.

The method we use to detect when the effects of an event-type e generated on a node n have finished consists of the following steps:

1. We register a new listener *l* to capture the event *e* in *n*. The code of the listener *l* invokes an asynchronous function specifying the callback function *cf*. What asynchronous function is actually invoked in l is mainly irrelevant; for instance, in Javascript, we can simply invoke *setTimeout(cf,0)*. Notice that as consequence of property 1 in section 2, it is guaranteed that *cf* will be executed after all the listeners triggered by the execution of *e* have finished. Therefore, if the listeners had not made any other asynchronous call, then the control arriving to *cf* would indicate that the effects of *e* had finished and the navigation sequence execution could continue. Nevertheless, since the listeners can actually execute other asynchronous calls, this is not enough.

2. To be notified of every asynchronous call executed by the listeners triggered by *e*, we redefine those asynchronous functions providing our own implementation of them (for instance, in Javascript we need to redefine *setTimeout, setInterval* and the functions used to execute AJAX requests such as *XMLHTTPRequest*). The template of our implementation of each function is shown in Fig 5. The function maintains a counter that is increased every time the function is invoked (the counter is maintained as a global variable initialized to zero for every emitted event). After increasing the counter, the function calls the former standard implementation of the asynchronous function provided by the browser but substituting the received callback function by a modified one (the *new_cf* function created in Figure 5). This new callback function invokes the original callback function and

then decreases the counter. This way, the counter always takes the value of the number of currently active calls.

3. When the callback function *cf* from step 1 is executed, it polls the counters associated to the asynchronous functions. When they are all 0, we know the asynchronous calls have finished and execution can proceed.

4. There may be some cases where the effects of *e* actually never finish. This is for instance the case when the *setInterval* function is used. This function executes the callback function at specified time intervals and, therefore, its effects last indefinitely unless the function *clearInterval* is used. In the generation-phase, if the *setInterval* calls are not cleared after a certain timeout, the system notifies it to the user so she/he can specify the desired action, which can be to wait a fixed time or wait for a certain number of intervals to complete.

```
old_asyncFunction = standardAsyncFunction;
new_asyncFunction = new function(param1,param2,…,paramn,cf) {
    counter++;        //counter is a global variable
    new_cf = new function() {
        result = cf();
        counter--;
        if (counter==0) {
            notifyEndAsyncFunctions();
        }
        return result;
    };
    old_asyncFunction(param1,param2,…,paramn,new_cf);
};
standardAsyncFunction = new_asyncFunction;
```

Fig. 5. Asynchronous Function Redefinition

In addition of the possible effects of an event in the current page, the event can also make the browser (or a frame inside the page) navigate to a new page. When the new page/frame is loaded (this can be detected using browser APIs), the load event is generated; this event has as target the body element of the page. Before continuing the execution of the navigation sequence, we need to wait until the end of the effects of the load event have finished, using the same technique used for the rest of events.

4 Evaluation

To evaluate the validity of our approach, we tested the implementation of our techniques with a wide range of AJAX-based web applications. We performed two kinds of experiments:

1. We selected a set of 75 real websites making extensive use of scripting code and AJAX technology. We used the prototype to record and reproduce one navigation sequence on each site. The navigation sequences automated the main purpose of the site. For instance, in electronic shops we automated the process of searching products; in webmail sites we automated the process required to access e-mails.

2. Some of the main APIs for generating AJAX-based applications such as Yahoo! User Interface Library (YUI) [16] and Google Web Toolkit (GWT) [4] include a set of example websites. At the time of testing, GWT included 5 web applications and YUI included 300 examples. We recorded and executed 12 navigation sequences in the web applications from GWT ensuring that every interface element from the applications was used at least once. In the case of YUI, we recorded 40 sequences in selected examples (choosing the more complex examples). This second group of tests is useful because many real websites use those toolkits.

Table 1. Experimental Results

Website	Played	Website	Played	Website	Played
www.a9.com/java	✓	www.fidelityasap.com	✓	www.optize.es	✓
www.abebooks.com	✓	www.fnac.es	✓	www.paginasamarillas.es	✓
www.accorhotels.com	✓	www.gmail.com	✓	www.penguin.co.uk	✓
www.addall.com	✓	www.gongdiscos.com	✓	people.yahoo.com	✓
www.voyages-sncf.com	✓	www.hotelopia.es	✓	code.jalenack.com/periodic	✓
www.alitalia.com/ES_ES/	✓	www.hotelsearch.com	✓	www.pixmania.com	✓
www.allbooks4less.com	✓	www.iberia.com	✓	www.planethome.de	✓
www.amadeus.net	✓	www.iit.edu	✓	www.priceline.com	✓
www.amazon.com	✓	www.imdb.com/search	✓	www.renault.es	✓
store.apple.com	✓	www.infojobs.net	✓	www.renfe.es	✓
www.atrapalo.com	✓	www.jet4you.com	✓	www.reuters.com	✓
autos.aol.com	✓	www.laborman.es	✓	www.rumbo.es	✓
www.balumba.es	✓	www.landrover.com	✓	www.shop-com.co.uk	✓
www.barnesandnoble.com	✓	www.es.lastminute.com	✓	www.sparkassen-immo.de	✓
www.bookdepository.co.uk	✓	www.marsans.es	✓	www.sterling.dk	✓
www.booking.com	✓	www.meridiana.it	✓	www.ticketmaster.com	✓
www.carbroker.com.au	✓	www.msnbc.msn.com	✓	tudulist.com	✓
www.casadellibro.com	✓	www.muchoviaje.com	✓	www.tuifly.com/es	✓
www.cervantesvirtual.com	✓	www.musicstore.com	✓	es.venere.com	✓
www.cia.gov	✓	www.myair.com	✓	www.viajar.com	✓
controlp.com	✓	www.mymusic.com	✓	www.vuelosbaratos.es	✓
www.digitalcamerareview.com	✓	www.es.octopustravel.com	✓	www.webpagesthatsuck.com	✓
www.ebay.es	✓	www.ofertondelibros.com	✓	news.search.yahoo.com/news/advanced	✓
www.edreams.es	✓	www.okipi.com	✓	news.yahoo.com	✗
www.elcorteingles.es	✓	vols.opodo.fr	✓	mail.yahoo.com	✓

The techniques proposed for the recording phase have been implemented using MSIE and the execution phase has been implemented using both MSIE and Firefox. In each group of experiments, we recorded the navigation sequences on MSIE and executed them using both MSIE and Firefox. The execution on MSIE allows us to measure the effectiveness of our techniques in both the recording and execution phases. We execute the sequences in Firefox to check that the algorithm presented in section 3.3 is valid in both browsers. Since MSIE and Firefox usually build different DOM trees for the same pages, in some cases the XPath expression generated by the recording in MSIE were manually modified to fit the DOM tree in Firefox. Notice that this is not a limitation of our approach: it only highlights the issue that the browser used for the recording and execution phase should be the same.

The results of the evaluation were encouraging (see Table 1). In the first set of experiments (real websites), 74 of 75 sequences were recorded and executed fine.

In the case of *news.yahoo.com*, the XPath expression generated to identify an element used an URL with a query parameter which changed every time the page was reloaded. This parameter is not a session identifier since it changes its value during the same session. If the recorded XPath expression is modified manually to ignore the value of this parameter, then the sequence works correctly. To solve problems like this, we could include redundant localization information; this way, if an element cannot be identified using the "minimal" expression, then we can still use the other information to search the nearest match in the page ([1] uses a similar idea that could be extended to deal with these cases, although they do not use other necessary information, such as hierarchical information). Another option is allowing the user to provide several examples of the same sequence for detecting those parameters.

The second group of tests was completely successful in GWT applications, while in the YUI case only one sequence could not be recorded. The problem was that the *blur* event was not being generated with the *setText* action. Once this was corrected, the sequence could be recorded.

5 Related Work

WebVCR [1] and WebMacros [11] were pioneer systems for web navigation sequences automation using the "recorder metaphor". Both systems were only able to record a reduced set of events (clicks and filling in form fields) on a reduced set of elements (anchors and form-related elements). In the execution phase they relied on HTTP clients that lacked the ability to execute scripting code or to support AJAX requests. Furthermore, the techniques they used for identifying the target elements of user actions were based on the text associated to the elements and the value of some specific pre-configured attributes (e.g. *href* for *A* tags and *src* for *FORM* tags).

Wargo [9] introduced using a commercial browser as execution component, thus supporting websites using scripting languages and guaranteeing that the websites will behave in the execution phase in the same way as when a human user accesses it. Nevertheless, it still showed the remaining previously mentioned problems.

Instead of using the "recorder" metaphor, in SmartBookmarks [6] the macros are generated retroactively; when the user reaches a page and bookmarks it, the system tries to automatically find the starting point of the macro. In order to do this, Smart-Bookmarks permanently monitors the user actions. As it was explained in section 3.1, recording user actions in the browser as the user navigates forces to either restrict the set of monitored events or suffering from an "event-flooding" problem. SmartBookmarks only supports the events click, load and change. Another drawback is that it relies on timeouts to determine when to continue executing the sequence. HtmlUnit [5] is an open-source tool for web applications unit testing. HtmlUnit does not provide a recording tool; instead, the user needs to manually create the navigation sequences using Java coding. In addition HtmlUnit uses its own custom browser instead of relying on conventional browsers. Although their browser has support for many Javascript and AJAX functionalities, this is vulnerable to small implementation differences that can make a web page to behave differently when accessed with the custom browser.

Selenium [13] is a suite of tools to automate web applications testing. Selenium uses the recorder metaphor through a toolbar installed in Firefox. It is only able to record a reduced set of events. To identify elements, Selenium uses a system based on the text or generates an XPath expression that does not try to be resilient to small changes. Another drawback is that Selenium does not detect properly the end of the effects caused by a user action in the recording process.

Sahi [12] is another open-source tool for automated testing of web applications. Sahi includes a navigation recording system and it allows the sequences to be executed in commercial browsers. To use Sahi, the user configures its navigator to use a proxy. Every time the browser requests a new page, the proxy retrieves it, adds listeners for monitoring user actions, and returns the modified page. Using a proxy makes the recording system independent of the web browser used. Nevertheless, using a proxy does not allow using approaches where the user explicitly indicates the actions to record; therefore, as discussed previously, it forces to choose between either monitoring only a reduced set of events or suffering from "event flooding". Sahi only supports recording events such as click and change. Other events such as *mouseover* can be used at the execution phase if the user manually codes the navigation scripts. Another drawback is that they do not detect the end of the effects caused by a user action, using timeouts instead.

In the commercial software arena, QEngine [10] is a load and functional testing tool for web applications. QEngine also uses the recorder metaphor through a toolbar installed in MSIE (also used as execution component). In addition of the most typical events supported by the previously mentioned systems, QEngine also supports a form of explicitly specifying *mouseover* events on certain elements, consisting in placing the mouse over the target element for more than a certain timeout (avoiding this way the "flooding" problem). Nevertheless, they do not capture other events such as *mouseout* or *mousemove*. To identify elements, they use a simple system based on the text, attributes and relative position of the element. While this may be enough for application-testing purposes where changes are controlled, it is not enough to deal with autonomous web sources. In addition, as previous systems, QEngine does not detect the end of the effects of an action. iOpus [7] is another web automation tool that uses the recorder metaphor. Their drawbacks with respect to our proposal are almost identical to those mentioned for QEngine.

Kapow [8] is yet another web automation tool oriented to the creation of mashups and web integration applications. Kapow uses its own custom browser. Therefore, in our evaluation it showed to be vulnerable to the formerly mentioned drawback: small implementation differences can make a web page to behave differently. For instance, from the set of 12 sequences from Google Web Toolkit we used in our tests, the Kapow browser could only successfully reproduce 1 of them. To identify the target elements, Kapow generates an XPath expression that tries to be resilient to small changes, although the details of the algorithm they use have not been published.

With respect to the algorithm to identify target elements, [2,14] have also addressed the problem of generating change-resilient XPath expressions. In those approaches, the user provides several example pages identifying the target element; and the system generalizes the expression by examining the differences between them. In our case, that would force the user to record the navigation sequence several times. We believe that process would be much more cumbersome to the user.

6 Conclusions

We have presented a set of new techniques to record and execute web navigation sequences in AJAX-based websites. Previous proposals show important limitations in the range of user actions that they can record and execute, the methods they use for identifying the target elements of user actions and/or how they wait for the effects of a user action to finish. Our techniques have been successfully implemented using both MSIE and Firefox. Our main contributions are a new method for recording navigation sequences able to scale to a wider range of events and a novel method to detect when the effects caused by a user action (including the effects of scripting code and AJAX requests) have finished, without needing to use inefficient timeouts. We have also evaluated our approach with more than 100 web applications, obtaining a high degree of effectiveness.

References

1. Anupam, V., Freire, J., Kumar, B., Lieuwen, D.: Automating web navigation with the WebVCR. In: Proceedings of WWW 2000, pp. 503–517 (2000)
2. Davulcu, H., Yang, G., Kifer, M., Ramakrishnan, I.V.: Computational Aspects of Resilient Data Extraction from Semistructured Sources. In: Proc. of ACM Symposium on Principles of Database Systems (PODS), pp. 136–144 (2000)
3. Document Object Model (DOM) Level 3 Events Specification, http://www.w3.org/TR/DOM-Level-3-Events/
4. Google Web Toolkit, http://code.google.com/webtoolkit/
5. HtmlUnit, http://htmlunit.sourceforge.net/
6. Hupp, D., Miller, R.C.: Smart Bookmarks: automatic retroactive macro recording on the web. In: Proc. of ACM Symposium on User Interface Software and Technology, UIST 2007 (2007)
7. iOpus, http://www.iopus.com
8. Kapow, http://www.openkapow.com
9. Pan, A., Raposo, J., Álvarez, M., Hidalgo, J., Viña, A.: Semi automatic wrapper-generation for commercial web sources. In: Proc. of IFIP WG8.1 Working Conference on Engineering Information Systems in the Internet Context 2002, pp. 265–283 (2002)
10. QEngine, http://www.adventnet.com/products/qengine/index.html
11. Safonov, A., Konstan, J., Carlis, J.: Beyond Hard-to-Reach Pages: Interactive, Parametric Web Macros. In: Proc. of the 7th Conference on Human Factors & the Web (2001)
12. Sahi, http://sahi.co.in/w/
13. Selenium, http://seleniumhq.org/
14. Lingam, S., Elbaum, S.: Supporting End-Users in the Creation of Dependable Web Clips. In: Proc. of WWW 2007, pp. 953–962 (2007)
15. XML Path Language (XPath), http://www.w3.org/TR/xpath
16. Yahoo! User Interface Library (YUI), http://developer.yahoo.com/yui

Supporting Personal Semantic Annotations in P2P Semantic Wikis

Diego Torres[1], Hala Skaf-Molli[2], Alicia Díaz[1], and Pascal Molli[2]

[1] LIFIA, Facultad de Informática,
Universidad Nacional de La Plata, Argentina
{diego.torres,alicia.diaz}@lifia.info.unlp.edu.ar
[2] LORIA – INRIA Nancy-Grand Est
Nancy Université, France
{skaf,molli}@loria.fr

Abstract. In this paper, we propose to extend Peer-to-Peer Semantic Wikis with personal semantic annotations. Semantic Wikis are one of the most successful Semantic Web applications. In semantic wikis, wikis pages are annotated with semantic data to facilitate the navigation, information retrieving and ontology emerging. Semantic data represents the shared knowledge base which describes the common understanding of the community. However, in a collaborative knowledge building process the knowledge is basically created by individuals who are involved in a social process. Therefore, it is fundamental to support personal knowledge building in a differentiated way. Currently there are no available semantic wikis that support both personal and shared understandings. In order to overcome this problem, we propose a P2P collaborative knowledge building process and extend semantic wikis with personal annotations facilities to express personal understanding. In this paper, we detail the personal semantic annotation model and show its implementation in P2P semantic wikis. We also detail an evaluation study which shows that personal annotations demand less cognitive efforts than semantic data and are very useful to enrich the shared knowledge base.

1 Introduction

Semantic Wikis [1,2,3,4] are one of the most successful Semantic Web applications. They are widely used for collaborative knowledge building. In semantic wikis [1,3], wikis pages are annotated with semantic data to facilitate the navigation, the information retrieving and ontology emerging. Semantic data represents the shared knowledge base which describes the common understanding of the community. The knowledge base is built collaboratively through an iterative and social process.

However, collaborative knowledge building is basically a spiraled process where knowledge first emerges at individual context and then is socialized [5,6]. This process [7,6] involves *externalization, publication, internalization* and *reaction*. Most of semantic wikis only support the knowledge socialization, but it is

S.S. Bhowmick, J. Küng, and R. Wagner (Eds.): DEXA 2009, LNCS 5690, pp. 317–331, 2009.

fundamental to support personal knowledge building too [8]. Personal Semantic Wikis [8,9] provide an easy way to manage personal knowledge often without collaborative functionality. However, to carry out a collaborative knowledge building activity, user needs to manage and combine both shared and personal knowledge.

Existing collaborative knowledge building systems support partially or completely this process. For instance, [10,11,12] are collaborative knowledge building systems, however, they are more oriented towards collaborative ontology development rather than ontology emerging. Other systems like semantic wikis [1,2,3,4,8] are more appropriate to support collaborative knowledge emerging, however, they do not provide functionalities to manage combined personal and shared understandings. For example, Semantic MediaWiki (SMW) only enables shared knowledge building. On the other hand, SemperWiki [9] only supports personal knowledge building. Currently, there are no semantic wikis that help people to combine and manage in a usable way both kind of knowledge.

The goal of this work is to propose an innovative semantic wiki approach that supports both personal and shared knowledge building. In this approach, the shared knowledge is unique and accessible to everyone, while the personal knowledge is only accessible by its owner and represents the user private view (perspective) of the shared one. Personal knowledge can differ from the shared one, but it can also have overlapped parts.

For the emerging of shared knowledge, we follow the same approach as SMW where *shared semantic annotations* are embedded in the wiki text by using a suitable syntaxis. For the personal knowledge, we propose *Personal Semantic Annotations* to externalize personal understanding. *Personal semantic annotations* are associated to the wiki page and they are only accessed by the owner user. For the end-user, the *personal semantic annotations* look like *tags*, however they are semantically richer: they support categories and individuals.

We believe that the addition of *Personal Semantic Annotations* to semantic wikis enables:

– To support the individual understanding in the collaborative knowledge building process [6]
– To provide personalized knowledge retrieving, structuring and navigation.
– To enable a combined personal and shared knowledge retrieving.
– To enrich the shared semantic annotations and to augment, therefore, the shared knowledge base.

Moreover, adding personal semantic annotations and shared ones involve complementary activities. Whereas adding a shared semantic annotation seems to be suitable during editing activity, adding a personal one seems to be more suitable during browsing activity. In order to validate these hypothesis, we have conducted an evaluation study.

In this paper, we introduce a peer to peer semantic wiki called *P-Swooki* that supports both personal and shared knowledge building. *P-Swooki* extends a peer-to-peer semantic wiki Swooki [3] by adding personal knowledge building.

We choose to validate our approach in a peer to peer semantic wiki because in a P2P architecture information dissemination is easily controlled *i.e.* shared annotations are broadcasted and integrated by all peers while personal semantic annotations remain local.

The paper is organized as follows. The next section 2 gives a brief background about collaborative knowledge building process. Section 3 introduces a P2P collaborative knowledge building process and discusses the personal annotations problematic. Sections 4 and 5 present the implementation and architecture of *P-Swooki*. Section 6 details a usage study which shows that personal annotations demand less cognitive efforts than the shared one and they are very useful to enrich the shared knowledge base. The last section concludes the paper and points further works.

2   Background: Collaborative Knowledge Building

The majority of works on knowledge management focus on *organizational knowledge management* [13,14]. Many of them follow the traditional KM approach [5] to creates large centralized knowledge repositories, in which corporate knowledge is collected, represented and organized, according to a single - shared - conceptual schema [15]. In [16], the authors noted that *"This centralized approach -and its underling objectivist epistemology- is one of the reasons why so many KM systems are deserted by users"*. In [13,16], the authors propose a P2P organizational knowledge management in order to make organizational memory more flexible. However, this approach is more suitable to *knowledge discovery and propagation* rather than collaborative and personal knowledge building.

Collaborative knowledge building focuses on understanding as a learning process where personal understanding can not be built internally without social interaction. People need to participate in a social process and create new knowledge collaboratively. Gerry Stahl in [6] proposes a conceptual collaborative knowledge building model which shows the *"mutual constitution of the individual and the social knowledge building as a learning process"*, as depicted in the figure 1. This process should be adapted to the P2P semantic wikis context.

Stahl's process starts with the description of the *personal understanding* by specifying personal beliefs, which are tacit. Then, they can be articulated in a *"language"* and enters into a social process of interaction with other people and their shared understanding. Later, this shared knowledge enters again in the personal understanding and provokes a change in personal beliefs, motivations and concerns. When this happens, these modifications become a new tacit understanding and will be the new starting point for future understanding and further learning. In [7], the authors reinterpreted this process and proposed a four steps spiral process for centralized knowledge sharing.

– knowledge *externalization* where knowledge goes from tacit to explicit. This is an individual activity.

– Knowledge *publication* where the knowledge goes from individual context to share context. This produces a new shared knowledge contribution.

– Knowledge *internalization* where knowledge goes from explicit to tacit and from the shared to the individual context.

– *Reaction* is the act of opening a discussion and argumentation linked to previous shared contribution to achieve a consensus. A reaction always involves a externalizations and an eventual publication.

In this work, we will adapt the simplified version of Stahl's process to the context of P2P semantic wikis.

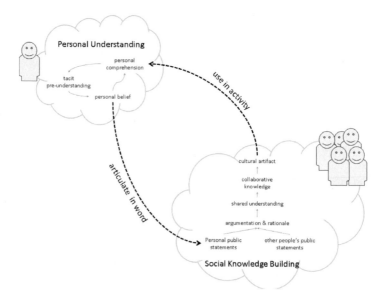

Fig. 1. Stahl's Collaborative Knowledge Building Process

3   P2P Collaborative Knowledge Building Approach

In this work, we extend P2P semantic wikis by supporting personal understanding building. In addition to shared semantic annotations embedded in the wiki text, users can also associate *personal semantic annotations* to semantic wiki pages. These private annotations express personal understanding of the users. For example, if a user was navigating to the semantic wiki page "Semantic Wiki" as it is shown in the figure 2, eventually, she would like to annotate this page as "Collaborative Tool", "Web" and "Semantic Wiki". If these annotations only express personal understanding, they should be private. Other annotations as "Semantic Web" or "Wiki" are shared, they could be defined by the same user or by other users. We can notice that users manage simultaneously both shared and personal semantic annotations.

We adapt the collaborative knowledge building process to P2P settings as it is detailed in the next section.

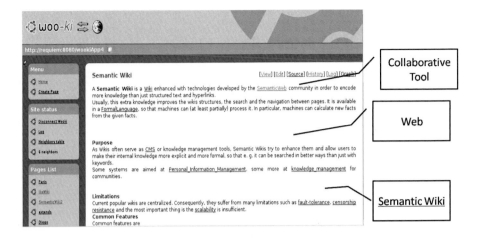

Fig. 2. Adding Personal Semantic Annotation in Semantic Wiki

3.1   P2P Collaborative Knowledge Building Process

A P2P collaborative knowledge building process is a continuous spiraled process which involves *externalization, publication, internalization* and *reaction*, where externalization and publications steps had to be redefined to support P2P settings. Internalization and Reactions are not modified.

Besides, users manage in a well-differentiated way both, personal and shared understandings. Every user needs to manage in separated spaces the personal and shared annotations. We define two repositories: the *personal understanding repository* and the *shared understanding repository* respectively. In a P2P setting, we consider that every user works in one peer and has both repositories. The *shared understanding repositories* will be eventually identical for all users due to the synchronization algorithms [17].

Our P2P collaborative knowledge building process redefines the externalization and publications steps as:

– *Externalization* where personal knowledge goes from tacit to explicit. Users use personal semantic annotations to externalize their own knowledge. This is an individual activity, this knowledge remains private in the context of the personal understanding space.

– *Publication* where the knowledge goes from the individual context to the shared one. As a result, a personal semantic annotation becomes a shared one. In other words, this involves to move a personal semantic annotation from a given user's personal understanding repository to the shared one. This step involves to replicate the annotation to every user as a shared annotation.

For example, in the figure 3 the "user1" externalizes "Collaborative Tool" personal annotation on her personal repository. Then, when she performs a publication this personal annotation should be disseminated to every user, even to herself. After the publication, the semantic annotation "Collaborative Tool" should appear in every shared understanding repository as a shared annotation.

Consequently, personal understanding building is achieved by supporting the separation of both knowledge repositories (personal and shared) and the externalization step.

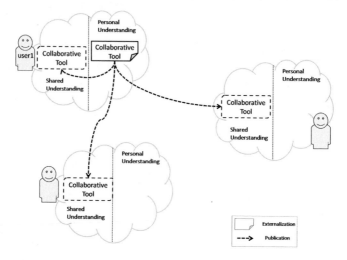

Fig. 3. P2P Collaborative Knowledge Building Process

This P2P collaborative knowledge building approach has several advantages:

– *Personal navigation*: the system allows the users to have simultaneously personal and shared navigation on the the same content. Shared navigation is the traditional navigation supported by any semantic wiki. Personal navigation is a new kind of navigation, it is personal and it is the consequence of the personal semantic annotations. The user has an instant gratification after adding personal semantic annotations.

– *Enrichment of shared knowledge*: the user can make public her personal semantic annotations. Consequently, the shared knowledge is enriched.

– *Improve system usability*: adding shared semantic annotations and personal semantic ones involves complementary activities. Whereas adding shared semantic annotations seems to be suitable during the editing activity, adding personal annotations seems to be more suitable during reading activity.

3.2   Personal Semantic Annotations: Individuals and Categories

Every semantic wiki page could be tagged with several personal semantic annotations as it was shown above. A personal semantic annotation can be a *category* or an *individual*.

Categories define a family of elements. For example, in the previous example (figure 2), the annotation "Semantic Wiki" was underlined in order to indicate that this wiki page is a *Semantic Wiki category definition*.

Individuals denote elements that fall at least in one category. *Semantic Mediawiki* is an individual that fall in the category Semantic Wiki. An Individual can belong to many categories.

A semantic wiki page can be annotated with many annotations. For example, a user personally would like to annotate the wiki page "Swooki" as a "Semantic Wiki" and as "P2P application".

Currently, the annotation model is simple, it only considers categories and individuals. In the near future, we will enrich it in order to support relationships and attributes.

4   P-Swooki: P2P Collaborative Knowledge Building System

We have developed *P-Swooki*, a P2P collaborative knowledge building system that extends the P2P semantic wiki *Swooki* with personal semantic annotations.

Shared semantic annotations are already supported by *Swooki* as detailed in the section 4.1. Therefore, we had only to add personal annotations functionalities to Swooki. In sections 4.2, 4.3 and 4.5, we detail the personal annotations management, the data model and its associated operations.

4.1   Shared Semantic Annotation Management

In Swooki every peer hosts a copy of all wiki pages and the *shared understanding repository*. When a peer updates its local copy of data, it generates a corresponding operation. This operation is processed in four steps:

1. It is executed immediately against the local replica of the peer,
2. it is broadcasted through the P2P network to all other peers,
3. It is received by the other peers,
4. it is integrated to their local replica. If needed, the integration process merges this modification with concurrent ones, generated either locally or received from a remote server.

To synchronize data, Swooki [18] implements a modified version of the P2P synchronization algorithm detailed in [17]. Swooki synchronization algorithm ensures the convergence on the wiki text and the *shared understanding repository* i.e. when the system is idle, all copies are identical.

4.2   Personal Semantic Annotations Management

In *P-Swooki*, personal semantic annotations are hosted locally. When a user updates her personal semantic annotations, she generates a corresponding operation. The operation is executed locally against the user *personal understanding repository*. This operation is *not* broadcasted to other peers.

The process to annotate a wiki page is simple as it was explained above. The system enables users to annotate a wiki page as a new *category* or as an *individual* of an existing category.

In order to handle personal semantic annotations, we extended Swooki's data model and defined new editing operations.

4.3   P-Swooki Data Model

The data model is an extension of Swooki [18,19] data model. Therefore, each semantic wiki peer has assigned a global unique identifier named *NodeID*.

As in any wiki system, the basic element is a wiki page, therefore every wiki page has assigned a unique identifier *PageID*, which is the name of the page. The name is set the page is created. If several servers create concurrently pages under the same name, their content will be directly merged by the synchronization algorithm. Notice that a *URI* can be used to unambiguously identify the concept described in the page. The *URI* must be global and location independent in order to ensure load balancing. For the sake of simplicity, in this paper, we use a string as page identifier.

The figure 4 describes the personal semantic annotations data model. This data model is described by the Ontology Definition Meta-model (ODM) [20].

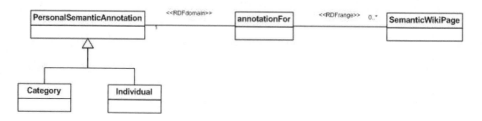

Fig. 4. Personal Semantic Annotation Data Model

4.4   Personal Semantic Annotation Storage Model

RDF is the standard data model for encoding semantic data. In P-Swooki, every peer has two local RDF repositories : *Personal Statements* and *Shared Statements*. They implement the *personal understanding repository* and the *shared understanding repository* respectively.

– The *Shared Statements* contains a set of RDF statements which were extracted from the wikis pages. A statement is defined as a triple (Subject, Predicate, Object) where the subject is the name of the page, the predicates (or properties) and the objects are related to the concept involved in the page.

– The *Personal Statements* contains personal semantic annotations which are represented as personal RDF statements. A personal RDF statement is defined as a triple (Subject, Predicate, Object) where the subject is the wiki page and the predicate annotates the page as a personal semantic annotation type as described in the next section.

We define two operations on the RDF repositories:

– insertRDF(R,t): adds a statement t to the *Personal Statements* or *Shared Statements* repository R.
– deleteRDF(R,t): deletes a statement t from the *Personal Statements* or *Shared Statements* repository R.

These operations are not manipulated directly by the end-user, they are called implicitly by the editing operations as it is shown in the following section.

4.5   Editing Operations

There are four editing operations for editing personal semantic annotations: *addIndividual, addCategory, delIndividual* and *delCategory*. An update is considered as a delete of old value followed by an insert of a new value.

1. *addCategory(PageID, CategoryName)*: where *PageID* is the identifier of the semantic wiki page. *CategoryName* is the name of the new category.
This operation sets the wiki page *PageId* as a category in the user personal repository. This operation calls the *insertRDF(Personal Statements,(PageId, RDF.Type, CategoryName))* function to add a new triplet into the personal RDF repository.

2. *addIndividual(PageID, CategoryName)*: sets the wiki page *PageID* as a member of the category *CategoryName*. If *CategoryName* does not exist, it is added automatically to the *Personal Statements* repository by calling the operation *addCategory* and then the operation automatically annotates the *PageId* as member of the *CategoryName*.
During this operation an RDF statement is added to the personal repository by calling *insertRDF(Personal Statements, (PageId, belongsTo, CategoryName))* where *belongsTo* is a predicate to associate an individual to a category.

3. *delIndividual(PageID, CategoryName*: eliminates the *PageID* as member of the category *CategoryName* from the personal RDF repository by calling *DeleteRDF(Personal Statements, (PageId, RDF.Type, CategoryName))*.

4. *delCategory(PageID,CategoryName*: first, calls the *delIndividual* operation for each member of the category *CategoryName*, and then deletes the category *CategoryName* from the personal RDF repository by calling the *DeleteRDF* operation.

5   *P-Swooki* Architecture

P-Swooki is implemented as an extension of Swooki. Swooki is a P2P semantic wiki which is implemented in Java as servlets in a Tomcat Server and uses Sesame 2.0 as RDF repository.

P-Swooki is developed over a Swooki architecture using one peer per user. A P-Swooki peer is compound by the following components (see figure 5). The

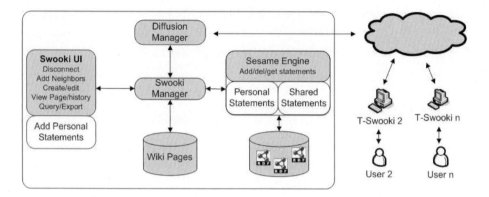

Fig. 5. P-Swooki Architecture

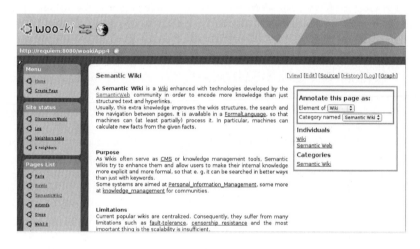

Fig. 6. P-Swooki Interface

grey boxes are Swooki components whereas the white ones are the P-Swooki components.

User Interface. The P-Swooki UI component is composed by the Swooki wiki editor and it incorporates the functionalities to make personal annotations. This basically divides the wiki page into two areas: the shared and private annotation spaces. The shared space is defined by a regular wiki editor supported by Swooki functionality. The private annotation one includes a box to add personal semantic annotations and to visualize them (see figure 6).

Swooki Manager. The Swooki manager implements the synchronizing algorithm.

Sesame Engine. We use a multi-set [18] extension of Sesame 2.0 [21] as RDF repository. Sesame is controlled by the Swooki manager for storing and retrieving

RDF statements. P-Swooki stores the private annotations using a different name space. This allows to reuse the storing and retrieving facilities already implemented by Swooki.

Diffusion Manager. The diffusion manager is in charge of maintaining the membership of the unstructured network and to implement a reliable broadcast for the shared repositories.

6   Evaluation

In this section, we present the evaluation of our approach. We have conducted two separate experiments, one in France and another one in Argentina. The total number of participants were 15 people. The participants ranged in age from 25 to 45. All participants were involved in computer science, all were familiar with wikis and 5 of them were familiar with semantic wikis and have some experience in ontology building. The participants were in different rooms and they were not allowed to communicate to each other during the experience.

We started the first experience in France by a short explanation about semantic wikis, shared knowledge and personal knowledge. We asked participants to develop a semantic wiki by using both kinds of annotations. They started with a non empty wiki. In fact, there were created in advance 2 semantic wikis pages; one about *Semantic Wiki* and another one about *Semantic Web*. We also suggested participants to use a special syntax in order to control vocabulary explosion as it occurs in folksonomies [22].

In order to consolidate the first experience, we have repeated the same experience in Argentina which confirmed the results previously obtained in France. These experiences show a preliminary evidence of the contribution of our approach regarding the usability of personal semantic annotations and their complementarity with the shared knowledge. In the following we show the results of these experiences. As both experiences showed the same outcome, we will only present the results from France.

The tables 1 and 2 show the type (individual or category) and the amount of personal semantic annotations that each each participant has added to the *Semantic Wiki* and *Semantic Web* wiki pages respectively.

The *Semantic Wiki* page was annotated by all the participants. They annotated this page as individual 17 times and as category 15 times. The most active participant added 11 personal semantic annotations to this page. The average number of annotations per participant was 4.5. The average without the most active participant was 3.5.

The *Semantic Web* page was annotated by all the participants. They annotated this page as individual 8 times and as category 9 times. The most active participant added 11 personal semantic annotations to this page. The average number of annotations per participant was 2.5. The average without the most active participant was nearly 1.

Table 1. Personal Semantic Annotation for *Semantic Wiki* Page

User	Individual	Category	Total
1	1	1	2
2	1	0	1
3	2	5	7
4	6	5	11
5	2	1	3
6	4	3	7
7	1	0	1

Table 2. Personal Semantic Annotation for *Semantic Web* Page

User	Individual	Category	Total
1	0	1	1
2	1	0	1
3	0	2	2
4	6	5	11
5	1	1	2
6	1	0	1
7	1	0	1

The results above confirm our initial hypothesis about the usefulness of the personal semantic annotations because all the participants have added personal annotations.

The table 3 shows the shared and personal semantic annotations used by the participants for the *Semantic Wiki* page.

Table 3. Personal and Shared Semantic Annotation for Semantic Wiki Page

Shared annotations	Individual	Category
Category: PersonalInformationManagement	ResearchTopic	SemanticWiki (4)
Category: KnowledgeManagement	SemanticWeb (4)	Wiki(2)
Category: SemanticWeb	CollaborativeTool(2)	SemanticWeb
Category:FormalLanguage	Web	Web (2)
KindOf: Wiki	NoDelete	CollaborativeTool
has: FactBox	Semantics	WebOfData
limitation : fault-tolerance	KnowledgeWeb	NoUndoTag
limitation : scalability	Something	WWW
limitation : censorship	Wiki (3)	CSCW
	Web	Semantic

The column *Shared annotations* regroups the shared semantic annotations. At the beginning of the experience, it was empty. This page could be annotate as a category such as *Category: PersonalInformationManagement* or as an object property such as *limitation : scalability*.

The second and third columns regroup all the personal semantic annotations. Notice that in some cases many users used the same semantic annotation. For instance, four users used *SemanticWeb* annotations and two users used *Wiki* annotations.

We can observe that the total number of semantic annotations is increased. Therefore, personal semantic annotations could be useful to augment the shared knowledge.

With this evaluation we learn the following lessons:

– Every participant used both personal and shared semantic annotations;
– Most participants said that it is easy to use personal semantic annotations, because it is not necessary to embed them into the text.
– Some participants had difficulties to distinguish between a category and an individual.
– All participants have manifested the importance to have a good user-interface to facilitate personal navigation.
– For some participants personal annotations were useful to structure their own navigational map according to their personal taxonomy.
– Personal annotations were easier for people not familiar with semantic wikis whereas someone familiar with semantic wikis did not see exactly the added value of personal semantic annotations.
– One participant did not understand the difference between personal and shared semantic annotations.
– For most participants, it was easier to add personal annotations when they were browsing and to add shared annotations when they were editing the wiki pages.
– Most participants found that combining both kind of annotations could help them to make better knowledge retrieving.

Although, it is premature, the average of personal annotations shows a tendency: people feel comfortable using personal annotations and adding personal annotations is a complementary activity in semantic wiki.

These first results encourage us to continue in this direction, however, we need to conduct large scale experiences to consolidate these results.

7 Conclusion and Further Work

In this paper, we have introduced an approach to manage personal and shared knowledge in P2P semantic wikis. Shared knowledge is managed as in any semantic wiki. On the other hand, personal knowledge is defined as personal semantic annotations.

We have designed a P2P collaborative knowledge building process by basically supporting personal understanding. Personal annotations is the mechanism we have proposed to support personal understanding. Personal semantic annotations are private and appear in the context of the wiki page. This approach involves personal, shared or personal and shared navigation and retrieving. In this research, we have adopted a P2P approach because it is the more suitable to control personal knowledge. We have implemented *P-Swooki* as an extension of the P2P semantic wiki, Swooki. As our approach is general, it could by applied to any semantic wiki, such as Semantic MediaWiki.

The evaluation of *P-Swooki* has confirmed our hypothesis that the usability of semantic wiki system can be improved by adding personal knowledge management. Most of the participants have used personal annotations. To consolidate these results, we plan to conduct more experimentations.

For instant, the personal annotation model is simple, it only considers categories and individuals. In the near future, we will enrich it in order to support relationships and attributes. Shared knowledge could be also enriched by publishing personal annotations. Therefore, it is needed to extend the current approach with a mechanism to "easily integrate" personal annotations in the shared repository and also in the wiki text.

From the experience, we also noticed that many users have defined the same personal semantic annotations. Although, these knowledge means a common understanding, they were not in the shared knowledge base. In the future, we plan to use discovering knowledge techniques to enrich the shared knowledge base.

Acknowledgments

This work was partially funded by the "P2P Semantic Wikis for large-scale distributed knowledge management and large community integration" project, which is sponsored by the MinCyT, Argentina and INRIA-CNRS, France.

References

1. Völkel, M., Krtözsch, M., Vrandecic, D., Haller, H., Studer, R.: Semantic wikipedia. Journal of Web Semantics 5(4) (2007)
2. Schaffert, S.: Ikewiki: A semantic wiki for collaborative knowledge management. In: WETICE, pp. 388–396. IEEE Computer Society, Los Alamitos (2006)
3. Rahhal, C., Skaf-Molli, H., Molli, P.: Swooki: A peer-to-peer semantic wiki. In: The 3rd Workshop: 'The Wiki Way of Semantics'-SemWiki, co-located with the 5th Annual European Semantic Web Conference (ESWC), Tenerife, Spain (2008)
4. Buffa, M., Erétéo, G., Faron-Zucker, C., Gandon, F., Sander, P.: SweetWiki: A Semantic Wiki. Journal of Web Semantics, special issue on Web 2.0 and the Semantic Web 6(1) (2008)
5. Nonaka, I., Takeuchi, H.: The Knowledge - Creating Company: How Japanese Companies Create the Dynamics of Innovation. Oxford University Press, Oxford (1995)
6. Stahl, G. (ed.): Group cognition: Computer support for building collaborative knowledge. MIT Press, Cambridge (2006)
7. Díaz, A., Baldo, G., Canals, G.: A framework for collaborative knowledge sharing with divergence. IADIS Int. Journal on WWW/Internet 5(2), 86–99 (2007)

8. Oren, E., Völkel, M., Breslin, J.G., Decker, S.: Semantic wikis for personal knowledge management. In: Bressan, S., Küng, J., Wagner, R. (eds.) DEXA 2006. LNCS, vol. 4080, pp. 509–518. Springer, Heidelberg (2006)
9. Oren, E.: Semperwiki: a semantic personal wiki. In: Proc. of 1st Workshop on The Semantic Desktop - Next Generation Personal Information Management and Collaboration Infrastructure, Galway, Ireland (2005)
10. Díaz, A., Baldo, G., Canals, G.: Co-protégé: Collaborative ontology building with divergences. In: DEXA Workshops, pp. 156–160. IEEE Computer Society, Los Alamitos (2006)
11. Tudorache, T., Noy, N.F., Tu, S.W., Musen, M.A.: Supporting collaborative ontology development in protégé. In: Sheth, A.P., Staab, S., Dean, M., Paolucci, M., Maynard, D., Finin, T., Thirunarayan, K. (eds.) ISWC 2008. LNCS, vol. 5318, pp. 17–32. Springer, Heidelberg (2008)
12. Sure, Y., Erdmann, M., Angele, J., Staab, S., Studer, R., Wenke, D.: OntoEdit: Collaborative ontology development for the semantic web. In: Horrocks, I., Hendler, J. (eds.) ISWC 2002. LNCS, vol. 2342, pp. 221–235. Springer, Heidelberg (2002)
13. Bonifacio, M., Cuel, R.: Knowledge nodes: the building blocks of a distributed approach to knowledge management. J. of Universal Computer Science 8 (2002)
14. Lee, H., Choi, B.: Knowledge management enablers, processes, and organizational performance: An integrative view and empirical examination. Journal of Management Information Systems 20(1), 179–228 (2003)
15. Malhotra, Y.: Why knowledge management systems fail? enablers and constraints of knowledge management in human enterprises. In: Holsapple, C. (ed.) Handbook on Knowledge Management. International Handbook on Information Systems, vol. 1. Springer, Heidelberg (2002)
16. Bonifacio, M., Bouquet, P., Mameli, G., Nori, M.: Peer-mediated distributed knowledge management. In: van Elst, L., Dignum, V., Abecker, A. (eds.) AMKM 2003. LNCS, vol. 2926, pp. 31–47. Springer, Heidelberg (2004)
17. Oster, G., Urso, P., Molli, P., Imine, A.: Data Consistency for P2P Collaborative Editing. In: Proceedings of the ACM Conference on Computer-Supported Cooperative Work - CSCW 2006, Banff, Alberta, Canada. ACM Press, New York (2006)
18. Skaf-Molli, H., Rahhal, C., Molli, P.: Peer-to-peer semantic wiki. In: Verlag, S. (ed.) DEXA 2009: 20th International Conference on Database and Expert Systems Applications (2009)
19. Weiss, S., Urso, P., Molli, P.: Wooki: a p2p wiki-based collaborative writing tool. In: Web Information Systems Engineering, Nancy, France. Springer, Heidelberg (2007)
20. Gaaevic, D., Djuric, D., Devedzic, V., Selic, B.: Model Driven Architecture and Ontology Development. Springer, Secaucus (2006)
21. Broekstra, J., Kampman, A., van Harmelen, F.: Sesame: A generic architecture for storing and querying RDF and RDF schema. In: Horrocks, I., Hendler, J. (eds.) ISWC 2002. LNCS, vol. 2342, p. 54. Springer, Heidelberg (2002)
22. Sen, S., Lam, S.K., Rashid, A.M., Cosley, D., Frankowski, D., Osterhouse, J., Harper, F.M., Riedl, J.: Tagging, communities, vocabulary, evolution. In: CSCW 2006: Proceedings of the 2006 20th anniversary conference on Computer supported cooperative work, pp. 181–190. ACM, New York (2006)

OrdPathX: Supporting Two Dimensions of Node Insertion in XML Data

Jing Cai and Chung Keung Poon

Department of Computer Science,
City University of Hong Kong
Tylor.Cai@student.cityu.edu.hk, ckpoon@cs.cityu.edu.hk

Abstract. We introduce a novel XML labeling scheme called OrdPathX which supports both leaf and internal node insertions for XML data. Dynamic XML labeling has been studied for years. However almost all labeling schemes allow only leaf node insertions. Inspired by the careting-in technique of OrdPath [7], we propose a new labeling algorithm which supports internal node insertions gracefully without relabeling. We will describe the labeling algorithm and the associated operations for various inter-node relationship determination. Experimental results show that OrdPathX can handle internal node insertions efficiently.

1   Introduction

Nowadays XML is a standard language for information representation and exchange over the Internet. An XML document comprises of hierarchically nested elements and can be naturally modeled as an ordered tree. Designing efficient labeling schemes for the nodes to support the determination of various structural relations has attracted much attention recently. In the past decade, many labeling schemes have been proposed, including *interval-based schemes* (e.g., [10,12,6,4]), *prefix-based schemes* (e.g. [8,7]) and others (e.g. [11]). However, most of them cannot handle internal node insertions efficiently.

In this paper we propose a new prefix-based scheme that can handle internal node insertion with a good performance. This is of particular interest as many academic and commercial applications employ prefix-based schemes as the underlying labeling scheme. They are used for XML maintenance [13,14], DBMS systems [15,16,7] and XML data indexing [17]. E.g., OrdPath is implemented in Microsoft SQL Server 2005 for execution plan optimization [9]. In the latest release of Microsoft SQL Server 2008, OrdPath is internally used by a new DBMS data type called *HierarchyID* which models the hierarchy of tree structured data [1,2].

Our labeling scheme, OrdPathX, is inspired by the "careting-in" technique of OrdPath and can handle both internal and leaf node insertions efficiently. Internal nodes updating is a common operation that may occur to an XML tree. Suppose we have the XML tree shown in Figure 1 which models a typical Linux based file directory structure. Any file/directory creation or deletion represents

S.S. Bhowmick, J. Küng, and R. Wagner (Eds.): DEXA 2009, LNCS 5690, pp. 332–339, 2009.

a change to this XML tree in which internal/leaf node insertion/deletion take place. Therefore, how to handle the internal nodes and leaf nodes updating efficiently is of great importance.

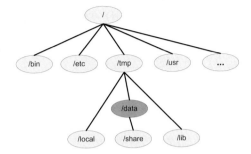

Fig. 1. Typical File Structure

The rest of this paper is organized as follows. We explain our new labeling scheme and the associated algorithms for inter-node relation determination in Section 2 and 3 respectively. In Section 4, we turn to the technical issue of encoding the labels efficiently and present our experimental evaluation as well. Section 5 concludes this paper.

2   The OrdPathX Labeling Scheme

We first describe the key ideas of the OrdPath labeling scheme. In the OrdPath scheme, the initial tree is labelled similar to a Dewey scheme but only positive odd numbers are used. To insert a node v as a child of u with label L, there are several cases. If v is the first child of u, then label v as $L.1$. Otherwise, if v is inserted as the rightmost (resp. leftmost) child of u and its left (resp. right) sibling v' has label $L.x$, then v has label $L.x + 2$ (resp. $L.x - 2$). Finally, if v is inserted between two consecutive children, v_1 and v_2, of u and their labels are $L.x$ and $L.x + 2$ respectively, then label v as $L.x + 1.1$. This is the essence of the "careting-in" technique in OrdPath.

Note that when x is an odd number, $x + 1$ is even. Thus, the label of a newly inserted node may contain sequences of even components but each such sequence will be followed by an odd component. More generally, an OrdPath label can be partitioned into a sequence of *chunks*, each chunk consisted of zero or more even components followed by an odd component. We can express the labels of v_1 and v_2 as $L.C_1$ and $L.C_2$ where C_1 (resp. C_2) is the last chunk of the label of v_1 (resp. v_2) and C_1 is lexicographically smaller than C_2. Then the new node v is labelled as $L.C'$ where $C' = caretin(C_1, C_2)$ is a label generated by the caret-in function that takes C_1 and C_2 as input and returns a label lexicographically in between C_1 and C_2.

Now, in the OrdPathX scheme, the label consists of an *Augmented OrdPath* *(AO)*, possibly followed by a *Parent Height (PH)*:

$$OrdPathX = AO.PH$$

where an AO is a sequence of *chunks* and between each pair of consecutive chunks there may exist an *Incremental Height* (IH). The IH and PH (to be explained) are themselves OrdPath labels. The initial assignment of labels to nodes are the same as that in the OrdPath scheme. Note that in the initial assignment, no labels will contain any IH and PH.

To facilitate our discussions, we call those nodes present in the initial tree *initial nodes* and those inserted as leaves and parents of existing nodes *leaf-insertion nodes* and *parent-insertion nodes* respectively.

2.1   An Illustrating Example

Suppose the initial tree contains a node v with parent v' (Figure 2(a)) and we are to insert nodes u_1, u_2, u_3 between them. At the end, we have created a sequence of parent-insertion nodes between v and v' and we call it a *Parent Insertion Chain* (PIC). See Figure 2(d).

First, consider the insertion of u_1. We label u_1 as $L.(1).C$ where "(1)" is an Incremental Height. See Figure 2(b). The presence of an IH between the last two chunks of a label indicates that u_1 is a parent-insertion node. We refer to this IH as the *Significant Incremental Height* (SIH). (The IHs between other chunks are not called SIH as their presence do not imply that the node is a parent-insertion node.) An SIH of "(1)" indicates that it is the first node inserted along the PIC between two non parent-insertion nodes (i.e., v' and v in this case). Moreover, we relabel v as $L.C.[1]$ where "[1]" is the Parent Height of v's label. It indicates that v has a parent-insertion node above it. In general, the PH of a node is set

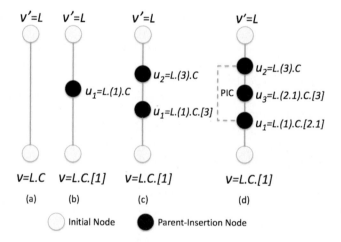

Fig. 2. An Example

to be identical to the SIH of its parent's label. The absence of PH implies that no node has been inserted as its parent.

Next, we insert u_2 as the new parent of u_1. Note that u_1 and u_2 are both parent-insertion nodes along the PIC between v' and v. We will make use of their SIH to encode their relative order along this PIC so that a higher node on the PIC will have a larger SIH. Thus, we label u_2 as $L.(3).C$. Since u_1 now has a parent-insertion node above it, we also relabel u_1 as $L.(1).C.[3]$. See Figure 2(c). Finally we insert node u_3 between u_1 and u_2. The concept of "Caret-in" is applied and so the SIH of u_3 is set as (2.1). See Figure 2(d).

In summary, the IH of the new node is assigned using the OrdPath scheme while the other parts of the AO are the same as its child.

2.2   Detail Procedures

Parent Node Insertions. Suppose we insert a node u as the parent of an existing node v. Let v' be the original parent of v.

(Case 1: both v' and v are not parent-insertion nodes.) Let their labels be L and $L.C$ respectively where C is the last chunk of v's label. Then node u is labelled as $L.(1).C$. As the parent of v is now u instead of v', we modify the label of v as $L.C.[1]$.

(Case 2: v is a parent-insertion node while v' is not.) Let the label of v' be L and that of v be $L.(ih).C$. Then label u as $L.(ih').C$ where ih' is the next OrdPath label larger than ih. Also, label v as $L.(ih).C.[ih']$.

(Case 3: v is not a parent-insertion node while v' is.) We need to distinguish whether v' was previously inserted as parent of v or v was appended as a child of v'. Let the label of v' be $L.(ih).C.[ph]$ where ph can be empty. (Subcase a) The label of v is $L.(ih).C.D$. Then the AO part of v' is a prefix of the label of v (and D is the last chunk generated when v was appended as a child of v'). In this case, u is the first parent-insertion node between v' and v. As in case 1, we label u as $L.(ih).C.(1).D$ and modify the label of v to $L.(ih).C.D.[1]$. (Subcase b) The label of v is $L.C.[ih]$. Then u is inserted as the last parent-insertion node along the PIC from some ancestor of v' to v. This PIC increases its length by 1 since u is inserted. So, we label u and v as $L.(ih').C.[ih]$ and $L.C.[ih']$ respectively where ih' is the next OrdPath label smaller than ih.

(Case 4: both v and v' are parent-insertion nodes.) Let the label of v' and v be $L.(ih2).C.[ph]$ and $L.(ih1).C.[ih2]$ respectively. Note: ph can be empty and $ih1$ is lexicographically smaller than $ih2$. Then label u and v as $L.(ih).C.[ih2]$ and $L.(ih1).C.[ih]$ respectively where $ih = caretin(ih1, ih2)$.

Leaf Node Insertions. Suppose we insert node u as a child of v so that u becomes a leaf. Basically, we will append a chunk to the label of v according to the OrdPath scheme. Specifically, let the label of v be $L.(ih).C.[ph]$ where ih and ph can be empty.

If v is not a parent-insertion node (i.e., ih is empty), then all its children are inserted after v and we label u as $L.C.D$ where D is computed using the OrdPath scheme, taking into account the relative position between u and the other child(ren) of v. Note that u does not inherit its parent's PH since that PH is merely used to keep the parent information of the parent of u and has nothing to do with u.

If v is a parent-insertion node (i.e., ih is non-empty), then one of its children, say w, was present before v was inserted. The label of w has the form $L.(ih1).C.[ih]$ where $ih1$ can be empty. If u is inserted on the left of w, we label u as $L.(ih).C.-0.D$ where D is a chunk computed using the OrdPath scheme so that it encodes the order of u among those siblings on the left of w. Similarly, if u is inserted on the right of w, we label u as $L.(ih).C.+0.D$ where D, computed by OrdPath, encodes the order of u among those siblings on the right of w. As we will see in Section 3, this $+/-0$ allows us to tell the relative order between u and w (as well as other siblings) easily. A "$+0$" (resp. '-0') component indicates that node is on the right (resp. left) side of w.

Figure 3 shows a sample tree with leaf insertions.

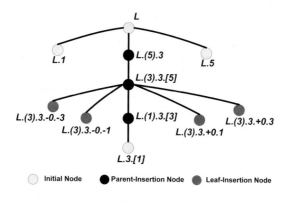

Fig. 3. Appending a child node

3   Determination of Inter-node Relationships

Parent-Child Relation. Consider a node u with label $L.(ih).C.[ph]$. We can deduce the AO part of u's parent as follows. If ph is non-empty, then the AO part is $L.(ph).C$. Otherwise, if ph is empty, then the AO part is L. Note that it is impossible to deduce the PH component of the parent.

To verify if a node v is the parent of u, we just apply the above algorithm on the label of u and see if the AO part thus generated matches the AO part of v's label.

Ancestor-Descendant Relation. To determine if two given nodes u and v are ancestor-descendant of each other, we consider two cases. (Case 1) If the AO part of v's label is a prefix of u's label, then v is ancestor of u. (Case 2) If u and v have the same sequence of components except for the SIH and PH, then v is an ancestor of u if and only if the SIH of v is lexicographically larger than the SIH of u. Note that in this case, v is a node inserted above u.

Sibling Relation. To determine if two nodes u and v are siblings, we check if they have the same parent. It suffices to deduce the AO part of their parent labels. To extend it to detect preceding/following sibling relation, we consider two cases.

(Case 1) If both u and v do not have a PH in their labels, then both do not have nodes inserted as their parents. In that case, we just compare their last chunks in lexicographic order.

(Case 2) If either u or v has a PH component, then the node (say u) with a PH component has one or more nodes inserted above it. We treat u as having "order 0" among its siblings. Note that v must not have a PH component. (Otherwise there would be another node inserted above v and this node is not the same as the node inserted above u.) If v has -0 in its first even component after the second last chunk, then v precedes u. Otherwise, v must have a +0 and so v follows u.

4 Implementation and Experimental Evaluation

Note that the use of dots and brackets in our labels are just for readers' easy reading. In the actual encoding, we employ the variable-length bit string representation (as used in OrdPath). Specifically, each component is encoded using the $I_i L_i O_i$ format where I_i is the component type indicating whether the component is an OrdPath, IH or PH. L_i is a prefix-free encoding [7] of the length of O_i, which in turn encodes the integer in binary. Note that we need 2 bits for I_i since we have three different types of label components.

We tested our scheme with both real world (Shakespeare's Play and TreeBank) and synthetic dataset (XMark of size 30MB and 336,244 nodes). The results show that OrdPathX behaves similarly on different datasets. Due to space limitation, we only present the results of XMark here.

In the experiment, we perform random parent insertions. We randomly pick a number of nodes (i.e., 10%, 20%, 40%, 60% and 80% of the total nodes) from

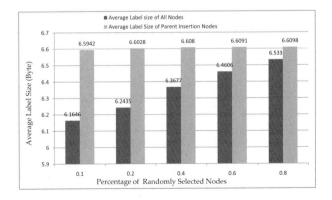

Fig. 4. Random Parent Inserion

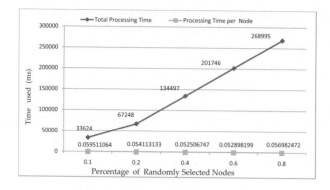

Fig. 5. Random Parent Inserion

the tree and insert one parent for each selected node. Figure 4 shows that the average label size of all nodes increases linearly as the proportion of selected nodes increases linearly while that of the inserted nodes always maintains at the same level regardless of how many parent insertions happen.

Figure 5 shows the total time as well as time per node to process parent insertions. Again, the total time increases linearly while the time per node remains relatively constant.

5   Conclusions and Future Work

Dynamic XML labeling has been studied for years. However, none of the existing labeling schemes avoid relabeling a large number of nodes when inserting a new parent node. In this paper, we extend the "careting-in" technique of OrdPath and propose a new labeling scheme called OrdPathX that supports both internal and leaf node insertions efficiently. In the future, we will investiagte the extension of OrdPathX to more generalized environment such as graph labeling. This could be useful in some fields such as Internet routing and graphical navigation.

References

1. http://www.microsoft.com/sqlserver/2008/en/us/whats-new.aspx
2. http://download.microsoft.com/download/3/4/C/
   34C25092-D2A1-4E56-83CF-923EF53BD390/SQLServer2008.pdf
3. Amagasa, T., Yoshikawa, M., Uemura, S.: A robust numbering scheme for XML documents. In: 19th International Conference on Data Engineering, pp. 705–707 (2003)
4. Cohen, E., Kaplan, H., Milo, T.: Labeling dynamic XML trees. In: Proceedings of the 21st Annual ACM Symposium on Principles of Database Systems, pp. 271–281 (2002)
5. Li, C., Ling, T.W., Hu, M.: Efficient updates in dynamic XML data: from binary string to quaternary string. The VLDB Journal 17, 573–601 (2008)

6. Li, Q., Moon, B.: Indexing and querying XML data for regular path expressions. In: Proceedings of the 27th International Conference on Very Large Data Bases, pp. 361–370 (2001)
7. O'Neil, P.E., O'Neil, E.J., Pal, S., Cseri, I., Schaller, G., Westbury, N.: ORDPATHs: Insert-friendly XML node labels. In: Proceedings of the 2004 ACM SIGMOD Conference on the Management of Data, pp. 903–908 (2004)
8. Tatarinov, I., Viglas, S., Beyer, K.S., Shanmugasundaram, J., Shekita, E.J., Zhang, C.: Storing and querying ordered XML using a relational database system. In: Proceedings of SIGMOD, pp. 204–215 (2002)
9. Sans, V., Laurent, D.: Prefix based numbering schemes for XML: techniques, applications and performances. In: Proceedings of the 35th International Conference on Very Large Data Bases, August 2008, pp. 23–28 (2008)
10. Santoro, N., Khatib, R.: Labeling and implicit routing in networks. The Computer Journal 28, 5–8 (1985)
11. Wu, X., Lee, M.L., Hsu, W.: A prime number labeling scheme for dynamic ordered XML trees. In: 20th International Conference on Data Engineering, pp. 66–78 (2004)
12. Yoshikawa, M., Amagasa, T., Shimura, T., Uemura, S.: XRel: a path-based approach to storage and retrieval of XML documents using relational databases. ACM Transactions on Internet Technology 1(1), 110–141 (2001)
13. Deschler, K., Rundensteiner, E.: MASS: A Multi-axis Storage structure for Large XML Documents. In: Proc. Conf. on Information and Knowledge Management, pp. 520–523 (2003)
14. Dang-Ngoc., T.T., Sans, V., Laurent, D.: Classifying XML Materialized views for their maintenance on distributed Web sources. In: Proc EGC Conf., RNTI, pp. 433–444 (2005)
15. Khaing, A., Thein, N.L.: A Persistent Labeling Scheme for Dynamic Ordered XML Trees. In: Proc. Conf. on Web Intelligence, pp. 498–501 (2006)
16. Gabillon, A., Fansi, M.: A persistent labelling scheme for XML and tree databases. In: Proc. SITIS Conf., pp. 110–115 (2005)
17. Duong, M., Zhang, Y.: LSDX: A New Labeling Schema for Dynamically Updating XML Data. In: Proc. ADC Conf., pp. 185–193 (2005)

XQSuggest: An Interactive XML Keyword Search System

Jiang Li and Junhu Wang

School of Information and Communication Technology,
Griffith University, Gold Coast, Australia
Jiang.Li@student.griffith.edu.au, J.Wang@griffith.edu.au

Abstract. Query suggestion is extensively used in web search engines to suggest relevant queries, which can help users better express their information needs. In this paper, we explore the application of query suggestion in XML keyword search and propose a novel interactive XML query system XQSuggest, which mainly targets non-professional users who roughly know the contents of the database. Our system extends conventional keyword search systems by instantly suggesting several understandable semantic strings after each keyword is typed in, so that the users can easily select their desired semantic string, which represents a specific meaning of the keyword, to replace the ambiguous keyword. We provide a novel algorithm to compute the final results. Experimental results are provided to verify the better effectiveness of our system.

1 Introduction

Keyword search has long been used to retrieve information from collections of text documents. Recently, keyword search in XML databases re-attracted attention of the research community because of the convenience it brings to users - there is no need for users to know the underlying database schema or complicated query language. Until now, a lot of research (e.g., XRank [2], SLCA [5], XSeek [3] and MaxMatch [4]) focuses on how to efficiently and meaningfully connect keyword match nodes and generate informative and compact results, but this only solves one side of the problem. The returned answers may be meaningful, but they may not be desired by the users. Therefore, the other side of the problem is how to accurately acquire the user's real intention, which is a difficult task because the keywords are inherently ambiguous. We use the following example to illustrate this problem.

Example 1. Suppose a user is interested in the population of all countries, and he submits the query {country, population} over the data tree in Fig. 1, which comes from Mondial data set [1]. However, every province and city element also has *country* and *population* attributes. In other words, the keywords *country* and *population* appear in different types of nodes. The systems can not know exactly which one is desired by the user just from the keywords. As a result, the irrelevant answers province (0.4) and city (0.4.4) will also be returned.

S.S. Bhowmick, J. Küng, and R. Wagner (Eds.): DEXA 2009, LNCS 5690, pp. 340–347, 2009.

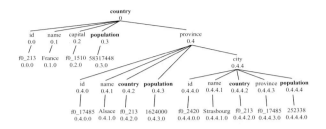

Fig. 1. Data tree t

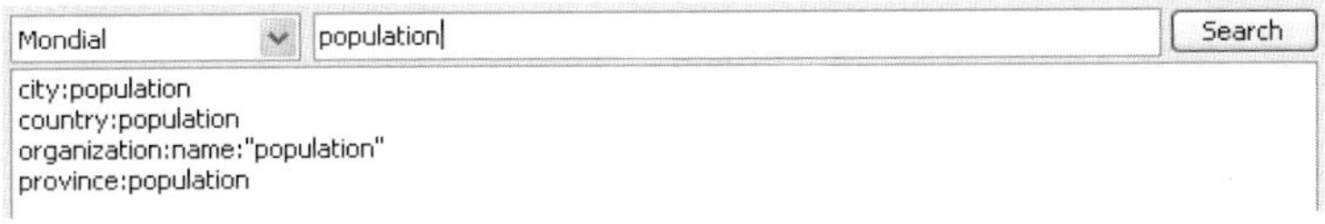

Fig. 2. A screen shot of query suggestion in XQSuggest

The example above shows the weakness of existing XML keyword search systems on returning relevant results when the submitted keywords have multiple meanings. In practice, most users of keyword search would roughly know the contents of the data and the meaning of each keyword they pose. They want to unambiguously express their needs but the keywords alone can not help them to do so. In this paper, we propose to solve the problem of ambiguity using query suggestion: When a keyword is typed in, the system can instantly suggest several *understandable* and *distinct* semantic strings. Then the user can select one to replace the ambiguous keyword. Fig. 2 shows a screen shot of query suggestion in our interactive XML keyword search system XQSuggest.

Our main contribution includes: (1) a fully implemented interactive XML keyword search system, XQSuggest, which allows for query suggestion and enables users to express their queries more clearly; (2) a novel algorithm for finding query results based on the semantic strings; (3) two optimization techniques that help to speed-up query processing; (4) experiments that verify the better effectiveness and usability of our system over conventional XML keyword search systems.

The rest of the paper is organized as follows. Section 2 provides background knowledge. Query suggestion technique is presented in Section 3. Section 4 then provides our new algorithm for finding the query results. Experimental studies are presented in Section 5. We conclude the paper in Section 6.

2   Preliminaries

An XML document is modeled as an unordered tree, called the *data tree*. Each internal node (i.e., non-leaf node) has a label, and each leaf node has a value. The internal nodes represent elements or attributes, while the leaf nodes represent the values of elements or attributes. Each node v in the data tree has a unique Dewey code, which represents the position of that node in the data tree. With this coding scheme, ancestor-descendant relationship can be easily identified: for any two nodes v_1, v_2 in data tree t, v_1 is an ancestor of v_2 iff v_1 is a prefix of v_2. Fig. 1 shows an example data tree.

Entity nodes. Most XML documents in reality are well designed and conform to a pre-defined schema. Therefore, even though an XML document is modeled as a tree, it is actually a container of related entities in the real world. Consider the data tree in Fig. 1, which is actually a collection of country, province and city entities. These entities are joined together through the ancestor-descendant relationship. We use an approach similar to that of [3] to identify entity nodes.

Definition 1. *Let t be a data tree. A node u in t is said to be a* simple node *if it is a leaf node, or has a single child which is a leaf node. A node u represents an* entity node *if: (1) it is root(t), or it corresponds to a *-node or +-node in the DTD (if DTD exists), or has siblings with the same label as itself (if DTD does not exist), and (2) it is not a simple node.*

Definition 2. *Let e_1 and e_2 be two entity nodes in t. If e_1 and e_2 have the same label, we say e_1 and e_2 are of the same entity-type. The entity-type will be referred to with the label name.*

Keyword query. A *keyword query* is a finite set of keywords $K = \{k_1, \ldots, k_n\}$. Given a keyword k and a data tree t, the search of k in t will check both the labels of internal nodes and values of leaf nodes for possible occurrence of k.

3   Query Suggestion

3.1   Semantic String

The basic idea of query suggestion is to replace each keyword with a *semantic string*, which has a more specific meaning than the keyword. Next we will precisely define the semantic strings associated with a keyword.

Definition 3. *Let t be the data tree and k be a keyword that occurs in t. A* semantic string *of k is a colon-separated sequence of labels $l_1 : l_2 : \cdots : l_n$ such that there is a path $u_1.u_2.\cdots.u_n$ in t, where u_n is a node whose label or value contains k, u_1 is the only entity node in the path (in other words, either u_n is an entity node and $u_1 = u_n$, or u_1 is the closest entity node above u_n), and l_i is the label of u_i for $i \in [1, n]$. The* length *of a semantic string is the number of labels in the path. The* type *of a semantic string is the same as the entity-type of its corresponding entity node.*

In some cases, a keyword can exist in both the element name and the value. In order to differentiate these two cases, if the keyword appears in the value, it will be double quoted in the semantic string. We call semantic strings that have double quoted keywords *predicates*.

Note: Each keyword occurring in the data tree has at least one semantic string. For the keyword in the label name of an entity node, its semantic string is the label name.

3.2   Boolean Operators

Sometimes, the user will specify more than one predicates within the same context. For the system, it is very difficult to judge the relationship between these predicates. Therefore, if the system allows the user to specify boolean operators (AND or OR) between semantic strings, the users will be able to express their queries more clearly. In our current implementation, the boolean operators can only be placed between the predicates of the same entity-type. If there are more than one predicates of the same entity-type in a query but no boolean operators are specified, the default boolean operator AND will apply.

3.3   Result of Query

When the ambiguous keywords are replaced with the semantic strings, each keyword has an exact meaning and is associated with a specific entity-type. We call the new query a *semantic string query*. Each semantic string will appear within some entity nodes, and we can join such entity nodes together as the result of the query. Note: the "join" here means concatenating entity nodes where the semantic strings appear and where an ancestor-descendant relationship exists. A more precise definition of a query result is given in Definition 4.

In the following, a subtree T of data tree t refers to a tree that can be obtained from t by erasing some entity nodes and all of their descendants. We say a semantic string s *appears in* an entity node $v \in T$ (and v *contains* s) if there is a path from v which is isomorphic to s. We say s appears in subtree T (and T *contains* s) if s appears in some entity node in T.

Definition 4. *A result of a semantic string query K is a subtree T of t with the following properties: (1) the root of T is an entity node in t, (2) every entity node in T contains at least one semantic string, (3) T contains all semantic strings in K, but no lower subtree of T contains all semantic strings in K, (4) if two predicates of the same entity-type have logic AND relationship, they must appear in the same entity node. We call such a result tree a* joined entity-node tree *(**JET**).*

4   Algorithm

4.1   Notations

Given a semantic string query $K = \{s_1, ..., s_n\}$, for each semantic string s_i, there is a stream, S_i, consisting of all the nodes which have the corresponding semantic string s_i. The nodes (i.e.,the Dewey codes) in each stream S_i are arranged in lexicographic order. The system needs to instantly suggest semantic strings after a keyword is typed in, so the performance of retrieving semantic strings of a keyword is very important. We build a B^+-tree on all of the words in t, each word k_i in the leaf of the B^+-tree points to the list of the corresponding semantic strings. It should be noted that there do not exist duplicates in a list.

The results of the algorithm are stored in the list RL. Each item in the list is a pointer which points to a list of entity nodes, which represents a JET to be returned to the user.

In order to facilitate keyword containment check and logical relationship check, we introduce four additional attributes $flag$, $pattern$, $isEntity$ and $logic$ for each semantic string during processing. The $flag$ attribute is a n-bit binary number, which indicates which keywords are contained in the sub-tree rooted at a node. The $pattern$ attribute is also a n-bit binary number, which presents which predicates of the same semantic string type are involved into a boolean formula. If the ith bit is set to 1, it means the semantic string s_i is involved into the boolean formula within the corresponding context.

We use the function $GetMinDewey(S_i)$ to get the smallest Dewey code (in lexicographic order) among the streams of semantic strings. The function $GetClosestEntityCode$ returns the Dewey code of the entity node associated with a semantic string, given the Dewey code and length of the semantic string.

4.2   The Stack-Based Algorithm

Our stack-based algorithm is shown in Algorithm 1.

Now, we explain the algorithm for evaluating queries. As mentioned earlier, we represent results using the joined tree of entity nodes (JET) associated with each semantic string, so after the minimal Dewey code of a semantic string is retrieved (line 5), the Dewey code of its entity node should be computed immediately for further processing (line 6), and this can be easily achieved with the length of the semantic string. After a node is popped up from the stack DS, if this node is an entity node, the algorithm first needs to perform a boolean logic check. If the desired boolean logic between the predicates of the same entity type is OR and any predicate is satisfied (line 12-13), the algorithm will modify the flag value to indicate all of the predicates are satisfied (line 14). After that, the algorithm will check whether it contains all the keywords (line 15). If a node is an entity node but does not contain all the keywords or it is not an entity node, the algorithm needs to copy the flag value to the top entry of the stack (line 20 and 22). Whether the unprocessed entity nodes exist is determined by the value of $DS.NumofEntityNode$, which is maintained by the algorithm to indicate how many entity nodes associated with the selected semantic strings exist in the stack DS (line 11 and 30).

4.3   Optimization Techniques

The performance of query evaluation can be seriously affected by the frequency of the keywords. If the keyword in the query is contained in the element names, there can be a large number of match nodes to be processed. First, given a query $K = \{s_1, ..., s_n\}$, if s_m is a sub-string of s_n, the s_m can be removed from K. This is because s_n can guarantee the satisfaction of s_m. Second, given a query $K = \{s_1, ..., s_n\}$, for the semantic strings with the same type, the algorithm can remove the semantic strings of the element or attribute, which at least occur

---

Algorithm 1. EvaluateQuery(K, S)

Input: Query $K = \{s_1, ..., s_n\}$, a set S of streams
Output: Result List RL

1: Create an empty result list RL
2: Create an empty list $ListItem$
3: Create an empty stack DS
4: **while** $DS \neq \emptyset$ **OR** $\neg isEnd(S_1) \wedge ... \wedge \neg isEnd(S_n)$ **do**
5:     $s_{min} = GetMinDewey(S_i)$
6:     $e_{min} = GetCloestEntityCode(s_{min}, S_{min}.SemanticStringSize)$
7:     Get longest common prefix lcp such that $DS[i].id = e_{min}[i]$ $(1 \leq i \leq lcp)$
8:     **while** $DS.size > lcp$ **do**
9:         $stackentry = DS.pop()$
10:         **if** $stackentry.isEntityNode = true$ **then**
11:             $DS.NumofEntityNode - -$
12:             **if** $stackentry.logic = OR$ **then**
13:                 **if** $stackentry.flag \& stackentry.pattern! = 0$ **then**
14:                     $stackentry.flag = stackentry.flag | stackentry.pattern$
15:             **if** $ContainAllKws(stackentry)$ **then**
16:                 Append current Dewey code in stack DS into the list $ListItem$
17:                 $RL.append(ListItem)$
18:                 Empty the list $ListItem$
19:             **else**
20:                 $CopyFlags(DS, stackentry)$
21:         **else**
22:             $CopyFlags(DS, stackentry)$
23:     **for** $i = lcp + 1$ to $e_{min}.length$ **do**
24:         $DS.push(e_{min}[i])$
25:     **if** DS is not empty **then**
26:         $SetFlag(DS[DS.size].flag, min)$
27:         $DS[DS.size].pattern = S_{min}.pattern;$
28:         $DS[DS.size].logic = S_{min}.logic$
29:     **if** $lcp! = e_{min}.length$ **then**
30:         $DS.numofEntityNode + +$
31: **procedure** COPYFLAGS($DS, child$)
32:     **if** $DS.NumofEntityNode > 0$ **then**
33:         **if** $child.flag \& child.pattern = child.pattern$ **then**
34:             $DS[DS.size].flag| = child.flag$
35:         **if** $child.isEntityNode = true$ **then**
36:             Append current Dewey code in stack DS into the list $ListItem$

---

once as a child of another element. This can be achieved by exploring the DTD file. The two optimization techniques work well because the time spent on processing semantic strings before query evaluation is much less than processing the keyword match nodes with a high frequency.

5 Experiments

All the experiments were performed on a 1.6GHz laptop with 1G RAM. We used the data sets WSU, SigmodRecord and Mondial obtained from [1], and selected five keyword queries for each data set. The queries are listed in Table 1. The corresponding semantic string queries are also listed in this table.

5.1 Reduction of Irrelevant Nodes

As we stated earlier, the multiple meanings of a keyword may cause a lot of irrelevant nodes to be processed. This can seriously influence the effectiveness

Table 1. Keyword queries and semantic string queries

WSU		
QW1	course,Lab	course,course:place:room:"lab"
QW2	CAC, 101	course:title:"cac" AND course:crs:"101"
QW3	instructor, MCELDOWNEY	course:instructor:"mceldowney"
QW4	ECON,bldg	course:title:"econ",course:place:bldg
QW5	ACCTG, times, place	course:prefix:"acctg",course:times,course:place
SigmodRecord		
QS1	Karen, title	article:authors:author:"karen",article:title
QS2	Anthony, Data	article:authors:author:"anthony" AND article:title:"data"
QS3	volume, 11, article	issue:volume:"11",article
QS4	article,data,John	article, article:title:"data" OR article:authors:author:"john"
QS5	database, volume,number	article:title:"database",issue:volume,issue:number
Mondial		
QM1	country,population	country:population
QM2	muslim,country	country:religions:"muslim", country
QM3	Belarus, population	country:name:"belarus", country:population
QM4	Ethnicgroups,Chinese,Indian,Capital	country:ethnicgroups:"chinese" OR country:ethnicgroups:"indian",country:capital
QM5	Turin, longitude, latitude	city:name:"turin", city:longitude,city:longitude

Table 2. Reduction of irrelevant nodes

Query	Keyword frequency	semantic string frequency	reduction	frequency after optimization	reduction
QW1	7925	3930	50.4%	6	99.9%
QW2	351	286	18.5%	286	0%
QW3	3932	6	99.8%	6	0%
QW4	4001	3942	1.5%	18	99.6%
QW5	7880	7880	0	32	99.6%
QS1	1506	1506	0	2	99.8%
QS2	185	185	0	185	0%
QS3	1603	1507	6.0%	1507	0
QS4	1713	1713	0	209	88.3%
QS5	481	478	1%	411	14.6%
QM1	18205	231	98.7%	231	0%
QM2	13476	324	97.6%	93	71.3%
QM3	4598	232	95.0%	2	99.1%
QM4	2070	281	86.4%	50	82.2%
QM5	2969	2793	5.9%	2793	0%

and efficiency of query. We compared the number of nodes that need to be processed before and after query suggestion (represented by keyword frequency and semantic string frequency), and calculated the reduction percentage. In addition, we also list the semantic string frequency of XQSuggest after applying the optimization techniques. The results are listed in Table 2.

5.2   Search Quality

We compared the search quality of XQSuggest with the most recent XML keyword search system MaxMatch [4], which claims better effectiveness than previous systems. The data sets we chose for comparison are Mondial and SigmodRecord. We evaluate the effectiveness of XQSuggest and MaxMatch based on *precision, recall* and *F-Measure* [4].

The comparisons of precision, recall and F-Measure over Mondial are illustrated in Fig. 3. As shown in the figure, XQSuggest achieves higher precision, recall and F-Measure than MaxMatch. This is mainly because the submitted

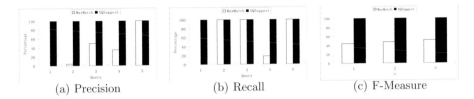

(a) Precision   (b) Recall   (c) F-Measure

Fig. 3. Precision, Recall and F-Measure on Mondial Data Set

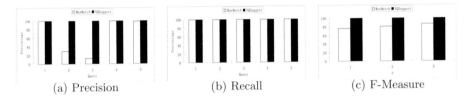

(a) Precision   (b) Recall   (c) F-Measure

Fig. 4. Precision, Recall and F-Measure on SigmodRecord Data Set

keywords exist in different types of nodes and the user can eliminate the irrelevant meanings with our system. Fig. 3 (c) presents the F-Measure of the queries with $\alpha = 0.5$, 1 and 2. It is shown that XQSuggest outperforms MaxMatch. The comparisons of Precision, Recall and F-Measure over SigmodRecord are illustrated in Fig. 4. On the SigmodRecord data set, the advantages of XQSuggest is not that obvious because most keywords in the data set have unique meanings.

6   Conclusion

In this paper, we explored the application of query suggestion in XML keyword search. Our system XQSuggest suggests several semantic strings after each keyword is typed in, which significantly reduces keyword ambiguity and facilitates the use of Boolean operators. We proposed an algorithm to find the results of the transformed query. Two optimization techniques were used in our algorithm in order to improve performance.

References

1. http://www.cs.washington.edu/research/xmldatasets
2. Guo, L., Shao, F., Botev, C., Shanmugasundaram, J.: Xrank: Ranked keyword search over XML documents. In: SIGMOD Conference, pp. 16–27 (2003)
3. Liu, Z., Chen, Y.: Identifying meaningful return information for XML keyword search. In: SIGMOD Conference, pp. 329–340 (2007)
4. Liu, Z., Chen, Y.: Reasoning and identifying relevant matches for xml keyword search. PVLDB 1(1), 921–932 (2008)
5. Xu, Y., Papakonstantinou, Y.: Efficient keyword search for smallest lcas in XML databases. In: SIGMOD Conference, pp. 527–538 (2005)

A Prüfer Based Approach to Process Top-k Queries in XML

Ling Li, Mong Li Lee, Wynne Hsu, and Han Zhen

School of Computing, National University of Singapore
{leeml,whsu,hanzhen}@comp.nus.edu.sg

Abstract. Top-k queries in XML involves retrieving approximate matching XML documents. Existing techniques process top-k queries in XML by applying one or more relaxations on the twig query. In this work, we investigate how Prüfer sequence can be utilized to process top-k queries in XML. We design a method called XPRAM that incorporates the relaxations into the sequence matching process. Experiment results indicate that the proposed approach is efficient and scalable.

1 Introduction

Top-k queries arise naturally in many database applications and has been extended to XML [2,3,7]. [2] introduces the notion of query relaxation. The structure of a query can be relaxed using edge generalization, leaf node deletion and subtree promotion and any combination of them. An XML document has both the content and structural aspects. For the former, an IR-based tf*idf paradigm is employed for text-search at each element level. For the latter, relaxation using edge generalization, leaf deletion and subtree promotion is applied (recursively). For example, the XML query in Figure 1(a) returns all the XML documents containing *person* element who has a phone and there is a profile to record his/her gender and age. The corresponding relaxations are shown in Figures 1(b) to 1(d).

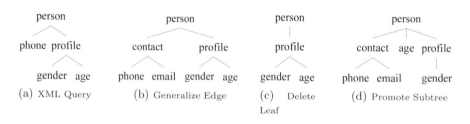

Fig. 1. Example Query and Its Relaxations

Two systems, FleXPath [3] and Whirlpool [7], have been developed for top-k queries in XML. FleXPath [3] gradually relax the query pattern until it returns at least k answers. Whirlpool [7] consider query relaxation with different join orders to prune the lower-ranked documents as early as possible.

S.S. Bhowmick, J. Küng, and R. Wagner (Eds.): DEXA 2009, LNCS 5690, pp. 348–355, 2009.
© Springer-Verlag Berlin Heidelberg 2009

In this work, we investigate another approach to answer top-k queries in XML. The works in [8] and [9] utilize Prüfer sequences for XML indexing and query processing. XML documents are transformed into labeled sequences. A twig query is also mapped to a sequence and subsequence matching is performed to find all the occurrences of a query. The advantage of using Prüfer sequence is that it allows holistic processing of a twig pattern without decomposing the twig into root-to-leaf paths. However, these works are focused on finding exact answers using Prüfer sequences and not top-k queries.

Based on the Prüfer sequence approach, we design a solution that relax the matching process instead of the naive way of generating exponential relaxations of a query. We call the proposed solution XPRAM (PRüfer sequences for Approximate Matching in XML). This approach incorporates edge generalization, leaf deletion and subtree promotion during the matching process to find top-k answers. We also extend Prüfer sequences with the preorder and postorder numbering of XML nodes to facilitate the containment relationship test. We carry out a set of experiments to demonstrate that the proposed approach is efficient and scalable, and outperforms existing approaches.

2   Preliminaries

The *Prüfer sequence* of an XML tree is constructed using a node removal method. Given a labeled tree T_n with n nodes labeled from 1 to n, usually in postorder, we delete the leaf node with the smallest label to form a smaller tree T_{n-1} and record a_1, which is the tag of the deleted node's parent. This process is repeated until only one node is left. The sequence $LPS = (a_1 \ a_2 \ \ldots \ a_{n-1})$ is called the Prüfer sequence of tree T_n. In this sequence, the tag of each leaf node appears once and the tag of each branch node appears as many times as the number of its child nodes in the tree. We further augment the Prüfer sequence with the preorder and postorder numbering of an XML tree to facilitate checking of the containment relationship between any two XML nodes and call it *Prüfer Sequence+ (LPS$^+$)*. Figure 2 shows an example XML tree and its corresponding augmented *Prüfer sequence*.

Definition 1 (Q-connected tree of T). *Given an XML tree $T(V_T, E_T)$ and a query pattern $Q(V_Q, E_Q)$, a Q-connected tree of T, denoted by $T_Q(V, E)$, is given by*

1. $V = V_T \cap V_Q$
2. $E = E' \cup E''$ *where* $E' = \{(n_i, n_j) \mid (n_i, n_j) \in E_T \text{ and } n_i, n_j \notin V_T - V\}$
   and $E'' = \{(parent(n), child(n)) \mid n \in V_T - V\}$

Consider the query Q in Figure 3. Figure 4 shows a Q-connected XML tree of $T1$ in Figure 2. Note that nodes with tags E, F and G do not occur in the Q-connected tree since they do not occur in Q. Further, the parent of each node that has been removed is connected to each of its child nodes with an AD axis. The Prüfer sequence+ of a Q-connected tree of T is called the Q-connected LPS+ of T, denoted by $LPS^+_{connect}(T, Q)$. It can be obtained directly from $LPS^+(T)$.

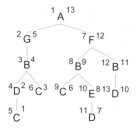

LPS(T₁) = C D B C B G A C B D E B F D B F A

LPS+(T₁) = C(5, 1) D(4,2) B(3,4) C(6,3) B(3,4) G(2,5) A(1,13) C(9,6) B(8,9)
D(11,7) E(10,8) B(8,9) F(7,12) D(13,10) B(12,11) F(7,12) A(1,13)

LPS(Q) = C B D B A

LPS(Q) = C(3,1) B(2,3) D(4,2) B(2,3) A(1,4)

Fig. 2. Example XML document $T1$ **Fig. 3.** Query pattern Q

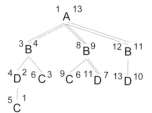

LPS_connect+(T₁, Q) = C(5, 1) D(4,2) B(3,4) C(6,3) B(3,4) A(1,13) C(9,6) B(8,9)
D(11,7) B(8,9) A(1,13) D(13,10) B(12,11) A(1,13)

Fig. 4. Q-connected XML tree of $T1$

Definition 2 (Q-conform sequence). *Given a tree T and a query Q, and their corresponding Prüfer sequence+ $LPS^+(T)$ and $LPS^+(Q)$, a subsequence s of $LPS^+(T)$ is a Q-conform sequence if s is a subsequence of $LPS^+(Q)$ and the corresponding tree structure of s can be obtained from Q by applying a combination of edge generalization, leaf node deletion and subtree promotion.*

A Q-conform sequence s is called a maximal Q-conform sequence of $LPS^+(T)$ if there does not exist another Q-conform sequence of $LPS^+(T)$, s', such that s' subsumes s. For example, both the subsequences $s_1 = (D\ (13,\ 10)\ B\ (12,\ 11)\ A\ (1,\ 13))$ and $s_2 = (C\ (6,\ 3)\ (D\ (13,\ 10)\ B\ (12,\ 11)\ A\ (1,\ 13))$ of $LPS^+(T1)$ conform to Q. However, only s_2 is a maximal Q-conform subsequence.

Definition 3 (Q-conform substring). *Given an XML document T and a query pattern Q, let s denote a substring of $LPS+(T)$, and s' be a string obtained by extracting only the tags from s. We say that s is a Q-conform substring, denoted by $LPS^+_{substr}(T, Q)$, if*

1. *s' is a substring of $s_{connect}$ where $s_{connect}$ is obtained from $LPS^+_{connect}(T, Q)$ by extracting only the tags.*
2. *s' is a substring of $LPS(Q)$.*

3. *each pair of nodes in V conforms to Q, where V is a set of nodes corresponding to the elements in s.*

A Q-conform substring s of $LPS^+(T)$ could be extended to the *longest Q-conform substring* of $LPS^+(T)$ if we insert the element which is the previous element of the first element of s in $LPS^+(T)$, in front of s, or if we append the element which is the next element of the last element of s in $LPS^+(T)$.

For example, a subsequence s_1, *(D (11,7) B (8,9) A (1,13))* of $LPS^+(T1)$ is a $LPS^+_{substr}(T1, Q)$, and another subsequence $s2$, *(C(9,6) B(8,9) D(11,7) B(8,9) A (1,13))*, which can be obtained from s_1 by inserting two elements in the front, is a longest Q-conform substring since it could not be further appended or inserted.

Given a Q-conform substring s and an element e, if we want to determine whether the string s + e is a Q-conform substring, we need to check whether the nodes corresponding to s and each element in s conform to Q. However, the following property allows us to restrict the test to just the nodes corresponding to e and the last element in s. The proof is provided in [10].

Property 1. Let T and Q denote an XML document and a query pattern respectively. Let $s = (e_i, \ldots, e_j)$ be a subsequence of $LPS+(T)$ such that s is a Q-conform substring. Let e_{j+1} denote the element after e_j in LPS+(T). If the nodes corresponding to the elements e_j and e_{j+1} conform to Q, then the nodes corresponding to the elements e_{j+1} and any element in s conform to Q.

3   XPRAM

The input to XPRAM is a set of XML documents and a query pattern Q. The output is a set of top-k documents. The main steps are:

1. Retrieve a set of longest Q-conform substrings of $LPS^+(T)$ for each input XML document T.
2. Concatenate the longest Q-conform substrings to get the maximal Q-conform sequences of $LPS^+(T)$.
3. Rank each document based on its maximal Q-conform sequences, and output the top-k documents.

3.1   Extract Longest Q-Conform Substrings

Given a query pattern Q issued over a set of XML documents, we need to retrieve the longest Q-conform substrings of $LPS^+(T)$ for each XML document T. Each element e in $LPS^+_{connect}(T, Q)$ is tested to see if it can be appended to s to form a substring of $LPS(Q)$ and a Q-conform substring. When the longest substring is obtained, an output is generated and the scanning process restarts from the next element in $LPS^+_{connect}(T, Q)$.

Suppose we issue query Q in Figure 3 over $T1$ in Figure 2. The set of longest Q-conform substrings of $LPS^+(T1)$ extracted by this step is $str_1 = (C(5,1))$, $str_2 = (C(6,3) B(3,4))$, $str_3 = (C(9,6) B(8,9) D(11,7) B(8,9) A(1,13))$, $str_4 = (D(4,2) B(3,4))$, $str_5 = (D(13,10) B(12,11) A(1,13))$, $str_6 = (A(1,13))$

3.2   Find Maximal Q-Conform Sequences

Next, we find the maximal Q-conform sequences of $LPS^+(T)$. We group the Q-conform substrings by its first element to avoid redundant computations, and use dynamic programming to process them. Algorithm 1 shows the details.

Algorithm 1. Find Maximal Q-conform Sequences

1: **Input:** set of longest Q-conform substrings S_{str} and query pattern Q
2: **Output:** S_{max} - a set of maximal Q-conform sequences of $LPS^+(T)$;
3: initialize $S_{max} = S_{str}$;
4: let A denote a sequence of tags obtained from elements in $LPS^+(Q)$ which are sorted by the postorder numbers in descending order;
5: **for** each tag a in A **do**
6:     **for** each sequence s_a in S_{max} such that the tag of the first element is a **do**
7:         let l = the tag of the last element in s_a;
8:         **for** each tag b from l to the last tag in A **do**
9:             let S = set of sequences in S_{max} where the tag of its first element is b;
10:             **for** each sequence s_b in S **do**
11:                 let s'_a = concatenate(s_a, s_b);
12:                 **if** s'_a != s_a **then**
13:                     replace s_a with s'_a in S_{max};
14:                 **end if**
15:             **end for**
16:         **end for**
17:     **end for**
18: **end for**
19: return S_{max};
    Function concatenate(s_a, s_b)
20: let e_b = first element of s_b and e_a = last element of s_a;
21: **if** e_a.tag == e_b.tag && e_b is identical to e_a **then**
22:     return s_a - e_a + s_b;
23: **else**
24:     let n_1 and n_2 be the elements of e_a and e_b respectively;
25:     **if** n_1 and n_2 conform to Q **then**
26:         return s_a + s_b;
27:     **else**
28:         let n_{anc} be the nearest ancestor of n_2 such that its element is in s_b;
29:         **while** n_{anc} != null **do**
30:             **if** n_1 and n_{anc} conform to Q **then**
31:                 return s_a + s_b;
32:             **else**
33:                 n_{anc} be the nearest ancestor of n_{anc} such that its element is in s_b;
34:             **end if**
35:         **end while**
36:     **end if**
37: **end if**
38: return s_a;

In this algorithm, maximal Q-conform sequences are gradually concatenated from the initial longest Q-conform substrings. Function *concatenate* is called to determine if two sequences can be concatenated. From the preorder and postorder numbers of the elements, we check if the corresponding nodes n_1 and n_2 conform to Q. Otherwise, without loss of generality, we check if n_1 and the ancestor of n_2 conform to Q. This tests for the case of subtree promotion. This process continues until all the longest Q-conform substrings have been processed.

To illustrate, let us consider the set of longest Q-substrings obtained in Section 3.1. We can concatenate str_4 with str_6 to form $s_1 = (D(4,2)\ B(3,4)\ A(1,13))$ since the corresponding nodes of elements $B(3,4)$ and $A(1,13)$ conform to Q. Similarly, str_2 and str_5 can be concatenated since the corresponding nodes of $B(3,4)$ and $A(1,13)$ conform to Q. Note that when we first check the nodes of $B(3,4)$ and $D(13,10)$, they do not conform to Q. The function *concatenate* will proceed to check the nodes of $B(3,4)$ and $B(12,11)$. Again, these nodes do not conform to Q too. The function continues to check the nodes of $B(3,4)$ and $A(1,13)$, and they conform to Q, hence str_2 and str_5 can be concatenated. When we examine the structure consisting of the nodes of the elements in str_2 + str_5, we find two XML fragments that can be mapped to Q with a subtree promotion. The result is $s_2 = (C(5,1)\ D(4,2)\ B(3,4)\ A(1,13))$, $s_3 = (C(5,1)\ D(13,10)\ B(12,11)\ A(1,13))$, $s_4 = (C(6,3)\ B(3,4)\ D(4,2)\ B(3,4)\ A(1,13))$, $s_5 = (C(6,3)\ B(3,4)\ D(13,10)\ B(12,11)\ A(1,13))$, $s_6 = (C(9,6)\ B(8,9)\ D(11,7)\ B(8,9)\ A(1,13))$. Note that s_1 is not in the final result, since it is subsumed by s_2 and s_4 and is not a maximal Q-conform sequence.

The XML documents retrieved can be ranked using existing ranking schemes that take into consideration the content [5] and/or path expressions [1,4,6]. Proofs of the soundness and completeness of XPRAM are given in [10].

4   Experiment Evaluation

We use the XMark data generator[1] to create the experimental dataset and vary three parameters to test the performance of XPRAM: *size of the input documents, value of k* and *query complexity*. The size of the dataset ranges from 1MB to 100MB. Edge generalization is provided by recursive nodes such as *parlist* in the XMark DTD. Leaf deletion is enabled by the optional nodes such as *incategory*, while subtree promotion is allowed by nodes such as *text*.

We run the same three queries as FleXPath [3] which have increasing complexity in terms of depth and fanout:

Q1: //item[./description/parlist]
Q2: //item[./description/parlist and ./mailbox/mail/text]
Q3: //item[./description/parlist/listitem and ./mailbox /mail /text[./bold and ./keyword and ./emph] and ./name and ./incategory]

All the experiments were carried out on a 2.58GHz Pentium 4 PC with 1.00 GB RAM, running WinXP. Each experiment is repeated 5 times, and the average time taken is recorded.

---

[1] http://monetdb.cwi.nl/xml/index.html

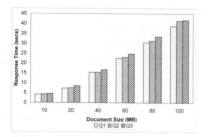

(a) E1. Varying document size and query complexity (k = 500)

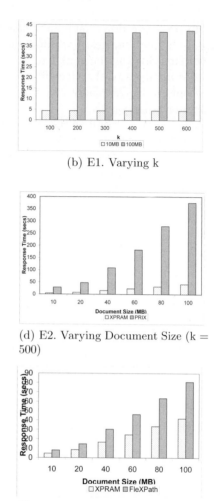

(b) E1. Varying k

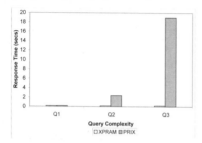

(c) E2. Varying query complexity (DocSize = 1MB)

(d) E2. Varying Document Size (k = 500)

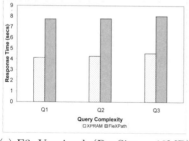

(e) E3. Varying k (DocSize = 10MB)

(f) E3. Varying Document Size (k = 500)

Fig. 5. XPRAM Experiments

The first set of experiments (E1) examines the time taken to process top-k queries. Figure 5(a) and Figure 5(b) show the result and confirm the scalability of XPRAM with increasing document size and k.

In the second set of experiments (E2), we modify the PRIX method to process top-k queries and compare it with XPRAM. The main modifications are:

1. For each document T and query pattern Q, we retrieve all the subsequences from LPS(T) which are subsequences of LPS(Q).
2. For each subsequence, we check the containment relationship specified in T.
3. The set of documents retrieved are ranked and the first k documents are returned.

In Figure 5(c) and Figure 5(d), we observe that as the query complexity or document size increases, the difference in the runtime of the two methods also increases. This is because the increase in the length of the Prüfer sequence results in a large number of subsequences retrieved for PRIX. The proposed XPRAM outperforms the PRIX approach.

The third set of experiments (E3) compares XPRAM with the Hybrid method in FleXPath [3]. Figure 5(e) shows the results of the time taken by both methods to process queries of increasing complexity, while Figure 5(f) shows the time taken as the sizes of the input documents vary. We observe that XPRAM outperforms FleXPath, indicating that relaxing the matching process instead of relaxing the query pattern is a faster approach.

5   Conclusion

Motivated by the growing importance of top k queries with the increasing XML repositories, we develop XPRAM, a Prüfer based approach for approximate matching in XML. The proposed solution increases the query performance in two ways. First, it is based on Prüfer sequence which provides for the holistic processing of twig queries. Second, instead of relaxing the query to retrieve approximate matches, XPRAM incorporates the relaxations (edge generalization, leaf node deletion and subtree promotion) into the matching process so that lower-ranked documents are quickly pruned off. Experiment results indicate that the proposed approach is efficient and scalable, and outperforms FleXPath.

References

1. Al-Khalifa, S., Yu, C., Jagadish, H.V.: Querying structured text in an xml database. In: ACM SIGMOD (2003)
2. Amer-Yahia, S., Koudas, N., Marian, A.: Structure and content scoring for xml. In: VLDB (2005)
3. Amer-Yahia, S., Lakshmanan, L.V.S., Pandit, S.: Flexpath: Flexible structure and full-text querying for xml. In: ACM SIGMOD (2004)
4. Botev, C., Shanmugasundaram, J., Amer-Yahia, S.: A texquery-based xml full-text search engine. In: ACM SIGMOD (2004)
5. Guo, L., Shao, F., Botev, C., Shanmugasundaram, J.: Xrank: Ranked keyword search over xml documents. In: ACM SIGMOD (2003)
6. Kaushik, R., Krishnamurthy, R., Naughton, J.F., Ramakrishnan, R.: On the integration of structure indexes and inverted lists. In: ACM SIGMOD (2004)
7. Marian, A., Amer-Yahia, S., Koudas, N., Srivastava, D.: Adaptive processing of top-k queries in xml. In: IEEE ICDE (2005)
8. Rao, P., Moon, B.: Prix: Indexing and querying xml using prufer sequences. In: IEEE ICDE (2004)
9. Tatikonda, S., Parthasarathy, S., Goyder, M.: Lcs-trim: Dynamic programming meets xml indexing and querying. In: VLDB (2007)
10. Zhen, H., Lee, M.L., Hsu, W.: Answering top-k queries in xml (submitted, 2009)

Bottom-Up Evaluation of Twig Join Pattern Queries in XML Document Databases

Yangjun Chen

Department of Applied Computer Science
University of Winnipeg
Winnipeg, Manitoba, Canada R3B 2E9
y.chen@uwinnipeg.ca

Abstract. Since the extensible markup language XML emerged as a new standard for information representation and exchange on the Internet, the problem of storing, indexing, and querying XML documents has been among the major issues of database research. In this paper, we study the twig pattern matching and discuss a new algorithm for processing *ordered twig pattern queries*. The time complexity of the algorithm is bounded by $O(|D|\cdot|Q| + |T|\cdot leaf_Q)$ and its space overhead is by $O(leaf_T\cdot leaf_Q)$, where T stands for a document tree, Q for a twig pattern and D is a largest data stream associated with a node q of Q, which contains the database nodes that match the node predicate at q. $leaf_T$ ($leaf_Q$) represents the number of the leaf nodes of T (resp. Q). In addition, the algorithm can be adapted to an indexing environment with XB-trees being used.

1 Introduction

The Extensible Markup Language (XML) is an emerging standard for data representation and exchange on the Internet. Tree pattern matching is one of the most important types of XML queries to extract information from XML sources. Normally, an XML document T is represented as a tree structure and typically a query Q specifies patterns of selection predicates on multiple elements that also have some specified tree structured relations. For instance, the XPath expression:

book[**title** = 'Art of Programming']*//author*[*fn* = 'Donald' and *ln* = 'Knuth']

asks for all those *author* elements that (i) have a child subelement *fn* with content 'Donald', (ii) have a child subelement *ln* with content 'Knuth', and are descendants of *book* elements that have a child *title* subelement with content 'Art of Programming'. It can be represented as a tree structure. So the query evaluation in XML document databases is essentially a tree matching problem.

We distinguish between two kinds of tree matchings. One is the so-called unordered tree matching, by which the order of siblings is not significant. The other is the ordered tree matching, by which the order of siblings should be taken into account. In the following definitions, $u \rightarrow v$ in Q stands for a child edge (/-edge) for a parent-child relationship; and $u \Rightarrow v$ for a descendant edge (//-edge) for an ancestor-descendant relationship. We also use *label*(v) to represent the name tag (i.e., symbol $\in \Sigma \cup \{*\}$) or the string associated with v.

S.S. Bhowmick, J. Küng, and R. Wagner (Eds.): DEXA 2009, LNCS 5690, pp. 356–363, 2009.
© Springer-Verlag Berlin Heidelberg 2009

Definition 1 (*unordered tree matching*). An embedding of a *twig* (small tree) pattern Q into an XML document T is a mapping $f: Q \to T$, from the nodes of Q to the nodes of T, which satisfies the following conditions:

(i) Preserve *node label*: For each $u \in Q$, $label(u) = label(f(u))$.

(ii) Preserve *parent-child/ancestor-descendant* relationships: If $u \to v$ in Q, then $f(v)$ is a child of $f(u)$ in T; if $u \Rightarrow v$ in Q, then $f(v)$ is a descendant of $f(u)$ in T.

If there exists a mapping from Q into T, we say, Q can be imbedded into T, or say, T contains Q. Notice that an embedding could map several nodes with the same tag name in a query to the same node in a database. It also allows a tree mapped to a path, by which the order of siblings is totally unconsidered. This definition is a little bit different from the ordered twig matching defined below.

Definition 2 (*ordered tree matching*). An embedding of a twig pattern Q into an XML document T is a mapping $f: Q \to T$, from the nodes of Q to the nodes of T, which satisfies the following conditions:

(i)   same as (i) in Definition 1.
(ii) same as (ii) in Definition 1.
(iii) Preserve *left-to-right order*: For any two nodes $v_1 \in Q$ and $v_2 \in Q$, if v_1 is to the left of v_2, then $f(v_1)$ is to the left of $f(v_2)$ in T.

v_1 is said to be to the left of v_2 if they are not related by the ancestor-descendant relationship and v_2 follows v_1. This kind of tree mappings is useful in practice. For example, an XML data model was proposed by Catherine and Bird [1] for representing interlinear text for linguistic applications, used to demonstrate various linguistic principles in different languages. For the purpose of linguistic analysis, it is essential to preserve the linear order between the words in a text [1]. In addition to interlinear text, the syntactic structure of textual data should be considered, which breaks a sentence into syntactic units such as noun clauses, verb phrases, adjectives, and so on. These are used by the language TreeBank [2] to provide a hierarchical representation of sentences. Therefore, by the evaluation of a twig pattern query against the Tree-Bank, the order between siblings should be considered [2, 3].

In 2003, Wang *et al.* [4] proposed a first index-based method, called *ViST*, for handling ordered twig pattern queries, by which the XML data are transformed into structure-encoded sequences and stored in a disk-based virtual *trie* using B^+-trees. One of the problems of this method is that the query processing strategy by straightforward sequence matching may result in false alarms. Another problem, as pointed out in [3], the size of indexes is higher than linear in the total number of elements in an XML document. Such problems are removed by a method, called *PRIX*, discussed in [3]. This method constructs two Prüfer sequences to represent an XML document: a numbered Prüfer sequence and a labeled Prüfer sequence. For all the labeled Prüfer sequences, a virtual trie is constructed, used as an index structure. In this way, the size of indexes is dramatically reduced to $O(|T|)$. But it suffers from very high *CPU* time overhead according to the following analysis. The method consists of a string matching phase and several so-called refinement phases, for which $O(k|Q|\log|Q|)$ time is needed (see page 328 in [3]), where k is the number of subsequences of a labeled Prüfer document sequence, which match Q's labeled Prüfer sequence. However, by the string matching defined in [3], a query pattern string can match non-consecutive segments within a document target string (see Definition 4.1 in [3], page 306). So in the worst case k is in the order of $O(|T|^{|Q|})$ since for each position i (in the target) matching the first element in the pattern string the second element of the pattern can match possibly at $|T| - i - 1$ positions; and for each position j matching the second element in the pattern, the third element in the pattern can possibly match at $|T| - j - 1$ positions, and so on. As an example, consider the following Prüfer string:

A … ab … bc … cd …d

in which each substring containing the same characters is of length $n/4$. Assume that the Prüfer string for a query is $abcd$. Then, there are $O(n^4)$ matching positions. For each of them, a tree embedding will be examined. (We note that if the string matching is restricted to consecutive segments, there is at most one matching for each position, at which the first element in the pattern matches. But it is not the case discussed in [3].)

In this paper, we propose a new method for processing ordered twig pattern queries. The main idea behind it is an algorithm for reconstructing tree structures from data streams as well as a new tree labeling technique for queries to represent *left-to-right relationships*. The new algorithm runs in $O(|D| \cdot |Q| + |T| \cdot leaf_Q)$ time and $O(leaf_T \cdot leaf_Q)$ space, where $leaf_T$ ($leaf_Q$) represents the number of the leaf nodes of T (resp. Q), and D is a largest data stream associated with a node q of Q, which contains the database nodes that match the node predicate at q.

The remainder of the paper is organized as follows. In Section 2, we restate the tree encoding [5], which can be used to facilitate the recognition of different relationships among the nodes of trees. In Section 3, we discuss our algorithm for evaluating ordered twig pattern queries. The paper concludes in Section 4.

2 Tree Labeling

In [5], an interesting tree encoding method was discussed, which can be used to identify different relationships among the nodes of a tree. Let T be a document tree. We associate each node v in T with a quadruple $\alpha(v) = (d, l, r, ln)$, where d is the document identifier (*DocId*), $l = LeftPos$, $r = RightPos$, and $ln = LevelNum$. Here, *LeftPos* and *RightPos* are generated by counting word numbers from the beginning of the document until the start and end of the element, respectively. By using such a data structure, the structural relationship between the nodes in an XML database can be simply determined [5]:

(i)   *ancestor-descendant*: a node v_1 associated with (d_1, l_1, r_1, ln_1) is an ancestor of another node v_2 with (d_2, l_2, r_2, ln_2) iff $d_1 = d_2$, $l_1 < l_2$, and $r_1 > r_2$.

(ii) *parent-child*: a node v_1 associated with (d_1, l_1, r_1, ln_1) is the parent of another node v_2 with (d_2, l_2, r_2, ln_2) iff $d_1 = d_2$, $l_1 < l_2$, $r_1 > r_2$, and $ln_2 = ln_1 + 1$.

(iii) *from left to right*: a node v_1 associated with (d_1, l_1, r_1, ln_1) is to the left of another node v_2 with (d_2, l_2, r_2, ln_2) iff $d_1 = d_2$, $r_1 < l_2$.

(See Fig. 3(a) for illustration.) In the rest of the paper, if for two quadruples $\alpha_1 = (d_1, l_1, r_1, ln_1)$ and $\alpha_2 = (d_2, l_2, r_2, ln_2)$, we have $d_1 = d_2$, $l_1 < l_2$, and $r_1 > r_2$, we say that α_2 is subsumed by α_1. For convenience, a quadruple is considered to be subsumed by itself. If no confusion is caused, we will use v and $\alpha(v)$ interchangeably. We can also assign LeftPos and RightPos values to the query nodes in Q for the same purpose as above. Finally we use $T[v]$ to represent a subtree rooted at v in T.

3   Main Algorithm

In this section, we describe our method. First, we discuss a kind of data stream transformation in 3.1, which provides the input to our main procedure. Then, in 3.2, the main algorithm is described in great detail.

3.1   Data Stream Transformation

As with *TwigStack* [4], each node q in a twig pattern (or say, a query tree) Q is associated with a data stream $B(q)$, which contains the positional representations (quadruples) of the database nodes v that match q (i.e., $label(v) = label(q)$). All the quadruples in a data stream are sorted by their (DocID, LeftPos) values. Therefore, iterating through the stream nodes in sorted order of their LeftPos values corresponds to access of document nodes in preorder. However, our algorithm needs to visit them in *postorder* (i.e., in sorted order of their RightPos values). For this reason, we maintain a global stack ST to make a transformation of data streams using the following algorithm. In ST, each entry is a pair (q, v) with $q \in Q$ and $v \in T$ (v is represented by its quadruple.)

Algorithm. *stream-transformation*($B(q_i)$'s)
input: all data streams $B(q_i)$'s, each sorted by LeftPos.
output: new data streams $L(q_i)$'s, each sorted by RightPos.
begin
1. **repeat until** each $B(q_i)$ becomes empty
2.   { identify q_i such that the first element v of $B(q_i)$ is of the minimal LeftPos
          value; remove v from $B(q_i)$);
3.   **while** ST is not empty and $ST.top$ is not v's ancestor **do**
4.   { $x \leftarrow ST.pop(\ )$; Let $x = (q_j, u)$;
5.   put u at the end of $L(q_i)$; }
7. $ST.push(q_i, v)$;
8.   }
end

In the above algorithm, ST is used to keep all the nodes on a path until we meet a node v that is not a descendant of $ST.top$. Then, we pop up all those nodes that are not v's ancestor; put them at the end of the corresponding $L(q_i)$'s (see lines 3 - 4); and push v into ST (see line 7.) The output of the algorithm is a set of data streams $L(q_i)$'s with each being sorted by RightPos values. However, we remark that the popped nodes are in postorder. So we can directly handle the nodes in this order without explicitly generating $L(q_i)$'s. But for ease of explanation, we assume that all $L(q_i)$'s are completely generated in the following discussion. We also note that the data streams associated with different nodes in Q may be the same. So we use $\boldsymbol{q}$ to represent the set of such query nodes and denote by $L(\boldsymbol{q})$ ($B(\boldsymbol{q})$) the data stream shared by them. Without loss of generality, assume that the query nodes in $\boldsymbol{q}$ are sorted by their RightPos values. We will also use $L(Q) = \{L(\boldsymbol{q}_1), ..., L(\boldsymbol{q}_l)\}$ to represent all the data streams with respect to Q, where each $\boldsymbol{q}_i$ ($i = 1, ..., l$) is a set of sorted query nodes that share a common data stream.

3.2 Main Procedure

First of all, we notice that iterating through $L(\boldsymbol{q}_1), ..., L(\boldsymbol{q}_l)$, i.e., the data streams sorted in increasing RightPos values, we navigate T in postorder. So, our algorithm works bottom-up. For the purpose of checking ordered tree embedding, we will first search Q in the breadth-first fashion, generating a number (called a *breadth-first number*) for each node q in Q, denoted as $bf(q)$, which can be used to represent the left-to-right order of siblings in a simple way (See Fig. 1(a) for illustration). Then, we use *interval(q)* to represent an interval covering all the breadth-first numbers of q's children. For example, for Q shown in Fig. 1(a), we have $interval(q_1) = [2, 3]$ and $interval(q_2) = [4, 5]$. In the following, we will use q and $bf(q)$ interchangeably.

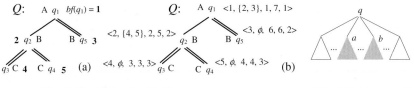

Fig. 1. Illustration for $L(q_i)$'s   **Fig. 2.** Subtrees in Q

Next, we associate each q with a tuple $g(q) = <bf(q), interval(q), LeftPos(q), RightPos(q), LevelNum(q)>$, as shown in Fig. 1(b). We say, a q is subsumed by a pair (L, R) if $L \leq LeftPos(q)$ and $R \geq RightPos(q)$. When checking the tree embedding of Q in T, we will associate each generated node v in T with a linked list A_v to record what subtrees in Q can be embedded in $T[v]$. Each entry in A_v is a quadruple $e = (q, interval, L, R)$, where q is a node in Q, $interval = [a, b] \subseteq interval(q)$ (for some $a \leq b$), $L = LeftPos(a)$ and $R = RightPos(b)$. Here, we use a and b to refer to the nodes with the breadth-first numbers a and b, respectively. Therefore, such a quadruple represents a set of subtrees (in $Q[q]$) rooted respectively at a, $a + 1$, ..., b (i.e., a set of subtrees rooted at a set of consecutive breadth-first numbers.) See Fig. 2 for illustration. In addition, the following two conditions are satisfied:

i) For any two entries e_1 and e_2 in A_v, $e_1.q$ is not subsumed by $(e_2.L, e_2.R)$, nor is $e_2.q$ subsumed by $(e_1.L, e_1.R)$. In addition, if $e_1.q = e_2.q$, $e_1.interval \not\subset e_2.interval$ and $e_2.interval \not\subset e_2.interval$.
ii) For any two entries e_1 and e_2 in A_v with $e_1.interval = [a, b]$ and $e_2.interval = [a', b']$, if e_1 appears before e_2, then $RightPost(e_1.q) < RightPost(e_2.q)$ or $RightPost(e_1.q) = RightPost(e_2.q)$ but $a < a'$. Condition (i) is used to avoid redundancy due to the following lemma.

Lemma 1. Let q be a node in Q. Let $[a, b]$ be an interval. If q is subsumed by $(LeftPos(a), RightPos(b))$, then there exists an integer $0 \leq i \leq b - a$ such that $bf(q)$ is equal to $a + i$ or q is an descendant of $a + i$.

Then, by imposing condition (i), A_v keeps only quadruples which represent pair-wise non-covered subtrees. Condition (ii) is met if the nodes in Q are checked along their increasing RightPos values. It is because in such an order the parents of the checked nodes must be non-decreasingly sorted by their RightPos values. Since we explore Q bottom-up, condition (ii) is always satisfied.

See Fig. 3 for a better understanding.

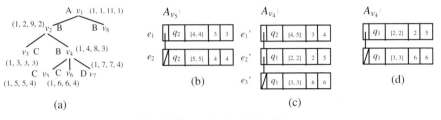

Fig. 3. Illustration for linked lists

In Fig. 3(a), we show a document tree T. Fig. 3(b) shows the linked list created for v_5 in T when it is generated and checked against q_3 and q_4 in Q shown in Fig. 1(a). Since both q_3 and q_4 are leaf nodes, $T[v_5]$ is able to embed either $Q[q_3]$ or $Q[q_4]$ and so we have two entries e_1 and e_2 in A_{v_5}. Note that $bf(q_3) = 4$ and $bf(q_3) = 5$. So we set their *intervals* to [4, 4] and [5, 5], respectively. In addition, each of them is a child of q_2. Thus, we have $e_1.q = e_2.q = q_2$. In Fig. 3(c), we show the linked list for v_4. It contains three entries e_1', e_2' and e_3'. Special attention should be paid to e_1'. Its *interval* is [4, 5], showing that $T[v_4]$ is able to embed both $Q[q_3]$ and $Q[q_4]$. In this case, $e_1'.L$ is set to 3 and $e_1'.R$ to 4. However, since $e_1'.q = q_2$ is subsumed by $(e_2'.L, e_2'.R) = (2, 5)$, the entry will be removed, and the linked list is reduced to a data structure shown in Fig. 3(d). With the linked lists associated with the nodes in T, the embedding of a sub-tree $Q[q]$ in $T[v]$ can be checked very efficiently. First, we define a simple operation over two intervals $[a, b]$ and $[a', b']$, which share the same parent:

$$[a, b] \, \Delta \, [a', b'] = \begin{cases} [a, b'] & \text{if } a \le a' \le b+1, b < b' \\ \text{undefined,} & \text{otherwise.} \end{cases}$$

For example, in A_{v_5}, we have an entry $(q_2, [4, 4], 3, 3)$. In A_{v_6} (which is exactly the same as A_{v_5}), we have an entry $(q_2, [5, 5], 4, 4)$. We can merge these two entries to form another entry $(q_2, [4, 5], 3, 4)$, which can be used to facilitate checking whether $T[v_4]$ embeds $Q[q_2]$.

The general process to merge two linked list is described below.

1. Let A_1 and A_2 be two linked list associated with the first two child nodes of a node v in T, which is being checked against q with $label(v) = label(q)$.
2. Scan both A_1 and A_2 from the beginning to the end. Let e_1 (from A_1) and e_2 (from A_2) be the entries encountered. We will perform the following checkings.
   - If RightPos($e_2.q$) > RightPos($e_1.q$), $e_1 \leftarrow next(e_1)$.
   - If RightPos($e_2.q$) < RightPos($e_1.q$), then $e_2' \leftarrow e_2$; insert e_2' into A_1 just before e_1; $e_2 \leftarrow next(e_2)$.

– If RightPos($e_2.q$) = RightPos($e_1.q$), then we will compare the intervals in e_1 and e_2. Let $e_1.interval = [a, b]$. Let $e_2.interval = [a', b']$.
If $a' > b + 1$, then $e_1 \leftarrow next(e_1)$.
If $a \leq a' \leq b + 1$ and $b < b'$, then replace $e_1.interval$ with $[a, b] \Delta [a', b']$ in A_1; e_1.RightPost $\leftarrow$ RightPos(b'); $e_1 \leftarrow next(e_1)$; $e_2 \leftarrow next(e_2)$.
If $[a', b'] \subseteq [a, b]$, then $e_2 \leftarrow next(e_2)$.
If $a' < a$, then $e_2' \leftarrow e_2$; insert e_2' into A_1 just before e_1; $e_2 \leftarrow next(e_2)$.

3. If A_1 is exhausted, all the remaining entries in A_2 will be appended to the end of A_1. The result of this process is stored in A_1, denoted as $merge(A_1, A_2)$. We also define

$$merge(A_1, ..., A_k) = merge(merge(A_1, ..., A_{k-1}), A_k),$$

where $A_1, ..., A_k$ are the linked lists associated with v's child nodes: $v_1, ..., v_k$, respectively. If in $merge(A_1, ..., A_k)$ there exists an e such that $e.interval = interval(q)$, $T[v]$ embeds $Q[q]$.

For the merging operation described above, we require that the entries in a linked list are sorted. That is, all the entries e are in the order of increasing RightPos($e.q$) values; and for those entries with the same RightPos($e.q$) value their intervals are 'from-left-to-right' ordered. Such an order is obtained by searching Q bottom-up (or say, in the order of increasing RightPos values) when checking a node v in T against the nodes in Q. Thus, no extra effort is needed to get a sorted linked list. Moreover, if the input linked lists are sorted, the output linked lists must also be sorted.

The above merging operation can be used only for the case that Q contains no $/$-edges. In the presence of both $/$-edges and $//$-edges, the linked lists should be slightly modified as follows.

i) Let q_j be a $/$-child of q with $bf(q_j) = a$. Let A_i be a linked list associated with v_i (a child of v) which contains an entry e with $e.interval = [c, d]$ such that $c \leq a$ and $a \leq d$.

ii) If $label(q_j) = label(v_i)$ and v_i is a $/$-child of v, e needn't be changed. Otherwise, e will be replaced with two entries:

    – ($e.q$, $[c, a - 1]$, LeftPos(c), LeftPos($a - 1$)), and
    – ($e.q$, $[a + 1, d]$, LeftPos($a + 1$), LeftPos(d)).

In terms of the above discussion, we give our algorithm for evaluating ordered twig pattern queries. In the process, we will generate *left-sibling* links from the current node v to the node u generated just before v if u is not a child (descendant) of v. However, in the following description, we focus on the checking of tree embedding and that part of technical details is omitted.

Algorithm. *tree-embedding(L(Q))*
Input: all data streams $L(Q)$.
Output: S_v's, with each containing those query nodes q such that $T[v]$ contains $Q[q]$.
begin
1. **repeat until** each $L(q)$ in $L(Q)$ become empty
2. {identify q such that the first element v of $L(q)$ is of the minimal RightPos value; remove v from $L(q)$;
3.   generate node v; $A_v \leftarrow$ f;
4.   let $v_1, ..., v_k$ be the children of v.

5. $B \leftarrow merge(A_{v_1}, ..., A_{v_k})$;
6. **for** each $q \in \boldsymbol{q}$ **do** { (*nodes in $\boldsymbol{q}$ are sorted.*)
7.   **if** q is a leaf **then** $\{S_v \leftarrow S_v \cup \{q\};\}$
8.   **else** (*q is an internal node.*)
9.   {**if** there exists e in B such that $e.interval = interval(q)$
10.   **then** $S_v \leftarrow S_v \cup \{q\};\}$
11. }
12. **for** each $q \in S_v$ **do** {
13.   append (q's parent, [$bf(q)$, $bf(q)$], q.LeftPos, q.RightPos to the end of A_v;}
14. $A_v \leftarrow merge(A_v, B)$; Scan A_v to remove subsumed entries;
15.   remove all A_{v_j}'s;}

16. }
end

In Algorithm *tree-embedding*( ), the nodes in T is created one by one. For each node v generated for an element from a $L(\boldsymbol{q})$, we will first merge all the linked lists of their children and store the output in a temporary variable B (see line 5). Then, for each $q \in \boldsymbol{q}$, we will check whether there exists an entry e such that $e.interval = inter\text{-}val(q)$ (see lines 8 - 9). If it is the case, we will construct an entry for q and append it to the end of the linked list A_v (see lines 12 - 13). The final linked list for v is established by executing line 14. Afterwards, all the A_{v_j}'s (for v's children) will be removed since they will not be used any more (see line 15).

Proposition 2. Algorithm *tree-embedding*( ) computes the entries in A_v's correctly.

4 Conclusion

In this paper, we have discussed a new method to handle the ordered tree matching in XML document databases. The main idea is the concept of intervals, which enables us to efficiently check from-left-to right ordering. The time complexity of the algorithm is bounded by $O(|D| \cdot |Q| + |T| \cdot leaf_Q)$ and its space overhead is by $O(leaf_T \cdot leaf_Q)$, where T stands for a document tree, Q for a twig pattern and D is a largest data stream associated with a node q of Q, which contains the database nodes that match the node predicate at q. $leaf_T$ ($leaf_Q$) represents the number of the leaf nodes of T (resp. Q).

References

[1] Catherine, B., Bird, S.: Towards a general model of Interlinear text. In: Proc. of EMELD Workshop, Lansing, MI (2003)
[2] Müller, K.: Semi-automatic construction of a question tree bank. In: Proc. of the 4th Intl. Conf. on Language Resources and Evaluation, Lisbon, Portual (2004)
[3] Rao, P., Moon, B.: Sequencing XML Data and Query Twigs for Fast Pat-tern Matching. ACM Transaction on Data base Systems 31(1), 299–345 (2006)
[4] Wang, H., Meng, X.: On the Sequencing of Tree Structures for XML Indexing. In: Proc. Conf., Data Engineering, Tokyo, Japan, April 2005, pp. 372–385 (2005)
[5] Zhang, C., Naughton, J., Dewitt, D., Luo, Q., Lohman, G.: on Supporting containment queries in relational database management systems. In: Proc. of ACM SIGMOD (2001)

Query Rewriting Rules for Versioned XML Documents

Tetsutaro Motomura[1], Mizuho Iwaihara[2],
and Masatoshi Yoshikawa[1]

[1] Department of Social Informatics, Kyoto University
[2] Graduate School of Information, Production, and Systems, Waseda University

Abstract. Shared and/or interactive contents such as office documents and wiki contents are often provided with both the latest version and all past versions. It is necessary to add version axes to XPath in order to trace version histories of fine-grained subdocuments of XML. Although research has been done on the containment and equivalence problems for XPath, which is a basic property of optimizing queries, there has been no research in the case for XPath extended with version axes. In this paper, we will propose query rewriting rules which can exchange between document axes and version axes, and prove that they are preserving query semantics. The rewriting rules enable us to swap path subexpressions between document axes and version axes to optimize queries.

1 Introduction

Structured documents having past versions are rapidly growing, especially among the area of wiki contents and office documents. Contents such as wiki are shared among many users and modified by them. Update history (change log) is also provided for examining creation processes and information sources. Old versions are compared with the latest version to check changed parts, and in cases of edit failures and malicious updates, an old version is recovered.

Version retrieval is essential for utilizing versioned contents. Through great contributions from a vast number of authors and by hundreds administrators, who can delete articles, Wikipedia[1] is rapidly growing and popular articles of Wikipedia contain more than one thousand of versions. Authors often browse version histories of their contributions to see how their articles have been augmented and updated by others. General users sometimes check version histories for data recency and provenance. However, version retrieval of Wikipedia offers quite limited functionalities. Since articles are the only unit of versions, users have to execute document-wise difference operation many times until desired results are obtained. Subversion[2] is a widely-used version control system for software resources. Subversion offers version retrieval, but users can specify only

---

[1] http://en.wikipedia.org/wiki
[2] http://subversion.tigris.org/

S.S. Bhowmick, J. Küng, and R. Wagner (Eds.): DEXA 2009, LNCS 5690, pp. 364–371, 2009.
© Springer-Verlag Berlin Heidelberg 2009

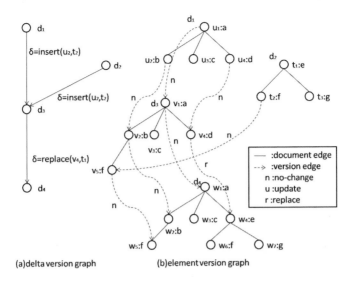

Fig. 1. (a) Delta version graph and (b) Element version graph

files as a unit of versions, so that again users need to take a number of file differences to track changes of specific parts of a file.

XML is a widely-accepted standard for structured documents, and used in many domains such as office documents and digital subscription. If we regard an office document or wiki content as an XML document tree, each paragraph corresponds to a node of the XML tree. XPath [1] is a W3C standard for querying XML documents. An XPath query is represented by a path expression and extracts a set of result nodes matching the expression from the node set of an XML document. Therefore, we can utilize XPath for retrieving and extracting information from XML documents. However, XPath provides no function for performing fine-grained version retrieval of tracing an update history of a document node. XVerPath [2] is a query language for versioned XML documents where XPath is extended with version axes, such as version parent/ancestor and version child/descendant. Fine-grained query functionalities over versioned XML have many applications, such as defining views on node-level history. Views on versioned XML can be defined based on users' interests and security requirements such as access control, where unnecessary or unauthorized subdocuments will be filtered out from the view.

Example 1. In Fig. 1, document d_3 is created from d_1 by inserted node t_2 as a child of u_2. Then document d_4 is created by replacing node v_4 with document d_2. Figure 1(a) is called a delta version graph, whose nodes represent documents and each edge corresponds to an update operation. Figure 1(b) is an element version graph, in which node-level version relationship is calculated from the delta version graph and represented by version edges, where each version edge is labeled with 'n', 'r', 'u' representing update types 'no-change', 'replace' and '(content) update', respectively. The document version d_4 is created by replacing v_4 in document d_3 with the subtree rooted at t_1. Let us now assume that we would like to retrieve the previous version of the nodes labeled as 'f' in d_4. We may first select all nodes labeled with 'f' in d_4, and then obtain the previous version of each node. On the other hand, we can obtain the same result by retrieving the nodes labeled with 'f' from d_3.

In the above example, the reason why there are two ways of retrieving version parents of the node is that there are multiple paths of version and document edges to reach the query targets. Such exchangability between document axis and version axis has not been discussed in the literature. For instance, research has been done on the containment and equivalence problem of XPath, which is a basic property for optimizing queries. A result in Milo and Suciu [3] showed that the containment problem for a fragment of XPath which consists of node tests, the child axis (/), the descendant axis (//), wildcards (*), which is denoted as $XP^{\{//,*\}}$, is in PTIME. The containment problem is co-NP complete for $XP^{\{//,*,[]\}}$, which is a fragment of XPath consisting of node tests, /, //, *, and predicates ([]) [4]. Since the size of versioned XML documents is very large because of its history information, it is necessary to optimize XVerPath queries in order to reduce search space and improve performance. However, we need to take a different approach from traditional XPath query processing to deal with multiple navigation paths of versioned XML documents.

The existence of multiple navigation paths allows us to swap evaluation order of document-axis and version-axis. However, such swapping of axes is not always possible. Thus, we need to investigate query rewriting rules and conditions for preserving query semantics under these rewriting rules.

In this paper, we consider queries in $XVP^{\{//,*,[],vpar\}}$, a fragment of XPath consisting of node tests, /, //, * and [], extended with version parent function *vpar*. In our versioned XML model, document versions are created by update operation delete, insert, replace and (content) update. We show a collection of query rewriting rules and their preconditions. For each update operation, a rule denoting exchangeability of *vpar* with either // or * is presented. Due to the space limitation, rewriting rules with regarded to insert, update, and replace, and all the proofs of this paper are omitted.

Versioned XML has been studied as space-efficient representation using references [5], lifespan [6] and deltas [7], and efficient retrieval of a specified version or a timestamp is discussed. For efficient detection of changes, structural difference of XML documents have been also studied [8] [9] [10], where structural difference is represented as a sequence of updates on an XML tree. While these works assume document-wise version retrieval or difference, our focus is on node-wise version retrieval. Version ancestors or descendants of a node or subdocument are queried in node-wise version retrieval, which is essential in handling large documents or views. We discuss basic tools for algebraic optimizations of queries involving intermixed navigation on document and version axes.

The rest of the paper is organized as follows: In Section 2, we define models of XML documents and version graphs, and query languages. In Section 3, we show a catalogue of query rewriting rules. In Section 4, we discuss formal properties of rewriting rules regarding delete. In Section 5, we show an application of elimination query redundancies using the rewriting rules shown in this paper. Section 6 is a conclusion.

2   Data Model and Query Languages

In this section, we introduce definitions of XML documents and version graphs, XPath and XVerPath queries, and their semantics.

In this paper, we model XML documents as unordered and labeled trees. An XML document is denoted as $d < N, E, r >$, where N is a labeled node set, E is a set of document edges, and $r \in N$ is the document root. Each node $n \in N$ has a content denoted by $content(n)$, which is a text containing element or attribute tag name from an infinite alphabet Σ. The label of node n from Σ is represented by $n.label$.

Let $d < N, E, r >$ be an XML document and let a and b in N, respectively. Then a is a *document ancestor* of b if a is an ancestor of b in the labeled tree of d.

2.1   Version Graphs

First, we give a definition of update operations for XML documents.

Definition 1. *(Update Operations). We define the following basic update operations on XML documents.*

1. *delete(x):delete the subtree with root node x from the document. Here, we represent the subtree with root x as $subtree(x) = (//*)(x)$, and represent the deleted set of nodes is represented as subtree(x).*
2. *insert(x, y):insert a new subtree(y) as a child of node x.*
3. *update(x, c):update the content of node x with $content(x) = c$.*
4. *replace(x, y):replace the subtree(x) by a new subtree(y).*

Next, we model a document-level update history satisfying the two properties: (1)a new document version is created by applying δ to a document, where δ is any update operation defined above, and (2)let D be a set of documents, then a document d_1 in D is obtained from another document d_2 in D by δ. This model is represented by a *delta version graph* $G_d < V_d, E_d >$, where is a directed acyclic graph such that the node set V_d corresponds to the document set D and the edge set E_d consists of the *version edges* $edge(d_1, \delta, d_2)$, such that $edge(d_1, \delta, d_2)$ exists if document d_2 is created from document d_1 by delta δ. We say that d_2 is a *child version* of d_1, and d_1 is a *parent version* of d_2.

Fig. 1 (a) shows an example of a delta version graph. For example, the version edge $edge(d_1, delete(u_3), d_3)$ means that the new document d_3 is generated by deleting the node u_3 from document d_1.

Now we define an *element version graph* for representing document node-level update history. An element version graph $G_e < V_e, E_e >$ is computed from a delta version graph G_d and a set of XML documents D, where V_e is a set of labeled document nodes and E_e is a set of labeled version edges. There are three edge labels n, u, r, meaning no-change, update, and replace, respectively. Let u and v be any nodes in V_e. Then a version edge (u, n, v) with label n means that v is a version child of u, but the content of v is not changed from u. A version edge (u, u, v) with label u means that v is a version child of u and v is updated from u by delta that created the document version of v. A version edge (u, r, v) with label r means that v is a version child of u and v is obtained by replacing the subtree rooted at u with the subtree rooted at v. The non-root nodes in a subtree newly introduced by an insert or replace operation does not have a version parent. There is a straightforward algorithm to create a unique element version graph from a delta version graph and document node set D, as described in [2]. Figure 1(b) illustrates an element version graph. For example, each node in the document d_1 except u_3 has a version child in the document d_3, because u_3 is deleted. Moreover, the content of v_1 is identical to the content of u_1 when d_3 created because u_1 is not change by the delete, and it is indicated by the label 'n' ('n' means no-change) of the version edge between u_1 and v_1.

Note that representing fully materialized versions are space-consuming and keeping only differences between versions are advantageous. Such difference-based representation of element version graphs is shown in [2]. However, the rewriting rules of this paper are independent from these physical representations of element version graphs.

2.2   Query Languages

In this section, we define query languages used in this paper. First, we define a fragment of XPath [1], the W3C standard query language for addressing parts of an XML document.

Definition 2. *(XPath Queries). We consider a subclass of XPath, consisting of expressions given by the following grammar.*

$$p \rightarrow \epsilon \mid l \mid * \mid p/p \mid p//p \mid p[q]$$

Here, ϵ is the empty path, 'l' is a label. '$*$' is a wildcard which selects an arbitrary label. The symbol '$/$' denotes the child axis selecting children of the context node, and '$//$' denotes the descendant-or-self axis that selects descendants of the context node and the context node itself. '$[]$' denotes a predicate which filters a node set and its grammar is given as follows:

$$q \rightarrow p \mid p \, \theta \, c$$

Here, c is a constant, p is an expression defined above, and θ is a comparison operator in $\{<, \leq, =, \geq, >\}$.

Let q be an XPath query and $d < N, E, r >$ be an XML document. We represent $q(r)$ the result of evaluating q on r. A query q is said to *contain* a query q', denoted by $q \sqsubseteq q'$, if $q(R) \sqsubseteq q'(R)$ holds for any set of nodes R. Two queries q and q' are *equivalent*, denoted by $q \equiv q'$, if and only if $q \subseteq q'$ and $q' \subseteq q$ hold.

Next, we define XVerPath [2] as $XP^{\{//,*,[]\}}$ extended with version axes.

Definition 3. *(XVerPath Queries). We define XVerPath$^{\{pc\}}$ as $XP^{\{//,*,[]\}}$ augmented with the following constructs:*

$$p \rightarrow vpar(el) \mid vchild(el)$$
$$el \rightarrow \epsilon \mid ell$$
$$ell \rightarrow n \mid u \mid r \mid ell\,, ell$$

Here, *vpar* stands for the version parent function and *vchild* stands for the version child function. The symbol *el* is a possibly empty list of version edge labels in $\{n, u, r\}$. If *el* is empty, it is regarded as an arbitrary label. Let G_e be an element version graph and v be a context node in G_e, and q be a query in $\{vpar(el), vchild(el)\}$. The semantics of $q(v)$ is defined as follows:

$$(vpar(el))(v) = \{\, u \mid edge(u, el, v) \in G_e \}$$
$$(vchild(el))(v) = \{\, w \mid edge(v, el, w) \in G_e \}$$

In the following, we often denote *el* as a subscript. Namely, $vpar_n$ is the same as $vpar(n)$. Note that the original XVerPath [2] includes the version ancestor-or-self axis (vanc) and the version descendant-or-self axis (vdesc). Query rewriting rules involving these axes are out of the scope of the paper. Let q be an arbitrary XVerPath$^{\{pc\}}$ query, G_d be a delta version graph, R be a subset of nodes in G_d. Then $q(R, G_d)$ is the result nodes of q evaluated over G_d and R. Here, since G_d contains all the versions and can be fixed, we omit G_d from the query expression and simply denote the query as $q(R)$. Containment and equivalence for XVerPath are defined in the same manner as for XPath.

3   Query Rewriting Rules

In this section, we present query rewriting rules which can be applied when an update operation δ is deletion. In Sections 4, we discuss formal properties of these rules.

In the following, let V be a node set belonging to a same version, and U be the node sets of the version parents of V. Also, let t_0 be the target node of an update operation δ which creates the version of V. For instance, if $\delta = delete(t_0)$, V is created by deleting the subtree rooted at t_0 from U.

In the following rewriting rule, rules are presented as a pair of a precondition and equivalent queries. Also the name of each rule is given in the format as 'update operation name – version axis – XPath operation'. For example, 'DEL-VPAR-*' means that the *vpar* function and a wildcard (*) can be interchanged when the update operation is delete.

In the following, q is either '$*$' or '$//$', and '$\backslash$' is the set difference.

Rule 1. *DEL-VPAR-$*$* and **Rule 2.** *DEL-VPAR-$//$*

  Precondition: $subtree(t_0)$ is deleted and V is created.

  Rewriting rule: $(/q/vpar_n)(V) = (/vpar_n/q)(V) \setminus subtree(t_0)$

4   Rewriting Rules Regarding Delete

In this section, we discuss how each query rewriting rule holds.

Let us assume that a deletion $\delta = delete(t_0)$ creates a document version T_v from T_u, and consider the possibility of swapping the evaluation order of an XVerPath query $q(T_v)$ and the version parent function $vpar$. Since the new XML document T_v is created by deleting the subtree rooted at t_0 from T_u, $T_v = T_u \setminus subtree(t_0)$ holds. We consider the situation whether q can be interchanged with the set difference '$\backslash$' so that $q(T_v) = q(T_u) \setminus q(subtree(t_0))$ holds. This property holds if q is either '$*$' or '$//$'.

Theorem 1. *(DEL-VPAR-$*$). Let G_e be an element version graph and V be a node set belonging to a same version, and $delete(t_0)$ be a delete operation that creates V. Then we have the following:*

$$(/*/vpar_n)(V) = (/vpar_n/*)(V) \setminus subtree(t_0)$$

The outline of the proof of Theorem 1 is as follows: first we show that the theorem holds for the case when the node set V is singleton, and extend the result using the following distribution law:

Theorem 2. *(Distribution). Let V_1, V_2 be any node sets and q be an arbitrary query in $XVP^{\{//,*,[],vpar,vchild\}}$. Then we have the following:*

$$q(V_1 \cup V_2) = q(V_1) \cup q(V_2)$$

The above query rewriting also holds for the descendant-or-self query '$//$'.

Theorem 3. *(DEL-VPAR-$//$). Let G_e be an element version graph, V be a node set belonging to a same version, $delete(t_0)$ be a delete operation that creates V, and q be any query in $XP^{\{//,*,[]\}}$. Then we have the following:*

$$(//vpar_n/q)(V) = (/vpar_n//q)(V) \setminus subtree(t_0)$$

Theorem 1 and Theorem 3 enable us to swap '$*$' (or '$//$') and $vpar$. If a deleted subtree and a node set applied by $vpar$ have a non-empty intersection $vpar_n(V) \cap subtree(t_0) \neq \emptyset$, the adjustment '$\setminus subtree(t_0)$' is required to remove the intersection. Otherwise, the equality holds in a simpler form without the adjustment: $(/*/vpar_n)(V) = (/vpar_n/*)(V)$.

5   Application of Rewriting Rules

In this section, we demonstrate elimination of redundant subqueries as an application of the query rewriting rules of this paper.

Example 2. Let us assume that document tree d_4 is created as the latest version by the update operation shown in Fig. 1. Also let us consider two queries such that $q_1 = /a/vchild_{n,r}$ and $q_2 = / * /vpar_{n,r,u}$.

Now we create a new query $q_3 = /a/vchild_{n,r}/ * /vpar_{n,r,u}$ by composing q_1 and q_2. Now consider evaluation of $q_3(v_1)$, where v_1 is the root node of document d_3: $(/a/vchild_{n,r}/*/vpar_{n,r,u})(v_1)$. Query q_3 can be transformed into the following equivalent query by Theorem 1: $(/a/vchild_{n,r}/vpar_{n,r,u}/*)(v_1) \backslash subtree(u_3)$. Since the subquery $vchild_{n,r}/vpar_{n,r,u}$ is an identity mapping (label 'u' is irrelevant here), the subquery can be eliminated, and we can obtain a query q_4: $(/a/*)(v_1) \backslash subtree(u_3)$. As a result, we can rewrite q_3 into q_4.

6   Conclusion

In this paper, we showed two query rewriting rules which can equivalently interchange between version axes and document axes for queries over versioned XML documents. We observe that more types of rules can be discovered by enumerating all possible interchangeable cases. The rewriting rules of the paper give us an interesting insight that querying over versions generated by a sequence of updates can be optimized by strategically choosing evaluation order of version and document axes.

References

1. Clark, J., DeRose, S.: XML Path Language(XPath) version 1.0 (1999), http://www.w3.org/TR/xpath/
2. Iwaihara, M., Hayashi, R., Chatvichienchai, S., Anutariya, C., Wuwongsue, V.: Relevancy-based access control and its evaluation on versioned XML documents. ACM Transactions on Information and System Security 10(1), 1–31 (2007)
3. Milo, T., Suciu, D.: Index structures for XPath expressions. In: ICDT, pp. 277–295 (1999)
4. Miklau, G., Suciu, D.: Containment and equivalence for a fragment of XPath. Journal of the ACM 51, 2–45 (2004)
5. Chien, S.Y., Tsotras, V.J., Zaniolo, C.: Efficient management of multiversion documents by object referencing. In: VLDB, pp. 291–300 (2001)
6. Chien, S.Y., Tsotras, V.J., Zaniolo, C., Zhang, D.: Supporting complex queries on multiversion XML documents. ACM Trans. Internet Techn. 6(1), 53–84 (2006)
7. Marian, A., Abiteboul, S., Cobena, G., Mignet, L.: Change-centric management of versions in an XML warehouse. In: VLDB, pp. 581–590 (2001)
8. Chawathe, S.S.: Comparing hierarchical data in external memory. In: VLDB, pp. 90–101 (1999)
9. Wang, Y., DeWitt, D.J., Cai, J.Y.: X-diff: An effective change detection algorithm for XML documents. In: ICDE, pp. 519–530 (2003)
10. Leonardi, E., Bhowmick, S.S.: Xanadue: a system for detecting changes to XML data in tree-unaware relational databases. In: SIGMOD, pp. 1137–1140 (2007)

Querying XML Data with SPARQL*

Nikos Bikakis, Nektarios Gioldasis, Chrisa Tsinaraki,
and Stavros Christodoulakis

Technical University of Crete, Department of Electronic and Computer Engineering
Laboratory of Distributed Multimedia Information Systems & Applications (TUC/ MUSIC)
University Campus, 73100, Kounoupidiana Chania, Greece
{nbikakis,nektarios,chrisa,stavros}@ced.tuc.gr

Abstract. SPARQL is today the standard access language for Semantic Web data. In the recent years XML databases have also acquired industrial importance due to the widespread applicability of XML in the Web. In this paper we present a framework that bridges the heterogeneity gap and creates an interoperable environment where SPARQL queries are used to access XML databases. Our approach assumes that fairly generic mappings between ontology constructs and XML Schema constructs have been automatically derived or manually specified. The mappings are used to automatically translate SPARQL queries to semantically equivalent XQuery queries which are used to access the XML databases. We present the algorithms and the implementation of SPARQL2XQuery framework, which is used for answering SPARQL queries over XML databases.

Keywords: Semantic Web, XML Data, Information Integration, Interoperability, Query Translation, SPARQL, XQuery, SPARQL2XQuery.

1 Introduction

The Semantic Web has to coexist and interoperate with other software environments and in particular with legacy databases. The *Extensible Markup Language (XML)*, its derivatives (*XPath*, *XSLT*, etc.), and the *XML Schema* have been extensively used to describe the syntax and structure of complex documents. In addition, XML Schema has been extensively used to describe the standards in many business, service, and multimedia application environments. As a result, a large volume of data is stored and managed today directly in the XML format in order to avoid inefficient access and conversion of data, as well as avoiding involving the application users with more than one data models. The database management systems offer today an environment supporting the XML data model and the XQuery access language for managing XML data. In the Web application environment the XML Schema acts also as a wrapper to relational content that may coexist in the databases.

Our working scenario assumes that users and applications of the Semantic Web environment ask for content from underlying XML databases using SPARQL. The

---

* An extended version of this paper is available at [14].

S.S. Bhowmick, J. Küng, and R. Wagner (Eds.): DEXA 2009, LNCS 5690, pp. 372–381, 2009.
© Springer-Verlag Berlin Heidelberg 2009

SPARQL queries are translated into semantically equivalent XQuery queries which are (exclusively) used to access and manipulate the data from the XML databases in order to return the requested results to the user or the application. The results are returned in RDF (N3 or XML/RDF) or XML [1] format. To answer the SPARQL queries on top of the XML databases, a mapping at the schema level is required. We support a set of language level correspondences (rules) for mappings between RDFS/OWL and XML Schema. Based on these mappings our framework is able to translate SPARQL queries into semantically equivalent XQuery expressions as well as to convert XML Data in the RDF format. Our approach provides an important component of any Semantic Web middleware, which enables transparent access to existing XML databases.

The framework has been smoothly integrated with the *XS2OWL* framework [9], thus achieving not only the automatic generation of mappings between XML Schemas and OWL ontologies, but also the transformation of XML documents in RDF format.

Various attempts have been made in the literature to address the issue of accessing XML data from within Semantic Web Environments [2, 4, 5, 6, 7, 8, 9, 10, 11, 12]. An extended overview of related work can be found at [13].

The rest of the paper is organized as follows: The mappings used for the translation as well as their encoding are described in Section 2. Section 3 provides an overview of the query translation process. The paper concludes in section 4.

2   Mapping OWL to XML Schema

The framework described here allows XML encoded data to be accessed from Semantic Web applications that are aware of some ontology encoded in OWL. To do that, appropriate mappings between the OWL ontology (O) and the XML Schema (XS) should exist. These mappings may be produced either automatically, based on our previous work in the *XS2OWL* framework [9], or manually through some mapping process carried out by a domain expert. However, the definition of mappings between OWL ontologies and XML Schemas is not the subject of this paper. Thus, we do not focus on the semantic correctness of the defined mappings. We neither consider what the mapping process is, nor how these mappings have been produced.

Such a mapping process has to be guided from language level correspondences. That is, the valid correspondences between the OWL and XML Schema language constructs have to be defined in advance. The language level correspondences that have been adopted in this paper are well-accepted in a wide range of data integration approaches [2, 4, 9, 10, 11]. In particular, we support mappings that obey the following language level correspondence rules: A class of O corresponds to a Complex Type of XS, a DataType Property of O corresponds to a Simple Element or Attribute of XS, and an Object Property of O corresponds to a Complex Element of XS.

Then, at the schema level, mappings between concrete domain conceptualizations have to be defined (e.g. the *employee* class is mapped to the *worker* complex type) following the correspondences established at the language level.

At the schema level mappings a mapping relationship between O and an XS is a binary association representing a semantic association among them. It is possible that for a single ontology construct more than one mapping relationships are defined. That is, a single source ontology construct can be mapped to more than one target XML

Schema elements (1:n mapping) and vice versa, while more complex mapping relationships can be supported.

The mappings considered in our work are based on the *Consistent Mappings Hypothesis*, which states that for each mapped property *Pr* of *O*:

   a. The domain classes of *Pr* have been mapped to complex types in *XS* that contain the elements or attributes that *Pr* has been mapped to.

   b. If *Pr* is an object property, the range classes of *Pr* have been mapped to complex types in *XS*, which are used as types for the elements that *Pr* has been mapped to.

2.1 Encoding of the Schema Level Mappings

Since we want to translate SPARQL queries into semantically equivalent XQuery expressions that can be evaluated over XML data following a given (mapped) schema, we are interested in addressing XML data representations. Thus, based on schema level mappings for each mapped ontology class or property, we store a set of XPath expressions (*"XPath set"* for the rest of this paper) that address all the corresponding instances (XML nodes) in the XML data level. In particular, based on the schema level mappings, we construct:

- A **Class XPath Set** X_C for each mapped class *C*, containing all the possible XPaths of the complex types to which the class *C* has been mapped to.
- A **Property XPath Set** X_{Pr} for each mapped property *Pr*, containing all the possible XPaths of the elements or/and attributes to which *Pr* has been mapped.

For ontology properties, we are also interested in identifying the property domains and ranges. Thus, for each property we define the X_{PrD} and X_{PrR} sets, where:

- The **Property Domains XPath Set** X_{PrD} for a property *Pr* represents the set of the XPaths of the property domain classes.
- The **Property Ranges XPath Set** X_{PrR} for a property *Pr* represents the set of the XPaths of the property ranges.

Example 1. Encoding of Mappings

   Fig. 1 shows the mappings between an OWL Ontology and an XML Schema.

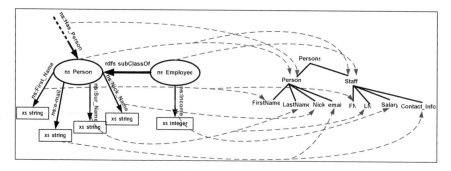

Fig. 1. Mappings Between OWL & XML

To better explain the defined mappings, we show in Fig. 1 the structure of the XML documents that follow this schema. The encoding of these mappings in our framework is shown in Fig.2.

Classes:	DataType Properties:
$X_{ns:Person}$={/Persons/Person, /Persons/Staff}	$X_{ns:First\_Name}$={/Persons/Person/FirstName, /Persons/Staff/FN}
$X_{ns:Employee}$={/Persons/Staff}	$X_{ns:Sur\_Name}$={/Persons/Person/LastName, /Persons/Staff/LN}
	$X_{ns:Nick\_Name}$={/Persons/Person/Nick }
Object Properties:	$X_{ns:e\text{-}mail}$ ={/Persons/Person/email, /Persons/Staff/Contact_Info}
$X_{ns:Has\_Person}$={/Persons/Person }	$X_{ns:Income}$={/Persons/Staff/Salary}

Fig. 2. Mappings Encoding

XPath Set Operators. For XPath Sets, the following operators are defined in order to formally explain the query translation methodology in the next sections:

- The unary ***Parent Operator*** ^{P}, which, when applied to a set of XPaths X (i.e. $(X)^P$), returns the set of the distinct parent XPaths (i.e. the same XPaths without the leaf node). When applied to the root node, the operator returns the same node.

Example 2. Let X={ /a , /a/b , /c/d , /e/f/g , /b/@f } then $(X)^P$={ /a , /a , /c , /e/f , /b }.

- The binary ***Right Child Operator*** ®, which, when applied to two XPath sets X and Y (i.e. X®Y), returns the members (XPaths) of the right set X, the parent XPaths of which are contained in the left set Y.

Example 3. Let X={ /a , /c/b } and Y={ /a/d , /a/c , /c/b/p , c/a/g } then
  X ®Y = { /a/d , /a/c , /c/b/p }.

- The binary ***Append Operator*** /, which is applied on an XPath set X and a set of node names N (i.e. X / N), resulting in a new set of XPaths Y by appending each member of N to each member of X.

Example 4. Let X={/a, /a/b} and N={c, d} then Y = X / N = {/a/c, /a/d, /a/b/c, a/b/d }.

XPath Set Relations. We describe here a relation among XPath sets that holds because of the *Consistent Mapping Hypothesis* described above. We will use this relation later on in the query translation process, and in particular in the variable bindings algorithm (subsection 3.1):

Domain-Range Property Relation: $\forall$ Property $Pr \Rightarrow X_{Pr\,R} = X_{Pr}$ and $X_{PrD} = \left(X_{Pr}\right)^P = \left(X_{Pr\,R}\right)^P$

The *Domain-Range Property Relation* can be easily understood taking into account the hierarchical structure of XML data as well as the *Consistent Mappings Hypothesis*. It describes that for a single property *Pr*:

- the XPath set of its ranges is equal to its own XPath set (i.e. the instances of its ranges are the XML nodes of the elements that this property has been mapped to).
- the XPath set of its domain classes is equal to the set containing its parent XPaths (i.e. the XPaths of the *CTs(Complex Types)* that contain the elements that this property has been mapped to).

3 Overview of the Query Translation Process

In this section we present in brief the entire translation process using a UML activity diagram. Fig. 3 shows the entire process which starts taking as input the given SPARQL query and the defined mappings between the ontology and the XML Schema (encoded as described in the previous sections). The query translation process comprises of the activities outlined in the following paragraphs.

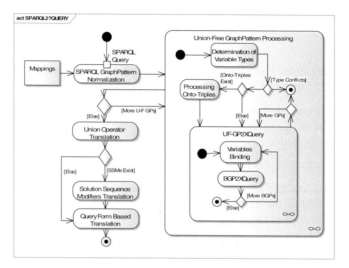

Fig. 3. Overview of the SPARQL Translation Process

SPARQL Graph Pattern Normalization. The *SPARQL Graph Pattern Normalization* activity re-writes the Graph-Pattern (*GP*) of the SPARQL query in an equivalent normal form based on equivalence rules. The SPARQL *GP* normalization is based on the *GP* expression equivalences proved in [3] and re-writing techniques. In particular, each *GP* can be transformed in a sequence *P1 UNION P2 UNION P3 UNION...UNION Pn*, where *Pi* (*1≤i≤n*) is a Union-Free *GP* (i.e. *GPs* that do not contain Union operators). This makes the *GP* translation process simpler and more efficient.

Union-Free Graph Pattern (UF-GP) Processing. The *UF-GP* processing translates the constituent *UF-GPs* into semantically equivalent XQuery expressions. The *UF-GP* Processing activity is a composite one, with various sub-activities. This is actually the step that most of the "real work" is done since at this step most of the translation process takes place. The *UF-GP processing* activity is decomposed in the following sub-activities:

– **Determination of Variable Types.** For every *UF-GP*, this activity initially identifies the types of the variables used in order to detect any conflict arising from the user's syntax of the input as well as to identify the form of the results for each variable. We define the following variable types: The *Class Instance Variable Type* (*CIVT*), The *Literal Variable Type* (*LVT*), The *Unknown Variable Type* (*UVT*), The

Data Type Predicate Variable Type (*DTPVT*), The *Object Predicate Variable Type* (*OPVT*), The *Unknown Predicate Variable Type* (*UPVT*).

We also define the following sets: The Data *Type Properties Set* (*DTPS*), which contains all the data type properties of the ontology. The *Object Properties Set* (*OPS*), which contains all the object properties of the ontology. The *Variables Set* (*V*), which contains all the variables that are used in the *UF-GP*. The *Literals Set* (*L*), which contains all the literals referenced in the *UF-GP*.

The determination of the variable types is based on a set of rules applied iteratively for each triple in the given *UF-GP*. Below we present a subset of these rules, which are used to determine the type (T_X) of a variable X:

Let S P O be a triple pattern.

1. If $P \in OPS$ and $O \in V \Rightarrow T_O = CIVT$. If predicate is an object property and object is a variable, then the type of the object variable is *CIVT*.

2. If $O \in L$ and $P \in V \Rightarrow T_P = DTPVT$. If the object is a literal value, then the type of the predicate variable is *DTPVT*.

– Processing Onto-Triples. *Onto-Triples* actually refer to the ontology structure and/or semantics. The main objective of this activity is to process *Onto-Triples* against the ontology (using SPARQL) and based on this analysis to bind (i.e. assigning the relevant XPaths to variables) the correct XPaths to variables contained in the *Onto-Triples*. These bindings are going to be used in the next steps as input to the *Variable Bindings* activity.

– UF-GP2XQuery. This activity translates the *UF-GP* into semantically equivalent XQuery expressions. The concept of a *GP*, and thus the concept of *UF-GF*, is defined recursively. The *BGP2XQuery* algorithm translates the basic components of a *GP* (i.e. Basic Graph Patterns - *BGPs* which are sequences of triple patterns and filters) into semantically equivalent XQuery expressions (see subsection 3.2). To do that a variables binding (see subsection 3.1) step is needed. Finally, *BGPs* in the context of a *GP* have to be properly associated. That is, to apply the SPARQL operators among them using XQuery expressions and functions. These operators are: *OPT*, *AND*, and *FILTER* and are implemented using standard XQuery expressions without any ad hoc processing.

Union Operator Translation. This activity translates the *UNION* operator that appears among *UF-GPs* in a *GP*, by using the *Let* and *Return* XQuery clauses in order to return the union of the solution sequence produced by the *UF-GPs* to which the Union operator applies.

Solution Sequence Modifiers Translation. This activity translates the SPARQL solution sequence modifiers using XQuery clauses (*Order By, For, Let,* etc.) and XQuery built-in functions (you can see the example in subsection 3.3.). The modifiers supported by SPARQL are *Distinct, Order By, Reduced, Limit,* and *Offset*.

Query Forms Based Translation. SPARQL has four forms of queries (*Select, Ask, Construct* and *Describe*). According to the query form, the structure of the final result is different. The query translation is heavily dependent on the query form. In particular, after the translation of any solution modifier is done, the generated XQuery is enhanced with appropriate expressions in order to achieve the desired structure of the results (e.g. to construct an RDF graph, or a result set) according to query form.

3.1   Variable Bindings

This section describes the variable bindings activity. In the translation process the term *"variable bindings"* is used to describe the assignment of the correct XPaths to the variables referenced in a given *Basic Graph Pattern (BGP)*, thus enabling the translation of *BGP* to XQuery expressions. In this activity, *Onto-Triples* are not taken into account since their processing has taken place in the previous step.

Definition 1 :   A triple pattern has the form (s,p,o) ϵ($I \cup B \cup V$) x ($I \cup V \cup B$) x ($I \cup B \cup L \cup V$), where I is a set of IRIs, B is a set of Blank Nodes, V is a set of Variables, and L the set of RDF Literals. In our approach, however, the individuals in the source ontology are not considered at all (either they do not exist, or they are not used in semantic queries).

Definition 2 :   A variable contained in a Union Free Graph Pattern is called a *Shared Variable* when it is referenced in more than one triple patterns of the same Union-Free Graph Pattern regardless its position in those triple patterns.

Variable Bindings Algorithm. When describing data with the RDF triples (s,p,o), subjects represent class individuals (RDF nodes), predicates represent properties (RDF arcs), and objects represent class individuals or data type values (RDF nodes). Based on that, and the *domain-range property* relation of Xpaths sets relations section we have: **a)** $X_s = X_{pD} = (X_{pR})^P = (X_p)^P$ **b)** $X_p = X_{pR}$ and **c)** $X_o = X_{pR}$.

Thus it holds that: $X_s = X_{pD} = (X_{pR})^P = (X_p)^P = (X_o)^P \Rightarrow X_s = (X_p)^P = (X_o)^P$ **(Subject-Predicate-Object Relation)**

This relation holds for every single triple pattern. Thus, the variable bindings algorithm uses this relation in order to find the correct bindings for the entire set of triple patterns starting from the bindings of any single triple pattern part (subject, predicate, or object).

In case of shared variables, the algorithm tries to find the maximum set of bindings (using the operators for XPath sets) that satisfy this relation for the entire set of triple patterns (e.g. the entire *BGP*). Once this relation holds for the entire *BGP* we have as a result that all the instances (in XML) that satisfy the *BGP* have been addressed.

The variable bindings algorithm in case of shared variables of *LVT* type it doesn't determine the XPaths for this kind of variable, since literal equality is independent of the XPaths expressions. Thus, the bindings for variables of this type cannot be defined at this step (mark as *"Not Definable"* at variable bindings rules). Instead, they will be handled by the *BGP2XQuery* (subsection 3.2) algorithm (using the mappings and the determined variables bindings).

The algorithm takes as input a *BGP* as well as a set of initial bindings and the types of variables as these are determined in the *"Determination of Variable Type"* activity. These initial bindings are the ones produced by the *Onto-Triple* processing activity and initialize the bindings of the algorithm. Then, the algorithm performs an iterative process where it determines, at each step, the bindings of the entire *BGP* (triple by triple). The determination of the bindings is based on the rules described below. This iterative process continues until the bindings for all the variables found in the successive iterations are equal. This means that no further modifications in the variable bindings are to be made and that the current bindings are the final ones.

Variable Bindings Rules. Based on the possible combinations of *S*, *P* and *O*, there are four different types of triple patterns (the ontology instance are not yet supported by our framework): **Type 1 :** $S \in V$, $P \in I$, $O \in L$. **Type 2 :** S, $O \in V$, $P \in I$. **Type 3 :** S, $P \in V$, $O \in L$. **Type 4 :** S, P, $O \in V$.

According to the triple pattern type, we have defined a set of rules for the variable bindings. In this section we present a sub-set of these rules due to space limitations.

In what follows the symbol $'$ in XPath sets denotes the new bindings assigned to the set at each iteration, while the symbol $\leftarrow$ denotes the assignment of a new value to the set. All the XPath sets are considered to be initially set to *null*. In that case, the intersection operation is not affected by the *null* set. E.g. X={ null } and Y= {/a/b , d/e} then X $\cap$ Y ={ /a/b , d/e }. The notation *"Not Definable"* is used for variables of type *LVT* as explained above. Consider the triple $S\ P\ O$:

- If the triple is of Type 1 $\Rightarrow X_S' \leftarrow X_{PD} \cap X_S$
- If the triple is of Type 2 $\Rightarrow X_S' \leftarrow X_{PD} \cap X_S \cap (X_O)^P$
  - If $P \in OPS \Rightarrow X_O' \leftarrow X_S' \circledR X_O$
  - If $P \in DTPS \Rightarrow X_O'$ *Non Definable* (as explained in previously)
- If the triple is of Type 3 $\Rightarrow X_S' \leftarrow X_{PD} \cap X_S$ and $X_P' \leftarrow X_S' \circledR X_P$
- If the triple is of Type 4 $\Rightarrow X_S' \leftarrow X_{PD} \cap X_S \cap (X_O)^P$ and $X_P' \leftarrow X_S' \circledR X_P$
  - If $T_O = CIVT$ or $T_O = UVT \Rightarrow X_O' \leftarrow X_P' \cap X_O$
  - If $T_O = LVT \Rightarrow X_O'$ *Non Definable* (as explained previously)

XPath Set Relations for Triple-Patterns. Among XPath sets of triple patterns there are important relations that can be exploited in the development of the XQuery expressions in order to correctly associate data that have been bound to different variables of triple patterns. The most important relation among XPath sets of triple patterns is that of extension:

Extension Relation: An XPath set A is said to be an extension of an XPath set B if all XPaths in A are descendants of the XPaths of B.

As an example of this relation, consider the XPath A' produced when applying the append (/) operator to an original XPath set A with a set of nodes.

The extension relation holds for the results of the variable bindings algorithm (*Subject-Predicate-Object Relation*) and implies that the XPaths bound to subjects are parents of the XPaths bound to predicates and objects of triple patterns.

3.2 Translating BGPs to XQuery

In this section we describe the translation of *BGPs* to semantically equivalent XQuery expressions. The algorithm manipulates a sequence of triple patterns and filters (i.e. a *BGP*) and translates them into semantically equivalent XQuery expressions, thus allowing the evaluation of a *BGP* on a set of XML data.

Definition 3 :   *Return Variables (RV)* are those variables for which the given SPARQL Query would return some information. The set of all *Return Variables* of a SPARQL query constitutes the set $RV \subseteq V$.

The BGP2XQuery Algorithm. We briefly present here the *BGP2XQuery* algorithm for translating *BGPs* into semantically equivalent XQuery expressions. The algorithm takes as input the mappings between the ontology and the XML schema, the *BGP*, the determined variable types, as well as the variable bindings. The algorithm is not executed triple-by-triple for a complete *BGP*. Instead, it processes subjects, predicates, and objects of all the triples separately. For each variable included in the *BGP*, the *BGP2XQuery* it creates a *For* or *Let* XQuery clause using the variable bindings, the input mappings, and the *Extension Relation* for triple-patterns (see subsection 3.1), in order to bound XML data into XQuery variables. The choice between the *For* and the *Let* XQuery clauses is based on specific rules so as to create a solution sequence based on the SPARQL semantics. Moreover, in order to associate bindings from different variables into concrete solutions, the algorithm uses the *Extension Relation*. For literals included in the *BGP*, the algorithm is using XPath predicates in order to translate them. Due to the complexity that a SPARQL filter may have, the algorithm translates all the filters into XQuery where clauses, although some "simple" of them (e.g. condition on literals) could be translated using XPath predicates. Moreover, SPARQL operators (Built-in functions) included in filter expressions are translated using built-in XQuery functions and operators. However, for some "special" SPARQL operators (like *same-Term, lang,* etc.) we have developed native XQuery functions that simulate them.

Finally, the algorithm creates an XQuery *Return* clause that includes the Return Variables (*RV*) that was used in the *BGP*.

There are some cases of share variables which need special treatment by the algorithm in order to apply the required joins in XQuery expressions. The way that the algorithm manipulates these cases depends on which parts (*subject-predicate-object*) of the triples patterns these shared variables refer to.

3.3 Example

We demonstrate in this example the use of the described framework in order to allow a SPARQL query to be evaluated in XML Data (based on Example 1). Fig. 4 shows how a given SPARQL query is translated by our framework into a semantically equivalent XQuery.

Consider the query: *"Return the Persons, their last name(s) and their nick name(s), whose first name is "John", whose last name begins with "A", and they have an e-mail address. The (existence of) nick name is optional. The query will return at most 30 solutions (LIMIT 30) ordered by last name value at descending order and skipping the first 5 solutions (OFFSET 5)".*

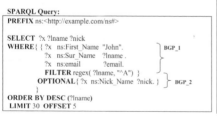

Fig. 4. SPARQL Query Translation Example

4   Conclusions

We have presented a framework and its software implementation that allows the evaluation of SPARQL queries over XML data which are stored in XML databases and accessed with the XQuery language. The framework assumes that a set of mappings between the OWL ontology and the XML Schema exists which obey to certain well accepted language correspondences.

The *SPARQL2XQuery* framework has been implemented as a software service which can be configured with appropriate mappings (between some ontology and XML Schema) and translates input SPARQL queries into semantically equivalent XQuery queries that are answered over the XML Database.

References

1. Beckett, D. (ed.): SPARQL Query Results XML Format. W3C Recommendation, January, 15, http://www.w3.org/TR/rdf-sparql-XMLres/
2. Bohring, H., Auer, S.: Mapping XML to OWL Ontologies. In: Leipziger Informatik-Tage 2005, pp. 147–156 (2005)
3. Pérez, J., Arenas, M., Gutierrez, C.: Semantics and complexity of SPARQL. In: Cruz, I., Decker, S., Allemang, D., Preist, C., Schwabe, D., Mika, P., Uschold, M., Aroyo, L.M. (eds.) ISWC 2006. LNCS, vol. 4273, pp. 30–43. Springer, Heidelberg (2006)
4. Rodrigues, T., Rosa, P., Cardoso, J.: Mapping XML to Exiting OWL ontologies. In: International Conference WWW/Internet 2006, Murcia, Spain, October 5-8 (2006)
5. Farrell, J., Lausenq, H.: Semantic Annotations for WSDL and XML Schema. W3C Recommendation, W3C (August 2007), http://www.w3.org/TR/sawsdl/
6. Groppe, S., Groppe, J., Linnemann, V., Kukulenz, D., Hoeller, N.: C.-t. Reinke: Embedding SPARQL into XQuery/XSLT. In: SAC 2008, pp. 2271–2278 (2008)
7. Akhtar, W., Kopecký, J., Krennwallner, T., Polleres, A.: XSPARQL: Traveling between the XML and RDF worlds – and avoiding the XSLT pilgrimage. In: Bechhofer, S., Hauswirth, M., Hoffmann, J., Koubarakis, M. (eds.) ESWC 2008. LNCS, vol. 5021, pp. 432–447. Springer, Heidelberg (2008)
8. Droop, M., Flarer, M., et al.: Embedding XPATH Queries into SPARQL Que-ries. In: Proc. of the 10th International Conference on Enterprise Information Systems
9. Tsinaraki, C., Christodoulakis, S.: Interoperability of XML Schema Applications with OWL Domain Knowledge and Semantic Web Tools. In: Proc. of the ODBASE (2007)
10. Cruz, I.R., Xiao, H., Hsu, F.: An Ontology-based Framework for XML Seman-tic Integration. In: Database Engineering and Applications Symposium (2004)
11. Christophides, V., Karvounarakis, G., et al.: The ICS-FORTH SWIM: A Powerful Semantic Web Integration Middleware. In: Proc. of the SWDB 2003, pp. 381–393 (2003)
12. Amann, B., Beeri, C., Fundulaki, I., Scholl, M.: Querying XML Sources Using an Ontology-Based Mediator. In: CoopIS/DOA/ODBASE 2002, pp. 429–448 (2002)
13. Bikakis, N., Gioldasis, N., Tsinaraki, C., Christodoulakis, S.: Semantic Based Access over XML Data. In: Proc. of 2nd World Summit on Knowledge Society 2009, WSKS 2009 (2009)
14. Bikakis, N., Gioldasis, N., Tsinaraki, C., Christodoulakis, S.: The SPARQL2XQuery Framework, http://www.music.tuc.gr/reports/SPARQL2XQUERY.PDF

Progressive Evaluation of XML Queries for Online Aggregation and Progress Indicator

Cheng Luo[1], Zhewei Jiang[2], Wen-Chi Hou[2], and Gultekin Ozsoyoglu[3]

[1] Department of Mathematics and Computer Science in Coppin State University,
2500 West North Avenue, Baltimore, MD, 21216, U.S.A.
cluo@coppin.edu

[2] Computer Science Department in Southern Illinois University Carbondale,
Carbondale, IL 62901, U.S.A.
{zjiang,hou}@cs.siu.edu

[3] Electrical Engineering and Computer Science Department in Case Western
Reserve University, Cleveland, OH 44106, U.S.A.
tekin@eecs.cwru.edu

Abstract. With the rapid proliferation of XML data, large-scale online applications are emerging. In this research, we aim to enhance the XML query processors with the ability to process queries progressively and report partial results and query progress continually. The methodology lays its foundation on sampling. We shed light on how effective samples can be drawn from semi-structured XML data, as opposed to flat-table relational data. Several innovative sampling schemes on XML data are designed. The proposed methodology advances XML query processing to the next level - being more flexible, responsive, user-informed, and user-controllable, to meet emerging needs and future challenges.

1   Introduction

The demand for an efficient and effective exploration method for large-scale XML repositories is more urgent than ever. There has been much research [1,17] devoted to designing efficient query processing techniques for XML data. However, due to the large size of the data and inherently complex nature of XML queries, many queries can still take hours or even days to complete [21]. Moreover, query results are generally reported after all the data (relevant to the queries) have been processed. No information about the query, such as partial results, approximate answers, or progress of query evaluation, is reported during the process, which may not be desirable for applications like OLAP, decision-support querying, and long-running transaction processing.

The deployments of OLAP, decision-support, and data mining systems in the future crucially hinge on their ability to provide timely feedbacks to users' queries. In this research, we aim to equip XML query processors with the power to timely display partial results, provide estimates, and report progresses of evaluation to meet the needs of emerging XML applications.

S.S. Bhowmick, J. Küng, and R. Wagner (Eds.): DEXA 2009, LNCS 5690, pp. 382–395, 2009.

Sampling presents an efficient and effective alternative to study data of large sizes. By processing only a small amount of sample data, approximate query answers can be derived quickly, responsively, and flexibly. It is well suited for applications like selectivity estimation, online analytical processing, decision-support querying, and long-running transaction processing. Indeed, there has been much research on sampling in relational databases during the past two decades [8,13,4,5]. While sampling has enjoyed so much success in relational databases, there has been virtually no or little work on sampling in XML databases. In this research, we show how informative samples can be drawn from XML documents.

Recently, research efforts have been put on continually providing estimates and reporting progress of evaluation in relational databases. For example, Hellerstein et al. [6] proposed processing aggregation queries incrementally and report estimates continually for online aggregation. Other work in progressive query evaluation includes [10,19]. There has also been some work on enhancing a system with a progress indicator by continually displaying the remaining query processing time [2,17,16]. Nevertheless, all this work is done in the context of relational databases rather than XML databases.

In this paper, we initiate the research of sampling on XML data and progressive evaluation of XML queries. We aim to enhance XML query processors with the ability to process both aggregation and non-aggregation queries progressively (or incrementally) and report results and progress of evaluation continually. The methodology lays its foundation on sampling. Several innovative sampling schemes on XML data are proposed. By continuously sampling from the database, queries are evaluated incrementally. For regular (non-aggregation) queries, the system continuously returns newly found query matches; for aggregation queries, it reports running estimates with confidence intervals. The progress of query processing is also reported continually. We have implemented and experimented with this methodology. The experimental results show that with only small amounts of sample data, we can accurately estimate aggregation queries and query processing time. The results also confirm that our sampling schemes are effective and well suited for progressive evaluation of XML queries.

The rest of the paper is organized as follows. Section 2 is the preliminaries. Section 3 introduces the framework for progressive evaluation of XML queries using sampling. Section 4 discusses the derivation of running estimates and their associated confidence intervals. Section 5 reports the experimental results and Section 6 concludes the paper.

2   Preliminaries

In this section, we introduce the notation and terminology that are used in the later parts of this paper.

XML Data Model. An XML document is generally modelled as a node-labelled, ordered tree $\mathcal{T} = (V_T, E_T)$, where V_T is the set of vertices in the tree and E_T the set of edges. Each node $v \in V_T$ corresponds to an element or an attribute in the document. Each directed edge $e = (u, v) \in E_T$ represents an element-subelement or element-attribute relationship between nodes u and v. If there is a directed edge from u to v, u is said to be a parent of v; if v is reachable from u via directed edges, u is said to be an ancestor of v.

Region coding has been adopted in much of the XML research [1,11,14]. In this scheme, each node in the tree is assigned a unique 3-ary tuple: (leftPos, rightPos, LevelNo), which represents the left, right positions, and level number of the node, respectively. The codes can be used easily to determine the structural relationships between nodes in the tree.

Twig Queries. Due to the tree-structured nature of XML data, XML queries are often expressed as path expressions, consisting of simple linear path expressions that describe linear paths without branches and complex path expressions that describe branched paths. The latter, commonly known as twig queries [1], represent a more general and complicated form of the XML queries.

Bruno, et al. [1] have designed an efficient twig query evaluation algorithm, called TwigStack. It has become the foundations of many other algorithms [14,15]. In this research, we will use it, with some minor modifications, as our underlying query evaluation algorithm.

XML Aggregation Queries. The XML query language provides syntax for computing summary information for a group of related document elements. Typical aggregation functions include COUNT, SUM, and AVERAGE.

Progress Indicators. Progress indicators report the progress of query evaluation on the fly [2,17]. The progress is usually measured by the fraction of work completed or, if possible, the remaining time. A progress bar is generally displayed to show the query execution progress. Users have the option to terminate the query evaluation when they feel that the estimation is sufficiently accurate.

3   Progressive Evaluation of Queries

3.1   Assumptions

TwigStack [1] is the underlying query processing algorithm for this research. In TwigStack, data nodes having the same label name are stored in ascending order by their left positions in a linear list, called a stream. Each stream is assumed to be indexed properly for direct access.

3.2   Progressive Evaluation Framework

The algorithm in Figure 1 outlines the progressive evaluation of an aggregation query. The process stops when a terminating condition is met, which can be a desired precision, a time constraint, or eof (i.e., all sampling units have been processed). For simplicity, the algorithm in Figure 1 assumes the terminating conditions are a desired precision represented by a specified confidence interval ϵ or eof. N is the population size. The function $SRWOR(1, N)$ returns a number $i, 1 \leq i \leq N$ or NULL (if all number are taken) using simple random sampling without replacement.

ProgressiveEvaluation()

```
begin
    confInterval=+∞;
    while ((confInterval > ε) and (i = SRWOR(1, N)! = NULL)) do
        sampleunit_range=PopulationTable[i];
        queryResults+=EvalTwig(sampleunit_range);
        (estimate,confInterval)=calEstimate(queryResults, ++sampleSize);
        Display(estimate,confInterval);
    end
end
```

Fig. 1. Progressive Evaluation of Queries

A sampling unit in this research is generally a sub-range of the document's region code domain, encompassing a sub-tree or a set of sub-trees of the data tree, depending upon the different sampling schemes (to be discussed later). Sampling units, represented by their region code ranges, are stored in a table, called the PopulationTable, for easy accesses. The function EvalTwig() uses TwigStack to find matches in the given sampling unit. Note that a slight modification to the TwigStack is needed so that each time it evaluates only a sub-range of the document's coding domain (i.e., only a sampling unit). This can be easily accomplished by using the range of a sample unit as the starting/terminating condition in the TwigStack. Besides displaying the newly found matches on the fly for non-aggregation queries, EvalTwig() also returns interesting aggregate values for the newly found matches. The computations of the running estimates and associated confidence intervals of the aggregation queries, indicated by the function calEstimate() in the algorithm, will be discussed in Section 4.

Desirable Properties of Progressive Evaluation. Sampling units are drawn continuously from the data tree and evaluated against the query in a progressive manner. Two basic properties are desirable of such a progressive evaluation.

Property 1. The progressive evaluation of a query must generate the complete set of query matches once the entire set of sampling units has been processed in some order.

Property 2. No duplicate query match is generated in the process of a progressive evaluation of sampling units.

3.3 Sampling from an XML Tree

There is a long history of using sampling for approximate aggregation query processing in relational databases [4,8,9,13]. Unfortunately, these techniques, applied generally to relational model's flat-table data, are not directly applicable to XML's semi-structured data. Naive sampling schemes, such as randomly sampling nodes from an XML tree or from individual node sets (i.e., groups of nodes of the same type) [20] can generally yield large variance in estimation, as

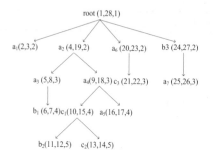

Fig. 2. An XML Data Tree

demonstrated by the multi-join estimation in relational databases. Consequently, one may have to resort to external index files for more effective matching [20], which unfortunately can be quite expensive. To the best of our knowledge, there has been no research that utilize the semi-structured nature of XML data to design effective XML sampling schemes.

We observe that to form informative and effective samples, relevant nodes must be drawn together as a unit, instead of independently like the naive sampling methods. In what follows, we discuss how to draw samples that preserve correlations of data nodes. Preserving correlations makes a sample informative and effective.

Before formally introducing the sampling schemes, let us first consider an example of what constitutes a sampling unit. For ease of exposition, we use the XML data tree and twig query in Figures 2 and 3, respectively, as an example. For a better demonstration of the structural relationships, Figure 4 shows the coverage of A nodes in the data tree.

For a given twig query, an XML data tree can be viewed as a set of sub-trees rooted at individual query root nodes. For example, the query root node of the query in Figure 3 is A and the XML data tree can be viewed as a set of sub-trees rooted at a_1, a_2, a_6, and a_7 as far as the query is concerned. The set of sub-trees constitutes the population from which individual sub-trees are drawn as sampling units. By continuously drawing sub-trees rooted at query root nodes, query evaluation can be accomplished progressively. We assume that simple random sampling without replacement [3] is the underlying technique used for drawing sub-trees from a sub-tree set.

Note that we do not materialize in any form the sub-trees. Drawing a sub-tree only means to retrieve nodes of the sub-tree from the respective streams. Since each sub-tree can be uniquely determined by its root node, we shall for simplicity represent a sub-tree by its root node. For instance, the aforementioned sub-tree set is represented by $\{a_1, a_2, a_6, a_7\}$.

Queries having the same root node type can use the same sub-tree set as their populations. Consequently, we only need to construct a sub-tree set for each node type (that could potentially become a query root node of some queries), and the total number of sub-tree sets to accommodate all possible queries is no greater than the number of node types in the data tree.

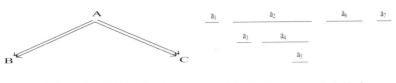

Fig. 3. A Sample XML Twig Query   **Fig. 4.** Coverage of A Nodes

3.4 Sampling Schemes

The set of sub-trees can be viewed as a partitioning of the data tree. The quantities of interesting statistics generally vary from sub-tree to sub-tree in a partitioning. If the distributions of the interested statistics are highly skewed among the sub-trees, estimates can have large variance. In this research, we design three partitioning schemes to form sub-tree sets that can be evaluated with ease and generate estimates with high accuracy.

We differentiate two types of query root nodes. As shown in Figure 2, there are A nodes like a_1, a_2, a_6, and a_7 that are not descendants of any other A nodes, and there are also A nodes like a_3, a_4, and a_5 that are descendants of some other A nodes. We shall call the former top query root (or top A) nodes and the latter non-top query root (or non-top A) nodes. Our goal is to construct sets of query root nodes (and thus sub-trees) that are suitable for progressive evaluation.

Top Query Root Node Sampling. In this scheme, only the top query root nodes are selected into the sub-tree set for sampling. That is, only those sub-trees rooted at top query root nodes are eligible for sampling. Consider Figure 2 again. The previously discussed sub-tree set $\{a_1, a_2, a_6, a_7\}$ is indeed the result of this partitioning scheme for all queries rooted at A. The rationale behind this scheme is that the top query root nodes cover disjoint ranges of the coding domain. Thus, no non-top query root nodes under a specific top query root node can form a query match with another top query root node. That is, no B and C nodes under a specific top A node can form a query match with another top A node. In addition, every non-top A node, e.g., a_3, a_4, or a_5, is covered by one and only one top A node. Consequently, data nodes will be retrieved from their respective streams at most once in the entire query processing process.

2	4	20	25
a_1	a_2	a_6	a_7

(2,3)	(4,19)	(9,18)	(20,26)
a_1	a_2	a_4	a_6+a_7

Fig. 5. Population Table for Top A Node

Fig. 6. Population Table for Hybrid Scheme

To facilitate sampling, we build a population table for each node type (that could potentially become a query root node type) to store the top nodes of the corresponding sub-trees. Figure 5 shows the population table for A nodes. Each cell in the table stores the left region code of a top A node for identification purpose. The right region code is not needed as it can be found once we locate the node in the respective stream.

Once a query root node is drawn from the population table, we evaluate the query against the sub-tree rooted at it. Evaluating a sub-tree is accomplished by examining the nodes of interest falling in the range of the sub-tree root node in the respective streams. For example, if a_2 is sampled, then we examine a_2, a_3, a_4, and a_5 from stream A, b_1 and b_2 from stream B, and c_1 and c_2 from stream C. In other words, all A, B, C nodes that fall in the range covered by a_2 are evaluated. Clearly, a progressive evaluation over this sub-tree set possesses the two desirable properties mentioned earlier.

Various index files can be used to locate and retrieve the desired nodes from the respective streams directly. But, for simplicity, we will not focus on this issue here. Interested readers are referred to [11] for details of such index files on XML data.

Again, we do not physically build any sample sub-tree. Instead, we just use the index files to locate the nodes of the sample sub-tree in the streams and direct them to TwigStack for evaluation in the normal way.

This method is very simple and should work well when the coverage of the query root nodes is roughly uniform and the tree structure is spread widely. However, if there are few top query root nodes, such as a_2, that cover large ranges of the coding domain, imbalanced workload and large variance in estimation can be expected. To remedy this problem, we devise two other sampling schemes in the following.

All Query Root Node Sampling. In this scheme, all query root nodes are included in the (sub-tree) set for sampling. For example, $\{a_1, a_2, a_3, a_4, a_5, a_6, a_7\}$ form the sub-tree set for all queries rooted at A. Since a non-top query root node, such as a_3, a_4, or a_5, falls in the ranges of its ancestors, in this case a_2, it may be evaluated multiple times. In order not to generate duplicate query matches, no query root node should be evaluated unless it is directly sampled. For example, when a_2 is drawn, we evaluate only a_2 from the A stream and, as usual, b_1 and b_2 from the B stream, and c_1 and c_2 from the C stream. Note that a_3 and a_4 are not evaluated here since they are not sampled directly. This schemes requires a little more modifications to the TwigStack than the previous sample scheme. Clearly, a progressive evaluation over this scheme generates the complete query results once all the nodes in the set are evaluated (Property 1), and no duplicate results are generated in the process (Property 2) due to the provision.

Note that some non-query-root nodes may be evaluated multiple times because they are covered by multiple query root nodes. For example, b_1 will be retrieved and evaluated twice if both a_2 and a_3 are sampled; similarly, c_1, c_2, and b_2 will also be retrieved twice if both a_2 and a_4 are sampled. This is due to the overlapping of the coding ranges of query root nodes, such as a_2 and a_3, and a_2 and a_4. This scheme can incur substantial overhead if the data are highly nested. In contrast, top query root node sampling scheme never retrieves or evaluates non-query root nodes more than once. However, this scheme does not require population tables. Samples can be drawn directly from the streams.

A Hybrid Sampling Scheme. The two aforementioned schemes each has its strengths and weaknesses. Here, we devise another sampling scheme to strike a balance between the variance and overhead. We attempt to combine (if possible) small consecutive query root nodes into a larger sampling unit, while extract (if possible) query root nodes nested in large query root nodes to form additional sampling units. Here, small/large nodes refer to nodes that have small/large coding ranges,respectively.

Two thresholds, an upper threshold and a lower threshold, are set up to guide the merge and extraction of the query root nodes. The upper threshold is used to constrain the size of the sample units, while the lower threshold is used to guard against forming too small sample units when extracting sampling units from large ones. There is no strict rule for setting up the upper/lower thresholds. Based on our experience, dividing the entire domain into several hundred subregions should be enough to yield good estimation even for a complex dataset like TreeBank. For example, one can set the upper bound to 0.5% of the entire range and lower bound to 0.25% (half the upper bound) to divide the data into roughly $1/((0.5\%+0.25\%)/2)=267$ subregions.

We attempt to combine as many consecutive small query root nodes that spread out over a range no greater than the upper threshold into a single sampling unit. On the other hand, if a query root node has a coverage greater than the upper bound, we look to extract its consecutive descendent query root nodes whose combined ranges are between the lower bound and the upper bound to form new sampling units. Note that if the consecutive nodes nested in a large node are too small (i.e., smaller than the lower threshold) to form sampling units, we will leave them alone as they are already covered by (or nested in) their ancestor nodes and will be sampled with them. By restraining the sizes of hopefully the majority of the sampling units between the lower and upper thresholds, this scheme is expected to yield estimates with small variance.

To ensure that no duplicate query matches are generated, we require that an extracted sampling unit be excluded from the evaluation of its ancestors. In addition, a non-top query root node is evaluated with its nearest ancestor query root node that is a sample unit. An example of using the hybrid sampling scheme can be found in the full version of the paper [12].

4   Running Estimates and Confidence Intervals

Because of the space limit, below is a brief discussion. A more detailed analysis can be found in [12].

4.1   Aggregation Queries

COUNT and SUM Queries. Let N be the total number of sampling units in the population table or the population size, and n the number of units drawn from the table or the sample size till now. Let $f = \frac{n}{N}$ be the sampling fraction. Let y_i, $1 \leq i \leq n$, be the quantity of interest associated with the ith randomly drawn sampling unit from the table. Let Y be the true result of the COUNT or SUM query, and $\hat{Y} = \frac{N}{n} \sum_{i=1}^{n} y_i$ an estimator of Y derived based on a simple random sampling without replacement sample of size n from the population table. The following properties hold following the theorems in [3].

Theorem 1. *$\hat{Y}$ is a consistent and unbiased estimator of Y.*

Theorem 2. *Let $s^2 = \frac{1}{n-1} \sum_{i=1}^{n} (y_i - \bar{y})^2$, where $\bar{y} = \frac{1}{n} \sum_{i=1}^{n} y_i$. $v(\hat{Y}) = \frac{N^2 s^2}{n}(1 - f)$ is an unbiased estimator of the variance of $\hat{Y}$.*

AVG Queries. AVG is computed as the ratio of the results of the respective SUM to COUNT queries, i.e., SUM / COUNT. Let y and x be the values of interest in the SUM and COUNT queries, respectively. Let $\hat{A} = \sum_{i=1}^{n} y_i / \sum_{i=1}^{n} x_i$ be an estimator of AVG. $\hat{A}$ is usually a slightly biased estimator for AVG when the sample size is small. In large samples, the distribution of $\hat{A}$ tends to normality and the bias becomes negligible [3].

An estimator for the variance of $\hat{A}$, denoted $v(\hat{A})$, can be derived [3] as

$$v(\hat{A}) = \frac{1-f}{n\bar{x}^2} \cdot \frac{\sum_{i=1}^{n} y_i^2 - 2\hat{A} \sum_{i=1}^{n} y_i x_i + \hat{A}^2 \sum_{i=1}^{n} x_i^2}{n-1} \tag{1}$$

where $\bar{x} = \sum_{i=1}^{n} x_i$.

If the sample size is large enough, for a given confidence probability p, based on the Central Limit Theorem, the associated confidence interval can be derived in the same way as those for the COUNT and SUM queries as $\hat{A} \pm \epsilon$, where $\epsilon = z_p \sqrt{v(\hat{A})}$.

5   Performance Evaluation

In this section, we report the experimental results of the proposed sampling schemes on progressive evaluation of queries. Three datasets, DBLP, XMark, and TreeBank, are used in the experiments to measure the performance of these methods in estimating aggregation queries and progress indicators.

5.1   Experimental Setting

The proposed sampling schemes are incorporated into the TwigStack algorithm [1] for progressive evaluation of queries. All algorithms are implemented in C++ and experiments are performed on a linux workstation with a 3.4GHz CPU and 1GB RAM.

We choose the datasets whose structural complexities range from simple to complex for experimental purposes. DBLP exhibits rather simple and uniform structures. It contains 3,332,130 elements, totaling 133 M bytes. XMark benchmark is a synthetic dataset generated based on an internet auction website database. It also exhibits very simple structures like DBLP but with a light degree of recursions. The dataset used in our experiments is around 116 MB. It has 79 labels and 1,666,315 elements. Another synthetic dataset TreeBank has complex and deep structures with a high degree of recursions. The dataset is around 86 MB with 250 labels and 2,437,666 elements. The depth of TreeBank document is as large as 36.

Queries with different structural characteristics and selectivity ratios are chosen for experiments. More than 50 twig queries are tested for each of the datasets.

The (absolute) relative error is used to measure the estimation errors of the methods for aggregation queries and progress indicators. We tested more than 50 different queries, each of which was run 5 times. The averages of the relative errors are reported.

5.2   Experimental Results

For simplicity, we shall call the top query root node sampling, the all query root node sampling, and the hybrid sampling the Sampling1, Sampling2, and Sampling3, respectively, in the following discussions.

DBLP Dataset Selectivity Estimation. Figure 7 shows the performance of the sampling methods on the DBLP dataset. As observed, the average relative errors of using only 5% samples are already less than 6% for Sampling1 and Sampling2 methods, and around 8% for the Sampling3. The errors decrease to around 3% when the sampling fractions increase to 25%.

The good performance of these methods is partly due to the uniformity of the dataset. Another factor that contributes to the good performance is the large numbers of sample units in the dataset, e.g., tens of thousands or hundreds of thousands in Sampling1 and Sampling2, for many queries. Consequently, even for a small sampling fraction, such as 5%, there are large numbers of sample units selected for evaluation, which help to form good representatives of the population and yield good estimates.

Sampling3 is able to form sample units of desired sizes by setting the upper/lower bounds. We have purposely constructed larger sample units (and thus less numbers of sample units) to observe the changes in estimation errors. In the experiments, the numbers of sample units in Sampling3 are around 10% of those in Sampling1 and Sampling2. As expected, Sampling1 and Sampling2 perform better than Sampling3 because their larger numbers of sample units tend to smooth out the variance better. In fact, Sampling3 could have performed just as well as the Sampling1 and Sampling2 if we had used smaller sample units. We will discuss more about Sampling3 when discussing the TreeBank dataset.

Progress Indicator. Due to the uniform tree structure of the dataset, all three sampling schemes are able to predict the execution time precisely (Figure 8). With 5% of the population sampled, the relative errors of the estimated execution time are 3.54%, 6.03%, and 4.63% for Sampling1, Sampling2, and Sampling3,

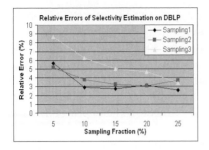

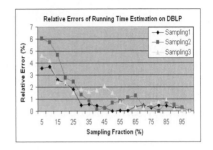

Fig. 7. Selectivity Estimation on DBLP

Fig. 8. Running Time Estimation on DBLP

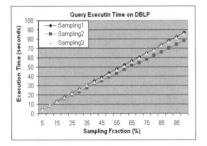

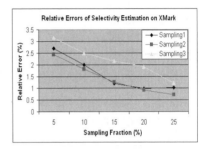

Fig. 9. Query Execution Time on DBLP

Fig. 10. Selectivity Estimation on XMark

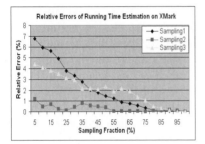

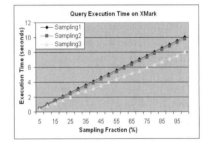

Fig. 11. Running Time Estimation on XMark

Fig. 12. Query Execution Time on XMark

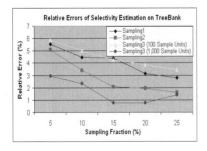

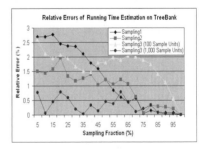

Fig. 13. Selectivity Estimation on TreeBank

Fig. 14. Running Time Estimation on TreeBank

respectively. The relative errors decease quickly to 1.78%, 2.43%, and 1.50% when 25% of the data sampled. The relatively steady system workload also contributes to the good performance.

Figure 9 shows the query processing time of the three methods. We measured the time after each batch of 5% sample units is processed. As observed, the processing time increases almost linearly with amount of samples processed. This implies that approximately the same amount of time is consumed on each batch of samples, which also explains the small errors of the execution time estimation in Figure 8. All three methods spend nearly the same amount of time in processing. The minor difference in execution time we believe is more related to the system loads than others. We also observe that the average overhead of a progressive evaluation is around 19.43%. That is, it takes 19.43% more time to evaluate the entire dataset in the progressive manner than the normal way.

Xmark Dataset Selectivity Estimation. The experimental results of XMark dataset are quite similar to those for DBLP because they both have quite uniform structures. In Figure 10, with only 5% samples taken, the relative errors of Sampling1, Sampling2, and Sampling3 are already as small as 2.70%, 2.43% and 3.16%, respectively. When the sampling fraction increases to 25%, the relative error decreases to 1.01%, 0.73%, 1.22%. Sampling1 and Sampling2 outperform Sampling3 because there are often tens or hundreds of thousands sampling units in Sampling1 and Sampling2 while there are only hundreds in Sampling3.

Progress Indicator. All three methods have very high accuracies. As shown in Figure 11, the relative errors of the execution time estimation with 5% samples are 6.76%, 1.21%, and 4.41% for Sampling1, Sampling2, and Sampling3, respectively. When the sampling fraction increase to 25%, the relative error decrease to 3.77%, 0.14% and 3.13%. Sampling2 seems to perform a little better than the other two. We attribute its success to the larger numbers of small sample units in the populations. The query processing time of three sampling methods on XMark dataset is shown in Figure 12. Again, due to the uniformity of the XMark dataset, the query processing time increases at a near-linear rate with sampling fraction.The average overhead of a progressive evaluation is 15.29%.

TreeBank Dataset Selectivity Estimation. The TreeBank dataset exhibits a complex structure and a high degree of recursions. As expected, Sampling1

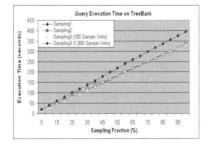

Fig. 15. Query Execution Time on TreeBank

does not perform nearly as well as Sampling2. This is due to the highly recursive nature of the structure that increases the variability of the number of query matches in Sampling1. But, still all these methods perform very well. No significant degradation is found when compared with the results of DBLP and XMark. We attribute this result to there being large enough numbers of sample units in these schemes to even out the variability.

We use two sets of upper and lower bounds in Sampling3 to construct sample units, one has 100 units and the other has 1,000 units. Note that these numbers are much smaller than the tens of thousands and hundreds of thousands sample units in the Sampling1 and Sampling2, respectively. As expected the 100-sample-unit Sampling3 does not perform as well as the others due to the too small number of sample units. But the 1,000-sample-unit Sampling3 outperforms the others, even though it also has a much smaller number of sample units (compared to the Sampling1 and Sampling2). It demonstrates the effectiveness of the measures taken to merge small subtrees and decompose large subtrees, which help smooth out the variability.

Progress Indicator. Again, all these methods provide accurate execution time estimation. As shown in Figure 14, with 5% samples, all methods yield estimates with less than 3% relative error. Sampling3 (with 1,000 sample units) is the best due to its "even-size" sample units and reasonably large numbers of sample units. Like other experiments, the processing time of all these methods increases almost linearly with sampling fraction, as shown in Figure 15. This indicates that all these schemes can smooth out the variability effectively and give good estimate of the execution time as demonstrated in Figure 14.

6   Conclusions

In this paper, we initiate the research of sampling on XML data and progressive evaluation of XML queries. We aim to enhance XML query processors with the ability to process both aggregation and non-aggregation queries progressively (or incrementally) and report the results and progress continually. Several innovative sampling schemes on XML data are proposed. Empirical studies show that all three sampling schemes are able to provide accurate estimation on aggregation query and progress indicators. The proposed methodology advances the XML query processing to the next level, being more flexible, responsive, and informative.

References

1. Bruno, N., Koudas, N., Srivastava, D.: Holistic twig joins: optimal xml pattern matching. In: Proceedings of the 2002 ACM SIGMOD international conf. on Management of data, pp. 310–321 (2002)
2. Chaudhuri, S., Narasayya, V., Ramamurthy, R.: Estimating progress of long running sql queries. In: Proc. ACM SIGMOD Conf., pp. 803–814 (2004)
3. Cochran, W.G.: Sampling Techniques. Wiley, Chichester (1977)

4. Ganguly, S., Gibbons, P., Matias, Y., Silberschatz, A.: Bifocal sampling for skew-resistant join size estimation. In: Proceedings of the 1996 ACM SIGMOD international conf. on Management of data, pp. 271–281 (1996)
5. Hass, P., Naughton, J., Seshadri, S., Stokes, L.: Sampling-based estimation of the number of distinct values of an attribute. In: Proc. 21st Intl. Conf. on Very Large Data Bases, pp. 311–322 (1995)
6. Hellerstein, J., Haas, P., Wang, H.: Online aggregation. In: Proc. ACM SIGMOD Conf., pp. 171–182 (1997)
7. Hoeffding, W.: Probability inequality for sums of bounded random variables. Journal of Amer. Statist. Assoc. (58), 13–30 (1964)
8. Hou, W.-C., Ozsoyoglu, G., Taneja, B.: Statistical estimators for relational algebra expression. In: Proc. 7th ACM Symp. on Principles of Database Systems, pp. 276–287 (1988)
9. Hou, W.-C., Ozsoyoglu, G., Taneja, B.: Processing aggregate relational queries with hard time constraints. In: Proc. ACM SIGMOD International Conf. on Management of Data, pp. 68–77 (1989)
10. Jermain, C., Dobra, A., Arumugam, S., Jashi, S., Pol, A.: A disk-based join with probabilistic guarantees. In: Proc. ACM SIGMOD Conf., pp. 563–574 (2005)
11. Jiang, H., Lu, H., Wang, W., Ooi, B.C.: Xr-tree: Indexing xml data for efficient structural joins. In: ICDE, pp. 253–264 (2003)
12. Jiang, Z., Luo, C., Hou, W., Ozsoyoglu, G.: Progressive Evaluation of XML Queries for Online Aggregation and Progress Indicator (Technical Report), http://www.cs.siu.edu/~zjiang/dexa08.pdf
13. Lipton, R.J., Naughton, J.F., Schneider, D.A.: Practical selectivity estimation through adaptive sampling. In: Proceedings 1990 ACM SIGMOD Intl. Conf. Managment of Data, pp. 1–11 (1990)
14. Lu, J., Chen, T., Ling, T.W.: Efficient processing of xml twig patterns with parent child edges: A look-ahead approach. In: Proceedings of CIKM, pp. 533–542 (2004)
15. Lu, J., Ling, T.W., Chan, C.-Y., Chen, T.: From region encoding to extended dewey: on efficient processing of xml twig pattern matching. In: Proceedings of the 31st international conf. on very large data bases (2005)
16. Luo, G., Naughton, J., Ellmann, C., Watzke, M.: Toward a progress indicator for database queries. In: Proc. ACM SIGMOD Conf., pp. 791–802 (2004)
17. Luo, G., Naughton, J., Ellmann, C., Watzke, M.: Increasing the accuracy and coverage of sql progress indicators. In: Proc. ACM SIGMOD Conf., pp. 853–864 (2005)
18. Ross, S.: Introduction to Probability Models, 2nd edn. Academic Press, London (1980)
19. Tan, K., Goh, C., Ooi, B.: Progressive evaluation of nested aggregate queries. VLDB Journal (9), 261–278 (2000)
20. Wang, W., Jiang, H., Lu, H., Yu, J.X.: Containment join size estimation: models and methods. In: Proceedings of the 2002 ACM SIGMOD International Conference on Management of Data, pp. 358–369 (2003)
21. Zhang, N., Ozsu, M.T., Aboulnaga, A., Ilyas, I.F.: Xseed: Accurate and fast cardinality estimation for xpath queries. In: Proc. 22nd Intl. Conf. on Data Engineering (2006)

Dynamic Query Processing for P2P Data Services in the Cloud

Pawel Jurczyk and Li Xiong

Emory University, Atlanta GA 30322, USA
{pjurczy,lxiong}@emory.edu

Abstract. With the trend of cloud computing, data and computing are moved away from desktop and are instead provided *as a service* from the cloud. Data-as-a-service enables access to a wealth of data across distributed and heterogeneous data sources in the cloud. We designed and developed DObjects, a general-purpose P2P-based query and data operations infrastructure that can be deployed in the cloud. This paper presents the details of the dynamic query execution engine within our data query infrastructure that dynamically adapts to network and node conditions. The query processing is capable of fully benefiting from all the distributed resources to minimize the query response time and maximize system throughput. We present a set of experiments using both simulations and real implementation and deployment.

1 Introduction

With the trend of cloud computing[1,2], data and computing are moved away from desktop and are instead provided *as a service* from the cloud. Current major components under the cloud computing paradigm include infrastructure-as-a-service (such as EC2 by Amazon), platform-as-a-service (such as Google App Engine), and application or software-as-a-service (such as GMail by Google). There is also an increasing need to provide data-as-a-service [1] with a goal of facilitating access to a wealth of data across distributed and heterogeneous data sources available in the cloud.

Consider a system that integrates the air and rail transportation networks with demographic databases and patient databases in order to model the large scale spread of infectious diseases (such as the SARS epidemic or pandemic influenza). Rail and air transportation databases are distributed among hundreds of local servers, demographic information is provided by a few global database servers and patient data is provided by groups of cooperating hospitals.

While the scenario above demonstrates the increasing needs for integrating and querying data across distributed and autonomous data sources, it still remains a challenge to ensure interoperability and scalability for such data services.

---

[1] http://en.wikipedia.org/wiki/Cloud_computing
[2] http://www.theregister.co.uk/2009/01/06/year_ahead_clouds/

S.S. Bhowmick, J. Küng, and R. Wagner (Eds.): DEXA 2009, LNCS 5690, pp. 396–411, 2009.

To achieve interoperability and scalability, data federation is increasingly becoming a preferred data integration solution. In contrast to a centralized data warehouse approach, a data federation combines data from distributed data sources into one single *virtual* data source, or a data service, which can then be accessed, managed and viewed as if it was part of a single system. Many traditional data federation systems employ a centralized mediator-based architecture (Figure 1). We recently proposed DObjects [2, 3], a P2P-based architecture (Figure 2) for data federation services. Each system node can take the role of either a mediator or a mediator and wrapper at the same time. The nodes form a virtual system in a P2P fashion. The framework is capable of extending cloud computing systems with data operations infrastructure, exploiting at the same time distributed resources in the cloud.

Fig. 1. Typical Mediator-Based Architecture   **Fig. 2.** P2P-Based Architecture

Contributions. In this paper we focus on the query processing issues of DObjects and present its novel dynamic query processing engine in detail. We present our dynamic distributed *query execution and optimization* scheme. In addition to leveraging traditional distributed query optimization techniques, our optimization is focused on dynamically placing (sub)queries on the system nodes (mediators) to minimize the query response time and maximize system throughput. In our query execution engine, (sub)queries are deployed and executed on system nodes in a dynamic (based on nodes' on-going knowledge of the data sources, network and node conditions) and iterative (right before the execution of each query operator) manner. Such an approach guarantees the best reaction to network and resource dynamics. We experimentally evaluate our approach using both simulations and real deployment.

2   Related Work

Our work on DObjects and its query processing schemes was inspired and informed by a number of research areas. We provide a brief overview of the relevant areas in this section.

Distributed Databases and Distributed Query Processing. It is important to distinguish DObjects and its query execution component from the many existing distributed database systems. At the first glance, distributed database systems have been extensively studied and many systems have been proposed. Earlier distributed database systems [4], such as R* and SDD-1, share modest targets for network scalability (a handful of distributed sites) and assume homogeneous databases. The focus is on encapsulating distribution with ACID guarantees. Later distributed database or middleware systems, such as Garlic [5], DISCO [6] or TSIMMIS [7], target large-scale heterogeneous data sources. Many of them employ a *centralized* mediator-wrapper based architecture (see Figure 1) to address the database heterogeneity in the sense that a single mediator server integrates distributed data sources through wrappers. The query optimization focuses on integrating wrapper statistics with traditional cost-based query optimization for single queries spanning multiple data sources. As the query load increases, the centralized mediator may become a bottleneck. More recently, Internet scale query systems, such as Astrolabe [8] and PIER [9], target thousands or millions of massively distributed homogeneous data sources with a peer-to-peer (P2P) or hierarchical network architecture. However, the main issue in such systems is how to efficiently route the query to data sources, rather than on integrating data from multiple data sources. As a result, the query processing in such systems is focused on efficient query routing schemes for network scalability.

The recent software frameworks, such as map-reduce-merge [10] and Hadoop[3], support distributed computing on large data sets on clusters of computers and can be used to enable cloud computing services. The focus of these solutions, however, is on data and processing distribution rather than on data integration.

While it is not the aim of DObjects to be superior to these works, our system distinguishes itself by addressing an important problem space that has been overlooked, namely, integrating large-scale heterogeneous data sources with both network and query load scalability without sacrificing query complexities and transaction semantics. In spirit, DObjects is a *distributed* P2P mediator-based system in which a federation of mediators and wrappers forms a virtual system in a P2P fashion (see Figure 2). Our optimization goal is focused on building effective sub-queries and optimally placing them on the system nodes (mediators) to minimize the query response time and maximize throughput.

The most relevant to our work are OGSA-DAI and its extension OGSA-DQP [11] introduced by a Grid community as a middleware assisting with access and integration of data from separate sources. While the above two approaches share a similar set of goals with DObjects, they were built on the grid/web service model. In contrast, DObjects is built on the P2P model and provides resource sharing on a peer-to-peer basis.

Data Streams and Continuous Queries. A large amount of efforts was contributed to the area of continuous or pervasive query processing [12, 8, 13, 14, 15, 16, 17]. The query optimization engine in DObjects is most closely related to SBON [18]. SBON presented a stream based overlay network for optimizing queries by carefully placing aggregation operators. DObjects shares a similar set of goals as SBON in distributing query operators based on on-going knowledge of network conditions. SBON uses a two step approach, namely, virtual placement and physical mapping for query placement based on a cost space. In contrast, we use a single cost metric with different cost features for easy decision making at

---

[3] http://hadoop.apache.org/core/

individual nodes for a local query migration and explicitly examine the relative importance of network latency and system load in the performance.

Load Balancing. Past research on load balancing methods for distributed databases resulted in a number of methods for balancing storage load by managing the partitioning of the data [19, 20]. Mariposa [21] offered load balancing by providing marketplace rules where data providers use bidding mechanisms. Load balancing in a distributed stream processing was also studied in [22] where load shedding techniques for revealing overload of servers were developed.

3   DObjects Overview

In this section we briefly describe DObjects framework. For further details we refer readers to [2, 3]. Figure 3 presents our vision of the deployed system. The system has no centralized services and thus allows system administrators to avoid the burden in this area. It also uses a *P2P resource sharing* substrate as a resource sharing paradigm to benefit from computational resources available in the cloud. Each node serves as a *mediator* that provides its computational power for a query mediation and results aggregation. Each node can also serve as a data adapter or wrapper that pulls data from data sources and transforms it to a uniform format that is expected while building query responses. Users can connect to any system node; however, while the physical connection is established between a client and one of the system nodes, the logical connection is between a client node and a virtual system consisting of all available nodes.

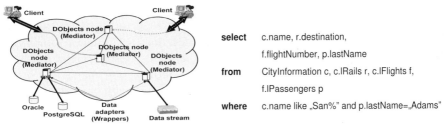

select	c.name, r.destination,
	f.flightNumber, p.lastName
from	CityInformation c, c.lRails r, c.lFlights f,
	f.lPassengers p
where	c.name like „San%" and p.lastName=„Adams"

Fig. 3. System architecture   **Fig. 4.** Query example

4   Query Execution and Optimization

In this section we focus on the query processing issues of DObjects, present an overview of the dynamic distributed query processing engine that adapts to network and resource dynamics, and discuss details of its cost-based query placement strategies.

4.1   Overview

As we have discussed, the key to query processing in our framework is to have a decentralized and distributed query execution engine that dynamically adapts to network and resource conditions. In addition to adapting "textbook" distributed query processing techniques such as distributed join algorithms and the learning curve approach for keeping statistics about data adapters, our query processing framework presents a number of innovative aspects. First, instead of generating a set of candidate plans, mapping them physically and choosing the best ones as in a conventional cost based query optimization, we create one initial abstract plan for a given query. The plan is a high-level description of relations between steps and operations that need to be performed in order to complete the query. Second, when the query plan is being executed, placement decisions and physical plan calculation are performed dynamically and iteratively. Such an approach guarantees the best reaction to changing load or latency conditions in the system.

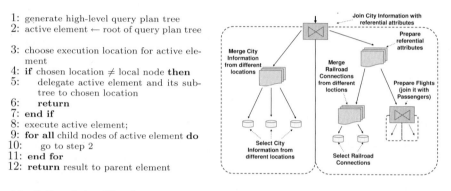

1: generate high-level query plan tree
2: active element ← root of query plan tree

3: choose execution location for active element
4: **if** chosen location ≠ local node **then**
5:     delegate active element and its subtree to chosen location
6:     **return**
7: **end if**
8: execute active element;
9: **for all** child nodes of active element **do**
10:     go to step 2
11: **end for**
12: **return** result to parent element

Alg. 1. Local algorithm for query processing

Fig. 5. Example of high-level query plan

It is important to highlight that our approach does not attempt to optimize physical query execution performed on local databases. Responsibility for this is pushed to data adapters and data sources. Our optimization goal is at a higher level focusing on building effective sub-queries and optimally placing those sub-queries on the system nodes to minimize the query response time.

Our query execution and optimization consists of a few main steps. First, when a user submits a query, a high-level query description is generated by the node that receives it. An example of such a query plan is presented in Figure 5. The plan corresponds to the query introduced in Figure 4 that queries for cities along with related referential attributes: railroad connections and flights. In addition, each flight will provide a list of passengers. Note that each type is provided by a different physical database. The query plan contains such elements as *joins*, *horizontal* and *vertical data merges*, and *select* operations that are performed on data adapters. Each element in the query plan has different algorithms of *optimization* (see Section 4.2).

Next, the node chooses active elements from the query plan one by one in a top-down manner for execution. Execution of an active element, however, can be delegated to any node in the system in order to achieve load scalability. If the system finds that the best candidate for executing current element is a remote node, the *migration of workload* occurs. In order to choose the best node for the execution, we deploy a network and resource-aware cost model that dynamically adapts to network conditions (such as delays in interconnection network) and resource conditions (such as load of nodes) (see Section 4.3). If the active element is delegated to a remote node, that node has a full control over the execution of any child steps. The process works recursively and iteratively, therefore the remote node could decide to move child nodes of submitted query plan element to other nodes or execute it locally in order to use the resources in the most efficient way to achieve good scalability. Algorithm 1 presents a sketch of the local query execution process. Note that our algorithm takes a greedy approach without guaranteeing the global optimality of the query placement. In other words, each node makes a local decision on where to migrate the (sub)queries.

4.2   Execution and Optimization of Operators

In previous section we have introduced the main elements in the high-level query plan. Each of the elements has different goals in the optimization process. It is important to note that the optimization for each element in the query plan is performed iteratively, just before given element is executed. We describe the optimization strategies for each type of operators below.

Join. Join operator is created when user issues a query that needs to join data across sites. In this case, join between main objects and the referenced objects have to be performed (e.g., join flights with passengers). The optimization is focused on finding the most appropriate join algorithm and the order of branch executions. The available join algorithms are nested-loop join (NLJ), semi-join (SJ) and bloom-join (BJ) [4]. In case of NLJ, the branches can be executed in parallel to speedup the execution. In case of SJ or BJ algorithms, the branches have to be executed in a pipeline fashion and the order of execution has to be fixed. Our current implementation uses a semi-join algorithm and standard techniques for result size estimations. There is also a lot of potential benefits in parallelization of the join operator execution using such frameworks as map-reduce-merge [10]. We leave this to our future research agenda.

Data Merge. Data merge operator is created when data objects are split among multiple nodes (horizontal data split) or when attributes of an object are located on multiple nodes (vertical data split). Since the goal of the data merge operation is to merge data from multiple input streams, it needs to execute its child operations before it is finished. Our optimization approach for this operator tries to maximize the parallelization of sub-branch execution. This goal is achieved by executing each sub-query in parallel, possibly on different nodes if such an approach is better according to our cost model that we will discuss later.

Select. Select operator is always the leaf in our high-level query plan. Therefore, it does not have any dependent operations that need to be executed before it finishes. Moreover, this operation has to be executed on locations that provide queried data. The optimization issues are focused on optimizing queries submitted to data adapters for a faster response time. For instance, enforcing an order (sort) to queries allows us to use merge-joins in later operations. Next, *response chunks* are built in order to support queries returning large results. Specifically, in case of heavy queries, we implement an iterative process of providing smaller pieces of the final response. In addition to helping to maintain a healthy node load level in terms of memory consumption, such a feature is especially useful when building a user interface that needs to accommodate a long query execution.

4.3   Query Migration

The key of our query processing is a greedy local query migration component for nodes to delegate (sub)queries to a remote node in a dynamic (based on current network and resource conditions) and iterative (just before the execution of each element in the query plan) manner. In order to determine the best (remote) node for possible (sub)query migration and execution, we first need a cost metric for the query execution at different nodes. Suppose a node migrate a query element and associated data to another node, the cost includes: 1) a transmission delay and communication cost between nodes, and 2) a query processing or computation cost at the remote node. Intuitively, we want to delegate the query element to a node that is "closest" to the current node and has the most computational resources or least load in order to minimize the query response time and maximize system throughput. We introduce a cost metric that incorporates such two costs taking into account current network and resource conditions. Formally Equation 1 defines the cost, denoted as $c_{i,j}$, associated with migrating a query element from node i to a remote node j:

$$c_{i,j} = \alpha * (DS/bandwidth_{i,j} + latency_{i,j}) + (1 - \alpha) * load_j \qquad (1)$$

where DS is the size of the necessary data to be migrated (estimated using statistics from data sources), $bandwidth_{i,j}$ and $latency_{i,j}$ are the network bandwidth and latency between nodes i and j, $load_j$ is the current (or most recent) load value of node j, and α is a weighting factor between the communication cost and the computation cost. Both cost terms are normalized values between 0 and 1 considering the potential wide variances between them.

To perform query migration, each node in the system maintains a list of candidate nodes that can be used for migrating queries. For each of the nodes, it calculates the cost of migration and compares the minimum with the cost of local execution. If the minimum cost of migration is smaller than the cost of local execution, the query element and its subtree is moved to the best candidate. Otherwise, the execution will be performed at the current node. To prevent a (sub)query being migrated back and forth between nodes, we require each node to execute at least one operator from the migrated query plan before further migration. Alternatively, a counter, or Time-To-Live (TTL) strategy, can be implemented to limit the number of migrations for the same (sub)query. TTL counter can be decreased every time a given (sub)tree is moved, and, if it reaches 0, the node has to execute at least one operator before further migration. The decision of a migration is made if the following equation is true:

$$min_j\{c_{i,j}\} < \beta * (1 - \alpha)load_i \qquad (2)$$

where $min_j\{c_{i,j}\}$ is the minimum cost of migration for all the nodes in the node's candidate list, β is a tolerance parameter typically set to be a value close to 1 (e.g. we set it to 0.98 in our implementations). Note that the cost of a local execution only considers the load of the current node.

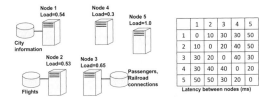

Fig. 6. Setup for Optimization Illustration

Illustration. To illustrate our query optimization algorithm, let us consider a query from Figure 4 with a sample system deployment as presented in Figure 6. Let us assume that a client submits his query to Node 5 which then generates a high-level query plan as presented in Figure 5. Then, the node starts a query execution. The operator at the root of the query plan tree is join. Using the equation 1 the active node estimates the cost for migrating the join operator. Our calculations will neglect the cost of data shipping for simplicity and will use $\alpha = 0.3$ and $\beta = 1.0$. The cost for migrating the query from Node 5 to Node 1 is: $c_{5,1} = 0.3 * (50/50) + (1 - 0.3) * 0.54 = 0.68$. Remaining migration costs are $c_{5,2} = 0.671$, $c_{5,3} = 0.635$ and $c_{5,4} = 0.33$. Using the equation 2 Node 5 decides to move the query to Node 4 ($c_{5,4} < 1.0 * (1 - 0.3) * 1.0$). After the migration, Node 4 will start execution of join operator at the top of the query plan tree. Let us assume that the node decides to execute the left branch first. CityInformation is provided by only one node, Node 1, and no data merge is required. Once the select operation is finished on Node 1, the right branch of join operation can be invoked. Note that Node 4 will not migrate any of the sub-operators (besides selections) as the cost of any migration exceeds the cost of local execution (the cost of migrations: $c_{4,1} = 0.558$, $c_{4,2} = 0.611$, $c_{4,3} = 0.695$ and $c_{4,5} = 0.82$; the cost of local execution: 0.21).

4.4   Cost Metric Components

The above cost metric consists of two cost features, namely, the *communication latency* and the *load* of each node. We could also use other system features (e.g. memory availability), however, we believe the load information gives a good estimate of resource availability at the current stage of the system implementation. Below we present techniques for computing our cost features efficiently.

Latency between Nodes. To compute the network latency between each pair of nodes efficiently, each DObjects node maintains a virtual coordinate, such that the Euclidean distance between two coordinates is an estimate for the communication latency. Storing virtual coordinates has the benefit of naturally capturing latencies in the network without a large measurement overhead. The overhead of maintaining a virtual coordinate is small because a node can calculate its coordinate after probing a small subset of nodes such as well-known landmark nodes or randomly chosen nodes. Several synthetic network coordinate schemes exist. We adopted a variation of Vivaldi algorithm [23] in DObjects. The algorithm uses a simulation of physical springs, where each spring is placed between any two nodes of the system. The rest length of each spring is set proportionally to current latency between nodes. The algorithm works iteratively. In every iteration, each node chooses a number of random nodes and sends a ping message to them and waits for a response. After the response is obtained, initiating node calculates the latency with remote nodes. As the latency changes, a new rest length of springs is determined. If it is shorter than before, the initiating node moves closer towards the remote node. Otherwise, it moves away. The algorithm always tends to find a stable state for the most recent spring configuration. An important feature about this algorithm is that it has great scalability which was proven by its implementation in some P2P solutions (e.g. in OpenDHT project [24]).

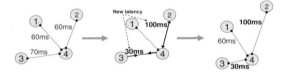

Fig. 7. Illustration of Virtual Coordinates Computation for Network Latency

Figure 7 presents an example iteration of the Vivaldi algorithm. The first graph on the left presents a current state of the system. New latency information is obtained in the middle graph and the rest length of springs is adjusted accordingly. As the answer to the new forces in the system, new coordinates are calculated. The new configuration is presented in the rightmost graph.

Load of Nodes. The second feature of our cost metric is the *load* of the nodes. Given our desired goal to support *cross-platform* applications, instead of depending on any OS specific functionalities for the load information, we incorporated a solution that assures good results in a heterogeneous environment. The main idea is based on time measurement of execution of a predefined test program that considers computing and multithreading capabilities of machines [25]. The program we use specifically runs multiple threads. More than one thread assures that if a machine has multiple CPUs, the load will be measured correctly. Each thread performs a set of predefined computations including a series of integer as well as floating point operations. When each of the computing threads finishes, the time it took to accomplish operations is measured which indicates current computational capabilities of the tested node. In order to improve efficiency of our load computation method, we can dynamically adjust the interval between consecutive measurements. When a node has a stable behavior, we can increase this interval. On the other hand, if we observe rapid change in the number of queries that reside on a given node, we can trigger the measurement.

After the load information about a particular node is obtained, it can be propagated among other nodes. Our implementation builds on top of a distributed event framework, REVENTS[4], that is integrated with our platform for an efficient and effective asynchronous communication among the nodes.

5   Experimental Evaluation

Our framework is fully implemented with a current version available for download[5]. In this section we present an evaluation through simulations as well as a real deployment of the implementation.

5.1   Simulation Results

We ran our framework on a discrete event simulator that gives us an easy way to test the system against different settings. The configuration of data objects relates to the configuration mentioned in Section 4 and was identical for all the experiments below. The configuration of data sources for objects is as follows: object CityInformation was provided by node1 and node2, object Flight by node3 and node4, object RailroadConnection by node1 and finally object Passenger by node2. All nodes with numbers greater than 4 were used as computing nodes. Load of a node affects the execution time of operators. The more operators were invoked on a given node in parallel, the longer the execution time was assumed. Different operators also had different impact on the load of nodes. For instance, a join operator had larger impact than merge operator. In order to evaluate the reaction of our system to dynamic network changes, the communication latency was assigned randomly at the beginning of simulation and changed a few times during the simulation so that the system had to adjust to new conditions in order to operate efficiently. The change was based on increasing or decreasing latency between each pair of nodes by a random factor not exceeding 30%. Table 1 gives a summary of system parameters (number of nodes and number of clients) and algorithmic parameter α with default values for different experiments.

Table 1. Experiment Setup Parameters

Test Case	Figure	# of Nodes (Mediators)	# of Clients	α
α vs. Query Workloads	8	6	14	*
α vs. # of Nodes	9	*	32	*
α vs. # of Clients	10	6	*	*
Comparison of Query Optimization Strategies	11	6	14	0.33
System Scalability	12, 13	20	*	0.33
Impact of Load of Nodes	14	*	256	0.33
Impact of Network Latencies	15	6	14	0.33

* - varying parameter

---

[4] http://dcl.mathcs.emory.edu/revents/index.php
[5] http://www.mathcs.emory.edu/Research/Area/datainfo/dobjects

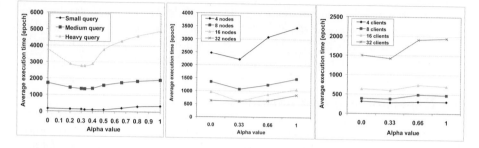

Fig. 8. Parameter Tuning - **Fig. 9.** Parameter Tuning - **Fig. 10.** Parameter Tuning -
Query Workloads   Number of Nodes   Number of Clients

Parameters Tuning - Optimal α. An important parameter in our cost metric
(introduced in equation 1) is α that determines the relative impact of load and
network latency in the query migration strategies. Our first experiment is an
attempt to empirically find optimal α value for various cases: 1) different query
workloads, 2) different number of nodes available in the system, and 3) different
number of clients submitting queries.

For the first case, we tested three query workloads: 1) small queries for City-
Information objects without referential attributes (therefore, no join operation
was required), 2) medium queries for CityInformation objects with two referential attributes (list of Flights and RailroadConnections), and 3) heavy queries
with two referential attributes of CityInformation of which Flight also had a
referential attribute. The second case varied the number of computational nodes
and used the medium query submitted by 32 clients simultaneously. The last
case varied a number of clients submitting medium queries.

Figure 8, 9 and 10 report average execution times for different query loads,
varying number of computational nodes, and varying number of clients respectively for different α. We observe that for all three test cases the best α value
is located around the value 0.33. While not originally expected, it can be explained as follows. When more importance is assigned to the load, our algorithm
will choose nodes with smaller load rather than nodes located closer. In this case,
we are preventing overloading a group of close nodes as join execution requires
considerable computation time. Also, for all cases, the response time was better
when only load information was used ($\alpha = 0.0$) compared to when only distance
information was used ($\alpha = 1.0$). For all further experiments we set the α value
to be 0.33.

Comparison of Optimization Strategies. We compare a number of varied
optimization strategies of our system with some baseline approaches. We give
average query response time for the following cases: 1) no optimization (a naive
query execution where children of current query operator are executed one by one
from left to right), 2) statistical information only (a classical query optimization
that uses statistics to determine the order of branch executions in join operations), 3) location information only ($\alpha = 1$), 4) load information only ($\alpha = 0$),
and 5) full optimization ($\alpha = 0.33$).

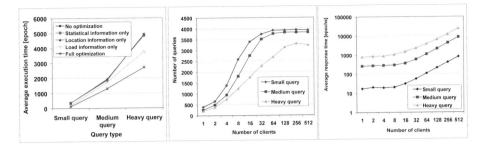

Fig. 11. Comparison of Different Query Optimization Strategies

Fig. 12. System Scalability (Throughput) - Number of Clients

Fig. 13. System Scalability (Response Time) - Number of Clients

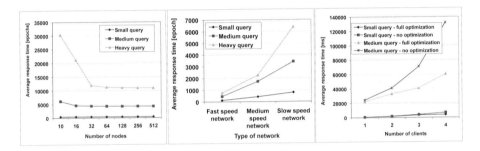

Fig. 14. Impact of Computational Resources

Fig. 15. Impact of Network Latency

Fig. 16. Average Response Time in Real System

The results are presented in Figure 11. They clearly show that, for all types of queries, the best response time corresponds to the case when full optimization is used. In addition, the load information only approach provides an improvement compared to the no optimization, statistical information only, and location information only approaches (the three lines overlap in the plot). The performance improvements are most manifested in the heavy query workload.

System Scalability. An important goal of our framework is to scale up the system for number of clients and load of queries. Our next experiment attempts to look at the throughput and average response time of the system when different number of clients issue queries. We again use three types of queries and a similar configuration to the above experiment.

Figures 12 and 13 present the throughput and average response time for different number of clients respectively. Figure 12 shows the average number of queries that our system was capable of handling during a specified time frame for a given number of clients. As expected, the system throughput increases as the number of clients increases before it reaches its maximum. However, when the system reaches a saturation point (each node is heavily loaded), new clients cannot obtain any new resources. Thus, the throughput reaches its maximum (e.g., around 3950 queries per specified time frame for the case of small queries at 64 clients). Figure 13 reports the average response time and shows a linear scalability. Please note that the figure uses logarithmic scales for better clarity.

Impact of Available Computational Resources. In order to answer the question how the number of nodes available in the system affects its performance, we measured the average response time for varying number of available system nodes with 256 clients simultaneously querying the system. The results are provided in Figure 14. Our query processing effectively reduces the average response time when more nodes are available. For small queries, 10 nodes appears to be sufficient as an increase to 16 nodes does not improve the response time significantly. For medium size queries 16 nodes appears to be sufficient. Finally, for heavy queries we observed improvement when we used 32 nodes instead of 16. The behavior above is not surprising and quite intuitive. Small queries do not require high computational power as no join operation is performed. On the other hand, medium and heavy queries require larger computational power so they benefit from a larger number of available nodes.

Impact of Network Latency. Our last experiment was aimed to find the impact of a network latency on the performance. We report results for three network speeds: a fast network that simulates a Fast Ethernet network offering speed of 100MBit/s, a medium network that can be compared to an Ethernet speed of 10MBit/s, and finally a slow network that represents speed of 1MBit/s.

The result is reported in Figure 15. The network speed, as expected, has a larger impact on the heavy query workload. The reason is that the amount of data that needs to be transferred for heavy queries is larger than medium and small queries, and therefore the time it takes to transfer this data in slower network will have much larger impact on the overall efficiency.

5.2   Testing of a Real Implementation

We also deployed our implementation in a real setting on four nodes started on general-purpose PCs (Intel Core 2 Duo, 1GB RAM, Fast Ethernet network connection). The configuration involved three objects, CityInformation (provided by node 1), Flight (provided by nodes 2 and 3) and RailroadConnection (provided by node 3). Node 4 was used only for computational purposes. We ran the experiment for 10,000 CityInformation, 50,000 Flights (20,000 in node 2 and 30,000 in node 3) and 70,000 RailroadConnections. The database engine we used was PostgreSQL 8.2. We measured the response time for query workloads including small queries for all relevant CityInformation and medium queries for all objects mentioned above. We used various number of parallel clients and $\alpha = 0.33$.

Figure 16 presents results for small and medium queries. It shows that the response time is significantly reduced when query optimization is used (for both small and medium queries). The response time may seem a bit high at the first glance. To give an idea of the actual overhead introduced by our system, we integrated all the databases used in the experiment above into one single database and tested a medium query from Java API using JDBC and one client. The query along with results retrieval took an average of 16s. For the same query, our system took 20s that is in fact comparable to the case of a local database. While the overhead introduced by DObjects cannot be neglected, it does not exceed reasonable boundary and does not disqualify our system as every middleware is expected to add some overhead. In this deployment, the overhead is mainly an effect of the network communication because data was physically distributed among multiple databases. In addition, the cost of distributed computing middleware and wrapping data into object representation also add to the overhead which is the price a user needs to pay for a convenient access to distributed data. However, for a larger setup with larger number of clients, we expect our system to perform better than *centralized* approach as the benefit from distributed computing paradigm and load distribution will outweigh the overhead.

6   Conclusion and Future Work

In this paper we have presented the dynamic query processing mechanism for our P2P based data federation services to address both geographic and load scalability for data-intensive applications with distributed and heterogeneous data sources. Our approach was validated in different settings through simulations as well as real implementation and deployment. We believe that the initial results of our work are quite promising. Our ongoing efforts continue in a few directions. First, we are planning on further enhancement for our query migration scheme. We are working on incorporating a broader set of cost features such as location of the data and dynamic adjustment of the weight parameter for each cost feature. Second, we plan to extend the scheme with a dynamic migration of active operators in real-time from one node to another if load situation changes. This issue becomes important especially for larger queries which last longer time in the system. Finally, we plan to improve the fault tolerance design of our query processing. Currently, if a failure occurs on a node involved in execution of a query, such query is aborted and error is reported to the user. We plan to extend this behavior with possibility of failure detection and allocation of a new node to continue execution of the operator that was allocated to the failed node.

Acknowledgement

We thank the anonymous reviewers for their valuable feedback. The research is partially supported by a Career Enhancement Fellowship by the Woodrow Wilson Foundation.

References

1. Logothetis, D., Yocum, K.: Ad-hoc data processing in the cloud. Proc. VLDB Endow. 1(2), 1472–1475 (2008)
2. Jurczyk, P., Xiong, L., Sunderam, V.: DObjects: Enabling distributed data services for metacomputing platforms. In: Proc. of the ICCS (2008)

3. Jurczyk, P., Xiong, L.: Dobjects: enabling distributed data services for metacomputing platforms. Proc. VLDB Endow. 1(2), 1432–1435 (2008)
4. Kossmann, D.: The state of the art in distributed query processing. ACM Comput. Surv. 32(4) (2000)
5. Carey, M.J., Haas, L.M., Schwarz, P.M., Arya, M., Cody, W.F., Fagin, R., Flickner, M., Luniewski, A.W., Niblack, W., Petkovic, D., Thomas, J., Williams, J.H., Wimmers, E.L.: Towards heterogeneous multimedia information systems: the Garlic approach. In: Proc. of the RIDE-DOM 1995, Washington, USA (1995)
6. Tomasic, A., Raschid, L., Valduriez, P.: Scaling Heterogeneous Databases and the Design of Disco. In: ICDCS (1996)
7. Chawathe, S., Garcia-Molina, H., Hammer, J., Ireland, K., Papakonstantinou, Y., Ullman, J.D., Widom, J.: The TSIMMIS project: Integration of heterogeneous information sources. In: 16th Meeting of the Information Processing Society of Japan, Tokyo, Japan (1994)
8. van Renesse, R., Birman, K.P., Vogels, W.: Astrolabe: A robust and scalable technology for distributed system monitoring, management, and data mining. ACM Trans. Comput. Syst. 21(2) (2003)
9. Huebsch, R., Chun, B.N., Hellerstein, J.M., Loo, B.T., Maniatis, P., Roscoe, T., Shenker, S., Stoica, I., Yumerefendi, A.R.: The architecture of pier: an internet-scale query processor. In: CIDR (2005)
10. Yang, H.c., Dasdan, A., Hsiao, R.L., Parker, D.S.: Map-reduce-merge: simplified relational data processing on large clusters. In: SIGMOD 2007: Proceedings of the 2007 ACM SIGMOD international conference on Management of data, pp. 1029–1040. ACM, New York (2007)
11. Alpdemir, M.N., Mukherjee, A., Gounaris, A., Paton, N.W., Fernandes, A.A.A., Sakellariou, R., Watson, P., Li, P.: Using OGSA-DQP to support scientific applications for the grid. In: Herrero, P., S. Pérez, M., Robles, V. (eds.) SAG 2004. LNCS, vol. 3458, pp. 13–24. Springer, Heidelberg (2005)
12. Madden, S., Franklin, M.J., Hellerstein, J.M., Hong, W.: Tag: A tiny aggregation service for ad-hoc sensor networks. In: OSDI (2002)
13. Yalagandula, P., Dahlin, M.: A scalable distributed information management system. In: SIGCOMM (2004)
14. Trigoni, N., Yao, Y., Demers, A.J., Gehrke, J., Rajaraman, R.: Multi-query optimization for sensor networks. In: DCOSS (2005)
15. Huebsch, R., Garofalakis, M., Hellerstein, J.M., Stoica, I.: Sharing aggregate computation for distributed queries. In: SIGMOD (2007)
16. Xiang, S., Lim, H.B., Tan, K.L., Zhou, Y.: Two-tier multiple query optimization for sensor networks. In: Proceedings of the 27th International Conference on Distributed Computing Systems, Washington, DC. IEEE Computer Society Press, Los Alamitos (2007)
17. Xue, W., Luo, Q., Ni, L.M.: Systems support for pervasive query processing. In: Proceedings of the 25th IEEE International Conference on Distributed Computing Systems (ICDCS 2005), Washington, DC, pp. 135–144. IEEE Computer Society, Los Alamitos (2005)
18. Pietzuch, P.R., Ledlie, J., Shneidman, J., Roussopoulos, M., Welsh, M., Seltzer, M.I.: Network-aware operator placement for stream-processing systems. In: ICDE (2006)
19. Aberer, K., Datta, A., Hauswirth, M., Schmidt, R.: Indexing data-oriented overlay networks. In: Proc. of the VLDB 2005, pp. 685–696 (2005)

20. Ganesan, P., Bawa, M., Garcia-Molina, H.: Online balancing of range-partitioned data with applications to peer-to-peer systems. Technical report, Stanford U. (2004)
21. Stonebraker, M., Aoki, P.M., Devine, R., Litwin, W., Olson, M.A.: Mariposa: A new architecture for distributed data. In: ICDE (1994)
22. Tatbul, N., Çetintemel, U., Zdonik, S.B.: Staying fit: Efficient load shedding techniques for distributed stream processing. In: VLDB, pp. 159–170 (2007)
23. Dabek, F., Cox, R., Kaashoek, F., Morris, R.: Vivaldi: A decentralized network coordinate system. In: Proceedings of the ACM SIGCOMM 2004 Conference (2004)
24. Sean Rhea, B.G., Karp, B., Kubiatowicz, J., Ratnasamy, S., Shenker, S., Stoica, I., Yu, H.: Opendht: A public dht service and its uses. In: SIGCOMM (2005)
25. Paroux, G., Toursel, B., Olejnik, R., Felea, V.: A java cpu calibration tool for load balancing in distributed applications. In: ISPDC/HeteroPar (2004)

A Novel Air Index Scheme for Twig Queries in On-Demand XML Data Broadcast

Yongrui Qin[1], Weiwei Sun[1,*], Zhuoyao Zhang[1], Ping Yu[2], Zhenying He[1], and Weiyu Chen[1]

[1] School of Computer Science, Fudan University
220 Handan Road, Yangpu District, Shanghai 200433, China
{yrqin,wwsun,zhangzhuoyao,zhenying,chwy}@fudan.edu.cn
[2] Distance Education College & e-Educational System Engineering Research Center,
East China Normal University
3363 Zhongshan Road (N.) Shanghai 200062, China
pyu@dec.ecnu.edu.cn

Abstract. Data broadcast is an efficient way for information dissemination in wireless mobile environments, and on-demand XML data broadcast is one of the most important research issues in this area. Indexing XML data on wireless channel is critical for this issue since energy management is very important in wireless mobile environments. Previous works have focused on air index schemes for single path queries. In this paper, we propose a novel air index scheme that builds concise air indexes for twig queries in on-demand XML data broadcast. We adopt the Document Tree structure as the basic air index structure for twig queries and propose to prune redundant structures of the basic Document Tree indexes to reduce the energy consumption. Then we propose to combine all the pruned indexes into one which can eliminate structure redundancy among the indexes to further reduce the energy consumption. Our preliminary experiments show that our air index scheme is very effective and efficient, as it builds concise air indexes and supports twig queries without losing any precision.

Keywords: XML, data broadcast, on-demand, air index, twig query.

1 Introduction

With the rapid development of wireless network technologies, users with mobile devices can access a large amount of information at anytime from anywhere. Data broadcast, as an efficient way for public information delivery to a large number of mobile users, offers great scalability, good power consumption, and efficient bandwidth utilization [1][2].

There are two typical broadcast modes for data broadcast [2]:

Broadcasting Mode. Data is periodically broadcast on the downlink channel. Clients only "listen" to that channel and download data they are interested in.

---

* Corresponding author.

S.S. Bhowmick, J. Küng, and R. Wagner (Eds.): DEXA 2009, LNCS 5690, pp. 412–426, 2009.
© Springer-Verlag Berlin Heidelberg 2009

On-Demand Mode. The clients send their requests to the server through uplink channel and the server considers all pending requests to decide the contents of next broadcast cycle.

Access efficiency and power conservation are two important issues in wireless data broadcast system. Accordingly, two critical metrics, *access time* and *tuning time* are used to measure the system's performance [1][2][3]. Air indexing techniques have also been studied in [1][2]. They introduce some auxiliary data structures in broadcast to indicate the arrival time of each data item. As a result, mobile clients know the arrival time of the requested data items in advance and can switch to the energy-saving mode (doze mode) during waiting. Therefore, the advantage of air index is reducing tuning time and thus a longer battery life can be attained. However, after introducing air index in broadcast, the broadcast cycles are lengthened and the access latency is increased. Therefore, concise indexes are always more preferable.

On-demand broadcast is an important topic in data broadcast research [2][4][5]. Most of the current broadcast researches are focusing on broadcasting data items with unique key values. The requests are key-based queries and the indexing methods are also key-based [1]. To retrieve information from the broadcast, users must know the item key in advance.

Besides the traditional structured information, such as records in relational databases, more and more information turn out to be semi-structured over the past few years. XML has rapidly gained popularity as a standard to represent semi-structured information, and also an effective format for data transmission and exchange.

XML data broadcast is a new research issue. In this paper, we focus on designing air index scheme for twig queries which are more powerful and accurate to express user interests in XML data than single path queries. Most of the air index schemes of traditional data broadcast are key-based [1] and are only suitable to index data items with unique key values. These index schemes can not be applied to XML data broadcast since XML data are semi-structured. Moreover, traditional XML index schemes for twig queries which are designed for efficient XPath/XQuery query processing are usually very large and even need to be stored in disks [7][8]. Since the bandwidth of a wireless channel is usually very limited and the storage capacity of mobile users is very low as well, these disk-based or large memory-based index schemes can not be applied to XML data broadcast as well.

In this paper, we adopt Document Tree structure as the basic air index structure. In order to reduce the size of the basic air index, we apply pruning technique on the Document Tree indexes of all XML documents to eliminate redundant structures and then combine the pruned indexes into one. In summary, the main contributions of this paper are:

- We adopt Document Tree structure as the basic air index structure for twig queries in on-demand XML data broadcast. This basic index structure can support twig queries in small space.
- We propose a novel index pruning technique to cut out redundant structures which do not satisfy any user requests. The total size of the air index is greatly reduced after pruned, while it still can support twig queries effectively without losing any precision.
- We put forward an efficient heuristic algorithm for the problem of pruning redundant structures of the air index.

- We propose to combine the pruned Document Tree indexes into one single index to further reduce the size of the air index.

We proceed with related works in Section 2. Section 3 describes our air index scheme that supports twig queries and the problems of pruning and combining Document Tree indexes. Section 4 presents an experimental study that evaluates the performance of our solution and Section 5 concludes this paper.

2   Related Works

Emerging as a new research issue, a lot of works dealing with the construction of air index for XML data broadcast have appeared recently. Some studies address the performance optimization of query processing of XML streams in wireless broadcast [9][10][11]. Several kinds of internal index structures are introduced so that clients can skip the irrelevant parts of data in evaluation. Firstly, based on these schemes, the client does not have any knowledge of how many documents actually satisfy his current request and has to monitor the stream all the while to retrieve the interesting parts of data. Secondly, these approaches index each document separately which prolongs the broadcast cycle and hence weakens the adaptability of on-demand mode. Moreover, they mainly focus on designing index schemes for single path queries.

Other works study the aggregation of the content of XML documents. Ref. [12] designs a new structure called RoxSum to describe the summary of the structure information of multiple XML documents. The VA-RoxSum [13] is also proposed to aggregate both structure and values information of multiple XML documents. These works consider only single path queries. Due to the loss of most branching information, they can not apply to twig queries directly.

3   Air Index Scheme for Twig Queries

In on-demand XML data broadcast, the broadcast server first collects mobile users' queries which request some XML data on the server, and then schedules the content to be broadcasted on the wireless channel according to the collected queries and builds an air index of the broadcasted content. In this section, we discuss the novel air index scheme for twig queries in detail.

3.1   The Basic Index Structure: Document Tree

Similar to DataGuide[6], the Document Tree structure keeps only the tree structure information of an XML document and removes all values and other content. The difference is Document Tree reserves all the original branching structure information which is necessary for processing twig queries while DataGuide keeps only single path structure information. Fig.1 shows examples of the Document Tree structure DT and its corresponding DataGuide DG.

Due to the loss of branching structure information, the DataGuide structure can not support twig queries. For example, given two twig queries {$q1$:/a/c[b]/b, $q2$:/a/b[c]/d}, according to DT, it is obvious that the original XML document of Fig.1 satisfies $q1$ but does not satisfy $q2$ since DT contains complete structure information of the original XML document. However, according to DG, $q1$ is not satisfied while

$q2$ is satisfied. Therefore, the DataGuide structure can not support twig queries and we should adopt Document Tree as the basic air index structure. As all the values and other contents have been removed, the Document Tree structure is usually much smaller than the original XML document and is much smaller than the index structures proposed in [7][8].

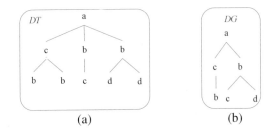

Fig. 1. Examples of document tree(*DT*) and DataGuide(*DG*)

3.2   Pruning Redundant Structures of Document Tree

The Document Tree index structure keeps all the branching structure information of the original XML document and can support twig queries effectively. However, according to a given set of mobile users' queries, there exist redundant structures in the complete Document Tree. The reasons are: 1) Every query only requests a small part of the original XML document and thus only matches a small part of structure of the Document Tree; 2) Since the twig queries contain branching structure constraints, the number of successful matching between the queries and the Document Tree decreases; 3) The queries has less possibility that they successfully match the deeper element nodes than that they successfully match the shallower element nodes. Therefore, we need to prune redundant structures of the Document Tree of an original XML document to reduce the size.

Fig.2 shows the examples of pruned Document Trees for two user queries. There is a set of two Document Trees with identifiers *DT1*, *DT2* in the figure. A set of two mobile users' queries {/a//b, /a/*[c]/c}, namely *Q* is also shown in the figure. According to query set *Q*, the pruning results are *PDT1* and *PDT2*. First, we prune redundant structures of *DT1* according to *Q*. Because *DT1* only satisfies query $q1$ in set *Q* but does not satisfy $q2$, we only reserve the necessary part of structure of *DT1* for $q1$. The pruning result is *PDT1* which reserves only the path prefix {/a/b} and prunes the unnecessary deep element nodes. Note that, the three label paths {/a/b/c, /a/b/d, /a/b/d} all satisfy $q1$ but we only reserve the necessary prefix that already satisfies $q1$. Furthermore, the two longer label paths {/a/c/b, /a/c/b} also satisfies $q1$, but in order to further reduce the size of the air index, we abandon the longer paths. Then we prune redundant structures of *DT2* according to *Q*. *DT2* satisfies both $q1$ and $q2$. We only reserve the left subtree of *DT2* that satisfies $q2$. Moreover, this structure already satisfies $q2$, thus we prune all the rest part of *DT2*. The pruning result is *PDT2*. The two pruning results only reserve the necessary part of structures of *DT1* and *DT2*. Obviously, they are concise and accurate since we can get the correct matching results from the pruned structures.

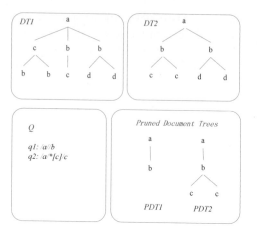

Fig. 2. Examples of pruned document trees

In general, each twig query matches more than one branch because the twig queries contain branching structure constraints. For a given set of twig queries, the queries may possibly share some branches that satisfy different branching structure constraints of them at the same time. For example, given two twig queries $q1$ and $q2$ and a Document Tree index has five branches $\{b1, b2, b3, b4, b5\}$. Suppose that branch groups $\{b1, b2\}$ and $\{b2, b3\}$ both can completely satisfy $q1$, while branch groups $\{b1, b3\}$ and $\{b4, b5\}$ both can completely satisfy $q2$. In this case, $\{b2, b3, b4, b5\}$ is a set of necessary branches that completely satisfies $q1$ and $q2$, because after removing any one branch in it, it can not completely satisfy both $q1$ and $q2$. However, it is not optimal, and the optimal set of necessary branches is $\{b1, b2, b3\}$ as it can completely satisfy $q1$ and $q2$ and includes the least branches.

We put forward a heuristic algorithm to prune redundant structures efficiently in the following. Our algorithm visits the element nodes in the Document Tree index DT in a *breadth-first-search* (BFS) way, and checks the path from root node to the current visiting node to see if the path can satisfy some single path queries or partially satisfy some branching structure constraints of some twig queries in Q. If some single path queries are satisfied by the current path, we mark all the element nodes of the current path as "*reserved*", which means all these nodes will not be pruned. And then we mark the satisfied queries as "*satisfied*" and will not process these queries any more. If some branching structure constraints of some twig queries are satisfied, we need to cache the current path first. The reason is that the twig queries contain several branching structure constraints and we can not check whether all the constraints have been satisfied only according to the current path. Then we check if any twig queries can be completely satisfied by the cached paths. If a twig query is completely satisfied by the cached paths, we mark all the element nodes of the corresponding paths as "*reserved*", and mark the satisfied twig query as "*satisfied*". After all element nodes in DT have been visited, all the element nodes that have been marked as "*reserved*" will not be pruned, while other element nodes are redundant and should be pruned. The pruning result is PDT. Similarly, the queries that have been marked as "*satisfied*" can be satisfied by DT, while other queries can not be satisfied by DT. Since all necessary structures are reserved by our algorithm, the pruning result *PDT* contains enough structure information to index the original XML document for the given query set *Q*. Particularly, because the algorithm visits and processes the element nodes in *DT* in a BFS way, we can guarantee that we have reserve the shortest paths that satisfy or partially satisfy some queries in *Q*. We also reserve the paths that satisfy the most branching structure constraints of twig queries (Step 10). Therefore, we always reserve the more desirable paths and thus can prune more redundant structures.

If the Document Tree index *DT* has *n* element nodes, then the computing complexity of our pruning algorithm is O(*n*).

Pruning Algorithm:

Input: Document Tree index *DT*; user query set *Q*.
Output: Document Tree index after pruned *PDT*.
Algorithm:
  1. visit nodes in *DT* in a *breadth-first-search* way;
  2. for each node *e* not marked "*reserved*" in *DT*
  3.   if the path *root*-to-*e* satisfies any single path queries in *Q* which have not been marked "*satisfied*", then
    4.   mark all nodes of the path as "*reserved*";
    5.   mark the satisfied single path queries as "*satisfied*";
    6.   if any twig queries in *Q* which have not been marked as "*satisfied*" are partially satisfied by the path *root*-to-*e*, then
      7.   cache all the path *root*-to-*e*;
      8.   check all cached paths;
      9.   if there are any twig queries are completely satisfied by the cached paths, then
      10.   mark all the element nodes of the corresponding path as "*reserved*"; if there are more than one cached paths that satisfy the same branching structure constraints of a twig query, mark all the element nodes of the path that satisfies the most branching structure constraints as "*reserved*"; otherwise, randomly mark all the element nodes of one of the paths as "*reserved*";
      11.   mark the satisfied twig queries as "*satisfied*";
  12. prune all nodes in *DT* that have not been marked as "*reserved*"; the pruning result is *PDT*.

3.3 Combining Pruned Document Tree Indexes

Note that, the Document Tree indexes of different original XML documents usually share many prefixes. As a result, if we broadcast these indexes one by one, mobile users need to process the same prefixes many times in different indexes to find out if their queries can be satisfied by any XML documents on the wireless channel. Based on this observation, we propose to combine the separate pruned indexes for each original XML document in order to eliminate the redundancy among the indexes.

Fig.3 shows an example of a combined index which combines the two pruned Document Tree indexes *PDT1* and *PDT2* shown in Fig.2. Since the two pruned indexes share the prefix {/a/b}, we should represent the prefix only one time in the combined index. Note that, as described in [12], identifying the leaf nodes of all the root-to-leaf paths with a document *ID* is enough to imply all internal path nodes of that document. We adopt the same labeling method to identify all the leaf nodes in the combined index in our solution. Suppose that *DT1* and *DT2* shown in Fig.2 are the indexes of the original XML documents *D1* and *D2*, and then the labeling result of the combined index can be shown in Fig.3. Obviously, the total number of document *IDs* in the combined index equals to the total number of leaf nodes in all separate pruned structure indexes.

The query $\{q1:/a//b\}$ shown in Fig.2 needs to process the prefix $\{/a/b\}$ in the combined index only one time in order to get the matching result. In this example, the prefix $\{/a/b\}$ satisfies query *q1*, therefore, according the labeling scheme of the combined index, both XML documents *D1* and *D2* satisfy query *q1*.

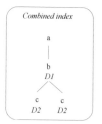

Fig. 3. Combined index of *PDT1* and *PDT2* in Fig.2

The combining process of pruned Document Tree indexes is different from that of RoxSum which is described in [12] because we must reserve all the branching structure contained in the separate pruned indexes to support twig queries. In this section, we discuss the problem of how to combine the pruned Document Tree indexes optimally.

First, in order to simplify our discussion, we only consider the simpler structures that only contain single paths from the root node to leaf nodes without any other branches, that is all nodes has at most one child except for the root node. We show that this kind of structure indexes can be equivalently converted to complete weighted bipartite graphs.

An example of this kind of special structure indexes is shown in Fig.4. Index *DT1* and *DT2* are two special structure indexes to be combined. They both have three single path branches, namely p_1, p_2, p_3 and p_4, p_5, p_6, respectively. The matching results between p_1, p_2, p_3 and p_4, p_5, p_6 can be converted to a weighted complete weighted bipartite graph. The weight of each edge of the bipartite graph equals to the number of matching nodes between the related single path branches. For example, the number of matching nodes between p_1 and p_5 is 3 and hence the weight of edge (p_1, p_5) in the bipartite graph is 3 as well. The right part of Fig.4 shows the result of the converted complete weighted bipartite graph. Its maximum matching of given weight is shown as the bold and grey edges. This maximum matching also indicates that the optimal combination of index *DT1* and *DT2* is: p_1 combining with p_5, p_2 combining with p_6, and p_3 combining with p_4.

The best known strongly polynomial time bound algorithm for weighted bipartite matching is the classical Hungarian method presented in [14], which runs in $O(|V|*(|E|+|V|*\log|V|))$ time. As to complete weighted bipartite graph, suppose the numbers of nodes in the two vertex sets are m, n, respectively, then we have $|V|=m+n$, $|E|=m*n$ and the maximum matching of given weight in complete bipartite graph has a computing complexity of $O(m^2*n+m*n^2)$. In other words, if two structure indexes which contains only single paths from root node to leaf nodes and the numbers of single paths in the two indexes are m, n, respectively, the optimal combination can be found in $O(m^2*n+m*n^2)$ running time. Particularly, if the numbers of single paths in the two indexes are both n, then the running time will be $O(n^3)$.

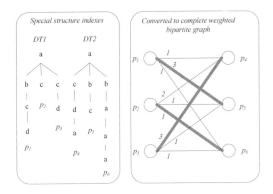

Fig. 4. Examples of single path structure indexes

Nevertheless, the general combining process of the pruned indexes is a little more complicated than the special cases. When combining single paths indexes, we can get the number of matching nodes of two single paths easily by comparing element nodes of them from the root node one by one. However, when combining some branching structures, we need to find out the largest number of matching nodes of their subtree structures first. For example, in Fig.5, if we want to find out the optimal combinations of the five subtrees of two root nodes "a" in the figure (these subtrees have been labeling with numbers from 1 to 5), we need to find out the optimal combinations of subtrees of the nodes in Layer2. In this example, the approach to find out the optimal combinations of subtrees of the nodes in Layer2 is exact the same as the single paths structures shown in Fig.4 because all nodes in Layer2 are single paths. In other words, we can use Hungarian method to find out the optimal combinations between nodes 1, 2, 3 and nodes 4, 5 in Layer2. According to these combining results, the root nodes in Layer1 now can use Hungarian method to find out the optimal combinations as well. The optimal combinations are: the two root nodes in Layer1 will be combined; and in Layer2, node 2 will be combined with node 4, and node 3 will be combined with node 5; finally, in Layer3, the child node c of node 2 will be combined with the child node c of node 4 and similarly, the child nodes d of node 3 will be combined with the child nodes of node 5.

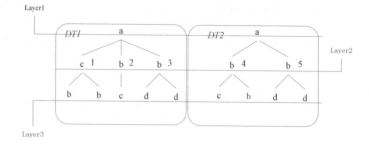

Fig. 5. Two Document Tree indexes to be combined

Therefore, we can find out the optimal combination of two given Document Tree indexes by recursively using the classical Hungarian method and then computing the optimal combinations of subtree structures in the Bottom-up way.

3.4 Two-Tier Structure of Air Index

Generally, traditional air indexes are one-tier structures. They index the corresponding data items based on key-based values. The keys and the data items are one-to-one relationships and thus the one-tier structures are efficient. However, in XML data broadcast, XML data are semi-structured and the air indexes are not key-based but structure-based. One-tier structures are inefficient because the combined air index and the XML documents on the wireless channel are one-to-many relationships. Based on this observation, we propose the two-tier structure to optimize the combined air index. Fig.6 shows examples of the one-tier structure and the two-tier structure.

As shown in the Fig.6(a), the one-tier structure uses the offsets to indicates the positions of XML documents on the wireless channel. $O1$ and $O2$ are the offsets of the two XML documents which will be broadcasted on the wireless channel. According to the air index and the offsets, mobile users can get position information of the required XML documents and then switch to *doze mode* until the documents arrive. Fig.6(b) shows an example of the two-tier structure of air index. In Fig.6(b), $D1$ and $D2$ are the document IDs of the XML documents on the wireless channel, and similarly, $O1$ and $O2$ are the offsets of the two XML documents. Mobile users can get the document IDs of the XML documents which satisfy their queries from the first tier. Then mobile users can get the offsets in the second tier according to the document IDs.

Note that, the length of the offset limits the maximum length of the XML documents which can be indexed by an air index. When the length of the offset is 3 bytes, an air index can index up to the maximum length of 16M bytes and when the length of the offset is 4 bytes, an air index can index up to 4G bytes which is quite enough for a wireless broadcast system. Therefore, the preferable length of the offset should be 4 bytes. On the other hand, when the length of document ID is 2 bytes, the maximum number of XML documents is up to 64K. Generally, this is enough to represents all the XML documents on the broadcast server. Based on this assumption, the total size of offsets of the air index in Fig.6(a) is 9*4=36(bytes) and the total size of offsets including the document IDs of the air index in Fig.6(b) is 9*2+2*(2+4)=30(bytes). Therefore, the two-tier structure can further reduce the total size of the air index and thus reduce energy consumption.

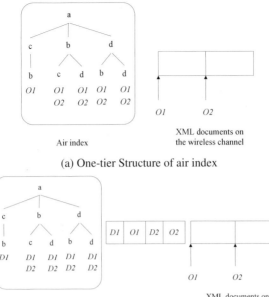

(a) One-tier Structure of air index

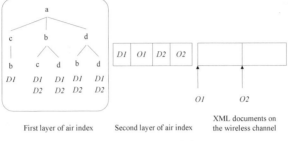

(b) Two-tier Structure of air index

Fig. 6. One-tier structure *vs.* Two-tier structure

In theory, in a wireless broadcast system, suppose there are N_{doc} XML documents that will be broadcasted on the wireless channel, the total number of leaf nodes in the pruned Document Tree indexes is N_{leaf_node}, the length of an offset is L_O bytes, and the length of a document *ID* is L_D bytes, then the total size of offsets in the one-tier structure is $N_{leaf\_node} * L_O$, and the total size of offsets including the document *ID*s in the two-tier structure is $N_{leaf\_node} * L_D + N_{doc} * (L_D + L_O)$. By comparing these two sizes we can infer that when we have

$$\frac{N_{leaf\_node}}{N_{doc}} > \frac{L_O + L_D}{L_O - L_D}$$

the total size of two-tier structure is smaller than one-tier structure. Generally, we have $L_O=4$(bytes), $L_D=2$(bytes), then as long as we have

$$\frac{N_{leaf\_node}}{N_{doc}} > 3$$

the total size of two-tier structure is smaller than one-tier structure. Note that, this condition is easily satisfied since in our experiments, we have

$$\frac{N_{leaf\_node}}{N_{doc}} > 20$$

in most cases. Hence, the two-tier structure is more efficient.

3.5 Accessing XML Data on Wireless Channel

On the mobile client end, the client first downloads the air index. Then it can find out which documents satisfy its queries and when it can download them. Since the air index supports the current submitted user queries without losing any precision, the mobile clients can correctly find out all the documents they require.

Due to the combining process, the matching process between the combined index on air and user queries on the mobile client ends is a little different from traditional query processing. First, for single path queries, the mobile clients need to confirm that which documents can satisfy their requests. The reason is that the internal path nodes of the combined index can only be identified by their corresponding leaf nodes and the mobile clients need to follow the prefixes of the combined index that satisfy their queries to finally get the corresponding document *ID*s. Second, for twig queries, the mobile clients need to check whether all the satisfying paths of a twig query belong to the same documents, because all branching structure constraints of a twig query should be satisfied by the branches from the same XML document; otherwise, the twig query has not been completely satisfied.

4 Experimental Evaluation

In this section, we first describe the experimental setup of our experiments and then study the performance of the pruning technique and the combining technique for the Document Tree index structure. We also study the comparison between two-tier structure and one-tier structure. Finally, we study the overall performance of our air index scheme.

4.1 Experimental Setup

In our experiments, synthetic XPath queries are generated using the generator in [15]. Experiments are run on a synthetic data set: News Industry Text Format (NITF) DTD, and 500 XML documents are generated. The average depth of all documents is about 8. Experiments on another synthetic data set on NASA DTD are also performed. We just report the previous ones, as the results for NASA are similar.

Two parameters are varied in the experiments: the number of queries (N_Q), and the probability of nested paths (P_{NP}). The descriptions of them are shown in the following Table 1.

Table 1. Workload parameters for our experiments

Parameter	Range	Default Value	Description
N_Q	100 to 1 000	500	Number of queries
P_{NP}	0 to 30%	10%	Probability of a nested path occurring at a location step

4.2   The Performance of Pruning Algorithm

We first define *Pruning Ratio(PR)* as follows:

$$PR = \frac{\text{size of original air index - size of air index after pruned}}{\text{size of original air index}} \times 100\%$$

therefore, if the *PR* is higher, we will prune more redundant structures and the performance of our pruning algorithm will be better; otherwise we will prune fewer redundant structures.

Fig.7 depicts the effect of N_Q. As the N_Q increases, the *PR* decreases because more and more structures are needed by the user queries. When N_Q =100, *PR*=80% and when N_Q =1 000, *PR*=63%. Therefore, our pruning algorithm can prune a large part of the original air index. Moreover, the *PR* decreases not so fast when N_Q is larger since the user queries share more and more structures with the other user queries.

Fig.8 shows the effect of P_{NP}. As the P_{NP} increases, the *PR* increases as well. This is because the selectivity of user queries decreases as P_{NP} increases and fewer user queries are satisfied and fewer structures are needed by the user queries. When P_{NP}=0%, *PR*=59%; and when P_{NP}=30%, *PR*=68%. As a result, our pruning algorithm can support twig queries efficiently.

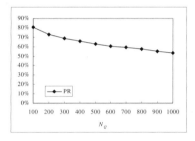

Fig. 7. Effect of N_Q

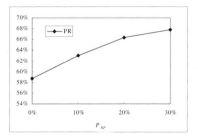

Fig. 8. Effect of P_{NP}

4.3   The Performance of Combining Algorithm

We first define *Combining Ratio(CR)* as follows:

$$CR = \frac{\text{size of pruned air index - size of combined air index}}{\text{size of pruned air index}} \times 100\%$$

therefore, if the CR is higher, we will combine more sharing prefixes and the performance of our combining algorithm is better; otherwise we will combine fewer sharing prefixes. The combining algorithm is applied to the pruned indexes directly.

Fig.9 depicts the effect of N_Q. Similar to Fig.7, as the N_Q increases, the CR decreases. More structures are needed as N_Q increases, and then they likely share more prefixes. Thus the CR decreases not as fast as the PR compared with Fig.7. When N_Q =100, CR=91% and when N_Q =1 000, CR=78%. Therefore, our combining algorithm can eliminate redundancy among the separate air indexes effectively. Similarly, the CR decreases not so fast when N_Q is larger since the user queries share more and more structures with the other user queries.

Fig.10 shows the effect of P_{NP}. As the P_{NP} increases, the CR increases as well. Similarly, this is because the selectivity of user queries decreases as P_{NP} increases. Thus, fewer user queries are satisfied and fewer structures are needed by the user queries. When fewer user queries are satisfied, it also indicates that the needed structures of air index share more prefixes. When P_{NP}=0%, CR=78%; and when P_{NP}=30%, CR=85%. Therefore, our combining algorithm can support twig queries efficiently.

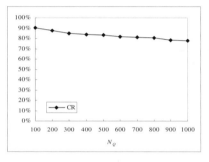

Fig. 9. Effect of N_Q

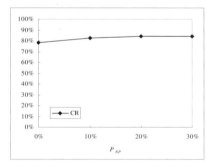

Fig. 10. Effect of P_{NP}

4.4 The Comparison between Two-Tier Structure and One-Tier Structure

Fig.11 shows the comparison between two-tier structure and one-tier structure. As the N_Q increases, more structures are needed and thus N_{leaf_node}, which has been discussed in Section 3.4, increases as well. As a result, the sizes of both structures increase. As can be seen from the figure, two-tier structure is smaller than one-tier structure. When N_Q =100, two-tier structure is 25% smaller than one-tier structure and when N_Q =1 000, two-tier structure is 30% smaller than one-tier structure. Therefore, two-tier structure is more preferable.

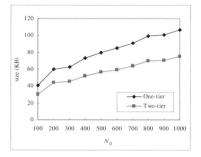

Fig. 11. Effect of N_Q

4.5 The Performance of Our Air Index Scheme

We have tested 500 XML documents in our experiments of which the total size is 6200K bytes. From Fig.11, we can see that our air index scheme builds only tens of K bytes indexes which are only about 1% of the original XML documents. Moreover, the total size of the complete Document Tree indexes of the original XML documents is about 665K bytes in our experiments. In other words, after using our air index scheme to prune redundant structures and to combine sharing prefixes, the final air index is only about 9% of the complete structure indexes. As a result, our air index scheme is quite effective and efficient and can reduce energy consumption significantly.

5   Conclusions

In this paper, we focus on designing effective and efficient air index scheme for twig queries in on-demand XML data broadcast. Twig queries are more powerful and more accurate to express user interests in XML data than single path queries. We have adopted the Document Tree structure as the basic air index structure for twig queries. We have proposed to prune redundant structures of the basic Document Tree indexes of all XML documents to reduce the energy consumption and have designed an efficient heuristic algorithm to prune redundant structures of indexes. We have also proposed to combine the sharing prefixes of the pruned indexes to further eliminate structure redundancy among the indexes.

In our experiments, the pruning algorithm can prune more than one half of the original Document Tree indexes in most cases and our combining algorithm shows great effectiveness as well since the combining ratio is usually larger than 80%. Our preliminary experiments show that our air index scheme is quite effective and efficient, as it builds comparatively small indexes and supports twig queries without losing any precision. In most cases, our scheme builds indexes that only reserve 9% of the complete structures information or only 1% of the original XML documents information.

Acknowledgments

This research is supported in part by the National Natural Science Foundation of China (NSFC) under grant 60503035, 60703093, and the National High-Tech Research and Development Plan of China under Grant 2006AA01Z234 and SRF for ROCS, SEM.

References

1. Imielinski, T., Viswanathan, S., Badrinath, B.R.: Data on Air: Organization and Access. IEEE Transactions on Knowledge and Data Engineering 9(3) (1997)
2. Xu, J., Lee, D., Hu, Q., Lee, W.C.: Data Broadcast. In: Handbook of Wireless Networks and Mobile Computing. John Wiley & Sons, Chichester (2002)
3. Acharya, S., Alonso, R., Franklin, M., Zdonik, S.: Broadcast disks: Data management for asymmetric communications environments. In: SIGMOD (1995)
4. Acharya, S., Muthukrishnan, S.: Scheduling On-Demand Broadcasts: New Metrics and Algorithms. In: MOBICOM 1998 (1998)
5. Sun, W., Shi, W., Shi, B., Yu, Y.: A Cost-Efficient Scheduling Algorithm of On-Demand Broadcasts. ACM Journal of Wireless Networks 9(3) (2003)
6. Goldman, R., Widom, J.: DataGuides: enabling query formulation and optimization in semistructured databases. In: VLDB 1997 (1997)
7. Kaushik, R., Bohannon, P., Naughton, J.F., Korth, H.F.: Covering indexes for branching path queries. In: SIGMOD Conference, pp. 133–144 (2002)
8. Wang, W., Wang, H., Lu, H., Jiang, H., Lin, X., Li, J.: Efficient Processing of XML Path Queries Using the Disk-based F&B Index. In: VLDB 2005, pp. 145–156 (2005)
9. Park, C., Kim, C., Chung, Y.: Efficient Stream Organization for Wireless Broadcasting of XML Data. In: Grumbach, S., Sui, L., Vianu, V. (eds.) ASIAN 2005. LNCS, vol. 3818, pp. 223–235. Springer, Heidelberg (2005)
10. Park, S., Choi, J., Lee, S.: An effective, efficient XML data broadcasting method in a mobile wireless network. In: Bressan, S., Küng, J., Wagner, R. (eds.) DEXA 2006. LNCS, vol. 4080, pp. 358–367. Springer, Heidelberg (2006)
11. Chung, Y., Lee, J.: An indexing method for wireless broadcast XML data. Information Sciences 177(9), 1931–1953 (2007)
12. Vagena, Z., Moro, M.M., Tsotras, V.J.: RoXSum: Leveraging Data Aggregation and Batch Processing for XML Routing. In: ICDE (2007)
13. Vagena, Z., Moro, M.M., Tsotras, V.J.: ValueAware RoXSum: Effective Message Aggregation for XMLAware Information Dissemination. In: WebDB 2007 (2007)
14. Kuhn, H.W.: The hungarian method for the assignment problem. Naval Research Logistics Quarterly, 83–97 (1955)
15. Diao, Y., Altinel, M., Franklin, M.J., Zhang, H., Fischer, P.: Path Sharing and Predicate Evaluation for High-Performance XML Filtering. TODS 28(4) (2003)

Semantic Fields: Finding Ontology Relationships

Ismael Navas-Delgado, Maria del Mar Roldán-García, and José F. Aldana-Montes

E.T.S.I. Informática. Computer Languajes and Computing Science Department,
Boulevard Louis Pasteur 35, 29071 Málaga, Spain
{ismael,mmar,jfam}@lcc.uma.es

Abstract. This paper presents the Semantic Field concept which enables the global comparison of ontologies. This concept uses the results obtained from ontology matching tools to estimate the global similarity between ontologies. It has been used in a tool called the Semantic Field Tool (SemFiT). A demo tool is provided (http://khaos.uma.es/SFD).

Keywords: Ontology Alignment, Ontology Distance.

1 Introduction

In the Semantic Web, the notion of ontology as a form of representing a particular universe of discourse (or some part of it) is very important. Ontology alignment is a key aspect of knowledge exchange in the Semantic Web; it allows organizations to model their own knowledge without having to stick to a specific standard. In fact, there are two good reasons why most organizations are not interested in working with a standard for modeling their knowledge: (a) it is very difficult or expensive for them to reach an agreement about a common standard, and (b) standards reached do not often fit in with the specific needs of the all participants in the standardization process.

Ontology alignment is perhaps the best way to solve the problems of heterogeneity. There are a lot of techniques for accurately aligning ontologies, but experience tells us that the complex nature of the problem to be solved makes it difficult for these techniques to operate satisfactorily for all kinds of data, in all domains, and as all users expect. This problem has been studied in several works [1][2][3].

Thus, interoperability relies on the ability to reconcile different existing ontologies, which may have overlapping or closely related domains. This reconciliation depends on the existence of ontology relationships, to relate terms in different possibly multiple ontologies in overlapping or related domains. However, to obtain the best results, we need to know how the ontologies overlap so that we can align them appropriately, or merge them into a new ontology.

In this context, we have proposed the concept of Semantic Field, which goes a step further in the search for ontology relationships. This concept uses the results obtained from ontology matching tools to estimate the global similarity between ontologies. Thus, users and applications can discover the global configuration of relationships between existing ontologies.

In this paper we present the Semantic Field concept through a framework for searching ontology relationships: the Ontology Matching and Alignment Framework

S.S. Bhowmick, J. Küng, and R. Wagner (Eds.): DEXA 2009, LNCS 5690, pp. 427–434, 2009.

(OMAF). This framework has been instantiated as a tool called the Semantic Field Tool (SemFiT, http://khaos.uma.es/SFD), to facilitate ontology alignment in the Semantic Web. It uses MaF (Matching Framework, http://khaos.uma.es/maf) to find mappings between ontologies and allows the insertion of existing matching algorithms. Finally, we have studied several application types that have been developed as real applications to validate this tool: the Ontology Search Engine (OSE) and the Semantic Web Services Automatic Composition and Matchmaking (SWS-ACoM) [4].

2   OMAF: The Ontology Mapping and Alignment Framework

The reuse of these ontologies enables the development costs of Semantic Web applications to be kept to a minimum. In this context it is necessary to use a mechanism to reduce the number of ontologies that a software developer, a software agent or a Semantic Web application have to review. In this section we present the concept of Semantic Field applied to the Semantic Web context, in which we will identify the ontologies related with our needs. Semantic Fields can be used in semantic Web applications to relate ontologies, and their application can help to reduce the search space required to find ontologies useful for solving user requests. Thus, users will be able to calculate which ontologies are useful in a specific domain and how they are related to enable semantic Web application interoperability.

2.1   Definitions

The **Semantic Neighbourhood** of an ontology, namely the pivot ontology, is not exactly a set of aligned ontologies. It is neither a set of integrated ontologies nor a new ontology obtained from the merging process of a set of ontologies. Rather, the Semantic Neighbourhood of an ontology is a set of ontologies, which is built based on the distance from the pivot ontology to the other known ontologies. Depending on the perspective (the pivot ontology) and the radio (maximum distance from the pivot ontology to all the ontologies in the semantic neighbourhood), the semantic neighbourhood will be composed of different ontologies (see example shown in Figure 1 right). Figure 1 (right) represents the neighbourhood, $SN_{O_X}^Y$, of the ontology X with a radio Y. Furthermore, ontologies can belong to different neighbourhoods, depending on their distance to pivot ontologies.

The **Semantic Field** is a more specific concept based on the user perspective and it is derived from the Semantic Neighbourhood. Ontologies from outside the Semantic Neighbourhood can be included in the user Semantic Field by means of concepts defined as relevant for a specific user (or user community). However, ontologies from the Semantic Neighbourhood cannot be excluded from the Semantic Field, because of the way in which the Semantic Field has been defined and then calculated. Thus, several ontologies that are not in the user's Semantic Neighbourhood are included in the Semantic Field if the concepts relevant for the user are related to these ontologies (see an example in Figure 1 left). This figure shows how an ontology can be part of a Semantic Field for a set of relevant concepts and/or roles of the ontology O_X (Q) and a Radio (Y), $_Q SF_{O_X}^Y$.

Therefore, the Semantic Field is a set of relevant ontologies for the user semantic focus.

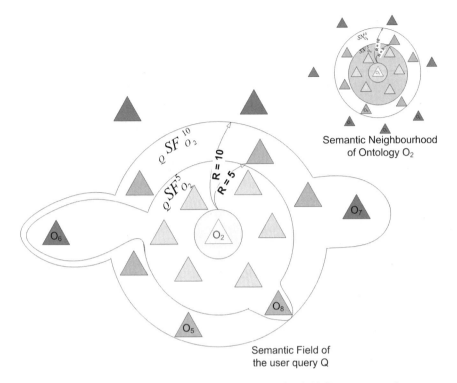

Semantic Neighbourhood
of Ontology O_2

Semantic Field of
the user query Q

Fig. 1. The Semantic Neighbourhood and Semantic Field Concept examples

2.2 SemFiT

This section presents the instantiation of the Ontology Matching and Alignment Framework (OMAF) as a final tool (the Semantic Field Tool, SemFiT). SemFiT has two main target users: ontology (and mapping) providers and final users. Ontology providers have as their main goal to make their ontologies publicly available. On the other hand, final users will make use of the registered information to locate ontologies and their relationship with other similar ontologies. Semantic Field can be calculated without determining which concepts are relevant for the user, but the best way to benefit from Semantic Fields is to set the relevant ontology and its relevant concepts.

In order to calculate the Semantic Field a matrix of mappings is required between registered ontologies (O_1, ... , O_s). Thus, both the implementations of the Semantic Field concept and its main interface require including mechanisms to calculate relationships between pairs of ontologies. The process of calculating these mappings could be manual, semi-automatic or automatic (it uses MaF [5]). Using this matrix the tool calculates the distance between pairs of ontologies. The target of the ontology distance is to find a global measurement to compare ontologies, based on concept to concept similarities.

The use of different formulas to calculate the Semantic Field will produce different results, from a detailed point of view (ontology to ontology distance), but there will be fewer differences when comparing a lot of ontologies, and when the user needs to locate ontologies by reducing the search-space. In our proposal the distance has been calculated using the following formulas (these formulas has produced promising results in previous tests [6]):

- $$DD(O_i, O_j) = \frac{\#Concepts(O_i)}{\displaystyle\sum_{c \in Concepts(O_i)} \max(mappings(c, O_j))}$$

- $$D(O_i, O_j) = \min(DD(O_i, O_j), DD(O_j, O_i))$$

The distance values provide hints of the mappings that can be found between ontologies, so the use of different formulas will affect the distance between two ontologies but it will not affect the global selection of ontologies (in as much as all the distances have been calculated in the same way).

At query time, it is possible for the user to define relevant concepts from the pivot ontology. In this case we can calculate the weight-based ontology distance, making use of ontology distance measurements which use this knowledge. Thus, the mappings involving these relevant concepts will reduce the distance between the registered ontologies. In this way the directed distance (DD) can be re-defined as:

- $$DD(O_i, O_j) = \frac{\#concepts(O_i)}{\displaystyle\sum_{c \in Concepts(O_i)} \max(R_F(c) * mappings(c, O_j))}$$

This formula is based on a relevance factor ($R_F(c)$), which indicates the relevance of each concept (set up by the user to indicate his/her preferences), and it should be noted that the distance formula is not changed. Non-relevant concepts will have $R_F(c)=1$ and relevant concepts will have $R_F(c)>1$. Thus, if we establish a factor value of 2, then the relevant concepts will be twice as important in the calculation. For example (Figure 2), given two ontologies twelve elements and three elements respectively, in which only one pair of concepts is similar (similarity value of 0.9), the directed distances are:

- $$DD(O_1, O_2) = \frac{3}{0.9} = 3.3$$

- $$DD(O_2, O_1) = \frac{12}{0.9} = 13.3$$

If we use Formula 2, then:

- $$D(O_1, O_2) = D(O_2, O_1) = 3.3$$

However, if one of the pairs of concepts that is similar is relevant ($R_F = 2$), then the directed distances are smaller:

- $$DD(O_1, O_2) = \frac{3}{1.8} = 1.6$$

- $$DD(O_2, O_1) = \frac{12}{1.8} = 6.6$$

And, for this case:

$$D(O_1, O_2) = D(O_2, O_1) = 1.6$$

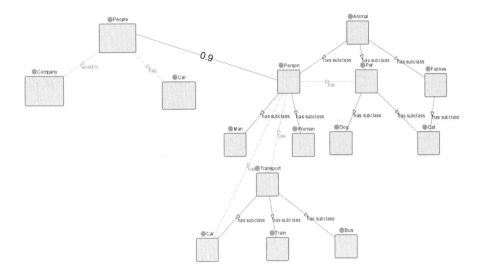

Fig. 2. Alignment Example

2.3 SemFiT Demo Tool

In order to show the capabilities of Semantic Field we have developed a demo tool for using the terms described in this section with a set of pre-calculated ontologies (available at http://khaos.uma.es/SFD/index.jsp).

Once registered (registration is free and is only used to trace the use of the demo tool), and logged into the system, users can view all the registered ontologies and their distances or select one of them as his/her relevant ontology and can also view all the others (Figure 3 top-left). Visualization of all the ontologies involves a view of the lines indicating distances between them (Figure 3).

However, the initial visualization of all the ontologies is not clear when we have a large set of ontologies. Thus, the visualization tool provides an option for selecting a threshold to create groups of ontologies (those that are at a distance less than a particular threshold), so it is easy to know which ontologies are more closely related (Figure 3 top-right).

The next step that a user can perform to locate an ontology is to select one of those registered as his/her relevant ontology, and filter ontologies outside its semantic neighbourhood. Once the relevant ontology has been selected (Figure 3) the user has a new item to be used, "View Neighbourhood", which will return a graph with a reduced number of ontologies.

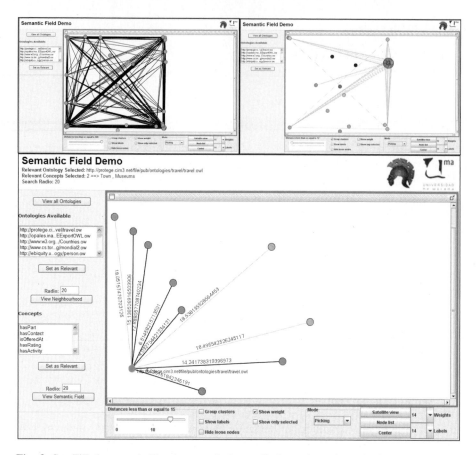

Fig. 3. SemFiT demo tool. The demo tool shows all the registered ontologies (top-left), and they can be organized in clusters (top-right). However, the main functionality is to calculate the semantic field for a given ontology, a radius and a set of relevant concepts (down).

The last step for selecting the ontologies related with the user preferences is to use the Semantic Field by selecting those concepts in the relevant ontology that are most important to the user. This will mean that the distance between some pairs of ontologies will be reduced (lower part of Figure 3). In this case the concepts chosen are related to all the ontologies, so all the distances have been reduced.

The example presented in this section provides a view of the potential use of Semantic Fields to locate ontologies, as a demo tool publicly available at http://khaos.uma.es/SFD/index.jsp. This tool shows how the use of Semantic Neighbourhood and Semantic Field concepts will produce a view of the available ontologies in which the user can easily locate ontologies.

3 Discussion

The use of ontologies for annotating information in Web documents and Semantic Web applications makes it necessary to locate existing ontologies in order to reuse them. The problem of finding a useful ontology for an application is that if such an ontology cannot be found, a new ontology will have to be developed (which is a costly process). Thus, several tools have been proposed for searching ontologies based on keyword searches, such as Swoogle and OntoKhoj [7]. However, the increasing number of available ontologies will produce longer lists of ontologies for each search (as for Google searches). In this way, other proposals like AKTiveRank [8], introduce the concept of ontology ranking into these searches. So, each search will produce a list of ontologies with a ranking number in order to help the user to decide which ontology is most useful for their requirements. The most sophisticated approach is Watson [9], which is a Semantic Web gateway that provides a single access to semantic information. It provides an API for finding Web documents, metadata, and entities inside these documents.

Ontology searches and ranking proposals do not take user knowledge in the domain into account. Thus, our proposal is to take advantage of the Semantic Fields to improve ontology searches. Semantic Fields can help in two contexts: when the user knows an ontology in the target domain and/or the user is able to express his/her knowledge as a simple ontology. In both cases the user can select the set of relevant concepts for his/her application.

In this paper we have also described how a dynamic, user-query centered concept (Semantic Field) has been implemented as a practical tool, SemFiT, for discovering ontology relationships in the Semantic Field of a user query related to a specific pivoting ontology. Semantic Fields can be used in different kinds of applications, such as ontology clustering and for ontology searches and location. SemFiT has been implemented to be parallelizable and with the aim of reducing the amount of calculus done at query time (when calculating the Semantic Field of a user query).

The calculation of a Semantic Field with a specific tool depends on being able to calculate mappings automatically. The use of different matching tools can provide different results. However, all the possible solutions will share the same characteristic: the Semantic Field provided is an indicator of the mappings that can be found. Furthermore, mappings that produce the Semantic Field could also be offered to the users.

Previously we asserted that using the Semantic Field benefits users by providing them with the ability to reduce the search-space in large sets of ontologies. In this context, the use of manual or semi-automatic mechanisms is not feasible because of the time required to produce or revise mappings. Thus, the methods proposed for registering ontologies assume that mappings will be automatically calculated by a mapping tool. However, in domains in which manual or semi-automatic mappings have been calculated previously, it is possible to add new methods to the interface to define the mappings for two registered ontologies (overwriting the automatic ones).

4   Conclusions

In this paper we have presented OMAF, a framework for developing tools based on the ontology relationships. This framework includes the use of the Semantic Field concept for measuring the ontology semantic distance (dissimilarity) between ontologies. The framework has been instantiated as a tool (SemFiT) that enables the global comparison of ontologies. This tool is available as a service for being used in application. Besides, we have provided an environment to test the Semantic Field concept using a graphical demo tool.

In SemFiT we have used MaF as a matching tool for calculating individual relationships between ontology concepts, and the information provided is used by the tool to calculate the ontology semantic distance (http://khaos.uma.es/maf).

The implementation of SemFiT is based on the use of a database for storing the information, so the calculation of Semantic Fields is a scalable process (thanks to the scalability of the relational database used).

Acknowledgments. Supported by TIN2008-04844 (Spanish Ministry of Education and Science), and the Junta de Andalucía project P07-TIC-02978.

References

1. Maedche, A., Staab, S.: Measuring similarity between ontologies. In: Gómez-Pérez, A., Benjamins, V.R. (eds.) EKAW 2002. LNCS, vol. 2473, pp. 251–263. Springer, Heidelberg (2002)
2. Ehrig, M., Sure, Y.: Ontology mapping - an integrated approach. In: Bussler, C.J., Davies, J., Fensel, D., Studer, R. (eds.) ESWS 2004. LCNS, vol. 3053, pp. 76–91. Springer, Heidelberg (2004)
3. Euzenat, J., et al.: State of the art on ontology alignment. Deliverable D2.2.3, Knowledge web NoE (2004)
4. Brogi, A., Corfini, S., Aldana, J.F., Navas, I.: Automated discovery of compositions of services described with separate ontologies. In: Dan, A., Lamersdorf, W. (eds.) ICSOC 2006. LNCS, vol. 4294, pp. 509–514. Springer, Heidelberg (2006)
5. MaF: the Matching Framework, http://khaos.uma.es/maf/
6. Navas, I., Sanz, I., Aldana, J.F., Berlanga, R.: Automatic generation of semantic fields for resource discovery in the semantic web. In: Andersen, K.V., Debenham, J., Wagner, R. (eds.) DEXA 2005. LNCS, vol. 3588, pp. 706–715. Springer, Heidelberg (2005)
7. Patel, C., et al.: Ontokhoj: a semantic web portal for ontology searching, ranking and classification. In: Proceedings of the 5th ACM international workshop on Web information and data management, pp. 58–61. ACM Press, New York (2003)
8. Alani, H., Brewster, C., Shadbolt, N.: Ranking ontologies with aktiverank. In: Proceedings of the International Semantic Web Conference, pp. 1–15 (2006)
9. Allocca, C., d'Aquin, M., Motta, E.: Finding equivalent ontologies in watson. In: Proceedings of the International Semantic Web Conference, Posters & Demos (2008)

Complete OWL-DL Reasoning Using Relational Databases

Maria del Mar Roldan-Garcia and Jose F. Aldana-Montes

University of Malaga, Departamento de Lenguajes y Ciencias de la Computacion
Malaga 29071, Spain
{mmar,jfam}@lcc.uma.es
http://khaos.uma.es

Abstract. Real Semantic Web applications, such as biological tools, use large ontologies, that is, ontologies with a large number (millions) of instances. Due to the increasing development of such applications, it is necessary to provide scalable and efficient ontology querying and reasoning systems. DBOWL is a Persistent and Scalable OWL reasoner which stores ontologies and implements reasoning using a relational database. In this paper we present an extension of DBOWL that implements all inference rules for OWL-DL. Furthermore, we describe briefly the reasoning algorithms and their completeness proofs.

1 Introduction

Semantic Web applications, such as biological tools, use large ontologies, that is, ontologies with a large number (millions) of instances. Description logic based tools, like Pellet [1] or RACER [2] allow us to manage OWL ontologies, but not very large ones. Reasoning algorithms are not scalable and are usually main memory oriented. These reasoners are highly optimized for reasoning on the ontology structure (Tbox reasoning in Description Logic nomenclature), but have problems when dealing with reasoning on instances (Abox reasoning). It is logical to think that applications in the Semantic Web will need to infer new knowledge from the explicit knowledge defined not only in the Tbox but especially in the Abox. The complex reasonings that should be implemented for the Semantic Web applications will need an optimal storage model, disk oriented, in order to be efficient and scalable. In the past few years there has been a growing interest in the development of systems for storing, querying and reasoning on large ontologies in the Semantic Web. Firstly, these systems were oriented to RDF storage [3] [4]. Nowadays, research is oriented to massive OWL storage. Several alternative approaches using relational technology have been presented. However, these proposals have some problems. On one hand, they are not complete with respect to OWL-DL reasoning. On the other hand, the performance and scalability of these tools is not satisfactory in some cases, particularly those which implement reasoning by means of datalog rules. Finally, the best tools are commercial tools, and users must pay to make use of them. In order to solve these problems, we have developed DBOWL, a persistent and

S.S. Bhowmick, J. Küng, and R. Wagner (Eds.): DEXA 2009, LNCS 5690, pp. 435–442, 2009.

scalable reasoner for very large OWL-DL ontologies. A preliminary version of DBOWL was presented in [5]. In this paper we present an updated version, which is more complete than the previous one. We also describe briefly the reasoning algorithms and their completeness proofs.

2   DBOWL

DBOWL [5] is a persistent and scalable OWL reasoner. It stores the OWL-DL ontologies in a relational database, and supports Tbox queries (queries on the ontology structure), Abox inferences (reasoning on the ontology instances) and ECQ (Extended Conjunctive Queries) queries [6]. Currently we are finishing a SPARQL query engine for DBOWL. In order to create the relational database for ontology storage, a Description Logic Reasoner is used. Thus, the consistency of the ontology as well as the inferences about the ontology structure is delegated to this reasoner and DBOWL focuses on reasoning on instances (large numbers of them). Both, Tbox queries and ECQ queries are implemented by translation to SQL [6]. Abox inferences are implemented by java functions and SQL views.

2.1   Storage in a Relational Database

DBOWL stores the OWL-DL ontologies in a relational database. We define a specific relational schema for our tool. Tables are categorized into 4 types: ontology information tables, Tbox tables, Abox tables, and descriptions tables.

Ontology information tables include *ontology_index* and *uri_index*, which store the ID and the URL of all ontologies and the ID and the URI of all instances in the database respectively. Tbox tables store the subclass, subproperty, equivalent class and equivalent property relationships, the disjoint classes and the properties characteristics. This information is provided by the Description Logic reasoner. The *hierarchy* table stores all class/subclass pairs while the *Hierarchyprops* table stores all property/subproperty pairs. On the other hand, the *equivalents* table stores all class/equivalent class pairs while *equivalentprops* stores all pair property/equivalent property pairs. Finally, the *disjoint* table stores all pairs of disjoint classes and the *propertytypes* table stores if the property is transitive, symmetric or functional, if it is the inverse of another property and also its domain and range. Using these tables we can easily obtain the class and property hierarchies and the equivalent classes/properties of a specific class/property.

Furthermore, all the information needed for evaluating Abox inferences is also in the database. There is at least one table for each kind of possible class description in the ontology: enumeration (*enumeration* table), value restrictions (*all*, *some*, *hasvalue* tables), cardinality restriction (*max* and *min* tables), intersection (*intersec* table), union (*union* table) and complement (*complement* table). These tables will be used to implement the Abox inference rules.

Finally, we create one Abox table for each class and each property in the ontology. The name of this table is the same as the class or property. These tables contain the IDs of instances explicitly defined as instances of the class or property. Tables representing classes contain only one column (*ID*) while tables representing properties contain two columns (*SUBJECT, OBJECT*).

2.2   Tbox Retrieval

Tbox retrieval can be evaluated directly using the query language. Currently, DBOWL supports all the Tbox queries implemented by RACER. In order to implement them, the information obtained from the DL reasoner is stored in the corresponding tables at load time. For example, we store the equivalent classes for each class in the database. Thus, we only need to query the database to evaluate the Tbox reasoning which evaluates if two classes are equivalent to each other or if they are the equivalent classes of a specific class. We also use the DL reasoner to obtain the properties domain and range, which are sometimes not explicitly asserted by the ontology, but they can be inferred. At query time, this information will be obtained by querying the database with a simple SQL query. Obviously, the performance of these Tbox reasonings, being sound and complete, is much better than in a description logic base reasoner which evaluate the reasoning each time in main memory.

2.3   Abox Inferences

The Abox inference rules currently supported by DBOWL cover OWL-DL completely. They are implemented as java functions using only the information stored in the database. We define views for each class and property in the ontology. These views define the set of instances of the corresponding class or property, i.e. instances explicitly asserted by the ontology plus instances inferred by the Abox inference rules.

In order to implement the Abox reasoning, we divided inferences rules into 4 groups, i.e. (1) rules which use instances of properties and produce instances of properties, (2) rules which use instances of properties and produce instances of classes, (3) rules which use instances of classes and properties and produce instances of classes, and (4) rules which use instances of classes and produce instances of properties. Examples of rules in group 1 are those for implementing the subpropertyOf and the transitivePropery rules. Group 2 contains those rules for inferences of domain and range. Rules for reasoning on descriptions are in group 3. The only rules in group 4 are the hasvalue2 and hasvalue3 rules. In order to create the views for each class and property in the ontology, rules in group 1 are evaluated in a fix-point algorithm. After that, rules in group 2 are evaluated. Rules in group 3 have the same treatment as rules in group 1. Finally, rules in group 4 are evaluated. In the case where these rules produce new results, all groups are reevaluated. This means that we evaluated a fix-point algorithm for all groups which is controlled using the group 4 rules.

When this algorithm finishes, views are created, which will be used by DBOWL queries in order to obtain complete results. Therefore, the computation of the inferred instances is doing at query time.

```
Procedure subClassOf                    Function ReasoningSubClassOf(Classname C):viewName
Variables                               Variables
    SetOfCassNames C                        SetOfClassNames D
    ClassName C_i                           ClassName D_i
Begin                                   Begin
    C ← getClassesDescOrderSub()            D ← getSubClasses(C)
    forall C_i ∈ C do                       if D = ∅ then
        ReasoningSubClassOf(C_i)                V_C ← π_id(lastviewfor(C))
    endfor                                  else
End                                             V_C ← π_id(lastviewfor(C)) ∪ π_id(lastviewfor(D_i))
                                            endif
                                        End
```

Fig. 1. subClassOf reasoning algorithm

2.4   Inference Rules Algorithms

Figure 1 shows, as an example, one algorithm for evaluating Abox inference rules, the subclassOf algorithm. It obtains a list of all classes in the ontology ordered according to its the level in the class hierarchy. This means that first classes in the list will be those without subclasses and all subclasses for a given class will be in a previous position in the list. The algorithms use the Oracle *connect by* clause to create this list. Then, views for each class are defined following the order in the list. Thus, a view for a class will be defined using the definition of the view for all its subclasses.

3   Related Works

DBOWL is an OWL reasoner. As OWL is based on DL, we must also study DL reasoners. Of these, RACER [2] is the most relevant and one of the most complete, and implements both Tbox and Abox reasoning. Furthermore, it provides its own query language, which allows simple conjunctive queries to be evaluated. It is not persistent however, and reasoning is implemented by reducing it to satisfiability. This means that large ontologies (with a large number of instances) cannot be loaded. PELLET [1] provides the same functionality as RACER but also has the same problems. In the past few years there has been a growing interest in the development of systems for storing large amounts of knowledge in the Semantic Web. Firstly, these systems were oriented to RDF storage [3] [4]. Nowadays, research is oriented to massive OWL storage. Several alternative approaches using relational technology have been presented. Instance Store [7] uses a DL reasoner for inferring Tbox information and storing it in a relational database. However, the ontology definition language does not allow the definition of binary relationships. From our point of view, this is an important expressiveness limitation. Moreover, Instance Store only evaluates subsumption of concepts and equivalent classes by reducing them to terminological reasonings and evaluates them using a DL reasoner. On the other hand, the QuONTO [8] system reduces the ontology definition language to DL-Lite [9], a description logic which is a subset of OWL-DL. Therefore, the soundness and completeness of the reasonings is ensured. It evaluates subsumption of concepts and conjunctive queries. The queries are rewritten using the Tbox information and they are translated to SQL. DLDB-OWL [10] extends a relational database with OWL inferences. This proposal uses a DL reasoner as Instance Store does, but the database schema is more complex. In its public distribution only the subsumption of concepts is implemented, but it is implemented using only the information stored in the database. BigOWLIM [11] is a commercial tool that stores DL-Lite and RDFS ontologies on disk by means of binary files. BigOWLIM reasoning is evaluated by TRREE, its own inference engine. This is the most complete current reasoner. Finally, Minerva [12] also stores the ontology in a relational database, but uses a DL reasoner for evaluating Tbox reasonings and a rule engine to evaluate Abox reasonings covering OWL-DL partially. In other words, it combines relational technology with logic rules. Minerva also evaluates SPARQL queries. Our proposal aims to subsume all these results, providing a persistent and scalable tool for querying and reasoning on OWL ontologies. To do this, we provide an optimized storage model which is efficient and scalable; we implement reasoning on top of a relational database and combine reasoning and querying.

4   DBOWL Performance and Completeness

In order to demonstrate the completeness and evaluate the performance of our tool we use UOB [13], a well known benchmark to compare repositories in the Semantic Web. This benchmark is intended to evaluate the performance of OWL repositories with respect to extensional queries over a large data set that commits to a single realistic ontology. Furthermore, the benchmark evaluates the system completeness and soundness with respect to the queries defined. This benchmark provides tree ontologies, i.e. a 20, 100 and 200 Megabyte ontologies and the queries results for each one. This first experiment is conducted on a PC with Pentium IV CPU of 2.13 GHz and 2G memory, running on a Windows XP service pack 2 with Java JRE 1.4.2 Release 16.

4.1   Performance

With respect to performance, figure 2 shows the response time of each benchmark query expressed in seconds for the 20MB and 100MB ontology. These ontology contains around 200.000 and 1.000.000 individuals respectively. Current results suggest that we are working in the right direction. Our highest response time is 28 miliseconds for the 100 MB ontology and the results also show that DBOWL scales well. We asume that real applications dealing with very big ontologies will require the use of better computers. We believe that in those environments DBOWL will return very good performance results.

4.2   Completeness

In order to check the completeness of DBOWL, two tests (empirical and theoretical) were carried out. First, we evaluated the UOB queries for the 100MG ontology in DBOWL and obtained the correct results for all queries. We also

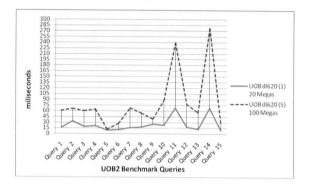

Fig. 2. DBOWL response times for the 20MG and 100MG ontologies

evaluated the UOB queries in several related tools, in order to prove that only DBOWL obtains the complete results for all of them. Figure 3 shows the number of instances returned by the tools and the UOB results. As we can see, DBOWL and UOB return different results for queries 11, 13 and 15. We check the UOB results for these queries and we believe that they are not correct, maybe because this part of the benchmark was never evaluated. For queries 11 and 15 DBOWL returns more results than UOB. This is because some synonymous of instances are not in the UOB result but instances are. For query 13 DBOWL returns less instances than UOB. This is because UOB result includes instances of all departments, but query 13 asks only for instances in the department0. Therefore, DBOWL results are the correct ones.

Some tools (Minerva, Jena2) cannot deal with the 100MB ontology because of its size (around 1.000.000 individuals), others, such as Instance Store do not support OWL-DL ontologies. As we can observe in Figure 3, BigOWLIM obtains good results, but it is not complete for OWL-DL. It only supports DL-Lite. Therefore, it cannot obtain complete results for several queries. With this test we demonstrate empirically that DBOWL can deal with very big ontologies obtaining complete results. On the other hand, we provide some theoretical proofs for every Abox inference rule algorithm, DBOWL returns the correct set of results. This is proof by contradiction. We present here the theoretical proof for the algorithms shown in section 2.4.

Lemma 1. Let D be a subclass of C. Let V_d be the last view created for the class D. Let V_c be the last view created for the class C. Let V_{c_S} be the view created by the subClassOf algorithm for the class C. If $< x >$ is a tuple in V_d, then $< x >$ is a tuple in V_{c_S}.

Proof. Suppose that $< x >$ is a tuple in V_d and $< x >$ is a NOT a tuple in V_{c_S}.

As D is a subclass of C, the subClassOf algorithm defines a view V_{c_S} for the class C as $\pi_{id}(lastviewfor(C)) \cup \pi_{id}(lastviewfor(D))$. That is $\pi_{id}(V_c) \cup \pi_{id}(V_d)$ (by premises).

As $< x >$ is a tuple in V_d then $< x >$ is a tuple in $\pi_{id}(V_d)$ (by definition of π in the relational algebra).

TOOL	Q1	Q2	Q3	Q4	Q5	Q6	Q7	Q8	Q9	Q10	Q11	Q12	Q13	Q14	Q15
DLDB-OWL	35	9152	666	386	0	757	3	67	0	22	-	37	239	0	0
Minerva	-	-	-	-	-	-	-	-	-	-	-	-	-	-	-
Sesame	35	5934	666	414	0	0	3	0	0	22	3264	0	0	0	0
Jena2	-	-	-	-	-	-	-	-	-	-	-	-	-	-	-
BigOWLIM	35	11305	666	414	200	757	139	342	1041	29	6180	37	414	6656	0
Kowari	35	0	0	0	0	0	0	0	0	22	58	0	0	0	0
Instance Store	-	-	-	-	-	-	-	-	-	-	-	-	-	-	-
DBOWL	35	11305	666	414	200	772	145	344	1041	29	6230	37	416	6913	73
UO Benchmark	35	11305	666	414	200	772	145	344	1041	29	6225*	37	10503*	6913	72*

Fig. 3. Number of instances returned by different tools (100MG ontology). Benchmark number of instances for marked (*) queries are explained in section 4.2.

Thus $< x >$ is a tuple in V_{c_S} (by definition of $\cup$ in the relational algebra). This produces a contradiction.

5   Conclusions and Future Work

This paper presents DBOWL, a tool for querying and reasoning on OWL-DL ontologies. It stores the ontologies in a relational database, using a description logic reasoner for pre-computing the class and property hierarchies, which are also stored in the database. DBOWL supports both Tbox and Abox reasoning and Extended Conjunctive Queries (ECQ). Reasonings are encapsulated by java functions making it possible to configure the tool according to the reasoning needs of the applications. In order to test our proposal we have used a benchmark and compared the results with several related tools. We also provided algorithms for evaluating Abox inference rules covering OWL-DL and their theoretical proof of completeness. The results obtained suggest that DBOWL is a promising OWL reasoner. Currently, DBOWL supports much bigger ontologies than traditional DL systems and it is being used in some applications in the System Biology domain, where particularly large ontologies are used, like the Gene Ontology (http://www.geneontology.org/) or TAMBIS (http://www.ontologos.org). It is also in use in some research projects such as the Spanish Ministry of Science and Innovation research project ICARIA (TIN2008-04844) or the Applied Systems Biology Project, P07-TIC-02978 (Innovation, Science and Enterprise Ministry of the regional government of the Junta de Andaluca). DBOWL is currently being used in real tools for Knowledge-Based Analysis in Systems Biology as KA-SB [14], obtaining very good results.

As future work, we plan to implement a SPARQL query engine for DBOWL. We also are studying some optimization techniques (like database indexes, parallel computation and incremental reasoning) in order to improve the response time of the queries.

Acknowledgements

This work has been funded by the Spanish MEC Grant (TIN2008-04844). We wish to thank Oscar Corcho for his suggestions about the theoretical proofs.

References

1. Sirin, E., Parsia, B., Grau, B.C., Kalyanpur, A., Katz, Y.: Pellet: A practical OWL-DL reasoner. Journal of Web Semantics 5(2) (2007)
2. Haarslev, V., Möller, R.: RACER system description. In: Goré, R.P., Leitsch, A., Nipkow, T. (eds.) IJCAR 2001. LNCS, vol. 2083, p. 701. Springer, Heidelberg (2001)
3. Broekstra, J., Kampman, A., van Harmelen, F.: Sesame: A generic architecture for storing and querying RDF and RDF schema. In: Horrocks, I., Hendler, J. (eds.) ISWC 2002. LNCS, vol. 2342, p. 54. Springer, Heidelberg (2002)
4. KAON. The Karlsruhe Ontology and Semantic Web Framework. Developer's Guide for KAON 1.2.7 (January 2004),
   http://km.aifb.uni-karlsruhe.de/kaon2/Members/rvo/KAON-Dev-Guide.pdf
5. Roldán-García, M.M., Aldana-Montes, J.F.: DBOWL: Towards a Scalable and Persistent OWL reasoner. In: The Third International Conference on Internet and Web Applications and Services. ICIW 2008, Athens, Greece, June 8-13 (2008)
6. Roldán-García, M.M., Molina-Castro, J.J., Aldana-Montes, J.F.: ECQ: A Simple Query Language for the Semantic Web. In: 7th International Workshop on Web Semantics, WebS 2008. DEXA 2008, Turin, Italy, September 1-5 (2008)
7. Horrocks, I., Li, L., Turi, D., Bechhofer, S.: The Instance Store: Description Logic Reasoning with Large Numbers of Individuals (2004)
8. Acciarri, A., Calvanese, D., De Giacomo, G., Lembo, D., Lenzerini, M., Palmieri, M., Rosati, R.: QuOnto: Querying Ontologies. In: Proceedings of the National Conference on Artificial Intelligence 2005, Part 4, vol. 20, pp. 1670–1671 (2005)
9. Calvanese, D., De Giacomo, G., Lenzerini, M., Rosati, R., Vetere, G.: DL-Lite: Practical reasoning for rich DLs. In: Proceedings of DL 2004. CEUR Electronic Workshop Proceedings (2004), http://ceur-ws.org/Vol-104/
10. Pan, Z., Heflin, J.: DLDB: Extending Relational Databases to Support Semantic Web Queries. In: Workshop on Practical and Scaleable Semantic Web Systems, ISWC 2003 (2003)
11. Kiryakov, A., Ognyanov, D., Manov, D.: OWLIM – a Pragmatic Semantic Repository for OWL. In: Proc. of Int. Workshop on Scalable Semantic Web Knowledge Base Systems (SSWS 2005), WISE 2005, New York City, USA, November 20 (2005)
12. Zhou, J., Ma, L., Liu, Q., Zhang, L., Yu, Y., Pan, Y.: Minerva: A scalable OWL ontology storage and inference system. In: Mizoguchi, R., Shi, Z.-Z., Giunchiglia, F. (eds.) ASWC 2006. LNCS, vol. 4185, pp. 429–443. Springer, Heidelberg (2006)
13. Ma, L., Yang, Y., Qiu, Z., Xie, G.T., Pan, Y., Liu, S.: Towards a complete OWL ontology benchmark. In: Sure, Y., Domingue, J. (eds.) ESWC 2006. LNCS, vol. 4011, pp. 125–139. Springer, Heidelberg (2006)
14. del Roldan-Garcia, M.M., et al.: KASBi: Knowledge-Based Analysis in Systems Biology. In: International Workshop on Semantic Web Applications and Tools for Life Sciences, SWAT4LS 2008 (2008)

FRESG: A Kind of Fuzzy Description Logic Reasoner

Hailong Wang, Z.M. Ma, and Junfu Yin

School of Information Science & Engineering, Northeastern University
Shenyang 110004, China
zongmin_ma@yahoo.com

Abstract. Based on the fuzzy description logic $F\text{-}ALC(G)$, we design and implement a fuzzy description logic reasoner, named FRESG1.0. FRESG1.0 can support the representation and reasoning of fuzzy data information with customized fuzzy data types and customized fuzzy data type predicates. We briefly introduce the reasoning services provided by FRESG1.0. Then, we particularize the overall architecture of FRESG1.0 and its design and implementation of the major components. In the paper, we pay more attention to illustrate the features of the reasoner as well as the algorithms and technologies adopted in the implementations.

Keywords: Fuzzy description logic, $F\text{-}ALC(G)$, reasoner, customized data type.

1 Introduction

In recent years, great progress have been made on the study of classic DL reasoners and many reasoners have been put forward, such as FaCT, RACER, Pellet, Jena, KAON2. For the goal to express and reason imprecise and uncertain information which widely exists in human knowledge and natural language, and as an attempt to process fuzzy data information, some fuzzy reasoners have been put forward, such as FiRE [1], GURDL [2], DeLorean [3], GERDS [4], YADLR [5], FuzzyDL [6]. All of these reasoners support fuzzy Description Logic. However, it has been pointed that most of these reasoners have limitations on expressing and reasoning fuzzy data type information. Furthermore, all of them, including FuzzyDL, cannot express and reason the customized data types and predicates [7].

For the reasons above, few efforts have been done for the representation and reasoning of customized fuzzy data types and predicates. In [8], the fuzzy data type group G is introduced into fuzzy description logic $F\text{-}ALC$ [9] and the fuzzy description logic $F\text{-}ALC(G)$ is prompted. This paper gives the *ABox* consistency checking algorithm of $F\text{-}ALC(G)$, introduces a reasoning framework [7] supporting the reasoning services of fuzzy data type information, and designs a reasoning algorithm with fuzzy data types. In order to verify the correctness of $F\text{-}ALC(G)$ reasoning algorithm and provide a testing platform for future extensions to fuzzy description logic $F\text{-}ALC(G)$ [8], this paper designs and implements a fuzzy description logic reasoner, named FRESG, which can provide reasoning services for customized fuzzy data types and customized fuzzy data type predicates.

S.S. Bhowmick, J. Küng, and R. Wagner (Eds.): DEXA 2009, LNCS 5690, pp. 443–450, 2009.

2   FRESG1.0 as a Fuzzy DL Reasoner

FRESG (**F**uzzy **R**easoning **E**ngine **S**upporting Fuzzy Data Type **G**roup) is a prototype reasoner, which is based on the *ABox* consistency checking algorithm of the fuzzy DL *F-ALC(G)* [8]. The current version of FRESG is named FRESG1.0. FRESG1.0 can support the consistency checking of a KB with empty *TBox*, and the future versions will provide supports for the consistency checking of a KB with simple *TBox* and general *TBox* restrictions. In fact, the consistency checking of a KB with empty *TBox* (equivalent to *ABox*) is the most basic reasoning problem in DL because the other inference services can be reduced to this case [9]. FRESG1.0 provides the following "standard" set of inference services, including:

- *ABox consistency checking*, which is used to check whether an *ABox* is "consistent" or "inconsistent";
- *Concept satisfiability*, which checks if it is possible for a concept to have any instances;
- *Entailment*. It is said that an interpretation I satisfies (is a model of) a KB K iff I satisfies each element in K [8]. A KB K entails an assertion α (denoted by $K \models \alpha$) iff every model of K also satisfies α. The problem of determining if $K \models \alpha$ holds is called *entailment* problem;
- *Retrieval*. According to the given query conditions, FRESG1.0 returns all the instances satisfying the conditions;
- *Realization*. Given an *ABox* A and a concept C, to find all the individuals a such that $\langle a: C, \bowtie, k \rangle \in A$.

Furthermore, the other important reasoning services, like *subsumption* and *classification*, can be reduced to the consistency checking of the corresponding *ABox*.

FRESG1.0 is a reasoner based on fuzzy DL *F-ALC(G)*, so its grammatical forms are in correspondence with the *F-ALC(G)* abstract syntax. Users can define their *ABox* according to the syntax of FRESG1.0 and then use FRESG1.0 reasoner to reason or query over the *ABox*. The detailed FRESG1.0 syntax can be referred to FRESG1.0 user manual. The following example 1 shows an *ABox* A_1 described in valid FRESG1.0 syntax.

Example 1.   (instance x (some R D) >= 0.7);
(instance x (all R C) >= 0.4);
(instance y C >= 0.2);
(instance y D >= 0.3);
(related x y R >= 0.5);
(instance x (some R (and C D)) < 0.5);
(instance y (dtsome T_1, (dtor p_1 p_2)) >= 0.7);
(instance y (dtatleast 3, T_2 T_3, p_3) >= 0.6);
(instance y (dtall T_2, p_1) >= 0.9);|

Its corresponding DL KB described in *F-ALC(G)* abstract syntax is as follows: A_1' = {$\langle x: \exists\ R.D, \geq, 0.7 \rangle$, $\langle x:\forall R.C, \geq, 0.4 \rangle$, $\langle y:C, \geq, 0.2 \rangle$, $\langle y:D, \geq, 0.3 \rangle$, $\langle (x,y):R, \geq, 0.5 \rangle$, $\langle x: \exists R.(C \sqcap D), <, 0.5 \rangle$, $\langle y: \exists T_1.p_1 \vee p_2, \geq, 0.7 \rangle$, $\langle y: \geq 3T_2, T_3.p_3, \geq, 0.6 \rangle$, $\langle y:\forall T_2.p_1, \geq, 0.9 \rangle$}.

Here x, y are individuals; C, D are concepts; R is an abstract role; T_1, T_2, T_3 are fuzzy data type roles; p_1, p_2, p_3 are fuzzy data type predicates based on base data type *xsd: integer*. Then, FRESG1.0 can provide reasoning services over the compiled A_1'.

3   FRESG1.0 Architecture and Design

The core of FRESG1.0 is a Tableaux reasoner and a fuzzy data type reasoner. The former is implemented based on the *F-ALC(G) ABox* consistency checking algorithm [8], and the latter is used to decide the satisfiability of fuzzy data type predicates conjunctions [8].

The main design goal of FRESG1.0 is that it has a small core reasoning engine, which is able to reason with customized fuzzy data type predicates and is suitable for extensions. Based on this goal, FRESG1.0 has the overall architecture shown in Fig.1. In the design, it requires that the fuzzy data type reasoner is independent of the Tableaux reasoner. On the other hand, if users want to add the construction operators of fuzzy data type expressions [8], they only need to modify the fuzzy data type manager, without having to change the Tableaux reasoner; If users want to add some basic data types supported by the reasoner, they only need to add the corresponding fuzzy data type checkers on the base of the current fuzzy data type reasoner. Such a design philosophy makes FRESG1.0 have highly modular structure and be easy extended. As a result, a solid foundation is laid for implementing more expressive DL reasoner and the further research. If the expression capacity of a certain part (e.g., the Tableaux reasoner, fuzzy data type manager or fuzzy data type checker) is amended and strengthened independently, a new fuzzy DL reasoner is formed, which is suitable for applications in different backgrounds.

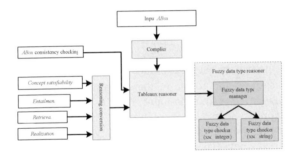

Fig. 1. The architecture of FRESG1.0 reasoner

The Tableaux reasoner checks the consistency of an *ABox*. The FRESG1.0 compiler translates the *ABox* described in FRESG1.0 syntax into the file described in abstract *F-ALC(G)* syntax which can be identified by the Tableaux reasoner. The Tableaux reasoner conducts the decision or query over the compiled file according to the "*ABox* consistency checking" command, or the query requirements converted by the unit of "Reasoning conversion". If there are some sub-queries about the customized fuzzy data type predicates in the query, the Tableaux reasoner invokes the fuzzy data type reasoner to decide the satisfiability of the fuzzy data type predicate conjunctions, and finally returns the corresponding results.

3.1   Tableaux Reasoner

The Tableaux reasoner has only one function: checking the consistency of an *ABox*. According to the model-theoretic semantics, an *ABox* is consistent if there is an interpretation that satisfies all the assertions in it. Such an interpretation is called a *model* of the *ABox*. The tableaux reasoner searches for such a model according to the *ABox* consistency checking algorithm [8].

In the process of implementation, the Tableaux reasoner firstly starts by constructing an initial forest $\mathcal{F}_{\mathcal{A}}$ for *ABox*. There are two kinds of nodes in the constructed forest $\mathcal{F}_{\mathcal{A}}$: abstract nodes (the normal labeled nodes) and data type nodes (unlabeled leaves of $\mathcal{F}_{\mathcal{A}}$). Each abstract node x is labeled by a set of triples and each triple is in the form of $\langle C, \bowtie, k \rangle$, which indicates the membership that the individual x belongs to $C \bowtie k$. The edges in the forest $\mathcal{F}_{\mathcal{A}}$ represent the relationship between nodes. An edge between two abstract nodes represents a fuzzy abstract role, and an edge between an abstract node and a data type node represents a fuzzy data type role. The initial forest $\mathcal{F}_{\mathcal{A}}$ is shown in "Initial Status" window in the form of tree.

According to the *ABox* consistency checking algorithm, the Tableaux reasoner repeatedly applies the tableaux expansion rules until a clash is detected in the label of a node, or until a clash-free forest is found, to which no more rules are applicable. FRESG1.0 applies *F-ALC(G)* Tableaux expansion rules in the following order:

1) For an abstract node, *F-ALC*-rules are used first and then *G*-rules are used [8];

2) *F-ALC*-rules are used according to the order of $\neg_{\bowtie}$-, $\sqcup_{\triangleright}$-, $\sqcap_{\triangleleft}$-, $\sqcap_{\triangleright}$-, $\sqcup_{\triangleleft}$-, $\exists_{\triangleright}$-, $\forall_{\triangleleft}$-, $\forall_{\triangleright}$-, and $\exists_{\triangleleft}$-rules;

3) *G*-rules are used according to the order of $\exists_{p\triangleright}$-, $\forall_{p\triangleleft}$-, $\geq_{p\triangleright}$-, $\leq_{p\triangleleft}$-, $\forall_{p\triangleright}$-, $\exists_{p\triangleleft}$-, $\leq_{p\triangleright}$-, and $\geq_{p\triangleleft}$-rules;

As a result of the use of uncertainty rules $\sqcup_{\triangleright}$- and $\sqcap_{\triangleleft}$-, a number of complete forest $\mathcal{F}_{\mathcal{A}1}, \mathcal{F}_{\mathcal{A}2},...., \mathcal{F}_{\mathcal{A}n} (n \geq 1)$ may be yielded. Finally, the Tableaux reasoner checks if these forests contain clashes or not. If one of these forests does not contain any clashes, there is a model about *ABox* and the input *ABox* is "consistent". If all these forests contain clashes, there is no model for *ABox* and the input *ABox* is "inconsistent". When the Tableaux reasoner checks the satisfiability of the forests, if there are some queries related to fuzzy data type, the Tableaux reasoner invokes the fuzzy data type reasoner to decide the satisfiability of the conjunction queries with fuzzy data type expression. The Tableaux reasoner returns the final results according to the results returned by fuzzy data type reasoner.

3.2 Fuzzy Data Type Reasoner

The fuzzy data type reasoner is used to decide the satisfiability of the conjunction query with fuzzy data type. In other words, the tableaux reasoner calls the fuzzy data type reasoner for the satisfiability of the fuzzy data type expression conjunction. The fuzzy data type reasoner includes one fuzzy data type manager and two fuzzy data type checkers. Among them, one checker is based on the basic data type *xsd: integer* and another one is based on the basic data type *xsd: string*.

The fuzzy data type manager works between the tableaux reasoner and the two fuzzy data type checkers. The design of the fuzzy data type manager is based on the algorithms in [8]. The Tableaux reasoner reduces the fuzzy data information in *ABox* to the form of fuzzy data type expressions conjunctions [8] using Tableaux expansion rules, and then call the fuzzy data type reasoner to check the satisfiability of the conjunction. The fuzzy data type manager decomposes the conjunction into disjunctions of fuzzy data type predicate conjunctions and then divides each of the predicates conjunction into two sub-conjunctions. One of the sub-conjunctions is based on *xsd: integer* and another one is based on the basic data type *xsd: string*. Then, the fuzzy data type manager sends these two sub-conjunctions to appropriate fuzzy data type checkers to decide their satisfiabilities. If the two sub-conjunctions of a fuzzy predicate conjunction are satisfiable, it is satisfiable. If any one of the fuzzy predicate conjunction is satisfiable, the input fuzzy data type expressions conjunctions is satisfiable.

The fuzzy data type checkers decide the satisfiability problem of fuzzy data type predicate conjunctions, where the fuzzy data type predicates are defined over a base data type in a fuzzy data type group. If some new base data types are needed to add into FRESG1.0, only corresponding fuzzy data type checkers are need to add. The design of the fuzzy data type checker is based on the algorithms in [8].

3.3   Reasoning Conversion

FRESG1.0 reasoning engine is designed to transform other reasoning tasks into *ABox* consistency checking tasks. The current version of FRESG provides not only *ABox* consistency checking services, but also the reasoning services such as concepts satisfaction checking, entailment reasoning function and querying. The FRESG1.0 provides these three kinds of reasoning services through transforming them into *ABox* consistency checking.

Firstly, it is easy to transform concepts satisfaction checking into *ABox* consistency checking. If users want to check the satisfaction of the $F\text{-}ALC(G)$-concept C, FRESG1.0 transforms it into checking consistency of the fuzzy *ABox* $\{(x: C) > 0\}$, and calls the Tableaux reasoner to decide its consistency (x is an abstract individual).

Secondly, the entailment reasoning function can also be transformed into *ABox* consistency checking task [9].

Thirdly, according to the query definition, FRESG1.0 returns all the individuals satisfying the query condition ($C \bowtie k$), so it is only required to do entailment reasoning for each individual x in the current *ABox* A. If $A \models \langle x: C, \bowtie, k \rangle$, x satisfies the querying condition and the result is displayed in the "Result Window". Otherwise, x does not satisfy the querying conditions. FRESG1.0 needs to decide every individual in *ABox* when querying, so the querying is inefficient. How to make the querying tasks more efficient is one of our future works.

4   Testing and Performance

All reasoning tasks that follow the FRESG1.0 syntax can be processed by the FRESG1.0. In order to verify the correctness of FRESG1.0, we provide two testing cases in this paper, which are designed to verify FRESG1.0 in two aspects, including *ABox* consistency checking and concepts satisfiability.

Example 2. In order to check the consistency of *ABox* A_1 in Example 1, we can run it in FRESG1.0, and the result is partly shown in Fig.2. According to the Tableaux expansion rules, the "Reasoning Details" window shows how the $\mathcal{F}_A$ is extended into four different kinds of completion-forests by the Tableaux reasoner. Then, the Tableaux reasoner calls the fuzzy data type reasoner, which checks the satisfiability of the conjunction query with customized fuzzy data type predicates. As a result, the first and second situations are satisfiable, but the third and fourth situations are unsatisfiable. So the "Final Result" window gives "The $ABox\ A_1$ is consistent" as a conclusion.

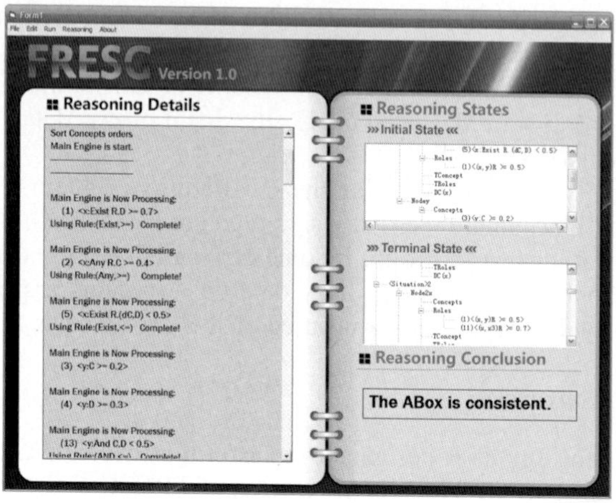

Fig. 2. The running results of FRESG1.0 reasoner

Example 3. Now we check the satisfiability of the concept (and (some R (all S (and C D))) (some S (dtsome T_1 T_2 T_3, (dtand E_1 E_2)))). Firstly, FRESG1.0 compiler transforms the concepts that follow the FRESG1.0 syntax into the concepts that follow the $F\text{-}ALC(G)$ syntax, i.e., $(\exists R.(\forall S.(C\sqcap D))) \sqcap (\exists S.(\exists T_1,T_2,T_3.(E_1\wedge E_2)))$. Secondly, using the unit of "reasoning conversion", FRESG 1.0 transforms the task that checks the satisfiability of concepts into the task that checks the consistency of the $ABox\ A_2$: $\{x : (\exists R.(\forall S.(C\sqcap D)))\sqcap(\exists S.(\exists T_1,T_2,T_3.(E_1\wedge E_2))) > 0\}$.

We can get useful information from the results: the initial forest that corresponds to A_2 is shown in the "Initial States" window. The "reasoning detail" window shows how $\mathcal{F}_A$ is expanded to a complete forest by the Tableaux reasoner. Then the Tableaux reasoner calls the fuzzy data type reasoner to check the satisfiability of the conjunction query with customized fuzzy data type predicates. Finally, because no clash is contained in the complete forest, A_2 is "consistent", and the "Final Result" window also gives the corresponding conclusion.

From the examples mentioned above, it can be seen that the customized fuzzy data type information can be expressed and reasoned in the FRESG1.0.

FRESG1.0 is open source (around 15000 lines of codes) and has complete documents and lots of testing cases. All of these can be obtained by emailing to the authors.

We have tested performance of the FRESG1.0 reasoner by checking the consistencies of different *ABoxes* with different sizes (as shown in Fig.3). The size of *ABox* is measured in instance number (both concept instances and role instances). In the logarithmic graph, the reasoning time is in scale to the *ABox* size.

	ABox1	ABox2	ABox3	ABox4	ABox5	ABox6	ABox7	ABox8	ABox9	ABox10
ABox size	100	200	300	400	500	600	700	800	900	1000
Time cost(s)	0.015	0.047	0.063	0.110	0.157	0.218	0.296	0.375	0.484	0.640

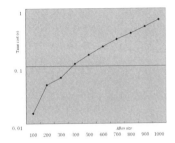

Fig. 3. Performance of the FRESG1.0 reasoner

5 Related Work

With the extensive and in-depth application of intelligent systems, the processing of fuzzy and uncertain information in the real world is becoming an essential part of intelligent systems [1-6, 8, 9]. It is imperative to develop fuzzy reasoners which can process the fuzzy Ontology and fuzzy DL. Table 1 lists some existing reasoners that can support the processing of fuzzy and uncertain information.

Table 1. The Comparisons of the Current Fuzzy Description Logic Reasoners

Reasoner	Supporting DL	Features
FiRE	f_{KD}-SHIN	A first fuzzy DL prototype reasoner, supporting GUI.
GURDL	f-ALC	expanding the optimization technology in classic DL, improving the performance of fuzzy reasoner
DeLorean	f-SHOIN	Converting the reasoning problems of f_{KD}-SHOIN to classic SHOIN to solve
GERDS	f-ALC	adding inverse role, top role, and bottom role on the base of f-ALC
YADLR	SLG algorithm	Not only supporting the membership in assertions is a constant, but also supporting it is a variable
FuzzyDL	f-SHIF(D)	Not only supporting fuzzy Lukasiewicz norms, but also supporting other fuzzy norms

Through comparison of the reasoners in Table 1, we find that all of the reasoners in Table 1 cannot support the representation and reasoning of customized fuzzy data type and predicates, which is very important in some intelligent systems [7, 8]. FRESG1.0 can support not only the representation and reasoning of fuzzy concept knowledge, but also fuzzy data information with customized fuzzy data types and predicates. The relationship of the reasoning capacity among these fuzzy reasoners (including FRESG1.0) is shown in Fig.4. In Fig.4, U means human knowledge. The dotted circle, solid circle and dotted & solid circle in Fig.4 denote FRESG1.0 reasoner and two fuzzy reasoner 1 & 2, respectively. In Fig.4, **C** stands for f-ALC that all fuzzy reasoners can support; **E** stands for the DLs that support fuzzy data type D [6], such as FuzzyDL in Table 5; **F** stands for the DLs that support customized fuzzy data type and customized fuzzy data type predicates; **B** stands for the remaining part with the same reasoning abilities of reasoner 1 and reasoner 2 in addition to f-ALC; **A** and **D** stands for the unique part of reasoner 1 and 2 in reasoning capacity, respectively. In Fig.4, **F** stands for the reasoning ability that the other fuzzy DL reasoners do not have, which is the unique part of FRESG1.0.

Fig. 4. The relationships among fuzzy description logic reasoners

6   Conclusion and Future Work

In this paper, we design and implement the fuzzy DL reasoner FRESG1.0 based on the fuzzy DL *F-ALC(G)*. It can support not only the representation and reasoning of fuzzy concept knowledge, but also fuzzy data information with customized fuzzy data types and customized fuzzy data type predicates.

FRESG1.0 reasoner is in constant update. In the near future, we are planning to extend the reasoner in three directions:

1) Improving the robustness and reliability through a large number of test examples and trying to improve its performance;
2) Further expanding its representation and reasoning abilities (e.g., the expansion of transitive role axioms (*S*), inverse roles (*I*), and number restrictions (*N*) etc.);
3) Expanding DIG standard interface so that it can be invoked by other reasoners.

References

1. Stoilos, G., Nikos, S., Stamou, G.: Uncertainty and the Semantic Web. J. IEEE Transaction on Intelligent Systems 21, 83–87 (2006)
2. Haarslev, V., Pai, H.I., Shiri, N.: Optimizing tableau reasoning in ALC extended with uncertainty. In: Proc. of the 2007 Int Workshop on Description Logics, pp. 307–314 (2007)
3. Bobillo, F., Delgado, M., Romero, J.: Optimizing the crisp representation of the fuzzy description logic SROIQ. In: Proc. of the 3rd ISWC Workshop on Uncertainty Reasoning for the Semantic Web (2007)
4. Habiballa, H.: Resolution strategies for fuzzy description logic. In: Proc. of the 5th Conference of the European Society for Fuzzy Logic and Technology, pp. 27–36 (2007)
5. Stasinos, K., Georgios, A.: Fuzzy-DL Reasoning over Unknown Fuzzy Degrees. In: Proc. of the 3rd Int Workshop on Semantic Web and Web Semantics, pp. 1312–1318 (2007)
6. Bobillo, F., Straccia, U.: fuzzyDL: An Expressive Fuzzy Description Logic Reasoner. In: Proc. of the 2008 IEEE Int. Conf. on Fuzzy Systems, pp. 923–930 (2008)
7. Pan, J.Z.: A Flexible Ontology Reasoning Architecture for the Semantic Web. J. IEEE Transaction on Knowledge and Data Engineering 19, 246–260 (2007)
8. Wang, H.L., Ma, Z.M.: A Decidable Fuzzy Description Logic F-ALC(G). In: Proc. of the 19th Int Conf on Database and Expert Systems Applications, pp. 116–123 (2008)
9. Straccia, U.: Reasoning within fuzzy description logics. J. Journal of Artificial Intelligence Research 14, 137–166 (2001)

Extracting Related Words from Anchor Text Clusters by Focusing on the Page Designer's Intention

Jianquan Liu, Hanxiong Chen, Kazutaka Furuse, and Nobuo Ohbo

Department of Computer Science, Graduate School of Systems and Information Engineering,
University of Tsukuba, 1-1-1 Tennodai, Tsukuba-shi, Ibaraki-ken, 305-8577, Japan
{ljq,chx,furuse,ohbo}@dblab.is.tsukuba.ac.jp

Abstract. Approaches for extracting related words (terms) by co-occurrence work poorly sometimes. Two words frequently co-occurring in the same documents are considered related. However, they may not relate at all because they would have no common meanings nor similar semantics. We address this problem by considering the page designer's intention and propose a new model to extract related words. Our approach is based on the idea that the web page designers usually make the correlative hyperlinks appear in close zone on the browser. We developed a browser-based crawler to collect "geographically" near hyperlinks, then by clustering these hyperlinks based on their pixel coordinates, we extract related words which can well reflect the designer's intention. Experimental results show that our method can represent the intention of the web page designer in extremely high precision. Moreover, the experiments indicate that our extracting method can obtain related words in a high average precision.

Keywords: Related word extraction, Anchor text, Clustering, Design intent.

1 Introduction

To search the Web, we must specify some keywords which can summarize our seeking purpose. Specifying appropriate keywords towards the seeking target makes the searching more effective. Apparently, extracting and suggesting related words, and analyzing similarity of words become important research topics.

Directly to survey the conventional methods, such as Google and Yahoo!, they suggest user related words by query log. However, their suggestions are less effective because search purposes are very different among the users. Then some studies use co-occurrence to extract and suggest related words. Nevertheless, high co-occurrence does not guarantee the semantic relatedness. Therefore, other studies follow contextual relation to analyze semantics, or derive context by analyzing the tag hierarchy of HTML page. However, the real context frequently changes with the dynamic layout, due to the widely use of CSS (Cascade Style Sheet). Two words in the same context of HTML structure will be divided by the real CSS layout. In other words, they are no longer related.

To address this problem of extracting related words, we challenge to extract the words that are considered correlative by the web page designers. For this purpose, we stand on the designer's side to consider the layout design. When designing a web page,

S.S. Bhowmick, J. Küng, and R. Wagner (Eds.): DEXA 2009, LNCS 5690, pp. 451–459, 2009.

the first thing coming up to their mind should be "How do I display the layout of the page?". Generally speaking, they would like to organize the related contents and put them in the same area on the page. Especially, hyperlink is one of the most important components of a page. Thus, they often put the related hyperlinks into the same zone on the page, which are identified by some associated anchor texts. It is natural to believe that all the hyperlinks appearing in the same area are certainly related. They would be related in contents, contexts, meanings or semantics in the designers' mind. Consequently, a method that can reflect or represent such a design intent based on the web page designer is highly desired. Analyzing such contents enables us to extract the related words and to provide the users with helpful search suggestion. Our novel approach is given in the following sections.

2 Related Works

Related works in this field can be roughly classified into two technical methods. One is to predict the related words by computing the similarity of words based on their co-occurrence frequencies in the documents. The other is to consider the semantic similarity by analyzing contextual texts, tag tree structures, or search results returned from search engines. Explanations of related works in each technical method, as well as their known problems, are presented as follows.

(1) Co-occurrence Similarity

Sato *et al.* [1] proposed a method of collecting a dozen terms that are closely related to a given seed term based on the occurrence frequencies. Their collection depends on the search results returned by search engine, without taking semantic similarity into consideration. Moreover, there are a number of attempts incorporating semantic knowledge into co-occurrence analysis [2], [3]. However, these methods are still highly dependent on co-occurrence frequency. In addition, Kraft *et al.* proposed a method for automatically generating refinements or related terms to the queries by mining the anchor texts for a large hypertext document collection [4]. This method is essentially dependent on occurrence frequency as well.

All these related works concentrate on the co-occurrence frequencies of related words and suffer from well known limitations, mostly due to their inability to exploit semantic similarity between terms: documents sharing terms that are different but semantically related will be considered as unrelated [5].

(2) Semantic Similarity

There is an approach which inferred the semantic similarity of a pair of single words from a corpus [5]. Sahami *et al.* introduced a novel method for measuring the similarity between short text snippets by leveraging web search results to provide greater context for the short texts [6]. Following this web-based kernel, Bollegala *et al.* [7] proposed a stronger method to compute semantic similarity using automatically extracted lexicon-syntactic patterns from text snippets. However, they strictly depend on contextual correlation of search results returned by search engines. Furthermore, there are a

```
<HTML>
<BODY width=400>
<DIV width=400>
    <DIV width=200 style="float:left; text-align:left" >
        <A href="#">Left text link1</A>
    </DIV>
    <DIV width=200 style="float:right;text-align:right">
        <A href="#">Right text link1</A>
    </DIV>
    <DIV width=200 style="float:left; text-align:left;
                    margin: 20px 0 0 -117px;">
        <A href="#">Left text link2</A>
    </DIV>
    <DIV width=200 style="float:right;text-align:right;
                    margin:20px -128px 0 0;">
        <A href="#">Right text link2</A>
    </DIV>
</DIV>
</BODY>
</HTML>
```

Fig. 1. An example: the HTML source (left) and its appearance on web browser (right)

number of researches focusing on semantic similarity based on analysis of contextual texts or tag tree structure [8], [9]. Additionally, the literature [10,11] give details on how to extract data information from HTML tag hierarchy. [10] tells us that the link blocks can help users to identify relevant zones on a multi-topic page. It indicates that the page designer often puts the relevantly semantic information together in the same link block. However, Hattori [12] pointed out that the content-distance based on the order of HTML tags does not always correspond to the intuitional distance between content elements on the actual layout of a web page. For instance, in Figure 1, analysis on HTML source may conclude that the four links are in the same link block, which is different from the real appearance. Rather, it is natural to consider that the four links belong to two blocks: the left one and the right one.

Observing the above related works, we mainly take on the following challenges in this paper: a) overcome the well known limitations of the methods based on co-occurrence frequency; b) keep consideration on semantic similarity of related words; c) solve the problems of page layout involved by CSS style in the page; d) extract related words by representing designer's intention.

3   Our Approach and Implementation

To address the challenges mentioned above, we propose a new analysis model named "BBCECI", as shown in Figure 2, which is composed of four modules. They are **B**rowser-**B**ased **C**rawling module, Link-based **E**xtracting module, **C**oordinate-based **C**lustering module and Cluster-based **I**ndexing module. The process flow is continuous in the BBCECI model. Its input are the pages crawled from the web space, and the output is an index database of related words.

Firstly, the Browser-based Crawling module, which is implemented as an extension to Mozilla Firefox, fetches the pages from the web space via a browser-based crawler. During the crawl process, the Link-based Extracting module extracts the values of the x-coordinate and y-coordinate, and the anchor text for each hyperlink from a page to compose a tuple *(x,y,anchortext)*. The extracting process launches after loading the page, which ensures that all the extracted information is based on the real appearance on the browser.

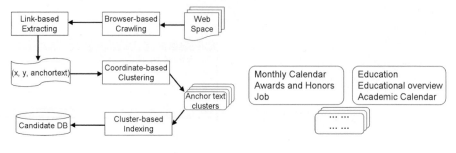

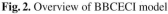

Fig. 2. Overview of BBCECI model   **Fig. 3.** Collection of anchor text clusters

Secondly, the Coordinate-based Clustering module treats each tuple *(x,y,anchortext)* as a point in the 2-dimensional Euclidean space, taking the string of *anchortext* as the label of each point. Then, it carries out the clustering for all points of each page using the well known DBSCAN algorithm [13]. After that, all the labels are assigned to certain clusters, where anchor texts (labels) in the same cluster are considered to be relevant.

Lastly, the Cluster-based Indexing module splits all anchor texts in the same cluster into single words, with stopword elimination and stemming. We consider all the words related to each other, so we weight them in each cluster respectively. In each cluster, the words are sorted by their weight values in descending order, and we create an entry of the index for each word. If there are duplicate entries of the same word, we merge them into one entry to make sure that all the entries in the database are unique.

This process results in a global index database. Consequently, we can provide users with related words while they search the Web. In the following subsections, we describe the details of implementation for each module.

3.1   Browser-Based Crawling

We propose a new browser-based crawling method to simulate the user browsing, without analyzing the HTML tag hierarchy at the background. The crawler is implemented as an extension to Mozilla Firefox browser. In order to achieve high-quality pages, we apply the **BFS** strategy to crawl the web [14]. Our browser-based crawler starts to visit URLs given by looking up the configuration file. Then it fetches all the accessible links in a page and extract their URL addresses embedded in the <A> tag by running a Javascript program. For example, against the following link, it fetches the value of *href* property as a new URL address. Here, the *href* is http://www.acm.org/, and "ACM Site" is the anchor text.

<A href="http://www.acm.org/">ACM Site</A>

The crawler will append the URL into the configuration file, as soon as it removes the duplicates. The crawler repeats the same extracting operation until it reaches the defined maximal number of URL addresses. During the crawling, the link information *(u,v)* denoting the hyperlink from page *u* to *v*, is stored as well.

3.2   Link-Based Extraction

The link-based extracting model runs after the crawling to extract the pixel coordinate information and anchor text of each link. Our extracting method is different from the conventional analysis based on HTML tag hierarchy. A simple Javascript function fetches all the DOM objects by the tag name <A> as follows.

```
var Aobjs = document.getElementsByTagName('A');
```

Using the variable **Aobjs**, we can achieve the coordinate value, the value of the *href* attribute, and the anchor text for each hyperlink. We define tuple *(x,y,anchortext)* to denote each extracted link, and store all the tuples.

3.3   Coordinate-Based Clustering

For this phase, we propose a coordinate-based clustering method to recognize the link blocks in a page, so that we can represent the intentions of page designers as clearly as possible. The well known DBSCAN algorithm [13] is introduced for clustering. After the clustering process, all the points in a page will be divided into some certain clusters. Then, we gather all the labels in the same cluster to form an anchor text cluster according to the result of clustering. When we complete the process for all the crawled pages, a large number of anchor text clusters will be formed as shown in Figure 3.

3.4   Cluster-Based Indexing

For each cluster shown in Figure 3, we split each anchor text into single words, eliminate the stopwords and process stemming. Then we compose all words from the same cluster into the same word set, believing that all the words in the same set are semantically related. We then weight the relatedness w_i for each word k_i in the word set S by the Formula 1.

$$w_i = \frac{Count(k_i)}{\sum_{k_i \in S}^{S} Count(k_i)} \tag{1}$$

$$w_i' = PR(p) \times w_i \quad (iff \ k_i \ is \ extracted \ from \ page \ p) \tag{2}$$

Referring to the anchor text clusters in Figure 3, we create an index table for each word set as shown in Table 1 and Table 2. All the related words associated with the indexed word are sorted in descending order of their relatedness values w_i.

Lastly, to make sure that all the index entries in the database are unique, we scan all the index tables and merge duplicates, weighting the word by Formula 2. To determine the relationship between two index entries of the same word, we cite the well known **PageRank** [15]. It is the merging idea that the word k_i extracted from page p is more important if the $PR(p)$ value is higher. In our study, for computing $PR(p)$, we determine the *damping factor* as a constant which is often assumed to be 0.85. For instance, "calendar" in Table 1 and Table 2 should be merged. Suppose the words in Table 1 are extracted from page u, and Table 2 from page v. As we have saved the link information (u,v) during the crawling, we can compute $PR(u)$ and $PR(v)$ immediately by those link information. Assume that *PR(u)=0.25*, *PR(v)=0.5* after computation, then the two entries of "calendar" are merged to one as shown in Table 3.

Table 1. Indexing table of set1

Entries	Related words
award	calendar($\frac{1}{5}$), honor($\frac{1}{5}$), job($\frac{1}{5}$), month($\frac{1}{5}$)
calendar	award($\frac{1}{5}$), honor($\frac{1}{5}$), job($\frac{1}{5}$), month($\frac{1}{5}$)
honor	award($\frac{1}{5}$), calendar($\frac{1}{5}$), job($\frac{1}{5}$), month($\frac{1}{5}$)
job	award($\frac{1}{5}$), calendar($\frac{1}{5}$), honor($\frac{1}{5}$), month($\frac{1}{5}$)
month	award($\frac{1}{5}$), calendar($\frac{1}{5}$), honor($\frac{1}{5}$), job($\frac{1}{5}$)

Table 2. Indexing table of set2

Entries	Related words
academic	education($\frac{2}{5}$), calendar($\frac{1}{5}$), overview($\frac{1}{5}$)
calendar	education($\frac{2}{5}$), academic($\frac{1}{5}$), overview($\frac{1}{5}$)
education	academic($\frac{1}{5}$), calendar($\frac{1}{5}$), overview($\frac{1}{5}$)
overview	education($\frac{2}{5}$), academic($\frac{1}{5}$), calendar($\frac{1}{5}$)

Table 3. Merged result

Entries	Related words
calendar	education(0.2), academic(0.1), overview(0.1), award(0.05), honor(0.05), job(0.05), month(0.05)

4   Experimental Evaluations

We performed experiments to evaluate the BBCECI model. The key idea of our approach is to represent the design intent of web page designer as much as possible, which guarantees that we can extract the relevant even semantically related words from the link blocks. Therefore, the precision of representing the designer's intention becomes an important evaluation criterion. We thus evaluated the precision in partitioning the link blocks based on different websites. Certainly, the most important one is that the precision of related words extracted by BBCECI model, therefore we evaluated it as well.

$$Precision = \frac{\sum_{i=1}^{K} \frac{|\{relevant\ links\}|}{|\{links \in Cluster_i\}|}}{K} \tag{3}$$

4.1   Evaluation Based on Different Websites

In this section, we performed the evaluation based on different websites. To apply DB-SCAN algorithm, we set *1-dist* as the appropriate *k-dist* condition to determine the paratemeters *MinPts* and *Eps* by experimental preparation. The precision is calculated using Formula 3 where K is the number of clusters divided by DBSCAN. To make the evaluation as fair as possible, the websites are collected from the Open Directory Project[1]. They are referred to four different categories, education, news, computer organization and data engineering, which are summarized in Table 4.

The browser-based crawler visited the websites in Table 4 following the BFS strategy to crawl 100 pages for each website. The DBSCAN algorithm was carried out to do clustering on the 100 pages for each website independently. According to the formula 3, the average precisions of partitioning link blocks for these websites were computed separately. After computation, the radar chart was plotted to visualize the results in the pellucid Figure 4. As shown in the radar chart, all the average precisions of partitioning link blocks are almost over 0.90 except the precision of Stanford-U. It is confident that

---

[1] http://www.dmoz.org/

Table 4. Summary of different websites

Short name	Description	URL address	Category
Tokyo-U	Tokyo University	http://www.u-tokyo.ac.jp/	Education
Stanford-U	Stanford University	http://www.stanford.edu/	Education
BBC	BBC News	http://news.bbc.co.uk/	News
CNN	Cable News Network	http://www.cnn.com/	News
BCS	British Computer Society	http://www.bcs.org/	Computer Org.
CRA	Computing Research Association	http://www.cra.org/	Computer Org.
SIGMOD	ACM Special Interest Group on MOD	http://www.sigmod.org/	Data Eng.
SIGKDD	ACM Special Interest Group on KDD	http://www.sigkdd.org/	Data Eng.

our BBCECI model can represent the link blocks in the extremely high precision, which is well reflecting the page designer's intention.

4.2 Precision of Related Words by BBCECI Model

Concerning to the focus of extracting related words, whether the words and the word of index entry are related should be examined manually. For this purpose, we performed experiments to evaluate the the precision of related words that are extracted by the BBCECI model, according to the results of examination.

Firstly, we drove the browser-based crawler to visit the website of Tsukuba University (http://www.tsukuba.ac.jp/english/), and crawled totally 200 pages following the BFS strategy. From the 200 pages, the link-based extracting module extracted 3,526 hyperlinks. Then we applied the DBSCAN algorithm to execute clustering for each page independently. As soon as the clustering process finished, all the hyperlinks were divided into 1,190 different anchor text clusters. Finally, we preprocessed all the anchor texts by doing tokenization, stemming and stopword elimination, achieving 4,647 single words as the output. By applying cluster-based indexing method, we totally extracted 801 related words and created an index table containing 801 index entries corresponding to their related words. All the index entries are unique by merging the duplicates

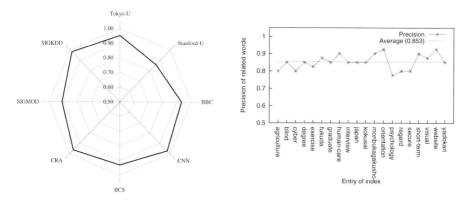

Fig. 4. Average precisions of partitioning link **Fig. 5.** Precision of related words extracted by blocks based on different websites   BBCECI model

Table 5. Output data of each process module

Process	Output	Quantity
Browser-based crawling	web pages	200
Link-based extraction	hyperlinks	3526
Coordinate-based clustering	clusters	1190
Cluster-based indexing	single words	4647
	related words	801
	index entries	801

using the method mentioned in the BBCECI model. The output data of each process module is summarized in Table 5.

As we known, if the size of experimental data is larger, the evaluation would be more accurate and reliable. However, manually evaluating the precisions on all the 801 entries in the index table is not impossible but an inefficient job. Therefore, we carried out the evaluation by sampling. We randomly selected 20 entries from the index table, and manually evaluated the precision of the top 20 related words for each entry.

The result of evaluation is shown in Figure 5. It is clear that all the precisions are almost over 0.8 except the entry of "psychology". It is more confident that the BBCECI model proposed in this paper can provide related words in a high average precision about 0.853. With such an index table of related words in hand, implementing a server to suggest related words to support the Web search will become an easy job.

5   Conclusions and Future Works

In this paper, we proposed the BBCECI model to extract related words, keeping consideration on their semantic similarities. The browser-based crawling module and link-based extracting module solve the problem arisen because of CSS style. The coordinate-based clustering module recognizes the link blocks in extremely high precision, which largely reflect the intention of page designer. The experimental evaluations indicate that the BBCECI model can extract related words in a high average precision over 0.90.

The main modules have been implemented, thus as the future work, we are planning to implement a keyword suggestion server, which provides users with related words to their queries when they search the Web. From an search engine combined with such a server, a user can receive helpful suggestion.

Besides, we are planing to carry out more experimental evaluations to compare our BBCECI model and the suggestion server with other related works that are focusing on the extraction of related words.

References

1. Sato, S., Sasaki, Y.: Automatic collection of related terms from the web. In: The Companion Volume to the Proceedings of 41st ACL, pp. 121–124 (2003)
2. Cristianini, N., Shawe-Taylor, J., Lodhi, H.: Latent semantic kernels. J. Intell. Inf. Syst. 18(2-3), 127–152 (2002)

3. Terra, E., Clarke, C.L.A.: Scoring missing terms in information retrieval tasks. In: Proceedings of the 13th CIKM, pp. 50–58 (2004)
4. Kraft, R., Zien, J.Y.: Mining anchor text for query refinement. In: Proceedings of the 13th WWW, pp. 666–674 (2004)
5. Kandola, J.S., Shawe-Taylor, J., Cristianini, N.: Learning semantic similarity. In: Proceedings of Advances in Neural Information Processing Systems 15 (NIPS), pp. 657–664 (2002)
6. Sahami, M., Heilman, T.D.: A web-based kernel function for measuring the similarity of short text snippets. In: Proceedings of the 15th WWW, pp. 377–386 (2006)
7. Bollegala, D., Matsuo, Y., Ishizuka, M.: Measuring semantic similarity between words using web search engines. In: Proceedings of the 16th WWW, pp. 757–766 (2007)
8. Pant, G.: Deriving link-context from html tag tree. In: Proceedings of the 8th SIGMOD workshop DMKD, pp. 49–55 (2003)
9. Hung, B.Q., Otsubo, M., Hijikata, Y., Nishida, S.: Extraction of anchor-related text and its evaluation by user studies. In: Proceedings Part I of the 12th HCI, pp. 446–455 (2007)
10. Chakrabarti, S.: In: Mining The Web: Discovering Knowledge from Hypertext Data, pp. 227–233. Elsevier, USA (2003)
11. Liu, B.: In: Web Data Mining: Exploring Hyperlinks, Contents, and Usage Data, pp. 323–379. Springer, Heidelberg (2007)
12. Hattori, G., Hoashi, K., Matsumoto, K., Sugaya, F.: Robust web page segmentation for mobile terminal using content-distances and page layout information. In: Proceedings of the 16th WWW, pp. 361–370 (2007)
13. Ester, M., Kriegel, H.P., Sander, J., Xu, X.: A density-based algorithm for discovering clusters in large spatial databases with noise. In: Proceedings of the 2nd KDD, pp. 226–231 (1996)
14. Najork, M., Wiener, J.L.: Breadth-first crawling yields high-quality pages. In: Proceedings of the 10th WWW, pp. 114–118 (2001)
15. Brin, S., Page, L.: The anatomy of a large-scale hypertextual web search engine. In: Proceedings of the 7th WWW, pp. 107–117 (1998)

Evolution of Query Optimization Methods: From Centralized Database Systems to Data Grid Systems*

Abdelkader Hameurlain

Institut de Recherche en Informatique de Toulouse IRIT, Paul Sabatier University
118, Route de Narbonne, 31062 Toulouse Cedex, France
Tel.: 33 (0) 5 61 55 82 48; Fax: 33 (0) 5 61 55 62 58
hameur@irit.fr

Abstract. The purpose of this talk is to provide a comprehensive state of the art concerning the evolution of query optimization methods from centralized database systems to data Grid systems through parallel, distributed and data integration systems. For each environment, we try to describe synthetically some methods, and point out their main characteristics.

Keywords: Query Optimization, Relational Database Systems Parallel and Distributed Database Systems, Data Integration Systems, Large Scale, Data Grid Systems.

1 Introduction

Query processing involves three steps: decomposition, optimization and execution. The first step decomposes a relational query (a SQL query) using logical schema into an algebraic query. During this step syntactic, semantic and authorization are done. The second step is responsible for generating an efficient execution plan for the given SQL query from the considered search space. The third step consists in implementing the efficient execution plan (or operator tree) [36]. In this talk, we focus only on query optimization methods. We consider multi-join queries without "group" and "order by" clauses.

Work related to the relational query optimization goes back to the 70s, and began mainly with the publications of Wong et al. [97] and Selinger et al. [80]. These papers motivated a large part of the database scientific community to focus their efforts on this subject. The optimizer's role is to generate, for a given SQL query, an optimal (or close to the optimal) execution plan from the considered search space. The optimization goal is to minimize response time and maximize throughput while minimizing optimization costs. An optimizer can be decomposed into three elements [26]: search space, search strategy, and cost model.

Several approaches, methods and techniques of query optimization have been proposed for various Database Management Systems DBMS (i.e. relational, deductive, distributed, object, parallel). The quality of query optimization methods depends

---

* An extended version of this paper will appear in the International Journal "Transactions on Large Scale Data and Knowledge Centered Systems", LNCS, Springer, Sept. 2009.

S.S. Bhowmick, J. Küng, and R. Wagner (Eds.): DEXA 2009, LNCS 5690, pp. 460–470, 2009.

strongly on the accuracy on the efficiency of cost models [1, 27, 28, 46, 70, 100]. There are two types of query optimization methods [17]: static and dynamic. Execution plans generated by a static optimizer can be sub-optimal, due to several causes [69]: estimation errors and unavailability of resources. In order to detect and modify the sub-optimal execution plans at run-time, dynamic optimization methods have been proposed in different environments: uni-processor, distributed, parallel and large scale [3, 7, 17, 38, 51, 53, 68, 75, 88, 99].

The purpose of this talk is to provide a state of the art concerning the evolution of query optimization methods in different environments. For each environment, we try to describe synthetically some methods, and to point out their main characteristics [69], especially, the nature of decision-making (centralized or decentralized), the type of modification (re-optimization or re-scheduling), the level of modification (intra-operator and/or inter-operator), and the type of event (estimation errors, delay, user preferences).

2   Uni-processor Relational Query Optimization

In the uni-processor relational systems, the query optimization process [13, 54, 63] consists of two steps: (i) logical optimization which consists in applying the classic transformation rules of the algebraic trees to reduce the manipulated data volume, and (ii) physical optimization which has roles of [63]: (a) determining an appropriate join method for each join operator, and (b) generating the order in which the joins are performed [48, 60] with respect to a cost model.

In the literature, we distinguish, generally, two classes of search strategies to solve the problem of the join scheduling: (i) Enumerative strategies, and (ii) Random strategies. Enumerative strategies are based on the generative approach. They use the principle of dynamic programming (e.g. optimizer of System R). The enumerative strategies are inadequate in optimizing complex queries because the number of execution plans quickly becomes too large [87]. To resolve this problem, random strategies are used. The transformational approach characterizes this kind of strategies. Several rules of transformation (e.g.; Swap, 3Cycle, Join commutativity) were proposed [48, 49, 85] where the validity depends on the nature of the considered search space [61].

The description of the principles of the search strategies leans on the generic search algorithms described in [60] and on the comparative study between the random algorithms proposed by [48, 49, 61, 85, 86]. Indeed, in [85, 86] and [48, 49], the authors concentrated their efforts on the performance evaluation of the random algorithms for Iterative Improvement and the Simulated Annealing. However, the difference of their results underlines the difficulty of such evaluation. In fact, for Swami and Gupta [85, 86], the Simulated Annealing algorithm is never superior to the Iterative Improvement whatever the time dedicated to the optimization is, while for Ioannidis and Cha Kong [48, 49], it is better than the Iterative Improvement algorithm after some optimization time.

According to the results in [47, 48, 49, 60, 85, 86], it is difficult to conclude about the superiority of a search strategy with regard to the one another.

The search strategies find the optimal solution more or less quickly according to their capacity to face the various problems. They must be adaptable to queries of diverse sizes (simple, medium, complex) and in various types of use (i.e. ad-hoc or repetitive)

[59]. A solution to this problem is the parameterization and the extensibility of query optimizers [9, 50, 59] possessing several search strategies, each being adapted for a type of queries.

3   Parallel and Distributed Query Optimization

Parallel relational query optimization methods [39] can be seen as an extension of relational query optimization methods developed for the centralized systems, by integrating the parallelism dimension [21, 37, 42, 63, 88, 94]. Indeed, the generation of an optimal parallel execution plan (or close to optimal), is based on either a two-phase approach [41, 44], or on a one-phase approach [15, 61, 79]. A two-phase approach consists in two sequential different steps: (i) generation of an optimal sequential execution plan, and (ii) resource allocation to this plan. The last step consists, at first, in extracting the various sources of parallelism, then, to assign the resources to the operations of the execution plan by trying to meet the allocation constraints (i.e. data locality, and various sources of parallelism). As far as the one-phase approach, the steps (i) and (ii) are packed into one integrated component [63].

We provide an overview of some static and dynamic query optimization methods in a parallel relational environment [39].

In a static context [39], the most advanced works are certainly those of Garofalakis et al. [29, 30]. They extend elegantly the propositions of [14, 26, 41] where the algorithms of parallel query are based on a uni-dimensional cost model. Furthermore, [30] tackle the scheduling problem (i.e. parallelism extraction) and the resource allocation in a context, which can be multi-query by considering a multidimensional model of used resources (i.e. preemptive, and non-preemptive). The proposals of [30] seem to be the richest in terms of categories of considered resources, exploited parallelisms, and various allocation constraints. In a dynamic context, the majority of proposed methods [12, 38, 57, 68, 77] point out the importance of: (i) the determination of the join parallelism degree and the resource allocation method, (ii) the introduction of a dynamic re-optimization algorithm which detects and corrects sub-optimality of the execution plan produced by the optimizer at compile time. The basic idea of this algorithm is founded on the collection of the statistics in some key-points during the query execution. The collected statistics correspond to the real values (observed during the execution), where the estimation is subject to error at compile time. These statistics are used to improve the resource allocation or by changing the execution plan of the remainder of the query (the part of the query, which is not executed yet). Obviously, the re-optimization process will be engaged only in case of estimation errors really bringing sub-optimality besides of the execution plan.

As far as the distributed query processing [58], the optimization process is composed of two steps: a global optimization step and a local optimization step [73]. The global optimization consists of: (i) determining the best execution site for each local sub-query considering data replication; (ii) finding the best inter-site operator scheduling, and (iii) placing these last ones. As for local optimization, it optimizes the local sub-queries on each site which are involved to the query evaluation.

There are two types of distributed query optimization: static and dynamic. Static query optimization methods [16, 64, 81, 84] are focused mainly on the optimization of inter-site communication costs, by reducing the data volume transferred between sites. The introduction of a new operator, semi-join based join [8, 16, 93], provides certainly more flexibility to optimizers. However, it increases considerably the size of search space.

Dynamic query optimization methods [22, 72] are based on dynamic scheduling (or re-scheduling) of inter-site operators to correct the sub-optimality due to the inaccuracies of estimations and variations of available resources.

4 Query Optimization in Large Scale Data Integration Systems

Data integration systems extend [89, 96, 98] the distributed database approach to multiple, autonomy, and heterogeneous data sources by providing uniform access. Heterogeneity and autonomous of data sources characterize data integration systems. Sources might be restricted due to the limitation of their query interface or certain attribute must be hidden due to privacy reasons. To handle the limited query capabilities of data sources, new mechanisms have been introduced [31, 66], such as, Dependant Join Operator which is asymmetric in nature. This asymmetric propriety causes the search space to be restricted and raises the issue of capturing valid execution plans [65, 66, 98].

As for the optimization methods, the community quickly noticed that the centralized optimization methods [3, 6, 11, 51, 52, 53, 56] could not be scaled up for the following reasons: (i) the number of messages which is relatively important on a network with low bandwidth and strong latency, and (ii) the bottleneck that forms the optimizer. Hence, the decentralized dynamic optimization methods [91, 92] correct the sub-optimality of execution plans by decentralizing the control.

Centralized and decentralized dynamic optimization methods can be classified according to the modification level of execution plans. This modification can be taken either on the intra-operator level, or on the inter-operator (sub-query) level.

Furthermore, in a large scale environment it becomes very convenient to make the query execution autonomous and self-adaptable. In this perspective, two close approaches have been investigated: the broker approach [18], and the mobile agent approach [5, 55, 97, 71, 78]. The second approach consists in using a programming model based on mobile agents [25], knowing that at present the mobile agent platforms supply only migration mechanisms, but they do not offer proactive migration decision policy.

5 Query Optimization in Data Grid Systems

Large scale and dynamicity of nodes characterize the grid systems. Large scale environment means [40]: (i) high numbers of data sources (e.g. databases, xml files), users, and computing resources (i.e. CPU, memory, network and I/O bandwidth) which are heterogeneous and autonomous, (ii) the network bandwidth presents, in average, a low bandwidth and strong latency, and (iii) huge volumes of data. Dynamicity of nodes means that a node can join, leave or fail at any time.

Recently, the grid computing [24], intended initially for the intensive computing, open towards the management of voluminous, heterogeneous, and distributed data on a large-scale environment. Grid data management [74] raises new problems and presents real challenges such as resource discovery and selection, query processing and optimization, autonomic management, security, and benchmarking. To tackle these fundamental problems [74], several methods have been proposed [4, 20, 33, 34, 45, 67, 82, 90]. A very good and complete overview addressing the most above fundamental problems is described in [74].

In this talk, we are interesting only in query processing and optimization methods proposed in data grid systems.

Several approaches have been proposed for distributed query processing (DQP) in data grid environments [2, 4, 33, 34, 35, 45, 82, 95]. Smith et al. [82] discuss the role of DQP within the Grid and determine the impact of using Grid for each step of DQP. The properties of grid systems such as flexibility and power make grid systems suitable platforms for DQP [82].

Convergence between grid technologies and web services leads researchers to develop standardized grid interfaces. Open Grid Services Architecture OGSA [23] is one of the most well known standards used in grids. OGSA-DQP [2] is a high level data integration tool for service-based Grids. It is built on a Grid middleware named OGSA-DAI [4] which provides a middleware that assists its users by accessing and integrating data from separate sources via the Grid. Wohrer et al. [95] describe the concepts that provide virtual data sources on the Grid and that implement a Grid data mediation service which is integrated into OGSA-DAI [4].

By analyzing the DQP approaches on the Grid, the research community focused on the current adaptive query processing approaches [6, 32, 43, 53] and proposed extensions in grid environments [19, 33, 35]. These studies achieve query optimization, by providing efficient resource utilization, without considering parallelization.

As far as parallelism dimension integration, many authors have re-studied DQP in order to be efficiently adopted by considering the properties (e.g.; heterogeneity) of grids. Several methods are proposed in this direction [10, 20, 34, 62, 76, 83] which define different algorithms for parallel query processing in grid environments. The proposed methods consider different forms of parallelism (e.g. pipelined parallelism), whereas all of them consider also resource discovery and load balancing.

6 Conclusion

Because of the importance and the complexity of the query optimization problem, the database community has proposed approaches, methods and techniques in different environments. In this talk, we wanted to provide a survey related to evolution of query optimization methods from centralized relational database systems to data grid systems through parallel and distributed database systems and data integration (mediation) systems. For each environment, we described some query optimization methods, and pointed out their main characteristics which allow comparing them.

Acknowledgement

We would like to warmly thank Professor Roland Wagner for his kind invitation to give this talk.

References

1. Adali, S., Candan, K.S., Papakonstantinou, Y., Subrahmanian, V.S.: Query Caching and Optimization in Distributed Mediator Systems. In: Proc. of ACM SIGMOD Intl. Conf. on Management of Data, pp. 137–148. ACM Press, New York (1996)
2. Alpdemir, M.N., Mukherjee, A., Gounaris, A., Paton, N.W., Fernandes, A.A.A., Sakellariou, R., Watson, P., Li, P.: Using OGSA-DQP to support scientific applications for the grid. In: Herrero, P., S. Pérez, M., Robles, V. (eds.) SAG 2004. LNCS, vol. 3458, pp. 13–24. Springer, Heidelberg (2005)
3. Amsaleg, L., Franklin, M., Tomasic, A.: Dynamic query operator scheduling for wide-area remote access. Distributed and Parallel Databases 6(3), 217–246 (1998)
4. Antonioletti, M., et al.: The design and implementation of Grid database services in OGSA-DAI. Concurrency and Computation: Practice & Experience 17, 357–376 (2005)
5. Arcangeli, J.-P., Hameurlain, A., Migeon, F., Morvan, F.: Mobile Agent Based Self-Adaptive Join for Wide-Area Distributed Query Processing. Jour. of Database Management 15(4), 25–44 (2004)
6. Avnur, R., Hellerstein, J.-M.: Eddies: Continuously Adaptive Query Processing. In: Proc. of the ACM SIGMOD, vol. 29, pp. 261–272. ACM Press, New York (2000)
7. Babu, S., Bizarro, P.: Adaptive Query Processing in the Looking Glass. In: Second Biennial Conf. on Innovative Data Systems Research, CIDR 2005, pp. 238–249 (2005)
8. Bernstein, P.A., Goodman, N., Wong, E., Reeve, C.L., Rothnie Jr.: Query Processing in a System for Distributed Databases (SDD-1). ACM Trans. Database Systems 6(4), 602–625 (1981)
9. Bizarro, P., Bruno, N., DeWitt, D.J.: Progressive Parametric Query Optimization. IEEE Transactions on Knowledge and Data Engineering 21(4), 582–594 (2009)
10. Bose, S.K., Krishnamoorthy, S., Ranade, N.: Allocating Resources to Parallel Query Plans in Data Grids. In: Proc. of the 6th Intl. Conf. on Grid and Cooperative Computing, pp. 210–220. IEEE CS, Los Alamitos (2007)
11. Bouganim, L., Fabret, F., Mohan, C., Valduriez, P.: A dynamic query processing architecture for data integration systems. Journal of IEEE Data Engineering Bulletin 23(2), 42–48 (2000)
12. Brunie, L., Kosch, H., Wohner, W.: From the modeling of parallel relational query processing to query optimization and simulation. Parallel Processing Letters 8, 2–24 (1998)
13. Chaudhuri, S.: An Overview of Query Optimization in Relational Systems. In: Symposium in Principles of Database Systems PODS 1998, pp. 34–43. ACM Press, New York (1998)
14. Chekuri, C., Hassan, W.: Scheduling Problem in Parallel Query Optimization. In: Sympo. in Principles of Database Systems PODS 1995, pp. 255–265. ACM Press, New York (1995)
15. Chen, M.S., Lo, M., Yu, P.S., Young, H.S.: Using Segmented Right-Deep Trees for the Execution of Pipelined Hash Joins. In: Proc. of the 18th VLDB Conf., pp. 15–26. Morgan Kaufmann, San Francisco (1992)
16. Chiu, D.M., Ho, Y.C.: A Methodology for Interpreting Tree Queries Into Optimal Semi-Join Expressions. In: Proc. of ACM SIGMOD, pp. 169–178. ACM Press, New York (1980)
17. Cole, R.L., Graefe, G.: Optimization of dynamic query evaluation plans. In: Proc. Of ACM SIGMOD, vol. 24, pp. 150–160. ACM Press, New York (1994)

18. Collet, C., Vu, T.-T.: QBF: A Query Broker Framework for Adaptable Query Evaluation. In: Christiansen, H., Hacid, M.-S., Andreasen, T., Larsen, H.L. (eds.) FQAS 2004. LNCS, vol. 3055, pp. 362–375. Springer, Heidelberg (2004)
19. Cybula, P., Kozankiewicz, H., Stencel, K., Subieta, K.: Optimization of distributed queries in grid via caching. In: Meersman, R., Tari, Z., Herrero, P. (eds.) OTM-WS 2005. LNCS, vol. 3762, pp. 387–396. Springer, Heidelberg (2005)
20. Da Silva, V.F.V., Dutra, M.L., Porto, F., Schulze, B., Barbosa, A.C., de Oliveira, J.C.: An adaptive parallel query processing middleware for the Grid. Concurrence and Computation: Practice and Experience 18, 621–634 (2006)
21. Dewitt, D.J., Gray, J.: The Future of High Performance Database Systems. Communication of the ACM, 85–98 (1992)
22. Evrendilek, C., Dogac, A., Nural, S., Ozcan, F.: Multidatabase Query Optimization. Journal of Distributed and Parallel Databases 5(1), 77–113 (1997)
23. Foster, I.: The Grid: A New Infrastructure for 21st Century Science. Physics Today 55(2), 42–56 (2002)
24. Foster, I., Kesselman, C.: The Grid: Blueprint for a New Computing Infrastructure. Morgan Kaufmann, San Francisco (2004)
25. Fuggetta, A., Picco, G.-P., Vigna, G.: Understanding Code Mobility. IEEE Transactions on Software Engineering 24(5), 342–361 (1998)
26. Ganguly, S., Hasan, W., Krishnamurthy, R.: Query Optimization for Parallel Execution. In: Proc. of the 1992 ACM SIGMOD, vol. 21, pp. 9–18. ACM Press, San Diego (1992)
27. Ganguly, S., Goel, A., Silberschatz, A.: Efficient and Accurate Cost Models for Parallel Query Optimization. In: Symposium in Principles of Database Systems PODS 1996, pp. 172–182. ACM Press, New York (1996)
28. Gardarin, G., Sha, F., Tang, Z.-H.: Calibrating the Query Optimizer Cost Model of IRO-DB, an Object-Oriented Federated Database System. In: Proc. of 22th VLDB, pp. 378–389. Morgan Kaufmann, San Francisco (1996)
29. Garofalakis, M.N., Ioannidis, Y.E.: Multi-dimensional Resource Scheduling for Parallel Queries. In: Proc. of ACM SIGMOD, pp. 365–376. ACM Press, New York (1996)
30. Garofalakis, M.N., Ioannidis, Y.E.: Parallel Query Scheduling and Optimization with Time- and Space - Shared Resources. In: Proc. of the 23rd VLDB Conf., pp. 296–305. Morgan Kaufmann, San Francisco (1997)
31. Goldman, R., Widom, J.: WSQ/DSQ: A practical approach for combined querying of databases and the web. In: Proc. of ACM SIGMOD, pp. 285–296. ACM Press, New York (2000)
32. Gounaris, A., Paton, N.W., Fernandes, A.A.A., Sakellariou, R.: Adaptive query processing: A survey. In: Eaglestone, B., North, S.C., Poulovassilis, A. (eds.) BNCOD 2002. LNCS, vol. 2405, pp. 11–25. Springer, Heidelberg (2002)
33. Gounaris, A., Paton, N.W., Sakellariou, R., Fernandes, A.A.A.: Adaptive Query Processing and the Grid: Opportunities and Challenges. In: Proc. of the 15th Intl. Dexa Workhop, pp. 506–510. IEEE CS, Los Alamitos (2004)
34. Gounaris, A., Sakellariou, R., Paton, N.W., Fernandes, A.A.A.: Resource Scheduling for Parallel Query Processing on Computational Grids. In: Proc. of the 5th IEEE/ACM Intl. Workshop on Grid Computing, pp. 396–401 (2004)
35. Gounaris, A., Smith, J., Paton, N.W., Sakellariou, R., Fernandes, A.A.A., Watson, P.: Adapting to Changing Resource Performance in Grid Query Processing. In: Pierson, J.-M. (ed.) VLDB DMG 2005. LNCS, vol. 3836, pp. 30–44. Springer, Heidelberg (2006)
36. Graefe, G.: Query Evaluation Techniques for Large Databases. ACM Computing Survey 25(2), 73–170 (1993)

37. Hameurlain, A., Morvan, F.: An Overview of Parallel Query Optimization in Relational Systems. In: 11th Intl Worshop on Database and Expert Systems Applications, pp. 629–634. IEEE Computer Society Press, Los Alamitos (2000)
38. Hameurlain, A., Morvan, F.: CPU and incremental memory allocation in dynamic parallelization of SQL queries. Journal of Parallel Computing 28(4), 525–556 (2002)
39. Hameurlain, A., Morvan, F.: Parallel query optimization methods and approaches: a survey. Journal of Computers Systems Science & Engineering 19(5), 95–114 (2004)
40. Hameurlain, A., Morvan, F., El Samad, M.: Large Scale Data management in Grid Systems: a Survey. In: IEEE Intl. Conf. on Information and Communication Technologies: from Theory to Applications, pp. 1–6. IEEE CS, Los Alamitos (2008)
41. Hasan, W., Motwani, R.: Optimization Algorithms for Exploiting the Parallelism - Communication Tradeoff in Pipelined Parallelism. In: Proc. of the 20th int'l Conf. on VLDB, pp. 36–47. Morgan Kaufmann, San Francisco (1994)
42. Hasan, W., Florescu, D., Valduriez, P.: Open Issues in Parallel Query Optimization. SIGMOD Record 25(3), 28–33 (1996)
43. Hellerstein, J.M., Franklin, M.J.: Adaptive Query Processing: Technology in Evolution. Bulletin of Technical Committee on Data Eng. 23(2), 7–18 (2000)
44. Hong, W.: Exploiting Inter-Operation Parallelism in XPRS. In: Proc. Of ACM SIGMOD, pp. 19–28. ACM Press, New York (1992)
45. Hu, N., Wang, Y., Zhao, L.: Dynamic Optimization of Sub query Processing in Grid Database, Natural Computation. In: Proc of the 3rd Intl Conf. on Natural Computation, vol. 5, pp. 8–13. IEEE Computer Society Press, Los Alamitos (2007)
46. Hussein, M., Morvan, F., Hameurlain, A.: Embedded Cost Model in Mobile Agents for Large Scale Query Optimization. In: Proc. of the 4th Intl. Symposium on Parallel and Distributed Computing, pp. 199–206. IEEE CS, Los Alamitos (2005)
47. Ioannidis, Y.E., Wong, E.: Query Optimization by Simulated Annealing. In: Proc. of the ACM SIGMOD Intl. Conf. on Management of Data, pp. 9–22. ACM Press, New York (1987)
48. Ioannidis, Y.E., Kang, Y.C.: Randomized Algorithms for Optimizing Large Join Queries. Proc. of ACM SIGMOD 19, 312–321 (1990)
49. Ioannidis, Y.E., Christodoulakis, S.: On the Propagation of Errors in the Size of Join Results. In: Proc. of the ACM SIGMOD, pp. 268–277. ACM Press, New York (1991)
50. Ioannidis, Y.E., Ng, R.T., Shim, K., Sellis, T.K.: Parametric Query Optimization. In: 18th Intl. Conf. on VLDB, pp. 103–114. Morgan Kaufmann, San Francisco (1992)
51. Ives, Z.-G., Florescu, D., Friedman, M., Levy, A.Y., Weld, D.S.: An adaptive query execution system for data integration. In: Proc. of the ACM SIGMOD Intl. Conf. on Management of Data, pp. 299–310. ACM Press, New York (1999)
52. Ives, Z.-G., Levy, A.Y., Weld, D.S., Florescu, D., Friedman, M.: Adaptive query processing for internet applications. Journal of IEEE Data Engineering Bulletin 23(2), 19–26 (2000)
53. Ives, Z.-G., Halevy, A.-Y., Weld, D.-S.: Adapting to Source Properties in Processing Data Integration Queries. In: Proc. of the ACM SIGMOD, pp. 395–406. ACM Press, New York (2004)
54. Jarke, M., Koch, J.: Query Optimization in Database Systems. ACM Comput. Surv. 16(2), 111–152 (1984)
55. Jones, R., Brown, J.: Distributed query processing via mobile agents (1997), http://www.cs.umd.edu/~rjones/paper.html (Found 14 November 2002)
56. Kabra, N., Dewitt, D.J.: Efficient Mid - Query Re-Optimization of Sub-Optimal Query Execution Plans. In: Proc. of the ACM, pp. 106–117. ACM Press, New York (1998)

57. Kosch, H.: Managing the operator ordering problem in parallel databases. Future Generation Computer Systems 16(6), 665–676 (2000)
58. Kossmann, D.: The State of the Art in Distributed Query Processing. ACM Computing Surveys 24(24), 422–429 (2000)
59. Lanzelotte, R.S.G.: OPUS: an extensible Optimizer for Up-to-date database Systems. Ph-D Thesis, Computer Science, PUC-RIO, avail. INRIA, Rocquencourt, n° TU-127 (1990)
60. Lanzelotte, R.S.G., Valduriez, P.: Extending the Search Strategy in a Query Optimizer. In: Proc. of the Int'l Conf. on VLDB, pp. 363–373. Morgan Kaufmann, San Francisco (1991)
61. Lanzelotte, R.S.G., Valduriez, P., Zaït, M.: On the Effectiveness of Optimization Search Strategies for Parallel Execution Spaces. In: Proc. of VLDB, pp. 493–504 (1993)
62. Liu, S., Karimi, H.A.: Grid query optimizer to improve query processing in grids. Future Generation Computer Systems 24(5), 342–353 (2008)
63. Lu, H., Ooi, B.C., Tan, K.-L.: Query Processing in Parallel Relational Database Systems. IEEE CS Press, Los Alamitos (1994)
64. Mackert, L.F., Lohman, G.M.: R* Optimizer Validation and Performance Evaluation for Distributed Queries. In: Proc. of the 12th Intl. Conf. on VLDB, pp. 149–159 (1986)
65. Manolescu, I.: Techniques d'optimisation pour l'interrogation des sources de données hétérogènes et distribuées, Ph-D Thesis, Versailles Univ., France (2001)
66. Manolescu, I., Bouganim, L., Fabret, F., Simon, E.: Efficient querying of distributed resources in mediator systems. In: Meersman, R., Tari, Z., et al. (eds.) CoopIS 2002, DOA 2002, and ODBASE 2002. LNCS, vol. 2519, pp. 468–485. Springer, Heidelberg (2002)
67. Marzolla, M., Mordacchini, M., Orlando, S.: Peer-to-Peer for Discovering resources in a Dynamic Grid. Jour. of Parallel Computing 33(4-5), 339–358 (2007)
68. Mehta, M., Dewitt, D.J.: Managing Intra-Operator Parallelism in Parallel Database Systems. In: Proc. of the 21th Intl. Conf. on VLDB, pp. 382–394 (1995)
69. Morvan, F., Hameurlain, A.: Dynamic Query Optimization: Towards Decentralized Methods. Intl. Jour. of Intelligent Information and Database Systems, Inderscience Publishers (in press, 2009)
70. Naacke, H., Gardarin, G., Tomasic, A.: Leveraging Mediator Cost Models with Heterogeneous Data Sources. In: Proc. of the 14th Intl. Conf. on Data Engineering, pp. 351–360. IEEE CS, Los Alamitos (1998)
71. Ozakar, B., Morvan, F., Hameurlain, A.: Mobile Join Operators for Restricted Sources. Mobile Information Systems: An International Journal 1(3), 167–184 (2005)
72. Ozcan, F., Nural, S., Koksal, P., Evrendilek, C., Dogac, A.: Dynamic query optimization in multidatabases. Data Engineering Bulletin CS 20(3), 38–45 (1997)
73. Özsu, M.T., Valduriez, P.: Principles of Distributed Database Systems, 2nd edn. Prentice-Hall, Englewood Cliffs (1999)
74. Pacitti, E., Valduriez, P., Mattoso, M.: Grid Data Management: Open Problems and News Issues. Intl. Journal of Grid Computing 5(3), 273–281 (2007)
75. Paton, N.W., Chávez, J.B., Chen, M., Raman, V., Swart, G., Narang, I., Yellin, D.M., Fernandes, A.A.A.: Autonomic query parallelization using non-dedicated computers: an evaluation of adaptivity options. VLDB Journal 18(1), 119–140 (2009)
76. Porto, F., da Silva, V.F.V., Dutra, M.L., Schulze, B.: An adaptive distributed query processing grid service. In: Pierson, J.-M. (ed.) VLDB DMG 2005. LNCS, vol. 3836, pp. 45–57. Springer, Heidelberg (2006)
77. Rahm, E., Marek, R.: Dynamic Multi-Resource Load Balancing in Parallel Database Systems. In: Proc. of the 21st VLDB Conf., pp. 395–406 (1995)

78. Sahuguet, A., Pierce, B., Tannen, V.: Distributed Query Optimization: Can Mobile Agents Help? (2000),
http://www.seas.upenn.edu/~gkarvoun/dragon/publications/sahuguet/ (Found December 11, 2003)

79. Schneider, D., Dewitt, D.J.: Tradeoffs in Processing Complex Join Queries via Hashing in Multiprocessor Database Machines. In: Proc. of the 16th VLDB Conf., pp. 469–480. Morgan Kaufmann, San Francisco (1990)

80. Selinger, P.G., Astrashan, M., Chamberlin, D., Lorie, R., Price, T.: Access Path Selection in a Relational Database Management System. In: Proc. of the 1979 ACM SIGMOD Conf. on Management of Data, pp. 23–34. ACM Press, New York (1979)

81. Selinger, P.G., Adiba, M.E.: Access Path Selection in Distributed Database Management Systems. In: Proc. Intl. Conf. on Data Bases, pp. 204–215 (1980)

82. Smith, J., Gounaris, A., Watson, P., Paton, N.W., Fernandes, A.A.A., Sakellariou, R.: Distributed query processing on the grid. In: Parashar, M. (ed.) GRID 2002. LNCS, vol. 2536, pp. 279–290. Springer, Heidelberg (2002)

83. Soe, K.M., New, A.A., Aung, T.N., Naing, T.T., Thein, N.L.: Efficient Scheduling of Resources for Parallel Query Processing on Grid-based Architecture. In: Proc. of the 6th Asia-Pacific Symposium, pp. 276–281. IEEE Computer Society Press, Los Alamitos (2005)

84. Stonebraker, M., Aoki, P.M., Litwin, W., Pfeffer, A., Sah, A., Sidell, J., Staelin, C., Yu, A.: Mariposa: A Wide-Area Distributed Database System. VLDB Jour. 5(1), 48–63 (1996)

85. Swami, A.N., Gupta, A.: Optimization of Large Join Queries. In: Proc. of the ACM SIGMOD Intl. Conf. on Management of Data, pp. 8–17. ACM Press, New York (1988)

86. Swami, A.N.: Optimization of Large Join Queries: Combining Heuristic and Combinatorial Techniques. In: Proc. of the ACM SIGMOD, pp. 367–376 (1989)

87. Tan, K.L., Lu, H.: A Note on the Strategy Space of Multiway Join Query Optimization Problem in Parallel Systems. SIGMOD Record 20(4), 81–82 (1991)

88. Taniar, D., Leung, C.H.C., Rahayu, J.W., Goel, S.: High Performance Parallel Database Processing and Grid Databases. John Wiley & Sons, Chichester (2008)

89. Tomasic, A., Raschid, L., Valduriez, P.: Scaling Access to Heterogeneous Data Sources with DISCO. IEEE Trans. Knowl. Data Eng. 10(5), 808–823 (1998)

90. Trunfio, P., et al.: Peer-to-Peer resource discovery in Grids: Models and systems. Future Generation Computer Systems 23(7), 864–878 (2007)

91. Urhan, T., Franklin, M.: XJoin: A reactively-scheduled pipelined join operator. IEEE Data Engineering Bulletin 23(2), 27–33 (2000)

92. Urhan, T., Franklin, M.: Dynamic pipeline scheduling for improving interactive query performance. In: Proc. of 27th Intl. Conf. on VLDB, pp. 501–510. Morgan Kaufmann, San Francisco (2001)

93. Valduriez, P.: Semi-Join Algorithms for Distributed Database Machines. In: Proc. of the 2nd Intl. Symposium on Distributed Databases, pp. 22–37. North-Holland, Amsterdam (1982)

94. Valduriez, P.: Parallel Database Systems: Open Problems and News Issues. Distributed and Parallel Databases 1, 137–165 (1993)

95. Wohrer, A., Brezany, P., Tjoa, A.M.: Novel mediator architectures for Grid information systems. Future Generation Computer Systems, 107–114 (2005)

96. Wiederhold, G.: Mediators in the Architecture of Future Information Systems. Journal of IEEE Computer 25(3), 38–49 (1992)

97. Wong, E., Youssefi, K.: Decomposition: A Strategy for Query Processing. ACM Transactions on Database Systems, 223–241 (1976)

98. Yerneni, R., Li, C., Ullman, J.D., Garcia-Molina, H.: Optimizing Large Join Queries in Mediation Systems. In: Beeri, C., Bruneman, P. (eds.) ICDT 1999. LNCS, vol. 1540, pp. 348–364. Springer, Heidelberg (1998)

99. Zhou, Y., Ooi, B.C., Tan, K.-L., Tok, W.H.: An adaptable distributed query processing architecture. Data & Knowledge Engineering 53(3), 283–309 (2005)

100. Zhu, Q., Motheramgari, S., Sun, Y.: Cost Estimation for Queries Experiencing Multiple Contention States in Dynamic Multidatabase Environments. Journal of Knowledge and Information Systems Publishers 5(1), 26–49 (2003)

Reaching the Top of the Skyline: An Efficient Indexed Algorithm for Top-k Skyline Queries

Marlene Goncalves and María-Esther Vidal

Universidad Simón Bolívar, Departamento de Computación, Apartado 89000
Caracas 1080-A, Venezuela
{mgoncalves,mvidal}@usb.ve

Abstract. Criteria that induce a Skyline naturally represent user's preference conditions useful to discard irrelevant data in large datasets. However, in the presence of high-dimensional Skyline spaces, the size of the Skyline can still be very large, making unfeasible for users to process this set of points. To identify the best points among the Skyline, the Top-k Skyline approach has been proposed. Top-k Skyline uses discriminatory criteria to induce a total order of the points that comprise the Skyline, and recognizes the best or top-k objects based on these criteria. Different algorithms have been defined to compute the top-k objects among the Skyline; while existing solutions are able to produce the Top-k Skyline, they may be very costly. First, state-of-the-art Top-k Skyline solutions require the computation of the whole Skyline; second, they execute probes of the multicriteria function over the whole Skyline points. Thus, if k is much smaller than the cardinality of the Skyline, these solutions may be very inefficient because a large number of non-necessary probes may be evaluated. In this paper, we propose the TKSI, an efficient solution for the Top-k Skyline that overcomes existing solutions drawbacks. The TKSI is an index-based algorithm that is able to compute only the subset of the Skyline that will be required to produce the top-k objects; thus, the TKSI is able to minimize the number of non-necessary probes. We have empirically studied the quality of TKSI, and we report initial experimental results that show the TKSI is able to speed up the computation of the Top-k Skyline in at least 50% percent w.r.t. the state-of-the-art solutions, when k is smaller than the size of the Skyline.

Keywords: Preference based Queries, Skyline, Top-k.

1 Introduction

Emerging technologies, such as the Semantic Web, Semantic Grid, Semantic Search and Peer to Peer, have made available a huge number of publicly very large data sources. For example, Google has indexed between ten and thirteen billion of Web pages by the time this paper has been written [19]. This exorbitant grow of data has impacted the performance of tasks whose complexity depends on the size of the input datasets, even when a large volume of these data may be irrelevant for solving some of these tasks. Particularly, the task of evaluating queries based on users preferences may be considerably affected by this situation. Thus, users have to be aware that a possibly

S.S. Bhowmick, J. Küng, and R. Wagner (Eds.): DEXA 2009, LNCS 5690, pp. 471–485, 2009.

large subset of input dataset may be useless, and criteria to efficiently discard irrelevant data need to be applied.

Skyline approaches have been successfully used to naturally express user's preference conditions useful to characterize relevant data in large datasets. However, in the presence of high-dimensional Skyline spaces, the size of the Skyline can still be very large, making unfeasible for users to process this set of points. To identify the best Skyline points, the Top-k Skyline has been proposed. Top-k Skyline uses discriminatory criteria to induce a total order of the Skyline points, and recognizes the top-k objects based on these criteria.

Several algorithms have been defined to compute the Top-k Skyline, but they may be very costly. First, they require the computation of the whole Skyline; second, they execute probes of the multicriteria function over the whole Skyline points. Thus, if k is much smaller than the cardinality of the Skyline, these solutions may be very inefficient because a large number of non-necessary probes may be evaluated, i.e., at least Skyline size minus k performed probes will be non-necessaries.

In this paper, we address the problem of computing Top-k Skyline queries efficiently. Given a multicriteria function m, a score function f and a set of objects $\mathcal{DO}$ which may be distributed among several data sources, we propose algorithms to identify the top-k objects from the Skyline, while probes of functions m and f are minimized.

Lets consider a government agency which offers a number of fellowships to the best three graduate students that apply to this grant. Applicants must submit their resumes, providing information on their academic and professional performance, which include recommendation letters and a statement of purpose. The summarized information is organized in the *Candidate* relational table, where fellowship candidates are described by an identifier, degrees, publications, GPA, and two ranking scores that denote the quality of the recommendation letters and the statement of purpose:

Candidate(Id,Degree,Publication,GPA,Recommendation, Statement)

Suppose the agency has received applications from ten candidates; Table 1 illustrates information of these candidates. Since the agency only can grant three fellowships, it has to select among the ten candidates, the top-3 that best meet its requirements.

According to the agency policy, all criteria are equally important and relevant; hence, either a weight or a score function cannot be assigned. A candidate can be chosen for granting a fellowship, if and only if, there is no other candidate with a higher degree, number of publications, and GPA. To nominate a candidate, the agency employees must identify the set of all the candidates that are non-dominated by any other candidate in terms of these criteria. Thus, tuples in table *Candidate* must be selected in terms of the values of: Degree, Publication, and GPA. For example, the candidate a dominates the candidates g and h because he has worse values in the Degree, Publication and GPA attributes. Following these criteria, the nominates are computed, and presented in Table 2.

To select the top-3 among the nominated graduate students, the average of the recommendation letter and statement of purpose score is used; therefore, three candidates are the new nominates: e, b and a.

Intuitively, to select the granted graduate students, queries based on user preferences have been posted against the table *Candidate*. There are several databases languages

Table 1. Candidates for three fellowships

Id	Degree	Publication	GPA	Recommendation	Statement
e	MsC	1	5.0	4.8	4.5
a	MsC	9	4.1	4.7	4.6
g	BEng	8	4.0	4.5	4.4
h	BEng	7	4.1	4.4	4.3
b	MsC	5	4.9	4.9	4.9
c	BEng	4	4.5	4.7	4.6
i	BEng	4	4.7	4.5	4.5
f	BEng	3	4.8	3.0	3.0
d	BEng	2	4.9	2.8	3.1
j	MsC	6	4.8	4.5	4.6

Table 2. Nominate Candidates for three fellowships

Id	Degree	Publication	GPA	Recommendation	Statement
e	MsC	1	5.0	4.8	4.5
a	MsC	9	4.1	4.7	4.6
b	MsC	5	4.9	4.9	4.9
j	MsC	6	4.8	4.5	4.6

to express preference queries. Skyline and Top-k are two user preference languages that could be used to identify some of the top-3 students. However, none of them will provide the complete set, and post-processing will be needed to identify all the students.

Skyline offers a set of operators to build an approximation of a Pareto curve or set of points that are non-dominated by any other point in the dataset. In consequence, by using Skyline, one could just obtain the nominated candidates.

Top-k approaches allow referees to implement a score function and filter some of the winners in terms of the score. In order to choose the top graduate students, top-k query engines compute the score for each tuple without checking dominance relationship among tuples in the dataset. However, it is not always possible to define such score function, because all criteria are equally important. Thus, the problem of selecting the granted students, corresponds to the problem of identifying the top-k elements in a partially ordered set.

Additionally, Skyline query engines construct a partially ordered set induced by the equally important criteria. Nevertheless, Top-k query engines select the top-k elements in terms of a score function that induces a totally ordered set. Therefore, to identify the granted students, a hybrid approach that combines the benefits of Skyline and Top-k is required. In this way, tuples in the answer will be chosen among the Skyline induced by a multiple criteria and then, ties will be broken using user-defined functions that eventually induce a total order.

Time complexity for answering preference queries is high and it depends on the dataset size and the number of probes performed. On one hand, in general, the problem of computing the Skyline is $O(n^2)$, where n is the number of tuples to be scanned. This is because all the n tuples need to be compared against themselves to probe the multicriteria function m[1]. On the other hand, the time complexity of selecting the top-k objects is $O(n \log n)$ because in the worst case, the whole input set has to be ordered[2]. Since a Top-k Skyline query engine needs to stratify the input data until the top-k objects are produced, time complexity of this task is $O(n^2)$. To decrease the processing time, Top-k Skyline query engines must implement efficient mechanisms to reduce the score and multicriteria function probes. To achieve this goal, we propose a query evaluation algorithm that minimize the number of non-necessary probes, i.e., this algorithm is able to identify the top-k objects in the Skyline, for which there are not k better Skyline objects in terms the score function f.

The paper is comprised of five sections. In Section 2 we define our Top-k Skyline approach; section 3 illustrates a description of state-of-the-art algorithms to compute Skyline queries. In Section 4 we propose the TKSI, an index based algorithm that is able to compute only the subset of the Skyline that will be required to produce the top-k objects. Section 5 reports the results of our experimental study. Section 6 summarizes existing related approaches. Finally, in Section 7, the concluding remarks and future work are pointed out.

2   Top-k Skyline

Given a set $\mathcal{DO} = \{o_1, \ldots, o_n\}$ of database objects, where each object o_j is characterized by p attributes $(A_1, \ldots, A_p)$; r different score functions $s_1, \ldots, s_q, \ldots, s_r$ defined over some of the p attributes, where each $s_i : O \to [0, 1]$, $1 \le i \le r$; a score function f defined on some scores s_i, which induces a total order of the objects in $\mathcal{DO}$; a multicriteria function m defined over some of the score functions $s_1, \ldots, s_q$, which rises a partial order of the objects in $\mathcal{DO}$; r sorted indexes $I_1, \ldots, I_r$ containing the identifier of all database objects in descending order by each score function s_i, respectively; and random access for each object from any index to the other indexes.

For simplicity, we suppose that scores related to the multicriteria function need to be maximized, and the score functions $s_1, \ldots, s_q, \ldots, s_r$ respect a natural ordering over p attributes. We define the Skyline S according to a multicriteria function m as follows:

$$S = \left\{ \begin{array}{l} o_i \in \mathcal{DO}/\neg \exists o_j \in \mathcal{DO} : (s_1(o_i) \le s_1(o_j) \wedge \cdots \wedge s_q(o_i) \le s_q(o_j) \wedge \\ \exists x \in \{1, ..., q\} : s_x(o_i) < s_x(o_j)) \end{array} \right\} \quad (1)$$

The conditions to be satisfied by the answers of a Top-k Skyline query w.r.t. to the functions m and f, are described as follows:

$$\xi_{<m,f,k>} = \left\{ o_j \in \mathcal{DO}/o_j \in S \wedge \neg(\exists^{k-|S|}o_l \in S : (f(o_l) > f(o_j))) \right\} \quad (2)$$

where, $\exists^t$ means that exists at most t elements in the set.

---

[1] A study of complexity Skyline problem is presented in [9].
[2] First the score function is probed for each instance, then data are ordered and finally, the top-k instances are returned.

Finally, the probes of the functions m and f required to identify the top-k objects in the Skyline correspond to necessary probes, i.e., a probe p of the functions m or f is necessary if and only if p is performed on an object $o \in \xi_{<m,f,k>}$. In this work, we define an algorithm that minimizes the number of non-necessary probes, while computing the Top-k Skyline objects with respect to the functions m and f.

3   Background

In this section we illustrate the functionalities of state-of-the-art Skyline algorithms, and how they can be used to compute the Top-k Skyline. For simplicity we assume that data are stored in relational databases. Particularly, each column of a table is accessed by means of an index structure and each index contains object identifiers and their scores. An object may be retrieved considering two types of accesses over the indexes: a sequential access retrieves an object o from a sorted index I, while a random access returns the score s from a given object identifier id.

First, we explain the Basic Distributed Skyline (BDS) defined by Balke et al. [2]. BDS computes a Skyline in two stages: in the first phase BDS constructs a Skyline superset in terms of a final object or the first object whose attributes have been completely recovered. In the second phase, BDS discards the dominated points from this superset.

Second, we describe the algorithm known as Basic Multi-Objective Retrieval (BMOR) proposed by Balke and Güntzer [1]. BMOR relies the computation of the Skyline on the construction of a virtual object which is comprised of the worst values of the multicriteria function seen so far; thus, a point in the Skyline corresponds to a seen object that dominates the virtual object. To compute the Top-k Skyline we extend BDS and BMOR in order to handle index structures, and we called these extensions Basic Distributed Top-k Skyline (BDTKS) and Basic Multi-Objective Retrieval for Top-k Skyline (BMORTKS), respectively.

Similarly to BDS and BMOR, the algorithms BDTKS and BMORTKS iteratively construct a superset of the Skyline S, and then, they probe the function f for the objects in this set. Elements are considered respecting the order induced by m.

BDTKS and BMORTKS assume that the values of each attribute A_i are indexed by I_i. All these indexes are scanned in a round robin fashion and only one object is recovered during each access. The virtual object is built with the worst seen values in each index I_i. If an object in the Skyline dominates the virtual object, then it is not necessary to continue looking for non-dominated objects. This is because the virtual object will dominate any unseen object, and any unseen object has worse value in each dimension of f than the virtual object. Similarly, the final object is built accessing each list I_i in a round robin fashion, and because the final object will dominate any unseen object, it is not required to look for more non-dominated objects.

To illustrate the behavior of BDTKS and BMORTKS, consider a projection of the relational table *Candidate*. Publication, GPA and Recommendation are collected from three different indexes as can be seen in Table 3. Indexes are sorted by Publication, GPA or Recommendation. Also, suppose that the agency is interested in those candidates with maximum number of publications and GPA.

To evaluate this query, BDTKS performs sorted access on the dataset in a round robin fashion in order to build a Skyline superset as it is shown in Table 4. Once BDTKS

Table 3. Datasets exported by indexes I_1, I_2 and I_3

I_1		I_2		I_3	
Id	Publication	Id	GPA	Id	Recommendation
a	9	e	5.0	b	4.9
g	8	b	4.9	e	4.8
h	7	d	4.9	a	4.7
j	6	j	4.8	c	4.7
b	5	f	4.8	i	4.5
c	4	i	4.7	g	4.5
i	4	c	4.5	j	4.5
f	3	h	4.1	h	4.4
d	2	a	4.1	f	3.0
e	1	g	4.0	d	2.8

completely sees the object j (the final object), it stops, and the Skyline superset is comprised of the objects a, e, g, b, h, d, j and f. BDTKS can stop at this point, because data are ordered in the indexes and any unseen object will be worse than j in each attribute. Then, BDTKS performs random access to retrieve the unseen scores from a, e, g, b, h, d, j and f; using all these values, it discards dominated objects from the Skyline superset. Finally, BDTKS outputs the Skyline which is composed of the objects a, e, j and b. Top-k objects are selected from there.

Table 4. Data accessed by BDTKS and BMORTKS

	BDTKS			BMORTKS		
id	Publication/GPA	Source	id	Publication/GPA	Source	Virtual Object
a	9	I_1	a	9	I_1	(9,_)
e	5.0	I_2	e	5.0	I_2	(9,5.0)
g	8	I_1	g	8	I_1	(8,5.0)
b	4.9	I_2	b	4.9	I_2	(8,4.9)
h	7	I_1	h	7	I_1	(7,4.9)
d	4.9	I_2	d	4.9	I_2	(7,4.9)
j	6	I_1	j	6	I_1	(6,4.9)
j	4.8	I_2	j	4.8	I_2	(6,4.8)
b	5	I_1	b	5	I_1	(5,4.8)
f	4.8	I_2	f	4.8	I_2	(5,4.8)
			c	4	I_1	(5,4.8)
			i	4.7	I_2	(5,4.8)

BMORTKS scans the same objects than BDTKS, but it constructs the virtual object. A virtual object contains the worst values seen in each sequential access. For each seen object, the BMORTKS performs random access for retrieving unseen values and it compares pair-wise the seen objects against the updated virtual object. Table 4 presents the list of virtual objects produced, where each pair represents the worst values seen for the indexes I_1 and I_2. At this point, the seen objects b and j dominates the last virtual object $(5, 4.8)$; thus, a Skyline superset is constructed because the objects b and j dominates any unseen object, and BMORTKS discards dominated objects producing the same Skyline than BDTKS.

Suppose that only one fellowship can be granted, i.e., a top-1 needs to be retrieved, then, both algorithms have performed ten and twelve probes to compute the Skyline, while only six of them are necessary. In the next section, we will describe the TKSI algorithm which is able to minimize the number of non-necessary probes.

4   TKSI – An Index-Based Algorithm to Compute Top-k Skyline Queries

In this section we propose the Top-k SkyIndex (TKSI) algorithm. TKSI is presented in the Figure 1. TKSI is an index-based solution able to compute a subset of the Skyline

---

INPUT:

- $\mathcal{DO}$: Data Set; m: Multicriteria Function; f: Score Function; k: Integer;
- $I = \{I_1, \ldots, I_q, \ldots, I_r\}$: Set of indexes on attributes in m and f;
- $s_1, \ldots, s_q, \ldots, s_r$: Score Functions;

OUTPUT:

- $\xi_{<m,f,k>}$: Top-k Skyline Objects

1) **INITIALIZE:**
    a) $\xi_{<m,f,k>} \leftarrow \emptyset;\ i \leftarrow 1;\ cont \leftarrow 1;$
    b) $min_1 \leftarrow 1.0, \ldots, min_r \leftarrow 1.0;$ /*The Highest Scores*/
2) **SEARCH TOP-K SKYLINE:** While ($cont < k$ and $\exists\, o \in \mathcal{DO}$)
    a) Select the following object o_t from the index I_r by sequential access;
    b) $min_r \leftarrow s_r(o_t);$
    c) Perform all random accesses to retrieve scores of object o_t using indexes $I - \{I_r\};$
    d) $i \leftarrow$ number of the index with the minimum value among the values:
        $(min_1 - s_1(o_t)), \ldots, (min_{r-1} - s_{r-1}(o_t));$
    e) If exists an object o between the first object and the object o_t in I_i, and o dominates to o_t
        i) Discard the object o_t
    f) If o_t is incomparable /*it is a Top-k Skyline Object*/
        i) Add o_t to $\xi_{<m,f,k>};$
        ii) $cont \leftarrow cont + 1;$
    g) $min_i \leftarrow s_i(o_t)$
3) **EXIT** Return the Top-k Skyline objects

Fig. 1. The TKSI Algorithm

Table 5. Normalized indexes and the TKSI Algorithm Execution

I_1		I_2		I_3	
Id	**Publication**	**Id**	**GPA**	**Id**	**Recommendation**
a	0.9	e	1.00	b	0.98
g	0.8	b	0.98	e	0.96
h	0.7	d	0.98	a	0.94
j	0.6	j	0.96	c	0.94
b	0.5	f	0.96	i	0.90
c	0.4	i	0.94	g	0.90
i	0.4	c	0.90	j	0.90
f	0.3	h	0.82	h	0.88
d	0.2	a	0.82	f	0.60
e	0.1	g	0.80	d	0.56

TKSI Execution		
Id	**Publication/GPA/Experience**	**Index**
b	0.98	I_3
e	1.00	I_2
b	0.98	I_2
d	0.98	I_2
j	0.96	I_2

required to produce the top-k objects; thus, TKSI minimizes the number of probes of the multicriteria and score functions and provides an efficient solution to the Top-k Skyline problem.

First, we illustrate the behavior of the TKSI with an example. Consider indexed values of Table 3 are normalized into range of [0,1] in Table 5 and the following query: the top-1 candidates with maximum experience among those candidates with maximum number of publications and GPA. To answer this query, Top-k SkyIndex accesses the objects from I_3 sequentially. For each accessed object o from I_3, TKSI verifies that o is a Top-k Skyline object. I_1 and I_2 contain the objects sorted descendantly. Because objects are sorted, it is very likely that any object with the higher values in each index of function m dominates the next objects in the indexes. For this reason, TKSI must select one of the indexes I_1 or I_2 in order to minimize the necessary probes over the multicriteria function m. The objects could be accessed in a round robin fashion. However, in order to speed up the computation, TKSI determines what is the index whose distance with respect to o is the lowest, i.e., the index that will avoid the access of more non-necessary objects. To do this, TKSI computes the distance D_1 as the difference between the last seen value from I_1 and the value for Publication of o ($min_1 - s_1(o)$), and D_2 as the difference between the last seen value from I_2 and the value for GPA of o ($min_2 - s_2(o)$). Next, TKSI selects the minimum value between D_1 and D_2. To compare the distances, the values of the attributes Publication and GPA are normalized.

Initially, TKSI accesses the first object b from I_3, and their values for Publication and GPA randomly (Step 2a-c)). Because of the objects from I_1 and I_2 have not been seen yet, TKSI assumes the last seen value is the maximum value possible for the attribute (Step 1c)). Therefore, the best distance between $D_1 = 1.0 - 0.5$ and $D_2 = 1.0 - 0.98$ is calculated (Step 2d)). In this case, I_2 has the minimum distance. Note that b is placed in the index I_2 in a lower position than the same object in I_1. Successively, the distances are calculated, and the objects e, b, d, and j from I_2 are accessed until the object j with a value lower in GPA is found. All these objects are compared against b to verify if some of them dominates it. Since, none of the objects e, b, d or j dominates b, the object b is a Top-k Skyline object (Step 2f)). If some object indexed by I_2 dominates b, a new object from I_3 is accessed. However, the algorithm decides to stop here because the objects behind j have worse values in Publication than b, and they may not dominate b. Moreover, the Top-1 Skyline object has been already found and only six of ten and twelve probes performed by BDTKS and BMORTKS were necessary. The detailed TKSI execution is showed in Table 5.

The TKSI algorithm is presented in Figure 1. In the first step (**INITIALIZE**), TKSI initializes the Top-k Skyline $\xi_{<m,f,k>}$; the minimum seen values for each index are in $min_1,\ldots,min_r$; finally, the variables i, and $cont$ are used as counters.

In step 2 (**SEARCH TOP-K SKYLINE**), TKSI identifies the Top-k Skyline objects. In the step 2a-c), the object o_t from I_r is accessed completely, and min_r is updated with the value of $s_r(o_t)$. In the step 2d), the index with the minimum distance to the object o_t is selected and scanned.

If o_t is dominated by some seen intermediate object in the selected index, then in step 2e) the object o_t is discarded. If the object o_t is non-dominated with respect the seen objects, then in step 2f) the object o_t is a Top-k Skyline object and it is inserted into $\xi_{<m,f,k>}$.

Thus, the algorithm continues until k objects have been found.

4.1   Properties of the TKSI Algorithm

The following property establishes that the TKSI algorithm is correct, i.e., TKSI computes the Top-k objects among the Skyline points.

Property 1 (The Best non-dominated objects). *Let b an object added by the TKSI algorithm to the $\xi_{<m,f,k>}$ set. Then, there exists not object $b' \in \mathcal{DO}$, s.t., b' is incomparable with b and $f(b') > f(b)$.*

Finally, the following theorem provides a lower-bound in the number probes that the algorithm TKSI can perform.

Theorem 1. *Let $\mathcal{DO}$ be a dataset. Let $I_i, \ldots, I_{r-1}$ be a set of indexes; each index is defined over one of the attributes of the multicriteria function m and it is ordered according to the m function. Let s be a score function defined over one attribute which is indexed by I_r and ordered according to m and s. To retrieve the top-k objects w.r.t. s in the Skyline induced by m, the TKSI algorithm performs at least $2k$ probes of the multicriteria function m.*

Proof
Suppose the Top-k Skyline objects are the top-k elements in the index I_r and that to compute them, each of these k objects only needs to be compared against the best of one index I_i on the attributes of the multicriteria function. To verify that an object o_t in I_r is a Top-k Skyline, o_t has to be pairwise compared against the best in one of the indexes I_i. Thus, the minimum number of probes required is $2k$.   ◇

5   Experimental Study

We have conducted an experimental study to empirically analyzed the quality of the TKSI algorithm w.r.t. the indexed-based algorithms BDTKS and BMORTKS.

5.1   Experimental Design

Datasets and Queries. The study was conducted on three relational tables populated with 100,000 tuples randomly generated. Each table contained an identifier and twelve columns that represent the score; values range from 0.0 to 1.0. The last six columns are highly correlated to the first six columns, i.e., the correlation between the i-th and (i+6)-th columns is higher than 90 %. A column may have duplicated values. The attribute values were generated following three data distributions:

 - *Uniform*: Attributes are independent of each other and their values are generated uniformly.
 - *Gaussian*:Attributes are independent of each other and their values are generated from five overlapping multidimensional Gaussian bells.
 - *Mixed*: Attributes are independent of each other. Data are divided into two groups of three columns: one group was generated using a Uniform distribution and the other, using a Gaussian distribution.

We randomly generated sixty queries characterized by the following properties: (a) only one table in the FROM clause; (b) the attributes in the multicriteria function and the score function were chosen randomly among the attributes of the table, following a Uniform distribution; (c) directives for each attribute of the multicriteria function were selected randomly considering only maximizing and minimizing criteria; (d) the number of attributes of the multicriteria function is six; (e) the number of attributes of the score function is one; (f) the argument of score function was chosen randomly following a Uniform distribution; and (g) k vary from 1%, 0.5%, 0.01%, 0.005%, 0.001% and 0.0001% of data size.

The average size of Skyline for sixty queries in each data distribution is reported in Table 6.

The experiments were evaluated on a SunFire V440 machine equipped with 2 processors Sparcv9 of 1.281 MHZ, 16 GB of memory and 4 disks Ultra320 SCSI of 73 GB running on SunOS 5.10 (Solaris 10). The BDTKS, BMORTKS, and TKSI algorithms were implemented in Java (64-bit JDK version 1.5.0 12).

TKSI implementation. The BDTKS, BMORTKS, and TKSI algorithms were developed on top of Oracle 9i. A set of sorted queries are executed for each criterion of the

Table 6. Average Skyline Size for each data distribution

Data Distribution	Average Skyline Size
Uniform	2,405
Gaussian	2,477
Mixed	2,539

multicriteria and the score functions, and the resultsets are stored on indexes. The resultsets are sorted ascendant or descendantly according to the criteria MIN or MAX of the multicriteria function. The resultset corresponding to the score function is sorted descendantly because the best k objects need to be retrieved. Each resultset is accessed on-demand.

Furthermore, a set of hash maps are built, one for each index. These hash maps are comprised of objects accessed by each index. Also, a variable for each index is updated with the last value seen in that index. Initially, these variables are set with the best values. Lately, they are updated according to the last object accessed by each index.

Thus, TKSI accesses the first object o from the index over the score function. It selects which is the index I_i that has the lowest gap with respect to o. The resultset of the selected index I_i is scanned until some object from I_i dominates o or none of the objects better than o in the attribute i dominates to o. If o is incomparable, then o is a Top-k Skyline object and it is added in the set of answers. This process continues until computing the top k objects.

Metrics. We report on the metrics: number of probes, seen-objects, sequential accesses, random accesses and total time. These metrics are defined as follows:

- Number of probes: Number of the multicriteria and the score function evaluations performed by the algorithm.
- Seen-objects: Number of different objects accessed by the algorithm.
- Sequential accesses: Number of sequential accesses performed by the algorithm.
- Random accesses: Number of random accesses performed by the algorithm.
- Accesses: Random accesses plus sequential accesses.
- Total time: Time of algorithm's evaluation. This metric is measured by using the time Solaris command.

5.2 Performance of the TKSI, BDTKS and BMORTKS Algorithms

We study the performance of the TKSI, BDTKS and BMORTKS algorithms. We computed the average of: number of probes, seen-objects, accesses, and total time, using the three algorithms. We selected and executed ten of the sixty generated queries in order to compare the three algorithms. This is because BMORTKS takes 8 hours approximately to finish the execution of ten queries.

Table 7 shows the results for the three algorithms. We confirm that the highest number of probes are performed by the BMORTKS algorithm. This may be because BMORTKS has to compare all the objects with the virtual object and then, once this condition is satisfied, the algorithm has to compare all these objects with themselves to determine which of them are non-dominated. Thus, probing with respect to the virtual object increases the number of comparisons of the BMORTKS algorithm. Additionally, the number of accesses and seen objects is similar among BMORTKS and BDTKS. Both results indicate that the number of probes affects evaluation time due to the number of probes was much higher for BMORTKS.

Finally, TKSI had the best performance because it does not build the Skyline completely, it only added necessary objects to the answer.

Table 7. Results for the algorithms TKSI, BDTKS and BMORTKS

	TKSI	**BDTKS**	**BMORTKS**
Number of Probes	412,049.8	23,749,796.4	27,201,876.5
Accesses	48,019.2	207,914.1	240,610.5
Seen-Objects	13,748.6	57,219.3	57,206.7
Total Time(sec)	185.88	339.47	27,870.81

Impact of the top-k in the performance of the TKSI algorithm. We study the performance of the TKSI and BDTKS, and the impact of k in the quality of TKSI. BMORTKS is not considered because as it was shown in the previous section, this algorithm is more expensive than TKSI and BDTKS. In this experiment, we executed sixty queries varying k.

Figure 2 reports the average of number of probes, the average of seen-objects, the average of sequential accesses, the average of random accesses and the average of total time used by TKSI and BDTKS algorithms. In general, we can observe that:

1) The number of probes and seen-objects is lower for the TKSI algorithm. This is because this algorithm builds the Skyline partially until finds the Top-k Skyline objects. On the other hand, the algorithm BDTKS constructs the Skyline completely and then, they perform non-necessary probes considering all objects in the superset of the Skyline.

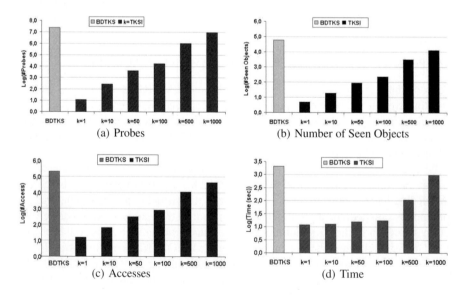

(a) Probes

(b) Number of Seen Objects

(c) Accesses

(d) Time

Fig. 2. Results for the algorithms TKSI and BDTKS

Table 8. t-test for average of Probes, Accesses, Seen Objects and Time

Domain	Probes		Sequential Accesses		Random Accesses		Seen-Objects		Total Time (sec)	
	TKSI	BDTKS	TKSI	BDTKS	TKSI	BDTKS	TKSI	BDTKS	TKSI	BDTKS
Uniform	1,904,018.55	23,749,796.40	5,688.45	80,804.90	14,324.15	335,023.30	2,864.83	57,219.30	209.06	2,036.78
Gaussian	1,726,737.80	25,703,896.00	5,251.67	83,788.80	13,210.13	342,259.90	2,642.03	58,447.50	186.21	2,162.61
Mixed	1,600,389.02	25,096,788.60	4,989.17	84,168.80	12,533.92	347,550.10	2,506.78	59,362.40	175.78	2,179.75
Average	1,743,715.12	24,850,160.33	5,309.76	82,920.83	13,356.06	341,611.10	2,671.21	58,343.07	190.35	2,126.38
t-test	*p-value=0.00039*		*p-value=0.00013*		*p-value=0.00008*		*p-value=0.00008*		*p-value=0.00040*	

2) The TKSI algorithm requires fewer sequential accesses and random accesses. This might be because, the algorithm avoids seeing unnecessary objects and does not create the Skyline completely when k is lower than the Skyline size.

3) Due to TKSI algorithm makes fewer probes and accesses, its time execution is the lowest.

4) TKSI executes a number of probes and accesses similar to BDTKS when $k = 1000$ because of this k value is close to Skyline size. In this case, TKSI builds the Skyline almost completely and therefore, it has similar performance with respect to BDTKS.

Impact of the data distribution in the performance of the TKSI algorithm. The objective of this experiment was to study the impact of the data properties on the performance of TKSI algorithm. We compare the values of the metrics: number of probes, accesses, seen-objects and total time for TKSI and BDTKS. Table 8 shows the results for the t-test. As this analysis shows, the differences for the number of probes, seen objects, accesses and total time are highly significant (at least 99.99% level).

6 Related Work

Different preference-based query languages have been defined to express user-preference criteria for a certain collection of data. These languages can be grouped into three paradigms: order-based, score-based and hybrid. The challenge of the order-based approaches is to identify the Skyline or set of objects that are non-dominated by any other object, and different techniques have been proposed to efficiently achieve this goal [1,2,3,9,14,15]. While these techniques could be used to construct the Skyline required to identify the top-k objects, they may be inefficient because they completely compute all the non-dominated points. Score-based techniques accomplish the problem of computing the top-k objects from an ordered set of points; although the number of seen objects is minimized during the top-k computation, the non-dominance relationship among the retrieved objects is not considered[5,6,8].

Finally, hybrid approaches compute the top-k objects among the ordering induced by a multicriteria function [4,7,10,11,12,13,16,17,18]; although existing solutions are able to identify the top-k objects, they may be inefficient because they rely on the construction of the complete Skyline. In [18] the top-k objects among a set of non-dominated points are identified in terms of the link-based metric PageRank that runs on a Skyline graph; this graph represents the dominance relationship between the points in the

Skyline when only sub-sets of the ranking parameters are considered. Similarly, the authors in [7,12,13] take into account the different dominance relationships among the Skyline points; users preference information and the Skyline frequency and the k representative Skyline metrics are used to distinguish the top-k objects, respectively. These approaches may be required if the score function to break the ties between the Skyline points is unknown. However, since [7,13] compute the whole Skyline and then, all the different Skyline sub-spaces, they may be very costly and inefficient when the difference between k and the size of the Skyline is large. Additionally, [12] only achieves an exact solution for a 2d-space. In [4,11] scanned-based approaches are proposed to compute the Top-k Skyline; the proposed techniques compute the top-k among a partially ordered set which is stratified into subsets of non-dominated objects, and are able to produce the top-k objects when the Skyline size is smaller than k. The main drawback of these approaches is that the skyline or strata have to be completely built. In this paper we also consider the Top-k Skyline problem, but our approach differs from the previous works in the following issues: first, we assume that a score function is provided by the users to break the ties between the non-dominated points; second, each of the attributes needed to compute the Skyline and the top-k objects are indexed and random access is available; third, the top-k score function is precomputed and stored in an index; and finally, Skyline size is greater or equal than k. These assumptions provided the basic of our TKSI algorithm, allowing to identify the top-k objects at the time the non-dominance relationship among some of the Skyline points is verified. Thus, the TKSI is able to minimize the number of non-necessary probes, when k is smaller than the size of the Skyline.

7   Conclusions and Future Work

In this work, the TKSI algorithm has been proposed and its performance has been empirically compare to extensions of the state-of-the-art algorithms: BDS and BMORS. The first, BDS, is based on the computation of a final object whose scores have been completely seen. The second, BMORS, computes a virtual object which is comprised of the minimum scores seen so far. Both algorithms are sound, but, they rely on the computation of a superset of the Skyline to identify the Top-k Skyline. On the other hand, TKSI builds the Skyline until it has computed the k objects. Initial experimental results show that TKSI computes the Top-k Skyline performing less number of probes and consuming less time, when k is smaller than the size of the Skyline.

We have not considered that there may exist several execution plans for a given Top-k Skyline query. Considering preference criteria during the query optimization might help to identify better execution plans. In the future, we plan to integrate Top-k Skyline techniques into a relational engine to select optimal execution plans that take into account decision criteria.

References

1. Balke, W.-T., Güntzer, U.: Multi-objective Query Processing for Database Systems. In: Proceedings of the International Conference on Very Large Databases (VLDB), Canada, pp. 936–947 (2004)

2. Balke, W.-T., Güntzer, U., Zheng, J.X.: Efficient distributed skylining for web information systems. In: Bertino, E., Christodoulakis, S., Plexousakis, D., Christophides, V., Koubarakis, M., Böhm, K., Ferrari, E. (eds.) EDBT 2004. LNCS, vol. 2992, pp. 256–273. Springer, Heidelberg (2004)

3. Börzönyi, S., Kossman, D., Stocker, K.: The Skyline operator. In: Proceedings of the International Conference on Data Engineering (ICDE), Germany, pp. 421–430 (2001)

4. Brando, C., Goncalves, M., González, V.: Evaluating top-k skyline queries over relational databases. In: Wagner, R., Revell, N., Pernul, G. (eds.) DEXA 2007. LNCS, vol. 4653, pp. 254–263. Springer, Heidelberg (2007)

5. Carey, M., Kossman, D.: On saying "Enough already!" in SQL. In: Proceedings of the ACM SIGMOD Conference on Management of Data, vol. 26(2), pp. 219–230 (1997)

6. Chang, K., Hwang, S.-W.: Optimizing access cost for top-k queries over Web sources: A unified cost-based approach. Technical Report UIUCDS-R-2003-2324, University of Illinois at Urbana-Champaign (2003)

7. Chan, C.-Y., Jagadish, H.V., Tan, K.-L., Tung, A.K.H., Zhang, Z.: On high dimensional skylines. In: Ioannidis, Y., Scholl, M.H., Schmidt, J.W., Matthes, F., Hatzopoulos, M., Böhm, K., Kemper, A., Grust, T., Böhm, C. (eds.) EDBT 2006. LNCS, vol. 3896, pp. 478–495. Springer, Heidelberg (2006)

8. Fagin, R.: Combining fuzzy information from multiple systems. Journal of Computer and System Sciences (JCSS) 58(1), 216–226 (1996); Proceedings of the Conference on Very Large Data Bases (VLDB), Norway, pp. 229–240 (2005)

9. Godfrey, P., Shipley, R., Gryz, J.: Maximal Vector Computation in Large Data Sets

10. Goncalves, M., Vidal, M.-E.: Preferred skyline: A hybrid approach between sQLf and skyline. In: Andersen, K.V., Debenham, J., Wagner, R. (eds.) DEXA 2005. LNCS, vol. 3588, pp. 375–384. Springer, Heidelberg (2005)

11. Goncalves, M., Vidal, M.E.: Top-k Skyline: A Unified Approach. In: Proceedings of OTM (On the Move) 2005 PhD Symposium, Cyprus, pp. 790–799 (2005)

12. Lee, J., You, G.-w., Hwang, S.-w.: Telescope: Zooming to interesting skylines. In: Kotagiri, R., Radha Krishna, P., Mohania, M., Nantajeewarawat, E. (eds.) DASFAA 2007. LNCS, vol. 4443, pp. 539–550. Springer, Heidelberg (2007)

13. Lin, X., Yuan, Y., Zhang, Q., Zhang, Y.: Selecting stars: The k most representative Skyline operator. In: Proceedings of the International Conference on Data Engineering (ICDE), Turkey, pp. 86–95 (2007)

14. Lo, E., Yip, K., Lin, K.-I., Cheung, D.: Progressive Skylining over Web-Accessible Database. Journal of Data and Knowledge Engineering 57(2), 122–147 (2006)

15. Papadias, D., Tao, Y., Fu, G., Seeger, B.: Progressive Skyline computation in database systems. ACM Transactions Database Systems 30(1), 41–82 (2005)

16. Pei, J., Jin, W., Ester, M., Tao, Y.: Catching the Best Views of Skyline: A semantic Approach Based on Decisive Subspaces. In: Proceedings of the Very Large Databases (VLDB), Norway, pp. 253–264 (2005)

17. Tao, Y., Xiao, X., Pei, J.: Efficient Skyline and Top-k Retrieval in Subspaces. IEEE Transactions on Knowledge and Data Engineering 19(8), 1072–1088 (2007)

18. Vlachou, A., Vazirgiannis, M.: Link-based ranking of Skyline result sets. In: Proc. of 3rd Multidiciplinary Workshop on Advances in Preference Handling (2007)

19. http://googleblog.blogspot.com/2008/07/we-knew-web-was-big.html

Energy Efficient and Progressive Strategy for Processing Skyline Queries on Air

JongWoo Ha[1], Yoon Kwon[2], Jae-Ho Choi[1], and SangKeun Lee[1]

[1] College of Information and Communication,
Korea University, Seoul, Republic of Korea
{okcomputer,redcolor25,yalphy}@korea.ac.kr
[2] Air Force Operations Command,
R.O.K.A.F, Republic of Korea
unikwon@korea.ac.kr

Abstract. Computing skyline and its variations is attracting a lot of attention in the database community, however, processing the queries in wireless broadcast environments is an uncovered problem despite of its unique benefits compared to the other environments. In this paper, we propose a strategy to process skyline queries for the possible expansion of current data broadcasting services. For the energy efficient processing of the skyline queries, the Sweep space-filling curve is utilized based on the existing DSI structure to generate broadcast program at a server side. The corresponding algorithms of processing skyline queries are also proposed for the mobile clients. Moreover, we extend the DSI structure based on a novel concept of Minimized Dominating Points (MDP) in order to provide a progressive algorithm of the queries. We evaluate our strategy by performing a simulation, and the experimental results demonstrate the energy efficiency of the proposed methods.

Keywords: Data broadcasting, skyline, energy efficient, progressive.

1 Introduction

In a multi-dimensional space, a data point p *dominates* another point p' if and only if, the attributes of p on all dimensions are smaller than or equal to those of p'. Given the notion of dominance relationship among data points, a skyline is defined as a set of data points, which is not *dominated* by any other data points. Computing skyline and its variations is attracting a lot of attention in the database community, mainly due to its importance for applications associated with multi-criteria decision making. Currently, the demand for the processing of skyline queries is expanding from conventional databases to various environments, including data streaming [13][19], the Web [15], mobile ad-hoc networks [7], and wireless sensor networks [10]. However, processing the queries in wireless broadcast environments is an uncovered problem despite of its unique benefits compared to the other environments; processing queries *on air* can be an attractive strategy for a server to support an arbitrary number of mobile clients without additional cost, as demonstrated in previous studies [12][20].

S.S. Bhowmick, J. Küng, and R. Wagner (Eds.): DEXA 2009, LNCS 5690, pp. 486–500, 2009.
© Springer-Verlag Berlin Heidelberg 2009

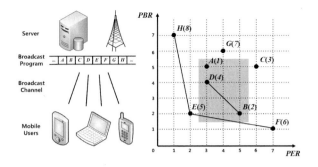

Fig. 1. Skyline Queries Processing in Wireless Broadcast Environments

In wireless broadcast environments, we focus on the problem of processing the *original* and *constrained skyline* queries[1], for which we suggest that three requirements should be satisfied: 1) the processing should terminate effectively within a broadcast cycle; 2) the processing of the constrained skyline query should minimize the energy consumption of the mobile clients; 3) the processing should be progressive. The first and second requirements consider the predominant issues associated with the access time and tuning time in wireless broadcast environments [8], while the third requirement originates from the desired properties of the processing methods for skyline queries. We describe the motivations for each of the suggested requirements as follows.

1. We can adopt the existing methods of disk-based processing of skyline queries, however, they will not comply with the first requirement. The methods are based on *random* access to the data, therefore, backtracking in a broadcast program, which leads mobile clients to tune in the next broadcast cycle, is inevitable. This fundamental disagreement with the *sequential* characteristics in wireless broadcast environments will introduce unbounded performance in terms of the access time [12][20].
2. A simple method to avoid this problem is to process skyline queries by tuning the entire data in a broadcast cycle and performing consecutive dominance test. In Figure 1, a client can obtain the original skyline, *E*, *F*, and *H*, by tuning into the data from *A* to *H*. However, in this case, the client would experience the worst-case performance in terms of the tuning time. Therefore, any strategies, which need to scan each of the data such as stream processing methods, does not satisfy the second requirement.
3. The progressive characteristic is desired in the context of processing skyline queries [9][15][17][18]. A progressive algorithm for processing skyline queries can immediately and effectively return skyline points as they are identified. Otherwise, the client has to wait until the processing terminates to obtain even a single skyline point. However, the progressive characteristic is not easy

---

[1] The original skyline query can be defined as an constrained skyline query by setting the entire space as the given constrained region.

to satisfy, since a data point can be either a skyline point or not according to the given constrained region. In Figure 1, for example, if a client sets the entire space to be a constrained region, the skyline will be identical to the *original* skyline. If another client sets the gray constrained region, then the skyline is B and D.

To deal with the problem of processing constrained skyline queries in a manner which satisfies all three requirements, we propose a strategy, referred to as *skyline on air*, for the possible expansion of current broadcasting services, including T-DMB [2] and 1seg [1]. First, we identify the benefits of the Sweep space-filling curve (SFC) [16] in the context of skyline query processing. Based on the Sweep SFC embedded in the distributed spatial index (DSI) [12], we design the *region-based pruning* and *point-based pruning* algorithms, which satisfy both the first and second requirements. Further, we observed that a processing algorithm can comply with the third requirement, if the index structure provides knowledge of the *minimized dominating points* (MDP). We propose the Extended-DSI to embed the MDP into our base index structure. Finally, we propose the corresponding *progressive point-based pruning* algorithm, which satisfies all three requirements.

The contributions of this paper include the following.

- We identify the benefits of the Sweep SFC and the MDP in the context of processing skyline queries.
- We propose an index structure, Extended-DSI, and processing algorithms, which allow our strategy to be energy efficient and progressive.
- Experiments are performed to evaluate the performance of *skyline on air*, and the results demonstrate the energy efficiency of the proposed methods.

To the best of our knowledge, this is the first attempt to process skyline queries in wireless broadcast environments. To make this initial study feasible, we assume that there is no update and no transmission errors of data.

The remainder of this paper is organized as follows. We discuss related research in Section 2, and Section 3 describes the proposed methods of processing constrained skyline queries in wireless broadcast environments. The experimental evaluation of the proposed methods is presented in Section 4, and Section 5 summarizes this paper and describes future works.

2   Related Work

In centralized databases, most of the existing methods of processing skyline queries share the same goal to improve the performance, i.e., minimizing the number of dominance tests in order to reduce the computational overhead and I/O cost. Block-Nested Loop (BNL) [3] conducts dominance tests for each data point in the dataset with every other data point. If it is not dominated by any other points, BNL returns the data as a skyline point. Divide-and-Conquer (D&C) [3] divides a given dataset into several regions each of which can be loaded in the main memory for processing. Then, it computes and merges the skyline candidates in each of the regions. Sort-Filter-Skyline (SFS) [5] introduces a monotonic scoring function which returns an ordering of dataset. This ordering guarantees that a data point is not dominated by other preceding data points and, therefore, it can safely ignore the corresponding dominance tests. However, the BNL, D&C, and SFS methods need to scan the whole dataset, and this will clearly lead to poor performance in terms of the tuning time in wireless broadcast environments.

To avoid the exhaustive scanning of the dataset, index-based methods were proposed. In [18], the first progressive strategy of computing skyline is introduced with Bitmap and Index. Nearest Neighbor [9] computes the skyline on an R-tree based on the observation that the nearest neighbor from the origin is always a skyline point. Branch-and-Bound [17] also utilizes an R-tree, and more importantly, this method guarantees the minimum I/O for a given R-tree. Based on the benefit of Z-order in processing skyline and k-dominant skyline queries, *ZSearch* is proposed in [11]. However, processing algorithms, which are based on *random* access to the data with tree indexing, will produce unbounded performances in terms of the access time, as demonstrated in [12][20].

As well as in centralized databases, current research of computing skyline deals with problems originating in various environments [7][10][13][15][19]. These studies also consider the characteristics of the environments and the progressive behavior of processing skyline queries. For example, the primary goal in [10] is to reduce the energy consumption of the wireless sensor nodes, and the importance of the progressive characteristic in the Web is also demonstrated in [15].

3   Proposed Methods

3.1   Sweep Space-Filling Curve with DSI

We first review the index structure, DSI, which is proposed for the efficient processing of location-dependent queries in wireless broadcast environments [12]. To take advantages of its linear and distributed characteristics, we adopt the DSI as our base index structure. Then, we identify the benefit of the Sweep SFC in the context of processing skyline queries.

In Figure 2, we show an example of DSI from the example dataset in Figure 1. In DSI, each data p has its own *order*, denoted by $p.order$, which is encoded based on the attributes of the data according to the Hilbert SFC [6]. For example, the data, E, which has attributes $(2, 2)$ in Figure 1, has an order 8. The dataset is divided into n_F *frames*, and an *index table* is stored together with the data in each frame. The index table consists of $\log_r n_F$ number of index table entries, where r is the exponential base called the *index base*. In the example, n_F and r are set to 8 and 2, respectively. Therefore, each index table contains 3 index table entries. Each ith index table entry has a pair of order and pointer for a data. The data pointed to by the pointer in the ith index table entry will be broadcast after an interval of r^i from the current frame, where $0 \leq i \leq \log_r n_F - 1$. For example, frame 8 in Figure 2 contains three index table entries to point to frames

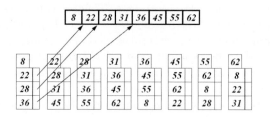

Fig. 2. An Example of DSI

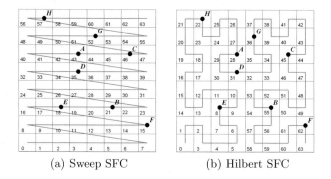

(a) Sweep SFC   (b) Hilbert SFC

Fig. 3. Example Dataset Varying Space-Filling Curves

22, 28, and 36, which are separated from the frame by intervals of 1, 2, and 4, respectively.

By default, the Hilbert SFC is adopted in DSI to assign an order to each data point. However, we employ the Sweep SFC to take advantages of its superiority compared to the Hilbert SFC for processing skyline queries, as descried below.

First, we can safely avoid a lot of pairwise dominance tests while guaranteeing the effectiveness of the query processing. In Figure 3(a), our example dataset is encoded in Sweep SFC. As shown in the figure, it is guaranteed that a data point p is not dominated by a data point q, if and only if, $p.order$ is less than or equal to $q.order$. Therefore, to test whether an example point D is a skyline point or not, we only need to conduct dominance tests against those data points with a smaller order value than that of D, i.e., F, E, and B.

Second, by adopting the Sweep SFC, it becomes an easy task to encode a multi-dimensional point into an order and to decode an order into a multi-dimensional point. Given an n-dimensional data point, $p = (a_1, a_2, ..., a_n)$, we can easily encode the data point, $p.order = \sum_{i=1}^{n}(a_i \times w^{i-1})$, where $0 \leq a_i < w$. The decoding is also done with a few arithmetic operations by taking the encoding process inversely. However, it is still a challenge to map multi-dimensional points efficiently with the Hilbert SFC. Moreover, the performance of the mapping becomes worse as the number of dimensions increases [4][14].

3.2 Region-Based Pruning Algorithm

Based on the Sweep SFC in DSI, we present the region-based pruning (RBP) algorithm in Algorithm 1. For the sake of simplicity, we assume that the attribute
values of all dimensions are integer types. To describe the RBP, a running example is illustrated in Figure 4. In this example, the constrained region is set to be
the entire space. Table 1 specifies the states of the variables at each stage of the
tuning buckets (the smallest logical unit of broadcast). For the sake of simplicity,
we assume that a bucket contains one frame in DSI, i.e., a single data and its
corresponding index structure is stored in a bucket, throughout the paper.

The basic idea of RBP is that we initially set the search space with the
given constrained region, and reduce the search space based on the *dominant
regions* and *empty regions*. In Figure 4, the dominant region of a data point
35 is represented with the gray region. Since any data points in the region is
not a skyline point, we safely remove the region from the search space. Basically,
removing dominant regions from the search space is for selectively tuning buckets
to reduce the tuning time. The empty region, induced by analyzing the broadcast
program, is a region in which there exists no data point. For example, RBP can
identify the two empty region, [36, 42] and [44, 45], from the index table of
bucket 35; there exists no data point between any two adjacent buckets in the
broadcast program. Since the RBP terminates when the search space becomes
empty, removing the empty regions is necessary to remove those regions, which
are not dominated by any data points in the search space, e.g., [0, 17].

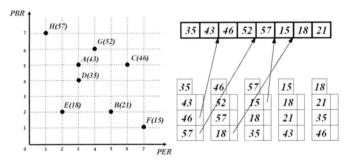

Fig. 4. An Example of Processing Skyline Query

Table 1. Running States of Region-Based Pruning Algorithm

Bucket	Program Analyzer	Skyline	Search Space
	(...)	∅	[0, 63]
35	(<u>35</u>, <u>43</u>, <u>46</u>, \0, <u>57</u>, ...)	35	[0, 34], [48, 50], [56, 57]
46	(35, 43, 46, <u>52</u>, 57, \0, 18, ...)	∅	[0, 18], [24, 25], [32, 33], [57, 57]
57	(35, 43, 46, 52, 57, <u>15</u>, 18, \0)	57	[15, 15], [18, 18], [24, 25], [32, 33]
15	(35, 43, 46, 52, 57, 15, 18, <u>21</u>)	57, 15	[18, 18]
18	(35, 43, 46, 52, 57, 15, 18, 21)	57, 15, 18	∅

Algorithm 1. Region-Based Pruning Algorithm

Input: *ConstrainedRegion*; **Output:** *Skyline*;
Variable: *searchSpace, bucket, pAnalyzer, newPoints, prunedRegion*;
01: *Skyline* = ∅;
02: *searchSpace* = getTargetSegmentsSet(*ConstrainedRegion*);
03: **while** (*searchSpace* != ∅)
04:      **if** (*bucket* == **null**)
05:          *bucket* = initialProbe();
06:      **else**
07:          *bucket* = getNextBucket(*pAnalyzer, searchSpace*);
08:      *newPoints* = getNewlyFoundPoints(*pAnalyzer, bucket, ConstrainedRegion*);
09:      *pAnalyzer* = updateProgramAnalyzer(*pAnalyzer, bucket, newPoints*);
10:      *searchSpace* −= getEmptyRegion(*pAnalyzer*);
11:      *prunedRegion* = getPrunedRegion(*searchSpace, newPoints*);
12:      **for** (*data* ∈ *Skyline*)
13:          **if** (*data.order* ∈ *prunedRegion*)
14:              *Skyline* −= *data*;
15:      *searchSpace* −= *prunedRegion*;
16:      **if** (*bucket.data* ∈ *searchSpace*)
17:          *Skyline* += *bucket.data*;
18:          *searchSpace* −= *bucket.data*;

We now describe the RBP in detail with the running example. First, RBP sets the *Skyline* to be an empty set (∅), and the search space to be the constrained region (line 01-02). When tuning the first bucket, RBP knows that the current bucket is 35, and that buckets 43, 46, and 57 will be broadcast after 1, 2, and 4 intervals, repectively. The bucket that will be broadcast after 3 interval is unknown, and is denoted by '\0' at the moment. This is maintained in a *program analyzer* to decide the next bucket to be tuned (line 04-09). By analyzing the index table, RBP discovers the two empty region, [36, 42] and [44, 45]. They are removed from the search space (line 10). Next, based on each of the newly found points from the current bucket (underlined in Table 1), it calculates the dominant region of each of the newly found data points (line 11). If the dominated region includes any data points in the *Skyline*, RBP drops the data from the *Skyline*, since it is actually not a skyline point (lines 12-14). After that, RBP removes the dominant region from the search space (lines 15). Note that data point 35 does not dominate itself. Since the reduced search space still includes the data in the current bucket, the algorithm adds the data to the *Skyline*. Then, it removes data point 35 from the search space (line 16-18).

RBP continues looping through lines 04-18 until the search space becomes empty. If not, it consults the program analyzer to decide the next bucket to be tuned (line 7). Since the search space does not include buckets 43 and 46 (They are included in the pruned region dominated by data point, 35), RBP does not need to tune these buckets. However, since there is an unknown data point between 46 and 57, it needs to tune bucket 46 to identify the unknown data point for effective processing. To get the *pointer* for the unknown bucket, the program

Table 2. Running States of Point-Based Pruning Algorithm

Bucket	Program Analyzer	Candidates	Skyline
	(...)	∅	∅
35	(<u>35</u>, <u>43</u>, <u>46</u>, \0, <u>57</u>, ...)	35, 57	35
46	(35, 43, 46, <u>52</u>, 57, \0, 18, ...)	57, 18	∅
57	(35, 43, 46, 52, 57, <u>15</u>, 18, \0)	57, 18, 15	57
15	(35, 43, 46, 52, 57, 15, 18, <u>21</u>)	57, 18, 15	57, 15
18	(35, 43, 46, 52, 57, 15, 18, 21)	57, 18, 15	57, 15, 18

analyzer decides to tune bucket 46. In our example, the processing terminates when it tunes bucket 18, and produces access and tuning time performances of 7 and 5 buckets, respectively.

The RBP algorithm provides an efficient way to selectively tune buckets, since it safely ignores pruned buckets. However, this algorithm may not be a feasible solution for mobile clients due to the extremely expensive cost of line 12 in algorithm 1, as we demonstrate in Section 4.2 by measuring the number of dominance tests. This is because of the fact that the algorithm applies dominance tests over the entire search space, and the search space grows *exponentially* as the dimension of the dataset increases.

3.3   Point-Based Pruning Algorithm

We propose an alternative point-based pruning (PBP) algorithm, to deal with the problem posed by RBP. The PBP significantly reduces the number of dominance tests so that the mobile client can reduce its energy consumption accordingly. To selectively tune buckets, RBP checks whether a bucket is included in the current search space or not. However, PBP algorithm utilizes a *shortcut* in order to have the same effect of tuning buckets selectively. The basic idea is that we can apply dominance tests *directly* to the data *points*. Based on pairwise dominance tests among every datas points found at the present moment, we can maintain a set of *candidates*, which is the target of the next tuning buckets. This process continues until every bucket containing candidates is tuned.

Algorithm 2 specifies the PBP algorithm. We describe PBP with the same running example of Figure 4, and depict only the differences between the PBP and RBP, due to space limitations. Table 2 specifies each stage of running the PBP algorithm. Note that the states of the program analyzer are identical to those of the previous RBP algorithm in Table 1, because they are identical in terms of the tuning of the next buckets. However, the PBP algorithm does not need to keep a search space. It only maintains a set of candidates in order to conduct dominance tests efficiently. When the algorithm initiates processing by tuning into bucket 35, it knows that data points 43 and 46 have no possibility being skyline points since they are dominated by point 35 (lines 06-09). Compared to the previous algorithm, PBP adds and drops Skyline based on the candidates (lines 10-14). PBP terminates when it discovers every point in the constrained

---

Algorithm 2. Point-Based Pruning Algorithm

Input: *ConstrainedRegion*; **Output:** *Skyline*;
Variable: *bucket, newPoints, pAnalyzer, candidates*;
01: *Skyline* = ∅;
02: *bucket* = initialProbe();
03: **while** (*bucket* != **null**)
04:      *newPoints* = getNewlyFoundPoints(*pAnalyzer, bucket, ConstrainedRegion*);
05:      *pAnalyzer* = updateProgramAnalyzer(*pAnalyzer, bucket, newPoints*);
06:      **for** (*order* ∈ *newPoints*)
07:          **if** (orderIsANewCandidate(*order, candidates*) == **true**)
08:              *candidates* −= getDropedCandidates(*candidates, order*);
09:              *candidates* += *order*;
10:      **for** (*data* ∈ *Skyline*)
11:          **if** (*data.order* ∉ *candidates*)
12:              *Skyline* −= *data*;
13:      **if** (*bucket.data* ∈ *candidates*)
14:          *Skyline* += *bucket.data*;
15:      *bucket* = getNextBucket(*pAnalyzer, candidates, ConstrainedRegion*);

---

region and tunes into every candidate. In the example of Figure 4, PBP produces access and tuning time performances of 7 and 5 buckets, respectively.

While the PBP significantly reduces the number of dominance tests, it does not provide the progressive characteristic when processing constrained skyline queries. For example, in Table 2, data 35 is set to be a skyline point when tuning bucket 35, however, it is not actually a skyline point. Next, we present our observation on how we can design a progressive algorithm to process constrained skyline queries.

3.4 Progressive Algorithm with Extended-DSI

A progressive algorithm for processing constrained skyline queries should be able to effectively detect whether a data in a currently tuned bucket is a skyline point or not, even if not every data point in the dataset has been found. This is not an easy problem, since a data point can be either a skyline point or not depending on the given constrained region.

We first present a naive solution with *full dominating points* (FDP). The FDP of a data point p, denoted by $p.fdp$ is defined as a set of data points, which dominate p. In Figure 5, for example, the $C(46).fdp$ is $\{E(18), B(21), D(35), A(43)\}$. Given an arbitrary constrained region, p is always a skyline point, if and only if, the constrained region includes p and does not include every data point in $p.fdp$. In Figure 5, we illustrate two example constrained regions, CR_1 and CR_2. The data point C is not a skyline point in the CR_1, since there exists those points, which dominate the C. In the CR_2, the data point C is a skyline point, because there are no such points, which dominates the C. Therefore, by checking the FDP of a data point, it is easily done to test that the data point is whether a skyline point or not.

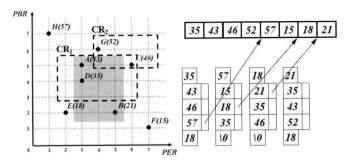

Fig. 5. An Example of Processing Constrained Skyline Query

Table 3. Running States of Progressive Point-Based Pruning Algorithm

Bucket	Program Analyzer	Candidates	Skyline
	(...)	∅	∅
35	(35, 43, 46, \0, 57, ...) (18:\0)	35	35
57	(35, 43, 46, \0, 57, 15, 18, \0)	35	35
18	(35, 43, 46, \0, 57, 15, 18, 21)	35, 21	35
21	(35, 43, 46, 52, 57, 15, 18, 21)	35, 21	35, 21

To enhance the above solution, we propose an alternative method with *minimized dominating points* (MDP). In the MDP, all the unnecessary data points for the progressive processing of skyline queries is removed from the PDP. The MDP of a data point p, denoted by $p.mdp$, is defined as a set of data points, which are included in $p.fdp$ and do not dominate other data points in $p.fdp$. For example, in Figure 5, the $C.mdp$ is $\{B, A\}$, in which E and D are removed from the $C.pdp$. Given an arbitrary constrained region, which contains p and does not contain any points in $p.mdp$, the data p is always a skyline point because there exist no such points that dominate p in the given constrained region. For example, we can see that a data point C is a skyline point for any constrained regions that include C and do not include both the B and A in Figure 5.

By embedding the MDP of a data point into the index structure, we can design a progressive algorithm for processing constrained skyline queries. However, our base index structure, DSI, has no slot to add the MDP. Therefore, we generalize the DSI and propose an index structure, Extended-DSI, which has sufficient additional space to store arbitrary information, called an *appendix*. This appendix is attached to each of the frames in the DSI. In this way, the original DSI [12] is an Extended-DSI with empty appendix. We show an example of the Extended-DSI on the right side of Figure 5. Note that we only need to insert the orders of the data points in $p.mdp$.

Finally, we propose the progressive point-based pruning (P-PBP) algorithm specified in Algorithm 3. We describe the P-PBP with an example in Figure 5. This example involves the processing of a constrained skyline query given the gray constrained region. As the algorithm tunes the first bucket, 35, it immediately

---

Algorithm 3. Progressive Point-Based Pruning Algorithm

Input: *ConstrainedRegion*; **Output:** *Skyline*;
Variable: *bucket, newPoints, pAnalyzer, candidates*;
01:   *Skyline* = ∅;
02:   *bucket* = initialProbe();
03:   **while** (*bucket* != null)
04:       **if** (*bucket.order* ∈ *ConstrainedRegion*)
05:           **if** (every point in *bucket.mdp* ∉ *ConstrainedRegion*)
06:               *Skyline* += *bucket.data*;
07:       *newPoints* = getNewlyFoundPoints(*pAnalyzer, bucket*);
08:       *pAnalyzer* = updateProgramAnalyzer(*pAnalyzer, bucket*);
09:       **for** (*order* ∈ *newPoints*)
10:           **if** (orderIsNewCandidate(*order, candidates*) == **true**)
11:               *candidates* -= getDropedCandidates(*candidates, order*);
12:               *candidates* += *order*;
13:       *bucket* = getNextBucket(*pAnalyzer, candidates, ConstrainedRegion*);

---

detects that the data in the bucket is a skyline point (lines 04-07). Note that the P-PBP does not need to drop false selection in the Skyline, because every data in the Skyline is guaranteed to be a skyline point. Since the appendix contains orders (without pointers) of data points for the corresponding MDP, the algorithm utilize them for pruning data points. As illustrated with (18:\0) in Table 3, the appendix of bucket 35 is stored without a pointer. This information is also used in the candidate tests in lines 09-12, and it will introduce another sequence of tuning buckets and another list of pruned points compared to PBP algorithm. Therefore, the tuning time performance of P-PBP, and generally, P-PBP performs better as demonstrated in Section 4.2.

After tuning bucket 35, the next bucket to be tuned is bucket 57. Since the given constrained region does not include any data points between 46 and 57, the algorithm safely ignores the unknown data points between 46 and 57 (line 13). Note that this is also applied in the previous RBP and PBP algorithms in line 8 and line 16, respectively.

4   Experimental Evaluation

4.1   Simulation Environment

To evaluate the proposed methods, we conducted a simulation by modeling a server, a broadcast channel, and an arbitrary number of mobile clients. Basically, our simulation environment is similar to that employed in [12]. Table 4 shows the default and detailed simulation parameters. We generate the datasets synthetically from $2D$ to $10D$. Each attribute is uniformly assigned in a $[0, 1023]$ space, which is similar to the *indep* dataset in [3]. ServerDBSize (number of buckets) is set to 1,000 similar to that in [8] for a feasible simulation with data broadcasting services (The server tend to keep the broadcast program as small as possible to increase the overall performance of the access time.) We specify the size of the

Table 4. Simulation Parameters

Parameters	Contents (Default)
ServerDBSize (N)	1000
Dimension (D)	2 to 10
Constrained Ratio (r_c)	1.0 to 0.5
Index Base (r)	2

Table 5. Analysis of Dataset

Dimension	2	3	4	5	6	7	8	9	10
Number of Skyline Points ($r_c = 1.0$)	11	30	87	181	274	422	484	699	739
Number of Skyline Points ($r_c = 0.5$)	7	20	20	15	12	9	4	3	1
Average Number of Points in FDP	245.0	124.7	62.4	30.9	15.1	7.5	4.1	1.7	1.2
Average Number of Points in MDP	5.0	12.0	16.1	14.7	10.3	6.2	3.7	1.6	1.1

skyline in the dataset in Table 5, because it has a clear relationship with the performances of the proposed methods. Table 5 also demonstrates that the size of MDP is quite smaller compared to the FDP. The object factor n_o is set to 1 in generating the Extended-DSI. In this setting, each bucket contains one frame, and this directly provides the knowledge of how many index tables are needed to process the query. The mobile clients set a constrained region to initiate query processing with the parameter r_c. When a clients set r_c to 0.5, for example, the attributes of the corresponding constrained region is set to [0, 511] for all D. If r_c is set to 1.o, then the client processes the original skyline queries. In the experiments, we measure the tuning time and access time [8] and the number of dominance test. The tuning time and the number of dominance tests imply the energy consumption originated from the communication and computational costs, respectively.

As well as the proposed algorithms, i.e., RBP, PBP, and P-PBP, we evaluate two additional algorithms, the Naive and Optimal, to acquire baseline performances. The Naive merely tunes every bucket in a broadcast cycle, conducting consecutive dominance tests, and it does not utilize the benefit of Sweep SFC for the dominance tests. On the other hand, Optimal is the ideal one. The Optimal knows every orders of the skyline points before processing, and it merely performs consecutive calls to *getNextBucket(pAnalyzer, Skyline.orders)* to acquire the index tables towards the given orders of the skyline points and to tune in the bucket containing the skyline points.

4.2   Experimental Results

We illustrate the tuning time performance in Figures 6 and 7. The tuning time of Naive is straightforward, as expected. The performance of Optimal has a strong relationship with the number of skyline points in the dataset. Optimal tunes buckets: 1) to acquire the actual skyline points; 2) to get the index tables towards the given orders of the skyline points. The difference in the tuning time

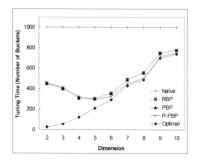

Fig. 6. Tuning Time ($r_c = 1.0$)

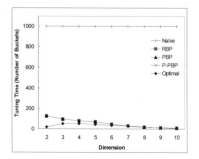

Fig. 7. Tuning Time ($r_c = 0.5$)

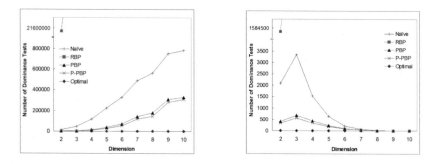

Fig. 8. No. Dominance Tests ($r_c = 1.0$)   **Fig. 9.** No. Dominance Tests ($r_c = 0.5$)

performances between Optimal and the proposed algorithms originates from tuning more buckets in order to: 3) resolve unknown data points during processing; 4) acquire candidates, which are not actually skyline points. As D increases, the number of tuning buckets for 3) and 4) is reduced, therefore, the differences in the performance of the proposed algorithms becomes smaller compared with the optimal algorithm. Note that the performance of P-PBP is close to that of Optimal beyond $6D$. The difference between Figures 6 and 7 originates from the volume of the constrained region set by r_c. When the constrained region does not include a data point, the proposed algorithms can safely ignore tuning the corresponding bucket for 3). Therefore, the tuning time performance gets better as r_c decreased.

Figures 8 and 9 illustrate the number of dominance tests during processing, which imply the energy consumptions for CPU operations. As discussed in section 3.2, RBP conducts a huge amount of dominance tests, due to its inefficiency of region-based pruning. The energy consumption of CPU operations is known to be less than that of the communication (i.e. tuning buckets), however, this huge amount of computation can reduce the battery life compared to that obtained with the alternative PBP algorithm. Note that even Naive performs moderately well compared to the RBP algorithm. Basically, the algorithms perform a larger number of dominance tests when the dataset has more skyline points and the

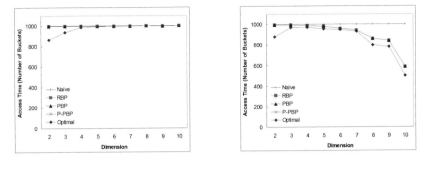

Fig. 10. Access Time ($r_c = 1.0$)   **Fig. 11.** Access Time ($r_c = 0.5$)

constrained region includes more data points. The difference in performance between Naive and PBP algorithms originates from: 5) the characteristics of the Sweep SFC; 6) the false selections of candidates. The difference in performance between PBP and P-PBP arises from the MDP, which gives more opportunity for pruning data points. In Figure 9, the performance in $4D$ is better that that in $3D$, although both datasets has the same number of skyline points. This is because of the fact that the impact of 6) is larger for $3D$ than for $4D$, since the constrained region in $3D$ includes more data points than that in $4D$.

Figures 10 and 11 illustrate the access time performance of the proposed methods. To process the original skyline query, the proposed algorithms should resolve every unknown data point in a broadcast cycle to guarantee the effectiveness of processing. Therefore, the access time is close to the length of the broadcast cycle in Figure 10. However, the proposed algorithms do not produce an access time of more than a broadcast cycle, thus satisfying the second requirements in the problem statement. Note that even Optimal also produces an access time performance close to the length of the broadcast cycle beyond $5D$, since the number of skyline points is large. In Figure 11, the access time steadily decreases as the constrained regions get smaller. This is because of the fact that the proposed algorithms only need to discover data points included in the constrained regions.

5   Conclusions

In this paper, we propose *skyline on air* as an energy efficient and progressive strategy for processing skyline queries for current data broadcasting services. We design an efficient algorithm in terms of the tuning time, based on the Sweep space-filling curve in DSI. With an efficient pruning method and the identified benefits of the Sweep SFC, we provide more opportunities for mobile clients to reduce their energy consumption by significantly reducing the number of dominance tests required for computing the skyline. Further, we propose Extended-DSI and the corresponding progressive point-based pruning algorithm, which enables the *skyline on air* to be a progressive strategy. The experimental evaluations demonstrate that the performance of proposed methods approaches that of the optimal one in terms of the tuning time as the dimension of the dataset increases. In future works, we plan to study the impact of transmission errors on the processing of skyline queries.

Acknowledgements. This study was supported by the Seoul Research and Business Development Program (10561), Seoul, Korea.

References

1. Digital broadcasting and 1seg, http://www.dpa.or.jp/
2. Terrestrial digital multimedia broadcasting (T-DMB), http://www.t-dmb.org/
3. Börzsönyi, S., Kossmann, D., Stocker, K.: The skyline operator. In: ICDE, pp. 421–430 (2001)
4. Chenyang, L., Hong, Z., Nengchao, W.: Fast n-dimensional hilbert mapping algorithm. In: ICCSA, pp. 507–513 (2008)
5. Chomicki, J., Godfrey, P., Gryz, J., Liang, D.: Skyline with presorting. In: ICDE, pp. 717–816 (2003)
6. Gotsman, C., Lindenbaum, M.: On the metric properties of discrete space-filling curves. IEEE Trans Image Process 5(5), 794–797 (1996)
7. Huang, Z., Jensen, C.S., Lu, H., Ooi, B.C.: Skyline queries against mobile lightweight devices in manets. In: ICDE, p. 66 (2006)
8. Imielinski, T., Viswanathan, S., Badrinath, B.R.: Data on air: Organization and access. IEEE Trans. Knowl. Data Eng. 9(3), 353–372 (1997)
9. Kossmann, D., Ramsak, F., Rost, S.: Shooting stars in the sky: An online algorithm for skyline queries. In: VLDB, pp. 275–286 (2002)
10. Kwon, Y., Choi, J.H., Chung, Y.D., Lee, S.: In-network processing for skyline queries in sensor networks. IEICE Transactions 90-B(12), 3452–3459 (2007)
11. Lee, K.C.K., Zheng, B., Li, H., Lee, W.C.: Approaching the skyline in z order. In: VLDB, pp. 279–290 (2007)
12. Lee, W.C., Zheng, B.: Dsi: A fully distributed spatial index for location-based wireless broadcast services. In: ICDCS, pp. 349–358 (2005)
13. Lin, X., Yuan, Y., Wang, W., Lu, H.: Stabbing the sky: Efficient skyline computation over sliding windows. In: ICDE, pp. 502–513 (2005)
14. Liu, X., Schrack, G.F.: Encoding and decoding the hilbert order. Softw., Pract. Exper. 26(12), 1335–1346 (1996)
15. Lo, E., Yip, K.Y., Lin, K.I., Cheung, D.W.: Progressive skylining over web-accessible databases. Data Knowl. Eng. 57(2), 122–147 (2006)
16. Mokbel, M.F., Aref, W.G., Kamel, I.: Analysis of multi-dimensional space-filling curves. GeoInformatica 7(3), 179–209 (2003)
17. Papadias, D., Tao, Y., Fu, G., Seeger, B.: An optimal and progressive algorithm for skyline queries. In: SIGMOD Conference, pp. 467–478 (2003)
18. Tan, K.L., Eng, P.K., Ooi, B.C.: Efficient progressive skyline computation. In: VLDB, pp. 301–310 (2001)
19. Tao, Y., Papadias, D.: Maintaining sliding window skylines on data streams. IEEE Trans. Knowl. Data Eng. 18(2), 377–391 (2006)
20. Zheng, B., Lee, W.C., Lee, D.L.: Spatial index on air. In: PerCom, pp. 297–304 (2003)

RoK: Roll-Up with the K-Means Clustering Method for Recommending OLAP Queries

Fadila Bentayeb and Cécile Favre

Université de Lyon (ERIC - Lyon 2)
5 av. Pierre Mendès-France
69676 Bron Cedex, France
bentayeb@eric.univ-lyon2.fr,
cecile.favre@univ-lyon2.fr
http://eric.univ-lyon2.fr

Abstract. Dimension hierarchies represent a substantial part of the data warehouse model. Indeed they allow decision makers to examine data at different levels of detail with On-Line Analytical Processing (OLAP) operators such as drill-down and roll-up. The granularity levels which compose a dimension hierarchy are usually fixed during the design step of the data warehouse, according to the identified analysis needs of the users. However, in practice, the needs of users may evolve and grow in time. Hence, to take into account the users' analysis evolution into the data warehouse, we propose to integrate personalization techniques within the OLAP process. We propose two kinds of OLAP personalization in the data warehouse: (1) adaptation and (2) recommendation.

Adaptation allows users to express their own needs in terms of aggregation rules defined from a child level (existing level) to a parent level (new level). The system will adapt itself by including the new hierarchy level into the data warehouse schema. For recommending new OLAP queries, we provide a new OLAP operator based on the K-means method. Users are asked to choose K-means parameters following their preferences about the obtained clusters which may form a new granularity level in the considered dimension hierarchy. We use the K-means clustering method in order to highlight aggregates semantically richer than those provided by classical OLAP operators. In both adaptation and recommendation techniques, the new data warehouse schema allows new and more elaborated OLAP queries.

Our approach for OLAP personalization is implemented within Oracle 10 g as a prototype which allows the creation of new granularity levels in dimension hierachies of the data warehouse. Moreover, we carried out some experiments which validate the relevance of our approach.

Keywords: OLAP, personalization, adaptative system, recommendation, schema evolution, clustering, K-means, analysis level, dimension hierarchy.

S.S. Bhowmick, J. Küng, and R. Wagner (Eds.): DEXA 2009, LNCS 5690, pp. 501–515, 2009.
© Springer-Verlag Berlin Heidelberg 2009

1 Introduction

Traditional databases aim at data management, i.e., they help organizing, structuring and querying data. Data warehouses have a very different vocation: analyzing data by exploiting specific multidimensional models (star, snowflake and constellation schemas). Data are organized around indicators called measures, and analysis axes called dimensions. Dimension attributes can form a hierarchy which compose various granularity levels. They allow users (decision makers) to examine data at different levels of detail by using On-Line Analytical Processing (OLAP) tools. Indeed, OLAP allows users to acquire a dynamic manipulation of the data contained in the data warehouse, in particular through hierarchies that provide navigational structures to get summarized or detailed data by rolling up or drilling down.

The main objective of data warehouses is to facilitate decision making. In order to satisfy the whole analysis needs of the majority of the users, a promising issue consists in considering a personalization process for OLAP analysis. By personalization, we mean considering the user to be in the center of the decision system, taking into account his or her own preferences, needs, etc. Research concerning personalization constitutes an emerging topic for the data warehouse domain [1].

In a previous work [2], we proposed an original approach to allow schema evolution in data warehouses independently from data sources. In this paper, we extend this approach to support users' analyses personalization in an interactive way following two main techniques, namely adaptation and recommendation. We propose then a general framework to integrate OLAP personalization in data warehouses. The originality of our framework consists in including additional information and/or knowledge into the data warehouse for further analysis. The solution we propose is implemented by creating new dimension hierarchies into the data warehouse model in order to get new OLAP queries.

In the adaptation technique, users define their additional information under the form of aggregation rules from a child level (existing level) to a parent level (new level). Then, the system adapts to the data warehouse schema by creating the new granularity level in a dimension hierarchy which allows the user to get his/her own personalized analysis.

In the recommendation technique, classical tools are designed to help users to find items within a given domain, according to their own preferences (user profile). The recommendation technique we propose is slightly different from classical ones since we use data mining techniques to extract relevant clusters. These latter possibly represent significant and more elaborated OLAP queries. Hence, users can fix the algorithm parameters in an interactive way until the suggestion of the system coincides with the users' objectives, validating, therefore, the suggestion. We define more precisely a new Roll-up operator based on K-means (RoK) method that creates a new (parent) level to which, a child level rolls up in a dimension hierarchy. Our RoK operator is indeed different from classical OLAP operators since it combines data mining and OLAP tools.

To integrate efficiently our proposition in the OLAP process, we implemented the K-means method inside the Oracle 10g Relational DataBase Management

System (RDBMS) under the form of a stored procedure. This allows treating efficiently large data sets directly inside the data warehouse, like an OLAP operator. In addition, we carried out some experiments which validate the relevance of our approach.

The rest of this paper is organized as follows. In Section 2, we present related work regarding personalization, combining OLAP and data mining and schema evolution in data warehouses. Then, in Section 3, we present our approach for personalized OLAP analysis in data warehouses. To illustrate our purpose, we provide an example from a real case study in Section 4. Section 5 details our data-mining based approach to recommend new OLAP queries and presents the data warehouse model evolution which supports our OLAP personalization approach. Section 6 presents the experiments we performed to validate our approach. We finally conclude this paper and provide some research perspectives in Section 7.

2   Related Work

Personalization in data warehouses is closely related to various research areas that we evoke in this section.

2.1   Personalization

Personalization has been studied since many years and constitutes always a hot topic in domains such as information retrieval (IR), databases (DB) and human-computer interaction (HCI). The general idea is to provide pertinent answers/adapted interfaces to the user according to his/her individual preferences [3]. Personalization is usually based on the concept of profile [4]. This profile is used to model the user himself, his/her needs, the group he/she belongs to and so on.

This profile is not defined in a standard way. In the context of HCI, the profile contains information that allows the adaptation of the interface according to preferences [5]. In the context of IR, the profile can be represented as a set of key words with ponderation [6] or a set of utility functions to express in a relative way domains of interest [7]. In the context of DB, the profile can contain the usual queries of a user i.e. usual predicates, or order in these predicates [8,9]. Thus, the system exploits these predicates to enrich queries and to provide more pertinent results.

Since data warehouses are characterized by voluminous data and are based on a user-centered analysis process, including personalization into the data ware-housing process becomes a new research issue [1]. Works in this domain are inspired from those proposed for personalization in IR, DB, and HCI. For example, selecting data for visualization, based on users' preferences [10] or facilitating the navigation into the data cube [11,12], or recommending some possible analyses according to navigation of other users [13].

2.2   Combining OLAP and Data Mining

OLAP operators have a powerful ability to organize and structure data allowing exploration and navigation into aggregated data. Data mining techniques are known for their descriptive and predictive power to discover knowledge from data. Thus OLAP and data mining are used to solve different kinds of analytic problems: OLAP provides summary data and generates rich calculations while data mining discovers hidden patterns in data. OLAP and data mining can complement each other to achieve, for example, more elaborated analysis.

In the context of data warehouses and OLAP, some data mining techniques can be used as aggregation operators. Thus many works are now focused on providing more complex operators to take advantages from the analysis capabilities of the data mining [14,15]. In our approach, we are going beyond these proposals by exploiting data mining not only at the final stage as OLAP operators but also to consider the data warehouse evolution and take into account users' preferences.

2.3 Data Warehouse Model Evolution

During OLAP analysis, business users often need to explore fact trends over the time dimension. This requires time-variant and non-volatile data. Thus, dimension updates and schema evolutions are logically prohibited because they can induce data loss or erroneous results. To deal with this problem, two categories of research emerged. The first category recommends extending the multidimensional algebra to update the model with a set of schema evolution operators [16,17] while the second category proposes temporal multidimensional data models [18,19]. These works manage and keep the evolutions history by time-stamping relations.

3   Personalization in Data Warehouses

3.1   General Approach

Generally, to carry out OLAP analysis, the user generates a data cube by selecting dimension level(s) and measure(s) which will satisfy his/her needs. Then, the user explores the obtained cube to detect similarities between data facts or dimension instances. For that, he/she explores different levels within a dimension. To help the user in this step, we propose to personalize his/her analysis according to his(her) individual needs and preferences. In this context, we provide a general framework for OLAP personalization shown in Figure 1.

To achieve OLAP personalization, our key idea consists in integrating new information or knowledge inside the data warehouse. Hence, we consider two kinds of knowledge: (1) explicit knowledge expressed by users themselves, and (2) knowledge extracted from the data.

In our framework, we identify four main processes: (1) knowledge acquisition which requires either explicit information or extracted information from the data using data mining techniques, (2) knowledge integration into the data warehouse, (3) data warehouse schema evolution, and (4) OLAP queries personalization.

In the following, we present our approach for OLAP personalization which is composed of two techniques, namely adaptation and recommendation. Each technique respects the four steps of our framework mentioned above.

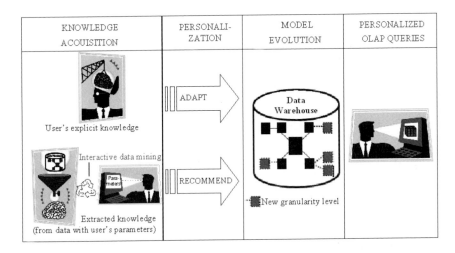

KNOWLEDGE ACQUISITION	PERSONALI-ZATION	MODEL EVOLUTION	PERSONALIZED OLAP QUERIES

Fig. 1. Framework for OLAP personalization

3.2 Adaptation-Based Personalization

Our adaptative data warehouse system aims to personalize analysis by integrating users's knowledge into the data warehouse, providing an answer to individual analysis needs. The user is asked to define his/her own knowledge in terms of if-then rules representing aggregations from a child level to a parent level. These rules are used to create a new granularity level in the considered dimension hierarchy. The if-clause, indeed, determines conditions on the attributes of the child level for grouping instances together forming a partition. The then-clause determines aggregates of the parent level, each one corresponds to a subset of the partition. In this case, the system is adaptative since it adapts itself by evolving the data warehouse schema to take into account new user's information.

3.3 Recommendation-Based Personalization

Classical OLAP operators are designed to create intuitive aggregates. However, to help users to find non expected and relevant aggregates expressing deep relations within a data warehouse, we propose to combine data mining techniques and OLAP. We choose to use the K-means clustering method, because of the format of its result, which is defined as a partition. The user is asked to fix the algorithms' parameters in an interactive way for obtaining relevant clusters. Then, the system recommends to the user the obtained clusters. If these latter are validated by the user, they are integrated into the data warehouse and a new hierarchy level is then created, allowing new OLAP queries which are proposed to the user.

To create a new level in a dimension hierarchy, we consider only classical hierarchies in both adaptation and recommendation techniques. In other words, each child occurrence in a child level is linked to a unique parent occurrence in

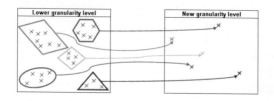

Fig. 2. Creation of a new granularity level

a parent level but each parent occurence can be associated with several child occurrences as showed in Figure 2.

4   Illustrative Example

To illustrate our approach for OLAP personalization in data warehouses, we use the example of the LCL company, which is a french bank we are collaborating with. We focus on an extract of the data warehouse concerning the management of accounts. We consider two measures which are the Net Banking Income (NBI) and the COST. The NBI is the profit obtained from the management of customers' accounts. As its name suggests, the second measure corresponds to the cost of customers' accounts. These measures are observed according to several dimensions: CUSTOMER, AGENCY and YEAR (Figure 3a). The dimension AGENCY is organized as a hierarchy which defines the geographical commercial structure of LCL, i.e. AGENCY is grouped into COMMERCIAL UNIT, which is grouped into DIRECTION.

Now, let us take the case of the person in charge of student products in the LCL french bank. He/she knows that there are three types of agencies: "student" for agencies which gather only student accounts, "foreigner" for agencies whose customers do not live in France, and "classical" for other agencies. However, this information is not stored in the data warehouse and therefore it cannot be used to carry out analysis about "student" agencies. Our adaptation-based personalization approach consists then in allowing the user to integrate his specific knowledge into the data warehouse. Then the system adapts itself according to this new user's knowledge by generating a new granularity level: AGENCIES GROUP that corresponds to the desired level in the AGENCY dimension (Figure 3b).

Fig. 3. a) Initial LCL data warehouse. b) Personalized LCL data warehouse.

Suppose now that the user wants to group agencies together according to the population of the city where the agency is located (Population) and the number of customers (CustNumber) but he/she doesn't kwow really how. To achieve this goal, our recommendation-based personalization approach consists then in extracting knowledge automatically from the data warehouse to provide possibly relevant clusters of agencies by using an unsupervised learning method, namely K-means. The system is then in charge of recommending to the user a new granularity level AGENCIES GROUP (Figure 3b) based on the obtained agencies clusters. The AGENCIES GROUP granularity level allows more elaborated OLAP queries. For instance, one may observe the evolution of ACCOUNTS MANAGEMENT (NBI) by CUSTOMER (Segmentation), YEAR (Year) and AGENCIES GROUP (AgenciesGroupName).

In the following, we detail our approach to recommend new OLAP queries based on the K-means method.

5 Framework for Recommending OLAP Queries

5.1 Basic Definitions

A data warehouse is a multidimensional database that can be defined as follows: $\mu = (\delta, \varphi)$, where δ is a set of **dimensions** and φ is a set of **facts** [17].

A **dimension schema** is a tuple $D = (L, \preceq)$ where L is a finite set of levels which contains a distinguished level named all, such that $dom(all) = \{all\}$ and $\preceq$ is a transitive and reflexive relation over the elements of L. The relation $\preceq$ contains a unique bottom level called l_{bottom} and a unique top level called all.

$$L = \{l_{bottom}, ..., l, ..., all \mid \forall\, l,\ l_{bottom} \preceq l \preceq all\}$$

Each level $l \in L$ is associated with a set of values $dom(l)$. For each pair of levels l and l' such that $l \preceq l'$, there exists a roll-up function f which is a partial function so that:

$$f_l^{l'} : dom(l) \longrightarrow dom(l')$$

A **fact table schema** F is defined as follows: $F = (I, M)$ where I is a set of dimension identifiers and M is a set of measures. A **fact table instance** is a tuple where the set of values for each identifier is unique.

To create data cubes, we use the *CUBE* operator [20] which is defined as follows: for a given fact table $F = (I = \{l_1 \in D_1, ..., l_p \in D_p\}, M)$, a set of levels $GL = \{l'_1 \in D_1, ..., l'_p \in D_p \mid l_i \preceq l'_i\ \forall i = 1..p\}$, and a set of measures m with $m \subset M$, the operation $CUBE(F, GL, m)$ gives a new fact table $F' = (GL, m')$ where m' is the result of aggregation (with roll-up functions $f_{l_1}^{l'_1}, ..., f_{l_p}^{l'_p}$) of the set of measures m from I to GL.

5.2 K-Means

K-means is known as a partitional clustering method that allows to classify a given data set X through k clusters fixed a priori [21,22]. The main idea is to define k centroids, one for each cluster, and then assign each point to one of the k clusters so as to minimize a measure of dispersion within the clusters. The algorithm is composed of the following steps:

1. *Place k initial points into the space represented by the data set X;*
2. *Assign each object x_i to the group that has the closest centroid c_j (the proximity is often evaluated with the euclidian metric);*
3. *Recalculate the positions of the k centroids when all objects have been assigned ;*
4. *Repeat Steps 2 and 3 until the centroids no longer move.*

The best grouping is the partition of the data set X that minimizes the sum of squares of distances between data and the corresponding cluster centroid.

We chose the K-means method for the following reasons: (1) its result format which is a partition that corresponds to the building process of the aggregation level in a dimension hierarchy, and (2) its low and linear algorithmic complexity which is crucial in the context of OLAP to provide the user with quick results.

5.3   Formalization

The K-means method enables us to classify instances of a level l, either on its own attributes, or on measure attributes in the fact table of the data warehouse. We exploit then the K-means clustering results to create a new level l_{new} and a roll-up function which relates instances of the child level l with the domain of the parent level l_{new}.

Roll-up with Generalize operator. An operator called **Generalize** is proposed in [17]. This operator creates a new level l_{new}, to which a pre-existent level l rolls up. A function f must be defined from the instance set of l, to the domain of l_{new}. We can summarize the formal definition of this operator as follows: given a dimension $D = (L = \{l_{bottom}, ..., l, ..., all\}, \preceq)$, two levels $l \in L$, $l_{new} \notin L$ and a function $f_l^{l_{new}} : instanceSet(l) \longrightarrow dom(l_{new})$. $Generalize(D, l, l_{new}, f_l^{l_{new}})$ is a new dimension $D' = (L', \preceq')$ where $L' = L \cup \{l_{new}\}$ and $\preceq' = \preceq \cup \{(l \to l_{new}), (l_{new} \to All)\}$, according to the roll-up function $f_l^{l_{new}}$.

Example. Consider the dimension AGENCY (Figure 3) and the roll-up function:

$f_{\text{AGENCY}}^{\text{POTENTIAL GROUP}} = ((\text{Charpennes, Big}), ..., (\text{Aubenas, Small}), ..., (\text{Lyon La Doua, Average}), ...)$.

Then, Generalize(AGENCY, AGENCY, POTENTIAL GROUP, $f_{\text{AGENCY}}^{\text{POTENTIAL GROUP}}$) adds a new level called POTENTIAL GROUP in the AGENCY dimension.

Hence, AGENCY $\to$ POTENTIAL GROUP constitutes another hierarchy for the AGENCY dimension.

Roll-up with RoK operator. In our case, the $f_l^{l_{new}}$ function is represented by our "RoK" (*Roll-up with K-means*) operator. Assume a positive integer k, a population $X = \{x_1, x_2, ..., x_n\}$ composed by n instances and a set of k classes $\mathcal{C} = \{\mathcal{C}_1, ..., \mathcal{C}_k\}$. By using the K-means algorithm described in section 5.2, $RoK(X, k)$ calculates the set $C = \{c_1, ..., c_k \mid \forall i = 1..k, c_i = barycenter(\mathcal{C}_i)\}$ and returns the roll-up function:

$$f_x^c = \{(x_j \rightarrow \mathcal{C}_i) \mid \forall j = 1..n \text{ and } \forall m = 1..k, dist(x_j, c_i) \leq dist(x_j, c_m)\}$$

Example. Let $X = \{x_1 = 2, x_2 = 4, x_3 = 6, x_4 = 20, x_5 = 26\}$ and $\mathcal{C} = \{\mathcal{C}_1, \mathcal{C}_2\}$. RoK(X, 2) returns the set $C = \{c_1 = 4, c_2 = 23\}$ with the roll-up function $f_x^c = \{(x_1 \rightarrow \mathcal{C}_1), (x_2 \rightarrow \mathcal{C}_1), (x_3 \rightarrow \mathcal{C}_1), (x_4 \rightarrow \mathcal{C}_2), (x_5 \rightarrow \mathcal{C}_2)\}$

Discussion. Comparing with the Generalize operator, our RoK operator generates automatically the new roll-up function. Our RoK operator is then more than a conceptual operator and provides a way to deal not only with the structure of the hierarchy, but also with the data of this hierarchy.

5.4   Algorithm

We present in the following the input parameters and the different steps of the personalization algorithm for the recommendation system.

 - A dimension $D = (L, \preceq)$, a level $l \in L$, a set of measure $m \in M$ (if required),
 - A level name $l_{new} \notin L$,
 - A positive integer $k \geq 2$ which will be the modality number of l_{new},
 - A variable *dataSource* that can take two values: 'F' (for *fact*) or 'D' (for *dimension*).

Step 1. Construction of the learning set X_l: This first step generates a learning set X_l from the instances of the pre-existing analysis level l. We consider a variable called *dataSource*. If the value of the variable equals to 'D', the population X_l is described by a part of attributes of the dimension D chosen by the user. Otherwise, X_l is generated by executing the operation $CUBE(F, l, m)$ whose parameters are also fixed by the user.

Example. Let us consider the two examples presented previously about the creation of the POTENTIAL GROUP and the COST GROUP levels.

Let us consider that one user needs to create a new level POTENTIAL GROUP from the AGENCY level. If the *dataSource* parameter equals to 'D', each agency will be described by a part of its descriptors in the data warehouse chosen by the user. For instance, the user can choose the CustNumber and the Population attributes for the reasons presented before (Figure 4a).

Now, let us suppose that the user needs to create a new level COST GROUP from the CUSTOMER level. If the *dataSource* parameter equals to 'F', our algorithm performs the operator CUBE (ACCOUNTS MANAGEMENT, CUSTOMER, COST) according to the choice of the user. Thus, we obtain the learning set described in Figure 4b.

AgencyName	CustNumber	Population
Charpennes	105	12 000
Aubenas	6	100
Lyon La Doua	60	6 000
Annonay	8	180

CUSTOMERid	COST
001	400
004	20
005	300
007	50

a)   b)

Fig. 4. a) The AGENCY analysis level described by a part of its own attributes. b) The CUSTOMER analysis level described by a measure.

Step 2. Clustering: During this step, the algorithm applies the *RoK* operator to the learning set X_l. If, for example, the parameter k equals to 2, the operation *RoK* on the Figure 4a gives the set $\mathcal{C} = \{\mathcal{C}_1(82.5; 9000), \mathcal{C}_2(7; 140)\}$ and the roll-up function:

$$f_{\text{AGENCY}}^{\text{POTENTIAL GROUP}} = ((\text{Charpennes}; \mathcal{C}_1), (\text{Aubenas}; \mathcal{C}_2),$$
$$(\text{Lyon La Doua}; \mathcal{C}_1), (\text{Annonay}; \mathcal{C}_2)).$$

Step 3. Creation of the new level: This step implements the new analysis level l_{new} in the data warehouse model. It is done after the validation of the user. To do this operation, our algorithm performs a *Generalize* operation on the dimension D, from the level l by using the roll-up function $f_l^{l_{new}}$ generated during the previous step.

Example. To materialize the POTENTIAL GROUP level in the AGENCY dimension, our algorithm performs the operator Generalize:

Generalize(AGENCY, AGENCY, POTENTIAL GROUP, $f_{\text{AGENCY}}^{\text{POTENTIAL GROUP}}$).

5.5   Feature Selection

To apply the K-means clustering method onto the data warehouse, we propose two strategies for the feature selection. The first one uses directly attributes that describe the level l to be classified while the second one uses measure attributes on the fact table aggregated over the level l. We are going to illustrate these two proposals with examples extracted from the LCL case study presented previously.

Proposal 1. K-means based on the dimension level features. Let us consider the next analysis objective: *Is it necessary to close agencies which make little income? And is it necessary to open new agencies in places which make a lot of income?*

To try to answer these questions, the user is going to study the NBI through the AGENCY dimension (Figure 3). To improve his/her analysis, the user can feel the need to aggregate agencies according to their potential. For that purpose, our operator allows the user to classify instances of the AGENCY level according to the population of the city where the agency is located in and the customer number of the agency. The objective is to create a new level which groups the instances of the AGENCY level in small, average or big potential (Figure 5).

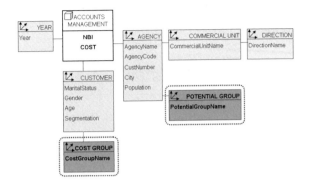

Fig. 5. LCL data warehouse model after addition of "COST GROUP" and "POTENTIAL GROUP" levels

Proposal 2. *K-means based on data fact measures.* Assume that the analysis objective of the user is to identify a customer grouping according to the costs. The idea is that a customer can cost much compared to an average cost but also bring much more than an average and vice versa. Thus it would be interesting to analyse the NBI according to groups of customer costs. With our proposal, the user can concretize this need with a new level in the CUSTOMER dimension. For that, our operator will summarize COST measure on the CUSTOMER level of the dimension. K-means is then performed to the result of this aggregation operation. After this clustering, the creation of the new level allows analysis according to groups of costs (Figure 5).

5.6 Data Warehouses Model Evolution for OLAP Personalization

Before the effective creation of the level, a validation phase by the user is required, since we are in a context of recommendation. The validation is given by the user only if the proposed level is an answer to his/her analysis needs. Note that, this personalization process provides the user with expressing his/her needs in terms of giving the value of the number of classes he/she wants and specifying the attributes involved in the K-means process.

Creating new granularity levels does not affect the integrity of existing data. The data warehouse is updated, allowing to share the new analysis possibilities with all decision makers, without requiring versions management.

6   Implementation and Experiments

We developped our approach inside the Oracle 10g RDBMS. Thus, we implemented the k-prototypes algorithm by using PL/SQL stored procedures. K-prototypes is a variant of the K-means method allowing large datasets clustering with mixed numeric and categorical values [23]. In our implementation, datasets are stored within a relational table. After the clustering process, the

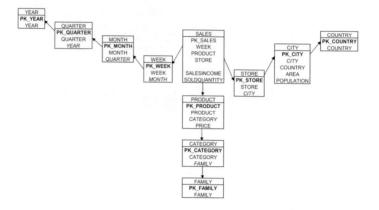

Fig. 6. Schema of the "Emode" data warehouse

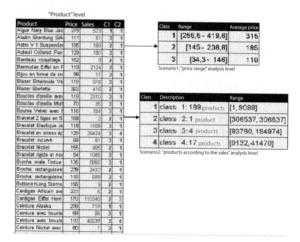

Fig. 7. Results of the two scenarios

model evolution is performed by using SQL operators: the new level is created with the *CREATE TABLE* command and the roll-up function is established with a primary key/foreign key association between the new and the existing levels.

We carried out some experiments under the "Emode" data warehouse. Emode is an e-trade data warehouse which is used as a demonstration database for the tool *"BusinessObject 5.1.6"*. We standardized the schema of this data warehouse compared to the diagram of Figure 6.

The *sales* fact table stores 89200 records and the *article* level of the *product* dimension contains 213 instances. According to our two proposals for "feature selection", we envisaged two scenarios:

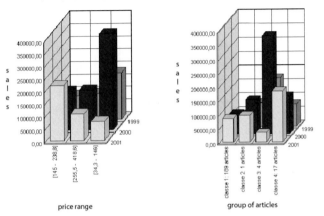

Fig. 8. Analysis results of the two scenarios

1. Creation of an *article price grouping* level which classifies the 213 articles according to their price,
2. Creation of another level *article sales grouping* which groups the articles according to the sales income.

Figure 7 shows the results of the two scenarios.

We created the *article price grouping* level with three possible values. With this level, we can analyse the influence of the prices on sales. Figure 8 shows the quantity sold for the 3 price categories. For instance, we can conclude that the products of the lower price (category 3) are those that are sold in larger quantities.

For the *article sales grouping* level, we obtain a level allowing to gather articles into four classes of sales income. Figure 8 shows the quantity sold according to the sales income information. Thus we can for instance affirm that the products that are the subject of the best sales (category 2) are not sold in the lowest quantities. Such a created level allows to confirm or deny the 80-20 rule.

We mention that a drill down allows to know more about the products composing the various created classes.

7   Conclusion and Perspectives

In this paper, we proposed a general framework to integrate knowlegde inside a data warehouse in order to allow OLAP personalization. Our personalization approach is supported by the data warehouse model evolution, independently of the data sources, and it provides to the users new analysis possibilities.

We exploit two types of knowledge: explicit knowledge which is directly expressed by users and implicit knowledge which is extracted from the data. In the first case, the system adapts itself by creating a new granularity level according to the user's needs. In the second case, the system recommends to the user a new analysis axis based on automatically extracted clusters from the data warehouse. If the user validates the proposition, a new granularity level is creatad in the dimension hierarchy.

Our recommendation system is based on a definition of a new OLAP operator, called RoK based on a combination between the K-means clustering method and a classical roll-up operator. RoK operator computes significant and more elaborated OLAP queries than the classical ones.

To validate our approach for OLAP personalization, we developed a proto-type within the Oracle 10g RDBMS and carried out some experiments which showed the relevance of our personalized data warehouse system. We mainly implemented the RoK operator in the form of a stored procedure using PL/SQL language.

This work opens several promising issues and presents new challenges in the domain of personalization in data warehouses. Firstly, instead of recommending to the user to create only one hierarchy level, we plan to generalize our recom-mendation approach to be able to recommend a fully dimension hierarchy by using for example the Agglomerative Hierarchical Clustering. The user will be asked to choose a number of classes after the learning process. In this case, the number of classes is not an input parameter. Secondly, we plan to refine the recommendation process. A promising issue consists in combining data mining and the concept of user profile for personalization. Hence, we suggest to consider users' analysis sessions which are composed of a set of queries. For each user pro-file, our key idea is to use frequent itemset mining methods for extracting the frequently asked queries. These latter are recommended to the user according to his/her profile. Finally, as a consequence of our data warehouse personalization and evolution approach, it is interesting to evaluate the performance of material-ized views maintenance. In other words, once a new level is created, how existing materialized views are updated and what is the cost?

References

1. Rizzi, S.: OLAP Preferences: A Research Agenda. In: DOLAP 2007, pp. 99–100 (2007)
2. Bentayeb, F., Favre, C., Boussaid, O.: A User-driven Data Warehouse Evolu-tion Approach for Concurrent Personalized Analysis Needs. Journal of Integrated Computer-Aided Engineering 15(1), 21–36 (2008)
3. Domshlak, C., Joachims, T.: Efficient and Non-Parametric Reasoning over User Preferences. User Modeling and User-Adapted Interaction 17(1-2), 41–69 (2007)
4. Korfhage, R.R.: Information storage and retrieval. John Wiley & Sons, Inc., Chich-ester (1997)
5. Manber, U., Patel, A., Robison, J.: Experience with personalization of yahoo! Com-munications of the ACM 43(8), 35–39 (2000)
6. Pretschner, A., Gauch, S.: Ontology Based Personalized Search. In: ICTAI 1999, Chicago, Illinois, USA, pp. 391–398 (1999)
7. Cherniack, M., Galvez, E.F., Franklin, M.J., Zdonik, S.B.: Profile-Driven Cache Management. In: ICDE 2003, Bangalore, India, pp. 645–656 (2003)

8. Chomicki, J.: Preference Formulas in Relational Queries. ACM Transactions on Database Systems 28(4), 427–466 (2003)

9. Koutrika, G., Ioannidis, Y.: Personalized Queries under a Generalized Preference Model. In: ICDE 2005, Tokyo, Japan, pp. 841–852 (2005)

10. Bellatreche, L., Giacometti, A., Marcel, P., Mouloudi, H., Laurent, D.: A Personalization Framework for OLAP Queries. In: DOLAP 2005, pp. 9–18 (2005)

11. Ravat, F., Teste, O.: Personalization and OLAP Databases. Annals of Information Systems, New Trends in Data Warehousing and Data Analysis (2008)

12. Jerbi, H., Ravat, F., Teste, O., Zurfluh, G.: Management of context-aware preferences in multidimensional databases. In: ICDIM 2008, pp. 669–675 (2008)

13. Giacometti, A., Marcel, P., Negre, E.: A Framework for Recommending OLAP Queries. In: DOLAP 2008, pp. 73–80 (2008)

14. BenMessaoud, R., Boussaid, O., Rabaseda, S.: A new OLAP aggregation based on the AHC technique. In: DOLAP 2004, pp. 65–72 (2004)

15. Kaya, M.A., Alhajj, R.: Extending OLAP with Fuzziness for Effective Mining of Fuzzy Multidimensional Weighted Association Rules. In: Li, X., Zaïane, O.R., Li, Z.-h. (eds.) ADMA 2006. LNCS, vol. 4093, pp. 64–71. Springer, Heidelberg (2006)

16. Blaschka, M., Sapia, C., Höfling, G.: On Schema Evolution in Multidimensional Databases. In: Mohania, M., Tjoa, A.M. (eds.) DaWaK 1999. LNCS, vol. 1676, pp. 153–164. Springer, Heidelberg (1999)

17. Hurtado, C., Mendelzon, A., Vaisman, A.: Maintaining Data Cubes under Dimension Updates. In: ICDE 1999, pp. 346–355 (1999)

18. Morzy, T., Wrembel, R.: Modeling a Multiversion Data Warehouse: A Formal Approach. In: ICEIS 2003, vol. 1, pp. 120–127 (2003)

19. Vaisman, A., Mendelzon, A.: Temporal Queries in OLAP. In: VLDB 2000, pp. 242–253 (2000)

20. Gray, J., Bosworth, A., Layman, A., Pirahesh, H.: Data Cube: A Relational Aggregation Operator Generalizing Group-By, Cross-Tab, and Sub-Total. In: ICDE 1996, pp. 152–159 (1996)

21. Forgy, E.: Cluster Analysis of Multivariate Data: Efficiency versus Interpretability of Classification. Biometrics 21

22. MacQueen, J.: Some Methods for Classification and Analysis of Multivariate Observations. In: Vth Berkeley Symposium, pp. 281–297 (1967)

23. Huang, Z.: Clustering Large Data Sets with Mixed Numeric and Categorical Values. In: PAKDD 1997 (1997)

On Index-Free Similarity Search
in Metric Spaces

Tomáš Skopal[1] and Benjamin Bustos[2]

[1] Department of Software Engineering, FMP, Charles University in Prague,
Malostranské nám. 25, 118 00 Prague, Czech Republic
skopal@ksi.mff.cuni.cz
[2] Department of Computer Science, University of Chile,
Av. Blanco Encalada 2120 3er Piso, Santiago, Chile
bebustos@dcc.uchile.cl

Abstract. Metric access methods (MAMs) serve as a tool for speeding similarity queries. However, all MAMs developed so far are index-based; they need to build an index on a given database. The indexing itself is either static (the whole database is indexed at once) or dynamic (insertions/deletions are supported), but there is always a preprocessing step needed. In this paper, we propose *D-file*, the first MAM that requires no indexing at all. This feature is especially beneficial in domains like data mining, streaming databases, etc., where the production of data is much more intensive than querying. Thus, in such environments the indexing is the bottleneck of the entire production/querying scheme. The idea of D-file is an extension of the trivial sequential file (an abstraction over the original database, actually) by so-called *D-cache*. The D-cache is a main-memory structure that keeps track of distance computations spent by processing all similarity queries so far (within a runtime session). Based on the distances stored in D-cache, the D-file can cheaply determine lower bounds of some distances while the distances alone have not to be explicitly computed, which results in faster queries. Our experimental evaluation shows that query efficiency of D-file is comparable to the index-based state-of-the-art MAMs, however, for zero indexing costs.

1 Introduction

The majority of problems in the area of database systems concern the efficiency issues – the performance of DBMS. For decades, the number of accesses to disk required by I/O operations was the only important factor affecting the DBMS performance. Hence, there were developed indexing structures [16,2], storage layouts [4], and also disk caching/buffering techniques [7]; all of these designs aimed to minimize the number of physical I/Os spent during a database transaction flow. In particular, disk caching is extremely effective in situations where repeated access to some disk pages happens within a single runtime session.

In some modern databases, like multimedia DBs (MMDBs), DNA DBs, time series DBs, etc., we need to use a similarity function $\delta(\cdot, \cdot)$ which serves as a relevance measure, saying how much a DB object is relevant to a query object. To

S.S. Bhowmick, J. Küng, and R. Wagner (Eds.): DEXA 2009, LNCS 5690, pp. 516–531, 2009.

speedup similarity search in MMDBs, there have been many indexing techniques developed, some being domain-specific and some others more general. The new important fact is that the performance of MMDBs is more affected by CPU costs than by I/O costs. In particular, in MMDBs community a single computation of similarity value $\delta(\cdot, \cdot)$ is employed as the logical unit for indexing/retrieval cost, because of its dominant impact on overall MMDB performance [18,5] (algorithms computing $\delta(\cdot, \cdot)$ are often super-linear in terms of DB object size). Thus, the I/O costs are mostly regarded as a minor component of the overall cost because of the computational complexity of similarity measures. The number of computations $\delta(\cdot, \cdot)$ needed to answer a query or to index a database is referred to as the *computation costs*.

1.1 Metric Access Methods

Among the similarity search techniques, *metric access methods* (MAMs) are suitable in situations where the similarity measure δ is a *metric* distance (in mathematical meaning). The metric postulates – reflexiveness, positiveness, symmetry, triangle inequality – allow us to organize the database within classes that represent some occupied partitions of the underlying metric space. The classes are usually organized in a data structure (either persistent or main-memory), called *index*, that is created during a preprocessing step (the indexing).

The index is later used to quickly answer typical similarity queries – either a *k nearest neighbors* (kNN) query like "return the 3 most similar images to my image of horse", or a *range query* like "return all voices more than 80% similar to the voice of nightingale". In particular, when issued a similarity query[1], the MAMs exclude many non-relevant classes from the search (based on metric properties of δ), so only several candidate classes of objects have to be sequentially searched. In consequence, searching a small number of candidate classes turns out in reduced computation costs of the query.

There were developed many MAMs so far, addressing various aspects – main-memory/database-friendly methods, static/dynamic indexing, exact/approximate search, centralized/distributed indexing, etc. (see [18,12,5,11]). Although various MAMs often differ considerably, they all share the two following properties:

1. MAMs are all *index-based*. For a given database, an index must exist in order to be able to process queries. Hence, the first query must be always preceded by a more or less expensive data preprocessing which results in an index (either main-memory or persistent).
2. Once its index is built, a MAM solves every query request *separately*, that is, every query is evaluated as it would be the only query to be ever answered. In general, no optimization for a *stream of queries* is considered by MAMs up to date. Instead, enormous research has been spent in "materializing" the data-pruning/-structuring knowledge into the index file itself.

---

[1] A range or kNN query can be viewed as a ball in the metric space (centered in query object Q with radius of the range/distance to kNN), so we also talk about *query ball*.

In the following, we consider three representatives out of dozens of existing MAMs – the *M-tree*, the *PM-tree*, and *GNAT*.

M-tree. The *M-tree* [6] is a dynamic (easily updatable) index structure that provides good performance in secondary memory, i.e., in database environments. The M-tree index is a hierarchical structure, where some of the data objects are selected as centers (also called references or local *pivots*) of ball-shaped regions, while the remaining objects are partitioned among the regions in order to build up a balanced and compact hierarchy of data regions.

PM-tree. The idea of PM-tree [13,14] is to enhance the hierarchy of M-tree by using information related to a static set of p global pivots P_i. In a PM-tree's non-leaf region, the original M-tree-inherited ball region is further cut off by a set of rings (centered in the global pivots), so the region volume becomes smaller. Similarly, the PM-tree leaf entries are extended by distances to the pivots, which are also interpreted as rings due to quantization. Each ring stored in a non-leaf/leaf entry represents a distance range bounding the underlying data with respect to a particular pivot. The combination of all the p entry's ranges produces a p-dimensional minimum bounding rectangle, hence, the global pivots actually map the metric regions/data into a "pivot space" of dimensionality p.

GNAT. The Geometric Near-Neighbor Access Tree (GNAT) [3] is a metric access method that extends the Generalized-Hyperplane Tree [15]. The main idea behind GNAT is to partition the space into zones that contain close objects. The root node of the tree contains m objects selected from the space, the so-called *split-points*. The rest of the objects is assigned to their closest split-point. The construction algorithm selects with a greedy algorithm the split-points, such that they are far away from each other. Each zone defined by the selected split-points is partitioned recursively in the same way (possibly using a different value for m), thus forming a search hierarchy. At each node of the tree, a $O(m^2)$ table stores the range (minimum and maximum distance) from each split-point to each zone defined by the other split-points.

1.2   Motivation for Index-Free Similarity Search

As mentioned earlier, the existing MAMs are all index-based. However, there emerge many real and potential needs for access methods that should provide index-free similarity search. We briefly discuss three cases where any data pre-processing (like indexing) is undesirable:

"Changeable" databases. In many applications, there are databases which content is intensively changing over time, like streaming databases, archives, logs, temporal databases, where new data arrives and old data is discarded frequently. Alternatively, we can view any database as "changeable" if the proportion of changes to the database exceeds the number of query requests. In highly changeable databases, the indexing efforts lose their impact, since the expensive indexing is compensated by just a few efficient queries. In the extreme case (e.g., sensory-generated data), the database could have to be massively updated in real time, so that any indexing is not only slow but even impossible.

Isolated searches. In complex tasks, e.g., in data mining, a similarity query over a single-purpose database is used just as an isolated operation in the chain of all required operations to be performed. In such case, the database might be established for a single or several queries and then discarded. Hence, index-based methods cannot be used, because, in terms of the overall costs (indexing+querying), the simple sequential search would perform better.

Arbitrary similarity function. Sometimes, the similarity measure is not defined a priori and/or can change over the time. This includes learning, user-defined or query-defined similarity, while in such case any indexing would lead to many different indexes, or it is not possible at all.

1.3 Paper Contribution

In this paper, we propose the D-file, an index-free MAM employing a main-memory structure – the D-cache. The D-cache (distance cache) stores distances computed during querying within a single runtime session. Hence, the aim of D-file is not to use an index, but to amortize the query costs by use of D-cache, similarly like I/O-oriented access methods amortize the I/O costs using disk cache. As in the case of simple sequential search, querying the D-file also means a sequential traversal of the entire database. However, whenever a DB object is to be checked against a query, instead of computing the DB object-to-query object distance, we request the D-cache for its tightest lower bound. This lower-bound distance is subsequently used to filter the DB object. Since many distances could have been computed during previous querying (for other query objects, of course), the lower bounds could be transitively inferred from the D-cache "for free", which results in reduced query costs.

2 Related Work

In this section, we briefly discuss existing index-free attempts to metric similarity search. In fact, to the best of our knowledge, there exist just one non-trivial approach applicable directly to the index-free metric search, as mentioned in Section 2.2. But first, in the following section we discuss the simple sequential scan – in a role of trivial index-free MAM.

2.1 Simple Sequential Scan

If no index structure is provided, the only way of answering a similarity query in metric spaces is to perform a sequential search of the database. By this approach, for both range and k-NN queries the search algorithm computes the distances between the query object and all objects in the database. With the computed distances it is trivial to answer both types of queries. This approach has, of course, the advantage that neither space nor preprocessing CPU time is required to start performing similarity queries. It follows that the cost of a sequential scan is linear in the size of the database. This, however, may be already prohibitively expensive, for example if the database is too large, e.g., it contains tens of millions of objects, or if the distance function is expensive to compute, e.g., the edit distance between strings is $O(nm)$ for sequences of lengths n and m.

For the particular case of vector spaces, the VA-file [17] is a structure that stores compressed feature vectors, providing thus an efficient sequential scan of the database. At query time, an approximation of the distances between query objects and compressed features are computed, discarding at this filtering step as many objects as possible. The search algorithm refines this result by computing the real distances to database objects only for the non-discarded vectors. While this approach could be a good alternative to the plain sequential scan, it only works in vector spaces and cannot be easily generalized to the metric case. Moreover, though the idea is based on sequential search, it is not index-free approach, because the compressed vectors form a persistent index – the VA-file.

2.2   Query Result Caching

Recently, the concept of *metric cache* for similarity search was introduced, providing a caching mechanism that prevents any underlying MAM (i.e., also simple sequential scan) to process as many queries as possible [8,9]. Basically, the metric cache stores a history of similarity queries and their answers (ids and descriptors of database objects returned by the query). When a next query is to be processed, the metric cache either returns the exact answer in case the same query was already processed in the past and its result still sits in the cache. Or, in case of a new query, such old queries are determined from the metric cache, that spatially contain the new query object inside their query balls. If the new query is entirely bounded by a cached query ball, a subset of the cached query result is returned as an exact answer of the new query. If not, the metric cache is used to combine the query results of spatially close cached queries to form an approximate answer of the new query. In case the approximated answer is likely to exhibit a large retrieval error, the metric cache gives up and forwards the query processing to the underlying retrieval system/MAM (updating the metric cache by the query answer afterwards).

We have to emphasize that metric cache is a higher-level concept that can be combined with any MAM employed in a content-based retrieval system. Hence, metric cache is just a standalone front-end subpart in the whole retrieval system, while the underlying MAM alone is not aware of the metric cache. On the other hand, the proposal of D-cache in the following text is a low-level concept that plays the role of integral part of a metric access method (the D-file, actually).

3   D-File

We propose an index-free metric access method, the *D-file*, which is a set of methods extending simple sequential search over the database. Unlike the

VA-file mentioned earlier, we emphasize that D-file is just an abstraction above the original database, hence, there is no additional "file" materialized along-side the database, that is, no additional *persistent* data structure is maintained, nor any preprocessing is performed. In other words, the D-file is the original database file equipped by a set of querying methods. Instead, the D-file uses a main-memory structure called D-cache (described in the next section). The D-cache has a simple objective – to gather (cache) distances already computed between DB and query objects within a single runtime session. Based on the stored distances, the D-cache can be asked to cheaply infer lower bound of some distance between a query object and a DB object. The D-file's query algorithms then use these lower bounds when filtering DB objects, see Algorithms 1 and 2.

Algorithm 1. (D-file kNN query)

---

set **kNNQuery**(Q, k) {
  Dcache.**StartQueryProcessing**(Q)
  let NN be array of k pairs $[O_i, \delta(Q, O_i)]$ sorted asc. wrt $\delta(Q, O_i)$, initialized to NN $= [[-, \infty], ..., [-, \infty]]$
  let r_Q denotes the actual distance component in NN$[k]$
  for each O_i in database **do**
    if Dcache.**GetLowerBoundDistance**$(Q, O_i) \leq r_Q$ **then**   // D-cache filtering
      compute $\delta(Q, O_i)$; Dcache.**AddDistance**$(Q, O_i, \delta(Q, O_i))$
      if $\delta(Q, O_i) \leq r_Q$ **then** insert $[O_i, \delta(Q, O_i)]$ into NN   // basic filtering
  return NN as result }

Algorithm 2. (D-file range query)

---

set **RangeQuery**(Q, r_Q) {
  Dcache.**StartQueryProcessing**(Q)
  for each O_i in database **do**
    if Dcache.**GetLowerBoundDistance**$(Q, O_i) \leq r_Q$ **then**   // D-cache filtering
      compute $\delta(Q, O_i)$; Dcache.**AddDistance**$(Q, O_i, \delta(Q, O_i))$
      if $\delta(Q, O_i) \leq r_Q$ **then** add O_i to the query result }   // basic filtering

---

4   D-Cache

The main component of D-file is the *D-cache* (distance cache) – a non-persistent (memory resident) structure, which tracks distances computed between query objects and DB objects, considering a single runtime session, i.e., contiguous sequence of queries. The track of distance computations is stored as a set of triplets, each of the form $[id(Q_i), id(O_j), \delta(R_i, O_j)]$, where Q_i is a query object, O_j is a DB object, and $\delta(Q_i, O_j)$ is their distance computed during the current session. We assume query as well as DB objects are uniquely identified.

Instead of considering a set of triplet entries, we can view the content of D-cache as a sparse matrix

$$
D = \begin{array}{c} \\ Q_1 \\ Q_2 \\ Q_3 \\ ... \\ Q_m \end{array}
\begin{array}{cccccc} O_1 & O_2 & O_3 & ... & O_n \\ & d_{12} & d_{13} & ... & \\ d_{21} & & & ... & d_{2n} \\ & & & ... & \\ ... & ... & ... & ... & ... \\ d_{m1} & & d_{m3} & ... & \end{array}
$$

where the columns refer to DB objects, the rows refer to query objects, and the cells store the respective query-to-DB object distances. Naturally, as new DB objects and query objects arrive into D-cache during the session, the matrix gets larger in number of rows and/or columns. Note that at the beginning of session the matrix is empty, while during the session the matrix is being filled. However, at any time of the session there can still exist entirely empty rows/columns.

Note that query objects do not need to be external, that is, a query object could originate from the database. From this point of view, an object can have (at different moments) the role of query as well as the role of DB object, however, the unique identification of objects ensures the D-cache content is correct.

Because of frequent insertions into D-cache, the matrix should be efficiently updatable. Moreover, due to operations described in the next subsection, we have to be able to quickly retrieve a cell, column, or row values. To achieve this goal, we have to implement the matrix by a suitable data structure(s).

4.1   D-Cache Functionality

The desired functionality of D-cache is twofold:

- First, given a query object Q and a DB object O on input (or even two DB objects or two query objects), the D-cache should quickly determine the *exact* value $\delta(Q, O)$, provided the distance is present in D-cache.
- The second functionality is more general. Given a query object Q and a DB object O on input, the D-cache should determine the tightest possible *lower bound* of $\delta(Q, O)$ without the need of an explicit distance computation.

Both of the functionalities allow us to filter some non-relevant DB objects from further processing, making the search more efficient. However, in order to facilitate the above functionality, we have to feed the D-cache with information about the involved objects. To exploit the first functionality, the current query object could have to be involved in earlier queries either as query object or as DB object the distance of which was computed against another query object.

To exploit the second functionality, we need to know distances to some past query objects DP_i^Q which are very close to the current query Q. Suppose for a while we know $\delta(Q, DP_1^Q), \delta(Q, DP_2^Q), \ldots$ – these will serve as *dynamic pivots* made-to-measure to Q. Since the dynamic pivots are close to Q, they should be very effective when pruning as they provide tight approximations of $\delta(Q, O_i)$. Having the dynamic pivots DP_i^Q, we can reuse some distances $\delta(DP_i^Q, O)$ still sitting in the D-cache matrix, where they were inserted earlier during the current session. Then, with respect to the pivots and available distances $\delta(DP_i^Q, O)$ in the matrix, the value $max_{DP_i^Q}\{\delta(DP_i^Q, O) - \delta(DP_i^Q, Q)\}$ is the tightest lower-bound distance of $\delta(Q, O)$. Similarly, $min_{DP_i^Q}\{\delta(DP_i^Q, O) + \delta(DP_i^Q, Q)\}$ is the tightest upper-bound distance of $\delta(Q, O)$. See the situation in Figure 1a.

4.2   Determining Dynamic Pivots

In principle, we consider two ways to obtain k dynamic pivots out of (all) previously processed queries:

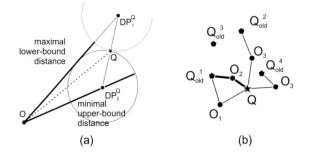

Fig. 1. (a) Lower/upper bounds to $\delta(Q,O)$. (b) Internal selection of dynamic pivot ($k = 1$) by closeness approximation (Q_{old}^1 is the winner).

(a) Recent. We choose k past query objects immediately, that is, before the current query processing actually starts. More specifically, we choose the k recently processed (distinct) query objects.

(b) Internal. When the current query processing is started, the first x distance computations of $\delta(Q, O_i)$ are used just to update D-cache, that is, the D-cache is still not used for filtering. After insertion of the respective x triplets into D-cache, the D-cache could be requested to select the k most promising dynamic pivots out of the past query objects Q_{old}^i.

The reason for computing x triplets is based on an expectation, that there could appear so-called *mediator objects* in D-cache, that is, objects O_m for which distances $\delta(Q, O_m)$ and $\delta(O_m, Q_j)$ will appear in D-cache. Such mediator objects provide an indirect distance approximation of $\delta(Q, Q_j)$ (remember that because of the triangle inequality $\delta(Q, Q_j) \le \delta(Q, O_m) + \delta(Q_m, R_j)$). The mediators can be used for selection of dynamic pivots as follows:

A dynamic pivot DP_j is chosen as $DP_j \in \{Q_{old}^i | \exists O_m(\delta(Q, O_m) + \delta(O_m, Q_{old}^i) \le k\text{-}min_{O_l}\{\delta(Q, O_l) + \delta(O_l, Q_{old}^i))\}\}$. The $k\text{-}min_{O_l}\{\cdot\}$ is the kth minimum value; with respect to any O_l for which D-cache stores distances $\delta(Q, O_l)$ and $\delta(O_l, Q_{old}^i)$. This means that an old query object will be selected as a dynamic pivot if its associated smallest "through-mediator approximation" of $\delta(Q, DP_j)$ is within the k smallest among all old query objects.

Figure 1b shows an example for $k = 1$. The really closest pivot Q_{old}^4 was not selected because the mediator O_3 is an outlier, while no better mediator for Q_{old}^4 was found in D-cache. Also note that for Q_{old}^3 there is no mediator at all.

After we determine the dynamic pivots, we compute their distances to Q. Note that this is the *only place* where we explicitly compute some extra distance computations (not computed when not employing D-cache).

4.3 D-Cache Implementation

The D-cache is initialized by D-file when the session begins. Besides this global initialization, the D-cache is also notified by D-file that a query has been started (method StartQueryProcessing). At that moment, a new query object is being processed so the current dynamic pivots have to be dismissed. Every time a distance $\delta(Q, O_i)$ value is explicitly computed, the triplet $[id(Q), id(O_i), \delta(Q, O_i)]$ is inserted into the D-cache (method AddDistance).

Besides the retrieval of the hopefully available exact distance between objects Q, O_i (method GetDistance(Q, O_i)), the main functionality is operated by method GetLowerBoundDistance, see Algorithm 3.

Algorithm 3. (GetLowerBoundDistance)

```
double GetLowerBoundDistance(Q, O_i) {
  let x be the number of computations ignored
  let k be the number of pivots to use
  let DP be the set of dynamic pivots and their distances to Q
  mComputed = mComputed + 1                              // mComputed=0 initialized at query start
  value = 0
  if mComputed ≤ x and determineMethod = internal then {     // internal pivot selection
    value = compute δ(Q, O_i)
    AddDistance(Q, O_i, value)
    if mComputed = x then
      DP = DeterminePivotsByMediatorDistances(Q, k)
  } else {                                                // lower bound construction/update
    for each P in DP do
      if cell(P, O_i) is not empty then
        value = max(value, cell(P, O_i) − δ(Q, P)) }
  return value }
```

The structure of D-cache itself is implemented by two substructures – the CellCache and the RowCache:

CellCache Structure. As the main D-cache component, the *CellCache* stores the distance matrix as a set of triplets $(id1, id2, \delta(Q_{id1}, O_{id2}))$ in a hash table, and provides retrieval of individual triplets. As a hash key, $(min(id1, id2),$ $max(id1, id2))$ is used. When applying the *recent* dynamic pivot selection, as defined in Section 4.2, the CellCache is the only D-cache component. Naturally, the hash-based implementation of CellCache is very fast (constant access time).

RowCache Structure. However, when applying *internal* selection of dynamic pivots, we need to retrieve rows from the distance matrix. This cannot be efficiently accomplished by CellCache, hence, we use the *RowCache* as a redundant data structure. In particular, the RowCache aims at efficiently computing the dynamic pivots to be used with the current query object. Thus, it must determine the mediator objects and compute the intersection between rows of the D-cache. It could also be used to obtain bounds of the distances between objects as with the CellCache, however the CellCache may be more efficient than the RowCache for this particular function.

The RowCache is implemented as a main-memory inverted file. Each row of this cache stores all the computed distances for a single query, and it is implemented as a linked list. Each time a distance between the query and an object from the database is computed, a node is added to the list. When a new query object is inserted in the cache, the algorithm creates a new row and stores there the computed distances to the new query object.

To compute the best dynamic pivots for a given query object, the RowCache determines firstly its mediator objects. That is, given two query objects, the current one and one from the cache, it returns the intersection of the corresponding rows in the RowCache. This is repeated for all query objects in the cache. Once the mediator objects are found, the algorithm determines the best dynamic pivots by selecting the k query objects with smallest indirect distance (i.e., obtained through a mediator) to the current query (see Algorithm 4). This algorithm can be efficiently implemented with a priority queue (max-heap), that keeps the k best pivots found so far, and replaces the worst of them when a better pivot is found. With our actual implementation of RowCache[2], in the worst case this algorithm takes $O(A * \log(A) + A * C + A * \log(k))$ time, where A is the number of rows in the RowCache, C is the maximum number of cells per row, and k is the number of pivots. In practice, however, there are usually only a few valid mediators per query objects, thus the average cost is closer to $O(\log(k))$.

Algorithm 4. (DeterminePivotsByMediatorDistances)

---

set **DeterminePivotsByMediatorDistances**(Q, k) {
  old = the set of past query objects Q_{old}^i
  winners = $\emptyset$ // set of k pairs [object, distance] ordered ASC on distance
  for each Q_{old}^i in old **do**
    // determine mediators, i.e. objects having distance to both Q and Q_{old}^i in D-cache
    // $row(\cdot) \cap row(\cdot)$ stands for all DB objects having defined both values on their position in the rows
    mediators = $row(Q_{old}^i) \cap row(Q)$
    for each M in mediators **do**
      update winners by $[Q_{old}^i, cell(Q, M) + cell(Q_{old}^i, M)]$
  return [winners, computed distances $\delta(Q, \text{winners}(i))$] }

The CellCache and RowCache are not necessarily synchronized, that is, both caches may not contain the distances for the same pair of objects. However, if a query is deleted from the CellCache this is also reflected in the RowCache.

Distance Replacement Policies. Since the storage capacity of D-cache is limited, the hash table of CellCache as well as the inverted list of RowCache are of user-defined size (in bytes, usually equally divided between CellCache and RowCache[3]). Once either of the structures has no space to accommodate a new distance, some old distance (near to the intended location) has to be released. We consider two policies for distance replacement:

LRU. A distance is released, which has been least recently used (read). The hypothesis is that keeping just the frequently used distances leads to tighter distance lower/upper bounds, thus to better pruning.

---

[2] The depicted implementation of the RowCache structure is not optimized. However, in the experimental evaluation we will show that the effectiveness of the "internal pivot determination" is worse than the simple "recent query objects", anyways.

[3] If the internal dynamic pivot selection (and RowCache) is not used, the entire D-cache capacity is given to CellCache.

Smallest distance. The hypothesis is an anticipation that small distances between queries and DB objects represent overlapped query and data regions; in such case (even the exact) small distance is useless for pruning, so we release it.

5   Experimental Evaluation

We have extensively tested the D-file, while we have compared its performance with M-tree, PM-tree, and GNAT. We have observed just the computation costs, that is, the number of distance computations spent by querying. For the index-based MAMs, we have also recorded the construction costs in order to give an idea about the indexing/querying trade-off.

5.1   The Testbed

We used 3 databases and 3 metrics (two continuous and one discrete):

– A subset of *Corel features* [10], namely 65,615 32-dimensional vectors of color moments, and the L_1 distance (the sum of the difference of coordinate values between two vectors). Note: L_2 is usually used with color histograms, but from the indexing point of view any L_p norm ($p \geq 1$) gives similar results.
– A synthetic *Polygons* set; 500,000 randomly generated 2D polygons varying in the number of vertices from 10 to 15, and the Hausdorff distance (maximum distance of a point set to the nearest point in the other set). This set was generated as follows: The first vertex of a polygon was generated at random; the next one was generated randomly, but the distance from the preceding vertex was limited to 10% of the maximum distance in the space. Since we have used the Hausdorff distance, one could view a polygon as a cloud of 2D points.
– A subset of *GenBank* file `rel147` [1], namely 50,000 protein sequences of lengths from 50 to 100, and the edit distance (minimum number of insertions, deletions, and replacements needed to convert a string into another).

Index-based MAM settings. The databases were indexed with M-tree and PM-tree, while GNAT was used to index just Corel and Polygons. For (P)M-tree, the node size was set to 2kB for Corel, and to 4kB for Polygons and GenBank databases (the node degree was 20–35). The PM-tree used 16 (static) pivots in inner nodes and 8 pivots in leaf nodes. Both M-tree and PM-tree used mM_RAD node splitting policy and the single-way object insertion. The GNAT arity (node degree) was set to 50 for Corel and to 20 for Polygons. For most querying experiments, we have issued 1,000 queries and averaged the results.

D-file settings. Unless otherwise stated, the D-cache used 100 MB of main memory and unlimited history of query objects' ids, i.e., we keep track of all the queries issued so far (within a session). The `recent` and `internal` dynamic pivot selection techniques were considered. Concerning `internal` selection, the number of initial fixed distance computations was set to $x = 1,000$. The smallest distance replacement policy was used in all tests. Furthermore, unless otherwise stated, the D-file used 100 dynamic pivots. The D-cache was reset/initialized before every query batch was started.

Table 1. Index construction costs (total distance computations)

index	Corel	Polygons	GenBank
M-tree	4,683,360	38,008,305	3,729,887
PM-tree	7,509,804	55,213,829	6,605,421
GNAT	60,148,055	497,595,605	n/a
D-file	0	0	0

5.2 Indexing

The first experiment was focused on indexing – the results are summarized in Table 1. Because of its static nature, note that GNAT is an order of magnitude more expensive than (P)M-tree. By the way, due to the expensive construction and the expensive edit distance, we could not index GenBank by GNAT.

5.3 Unknown Queries

The second set of tests was focused on the impact of D-file on querying when considering "unknown" queries, that is, query objects outside the database. It has to be emphasized that for unknown queries the D-file cannot take advantage of the trivial method GetDistance, because for an arriving query object there cannot be any record stored within D-cache at the moment the query starts. Thus, D-file can effectively use just the D-cache's non-trivial method GetLower-BoundDistance.

First, for the Corel database we have sampled queries with "snake distribution" – for an initially randomly sampled query object, its nearest neighbor was found, then the nearest neighbor's nearest neighbor, and so on. The intention for snake distribution was led by an anticipation that processing of a single "slowly moving" query will benefit from D-cache (which might provide tighter lower bounds). Because the D-cache should benefit from a long sequence of queries, we were interested in the impact of growing query batch size, see Figure 2a. As we can see, this assumption was not confirmed, the D-file costs generally follow

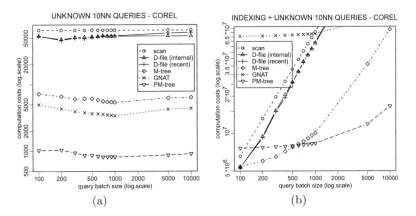

Fig. 2. Unknown 10NN queries on Corel: (a) queries only (b) indexing + queries

the other MAMs – the performance gain of D-file queries is rather constant. Although in this test the query performance of D-file is not very good when compared to the other MAMs, in Figure 2b the situation takes into account also indexing costs. When just a small- or medium-sized query batch is planned for a database, the costs (total number of distance computations spent on indexing + all queries) show the D-file beats GNAT and PM-tree considerably. Note that in Figure 2b the costs are not averaged per query but total (because of correct summing with indexing costs).

The second "unknown" set of queries (randomly sampled range queries having radius $r = 5$) was performed on the GenBank database, considering growing number of D-cache's dynamic pivots, see Figure 3a. The costs decrease with growing number of pivots, the D-file shows a reduction of costs by 45%, when compared to the sequential search. The `recent` dynamic pivot selection is the winning one. In Figure 3b the situation is presented in indexing + querying costs (total costs for 100 queries were considered).

5.4   Database Queries

For the third set of experiments, we have considered database (DB) queries. As opposed to unknown queries, the DB queries consisted of query objects randomly sampled from the databases and having also the same ids as in the database. Although not as general as unknown queries, the DB queries are legitimate queries – let us bring some motivation. In a typical general-purpose image retrieval scenario, the user does not have the perfect query image (s)he wants. Instead, (s)he issues any available "yet-satisfactory" image query (being an unknown query) and then iteratively browses (navigates) the database by issuing subsequent DB queries given by an image from the previous query result.

The browsing is legitimate also for another reason. Even if we have the right query image, the similarity measure is often imperfect with respect to the specific user needs, and so the query result may be unsatisfactory. This could be compensated by further issuing of DB queries matching the user's intent better.

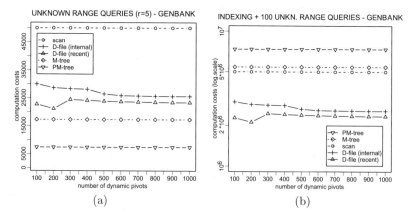

Fig. 3. Unknown range queries on GenBank: (a) queries only (b) indexing + queries

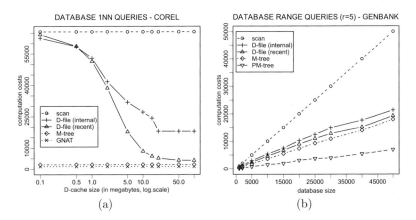

Fig. 4. (a) DB 10NN queries on Corel. (b) DB range queries on GenBank.

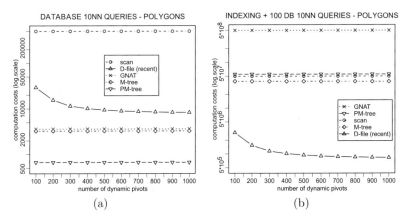

Fig. 5. DB 10NN queries on Polygons: (a) queries only (b) indexing + queries

For DB queries, we have anticipated much greater impact of D-cache, because distances related to a newly arriving query (being also DB object) could reside within D-cache since previous query processing. Consequently, for DB queries also the trivial GetDistance method can effectively take advantage, so that we could obtain an exact distance for filtering, rather than only a lower bound.

In Figure 4a, the results of 1NN queries on Corel are presented, considering varying D-cache storage capacity (we actually performed 2NN search, because the query object is DB object). We observe that a small D-cache is rather an overhead than a benefit, but for growing D-cache size the D-file costs dramatically decrease (to 6% of seq. search). At 10–20 MB the D-cache is large enough to store all the required distances, so beyond 20 MB the performance gain stagnates. However, note that there is a significant performance gap between D-files employing `recent` and `internal` pivot selection. This should be an evidence that exact distances retrieved from D-cache are not the dominant pruning factor even for DB queries, because the GetDistance method is pivot-independent. Hence, the effect of non-trivial "lowerbounding" is significant also for DB queries.

In the fourth experiment, see Figure 4b, the growing GenBank database was queried on range. Here, the D-file performance approaches M-tree, while the performance scales well with the database growth. In the last experiment, see Figure 5a, we tested the effect of the number of dynamic pivots on the largest database – Polygons. For D-file, the costs fell down to 1.7% of sequential search, while they were decreasing with the increasing number of D-cache's dynamic pivots. In Figure 5b, the total indexing + querying costs are presented for 100 10NN queries, beating the competitors by up to 2 orders of magnitude.

6   Conclusions

In this paper, we presented the D-file – the first index-free metric access method for similarity search in metric spaces. The D-file operates just on the original database (i.e., it does not use any indexing), while, in order to support efficient query processing, it uses lower bound distances cheaply acquired from the D-cache structure. The D-cache is a memory-resident structure which keeps track of distances already computed during the actual runtime session.

The experiments have shown that D-file can compete with the state-of-the-art index-based MAMs (like M-tree, PM-tree, GNAT). Although in most cases the separate query costs are higher for D-file, if we consider the total indexing+querying costs, the D-file performs better for small- and middle-sized query batches. Thus, the usage of D-file could be beneficial either in tasks where only a limited number of queries is expected to be issued, or in tasks where the indexing is inefficient or not possible at all (e.g., highly changeable databases).

Future Work. In the future, we would like to employ the D-cache also by the index-based MAMs, since its ability to provide lower/upper bounds to distances is not limited just to D-file. Moreover, by using D-cache the index-based MAMs could take advantage not only from the improved query processing, but also from an increased performance of indexing.

Acknowledgments

This research was supported by Czech Science Foundation project no. 201/09/0683 (first author), and by FONDECYT (Chile) Project 11070037 (second author).

References

1. Benson, D.A., Karsch-Mizrachi, I., Lipman, D.J., Ostell, J., Rapp, B.A., Wheeler, D.L.: Genbank. Nucleic Acids Res. 28(1), 15–18 (2000)
2. Böhm, C., Berchtold, S., Keim, D.: Searching in High-Dimensional Spaces – Index Structures for Improving the Performance of Multimedia Databases. ACM Computing Surveys 33(3), 322–373 (2001)

3. Brin, S.: Near neighbor search in large metric spaces. In: Proc. 21st Conference on Very Large Databases (VLDB 1995), pp. 574–584. Morgan Kaufmann, San Francisco (1995)

4. Carson, S.D.: A system for adaptive disk rearrangement. Software - Practice and Experience (SPE) 20(3), 225–242 (1990)

5. Chávez, E., Navarro, G., Baeza-Yates, R., Marroquín, J.L.: Searching in metric spaces. ACM Computing Surveys 33(3), 273–321 (2001)

6. Ciaccia, P., Patella, M., Zezula, P.: M-tree: An Efficient Access Method for Similarity Search in Metric Spaces. In: VLDB 1997, pp. 426–435 (1997)

7. Effelsberg, W., Haerder, T.: Principles of database buffer management. ACM Transactions on Database Systems (TODS) 9(4), 560–595 (1984)

8. Falchi, F., Lucchese, C., Orlando, S., Perego, R., Rabitti, F.: A metric cache for similarity search. In: LSDS-IR 2008: Proceeding of the 2008 ACM workshop on Large-Scale distributed systems for information retrieval, pp. 43–50. ACM Press, New York (2008)

9. Falchi, F., Lucchese, C., Orlando, S., Perego, R., Rabitti, F.: Caching content-based queries for robust and efficient image retrieval. In: EDBT 2009: Proceedings of the 12th International Conference on Extending Database Technology, pp. 780–790. ACM Press, New York (2009)

10. Hettich, S., Bay, S.: The UCI KDD archive (1999), http://kdd.ics.uci.edu

11. Hjaltason, G.R., Samet, H.: Index-driven similarity search in metric spaces. ACM Trans. Database Syst. 28(4), 517–580 (2003)

12. Samet, H.: Foundations of Multidimensional and Metric Data Structures. Morgan Kaufmann, San Francisco (2006)

13. Skopal, T.: Pivoting M-tree: A Metric Access Method for Efficient Similarity Search. In: Proceedings of the 4th annual workshop DATESO, Desná, Czech Republic, ISBN 80-248-0457-3, also available at CEUR, vol. 98, pp. 21–31 (2004) ISSN 1613-0073, http://www.ceur-ws.org/Vol-98

14. Skopal, T., Pokorný, J., Snášel, V.: Nearest Neighbours Search Using the PM-Tree. In: Zhou, L.-z., Ooi, B.-C., Meng, X. (eds.) DASFAA 2005. LNCS, vol. 3453, pp. 803–815. Springer, Heidelberg (2005)

15. Uhlmann, J.: Satisfying general proximity/similarity queries with metric trees. Information Processing Letters 40(4), 175–179 (1991)

16. Vitter, J.S.: External memory algorithms and data structures: dealing with massive data. ACM Computing Surveys 33(2), 209–271 (2001)

17. Weber, R., Schek, H.-J., Blott, S.: A quantitative analysis and performance study for similarity-search methods in high-dimensional spaces. In: VLDB 1998: Proceedings of the 24rd International Conference on Very Large Data Bases, pp. 194–205. Morgan Kaufmann Publishers Inc., San Francisco (1998)

18. Zezula, P., Amato, G., Dohnal, V., Batko, M.: Similarity Search: The Metric Space Approach (Advances in Database Systems). Springer, Secaucus (2005)

An Approximation Algorithm for Optimizing Multiple Path Tracking Queries over Sensor Data Streams

Yao-Chung Fan[1] and Arbee L.P. Chen[2,*]

[1] Department of Computer Science, National Tsing Hua University, Taiwan, R. O. C.
[2] Department of Computer Science, National Chengchi University, Taiwan, R. O. C.
d938318@oz.nthu.edu.tw, alpchen@cs.nccu.edu.tw

Abstract. Sensor networks have received considerable attention in recent years and played an important role in data collection applications. Sensor nodes have limited supply of energy. Therefore, one of the major design considerations for sensor applications is to reduce the power consumption. In this paper, we study an application that combines RFID and sensor network technologies to provide an environment for moving object path tracking, which needs efficient join processing. This paper considers multi-query optimization to reduce query evaluation cost, and therefore power consumption. We formulate the multi-query optimization problem and present a novel approximation algorithm which provides solutions with suboptimal guarantees. In addition, extensive experiments are made to demonstrate the performance of the proposed optimization strategy.

Keywords: Query Processing, Sensor Network, Query Optimization, and Distributed Database.

1 Introduction

Sensor networks have received considerable attention in recent years. A sensor network consists of a large number of sensor nodes. In general, sensor nodes are equipped with the abilities of sensing, computing, and communicating.

One of the features for sensor networks is resource limitation. Sensor nodes typically are limited in computing power, network bandwidth, storage capability, and energy supply. Resource conservation therefore becomes a major consideration when devising sensor applications.

Sensor networks provide a new way of data collection and create new needs for information processing in a variety of scenarios. In this paper, we introduce a new application that combines RFID (Radio Frequency Identification) and sensor network technologies for customer path tracking. In the application, efficient join processing is critical to its performance. We consider multi-query optimization to reduce the cost of join processing, and therefore power consumption.

---

* To whom all correspondence should be sent.

S.S. Bhowmick, J. Küng, and R. Wagner (Eds.): DEXA 2009, LNCS 5690, pp. 532–546, 2009.

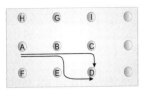

Fig. 1. Motivating Scenario

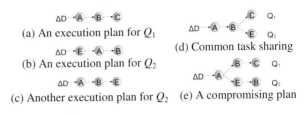

(a) An execution plan for Q_1

(b) An execution plan for Q_2

(c) Another execution plan for Q_2

(d) Common task sharing

(e) A compromising plan

Fig. 2. Execution Plans

Application. Consider a set of sensor nodes deployed in a shopping mall, as shown in Figure 1. The sensor nodes are equipped with an RFID reader. Moreover, the customers who enter the monitoring area are equipped with a RFID tag. When a customer passes through a sensor node, the sensor node detects the event and generates a tuple of data. This tuple of data contains the customer identification number (CID), the timestamp when the tuple was generated (TID), and the sensor identification number (SID). The sensor nodes inherently generate data streams if customers keep entering the area.

Such RFID and sensing platform provides an environment to infer customer activities from the sensor readings. For example, the shopping mall manager may want to know the customers who walk through nodes A, B, C, and D in the mall. Having such information, the manager may be able to infer that the customers are looking for something, and then make suitable assistances or recommendations to the customers. The manager may issue the following query Q_1:

Q_1: Select CID
   From A, B, C, D
   Where A.CID = B.CID = C.CID = D.CID
   AND A.TID < B.TID < C.TID < D.TID
   Window = 10 minutes
   Action = *Service1* (UID)

Q_2: Select CID
   From A, B, D, E
   Where A.CID = B.CID = E.CID = D.CID
   AND A.TID < B.TID < E.TID < D.TID
   Window = 10 minutes
   Action = *Service2* (CID)

In the query, "Select" clause specifies the customer identification numbers, "From" clause indicates the sensor data streams, "Where" clause joins the readings among the indicated streams with suitable predicates, "Window" clause specifies the valid time constraint for the customers along the path, and "Action" clause indicates the proper action of providing some service to the qualified customers. We refer to such a query as a *path tracking* query.

To process the query, a naïve approach is to ask nodes A, B, C, and D to send the tuples within the window to a base station to perform the evaluations [1]. However, this approach suffers from high communication cost as each node transmits all the data to the base station regardless of whether the data contribute to the query results. This approach will be inefficient when the join selectivity is low. Therefore, in this paper we consider *in-network query processing strategy* as follows.

Take Q_1 as an example. Each node keeps a buffer to store tuples generated in the past ten minutes. Since we require D.TID to be the largest among the four timestamps, the query processing starts at node D. When a new tuple ΔD of node D is generated, node D routes ΔD to the other nodes to probe for matching according to a given execution plan, which describes the order the probing of ΔD is perfomed in the other nodes. For example, Figure 2(a) shows an execution plan $\Delta D \rightarrow A \rightarrow B \rightarrow C$ for Q_1, which indicates that ΔD is firstly sent to node A, and if ΔD fails to match, the processing terminates. Otherwise, $\Delta D \bowtie A$ is processed and the result routed to B for further probing. A tuple that matches all the nodes produces a join result, which is then returned to the base station.

Note that the efficiency of the query processing depends on the order of the probing. This is because a newly generated tuple that is dropped by one of the nodes cannot produce a join result. Therefore, a worst case of the processing will be that a tuple matches all the nodes except the last one. In this case, no result is produced and all prior processing efforts are wasted. However, if we swap the last node with the first one for the probing, the tuple will get dropped in the first probe rather than in the last probe. Therefore, choosing a probing order is critical to the query processing efficiency. In the following discussion, we refer to the strategy that *optimizes query processing by choosing probing orders* as the *join order optimization* (JOO) strategy.

In addition to the JOO strategy, another opportunity for improving the efficiency is the *common-task-sharing optimization* (CTSO) strategy. The basic idea behind the CTSO strategy is to share common tasks among queries when multiple queries are present. For example, assume now that in addition to Q_1, the manager poses another query Q_2. In such a case, the two queries have two tasks in common, i.e. probing for node A and node B. To avoid redundant execution, we can plan Q_1 to be $\Delta D \rightarrow A \rightarrow B \rightarrow C$ and Q_2 to be $\Delta D \rightarrow A \rightarrow B \rightarrow E$ to share the probing for node A and node B. Figure 2(d) shows the resulting plan.

Note that we refer to routing a tuple or an intermediate result to some node to probe for matching as a *task*. For example, the processing of Q_1 has three tasks, i.e., probing ΔD with node A, node B, and node C for matching. For ease of discussion, we denote the tasks by {A, B, C}.

Challenge. Both the JOO strategy and the CTSO strategy reduce the cost for processing queries. However, the combination of the two strategies brings an interesting tradeoff to study. In some cases, the two strategies *complement* each other. For example, assume the plans in Figure 2(a) and Figure 2(c) are suggested by the JOO strategy, we can further share the common tasks between the two plans to have a combined plan, as shown in Figure 2(d).

Nevertheless, the two strategies generally are *conflicting* to each other. For example, assume the JOO strategy suggests the plan in Figure 2(b) instead of the one in Figure 2(c). In this case, when planning Q_1 and Q_2, if we choose to use the JOO strategy, we lose the opportunities of further using the CTSO strategy.

On the other hand, if we choose CTSO, we need to adopt the plan in Figure 2(c). If the cost saved by the CTSO strategy is less than the cost incurred from the violation of the JOO strategy, sharing common tasks will be a bad decision, and vice versa.

An important thing to note about is that there are many options for combining the two strategies. For example, we can plan Q_2 to be $\Delta D \rightarrow A \rightarrow E \rightarrow B$ to share {A} with Q_1. Figure 2(e) shows the resulting plan. In this case, the plan for Q_2, i.e. $\Delta D \rightarrow A \rightarrow E \rightarrow B$, may be a suboptimal result in terms of the JOO strategy. However, using this plan may be more beneficial to the overall cost than using the plan that exclusively uses the JOO strategy or the CTSO strategy. Therefore, the challenge lies in how to balance the two optimization strategies such that the overall cost for query processing can be minimized.

This paper presents a framework for efficiently combining the two strategies. We first formulate the search space of finding an optimal combination, and then show the complexity of finding the optimal combination. Furthermore, we propose a novel approximation algorithm, which provides solutions with sub-optimal guarantees.

The rest of the paper is organized as follows. Section 2 discusses the cost model for the query processing. Section 3 presents the optimization strategies. The experiment results are provided in Section 4. Section 5 provides related works. Finally, we conclude the paper in Section 6 and give some directions for future research.

2   Cost Model

We first provide the terminology and the assumptions we use through this paper. We consider a sensor network as one which consists of a set of sensor nodes $\{N_1, N_2, ..., N_n\}$ and a base station which has no energy and memory limitations. The sensor nodes are well-synchronized. The base station keeps the network topology and there are no communication delays in the sensor network. The sensor nodes generate a set of data streams $\Phi = \{S_1, S_2, ..., S_n\}$.

A path-tracking query Q joins stream S_j with some other streams in Φ. We denote this subset of Φ as θ, and use the notation $Q(O)$ to refer to the execution of Q following a given join order O. A join order O is a permutation $O = <S_{O1}, S_{O2}, ..., S_{O|\theta|}>$ of the elements in θ. We use $Cost(Q(O))$ to denote the expected cost for processing ΔS_j using $Q(O)$.

All path-tracking queries are posed at the base station, from which the queries are planned and disseminated into the sensor network. All join results are collected at the base station. In this paper, we focus on the in-network join processing of the queries. The further processing of join results, such as making recommendations, take places at the base station, and are beyond the scope of this paper. In this study, we assume that all path-tracking queries have the same window specification. Further Relaxation of this assumption can be a future direction to proceed.

Cost Model. In the following, we elaborate on the mechanism of processing a path-tracking query Q in a sensor network, and present a cost model for the processing of the path-tacking query. The model helps us to make a decision on choosing a good join order. We use ΔS_j to denote the newly generated tuple from S_j.

In processing ΔS_j using a given join order $O = <S_{O1}, S_{O2}, ..., S_{O|\theta|}>$, the expected cost of the query processing is given by

$$Cost(Q(O)) = \sum_{i=1}^{|\theta|}(\gamma \cdot \sigma_i \cdot \sum_{j=1}^{i}d_j),$$

where σ_i stands for the probability of $\Delta S_j \bowtie ... \bowtie S_{Oi-1}$ being dropped by S_{Oi}, d_i denotes the minimum number of hop distance between N_{Oi} and N_{Oi-1}, and γ is the cost for one hop transmission. Note that $N_{O0}=N_j$.

Example 1. We use Q_1 to illustrate the cost model. Assume Q_1 uses $O = < B, C, A>$, and the routing topology is as shown in Figure 3. Assume that $\gamma = 100$, $\sigma_1 = 0.3$, $\sigma_2 = 0.1$, and $\sigma_3 = 0.5$. In processing ΔD using $O = < B, C, A>$, there are four cases to consider:

Case 1: ΔD is dropped by node B.

In this case, the cost for processing ΔD is $0.3 \cdot (1 \cdot 100) = 30$ since if ΔD dropped by node B, only one transmission is performed.

Case 2: $\Delta D \bowtie B$ is dropped by node C.

In this case, the cost for the processing is $0.1 \cdot ((1+1) \cdot 100) = 20$.

Case 3: $\Delta D \bowtie B \bowtie C$ is dropped by node A.

In this case, the cost for the processing is $0.5 \cdot ((1+1+2) \cdot 100) = 200$, since four transmissions has been performed.

Case 4: $\Delta D \bowtie B \bowtie A \bowtie C$ is produced.

The processing cost is $0.1 \cdot (4 \cdot 100) = 40$.

Thus, the total cost is $30 + 20 + 200 + 40 = 290$.   ∎

Optimization Goal. Given a set of queries $W = \{Q_1, Q_2, ..., Q_m\}$ at node N_j, the goal is to minimize $\sum_{i=1}^{m} Cost\ (Q_i(O))$.

Toward this goal, the JOO strategy aims to reduces $Cost(Q_i(O))$ for each query, and the CTSO optimization aims to avoid redundant executions among multiple queries. Our goal is to combine and balance the two strategies for maximal benefits.

3   Multi-query Optimization

This section presents optimization strategies for processing multiple path tracking queries. In the following discussion, we refer to the plan for processing a query individually as a *local plan*, and use the term *global plan* to refer to a plan that provides a way to compute results for multiple queries.

3.1   Join Order Optimization

The first option for improving efficiency is to optimize the queries individually by choosing their join order. A naive method for finding an optimal join order is to enumerate all possible orders. Nevertheless, such approach results in the complexity of $O(|\theta|!)$. In general, finding an optimal join order without *independent assumptions* [2] is a NP-hard problem. Therefore, heuristic approaches are often adopted to find good join orders. There are many interesting works, and we refer the reader to the survey in [2].

In this study, we greedily choose the stream that drops the maximum number of tuples with a minimum cost at each step to determine a join order. The basic idea is that if a tuple ΔS_j eventually gets dropped, it is better that it gets dropped as early as possible. Therefore, when choosing S_{Oi}, $1 < i < |\theta|$, the stream which most likely drops $\Delta S_j \bowtie ... \bowtie S_{Oi-1}$ should be chosen. However, if the chosen stream is far away from N_{Oi-1}, the execution will instead incur a great deal of message traffics. Therefore, we further consider the communication cost. That is, the join orders are chosen by balancing the two factors. That is, we choose the stream that most likely drops $\Delta S_j \bowtie ... \bowtie S_{Oi-1}$ and that needs as few hop transmissions as possible to reach. Our idea is best seen by an example. Again, we use Q_1 to illustrate how we choose the join orders.

Example 2. Assume that the probability that ΔD is dropped by A is 0.5, the probability of ΔD is dropped by B is 0.3, and the probability of ΔD is dropped by C is 0.1. The routing topology is as shown in Figure 3. For ease of illustration, we assume the probabilities are independent.

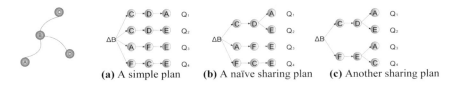

(a) A simple plan   **(b)** A naïve sharing plan   **(c)** Another sharing plan

Fig. 3. A Topology   **Fig. 4.** Execution Plans

Our approach for choosing join orders proceeds as follows. First, node B is chosen to be S_{O1} because it provides an expectation of 0.3/1=0.3 (the expected number of ΔD to be dropped per number of hop transmissions) while node A only provides an expectation of 0.5/2=0.25 and node C only provides an expectation of 0.1/2=0.05. Then, node A is chosen to be S_{O2} since it provides an expectation of 0.5 to drop $\Delta D \bowtie B$ while node C only provides an expectation of 0.1. Finally, node C is chosen to be S_{O3} and the order $O = <B, A, C>$ is outputted. ∎

3.2 Naïve Sharing Strategy

Having described the JOO strategy, we now turn to the problem of optimizing multiple queries. Let us first consider an example for optimizing multiple queries.

Example 3. Given a set of queries $W = \{Q_1, Q_2, Q_3, Q_4\}$ at node B, with $\theta_1 = \{A, C, D\}$, $\theta_2 = \{C, D, E\}$, $\theta_3 = \{A, E, F\}$, and $\theta_4 = \{F, C, E\}$. For ease of discussion, we assume that the JOO strategy suggests the four locally optimized plans, as shown in Figure 4(a). Note that no common tasks between the four plans are shared in Figure 4(a).

One straightforward method to further improve the global plan in Figure 4(a) is to combine the locally optimized plans which have common prefixes. For example, we can combine the plans for Q_1 and Q_2 in Figure 4(a) to produce a global execution plan as shown in Figure 4(b). This execution plan is more efficient than the plan in Figure 4(a), because the redundant tasks are avoided and the plan does not violate the optimized join orders. We refer to this strategy as the *naïve sharing strategy*.

However, the naïve sharing strategy restricts the scope of optimizations. The main problem with the naïve sharing strategy is that more than one possible local plan can be used to process a query. The naïve sharing strategy, however, only considers the locally optimized plans. For example, although there is no common prefix between the locally optimized plans of Q_3 and Q_4, we still can plan Q_3 to be <F, E, A> and Q_4 to be <F, E, C> such that the common tasks {F, E} can be shared. Figure 4(c) shows the resulting plan, which may be more efficient than the one shown in Figure 4(b). In fact, in addition to sharing {F, E}, the queries have other tasks in common, and the naïve sharing strategy does not consider them to be shared, potentially missing further optimization opportunities.

3.3 Hybrid Optimization Strategy

In this section, we study the problem of combining the JOO strategy and the CTSO strategy for maximal benefits. We first discuss the construction of the search space for combining the two strategies (Section 3.3.1), and then propose a randomized algorithm for finding the optimal combination (Section 3.3.2).

3.3.1 Search Space

We first state two definitions for the subsequent discussions. The definitions are used to describe elements in the search space.

Given a set of queries $W = \{Q_1, Q_2, ..., Q_m\}$ at node N_j. In the following discussion, we refer a set of tasks as a *task-set*. We say a query Q *contains* a task-set if the task-set is contained in θ. The *count* of a task-set is the number of queries of W having the task-set. For example, $\{C, D\}$ is a task-set of Q_1 and the count for $\{C, D\}$ is two in Example 3.

Definition 1. We say a task-set is a *common task-set*, CTS, among $W = \{Q_1, Q_2, ...,Q_m\}$ if and only if the count of the task-set is equal to or larger than two, i.e. at least two queries in W contain the task-set.

Definition 2. For a given CTS, we call a subset w_{CTS} of W whose elements all contain the CTS as the *sharing set* of the CTS.

In Example 3, $\{C, D\}$ is a CTS and $w_{\{C, D\}}= \{Q_1, Q_2\}$ is the sharing set of $\{C, D\}$. An important thing to note is that a sharing set w_{CTS} represents some execution way to share the CTS for the queries in the sharing set. For example, $w_{\{C,D\}}=\{Q_1, Q_2\}$ represents the execution that shares $\{C, D\}$ for Q_1 and Q_2, i.e. the upper plan shown in Figure 4(b).

The first step for finding a global optimal plan is to enumerate all possible executions for processing queries. Note that all the enumerated executions form the search space for a global optimal plan. For a given set of queries W, the algorithm for constructing the search space proceeds as follows.

The first step is to derive all CTS from W. The process of finding CTS amounts to the process of deriving the frequent item-set [4] with count ≥ 2 from $\{\theta_1, \theta_2, ..., \theta_m\}$ because only the task-set with count ≥ 2 are the CTS for queries in W.

Second, for each CTS derived from W, we enumerate the associated sharing sets. Note that there is more than one sharing set for a CTS. By definition, any subset S of a sharing set with $|S| \geq 2$ is also a sharing set. In Example 3, the subsets $\{Q_1, Q_2\}$, $\{Q_1, Q_4\}$ and $\{Q_2, Q_4\}$ of $w_{\{C\}} = \{Q_1, Q_2, Q_4\}$ are also sharing sets, because if the queries can share $\{C\}$, then any combination of the queries definitely can share the common task $\{C\}$.

Third, for each sharing set w_{CTS}, we represent the w_{CTS} by a vector $\mathbf{a}_{CTS} \in \{0, 1\}^m$ whose component a_i equals to one if $Q_i \in w_{CTS}$, and zero otherwise. In Example 3, $\mathbf{a}_{\{C\}} = (1, 1, 0, 1)$ describes that $Q_1, Q_2,$ and Q_4 share $\{C\}$.

Fourth, the above steps find all possible executions sharing common tasks for queries. However, there are also possibilities that queries share nothing. That is, the queries use their locally optimized plans given by the JOO strategy. Therefore, we also include the options that the queries are individually executed.

To establish the search space, we combine the vectors, which represent the sharing sets, into a matrix $\mathbf{A}$ (column wise) and their associated sharing costs into a cost vector $\mathbf{c}$.

The search space for the problem can be then formulated as

$$
\begin{aligned}
&\text{Minimize} \quad \mathbf{c} \cdot \mathbf{x} \\
&\text{Subject to } \mathbf{A} \cdot \mathbf{x} = \mathbf{e}, \\
&\qquad\quad x_i \in \{1,0\}, \quad i = 1, ..., n
\end{aligned}
$$

The vector **x** contains components x_i, $i = 1, ..., n$, which equals to 1 if A_i is chosen, and 0 otherwise, and **e** denotes the vector whose components all equal to one. A vector **x** satisfying $A \cdot x = e$ is called a *feasible solution*. Note that each feasible solution corresponds to a global execution plan for processing W. Therefore, our goal now is to find a vector x^* with subject to A with minimal cost $c \cdot x^*$. Theorem 1 shows the complexity of the problem.

Theorem 1. *Given a set of queries* $W = \{Q_1, Q_2, ..., Q_m\}$, *the problem of selecting one join order for each query with the goal of minimizing* $\sum_{i=1}^{m} Cost(Q_i(O))$ *is NP-Complete.*

Proof. By showing the equivalence between a set partition [3], a well-known NP-Complete problem, and our problem, we shows that no polynomial time algorithm is expected to exist. ∎

$$
\begin{bmatrix}
0 & 0 & 1 & 1 & 1 & 1 & 1 & 0 & 1 & 0 & 0 & 0 & 0 & 0 & 1 & 0 & 0 & 0 \\
1 & 0 & 1 & 0 & 1 & 1 & 0 & 1 & 1 & 1 & 1 & 1 & 0 & 0 & 0 & 1 & 0 & 0 \\
0 & 1 & 0 & 1 & 0 & 0 & 0 & 0 & 0 & 1 & 1 & 0 & 1 & 1 & 0 & 0 & 1 & 0 \\
1 & 1 & 0 & 0 & 1 & 0 & 1 & 1 & 0 & 1 & 0 & 1 & 1 & 1 & 0 & 0 & 0 & 1 \\
\{C,E\} & \{E,F\} & \{C,D\} & \{A\} & \{C\} & \{C\} & \{C\} & \{C\} & \{D\} & \{E\} & \{E\} & \{E\} & \{E\} & \{F\} & \{\} & \{\} & \{\} & \{\} \\
c_1 & c_2 & c_3 & c_4 & c_5 & c_6 & c_7 & c_8 & c_9 & c_{10} & c_{11} & c_{12} & c_{13} & c_{14} & c_{15} & c_{16} & c_{17} & c_{18}
\end{bmatrix}
$$

Fig. 5. The search space for W

Example 4. The search space for W in Example 3 is constructed as follows.

Step1: Find *CTS*s from W. We have $\{C, E\}$, $\{E, F\}$, $\{C, D\}$, $\{A\}$, $\{C\}$, $\{D\}$, $\{E\}$ and $\{F\}$.

Step2: For each *CTS*, we enumerate the associated sharing sets and then generate a vector and a sharing cost to represent each enumerated sharing set. For $\{C, E\}$, we have the sharing vector $a_1 = (0, 1, 0, 1)$ and the sharing cost c_1 with respect to the sharing set $w_{\{C, E\}} = \{Q_2, Q_4\}$. Likewise, we have $a_2 = (0, 0, 1, 1)$ and c_2 for $\{E, F\}$, $a_3 = (1, 1, 0, 0)$ and c_3 for $\{C, D\}$, and $a_4 = (1, 0, 1, 0)$, and c_4 for $\{A\}$. Note that, for $\{C\}$, we will have $a_5 = (1, 1, 0, 1)$, $a_6 = (1, 1, 0, 0)$, $a_7 = (1, 0, 0, 1)$, and $a_8 = (0, 1, 0, 1)$ with respect to the combinations of the sharing set $w_{\{C\}}$, i.e. $\{Q_1, Q_2, Q_4\}$ $\{Q_1, Q_2\}$, $\{Q_1, Q_4\}$ and $\{Q_2, Q_4\}$. Likewise, for $\{D\}$, we have $a_9 = (1, 1, 0, 0)$. For $\{E\}$, we have $a_{10} = (0, 1, 1, 1)$, $a_{11} = (0, 1, 1, 0)$, $a_{12} = (0, 1, 0, 1)$, and $a_{13} = (0, 0, 1, 1)$ with respect to the combinations of the sharing set $w_{\{E\}}$. For $\{F\}$, we have $a_{14} = (0, 0, 1, 1)$

Step3: For the options of the individual executions for the queries, we have $a_{15} = (1, 0, 0, 0)$, $a_{16} = (0, 1, 0, 0)$, $a_{17} = (0, 0, 1, 0)$, and $a_{18} = (0, 0, 0, 1)$, and the associated sharing costs.

Step4: Finally, we have a 4×18 matrix A and the corresponding cost vector c, where $A_i = a_i$ and $c_i = c_i$, $i = 1, ..., 18$.

Figure 5 shows the search space for W, where columns describe all possible executions and the associated *CTS*s. For example, the first column describes the execution that shares $\{C, E\}$ between Q_2 and Q_4 with cost c_1. ∎

3.3.2 ε-Approximation Solution

We consider approximation solutions for combining the JOO strategy and the CTSO strategy due to the prohibitive complexity of finding an optimal combination. We propose a novel randomized algorithm, named **RARO**, based on the linear programming rounding technique [5].

Definition 4. We say an algorithm is an ε-approximation algorithm for a minimization problem with optimal cost *opt*, if the algorithm runs in polynomial time and returns a feasible solution with cost opt^+, such that $opt^+ \leq (1 + \varepsilon) \cdot opt$, for any $\varepsilon > 0$. We refer the solution as an ε-approximation solution.

For solving the problem: minimizing $\mathbf{c} \cdot \mathbf{x}$ subject to $\mathbf{A} \cdot \mathbf{x} = \mathbf{e}$ and $\mathbf{x} \in \{0, 1\}^n$ formulated in Section 3.3.1, the basic idea behind Algorithm **RARO** is as follows.

First, we solve the relaxed version of the given problem, in which the original integrality constraint ($x_i \in \{0, 1\}$, $i = 1, ..., n$) is replaced by ($x_i = [0, 1]$, $i = 1, ..., n$). That is, we allow x_i to take real values between 0 and 1. The relaxed problem can be solved in polynomial time [5]. Let $\mathbf{x}^+ = \{x^+_1, ..., x^+_n\}$ be the solution of the relaxed problem. Note that the components of $\mathbf{x}^+$ might not be integral but fractional, meaning that $\mathbf{x}^+$ is not a feasible solution for our problem.

To obtain a feasible solution, we then randomly round these fractional values. More precisely, we treat $\mathbf{x}^+$ as a probability vector, and, for $i = 1, ..., n$, set x_i to one with probability x^+_i. After this rounding process, we obtain a solution whose components take values from $\{0, 1\}$, which probably is a feasible solution for the problem. We then verify whether the solution is feasible. The condition whether a solution is feasible can be verified in polynomial time. In case the solution is not fulfilled, we repeat the rounding process to obtain another one.

The probability to obtain a feasible solution can be small. However, if we keep repeating the rounding process, the possibility to obtain a feasible solution grows. Algorithm **RARO** is based on such principle. More specifically, Algorithm **RARO** runs Procedure **Rounding** $g(\varepsilon)$ times and returns $\mathbf{x}_{min}$ with a minimal cost $\mathbf{c} \cdot \mathbf{x}_{min}$ as the final solution to the given problem, where $g(\varepsilon) = \ln 10^{-3} / \ln(1 - (1 - 1/e)^m (1 - 1/\varepsilon))$.

Theorem 2 shows the probability for Procedure **Rounding** in Figure 6 to output a feasible solution satisfying approximation guarantees. Theorem 3 shows the probability that Algorithm **RARO** fails to output an ε-approximation solution is at most 0.001.

Lemma 1. *With probability at least* $(1 - 1/e)^m$, *Procedure* **Rounding** *returns a feasible solution.*

Proof. In a rounding process, the probability that a query Q_i is not planned is: $\Pr[Q_i$ is not planned$] = \prod_{k:A_k^i=1} (1 - x_k^+)$. Let n_i be the number of ways to execute Q_i, that is, the number of **A**'s column whose the i-th component equals to 1. Since $\sum_{k:A_k^i=1} x_k^+ = 1$, it is clear that $\Pr[Q_i$ is not planned$]$ is maximized when $x_k^+ = 1 / n_i$.

Therefore, we get $\Pr[Q_i$ is not planned$] = \prod_{k:A_k^i=1} (1 - x_k^+) \leq (1 - 1/n_i)^{n_i} \cong 1/e$.

Conversely, $\Pr[Q_i$ is planned$] = 1 - 1/e$.

Therefore, the probability of the event all queries are planned is $(1 - 1/e)^m$. ∎

```
Algorithm RARO
Input: A search space A and the ε paramter
Output: A solution x_min for our problem
1.    Solve A·x = e, where x_j = [0, 1], ∀j∈n
2.    Treat the solution x' = {x'_1, ..., x'_n} as a probability vector
3.    Set G = ln 10^{-3} / ln(1 - (1 - 1/e)^m (1 - 1/ε))
4.    Set x_min and mincost = ∞
5.    For i = 1 to G
6.        Set x = 0
7.        x = Rounding ( x' )
8.        If ( c·x_min < mincost and A·x = e )
9.            x_min = x
10.       EndIF
11.   End
12.   Return x_min

Procedure Rounding
Input: a probability vector x'
Output: a decision vector x, x∈ {0, 1}^n, initially are all zero.
1.        For i = 1 to n
2.            Set x_i to 1 with probability x'_i
3.        EndFor
4.        Return x
```

Fig. 6. Algorithm *RARO*

Lemma 2. *For* $\varepsilon > 1$, *with probability at least* $1 - 1/\varepsilon$, *Procedure* **Rounding** *returns a solution with a cost lower than* $\varepsilon \cdot opt$, *where opt denotes the optimal cost for our optimization problem.*

Proof. Let $\mathbf{x}^+$ be the solution returned by Procedure **Rounding**. For the expected cost of $\mathbf{c} \cdot \mathbf{x}^+$, we have $E[\mathbf{c} \cdot \mathbf{x}^+] = \sum_{i=1}^n c_i \cdot x_i^+$, which is the same as the optimal cost to the relaxed problem. We denote the optimal cost for the relaxed problem as opt_f. Note that $opt \geq opt_f$.

Applying the Markov's Inequality to $\mathbf{c} \cdot \mathbf{x}^+$, we have

$\Pr[\mathbf{c} \cdot \mathbf{x}^+ \geq \varepsilon \cdot opt_f] \leq (opt_f / \varepsilon \cdot opt_f) \leq 1/\varepsilon$.

Equivalently, $\Pr[\mathbf{c} \cdot \mathbf{x}^+ < \varepsilon \cdot opt_f] > 1 - 1/\varepsilon$. As we note that $opt \geq opt_f$, we have $\Pr[\mathbf{c} \cdot \mathbf{x} < \varepsilon \cdot opt_f \leq \varepsilon \cdot opt] > 1 - 1/\varepsilon$. ∎

Theorem 2. *With probability at least* $(1 - 1/e)^m (1 - 1/\varepsilon)$, *Procedure* **Rounding** *produces an* ε-*approximation solution to the given problem.*

Proof. By Lemma 1 and Lemma 2. ∎

Theorem 3. *The probability that Algorithm* **RARO** *fails to output an* ε-*approximation solution is at most 0.001.*

Proof. The event S that Algorithm **RARO** fails to output an ε-approximation solution occurs only when all the calls of Procedure **Rounding** fails to produce an ε-approximation solution. Therefore, we proceed with the proof by showing that the probability of S is at most 0.001.

First, let us consider the probability of the event that calling Procedure **Rounding** x times and none of them produce an ε-approximation solution. Obviously, this probability is $(1 - (1 - 1/e)^m (1 - 1/\varepsilon))^x$ by Theorem 2.

Then, note that Procedure **Rounding** is called $\ln 10^{-3} / \ln(1 - (1 - 1/e)^m (1 - 1/\varepsilon))$ times in Algorithm **RARO**. Therefore, the probability of the event S is given by

$$\text{Prob}(S) = (1 - (1 - 1/e)^m (1 - 1/\varepsilon))^{\ln 10^{-3} / \ln(1 - (1 - 1/e)^m (1 - 1/\varepsilon))} .$$

By taking the natural logarithm of both sides of the above equation, we have

$$\ln(\text{Prob}(S)) = \ln(10^{-3}) / \ln(1 - (1 - 1/e)^m (1 - 1/\varepsilon)) \cdot (\ln(1 - (1 - 1/e)^m (1 - 1/\varepsilon))) .$$

Rewriting the equation, we obtain

$$\text{Prob}(S) = 0.001. \hspace{4cm} \blacksquare$$

4   Performance Evaluation

4.1   Prototyping Experiences

We have completed an initial implementation for the path tracking application by adopting Tmotesky nodes and SkyeRead-M1-Mini RFID reader. All programs in the base station are coded using Java, and the sensor network is programmed by Maté [6]. The plans of the given queries are optimized at the base station (BS), and then are disseminated into the network by Maté. The dissemination begins with the broadcast of Maté capsules from the root of the network. As each node hears the capsules, it decides if the Maté capsules should be installed locally or need to be flooded to its children or both. Our queries successfully run on this environment. We have nine Tmotesky nodes equipped with M1-Mini Reader in our implementation. However, this implementation is too small in its scale to measure the performances of our strategies. Consequently, the experiment results in this section are based on a simulation.

4.2   Experiment Setup

In J-sim, a sensor network consists of three types of nodes: *sensor nodes* (detect events), *target nodes* (generate events), and *sink nodes* (utilize and consume the senor information). In our setting, we create a monitoring area in which a regular 13*14 grid with 36 sensor nodes is simulated. The sensor nodes are assumed to have a 2*2 grid of monitor radius. The *sink node* is on the left-upper corner of the grid. Figure 7 shows this setting. In addition, we generate 6000 target nodes that simulate customers with RFID tags. The target nodes are assumed to enter the area from the entrance and leave from the exits of the area. We use *random walk* [5] to simulate the customers' moving behavior: the target nodes on the area move to every possible direction with equal probability, and stop its move process when it reaches the exits. When a target node enters the monitor region of a sensor node, the sensor node generates a tuple (5bytes) containing the sensor ID (1byte), the target node ID (3bytes), and the time-stamp (1byte). As mentioned earlier, if target nodes keep entering the area, the sensor nodes generate data streams. We process the streams by the proposed strategies, and compare the performance of the strategies in this environment. In all experiments we show the average value of 100 runs.

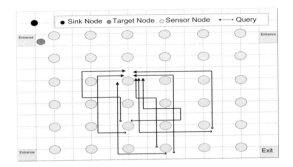

Fig. 7. Experiment setup

4.3 Evaluation

Basic Comparison. In the first set of experiments, we evaluate the performance of the presented optimization techniques. Join-order optimizations (JOO): all queries are individually optimized by the join-order optimizations. Common-task-sharing optimizations (CTSO): the queries are optimized by only considering share common tasks among queries. Naïve sharing strategy (NSS): all the queries are firstly optimized by JOO and then share the common prefix among queries. Hybrid Optimization (HO): the queries are optimized by the HO optimization with parameter $\varepsilon = 2$.

We pose eight queries on the simulated environment, as shown in Figure 7. The average selectivities for the queries are 0.2104. Figure 8 shows the experiment results, where y-axis shows the energy consumption for processing the given queries. We can see that HO significantly outperforms the other strategies.

Effect of In-network Processing. This experiment demonstrates the benefits of in-network processing against the centralized processing (all data are collected and evaluated in a BS). We mainly compare the centralized processing (CP) with JOO (as a representative of in-network processing). Two query sets are used to evaluate the performance, one with higher selectivities (average selectivity = 0.6115) of joining three streams (n=3) and the other with lower selectivities of joining five streams (0.3116, n=5). We vary the level of nodes at which the execution plans are installed to observe the performance of the strategies. Figure 9 shows the results, where x-axis is the level of nodes (the sink node at level 0) and y-axis is the number of transmissions involved in the query processing.

Note that the poor performance of CP comes from the fact that CP transmits all data to BS regardless of whether the data contribute to the query results. This effect can be observed from the selectivity factor. When the selectivity is 0.3116, JOO obviously outperforms CP, because most of tuples get dropped during execution. From this experiment, we know that CP is only good for the query processing that is close to the sink, with very high selectivity, and few streams involved in queries.

Effect of Low Sharing Opportunity. This experiment studies the effect of low sharing opportunity. We consider a query set where each query only has one common task with others. Even under this restriction, HO still has good results, as shown in Figure 10. This is because if no common task exists, HO will degenerate to JOO, i.e., all queries will be optimized individually. One thing to note in this experiment is that JOO outperforms CTSO. As we mentioned, blindly sharing the common tasks is not always beneficial to the query processing. In this experiment with low sharing opportunity, the poor performance of CTSO comes from the fact that the tasks saved by sharing are less than the cost incurred from the violation of the individual optimizations.

Effect of Number of Queries. This experiment studies the effect of the number of queries involved in the processing. Figure 11 shows our results, where x-axis is the number of the queries involved in the processing and y-axis is the energy consumption. We vary the number of queries to observe the performance of the strategies. We can see that HO always outperforms the other strategies and shows advantages as the number of queries increases, since when the number of queries increases, the sharing opportunities also increase.

5   Related Work

The problem of processing multiple queries in a data stream environment has been studied in several fronts [7][9][10][11][12][13].

Hammad et al.[7] study the problem of sharing executions of queries that that have different join window specification, where the proposed techniques are only capable of handling two-way join queries. The approach proposed in [9] address the problem of optimizing multiple group-by queries in GigaScope [8], where the queries can only differ in the grouping attributes.

In addition to sharing query plans, another alternative for processing multiple queries is to use the *query predicate index*. The main idea for query predicate index [10][11][12] is to decompose queries into operators and use an index structure to simultaneously evaluate multiple queries. One concern for these approaches is the intuition that the sharing among queries always benefits the overall query processing. However, this intuition does not always hold. In order to enable sharing, some executions, such as selection and projection, may need to be postponed, which can result in a significant increase in the size of intermediate results. Moreover, if the operators are with high selectivity, executing the queries individually can be more beneficial to the overall cost.

Krishnamurthy et al.[12] address sharing executions of the multiple aggregates with different window specification and different selection predicates over single data stream. The idea is to fragment input tuples into disjoint sets with respect to the query predicates and then on-the-fly answer queries with the associated disjoint sets.

All the above-mentioned works [7][9][10][11][12] use the same framework where massive data streams are collected and sent to a central processing engine where the data are processed. As reported in our experiments, such frameworks are not efficient for processing the streams formed by a sensor network due to power considerations.

Huebsch et al. [13] extends the idea of fragmenting tuples [12] to processing multiple aggregation-queries over distributed data streams, where the queries are assumed to differ in their selection predicates. Both [12][13] focus on the aggregate queries, but none of them permit join processing.

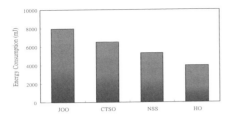

Fig. 8. Performance comparison

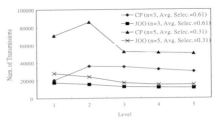

Fig. 9. Effect of in-network processing

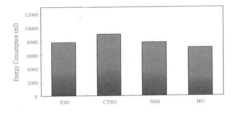

Fig. 10. Effect of low sharing opportunity

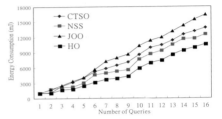

Fig. 11. Effect of number of queries

The in-network query processing systems for sensor networks, including Tinydb [16], TAG [15], and Cougar [14], support simultaneous multiple query executions. However, none of them addresses the problem of optimizing multiple join-queries.

The recent works [17][18][19] propose multiple query optimization techniques for sensor networks. However, the proposed techniques all focus on optimizing multiple aggregation or selection queries, which are orthogonal to the problem we solve. Müller et al. [18] considers the problem of optimizing multiple aggregation queries for sensor networks, while the work [17][19] addresses the problem of merging multiple selection queries which request different data at different acquisition rates.

6 Conclusion and Future Work

In this paper, we investigate the problem of efficiently evaluating path tracking queries in a sensor network application. We consider exploiting multi-query optimizations. We formulate the problem of the multi-query optimization and show the complexity of finding the optimal solution. We propose an efficient algorithm, which provide solutions with sub-optimal guarantees. The experiment result shows that our strategies ensure scalability, minimize the message traffic, and therefore reduce energy consumption.

In this study, we consider scalability issues of processing path-tracking queries. However, in stream environments, the adaptivity issues are also critical when dealing with the dynamic nature of data streams. If the data distribution of streams fluctuates over time, an adaptive approach to execute the queries is therefore essential for a good performance.

In addition, we are currently implementing the path tracking application in a large sensor network. This raises some practical problems at the system level, such as communication scheduling, etc. We plan to explore these issues as a part of our future work.

References

[1] Viglas, S., Naughton, J.F., Burger, J.: Maximizing the output rate of multi-way join queries over streaming information sources. In: Proc. of the Intl. Conf. on Very Large Data Bases, pp. 285–296 (2003)

[2] Babu, S., et al.: Adaptive ordering of pipeline stream filters. In: Proc. of the ACM SIGMOD Conf. on Management of Data, pp. 407–418 (2004)

[3] Balas, E., Padberg, M.: Set partition: a survey. SIAM review (18), 710–760 (1976)

[4] Han, J., Kamber, M.: Data Mining Concepts and Techniques. Morgan Kaufmann, San Francisco (2001)

[5] Motwani, R., Raghavan, P.: Randomized Algorithms. Cambridge Press (1995)

[6] Levis, P., Gay, D.: Maté: a tiny virtual machine for sensor networks. In: Proc. of Intl. Conf. on Architectural Support for Programming Languages and Operating Systems, pp. 85–95 (2002)

[7] Hammad, M.A., et al.: Scheduling for shared window joins over data streams. In: Proc. of the Intl. Conf. on Very Large Data Bases, pp. 297–308 (2003)

[8] Cranor, C.D., et al.: Gigascope: a stream database for network application. In: Proc. of the ACM SIGMOD Conf. on Management of Data, pp. 647–651 (2003)

[9] Srivastava, D., et al.: Multiple aggregations over data streams. In: Proc. of the ACM SIGMOD Conf. on Management of Data, pp. 299–310 (2005)

[10] Chandrasekaran, S., Franklin, M.J.: Streaming Queries over Streaming Data. In: Proc. of the Intl. Conf. on Very Large Data Bases, pp. 203–214 (2002)

[11] Madden, S., et al.: Continuously Adaptive Continuous Queries over Streams. In: Proc. of the ACM SIGMOD Conf. on Management of Data, pp. 49–60 (2002)

[12] Krishnamurthy, S., et al.: On-the-fly sharing for streamed aggregation. In: Proc. of the ACM SIGMOD Conf. on Management of Data, pp. 623–634 (2006)

[13] Huebsch, R., et al.: Sharing aggregate computation for distributed queries. In: Proc. of the Intl. Conf. on Very Large Data Bases (2007)

[14] Yao, Y., Gehrke, J.: Query processing in sensor networks. In: Proc. of Intl. Conf. on Innovative Data Systems Research (2003)

[15] Madden, S., et al.: TAG: a tiny aggregation service for ad-hoc sensor networks. In: Proc. of Annual Symps. on Operating System Design and Implementation, pp. 131–146 (2002)

[16] Madden, S., et al.: TinyDB: an acquisitional query processing system for sensor networks. ACM Trans. on Database Systems 30(1), 122–173 (2005)

[17] Trigoni, N., et al.: Multi-query optimization for sensor networks. In: Proc. of Intl. Conf. on Distributed Computing in Sensor Systems, pp. 301–321 (2005)

[18] Müller, R., Alonso, G.: Efficient sharing of sensor networks. In: Proc. of Intl. Conf. on Mobile Ad-hoc and Sensor Systems, pp. 101–118 (2005)

[19] Xian, S., et al.: Two-Tier Multiple query optimization for sensor networks. In: Proc. of the IEEE Intl. Conf. Distributed Computing System, pp. 39–47 (2007)

A Versatile Record Linkage Method by Term Matching Model Using CRF

Quang Minh Vu, Atsuhiro Takasu, and Jun Adachi

National Insitute of Informatics, Tokyo 101-8430, Japan
{vuminh,takasu,adachi}@nii.ac.jp

Abstract. We solve the problem of record linkage between databases where record fields are mixed and permuted in different ways. The solution method uses a conditional random fields model to find matching terms in record pairs and uses matching terms in the duplicate detection process. Although records with permuted fields may have partly reordered terms, our method can still utilize local orders of terms for finding matching terms. We carried out experiments on several well-known data sets in record linkage research, and our method showed its advantages on most of the data sets. We also did experiments on a synthetic data set, in which records combined fields in random order, and verified that it could handle even this data set.

1 Introduction

Information on the web is growing at an explosive rate [10], and information about an object may appear in several places. To retrieve such information effectively, we need to merge all the spread out information on the same object. Some of the previous studies have dealt with collecting information about companies, people, etc. [16,1,3]. In this study, we tried to collect information about journal papers from different data resources. An application is extraction of information about papers from publication lists posted on the web and matching them with records in research paper databases. Since databases and publication lists are different resources, the field orders and field permutations might be different; this makes linkage a more challenging task. We devised a versatile method for linking records that can work well with field-reordered records.

Most of the previous studies on record linkage targeted databases with records that have similar field orders [4,13,8]. For example, some focused on field segmentation records, where two sets of fields in two databases are the same. The methods that were developed in these studies work well with long string combinations from the same set of fields and when the fields are combined in the same order. Hereafter, we call such field segmentation records and field combination string records *segmentation records* and *string records*, for short. For these kinds of records, the previous studies built matching models based on the string edit distance to find common information between two records and to find different information that is inserted/deleted to/from one record. Although methods based on the string edit distance are effective for records with the same field

S.S. Bhowmick, J. Küng, and R. Wagner (Eds.): DEXA 2009, LNCS 5690, pp. 547–560, 2009.

order, they are of limited benefit when the records are from different databases and have different field orders. For example, in these two records: *"Four Seasons Grill Room, 854 Seventh Ave., New York City"* and *"854 Seventh Ave., New York City, Four Seasons Grill Room"*, the common text "Four Seasons Grill Room" appear at the head of one record but at the tail of the other record, and the string edit distance method will regard this text as being deleted from both records. This fault may degrade the duplicate detection performance.

There is another record linkage method that builds bags of words for records and measures the weights of common terms [4]. This method can find record pairs that have different field orders, but it neglects the term orders in a string. Therefore, it is difficult to detect overlapping phrases, and this drawback may degrade the duplicate detection performance.

To detect duplications in records that have different field orders, we tried to find matching information at the term level and combine the matching information for the duplicate detection process. Our approach uses a labeling method to find matching terms. Terms in one record are used as labels for matching terms in another record. We solve this labeling problem by using a conditional random fields (CRF)[11,15] model. CRF is usually used to solve the sequence labeling problem. For example, we applied CRF to bibliographic information extraction from title pages of academic articles where term sequences appearing in title pages are labeled with bibliographic components [14]. Unlike standard CRF applications including our previous study [14], here we use CRF to align terms in a pair of bibliographic records. Our way of CRF application is similar to the study [13], where CRF was used to measure the costs of a learnable string edit distance. However, their model suffers when the term orders of the compared strings are different. Our CRF model is analogous to a CRF model which aligns terms in machine translation [5]. Our model detects reordered terms in different records, while their model [5] detects terms having the same meaning in two languages.

The rest of this paper is organized as follows. Section 2 summarizes the related studies on record linkage. Section 3 presents our term matching model that is based on the CRF model and our duplicate detection classifier that is based on the support vector machine (SVM) method [7]. Section 4 shows experimental results on several well-known data sets and compares them with the results of previous research. Section 5 discusses the advantages of our model and compares it with other CRF models in other applications so that the reader can better understand the characteristics of our approach. Finally, Section 6 concludes this research.

2   Related Work

In [9,4,13], the string edit distance method is used to detect duplicate parts and different parts of two records. The authors carefully considered edit operations, so that they could be used to recognize important text changes that help to differentiate two distinct records. To recognize important text changes, they used an SVM model in [4] and a CRF model in [13]. However, these approaches

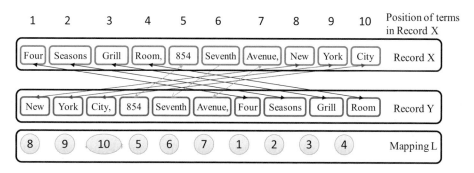

Fig. 1. Example of term matching using the proposed model

have trouble handling records with different field orders. If the fields of the two records are not aligned or their field combinations are in different orders, there are no reasonable edit operations to transform one record into another record.

In [4], the authors also used bags of words of records to find duplications. They improved the term frequency-inverse document frequency (tf-idf) weighting scheme [2,12] in the traditional vector space model by using a model to learn important words that are strongly related to a specific database. This method can be applied to records that have different data field orders. However, since the bag-of-words approach ignores term orders in a string, it can not recognize terms' consecutiveness, and therefore, it can not utilize matching phrases in records.

3   Term Matching CRF Model

3.1   Application of CRF to Term Matching Problem

Let us define the notations used in this paper. For a sequence X, $|X|$ denote the length of the sequence. For a set S and integer l, S^l denotes a set of sequences of elements of S whose length is l. We detect identical record pairs by performing the following two steps:

1. *term mapping*: map terms in a pair of records using CRF, and
2. *resolution*: decide the identity of the pair based on the term mapping using SVMs.

Let us consider the pair of records in Section 2. Figure 1 shows a term mapping between the pair of records X and Y. Note that the mapping does not preserve the order of terms.

A record is denoted by a sequence $X = (x_1, x_2, ..., x_n)$, where each element x_i represents a term in the record. For a segmentation record, we encode the field type and field ID together with the spelling of the term in order to utilize them as features in CRF. Therefore, each term x_i contains three pieces of information: $x_i = (x_i^{spell}, x_i^{field}, x_i^{id})$, where x_i^{spell} is the term spelling, x_i^{field} type of field where x_i exists, and x_i^{id} the field ID. For example, the field type of the first to third terms of the record X in Fig. 1 is restaurant name, whereas the field type of the 8th to the 10th terms is city name. Records often contain multiple values for a field such as the authors of the article. The field ID is used to discriminate fields of the same type. For example, regarding a term x_i for the first author and a term x_j for the second author, their field types are the same, but their field IDs are different: i.e., $x_i^{field} = x_j^{field}$ and $x_i^{id} \neq x_j^{id}$. Since a string record does not have any field information, we introduce an imaginary field type $string$ and assign it to all terms, i.e., $\hat{x}_i = (x_i^{spell}, string, 1)$ for any ith term in a string record, where the field ID of all terms is 1.

We denote a mapping between terms in record X and Y as a list $m \equiv (m_1, m_2, \cdots, m_{|Y|})$ of positions of terms in X where m_i denotes the position of the term in X that is mapped to the ith term in Y. For example, the mapping in Fig. 1 is represented with $(8, 9, 10, 5, 6, 7, 1, 2, 3, 4)$ by which, for instance, the first term "New" in record Y is mapped to the 8th term "New" in record X.

To handle the term mapping by using a linear chain CRF model, for each pair of records (X, Y), we use the terms in one record as labels and assign one of them to each term in the other record. Formally, we initially set a null position 0 that means that there is no corresponding term in X. For a pair of records (X, Y), we use $L_X \equiv \{0, 1, 2, \cdots, |X|\}$ as the set of labels for terms in Y. The term mapping between a record X and Y is defined as a label sequence $m \in L_X^{|Y|}$.

As in the linear chain CRF model, we assume that the mapping of term y_i is determined by the mapping of y_{i-1} and y_i itself. Then, the probability that $m = (m_1, m_2, \cdots, m_{|Y|}) \in L_X^{|Y|}$ is a mapping between a record X and Y is given by

$$p(m \mid X, Y, \theta)$$

$$\propto \exp \left(\sum_{i=1}^{|Y|} \sum_k \lambda_k f_k(x_{m_i}, y_i) + \sum_{i=2}^{|Y|} \sum_h \mu_h g_h(x_{m_{i-1}}, x_{m_i}, y_{i-1}, y_i) \right) \quad (1)$$

where $f_k(x_{m_i}, y_i)$ and $g_h(x_{m_{i-1}}, x_{m_i}, y_{i-1}, y_i)$ are feature functions discussed in the next section; λ_k and μ_h are parameters of the proposed term matching CRF model; and θ denotes the parameters $\{\lambda_k\}_k \cup \{\mu_h\}_h$. The optimal term mapping for a record X and Y is obtained by solving the optimization problem

$$\underset{m \in L_X^{|Y|}}{\operatorname{argmax}} \; p(m \mid X, Y, \theta) . \quad (2)$$

3.2   Parameter Learning Algorithm

Given a set of training data $\{(X_1, Y_1, m_1), (X_2, Y_2, m_2), \cdots, (X_n, Y_n, m_n)\}$, we have to find an optimal set of parameters $\theta = \{\lambda_k\}_k \cup \{\mu_l\}_l$ that best models the learning data. The object function is the summary of likelihoods of all training data, as follows.

$$\Phi(\boldsymbol{\theta}) = \sum_{i=1}^{n} \log p(\boldsymbol{m}_i | \boldsymbol{Y}_i, \boldsymbol{X}_i) \tag{3}$$

CRF uses regularization to avoid overfitting: it gives a penalty to weight vectors whose norm is too large. The penalty used is based on the Euclidean norm of $\boldsymbol{\theta}$ and on a regularization parameter $1/2\sigma^2$ that determines the strength of the penalty. Therefore, the object function becomes.

$$\Phi(\boldsymbol{\theta}) = \sum_{i=1}^{n} \log p(\boldsymbol{m}_i | \boldsymbol{Y}_i, \boldsymbol{X}_i) - \sum_{k} \frac{\lambda_k^2}{2\sigma^2} - \sum_{l} \frac{\mu_l^2}{2\sigma^2} \tag{4}$$

To maximize $\Phi(\boldsymbol{\theta})$, CRF uses the LBFGS algorithm[6] to update parameters $\boldsymbol{\theta}$ iteratively so that $\Phi(\boldsymbol{\theta})$ approaches the global maximum point. At each iteration step, the LBFGS algorithm uses partial differential coefficients $\frac{\partial \Phi}{\partial \lambda_k}$, $\frac{\partial \Phi}{\partial \mu_l}$ to update parameters. See [15] for details about calculating these coefficients.

3.3   Label Assigning Algorithm

To find term associations between two records, we have to maximize $p(\boldsymbol{m} \mid \boldsymbol{X}, \boldsymbol{Y}, \boldsymbol{\theta})$ which is equivalent to the following optimization problem.

$$\boldsymbol{m} = \operatorname*{argmax}_{\boldsymbol{m}} \left(\sum_{i=1}^{|\boldsymbol{Y}|} \sum_{k} \lambda_k f_k(\boldsymbol{x}_{m_i}, \boldsymbol{y}_i) + \sum_{i=2}^{|\boldsymbol{Y}|} \sum_{h} \mu_h g_h(\boldsymbol{x}_{m_{i-1}}, \boldsymbol{x}_{m_i}, \boldsymbol{y}_{i-1}, \boldsymbol{y}_i) \right)$$

$$= \operatorname*{argmax}_{\boldsymbol{m}} \left(\sum_{i=1}^{|\boldsymbol{Y}|} \sum_{k} \phi(\boldsymbol{x}_{m_{i-1}}, \boldsymbol{x}_{m_i}, \boldsymbol{y}_{i-1}, \boldsymbol{y}_i) \right) \tag{5}$$

where $\phi(\boldsymbol{x}_{m_{i-1}}, \boldsymbol{x}_{m_i}, \boldsymbol{y}_{i-1}, \boldsymbol{y}_i) = \sum_k \lambda_k f_k(\boldsymbol{x}_{m_i}, \boldsymbol{y}_i) + \mu_h g_h(\boldsymbol{x}_{m_{i-1}}, \boldsymbol{x}_{m_i}, \boldsymbol{y}_{i-1}, \boldsymbol{y}_i)$
Let $\psi(l, \boldsymbol{m}_l)$ be $\sum_{i=1}^{l} \phi(\boldsymbol{x}_{m_{i-1}}, \boldsymbol{x}_{m_i}, \boldsymbol{y}_{i-1}, \boldsymbol{y}_i)$, where $\boldsymbol{m}_l = (m_1, m_2, \cdots, m_l)$.
Then, we have $\psi(l+1, \boldsymbol{m}) = \psi(l, \boldsymbol{m}) + \phi(\boldsymbol{x}_{m_l}, \boldsymbol{x}_{m_{l+1}}, \boldsymbol{y}_l, \boldsymbol{y}_{l+1})$. We also have

$$\max_{\boldsymbol{m}_{l+1}} \psi(l+1, \boldsymbol{m}_l, m_{l+1}) = \max_{m_{l+1}} \left(\max_{\boldsymbol{m}_{l-1}} \psi(l, \boldsymbol{m}_{l-1}, m_l) + \phi(\boldsymbol{x}_{m_l}, \boldsymbol{x}_{m_{l+1}}, \boldsymbol{y}_l, \boldsymbol{y}_{l+1}) \right) \tag{6}$$

Using Eq. (6), we can solve $\max_{m_{l+1}} \psi(l+1, \boldsymbol{m}_l, m_{l+1})$ consecutively in a dynamic programming manner, which starts at $\max_{m_0} \psi(0, \cdot, m_0) = 0$, where m_0 is a dummy mapping. When we finish at $l = |\boldsymbol{Y}|$, we get the optimal solution of Eq. (5).

3.4   Feature Functions for Term Matching Model

We use a linear chain graph for our term matching model and build feature functions for nodes and edges as follows.

Node Feature Functions. A node feature function $f_k(\boldsymbol{x}, \boldsymbol{y})$ is defined for each field type t and measures the similarity of terms $\boldsymbol{x}$ and $\boldsymbol{y}$. It is defined as

$$f_k(\boldsymbol{x}, \boldsymbol{y}) = \begin{cases} \sigma(x^{spell}, y^{spell}) & \text{if } x^{field} = y^{field} = t \\ 0 & \text{otherwise} \end{cases} \qquad (7)$$

where $\sigma(x^{spell}, y^{spell})$ is one of the following string similarities:

- *Full matching function:* $\sigma(x^{spell}, y^{spell}) = 1$ if $x^{spell} = y^{spell}$; otherwise, it is 0.
- *Abbreviation matching function:* $\sigma(x^{spell}, y^{spell}) = 1$ if x^{spell} is the abbreviated form of y^{spell} or y^{spell} is the abbreviated form of x^{spell}. Otherwise it is 0.
- *Close string matching function:* $\sigma(x^{spell}, y^{spell})$ is the edit distance between x^{spell} and y^{spell}.
- *Close number matching function:* $\sigma(x^{spell}, y^{spell}) = 1$ if both x^{spell} and y^{spell} are numeric and $|x^{spell} - y^{spell}|$ is less than a threshold. This function is used to measure the similarity of numeric fields, such as year published in bibliographic records.

Since the string record does not have any field information, we introduce a wild card *General* of the field type that matches any field x^{field} and y^{field} in eq. (7);, i.e., a node feature function for the type *General* returns $\sigma(x^{spell}, y^{spell})$ independent of the field types x^{field} and y^{field}.

Edge Feature Functions. An edge feature function $g_h(\boldsymbol{x}_{m_{i-1}}, \boldsymbol{x}_{m_i}, \boldsymbol{y}_{i-1}, \boldsymbol{y}_i)$ is defined for each field type *type* and a mapping $\boldsymbol{m}$. It measures the similarity of two consecutive terms. It is defined as

$$g_h(\boldsymbol{x}_{m_{i-1}}, \boldsymbol{x}_{m_i}, \boldsymbol{y}_{i-1}, \boldsymbol{y}_i)$$
$$= \begin{cases} \sigma(x_{m_{i-1}}^{spell}, y_{i-1}^{spell}) \cdot \sigma(x_{m_i}^{spell}, y_i^{spell}) & \text{if } x_{m_{i-1}}^{field} = x_{m_i}^{field} = y_{i-1}^{field} = y_i^{field} = type, \\ & \qquad x_{m_{i-1}}^{id} = x_{m_i}^{id}, y_{i-1}^{id} = y_i^{id} \\ 0 & \text{otherwise} \end{cases}$$
$$(8)$$

where $\sigma(x^{spell}, y^{spell})$ is same as the node feature function.

3.5   Feature Vectors for Resolution by SVM

For resolution by SVM, we have to build feature vectors to represent records' similarity and use these feature vectors to separate duplicate pairs from non-duplicates. We create two feature values as follows.

Feature Values Derived from CRF Feature Functions. For each feature function in the CRF model, we summarize its values across all nodes and normalize it by the record length. Let $\boldsymbol{m}^* = (m_1^*, m_2^*, \cdots, m_{|\boldsymbol{Y}|}^*)$ denote the optimal mapping obtained by solving the problem (2). For records X and Y, the feature $\hat{f}_k$ derived from the node feature function $f_k(\cdot, \cdot)$ is

$$\hat{f}_k = \frac{1}{|Y|} \sum_{i=1}^{|Y|} f_k(x_{m_i^*}, y_i) \ . \tag{9}$$

Similarly, for a record X and Y, the feature $\hat{g}_h$ derived from the edge feature function $g_h(\cdot, \cdot, \cdot, \cdot)$ is

$$\hat{g}_h = \frac{1}{|Y|} \sum_{i=1}^{|Y|-1} g_h(x_{m_{i-1}^*}, x_{m_i^*}, y_{i-1}, y_i) \ . \tag{10}$$

Heuristic Feature Values. In addition to feature values described in the previous subsection, we also create the following heuristic features from matching terms that are useful for duplicate detection.

1. Number of terms to be deleted/inserted: Since the CRF model only calculates features from matching terms, we created this feature to take into account different terms for the duplicate detection process.
2. Number of consecutive terms to be deleted/inserted: this feature can put more penalty points on deleted/inserted phrases.
3. Position of terms to be deleted/inserted: For journal citation records, information such as author name often appears at the beginning of strings, whereas venue information often appears at the end. We use these features to differentiate term importance on the basis of position.

4 Experiments

4.1 Experimental Methods

Data sets. We carried out experiments on three well-known data sets that have been used in previous studies. The first data set contains records on restaurant information. We use four fields in this data set: name, address, city and cuisine. The second and the third data sets are the *Cora* and *Citeseer* data sets, and

Table 1. Number of records and duplications in the data sets

Data set	Number of records	Duplications
Restaurant	864	112 duplicate pairs
Cora	1295	122 unique papers
Citeseer Reasoning	514	196 unique papers
Citeseer Face	349	242 unique papers
Citeseer Reinforcement	406	148 unique papers
Citeseer Constraint	295	199 unique papers

they contain citations of papers. In the *Cora* dataset, citations are segmented into fields, and we used five fields in our experiments: author, title, venue, year, and page number. As for the *Citeseer* data set, we used the same subset as in [13] that consists of papers about four topics: *Reasoning, Face, Reinforcement,* and *Constraint*. Citations in this data set are long string records whose fields are not segmented. Details regarding the number of records and the number of duplications are shown in Table 1.

Table 2. Feature functions for segmentation records of citations in CRF term matching model

Feature type	Field type	Label difference	Matching type 1	Matching type 2
Edge	Author	1	Full	Full
Edge	Author	1	Abbreviation string	Full
Edge	Author	1	Full	Abbreviation string
Edge	Author	1	Abbreviation string	Abbreviation string
Edge	Author	-1	Full	Full
Edge	Author	-1	Abbreviation string	Full
Edge	Author	-1	Full	Abbreviation string
Edge	Author	-1	Abbreviation string	Abbreviation string
Node	Author		Full	
Edge	Title	1	Full	Full
Edge	Title	1	Full	Abbreviation string
Edge	Title	1	Abbreviation string	Full
Edge	Title	1	Full	Close string
Edge	Title	1	Close string	Full
Node	Title		Full	
Edge	Venue	1	Full	Full
Edge	Venue	1	Abbreviation string	Full
Edge	Venue	1	Full	Abbreviation string
Node	Venue		Full	
Node	Page		Full	
Node	Page		Close number	
Node	Year		Full	
Node	Year		Close number	

Feature Functions in the CRF Term Matching Model. In our CRF term matching model, feature functions are used to find the best way to match terms between record pairs. Tables 2 and 3 list examples of feature functions used for citation segmentation records and restaurant string records, respectively. In these tables, the "label difference" column means the difference between the positions of two labels. The "matching type 1" and "matching type 2" columns mean the string matching functions used to match node terms and label terms at the current position and at the previous position, respectively. The notation "Full", "Abbreviation string", "Close string", and "Close number" mean two terms match exactly, one term is abbreviation of another, two terms have small string edit distance, and two term numbers have a small difference in value.

Table 3. Feature functions for string records of restaurant information in CRF term matching model

Feature type	Field type	Label difference	Matching type 1	Matching type 2
Edge	General	1	Full	Full
Edge	General	1	Abbreviation string	Full
Edge	General	1	Full	Abbreviation string
Edge	General	1	Close string	Full
Edge	General	1	Full	Close string
Node	General		Full	
Node	General		Abbreviation string	

Parameter Learning for the CRF Term Matching Model. Our CRF term matching model requires the parameters of the feature functions to be tuned. To find an optimal set of parameters, we prepared a set of duplicate records and annotated matching terms in record pairs. These duplicate records are from a neutral resource other than the restaurant data set, the Citeseer data set, and the Cora data set. We then ran the traditional CRF parameter learning algorithm to find the optimal parameters.

SVM Classifier Learning. We used an SVM classifier to decide the identity of each pair of records. In this experiment, we used the SVM*light* tool.[1] We created training and test data for the SVMs as follows. First, we group records into clusters of duplicates. When making pairs of records even from these clusters, the data is imbalanced; i.e., it contains only a few positive pairs and many negative pairs. Therefore, we do sampling to prepare training data. The sampling method affects the observed. Hence, we used the following sampling methods that are similar to those of the previous studies [4,13].

1. **Selection of negative pairs from the top**
   In [4], the authors roughly grouped records into overlapping clusters and selected all pairs of records in the same cluster for their experiments. This way of sampling resulted in most of the positive pairs and negative pairs with high similarity being selected. Our first sampling method was similar to this method. For positive pairs, we selected all positive pairs in the data set. For negative pairs, we first measured their similarities using the tf-idf vector space model and ranked pairs by their similarities. We then selected negative pairs from the top so as to get k times more negative pairs than positive pairs. We call this sampling method *similarity sampling*.

2. **Selection of negative pairs by random sampling**
   In [13], the authors selected all positive pairs in the data set. Next, they removed negative pairs which were too different and sampled the rest of the negative pairs randomly. They selected ten times more negative pairs than positive pairs. Our second method of pair selection is similar to this one. For positive pairs, we select all positive pairs in the data set. For negative

---

[1] http://svmlight.joachims.org

pairs, we first measure the similarities of the record pairs by using the tf-idf vector space model. Then, we choose negative pairs with the top similarities to get $2k$ times as many negative pairs as positive pairs. From these negative pairs, we randomly sampled pairs to get half of them. In the end, the number of remaining negative pairs was k times larger than the number of positive pairs. We call this sampling method *random sampling*.

These two ways of sampling create two data sets that have different characteristics. We set $k = 10$ in the experiments with the restaurant data set and the four subsets in the *Citeseer* data set. Regarding the *Cora* data set, the duplicate clusters are large, so the number of positive pairs is also large. Therefore, we set $k = 3$ in the experiments with the *Cora* data set.

Evaluation Metrics. As in the previous studies, we sorted the record pairs by referring to the scores obtained by the SVM classifier and calculated the precision, recall, and f-measure values from the sorted results. We recorded the maximum f-measure value in each test. We used a 50/50 training/test split of data and repeated the random split process 10 times and did cross validations. Then, we took the average of the maximum f-measure values across all tests.

4.2   Experiments on Traditional Data Sets

We carried out experiments on the restaurant data set, *Cora* data set, and *Citeseer* data set. We carried out three experiments on the restaurant data set, using name only, address only, and four fields of name, address, city, and cuisine. We carried out one experiment on the *Cora* data set by using five fields of author, title, venue, year, and page and four experiments on four subsets of the *Citeseer* data: *Reasoning, Face, Reinforcement,* and *Constraint.*

Comparison with Bilenko's Method. We generated the training data and test set by using similarity sampling. Since it is similar to the selection method in [4], it allows us to compare Bilenko's approach directly. The results are shown in Table 4, where Bilenko's results are copied from [4]. As can be seen in Table 4, our approach outperforms Bilenko's approach on six of the eight sets.

Fig. 2 shows the precisions and f-measures for each recall. Graphs (a), (b) and (c) respectively show the performances for the restaurant data set when using the field name only, the field address only, and four fields. Graph (d) shows the performance for Cora data set. As shown in this graph, the proposed method keeps high precision until high recall.

Comparison with McCallum's Method. In this experiment, we generated training and test data by random sampling. Since it is similar to the selection method in [13], and it allows us to compare our method with McCallum's approach directly. The results are shown in Table 5, where McCallum's results are copied from [13]. As can be seen, our approach outperforms McCallum's approach on all six sets.

Table 4. Comparison with Bilenko's approach

Data set	Restaurant			Cora	Citeseer			
Fields / topic	Name	Address	Name, address, city, cuisine	Author, title, venue, page, year	Reason-ing	Face	Reinforce-ment	Constr-aint
Bilenko's approach	43.3%	71.2%	**92.2%**	86.7%	93.8%	**96.6%**	90.7%	94.1%
Our approach	**86.2%**	**74.7%**	90.16%	**87.4%**	**95.6%**	94.4%	**94.9%**	**96.9%**

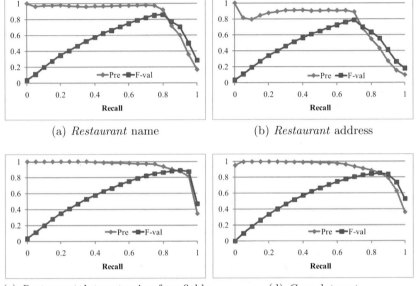

(a) *Restaurant* name

(b) *Restaurant* address

(c) *Restaurant* data set using four fields

(d) *Cora* data set

Fig. 2. Relationship between recall and precision

4.3   Experiments on Synthetic Data Sets

We carried out two experiments on synthetic data sets whose records had their fields permuted. Records in the *Cora* data set are segmented into fields. We combined fields in random order to create record pairs with different field orders. The two experiments are as follows. In the first experiment, we combined fields in only one record to create a pair between one segmentation record and one string record. In the second experiment, we combined fields in both records to create a pair of string records. These two combinations created record pairs with permuted orders, and they have not been used in previous research. The results are listed in Table 6. As can be seen, the first experiment produced results that are equivalent to those for the records with same field orders. This outcome can be explained by arguing that the term matching results are the equivalent to those in the previous experiment and information about field types can be

Table 5. Comparison with McCallum's approach

Data set	Restaurant		Citeseer			
Field / topic	Name	Address	Reasoning	Face	Reinforcement	Constraint
McCallum's approach	44.8%	78.3%	96.4%	91.8%	91.7%	97.6%
Our approach	**88.4%**	**79.6%**	**96.5%**	**95.4%**	**96.6%**	**97.8%**

Table 6. Experimental results on the synthetic data set created from the *Cora* data set

Experiment method	Performance
Pairs of one field record and one string record	**87.5%**
Pairs of two string records	83.9%

exploited from the field segmentation records in this experiment. In the second experiment, the performance slightly deteriorates, but the result is reasonable. In this experiment, the information on field types was removed from both records in each pair, and this was the main cause of degradation. Fig. 3 is an example alignment output of a pair of re-ordered records. As can be seen, our approach can detect consecutive matched terms and calculate a good mapping result.

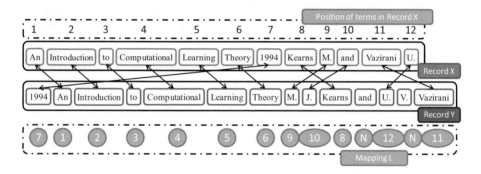

Fig. 3. An alignment result on a pair of string records

5   Discussion

Duplicate detection of records that have different field orders is more difficult than duplicate detection of records that have the same field order because words are partly reordered in records. To detect duplicate records effectively, the linkage method must be robust to this reordering. Furthermore, although record fields are in different orders, the terms in the same field still keep their order across duplicate records. Therefore, an effective linkage method should be able to recognize this local order for the matching process. Neither the string edit distance approach, nor the bag-of-words approach satisfies both requirements. The string edit distance can recognize local orders but it is weak when faced with field reordering. On the other hand, the bag of words approach is robust to field reordering but it is weak in regard to recognition of terms' local orders. Our approach, on the other hand, satisfies both requirements. It encodes each term by one node in a chain graph, and the algorithm to label pairs of matching terms is robust to field reordering. Our approach can also capture the local order of terms, since it creates feature functions on edges between two consecutive nodes and outputs matching weights for consecutive matching terms. The improvements on most data sets, in particular, on the restaurant name data set and the citation data sets, confirm the advantages of our approach. The results on synthetic records whose fields were randomly ordered are the same as on records that have the same field order. This fact shows that although the fields are randomly reordered, our approach can still utilize terms' local orders to detect consecutive matching terms between duplicate records. This is the main advantage of our approach in comparison with the previous ones. Compared with previous approaches, our method is little more expensive: There is a small cost to prepare term alignments in the training phase. The training dataset is independent from test datasets, so the trained model can be used with different test datasets. In our experiment, we used a training set of 58 aligned pairs. We updated the parameters 1000 times by using the LBFGS algorithm, and it took 32 minutes on a four 3.2Ghz CPU, 8GB memory machine.

In [4,13], the authors consider the details of string edit operations by determining which letters or terms are deleted/inserted. Our approach, on the other hand, only considers the weights of matching terms. However, it can be extended to consider deleted/inserted terms in detail. For example, from the output of the CRF term matching model, we can create features on deleted/inserted terms and use an SVM model to differentiate the importance of deleted/inserted terms.

The proposed term matching model allows many-to-one mappings between a record X and Y, as shown in the definition of the mapping. That is, different terms in the record Y may be mapped to the same term in the record X. This feature of the term matching model is not preferable. However, each term in the record Y tends to mapped to different term in the record X because of the edge feature functions.

6   Conclusions

We proposed a new method for solving the problem of record linkage between different databases. Our method can find duplicate records from databases that have permuted field orders. We built a term matching model based on the CRF method to find matching terms in records. After extracting the matching terms, we built matching vectors for the record pairs and used an SVM classifier to separate duplicate pairs from non-duplicate pairs.

We experimented on traditional data sets, and our approach showed improvements in comparison with the previous approaches that used either the string edit distance method or the vector space model method. Our method has the good point of the string edit distance method as well as the good points of the vector space model method. That is, it can utilize term orders inside the same field, and it can cope well with field reordering among databases. We also created a synthetic data set by reordering record fields in a random manner, and the results on this synthetic data set were equivalent to those for well-aligned record fields.

References

1. Asano, Y., Nishizeki, T., Toyoda, M., Kitsuregawa, M.: Mining communities on the web using a max-flow and a site-oriented framework. IEICE - Trans. Inf. Syst. E89-D(10), 2606–2615 (2006)
2. Baeza-Yates, R., Ribeiro-Neto, B.: Modern Information Retrieval. Addison Wesley Longman Publishing (1999)
3. Bhattacharya, I., Getoor, L.: A latent dirichlet model for unsupervised entity resolution. In: SDM (2006)
4. Bilenko, M., Mooney, R.J.: Adaptive duplicate detection using learnable string similarity measures. In: KDD 2003: Proceedings of the ninth ACM SIGKDD international conference on Knowledge discovery and data mining, pp. 39–48. ACM Press, New York (2003)
5. Blunsom, P., Cohn, T.: Discriminative word alignment with conditional random fields. In: ACL-44: Proceedings of the 21st International Conference on Computational Linguistics and the 44th annual meeting of the Association for Computational Linguistics, Morristown, NJ, USA, pp. 65–72. Association for Computational Linguistics (2006)
6. Byrd, R.H., Nocedal, J., Schnabel, R.B.: Representations of quasi-newton matrices and their use in limited memory methods. Math. Program. 63(2), 129–156 (1994)
7. Cristianini, N., Shawe-Taylor, J.: An Introduction to Support Vector Machines and Other Kernel-based Learning Methods. Cambridge University Press, Cambridge (2000)
8. Hernandez, M.A., Stolfo, S.J.: Real-world data is dirty: Data cleansing and the merge/purge problem. Data Mining and Knowledge Discovery 2(1), 9–37 (1998)
9. Jaro, M.A.: Advances in record-linkage methodology as applied to matching the 1985 census of tampa, florida. Journal of the American Statistical Association 84(406), 414–420 (1989)
10. Kitsuregawa, M.: 'Socio Sense' and 'Cyber Infrastructure' for information explosion era': Projects in japan. In: DASFAA, pp. 1–2 (2007)
11. Lafferty, J., Mccallum, A., Pereira, F.: Conditional random fields: Probabilistic models for segmenting and labeling sequence data. In: Proc. 18th International Conf. on Machine Learning, pp. 282–289. Morgan Kaufmann, San Francisco (2001)
12. Manning, C.D., Schutze, H.: Foundations of Statistical Natural Language Processing. MIT Press, Cambridge (2003)
13. Mccallum, A., Bellare, K., Pereira, F.: A conditional random field for discriminatively-trained finite-state string edit distance. In: Conference on Uncertainty in AI, UAI (2005)
14. Ohta, M., Yakushi, T., Takasu, A.: Bibliographic element extraction from scanned documents using conditional random fields. In: Proc. 3rd International Conf. on Digital Information Management, pp. 99–104 (2008)
15. Sutton, C., Mccallum, A.: An introduction to conditional random fields for relational learning. In: Getoor, L., Taskar, B. (eds.) Introduction to Statistical Relational Learning. MIT Press, Cambridge (2007)
16. Vu, Q.M., Takasu, A., Adachi, J.: Improving the performance of personal name disambiguation using web directories. Inf. Process. Manage. 44(4), 1546–1561 (2008)

On-the-Fly Integration and Ad Hoc Querying of Life Sciences Databases Using LifeDB[*]

Anupam Bhattacharjee[1], Aminul Islam[1], Mohammad Shafkat Amin[1],
Shahriyar Hossain[1], Shazzad Hosain[1], Hasan Jamil[1], and Leonard Lipovich[2]

[1] Department of Computer Science, Wayne State University, USA
[2] Center for Molecular Medicine and Genetics, Wayne State University, USA
{anupam,aminul,shafkat,shah_h,shazzad,hmjamil,llipovich}@wayne.edu

Abstract. Data intensive applications in Life Sciences extensively use the hidden web as a platform for information sharing. Access to these heterogeneous hidden web resources is limited through the use of predefined web forms and interactive interfaces that users navigate manually, and assume responsibility for reconciling schema heterogeneity, extracting information and piping, transforming formats and so on in order to implement desired query sequences or scientific work flows. In this paper, we present a new data management system, called *LifeDB*, in which we offer support for currency without view materialization, and autonomous reconciliation of schema heterogeneity in one single platform through a declarative query language called *BioFlow*. In our approach, schema heterogeneity is resolved at run time by treating the hidden web resources as a virtual warehouses, and by supporting a set of primitives for data integration on-the-fly, extracting information and piping to other resources, and manipulating data in a way similar to traditional database systems to respond to application demands.

1 Introduction

Data and application integration in Life Sciences play an important and essential role. In traditional approaches, data and tools for interpreting them from multiple sources are warehoused in local machines, and applications are designed around these resources by manually resolving any existing schema heterogeneity. This approach is reliable, and works well when the application's resource need, or the data sources do not change often, requiring partial or full overhauling. The disadvantage is that the warehouse must be synchronized constantly with the sources to stay current leading to huge maintenance overhead. The alternative has been to write applications by dedicated communication with the data sources, again manually mediating the schema. While this approach removes the physical downloading of the source contents and buys currency, it still requires manual mediation, coping with changes in the source, and writing source specific

---

[*] Research supported in part by National Science Foundation grants CNS 0521454 and IIS 0612203, and National Institutes of Health NIDA grant 1R03DA026021-01.

S.S. Bhowmick, J. Küng, and R. Wagner (Eds.): DEXA 2009, LNCS 5690, pp. 561–575, 2009.

glue codes that cannot be reused. The basic assumption here is that the sources are autonomous and offers a "use as you see" and hands off support. That means that the application writer receives no support in any form or manner from the sources other than the access.

There has been a significant effort to alleviate the burden on the application writers for this alternative approach by developing libraries in popular scripting languages such as Perl and PHP for accessing and using popular resources such as GenBank, UCSC, PDB, etc. These sources can change their structures without invalidating these libraries, and we have to necessarily write applications using the sources for which tested scripts are available. Consequently, applications that demand change, access to new resources, are transient or ad hoc, and are not ready to commit to significant maintenance overhead as in the former approach still remain ill served. In this paper, we propose a new approach to on-the-fly autonomous information integration that removes several of the hurdles in accessing Life Sciences resources at a throw away cost, and without any need for strict coupling or dependence among the sources and the applications.

We have developed a new data management system called *LifeDB* offering a third alternative that combines the advantages of the previous two approaches – currency and reconciliation of schema heterogeneity, in one single platform through a declarative query language called *BioFlow*. In our approach, schema heterogeneity is resolved at run time by treating the hidden web resources as a virtual warehouse, and by supporting a set of primitives for data integration on-the-fly, to extract information and pipe to other resources, and to manipulate data in a way similar to traditional database systems. At the core of this system are the schema matching system *OntoMatch* [3], the wrapper generation system *FastWrap* [1], and a visual editor called *VizBuilder* [10], using which users are able to design applications using graphical icons without the need for ever learning BioFlow for application design. In BioFlow, we offer several language constructs to support mixed-mode queries involving XML and relational data, workflow materialization as processes and design using ordered process graphs, and structured programming using process definition and reuse. We introduce LifeDB's architecture, components and features in subsequent sections. However, we will not discuss FastWrap, OntoMatch and VizBuilder in any detail although they are integral components of LifeDB. We refer interested readers to the respective articles in the literature that are now published.

1.1   A Motivating Application: BioFlow by Example

To illustrate the capabilities of LifeDB, we adapt a real life Life Sciences application discussed in [8] which has been used as a use case for many other systems and as such can be considered a benchmark application for data integration. A substantial amount of glue codes were written to implement the application in [8] by manually reconciling the source schema to filter and extract information of interest. Our goal in this section is to show how simple and efficient it is to develop this application in LifeDB.

(a) genes

miRNA	chromosome
hsa-mir-10a	ch 17
hsa-mir-205	ch 1

(b) sangerRegulation

microRNA	geneName	pValue
hsa-mir-10a	FLJ36874	0.004
hsa-miR-196b	MYO16	0.009

(e) regulation

geneID	miRNA	targetSites	pValue
FLJ36874	hsa-mir-10a	10	0.004
FLJ36874	hsa-mir-10b	3	null
RUNDC2C	hsa-mir-205	8	null
MYO16	hsa-miR-196b	null	0.009

(c) micrornaRegulation

geneID	miRNA	targetSites
FLJ36874	hsa-mir-10a	10
FLJ36874	hsa-mir-10b	3
RUNDC2C	hsa-mir-205	8

(d) proteinCodingGene

Gene	p63Binding
FLJ36874	Y
RUNDC2C	Y
MYO16	N

(f) proteinCodingGeneRegulation

geneID	miRNA	targetSites	pValue	p63Binding
FLJ36874	hsa-mir-10a	10	0.004	Y
FLJ36874	hsa-mir-10b	3	null	Y
RUNDC2C	hsa-mir-205	8	null	Y
MYO16	hsa-miR-196b	null	0.009	N

Fig. 1. User tables and data collected from microRNA.org and microrna.sanger.ac.uk

The query, or workflow, the user wants to submit is the hypothesis: *"the human p63 transcription factor indirectly regulates certain target mRNAs via direct regulation of miRNAs"*. If positive, the user also wants to know the list of miRNAs that indirectly *regulate* other target mRNAs with high enough confidence score (i.e., $pValue \leq 0.006$ and $targetSites \geq 2$), and so he proceeds as follows. He collects 52 genes along with their chromosomal locations (shown partially in figure 1(a) as the table *genes*) from a wet lab experiment using the host miRNA genes and maps at or near genomic p63 binding sites in the human cervical carcinoma cell line ME180. He also has a set of several thousand direct and indirect protein-coding genes (shown partially in figure 1(d) as the table *proteinCodingGenes*) which are the targets of p63 in ME180 as candidates. The rest of the exploration thus proceeds as follows.

He first collects a set of genes (*geneIDs*) for each of the miRNAs in the table *genes*, from the web site www.microrna.org by submitting one miRNA at a time in the window shown in figure 2(a), that returns for each such gene, a set of gene names that are known to be targets for that miRNA. The site returns the response as shown in figure 2(b), from which the user collects the *targetSites* along with the gene name partially shown as the table *micrornaRegulation* in figure 1(c). To be certain, he also collects the set of gene names for each miRNA in table *genes* from microrna.sanger.ac.uk in a similar fashion partially shown in table *sangerRegulation* in figure 1(b). Notice that this time the column *targetSites* is not available, so he collects the *pValue* values. Also note that the scheme for each of the tables are syntactically heterogeneous, but semantically they are similar (i.e., *miRNA* $\equiv$ *microRNA*, *geneName* $\equiv$ *geneID*, and so on). He does so because the data in the two databases are not identical, and there is a chance that querying only one site may not return all possible responses. Once these two tables are collected, he then takes a union of these two sets of gene names (in *micrornaRegulation* and *sangerRegulation*), and finally selects the genes from the intersection of the tables *proteinCodingGene* (that bind to p63, i.e., *p63Binding* = 'N') and *micrornaRegulation* $\cup$ *sangerRegulation* as his response.

To compute his answers in BioFlow using LifeDB, all he will need to do is execute the following script that fully implements the application. It is interesting to note that in this application, the total number of data manipulation statements used are only seven (statements numbered (2) through (8)). The rest of

(a) microRNA.org input form.   (b) microRNA.org returned page.

Fig. 2. Typical user interaction interface at microRNA.org site

the statements are data definition statements needed in any solution using any other system. We will describe shortly what these data manipulation sentences mean in this context. For now, a short and intuitive explanation is in order while we refer interested readers to [12,11] for a more complete exposition.

```
process compute_mirna                                            (1)
{
  open database bioflow_mirna;
  drop table if exists genes;
  create datatable genes {
    chromosome varchar(20), start int, end int, miRNA varchar(20) };
  load data local infile '/genes.txt'
    into table genes fields terminated by '\t'
    lines terminated by '\r\n';

  drop table if exists proteinCodingGene;
  create datatable proteinCodingGene {
    Gene varchar(200), p63binding varchar(20) };
  load data local infile '/proteinCodingGene.txt'
    into table proteinCodingGenes fields terminated by '\t'
    lines terminated by '\r\n';

  drop table if exists micrornaRegulation;
  create datatable micrornaRegulation {
    mirna varchar(200), targetsites varchar(200), geneID varchar(300) };

  define function getMiRNA
    extract mirna varchar(100), targetsites varchar(200),
    geneID varchar(300)
    using wrapper mirnaWrapper in ontology mirnaOntology
    from "http://www.microrna.org/microrna/getTargets.do"
    submit( matureName varchar(100), organism varchar(300) );    (2)
```

```
insert into micrornaRegulation
   call getMiRNA select miRNA, '9606' from genes ;                    (3)

drop table if exists sangerRegulation;
create datatable sangerRegulation {
   microRNA varchar(200), geneName varchar(200), pvalue varchar(200) };

define function getMiRNASanger
   extract microRNA varchar(200), geneName varchar(200),
   pvalue varchar(30)
   using wrapper mirnaWrapper in ontology mirnaOntology
   from "http://microrna.sanger.ac.uk/cgi-bin/targets/v5/hit_list.pl/"
   submit( mirna_id varchar(300), genome_id varchar(100) );          (4)

insert into sangerRegulation
   call getMiRNASanger select miRNA, '2964' from genes ;             (5)

create view regulation as
combine micrornaRegulation, sangerRegulation
   using matcher OntoMatch identifier gordian;                        (6)

create view proteinCodingGeneRegulation as
link regulation, proteinCodingGene
   using matcher OntoMatch identifier gordian;                        (7)

select *
   from proteinCodingGeneRegulation
   where pValue <= 0.006 and targetSites >= 2 and p63binding='N';     (8)

close database bioflow_mirna;
}
```

In the above script, the statements numbered (1) through (7) are most interesting and unique to BioFlow. The define function statements (2) and (4) essentially declare an interface to the web sites at URLs in the respective from clauses, i.e., microrna.org and microrna.sanger.ac.uk. The extract clause specifies what columns are of interest when the results of computation from the sites are available, whereas the submit clauses say what inputs need to be submitted. In these statements, it is not necessary that the users supply the exact variable names at the web site, or in the database. The wrapper (FastWrap) and the matcher (OntoMatch) named in the using clause and available in the named ontology mirnaOntology, actually establish the needed schema correspondence and the extraction rules needed to identify the results in the response page. Essentially, the define function statement acts as an interface between LifeDB and the web sites used in the applications. This statement was first introduced in [5] as the so called *remote user defined function* for databases where the input to the function is a set of tuples to which the function returns a table. However, the construct in [5] was too rigid and too mechanistic with the user needing to supply all the integration instructions. Actually, it could not use a wrapper or a schema matcher. The user needed to supply the exact scheme and exact data extraction rules. In BioFlow, it is now more declarative and intuitive.

To invoke the form functions and compute queries at these sites, we use `call` statements at (3) and (5). The first statement calls `getMiRNA` for every tuple in table `genes`, while the second call only sends one tuple to `getMiRNASanger` to collect the results in tables `micrornaRegulation` and `sangerRegulation`. The statements (6) and (7) are also new in BioFlow. They capture respectively the concepts of *vertical* and *horizontal* integration in the literature. The `combine` statement collects objects from multiple tables possibly having conflicting schemes into one table. To do so, it also uses a key identifier (such as `gordian` [15]) to recognize objects across tables. Such concepts have been investigated in the literature under the titles record linkage or object identification. For the purpose of this example, we adapted GORDIAN [15] as one of the key identifiers in BioFlow. The purpose of using a key identifier is to recognize the fields in the constituent relations that essentially make up the object key[1], so that we can avoid collecting non-unique objects in the result. The `link` statement, on the other hand, extends an object in a way similar to join operation in relational algebra. Here too, the schema matcher and the key identifier play an important role. Finally, the whole script can be stored as a named *process* and reused using BioFlow's `perform` statement. In this example, line (1) shows that this process is named `compute_mirna` and can be stored as such for later use. We will take up these issues again in section 2.2 when we discuss BioFlow.

1.2   Related Research

Before we introduce LifeDB more formally, we would like to mention how LifeDB stands relative to its predecessors. There are several well known data integration systems in the literature. We single out only a few more recent ones, simply because they improve upon many older ones, and are close to LifeDB in spirit than many others. BioKleisli [7] and Biopipe [9] are two such data integration systems for Life Sciences that help define workflows and execute queries. While they are useful, they actually fall into the second category of application design approaches that we discussed in section 1. As such, they are very tightly coupled with the source databases because they directly interact with them using the services supported by the sources. The scripting language just eases the tediousness of developing applications. Schema mediation remains user responsibility and they do not deal with hidden web resources, or ad hoc data integration. One of the systems, ALADIN [2] (ALmost Automatic Data INtegration), falls into the first category and supports integration by locally creating a physical relational database warehouse manually. However, it uses technologies such as key identification and record linkage to mediate schema heterogeneity, applying domain and application specific knowledge and thus having limited application.

---

[1] Note that object key in this case is not necessarily the primary keys of the participating relations.

In LifeDB, integration is declarative, fully automatic and does not rely on a local copy of the resources (it uses virtual warehousing). Neither does it depend on application or domain specific knowledge. For all queries, workflows and data integration requests, the schema mediation, wrapper generation and information extraction are carried out in real time. One of the recent integration systems, MetaQuerier [4], has features similar to LifeDB. It integrates hidden web sources, and uses components similar to LifeDB. But MetaQuerier integrates the components using glue codes making it resistant to change. In LifeDB on the other hand, we focus on a framework where no code writing will be necessary for the development of any application regardless of the sites or resources used. In addition, the application will never require specialized site cooperation to function.

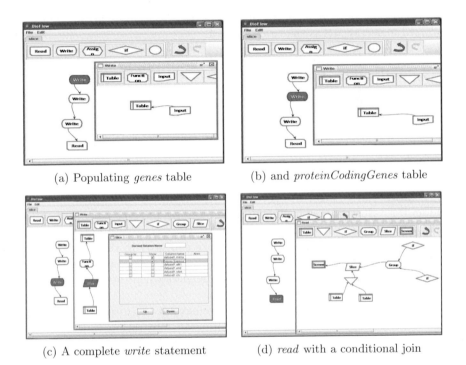

(a) Populating *genes* table   (b) and *proteinCodingGenes* table

(c) A complete *write* statement   (d) *read* with a conditional join

Fig. 3. Partial miRNA application script development using VizBuilder

LifeDB stands out among all the systems in two principal ways. First, it extends SQL with automatic schema mediation, horizontal and vertical integration, hidden web access, and process definition primitives. Apart from the fact that for the purpose of programming convenience, it retains a few non-declarative language constructs such as assignment and loop statements, it is almost entirely declarative. It separates system aspects from the language, making it possible to change the underlying system components as newer and better technologies become available, and improves its functionality without compromising the semantics of its query language, BioFlow. Second, it successfully avoids physical warehousing without sacrificing autonomy of the sources – in other words, it does not depend on specialized services from the source databases such as option to run stored scripts, access to the server to submit scripts for execution and so on, the way BioKleisli or Biopipe require. Yet, it supports declarative integration at a throw away cost. Unlike ALADIN and MetaQuerier, it is also fully automatic, and no user intervention or glue code writing is required to enable integration and access.

Finally, LifeDB goes one step further by supporting a graphical query builder front-end which interested readers may find in [10]. Using this query builder, it is possible to write the miRNA application just described, visually (without even knowing BioFlow) and execute it right from the front end. Almost all the sequence of VizBuilder panels used to write the miRNA application script discussed in section 1.1 is shown in figure 3.

2   LifeDB and BioFlow

2.1   Integration Model

The successful development of a declarative statement for hidden web access, schema mediation, and horizontal and vertical integration can be largely attributed to the integration model adopted in LifeDB, as shown in figure 4. In this abstract model, web forms are viewed roughly as functions to which a set of arguments can be passed to get a table as a returned value. In the process of sending the values, a match function μ determines the schema correspondence to map arguments accordingly. An extraction function η also isolates the table from the response page. These attributes are captured in the using clause of the define function statement discussed in section 1 as wrapper and matcher options. The submission and extraction commences when the call statement is executed (i.e., statement (3)).

In our abstract model, we assume that we have at our disposal a set of functions for a set of specific operations. For automatic integration, we need to somehow transform a hidden web form into a relation. We also recognize that a web form itself is a function φ that returns a relation in response to a set of submitted parameter, a mapping in essence. The *transform* function thus takes three different functions such as matching function ($\mu \in \Sigma$), form function ($\varphi \in \Upsilon$), extract function ($\eta \in \Xi$) and the output schema S as parameters to convert the hidden web form into a relation. The whole transformation can be written as

$$\tau^{\mu}_{\varphi,\eta,S}(r) = \pi^{\mu}_{S}(\eta(\varphi^{\mu}(r)))$$

Another feature the model supports is vertical and horizontal integration. In this view, data sets (or tables) from various sources can be extracted using η and collected in a set using a set union type operation, called *combine* (χ) and expressed as $r \leftarrow \eta(d_i)\chi^\mu_\kappa \eta(d_j)$, or a join type operation, called $link^2$ (λ) and written analogously as $r \leftarrow \eta(d_i)\chi^\mu_\kappa \eta(d_j)$. Both these operations require reconciliation of schema heterogeneity using a match function μ, and object identification using a key discovery/identifier function κ. In the BioFlow statements (6) and (7), these two implementations are shown where a matcher has been used to map the schema, and additionally, a key identifier has been used to disambiguate the objects in the data sets. Notice that unlike the **define function** statements, the **wrapper** option is missing. The reason for this is that wrappers have already been applied to these data sets during the **call** statement execution that possibly created these sets. Given that these constructs are modular, it is now possible to mix and match them to create arbitrarily complex sequence of statements that can mimic any application within the model's semantic scope.

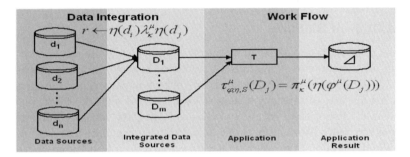

Fig. 4. BioFlow abstract data integration model

2.2 The BioFlow Language Basics

Like many others [6,14], we too abstract the web as a set of relations (flat or nested). This view establishes a clean correspondence between web documents containing tables, and database relations[3]. This view also makes it possible to treat web documents the same way as relations once they are converted (such as using the extraction function η) into tables[4]. Thus, from a single database point of view, the classical theories of relational data model hold and the notion of functional dependencies and keys carry over immediately to our data model. However, due to the presence of heterogeneity and the need for semantic integration, we need to equip the traditional model with additional machineries.

Semantic Equivalence. To model schema mapping, and the notions of horizontal and vertical integration, and to understand what to expect from BioFlow, we introduce the concept of semantic equivalence using *term similarity* ($\sim$),

---

[2] In the literature it is also known as data fusion or aggregation operation.
[3] Documents that do not contain tables, are considered empty relations.
[4] By making this assumption we are in a way assuming that for the rare set of hidden web databases that produce information in a form other than tables, our model will not be appropriate.

attribute equivalence ($\simeq$), *value* and *object equivalence* that are part of our database functions μ (match)[5], κ (key identifier)[6] and η (extraction)[7].

For example, for two attribute names *miRNA* and *microRNA*, and two values *ch17* and *ch-17*, the relationships *miRNA* $\sim$ *microRNA* (attribute name similarity) and *ch17* $\sim$ *ch-17* (value similarity) hold. In this example, *miRNA* $\simeq$ *microRNA* holds since *miRNA* $\sim$ *microRNA* holds and *type(miRNA)* = *type(microRNA)* holds. The value *hsa-mir-10a* in figure 1(b) is equivalent to *hsa-mir-10a* in figure 1(c) since *miRNA* $\simeq$ *microRNA* and *hsa-mir-10a* $\sim$ *hsa-mir-10a* hold. Finally, for the entity sets *micrornaRegulation* in figure 1(b) and *sangerRegulation* in figure 1(c), *microRNA* or *miRNA* could be considered as their candidate keys. So, objects *<FLJ36874, hsa-mir-10a, 10>* and *<hsa-mir-10a, FLJ36874, 0.004>* are equivalent even though they have a heterogeneous set of attributes.

2.3   A Tour of BioFlow

We have already discussed some of the features and statements of BioFlow in section 1 in the context of the miRNA application in [8]. Since a complete discussion on BioFlow is not within the scope of this paper, we only discuss a few interesting features of this language in this section and refer the reader to [12,11] for a complete exposition. At a very high level, BioFlow consists of five types of statements: resource description, control and structuring, data integration, workflow and data manipulation statements. Resource description statements include table definitions and remote user defined functions such as `define function` statements. Tables in BioFlow can be relational tables or XML documents which are stored, managed and processed in their native form without translation in local or remote storage. It is easy to see that the `define function` construct can also be used to include application tools (not only databases) in the workflows or queries in the same way we use and access hidden web databases. Largely, the semantics of tables resemble the semantics of approximate or inaccurate databases in the literature with the additional property that in BioFlow, attributes are also inaccurate or approximate. This is so because we also deal with schema heterogeneity. The notions of various types of similarities introduced in section 2.2 play a major role in the BioFlow semantics. For example, consider the query below in the context of the table `genes` in figure 1(a).

```
select microRNA from genes where allele="ch-17";
```

This query will return a relation with a single column called *microRNA* with only one row having the value "hsa-mir-10a". Although the table `genes` does not have

---

[5] The match function μ takes two relational schemes as input and returns the schema correspondences as a list of pairs of equivalent attributes.

[6] Given a relation instance as input, the κ function discovers all candidate keys in a relation that hold on that instance.

[7] The extraction function η converts a given web page to a set of relations if, and only if, the page contains regular patterns that can be identified as objects. In a sense, this function implements the definition of wrappers after Laender [13].

a column named *microRNA*, the query succeeds because *microRNA* $\simeq$ *miRNA* (edit distance similarity), *chromosome* $\simeq$ *allele* (synonym similarity), and *"ch-17"* $\simeq$ *"ch 17"* (value similarity), hold. The consequence of such a powerful feature is that users no longer have to be aware of exact terms in BioFlow and can design their applications without the full knowledge of the underlying databases, a feature that comes handy in the presence of schema heterogeneity in large and open databases on the internet. The match functions defined earlier make all these possible.

Control and structuring statements in BioFlow include assignment statements, branching statements (`if then else`), looping statements (`repeat until`), and stored process reuse statement `include`. Their semantics parallel the traditional semantics in programming languages and hence require no additional discussion. These programming statements are incorporated in BioFlow for the completeness of workflow definitions and programming ease even though some of these constructs are non-declarative in nature. Data manipulation statements on the other hand, are purely declarative and include `call` statements, `select from where` statements, and update statements such as `insert` and `delete`. The only difference from traditional SQL is that `call` statements behave like a `select` statement and can be used where a `select` can be used. Also, in `select` statements, we allow nested data item expressions such as in XQuery or XPath to be able to refer to nested attributes in a table[8].

The last two categories of statements are particularly unique to BioFlow. The data integration statements `link` and `combine`, and the dual role of `call` as a data integration statement deserves a short discussion. Two tables in BioFlow can be combined into one (vertical integration) table if they share a candidate key. For example, the tables *micrornaRegulation* and *sangerRegulation* are combine compatible since they share the keys *miRNA* and *microRNA* (and *microRNA* $\simeq$ *miRNA* holds). The resultant scheme of a combine operation is determined according to the following definition.

Definition 2.1 (Combined Schema). Let r and s be two relations over the schemes $R(A_1, A_2, \ldots, A_m)$, and $S(B_1, B_2, \ldots, B_n)$ respectively, and μ be any match function. Also let $A = A_1, A_2 \ldots, A_p$, and $B = B_1, B_2 \ldots, B_p$, where $p \leq min(m, n)$, be all the attributes in R and S such that $< A_i, B_i > \in \mu(R, S), i = 1, \ldots, p$ holds. Then the combined schema of R and S will be $(R \cup (S - B))$.

Consider, the following `combine` statement in the context of the relations in figure 1.

```
combine sangerRegulation, micrornaRegulation
  using matcher OntoMatch, identifier gordian;
```

---

[8] In the current version of BioFlow, although we allow relational tables and XML documents to be intermixed, we do not allow nested relational tables. All nested relations are stored in XML documents by default. But the `select` statements uniformly use nested expressions to refer to data items and distinguishes the table types based on the context it is used.

In this example, $\mu(micrornaRegulation,\ sangerRegulation)$ returns $\{<geneID,\ geneName>,\ <miRNA,\ microRNA>\}$ as match pairs, and hence the combined scheme $\{geneID,\ miRNA,\ targetSites,\ pValue\}$ may be produced for the table *regulation* (figure 1(e)). However, the content of *regulation* will be the set of distinct "objects" that appear in any one of these relations. In this example, the matcher OntoMatch establishes the schema correspondence ($\{<geneID,\ geneName>,\ <miRNA,\ microRNA>\}$) and the key identifier function GOR-DIAN returns $miRNA$ and $microRNA$ as candidate keys. Consequently, the *regulation* instance will have four objects because one object, *hsa-mir-10a*, is common. Since each of these source tables is missing a few attributes of the other tables, these columns are padded with *null*s as shown. The important point here is that the two schemes need not be union compatible as in relational model counterpart. It is also distinct from outer union operations since objects are selected using the notion of keys in combine and generally are not equivalent.

Differently from `combine`, `link` requires that two relations share a candidate key and foreign key relationship in a one to many fashion, giving an impression of a one to many join in traditional databases. But unlike join, and similar to combine, the link operation is dependent upon a match function μ and key discovery function κ, to establish correspondence and identify objects for join. The results of the link operation below is shown in figure 1(f).

```
link regulation, proteinCodingGene
    using matcher OntoMatch, identifier gordian;
```

Finally, workflow statements in BioFlow include the process definition statement, `perform` statement and `wait` statement. A process in BioFlow is a self contained unit of computation. In other words, a complete BioFlow program can be compiled, executed and stored separately, and when needed, can be retrieved or included in other programs or processes within the same database. Thus, each process is uniquely named, has its own associated resource description and data manipulation components, and executes in separate independent process threads[9]. However, they follow BioFlow's global resource definition principle and so duplicate resource definition, and variable naming or block naming are not allowed[10]. It should be apparent that complex processes can be created using the constructs allowed in BioFlow. The general structure of a process is as follows.

```
process P statements/blocks;
```

Processes cannot be nested, and thus, a process cannot contain another process. However, a process may include another process and use it as part of its

---

[9] Note that process descriptions are part of the resource description component of each BioFlow program. When a process is included as part of another program, it is assumed that all resources needed to execute the process is already included in the parent program's resource description section. Failure to do so may result in a run time error.

[10] The difference between a statement block and a process should be clear. A process is a self contained reusable BioFlow program, whereas statement blocks are not.

computational task. This also means, processes may call each other in mutually recursive manner, but cannot include self recursion. The execution of processes can be fairly sophisticated, and using the powerful constructs, extremely expressive workflows can be described and executed. In BioFlow, processes can be executed in a graph like fashion using the following constructs. A process execution statement can appear anywhere a statement can appear in a BioFlow program.

perform [parallel] $p_1 \ldots, p_n$ [after $q_1 \ldots, q_k$] [leave];

The parallel option creates n number of processes to execute the named processes in the list simultaneously. This feature becomes handy when submitting several processes to multiple sites is needed, and running them in tandem without waiting for any of the sites in a serial fashion to save time is necessary. The after option allows the processes to be performed only after the named processes are all completed, thus giving a *wait on* feature to sequence processes. The leave option allows the system to move to the next statement once the perform statement is scheduled to be executed, without waiting for its completion. In order to check if a process scheduled has already completed execution, we provide a function called $pending(p)$ where p is a process name. This function returns true if the process is still executing, otherwise it returns false. The wait on statement below also uses $pending(p)$ to check the status of processes in order to halt processing of the subsequent statements until the list of processes finished execution.

wait on $p_1 \ldots, p_k$;

2.4   LifeDB Architecture

The architecture of the current implementation of LifeDB is shown in figure 5 in which we show its essential components. Aside from the usual components, the following deserve special mention – the visual query interface VizBuilder,

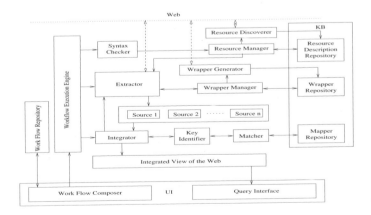

Fig. 5. LifeDB system architecture

wrapper generator FastWrap, and schema matcher OntoMatch. In the current implementation, we did not include the *Resource Discoverer* component, and the knowledgebase component for the storage of wrappers and schema maps. The purpose of the knowledgebase, called the ontology, is to reuse these information when available and not redo the extraction function generation or matching schema at run time. We have used monetDB [16] as our storage system and underlying query processor due to its support for both XML and relational data. As discussed before, we have also used GORDIAN as our key identification function for link and combine operators.

3   Summary and Future Plans

Our goal in this paper was to present LifeDB, a highly adaptive middleware system for Life Sciences data integration *on-the-fly*. It combined the existing integration technologies automatically using a declarative language called BioFlow. The BioFlow language continues to evolve as we gain more experience with LifeDB and BioFlow, and we expect it to be so for the near future. In our future release, we plan to include automatic resource discovery, which will be used as a mechanism to generate ontologies involving the wrapper definitions and schema matching information per site to facilitate reuse and improve efficiency. Once implemented, this module will also help separate the user model of application from the real world scenarios making the user more independent of the changes or disparities in the real world.

Current implementation of FastWrap does not generate column names of extracted table, forcing a compilation step to generate the attribute names through user intervention when the column names are missing in the source pages. Although we have a fairly good method for addressing this issue in such situations, it is still manual. We plan to remove this limitation in our next release. Although in our current release, we support a combination of data types – tables and XML documents – they are processed separately. This is because the back end data management system monetDB does not allow mixing relational data with XML documents in the same query. We are working to remove this limitation in our future versions of LifeDB.

References

1. Amin, M.S., Jamil, H.: FastWrap: An efficient wrapper for tabular data extraction from the web. In: IEEE International Conference on Information Reuse and Integration, Las Vegas, Nevada (August 2009)
2. Bauckmann, J.: Automatically Integrating Life Science Data Sources. In: VLDB PhD Workshop (2007)
3. Bhattacharjee, A., Jamil, H.: OntoMatch: A monotonically improving schema matching system for autonomous data integration. In: IEEE International Conference on Information Reuse and Integration, Las Vegas, Nevada (August 2009)
4. Chang, K., He, B., Zhang, Z.: Toward large scale integration: Building a MetaQuerier over databases on the web. In: CIDR Conference (2005)

5. Chen, L., Jamil, H.M.: On using remote user defined functions as wrappers for biological database interoperability. International Journal of Cooperative Information Systems 12(2), 161–195 (2003)
6. Chu, E., Baid, A., Chen, T., Doan, A., Naughton, J.F.: A relational approach to incrementally extracting and querying structure in unstructured data. In: VLDB 2007, Vienna, Austria, pp. 1045–1056 (2007)
7. Davidson, S.B., Overton, G.C., Tannen, V., Wong, L.: BioKleisli: A digital library for biomedical researchers. International Journal on Digital Libraries 1(1), 36–53 (1997)
8. Gusfield, D., Stoye, J.: Relationships between p63 binding, DNA sequence, transcription activity, and biological function in human cells. Mol. Cell. 24(4), 593–602 (2006)
9. Hoon, S., Ratnapu, K.K., Chia, J.-M., Kumarasamy, B., Juguang, X., Clamp, M., Stabenau, A., Potter, S., Clarke, L., Stupka, E.: Biopipe: A flexible framework for protocol-based bioinformatics analysis. Genome Research 13(8), 1904–1915 (2003)
10. Hossain, S., Jamil, H.: A visual interface for on-the-fly biological database integration and workflow design using VizBuilder. In: 6th International Workshop on Data Integration in the Life Sciences, Manchester, UK (July 2009)
11. Jamil, H., El-Hajj-Diab, B.: BioFlow: A web-based declarative workflow language for Life Sciences. In: 2nd IEEE Workshop on Scientific Workflows, Honolulu, Hawaii, pp. 453–460. IEEE Computer Society Press, Los Alamitos (2008)
12. Jamil, H., Islam, A.: The power of declarative languages: A comparative exposition of scientific workflow design using BioFlow and Taverna. In: 3rd IEEE Workshop on Scientific Workflows, Los Angeles, CA, July 2009, IEEE Computer, Los Alamitos (2009)
13. Laender, A., Ribeiro-Neto, B., da Silva, A.S.: DEByE - date extraction by example. Data Knowl. Eng. 40(2), 121–154 (2002)
14. Minton, S.N., Nanjo, C., Knoblock, C.A., Michalowski, M., Michelson, M.: A heterogeneous field matching method for record linkage. In: ICDM, November 2005, vol. 27 (2005)
15. Sismanis, Y., Brown, P., Haas, P.J., Reinwald, B.: GORDIAN: efficient and scalable discovery of composite keys. In: VLDB 2006, pp. 691–702 (2006)
16. Zhang, Y., Boncz, P.: XRPC: interoperable and efficient distributed XQuery. In: VLDB, pp. 99–110 (2007)

Analyses and Validation of Conditional Dependencies with Built-in Predicates

Wenguang Chen[1], Wenfei Fan[2,3], and Shuai Ma[2]

[1] Peking University, China
[2] University of Edinburgh, UK
[3] Bell Laboratories, USA

Abstract. This paper proposes a natural extension of conditional functional dependencies (CFDs [14]) and conditional inclusion dependencies (CINDs [8]), denoted by CFD^Ps and CIND^Ps, respectively, by specifying patterns of data values with $\neq, <, \leq, >$ and $\geq$ predicates. As data quality rules, CFD^Ps and CIND^Ps are able to capture errors that commonly arise in practice but cannot be detected by CFDs and CINDs. We establish two sets of results for central technical problems associated with CFD^Ps and CIND^Ps. (a) One concerns the satisfiability and implication problems for CFD^Ps and CIND^Ps, taken separately or together. These are important for, *e.g.*, deciding whether data quality rules are dirty themselves, and for removing redundant rules. We show that despite the increased expressive power, the static analyses of CFD^Ps and CIND^Ps retain the same complexity as their CFDs and CINDs counterparts. (b) The other concerns validation of CFD^Ps and CIND^Ps. We show that given a set Σ of CFD^Ps and CIND^Ps on a database D, a set of SQL queries can be automatically generated that, when evaluated against D, return all tuples in D that violate some dependencies in Σ. This provides commercial DBMS with an immediate capability to detect errors based on CFD^Ps and CIND^Ps.

1 Introduction

Extensions of functional dependencies (FDs) and inclusion dependencies (INDs), known as *conditional functional dependencies* (CFDs [14]) and *conditional inclusion dependencies* (CINDs [8]), respectively, have recently been proposed for improving data quality. These extensions enforce patterns of semantically related data values, and detect errors as violations of the dependencies. Conditional dependencies are able to capture more inconsistencies than FDs and INDs [14,8].

Conditional dependencies specify constant patterns in terms of equality ($=$). In practice, however, the semantics of data often needs to be specified in terms of other predicates such as $\neq, <, \leq, >$ and $\geq$, as illustrated by the example below.

Example 1. An online store maintains a database of two relations: (a) item for items sold by the store, and (b) tax for the sale tax rates for the items, except artwork, in various states. The relations are specified by the following schemas:

> item (id: string, name: string, type: string, price: float, shipping: float,
>   sale: bool, state: string)
> tax (state: string, rate: float)

S.S. Bhowmick, J. Küng, and R. Wagner (Eds.): DEXA 2009, LNCS 5690, pp. 576–591, 2009.

id	name	type	price	shipping	sale	state
t_1:	b1	Harry Potter	book	25.99	0	T	WA
t_2:	c1	Snow White	CD	9.99	2	F	NY
t_3:	b2	Catch-22	book	34.99	20	F	DL
t_4:	a1	Sunflowers	art	5m	500	F	DL

(a) An item relation

	state	rate
t_5:	PA	6
t_6:	NY	4
t_7:	DL	0
t_8:	NJ	3.5

(b) tax rates

Fig. 1. Example instance D_0 of item and tax

where each item is specified by its id, name, type (*e.g.*, book, CD), price, shipping fee, the state to which it is shipped, and whether it is on sale. A tax tuple specifies the sale tax rate in a state. An instance D_0 of item and tax is shown in Fig. 1.

One wants to specify dependencies on the relations as data quality rules to detect errors in the data, such that inconsistencies emerge as violations of the dependencies. Traditional dependencies (FDs, INDs; see, *e.g.*, [1]) and conditional dependencies (CFDs, CINDs [14,8]) on the data include the following:

> cfd$_1$: item (id $\rightarrow$ name, type, price, shipping, sale)
> cfd$_2$: tax (state $\rightarrow$ rate)
> cfd$_3$: item (sale $=$ 'T' $\rightarrow$ shipping $= 0$)

These are CFDs: (a) cfd$_1$ assures that the id of an item uniquely determines the name, type, price, shipping, sale of the item; (b) cfd$_2$ states that state is a key for tax, *i.e.*, for each state there is a unique sale tax rate; and (c) cfd$_3$ is to ensure that for any item tuple t, if $t[\text{sale}] =$ 'T' then $t[\text{shipping}]$ must be 0; *i.e.*, the store provides free shipping for items on sale. Here cfd$_3$ is specified in terms of patterns of semantically related data values, namely, sale $=$ 'T' and shipping $= 0$. It is to hold only on item tuples that match the pattern sale $=$ 'T'. In contrast, cfd$_1$ and cfd$_2$ are traditional FDs without constant patterns, a special case of CFDs. One can verify that no sensible INDs or CINDs can be defined across item and tax.

Note that D_0 of Fig. 1 satisfies cfd$_1$, cfd$_2$ and cfd$_3$. That is, when these dependencies are used as data quality rules, no errors are found in D_0.

In practice, the shipment fee of an item is typically determined by the price of the item. Moreover, when an item is on sale, the price of the item is often in a certain range. Furthermore, for any item sold by the store to a customer in a state, if the item is *not* artwork, then one expects to find the sale tax rate in the state from the tax table. These semantic relations cannot be expressed as CFDs of [14] or CINDs of [8], but can be expressed as the following dependencies:

> pfd$_1$: item (sale $=$ 'F' & price $\leq 20 \rightarrow$ shipping $= 3$)
> pfd$_2$: item (sale $=$ 'F' & price > 20 & price $\leq 40 \rightarrow$ shipping $= 6$)
> pfd$_3$: item (sale $=$ 'F' & price $> 40 \rightarrow$ shipping $= 10$)
> pfd$_4$: item (sale $=$ 'T' $\rightarrow$ price ≥ 2.99 & price < 9.99)
> pind$_1$: item (state; type $\neq$ 'art') $\subseteq$ tax (state; nil)

Here pfd$_2$ states that for any item tuple, if it is not on sale and its price is in the range $(20, 40]$, then its shipment fee must be 6; similarly for pfd$_1$ and pfd$_3$. These dependencies extend CFDs [14] by specifying patterns of semantically related data values in terms of predicates $<, \leq, >$, and $\geq$. Similarly, pfd_4 assures that for any item tuple, if it is on sale, then its price must be in the range $[2.99, 9.99)$. Dependency pind_1 extends CINDs [8] by specifying patterns with $\neq$: for any item tuple t, if $t[\mathsf{type}]$ is *not* artwork, then there must exist a tax tuple t' such that $t[\mathsf{state}] = t'[\mathsf{state}]$, *i.e.*, the sale tax of the item can be found from the tax relation.

Using pfd_1–pfd_4 and pind_1 as data quality rules, we find that D_0 of Fig. 1 is *not* clean. Indeed, (a) t_2 violates pfd_1: its price is less than 20, but its shipping fee is 2 rather than 3; similarly, t_3 violates pfd_2, and t_4 violates pfd_3. (b) Tuple t_1 violates pfd_4: it is on sale but its price is not in the range $[2.99, 9.99)$. (c) The database D_0 also violates pind_1: t_1 is not artwork, but its state cannot find a match in the tax relation, *i.e.*, no tax rate for WA is found in D_0.   □

None of pfd_1–pfd_4 and pind_1 can be expressed as FDs or INDs [1], which do not allows constants, or as CFDs [14] or CINDs [8], which specify patterns with equality $(=)$ only. While there have been extensions of CFDs [7,18], none of these allows dependencies to be specified with patterns on data values in terms of built-in predicates $\neq, <, \leq, >$ or $\geq$. To the best of our knowledge, no previous work has studied extensions of CINDs (see Section 6 for detailed discussions).

These highlight the need for extending CFDs and CINDs to capture errors commonly found in real-life data. While one can consider arbitrary extensions, it is necessary to strike a balance between the expressive power of the extensions and their complexity. In particular, we want to be able to reason about data quality rules expressed as extended CFDs and CINDs. Furthermore, we want to have effective algorithms to detect inconsistencies based on these extensions.

Contributions. This paper proposes a natural extension of CFDs and CINDs, provides complexity bounds for reasoning about the extension, and develops effective SQL-based techniques for detecting errors based on the extension.

(1) We propose two classes of dependencies, denoted by CFD^ps and CIND^ps, which respectively extend CFDs and CINDs by supporting $\neq, <, \leq, >, \geq$ predicates. For example, all the dependencies we have encountered so far can be expressed as CFD^ps or CIND^ps. These dependencies are capable of capturing errors in real-world data that cannot be detected by CFDs or CINDs.

(2) We establish complexity bounds for the satisfiability problem and the implication problem for CFD^ps and CIND^ps, taken separately or together. The satisfiability problem is to determine whether a set Σ of dependencies has a nonempty model, *i.e.*, whether the rules in Σ are consistent themselves. The implication problem is to decide whether a set Σ of dependencies entails another dependency φ, *i.e.*, whether the rule φ is redundant in the presence of the rules in Σ. These are the central technical problems associated with any dependency language.

We show that despite the increased expressive power, CFD^ps and CIND^ps do not increase the complexity for reasoning about them. In particular, we show that the satisfiability and implication problems remain (a) NP-complete and coNP-complete for CFD^ps, respectively, (b) in $O(1)$-time (constant-time) and EXPTIME-complete for CIND^ps, respectively, and (c) are undecidable when CFD^ps and CIND^ps are taken together. These are *the same as* their CFDs and CINDs counterparts. While data with linearly ordered domains often makes our lives harder (see, *e.g.*, [21]), CFD^{p}s and CIND^{p}s do not complicate their static analyses.

(3) We provide SQL-based techniques to detect errors based on CFD^{p}s and CIND^{p}s. Given a set Σ of CFD^{p}s and CIND^{p}s on a database D, we automatically generate a set of SQL queries that, when evaluated on D, find all tuples in D that violate some dependencies in Σ. Further, the SQL queries are independent of the size and cardinality of Σ. No previous work has been studied error detection based on CINDs, not to mention CFD^{p}s and CIND^{p}s taken together. These provide the capability of detecting errors in a single relation (CFD^{p}s) and across different relations (CIND^{p}s) within the immediate reach of commercial DBMS.

Organizations. Sections 2 and 3 introduce CFD^{p}s and CIND^{p}s, respectively. Section 4 establishes complexity bounds for reasoning about CFD^{p}s and CIND^{p}s. Section 5 provides SQL techniques for error detection. Related work is discussed in Section 6, followed by topics for future work in Section 7.

2   Incorporating Built-in Predicates into CFDs

We now define CFD^{p}s, also referred to as *conditional functional dependencies*, by extending CFDs with predicates ($\neq, <, \leq, >, \geq$) in addition to equality ($=$).

Consider a relation schema R defined over a finite set of attributes, denoted by attr(R). For each attribute $A \in$ attr(R), its domain is specified in R, denoted as dom(A), which is either finite (*e.g.,* bool) or infinite (*e.g.,* string). We assume *w.l.o.g.* that a domain is totally ordered if $<, \leq, >$ or $\geq$ is defined on it.

Syntax. A CFD^p φ on R is a pair $R(X \rightarrow Y,\ T_p)$, where (1) X, Y are sets of attributes in attr(R); (2) $X \rightarrow Y$ is a standard FD, referred to as the FD *embedded in φ*; and (3) T_p is a tableau with attributes in X and Y, referred to as the *pattern tableau* of φ, where for each A in $X \cup Y$ and each tuple $t_p \in T_p$, $t_p[A]$ is either an unnamed variable '_' that draws values from dom(A), or 'op a', where op is one of $=, \neq, <, \leq, >, \geq$, and '$a$' is a constant in dom($A$).

If attribute A occurs in both X and Y, we use A_L and A_R to indicate the occurrence of A in X and Y, respectively, and separate the X and Y attributes in a pattern tuple with '$\|$'. We write φ as $(X \rightarrow Y,\ T_p)$ when R is clear from the context, and denote X as LHS(φ) and Y as RHS(φ).

Example 2. The dependencies cfd1–cfd3 and pfd1–pfd4 that we have seen in Example 1 can all be expressed as CFD^{p}s. Figure 2 shows some of these CFD^{p}s: φ_1 (for FD cfd$_2$), φ_2 (for CFD cfd$_3$), φ_3 (for pfd$_2$), and φ_4 (for pfd$_4$). □

Semantics. Consider CFD^p $\varphi = (R : X \rightarrow Y,\ T_p)$, where $T_p = \{t_{p1}, \ldots, t_{pk}\}$.

A data tuple t of R is said to *match* LHS(φ), denoted by $t[X] \asymp T_p[X]$, if *for each tuple t_{pi} in T_p and each attribute A in X*, either (a) $t_{pi}[A]$ is the wildcard '_' (which matches any value in dom(A)), or (b) $t[A]$ op a if $t_{pi}[A]$ is 'op a', where the operator op ($=, \neq, <, \leq, >$ or $\geq$) is interpreted by its standard semantics. Similarly, the notion that t *matches* RHS(φ) is defined, denoted by $t[Y] \asymp T_p[Y]$.

Intuitively, each pattern tuple t_{pi} specifies a condition via $t_{pi}[X]$, and $t[X] \asymp T_p[X]$ if $t[X]$ satisfies the *conjunction* of all these conditions. Similarly, $t[Y] \asymp T_p[Y]$ if $t[Y]$ matches all the patterns specified by $t_{pi}[Y]$ for all t_{pi} in T_p.

(1) $\varphi_1 = $ tax (state $\rightarrow$ rate, T_1)

state	rate
T_1: | - | - |

(2) $\varphi_2 = $ item (sale $\rightarrow$ shipping, T_2)

sale	shipping
T_2: | = T | = 0 |

(3) $\varphi_3 = $ item (sale, price $\rightarrow$ shipping, T_3)

sale	price	shipping
= F	> 20	= 6
T_3: | = F | ≤ 40 | = 6 |

(4) CFD^P $\varphi_4 = $ item (sale $\rightarrow$ price, T_4)

sale	price
= T	≥ 2.99
T_4: | = T | < 9.99 |

Fig. 2. Example CFD^{P}s

An instance I of R *satisfies* the CFD^P φ, denoted by $I \models \varphi$, if for *each pair* of tuples t_1, t_2 in the instance I, if $t_1[X] = t_2[X] \asymp T_p[X]$, then $t_1[Y] = t_2[Y] \asymp T_p[Y]$. That is, if $t_1[X]$ and $t_2[X]$ are equal and in addition, they both match the pattern tableau $T_p[X]$, then $t_1[Y]$ and $t_2[Y]$ must also be equal to each other and they both match the pattern tableau $T_p[Y]$.

Observe that φ is imposed only on the subset of tuples in I that match LHS(φ), rather than on the entire I. For all tuples t_1, t_2 in this subset, if $t_1[X] = t_2[X]$, then (a) $t_1[Y] = t_2[Y]$, *i.e.*, the semantics of the embedded FDs is enforced; and (b) $t_1[Y] \asymp T_p[Y]$, which assures that the *constants* in $t_1[Y]$ match the *constants* in $t_{pi}[Y]$ for all t_{pi} in T_p. Note that here tuples t_1 and t_2 can be the same.

An instance I of R satisfies a set Σ of CFD^{P}s, denoted by $I \models \Sigma$, if $I \models \varphi$ for each CFD^P φ in Σ.

Example 3. The instance D_0 of Fig. 1 satisfies φ_1 and φ_2 of Fig. 2, but neither φ_3 nor φ_4. Indeed, tuple t_3 violates (*i.e.*, does not satisfy) φ_3, since $t_3[\text{sale}] = $ 'F' and $20 < t_3[\text{price}] \leq 40$, but $t_3[\text{shipping}]$ is 20 instead of 6. Note that t_3 matches LHS(φ_3) since it satisfies the condition specified by the *conjunction* of the pattern tuples in T_3. Similarly, t_1 violates φ_4, since $t_1[\text{sale}] = $ 'T' but $t_1[\text{price}] > 9.99$. Observe that while it takes two tuples to violate a standard FD, a single tuple may violate a CFD^P. □

Special Cases. (1) A standard FD $X \rightarrow Y$ [1] can be expressed as a CFD $(X \rightarrow Y, T_p)$ in which T_p contains a single tuple consisting of '_' only, without constants. (2) A CFD $(X \rightarrow Y, T_p)$ [14] with $T_p = \{t_{p1}, \ldots, t_{pk}\}$ can be expressed as a set $\{\varphi_1, \ldots, \varphi_k\}$ of CFD^{P}s such that for $i \in [1, k]$, $\varphi_i = (X \rightarrow Y, T_{pi})$, where T_{pi} contains a single pattern tuple t_{pi} of T_p, with equality (=) only. For example, φ_1 and φ_2 in Fig. 2 are CFD^{P}s representing FD cfd2 and CFD cfd3 in Example 1, respectively. Note that all data quality rules in [10,18] can be expressed as CFD^{P}s.

3   Incorporating Built-in Predicates into CINDs

Along the same lines as CFD^{P}s, we next define CIND^{P}s, also referred to as *conditional inclusion dependencies*. Consider two relation schemas R_1 and R_2.

Syntax. A CIND^P ψ is a pair $(R_1[X; X_p] \subseteq R_2[Y; Y_p], T_p)$, where (1) X, X_p and Y, Y_p are lists of attributes in attr(R_1) and attr(R_2), respectively; (2) $R_1[X] \subseteq R_2[Y]$ is a standard IND, referred to as the IND *embedded* in ψ; and (3) T_p is a tableau, called the *pattern tableau* of ψ defined over attributes $X_p \cup Y_p$,

(1) $\psi_1 = ($item $[$state; type$] \subseteq$ tax $[$state; nil$], \; T_1)$,

(2) $\psi_2 = ($item $[$state; type, state$] \subseteq$ tax $[$state; rate$], \; T_2)$

$$T_1: \frac{\text{type} \; \| \; \text{nil}}{\neq \text{art} \; \| \;}$$

$$T_2: \frac{\text{type} \; | \; \text{state} \; \| \; \text{rate}}{\neq \text{art} \; | = \text{DL} \; \| = 0}$$

Fig. 3. Example CIND^{p}s

and for each A in X_p or Y_p and each pattern tuple $t_p \in T_p$, $t_p[A]$ is either an unnamed variable '_' that draws values from $\mathsf{dom}(A)$, or 'op a', where op is one of $=, \neq, <, \leq, >, \geq$ and 'a' is a constant in $\mathsf{dom}(A)$.

We denote $X \cup X_p$ as $\mathsf{LHS}(\psi)$ and $Y \cup Y_p$ as $\mathsf{RHS}(\psi)$, and separate the X_p and Y_p attributes in a pattern tuple with '$\|$'. We use nil to denote an *empty* list.

Example 4. Figure 3 shows two example CIND^{p}s: ψ_1 expresses pind_1 of Example 1, and ψ_2 refines ψ_1 by stating that for any item tuple t_1, if its type is not art and its state is DL, then there must be a tax tuple t_2 such that its state is DL and rate is 0, *i.e.*, ψ_2 assures that the sale tax rate in Delaware is 0. □

Semantics. Consider CIND^p $\psi = (R_1[X; \; X_p] \subseteq R_2[Y; \; Y_p], \; T_p)$. An instance (I_1, I_2) of (R_1, R_2) *satisfies* the CIND^p ψ, denoted by $(I_1, I_2) \models \psi$, iff for *each* tuple $t_1 \in I_1$, if $t_1[X_p] \asymp T_p[X_p]$, then there *exists* a tuple $t_2 \in I_2$ such that $t_1[X] = t_2[Y]$ and moreover, $t_2[Y_p] \asymp T_p[Y_p]$.

That is, if $t_1[X_p]$ matches the pattern tableau $T_p[X_p]$, then ψ requires the existence of t_2 such that (1) $t_1[X] = t_2[Y]$ as required by the standard IND embedded in ψ; and (2) $t_2[Y_p]$ must match the pattern tableau $T_p[Y_p]$. In other words, ψ is "conditional" since its embedded IND is applied only to the subset of tuples in I_1 that match $T_p[X_p]$, and moreover, the pattern $T_p[Y_p]$ is enforced on the tuples in I_2 that match those tuples in I_1. As remarked in Section 2, the pattern tableau T_p specifies the *conjunction* of patterns of all tuples in T_p.

Example 5. The instance D_0 of item and tax in Fig. 1 violates CIND^p ψ_1. Indeed, tuple t_1 in item *matches* $\mathsf{LHS}(\psi_1)$ since $t_1[$type$] \neq$ 'art', but there is no tuple t in tax such that $t[$state$] = t_1[$state$] =$ 'WA'. In contrast, D_0 satisfies ψ_2. □

We say that a database D satisfies a set Σ of CINDs, denoted by $D \models \Sigma$, if $D \models \varphi$ for each $\varphi \in \Sigma$.

Safe CIND^{p}s. We say a CIND^p $(R_1[X; \; X_p] \subseteq R_2[Y; \; Y_p], \; T_p)$ is *unsafe* if there exist pattern tuples t_p, t'_p in T_p such that either (a) there exists $B \in Y_p$, such that $t_p[B]$ and $t'_p[B]$ are not satisfiable when taken together, or (b) there exist $C \in Y, A \in X$ such that A corresponds to B in the IND and $t_p[C]$ and $t'_p[A]$ are not satisfiable when taken together; *e.g.*, $t_p[$price$] = 9.99$ and $t'_p[$price$] \geq 19.99$.

Obviously unsafe CIND^{p}s do not make sense: there exist no nonempty database that satisfies unsafe CIND^{p}s. It takes $O(|T_p|^2)$-time in the size $|T_p|$ of T_p to decide whether a CIND^p is unsafe. Thus in the sequel we consider safe CIND^p only.

Special Cases. Observe that (1) a standard CIND $(R_1[X] \subseteq R_2[Y])$ can be expressed as a CIND^p $(R_1[X; \; \text{nil}] \subseteq R_2[Y; \; \text{nil}], \; T_p)$ such that T_p is simply a *empty* set; and (2) a CIND $(R_1[X;\ X_p] \subseteq R_2[Y;\ Y_p],\ T_p)$ with $T_p = \{t_{p1}, \ldots, t_{pk}\}$ can be expressed as a set $\{\psi_1, \ldots, \psi_k\}$ of CIND^{p}s, where for $i \in [1, k]$, $\psi_i = (R_1[X;\ X_p] \subseteq R_2[Y;\ Y_p],\ T_{pi})$ such that T_{pi} consists of a single pattern tuple t_{pi} of T_p defined in terms of equality (=) only.

4   Reasoning about CFD^{p}s and CIND^{p}s

The satisfiability problem and the implication problem are the two central technical questions associated with any dependency languages. In this section we investigate these problems for CFD^{p}s and CIND^{p}s, separately and taken together.

4.1   The Satisfiability Analysis

The satisfiability problem is to determine, given a set Σ of constraints, whether there exists a *nonempty* database that satisfies Σ.

The satisfiability analysis of conditional dependencies is not only of theoretical interest, but is also important in practice. Indeed, when CFD^{p}s and CIND^{p}s are used as data quality rules, this analysis helps one check whether the rules make sense themselves. The need for this is particularly evident when the rules are manually designed or discovered from various datasets [10,18,15].

The Satisfiability Analysis of CFD^{p}s. Given any FDs, one does not need to worry about their satisfiability since any set of FDs is always satisfiable. However, as observed in [14], for a set Σ of CFDs on a relational schema R, there may not exist a *nonempty* instance I of R such that $I \models \Sigma$. As CFDs are a special case of CFD^{p}s, the same problem exists when it comes to CFD^{p}s.

Example 6. Consider CFD^p $\varphi = (R : A \to B,\ T_p)$ such that $T_p = \{(\_ \parallel = a), (\_ \parallel \neq a)\}$. Then there exists no *nonempty* instance I of R that satisfies φ. Indeed, for any tuple t of R, φ requires that both $t[B] = a$ and $t[B] \neq a$.   □

This problem is already NP-complete for CFDs [14]. Below we show that it has the same complexity for CFD^{p}s despite their increased expressive power.

Proposition 1. *The satisfiability problem for CFD^{p}s is NP-complete.*   □

Proof sketch: The lower bound follows from the NP-hardness of their CFDs counterparts [14], since CFDs are a special case of CFD^{p}s. The upper bound is verified by presenting an NP algorithm that, given a set Σ of CFD^{p}s defined on a relation schema R, determines whether Σ is satisfiable.   □

It is known [14] that the satisfiability problem for CFDs is in PTIME when the CFDs considered are defined over attributes that have an infinite domain, *i.e.,* in the absence of finite domain attributes. However, this is no longer the case for CFD^{p}s. This tells us that the increased expressive power of CFD^{p}s does take a toll in this special case. It should be remarked that while the proof of Proposition 1 is an extension of its counterpart in [14], the result below is new.

Theorem 2. *In the absence of finite domain attributes, the satisfiability problem for CFD^{p}s remains NP-complete.*   □

Proof sketch: The problem is in NP by Proposition 1. Its NP-hardness is shown by reduction from the 3SAT problem, which is NP-complete (cf. [17]).   □

The Satisfiability Analysis of CIND^{p}s. Like FDs, one can specify arbitrary INDs or CINDs without worrying about their satisfiability. Below we show that CIND^{p}s also have this property, by extending the proof of its counterpart in [8].

Proposition 3. *Any set Σ of* CIND^{p}s *is always satisfiable.*   □

Proof sketch: Given a set Σ of CIND^{p}s over a database schema $\mathcal{R}$, one can always construct a *nonempty* instance D of $\mathcal{R}$ such that $D \models \Sigma$.   □

The Satisfiability Analysis of CFD^{p}s and CIND^{p}s. The satisfiability problem for CFDs and CINDs taken together is undecidable [8]. Since CFD^{p}s and CIND^{p}s subsume CFDs and CINDs, respectively, from these we immediately have:

Corollary 4. *The satisfiability problem for* CFD^{p}s *and* CIND^{p}s *is undecidable.*□

4.2   The Implication Analysis

The implication problem is to determine, given a set Σ of dependencies and another dependency ϕ, whether or not Σ entails ϕ, denoted by $\Sigma \models \phi$. That is, whether or not for all databases D, if $D \models \Sigma$ then $D \models \phi$.

The implication analysis helps us remove redundant data quality rules, and thus improve the performance of error detection and repairing based on the rules.

Example 7. The CFD^{p}s of Fig. 2 imply CFD^{p}s $\varphi =$ item (sale, price $\rightarrow$ shipping, T), where T consists of a single pattern tuple (sale =‘F’, price $= 30 \parallel$ shipping $= 6$). Thus in the presence of the CFD^{p}s of Fig. 2, φ is redundant.   □

The Implication Analysis of CFD^{p}s. We first show that the implication problem for CFD^{p}s retains the same complexity as their CFDs counterpart. The result below is verified by extending the proof of its counterpart in [14].

Proposition 5. *The implication problem for* CFD^{p}s *is* coNP-*complete.*   □

Proof sketch: The lower bound follows from the coNP-hardness of their CFDs counterpart [14], since CFDs are a special case of CFD^{p}s. The coNP upper bound is verified by presenting an NP algorithm for its complement problem, *i.e.*, the problem for determining whether $\Sigma \not\models \varphi$.   □

Similar to the satisfiability analysis, it is known [14] that the implication analysis of CFDs is in PTIME when the CFDs are defined only with attributes that have an infinite domain. Analogous to Theorem 2, the result below shows that this is no longer the case for CFD^{p}s, which does not find a counterpart in [14].

Theorem 6. *In the absence of finite domain attributes, the implication problem for* CFD^{p}s *remains* coNP-*complete.*   □

Proof sketch: It is in coNP by Proposition 5. The coNP-hardness is shown by reduction from the 3SAT problem to its complement problem, *i.e.*, the problem for determining whether $\Sigma \not\models \varphi$.   □

Table 1. Summary of complexity results

Σ	General setting		Infinite domain only	
	Satisfiability	Implication	Satisfiability	Implication
CFDs [14]	NP-complete	coNP-complete	PTIME	PTIME
CFD^{p}s	NP-complete	coNP-complete	NP-complete	coNP-complete
CINDs [8]	$O(1)$	EXPTIME-complete	$O(1)$	PSPACE-complete
CIND^{p}s	$O(1)$	EXPTIME-complete	$O(1)$	EXPTIME-complete
CFDs + CINDs [8]	undecidable	undecidable	undecidable	undecidable
CFD^{p}s + CIND^{p}s	undecidable	undecidable	undecidable	undecidable

The Implication Analysis of CIND^{p}s. We next show that CIND^{p}s do not make their implication analysis harder. This is verified by extending the proof of their CINDs counterpart given in [8].

Proposition 7. *The implication problem for* CIND^{p}s *is* EXPTIME-*complete.*   □

Proof sketch: The implication problem for CINDs is EXPTIME-hard [8]. The lower bound carries over to CIND^{p}s since CIND^{p}s subsume CINDs. The EXPTIME upper bound is shown by presenting an EXPTIME algorithm that, given a set $\Sigma \cup \{\psi\}$ of CIND^{p}s over a database schema $\mathcal{R}$, determines whether $\Sigma \models \psi$.   □

It is known [8] that the implication problem is PSPACE-complete for CINDs defined with infinite-domain attributes. Similar to Theorem 6, below we present a new result showing that this no longer holds for CIND^{p}s.

Theorem 8. *In the absence of finite domain attributes, the implication problem for* CIND^{p}s *remains* EXPTIME-*complete.*   □

Proof sketch: The EXPTIME upper bound follows from Proposition 7. The EXPTIME-hardness is shown by reduction from the implication problem for CINDs in the general setting, in which finite-domain attributes may be present; the latter is known to be EXPTIME-complete [8].   □

The Implication Analysis of CFD^{p}s and CIND^{p}s. When CFD^{p}s and CIND^{p}s are taken together, their implication analysis is beyond reach in practice. This is not surprising since the implication problem for FDs and INDs is already undecidable [1]. Since CFD^{p}s and CIND^{p}s subsume FDs and INDs, respectively, from the undecidability result for FDs and INDs, the corollary below follows immediately.

Corollary 9. *The implication problem for* CFD^{p}s *and* CIND^{p}s *is undecidable.*   □

Summary. The complexity bounds for reasoning about CFD^{p}s and CIND^{p}s are summarized in Table 1. To give a complete picture we also include in Table 1 the complexity bounds for the static analyses of CFDs and CINDs, taken from [14,8]. The results shown in Table 1 tell us the following.

(a) Despite the increased expressive power, CFD^{p}s and CIND^{p}s do not complicate the static analyses: the satisfiability and implication problems for CFD^{p}s and CIND^{p}s have the same complexity bounds as their counterparts for CFDs and CINDs, taken separately or together.

(b) In the special case when CFD^{p}s and CIND^{p}s are defined with infinite-domain attributes only, however, the static analyses of CFD^{p}s and CIND^{p}s do not get simpler, as opposed to their counterparts for CFDs and CINDs. That is, in this special case the increased expressive power of CFD^{p}s and CIND^{p}s comes at a price.

5   Validation of CFD^{p}s and CIND^{p}s

If CFD^{p}s and CIND^{p}s are to be used as data quality rules, the first question we have to settle is how to effectively detect errors and inconsistencies as violations of these dependencies, by leveraging functionality supported by commercial DBMS. More specifically, consider a database schema $\mathcal{R} = (R_1, \ldots, R_n)$, where R_i is a relation schema for $i \in [1, n]$. The error detection problem is stated as follows.

The *error detection problem* is to find, given a set Σ of CFD^{p}s and CIND^{p}s defined on $\mathcal{R}$, and a database instance $D = (I_1, \ldots, I_n)$ of $\mathcal{R}$ as input, the subset $(I_1', \ldots, I_n')$ of D such that for each $i \in [1, n]$, $I_i' \subseteq I_i$ and each tuple in I_i' violates at least one CFD^p or CIND^p in Σ. We denote the set as $\mathsf{vio}(D, \Sigma)$, referred to it as *the violation set* of D w.r.t. Σ.

In this section we develop SQL-based techniques for error detection based on CFD^{p}s and CIND^{p}s. The main result of the section is as follows.

Theorem 10. *Given a set Σ of CFD^{p}s and CIND^{p}s defined on $\mathcal{R}$ and a database instance D of $\mathcal{R}$, where $\mathcal{R} = (R_1, \ldots, R_n)$, a set of SQL queries can be automatically generated such that (a) the collection of the answers to the SQL queries in D is $\mathsf{vio}(D, \Sigma)$, (b) the number and size of the set of SQL queries depend only on the number n of relations and their arities in $\mathcal{R}$, regardless of Σ.*   □

We next present the main techniques for the query generation method. Let $\Sigma^i_{\mathsf{cfd^p}}$ be the set of all CFD^{p}s in Σ defined on the same relation schema R_i, and $\Sigma^{(i,j)}_{\mathsf{cind^p}}$ the set of all CIND^{p}s in Σ from R_i to R_j, for $i, j \in [1, n]$. We show the following. (a) The violation set $\mathsf{vio}(D, \Sigma^i_{\mathsf{cfd^p}})$ can be computed by *two* SQL queries. (b) Similarly, $\mathsf{vio}(D, \Sigma^{(i,j)}_{\mathsf{cind^p}})$ can be computed by a *single* SQL query. (c) These SQL queries encode pattern tableaux of CFD^{p}s (CIND^{p}s) with data tables, and hence their sizes are independent of Σ. From these Theorem 10 follows immediately.

5.1   Encoding CFD^{p}s and CIND^{p}s with Data Tables

We first show the following, by extending the encoding of [14,7]. (a) The pattern tableaux of all CFD^{p}s in $\Sigma^i_{\mathsf{cfd^p}}$ can be encoded with *three data tables*, and (b) the pattern tableaux of all CIND^{p}s in $\Sigma^{(i,j)}_{\mathsf{cind^p}}$ can be represented as *four data tables*, no matter how many dependencies are in the sets and how large they are.

Encoding CFD^{p}s. We encode all pattern tableaux in $\Sigma^i_{\mathsf{cfd^p}}$ with three tables enc_L, enc_R and $\mathsf{enc}_{\neq}$, where enc_L (resp. enc_R) encodes the non-negation $(=, <, \leq, >, \geq)$ patterns in LHS (resp. RHS), and $\mathsf{enc}_{\neq}$ encodes those negation $(\neq)$ patterns. More specifically, we associate a unique id cid with each CFD^{p}s in $\Sigma^i_{\mathsf{cfd^p}}$, and let enc_L consist of the following attributes: (a) cid, (b) each attribute

(1) encL				
cid	sale	price	price>	price<
2	T	null	null	null
3	F	-	20	40
4	T	null	null	null

(2) encR				
cid	shipping	price	price>	price<
2	0	null	null	null
3	6	null	null	null
4	null	-	2.99	9.99

(3) enc≠		
cid	pos	att	val

Fig. 4. Encoding example of CFD^{P}s

A appearing in the LHS of some CFD^{P}s in $\Sigma^i_{\text{cfd}^P}$, and (b) its four companion attributes $A_>$, $A_\geq$, $A_<$, and $A_\leq$. That is, for each attribute, there are five columns in encL, one for each non-negation operator. Similarly, encR is defined. We use an enc≠ tuple to encode a pattern $A \neq c$ in a CFD^P, consisting of cid, att, pos, and val, encoding the CFD^P id, the attribute A, the position ('LHS' or 'RHS'), and the constant c, respectively. Note that the arity of encL (encR) is bounded by $5*|R_i|+1$, where $|R_i|$ is the arity of R_i, and the arity of enc≠ is 4.

Before we populate these tables, let us first describe a preferred form of CFD^{P}s that would simplify the analysis to be given. Consider a CFD^P $\varphi = R(X \to Y, T_p)$. If φ is not satisfiable we can simply drop it from Σ. Otherwise it is equivalent to a CFD^P $\varphi' = R(X \to Y, T'_p)$ such that for any pattern tuples t_p, t'_p in T'_p and for any attribute A in $X \cup Y$, (a) if $t_p[A]$ is op a and $t'_p[A]$ is op b, where op is not $\neq$, then $a = b$, (b) if $t_p[A]$ is '_' then so is $t'_p[A]$. That is, for each non-negation op (resp. _), there is a *unique* constant a such that $t_p[A] = $ 'op a' (resp. $t_p[A] = $ _) is the only op (resp. _) pattern appearing in the A column of T'_p. We refer to $t_p[A]$ as $T'_p(\text{op}, A)$ (resp. $T'_p(\_, A)$), and consider w.l.o.g. CFD^{P}s of this form only. Note that there are possibly multiple $t_p[A] \neq c$ patterns in T'_p,

We populate encL, encR and enc≠ as follows. For each CFD^P $\varphi = R(X \to Y, T_p)$ in $\Sigma^i_{\text{cfd}^P}$, we generate a distinct cid id$_\varphi$ for it, and do the following.

– Add a tuple t_1 to encL such that (a) $t[\text{cid}] = $ id$_\varphi$; (b) for each $A \in X$, $t[A] = \_$ if $T'_p(\_, A)$ is '_', and for each non-negation predicate op, $t[A_{\text{op}}] = $ 'a' if $T'_p(\text{op}, A)$ is 'op a'; (c) we let $t[B] = $ 'null' for all other attributes B in encL.

– Similarly add a tuple t_2 to encR for attributes in Y.

– For each attribute $A \in X \cup Y$ and each $\neq a$ pattern in $T_p[A]$, add a tuple t to enc≠ such that $t[\text{cid}] = $ id$_\varphi$, $t[\text{att}] = $ 'A', $t[\text{val}] = $ 'a', and $t[\text{pos}] = $ 'LHS' (resp. $t[\text{pos}] = $ 'RHS') if attribute A appears in X (resp. Y).

Example 8. Recall from Fig. 2 CFD^{P}s φ_2, φ_3 and φ_4 defined on relation item. The three CFD^{P}s are encoded with tables shown in Fig. 4: (a) encL consists of attributes: cid, sale, price, price> and price<; (b) encR consists of cid, shipping, price, price> and price<; those attributes in a table with only 'null' pattern values do not contribute to error detection, and are thus omitted; (c) enc≠ is empty since all these CFD^{P}s have no negation patterns. One can easily reconstruct these CFD^{P}s from tables encL, encR and enc≠ by collating tuples based on cid. □

Encoding CIND^{P}s. All CIND^{P}s in $\Sigma^{(i,j)}_{\text{cind}^P}$ can be encoded with four tables enc, encL, encR and enc≠. Here encL (resp. encR) and enc≠ encode non-negation

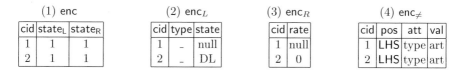

(1) enc		
cid	state$_L$	state$_R$
1	1	1
2	1	1

(2) enc$_L$		
cid	type	state
1	_	null
2	_	DL

(3) enc$_R$	
cid	rate
1	null
2	0

(4) enc$_{\neq}$			
cid	pos	att	val
1	LHS	type	art
2	LHS	type	art

Fig. 5. Encoding example of CIND^{P}s

patterns on relation R_i (resp. R_j) and negation patterns on relations R_i or R_j, respectively, along the same lines as their counterparts for CFD^{P}s. We use enc to encode the INDs *embedded* in CIND^{P}s, which consists of the following attributes: (1) cid representing the id of a CIND^P, and (2) those X attributes of R_i and Y attributes of R_j appearing in some CIND^{P}s in $\Sigma_{\text{cind}^P}^{(i,j)}$. Note that the number of attributes in enc is bounded by $|R_i| + |R_j| + 1$, where $|R_i|$ is the arity of R_i.

For each CIND^P $\psi = (R_i[A_1 \ldots A_m;\ X_p] \subseteq R_j[B_1 \ldots B_m;\ Y_p],\ T_p)$ in $\Sigma_{\text{cind}^P}^{(i,j)}$, we generate a distinct cid id$_\psi$ for it, and do the following.

- Add tuples t_1 and t_2 to enc$_L$ and enc$_R$ based on attributes X_p and Y_p, respectively, along the same lines as their CFD^P counterpart.
- Add tuples to enc$_{\neq}$ in the same way as their CFD^P counterparts.
- Add tuple t to enc such that $t[\text{cid}] = \text{id}_\psi$. For each $k \in [1, m]$, let $t[A_k] = t[B_k] = k$, and $t[A] = $ 'null' for the rest attributes A of enc.

Example 9. Figure 4 shows the coding of CIND^{P}s ψ_1 and ψ_2 given in Fig. 3. We use state$_L$ and state$_R$ in enc to denote the occurrences of attribute state in item and tax, respectively. In tables enc$_L$ and enc$_R$, attributes with only 'null' patterns are omitted, for the same reason as for CFD^{P}s mentioned above. □

Putting these together, it is easy to verify that at most $O(n^2)$ data tables are needed to encode dependencies in Σ, regardless of the size of Σ. Recall that n is the number of relations in database $\mathcal{R}$.

5.2   SQL-Based Detection Methods

We next show how to generate SQL queries based on the encoding above. For each $i \in [1, n]$, we generate *two* SQL queries that, when evaluated on the I_i table of D, find vio$(D, \Sigma_{\text{cfd}^P}^i)$. Similarly, for each $i, j \in [1, n]$, we generate a *single* SQL query $Q_{(i,j)}$ that, when evaluated on (I_i, I_j) of D, returns vio$(D, \Sigma_{\text{cind}^P}^{(i,j)})$. Putting these query answers together, we get vio(D, Σ), the violation set of D w.r.t. Σ.

Below we show how the SQL query $Q_{(i,j)}$ is generated for validating CIND^{P}s in $\Sigma_{\text{cind}^P}^{(i,j)}$), which has not been studied by previous work. For the lack of space we omit the generation of detection queries for CFD^{P}s, which is an extension of the SQL techniques for CFDs discussed in [14,7].

The query $Q_{(i,j)}$ for the validation of $\Sigma_{\text{cind}^P}^{(i,j)}$ is given as follows, which capitalizes on the data tables enc, enc$_L$, enc$_R$ and enc$_{\neq}$ that encode CIND^{P}s in $\Sigma_{\text{cind}^P}^{(i,j)}$.

> **select** R_i.*
> **from** R_i, enc$_L$ L, enc$_{\neq}$ N

where $R_i.X \asymp L$ **and** $R_i.X \asymp N$ **and not exists** (
    select $R_j.*$
    from R_j, enc H, enc$_R$ R, enc$_{\neq}$ N
    where $R_i.X = R_j.Y$ **and** L.cid $= R$.cid **and** L.cid $= H$.cid **and**
        $R_j.Y \asymp R$ **and** $R_j.Y \asymp N$)

Here (1) $X = \{A_1, \ldots, A_{m1}\}$ and $Y = \{B_1, \ldots, B_{m2}\}$ are the sets of attributes of R_i and R_j appearing in $\Sigma_{\text{cind}^p}^{(i,j)}$, respectively; (2) $R_i.X \asymp L$ is the conjunction of

$L.A_k$ **is null or** $R_i.A_k = L.A_k$ **or** $(L.A_k = \text{'\_'}$
    and $(L.A_{i_>}$ **is null or** $R_i.A_k > L.A_{i_>})$ **and** $(L.A_{i_{\geq}}$ **is null or** $R_i.A_k \geq L.A_{k_{\geq}})$
    and $(L.A_{k_<}$ **is null or** $R_i.A_k < L.A_{k_<})$ **and** $(L.A_{i_{\leq}}$ **is null or** $R_i.A_k \leq L.A_{i_{\leq}}))$

for $k \in [1, m_1]$; (3) $R_j.Y \asymp R$ is defined similarly for attributes in Y; (4) $R_i.X \asymp N$ is a shorthand for the conjunction below, for $k \in [1, m_1]$:

not exists (select $*$ **from** N **where** L.cid $= N$.cid **and** N.pos $=$ 'LHS' **and**
        N.att $=$ 'A_k' **and** $R_i.A_k = N$.val);

(5) $R_j.Y \asymp N$ is defined similarly, but with N.pos $=$ 'RHS' ; (6) $R_i.X = R_j.Y$ represents the following: for each A_k ($k \in [1, m_1]$) and each B_l ($l \in [1, m_2]$), $(H.A_k$ **is null or** $H.B_l$ **is null or** $H.B_l \neq H.A_k$ **or** $R_i.A_k = R_j.B_l)$.

Intuitively, (1) $R_i.X \asymp L$ **and** $R_i.X \asymp N$ ensure that the R_i tuples selected match the LHS patterns of some CIND^{p}s in $\Sigma_{\text{cind}^p}^{(i,j)}$; (2) $R_j.Y \asymp R$ **and** $R_j.Y \asymp N$ check the corresponding RHS patterns of these CIND^{p}s on R_j tuples; (3) $R_i.X = R_j.Y$ enforces the *embedded* INDs; (4) L.cid $= R$.cid **and** L.cid $= H$.cid assure that the LHS and RHS patterns in the same CIND^p are correctly collated; and (5) **not exists** in Q ensures that the R_i tuples selected violate CIND^{p}s in $\Sigma_{\text{cind}^p}^{(i,j)}$.

Example 10. Using the coding of Fig. 5, an SQL query Q for checking CIND^{p}s ψ_1 and ψ_2 of Fig. 3 is given as follows:

select $R_1.*$ **from** item R_1, enc$_L$ L, enc$_{\neq}$ N
where $(L$.type **is null or** R_1.type $= L$.type **or** L.type $=$ '_') **and not exist** (
  select $*$ **from** N
  where N.cid $= L$.cid **and** N.pos $=$ 'LHS' **and** N.att $=$ 'type')
  and $(L$.state **is null or** R_1.state $= L$.state **or** L.state $=$ '_') **and not exist** (
  select $*$ **from** N
  where N.cid $= L$.cid **and** N.pos $=$ 'LHS' **and** N.att $=$ 'state' **and** R_1.state $=N$.val)

  and not exists (
  select $R_2.*$ **from** tax R_2, enc H, enc$_R$ R
  where $(H$.state$_L$ **is null or** H.state$_R$ **is null or** H.state$_L$! $= H$.state$_R$ **or**
    R_2.state $= R_1$.state) **and** L.cid $= H$.cid **and** L.cid $= R$.cid **and**
    $(R$.rate **is null or** R_2.rate $= R$.rate **or** R.rate $=$ '_') **and not exist** (
    select $*$ **from** N
    where N.cid $= R$.cid **and** N.pos$=$ 'RHS' **and** N.att $=$ 'rate' **and** R_2.rate $=N$.val))

The SQL queries generated for error detection can be simplified as follows. As shown in Example 10, when checking patterns imposed by enc, enc$_L$ or enc$_R$, the queries need not consider attributes A if $t[A]$ is 'null' for each tuple t in the table. Similarly, if an attribute A does not appear in any tuple in $enc_{\neq}$, the queries need not check A either. From this, it follows that we do not even need to generate those attributes with only 'null' patterns for data tables enc, enc_L or enc_R when encoding CIND^{p}s or CFD^{p}s. $\square$

6   Related Work

Constraint-based data cleaning was introduced in [2], which proposed to use dependencies, *e.g.*, FDs, INDs and denial constraints, to detect and repair errors in real-life data (see, *e.g.*, [11] for a comprehensive survey). As an extension of traditional FDs, CFDs were developed in [14], for improving the quality of data. It was shown in [14] that the satisfiability and implication problems for CFDs are NP-complete and coNP-complete, respectively. Along the same lines, CINDs were proposed in [8] to extend INDs. It was shown [8] that the satisfiability and implication problems for CINDs are in constant time and EXPTIME-complete, respectively. SQL techniques were developed in [14] to detect errors by using CFDs, but have not been studied for CINDs. This work extends the static analyses of conditional dependencies of [14,8], and has established several new complexity results, notably in the absence of finite-domain attributes (*e.g.*, Theorems 2, 6, 8). In addition, it is the first work to develop SQL-based techniques for checking violations of CINDs and violations of CFD^{p}s and CIND^{p}s taken together.

Extensions of CFDs have been proposed to support disjunction and negation [7], cardinality constraints and synonym rules [9], and to specify patterns in terms of value ranges [18]. While CFD^{p}s are more powerful than the extension of [18], they cannot express disjunctions [7], cardinality constraints and synonym rules [9]. To our knowledge no extensions of CINDs have been studied. This work is the first full treatment of extensions of CFDs and CINDs by incorporating built-in predicates ($\neq, <, \leq, >, \geq$), from static analyses to error detection.

Methods have been developed for discovering CFDs [10,18,15] and for repairing data based on either CFDs [13], traditional FDs and INDs taken together [5], denial constraints [4,12], or aggregate constraints [16]. We defer the treatment of these topics for CFD^{p}s and CIND^{p}s to future work.

A variety of extensions of FDs and INDs have been studied for specifying constraint databases and constraint logic programs [3,6,19,20]. While the languages of [3,19] cannot express CFDs, constraint-generating dependencies (CGDs) of [3] and constrained tuple-generating dependencies (CTGDs) of [20] can express CFD^{p}s, and CTGDs can also express CIND^{p}s. The increased expressive power of CTGDs comes at the price of a higher complexity: both their satisfiability and implication problems are undecidable. Built-in predicates and arbitrary constraints are supported by CGDs, for which it is not clear whether effective SQL queries can be developed to detect errors. It is worth mentioning that Theorems 2 and 6 of this work provide lower bounds for the consistency and implication analyses of CGDs, by using patterns with built-in predicates only.

7   Conclusions

We have proposed CFD^{p}s and CIND^{p}s, which further extend CFDs and CINDs, respectively, by allowing patterns on data values to be expressed in terms of $\neq, <, \leq, >$ and $\geq$ predicates. We have shown that CFD^{p}s and CIND^{p}s are more powerful than CFDs and CINDs for detecting errors in real-life data. In addition, the satisfiability and implication problems for CFD^{p}s and CIND^{p}s have the same complexity bounds as their counterparts for CFDs and CINDs, respectively. We have also provided automated methods to generate SQL queries for detecting errors based on CFD^{p}s and CIND^{p}s. These provide commercial DBMS with an immediate capability to capture errors commonly found in real-world data.

One topic for future work is to develop a dependency language that is capable of expressing various extensions of CFDs (*e.g.*,CFD^{p}s, eCFDs [7] and CFD^{c}s [9]), without increasing the complexity of static analyses. Second, we are developing effective algorithms for discovering CFD^{p}s and CIND^{p}s, along the same lines as [10,18,15]. Third, we plan to extend the methods of [5,13] to repair data based on CFD^{p}s and CIND^{p}s, instead of using CFDs [13], traditional FDs and INDs [5], denial constraints [4,12], and aggregate constraints [16].

Acknowledgments. Fan and Ma are supported in part by EPSRC E029213/1. Fan is a Yangtze River Scholar at Harbin Institute of Technology. Chen is sponsored by Chinese grants 006BAH03B03 and 2007AA01Z159.

References

1. Abiteboul, S., Hull, R., Vianu, V.: Foundations of Databases. Addison-Wesley, Reading (1995)
2. Arenas, M., Bertossi, L.E., Chomicki, J.: Consistent query answers in inconsistent databases. In: PODS (1999)
3. Baudinet, M., Chomicki, J., Wolper, P.: Constraint-Generating Dependencies. J. Comput. Syst. Sci. 59(1), 94–115 (1999)
4. Bertossi, L.E., Bravo, L., Franconi, E., Lopatenko, A.: The complexity and approximation of fixing numerical attributes in databases under integrity constraints. Inf. Syst. 33(4-5), 407–434 (2008)
5. Bohannon, P., Fan, W., Flaster, M., Rastogi, R.: A cost-based model and effective heuristic for repairing constraints by value modification. In: SIGMOD (2005)
6. Bra, P.D., Paredaens, J.: Conditional dependencies for horizontal decompositions. In: ICALP (1983)
7. Bravo, L., Fan, W., Geerts, F., Ma, S.: Increasing the expressivity of conditional functional dependencies without extra complexity. In: ICDE (2008)
8. Bravo, L., Fan, W., Ma, S.: Extending dependencies with conditions. In: VLDB (2007)
9. Chen, W., Fan, W., Ma, S.: Incorporating cardinality constraints and synonym rules into conditional functional dependencies. IPL 109(14), 783–789 (2009)
10. Chiang, F., Miller, R.J.: Discovering data quality rules. In: VLDB (2008)
11. Chomicki, J.: Consistent query answering: Five easy pieces. In: ICDT (2007)
12. Chomicki, J., Marcinkowski, J.: Minimal-change integrity maintenance using tuple deletions. Inf. Comput. 197(1-2), 90–121 (2005)

13. Cong, G., Fan, W., Geerts, F., Jia, X., Ma, S.: Improving data quality: Consistency and accuracy. In: VLDB (2007)
14. Fan, W., Geerts, F., Jia, X., Kementsietsidis, A.: Conditional functional dependencies for capturing data inconsistencies. TODS 33(2) (2008)
15. Fan, W., Geerts, F., Lakshmanan, L.V., Xiong, M.: Discovering conditional functional dependencies. In: ICDE (2009)
16. Flesca, S., Furfaro, F., Parisi, F.: Consistent query answers on numerical databases under aggregate constraints. In: Bierman, G., Koch, C. (eds.) DBPL 2005. LNCS, vol. 3774, pp. 279–294. Springer, Heidelberg (2005)
17. Garey, M., Johnson, D.: Computers and Intractability: A Guide to the Theory of NP-Completeness. W. H. Freeman and Company, New York (1979)
18. Golab, L., Karloff, H.J., Korn, F., Srivastava, D., Yu, B.: On generating near-optimal tableaux for conditional functional dependencies. In: VLDB (2008)
19. Maher, M.J.: Constrained dependencies. TCS 173(1), 113–149 (1997)
20. Maher, M.J., Srivastava, D.: Chasing Constrained Tuple-Generating Dependencies. In: PODS (1996)
21. Van der Meyden, R.: The complexity of querying indefinite data about linearly ordered domains. JCSS 54(1) (1997)

Discovering Sentinel Rules
for Business Intelligence

Morten Middelfart[1] and Torben Bach Pedersen[2]

[1] TARGIT A/S*
Aalborgvej 94, 9800 Hjørring, Denmark
morton@targit.com
[2] Aalborg University – Department of Computer Science
Selma Lagerløfs Vej 300, 9220 Aalborg Ø, Denmark
tbp@cs.aau.dk

Abstract. This paper proposes the concept of *sentinel rules* for multi-dimensional data that warns users when measure data concerning the external environment changes. For instance, a surge in negative blogging about a company could trigger a sentinel rule warning that revenue will decrease within two months, so a new course of action can be taken. Hereby, we expand the window of opportunity for organizations and facilitate successful navigation even though the world behaves chaotically. Since sentinel rules are at the schema level as opposed to the data level, and operate on data *changes* as opposed to absolute data values, we are able to discover strong and useful sentinel rules that would otherwise be hidden when using sequential pattern mining or correlation techniques. We present a method for sentinel rule discovery and an implementation of this method that scales linearly on large data volumes.

1   Introduction

The Computer Aided Leadership and Management (CALM) concept copes with the challenges facing managers that operate in a world of chaos due to the globalization of commerce and connectivity [7]; in this chaotic world, the ability to continuously act is far more crucial for success than the ability to long-term forecast. The idea in CALM is to take the Observation-Orientation-Decision-Action (OODA) loop (originally pioneered by "Top Gun"[1] fighter pilot John Boyd in the 1950s [6]), and integrate business intelligence (BI) technologies to drastically increase the speed with which a user in an organization cycles through the OODA loop. Using CALM, any organization can be described as a set of OODA loops that are continuously cycled to fulfill one or more Key Performance Indicators (KPI's). One way to improve the speed from observation to action is to expand the "horizon" by allowing the user to see data from the external environment,

---

* This work was supported by TARGIT A/S. The experiments were assisted by lead-programmers Jan Krogsgaard and Jakob Andersen.

[1] Colonel John Boyd was fighter instructor at Nellis Air Force Base in Nevada, the predecessor of U.S. Navy Fighter Weapons School nicknamed "Top Gun".

S.S. Bhowmick, J. Küng, and R. Wagner (Eds.): DEXA 2009, LNCS 5690, pp. 592–602, 2009.

and not only for the internal performance of the organization. Another way is to give early warnings when factors change that might influence the user's KPI's, e.g., revenue. Placing "sentinels" at the outskirts of the data available harness both ways of improving reaction time and thus organizational competitiveness.

A sentinel rule is a relationship between two measures, A and B, in an OLAP database where we know, that a change in measure A at one point in time affects measure B within a certain *warning period*, with a certain confidence. If such a relationship exists, we call measure A the *source measure*, and measure B the *target measure*. Usually, the target measure is, or contributes to, a KPI. The source measure ideally represents the external environment, or is as close to the external environment as possible.

The idea that some actions or incidents are interlinked has been well explored in *association rules* [1]. In general, association rule mining seeks to find co-occurrence patterns within *absolute data values*, whereas our solution works on the *relative changes in data*. In addition, association rule mining typically works on *categorical data*, i.e., dimension values, whereas our solution works on *numerical data* such as measure values. *Sequential pattern mining* introduces a sequence in which actions or incidents take place, with the intention of predicting one action or incident based on knowing another one. This adds to the complexity of association rules which makes the Apriori approach even more costly [3], thus new approaches to improving the performance of mining sequential patterns have emerged [5,9,12,10], and have also given rise to *multi-dimensional pattern mining* [11]. Sequential pattern mining allows a time period to pass between the premise and the consequent in the rule, but it remains focused on co-occurrence patterns within absolute data values for categorical data. Furthermore, our solution generates rules at the *schema level*, as opposed to the *data level*, using a contradiction elimination process. The combination of schema-level rules based on relative changes in data allows us to generate fewer, more general, rules that cannot be found with neither association rules nor sequential pattern mining. In the full paper [8] we specifically demonstrate why sequential pattern mining does not find any meaningful rules in our running example presented in Section 2.

Other approaches to interpreting the behavior of data sequences are various regression [2] and correlation [4,13] techniques which attempt to describe a functional relationship between one measure and another. In comparison, we can say that sentinel rules are a set of *"micro-predictions"* that are complementary to regression and correlation techniques. Sentinel rules are useful for discovering strong relationships between a smaller subset within a dataset, and thus they are useful for detecting warnings whenever changes (that would otherwise go unnoticed) in a relevant source measure occur. In addition, regression and correlation techniques do not support uni-directional relationships such as our solution. Regression and correlation based techniques, on the other hand, are useful for describing the overall trends within a dataset. In the full paper [8], we specifically provide a concrete, realistic example where nothing useful is found using correlation, while sentinel rules *do* find an important relationship within a subset of the data.

The novel contributions in this paper include the sentinel rule concept, and an algorithm that discover sentinel rules on multi-dimensional data that scales linearly on large volumes of synthetic and real-world data. We give a formal definition of sentinel rules, and we define the indication concept for rules and for source and target measures. In this context, we provide a contradiction elimination process that allows us to generate more general rules that are easy to interpret. We also provide a useful notation for sentinel rules. We believe that we are the first to propose the concept of sentinel rules, and to provide an algorithm and implementation for discovering them.

The next section presents the formal definition, Section 3 presents an algorithm for discovering sentinel rules. Section 4 presents implementation and experiments, and Section 5 presents conclusion and future work.

2   Problem Definition

Running Example: Imagine a company that sells products world-wide, and that we, in addition to the traditional financial figures such as revenue, *Rev*, have been monitoring the environment outside our organization and collected that information in three measures. The measure *NBlgs* represents the number of times an entry is written on a blog where a user is venting a negative opinion about our company or products. The measure *CstPrb* represents the number of times a customer contacts our company with a problem related to our products. The measure *WHts* represents the number of hits on our website, and this figure has been cleansed in order to represent human contact exclusively, eliminating traffic by robots etc. In Table 1 we see a subset from our database, representing each quarter in year 2007 across three geographical regions. It should be noted that a subset like Table 1 can easily be extracted from a multi-dimensional database, i.e., if the desired data are the base level of the database no processing is needed, if the desired levels are higher than the base level, the data might or might not be preaggregated. However, both extraction and aggregation are typically basic built

Table 1. Example dataset

T: Time	D_2: Region	M_1: NBlgs	M_2: CstPrb	M_3: WHts	M_4: Rev
2007-Q1	Asia	20	50	1,000	10,000
2007-Q2	Asia	21	45	1,500	9,000
2007-Q3	Asia	17	33	2,000	11,000
2007-Q4	Asia	15	34	2,500	13,000
2007-Q1	EU	30	41	3,000	20,000
2007-Q2	EU	25	36	3,500	25,000
2007-Q3	EU	22	46	4,000	28,000
2007-Q4	EU	19	37	4,500	35,000
2007-Q1	USA	29	60	5,000	50,000
2007-Q2	USA	35	70	5,500	55,000
2007-Q3	USA	40	72	6,500	45,000
2007-Q4	USA	39	73	7,500	40,000

in functions of any multi-dimensional database. The three measures: NBlgs, CstPrb and WHts, representing the external environment around our company, have been presented along with the internal measure, Rev, representing our Revenue.

We are interested in discovering whether we can use any of the external measures to predict a future impact on the internal Revenue measure; in other words we are looking for sentinel rules where one of the measures $M_1...M_3$ can give us an early warning about changes to M_4. To distinguish between which measures are "causing" the other, we call the measures $M_1...M_3$ *source measures* and the measure M_4 is called the *target measure*.

Formal Definition: Let C be a multi-dimensional data cube containing a set of dimensions: $D = \{D_1, D_2...D_n\}$ and a set of measures: $M = \{M_1, M_2...M_p\}$. We denote the members of the dimensions in D by $d_1, d_2...d_n$ and we denote the corresponding *measure values* for any combination of dimension members by $m_1, m_2...m_p$. A measure value is a function, M_i, that returns the value of a given measure corresponding to the dimension members it is presented with. We will now provide a series of definitions that define a source measure, A, is a *sentinel* for a target measure, B, i.e., a guarded watchtower from which we monitor A in order to know about changes ahead of time to B. The sentinel rule between A and B is denoted $A \rightsquigarrow B$. We assume, without loss of generality, that there is only one time dimension, T, in C, and that $T = D_1$, and subsequently $t = d_1$. The formal definitions are listed in Formulae 1 to 10 below.

A fact, f, in C is defined in Formula (1). Given a fact f, the measure M_i is a function $M_i(t, d_2, d_3...d_n) = m_i$. The "dimension" part of f, $(t, d_2, d_3...d_n)$, is called a cell. The *shifting* of a fact f, f', is a fact with the same non-time dimension values $(d_2...d_n)$ as f, but for time period $t + o$, if it exists in C, i.e., a

$$f = (t, d_2, d_3...d_n, m_1, m_2...m_p) \tag{1}$$

$$Shift(C, f, o) = f' = (t + o, d_2, d_3...d_n, m'_1, m'_2...m'_p) \text{ if } f' \in C \tag{2}$$

$$Diff(C, f, o) = (t, d_2, d_3...d_n, \frac{m'_1 - m_1}{m_1}, \frac{m'_2 - m_2}{m_2}...\frac{m'_p - m_p}{m_p})$$
$$\text{where } f = (t, d_2, d_3...d_n, m_1, m_2...m_p) \wedge f \in C \wedge$$
$$f' = Shift(C, f, o) = (t + o, d_2, d_3...d_n, m'_1, m'_2...m'_p) \wedge f' \in C \tag{3}$$

$$x = (t, d_2, d_3...d_n, \frac{m'_1 - m_1}{m_1}, ..., \frac{m'_i - m_i}{m_i}, ..., \frac{m'_p - m_p}{m_p}) \wedge |\frac{m'_i - m_i}{m_i}| \geq \alpha \tag{4}$$

$$ST(C, o, w) = \{(Diff(C, f, o), Diff(C, Shift(C, f, w), o))|f \in C\} \tag{5}$$

$$ContraRule(IndRule) = IndPrem(IndRule) \rightarrow \overline{IndCons(IndRule)} \tag{6}$$

$$ElimSupp(IndRule) = IndSupp_{IndRule} - IndSupp_{ContraRule(IndRule)} \tag{7}$$

$$MaxRule = \begin{cases} \{IndRule_i \mid IndRule_i \in A \rightarrow B \ \wedge \ ElimSupp(IndRule_i) > 0\} \\ \qquad \text{if } IndSupp_{A \rightarrow B} >= IndSupp_{A \rightarrow inv(B)}, \\ \{IndRule_i \mid IndRule_i \in A \rightarrow inv(B) \ \wedge \ ElimSupp(IndRule_i) > 0\} \\ \qquad \text{if } IndSupp_{A \rightarrow B} < IndSupp_{A \rightarrow inv(B)}. \end{cases} \tag{8}$$

$$SentSupp_{A \rightsquigarrow B} = \begin{cases} IndSupp_{A\blacktriangle} & \text{if } A\blacktriangledown \rightarrow B* \notin MaxRule, \\ IndSupp_{A\blacktriangledown} & \text{if } A\blacktriangle \rightarrow B* \notin MaxRule, \\ IndSupp_{A\blacktriangle} + IndSupp_{A\blacktriangledown} & \text{otherwise.} \end{cases} \tag{9}$$

$$Conf_{A \rightsquigarrow B} = \frac{\sum_{IndRule_i \in MaxRule} ElimSupp(IndRule_i)}{SentSupp_{A \rightsquigarrow B}} \tag{10}$$

period of o members later on the time dimension. We denote the *offset*, o, and define Shift as shown in Formula (2). Since we are interested in the change in data, we introduce the *measure difference function, Diff*. With *Diff*, we find the relative changes to each of the measures during the time period specified by the offset. *Diff* is defined as shown in Formula (3). Given a threshold, α, we say that $x \in Diff(C, f, o)$ is an *indication* on a measure, M_i, if Formula (4) holds. We say that an indication on M_i, x, is *positive*, denoted $M_i\blacktriangle$, when $\frac{m_i' - m_i}{m_i} > 0$ and consequently that an indication, x, is *negative*, denoted $M_i\blacktriangledown$, when $\frac{m_i' - m_i}{m_i} < 0$. We define a wildcard, $*$, meaning that M_i* can be either $M_i\blacktriangle$ or $M_i\blacktriangledown$.

In our running example, when assessing whether a relationship exists, we are not concerned with minor fluctuations, so we define a threshold of 10%, meaning that a measure has to change at least 10% up or down in order to be of interest. Furthermore, given the dataset we have, we are interested in seeing the changes that occur over *quarters* as presented in Table 1. This means that we set the threshold $\alpha = 10\%$ and then the offset $o = 1$ *Quarter*. In Table 2, we have calculated the changes from each quarter to the next and subjected each change to an evaluation against the threshold of 10% change. We denote positive indications by $\blacktriangle$ and subsequently negative by $\blacktriangledown$, if a change is less than 10% in either direction it is deemed "neutral". Please note that since we are dealing with *changes* between periods, we naturally get one less row for each region.

A Source-Target Set, ST as seen in Formula (5), is defined as paired indications of changes over time, where the source and target measures have been shifted with the offset, o. The target measures have additionally been shifted with a *warning period*, w, which is the timeframe after which we should expect a change on a target measure, after an indication on a source measure has occurred. We say that $(x, x') \in ST(C, o, w)$ *supports* the *indication rule* $A\blacktriangle \rightarrow B\blacktriangle$ if x is an indication of $A\blacktriangle$ and x' is an indication of $B\blacktriangle$. In this case, we also say that x supports $A\blacktriangle$ and x' supports $B\blacktriangle$. The *support* of an indication rule is the number of $(x, x') \in ST(C, o, w)$ which supports the rule. The support of indication rules $A\blacktriangledown \rightarrow B\blacktriangledown$, $A\blacktriangle \rightarrow B\blacktriangledown$ and $A\blacktriangledown \rightarrow B\blacktriangle$ as well as the support for indications $A\blacktriangledown$ and $B\blacktriangledown$ are defined similarly. We denote the support of an indication and an indication rule by *IndSupp* followed by the name of the indication or indication rule, respectively, e.g., $IndSupp_{A\blacktriangle}$ and $IndSupp_{A\blacktriangle \rightarrow B\blacktriangle}$.

A sentinel rule is an *unambiguous* relationship between A and B, thus we must first eliminate contradicting indication rules, if such exist, before we have

Table 2. Indications between quarters

T: Time	D_2: Region	M_1: NBlgs	M_2: CstPrb	M_3: WHts	M_4: Rev
'07:Q1→Q2	Asia	neutral	$M_2\blacktriangledown$	$M_3\blacktriangle$	$M_4\blacktriangledown$
'07:Q2→Q3	Asia	$M_1\blacktriangledown$	$M_2\blacktriangledown$	$M_3\blacktriangle$	$M_4\blacktriangle$
'07:Q3→Q4	Asia	$M_1\blacktriangledown$	neutral	$M_3\blacktriangle$	$M_4\blacktriangle$
'07:Q1→Q2	EU	$M_1\blacktriangledown$	$M_2\blacktriangledown$	$M_3\blacktriangle$	$M_4\blacktriangle$
'07:Q2→Q3	EU	$M_1\blacktriangledown$	$M_2\blacktriangle$	$M_3\blacktriangle$	$M_4\blacktriangle$
'07:Q3→Q4	EU	$M_1\blacktriangledown$	$M_2\blacktriangledown$	$M_3\blacktriangle$	$M_4\blacktriangle$
'07:Q1→Q2	USA	$M_1\blacktriangle$	$M_2\blacktriangle$	$M_3\blacktriangle$	$M_4\blacktriangle$
'07:Q2→Q3	USA	$M_1\blacktriangle$	neutral	$M_3\blacktriangle$	$M_4\blacktriangledown$
'07:Q3→Q4	USA	neutral	neutral	$M_3\blacktriangle$	$M_4\blacktriangledown$

Table 3. Target and source measure comparison

T: Time	D_2: Region	M_1: NBlgs	M_2: CstPrb	M_3: WHts	M_4': Rev
'07:Q1→Q2	Asia	neutral	$M_2\blacktriangledown$	$M_3\blacktriangle$	$M_4'\blacktriangle$
'07:Q2→Q3	Asia	$M_1\blacktriangledown$	$M_2\blacktriangledown$	$M_3\blacktriangle$	$M_4'\blacktriangle$
'07:Q1→Q2	EU	$M_1\blacktriangledown$	$M_2\blacktriangledown$	$M_3\blacktriangle$	$M_4'\blacktriangle$
'07:Q2→Q3	EU	$M_1\blacktriangledown$	$M_2\blacktriangle$	$M_3\blacktriangle$	$M_4'\blacktriangle$
'07:Q1→Q2	USA	$M_1\blacktriangle$	$M_2\blacktriangle$	$M_3\blacktriangle$	$M_4'\blacktriangledown$
'07:Q2→Q3	USA	$M_1\blacktriangle$	neutral	$M_3\blacktriangle$	$M_4'\blacktriangledown$

a sentinel rule. We refer to this process as the *contradiction elimination process*, and we use it to remove indication rules with the same premise, but a different consequent, and vice versa, e.g., if both $A\blacktriangle \rightarrow B\blacktriangle$ and $A\blacktriangle \rightarrow B\blacktriangledown$ or if both $A\blacktriangle \rightarrow B\blacktriangle$ and $A\blacktriangledown \rightarrow B\blacktriangle$ are supported. To eliminate such contradictions, we pair the indication rules in two sets that do not contradict each other, and we denote these sets by $A \rightarrow B$ and $A \rightarrow inv(B)$, as follows: $A \rightarrow B = \{A\blacktriangle \rightarrow B\blacktriangle, A\blacktriangledown \rightarrow B\blacktriangledown\}$ and $A \rightarrow inv(B) = \{A\blacktriangle \rightarrow B\blacktriangledown, A\blacktriangledown \rightarrow B\blacktriangle\}$. Here *inv* indicates an inverted relationship between the indications on A and B, e.g. if $A\blacktriangle$ then $B\blacktriangledown$, and vice versa. For the purpose of being able to deduct the support of the indication rule(s) we eliminate, we define functions for returning the premise and the consequent indication, *IndPrem* and *IndCons*, from an indication rule $A\blacktriangle \rightarrow B\blacktriangle$ as follows: $IndPrem(A\blacktriangle \rightarrow B\blacktriangle) = A\blacktriangle$ and $IndCons(A\blacktriangle \rightarrow B\blacktriangle) = B\blacktriangle$. Furthermore, we define the complement of an indication as follows: $\overline{A\blacktriangle} = A\blacktriangledown$ and $\overline{A\blacktriangledown} = A\blacktriangle$. We can now define a contradicting indication rule as a function, *ContraRule*, for an indication rule, *IndRule*, as shown in Formula (6). The support after elimination, *ElimSupp*, of an indication rule, *IndRule*, where the support of the contradicting indication rule, *ContraRule(IndRule)*, has been eliminated is calculated as shown in Formula (7).

MaxRule is the set of indication rule(s), $IndRule_i$, in the set ($A \rightarrow B$ or $A \rightarrow inv(B)$) with the highest *IndSupp* and where $ElimSupp(IndRule_i) > 0$. With *MaxRule*, we have identified the best indication rule(s) for a sentinel rule that represents an unambiguous relationship between A and B, i.e., the non-contradicting indication rules with the highest *ElimSupp*. In other words, we have eliminated the *contradicting* indication rules where the premise contradicts the consequent, as well as the *orthogonal* indication rules where different premises have the same consequent. If the *MaxRule* set consists of only one indication rule, we refer to the sentinel rule based on this as a *uni-directional* rule.

We denote the support of a sentinel rule by *SentSupp*, followed by the name of the sentinel rule, e.g., $SentSupp_{A \rightsquigarrow B}$. For a potential sentinel rule, $A \rightsquigarrow B$, we define *SentSupp* as the sum of the support of source measure indications for the indication rule(s) contained in the sentinel rule as shown in Formula (9). We note the difference between the support of an indication rule, *IndSupp*, and a sentinel rule, *SentSupp*. Specifically, when calculating the support of a sentinel rule, $SentSupp_{A \rightsquigarrow B}$, we only consider the support of indications on the source measure (the premise), $A\blacktriangle$ and $A\blacktriangledown$. With indication rules, both indications on the source and target measure needs to occur. The reason is, that the consequential support of indications on the target measure, $B\blacktriangle$ or $B\blacktriangledown$, is taken into consideration when calculating the confidence of the sentinel rule in Formula (10). In the case of a uni-directional rule (the two first cases) we only consider the support of indications on the source measure that have the same direction as the one indication rule in *MaxRule*; this is done in order not to penalize otherwise good uni-directional rules in terms of confidence. We denote confidence by *Conf*, and define the confidence for a sentinel rule, $A \rightsquigarrow B$, as follows shown in Formula (10). The minimum threshold for *SentSupp* is denoted β, and the minimum threshold for *Conf* is denoted γ. With these definitions, we say that a sentinel rule, $A \rightsquigarrow B$, with an offset, o, and a warning period, w, exists in C when $SentSupp_{A \rightsquigarrow B} \geq \beta$ and $Conf_{A \rightsquigarrow B} \geq \gamma$. α, β, γ, o, and w are provided by the user, and typically set iteratively based on the user's experience.

To express sentinel rules with easy readability, we use $\rightsquigarrow$ to show that there is a sentinel rule between a source measure, A, and a target measure, B. In the case, where a bi-directional rule represents an inverted relationship between the source and the target measure, we add *inv* to the target measure. In the case where the rule is uni-directional, we add ▲ or ▼ to both the source and the target measure to express the direction of the sentinel rule.

In our running example, we limit ourselves to investigating whether sentinel rules exist between any of the source measures $M_1...M_3$ and the target measure M_4. We now need to compare the changes in $M_1...M_3$ to changes in M_4 at a later time. In this case, we choose the timeframe of 1 quarter again, meaning that warning period $w = 1$ *Quarter*. In Table 3, we show the comparison between the source measure indications and the target measure indication one quarter later. The measure M_4 is basically moved one line up -or as shown in Table 3; one quarter back. This means that all source measures for Asia changing 2007: Q2→Q3 as shown in the left column are now compared on the same line, within the same row, to the change on the target measure, M_4, for Asia changing 2007: Q3→Q4 and so on. The shift of M_4 shown in the row with data for the period one quarter earlier is denoted M_4'. Please note that we get one less row for each geographical region since we are looking at changes between the periods.

Based on Table 3, we count the support for each combination of indication changes, the indication rules, for each potential sentinel rule; in addition, we can count the support of the relationship overall, basically the support means counting all rows that do not have a "neutral" change on the source measure since we

Indication rule and sentinel rule support

Table 4. $M_1 \rightsquigarrow M_4$

M_1	M_4'	$IndSupp$
M_1▲	M_4'▲	0
M_1▼	M_4'▼	0
M_1▲	M_4'▼	2
M_1▼	M_4'▲	3
$SentSupp_{M_1 \rightsquigarrow M_4} = 5$		

Table 5. $M_2 \rightsquigarrow M_4$

M_2	M_4'	$IndSupp$
M_2▲	M_4'▲	1
M_2▼	M_4'▼	0
M_2▲	M_4'▼	1
M_2▼	M_4'▲	3
$SentSupp_{M_2 \rightsquigarrow M_4} = 3$		

Table 6. $M_3 \rightsquigarrow M_4$

M_3	M_4'	$IndSupp$
M_3▲	M_4'▲	4
M_3▼	M_4'▼	0
M_3▲	M_4'▼	2
M_3▼	M_4'▲	0
$SentSupp_{M_3 \rightsquigarrow M_4} = 6$		

Table 7. $M_1 \rightsquigarrow M_4$

M_1	M_4'	$ElimSupp$
M_1▲	M_4'▼	2
M_1▼	M_4'▲	3
$SentSupp_{M_1 \rightsquigarrow M_4} = 5$		
$Conf_{M_1 \rightsquigarrow M_4} = \frac{5}{5} = 100\%$		
Conformance: ok		

Table 8. $M_2 \rightsquigarrow M_4$

M_2	M_4'	$ElimSupp$
M_2▼	M_4'▲	3
$SentSupp_{M_2 \rightsquigarrow M_4} = 3$		
$Conf_{M_2 \rightsquigarrow M_4} = \frac{3}{5} = 100\%$		
Conformance: ok		

Table 9. $M_3 \rightsquigarrow M_4$

M_3	M_4'	$ElimSupp$
M_3▲	M_4'▲	2
$SentSupp_{M_3 \rightsquigarrow M_4} = 6$		
$Conf_{M_3 \rightsquigarrow M_4} = \frac{2}{6} = 33\%$		
Conformance: failed		

define indications as being either positive or negative. For example, we see summarized in Table 4, that the indication rule $M_1\blacktriangledown \rightarrow M_4'\blacktriangle$ is supported 3 times in the dataset shown in Table 3; we say that the indication rule $M_1\blacktriangledown \rightarrow M_4'\blacktriangle$ has a support of 3, and the sentinel rule $M_1 \rightsquigarrow M_4$ has a support of all indication rule combinations which in this case is 5. Table 4 through 6 lists the indication rules for each potential sentinel rule with their respective support (Formula (9)).

As mentioned earlier, the ideal sentinel rule describes changes bi-directionally so that it can "predict" both positive and negative changes on the target measure. However, the relationship also needs to be non-contradictory in order to be useful as a sentinel rule. To do this, we eliminate the indications that contradict each other as described in Formulae 6 and 7. In Table 5 we find the a uni-directional rule where the two contradicting indication rules have equal support, thus we disregard these indications completely (Formula (9)) and therefore $SentSupp_{M_2 \rightsquigarrow M_4}=3$. In Table 6 the contradiction elimination process does not eliminate both indication rules, it reduces the two indication rules to one and decreases $ElimSupp$ (Formula (7)) in the calculation of confidence. In order to identify the best sentinel rules, we set the thresholds $\beta = 3$ and $\gamma = 60\%$. Table 7 through 9 show the sentinel rules from our running example and their respective conformance to the thresholds we have set. As seen in Table 8 and 9, we end up having uni-directional sentinel rules, since the indication rules $M_2\blacktriangle \rightarrow M_4'\blacktriangle$ and $M_2\blacktriangle \rightarrow M_4'\blacktriangledown$, as shown in Table 5, contradict each other and have equal support. In addition, the indication rules $M_3\blacktriangle \rightarrow M_4'\blacktriangle$ and $M_3\blacktriangle \rightarrow M_4'\blacktriangledown$ contradict each other in Table 6. Of these, $M_3\blacktriangle \rightarrow M_4'\blacktriangle$ is strongest and "wins" the elimination process (Formula (8)) as seen in Table 9.

In this example, we have found two sentinel rules that can provide our company with an early warning. If we monitor the changes to M_1, the number of negative blog entries, we will know one quarter in advance whether to expect an increase or a decrease in M_4 Revenue. If we monitor the number of times a customer contacts our company with a problem related to our products, M_2, we will know one quarter ahead whether to expect an increase in Revenue. Using the notation defined earlier in this section, we can express the rules found in our running example as follows: $NBlgs \rightsquigarrow inv(Rev)$ and $CstPrb\blacktriangledown \rightsquigarrow Rev\blacktriangle$.

3   The FindSentinels Algorithm

The following algorithm has been implemented in SQL on a Microsoft SQL Server 2005. The actual SQL code can found in the full paper [8]. We assume without loss of generality that of the p measures in the dataset, C, $M_1...M_{p-1}$ are the source measures and M_p is the target measure.

Step 1 creates a temporary table where each unique value of t, is sorted in ascending order and assigned an integer, Id, growing by 1 for each t. This temporary table will allow us to select values of t for comparison with a given distance in periods, regardless of the format of the period field, t, in the database. To optimize performance, we create an index on the period table. By joining 4 copies of each of the original dataset and the period table (one for each of the periods: $t, t + o, t + w$, and $t + w + o$), we create a Source-Target set (Formula (5)) and calculate indications (Formulae (3) and (4)) for our selected p-1 source measures and one target measure. We calculate these indications for each cell (dimension combination) in the dataset, and return -1, 0, or 1 depending on whether the indication is negative, neutral or positive against the threshold α.

Step 2 counts the number of positive and negative indications on the source measure, and for each of these source measure indications, it summarizes the indications on the target measure. Since the indications are expressed as -1, 0 or 1, our contradiction elimination process can be carried out using sum.

Table 10. The *FinalResult* table

SentinelRule	SentSupp	Conf
NBlgs->inv(Rev)	5	100
CstPrb_dec->Rev_inc	3	100

Step 3 retrieves the potential rules from previous output, meaning that a source measure needs to have at least one indication with a consequential indication on the target measure, i.e., *ElimSupp*<> 0. For each of these rules, we calculate the sum of the support of source measure indications, *SentSupp*, the sum of absolute indications on the target measure, *AbsElimSupp*, as well as *MaxElimSupp* which is max(*ElimSupp*). In addition, we calculate the *Direction* of the relationship between source and target measure where 1 is straightforward and -1 is inverted. The nature of *Direction* also helps us eliminate orthogonal rules since these will always have *Direction=0*. This is true because an orthogonal relationship means that both positive and negative indications on the source measure leads to only one type of indication on the target measure. Finally,

Algorithm FindSentinels

Input: A dataset, C, an offset, o, a warning period, w, a threshold for indications, α, a minimum *SentSupp* threshold, β, and a minimum *Conf* threshold, γ.

Output: Sentinel rules with their respective *SentSupp* and *Conf*.

Method: Sentinel rules are discovered as follows:

1. Scan the dataset C once and retrieve unique values of t into an indexed subset. Use the subset to reference each cell $(t, d_2, \dots, d_n) \in C$ with the corresponding cells for $\{t + o, t + w, t + w + o\} \in C$. Output a Source-Target set (Formula (5)) for each cell, $(t, d_2, \dots, d_n)$, where the indications (Formulae (3) and (4)) on source measures, $M_1 \dots M_{p-1}$, are calculated using $\{t, t + o\}$ and the indications on target measure, M_p, is calculated using $\{t + w, t + w + o\}$.

2. For each positive and negative source measure indication, $M_i Ind$, in the output from Step 1, count the number of source measure indications as $IndSupp_i$ and sum the target measure indications as $ElimSupp_i$.

3. Retrieve from the output from Step 2, each source measure, $M_i \in M_1 \dots M_{p-1}$, where $ElimSupp <> 0$. For each of these source measures, calculate: $SentSupp = \text{sum}(IndSupp)$, $AbsElimSupp = \text{sum}|ElimSupp|$, $MaxElimSupp = \text{max}(ElimSupp)$, $Direction = \text{avg}(\text{sign}(M_i Ind)^* \text{sign}(ElimSupp))$, and $IndRuleCount$ as the number of different indications (positive, negative). Output the rules where $SentSupp >= \beta$ and $Conf = \frac{AbsElimSupp}{SentSupp} >= \gamma$, use $IndRuleCount=2$ to identify bi-directional rules and $Direction$ to describe whether the relationship is straight-forward or inverted. For uni-directional rules ($IndRuleCount = 1$) use the combinations of $Direction$ and sign($MaxElimSupp$) to describe the relationship.

we calculate the number of indication rules, *IndRuleCount*, in the potential sentinel rule. This information is used to distinguish between bi- and uni-directional rules. Using this information, we can now identify the sentinel rules that comply with the criteria of *SentSupp* $>= \beta$ and *Conf* $>= \gamma$. In addition, we can use the values of *IndRuleCount*, *Direction*, and *MaxElimSupp* to describe the sentinel rule in accordance with our notation. We store the output in a table called *FinalResult*.

Upon execution of the algorithm, FindSentinels, with the dataset from our running example as *C*, we get the output table named *FinalResult* as seen in Table 10. We note that the result is similar to that of Tables 7 & 8.

4 Implementation and Experiments

In the full paper [8] we conclude that the FindSentinels algorithm has $\mathcal{O}(n)$ computational complexity, where n is the size of *C*, and the algorithm thus scales linearly. We implemented the algorithm on a Microsoft SQL Server and experimentally validated that our SQL implementation does indeed scale linearly up to 10 million rows of realistic data that was synthetically generated, modeled after our running example.

5 Conclusion and Future Work

We have proposed a novel approach for discovering so-called sentinel rules in a multi-dimensional database for business intelligence. The sentinel rules were generated at schema level, which means that they are more general and cleansed for contradictions, and thus easy to interpret. We provided an algorithm for sentinel discovery that scales linearly on large volumes of data. With regards to novelty, we specifically demonstrated that sentinel rules are different from sequential pattern mining, since sentinel rules operate at the schema level and use a contradiction elimination process to generate fewer, more general rules. Furthermore, we found sentinel rules to be complementary to correlation techniques by discovering strong relationships between a smaller subset within a dataset that would otherwise be "hidden in the average" using correlation techniques.

For future work, the algorithm could be extended with the ability to automatically fit α, β, γ, o, and w, and to seek for rules with multiple source measures In addition, the algorithm could be extended to exploit multi-dimensionality.

References

1. Agrawal, R., Imielinski, T., Swami, A.: Mining association rules between sets of items in large databases. In: Proc. of ACM SIGMOD, pp. 207–216 (1993)
2. Agrawal, R., Lin, K.I., Sawhney, H.S., Shim, K.: Fast similarity search in the presence of noise, scaling, and translation in timeseries databases. In: Proc. of VLDB, pp. 490–501 (1995)

3. Agrawal, R., Srikant, R.: Mining Sequential Patterns. In: Proc. of ICDE, pp. 3–14 (1995)
4. Han, J., Kamber, M.: Data Mining Concepts and Techniques, 2nd edn. Morgan Kaufmann Publishers, San Francisco (2006)
5. Han, J., Pei, J., Mortazavi-Asl, B., Chen, Q., Dayal, U., Hsu, M.: FreeSpan: frequent pattern-projected sequential pattern mining. In: Proc. of KDD, pp. 355–359 (2000)
6. Lind, W.S.: Maneuver Warfare Handbook. Westview Press (1985)
7. Middelfart, M.: CALM: Computer Aided Leadership & Management - How Computers can Unleash the Full Potential of Individuals and Organizations in a World of Chaos and Confusion. iUniverse (2005)
8. Middelfart, M., Pedersen, T.B.: Discovering Sentinel Rules for Business Intelligence. DB Tech Report no. 24, dbtr.cs.aau.dk
9. Pei, J., Han, J., Mortazavi-Asl, B., Pinto, H., Chen, Q., Dayal, U., Hsu, M.C.: PrefixSpan: Mining Sequential Patterns by Prefix-Projected Growth. In: Proc. of ICDE, pp. 215–224 (2001)
10. Pei, J., Han, J., Mortazavi-Asl, B., Wang, J., Pinto, H., Chen, Q., Dayal, U., Hsu, M.: Mining Sequential Patterns by Pattern-Growth: The PrefixSpan Approach. IEEE TKDE 16(11), 1424–1440 (2004)
11. Pinto, H., Han, J., Pei, J., Wang, K., Chen, Q., Dayal, U.: Multi-Dimensional Sequential Pattern Mining. In: Proc. of CIKM, pp. 81–88 (2001)
12. Srikant, R., Agrawal, R.: Mining Sequential Patterns: Generalizations and Performance Improvements. In: Apers, P.M.G., Bouzeghoub, M., Gardarin, G. (eds.) EDBT 1996. LNCS, vol. 1057, pp. 3–17. Springer, Heidelberg (1996)
13. Zhu, Y., Shasha, D.: StatStream: Statistical Monitoring of Thousands of Data Streams in Real Time. In: Proc. of VLDB, pp. 358–369 (2002)
14. Zurawski, R., Jensen, C.S., Pedersen, T.B. (eds.): The Industrial Information Technology Handbook. Multidimensional Databases and OLAP. CRC Press, Boca Raton (2005)

Discovering Trends and Relationships among Rules

Chaohai Chen, Wynne Hsu, and Mong Li Lee

School of Computing, National University of Singapore
{chaohai,whsu,leeml}@comp.nus.edu.sg

Abstract. Data repositories are constantly evolving and techniques are needed to reveal the dynamic behaviors in the data that might be useful to the user. Existing temporal association rules mining algorithms consider time as another dimension and do not describe the behavior of rules over time. In this work, we introduce the notion of trend fragment to facilitate the analysis of relationships among rules. Two algorithms are proposed to find the relationships among rules. Experiment results on both synthetic and real-world datasets indicate that our approach is scalable and effective.

1 Introduction

With the rapid proliferation of data, applying association rule mining to any large dataset would lead to the discovery of thousands of associations, many of which are neither interesting nor useful. In a dynamic environment where changes occur frequently, often over short periods, it is important to discover the evolving trends.

Table 1 lists the association rules discovered for a sample dataset. At first glance, none of the rules seems interesting. However, on closer examination of the rules, we observe that the confidence of the rule "beer $\Rightarrow$ chip" is 20% in 1997, 40% in 1998, and 80% in 1999. In other words, there is an increasing trend in the confidence values of "beer $\Rightarrow$ chip" from 1997 to 1999. This could be a useful piece of information.

Further, when we examine the rules "toothbrush A $\Rightarrow$ toothpaste C" and "toothbrush B $\Rightarrow$ toothpaste C" over each individual year, we observe that the confidence of the rule "toothbrush A $\Rightarrow$ toothpaste C" decreases from 1997 to 1999, that is, it drops from 100% to 80% to 60%. In contrast, the confidence of the rule "toothbrush B $\Rightarrow$ toothpaste C" increases from 60% in 1997 to 80% in 1998 and 100% in 1999. These two rules seem to exhibit a negative correlation. This may indicate that they have a competing relationship. That is, customers who buy toothbrush A or B tend to buy toothpaste C; however, over the years, customers who buy toothbrush B are more likely to buy toothpaste C, whereas customers who buy toothbrush A are less likely to buy toothpaste C. As such, if toothpaste C is the key product and the company wants to increase the sale of toothpaste C, it could produce more toothbrush B rather than A as a promotion for buying toothpaste C.

On the other hand, suppose the confidence of "toothbrush A $\Rightarrow$ toothpaste C" decreases from 60% in 1997 to 50% in 1998 to 40% in 1999, and the confidence of "toothbrush B $\Rightarrow$ toothpaste C" also decreases from 70% in 1997 to 60% in 1998 to 50% in 1999. However, the confidence of "toothbrush A, toothbrush B $\Rightarrow$ toothpaste C"

S.S. Bhowmick, J. Küng, and R. Wagner (Eds.): DEXA 2009, LNCS 5690, pp. 603–610, 2009.

increases from 50% in 1997 to 70% in 1998 to 90% in 1999. The relationship among these three rules could be interesting since it is counter-intuitive. It could indicate that the combined effect of toothbrush A and toothbrush B is opposite to that of toothbrush A and B individually. As such, the company could sell toothbrush A and B together rather than individually if it wants to increase the sell of toothpaste C.

Motivated by the above observations, we investigate the dynamic aspect of association rules. There has been some research to analyze the dynamic behavior of association rules over time and detect emerging pattern or deviation between two consecutive datasets. Baron et al. [1] consider a rule as a time object, and design a model to capture the content, statistical properties and time stamp of rules. Subsequent works [2-4] use the model to monitor the statistical properties of a rule at different time points in order to detect interesting or abnormal changes in a rule. Liu et al. [5] also examine the temporal aspect of an association rule over time to discover trends. Statistical methods such as the chi-square tests are used to analyze the interestingness of an association rule over time, and a rule is classified as a stable rule, or exhibits increasing or decreasing trend. In Chen et al. [6], the authors identify contiguous intervals during which a specific association rule holds, as well as the periodicity of the rule. Dong et al. [7] design an algorithm to discover emerging patterns whose supports increase significantly over two time points.

To date, no work has been done to discover the relationships about the changes of the rules over time. In this paper, we propose four types of relationships among rules based on the correlations in their statistical properties. The contributions of this work are summarized as follows:

1. Propose four types of relationships among rules over time
2. Design algorithms to discover such relationships
3. Verify the efficiency and effectiveness of the proposed approaches with both synthetic and real-world datasets.

Table 1. Association Rules Discovered in a Sample Dataset

Rule	Confidence
beer ⇒ chip	50%
chip ⇒ beer	63%
beer ⇒ toothpaste C	85%
cake ⇒ toothpaste C	77%
chip ⇒ toothpaste C	72%
toothbrush A ⇒ toothpaste C	76%
toothbrush B ⇒ toothpaste C	76%
toothpaste C ⇒ toothbrush A	55%
toothpaste C ⇒ toothbrush B	55%
toothbrush A, toothbrush B ⇒ toothpaste C	66%
toothpaste C => beer	52%
toothbrush A => cake	66%
toothpaste C => chip	78%
toothbrush B => beer	62%
....	....

beer ⇒ chip

Year	Confidence
1997	20%
1998	40%
1999	80%

toothbrush A ⇒ toothpaste C

Year	Confidence
1997	100%
1998	80%
1999	60%

toothbrush B ⇒ toothpaste C

Year	Confidence
1997	60%
1998	80%
1999	100%

2 Preliminaries

We model a rule's confidence over time as a time series, denoted as $\{y_1, y_2, ..., y_n\}$.

Definition 1 (Strict Monotonic Series). A time series $\{y_1, y_2, ..., y_n\}$ is a strict monotonic series if

   a) $y_i - y_{i+1} > 0 \ \forall \ i \in [1, n\text{-}1]$ (monotonic decreasing) or
   b) $y_i - y_{i+1} < 0 \ \forall \ i \in [1, n\text{-}1]$ (monotonic increasing)

Definition 2 (Constant Series). A time series $\{y_1, y_2, ..., y_n\}$ is a constant series if $\forall$ $i \in [1, n\text{-}1]$, $|y_i - y_{i+1}| < \varepsilon$ where $0 < \varepsilon \ll 1$.

Definition 3 (Inconsistent Sub-Series). Given a time series $\{y_1, y_2, ..., y_n\}$, we say that $\{y_i, ..., y_j\}$, $1 \le i < j \le n$, is an inconsistent sub-series in $\{y_1, y_2, ..., y_n\}$ if by removing $\{y_i, ..., y_j\}$, we obtain a strict monotonic or constant time series $\{y_1,..., y_{i-1}, y_{j+1}, ...,y_n\}$.

Definition 4 (Trend Fragment). Suppose $T = \{y_1, y_2, ..., y_n\}$ is a time series with k inconsistent sub-series $S_1, S_2, ..., S_k$. Let $|S_i|$ denote the number of time points in sub-series S_i. We call T a trend fragment if

   a) $|S_i| <$ max_inconsistentLen, $1 \le i \le k$;
   b) $n - \sum_i |S_i| >$ min_fragmentLen

where min_fragmentLen and max_inconsistentLen are user-specified parameters denoting the minimum length of a trend fragment and the maximum length of an inconsistent series respectively.

Based on the definition of trend fragment, we define the relationships among rules over time. These rules have the same consequent C and their relationships are based on the confidence correlations. We use the Pearson correlation coefficient [8] to measure the confidence correlation:

$$\rho_{X,Y} = \frac{E(XY) - E(X)E(Y)}{\sqrt{E(X^2) - E^2(X)}\sqrt{E(Y^2) - E^2(Y)}} \tag{1}$$

where X and Y are two confidence series and E is the expected value operator.

Suppose we have three rules: $R_1: \alpha \Rightarrow C$, $R_2: \beta \Rightarrow C$, $R_3: \alpha \cup \beta \Rightarrow C$ with the same consequent C, $\alpha \not\subset \beta$, $\beta \not\subset \alpha$. Let CS_1, CS_2, CS_3 be the confidence values of R_1, R_2, R_3 over the period $[t_1, t_2]$ in which CS_1, CS_2, CS_3 are trend fragments and let δ be a user-defined tolerance. Depending on the values of ρ_{CS_i, CS_j} we have the following types of relationships.

Definition 5 (Competing Relationship). Suppose CS_1 and CS_2 are monotonic trend fragments. We say $R_1 : \alpha \Rightarrow C$ and $R_2 : \beta \Rightarrow C$ $(\alpha \cap \beta = \phi)$ have a competing relationship in $[t1, t2]$ if $\rho_{CS_1, CS_2} < -1 + \delta$.

Definition 6 (Diverging Relationship). Suppose CS_1, CS_2 and CS_3 are monotonic trend fragments. We say $R_1 : \alpha \Rightarrow C$ and $R_2 : \beta \Rightarrow C$ have a diverging relationship with $R_3: \alpha \cup \beta \Rightarrow C$ in $[t1, t2]$ if

   a) $\rho_{CS_1, CS_2} > 1 - \delta$,
   b) $\rho_{CS_1, CS_3} < -1 + \delta$, or $\rho_{CS_2, CS_3} < -1 + \delta$

Definition 7 (Enhancing Relationship). Suppose CS_1 and CS_3 are monotonic trend fragments while CS_2 is a constant trend fragments. We say $R_1 : \alpha \Rightarrow C$ and $R_2 : \beta \Rightarrow C$ have an enhancing relationship with $R_3 : \alpha \cup \beta \Rightarrow C$ in [t1, t2] if

   a) $\rho_{CS1, CS3} < -1 + \delta$
   b) CS_1 is monotonic decreasing and CS_3 is monotonic increasing

Definition 8 (Alleviating Relationship). Suppose CS_1 and CS_3 are monotonic series while CS_2 is a constant series. We say $R_1 : \alpha \Rightarrow C$ and $R_2 : \beta \Rightarrow C$ have an alleviating relationship with $R_3 : \alpha \cup \beta \Rightarrow C$ in [t1, t2] if

   a) $\rho_{CS1, CS3} < -1 + \delta$
   b) CS_1 is monotonic increasing and CS_3 is monotonic decreasing

3   Proposed Approach

We first partition the original dataset according to some time granularity and mine the association rules for each partition. Next, we scan the confidence series of a rule from left to right and group the values into consistent sub-series. We merge the adjacent sub-series if the gap between the two series is less than max_inconsistentLen and the merged series is strictly monotonic or constant. The merged sub-series, with lengths greater than min_fragmentLen, are identified as trend fragments. Finally, we find the relationships among rules over time.

Algorithm 1. FindRelAmongRules RBF

Input: all rules with trend fragments; Output: relationships among rules
1.   For each pair rules r_i and r_j
2.   Let TFS_i and TFS_j be the sets of trend fragments of r_i and r_j respectively.
3.   For each $<f_i, f_j> \in TFS_i \times TFS_j$ do
4.   If (overlap(f_i, f_j) > minRatio)
5.   corr = calculateCorrelation (f_i, f_j, overlap(f_i, f_j))
6.   If (corr < -1 + δ and r_i, r_j have no common items in the antecedent)
7.   Output competing relationship (r_i, r_j, overlap(f_i, f_j))
8.   Find the combined rule r_k
9.   If (r_k exists)
10.   Let TFS_k be the set of trend fragments of r_k
11.   For each $<f_i, f_j, f_k> \in TFS_i \times TFS_j \times TFS_k$ do
12.   If (overlap(f_i, f_j) > minRatio and overlap(f_i, f_k) > minRatio)
13.   If (both f_i and f_j are not stable)
14.   corr = calculateCorrelation (f_i, f_j, overlap(f_i, f_j))
15.   If (corr > 1 - δ) corr = calculateCorrelation (f_i, f_k, overlap(f_i, f_k))
16.   If (corr < -1 + δ) Output diverging relationship (r_i, r_j, r_k, overlap(f_i, f_j, f_k))
17.   Else if either f_i or f_j is stable but not both
18.   WLOG, suppose f_i is not stable and f_j is stable
19.   corr = calculateCorrelation (f_i, f_k, overlap(f_i, f_k))
20.   If (corr < -1 + δ)
21.   If (f_k is increasing) Output enhancing relationship (r_i, r_j, r_k, overlap(f_i, f_j, f_k))
22.   Else output alleviating relationship (r_i, r_j, r_k, overlap(f_i, f_j, f_k))

Algorithm 2. FindRelInGroups GBF

Input: groups of comparable trend fragments
Output: relationships among rules in each group
1. For each group of comparable trend fragment G
2.   For each $<f_i, f_j> \in G \times G$ do
3.   Let r_i and r_j be the rules associated with f_i and f_j respectively.
4.   If (r_i is a sub-rule of r_j or vice versa)
5.   WLOG, let r_i be a sub-rule of r_j
6.   Find the other sub-rule r_k such that r_i is the combined rule of r_j and r_k
7.   Rename the combined rule as r_c and the two sub-rules as r_i and r_j.
8.   Else find the combined rule r_c of r_i and r_j
9.   Find the corresponding fragment f_c for r_c in G
10.   corr = calculateCorrelation ($f_i, f_j,$ overlap(f_i, f_j))
11.   If (corr $< -1 + \delta$ and r_i, r_j have no common items in the antecedent)
12.   Output competing relationship ($r_i, r_j,$overlap(f_i, f_j))
13.   If (f_c is not stable)
14.   If $f_i,$ and f_j are not stable
15.   corr = calculateCorrelation ($f_i, f_j,$ overlap(f_i, f_j))
16.   If (corr $> 1 - \delta$)
17.   corr = calculateCorrelation ($f_i, f_c,$ overlap(f_i, f_c))
18.   If (corr $< -1 + \delta$)
19.   Output diverging relationship ($r_i, r_j, r_c,$ overlap(f_i, f_j, f_c))
20.   Else if either $f_i,$ or f_j is stable but not both
21.   WLOG, suppose $f_i,$ is not stable and f_j is stable
22.   corr = calculateCorrelation ($f_i, f_c,$ overlap(f_i, f_c))
23.   If (corr $< -1 + \delta$)
24.   If (f_c is increasing) output enhancing relationship ($r_i, r_j, r_c,$ overlap(f_i, f_j, f_c))
25.   Else output alleviating relationship ($r_i, r_j, r_c,$ overlap(f_i, f_j, f_c))

A naïve approach is to perform pair-wise comparisons of the rules by comparing their trend fragments. Two trend fragments are comparable if their time intervals overlap significantly, i.e., the length of the overlap time intervals is greater than min-Ratio. With this, we confine the correlation computation to the overlap time interval of comparable trend fragments.

Definition 9 (Combined Rule). Suppose we have three rules $r_i : \alpha \Rightarrow C$, $r_j : \beta \Rightarrow C$, $r_k : \gamma \Rightarrow C$. If $\alpha \cup \beta = \gamma$, $\alpha \not\subset \beta$ and $\beta \not\subset \alpha$, we say r_k is the combined rule of r_i and r_j.

Definition 10 (Sub-Rule). Given rules $r_i : \alpha \Rightarrow C$, $r_k : \gamma \Rightarrow C$, if $\alpha \subset \gamma$, then r_i is a sub-rule of r_k.

Algorithm 1 (RBF) gives the details to find relationships among rules. For each pair of rules, we find their trend fragments (Line 2). For each pair of comparable trend fragments, we determine their correlation (lines 4-5). If they have a competing relationship, we output the rules and their overlap intervals (lines 6-7). Line 8 finds the combined rule. Correlations are computed to determine whether the rules have a diverging, enhancing or alleviating relationship (line 11-22).

Note that the naive approach requires scanning all the rules even when they do not have any comparable trend fragments. Algorithm 2 (GBF) shows an optimization that groups comparable trend fragments and process only those rules whose trend fragments are in the same group. The grouping of trend fragments proceeds by first sorting the fragments in ascending order of their start times, followed by their end times. Trend fragments with significant overlap of time intervals are placed in the same group. For each pair of comparable trend fragments, we find the corresponding rules as well as their combined rule (lines 2-9). Then we compute the correlation between the rules to determine their relationships (lines 10-25).

4   Performance Study

We implemented the algorithms in C++ and carried out the experiments on a 2.33 GHz PC with 3.25 GB RAM, running Windows XP. We extend the synthetic data generator in [10] to incorporate time and class information.

4.1   Experiments on Synthetic Dataset

We first evaluate the performance of algorithms RBF and GBF on synthetic datasets. to find the relationships among rules. Fig.1 shows the running time of RBF and GBF when the number of rules increases from 1000 to 10 000 and min_ratio is 0.85. We observe that GBF outperforms RBF and is scalable. As the number of rules increases, the running time of RBF increases faster than GBF. This is because RBF performs pair-wise comparisons among rules, while GBF groups comparable fragments and performs pruning to avoid unnecessary comparisons.

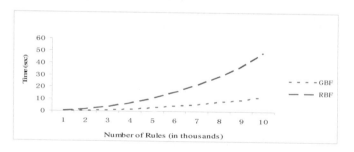

Fig. 1. Running Time of GBF and RBF

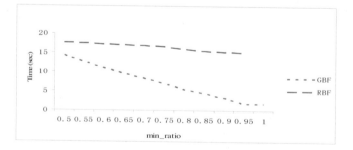

Fig. 2. Varying min_ratio in GBF and RBF

We also evaluate the sensitivity of RBF and GBF when min_ratio varies from 0.55 to 1. The number of rules is fixed at 5000. The performance of RBF and GBF are shown in Fig.2. We observe that GBF is faster than RBF. As the min_ratio increases from 0.55 to 1, the running time of GBF decreases rapidly, while the running time of RBF remains relatively constant. The reason is that when min_ratio is large, many combined rules do not have comparable fragments with the sub-rules and there is no relationship among them. GBF finds pairs of combined rule and its sub-rules only if they have fragments in the same group of comparable fragments. However, RBF finds each pair of combined rule and its sub-rules even when the rules do not have comparable fragments.

4.2 Experiments on Real World Dataset

Finally, we demonstrate the applicability of the algorithms to discover meaningful relationships among rules in a real-world dataset. The dataset is the currency exchange rate dataset from Duke Statistics Data Set Collection [9]. It contains the prices of 12 currencies relative to the US dollar from 10/9/1986 to 8/9/1996.

We transform the changes of the currency prices for each day into a transaction as follows. We compare the price of each currency for each day with its price for the previous day. Each increase or decrease of the price is associated with a corresponding Integer item in the transaction.

Table 2 shows the number of relationships found when we target the increase of five different currencies. Table 3 gives a sample of the relationships discovered.

Table 2. Number of Relationships in Different Categories

Currency	Diverging	Enhancing	Alleviating	Competing
France Franc	31	0	31	755
German Mark	9	0	0	548
New Zealand	286	4	0	311
SpainPeseta	107	1	78	807
Sweden Krone	319	0	28	317

Table 3. Examples of Relationships

	Relationship	Rules	Period
1	Competing	NLG+, DEM- => ESP+↑ NZD+, JPY- => ESP+ ↓	1987-1991
2	Diverging	AUD-,CAD-,FRF-,GBP-=> ESP+↑ AUD-, FRF-,GBP- => ESP+↓ CAD- => ESP+ ↓	1990-1992
3	Diverging	FRF+,ESP+,AUD-,CAD- => SEK+↓ AUD-,CAD- => SEK+↑ FRF+,ESP+ => SEK+ ↑	1991-1994
4	Enhancing	AUD-,FRF-,JPY-,SEK-, CHE- =>ESP+↑ AUD-,FRF-,CHE- =>ESP+↓ JPY-,SEK- =>ESP+ −	1990-1992

The symbols in Table 3 " ↑ , " ↓ " , and " – " denote that the confidence of the rule increases, decreases and remain stable respectively. The symbols "+" and "-" on the right to each currency symbol denote that the price of the currency increase and decrease respectively. The diverging relationship among the three rules "UD-,CAD-,FRF-,GBP-=> ESP+↑" "UD-, FRF-,GBP- => ESP+↓ " and "CAD- => ESP+ ↓" in Table 3 indicates that the confidence of "AUD-, FRF-,GBP- => ESP+" and "CAD- => ESP+" decrease over time while the confidence of their combined rule "AUD-,CAD-,FRF-,GBP- => ESP+" increases. This could be an important piece of information to currency traders.

5 Conclusion

In this work, we have analyzed the dynamic behavior of association rules over time and proposed four types of relationships among association rules. These relationships could reveal the correlations about the effect of the conditions on the consequent over time. We have designed two algorithms to discover the relationships among rules. Experiments on both synthetic and real-world datasets show that our approaches are efficient and effective.

References

1. Baron, S., Spiliopoulou, M.: Monitoring Change in Mining Results. In: 3rd International Conference on Data Warehousing and Knowledge Discovery, pp. 51–60 (2001)
2. Baron, S., Spiliopoulou, M., Gunther, O.: Efficient Monitoring of Patterns in Data Mining Environments. In: 7th East European Conference on Advances in Databases and Information Systems, pp. 253–265 (2003)
3. Baron, S., Spiliopoulou, M.: Monitoring the Evolution of Web Usage Patterns. In: 1st European Web Mining Forum Workshop, pp. 181–200 (2003)
4. Spiliopoulou, M., Baron, S., Giinther, O.: Temporal Evolution and Local Patterns. In: International Seminar on Local Pattern Detection, pp. 190–206 (2005)
5. Liu, B., Ma, Y., Lee, R.: Analyzing the Interestingness of Association Rules from the Temporal Dimension. In: 1st IEEE International Conference on Data Mining, pp. 377–384 (2001)
6. Chen, X., Petrounias, I.: Mining Temporal Features in Association Rules. In: 3rd European Conference on Principles of Data Mining and Knowledge Discovery, pp. 295–300 (1999)
7. Dong, G., Li, J.: Efficient mining of emerging patterns: discovering trends and differences. In: 5th International Conference on Knowledge Discovery and Data Mining, pp. 43–52 (1999)
8. Mann, P.S.: Introductory Statistics. John Wiley & Sons, Chichester (2003)
9. Duke Statistics Dataset.
   http://www.stat.duke.edu/
   data-sets/mw/ts_data/all_exrates.html
10. Agrawal, R., Srikant, R.: Fast algorithms for mining association rules. In: 20th International Conference on Very Large Databases, pp. 487–499 (1994)

Incremental Ontology-Based Extraction and Alignment in Semi-structured Documents

Mouhamadou Thiam[1,3], Nacéra Bennacer[2], Nathalie Pernelle[1], and Moussa Lô[3]

[1] LRI, Université Paris-Sud 11, INRIA Saclay Ile de France
2-4 rue Jacques Monod, F-91893 Orsay Cedex, France
[2] SUPELEC, 3 rue joliot- curie, F-91192 Gif-sur-Yvette cedex, France
[3] LANI, Université Gaston Berger, UFR S.A.T, BP 234 Saint-Louis, Sénégal
{mouhamadou.thiam,nathalie.pernelle}@lri.fr,
nacera.bennacer@supelec.fr,
lom@ugb.sn
http://www.lri.fr/~thiam

Abstract. *SHIRI*[1] is an ontology-based system for integration of semi-structured documents related to a specific domain. The system's purpose is to allow users to access to relevant parts of documents as answers to their queries. *SHIRI* uses RDF/OWL for representation of resources and SPARQL for their querying. It relies on an automatic, unsupervised and ontology-driven approach for extraction, alignment and semantic annotation of tagged elements of documents. In this paper, we focus on the *Extract-Align* algorithm which exploits a set of named entity and term patterns to extract term candidates to be aligned with the ontology. It proceeds in an incremental manner in order to populate the ontology with terms describing instances of the domain and to reduce the access to extern resources such as Web. We experiment it on a HTML corpus related to call for papers in computer science and the results that we obtain are very promising. These results show how the incremental behaviour of *Extract-Align* algorithm enriches the ontology and the number of terms (or named entities) aligned directly with the ontology increases.

Keywords: Information Extraction, Semantic Annotation, Alignment, Ontology, Semi-structured documents, OWL, RDF/RDFS.

1 Introduction

Information available on the Web is mostly in HTML form and thus is more or less syntactically structured. The need to automate these information processing, their exploitation by applications and their sharing justify the interest that research carries on the semantic Web. Because of the lack of semantic, the querying over these resources are generally based on keywords. This is not satisfying because it does not ensure answer relevance and the answer is a whole

---

[1] Système Hybride d'Intégration et de Recherche d'Information, Digiteo labs project.

S.S. Bhowmick, J. Küng, and R. Wagner (Eds.): DEXA 2009, LNCS 5690, pp. 611–618, 2009.
© Springer-Verlag Berlin Heidelberg 2009

document. The annotation of web resources with semantic metadata should allow better interpretation of their content. The metadata semantics are defined in a domain ontology through domain concepts and their relations. Nevertheless, manual annotation is time-consuming and the automation of annotation techniques is a key factor for the future web and its scaling-up.

Many works belonging to complementary research fields such as machine learning, knowledge engineering and linguistics investigate the issue of annotation of such documents. Some works are based on supervised approaches or on the existence of structure models in the input documents as in [7], [8], [10] or in text as in [3], [12]. Generally, the assumed hypotheses are incompatible with the heterogeneity and the great number of documents. Now, one information may appear in different kinds of structure depending on the document forms. Some unsupervised approaches are specialized in structured parts such as tables [15]. Moreover, one document may contain both structured and unstructured parts.

Except for named entities, instances are often drowned in text, so they are not easily dissociable. Even advanced Natural Language Processing techniques often adapted to very specific corpora could not succeed.

Named Entities Recognition (NER) aims to locate and classify elements in text into predefined categories such as the names of persons, organizations, locations, dates, etc. Some unsupervised Named-entity recognition systems are based on lexical resources ([9]), or on lexical resources built thanks to data available on the web ([12], [2]). Some approaches use the Web as a possible corpora to apply pattern and find terms to annotate a named entity of a resource [3]. Because this method is time-consuming, it has to be applied when other strategies fail.

The automation of heterogeneous documents annotation can also be based on terms that describe concepts that are not named entities. The different extraction techniques can be categorized as linguistic, statistic or hybrid ([14], [13]).

Once a term or a named entity is extracted, it has to be compared to the set of terms that belongs to the Ontology (concept labels or named entities). Similarity measures that can be used to estimate a semantic similarity between named entities or terms have been extensively studied ([6]).

SHIRI[1] can be introduced as an ontology-based integration system for semi-structured documents related to a specific domain. The system purpose is to allow users to access to relevant parts of HTML documents as answers to their queries. *SHIRI* uses RDF/OWL standard W3C languages for representation of resources and SPARQL for their querying. The system relies on an automatic, unsupervised and ontology-driven approach for extraction, alignment and semantic annotation of documents tagged elements. The extraction of term candidates to be aligned with the ontology relies on a set of named entity and term patterns. It proceeds in an incremental manner in order to populate the ontology with terms describing domain instances and to reduce the access to extern resources such as Web. The annotation of these terms is associated to tagged element of the HTML document (named structural unit) [1]. Actually, terms are generally not located in an accurate manner and may be drowned in a same structural unit. In this paper we focus on the algorithm defined for the extraction and the alignment named *Extract-Align* algorithm. We experiment and validate it on a HTML corpus related to call for papers in computer science and the results that we obtain are very promising. These results show how the incremental behaviour of *Extract-Align* algorithm enriches the ontology and how the number of terms (or named entities) aligned directly with the ontology increases. In section 2, we detail the extraction and alignment approach. In section 3, we present the results of the experiments made on a corpus related to call for papers. In section 4, we conclude and give some perspectives.

2   Incremental and Semantic Alignment Approach

In this section, we focus on the terms extraction and their alignment with the ontology. The extraction method applies a set of patterns to extract term candidates. It distinguishes the named entity patterns and the term patterns. The term candidates are to be aligned with the concepts of the domain ontology. This alignment is either directly done with the ontology or indirectly thanks to the Web. The ontology is then populated with the aligned terms that are exploited for the next alignments.

2.1   Ontology Description

Let $\mathcal{O}(\mathcal{C}, \mathcal{R}, \preceq, \mathcal{S}, \mathcal{A}, \mathcal{L_{EX}})$ be the domain ontology where $\mathcal{C}$ is the set of concepts, $\mathcal{R}$ is the set of relations between concepts, $\preceq$ denotes the subsumption relation between concepts and between relations. $\mathcal{S}$ defines the domain and the range for each relation and $\mathcal{A}$ is a set of axioms and rules defined over concepts and relations.

$\mathcal{L_{EX}}(\mathcal{L}, \mathcal{T}, prefLabel, altLabel, hasTerm, hastermNe)$ defines the set $\mathcal{L}$ of concept labels and the set $\mathcal{T}$ of terms or named entites describing the concepts of the domain. Each concept $c \in \mathcal{C}$ is related to a preferred label via *prefLabel* property and to alternate labels via *altLabel* [2] belonging to $\mathcal{L}$. Each concept $c \in \mathcal{C}$ is related to terms via *hasTerm* property and to named entities via *hastermNe* belonging to $\mathcal{T}$. We assume that the sets $\mathcal{L}$ and $\mathcal{T}$ are initialized respectively by a set of labels and a set of terms selected by the domain expert.

Example 1. Labels and terms selected for the *Topic* concept c of computer science domain include the following:

prefLabel(c, 'Topic'), altLabel(c, 'field'), altLabel(c, 'area'), altLabel(c, 'theme'), hasTerm(c, 'communications protocol'), hasTerm(c, 'data encryption'), hasTerm(c, 'information'), hasTerm(c, 'object-oriented programming language')

The set of terms $\mathcal{T}$ is enriched by extracted terms as documents are processed. Since this enrichment is automatic, some terms may be irrelevant, that's why we distinguish them from labels. If the expert decides to validate the ontology, it is possible that some of them become labels.

---

[2] Properties defined in SKOS: Simple Knowledge Organization System.

2.2   Extract-Align Algorithm

The *SHIRI* extraction and alignment approach proceeds in an incremental manner. Each Extract-Align invocation processes a subset of documents. More precisely, at each invocation, the algorithm is applied to a subset of documents D belonging to the same domain, to the ontology of this domain $\mathcal{O}$, to a set of patterns P, to a set *Processed* of terms handled in previous steps. The algorithm distinguishes two types of patterns : syntactic named entity patterns and syntactic term patterns. These two types of patterns are used to extract a set of term candidates denoted $\mathcal{I}$ (see example in table 1). Each $t \in \mathcal{I}$ is identified by the sequence of the numbered words according to their occurrence order in the document. These terms are to be aligned wih the set of labels $\mathcal{L}$ and the set of terms $\mathcal{T}$ defined in the ontology $\mathcal{O}$.

At each step, the algorithm attempts to directly align terms of $\mathcal{I}$ with the ontology, otherwise by using the web. Besides, each step enriches the set $\mathcal{T}$ of domain terms and named entities, so the number of web invocations should be reduced when the next documents will be processed. That is also the reason why the set of unaligned processed terms are kept in *Processed*.

The function $alignTerm(t)$ is applied to each $t \in \mathcal{I}$ and returns a set of concepts $C_t \subset \mathcal{C}$ if it succeeds. Then, t is added to $\mathcal{T}$ and related to each $c \in C_t$ via *hasTerm* or via *hastermNe* relations depending on the matched pattern (see example below). The invoked $alignTerm(t)$ function uses similarity measures that are appropriate to compare two named entities or two terms.

The unaligned terms are submitted to the Web like in CPankow approach [3]: lexico-syntactic Hearst patterns for hyponymy [5] are used to construct queries containing the unaligned term t. These queries are submitted to a search engine in order to find a set of label candidates L_t. For each $l \in L_t$, the function $webAlign(l)$ is applied and returns a set of concepts $C_t \subset \mathcal{C}$. If $webAlign(l)$ succeeds, then, l and t are added to $\mathcal{T}$. t is related to each $c \in C_t$ via *hasTerm* or via *hastermNe* relations depending on the matched pattern. l is related to each $c \in C_t$ via *hasTerm* relation. Since l is extracted automatically it is considered as a term.

In addition, the term candidates $\mathcal{I}$ are also processed in an incremental manner from the longest to the shortest. We assume that a term is more precise and meaningful than the terms it contains. For example *distributed databases* is more precise than *databases*. But for a term such as *Interoperability of data on the Semantic Web*, the alignment will fail very probably. We denote a term of length k occurring at position i in the document as a sequence of k words: $t_i^k = w_i w_{i+1}...w_{i+k-1}$, where w_{i+j} denote the word at position $i+j$, $j \in [0, k-1]$. We note $\mathcal{I}^k = \{t_i^k, i \in [1, N]\}$ the set of extracted terms of length k varying from *len* to 1 (*len* is the maximal length).

At iteration k, the algorithm proceeds terms of $\mathcal{I}^k$ and $\mathcal{I} = \bigcup_{i=1}^{k} \mathcal{I}^k$. We say that $t_{i_2}^{k_2}$ is included in $t_{i_1}^{k_1}$ if $k_2 < k_1$ and $i_2 \in [i_1, i_1 + k_1 - 1]$. When the system aligns a term $x \in \mathcal{I}^k$ then $\forall y \in \bigcup_{i=1}^{k-1} \mathcal{I}^i$ such that y is included in x, y is deleted from $\mathcal{I}$.

Table 1. An example of extracted terms

Original Text	Extracted Terms
..Areas$_{71}$ of$_{72}$ interest$_{73}$ are$_{74}$ distributed$_{75}$ databases$_{76}$ and$_{77}$ artificial$_{78}$ intelligence$_{79}$. The$_{80}$ workshops$_{81}$ SEMMA$_{82}$ focuses$_{83}$ also$_{84}$ on$_{85}$ databases$_{86}$. Intelligence$_{87}$ areas$_{88}$..	..[Areas$_{71}$] of$_{72}$ interest$_{73}$ are$_{74}$ [distributed$_{75}$ [databases$_{76}$]] and$_{77}$ [artificial$_{78}$ [intelligence$_{79}$]]. The$_{80}$ [workshops$_{81}$] SEMMA$_{82}$ focuses$_{83}$ also$_{84}$ on$_{85}$ [databases$_{86}$]. [Intelligence$_{87}$] [areas$_{88}$]..

Example: Given the text in table 1 and the two patterns $P_t^1 = JN$ and $P_t^2 = N$ where J denotes an adjective and N a name, the extracted terms are the following: In this example, $\mathcal{I}^1 = \{$Areas$_{71}$, databases$_{76}$, intelligence$_{79}$, workshops$_{81}$, databases$_{86}$, Intelligence$_{87}$, areas$_{88}\}$ and $\mathcal{I}^2 = \{$distributed$_{75}$ databases$_{76}$, artificial$_{78}$ intelligence$_{79}\}$. The terms [*distributed$_{75}$ databases$_{76}$*] and [*artificial$_{78}$ intelligence$_{79}$*] are aligned with the concept [*Topic*]. So, we delete *databases$_{76}$*, *Intelligence$_{79}$* from $\mathcal{I}^1$.

Three kinds of outputs result from the Extract-Align invocation: (1) rdf triples which enrich the ontology with terms or named entities describing the concepts (*hasTerm* and *hastermNe* relations), (2) rdf triples refering the structural units of the documents, the concepts these units contain (*containInstanceOf*) and the values of corresponding terms or named entities (*hasValueInstance*) and (3) the set of all processed terms.

3   Validation of Extract-Align Algorithm

Let $\mathcal{O}$ be the domain ontology of *Call for Papers for Computer Science Conferences*. The named entities are the events (i.e. conferences, workshops), the persons, their affiliations (i.e. team, laboratory and/or university), and the locations (university, city or country) of the events. Each concept is described by a preferred label and a set of alternative labels. For example, *scientist* and *people* are related to the *Person* concept. The expert has also exploited Wordnet to select a set of 353 domain terms such as {*Communications protocol, data encryption, information, ...*} that are related to *Topic* concept via *hasTerm* property. The corpus we collect is composed of a set of 691 HTML documents (250542 words after pre-processing).

Named entities are automatically extracted from the document collection using Senellart specialized technique [2] which exploits DBLP (Digital Bibliography and Library Project) to identify accurately person names and dates. We also use the set of C-Pankow syntactic patterns to extract other named entities instances. Terms are extracted using the patterns defined in [4].

To retrieve web label candidates, a set of queries is constructed for each term or named entity. The queries are constructed like in C-Pankow approach using : hearst patterns, copula and definites (noun phrase introduced by the definite determiner *The*). The web labels are selected when the confidence measure value is over 0.2. For the alignment of the term candidates or web labels, we use Taxomap tool [11]. Its aim is to discover correspondences between concepts of two taxonomies. It relies on terminological and structural techniques applied sequentially and performs an oriented alignment from a source ontology to a target ontology. We only exploit the terminological strategies of Taxomap i.e. syntactic-based similarity measures applied on concept labels and terms (term inclusion, n-gram similarity,...). For the alignment of named entities, the similarity is strictly the equality between terms.

For example, in our experiments, the term *reinforcement learning* is directly aligned with the *Topic* concept thanks to the term *learning*. The term *World Wide Web* has not been aligned directly with $\mathcal{O}$. One of the web label candidate is *information ressources*, Taxomap aligns it with *information* term related to *Topic* concept. *Reinforcement learning, information ressources, World Wide Web* are then added to $\mathcal{T}$ and related to *Topic* concept via *hasTerm* property.

Table 2. Results for Named Entity and Term Pattern

	Named Entity Patterns					Term Patterns		
	Aligned With $\mathcal{O}$	Using the Web	Precision	Precision with incomplete NE	Recall	Aligned With $\mathcal{O}$	Precision	Recall
Affiliation	0	1317	84.18%	86.07%	70.83%	165	96.97%	91.95%
Location	0	1097	98.53%	99.02%	91.86%	143	80.42%	78.77%
Person	745	362	89.47%	90.85%	79.5%	206	63.59%	59.01%
Event	0	741	64.35%	86.13%	84.47%	80	65.00%	65.00%
Date	456	0	97.58%	97.58%	74.17%	-	-	-
Topic	-	-	-	-	-	276	65.58%	59.34%

Table 2 shows the results we obtain for named entity patterns. We present the precision and recall measures for named entites that are aligned either directly or using the Web label candidates. Since the granularity of the *Shiri-Annot* annotation is the structural unit, we consider that a named entity is incomplete but correct if the complete name appears in the structural unit where it is extracted. For instance, *International Conference* is incomplete but the structural unit contains the whole name which is *International Conference on Web Services*. The Web allows to align approximatively 74 % of all the aligned named entities. All the affiliations, events and locations are found thanks to the Web. Furthermore, the table shows that by taking into account incomplete named entities the precision increases especially for events. These named entities are often partially extracted due to their length and their complexity. Table 2 shows that thanks to term patterns, the concept *Topic* is enriched of 78% de terms. Other named entities have been also found with good precision and recall for affiliations which are often described using complex terms. Table 3 shows that a lot of terms occurs many times since all documents talk about the same domain. Obviously, most of them are not aligned with the ontology. Moreover, those which are included in aligned terms are not processed.

Table 3. Number of Extracted Terms by Length

Length	1	2	3	4	5	7	**Sum**
Extracted Terms	101430	32712	17912	5704	966	104	158828
Extracted Terms (distinct)	14413	15806	10020	3797	602	48	44680

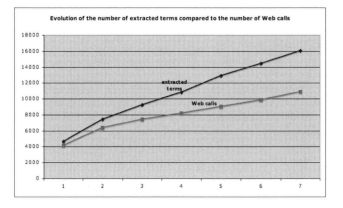

Fig. 1. Extracted terms, Web calls versus the number of documents (by ten)

Figure 1 shows: (1) the evolution of the number of terms extracted according to the number of documents (by ten) (2) the evolution of the number of web calls according to the number of documents (by ten). The results show that the number of Web invocations decreases with the number of processed documents. This explains by the incremental behaviour of Extract-Align algorithm : (1) the more the ontology is populated by new terms, the more a term candidate can be directly aligned and (2) all term web alignements which fail are stored (*Processed* data).

4   Conclusions and Future Works

In this paper, we have presented an automatic, unsupervised and ontology-driven approach for extraction, alignment and semantic annotation of tagged elements of documents. The *Extract-Align* algorithm proceeds in an incremental manner in order to populate the ontology with terms describing instances of the domain and to reduce the access to extern resources such as Web.

We experiment and validate our approach on a HTML corpus related to call for papers in computer science and the results are promising. These results show how the ontology is enriched and how the number of terms (or named entities) aligned directly with the ontology increases. The constructed ontology can be validated by a domain expert in order to select among the terms those to be removed or those to become concept labels.

A short-term perspective is the exploitation of the annotation model to reformulate domain queries in order to adapt them to the various levels of precision of the annotations. A further perspective is to study how a quality measure can be associated to each annotation triple. We also plan to apply our approach to other domains like e-commerce web sites.

References

1. Thiam, M., Pernelle, N., Bennacer, N.: Contextual and Metadata-based Approach for the Semantic Annotation of Heterogeneous Documents. In: ESWC-SeMMA workshop, Tenerife, Spain (2008)
2. Senellart, P.: Understanding the Hidden Web. PHD Thesis, University of Paris (December 11, 2007)
3. Cimiano, P., Handschuh, S., Staab, S.: Gimme'The Context: Context Driven Automatic Semantic Annotation With C-PANKOW. In: WWW conference (2005)
4. Arppe, A.: Term Extraction from unrestricted Text. In: The Nordic Conference on Computational Linguistics, NoDaLiDa (1995)
5. Hearst, M., Marti, A.: Automatic acquisition of hyponyms from large text corpora. In: Proceedings of the 14th International conference on Computational linguistics, 1992, France, pp. 539–545 (1992)
6. Cohen, W.W., Ravikumar, P., Fienberg, S.E.: A comparison of string distance metrics for name-matching tasks. In: IIWeb, pp. 73–78 (2003)
7. Crescenzi, V., Mecca, G., Merialdo, P.: RoadRunner: Towards Automatic Data Extraction from Large Web Sites. In: Very Large Data Bases Conference, VLDB (2001)
8. Davulcu, H., Vadrevu, S., Nagarajan, S.: OntoMiner: Automated Metadata and instance Mining from News Websites. The International Journal of Web and Grid Services (IJWGS) 1(2), 196–221 (2005)
9. Borislav, P., Atanas, K., Angel, K., Dimitar, M., Damyan, O., Miroslav, G.: KIM - Semantic Annotation Platform. Journal of Natural Language Engineering 10(3-4), 375–392 (2004)
10. Baumgartner, R., Flesca, S., Gottlob, G.: Visual Web Information Extraction with Lixto. The VLDB Journal, 119–128 (2001)
11. Hamdi, F., Zargayouna, H., Safar, B., Reynaud, C.: TaxoMap in the OAEI 2008 alignment contest. In: Ontology Alignment Evaluation Initiative (OAEI) 2008 Campaign - Int. Workshop on Ontology Matching (2008)
12. Etzioni, O., Cafarella, M., Downey, D., Kok, S., Popescu, A., Shaked, T., Soderland, S., Weld, D., Yates, A.: Unsupervised named-entity extraction from the web: An experimental study. Artificial Intelligence 165(1), 91–134 (2005)
13. Navigli, R., Velardi, P.: Learning Domain Ontologies from Document Warehouses and Dedicated Web Sites. Computational Linguistics 30(2), 151–179 (2004)
14. Drouin, P.: Term extraction using non-technical corpora as a point of leverage. Terminology 9(1), 99–117 (2003)
15. Cafarella, M.J., Halevy, A., Zhe Wang, D.: Uncovering the relational web. In: Proceedings of WebDB, Canada, (2008)

Tags4Tags: Using Tagging to Consolidate Tags

Leyla Jael Garcia-Castro[1], Martin Hepp[1], and Alexander Garcia[2]

[1] E-Business and Web Science Research Group, Universität der Bundeswehr München,
D-85579 Neubiberg, Germany
leyla.garcia@unibw.de, mhepp@computer.org
[2] Department of Computational Linguistics, University of Bremen,
D-28359 Bremen, Germany
cagarcia@uni-bremen.de

Abstract. Tagging has become increasingly popular and useful across various social networks and applications. It allows users to classify and organize resources for improving the retrieval performance over those tagged resources. Within social networks, tags can also facilitate the interaction between members of the community, *e.g.* because similar tags may represent similar interests. Although obviously useful for straightforward retrieval tasks, the current meta-data model underlying typical tagging systems does not fully exploit the potential of the social process of finding, establishing, challenging, and promoting symbols, *i.e.* tags. For instance, the social process is not used for establishing an explicit hierarchy of tags or for the collective detection of equivalencies, synonyms, morphological variants, and other useful relationships across tags. This limitation is due to the constraints of the typical meta-model of tagging, in which the subject must be a Web resource, the relationship type is always *hasTag,* and the object must be a tag as a literal. In this paper, we propose a simple yet effective extension for the current meta-model of tagging systems in order to exploit the potential of collective tagging for the emergence of richer semantic structures, in particular for capturing semantic relationships between tags. Our approach expands the range of the object of tagging from Web resources only to the union of (1) Web resources and (2) pairs of tags, i.e., users can now use arbitrary tags for expressing typed relationships between a pair of tags. This allows the user community to establish similarity relations and other types of relationships between tags. We present a first prototype and the results from an evaluation in a small controlled setting.

Keywords: Social Web, folksonomy, tagging, meta-model, emergent semantics, conceptual graphs, Semantic Web, Web 2.0.

1 Introduction

Nowadays social tagging systems (STS), and the resulting knowledge structures known as folksonomies [1], are widely used on the Web. Tagging typically works by assigning short lexical elements to resources in a collaborative environment, mainly for document retrieval. Popular sites focus on tagging Web resources (*e.g.* Delicious, http://www.delicious.com/, and Connotea, http://www.connotea.org/), images (*e.g.* Flickr, http://www.flickr.com/), or blogs and other user-generated content

S.S. Bhowmick, J. Küng, and R. Wagner (Eds.): DEXA 2009, LNCS 5690, pp. 619–628, 2009.

(*e.g.* Technorati, http://www.technorati.com/). Recently, respective technology has also been used in corporate networks such as the Electricité de France Intranet [2], where tags were used in blogs to promote knowledge sharing inside the organization.

It can be assumed that the popularity of tagging is not only due to the simplicity of the tagging operation itself, but also because tags effectively facilitate search and navigation over tagged resources [3]. From the technical perspective, there are several attractive features of tagging systems that create added value for users: First, the use of URIs for resources provides reliable, unique identifiers for documents, which allows for the consolidation of meta-data. Second, the sites provide a collaborative environment with an explicit representation of users, which allows discovering implicit relationships, *e.g.* networks of users with similar skills, tasks, or interests. Third, tagging systems provide simple yet effective support for the emergence of consensus on (i) the exact lexical form of a tag and (ii) the appropriateness of a tag for a certain resource based on collaborative filtering and recommendations. This helps to avoid orphaned tags and reduces lexical or morphological variations; at the same time, it keeps up with the high agility and the good coverage of rare but still relevant elements, *i.e.* such that are on the long tail. Centralized approaches, including classical ontology-based solutions often lag behind in their coverage of novel or specific domain elements [4]. Furthermore, as tags work like bookmarks or indexes, they help to reduce spam-induced noise in search engines and enable text-based queries over elements like images [5]. Moreover, tagging does not impose rigid categories or a controlled vocabulary on users but gives to users the possibility to freely create and associate terms, i.e., descriptors, to resources.

Although tagging has proven to provide significant benefits, there are also relevant limitations of the current state of technology. Typical problems are (i) tag ambiguity, (ii) missing links between multiple synonyms, spelling variants, or morphological variants, and (iii) variation in the level of granularity and specificity of the tags used caused by differences in the domain expertise of agents [2, 3, 6, 7]. These limitations are problematic for *e.g.* (i) developing intelligent user interfaces for annotations, (ii) improving navigation and querying based on annotations, and (iii) integrating content from diverse and heterogeneous data sources [6].

Additional formal structures may help to overcome some of problems mentioned above [5, 8, 9]. A main question, however, is whether such formal structures are imposed explicitly in the tagging stage or derived implicitly by mining tagging data.

While there exist many proposals for the latter approach, we propose to expand the underlying meta-model of tagging systems from attaching tags to resources only to attaching tags to resources and *arbitrary pairs of tags,* i.e., pairs of the form (tag, tag). Our motivation is to exploit the positive technical and social effects of tagging for the construction and the management of more powerful conceptual structures in information systems and on the Web.

Some of the expected advantages of this model are: (i) supporting the emergence of explicit relationships between tags, (ii) adding meaning to numerical tags, (iii) building a "tagsonomy", *i.e.,* a conceptual graph of tags, including relationships between them, and (iv) improving the basis for adding formal semantics to tags by mining techniques.

We expect this to improve the retrieval performance of tagging systems and to help building conceptual graphs from a set of tags. This may turn STS into true sources of collective intelligence. Such likely requires the aggregation and recombination of data collected from annotations in order to create new knowledge and new ways of learning that individual humans cannot achieve by themselves [10].

This paper is organized as follows. Section 2 introduces our model and motivates our approach. Section 3 summarizes our preliminary evaluation by means of a controlled experiment on establishing relations between tags. Section 4 summarizes related work and discusses our contribution. Section 5 sketches future directions for research.

2   A Vision to Expand the Scope of Taggable Objects

Currently, STS allow agents (A), i.e. users, to add tags (T) to resources (R); each respective activity is called tagging (TA). This simple setting already allows a high degree of variation as presented in Fig. 1.

Some of the strengths of STS arise from those combinations such as promoting serendipity, facilitating convergence, and supporting collaboration by means of filters and recommendations based on existing tags. Since users can share their tags with the community, they are building not only a knowledge representation for themselves but are also helping others to discover associations that were not previously known. This increases the collectively available expertise [11] and supports the social reinforcement by means of enabling social connections based on common interests [8]. In other words, the aggregation of many individual user contributions can by itself create an added value in STS [10].

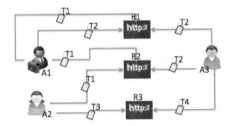

Fig. 1. Variations in tagging resources

A STS can be represented as a graph where agents (A), resources (R), and tags (T) are the vertices (V) and tagging activities (TA) are the edges. The predominant tagging meta-model relies mainly on a triple of (i) agents, i.e. users, (ii) resources, and (iii) tags. It is also possible to include the system where the tagging took place and a polarity to assign negative or positive values to tags as Gruber proposes [5]. Some authors add relations between tags such us *relatedTo* and *equivalentTo* which are taken from Newman's model [12]. The typical meta-model used today can be formalized as follows:

$$STS = <V, TA > | V = A \cup T \cup R \text{ and } TA \subset \{(A, T, R)\} \qquad (1)$$

Although tagging has proven to provide significant benefits, there are also important limitations. Several of these limitations are caused by the constraints in the current meta-model, as the subject of the tagging must be a Web resource, the relationship type is always *hasTag,* and the object must be a tag.

Often, social tagging suffers from a lack of structure and contextualization. For example, we often do not know for which purpose a particular user attached a certain tag to a certain resource. The assignment may *e.g.* be relevant or valid only in the context of a particular task. Since all tags are organized in a shallow way, the navigation, querying, and retrieval of resources is limited [2, 6] and free relations between tags cannot be established by the same social process. Also, it has been observed that people need to contextualize communication with other people because that fosters the creation of new knowledge [10]. Being able to tag other agents' previous tagging activities would support that.

The key motivation for our approach is to keep the simplicity and popularity of tagging while using them on richer conceptual structures, instead of solely trying to derive those structures from tagging data by mining techniques. In our opinion, the current meta-model does not fully exploit the potential of tagging for finding, establishing, challenging and promoting symbols in a community. For instance, the social convergence in STS is not used for establishing an explicit hierarchy of tags or for the collective detection of equivalencies, synonyms, morphological variants, and other useful relationships across tags.

We propose to expand the current meta-model by opening up the range of taggable objects *from resources only to resources and relations between tags*. We aim at relating entities from this expanded set in a semantic manner by means of that approach. Our model is shown in Fig 2.

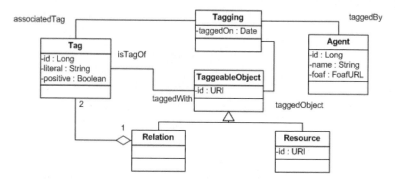

Fig. 2. Expanding the scope of taggable objects

A STS to support our model can be still represented as a graph. As before, agents (A), Web resources (R), and tags (T) are the vertices (V) and tagging activities (TA) are the edges. One new vertex is required in order to represent relations between tags (RT). It can be formalized as follows:

$$STS = <V, TA> \mid V = A \cup T \cup R \cup RT \text{ and } RT \subset \{T \times T\} \qquad (2)$$

Our model intends to widen the scope of social tagging. The subject of the tagging remains an agent and the predicates remains a tag, but the object now can be either a Web resource or a type of relation between tags. The relationship type can still be *hasTag* but it is also possible to add new relation types between any pair of tags. This minimal change should facilitate the usage of our new model by users familiar with

Table 1. Summary of possible tagging activities with Tags4Tags

Subject	Predicate	Object	Example
Agent	Tag	Resource	(agent1, travel, http://vacations.com)
		(Tag, Tag)	(agent1, englishToSpanish, (tag:travel, tag:viaje))

traditional tagging. A summary of possible tagging scenarios with our extended model is shown in Table 1.

Since our model is based on the current meta-model, it will likely be possible to adapt and apply (i) existing approaches to derive formal structures from tagging data such as FLOR [6] and SCARLET [13], (ii) normalization and disambiguation techniques such as [14-16], (iii) the addition of meaning to tags by using URIs [9], and (iv) techniques and tools for the tag data consolidation among platforms [8].

With the proposed extension, we basically allow people to build and maintain conceptual graphs based on tagging and complement this by social mechanisms for convergence. The resulting networks can be an important starting point to allow better retrieval and more sophisticated processing, and will likely allow more powerful approaches for deriving formal structures.

3   Prototype and Implementation

In order to evaluate our model, we developed a first prototype with the main goal of analyzing how well people are able to relate objects by tags representing the type of relationship. The model was initially populated with resources, agents, and tags related to "travel", which we collected via the Connotea API (http://www.connotea.org/wiki/WebAPI), and relationships which we collected from participants by means of a Java Web-based application. The architecture of our prototype is presented in Fig. 3.

In order to capture relationships between taggable objects, we provide two columns: The left-hand side corresponds to the subject of the relationship and right-hand side corresponds to the object. The prototype was presented as a game where participants had to find as many relations as they could in a given period of time. The tool provides a set of recommended tags for likely relationships as well as suggestions based on existing tags in the system, see Fig. 4. Note that the predefined relations like isPartOf are also just tags. A formal meaning can be associated to those based on the outcome of the social tagging process.

The prototype was implemented as a Web-based application using Java 1.6 as the development language, the Spring Framework (http://www.springsource.org/) and Velocity 1.4 (http://velocity.apache.org/) to manage the Model-View-Controller architecture; ExtJS (http://extjs.com/), BoxOver (http://boxover.swazz.org/ example.html), and Autosuggest BSN (http://www.brandspankingnew.net/specials/ ajax_autosuggest/ajax_autosuggest_autocomplete.html) libraries for the user interface, Direct Web Remoting (http://directwebremoting.org/) for AJAX, and Jena (http://jena.sourceforge.net/) as the underlying Semantic Web framework.

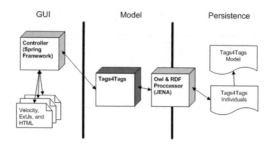

Fig. 3. Prototype architecture

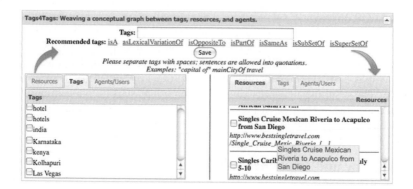

Fig. 4. A screenshot of the prototype to relate taggable objects

4   Evaluation

The prototype described in the previous section was used to test whether the T(T, T) pattern, *i.e.* tags attached to pairs of tags, can be used to consolidate tag sets and to elicit useful relationships between tags. Those could be used for more powerful tag-based retrieval. Thus the main goal was to find out whether our approach is a feasible way of gathering user input to capture equivalencies and relationships, such as narrower and broader relations. We wanted to evaluate whether (i) average computer users are able to grasp our idea and employ it with minimal instructions, and whether (ii) the collected data is of sufficient quality to be useful.

4.1   Methodology

We recruited ten individuals from our university, both employees and students, and asked them to spot and enter as many relations as they could in a predefined period of time. All participants, seven bachelor students, one PhD student, and two researchers, had experience using the Web and some of them using tagging systems.

A set of 92 tags related to "travel" was randomly selected via the Connotea API as the initial data. We only considered tags longer than four characters and resources with titles longer than five characters. Once our model was populated with the data, we developed the Web-based prototype to allow people to relate tags easily. The prototype, already presented in the previous section, was build with Eclipse (http://www.eclipse.org) and was tested manually.

Participants were asked independently to take part in the experiment and they received brief oral instructions only. The experiment was conducted in a sequential order on a single machine. All data was collected in an RDF file, and relations established by the participants were automatically loaded as part of the initial data for the following participants. This happens because relations are established by means of tags, thus they become part of the set of tags as well.

4.2 Results

All participants used the application with ease and established multiple different relations between tags. We observed that the task was harder for the first participants since they had only the initial data, and it got increasingly easier for the later participants since they could reuse relationships previously collected.

The relations between similar tags were quite consistent among users instead of the use of different tags, see Fig. 5. For instance, people attached similar tags such as "canbereachedby" and "canbevisitedby" to relate places like Berlin and Antigua to the tag "airline". Another example is the relation between "vacations" and "vacation" with tags such as "singular" and "isLexicalVariationOf". Someone with more experience in triples construction could have added a tag "isSubSetOf" between "singular" and "isLexicalVariationOf", *i.e.* ("isSubSetOf", (singular, isLexicalVariationOf)), in order to consolidate data.

Fig. 5. Some of the results found

In order to achieve a better use of some of the recommended relationships such as "isA" and "isPartOf" as well as others proposed by users themselves, it could be necessary to identify the domain or context of taggable objects. It could also be useful to understand the meaning of some tags such as "avianFlu". This particular tag was related to the tag "airline" with the tag "isA", *i.e.* ("isA", (avianFlu, airline)), by one of the participants; however this tag was initially attached to a resource related to a virus known as "avian influenza". Additionally, background and education are also important to understand some tags; a biologist or a medical doctor would hardly have misunderstood the meaning of "avianFlu". We assume that in Web-wide tagging systems based on our approach, the increased mass of tagging data will simplify filtering out noise and contradictions more easily.

5 Discussion and Related Work

5.1 Related Work

There are different approaches to improve social tagging by means of structured meta-models and ontologies. Approaches in this vein can be classified into five main groups: (i) modeling tagging [12], (ii) augmenting the user-contributed data [5], (iii) adding meaning to tags [9, 17], (iv) adding meta-data in order to improve retrieval, information exchange and knowledge reuse [18], and (v) enhancing sharing and reuse of social tagging data through different platforms [8].

Newman [12] proposes an ontology to "model the relationship between an agent, an arbitrary resource, and one or more tags". His model represents tags, agents, resources and tagging activities as classes and relates them via object properties; some relations between tags such as *relatedTo* and *equivalentTo* are also modeled. A tagging activity is defined as a triple, which corresponds to a resource tagged by an agent with an associated tag. The conceptual approach of Newman's ontology is based on the theoretical work by Gruber [5] and is taken as the baseline for other models because of its simple but comprehensive nature. Gruber's approach [5] is broader than Newman's because it includes other elements rather than tags, agents, and resources. Hence Gruber works with a quintuple; this quintuple incorporates information about (i) the system where the tagging took place and (ii) a polarity to represent positive and negative tags. Gruber's ontology is part of the TagCommon project (http://tagcommons.org/).

MOAT [9] is one of the models that extends Newman's ontology. Its aim is to semantically enrich content from free tagging by means of providing a way for users to attach meanings to their tags. A MOAT meaning refers to a Web resource and is part of the tagging; for instance, users could attach a meaning from DBPedia (http://dbpedia.org/) or any other resource they choose. Another approach to add meanings to tags, Extreme Tagging, is presented in [17]. Since a tag can have different meanings in different contexts, tagging tags is used in that approach to disambiguate the respective contexts. The underlying meaning of a tag can be revealed by means of another tag. Kim *et al.* [8] propose a system named SCOT, a semantic cloud of tags to represent the structure and semantics of tagging data and to allow the import and export among different platforms. Tagging activities are represented as a tag cloud, which includes user-tag and tag-tag relations. SCOT uses SIOC to describe site information, FOAF to represent agents (both humans and machines), and SKOS to represent tags and allow semantic relations among them.

Oren *et al.* [18] explore the meaning of semantic annotations but do not propose an explicit model. According to them, tagging expresses an unspecified relation between the resource and the tag. Thus, according to their position, making a complex statement about the real world is not possible, but only assigning tags, because of the lack of context.

5.2 Findings and Results

Our proposal builds mainly upon the work of Gruber [5] and Passant & Laublet [9]. While our approach is different from all other ones we are aware of, it remains widely compatible with existing algorithms and tools. Currently, we do not yet explore the problem of import and export of tagging data across platforms as done by Kim et al.

[8]. Our proposal allows tagging tags same as in the work by Tanasescu & Streibel [17], and defining hierarchical relationships as e.g. offered by Bibsonomy (http://www.bibsonomy.org/). In contrast to existing approaches, our model expands the scope of taggable objects to a much broader set than described in any previous work, and uses this expansion to support the collective construction of conceptual graphs involving relationships between tags. Furthermore, by means of these networks, we expect to facilitate the disambiguation of tags in a similar way than Yeung, Gibbins & Shadbolt [16] propose.

Through the experiments described in the evaluation section, we gained preliminary evidence that our approach can be used with minimal instruction by average users familiar with traditional tagging systems. Also, we can see that a relatively simple expansion of the current tagging meta-model facilitates (i) the construction of conceptual graphs and (ii) the inference of a hierarchy of tags and other meaningful relationships such as synonyms and antonyms based on standard mining techniques, which will be immediately useful for query expansion and disambiguation.

6   Conclusions and Future Work

We have proposed a minimal yet fundamental expansion of the meta-model of tagging in order to empower the construction of richer conceptual structures while keeping the ease and popularity of free tagging. Our model adds a semantic level to free-tagging in order to improve search and retrieval by means of the addition of new elements in the tagging operation. Such can be used to (i) build complex conceptual graphs that represent the underlying relations between tags, (ii) improve those networks by social mechanisms for convergence, (iii) use those networks to disambiguate tag meanings and for query expansion, and (iv) reduce the gap for deriving formal structures from tagging data for other purposes.

Our model proved to be feasible, even though the first prototype needs to be improved and complemented in order to be able to collect more data about tagging activities in agents, tags, and relationships other than just pairs of tags. Additionally, more tests are required to determine more precisely how our model can be used (i) to improve search and retrieval in STS and social convergence mechanisms, and (ii) to derive formal structures from tagging.

According to the results of the evaluation, some improvements are needed and would be useful to derive formal structures and improve search and retrieval:

- Semi-automatic consolidation of data, *i.e.,* tags and relationships, by means of normalization and disambiguation techniques to reduce lexical variations and suggest hypernyms and synonyms.
- Contextualization of taggable objects to avoid misinterpretations such as taking "avianFlu" as an airline instead of a disease. It would be also useful to collect more meaningful relationships.
- Allowing the characterization of relationships (such as symmetry and transitivity) could be very useful. However, instructing average users to use this feature properly could be very difficult as well.

In a nutshell, we hope that our expanded model will help to improve the performance of tagging systems while keeping up with their popularity and ease of use on a Web scale.

References

1. VanDerWal, T., Folksonomy.: (2007),
   http://www.vanderwal.net/folksonomy.html (retrieved April 2, 2009)
2. Passant, A.: Using Ontologies to Strengthen Folksonomies and Enrich Information Retrieval in Weblogs: Theoretical background and corporate use-case. In: International Conference on Weblogs and Social Media, USA (2007)
3. Specia, L., Motta, E.: Integrating folksonomies with the semantic web. In: Franconi, E., Kifer, M., May, W. (eds.) ESWC 2007. LNCS, vol. 4519, pp. 624–639. Springer, Heidelberg (2007)
4. Hepp, M.: E-business vocabularies as a moving target: Quantifying the conceptual dynamics in domains. In: Gangemi, A., Euzenat, J. (eds.) EKAW 2008. LNCS, vol. 5268, pp. 388–403. Springer, Heidelberg (2008)
5. Gruber, T.: Ontology of Folksonomy: A Mash-up of Apples and Oranges. International Journal on Semantic Web & Information Systems 3(2) (2007)
6. Angeletou, S., Sabou, M., Motta, E.: Semantically Enriching Folksonomies with FLOR. In: Proceedings of the European Semantic Web Conference - Worshop on Colletive Intelligence and the Semantic Web, Spain (2008)
7. Golder, S.A., Huberman, B.A.: Usage patterns of collaborative tagging systems. Journal of Information Science 32(2), 198–208 (2006)
8. Kim, H.-L., Breslin, J.G., Yang, S.-K., Kim, H.-G.: Social semantic cloud of tag: Semantic model for social tagging. In: Nguyen, N.T., Jo, G.-S., Howlett, R.J., Jain, L.C. (eds.) KES-AMSTA 2008. LNCS, vol. 4953, pp. 83–92. Springer, Heidelberg (2008)
9. Passant, A., Laublet, P.: Meaning Of A Tag: A Collaborative Approach to Bridge the Gap Between Tagging and Linked Data. In: International World Wide Web Conference - Linked Data on the Web Workshop, China (2008)
10. Gruber, T.: Collective Knowledge Systems: Where the Social Web meets the Semantic Web. Web Semantics: Science, Services and Agents on the World Wide Web (2007)
11. Lemieux, S.: Social Tagging and the Enterprise: Does Tagging Work at Work?,
    http://www.semanticuniverse.com/
    articles-social-tagging-and-enterprise-does-tagging-work-work.
    html (retrieved February 10, 2009)
12. Newman, R.: Tag Ontology Design (2004),
    http://www.holygoat.co.uk/projects/tags/
    (retrieved February 16, 2009)
13. Sabou, M., d'Aquin, M., Motta, E.: SCARLET: SemantiC relAtion discoveRy by harvesting onLinE onTologies. In: Bechhofer, S., Hauswirth, M., Hoffmann, J., Koubarakis, M. (eds.) ESWC 2008. LNCS, vol. 5021, pp. 854–858. Springer, Heidelberg (2008)
14. Gracia, J., et al.: Querying the Web: A Multi-ontology Disambiguation Method. In: International Conference on Web Engineering, USA (2006)
15. Sabou, M., d'Aquin, M., Motta, E.: Using the Semantic Web as Background Knowledge for Ontology Mapping. In: International Semantic Web Conference - Workshop on Ontology Matching, Grecia (2006)
16. Yeung, C.A., Gibbins, N., Shadbolt, N.: Understanding the Semantics of Ambiguous Tags in Folksonomies. In: International Workshop on Emergent Semantics and Ontology Evolution, Korea (2007)
17. Tanasescu, V., Streibel, O.: Extreme Tagging: Emergent Semantics through the Tagging of Tags. In: International Workshop on Emergent Semantics and Ontology Evolution, Korea (2007)
18. Oren, E., et al.: What are Semantic Annotations (2006)

Detecting Projected Outliers in High-Dimensional Data Streams

Ji Zhang[1], Qigang Gao[2], Hai Wang[3], Qing Liu[1], and Kai Xu[1]

[1] CSIRO ICT Center, Hobart, TAS, Australia
{ji.zhang,q.liu,kai.xu}@csiro.au
[2] Dalhousie University, Halifax, NS, Canada
qggao@cs.dal.ca
[3] Saint Mary's University, Halifax, NS, Canada
hwang@smu.ca

Abstract. In this paper, we study the problem of projected outlier detection in high dimensional data streams and propose a new technique, called Stream Projected Ouliter deTector (SPOT), to identify outliers embedded in subspaces. Sparse Subspace Template (SST), a set of subspaces obtained by unsupervised and/or supervised learning processes, is constructed in SPOT to detect projected outliers effectively. Multi-Objective Genetic Algorithm (MOGA) is employed as an effective search method for finding outlying subspaces from training data to construct SST. SST is able to carry out online self-evolution in the detection stage to cope with dynamics of data streams. The experimental results demonstrate the efficiency and effectiveness of SPOT in detecting outliers in high-dimensional data streams.

1 Introduction

Outlier detection is an important research problem in data mining that aims to find objects that are considerably dissimilar, exceptional and inconsistent with respect to the majority data in an input database [9]. In recent years, we have witnessed a tremendous research interest sparked by the explosion of data collected and transferred in the format of streams. Outlier detection in data streams can be useful in many fields such as analysis and monitoring of network traffic data, web log, sensor networks and financial transactions, etc.

A key observation that motivates our work is that outliers existing in high-dimensional data streams are embedded in some lower-dimensional subspaces. Here, a subspace refers to as the data space consisting of a subset of attributes. These outliers are termed *projected outliers* in the high-dimensional space. The existence of projected outliers is due to the fact that, as the dimensionality of data goes up, data tend to become equally distant from each other [1]. As a result, the difference of data points' outlier-ness will become increasingly weak and thus undistinguishable. Only in moderate or low dimensional subspaces can significant outlier-ness of data be observed.

The problem of detecting projected outliers from high-dimensional data streams can be formulated as follows: given a φ-dimensional data stream $\mathcal{D}$, for each data point $p_i = \{p_{i1}, p_{i2}, \ldots, p_{i\varphi}\}$ in $\mathcal{D}$, the projected outlier detection method performs a mapping as

S.S. Bhowmick, J. Küng, and R. Wagner (Eds.): DEXA 2009, LNCS 5690, pp. 629–644, 2009.

$f : p_i \rightarrow (b, S_i, Score_i)$. b is a Boolean variable indicating whether or not p_i is a projected outlier. p_i is a project outlier (*i.e.*, $b = true$) if there is one or more subspaces where p_i is an outlier. These subspaces are called the outlying subspaces of p_i. In this case, S_i is the set of outlying subspaces of p_i and $Score_i$ is the corresponding outlier-ness score of p_i in each subspace of S_i. The users have the discretion to pick up the top k projected outliers that have the highest outlier-ness. In contrast, the traditional definition of outliers does not explicitly present outlying subspaces of outliers in the final result as outliers are detected in the full or a pre-specified data space that is known to users before outliers are detected.

In this paper, we present a new technique, called Stream Projected Outlier deTector (SPOT), to approach the problem of outlier detection in high-dimensional data streams. The major technical contributions of this paper can be summarized as follows:

- SPOT constructs the novel Sparse Subspace Template (SST) to detect projected outliers. SST consists of a number of mutually supplemented subspace groups that contribute collectively to an effective detection of projected outliers. SPOT is able to perform supervised and/or unsupervised learning to construct SST, providing a maximum level of flexibility to users. The strategy of self-evolution of SST has also been incorporated into SPOT to greatly enhance its adaptability to dynamics of data streams;
- Unlike most other outlier detection methods that measure outlier-ness of data points based on a single criterion, SPOT adopts a more flexible framework allowing for the use of multiple measures for outlier detection. Employing multiple measures is generally more effective than a single measure. SPOT utilizes the Multi-Objective Genetic algorithm (MOGA) as an effective search method to find subspaces that are able to optimize these criteria for constructing SST;
- We show that SPOT is efficient and outperforms the existing methods in terms of effectiveness through experiments on both synthetic and real-life data sets.

The rest of this paper is organized as follows. The basic concepts and definitions used in SPOT will be presented in Section 2. In Section 3, we dwell on algorithms of SPOT, with an emphasis on the learning and detection stages of SPOT. Experimental results of SPOT are reported in Section 4. The final section concludes this paper.

2   Concepts and Definitions

2.1   Data Space Partitioning

To facilitate the quantification of data synopsis for outlier detection, a hypercube is superimposed and equi-width partition of domain space is performed. Each attribute of the data is partitioned into a few number of non-overlapping equal-length intervals. The cells in hypercube can be classified into two categories, *i.e.*, the base and projected cells. A *base cell* is a cell in the full data space (with the finest granularity in hypercube). The dimensionality (*i.e.*, number of attributes) of a base cell is equal to φ, where φ is the dimension of the data stream. A *projected cell* is a cell that exists in a particular subspace. The dimensionality of a projected cell is smaller than φ.

2.2 Data Synopsis

Based on the hypercube structure, we employ two data synopsis, called *Base Cell Summary* (BCS) and *Projected Cell Summary* (PCS), to capture the major underlying characteristics of the data stream for detecting projected outliers. They are defined as follows.

Definition 1. Base Cell Summary (BCS): The Base Cell Summary of a base cell c in the hypercube is a triplet defined as $BCS(c) = \{D_c, \vec{LS_c}, \vec{SS_c}\}$, where D_c, $\vec{LS_c}$ and $\vec{SS_c}$ denote the number of points in c, the sum and squared sum of data values in each dimension of points in c, respectively, *i.e.*, $\vec{LS_c} = \sum \vec{p_i}$ and $\vec{SS_c} = \sum \vec{p_i}^2$, for p_i located in c, $1 \le i \le \varphi$.

BCS features two desirable properties, *i.e.*, additivity and incremental maintainability [21], that can be used to compute data synopsis for projected cells in subspaces.

Definition 2. Projected Cell Summary (PCS): The Projected Cell Summary of a cell c in a subspace s is a triplet defined as $PCS(c, s) = (RD, IRSD, IkRD)$, where RD, $IRSD$ and $IkRD$ are the Relative Density, Inverse Relative Standard Deviation and Inverse k-Relative Distance of data points in c of s, respectively.

RD, $IRSD$ and $IkRD$ are three effective measures to represent the overall data sparsity of each projected cell from different perspectives. They are used together in SPOT to achieve a good measurement of data outlier-ness. They are all defined as ratio-type measures in order to achieve statistical significance in measurement. The outlier-ness thresholds defined based on ratio-type measures are also intuitive and easy to specify.

Definition 3. Relative Density (RD): Relative Density of a cell c in subspace s measures the relative density of c w.r.t the expected level of density of non-empty cells in s. If the density of c is significantly lower than the average level of cell density in the same subspace, then the data in c can be labeled as outliers. RD is calculated as $RD(c, s) = \frac{D_c}{E(D_s)}$, where D_c and $E(D_s)$ represent the density (*i.e.*, number of points) in c and the expected density of all the cells in s. Since $E(D_s) = \frac{N}{\delta^{|s|}}$, where N corresponds to the effective stream length (the decayed total number of data points at a certain time), thus, $RD(c, s) = \frac{D_c \cdot \delta^{|s|}}{N}, p \in c$.

Definition 4. Inverse Relative Standard Deviation (IRSD): Inverse Relative Standard Deviation of a cell c in subspace s is defined as inverse of the ratio of standard deviation of c in s against the expected level of standard deviation of non-empty cells in s. Under a fixed density, if the data in a cell features a remarkably high standard deviation, then the data are distributed more sparsely in the cell and the overall outlier-ness of data in this cell is high. $IRSD(c, s)$ is computed as $IRSD(c, s) = \left[\frac{\sigma_c}{E(\sigma_s)}\right]^{-1}$, where σ_c denotes the standard deviation of c and $E(\sigma_s)$ denotes the expected standard deviation of cells in subspace s. Since σ_c is larger than 0 but does not exceed the length of the longest diagonal of the cell in subspace s, which is $\sqrt{|s|}l$, where $|s|$ is the dimensionality of s and l is the side length of each interval, thus $E(\sigma_s)$ can be estimated as $E(\sigma_s) = \frac{0 + \sqrt{|s|}l}{2} = \frac{\sqrt{|s|}l}{2}$.

Definition 5. Inverse k-Relative Distance (IkRD): Inverse k-Relative Distance for a cell c in a subspace s is the inverse of ratio of the distance between the centroid of c and its nearest representative points in s against the average level of such distance in s for all the non-empty cells. A high IkRD value of c indicates that c is noticeably far from the dense regions of the data in s, thus the outlier-ness of data in c is high. The IkRD is defined as $IkRD(c,s,k) = \left[\frac{k\_dist(c,s)}{average\_k\_dist(c_i,s)} \right]^{-1}$, $k\_dist(c,s)$ is the sum of distances between the centroid of c and its k nearest representative points in s and $average\_k\_dist(c_i,s)$ is the average level of $k\_dist(c_i,s)$ for all the non-empty cells in s.

3   Stream Projected Outlier Detector (SPOT)

An overview of SPOT is presented in Figure 1. SPOT can be broadly divided into two stages: the learning stage and the detection stage. SPOT can further support two types of learning, namely offline learning and online learning. In the offline learning, Sparse Subspace Template (SST) is constructed using either the unlabeled training data (*e.g.*, some available historic data) and/or the labeled outlier examples provided by domain experts. SST is a set of subspaces that features a higher data sparsity/outlier-ness than others. It casts light on where projected outliers are likely to be found in the high-dimensional space. SST consists of three groups of subspaces, *i.e.*, Fixed SST Subspaces ($\mathcal{FS}$), Unsupervised SST Subspaces ($\mathcal{US}$) and Supervised SST Subspaces ($\mathcal{SS}$), where $\mathcal{FS}$ is a compulsory component of SST while $\mathcal{US}$ and $\mathcal{SS}$ are optional components. SST is mainly constructed in an unsupervised manner where no labeled examples are required. However, it is possible to use the labeled outlier exemplars to further improve SST. As such, SPOT is very flexible and is able to cater for different practical applications that may or may not have available labeled exemplars. Multiobjective Genetic Algorithm (MOGA) is used for outlying subspace search in constructing $\mathcal{US}$ and $\mathcal{SS}$.

When SST is constructed, SPOT can start to screen projected outliers from constantly arriving data in the detection stage. The arriving data will be first used to update the data summaries (*i.e.*, PCSs) of the cell it belongs to in each subspace of SST. This data will then be labeled as a projected outlier if PCS values of the cell where it belongs to are lower than some pre-specified thresholds. The detected outliers are archived in the so-called Outlier Repository. Finally, all or only a specified number of the top outliers in Outlier Repository will be returned to users when the detection stage is finished.

During the detection stage, SPOT can perform online learning periodically. The online learning involves updating SST with new sparse subspaces that SPOT finds based on the current data characteristics and the newly detected outliers. Online learning improves SPOT's adaptability to dynamic of data streams.

3.1   Learning Stage of SPOT

Since the number of subspaces grows exponentially with regard to data dimensionality, evaluating each streaming data in each possible subspace becomes prohibitively expensive. As such, we only evaluate each point in SST alternatively, in an effort to render this problem tractable. The central task of the offline learning stage is to construct SST.

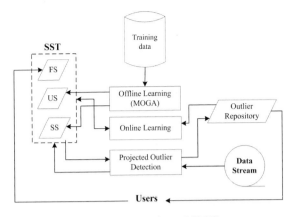

Fig. 1. An overview of SPOT

Construction of SST. It is desired that SST can contain one or more outlying subspaces for as many projected outliers in the streams as possible. To do this, SST is designed to contain a few groups of subspaces that are generated by different underlying rationales. Different subspace groups supplement each other towards capturing the right subspaces where projected outliers are hidden. This helps enable SPOT to detect projected outliers more effectively. Specifically, SST contains the following three subspace groups, *Fixed SST Subspaces ($\mathcal{FS}$), Unsupervised SST Subspaces ($\mathcal{US}$)* and *Supervised SST Subspaces ($\mathcal{SS}$)*, respectively.

• *Fixed SST Subspaces ($\mathcal{FS}$)*
Fixed SST Subspaces ($\mathcal{FS}$) contains all the subspaces in the full lattice whose maximum dimension is $MaxDimension$, where $MaxDimension$ is a user-specified parameter. In other words, $\mathcal{FS}$ contains all the subspaces with dimensions of $1, 2, \cdots,$ $MaxDimension$. $\mathcal{FS}$ satisfies that $\forall s$, we have $|s| \leq MaxDimension$ if and only if $s \in \mathcal{FS}$. Construction of $\mathcal{FS}$ does not require any learning process.

• *Unsupervised SST Subspaces ($\mathcal{US}$)*
Unsupervised SST Subspaces ($\mathcal{US}$) are constructed through an unsupervised offline learning process. We assume that a set of historical data is available for this unsupervised learning. All the training data are scanned and assigned into one (and only one) cell in the hypercube. The BCS of each occupied cell in H are maintained during this data assignment process. Multi-Objective Genetic Algorithm (MOGA) is then applied on the whole training data to find the subspaces that feature a higher number of outliers. These subspaces will be added to $\mathcal{US}$.

Once we have obtained the initial $\mathcal{US}$, we can further procure more useful subspaces for $\mathcal{US}$. We can find the outlying subspaces of the top training data that have the highest overall outlying degree. The selected training data are more likely to be considered as outliers and they can be potentially used to detect more similar outliers in the stream. The overall outlying degree of the training data is computed in an unsupervised manner by employing clustering analysis.

As the distance between two data points may vary significantly in different subspaces, we therefore expect the distance metric used in the clustering to be able to well reflect the overall distance of data points in difference subspaces, especially those where outliers are likely to be detected. To achieve this, we employ a novel distance metric, called *Outlying Distance(OD)*, for clustering training data. It is defined as the average distance between two points in top sparse subspaces of the whole training data obtained by MOGA, *i.e.*, $OD(p_1, p_2) = \frac{\sum_{i=1}^{m} dist(p_1, p_2, s_i)}{m}$, where m is the number of top sparse subspaces returned by MOGA and s_i is the ith subspaces in this set.

We utilize *lead clustering method*, also called the fixed-width clustering, to cluster the whole training data into a few clusters. Lead clustering method is a highly efficient clustering method that adopts an incremental paradigm to cluster data. For each data p in the data set that has yet been clustered, it will be assigned to the cluster c' such that $OD(p, c') < d_c$ and $\forall c_i \neq c', OD(p, c') \leq OD(p, c_i)$. The centriod of c' will be updated upon the cluster assignment of p. If $\forall c_i$, we have $OD(p, c_i) \geq d_c$, then a new cluster is initiated and p becomes the centroid of this new cluster. These steps will be repeated until all the data in the data set have been clustered.

Due to its incremental nature, lead clustering method features a promising linear scalability with regard to number and dimensions of training data. However, its result is sensitive to the order in which the training data are clustered. To solve this problem, we perform lead clustering in multiple runs under different data orders. Even though an outlier may be assigned into different clusters in different runs, the chance that it is assigned to a small cluster is relatively high. Hence, the outlying degree of a training data can be measured by the average size of clusters it belongs to in different runs.

● *Supervised SST Subspaces* ($\mathcal{SS}$)
In some applications, a small number of outlier exemplars may be provided by domain experts or are available from some earlier detection process. These outlier exemplars can be considered as carriers of domain knowledge that are potentially useful to improve SST for better a detection effectiveness. MOGA is applied on each of these outlier exemplars to find their top sparse subspaces. There subspaces are called Supervised SST Subspaces ($\mathcal{SS}$). Based on $\mathcal{SS}$, example-based outlier detection [22] can be performed that effectively detects more outliers that are similar to these outlier examples.

Computing PCS of a Projected Cell. In MOGA, PCS of projected cells need to be computed. The central components required for quantifying RD, IRSD, IkRD of PCS for a projected cell c are the density (D_c), mean (μ_c) and standard deviation (σ_c) of data points in c and the representative points in the subspace where c exists. Fortunately, as shown in [21], D_c, μ_c and σ_c can be efficiently obtained using the information maintained in BCS. Thus, we only show how representative points can be obtained for computing IkRD here. The representative points of a subspace s are the centroids of a selected set of non-empty cells in s. This selected set of cells are termed *coverage cells*. A rule-of-thumb in selecting coverage cells in a subspace s is to select a specified number of the most populated cells in s such that they cover a majority of data in s, such as 90%. It is noted that the initial representative data for different subspaces are obtained offline before outlier detection is conducted, thus it is not subject to the one-pass scan and time-criticality requirements of data stream applications.

Figure 2 and 3 present the algorithms of the unsupervised and the supervised learning in SPOT.

Algorithm. SPOT_Unsupervised_Learning (D_T, d_c, N_{runs})

Input: Training data D_T, clustering distance threshold d_c, number of clustering runs N_{runs};
Output: Unsupervised SST subspaces;
1. $US \leftarrow \emptyset$;
2. **MOGA**(D_T);
3. $US \leftarrow US \cup$ top sparse subspaces of D_T;
4. FOR i=1 to N_{runs} DO
5.   **Cluster**(D_T, US, d_c);
6.   **ComputeOF**(D_T);
7.   FOR each top training data p DO {
8.   **MOGA**(p);
9. $US \leftarrow US \cup$ top sparse subspaces of p; }
10. $SST \leftarrow US$;
11. **Return**(SST);

Fig. 2. Unsupervised learning algorithm of SPOT

Algorithm. SPOT_Supervised_Learning (OE)

Input: Set of outlier examplars OE;
Output: Supervised SST subspaces;
1. $SS \leftarrow \emptyset$;
2. FOR each outlier examplar o in OE DO {
3.   **MOGA**(o);
4. $SS \leftarrow SS \cup$ top sparse subspaces of o; }
5. $SST \leftarrow SS$;
6. **Return**(SST);

Fig. 3. Supervised learning algorithm of SPOT

3.2 Multi-Objective Genetic Algorithm (MOGA)

In the offline learning stage, we employ Multi-Objective Genetic Algorithm (MOGA) to search for subspaces whose RD, IRSD and IkRD objectives can be minimized in construction of SST. MOGA conducts search of good subspaces through a number of generations each of which has a population containing a specific number of individuals (*i.e.*, subspaces). The subspaces in the first generation are typically generated randomly, while the subspaces in the subsequent generations are generated by applying search operators such as crossover and mutation on those subspaces in their preceding generation. In a multi-objective minimization problem, subspaces in the population of each generation can be positioned on different trade-off surfaces in the objective function space. The subspaces located on a surface closer to the origin is better than the one far from the origin. The superiority (or inferiority) of the subspaces on the same surface are not distinguishable. The surface where the optimal subspaces are located is called *Pareto Front*. The goal of MOGA is to gradually produce an increasing number of optimal subspaces, located in the Pareto Front, from non-optimal subspaces as evolution proceeds. MOGA provides a good general framework for dealing with multi-objective search problems. However, we still need to perform ad-hoc design of MOGA in SPOT for outlying subspace search, including individual representation, objective functions, fitness function, selection scheme and elitism.

Individual Representation. In SPOT, all individuals are represented by binary strings with fixed and equal length φ, where φ is the number of dimensions of the dataset. Each bit in the individual will take on the value of "0" and "1", respectively, indicating whether or not its corresponding dimension is selected for a particular subspace.

Objective Functions. We need to define objective functions for subspaces w.r.t two types of data. The first type of data is *a single data point*. This applies to each top training data and each outlier exemplar. For a single data point, we have $f(p, s) = f(c, s)$, meaning that the objective function of subspace s w.r.t data point p is the data sparsity (measured by RD, IRSD, IkRD) of the cell c in s where p belongs to. The second type of data is *the whole training data* D_t. The objective function of s for D_t is defined as the percentage of data points in D_t with low PCS in s. Its calculation involves summing up the densities of projected cells in s with lower PCS. Note that the objective functions for both types of data are 3-dimensional function vectors.

Fitness Function. The fitness of a subspace s in SPOT is defined based upon its Pareto-count in the whole population. The Pareto-count of s is the number of subspaces that are inferior to s in the population, *i.e.*, $fitness(s) = |\{s_i : s \succ s_i\}|$. s_i is inferior to s, denoted as $s \succ s_i$, if none of the objective function values of s_i is better than or as good as s.

Selection Scheme. *Pareto-based selection scheme* is used to select fitter solutions in each step of evolution. It is a stochastic selection method where the selection probability of a subspace is proportional to its fitness value, *i.e.*, $Pr(s) = \frac{fitness(s)}{\sum_{i=1}^{P} fitness(s_i)}$, where P is the population size.

Elitism. Elitism is the effort to address the problem of losing those potentially good solutions during the optimization process because of the randomness of MOGA. If no special measures are taken, MOGA cannot guarantee the individuals obtained in a newer generation always outperform those in the older one. In SPOT, we use the *elitism* method that directly copies the best or a few best subspaces in the population of one generation to the next one, in an attempt to achieve a constantly improving population of subspaces.

3.3   Detection Stage of SPOT

The detection stage of SPOT performs outlier detection for arriving streaming data. As streaming data arrive continuously, the data synopsis PCS of the projected cell where the streaming data belongs to in each subspace of SST are first updated in order to capture new information of the arriving data. Hash function is employed here to quickly map a data into the cell it is located in any subspace. Then, the data is labeled as a projected outlier on the fly if RD, IRSD or IkRD of the cell it belongs to in one or more

SST subspaces falls under the pre-specified outlier-ness thresholds. These subspaces are the outlying subspaces of this outlier. All the outliers, together with their outlying subspaces and PCS of the cell they belong to in these outlying subspaces, are output to the Outlier Repository. All or a specified number of the top outliers in Outlier Repository are retuned to users in the end.

Updating PCS of a Projected Cell. The detection stage of SPOT involves the update of PCS of projected cells as data in the stream are being processed. In this subsection, we will demonstrate the incremental property of PCS, based on which PCS can be self-maintainable without using BCS.

• *Update RD of PCS.* Suppose that the latest snapshot of PCS of c is created at time T with an effective stream length N, then we can update RD in PCS of c incrementally when a new data point arrives at c at the time T' ($T \leq T'$) with an effective stream length N' as follows:

$$RD(c,s)^{T'} = \frac{[df(T'-T)\frac{RD(c,s)^T \cdot N}{\delta^{|s|}} + 1] \cdot \delta^{|s|}}{N'} \tag{1}$$

where $\frac{RD(c,s)^T \cdot N}{\delta^{|s|}}$ is the density of c at time T and $df(T'-T)\frac{RD(c,s)^T \cdot N}{\delta^{|s|}} + 1$ is thus the new density of c at time T'. After simplifying Eq. (1), we can get

$$RD(c,s)^{T'} = df(T'-T)\frac{N}{N'}RD(c,s)^T + \frac{\delta^{|s|}}{N'} \tag{2}$$

From Eq. (2) we can see that, for incremental maintenance of RD, we need to only additionally maintain, for each non-empty cell in the subspace, the effective stream length N when PCS of this cell is updated last time.

• *Update IRSD of PCS.* Suppose the density of cell c is m and let $IRSD(c,s)^T$ be the IRSD of cell c in subspace s at time T, which can be computed as follows based on the definition of $IRSD(c,s)^T$:

$$IRSD(c,s)^T = \frac{\sqrt{|s|l}}{2\sigma(c)} = \frac{\sqrt{|s|l}}{2}\sqrt{\frac{m-1}{\sum_{i=1}^m Dist(p_i,\mu_c)^2}} \tag{3}$$

where p_i is located in c. Based on Eq. (3), we can get

$$\sum_{i=1}^m Dist(p_i,\mu_c)^2 = \frac{|s|l^2(m-1)}{4(IRSD(c,s)^T)^2} \tag{4}$$

The $IRSD(c,s)$ after the $(m+1)^{th}$ point is assigned into c at time T' ($T \leq T'$) is computed as

$$IRSD(c,s)^{T'} = \frac{\sqrt{|s|l}}{2}\sqrt{\frac{df(T'-T)(m-1)+1}{df(T'-T)\sum_{i=1}^m Dist(pi,\mu'_c)^2 + Dist(p_{m+1},\mu'_c)^2}} \tag{5}$$

where μ'_c denotes the *new* mean of points in c when the $(m+1)^{th}$ point is inserted into c.

Algorithm. SPOT_Detection (SST, t, N_{can}, top_k)

Input: SST, self-evolution time period t and number of new subspaces generated N_{can} in self-evolution of SST;

Output: Outlier_Repository where outliers detected are stored.

1. SST_Candidate← ∅;
2. WHILE NOT (end of stream) DO {
3. IF a new data p in the stream arrives THEN {
4.   **Update_BCS**(p);
5.   **Update_PCS**(p, SST, SST_Candidate);
6.   IF (**Outlier_Detection**(p, SST)=True) THEN
7.   **Insert**(Outlier_Repository, p); }
8. IF ((Curent_time–Start_time) mod t=0) THEN {
9.   SST← **SST_Evolution**(SST, SST_Candidate);
10.   SST_Candidate ← **Generate_SST_Candidate**(SST, N_{can});
11.   For each new outliers o added to Outlier_Repository DO{
12.   **MOGA**(o);
13.   SST← SST ∪ top sparse subspaces of o; } } }
14. **Return**(**top_k_outliers**(Outlier_Repository, top_k));

Fig. 4. Detection algorithm of SPOT

● *Update Representative Points in a Subspace.* To ensure a quick computation of IkRD for a new data in the stream, we need to obtain the current set of coverage cells efficiently. To solve this problem, we devise the following heuristics to minimize the number of re-sorting of projected cells:

1. If a new data falls into one of the coverage cells in a subspace, then there is no need to update the current set of coverage cells in this subspace;

2. Both the total coverage and the minimum density of the current set of coverage cells in each subspace s of SST are maintained, denoted by Cov and Den_{min}, respectively. If a new data falls into a non-coverage cell c' in s, then there is no need to update the current set of coverage cells in s either if we have $Cov' > q$ and $den(c') \leq Den'_{min}$, where Cov' and Den'_{min} correspond respectively to the decayed Cov and Den_{min} after the new data is processed, and q denotes the coverage ratio required for the coverage cells. Both Cov' and Den'_{min} can be updated efficiently.

The detection algorithm of SPOT is presented in Figure 4.

4   Experimental Results

We use both synthetic and real-life datasets for performance evaluation. All the experimental evaluations are carried out on a Pentium 4 PC with 512MB RAM.

Synthetic Data Sets. The synthetic data sets are generated by two high-dimensional data generators. The first generator SD1 is able to produce data sets that generally exhibit remarkably different data characteristics in different subspaces. The number, location, size and distribution of the data in different subspaces are generated randomly. This generator has been used to generate high dimensional data sets for outlying subspace detection [20][18][19]. The second synthetic data generator SD2 is specially designed for comparative study of SPOT and the existing methods. Two data ranges are defined as $R_1 = (a, b)$ and $R_2 = (c, d)$, where $b + l < c$, l is the length of a partitioning interval in each dimension. This ensures that the data points in R_1 and R_2 will not fall into the same interval for each dimension. In SD2, we first generate a large set of normal data points D, each of which will fall into R_1 in $\varphi - 1$ dimensions and into R_2 in only one dimension. We then generate a small set of projected outliers O. Each projected outlier will be placed into R_2 for all the φ dimensions. Given the large size of D relative to O, no projected outliers will exist in D. An important characteristic of SD2 is that the projected outliers appear perfectly normal in all 1-dimensional subspaces.

Real Data Sets. We also use 4 real-life multi- and high-dimensional datasets in our experiments. The first two data sets come from the UCI machine learning repository. They are called *Letter Image* (RD1, 17-dimensions) and *Musk* (RD2, 168-dimensions), respectively. The third real-life data set is the *KDD-CUP'99 Network Intrusion Detection stream data set* (RD3, 42 dimensions). RD3 has been used to evaluate the clustering quality for several stream clustering algorithms [4][5]. The fourth real-life data set is the MIT wireless LAN (WLAN) data stream (RD4, 15 dimensions), which can be downloaded from *http://nms.lcs.mit.edu/ mbalazin/wireless/*.

4.1 Scalability Study

The scalability study investigates the scalability of SPOT (both learning and detection processes) w.r.t length N and dimensionality φ of data streams. The learning process we study here refers only to the unsupervised learning that generates $\mathcal{US}$ of SST. Due to its generality, SD1 with different N and φ is used in all scalability experiments.

Scalability of Learning Process w.r.t N. Figure 5 shows the scalability of unsupervised learning process w.r.t number of training data N. The major tasks involved in the unsupervised learning process are multi-run clustering of training data, selection of top training data that have highest outlying degree and application of MOGA on each of them to generate $\mathcal{US}$ of SST. The lead clustering method we use requires only a single scan of the training data, and the number of top training data we choose is usually linearly depended on N. Therefore, the overall complexity of unsupervised learning process scales linearly w.r.t N.

Scalability of Learning Process w.r.t φ. Since the construction of $\mathcal{FS}$ in SST does not need any leaning process, thus the dimension of training data φ will only affect the complexity of learning process in a linear manner, *provided that a fixed search workload is specified for MOGA*. As confirmed in Figure 6, we witness an approximately linear growth of execution time of learning process when φ is increased from 20 to 100 under a fixed search workload in MOGA.

Scalability of Detection Process w.r.t N. In Figure 7, we present the scalability result of detection process w.r.t stream length N. In this experiment, the stream length is set much larger than the number of training data in order to study the performance of SPOT in coping with large data streams. Figure 7 shows a promising linear behavior

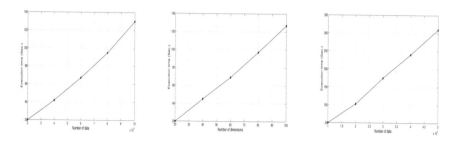

Fig. 5. Scalability of learning process w.r.t data number   **Fig. 6.** Scalability of learning process w.r.t data dimension   **Fig. 7.** Scalability of detection process w.r.t data number

of detection process when handing an increasing amount of streaming data. This is because that the detection process needs only one scan of the arriving streaming data. In addition, since BCS and PCS are both incrementally maintainable, detection process of SPOT thus becomes very efficient. This leads to a high throughput of SPOT and enables it to deal with fast data streams.

Scalability of Detection Process w.r.t φ. φ affects the size of $\mathcal{FS}$ that is used in detection process. When $MaxDimension$ is fixed, the size of $\mathcal{FS}$ is in an exponential order of φ, which is usually much larger than that of $\mathcal{US}$ and OS. This causes $\mathcal{FS}$ to dominate the whole SST. As such, the execution time of detection process is expected to grow exponentially w.r.t φ. We typically set lower $MaxDimension$ values for data streams with higher dimensionality to prevent an explosion of $\mathcal{FS}$. We first use $MaxDimension = 3$ for data streams of different dimensions and we can see an exponential behavior of the detection process. Then, we use variable values for $MaxDimension$ to adjust the size of $\mathcal{FS}$. We set $MaxDimension = 4$ for data sets with dimension of 20 and 40, set $MaxDimension = 3$ for data sets with dimension of 60 and finally set $MaxDimension = 2$ for data sets with dimension of 80 and 100. If this strategy is used, we will witness an irregularly-shaped, rather than an exponential, curve of the detection process. The results are presented in Figure 8.

4.2   Effectiveness Study

Convergence Study of MOGA. We first study the convergence of MOGA in terms of optimizing RD, IRSD and IkRD. Convergence of MOGA is crucial to outlying subspaces search in SPOT. We investigate the average of RD, IRSD and IkRD of the top 10 subspaces in the population of each generation of MOGA. This experiment is conducted on SD1, RD1, RD2, RD3 and RD4. Only the results of RD3 (KDD-CUP'99 data stream) are presented (see Figure 9). Similar results are obtained for other datasets. Generally speaking, the criteria are being improved (minimized) as more generations are performed in MOGA. This indicates a good convergence behavior of MOGA in searching outlying subspaces. However, there are some abrupt increase of optimizing criteria values in the search process. The underlying reason is that, when elitism is not used, there is a higher chance of occurrence of *degrading generations* in the evolution. This is due to the randomized nature of MOGA that likely renders good solutions in one

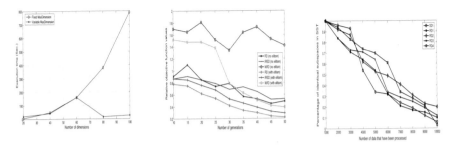

Fig. 8. Scalability of detection process w.r.t data dimension

Fig. 9. Convergence study of MOGA

Fig. 10. Evolution of SST

generation to be lost in the next one by crossover or mutation operations. When elitism is employed, we can achieve a better optimization of RD, IRSD and IkRD.

Detection Performance of SST. The three sub-groups of SST work collectively to detect outliers from data streams. Their respective contribution in outlier detection is not identical though. This experiment aims to study the percentage of outliers that can be detected by using each of the sub-groups of SST alone. The experimental results indicate that by using $\mathcal{FS}$ alone, which covers the full low-dimensional space lattice, we can detect as many as 80-90% of the total outliers that exist, while $\mathcal{US}$ and $\mathcal{SS}$ can further help the system detect another 10-20% of the outliers.

Evolution of SST. One of the important features of SPOT is its capability of self-evolution. This is a very useful feature of SPOT to cope with the fast-changing data streams. In this experiment, we investigate the evolution of SST as an informative indicator of concept drift of data streams. The setup is as follows. We keep the initial version of SST (*i.e.*, the SST obtained after 1000 data points are processed) and record the versions of SST when every 1000 data (up to 10,000) are processed afterwards. Self-evolution is activated at every 1000-data interval. We compare different SST versions with the initial one and calculate the percentage of identical subspaces. SD1, RD1, RD2, RD3 and RD4 are used for this experiment. The results are shown in Figure 10. We find that an increasing number of subspaces in the initial SST have disappeared in the later SST versions as more data points are processed. We use the same seeds in MOGA, ruling out the randomness in individual generation for different self-evolution sessions. Therefore, the change of SST is due to the varying characteristics of the data stream and outliers we detect in different stages.

Comparative Study. Since there is little research conducted in projected outlier detection for high-dimensional data streams, we cannot find the techniques tackling exactly the same problem as SPOT does for comparison. However, there are some related existing approaches for detecting outlier detection from data streams that we can compare SPOT with. They can be broadly categorized as methods using *histogram, Kernel density function, distance-based function* and *clustering analysis*, respectively.

Histogram technique creates bins for each dimension of the data stream. The density of each bin of the histogram are recorded. The sparsity of a bin in the histogram can be quantified by computing the ratio of the density of this bin against the average density

Methods	SD2		RD3		RD4	
	DR	FPR	DR	FPR	DR	FPR
Histogram	0%	32.1%	43.3%	24.7%	48.9%	19.2%
Kernel density function (random single subspace)	3.2%	1.1%	7.3%	0%	2.8%	0.3%
Kernel density function (random multi subspaces)	87.1%	1.5%	79.3%	8.6%	83.8%	7.4%
Kernel density function (SST)	100%	0%	84.3%	4.6%	91.9%	7.5%
Incremental LOF (random single subspace)	6.2%	0%	8.9%	0.1%	4.5%	0%
Incremental LOF (random multi subspaces)	85.2%	2.5%	79.2%	3.2%	81.7%	4.9%
Incremental LOF (SST)	95%	0.7%	91.4%	7.9%	90.7%	8.9%
HPStream	8.3%	1.2%	7.3%	1.1%	3.1%	0.8%
SPOT	100%	0%	89.5%	5.3%	92.9%	6.5%

Fig. 11. Comparative study results

of all the bins in the histogram for this dimension. A data point is considered as outlying in a dimension if it falls into an excessively sparse bin. Kernel density function models local data distribution in a single or multiple dimensions of space. A point is detected as an outlier if the number of values that have fallen into its neighborhood is less than an application-specific threshold. The number of values in the neighborhood can be computed by the kernel density function. Distance-based function draws on some distance metrics to measure the local density of data in order to find outliers. A major recent distance-based method for data stream is called *Incremental LOF* [15]. Clustering analysis can also be used to detect outliers from those data that are located far apart from data clusters. *HPStream* [5] is the representative method for finding subspace clusters in high-dimensional data streams.

The performance of all the methods are measured by *detection rate (DR)* and *false positive rate (FPR)*. The results are summarized in Figure 11. To facilitate result analysis, two selection rules are devised based on DR and FPR. They are $R_1 : DR > 90\%$ and $FPR < 5\%$, $R_2 : DR > 85\%$ and $FPR < 10\%$. Obviously, R_1 is more desirable than R_2 in terms of detection performance. We highlight the cells in the table that satisfy R_1 using red colour and those that satisfy R_2 using purple colour, respectively. After colouring, it becomes much easier for us to see, from a macro scale, that SPOT and Incremental LOF (SST) achieve the best performance overall for the three data sets, followed by Kernel density function (SST), Kernel density function (random multi subspaces) and Incremental LOF (random multi subspaces). Histogram, Kernel density function (random single subspace), Incremental LOF (random single subspace) and HPStream bottom the list, which does not contain any colored (DR, FPR) pairs. Compared with other competitive methods, SPOT is advantageous that it is equipped with subspace exploration capability, which contributes to a good detection rate. Moreover, using multiple criteria enables SPOT to deliver much more accurate detection which helps SPOT to reduce its false positive rate.

5   Related Work

There have been abundant research in outlier detection in the past decade. Most of the conventional outlier detection techniques are only applicable to relatively low dimensional static data [7][10][11][13][17][14]. Because they use the full set of attributes for outlier detection, thus they are not able to detect projected outliers. They cannot handle data streams either. Recently, there are some emerging work in dealing with outlier detection either in high-dimensional static data or data streams. However, there have not been any reported concrete research work so far for exploring the intersection of these two active research directions. For those methods in projected outlier detection in high-dimensional space [3][22][20][18][19], they can detect projected outliers that are embedded in subspaces. However, their measurements used for evaluating points' outlier-ness are not incrementally updatable and many of the methods involve multiple scans of data, making them incapable of handling data streams. For instance, [3][22] use the Sparsity Coefficient to measure data sparsity. Sparsity Coefficient is based on an equi-depth data partition that has to be updated frequently in data stream. This will be expensive and such updates will require multiple scan of data. [20][18][19] use data sparsity metrics that are based on distance involving the concept of kNN. This is not suitable for data streams either as one scan of data is not sufficient for retaining kNN information of data points. One the other hand, the techniques for tackling outlier detection in data streams [12][2][16] rely on full data space to detect outliers and thus projected outliers cannot be discovered by these techniques.

6   Conclusions

In this paper, we propose SPOT, a novel technique to deal with the problem of projected outlier detection in high-dimensional data streams. SPOT is equipped with incrementally updatable data synapses (BCS and PCS) and is able to deal with fast high-dimensional streams. It is flexible in allowing for both supervised and unsupervised learning in generating the detecting template SST. It is also capable of handling dynamics of date streams. The experimental results demonstrate the efficiency and effectiveness of SPOT in detecting outliers in high-dimensional data streams.

References

1. Aggarwal, C.C., Yu, P.S.: An effective and efficient algorithm for high-dimensional outlier detection. VLDB Journal 14, 211–221 (2005)
2. Aggarwal, C.C.: On Abnormality Detection in Spuriously Populated Data Streams. In: SDM 2005, Newport Beach, CA (2005)
3. Aggarwal, C.C., Yu, P.S.: Outlier Detection in High Dimensional Data. In: SIGMOD 2001, Santa Barbara, California, USA, pp. 37–46 (2001)
4. Aggarwal, C.C., Han, J., Wang, J., Yu, P.S.: A Framework for Clustering Evolving Data Streams. In: VLDB 2003, Berlin, Germany, pp. 81–92 (2003)
5. Aggarwal, C.C., Han, J., Wang, J., Yu, P.S.: A Framework for Projected Clustering of High Dimensional Data Streams. In: VLDB 2004, Toronto, Canada, pp. 852–863 (2004)
6. Angiulli, F., Pizzuti, C.: Fast outlier detection in high dimensional spaces. In: Elomaa, T., Mannila, H., Toivonen, H. (eds.) PKDD 2002. LNCS, vol. 2431, pp. 15–26. Springer, Heidelberg (2002)
7. Breuning, M., Kriegel, H.-P., Ng, R., Sander, J.: LOF: Identifying Density-Based Local Outliers. In: SIGMOD 2000, Dallas, Texas, pp. 93–104 (2000)
8. Guttman, A.: R-trees: a Dynamic Index Structure for Spatial Searching. In: SIGMOD 1984, Boston, Massachusetts, pp. 47–57 (1984)
9. Han, J., Kamber, M.: Data Mining: Concepts and Techniques. Morgan Kaufman Publishers, San Francisco (2000)

10. Knorr, E.M., Ng, R.T.: Algorithms for Mining Distance-based Outliers in Large Dataset. In: VLDB 1998, New York, NY, pp. 392–403 (1998)
11. Knorr, E.M., Ng, R.T.: Finding Intentional Knowledge of Distance-based Outliers. In: VLDB 1999, Edinburgh, Scotland, pp. 211–222 (1999)
12. Palpanas, T., Papadopoulos, D., Kalogeraki, V., Gunopulos, D.: Distributed deviation detection in sensor networks. SIGMOD Record 32(4), 77–82 (2003)
13. Ramaswamy, S., Rastogi, R., Kyuseok, S.: Efficient Algorithms for Mining Outliers from Large Data Sets. In: SIGMOD 2000, Dallas Texas, pp. 427–438 (2000)
14. Papadimitriou, S., Kitagawa, H., Gibbons, P.B., Faloutsos, C.: LOCI: Fast Outlier Detection Using the Local Correlation Integral. In: ICDE 2003, Bangalore, India, p. 315 (2003)
15. Pokrajac, D., Lazarevic, A., Latecki, L.: Incremental Local Outlier Detection for Data Streams. In: CIDM 2007, Honolulu, Hawaii, USA, pp. 504–515 (2007)
16. Subramaniam, S., Palpanas, T., Papadopoulos, D., Kalogeraki, V., Gunopulos, D.: Online Outlier Detection in Sensor Data Using Non-Parametric Models. In: VLDB 2006, Seoul, Korea, pp. 187–198 (2006)
17. Tang, J., Chen, Z., Fu, A.W.-c., Cheung, D.W.: Enhancing effectiveness of outlier detections for low density patterns. In: Chen, M.-S., Yu, P.S., Liu, B. (eds.) PAKDD 2002. LNCS, vol. 2336, p. 535. Springer, Heidelberg (2002)
18. Zhang, J., Lou, M., Ling, T.W., Wang, H.: HOS-Miner: A System for Detecting Outlying Subspaces of High-dimensional Data. In: VLDB 2004, Toronto, Canada, pp. 1265–1268 (2004)
19. Zhang, J., Gao, Q., Wang, H.: A Novel Method for Detecting Outlying Subspaces in High-dimensional Databases Using Genetic Algorithm. In: ICDM 2006, Hong Kong, China, pp. 731–740 (2006)
20. Zhang, J., Wang, H.: Detecting Outlying Subspaces for High-dimensional Data: the New Task, Algorithms and Performance. In: Knowledge and Information Systems (KAIS), pp. 333–355 (2006)
21. Zhang, T., Ramakrishnan, R., Livny, M.: BIRCH: An Efficient Data Clustering Method for Very Large Databases. In: SIGMOD 1996, Montreal, Canada, pp. 103–114 (1996)
22. Zhu, C., Kitagawa, H., Faloutsos, C.: Example-Based Robust Outlier Detection in High Dimensional Datasets. In: ICDM 2005, Houston, Texas, pp. 829–832 (2005)

Significance-Based Failure and Interference Detection in Data Streams

Nickolas J.G. Falkner and Quan Z. Sheng

School of Computer Science, The University of Adelaide
Adelaide, SA 5005, Australia
{jnick,qsheng}@cs.adelaide.edu.au

Abstract. Detecting the failure of a data stream is relatively easy when the stream is continually full of data. The transfer of large amounts of data allows for the simple detection of interference, whether accidental or malicious. However, during interference, data transmission can become irregular, rather than smooth. When the traffic is intermittent, it is harder to detect when failure has occurred and may lead to an application at the receiving end requesting retransmission or disconnecting. Request retransmission places additional load on a system and disconnection can lead to unnecessary reversion to a checkpointed database, before reconnecting and reissuing the same request or response. In this paper, we model the traffic in data streams as a set of significant events, with an arrival rate distributed with a Poisson distribution. Once an arrival rate has been determined, over-time, or lost, events can be determined with a greater chance of reliability. This model also allows for the alteration of the rate parameter to reflect changes in the system and provides support for multiple levels of data aggregation. One significant benefit of the Poisson-based model is that transmission events can be deliberately manipulated in time to provide a steganographic channel that confirms sender/receiver identity.

1 Introduction

The extensive use of sensor networks and distributed data gathering systems has increased both the rate and quantity of data that is delivered to receiving and processing nodes. Rather than processing a finite number of static records at a computationally-convenient time, data streams represent the fluctuating and potentially continuous flow of data from dynamic sources [1,2,3]. Data streams provide a rich and challenging source of data, with pattern identification and structure extraction providing important business knowledge for scientific, financial and business applications.

Several challenges occur when processing a data stream. Current data stream management techniques focus on a continuously generated stream of information and project analysis windows onto this stream, based on time intervals. While this works for a large number of applications, applications that wish to minimise onwards transmission based on the importance of data values may produce a data stream that appears discontinuous or fragmented. While a stream may be seen to never terminate, with a continuous flow of data, at a given time there may be no transmission activity within the data stream. Although the data stream may represent an abstract continuous data feed, it is implemented

S.S. Bhowmick, J. Küng, and R. Wagner (Eds.): DEXA 2009, LNCS 5690, pp. 645–659, 2009.
© Springer-Verlag Berlin Heidelberg 2009

as a flow of data packets. Given that a data stream is composed of individual data packets, all data streams are, at a network level, non-continuous but, within the database and sensing world, the *stream* is considered to be a continuous source of data from establishment to disconnection. Analysing this stream often requires a Data Stream Management System (DSMS) [4] that will place a set of expectations upon the performance characteristics of the data stream. The DSMS then provides an ability to query the stream, in an irregular, periodic or continuous manner. While data traffic on two data streams can be irregularly spaced or presented out-of-order [5,6], there is, however, an assumption that packets in a single stream will arrive at an, explicitly or implicitly defined, arrival rate that meets the expectations of both producer and consumer. Some challenges that we must address to provide the widest applicable model for data streams include:

- **Relaxing rigid time constraints:** The assumption that a data stream is both continuous and regularly periodic is a useful assumption, except where it is incorrect. For wireless sensor networks, with low energy budgets and aperiodic transmission profiles, a requirement to conform to a near-continuous transmission schedule will drain their resources rapidly. A requirement to conform to centralised time constraints places a synchrony burden on the system that requires specialised hardware or time poll updates to minimise drift.
- **Managing significant data as a separate class of values:** In any stream of data, some values will be of more significance than others. We wish to be able to process significant events in a timely fashion, potentially in a large collection of insignificant values. Without a classification model that allows the separation of the values in a data stream into distinct sets of significant events, we must deal with all values in the stream at the same level of priority for analysis and onwards transmission.
- **Failure detection without constant polling:** We wish to be able to detect failure and, preferably, interference or tampering without having to send a continuous set of data packets. Unless they are significant, continuous data packets only add to the amount of traffic that must be discarded at the sink, reducing effective bandwidth and consuming resources.

Consider two important applications for data stream interpretation: determining similar subsequences between data streams and determining skyline elements within a data stream. The subsequence problem requires the identification of elements of the data stream that are sufficiently characteristic to be identified in another stream, and the skyline problem requires us to find dominant elements in the stream, either across the entire stream or within a sliding window. In both cases, we are moving beyond a naive description of the stream it is now a stream of significant events, where this significance and its definition are key to our ability to mine the data stream while it is still being produced. Without an ongoing, and potentially evolvable, definition of significance, we cannot begin processing until we are sure that we have received all of the data and can now conduct our comparisons or selections with absolute certainty. This is, obviously, not a workable solution.

Our contribution is to provide a statistically-based model that can provide multiple views of the same data stream, with significance thresholds and arrival expectations defined for each view. By communicating expected rate information and comparing with actual rate information, we can detect failure and interference, and, in addition, exploit the statistical characteristics to provide an out-of-band communications channel. In this paper, we will provide a description of a significant event data stream that defines the nature of significance, the impact of the significance on the interpretation of the stream, and the expectation of the inter-arrival time in a way that allows the consumer and producer to agree upon a given rate, without being required to enforce a synchronous schedule.

The remaining sections of the paper are organized as follows. Section 2 presents the related work in this area. Section 3 introduces our model and provides the key definitions. Section 4 presents the failure detection model. The Out-of-band encoding mechanism is described in Section 5, with the results of experiments conducted to verify the work presented in Section 6. Finally, Section 7 provides some concluding remarks.

2 Related Work

Within data stream research, there are many examples of the application of an implied significant event stream over a continuous data stream. Dynamic Time Warping (DTW) [5] allows for the measurement of the similarity of two data sequences by allowing for a non-linear warping of the time scale on the measurements to provide a measurement of similarity between distinct events, regardless of their position in the data sequence. This does not, however, provide any estimates as to the constraining window that contains events of likely similarity. Other work has used Poisson-distributed test data to measure performance with well-established characteristics [7,8], but does not separate these Poisson streams for analysis, nor take the cumulative stream characteristics into account when multiple streams merge.

Streaming pattern discovery in multiple time-series (SPIRIT) [6] is an approach that employs principal component analysis to allow the location of correlations and hidden variables that are influencing the reported readings. SPIRIT is designed to allow the rapid incorporation of received events into dynamic weights, rather than detecting when a predicted event has not been received. SPIRIT can provide a reliable estimate by applying forecasting techniques to the hidden variables derived from the data stream and can use these to estimate missing data in the x_t data set, based on the x_{t-1} data set. However, this assumes continuous and regular data intervals. SPIRIT assumes periodicity in the data stream, rather than an irregular data stream or a data stream where the significant events are not spaced regularly in time.

StatStream [9] allows for the real-time analysis of large volumes of data, employing the Discrete Fourier Transform and a three-tiered time interval hierarchy. StatStream does assume that the data stream cannot be regarded as having a terminating point, and works on the data stream continuously, in order to meet real-time requirements. To be explicit, a data stream is regarded as a sequence, rather than a set. StatStream provides a basis for stating that any statistic present in the data stream at time t will be reported at time $t+v$, where v is a constant and is independent of the size and duration of the stream [9]. StatStream establishes three time periods: i) timepoints - the system quantum, ii) basic window - a consecutive subsequence of time points which comprise a digest, and iii) sliding window - a user-defined consecutive subsequence of basic windows that will form the basis for the time period over which a query may be executed.

While this provides a great deal of flexibility in dealing with intervals, StatStream expects to have at least one value per timepoint and, if one is not present, an interpolated value is used. This does not accommodate two possibilities. The first is that there is a reading but it is not sufficiently important to report at that time, where the system is filtering the result. The second is that there is no sensor reading to report at that point because the size of the time interval and the expected number of events leads to an event/interval ratio less than 1. More importantly, interpolation in the face of missing data may insert a false reading into the network. To explain this, it is first important to realise that, in the event of multiple values being reported in one timepoint, StatStream will provide a summary value. The reported arrival of an event in an adjacent interval can result in the preceding time interval being reported incorrectly, with the next interval providing a summary value that is also artificially high. While the synthesis of summary values will, over time, produce the same stream characteristics, there is no clear indication that an irregularity has occurred, nor can action take place to rectify the mistake. This summarisation can also obscure the point where a value, or set of values, has crossed the significance threshold.

There is a great deal of existing work in the field of sensor networks pertaining to more efficient use of resources through filtering, the discarding of insignificant data, and aggregation of significant data for more efficient upstream transmission. There is very little work that addresses the modelling of the implicit stream of significant events that these networks generate. Gu et al. [10] discuss a lightweight classification scheme for wireless sensor networks employing a hierarchical classification architecture but do not address statistical detection of node failure. Solis and Obraczka [11] discuss several aggregation mechanisms for power-efficiency but use temporal aggregation mechanisms and do not address node failure. Ye et al. [12] propose a statistical filtering process for injected false data but do not address statistical mechanisms for determining sensor operation.

3   Traffic Modelling in Intermittent Data Streams

Formally, a data stream is an ordered pair (s, δ) where s is a sequence of tuples of the form $(\xi_0, \xi_1...\xi_N)$, representing the data values contained in a single data packet, and δ is a sequence of time intervals where $\forall i, \delta_i > 0$.

In a regular, continuous data stream, the deviation from the average arrival time for each i is much smaller than the average arrival time (Equation 1). When data starts to arrive in an irregular fashion, the deviation can be of the same order of magnitude as the average time with a multiplier k, where k may be a fixed bound or may vary with time (Equation 2). The interpacket time interval, δ_i, must be sufficiently small that the client's expectation of continuity is not contradicted, otherwise the stream no longer appears continuous, i.e., it appears to be composed of irregular packets.

$$|\delta_{average} - \delta_i| < \epsilon, \forall i, \epsilon \ll \delta_{average} \tag{1}$$

$$|\delta_{average} - \delta_i| <= k\epsilon, \forall i, \epsilon \approx |\delta_{average}| \tag{2}$$

A user's expectation of continuity of a stream is only satisfied if k is sufficiently small that the δ_i values have an acceptable upper bound, δ_{max} (Equation 3). However, determining a reasonable expectation for δ_{max} is difficult, where data inter-arrival is irregular, as it may require a great deal of observation that is potentially only valid for one producer and consumer pair, or for one connection at a given time. This problem becomes more complex when we seek to place additional structure into the ongoing interpretation of a data stream.

$$\forall i, \delta_i < \delta_{max}, \delta_{max} >= \delta_{average} \tag{3}$$

3.1 Definition of the Model

Definition 1. *A significant event data stream, SED, is an ordered tuple* $(\mathcal{L}, \mathcal{E}, \Delta_\mathcal{L}, \lambda_\mathcal{L})$ *where:*

- *A tuple* $(\xi_0, \xi_1...\xi_N)$ *in the sequence* s *is significant if the removal or alteration of the tuple will have a significant and discrete impact on the result of a specified computation that depends on* s.
- $\mathcal{L}$ *is a unique label, identifying this SED.*
- $\mathcal{E}$ *is a sequence of tuples, such that* $\mathcal{E} \subseteq s$ *and all tuples in* $\mathcal{E}$ *are significant.*
- $\Delta_\mathcal{L}$ *is a sequence of time intervals where* $\forall i, \Delta_\mathcal{L} > 0$ *and* $\Delta_\mathcal{L} \sim Pois(\lambda_\mathcal{L})$.
- $\lambda_\mathcal{L}$ *is the expected arrival rate of* $\mathcal{E}_i \in \mathcal{E}$.
- *Any SED is a subsequence of an existing data stream,* $\mathcal{S} : (s, \delta)$ *where* $(s_i, \delta_i) \in (\mathcal{E}, \Delta) \not\Rightarrow (s_{i+1}, \delta_{i+1}) \in (\mathcal{E}, \Delta)$. □

Definition 2. *A SED Implementing Model (SEDIM) for a data stream is defined as follows:*

- *The data stream* $\mathcal{S}$ *is an ordered pair* (s, δ) *as defined previously,*
- *Within* $\mathcal{S}$, *there exists a set of significant event data streams,* $\hat{S} : (\hat{S}_0...\hat{S}_i), i \geq 0$.
- *A significant event* $\mathcal{E}$ *in* $(\hat{S}_i)$ *is defined such* $\forall \mathcal{E}_i \in \mathcal{E}, V_{lower_i} \leq \mathcal{E}_i \leq V_{upper_i}$, *where* V_{upper_i} *and* V_{lower_i} *represent the bounds of significance for the SED* $\hat{S}_i$.
- *The set* $\Lambda : (\lambda_{\mathcal{L}_0}...\lambda_{\mathcal{L}_n})$ *is a set of independent rate parameters for the expected arrival rate of significant data in the SEDs* $\hat{S}$.
- t_{now} *is defined as the current time in the system.*
- t_{window} *is defined as the time to the expiry of the current sliding window associated with a user query. An ongoing query may generate many subqueries, all of which have their own sliding window.*
- *The received set* $\mathcal{R}$ *is the set of all events that have been consumed in order to meet the requirements of the current query. The event* $\mathcal{E}_\mathcal{R}$ *is the last event that has been received and the event* $\mathcal{E}_{\mathcal{R}+1}$ *is the next event that will occur, regardless of the SED or stream that generates it. All events* $\mathcal{E}_{\mathcal{R}+1}$ *have an associated time interval* $\Delta_{\mathcal{R}+1}$ *and, by definition,* $\sum_{j=0}^{\mathcal{R}+1} \Delta_j > t_{now}$. □

3.2   Rationale

We must first justify that a data stream can be regarded as continuous, and still have the potential for irregular and insignificant data packet transfer. Many data stream applications make a similar assumption to StatStream, namely that there exists a system quantum and that events will arrive, at a roughly equal spacing, at a rate determined by the size of this quantum. For example, if the quantum is a second, the expected arrival rate is one event per second. This does, however, immediately provide the possibility of a continuous stream that can be viewed in such a way as to appear non-continuous in transmission. Consider a system with a one second timepoint and then, without changing the rate of arrival, change the timepoint to 0.1 seconds. Now the arrival rate is 1 event in every 10 timepoints and this, originally continuous and 100% utilised, data stream is now occupied 10% of the time. Considering the impact of pragmatic networking considerations on the transmission of data, given that it takes a finite non-zero time to transfer a network packet, there must exist a timepoint, T_ϵ, such that any data stream may be regarded non-continuous, regardless of the regularity of data transmission.

A more complex problem arises when, rather than managing regular transmission in a non-continuous data stream, we must consider the effect of irregularly spaced data, whether this is due to an absence of data or the insignificance of the data being transmitted. In the previous example, non-continuous data only constitutes a problem if the receiver fails to define their own aggregation or summary operations correctly for a given query in a sliding window. For example, if the value 10 is sent once per 10 timepoints, is the average over time 10 or 1? This will have an impact on the final answer delivered in response to the query "What is the average value over the sliding window from t_i to t_{i+x}?". As previously illustrated in the discussion of StatStream, a value that is incorrectly placed into a different timepoint can have a significant impact on the result of queries that span a subsequence, and this extends to the boundaries of sliding windows if they are purely timebased, rather than taking into account the possibility that an interval does not contain the same number of events. The query "What is the average value of the significant events received over the sliding window from t_i to t_{i+x}?" is an unambiguous query and, in the example above, would result in the value 10.

A data stream that is intermittently active may or may not have an associated sequence of significant events, given that the definition of significance is associated primarily with the consumer. However, if a significant event sequence exists within a data stream, the sequence may have a regular period that allows simple prediction of inter-event time, as the data stream can be composed of ongoing data combined with a regularly inserted significant element. It is, however, far more likely that events of significance will be more randomly distributed, unless what is being monitored for significance is naturally periodic or has been defined to be so.

We have employed the Poisson distribution to model the behaviour of significant event sequences arriving over data streams. The Poisson distribution is used where a number of discrete events occur within a given time-interval. Where an approximation can be made to the arrival rate, the Poisson distribution can be used to establish the inter-arrival time and also determine when it is likely that an event is lost or non-existent, rather than late. A significant advantage of the Poisson distribution is that, among other benefits, the combination of Poisson processes is another Poisson process with a rate parameter that is the sum of the composing processes. Changing the arrival rate allows the immediate alteration of the expectation of the arrival rate and inter-arrival rate of future events, without requiring the storage of previous event and time pairs.

The Poisson distribution, for rate λ, has a mean of λ. Thus, once an arrival rate has been established for a given interval, λ, we would expect λ events per interval. Given that we wish to be able to predict arrival and, hence, failure to arrive, we need to be able to predict the inter-arrival time. If the number of arrivals in a time interval $[t_i...t_{i+x}]$ follows the Poisson distribution, with rate parameter λ, then the lengths of the inter-arrival times are defined as $\mathcal{I} \sim Exponential(\lambda)$, with a mean inter-arrival time of $\frac{1}{\lambda}$. This is an important result as it clearly shows that the Poisson distribution will correctly model a continuous, regularly spaced data stream but it also gives us the ability to model a non-continuous, irregularly spaced data stream.

3.3   An Example

Wireless sensor network applications may employ multiple sensors in an observation domain, on the same node or by combining the results of several nodes [13,14]. This example illustrates a surveillance application, with a widespread WSN constructed of nodes that employ vibration sensors, photosensors and acoustic sensors. This WSN is spread over a large geographical range and has a lifespan measured in months. Maintenance is limited, due to the cost and time involved in travelling to the sensor nodes. With the sensor nodes ground-mounted, the vibration sensors report ground movement in the vicinity of the sensor, photosensors report an interruption to a light beam projected at the node from a nearby position and the acoustic sensors provide a measure of the acoustic pressure in the region. These three sensors display the range of possible sensor event generation. The vibration sensor will be continuously reading small vibrations in the ground, the photosensor is effectively boolean in that the light beam is either striking it or it is interrupted, and the acoustic sensor is more likely to manifest a combination of the previous two, as sound pressures can easily drop below the detectable level but have a continuous distribution once detected.

At each sensor node, a sequence of tuples, s, is generated each time transmission occurs. The sensor readings are only a subset of this tuple as the $(\xi_0...\xi_N)$ in s also include information such as source and destination, as well as any other system-specific information. The sensor reading tuple takes the form $(\eta_0, \eta_1, \eta_2, \eta_3)$, where:

- η_0 is the timepoint at which the data was sensed,
- η_1 is the vibration reading,
- η_2 is the photosensor reading,
- η_3 is the acoustic pressure

The sensor nodes employ both filtering and aggregation to limit upstream transmission but, given their remote location, each node must transmit sufficiently often to prevent unneeded maintenance visits [11]. In this example, the nodes employ local significance filtering, but no aggregation. Aggregation is carried out at a regional level, with filtering also employed on the aggregates where necessary, and may take a number of forms [15,16]. Such aggregation and filtering is vital because there is a maximum capacity

SED	t_0	t_1	t_2	t_3	t_4	t_5	t_6	t_7	t_8	t_9	t_{10}	t_{11}	t_{12}	t_{13}	t_{14}	t_{15}
VIB:	10	10	4	6	4		100		10	10	4	6	4		100	
PS:							1		1						1	1
AP:					120			120								
INSIG:	2	2	2	2	1	3	1	3	1	2	2	1	2	3	1	2

Fig. 1. Sample data stream decomposition

for a given sensor network, not just because of the information that each node wishes to send, because of the requirement for nodes to route information for other nodes. Ultimately, any sensor network has a maximum throughout, based on the number of nodes and their available bandwidth [17]. It is essential to keep bandwidth use below this threshold.

The significance thresholds for instantaneous readings on each sensor are Sig_{vib}, Sig_{ps} and Sig_{ap}, reflecting the level that separates filtered events from unfiltered events.

Defining Significant Event Streams. Within this SEDIM, there are four SEDs. These are:

- $(VIB, \mathcal{E} : \eta_1 >= Sig_{vib}, \Delta_{Vib}, \lambda_{Vib})$
- $(PS, \mathcal{E} : \eta_2 >= Sig_{ps}, \Delta_{ps}, \lambda_{ps})$
- $(AP, \mathcal{E} : \eta_3 >= Sig_{ap}, \Delta_{ap}, \lambda_{ap})$
- $(INSIG, \mathcal{E} \notin [VIB, PS, AP], \Delta_{insig}, \lambda_{insig})$

and the corresponding data model is $\mathcal{S}_{\mathcal{E}}$, where

- $\mathcal{S}_{\mathcal{E}} : [VIB, PS, AP, INSIG]$
- $\Lambda : [\lambda_{Vib}, \lambda_{ps}, \lambda_{ap}, \lambda_{insig}]$

Given that the events in $\mathcal{S}_{\mathcal{E}}$ are the tuples $(\xi_0...\xi_N)$, where each ξ is composed of transmission headers and footers and the sensor reading tuple $(\eta_0, \eta_1, \eta_2, \eta_3)$, we can now model an individual SED as a set of events that meet the SED criteria. This allows us to provide sample data streams for each SED, which would be interspersed in the true data stream, and analyse them separately at any node that can carry out filtering, aggregation and analysis.

The Sample Streams. This example presents tuples containing the sample data detected by the sensor node. Each tuple entry represents one poll of the sensors and all tuples are defined to be in synchrony. Each tuple entry is a timestamp (η_0) and the associated value (i.e., η_1, η_2, or η_3). The INSIG SED contains the cardinality of insignificant values, rather than the values themselves, although this is an implementation decision. Figure 1 shows a decomposed data stream, with the values allocated to the SEDs and insignificant values shown as cardinalities in the INSIG SED.

If we define t_0 as the time at the start of interval Δ_i and t_{now} as t_{15}, then we can regard each t_i as a discrete tick within the time interval. For the purposes of the example, the range $[t_0..t_{now}]$ is divided between time intervals Δ_i and Δ_{i+1}.

The estimated arrival rates (EAR), defined by previous observation and established as system baselines, for each SED are given in Figure 2 along with the Time Interval

SED	EAR	TIAAR (Δ_i, Δ_{i+1})	AAR (Δ_i, Δ_{i+1})	Interval
VIB	6	(6,6)	(6,6)	8 ticks
PS	2	(1,3)	(2,2)	8 ticks
AP	1	(2,0)	(1,1)	8 ticks

Fig. 2. Estimated arrival rates

SED	Transmission times (ticks)
VIB	t_8 and t_{16}
PS	t_9 and t_{16}
AP	t_{16} (first transmission)

Fig. 3. Transmission times

Aligned Arrival Rate (TIAAR), the Actual Arrival Rate (AAR) and the interval over which the rate is measured. TIAAR shows the rate that would be returned by a naive interpretation of the data stream on strict time boundaries, while AAR gives the arrival rate adjusted for the relaxation implicit in accepting data within the Poisson noise. Both TIAAR and AAR are given as a pair of values, one for each time interval.

The INSIG stream is continuous, as the cardinality of events is continuously generated. However, this SED is for internal reference only, and is not transmitted to other nodes (although a summary may be requested by a superior node).

The AAR values match the EAR values because 'late' events can be accepted within $\sqrt{EAR}$, in terms of the fraction of the previous interval that the collection window is still considered to be opened. Importantly, events accepted as part of a previous window cannot be counted as part of a subsequent window and, in the case of SED_{PS}, the second interval has the range $[t_9..t_{15}]$, as the t_8 tick has been included in the previous window. While not shown in Figure 1, t_{16} is the next tick that will occur after t_{15} and denotes the start of the next interval.

The delay introduced by this scheme can be seen by displaying the transmission times, as a given tick t_i, where the tuple is encapsulated and sent to the network as a significant packet. These transmissions times are shown in Figure 3.

The AP delay is the most significant, as we must wait an entire interval to ensure that the event at t_7 is an advanced event from $\Delta_i + 1$ (within the noise parameter), rather than an additional event from interval Δ_i. As the λ_{AP} is 1, and $\sqrt{1}$ is 1, we can wait an entire interval before receiving the packed that was supposed to have occurred in Δ_i. However, this is the worst case situation - the maximum delay inserted is one interval length in the rare situation that we are only expecting one event per interval.

4   Failure Detection

Failure detection in a continuous, regular data stream is relatively straight forward: the data stops. In a SED, a missing event poses a more complex problem as there are different possibilities:

1. Events are still arriving but are below the significance threshold of a given SED.
2. An event will be sent but it will arrive slightly after the deadline.
3. The event is actually lost.

In terms of the data stream $\mathcal{S}$, we can describe each of these possibilities as equations.

$$\exists \hat{\mathcal{S}}_i : \forall \mathcal{E}_i \in \mathcal{E}, \mathcal{E}_i \notin [\mathcal{V}_{lower_i}...\mathcal{V}_{upper_i}], \Delta_i >= t_{now} \tag{4}$$

$$\exists \hat{\mathcal{S}}_i : \exists (\mathcal{E}_{\mathcal{R}+1}, \Delta_{\mathcal{R}+1}), \Delta_{\mathcal{R}+1} > t_{window} \tag{5}$$

$$\nexists \hat{\mathcal{S}}_i : \exists (\mathcal{E}_{\mathcal{R}+1}, \Delta_{\mathcal{R}+1}) \tag{6}$$

In $\mathcal{S}$, equation 4 only constitutes failure in the SED that has the restrictive range of significance. In this case, it does not constitute a stream failure but indicates that no significant events are arriving. This may indicate that there are no significant events to report or that there is a mis-reporting of events. There are several possible reactions:

1. After a given time, adjust the range of significance to reflect increased knowledge of the data stream contents. This is referred to as *rate relaxation*.
2. The producer can generate a system message that provides evidence that, should a significant event occur, it will be generated and passed on. We refer to this as *significance exchange*.
3. If test equipment is in place, an artificially significant event is generated and sent upstream, to be discarded by the consumer. We refer to this as *significance verification*.
4. Ignore it and drop this SED from $\mathcal{S}$. This may also be considered equivalent to a rate relaxation to a parameter of zero - we expect nothing significant to occur in an interval.
5. Report it.

Rate relaxation increases the time over which events may be detected. As the Poisson distribution may only take integer valued parameters, the minimum non-zero rate is one event per interval. Once the rate has been relaxed to one event per interval, the only further relaxation possible is the extension of the interval and this is carried out by doubling the interval size, to a maximum value of 1 calendar year, although it may be smaller for a more short-lived system. Once relaxation has occurred to the occurrence and time limit, any further relaxation will set the rate parameter to zero, effectively terminating the expectation of arrivals in this SED. This is equivalent to reaction 4.

Significance exchange requires both producer and consumer to be able to exchange meta-values that describe the context or value ranges expected for the values. XML, provided that there is a contextual basis such as RDF or OWL-XML in place describing the shared context, may be used to exchange system messages that are not interpreted as events but contain information that confirm what both parties consider significance to be. These messages can piggy back onto events, if the events are wrapped in XML and are tagged by the producer as significant or insignificant. However, by choosing the lowest value $x : x > \mathcal{V}_{lower_i}$ and sending this as a test, the producer clearly indicates where the threshold is.

Significance verification requires that the producer be capable of injecting a significant event into the data stream and reporting on it, while labelling it in such a way that it is not treated as a significant event elsewhere. This is also a system, or test, message but, instead of simulating event handling, the test event is injected prior to detection. This requires a comprehensive test harness if physical sensors are being employed. Where data is not being acquired directly through physical sensors, we can separate the network reading component and data processing component of the producer and insert a data injection mechanism between the two. This also relies upon the ability of both producer and consumer to agree upon what constitutes a test message, and to be able to send and receive meta-values, rather than a composite stream of values, parsed from a purely structural perspective into a value stream with no type or context information.

Equation 5 may cause problems in a system with a synchronous time requirement, from the movement of values into adjacent cells potentially leading to incorrect summarisation, but is manageable within a SED. This is due to key properties of the Poisson distribution. The Poisson distribution has a mean of λ but also has a variance of λ. For a given interval, the number of observed arrivals fluctuates about the mean λ with standard deviation of $\sqrt{\lambda}$, where these fluctuations are referred to as the *Poisson noise*.

Poisson noise, or *shot noise*, describes the statistical fluctuation observed from the arrival of finite, discrete events. This effect is seen in electronic applications, photon counters and particle simulations. The Poisson noise increases as the rate of arrival increases, but at a slower rate. For any number of events sampled, where the sample has a Poisson distribution, the average number of events does not reflect that true number of events detected in that interval but the actual result will be distributed about the average with a Poisson distribution.

Similar to StatStream, we now have a time interval within which we will have been able to detect the vast majority of failures and it is a fixed time, given by the arrival rate λ and the time interval, $\mathcal{T}_{\mathcal{I}}$. We declare likely failure in Δ_i for a rate λ if, for a given interval $[\Delta_i...(\Delta_{i+1}/\sqrt{\lambda})]$, the cardinality of events $\mathcal{E}_i$ in the interval is less than λ.

$$Fail(\Delta_i, \lambda) : card(\mathcal{E}_i \in [\Delta_i...(\Delta_{i+1}/\sqrt{\lambda})]) < \lambda \qquad (7)$$

We have now presented the way of dealing with possibilities contained in equations 5 and 6. If an event is merely delayed, waiting for a pre-determined period beyond the original deadline will capture the event and there is no need to handle the event as lost or carry out any adjustments to the rate parameters. However, if the event doesn't arrive, even within the noise interval, then we have successfully detected a failure event and we can now take actions as outlined in the solution to equation 4.

A useful result of equation 7 is that it is immediately apparent that the higher the rate of arrival, the shorter the proportion of an interval that is required to detect failure. In a system with a high rate of arrival and short interval, this means that failure can be detected in very short time. Conversely, a system with low rate of arrival has a correspondingly long time to failure detection. This immediately motivates the need for the use of an artificially high significance rate, employing test data to keep the rate high, while not requiring a high rate of actual events. Significance verification or significance testing can both be used to achieve these aims.

Where interest is primarily on the waiting time to a given event somewhere in the stream, the Gamma (Γ) distribution is a family of continuous probability distributions with two paramers, k and θ, that is used for modelling waiting times. With integer k, the Γ distribution is the *Erlang* distribution and is the probability distribution of the waiting time until the k-th event in a Poisson process with rate $1/\theta$. Rather than monitoring every event to determine failure, which is energy intensive, we can now observe a k-th event to determine if the waiting times are meeting our estimates.

5   Encoding Information within the Poisson Noise

One of the advantages of allowing events to be legitimately "placed" within an interval, or within the Poisson noise of the interval, is that this information can be used as an additional communications channel. In a continuous, regular data stream, varying the regularity deliberately can be used to indicate out-of-band information that allows communication between producer and consumer, without placing an additional data burden on the main channel. In a sensor network this is vital as the smaller the data stream is, the less power is consumed in all parts of the network for transmission and processing. Examples of out-of-band channel use in the real world range from the use of tone in spoken language to signify additional meaning, such as questioning, sarcasm or dubiety, when the semantic content of the written form does not need to capture this.

Encoding information within the Poisson noise requires either that the producer and consumer have a synchronous communication channel, where deliberate movement is detectable, or have the ability to embed timestamps into their data streams to indicate the point at which they planned to send the data.

In a continuous, synchronous channel environment, producer and consumer will exchange λ events per interval. The simplest encoding available in the Poisson noise is to vary the arrival time of the final event and to reduce or increase the inter-arrival time. If we encode 0 as an unlikely reduction in arrival time and 1 as an unlikely increase in inter-arrival time, then it is possible to send binary messages from producer to consumer at the rate of 1 bit per event. This, however, does rely upon the channel in question being highly reliable, with a well-defined rate of arrival.

Where we cannot assume regularity, we must use timestamps, to allow the producer to indicate to the consumer that they had planned to send the data at time t, but actually did so at time $t+x$. Whatever x is, it must still fall within the Poisson noise interval but, in this case, we now have a greater range of possible value representations available, as the reference timestamp is within the stream.

One application of this is in non-continuous, irregularly reporting low-power sensors such as wireless surveillance sensor nodes. If these nodes only report periodically, how do we know that they haven't been tampered with in the interim? One approach is to provide a pseudo-random number generator of known and very large period, or a set of pre-generated pseudo-random numbers, to both producer and consumer and to offset the producer's messages by an interval based on these numbers. This reduces both the predictability of the event transmission and provides a low-power identification mechanism for a node. By choosing a generator with a large period, determining the sequence of numbers by observations is infeasible. If the seed, or the pre-generated sequence, are physically protected within the node, then an intruder is limited in how they can bypass a node, as they cannot replace it without losing the identification out-of-band channel, and disabling the sensor will, after some interval, generate a $Fail(\Delta_i, \lambda)$ event, which will also constitute a reportable warning event.

6 Experimental Results

We used a number of simulated test environments to test classification, failure detection reliability and the time to detection of our approach. Due to space constraints, we discuss one here, a statistical simulation based on event queues, with randomised failure.

In the experiment, a SEDIM entity (SEDIMent) was constructed as a set of simulated data streams, composed of data from three sensors. The time interval was set to 3600 seconds and λ increased from 1 event per interval to 30 events per interval. The SED-based classification was used to classify the cumulative data stream into VIB, PS, AP and INSIG SEDs. Stream transmission rates were monitored at the transmitting node and at upstream nodes. The experiment was designed to test: i) the correct detection of event loss when events fell outside of the Poisson noise interval, ii) the correct estimates of the rates of individual SEDs for determining failure of an individual sensor, iii) the correct estimates of total node failure in upstream nodes, iv) maintaining node liveness through the use of injected test data, implementing significance exchange and significance verification, as described in Section 4, and finally v) node identification using variation in the Poisson noise.

Experiments were run with failure rates ranging from 0 to 50% of nodes, with 1000 iterations of each experimental configuration. The failure of an individual event to arrive was detected within one time interval with a cumulative success rate 89.2%. Where no errors occurred in contiguous intervals, the probability of success increased to 99.6%. This established successful detection of event loss. The time taken to detect failure was proportional to $\frac{1}{\lambda}$, as expected.

The removal of INSIG data streams from transmission, and the calculation of individual arrival rates for all other SED, was tested at the event injecting node and at the simulated sink node. Transmitted events maintained their TIAAR across the system and measurements of failure detection and arrival rate were consistent with the EAR for the decomposed streams across all experiments.

The simulated sink node calculated the cumulative arrival rate, Λ_{EAR}, for all simulated sensor nodes. Individual nodes maintained a counter of the number of events where INSIG was the only active SED. In any interval where INSIG was the only active SED for the entirety of the interval, a test packet was injected into the data stream, giving the EAR for the node's active SEDs and an example classification of significance. If received at the sink, the node continued to be marked live and the classification condition was checked. Total node failure was simulated by setting all significance thresholds to the maximum value and disabling test packet injection. In this case, total node failure was detected at sink nodes within $((1 + (\frac{1}{\lambda}))\delta_i)$.

Finally, node identification was tested by dividing the first Poisson noise interval within a time interval into millisecond intervals, and encoding the transmission time

Failure rate	0.00	0.05	0.10	0.15	0.20	0.25
MoS	0.00	0.724	2.641	15.277	99.142	194.905

Failure rate	0.30	0.35	0.40	0.45	0.50	
MoS	271.944	337.732	393.982	446.978	498.445	

Fig. 4. Measure of Suspicion metrics for increasing failure rates

as marked-up XML data, accompanying the data stream. This transmission time was moved within the available slots by employing a fixed-period rotating set of random numbers, with the sequence known to both an individual node and the sink. In the event of an event being labelled as failed, due to falling outside of the expected range, the random number that would have matched the offset is discarded at the sink. This automatically causes the rejection of the packet, should it arrive late. A *Measure of Suspicion* (MoS) is kept at the sink node, increasing monotonically for every packet that either fails to arrive, or arrives with an unexpected offset. The MoS is decreased by one for every 10 packets that arrive with the expected offset. In testing, the average MoS for the system was 0.00 for experiments without synthetic failure and all events arriving within the intervals, as expected. Figure 4 shows the MoS for higher failure rates with a thousand individual trials of one thousand events.

It is of little surprise that higher failure rates have higher levels of uncertainty as a large number of the confirmation numbers will be dropped from the queue. At failure rates above 10%, the reduction in MoS is dominated by the ongoing increase, and approaches the failure rate multiplied by the number of events, 1000 in this case. However, for low failure rate data stream producing networks, a MoS threshold of 1 allows for the detection of increasing failure rates and the possibility of compromised nodes.

7   Conclusions

We have provided a model for the flow of significant events in data streams, in terms of the rate of arrival of these events, and the distribution of these events. This model is suitable for the modelling of both regular and irregular data streams, and is event-focused, rather than time-focused. By employing this model, it is possible to detect failure in the data stream, where this failure is a failure of transmission, or an absence of significant events.

Our experimental results clearly show that this model adapts to change, as well as reducing network overhead due to i) a minimisation of polling or liveness information, and ii) a well-defined expectations of network behaviour. We have also shown a simple application of our approach, which allows additional use of a channel without having to alter time boundaries or expected arrival rates.

We have already developed three simulation models, one statistical, one software based as a node level simulation and one grounded in the TinyOS Simulation (TOSSIM) [18] environment for power consumption. We are further developing an implementation of SEDIM in a WSN environment, and ad-hoc networking environment on hand-held computing devices. The power consumption in a mobile and distributed environment is further constrained by the requirement to support ad-hoc routing protocols and route discovery. In this role, the SEDIM approach will provide significant power savings, that will allow more aggressive route discovery and maintenance, supporting stream-based communication over ad-hoc networks, as well as packet-based communication.

References

1. Golab, L., Özsu, M.T.: Issues in data stream management. SIGMOD Rec. 32(2), 5–14 (2003)
2. Gaber, M.M., Zaslavsky, A., Krishnaswamy, S.: Mining Data Streams: A Review. SIGMOD Rec. 34(2), 18–26 (2005)
3. Babcock, B., Babu, S., Datar, M., Motwani, R., Widom, J.: Models and Issues in Data Stream Systems. In: PODS 2002: Proc. of the 21st ACM SIGMOD-SIGACT-SIGART Symposium on Principles of Database Systems, pp. 1–16. ACM, New York (2002)
4. Babu, S., Widom, J.: Continuous Queries over Data Streams. SIGMOD Rec. 30(3), 109–120 (2001)
5. Berndt, D.J., Clifford, J.: Using Dynamic Time Warping to Find Patterns in Time Series. In: AAAI 1994 Workshop on Knowledge Discovery in Databases, pp. 359–370. AAAI Press, Menlo Park (1994)
6. Papadimitriou, S., Sun, J., Faloutsos, C.: Streaming Pattern Discovery in Multiple Time-series. In: VLDB 2005: Proc. of the 31st Intl. Conference on Very Large Data Bases, pp. 697–708. ACM, New York (2005)
7. Bai, Y., Wang, F., Liu, P.: Efficiently filtering RFID data streams. In: CleanDB: The First International VLDB Workshop on Clean Databases, pp. 50–57. ACM, New York (2006)
8. Wei, Y., Son, S.H., Stankovic, J.A.: RTSTREAM: Real-Time Query Processing for Data Streams. In: 9th IEEE International Symposium on Object/component/service-oriented Real-Time Distributed Computing, pp. 141–150 (2006)
9. Zhu, Y., Shasha, D.: StatStream: Statistical Monitoring of Thousands of Data Streams in Real Time. In: VLDB 2002: Proc. of the 28th Intl. Conference on Very Large Data Bases, VLDB Endowment, pp. 358–369 (2002)
10. Gu, L., Jia, D., Vicaire, P., Yan, T., Luo, L., Tirumala, A., Cao, Q., He, T., Stankovic, J.A., Abdelzaher, T., Krogh, B.H.: Lightweight Detection and Classification for Wireless Sensor Networks in Realistic Environments. In: SenSys 2005: Proc. of the 3rd Intl. Conference on Embedded Networked Sensor Systems, pp. 205–217. ACM, New York (2005)
11. Solis, I., Obraczka, K.: In-Network Aggregation Trade-offs for Data Collection in Wireless Sensor Networks. Intl. Journal of Sensor Networks 1(3–4), 200–212 (2007)
12. Ye, F., Luo, H., Lu, S., Zhang, L.: Statistical En-Route Filtering of Injected False Data in Sensor Networks. IEEE Journal on Selected Areas in Communications 23(4), 839–850 (2005)
13. Pottie, G.J., Kaiser, W.J.: Wireless Integrated Network Sensors. Commun. ACM 43(5), 51–58 (2000)
14. Feng, J., Koushanfar, F., Potkonjak, M.: Sensor Network Architecture. Number 12 in III. In: Handbook of Sensor Networks. CRC Press, Boca Raton (2004)
15. Madden, S., Franklin, M.J., Hellerstein, J.M., Hong, W.: TAG: a Tiny AGgregation Service for Ad-hoc Sensor Networks. SIGOPS Oper. Syst. Rev. 36(SI), 131–146 (2002)
16. Petrovic, M., Burcea, I., Jacobsen, H.A.: S-ToPSS: Semantic Toronto Publish/Subscribe System. In: VLDB 2003: Proc. of the 29th Intl. Conference on Very Large Data Bases, VLDB Endowment, pp. 1101–1104 (2003)
17. Gupta, P., Kumar, P.R.: The Capacity of Wireless Sensor Networks. IEEE Trans. Info. Theory 46(2) (2000)
18. Levis, P., Lee, N., Welsh, M., Culler, D.: TOSSIM: Accurate and Scalable Simulation of Entire TinyOS Applications. In: SenSys '03: Proc. of the 1st Intl. Conference on Embedded Networked Sensor Systems, pp. 126–137. ACM, New York (2003)

Incremental and Adaptive Clustering Stream Data over Sliding Window

Xuan Hong Dang[1], Vincent C.S. Lee[1], Wee Keong Ng[2], and Kok Leong Ong[3]

[1] Monash University, Australia
{xhdang,vincent.lee}@infotech.monash.edu
[2] Nanyang Technological University, Singapore
awkng@ntu.edu.sg
[3] Deakin University, Australia
leong@deakin.edu.au

Abstract. Cluster analysis has played a key role in data stream understanding. The problem is difficult when the clustering task is considered in a sliding window model in which the requirement of outdated data elimination must be dealt with properly. We propose SWEM algorithm that is designed based on the Expectation Maximization technique to address these challenges. Equipped in SWEM is the capability to compute clusters incrementally using a small number of statistics summarized over the stream and the capability to adapt to the stream distribution's changes. The feasibility of SWEM has been verified via a number of experiments and we show that it is superior than Clustream algorithm, for both synthetic and real datasets.

1 Introduction

In recent years, we are seeing a new class of applications that changed the traditional view of databases as a static store of information. These applications are commonly characterized by the unbounded data streams they generate (or receive), and the need to analyze them in a continuous manner over limited computation resources [10, 3, 8]. Further, stream data can be lost under high speed conditions, become outdated in the analysis context, or intentionally dropped through techniques like sampling [4] or load shedding [16]. This makes it imperative to design algorithms that compute answers in a continuous fashion with only one scan over stream data whilst operating under the resource limitations.

Among various data mining tasks, clustering is one of the most important tasks that widely helps to analyze and uncover structures in the data. Research in data stream clustering reported so far has mostly focused on two mining models, the landmark window [12,2,18] and the forgetful window [7,8]. In the former one, clustering results are computed based on the entire data elements generated so far in the stream. In the latter model, they are also discovered from the complete stream history; however, the weight (or importance) of each data instance is decreased with time by using a fading function. While these two mining models are useful in some data stream applications, there is a strong demand to devise

S.S. Bhowmick, J. Küng, and R. Wagner (Eds.): DEXA 2009, LNCS 5690, pp. 660–674, 2009.

novel techniques that are able to cluster the data stream in a sliding window model. For example, in network intrusion monitoring, the changes of stream characteristics in the past several hours are much more valuable compared to those happened in days ago [11]. Moreover, it has also been shown that [19] performing a stream mining or querying task in the sliding window model is the most general and the most challenging work since it further deals with the problem of outdated data elimination.

In this paper, we propose SWEM algorithm (clustering data streams in a time-based Sliding Window with Expectation Maximization technique) to address the above challenges. Compared to other stream clustering algorithms relying on k-means [13,12,2,1], an EM-based algorithm is a soft clustering method and thus, it has some properties that are desirable for stream environments such as robustness to noise or the ability to handle missing data (as clearly shown in [18]). Furthermore, it has a strong statistical basis and theoretically guarantees optimal convergence. The SWEM algorithm consists of two stages which are strictly designed to address the problem of constrained memory usage and one-pass processing over data streams. In the first stage, SWEM scans data records arriving in the stream and summarizes them into a set of micro components where each one is characterized by a small number of statistics. These small amount of information in turns are effectively used in the second stage of SWEM to approximate the set of global clusters. Most importantly, in order to address the problem of characteristics changing in stream environments, we develop a method to adaptively split and merge micro components. Such an approach provides a flexible technique to re-distribute micro components across the data space and thus efficiently approximates the stream's evolving distribution. Experiments on various data streams have empirically shown that SWEM is very effective in finding clusters from a time-based sliding window. It is not only able to process data streams in an incremental fashion with confined memory space, its clustering qualities are high and close to those of a conventional EM (working without any stream's limits). In addition, the algorithm also outperforms the well-known CluStream algorithms [2] both in time performance and clustering qualities.

2   Related Work

Previous work on clustering data streams focused on developing space-efficient and one-pass algorithms. STREAM is one of the first algorithms that addresses the problem in a landmark window model [13]. It is a divide-and-conquer algorithm that builds clusters incrementally and hierarchically using bicriterion approximation algorithms. In [6,15], this approach is extended in that both the approximation factor and the storage memory are improved. It is also expanded for stream clustering in a sliding window model [5]. Unfortunately, in order to deal with the issue of finding clusters in window sliding mode, their algorithm simply re-clusters all summarized data points. Hence, there is no incremental work for the process of computing global clusters. Also, there is no experimental work to report the accuracy and effectiveness of the algorithm. Furthermore, by adopting the exponential histogram [9] as the framework for summarizing data, their algorithm can only work with count-based sliding window where the number of data records arriving in the window must be fixed in advance.

Recently, CluStream [2], DenStream [7] and D-Stream [8] are representative algorithms focusing on evolving data stream clustering. In CluStream [2], it divides the clustering process into online and offline components. The online component is designed to quickly update raw data into a set of micro clusters and the offline component is developed to compute clustering results based on snapshots captured at different times on the micro clusters' set. To address the issue of confined memory space, CluStream stores snapshots at various levels of granularity with recent ones stored at finer granularity. This technique is then later extended in HPStream [1] where the concept of projected clusters is introduced to cluster high dimensional data streams. DenStream [7] is another algorithm proposed to address the issue of clustering evolving data streams. The authors extended the DBSCAN method to the online-offline framework proposed in Clustream [2]. A fading function is also used to gradually reduce the weight of each micro cluster with time. The DBSCAN with the concept of *density connectivity* is then applied to the set of micro clusters to derive global clustering results. D-Stream [8] is another density-based algorithm for stream clustering. It divided the data space into a set of disjointed grids. Then, data instances are mapped into corresponding grids based on their dimensional values. Similar to DenStream, a fading function is used in D-Stream to reduce the weight of each grid over time. The final clustering results are obtained based on the density and the grids' connectivity. Motivated by the drawbacks of the hard clustering techniques (e.g., based on k-means) applied on various stream environments, CluDistream [18] has recently been proposed to cluster data streams based on the Expectation Maximization technique. Although this algorithm has been shown to provide significant results over other methods (especially in distributed stream environments where data could be noised or missed due to transmission), it only addressed the clustering problem in a *landmark* window where each EM algorithm is simply implemented at each node of the distributed network.

3   Problem Formulation

We consider a data stream as a time ordered series of data points $\mathcal{DS} = \{x_1, x_2, ..., x_n, ...\}$ where each x_i has the form $x_i = \{x_i^1, x_i^2, ..., x_i^d\}$ in d-dimensional space. We focus on the *time-based* sliding window model. In this model, let $TS_0, ..., TS_{i-b+1}, ..., TS_i$ denote the time periods elapsed so far in the stream. Each time period contains multiple data records arriving in that interval. Given an integer b, the time-based sliding window is defined as the set of data records arriving in the last b time periods and denoted by $\mathcal{SW} = \{TS_{i-b+1}, ..., TS_{i-1}, TS_i\}$. TS_i is called the latest time slot and TS_{i-b} is called the expiring one in the sliding window. When the time shifts to the new slot TS_i, the effect of all records in TS_{i-b} will be eliminated from the clustering model.

In our sliding window model, once data points in a time slot are processed, they cannot be retrieved for further computation at a later time (unless their summarized statistics are explicitly stored in memory). The amount of memory available is assumed to be limited. In particular, it is sub-linear in the window's size. As such, approaches that require storing the entire data points in the sliding window are not acceptable in this model. We also assume that streaming data evolve with time and records are generated as a result of a dynamic statistical process that consists of k mixture models (or components). Each model corresponds to a cluster that follows a multivariate normal distribution. Consequently, any cluster $C_h, 1 \le h \le k$, is characterized by a parameter: $\phi_h = \{\alpha_h, \mu_h, \Sigma_h\}$ where $\alpha_h, \mu_h, \Sigma_h$ are respectively the weight, vector mean(determining the center), and covariance matrix (determining the shape) of the cluster.

Accordingly, we define our problem as follows: Given the number of time slots b of the sliding window, our goal is to cluster the data stream incrementally by identifying a set of parameters $\Phi_G = \{\phi_1, ..., \phi_k\}$ that optimally fit the current set of data points arriving in the last b time periods of the stream.

4   Algorithm Description

4.1   Initial Phase

We compute m distributions (also called micro components) modelling the data within each time slot of the sliding window. Let Φ_L be the set of parameters of these local components, i.e., $\Phi_L = \{\phi_1, ..., \phi_m\}$, where each micro component $MC_\ell, 1 \le \ell \le m$, is also assumed to follow a multivariate Gaussian distribution characterized by three parameters $\phi_\ell = \{\alpha_\ell, \mu_\ell, \Sigma_\ell\}$. For the initial phase where $SW = \{TS_0\}$, the initial values for these parameters will be randomly chosen. We clarify the following concepts and assumptions.

In our model, each data point belongs to all components yet with different probability. Given x, its probability in component ℓ^{th} is computed based on the Bayes rule: $p(\phi_\ell|x) = \alpha_\ell \times p_\ell(x|\phi_\ell)/p(x) = \alpha_\ell \times p_\ell(x|\phi_\ell)/\sum_{i=1}^{m} \alpha_i \times p_i(x|\phi_i)$, in which $p_\ell(x|\phi_\ell) = \frac{1}{(2\pi)^{d/2}|\Sigma_\ell|^{1/2}} \exp\left[-\frac{1}{2}(x - \mu_\ell)^T \Sigma_\ell^{-1}(x - \mu_\ell)\right]$, where μ_ℓ is the d-dimensional mean vector and Σ_ℓ is the covariance $d \times d$ matrix. $|\Sigma_\ell|$ and $(\Sigma_\ell)^{-1}$ are respectively the determinant and inverse matrix of Σ_h.

Since we assume each attribute is independent of one another, Σ_ℓ is a diagonal variance matrix. Its determinant and inverse matrix can thus be easily computed: $|\Sigma_\ell| = \prod_{i=1}^{d}(\sigma_\ell^i)^2$ and $\Sigma_\ell^{-1} = \left(1/(\sigma_\ell^1)^2 \quad 1/(\sigma_\ell^2)^2 \quad \cdots \quad 1/(\sigma_\ell^d)^2\right) \mathbf{I}$, where σ_ℓ^i is the variance at dimension i of MC_ℓ and $\mathbf{I}$ is the identity matrix.

As we assume data points in the stream are generated independently, the probability of n data records arriving in TS_0 is computed by the product:

$$p(TS_0|\Phi_L) = \prod_{x_i \in TS_0} p(x_i|\Phi_L) = \prod_{i=1}^{n}\sum_{\ell=1}^{m} \alpha_\ell \times p_\ell(x_i|\phi_\ell)$$

Since this probability is small, we typically work with its log likelihood form. We define the average log likelihood measure which is used to evaluate how well the set of micro components approximates the stream within this time period by: $Q(\Phi_L) = \frac{1}{|TS_0|} \log \prod_{x \in TS_0} \sum_{h=1}^{m} \alpha_\ell \times p_\ell(x_i|\phi_\ell)$.

In its first stage, SWEM maximizes this quantity by beginning with an initial estimation of Φ_L and iteratively updates it until $|Q(\Phi_L^{t+1}) - Q(\Phi_L^t)| \leq \epsilon$. Specifically, SWEM updates Φ_L, it at iteration $t+1$ as follows:

In the E-step:
$$p(\phi_\ell|x) = \frac{\alpha_\ell^{(t)} \times p_\ell(x|\mu_\ell^{(t)}, \Sigma_\ell^{(t)})}{\sum_i \alpha_i^{(t)} \times p_i(x|\mu_i^{(t)}, \Sigma_i^{(t)})}$$

In the M-step:
$$\alpha_\ell^{(t+1)} = \frac{1}{n}\sum_{x \in TS_0} p(\phi_\ell|x); \quad \mu_\ell^{(t+1)} = \sum_{x \in TS_0} \frac{p(\phi_\ell|x)}{n_\ell} \times x;$$
$$\Sigma_\ell^{(t+1)} = \sum_{x \in TS_0} \frac{p(\phi_\ell|x)}{n_\ell} \times (x - \mu_\ell^{(t+1)})(x - \mu_\ell^{(t+1)})^T$$

where $n_\ell = \sum_{x \in TS_0} p(\phi_\ell|x)$.

When convergence, these micro components are approximated by only keeping a triple $T_\ell = \{N_\ell, \theta_\ell, \Gamma_\ell\}$ for each MC_ℓ. Essentially, let S_ℓ be the set of records assigned to MC_ℓ (to which they have the highest probability), then $N_\ell = |S_\ell|$ is the cardinality of the set S_ℓ; $\theta_\ell = \sum_{x_i \in S_\ell} x_i$ is the linear summation of the data points in the set S_ℓ; and $\Gamma_\ell = \sum_{x_i \in S_\ell} x_i x_i^T$ is the squared summation of these points.

The important property of T_ℓ is that it is sufficient to compute the mean and covariance of MC_ℓ. Concretely, $\mu_\ell = N_\ell^{-1}\theta_\ell$ and $\Sigma_\ell = N_\ell^{-1}\Gamma_\ell - N_\ell^{-2}\theta_\ell \times \theta_\ell^T$. Furthermore, its additive property ensures that the mean and covariance of a merged component can be computed from the values of each member component: $T_\ell = \{N_{\ell_1} + N_{\ell_2}, \theta_{\ell_1} + \theta_{\ell_2}, \Gamma_{\ell_1} + \Gamma_{\ell_2}\}$ given two member components $T_{\ell_1} = \{N_{\ell_1}, \theta_{\ell_1}, \Gamma_{\ell_1}\}$ and $T_{\ell_2} = \{N_{\ell_2}, \theta_{\ell_2}, \Gamma_{\ell_2}\}$.

Therefore, SWEM treats these triples as the sufficient statistics summarizing the information regarding the data points within the first time slot of the stream. They are then used in the second stage of the algorithm to compute the k global clusters. For each $\phi_h \in \Phi_G$, SWEM updates its parameters as follows:

E-step:
$$p(\phi_h|T_\ell) = \frac{\alpha_h^{(t)} \times p_h(\frac{1}{N_\ell}\theta_\ell|\mu_h^{(t)}, \Sigma_h^{(t)})}{\sum_{i=1}^k \alpha_i^{(t)} \times p_i(\frac{1}{N_\ell}\theta_\ell|\mu_i^{(t)}, \Sigma_i^{(t)})}$$

M-step:
$$\alpha_h^{(t+1)} = \frac{1}{n}\sum_{\ell=1}^m N_\ell \times p(\phi_h|T_\ell); \quad \mu_h^{(t+1)} = \frac{1}{n_h}\sum_{\ell=1}^m p(\phi_h|T_\ell) \times \theta_\ell;$$
$$\Sigma_h^{(t+1)} = \frac{1}{n_h}\left[\sum_{\ell=1}^m p(\phi_h|T_\ell)\Gamma_\ell - \frac{1}{n_h}\sum_{\ell=1}^m \left(p(\phi_h|T_\ell)\theta_\ell\right)\left(p(\phi_h|T_\ell)\theta_\ell\right)^T\right]$$

where $n_h = \sum_{\ell=1}^m N_\ell \times p(\phi_h|T_\ell)$.

4.2 Incremental Phase

This phase is incrementally applied when data in a new time slot arrive at the system. Nonetheless, different from the initial phase where SWEM has to randomly choose initial parameters for micro components, in this phase it utilizes the converged parameters in the previous time slot as the initial values for the mixture models. The rationale behind is that we expect the stream's characteristics between two consecutive time slots only change slightly. Thus, inheriting the converged parameters of the previous time slot can minimize the number of iterations for the next one.

Essentially, after the first iteration on the data points arriving in the new time slot, SWEM compares the quantity $Q(\Phi_L)$ with the converged one in the previous time slot. If two values are not much different, it is safe to say that the distribution in the new time interval is similar to the previous one. Otherwise,

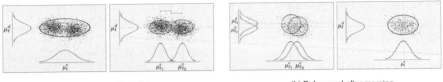

(a) Before and after splitting   (b) Before and after merging

Fig. 1. Split and merge components

the stream distribution has changed and it is necessary to adapt the set of micro components accordingly in order to avoid the local maximal problem [17]. We therefore develop in SWEM two split and merge operations to *discretely* redistribute components across the entire data space.

Figure 1(a) illustrates the idea of SWEM's split operation. A micro component is chosen for dividing if it is large enough and has the highest variance sum (i.e., its data are mostly spread). An MC_ℓ is considered large if its weight $\alpha_\ell > \frac{2}{m}\sum_i \alpha_i$ and the dimension for the split is the one having the maximum variance. Let e be such dimension, then the new means and variances for e in two new micro components MC_{ℓ_1} and MC_{ℓ_2} are approximated as follows:

$$\mu_{\ell_1}^e = \int_{\mu_\ell^e - 3\sigma_\ell^e}^{\mu_\ell^e} x \times p_{\ell,e}(x|\phi_\ell)dx; \qquad \mu_{\ell_2}^e = \int_{\mu_\ell^e}^{\mu_\ell^e + 3\sigma_\ell^e} x \times p_{\ell,e}(x|\phi_\ell)dx$$

$$(\sigma_{\ell_1}^e)^2 = \int_{\mu_\ell^e - 3\sigma_\ell^e}^{\mu_\ell^e} x^2 \times p_{\ell,e}(x|\phi_\ell)dx - (\mu_{\ell_1}^e)^2; \ (\sigma_{\ell_2}^e)^2 = \int_{\mu_\ell^e}^{\mu_\ell^e + 3\sigma_\ell^e} x^2 \times p_{\ell,e}(x|\phi_\ell)dx - (\mu_{\ell_2}^e)^2$$

For other dimensions, their means and variances are kept unchanged. The above computations are derived given the observation that each dimension is to follow a Gaussian distribution in which 99.7% of data are within $\mu \pm 3\sigma$. Therefore, the integrals with lower and upper bounds chosen in this range can approximately cover all component's data points projecting on that dimension.

On the other hand, as illustrated in Figure 1 (b), two components are selected for merging if they are small and close enough. An MC_ℓ is considered small if its weight $\alpha_\ell < \frac{1}{2m}\sum_i \alpha_i$ and the distance, which is measured based on the Mahalanobis distance, between two components are closed enough. Notice that compared to the Euclidean distance, the Mahalanobis works more effectively in SWEM since it takes into account the covariance matrix (actually describing the spread of component). We define the average squared Mahalanobis distance between two components as follows: $Avg(D_{i,j}) = \frac{1}{2}[(\mu_i - \mu_j)^T \Sigma_j^{-1}(\mu_i - \mu_j) + (\mu_j - \mu_i)^T \Sigma_i^{-1}(\mu_j - \mu_i)]$.

Given MC_{ℓ_1} and MC_{ℓ_2} to be merged, SWEM computes parameters for their merging component MC_ℓ based on the additive property:

$$\alpha_\ell = \alpha_{\ell_1} + \alpha_{\ell_2}; \quad \mu_\ell = \frac{\alpha_{\ell_1}}{\alpha_{\ell_1}+\alpha_{\ell_2}} \times \mu_{\ell_1} + \frac{\alpha_{\ell_2}}{\alpha_{\ell_1}+\alpha_{\ell_2}} \times \mu_{\ell_2}; \quad \Sigma_\ell = \frac{\Gamma_\ell}{n(\alpha_{\ell_1}+\alpha_{\ell_2})} -$$

$\frac{\theta_\ell \times \theta_\ell^T}{(n(\alpha_{\ell_1}+\alpha_{\ell_2}))^2}$. In that,

$\theta_\ell = n \times (\alpha_{\ell_1} \times \mu_{\ell_1} + \alpha_{\ell_2} \times \mu_{\ell_2}); \quad \Gamma_\ell = n[\alpha_{\ell_1}(\Sigma_{\ell_1} + \mu_{\ell_1}\mu_{\ell_1}^T) + \alpha_{\ell_2}(\Sigma_{\ell_2} + \mu_{\ell_2}\mu_{\ell_2}^T)].$

When SWEM converges to the optimal solution, m micro components are again summarized into m triples. Then, k global clusters are updated with these new statistics. This second stage of the algorithm is analogous to the second stage of the initial phase. Nonetheless, SWEM derives the k global models from only $(m + k)$ components in this incremental phase.

4.3   Expiring Phase

We apply this phase of the SWEM algorithm when the window slides and the oldest time slot is expiring from the mining model. As such, it is necessary to update $\Phi_G = \{\phi_1, \ldots, \phi_k\}$ by subtracting the statistics summarized in the local model $\Phi_L = \{\phi_1, \phi_2, \ldots, \phi_m\}$ of the expiring time slot. SWEM controls this process by using a fading factor λ $(0 < \lambda < 1)$ to gradually remove these statistics. The closer to 1 the λ is, the smaller the amount of the reducing weights being eliminated at each iteration of the algorithm. Thus, this factor provides us a method to prevent the effect of each expiring component from reducing too fast or too slow, which would cause the local optimal convergence in SWEM. Essentially, at each iteration t of the algorithm, SWEM reduces the weight of each expiring MC_ℓ by $N_\ell^{(t)} = \lambda^{(t)} N_\ell$. This also means that the reducing amount, denoted by $r_\ell^{(t)}$, is:

$$r_\ell^{(t)} = (1 - \lambda)\lambda^{(t-1)} N_\ell \tag{1}$$

The following theorem guarantees that the number of iterations t can be any arbitrary integer while the total reducing weights on each expiring component approaches (but never exceeds) its original value.

Theorem 1. *Let t be an arbitrary number of iterations used by the SWEM algorithm. Then for each expiring micro component MC_ℓ: $\lim_{t \to \infty} \sum_t r_\ell^{(t)} = N_\ell$.*

Proof. At the first iteration, $N_\ell^{(1)} = \lambda N_\ell$. Thus the reducing amount is $r_\ell^{(1)} = N_\ell - \lambda N_\ell = (1 - \lambda)N_\ell$. At the second iteration, $N_\ell^{(2)} = \lambda N_\ell^{(1)} = \lambda^2 N_\ell$ and $r_\ell^{(2)} = N_\ell^{(1)} - \lambda N_\ell^{(1)} = (1 - \lambda)\lambda N_\ell$.

By induction, at the iteration t, the reducing weight is $r_\ell^{(t)} = (1 - \lambda)\lambda^{(t-1)} N_\ell$. Therefore, the total reducing amount so far is:

$$\sum_t r_\ell^{(t)} = (1 - \lambda)N_\ell + (1 - \lambda)\lambda N_\ell + \ldots + (1 - \lambda)\lambda^{(t-1)} N_\ell$$

$$= (1 - \lambda)N_\ell[1 + \lambda + \ldots + \lambda^{t-1}]$$

It is clear that:
$$\lim_{t \to \infty}(1 - \lambda)[1 + \lambda + \ldots + \lambda^{t-1}] = \lim_{t \to \infty}(1 - \lambda^t) = 1 \text{ since } \lambda < 1. \qquad \square$$

Given the factor λ to progressively remove out-dated statistics, the E-step computes the posterior probability for each expiring component by:

$$p(\phi_h|T_\ell) = \frac{\alpha_h^{(t)} \times p_h(\frac{1}{N_\ell}\theta_\ell|\mu_h^{(t)}, \Sigma_h^{(t)})}{\sum_{i=1}^k \alpha_i^{(t)} \times p_i(\frac{1}{N_\ell}\theta_\ell|\mu_i^{(t)}, \Sigma_i^{(t)})}$$

and at the M-step, these posterior probabilities is subtracted from the global models. Specifically, the number of data points in the sliding window n_G is updated by: $n_G^{(t+1)} = n_G^{(t)} - \sum_{\ell=1}^{m} r_\ell^{(t+1)}$, in that the amount $\sum_{\ell=1}^{m} r_\ell^{(t+1)}$ is the total number of data points removed when applying Equation 1 at $t+1$. Subsequently, the weight of each global model ϕ_h is simply updated by:

$$\alpha_h^{(t+1)} = \frac{n_h^{(t+1)}}{n_G^{(t+1)}} \quad \text{where} \quad n_h^{(t+1)} = \alpha_h^{(t)} \times n_G^{(t)} - \sum_{\ell=1}^{m} p(\phi_h|T_\ell) \times r_\ell^{(t+1)}$$

The first factor $\alpha_h^{(t)} \times n_G^{(t)}$ is the estimated number of data points belonging to the global model ϕ_h while the second factor is its total number of data points (weighed by the posterior probability) now expiring at $t+1$.

To update new value for μ_h and Σ_h of each global model ϕ_h, SWEM first computes: $\theta_h^{(t+1)} = \theta_h^{(t)} - \sum_{\ell=1}^{m} p(\phi_h|T_\ell) \times r_\ell^{(t+1)} \times \theta_\ell/N_\ell$, in that θ_ℓ/N_ℓ actually is the vector mean of the expiring micro component (which does not change during this process). Then $\mu_h^{(t+1)}$ and $\Sigma_h^{(t+1)}$ are computed by: $\mu_h^{(t+1)} = \frac{\theta_h^{(t+1)}}{n_h^{(t+1)}}$; $\quad \Sigma_h^{(t+1)} = \frac{1}{n_h^{(t+1)}} \left[\Gamma_h^{(t+1)} - \frac{1}{n_h^{(t+1)}} \theta_h^{(t+1)} \theta_h^{(t+1)^T} \right]$

where $\Gamma_h^{(t+1)} = \Gamma_h^{(t)} - \sum_{\ell=1}^{m} \left[p(\phi_h|T_\ell) \times r_\ell^{(t+1)} \times \frac{\theta_\ell}{N_\ell} \right] \left[p(\phi_h|T_\ell) \times r_\ell^{(t+1)} \times \frac{\theta_\ell}{N_\ell} \right]^T$

and $\quad \Gamma_h^{(t)} = n_h^{(t)} \left[\Sigma_h^{(t)} + \mu_h^{(t)} \mu_h^{(t)^T} \right]$

During this process, whenever any global model has weight becoming too small, it is safe to remove it from the mining results. This happens if a global model was formed by only the data points arriving in the expiring time slot. Thus, when this time interval is beyond the window, such a global model is eliminated as well. When a global model is deleted, another one which has the highest summation on variances should be split in order to keep the number of global clusters unchanged. The computation for a global model splitting is similar to the case with splitting micro components presented in Section 4.2.

Finally, the memory space of SWEM is guaranteed by the following theorem.

Theorem 2. *The memory space consumption of SWEM is $O(n + (d^2 + d + 1)(mb + k))$, where n is the maximal number of data points arriving in a time slot and b is the number of time slots within a sliding window.*

Proof. The memory consumption includes two parts. The first part is the data buffer storing the new data points. We assume that n is the maximal number of data points that can arrive in one time slot. The second part is the space for recording the parameters of the mixture Gaussian models at each time slot as well as the entire sliding window. For a Gaussian mixture model of k global components, the memory consumption includes k weights α_h, k d-dimensional mean vectors μ_h, and k $d \times d$ covariance matrices. At each time slot of the sliding window, the memory space needs to maintain m micro components, each characterized by a weight N_ℓ, a d-dimensional vector sum θ_ℓ and a $d \times d$ matrix of squared sum Γ_ℓ. There are b time slots in the sliding window and therefore in total, the memory consumption of SWEM is $O(n + (d^2 + d + 1)(mb + k))$. □

Table 1. Average log likelihood returned by stdEM, SWEMw/oG and SWEM

TS	D2.K10.N100k			D4.K5.N100k			D10.K4.N100k		
	stdEM	w/oG	SWEM	stdEM	w/oG	SWEM	stdEM	w/oG	SWEM
2	-10.436	-10.512	-10.512	-19.252	-19.276	-19.276	-47.846	-47.869	-47.869
4	-10.427	-10.446	-10.446	-19.192	-19.215	-19.215	-47.933	-48.010	-48.010
6	-10.451	-10.604	-10.716	-19.164	-19.220	-19.326	-47.702	-47.712	-47.726
8	-10.444	-10.700	-10.735	-19.188	-19.226	-19.245	-47.859	-47.884	-47.886
10	-10.439	-10.523	-10.579	-19.202	-19.247	-19.258	-47.759	-47.820	-47.873

5   Experimental Results

5.1   Experimental Setup

We implement three algorithms, our SWEM technique, a standard EM algorithm denoted by stdEM[1], and CluStream algorithm [2] using Microsoft Visual C++ version 6.0. All experiments are conducted on a 1.9GHz Pentium IV PC with 1GB memory space running on the Windows XP platform. In the following, to describe a dataset we use D to denote its dimensions, K to denote its number of clusters, and N to its size in terms the number of data records. We first evaluate the clustering quality of SWEM based on the results returned by the standard EM technique. Then, its sensibility to various parameter settings is verified. Finally, the performance of SWEM is compared with that of CluStream algorithm.

5.2   Clustering Quality Evaluation

Using the method described in [18], we generate three datasets each with 100,000 data records and the number of dimensions and clusters are varied from 2 to 10. The data points of each dataset follow a series of Gaussian distributions. To simulate the evolution of the stream over time, we generate new Gaussian distribution for every 20k points by probability of 10%. With the above notations, three datasets are respectively denoted $D2.K10.N100k$, $D4.K5.N100k$ and $D10.K4.N100k$. Unless otherwise indicated, we set the sliding window equal to 5 time slots and use the following parameters: the number of micro components $m = 6K$ (where K is the number of global clusters), the error bound on average log likelihood $\epsilon = 0.001$, the merging threshold based on Mahalanobis distance $Avg(D_{i,j}) = 1$ and the fading factor $\lambda = 0.8$. Similar to other clustering algorithms relying on the EM technique [17, 18], the clustering qualities of SWEM and stdEM are evaluated by using the average log likelihood measure.

Table 1 shows the results returned by our algorithms where datasets are divided into 10 time intervals, each with 10k of data points. The results are reported after each of 2 time slots. This table also presents the outputs of SWEMw/oG, a variation of SWEM[2]. It can be seen that in all cases, our SWEM

---

[1] stdEM works without any constraint of stream environments.

[2] SWEMw/oG differs from SWEM in the expiring phase where it derives the global models by simply re-clustering all sets of micro components maintained.

Table 2. Clustering means on $D4.K5.N100k$ and $D4.K5.N100k$ with 5% noise

			D4.K5.N100k			D4.K5.N100k with noise	
		TM	stdEM	w/oG	SWEM	stdEM	w/oG	SWEM
C1	Dim 1	-165	-165.861	-165.618	-165.930	-163.054	-161.716	-162.050
	Dim 2	281	282.269	282.597	282.797	280.152	279.594	276.758
	Dim 3	-114	-114.070	-113.65	-113.800	-112.744	-110.741	-109.878
	Dim 4	175	175.863	176.609	176.471	172.770	172.969	172.368
C2	Dim 1	124	122.275	122.365	121.860	123.955	123.915	123.539
	Dim 2	-127	-125.064	-125.412	-125.454	-115.209	-116.539	-122.902
	Dim 3	188	188.376	188.3	188.527	179.520	177.993	186.276
	Dim 4	92	91.753	91.9252	91.523	89.571	91.093	97.919
C3	Dim 1	-3	-1.918	-1.90395	-1.745	-3.732	-2.686	-2.377
	Dim 2	224	223.657	223.699	223.446	222.349	222.635	222.477
	Dim 3	-53	-52.288	-52.2454	-52.113	-52.760	-51.036	-50.682
	Dim 4	-176	-175.382	-175.045	-175.102	-175.299	-174.635	-174.607
C4	Dim 1	295	297.043	297.839	297.536	295.111	294.120	296.555
	Dim 2	155	155.647	155.704	156.406	154.611	153.671	154.623
	Dim 3	276	275.964	275.681	275.236	275.875	274.569	274.624
	Dim 4	-73	-72.912	-73.3848	-73.182	-73.159	-75.363	-77.620
C5	Dim 1	245	246.302	246.922	246.851	245.685	245.970	243.827
	Dim 2	11	10.4482	8.990	9.525	11.182	14.011	9.430
	Dim 3	-154	-152.044	-152.077	-152.012	-155.230	-153.562	-152.924
	Dim 4	153	152.852	153.947	153.555	153.428	153.834	152.462

and SWEMw/oG algorithms almost obtain the accuracy close to that of the st-dEM. It is important to note that SWEM and SWEMw/oG process these data streams incrementally whilst stdEM clusters entire data points appearing in the sliding window at once and without any streaming constraints. It is further to observe that the clustering results of SWEM are not much different from SWEMw/oG which clearly indicating that the technique of gradually reducing weights of expiring components from global clusters works effectively in SWEM. Notice that the number of data points SWEM has to compute at the expiring phase is fixed and only $(k+m)$ while that of SWEMw/oG is $b \times m$ and dependent on the window's size b.

In Table 2, we provide more details on our algorithms where the true means (TM column) and the approximate ones of each cluster are reported for the dataset $D4.K5.N100k$ (at its last time slot). We also further add 5% of random noise to this dataset and the corresponding results are reported in the last three columns of the table. It is observed that the approximate means returned by SWEM and SWEMw/oG remain very close to those of the true clusters despite the noise appearance.

To further simulate significant changes in the stream distribution, we randomly generate two completely different distributions ($D = 4$, $K = 5$, and random noise remains 5%), each with 50k data points. Consequently, the dataset $D4.K5.N100k.AB$ is formed by merging two distributions. In the following

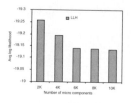
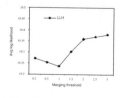
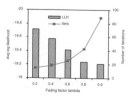

Fig. 2. Micro Components vs. Accuracy

Fig. 3. Merging threshold $Avg(D_{i,j})$ vs. Accuracy

Fig. 4. Fading Factor vs. Accuracy

sections, we test the sensitivity of SWEM on various parameter settings by using this dataset.

Varying Number of Micro Components: It is obvious that the number of micro components should be larger than the number of natural clusters in order to obtain a clustering of good quality. Nevertheless, a very large number of micro components is inefficient in terms of running time and memory storage since the complexity of the SWEM's first stage increases linearly with the number of micro components maintained. In order to observe how this parameter affects the accuracy of our algorithm, we run SWEM with $D4.K5.N100k.AB$ where the ratio between the number of micro components and the natural clusters is varied from 2 to 10. Figure 2 reports the experimental results where we compute the average accuracy on all time slots. From the figure we observe that if the number of micro components is close to the number of global clusters, the clustering quality is poor. This is because a very small number of micro components is harder to approximate the characteristics of the stream, especially in the situation where the distribution changes considerably. A poor approximation often causes worse in later phases where the SWEM needs to remove expiring information. When the number of micro components increases, the average log likelihood increases as well. However, we realize that this value becomes stable when the number of micro components is set around $m = 6K$. This indicates that we do not need to set the number of micro components too large, yet still able to achieve a high clustering accuracy.

Varying Merging Threshold: Recall that the merging and splitting operations are invoked to re-distribute micro components when SWEM detects a significant change happened in the stream's distribution. Furthermore, since SWEM always maintains a set of m micro components at each time slot, when two components are merged, another one with biggest summation in variances will be split. We report the clustering results when the merging threshold $Avg(D_{i,j})$ is varied. Figure 3 reports our results at the time slot 6 at which the algorithm detects a significant change in the data distribution. From the figure, it is observed that when the merging threshold is either too small or too large, the average log likelihood results are poor. The reason is that when $Avg(D_{i,j})$ is set too small, there are almost no micro components being merged (although they are close to each other). In the other extreme when $Avg(D_{i,j})$ is set too large, many

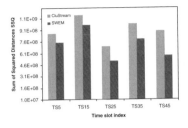 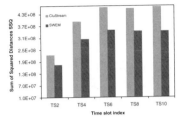

Fig. 5. Clustering quality comparison on real-world Network Intrusion dataset

Fig. 6. Clustering quality comparison on D10.K4.N100k dataset

micro components are frequently merged and split; this causes a poor result on clustering analysis. The highest average log likelihood is achieved when this merging threshold is set to be round 1.

Varying Fading Factor: The last parameter that may impact the clustering quality of SWEM is the fading factor λ. Figure 4 shows the relationship between the fading factor, the number of iterations used in the expiring phases, and the average log likelihood quality. The results are reported at the time slot 10 where the last interval of the first distribution in $D4.K5.N100k.AB$ is eliminated from the sliding window. As expected, when λ is set closer to 1, the algorithm reduces the weight of each expiring micro component slowly and needs more iterations. Accordingly, the quality of the clustering results is better (indicating by the larger average log likelihood value). In order to achieve high quality of clustering results, the best value of λ can be set between 0.8 and 0.9. It is also worth noting that since this expiring step is executed in the second stage where SWEM works only with micro components and global clusters (which actually are only a small number of weighted data points), the large number of iterations utilizing in this stage is therefore generally acceptable.

5.3   Comparison with CluStream

As presented in Section 2, CluStream [2] is an algorithm proposed to cluster entire data streams. Nonetheless, it is possible to modify CluStream for working in the window model. Specifically, instead of storing the snapshots at different levels of granularity, we let CluStream maintain each snapshot precisely at every time slot and those expiring from the sliding window will be deleted immediately. The time horizon in CluStream is chosen equal to the size of the sliding window. Furthermore, we keep the factor 2 for the root mean square (RMS) deviation as indicated in [2], this value produces the best CluStream's results.

It is also important to note that, in CluStream, only data points arriving in the first time slot are clustered until the k-means method converges. For the rest of the stream, each arriving point is clustered online by absorbing it to one micro cluster or forming itself as a new micro cluster. Hence, the convergence of CluStream is not guaranteed after the first time slot. We experimentally realize that such a process is only effective when the stream distribution is relatively stable over time. When the stream distribution remarkably changes, CluStream's qualities reduce considerably. Thus in the following experiments, we let CluStream run until its convergence at each time slot. This compromises the processing time but improves CluStream's clustering. We compare SWEM and CluStream on their clustering quality (measured in sum of squared distances SSQ) and execution time. Figures 5 and 6 report the SSQ values of two algorithms on the KDD-CUP'99 Network Intrusion Detection dataset and D10.K4.N100k (with 5% of random noise). As observed from the figures, the SSQ value of SWEM is always smaller than that of CluStream in both these real-world and artificial datasets. For example, at time slot 25 of the network intrusion dataset, the SSQ of SWEM is 35% smaller than that of CluStream. This explicitly indicates that the data points in each of the clusters obtained by SWEM are more similar and compact than those obtained by CluStream. To explain for these results, note that CluStream computes micro clusters based on Euclidean distance and it does not take into account clusters' shapes in identifying closest center for each data point. Furthermore, the hard assignment (due to using K-means) is highly sensitive to the noise and outliers since a small number of noise data can influence the computation of the clusters' means substantially [14, 18]. On the other hand, both micro clusters' centers and shapes are effectively utilized in SWEM. The using of a Mahalanobis-based distance has improved the SWEM's capability in identifying correct cluster centers. Additionally, SWEM is less affected by the noise data due to the advantage of soft assignment of the EM technique. Its approximate micro components are therefore usually produced in better quality and consequently the global clusters are also derived more accurately.

In order to provide more insights, we observe the clustering quality of two algorithms on $D4.K5.N100k.AB$ dataset. Figure 7 shows the clustering results in which we set the window's size equal to 4. As observed, the SSQ values of both SWEM and CluStream are linearly increased in the first four time slots and slightly changed in the fifth one (since data are generated from the same distribution). However, the clustering quality of CluStream is remarkably worse than that of SWEM at time slot 6 and subsequent ones. This can be understood by the design of CluStream, when a new instance is determined too far from the set of micro clusters and cannot be absorbed by any, CluStream simply creates a new micro cluster for it and merges other ones. As such, some new micro clusters have only one or a few points whereas the others have many. This causes the clusters' weights very imbalance and usually leads to poor approximation on new changing distribution. On the contrary, SWEM re-distributes the set of micro components by applying the merging and splitting operations. A large component having the highest variance sum will be split whilst two small ones will be merged if they are sufficiently close. Such operations not only *discretely* re-distribute micro components in the entire data space but also manage to make the weight of each component not much different from one another. Consequently, new evolving changes in the stream can be essentially approximated by SWEM. As in Figure 7, the SSQ value reported at the last time slot of SWEM is only $180k$ whereas that of CluStream is $233k$.

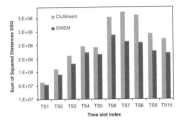
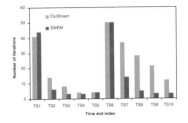

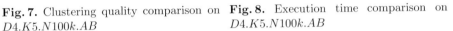

Fig. 7. Clustering quality comparison on $D4.K5.N100k.AB$

Fig. 8. Execution time comparison on $D4.K5.N100k.AB$

In order to evaluate the execution time of two algorithms, we continue using $D4.K5.N100k.AB$. The maximal number of iterations of both algorithms is set to 50 and the execution time is measured in terms of this value. Figure 8 reports the performance of two algorithms at each time slot. At the first time interval, SWEM uses a slightly more number of iterations than Clustream. For the next four ones, the number of iterations of both algorithms reduces considerably since the dataset's distribution is steady. At the critical time slot 6, both algorithms reach the maximum value due to the change in distribution. However, it is interesting to observe the results in the subsequent intervals. While SWEM's iterations reduces sharply, that number of CluStream remains very high. This is attributed to the fact that SWEM computes a set of micro components for each time slot separately; yet it always uses the converged models of the previous time slot as the initial parameters for the next one. This approach usually makes the converged parameters to be quickly achieved if the stream distribution does not significantly change between two consecutive time slots. On the other hand, CluStream always tries to update new data instances into a set of micro clusters maintained so far in the stream (regardless of how much the current distribution is different or similar from the past one). Consequently, CluStream often needs more time to converge even in the case the distribution is stable between time intervals. As observed from Figure 8, the difference in execution time of two algorithms is clearly reflected in the last four time slots of the stream.

6   Conclusions

In this paper, we have addressed the problem of clustering data streams in a sliding window, one of the most challenging mining model. We proposed SWEM algorithm that is able to compute clusters in a strictly single scan over the stream and work within confined memory space. Importantly, two techniques of splitting and merging components have been developed in SWEM in order to address the problem of time-varying data streams. The feasibility of our proposed algorithm was also verified via a number of experiments. Moreover, SWEM has a solid mathematical background as it is designed based on the EM technique. Such a mathematically sound tool has been shown to be stable and effective in many domains despite the mixture models it employs being assumed to follow multivariate Gaussian distributions.

References

1. Charu, C.A., Jiawei, H., Yu, P.S.: A framework for projected clustering of high dimensional data streams. In: VLDB conference, pp. 852–863 (2004)
2. Aggarwal, C.C., Han, J., Wang, J., Yu, P.S.: A framework for clustering evolving data streams. In: VLDB Conference, pp. 81–92 (2003)
3. Babcock, B., Babu, S., Datar, M., Motwani, R., Widom, J.: Models and issues in data stream systems. In: PODS, pp. 1–16 (2002)
4. Babcock, B., Datar, M., Motwani, R.: Sampling from a moving window over streaming data. In: SODA, pp. 633–634 (2002)
5. Babcock, B., Datar, M., Motwani, R., O'Callaghan, L.: Maintaining variance and k-medians over data stream windows. In: PODS (2003)
6. Moses, C., Liadan, O., Better, R.P.: Streaming algorithms for clustering problems. In: ACM symposium on Theory of computing, pp. 30–39 (2003)
7. Cao, F., Ester, M., Qian, W., Zhou, A.: Density-based clustering over an evolving data stream with noise. In: SDM (2006)
8. Chen, Y., Tu, L.: Density-based clustering for real-time stream data. In: SIGKDD Conference, pp. 133–142 (2007)
9. Datar, M., Gionis, A., Indyk, P., Motwani, R.: Maintaining stream statistics over sliding windows. In: SODA, pp. 635–644 (2002)
10. Garofalakis, M., Gehrke, J., Rastogi, R.: Querying and mining data streams: you only get one look a tutorial. In: SIGMOD Conference (2002)
11. Giannella, C., Han, J., Pei, J., Yan, X., Yu, P.S.: Mining Frequent Patterns in Data Streams at Multiple Time Granularities. Next Generation Data Mining (2003)
12. Guha, S., Meyerson, A., Mishra, N., Motwani, R., O'Callaghan, L.: Clustering Data Streams: Theory and Practice. IEEE TKDE 15 (2003)
13. Guha, S., Mishra, N., Motwani, R., O'Callaghan, L.: Clustering Data Streams. In: Proc. Symp. on Foundations of Computer Science (November 2000)
14. Han, J., Kamber, M.: Data mining: concepts and techniques (2001)
15. Liadan, O., Nina, M., Sudipto, G., Rajeev, M.: Streaming-data algorithms for high-quality clustering. In: ICDE (2002)
16. Nesime, T., Ugur, Ç., Stanley, B.Z., Michael, S.: Load shedding in a data stream manager. In: VLDB, pp. 309–320 (2003)
17. Ueda, N., Nakano, R.: Deterministic annealing em algorithm. Neural Netw. 11(2), 271–282 (1998)
18. Aoying, Z., Feng, C., Ying, Y., Chaofeng, S., Xiaofeng, H.: Distributed data stream clustering: A fast em-based approach. In: ICDE Conference, pp. 736–745 (2007)
19. Zhu, Y., Shasha, D.: Statstream: Statistical monitoring of thousands of data streams in real time. In: VLDB Conference (2002)

Alignment of Noisy and Uniformly Scaled Time Series

Constanze Lipowsky, Egor Dranischnikow, Herbert Göttler, Thomas Gottron,
Mathias Kemeter, and Elmar Schömer

Institut für Informatik, Johannes Gutenberg-Universität Mainz
55099 Mainz, Germany
{lipowsky,dranisch,goettler,gottron,
schoemer}@informatik.uni-mainz.de, kemeter@gmail.com

Abstract. The alignment of noisy and uniformly scaled time series is an important but difficult task. Given two time series, one of which is a uniformly stretched subsequence of the other, we want to determine the stretching factor and the offset of the second time series within the first one. We adapted and enhanced different methods to address this problem: classical FFT-based approaches to determine the offset combined with a naïve search for the stretching factor or its direct computation in the frequency domain, bounded dynamic time warping and a new approach called shotgun analysis, which is inspired by sequencing and reassembling of genomes in bioinformatics. We thoroughly examined the strengths and weaknesses of the different methods on synthetic and real data sets. The FFT-based approaches are very accurate on high quality data, the shotgun approach is especially suitable for data with outliers. Dynamic time warping is a candidate for non-linear stretching or compression. We successfully applied the presented methods to identify steel coils via their thickness profiles.

Keywords: Time series, linear time warping, alignment, stretching factor, offset, FFT, bounded dynamic time warping, shotgun analysis, linear regression.

1 Introduction

Given two time series X and Y, where Y is a subsequence of X in the sense that, for a human observer, Y looks similar to a part of X, we want to find an alignment of X and Y. However, compared to X the values of Y are more or less distorted. What makes the problem worse is the fact that Y is either stretched or compressed relative to its corresponding part of X. Now, our aim is to align the two data series automatically, that means to find the counterpart of Y within X in spite of distortion and stretching or compression. To achieve this, we need to determine the offset and the stretching or compression factor of Y relative to X. Figure 1 illustrates this situation: The two plotted data series look similar but are not easy to compare by a computer because the thinner, brighter one is stretched relative to the thicker, darker one.

This problem is highly relevant in practice, e.g., in steel production where the problem was posed and our test data comes from. All our methods can also deal with overlapping data series. The assumption that the second time series is part of the first one is not necessary but is true for the following practical example.

S.S. Bhowmick, J. Küng, and R. Wagner (Eds.): DEXA 2009, LNCS 5690, pp. 675–688, 2009.
© Springer-Verlag Berlin Heidelberg 2009

In the production of steel, coils are important (semi-finished) goods. A coil is a flat wound up steel strip of a certain width and thickness and varying length from a few hundred up to several thousand meters. During the production process a coil passes through different machines and steps. It is repeatedly unwounded, lumbered, cut into pieces, welded together with parts of other coils and wound up again. In between the production steps the coils are stored in the company's interim storage facility. In order to control the whole process and to retrace the origin of a certain piece of steel, it is important to be able to follow a coil and/or pieces of it throughout the whole production process. Therefore, the company *iba AG* in Fürth, Germany came up with the idea to derive a unique identification of each coil via its thickness profiles, the so called "fingerprints" [1], similar to the fingerprints which are unique for humans. The thickness of each coil is measured in certain fixed time intervals before and after each production step (e.g. every 10 ms). The *varying* throughput speed during this process is measured through the varying velocity of the rollers. So, it is possible to convert the time based data to locations on the steel strip – an information which is more interesting in the context of quality control. Based on these measures the thickness of the coil at certain positions or in certain discrete distances (typically every ten centimeters) is calculated by linear interpolation. Two fingerprints of the same coil after one and before the next production step are always similar but not identical. Differences arise due to different measuring devices, measuring inaccuracies, failure of measuring devices and the generally difficult circumstances of the production process (dirt, steam, large temperature differences and changes in the material). Surprisingly, some particular production steps (e.g. galvanization) do not change the fingerprint too much, so it is still possible to recognize the coil afterwards with our techniques.

Inaccuracies in thickness measuring lead to vertical errors, inaccuracies in speed measurement cause horizontal displacements. Because of the latter ones, it is not possible to find the optimal starting position of the second fingerprint within the first one by simply minimizing the mean squared error between the values of the two data series. As can be seen in figure 1, it is necessary to stretch one of the two fingerprints like an elastic band before it is possible to calculate a good alignment of the two data series.

Our main aim was to develop algorithms to align two fingerprints of the same coil after one and before the following production step. Therefore, we have to deal with vertical and horizontal errors as described above. In order to be able to compare the two data series, we consider one fingerprint as fixed (we will refer to it as the "fixed coil") and transform the other coil (the "align coil") onto the same scale. The necessary stretching/compression of the align coil corresponds to a horizontal scaling. An additional difficulty is that in most production steps short pieces of the coil are cut off at the beginning and at the end, because they have been damaged or are inhomogeneous (head and tail scrap). Hence, we also have to find the starting position of the align coil within the fixed coil. This corresponds to a positive offset. So, the assumption holds that the second data series is an inner part of the first one.

Even though all described methods have been developed, adapted or/and chosen to solve the described problem for thickness profiles of steel coils, they can also be applied to other data sets. We successfully used them on width profiles of steel coils and made some promising experiments on sea shell data. Like trees, sea shells form

a)

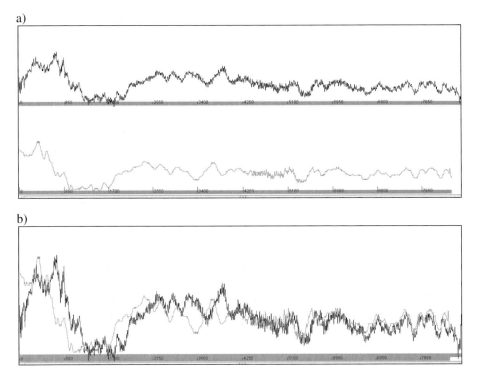

b)

Fig. 1. Two fingerprints of the same coil after one and before the next production step: a) the two profiles below each other, b) the two profiles in the same window: Obviously, there are similarities but those similarities are not easy to detect automatically when comparing the two series straight away because of the horizontal stretching/compression

annual "rings"[1] which can be compared in order to recognize particular environmental influences (e.g., extremely warm and cold years or the eruption of a volcano) or to reconstruct a chronology.

In general, there are two different approaches: Either to calculate/estimate the horizontal scaling first and figure out the offset in a second step or to calculate these two parameters simultaneously. We developed methods for both variants, implemented them in Java and examined them on different real and synthetic data sets.

The rest of the paper is organized as follows: In section 2 we briefly deal with related work. In particular, we will explain some algorithms and ideas which motivated our approaches to solve the task of estimating a horizontal scaling and an offset in time series. We then formally describe the problem in section 3 and describe some aspects of the data we worked with. In section 4 we explain the different algorithms and approaches we used. The results and evaluation methodology of our experiments are listed in section 5, before we conclude the paper in section 6 with a discussion of our findings and a look at future work.

---

[1] In sea shells, these structures are not circular but differences in the annual rates of shell growth show up as lines with different distances and can be measured, too.

2   Related Work

Our problem is a special form of time series analysis. There are similar problems in image processing, automatic speech recognition, dendrochronology and bioinformatics. We took the following approaches into account, modified and enhanced them where necessary and applied them to our data.

2.1   Calculation of Offset and Scaling Based on Fast Fourier Transform (FFT)

To find an alignment of two data series with the same scaling, a naïve approach is to calculate the mean squared error for each possible offset. More precisely, we place the align coil at the starting position of the fixed coil first and slide it point by point to the right afterwards. The mean squared error between the time series is calculated for each possible position. The best match comes with the least error and its starting position is the offset. The whole procedure can be accelerated by carrying out the necessary operations via FFT, which reduces the quadratic run time to $O(n \cdot \log n)$ for the calculation of the correlation (for details see section 4.1).

We can repeat this process for every possible scaling and finally pick the combination of scaling and offset which delivers the least mean squared error over all possible combinations (see section 3.1 and section 4.1).

Furthermore, it is even possible to determine the scaling factor directly by using FFT [2]. It shows up as an impulse in the frequency domain (see section 4.2).

2.2   Dynamic Time Warping (DTW)

Dynamic time warping [3,4] is used for pattern-recognition through comparison of two data series, e.g., for image retrieval, handwriting retrieval, speech recognition and to determine the age of a given piece of wood by comparing its annual rings' structure to the reconstructed dendrochronology of wood for thousands of years [5,6,7]. The order of the data points stays unchanged, but each value in the second sequence is associated with the "best fitting value" at an allowed position in the first one. Several points in the second data series can be mapped onto the same point within the first sequence and vice versa, as long as the order within each series remains unchanged. Another restriction is that each point of the second data series has to be mapped on a point of the first one. The entire process corresponds to finding a path in the matrix of all possible point assignments, which minimizes the squared error. This optimization problem can be solved via dynamic programming.

Run time and space of the so far described algorithm are quadratic. It is possible to adapt the Hirschberg algorithm [8] to this task so that space becomes linear by only doubling the run time. However, methods with quadratic run time are not suitable for long data series as in our steel coil example where the series can have up to 30,000 data points. Since we figured out that the horizontal scaling of our data series is always between 0.9 and 1.1, we can restrict the matrix to a corridor around the diagonal, which reduces the run time to $O(b \cdot n)$ where b is the width of the corridor and thus, speed up the process a lot [5,9]. This variation of DTW is called Bounded Dynamic Time Warping (BDTW, see section 4.3).

2.3 Shotgun Alignment

Bioinformatics deals with alignment problems, too [10,11]. At first glance, these problems are quite different from our problem because there it is the aim to calculate an optimal or at least a good alignment of two DNA or protein sequences which are represented as strings over a finite alphabet. But, when a coil consists of parts which were welded together, our problem looks in a way similar to alignment-problems of pieces of DNA in bioinformatics. So, we had the idea to get inspiration from this field of research. Instead of calculating an alignment of two DNA or protein sequences, we have to calculate an alignment of two different but similar discrete data series. An important difference is that, in our case, we have no fixed character set but different discrete numbers/thicknesses. We can deal with this difference by comparing the interpolated thicknesses at certain positions itself instead of using weights for each possible pair of characters. A second difference is that gaps in the middle of a coil do not occur or are at least extremely rare.

An additional problem is to calculate the horizontal scaling, which can be addressed by the following idea: Venter et al. [12] were the first to use the shotgun sequencing method to figure out the DNA-sequence of the human genome. Their idea was to produce several copies of the human genome, to cut each copy of the huge human chromosomes randomly into very small pieces, to sequence these pieces and to reassemble them automatically by using the overlaps. This process was much faster than the techniques used in the public Human Genome Project and only in highly repetitive regions less accurate (see section 4.4).

3   The Data Series

3.1 Definition of the Alignment-Problem

Given two data series $X = x_0, x_1, x_2, ..., x_{n-1}$ and $Y = y_0, y_1, y_2, ..., y_{m-1}$ we want to minimize the mean squared error of the overlapping part.

The problem is that it is not possible to compare the given data points directly because of a different horizontal scaling. This scaling is assumed to be constant for each pair of data series. According to our studies, this delivers good results for the alignments of coil fingerprints. In other practical examples, we might have to cut the data series into smaller pieces and to calculate a scaling for each piece separately.

If we calculate the scaling s first, keep one data series unchanged (the fixed coil) and adapt the other data series (the align coil) by linear interpolation

$$\tilde{y}_i = y_{\lfloor s \cdot i \rfloor} + (s \cdot i - \lfloor s \cdot i \rfloor) \cdot (y_{\lfloor s \cdot i \rfloor + 1} - y_{\lfloor s \cdot i \rfloor})$$

we can calculate the mean squared error for each possible starting position of the modified second data series $\tilde{Y} = \tilde{y}_0, \tilde{y}_1, \tilde{y}_2, ..., \tilde{y}_{r-1}$ with $r = \lfloor s \cdot m \rfloor$ within the first data series X.

As explained before, in steel production a coil is at most trimmed between two production steps. So, the second fingerprint should be found completely inside the first one. Hence, we can use the following formula to calculate the mean squared error

$$MSE = \frac{1}{r} \cdot \sum_{i=0}^{r-1} (x_{t+i} - \tilde{y}_i)^2$$

where $t \geq 0$ is the offset and $r \geq 0$ is the length of the corresponding part of Y within X.

It is necessary to transform the second data series vertically first, so that the mean values of both data series are equal, because MSE as distance measure is influenced by different values.

3.2 Uniqueness of the Fingerprints

Before actually aligning thickness profiles, we wanted to make sure that such a profile is really characteristic for a certain coil/piece of steel in a sense that it satisfies the fingerprint idea. Therefore, we wanted to know, how long a piece has to be to determine its origin or – in other words – how many values we need to make a decision. Since we had only a limited set of real data, we additionally generated and examined synthetic data to get a more general idea. As we figured out, our example data series can be modeled as a damped random walk, in our case as a discrete Ornstein-Uhlenbeck-process [13]. The measuring point x_{i+1} can be calculated as follows

$$x_{i+1} = \alpha \cdot x_i + z$$

where α is a damping factor and the random variable $z \overset{d}{=} N(0, \sigma^2)$ is normally distributed with mean 0 and standard deviation σ.

We estimated the parameters α and σ of this damped random walk from our data sets to generate synthetic data series. Visually those artificial data series could not be distinguished from a real data series by experts from the *iba AG* ("Turing test"). Then, we generated random walk series with 1,000,000 values, copied pieces of a given length by chance, added some realistic noise to those pieces and tried to locate their starting position within the original data series. We figured out that the noise is normally distributed and, thus, can be modeled as white noise (with different intensities in decibel).

Our experiments showed that it is extremely unlikely to find two corresponding fingerprints by chance: Pieces of 1,500 points were located correctly in 99.8% of all test. Data series of 3,000 points or more can be considered as unique. As for the steel coils we usually have more than 8,000 data points, their thickness profiles can definitely be seen as fingerprints.

4 Algorithms

In this section we describe several algorithms we developed, adapted or simply applied to solve the task of aligning noisy and uniformly scaled time series.

4.1 Naïve Alignment (Without and With FFT)

The simplest approach is to calculate the optimal position of the align coil within the fixed coil by directly minimizing the mean squared error. As we found out empirically, the horizontal scaling always varies only between 0.9 and 1.1, we try every possible scaling in between (the number is limited because our data sets are discrete, so it is sufficient to try each scaling that maps at least one point of the align coil onto a different one of the fixed coil). We then choose the parameters for scaling and offset that produced the smallest mean squared error. Obviously, this process is quadratic for each possible scaling and therefore quite time consuming, but it can be accelerated through the use of a fast Fourier transform as follows. The calculation of the mean squared error can be split into three sums

$$MSE = \frac{1}{r} \cdot \sum_{i=0}^{r-1} (x_{t+i} - \tilde{y}_i)^2 = \frac{1}{r} \cdot \left(\sum_{i=0}^{r-1} x_{t+i}^2 - 2 \cdot \sum_{i=0}^{r-1} x_{t+i} \cdot \tilde{y}_i + \sum_{i=0}^{r-1} \tilde{y}_i^2 \right)$$

where the first one varies only by one value when the offset is moved one step further, the second one is the correlation and can be calculated simultaneously for all possible offsets in the frequency domain with a variation of the common FFT-based calculation of the convolution [14] and the third one is constant for all offsets.

4.2 Calculation of the Scaling in the Frequency Domain

It is also possible to calculate the scaling directly in the frequency domain. The idea comes from the field of image processing. We followed the course of action proposed in [2] and slightly optimized it for the one dimensional data.

Given a function $f_1(t)$ and its scaled and translated replica $f_2(t) = f_1(st + h)$, their corresponding Fourier transforms F_1 and F_2 will be related by

$$F_2(x) = \frac{1}{s} F_1(s^{-1}x) e^{-i2\pi h s^{-1} x}$$

Therefore the following relation for the magnitudes of F_1 and F_2 (m_1 and m_2 respectively) holds after converting the x-axis to the logarithmic scale

$$m_2(y) = \frac{1}{|s|} m_1(y - a)$$

where $y = \log x$ and $a = \log s$.

Thus the scaling is reduced to a translational movement and can be found by the phase correlation technique, which uses the cross-power spectrum

$$C(z) = \frac{M_1(z) \cdot M_2(z)^*}{\left| M_1(z) \cdot M_2(z)^* \right|} = e^{-i2\pi a z}$$

where M_1 denotes the Fourier transform of m_1 and $M_2{}^*$ denotes the complex conjugate of the Fourier transform of m_2.

By taking inverse Fourier transform on the cross-power spectrum, we will have an impulse, which is approximately zero everywhere except at the sought-after displacement a. After scaling the data, we can use the same phase correlation technique for finding the offset h.

Although the described theory can be applied to our problem, there are some subtleties which must be considered in order to get good results in practice. The problems arise because the assumption that one function is a replica of another is not entirely true due to noise and other errors. On the other hand, we use the discrete Fourier transform for concrete computation and this can be a source of further errors due to aliasing and other effects.

The following course of action seems to yield the best results:

1. Not the magnitude spectra M_i but log-magnitude spectra $\log M_i$ should be used.
2. Because of the fact, that the Fourier spectrum is conjugate symmetric for real sequences, only one half of the spectrum should be used.
3. Only a small window in the middle of the logarithmic scale can be used for our purposes. This is due to the fact, that the first points on the logarithmic scale are calculated by means of the linear interpolation from only very few data points in the original lattice and thus contain not much information. On the other hand, it seems as if the higher frequencies were the consequence of the noise and do not comprise any useful information. Experiments have shown that the choice of the right window is the most crucial.

The complexity of this approach is dominated by the costs of getting the Fourier transforms and therefore is of time complexity $O(n \cdot \log n)$ when using the fast Fourier transform.

The sensitive spot of this technique is the determination of the right scaling. The noise in the data has a negative impact on the correctness of the result yielded by the algorithm. Using the FFT accelerated approach described in 4.1 to determine the horizontal scaling gives better results, but is also susceptible to extreme noise.

This technique is a good choice for time series with little noise, since the probability of an incorrect matching rises with the level of the noise.

4.3 Bound Dynamic Time Warping with Regression Analysis

As mentioned before, BDTW can be used to align two given coils, as well. Each point of the align coil is mapped on the best fitting point of the fixed coil while maintaining the order in both series. Then, each pair of mapped indices $(k_i \mid l_i)$, i.e., data point x_{k_i} of the fixed coil has been mapped on data point y_{l_i} of the align coil, is interpreted as a point in a two dimensional coordinate system. Then, the best linear approximation $f(k) = m \cdot k + b$ of all those points is calculated by regression analysis. Therefore, the squared error

$$SE = \sum_{i=0}^{p-1} (l_i - f(k_i))^2$$

is minimized which can be easily done by solving a system of linear equations.

Above, m is an estimation for the horizontal scaling of the align coil, b is an estimation for the offset. Figure 2 shows a mapping illustrating the alignment.

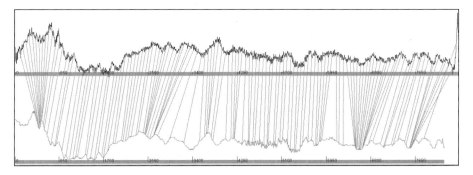

Fig. 2. BDTW shown at the example of two coils

4.4 Shotgun Analysis (Without and With Regression Analysis)

The method of shotgun sequencing and reassembling the parts afterwards was invented by Celera Genomics in the Human Genome Project [12] and inspired us for another approach: The shorter the align coil, the less impact has the usually small horizontal scaling on the alignment. Hence, our idea was to cut the align coil into pieces of a certain length m_s (e.g., of 50 meters = 500 values) and to locate these snippets within the fixed coil by a normal naïve alignment where no horizontal scaling is taken into account. Figure 3 gives an impression how the method works.

The median of the differences between the calculated starting positions of every pair of subsequent snippets divided by the length of the snippets can be used as horizontal scaling between the two data series. After an interpolation step as described in 3.1 the offset of the whole align coil can be calculated naïvely by minimizing the mean squared error.

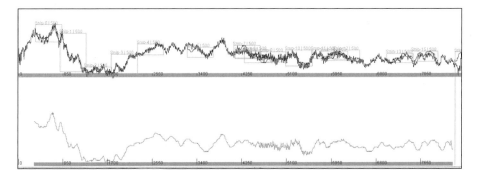

Fig. 3. The "best" positions of the different snippets of the second coil within the first one are shown by boxes. As you can see, the first four snippets are positioned extremely good, while the positions of some other snippets are inconsistent.

The naïve alignment at the end is not necessary. It is possible to calculate scaling and offset simultaneously by linear regression: The expected and the found starting position of each snippet is interpreted as a point in a two dimensional coordinate system (k_i is the expected starting position, l_i is the found starting position). Again the best linear approximation $f(k) = m \cdot k + b$ of all those points is calculated, where m is an estimation for the horizontal scaling of the align coil and b is an estimation for the offset. The computation time is quadratic or more precisely $O(c \cdot m_s \cdot n)$ where c is the number of the snippets and m_s is the length of each snippet.

One problem is that the order of the pieces can be inconsistent, especially when a piece seems to fit well on different positions inside the fixed coil, what can be true for quite a number of them. To avoid this, we used only characteristic snippets to construct the regression line. A snippet is assumed to be characteristic, if the error at its best fitting position is considerably lower than at any other position.

Obviously, this is a heuristic approach. On high quality data, it might be less accurate than the algorithms described before (even though it still produces good results that cannot be distinguished from the other results visually in most cases). Its main advantage is that it is very tolerant to noise and local outliers. This occurs quite often in practice because of measurement failures and/or dirt on a coil. Extremely bad fitting snippets can simply be ignored.

4.5 Alternative Quality Criteria

Instead of the mean squared error two other quality criteria can be used, the (ordinary) correlation or the coefficient of parallel variation[2] [15]. The latter can be seen as a signum function on the differences between data points. An advantage of the correlation is that it implies a normalization of the data. It is tolerant to different means and to a vertical scaling of the data. The coefficient of parallel variation is even more tolerant to noise, as it merely considers the direction of a change what makes it especially helpful for the naïve approach on bad data. It can also be calculated via our FFT algorithm.

5 Experiments

To analyze the quality of the different algorithms which we described in section 4, we ran several experiments. They had different aims, provided insights in different aspects of the algorithms' performance and reflected different tasks in our application scenario of steel coil tracking.

The time complexity of all algorithms has been explained above. In practice, they all need 0.5 to 3 seconds to construct an alignment of two data series with 5,000 to 8,000 measuring points. It is hard to compare their run time exactly because it depends on length and structure of the data series.

5.1 Determination of the Horizontal Scaling

Since the horizontal scaling was unknown for all practical examples, we used synthetic data to show that the developed methods are able to determine it correctly. Therefore, we created "coil data" with head and tail scrap artificially: We first created

---

[2] In related literature, we also encountered the German technical term "Gleichläufigkeitskoeffizient" quite often.

1,000 time series modeled by a damped random walk (the "fixed coils"). Then, we scaled and trimmed a copy of these data and added some white noise to produce corresponding "align coils". Since, for these data, we know the scaling (and the offset), they are perfectly suitable for our tests (a gold standard). The average difference between real and calculated scaling was less then $5 \cdot 10^{-4}$ for all our methods, which means that all methods have an average accuracy of more than three decimal places.

5.2 Recognition of Head and Tail Scrap

The intention of this experiment was to analyze the capability of the algorithms to estimate the parameters for offset and scaling. If head and tail scrap are calculated correctly, also the scaling must have been determined correctly. This is clear because our methods calculate only scaling and offset. The ending position is computed from these two parameters afterwards.

We used different data series for this test. On the one hand, we used real world data of coil thickness profiles recorded during steel production at the end of the first and before the second production step. On the other hand, we used again synthetic data.

5.2.1 Real World Data

As mentioned before, the real world data is noisy under several aspects: First of all, the devices measuring the thickness are based on different technologies and, therefore, have a different resolution and accuracy. Second, the speed of the steel strip while passing through the production plant is measured indirectly and at a different point than the thickness. This setup causes the observed differences in the horizontal scaling in the data. In the beginning of the research project the steel mill *voestalpine AG* in Linz, Austria provided us with information on 20 coils. We had two data series for each coil, one measured after the hot rolling mill and one right before the next production step, the cold rolling mill. In addition, we knew the real length of the coil along with the amount of steel cut off at the beginning and at the end of the coil before the second production step (head and tail scrap). This information should have been sufficient to use it as a gold standard to compare our calculated results – assumed the data were precise enough. Unfortunately, the production conditions did not allow a correct measurement of the parts that had been cut off, so the values given to us finally turned out to be only rough estimates. A visual analysis of the alignments revealed that our methods delivered far more accurate results. Such an alignment is shown in figure 4.

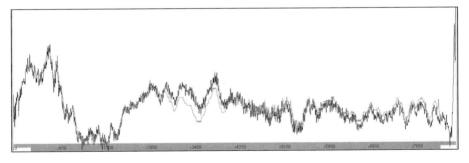

Fig. 4. An alignment of two fingerprints of the coil already shown in figure 1

5.2.2 Artificial Data

To run tests on a larger scale and to be able to evaluate the results, we again used the artificial data described in 5.1. Apart from white noise, scaling and trimming of a coil at the beginning and the end we observed other, more rare and particular noise in real world coil data. The devices to measure and record the thickness might temporarily fail. In this case the data series contains a long period (several hundred data points) of zero values. Another problem is dirt, pieces of steel or holes in the coil. They cause extremely short (1 to 5 data points) peaks in the time series. This kind of noise is difficult to handle.

To analyze how robust the algorithms are towards these phenomena we created further artificial data which contained also this kind of noise. The results of our test on synthetic data are shown in table 1 and 2.

Table 1. Accuracy of the different methods using MSE as quality criterion: Fraction of correctly aligned data series: An alignment was classified as correct when the real and the calculated starting and ending position of the second data series within the first one where no more than 5 points away from each other (equal to half a meter)

method	normal	with 3 peaks	with 3 zero lines	with both
Alignment with FFT	1.0	0.659	1.0	0.634
BDTW with Regression	1.0	0.117	0.546	0.107
Shotgun Analysis	0.952	0.953	0.96	0.945

Table 1 shows e.g., that in all "simple" cases the second data series was aligned correctly to the first one with the FFT-based approach described in section 4.1 and with the BDTW method described in 4.3, whereas the Shotgun Analysis described in 4.4 achieves comparably better results on data series with errors (e.g., still 94.5% correct alignments of all data series with zero lines and peaks).

Table 2. Accuracy of the different methods using MSE as quality criterion: Average difference (in data points) between real and found starting and ending position of the second data series within the first

Method	normal	with 3 peaks	with 3 zero lines	with both
Alignment with FFT	0.51|0.20	17.95|17.89	0.51|0.19	19.73|19.65
BDTW with Regression	1.62|1.15	119.85|119.88	7.79|7.72	115.20|115.26
Shotgun Analysis	1.64|1.55	1.72|1.64	1.72|1.61	6.60|6.54

In table 2, you can see the average difference of "real" and calculated head and tail scrap. The first number stands for the average difference in head scrap, the second one for the average difference in tail scrap. Here, it becomes clear that the Shotgun Analysis is only slightly less accurate than for example BTDW with Regression on good data. The result in table 1 seems worse because there, an alignment is categorised as wrong if its starting or ending position differs 6 or 7 instead of the allowed maximal 5 points from the "real" starting or ending position. This difference can hardly be noticed by visual inspection.

5.3 Searching a Database for Fitting Pairs of Fingerprints

The last test comprised again the fingerprint idea. For this analysis, we were provided with data of about 1,000 hot rolled strips and 191 cold rolled strips. The latter ones were the values of 191 coils among the 1,000 whose thickness was measured again at the beginning of the next production step. Here, we had a gold standard from the steel producers database, providing unique numerical identifiers for the coils. Our task was to identify the 191 among the 1000. Even with the simplest approach described in 4.1, we achieved the very high accuracy of 98%, that means, given the fingerprint of a coil from the beginning of the second production step, we where able to identify the corresponding one from the end of the first production step in most cases.

To accelerate the process and as for this matching task we do not need highly precise alignments, it is possible to compress the data by a method called piecewise aggregate approximation (PAA). Here, e.g. 30 consecutive points are mapped onto one new point by calculating their mean value. This way one gets a compression factor of 30. This gives a considerable speed-up since many candidate coils can be excluded fast. On the compressed data, we got an even higher accuracy of 100% because small inaccuracies are smoothened through the averaging. In this way, it is possible to identify the corresponding hot rolled strip to a given cold rolled strip within 1,000 candidates in less than 5 seconds.

6 Conclusions and Future Work

As illustrated above, the different methods have different strengths and weaknesses: Exact methods are more precise on good data sets but less robust to noise and errors. Therefore, for the practical use, it would be a good idea to have an automated choice of the alignment method based on the peculiarities of the involved data series. Another idea is to manipulate the data in a way that extreme outliers are "smoothed" before the alignment. We already made some experiments and got promising results but have not yet tested them systematically.

In some (exceptional) case, there can be gaps within a data series (e.g., because an erroneous part has been cut out). It is already possible to deal with those cases by cutting such a data series into two pieces and then aligning each piece. The shotgun method is in principle able to ignore the missing parts and to calculate the scaling only for the segments that can be found in both data series. An additional task would be to detect such gaps automatically and to calculate two different offsets and scalings for the two parts.

So far, our methods are restricted to a linear/constant horizontal scaling, which might not be true in all practical examples. Some of our methods (especially the ideas described in 4.3 and 4.4) have the potential to overcome this limitation, e.g., by replacing the linear regression by more sophisticated methods.

An alternative in practice is to compute alignments based on width profiles instead of thickness profiles which is also possible with the described methods. Besides, the described methods can basically be used to align any two data series of numerical values. Another practical example we are dealing with is the alignment of annual sea shell growth measurements to obtain a chronology as described above (see section 1). Therefore, it would be good to allow the alignment of more than two data series.

Acknowledgements

We want to thank the *voestalpine AG* in Linz, Austria for the allocation of real world fingerprint data from steel coils and the *iba AG* in Fürth, Germany for good and interesting cooperation.

Remarks

A selection of the methods described in this paper has been integrated into the freely available tool iba-Analyzer which can be found at the website of the *iba AG* at http://www.iba-ag.com/.

References

[1] Anhaus, H.: Verfahren und Vorrichtung zur Identifizierung eines Teilstücks eines Halbzeugs, Patentblatt DE102006006733B3 (August 23, 2007)

[2] Reddy, B.S., Chatterji, B.N.: An FFT-Based Technique for Translation, Rotation, and Scale-Invariant Image Registration. IEEE Transaction on Image Processing 5(8) (1987)

[3] Toyoda, M., Sakurai, Y., Ichikawa, T.: Identifying similar subsequences in data streams. In: Bhowmick, S.S., Küng, J., Wagner, R. (eds.) DEXA 2008. LNCS, vol. 5181, pp. 210–224. Springer, Heidelberg (2008)

[4] Chu, S., Keogh, E., Hart, D., Pazzani, M.: Iterative Deepening Dynamic Time Warping for Time Series. In: Proceedings of the Second SIAN International Conference on Data Mining (2002)

[5] Ratanamahatana, C.A., Keogh, E.: Everything you know about Dynamic Time Warping is Wrong. In: 3rd Workshop on Mining Temporal and Sequential Data, in conjunction with 10th ACM SIGKDD Int. Conf. Knowledge Discovery and Data Mining (KDD 2004), Seattle, WA (2004)

[6] Euachongprasit, W., Ratanamahatana, C.A.: Efficient multimedia time series data retrieval under uniform scaling and normalisation. In: Macdonald, C., Ounis, I., Plachouras, V., Ruthven, I., White, R.W. (eds.) ECIR 2008. LNCS, vol. 4956, pp. 506–513. Springer, Heidelberg (2008)

[7] Wenk, C.: Algorithmen für das Crossdating in der Dendrochronologie, diploma thesis, Freie Universität Berlin (1997)

[8] Hirschberg, D.S.: A Linear Space Algorithm for Computing Maximal Common Subsequences. Commun. ACM 18(6), 341–343 (1975)

[9] Salvador, S., Chan, P.: Fast DTW: Toward Accurate Dynamic Time Warping in Linear Time and Space. Intelligent Data Analysis 11(5), 561–580 (2007)

[10] Needleman, S.B., Wunsch, C.D.: A general method applicable to the search for similarities in the amino acid sequence of two proteins. Journal of Molecular Biology 48, 443–453 (1970)

[11] Smith, T.F., Waterman, M.S.: Identification of common molecular subsequence. J. Mol. Biol. 147, 195–197 (1981)

[12] Venter, L.C., et al.: The Sequence of the Human Genome. Science 291, 1304–1351 (2001)

[13] Uhlenbeck, G.E., Ornstein, L.S.: On the theory of Brownian Motion. Phys. Rev. 36, 823–841 (1930)

[14] Vetterling, W.T., Teukolsky, S.A., Press, W.A., Flannery, B.P.: Numerical Recipes in C, 2nd edn. Cambridge Univ. Press, Cambridge (1999)

[15] Kemeter, M.: Effizientes Alignment von Stahlband-Fingerprints, diploma thesis, Johannes Gutenberg-Universität Mainz (2008)

Extracting Decision Correlation Rules

Alain Casali[1] and Christian Ernst[2]

[1] Laboratoire d'Informatique Fondamentale de Marseille (LIF),
CNRS UMR 6166, Aix Marseille Université
IUT d'Aix en Provence, Avenue Gaston Berger,
13625 Aix en Provence Cedex, France
`alain.casali@lif.univ-mrs.fr`
[2] Ecole des Mines de St Etienne, CMP - Georges Charpak
880 avenue de Mimet, 13541 Gardanne
`ernst@emse.fr`

Abstract. In this paper, two concepts are introduced: decision correlation rules and contingency vectors. The first concept results from a cross fertilization between correlation and decision rules. It enables relevant links to be highlighted between sets of patterns of a binary relation and the values of target items belonging to the same relation on the twofold basis of the Chi-Squared measure and of the support of the extracted patterns. Due to the very nature of the problem, levelwise algorithms only allow extraction of results with long execution times and huge memory occupation. To offset these two problems, we propose an algorithm based both on the lectic order and contingency vectors, an alternate representation of contingency tables.

1 Introduction and Motivation

An important field in data mining is related to the detection of links between values in a binary relation with reasonable response times. In order to solve this problem, Agrawal et al. (1996) have introduced levelwise algorithms which allow the computation of association rules. Those express directional links ($X \rightarrow Y$ for example), based on the support-confidence platform. While introducing literalsets, Wu et al. (2004) propose computing positive and/or negative association rules, with the aim of extracting rules such as $\neg X \rightarrow Y$ or $\neg X \rightarrow \neg Y$ or $X \rightarrow \neg Y$.

A literal is a pattern $X\overline{Y}$ in which X is also called the positive part and $\overline{Y}$ the negative part. To compute such rules, the authors always use the support-confidence platform by redefining the support of a literal: the number of transactions of the binary relation including X and containing no 1-item (item of cardinality 1) of Y. Brin et al. (1997) propose the extraction of correlation rules, where the platform is no longer based on the support or the confidence of the rules, but on the Chi-Squared statistical measure, written χ^2. The use of χ^2 is well-suited for several reasons:

S.S. Bhowmick, J. Küng, and R. Wagner (Eds.): DEXA 2009, LNCS 5690, pp. 689–703, 2009.

1. It is a more significant measure in a statistical way than an association rule;
2. The measure takes into account not only the presence but also the absence of the items; and
3. The measure is non-directional, and can thus highlight more complex existing links than a "*simple*" implication.

The crucial problem, when computing correlation rules, is the memory usage required by levelwise algorithms. For a pattern X, the computation of the χ^2 function is based on a table including $2^{|X|}$ cells. Thus, at level i, we have to generate the $C^i_{|\mathcal{I}|}$ candidates, in the worst case scenario, and to store them, as well as the equivalent number of tables. If each cell is encoded over 2 bytes, at the 3rd level, a levelwise algorithm thus requires for the storage of the tables, still in the worst case scenario, 2.5 GB of memory, and 1.3 TB at the 4th level. This is why Brin et al. (1997) compute only correlations between two values of a binary relation. Using a given threshold $MinCor$, Grahne et al. (2000) show that the constraint "$\chi^2(X) \geq MinCor$" is a monotone constraint. Consequently, the set of rules resulting is a convex space (Vel, 1993). This set can be represented by its minimal border, noted L (Hirsh, 1994). In the latter paper, the author proposes a levelwise algorithm to compute this border. The deduction of the χ^2 value for a pattern which belongs to the convex space is carried out by using an approximation according to the values of the χ^2 of the patterns for L included in the current pattern.

On the other hand, when applying APC[1] approaches in semiconductor manufacturing, it is important to highlight correlations between parameters related to production in order to rectify possible drifts of the associated processes. Within this framework, in collaboration with STMicroelectronics and ATMEL, our current work is focused on the detection of the main parameters having an impact on the yield. The analysis is based on CSV files of measurements associated with production lots, whose characteristics are to have a huge number of columns (nature of the measurements) with regard to the number of rows (measures). We want to highlight correlations between the values of some columns and those of a target column (a particular column of the file, the yield).

To solve these problems, we introduce in this article the concept of decision correlation rules, a restriction of correlation rules containing a value of one target column. In order to compute these rules:

1. We use the lectic order (Ganter and Wille, 1999) to browse the powerset lattice;
2. We propose the concept of contingency vector: a new approach to the contingency tables;
3. We show how to build the contingency vector of a pattern with a cardinality i with the contingency vector of one of its subsets with a cardinality $i-1$ (which is impossible with contingency tables); and
4. We take advantage of the lectic order, the contingency vectors and the recursing mechanisms of construction to propose the LHS-CHI2 Algorithm.

---

[1] Advanced Process Control.

Finally, we carry out experiments on relations provided by the above mentioned manufacturers, and compare our results with a levelwise approach proposed in (Casali and Ernst, 2007).

The paper is organized as follows: in Section 2, the bases of correlation rules and of the lectic order are recalled. Section 3 describes the concepts used for mining decisional correlation rules and our algorithm. Experiments are detailed in Section 4. As a conclusion, we summarize our contributions and outline some research perspectives.

2   Related Work

In this section, the concepts of correlation rules (Brin et al., 1997) and of lectic order (Ganter and Wille, 1999) are first recalled. We also introduce the Ls Algorithm (Laporte et al., 2002), which allows the powerset lattice for a given set of items to be enumerated, according to the lectic order.

2.1   Correlation Rules

Let r be a binary relation (a transaction database) over a set of items $\mathcal{R} = \mathcal{I} \cup T$. In our approach, $\mathcal{I}$ represents the values (or the items) of the binary relation used as analysis criteria and T is a target attribute. For a given transaction, the target attribute does not necessarily have a value. The computation of the value for the χ^2 function for an item $X \subseteq \mathcal{R}$ is based on its contingency table. In order to simplify the notation, we introduce, in a first step, the lattice of the literalsets associated with a pattern $X \subseteq \mathcal{R}$. This set contains all the literalsets that can be built up given X, and with a cardinality $|X|$.

Definition 1 (Literalset Lattice). *Let $X \subseteq \mathcal{R}$ be a pattern, we denote by $\mathbb{P}(X)$ the literalset lattice associated with X. This set is defined as follows: $\mathbb{P}(X) = \{Y\overline{Z}$ such that $X = Y \cup Z$ and $Y \cap Z = \emptyset\} = \{Y\overline{Z}$ such that $Y \subseteq X$ and $Z = X \backslash Y\}$.*

Example 1. The literalset lattice associated with $X = \{A, B, C\}$ contains the following elements: $\{ABC, AB\overline{C}, AC\overline{B}, BC\overline{A}, A\overline{BC}, B\overline{AC}, C\overline{AB}, \overline{ABC}\}$.

Definition 2 (Contingency Table). *For a given pattern X, its contingency table, noted $CT(X)$, contains exactly $2^{|X|}$ cells. Each cell yields the support of a literalset $Y\overline{Z}$ belonging to the literalset lattice associated with X, i.e. this number represents the number of transactions of the relation r including Y and containing no 1-item of Z.*

Example 2. With the relation example r given in Table 1, Table 2 shows the contingency table of pattern BC.

In a second step, for each cell $Y\overline{Z}$ of the contingency table associated with X, we compute its expectation value (or average). In other words, we measure the

A. Casali and C. Ernst

Table 1. Relation example r

Tid	Item	Target
1	BCF	t_1
2	BCE	t_1
3	BCF	t_2
4	BC	-
5	BD	t_1
6	B	-
7	ACF	t_1
8	AC	-
9	AE	t_1
10	F	t_2

Table 2. Contingency table of pattern BC

	B	$\overline{B}$	$\sum_{raw}$
C	4	2	6
$\overline{C}$	2	2	4
$\sum_{column}$	6	4	10

theoretical frequency in case of independence of the 1-items included in $Y\overline{Z}$, see formula (1).

$$E(Y\overline{Z}) = |r| * \prod_{y \in Y} \frac{Supp(y)}{|r|} * \prod_{z \in Z} \frac{Supp(\overline{z})}{|r|} \tag{1}$$

In order to compute the value of the χ^2 function for a pattern X, for each item $Y\overline{Z}$ belonging to its literalset lattice, we measure the difference between the square of the support of $Y\overline{Z}$ and its expectation value, and divide by the average of $Y\overline{Z}$. Finally, all these values are summed (see formula (2)).

$$\chi^2(X) = \sum_{Y\overline{Z} \in \mathbb{P}(X)} \frac{(Supp(Y\overline{Z}) - E(Y\overline{Z}))^2}{E(Y\overline{Z})} \tag{2}$$

Brin et al. (1997) show that there is a single degree of freedom between the items. A table giving the centile values in function of the χ^2 value for X (Spiegel, 1990) can be used in order to obtain the correlation rate for X.

Example 3. Continuing our example, the χ^2 value for the pattern BC can be developed in a simple way: $\chi^2(BC) = \frac{(4-6*\frac{6}{10})^2}{6*\frac{6}{10}} + \frac{(2-4*\frac{6}{10})^2}{4*\frac{6}{10}} + \frac{(2-6*\frac{4}{10})^2}{6*\frac{4}{10}} + \frac{(2-4*\frac{4}{10})^2}{4*\frac{4}{10}} \simeq$ 0.28. This value corresponds to a correlation rate of about 45%.

Unlike association rules, a correlation rule is not represented by an implication but by the patterns for which the value of the χ^2 function is larger than or equal to a given threshold.

Definition 3 (Correlation Rule). *Let $MinCor$ (≥ 0) be a threshold given by the end-user and $X \subseteq \mathcal{R}$ a pattern. If the value for the χ^2 function for X is larger than or equal to $MinCor$, then this pattern represents a valid correlation rule.*

Moreover, in addition to the previous constraint, many authors have proposed some criteria to evaluate whether a correlation rule is semantically valid (Moore,

1986). Generally, the Cochran criteria are used: all literalsets of a contingency table must have an expectation value not equal to zero and 80% of them must have a support larger than 5% of the whole population. This last criterium has been generalized by Brin et al. (1997) as follows: $MinPerc$ of the literalsets of a contingency table must have a support larger than $MinSup$, where $MinPerc$ and $MinSup$ are thresholds specified by the user.

Example 4. Let $MinCor = 0.25$, then the correlation rule materialized by the pattern BC is valid ($\chi^2(BC) \simeq 0.28$). However, the correlation rule represented by the pattern Bt_1 is not valid ($\chi^2(Bt_1) \simeq 0.1$).

2.2   Lectic Order

The lectic order, denoted by $<_{lec}$, makes it possible to enumerate all the subsets of an itemset $\mathcal{I}$. This order allows the closed lattice of a binary relation (Ganter and Wille, 1999) to be computed, or to serve as a basis for the computation of the Partition Cube (Laporte et al., 2002): a lossless reduction of the (iceberg) data cube. The definition of the lectic order is given below:

Definition 4 (Lectic Order). *Let $\mathcal{I}$ be a set of items. Let us suppose that the items are totally ordered and therefore comparable two by two via an order denoted by $\preceq$. If X and $Y \subseteq \mathcal{I}$, then we have: $X <_{lec} Y \Leftrightarrow max_{\preceq}(X \backslash (X \cap Y)) \preceq max_{\preceq}(Y \backslash (X \cap Y))$. In other words, we deprive X and Y of their common part, and we see whether the last element of X is smaller (according to the $\preceq$ order) than the last one of Y.*

Example 5. Let us consider the set $\mathcal{I} = \{A, B, C\}$ totally ordered according to the lexicographic order. The enumeration of the subsets of $\mathcal{I}$, according to the lectic order, produces the following result: $\emptyset <_{lec} A <_{lec} B <_{lec} AB <_{lec} C <_{lec} AC <_{lec} BC <_{lec} ABC$.

In order to enumerate all the subsets of $\mathcal{I}$ according to the lectic order, the Lectic Subset Algorithm, noted Ls (Laporte et al., 2002), is used. The associated execution tree is a balanced tree, and is based on a double recursive call. Being given a node of the tree (representing a pattern $X \subseteq \mathcal{I}$,) the left sub-tree generates sub-patterns of X not containing $max_{\preceq}(X)$, whereas the right sub-tree leads to sub-patterns of X containing $max_{\preceq}(X)$.

   Algorithm 1 provides a pseudo code for Ls. The first call to this algorithm is done with $X = \emptyset$ and $Y = \mathcal{I}$.

Example 6. Figure 1 shows the execution tree of the Ls Algorithm for $\mathcal{I} = \{A, B, C\}$.

The following proposition expresses the fact that the lectic order is compatible with the anti-monotone constraints. Consequently, we can modify the Ls Algorithm to take into account a conjunction of anti-monotone constraints.

Proposition 1. *Let be X, $Y \subseteq \mathcal{I}$ two itemsets. If $X \subset Y$, then $X <_{lec} Y$ (Ganter and Wille, 1999).*

Alg. 1. Algorithm Ls

Input: X and Y two itemsets
Output: Powerset lattice of X
 1. **if** $Y = \emptyset$ **then Output** X
 2. $A := max(Y)$
 3. $Y := Y \backslash \{A\}$
 4. $LS(X, Y)$
 5. $Z := X \cup \{A\}$
 6. $LS(Z, Y)$

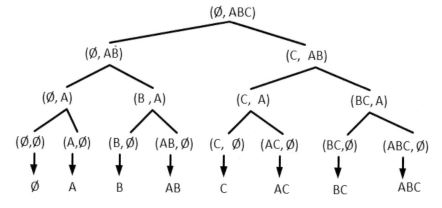

Fig. 1. Execution tree of Ls for $\mathcal{I} = \{A, B, C\}$

3   The LHS-Chi2 Algorithm

The contingency vectors are another representation of the contingency tables. We show that, for a given pattern $X \cup A$ ($X \subseteq \mathcal{R}, A \in \mathcal{R} \backslash X$), the computation of its contingency vector is possible using the contingency vector of X and the list of the row identifiers of the relation containing A. Then, we introduce the concept of decision correlation rule: a restriction of correlation rules, in such a way that the only rules containing a value of the target attribute are kept. Finally, the LHS-CHI2 Algorithm is used in order to compute the decision correlation rules based on the contingency vectors.

3.1   Contingency Vectors

A literal $Y\overline{Z}$, belonging to the literalset lattice associated with a pattern X, is represented in machine with vectors of $|X|$ bits. For a 1-item $x \in X$, the value of the vector of bits has a value of 1 if $x \in Y$ (the 1-item belongs to the positive part of the literal), and 0 otherwise. Thus, comparing two literals $Y_1\overline{Z_1}$ and $Y_2\overline{Z_2}$ belonging to the literalset lattice associated with the pattern X, consists in comparing each integer corresponding to the binary value of the associated vector of bits. Such a comparison is equivalent to extending the definition of the lectic order to the literalset one. A literalset precedes another literalset, according to the lectic order, if and only if the positive part of the first literal precedes, according to the lectic order, the positive part of the second literal. This order makes it possible to totally order the whole literalset lattice associated with the pattern X.

Definition 5 (Lectic order for a literalset). *Let $X \subseteq \mathcal{R}$ be a pattern, $Y_1\overline{Z_1}$ and $Y_2\overline{Z_2}$ two elements of the literalset lattice associated with the pattern X. The definition of the lectic order is extended over the literalsets as follows: $Y_1\overline{Z_1} <_{lec} Y_2\overline{Z_2}$ if and only if $Y_1 <_{lec} Y_2$.*

Example 7. The literalset lattice associated with the pattern $X = \{A, B, C\}$ according to the lectic order is the following: $\overline{ABC} <_{lec} \overline{AB}C <_{lec} \overline{BA}C <_{lec} AB\overline{C} <_{lec} C\overline{AB} <_{lec} AC\overline{B} <_{lec} BC\overline{A} <_{lec} ABC$.

Definition 6 (Equivalence Class associated with a literal). *Let $Y\overline{Z}$ be a literal. Let us denote by $[Y\overline{Z}]$ the equivalence class associated with the literal $Y\overline{Z}$. This class contains the set of transaction identifiers of the relation including Y and containing no value of Z (i.e., $[Y\overline{Z}] = \{i \in Tid(r)$ such that $Y \subseteq Tid(i)$ and $Z \cap Tid(i) = \emptyset\}$).*

Example 8. With our relation example (see Table 1), we have $[B\overline{C}] = \{5, 6\}$.

Proposition 2. *Let $X \subseteq \mathcal{R}$ be a pattern. The union of the equivalence classes $[Y\overline{Z}]$ of the literalset lattice associated with X is a partition (Laurent and Spyratos, 1988) of the identifiers of relation r. In other words:*

$$\bigcup_{Y\overline{Z} \in \mathbb{P}(X)} [Y\overline{Z}] = Tid(r)$$

Proof. Let $[Y_1\overline{Z_1}]$ and $[Y_2\overline{Z_2}]$ two equivalence classes belonging to the same literalset lattice associated with X. We show that a single transaction i can only belong to a single equivalence class. Let us suppose that i belongs to both equivalence classes. From the definition of an equivalence class, we have:

1. $\forall y_1 \in Y_1, \ y \in Tid(i)$
2. $\forall y_2 \in Y_2, \ y \in Tid(i)$

Since $Y_1\overline{Z_1}$ and $Y_2\overline{Z_2}$ belong to the same literalset lattice associated with X, we have two possibilities to build $Y_2\overline{Z_2}$ up from $Y_1\overline{Z_1}$:

1. There is at least one 1-item $y_1 \in Y_1$ such that $y_1 \in Z_2$, then $Z_2 \cap Tid(i) \neq \emptyset$ and, as a consequence: $i \notin [Y_2\overline{Z_2}]$
2. There is at least one 1-item $y_2 \in Y_2$ such that $y_2 \in Z_1$, then $Z_1 \cap Tid(i) \neq \emptyset$ and, as a consequence: $i \notin [Y_1\overline{Z_1}]$

We deduce that a single transaction i can only belong to a single equivalence class. □

Definition 7 (Contingency Vector). *Let $X \subseteq \mathcal{R}$ be a pattern. The contingency vector of X, denoted $CV(X)$, regroups the set of the literalsets equivalence classes belonging to $\mathbb{P}(X)$ and ordered according to the lectic order.*

Proposition 2 ensures that a single transaction identifier belongs only to one single equivalence class. Consequently, for a given pattern X, its contingency vector is an exact representation of its contingency table. To derive the contingency table from a contingency vector, it is enough to compute the cardinality of each of its equivalence classes. If the literalsets, related to the equivalence classes of a contingency vector, are ordered according to the lectic order, it is possible to know, because of the binary coding used, the literal relative at a position i of a contingency vector ($i \in [0; |X| - 1]$). This is because the literal and the integer i have the same binary coding.

Example 9. With our example relation (see Table 1), the contingency vector associated with the pattern BC is the following: $CV(BC) = \{[\overline{BC}], [B\overline{C}], [C\overline{B}], [BC]\} = \{\{9, 10\}, \{5, 6\}, \{7, 8\}, \{1, 2, 3, 4\}\}$.

The following proposition is the main result of our paper. It shows how to compute the contingency vector of the pattern $X \cup A$ given the contingency vector of X and the set of identifiers of the relation containing pattern A.

Proposition 3. *Let $X \subseteq \mathcal{R}$ be a pattern and $A \in \mathcal{R} \backslash X$ a 1-item. The contingency vector of the pattern $X \cup A$ can be computed given the contingency vectors of X and A as follows:*

$$CV(X \cup A) = (CV(X) \cap [\overline{A}]) \cup (CV(X) \cap [A]) \tag{3}$$

However, for a 1-item A, and by definition, the contingency vector of A contains respectively the set of the identifiers of the relation which does not contain A and the set of the identifiers of the relation which contains A. Moreover, we have: $Tid(\overline{A}) = Tid(r) \backslash Tid(A)$; as a consequence, formula (3) can be rewritten as follows:

$$CV(X \cup A) = (CV(X) \cup (Tid(r) \backslash Tid(A))) \cup (CV(X) \cap Tid(A)) \tag{4}$$

Proof. From the definition of a contingency vector, we have:

$$CV(X \cup A) = \bigcup_{Y\overline{Z} \in \mathbb{P}(X \cup A)} [Y\overline{Z}] = \bigcup_{Y\overline{Z} \in \mathbb{P}(X)} [Y\,\overline{AZ}] \cup \bigcup_{Y\overline{Z} \in \mathbb{P}(X)} [AY\overline{Z}] \ (a)$$

Moreover:

$$- \ [Y\overline{AZ}] = \{i \in Tid(r) \text{ such that } Y \subseteq Tid(i) \text{ and } ZA \cap Tid(i) = \emptyset\}$$
$$= \{i \in Tid(r) \text{ such that } Y \subseteq Tid(i) \text{ and } Z \cap Tid(i) = \emptyset\} \cap Tid(\overline{A})$$
$$= [Y\overline{Z}] \cap [\overline{A}] \ (b)$$
$$- \ [AY\overline{Z}] = \{i \in Tid(r) \text{ such that } AY \subseteq Tid(i) \text{ and } Z \cap Tid(i) = \emptyset\}$$
$$= \{i \in Tid(r) \text{ such that } Y \subseteq Tid(i) \text{ and } Z \cap Tid(i) = \emptyset\} \cap Tid(A)$$
$$= [Y\overline{Z}] \cap [A] \ (c)$$

By pushing intermediate results (b) and (c) into (a), we get:

$$CV(X \cup A) = (\bigcup_{Y\overline{Z} \in \mathbb{P}(X)} [Y\overline{Z}] \cap [\overline{A}]) \cup (\bigcup_{Y\overline{Z} \in \mathbb{P}(X)} [Y\overline{Z}] \cap [A])$$

Which can be simplified as follows: $CV(X \cup A) = (CV(X) \cap [\overline{A}]) \cup (VC(X) \cap [A])$.   □

Example 10. With the example relation (see Table 1), we have $CV(B) = \{\{7, 8, 9, 10\}, \{1, 2, 3, 4, 5, 6\}\}$ and $CV(C) = \{\{5, 6, 9, 10\}, \{1, 2, 3, 4, 7, 8\}\}$. By applying Proposition 3, and by ordering the literalsets of the literalset lattice associated with BC according to the lectic order, the contingency vector of BC is the following: $CV(BC) = \{\{9, 10\}, \{5, 6\}, \{7, 8\}, \{1, 2, 3, 4\}\}$. Thus, we retrieve the result of Example 9.

Algorithm 2 is used, given the contingency vector of a pattern X and the set of the transaction identifiers containing a 1-item A, to build the contingency vector of the pattern $X \cup A$ sorted according to the lectic order over the literalset lattice $\mathbb{P}(X \cup A)$.

Alg. 2. CREATE_CV Algorithm

Input: $CV(X)$ contingency vector of X, $Tid(A)$
Output: contingency vector of $X \cup A$ sorted according to the lectic order
1. $CV(Z) := \{\emptyset\}$
2. **for all** Equivalence class $[Y\overline{Z}] \in \mathbb{P}(X)$ according to the lectic order **do**
3. $CV(Z) := CV(Z) \cup ([Y\overline{Z}] \cap (Tid(r)\backslash(Tid(A)))) \cup ([Y\overline{Z}] \cap Tid(A))$
4. **end for**
5. **return** $CV(Z)$

The computation of a CV needs one database scan, and the following transition to the associated CT another one (overheads are ignored). This leads to a complexity of $2*|r|$, or $\mathcal{O}(|r|)$ whatever the number of cells of the CT. A classical computation of a CT at level i also needs one database scan; but here, in the worst case, each of the CT's cells is involved in one operation, which globally forces $2^i * |r|$ operations. Because 2^i is generally much smaller in comparison to $|r|$, the complexity is also of $\mathcal{O}(|r|)$. But when going into detail, the difference between the two methods is $2^{i-1} * |r|$ operations. With a database containing for example 500 transactions, this difference for a single CT computation is 4000 operations at level 4.

3.2   Decision Correlation Rules

Definition 8 (Decision Correlation Rules). *Let $X \subseteq \mathcal{R}$ be a pattern, and $MinCor$ a given threshold. The pattern X represents a valid decisional correlation rule if and only if:*

1. X contains a value of the target attribute $\mathcal{T}$;
2. $\chi^2(X) \geq MinCor$.

Example 11. With our relation example (see Table 1), if $MinCor = 0.25$, the decision correlation rule materialized by the pattern BCt_1 is a valid rule because:

1. $t_1 \in \mathcal{T}$ and $t_1 \in BCt_1$;
2. $\chi^2(BCt_1) \simeq 0.28$ ($\geq MinCor$).

The Lectic Hybrid Subset-Chi2 Algorithm, or LHS-CHI2, makes it possible to extract the whole set of decision correlation rules for a relation r satisfying the threshold constraint $MinCor$ for the χ^2 function. This algorithm is an adaptation of the Ls Algorithm to our context. This adaptation helps, among other things, to take into account contingency vectors. Moreover, we have added several monotone and anti-monotone constraints in order to prune the search space (Grahne et al., 2000). These constraints are:

1. A value of the target attribute must be present in the extracted pattern (monotone constraint);
2. As the χ^2 computation has no significance for a 1-item, we make sure that the cardinality of an examined pattern is larger than or equal to two (monotone constraint);
3. Since the χ^2 function is an increasing function, we impose a maximum cardinality, noted $MaxCard$, on the number of patterns to examine which, usually, does not exceed 8 (anti-monotone constraint);
4. All literalsets of a contingency table must have an expectation value not equal to zero (anti-monotone constraint).
5. Because the obtained rules must have a semantics on the relation, at least $MinPerc$ of the cells of the contingency table must have a support larger than or equal to $MinSup$. This constraint is expressed by the predicate Ct-$Perc$ in our algorithm. This predicate has three parameters: the contingency vector, $MinPerc$ and $MinSup$ (anti-monotone constraint).

Laporte et al. (2002) have modified the Ls Algorithm in order to compute iceberg data cubes. The authors take into account an anti-monotone constraint threshold from which the satisfiability is evaluated before the second recursive call of the Algorithm Ls (line 6). The authors use a pruning step with the negative border (Mannila and Toivonen, 1997) in order to only examine the most *"interesting"* cuboids (patterns in our context). In the same spirit, we modify the Ls Algorithm in order to take into account the five constraints above and to compute the χ^2. The result is an algorithm requiring, in the worst case, $2 * |\mathcal{R}| + 1$ contingency vectors in memory: we need $|\mathcal{R}|$ contingency vectors for the 1-items, the height of our tree is bounded by $|\mathcal{R}|$, and we need an additional contingency vector for the current node computation. This number has to be compared to the number of contingency tables to be computed at each level using a levelwise algorithm ($C^i_{|\mathcal{R}|}$).

Proposition 1 justifies the integration of these constraints in our algorithm. However, we do not carry out pruning using the negative border. Instead, we use the positive one (Mannila and Toivonen, 1997) relating to the predicate *CtPerc*. The use of the positive border is justified on the basis of the experiments carried out by Flouvat et al. (2005). The authors show that the positive border is of highly reduced cardinality by comparison with the negative border. As a consequence, the satisfiability tests of the anti-monotone constraints are faster when the positive border is used. In our context, we make sure that the pattern Z, used as a parameter within the second recursive call of the algorithm, has all its direct subsets included in one of the elements of the positive border (line 8). Let us emphasize that this test is carried out in the *AprioriGen* function (Agrawal et al., 1996) during the generation of the candidates of level $i+1$ using the frequent patterns for level i. If pattern Z is a candidate, then we compute its contingency vector by making sure that the literalsets relating to the classes of equivalence are sorted according to the lectic order (line 9) by calling Algorithm 2. If the pattern satisfies the anti-monotone constraints (line 10), we update the positive border (line 11), and then carry out the second recursive call of the algorithm (line 12). The set of the monotone constraints is evaluated on the leaves of the execution tree (line 1). By convention, we consider that we have $CV(\emptyset) = \{Tid(R), \emptyset\}$. The positive border is initialized with $\{\emptyset\}$. The pseudo code of LHS-CHI2 is provided in Algorithm 3. The first recursive call to LHS-CHI2 is carried out with $X = \emptyset$ and $Y = \mathcal{R}$.

Alg. 3. LHS-CHI2 Algorithm

Input: X and Y two patterns
Output: $\{itemset\ Z \subseteq X\ such\ that\ \chi^2(Z) \geq MinCor\}$

1. **if** $Y = \emptyset$ **and** $|X| \geq 2$ **and** $\exists c \in \mathcal{C} : c \in X$ **and** $\chi^2(X) \geq MinCor$ **then**
2.   Output $X, \chi^2(X)$
3. **end if**
4. $A := max(Y)$
5. $Y := Y \backslash \{A\}$
6. LHS-CHI2(X,Y)
7. $Z := X \cup \{A\}$
8. **if** $\forall z \in Z, \exists W \in BD^+ : \{Z \backslash z\} \subseteq W$ **then**
9. $VC(Z) := CREATE\_CV(CV(X), Tid(A))$
10.   **if** $|Z| \leq MaxCard$ **and** $CtPerc(CV(Z), MinPerc, MinSup)$ **then**
11. $BD^+ := max_\subseteq(BD^+ \cup Z)$
12.   LHS-CHI2(Z,Y)
13.   **end if**
14. **end if**

Example 12. The results of LHS-CHI2 with the thresholds $MinSup = 0.2$, $MinPerc = 0.25$ and $MinCor = 0.25$ for our relation example (see Table 1) are shown in Table 3.

Table 3. Results of the LHS-CHI2 algorithm over Table1 with the thresholds of Example 12

Decision Correlation Rule	χ^2 Value
At_1	0.48
BCt_1	0.28
BFt_1	0.28

4   Experimental Analysis

Some representative results of the LHS-CHI2 algorithm are presented below. The comparison is made with a standard levelwise algorithm, hereafter called LEVELWISE, based on the same monotone and anti-monotone constraints as those used in LHS-CHI2. The main difference is that the LEVELWISE method does not use contingency vectors but uses standard computation of contingency tables. Such an implementation of LEVELWISE was presented in (Casali and Ernst, 2007), as well as the *"cleaning"* aspects of the files analyzed in input, summarized first of all for the sake of clarity.

As emphasized in Section 1, the experiments were done on different CSV files of real value measures supplied by STMicroelectronics (STM) and ATMEL (ATM). These files have between 800 and 1500 columns, about 300 rows (containing null values), and one or more target columns. To carry out pre-processing and transformation of these files in the form of a base of transactions, we have, in a practical way (Pyle, 1999):

1. Launched a stage of column pre-processing by eliminating the ones poorly significant through adequate thretholds: doubles, columns having few different values or presenting too many null values, etc; then
2. Discretized the values of the residual columns: for each one, we standardized its values, then cut out the obtained values in intervals, and finally allocated a code interval to each value: 2 identical values of 2 different columns are thus differently coded out, and cannot interfere in the analysis. As with other items, absent values are treated as an item, but are not looked for correlations.

All experiments were conducted on an HPWorkstation (1.8 GHz processor with a RAM of 2Go), and the software was developed in C language.

The results of the experiments are presented on Figures 2 through 4. They are associated with an analysis of 2 files supplied by the two manufacturers. The first (STM) contains 1241 columns, 296 lines and a target attribute. The second (ATM) consists of 749 columns, 213 lines, and 3 target columns.

Figures 2 and 3 show the evolution of the execution times for both methods for the two files when $MinSup$ varies and $MinPerc$ and $MinCor$ are fixed. The STM (resp. ATM) file contains 3384 (resp. 1136) items after carrying out cleanings. As the graphs show, the answer times of the LHS-CHI2 method are between 30% and 70% better than LEVELWISE, even if they remain very large

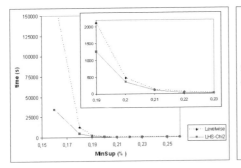 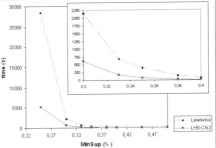

Fig. 2. Execution time with $MinPerc =$ 0.34, $MinCor = 1.6$ (STM file - target1)

Fig. 3. Execution time with $MinPerc =$ 0.24, $MinCor = 2.8$ (ATM file - target2)

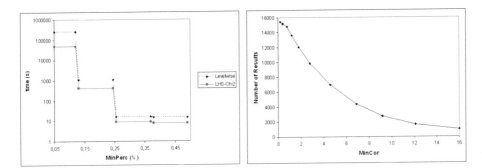

Fig. 4. Execution time with $MinSup =$ 0.24, $MinCor = 6.9$ (STM file - target1)

Fig. 5. Results with $MinSup = 0.38$, $MinPerc = 0.24$ (ATM file - target3)

when using very small thresholds. In each of the cases, an increasing windowing of the results is provided for consequent sub-intervals of $MinSup$.

Figure 4 shows the execution times for the STM file (using the same configuration as the experiment of Figure 2) when $MinSup$ and $MinCor$ are constant and when $MinPerc$ varies. The curve in staircases thus explains: a CT associated with a i-pattern containing 2^i cells, to say that $MinPerc$ of its cells must have the support means that $\lceil 2^i * MinPerc \rceil$ cells must have it. So for a 3-pattern, defining a value for $MinPerc$ varying between 0% and 12.49% means specifying that one single cell of the CT has to have the support, etc.... The proposed scale is logarithmic, because answer times for small values of $MinPerc$ are very large (more than 13 hours for LHS-CHI2, and about 69 hours for LEVELWISE with $MinPerc = 0.12$). No result is returned when $MinPerc$ is defined beyond 50%.

Figure 5 shows the number of extracted rules (identical for both methods) after the search when $MinPerc$ and $MinSup$ are fixed with suitable values and when $MinCor$ varies. In that particular case (an ATM file with a different target from the one used in the first test), execution times are identical whatever the value of $MinCor$, but are of the order of 2 minutes with LHS-CHI2, and about 17 minutes for LEVELWISE. This means that the $MinCor$ threshold only has small effect on performance. We recall that a value of 2.71 for MinCor corresponds to a correlation of 90%.

Finally, let us emphasize that the experimental sets used in the first three cases produce decision correlation rules with a cardinality of 4. This is the kind of information that is of interest for semiconductor manufacturers, as well as the possible crossings between rules of cardinality 3 and 4.

5   Conclusion and Future Work

In this paper, we have introduced two concepts:

- Decision correlation rules, i.e. restricted correlation rules containing a value of a target attribute, and
- Contingency vectors, i.e. an alternative representation of contingency tables, which are more concise and offer better properties related to performance. We have also proposed an algorithm based on the lectic order to go through the literalset lattice.

The LHS-CHI2 algorithm uses the inference property of the contingency vector of a pattern given the contingency vector of one of its direct subsets. The experiments show that the proposed method computes rules faster than those offered by levelwise algorithms. The implementation approach enabled us to discover new correlations between the parameters of the files that have been studied: approximately 25% of the correlation rules determined by the first experiment were unknown to STM, and the quasi-totality of the results obtained has been experimentally validated.

To continue our work, we intend: To enlarge done experimentations in order to produce more relevant results both considering space occupation and algorithmic complexity; To optimize the processing stages upstream of the algorithm (aggregation of attributes, merging of intervals) while safeguarding the context in order to obtain a greater number of rules and / or more significant results; and To widen the correlation rules extraction problem on items to that of computation of correlation rules for literalsets.

Acknowledgments

This work was supported by Research Project "Rousset 2003-2008", financed by the Communauté du Pays d'Aix, Conseil Général des Bouches du Rhône and Conseil Régional Provence Alpes Côte d'Azur.

References

Agrawal, R., Mannila, H., Srikant, R., Toivonen, H., Verkamo, A.I.: Fast Discovery of Association Rules. In: Advances in Knowledge Discovery and Data Mining, pp. 307–328 (1996)

Brin, S., Motwani, R., Silverstein, C.: Beyond market baskets: generalizing association rules to correlations. In: Proceedings of the International Conference on Management of Data, SIGMOD, pp. 265–276 (1997)

Casali, A., Ernst, C.: Extracting correlated sets using the chi-squared measurement within n-ary relations: An implementation. In: 8th European Advanced Equipement Control / Advanced Process Control, AEC/APC, 4 p. (2007)

Flouvat, F., De Marchi, F., Petit, J.M.: A thorough experimental study of datasets for frequent itemsets. In: Proceedings of the 5th International Conference on Data Mining, ICDM, pp. 162–169 (2005)

Ganter, B., Wille, R.: Formal Concept Analysis: Mathematical Foundations. Springer, Heidelberg (1999)

Grahne, G., Lakshmanan, L., Wang, X.: Efficient Mining of Constrained Correlated Sets. In: Proceedings of the 16th International Conference on Data Engineering, ICDE, pp. 512–524 (2000)

Hirsh, H.: Generalizing version spaces. Machine Learning 17(1), 5–46 (1994)

Laporte, M., Novelli, N., Cicchetti, R., Lakhal, L.: Computing full and iceberg datacubes using partitions. In: Hacid, M.-S., Raś, Z.W., Zighed, D.A., Kodratoff, Y. (eds.) ISMIS 2002. LNCS, vol. 2366, pp. 244–254. Springer, Heidelberg (2002)

Laurent, D., Spyratos, N.: Partition semantics for incomplete information in relational databases. In: Proceedings of the International Conference on Management of Data, SIGMOD, pp. 66–73 (1988)

Mannila, H., Toivonen, H.: Levelwise Search and Borders of Theories in Knowledge Discovery. Data Mining and Knowledge Discovery 1(3), 241–258 (1997)

Moore, D.S.: Tests of chi-squared type. In: D'Agostino, R.B., Stephens, M.A. (eds.) Goodness-of-Fit Techniques, pp. 63–95. Marcel Dekker, New York (1986)

Pyle, D.: Data Preparation for Data Mining. Morgan Kaufmann, San Francisco (1999)

Spiegel, M.R.: Théorie et applications de la statistique, Schaum (1990)

Vel, M.: Theory of Convex Structures. North-Holland, Amsterdam (1993)

Wu, X., Zhang, C., Zhang, S.: Efficient mining of both positive and negative association rules. ACM Trans. Inf. Syst. 22(3), 381–405 (2004)

Maintaining the Dominant Representatives on Data Streams[*]

Wenlin He[1,2], Cuiping Li[1,2], and Hong Chen[1,2]

[1] Key Labs of Data and Knowledge Engineering, Ministry of Education, China
[2] School of Information, Renmin University of China, Beijing, China
{hewl,licuiping,chong}@ruc.edu.cn

Abstract. It is well known that traditional skyline query is very likely to return over many but less informative data points in the result, especially when the querying dataset is high-dimensional or anti-correlated. In data stream applications where large amounts of data are continuously generated, this problem becomes much more serious since the full skyline result cannot be obtained efficiently and analyzed easily. To cope with this difficulty, in this paper, we propose a new concept called *Combinatorial Dominant* relationship to abstract dominant representatives of stream data. Based on this concept, we propose three novel skyline queries, namely *basic convex skyline query (BCSQ)*, *dynamic convex skyline query (DCSQ)*, and *reverse convex skyline query (RCSQ)*, combining the concepts of convex derived from geometry and the traditional skyline for the first time. These queries can adaptively abstract the contour of skyline points without specifying the size of result set in advance and promote information content of the query result. To efficiently process these queries and maintain their results, we design and analyze algorithms by exploiting a memory indexing structure called DCEL which is used to represent and store the arrangement of data in the sliding window. We convert the problems of points in the primal plane into those of lines in dual plane through dual transformation, which helps us avoid expensive full skyline computation and speeds up the candidate set selection. Finally, through extensive experiments with both real and synthetic datasets, we validate the representative capability of CSQs, as well as the performance of our proposed algorithms.

1 Introduction

Skyline is based on a relationship called dominate whose definition is as follows.

Definition 1. *Dominate* ($p \prec q$). *Assume $D = \{D_1,...,D_n\}$ is an N-dimensional set. For each of the dimension D_i, we define an order $\prec D_i$. We say that p is better than q in dimension D_i (denoted as $p_i \prec q_i$) if p_i comes before q_i based on $\prec D_i$ or conversely, q_i is worse than p_i (also denoted as $q_i \prec p_i$). If p_i and q_i are equal, we*

---

[*] This research is supported by National 863 Hi-Tech R & D Plan of China under Grant No. 2008AA01Z120, NSFC under Grant Nos. 60673138, 60603046, Program for New Century Excellent Talents in University and Union Project with BMEC on industry-study-research.

S.S. Bhowmick, J. Küng, and R. Wagner (Eds.): DEXA 2009, LNCS 5690, pp. 704–718, 2009.
© Springer-Verlag Berlin Heidelberg 2009

denote them as $p_i = q_i$. *A point p is said to dominate q if p is better or equal to q in all dimensions, and is better than q in at least one of the dimensions.*

Under this concept, skyline query is used to retrieve a set of multi-dimensional data points not dominated by any other in the querying dataset. For example, a CarDB storing multi-dimensional car instances is available. Eight tuples with two dimensions, mileage and price, are selected as Fig. 1(a) shows. Thus the traditional (full) skyline points of these tuples are { p_1, p_2, p_3, p_5, p_6 } (Fig. 1(b)). Most existing methods of computing skyline focus on the applications with relatively static environments. In highly dynamic applications like data stream, it is challenging to evaluate skyline queries due to: (i) the quantity of the stream data is considerably large; (ii) stream data are generated with continuous, unpredictable manner.

ID	Mileage	Price
p_1	27k	20k
p_2	54k	8k
p_3	68k	2k
p_4	90k	7.5k
p_5	16k	25k
p_6	32k	13k
p_7	68k	25k
p_8	80k	16k

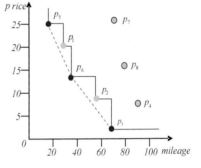

(a) Dataset Example   (b) Skyline of 2D Points

Fig. 1. Example of Skyline

A big problem of traditional skyline queries is that it is very likely to retrieve over many but less informative points. Referring back to the example above, we can easily find that 5/8 tuples of the dataset are skyline points. Theoretically, in the cases where dataset is anti-correlated or high dimensional, the amount of skyline points could be very large. The problem becomes much more serious in data stream applications, since the real-time characteristic and effectiveness will be affected to retrieve full skyline result. Thus a concise and more informative result is desired. Paper [12] investigated the problem of computing k skyline points such that the total number of (distinct) data points dominated by one of the k skyline points is maximized. The points in the result are called "top-k representative skyline points". Similar problems are studied in paper [7]. In this paper, we study the problem of reducing the amount of skyline points and abstracting the representative ones. The primary difference between our work and the existing ones is that they quantify the number of skyline points, say K, in advance, while we don't specify this value in query but adaptively select sufficient points to represent the dominant situation of querying dataset. For example, the set { p_3, p_5, p_6 } (the dark colored points set in Fig.1(b)) are our desired dominant representatives. They are better than full skyline points, since they are more succinct and stand to lose little dominant information compared to full skyline result. And we will adaptively get sufficient representatives rather then specify the amount, e.g. 3 in the example. From geometry view, these representative skyline points comprise the low part of convex hull. Based on this observation, this query is denoted as **Basic Convex Skyline Query (BCSQ)**. For another two significant skyline queries, namely **Dynamic Skyline Query** [10] and **Reverse Skyline Query** [5], we present their corresponding convex skyline queries as well. This paper focus on efficiently solving these novel skyline queries over the sliding window in data stream applications, and its contributions are summarized as follows:

1. We propose a novel dominate relationship called *combinatorial dominate*. Based on it, Convex Skyline Queries (CSQs) are formally stated to get the dominant representatives for the first time, making the general skyline result more concise and informative;

2. We present approaches to index points in the sliding window for data stream applications and give CSQs' geometric representations; We design algorithms and make use of pruning technique to efficiently solve CSQs by employing an arrangement representation called DCEL;

3. We present elaborative experiments to evaluate the performance and efficiency of our algorithms to answer CSQs.

The rest of this paper is organized as follows. In section 2, we give problem definitions of new dominate relationships and CSQs; section 3 introduces preliminaries about arrangement and DCEL employed in our algorithms. In section 4, we propose and analyze algorithms to efficiently answer BCSQ by using DCEL. In section 5, we first introduce an algorithm to retrieve globally convex skyline (GCS) to obtain a candidate set, then we develop algorithms to answer DCSQ and RCSQ based on GCS. Section 6 conducts a series of experiments to verify the efficiency of proposed algorithms. Section 7 surveys the related work and finally in section 8, we conclude this paper and give the research direction of our future work.

2   Background

2.1   Problem Definition

To obtain the dominant representatives of dataset, we introduce concepts called combinatorial dominate and dynamic combinatorial dominate for the first time. Based on these concepts, we state CSQs problems.

Definition 2. Combinatorial Dominate ($p_1, \ldots, p_d \prec_c q$). *Given a d-dimensional data set D, if there exists at least one combination $p^* = \alpha_1 p_1 + \cdots + \alpha_d p_d$, where $0 \le \alpha_1 \le 1, \ldots, 0 \le \alpha_d \le 1, \alpha_1 + \cdots + \alpha_d = 1$, dominating q. Then we say that $p_1, \ldots, p_d$ combinatorial dominate q, or q is combinatorial dominated by $p_1, \ldots, p_d$.*

Problem 1: ***Basic Convex Skyline Query (BCSQ).*** *Given a d-dimensional data set D, BCSQ retrieves all data points that are not combinatorial dominated by any combination of other points in D. Its result is denoted as* **BCS.**

In 2D space where each point contains two attributes, all the possible combinations of two distinct points constitute a line connecting them; in 3D space, the combinations of three distinct points constitute a triangular area; More general, for d-dimensional (d>3) dataset, the combinations of d points constitute a d-vertices region in hyperspace. Figure 2 (a) shows an example of *Combinatorial Dominate*. In traditional skyline query, $A1 \cup A2$ is cs_1's dominant area, namely any point locating in this area is dominated by cs_1; $A1 \cup A4$ is that of cs_2's. Having the concept of combinatorial dominate, any point besides traditional skylines locating in A3 is combinatorial dominated by cs_1, cs_2. Thus in Figure 1(b), p_1 and p_2 are skyline points not BCS, since they are combinatorial dominated by p_5, p_6 and p_3, p_6 respectively.

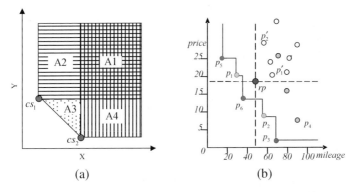

(a)   (b)

Fig. 2. (a) Combinatorial Dominate (b) Dynamic Convex Skyline

Definition 3. ***Dynamically Combinatorial Dominate*** *($p_1,..., p_d \prec_{DC} q$). Given a d-dimensional data set D and a reference point rp, assume p' is the mapping point of point p to rp's 1st quadrant. If $p'_1,..., p'_d$ combinatorial dominate q , then we say that $p_1,..., p_d$ dynamically combinatorial dominate q , or q is dynamically combinatorial dominated by $p_1,..., p_d$ .That means there exists at least one combination $p^* = \alpha_1 p_1 + \cdots + \alpha_d p_d$, where $0 \le \alpha_1 \le 1,..., 0 \le \alpha_d \le 1, \alpha_1 + \cdots + \alpha_d = 1$, which satisfies*

1). for all $i \in \{1,...,d\}$: $| p^{*i} - rp^i | \le | q^i - rp^i |$

2). at least one $j \in \{1,...,d\}$: $| p^{*j} - rp^j | < | q^j - rp^j |$.

Problem 2: ***Dynamic Convex Skyline Query (DCSQ).*** *Given a d-dimensional data set D and a reference point rp, DCSQ relative to rp retrieves all data points in D that are not dynamically combinatorial dominated by any combination of other points. Its result is denoted as* **DCS.**

Figure 2 (b) demonstrates rp's DCS comprised of p_1 and p_2, while its dynamic skyline points are $\{p_1, p_2, p_6\}$.

Problem 3: *Reverse Convex Skyline Query (RCSQ).* *Given a d-dimensional data set D and a reference point rp, RCSQ according to rp retrieves a set denoted as P, that for each point $p \in P$, rp is a dynamic convex skyline point of p. Its result is denoted as **RCS**.*

For example in figure 2(b), p_1 is one of rp's RCS since rp is p_1's DCS. Note RCSQ could return 0 point, which means the reference point is not any point's DCS of the given dataset.

2.2 Challenges for CSQs

A naïve brute-force approach to solve BCSQ could include two steps: 1) full skyline calculation; 2) BCS refining. The first step can be simply accomplished by any existing techniques; the second step checks whether skyline point is combinatorial dominated by any combination of other distinct two points (actually, they must be skyline points too). However, this approach is too expensive for data stream application. In term of space cost, it solely depends on the adopted technique. Moreover, we have to take into consideration the update of utilized index.

To solve problems 2 and 3, similar methods could be considered. For example, to answer DCSQ, according to its definition, all points will firstly be mapped to the reference point's 1st quadrant in a transformed space where the reference point becomes the new origin. After this transformation, DCSQ is converted to BCSQ, and we can issue a BCSQ to get the answer. The primary weakness of this approach is that since a transformed space with the same size as the original space is needed to construct for each ad hoc query, the space cost is unbearable huge.

3 Preliminaries

In this section, we briefly present duality, arrangement and DCEL, which will be employed in our algorithms to evaluate CSQs.

3.1 Duality and Arrangement

Duality and arrangement often combined to solve complex computational geometry problems and applied in many scientific areas such as Robotics and Computer Graphics etc [1, 2, 3]. In 2D space, point-line duality is a symmetric transformation for exchanging points and lines. Through the dual transformation, a point in the original plane (primal plane) is converted into a line in another plane called dual plane. Properties in the dual plane can be used to solve those in primal plane. Based on duality transformation, we present the definition of arrangement in dual plane.

Definition 4. *Arrangement(L).* *Let L is a set of n line in $\mathbb{R}^2$, L subdivides $\mathbb{R}^2$ into several regions, which are called faces or cells. The edges of subdivision are line segments or half edge; the vertices are the intersecting points of two lines of L. Under* *this subdivision, arrangement of L is the adjacent relation between vertices, edges and cells, denoted A(L).*

Fig. 3(a) illustrates a 2D arrangement including 3 lines, 6 points and 12 faces (1 outer face and 11 inner faces). The combinatorial complexity of $A(L)$ is the overall number of vertices, edges and faces in $A(L)$. Further, it can be proved that an arrangement of n Lines contains at most $O(n^2)$ vertices, $O(n^2)$ edges and $O(n^2)$ faces [3]. Thus the total complexity of an arrangement is $O(n^2)$.

One of the most important theorems in arrangement study to our study is zone theory [4]. It guarantees that the time complexity of inserting a line into (or deleting a line from) an arrangement with n existing lines is $O(n)$.

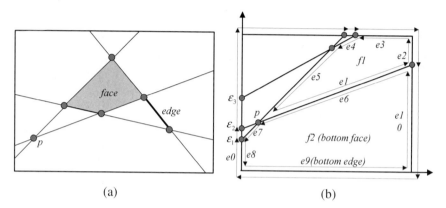

(a)   (b)

Fig. 3. (a) 2D Arrangement (b) DCEL

3.2 Double-Connected-Edge-List (DCEL)

DCEL is a data structure of representing and storing arrangement which was first introduced in [2]. Fig. 3 (b) illustrates a DCEL constructed by three dual lines $\varepsilon_1, \varepsilon_2, \varepsilon_3$ whose corresponding points in primal plane are $p_1(1,4), p_2(3,1), p_3(4,2)$ respectively. DCEL is comprised of three core sub structures as follows.

- *halfedge.* It's a directed segment of dual line, and maintains a pointer to its twin halfedge. A halfedge and its twin edge share the same end vertices but have opposite directions, e.g. $e1$ and $e6$. A halfedge maintains pointers to its end vertices and associative incident face.
- *vertex.* Its incident edges are circularly maintained.
- *face.* It maintains the halfedge list constructing itself (e.g. edge list of face f1 is $e_1 \rightarrow e_2 \rightarrow e_3 \rightarrow e_4 \rightarrow e_5 \rightarrow e_1$). The halfedges in the lists constitute a doubly connected circular, which make it feasible to traverse the face by both clockwise and counter-clockwise patterns.

Some other useful structures used to explain our algorithms are listed as follows:

- *bounding box.* It is the minimum rectangle including all possible intersecting vertices of dual lines. Notice we only consider the situation in 1st quadrant of the plane.

- *bottom edge.* It is the half edge on X axes, whose source vertex is the origin point. In the arrangement figure 3(b) shows, the bottom edge is e_9.
- *bottom face.* It is the face containing the bottom edge, e.g., *f2* in fig. 3 (b).
- *the first outer edge.* 1st half edge on Y axes in the outer face, i.e., e_0 in fig. 3(b).

4   Using DCEL to Answer BCSQ

In this section, we firstly introduce the relationship between skyline and top-k queries, and then explain how to index points in the sliding window by using DCEL; based on this structure, we describe and analyze our first algorithm to answer BCSQ.

4.1   Skyline vs. Top-k

A Top-k query retrieves the k highest scoring tuples from a dataset with respect to a scoring function defined on the attributes of a tuple. Some techniques for solving top-k problem can be utilized to answer skyline queries [6]. Here, we present a meaningful property about their relationship, which helps answer CSQs.

Property 1. For any linear scoring function $f = w_1 x_1 + \cdots + w_d x_d$, where d is the dimensionality of data set, $w_i, 1 \le i \le d$ is the weight of dimension i, all top-1 points must lie on the skyline. However, not all skyline points can be top-1 under certain weight $W = \{w_1 \cdots, w_d\}$.

Based on property 1, we have the following important corollary:

Corollary 1. A convex skyline point is a top-1 point with respect to certain $W = \{w_1 \cdots, w_d\}$. Any non top-1 skyline point cannot be top-1 point to any W and is combinatorial dominated by the combination of convex skyline points.

Inspired by this corollary and the approach to answer top-k queries in [6], we use arrangement to handle BCSQ. More specifically, we first construct and maintain DCEL indexing of data in the sliding window; at any instances when the window update, we retrieve all top-1 points to answer BCSQ.

4.2   Construct Arrangement to Index Points in the Sliding Window

The dual transformation is not unique, and could be determined according to the user's purpose. In our algorithms, we use $y = Yx + X$ to map a point in primal plane to a line in dual plane, where X and Y are the horizontal and vertical coordinates respectively. To adapt data stream environment, we employ the method to incrementally construct and maintain DCEL from [4]. Figure 3 (b) depicts a DCEL including three point's dual lines. Now we assume the forth point enters. It is first mapped to a dual line (denoted l_{new}). The insertion starts from e_0 and continues until one edge on X axis intersecting with l_{new}. Then the traverse turn to the intersecting edge's twin edge. The first intersecting vertex iv_1 is recorded for later face split processing. Then we enter into the face which the twin edge belongs to and continues traversal until l_{new} intersects with a new edge. Now the second interesting vertex iv_2 is obtained. We connect iv_1 and iv_2 to generate a new halfedge twin. Besides, two new faces f_{new1}, f_{new2} are created to replace the old one f_{old}. The original half edges of f_{old} and the new created half edges are reallocated to f_{new1}, f_{new2}. Other sub structures of DCEL are adjusted if necessary. The traversal and insertion terminate until the boundary of the bounding box is reached. When a point is removed, its dual line needs to be deleted in the arrangement. Since the deletion processing is similar to the insertion, we omit the description for simplicity. The time complexity of incrementally constructing DCEL including n lines is $O(n^2)$ [4].

4.3 Answer Basic Convex Skyline Query

Theorem 1. Given the arrangement with the dual transformation: $y = Yx + X$, the points whose dual line shares face with the bottom half edge belong to BCS.

Proof. Assume the point q satisfies the condition but it is not a convex skyline. According to the definition of BCS, there must be at least two points (denoted p_1, p_2) combinatorial dominating q. However, through the transformation: $y = Yx + X$, the dual lines of p_1 and p_2 will prevent that of q from sharing face with the bottom face edge. Thus it conflicts with the assumption.

Thus, we just need to abstract these half edges of the bottom face, and use their original points to answer BCS. Algorithm 1 represents the complete processing: lines 3-5 shows the processing of counter-clockwise traversing from the half edges of bottom face on the bounding box. Lines 6-9 add dual lines' corresponding points into the result. In figure 4(b), to answer BCSQ, e_9 and e_{10} are firstly visited and ignored; then e_6 and e_7 are visited, meanwhile their corresponding points are added to BCS. When e_8 , a half edge on X axis also on bounding box, is reached, the algorithm terminates and final BCS are obtained.

Algorithm 1. Convex skyline retrieval algorithm **BCSQ(A)**

```
Input:     the arrangement indexing the points in the sliding
window A
Output:  BCS of data points in the sliding window
01: BCS ={}
02: edge = e_bottom
03: while edge.onBoundingBox() == true do
04:      edge=edge.next()
05: end while
06: while edge.onBoundingBox() == false do
07:      BCS.add( edge.getPoint() )
08:      edge=edge.next()
09: end while
10: return BCS;
```

Since in the extreme case, all points belong to BCS, the time complexity of algorithm 1 is $O(n)$.

5 Handle DCSQ and RCSQ

In this section, we cope with DCSQ and RCSQ. *Globally Convex Skyline (GCS)* is introduced for obtaining a candidate set of DCSQ and RCSQ; then we develop algorithms to answer DCSQ and RCSQ based on GCS.

5.1 Global Convex Skyline Query Processing

To overcome difficulties mentioned in section 2, we introduce a new concept called global convex skyline inspired by a notation called *global skyline* introduced in [5]. The authors proved the global skyline is the super set of the reverse skyline. This property helps to prune unqualified points and thus decrease the computation cost. However, the approach in [5] cannot be directly applied to our problems, since their index structure is R-tree, different from ours. Interestingly, we find that the *GCS* can be used to prune unqualified points effectively and easily evaluated by using arrangement.

Definition 6. *Globally Combinatorial Dominate*($p_1, ..., p_d \prec_{GC} q$). *In a d-dimensional dataset, given a reference point rp. If $q, p_1, ..., p_d$ locate in the same quadrant according to rp, and q is combinatorial dominated by $p_1, ..., p_d$ then we say $p_1, ..., p_d$ globally combinatorial dominate q relative to rp, or q is globally combinatorial dominated by $p_1, ..., p_d$ relative to rp.*

Based on this definition, given a d-dimensional dataset and a reference point rp, **Global Convex Skyline Query (GCSQ)** retrieves a set of points which are not globally combinatorial dominated by any combination of other points relative to rp (denoted as GCS(rp)). Figure 4(a) illustrates an example: p_3 and p_5 globally convex dominate p_4 in rp 's 3rd quadrant; p_6 and p_9 globally combinatorial dominate p_7 and p_8 in rp 's 4th quadrant. The global convex skyline of rp is the union of partial results in four quadrants, i.e. { p_1 , p_3 , p_5 , p_6 , p_9 , p_{10} , p_{11} , p_{12} }.

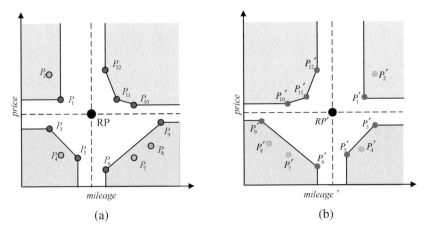

(a)   (b)

Fig. 4. (a) Globally Combinatorial Dominate (b) SST Example

To answer DCSQ and RCSQ, we present two important lemmas as follows.

Lemma 1. *Let rp be the reference point and GCS(rp) be rp's global convex skyline points, DCS(rp) be rp's dynamic convex skyline, and RCS(rp) be the set of reverse convex skyline, then* $DCS(rp) \subseteq GCS(rp)$ *and* $RCS(rp) \subseteq GCS(rp)$.

Lemma 2. *Given a reference point rp and dual transformation* $y = Yx + X$, *GCS(rp) of rp's* 1st *and* 3rd *quadrants are such points that their dual lines share face with rp's dual line.*

In Figure 4 (a), p_3, p_4 and p_5 all locate in 3rd quadrant of the reference point. p_4 doesn't belong to GCS since it is globally combinatorial dominated by p_3 and p_5. Figure 5(a) shows their dual line relationship. Both intercept and slope of $\varepsilon(p_4)$ are between those of $\varepsilon(p_3)$ and $\varepsilon(p_5)$, which prevent $\varepsilon(p_4)$ from sharing face with $\varepsilon(rp)$ in the arrangement. Besides, $\varepsilon(p_3)$ and $\varepsilon(p_5)$ share face with $\varepsilon(rp)$ as the shadowed area indicates, therefore GCS(rp) in 3rd quadrant.

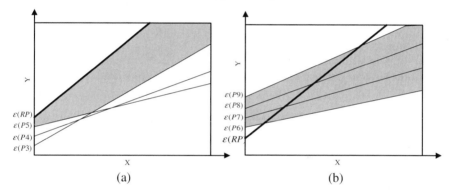

(a)   (b)

Fig. 5. (a) Dual lines of points in Q3 (b) Dual lines of points in Q4

However, this lemma is not applicable for 2nd and 4th quadrants. For example, though dual lines of p_7 and p_8, i.e. $\varepsilon(p_7)$ and $\varepsilon(p_8)$, share face with $\varepsilon(rp)$ (Fig.5 (b)), but p_7 and p_8 are globally combinatorial dominated by p_6 and p_9 (Fig.4 (a)).To make use of lemma 2, we introduce Symmetric and Shifting Transformation as follow.

Definition 7. *Symmetric and Shifting Transformation (SST).* *Given a point* $p(x, y)$, *SST generates a mapped point* $p'(MAX_i - x, y)$ *for* p, *where* MAX_i *is the maximum value of dimension i.*

Figure 4(b) indicates the result after SST is applied to original points in figure 4 (a). After SST, we convert the problem to get GCS(rp) in 2nd and 4th quadrants in the original space to getting GCS(rp) in 1nd and 3rd quadrants in SST space where lemma 2 is in effect. Based on lemma 1 and SST, we develop an algorithm called DA_GCSQ using double arrangements to evaluate GCSQ. These arrangements index the original and SSTed points respectively. Through face-sharing checking in two arrangements, we get partial GCS(rp) in different quadrants and merge them to form the result of GCS. DA_GCSQ takes several more steps than simple inserting a dual line, but they have the same order, i.e., $O(n)$. For space limitation, the algorithm is omitted.

5.2 Answer DCSQ

After the reference point's GCS is obtained, points belonging to GCS are uniformly mapped into the 1st quadrant of the reference point; then face-sharing checking is invoked to determine which points belong to DCS. These processes are straightforward and simple. However, a big problem arises: where to place these new dual lines of mapped points? It's not a reasonable approach to place them in the original arrangement for two reasons: 1) data in the sliding window are changing in a rapid rate, which makes the update of original arrangements is frequent; 2) CSQs are ad hoc. Hence this approach would complicate arrangements maintenance and affect the independence of ad-hoc queries.

To overcome this difficulty, our method is to use *partial sub space transformation*: we create an empty sub arrangement for each reference point, and insert points in GCS into it rather on the original full arrangements. This is much clear since the operations of arrangement update caused by data and queries are distinguished. Moreover, having the lemma 1, the search space of DCS and RCS are greatly reduced through GCS evaluation. The space and computation costs of partial space transformation are therefore relatively low. Algorithm 2 gives the complete processes.

Algorithm 2. Dynamic convex skyline retrieval algorithm **SA_DCSQ(A_1 , A_2 , rp)**

```
Input :    arrangements indexing the original and SSTed points
Output:    DCS of rp in the sliding window
01: SA=createDCEL()   //create an empty arrangement for rp
02: GCS=DA_GCSQ( A₁,A₂,rp)      // get the rp's GCS first
03: for each point p in GCS do
04:     mp=map(p, rp)            // map point into rp's 1ˢᵗ quadrant
05:        SA.insert(mp)
06: end for
07: DCS=BCSQ(SA)                // retrieve the BCS of SA
08: return DCS
```

Since the time complexity of DA_GCSQ is $O(n)$ and constructing an arrangement takes $O(n^2)$ running time, the overall time complexity of SA_DCSQ is $O(n+|GCS(rp)|^2)$. Moreover, since the maximum of $|GCS(rp)|$ is n, the time complexity of SA_DCSQ is $O(n^2)$.

5.3 Answer Reverse Convex Skyline Query

Similar to SA_DCSQ, we design an algorithm called SA_RCSQ to retrieve RCS of a reference point rp in algorithm 3. The algorithm is based on lemma 2. When a RCSQ is issued, GCS of the reference point is firstly retrieved (line 2). Then for each point p in GCS(rp), its own GCS denoted GCS(p) is similarly calculated(line 4). With the guarantee of lemma 2, we can make a preliminary pruning (line 5). The point is pruned if rp doesn't belong to its GCS. Otherwise, the point's DCS is calculated to inspect whether rp is in this result set (lines 5-10). If yes, the checking point p is added to RCS(rp)(lines 7-9).

Algorithm 3. Reverse convex skyline retrieval algorithm **SA_RCSQ(A_1, A_2, rp)**

```
Input  :  arrangements indexing the original and SSTed points
Output:   RCS dynamic convex skyline points relative to rp in the
          sliding window
01: RCS={}
02: GCS(rp)= DA_GCSQ(A₁, A₂,rp)              // get GCS(rp) first
03: for each point p∈ GCS do
04:     GCS(p)= DA_GCSQ(A₁,A₂,p)
05:         if rp∈ GCS(p) then      // prune p if rp is not its GCS
06:             DCS(p)=SA_DCSQ(A₁, A₂,p)
07:                 if  rp belongs to DCS(p) then
08:                     RCS.add(p)
09:                 end if
10:         end if
11: end for
12: return RCS
```

The running time of SA_RCSQ mainly depends on the following three factors: 1) the size of GCS(rp); 2) the time complexity of DA_GCSQ; 3) the time complexity of SA_DCSQ. The maximum value of GCS(rp) is equal to the window size, in the case when all points in the sliding window belong to GCS(rp). As discussed above, the time complexities of DA_GCSQ and SA_DCSQ are $O(n)$ and $O(n^2)$ respectively, and $\max(|GCS(rp)|) = n$, the overall time complexity of SA_RCSQ is thereafter $O(n^3)$. However, since in common cases $|GCS(rp)|$ is small, the overall time cost is relatively low.

6   Performance Evaluation

In this section, we introduce experiments to evaluate our proposed algorithms. They were implemented by using MS Visual C++ 6.0, and conducted on a PC with Intel Pentium 4 2.4GHz CPU, 1G main memory and 160G hard disk, running MS Windows XP Professional Edition. We conducted experiments on both synthetic and real life datasets. However, due to space limitation, we only report results on synthetic datasets here. Results from real life datasets mirror the result of the synthetic datasets closely. Three synthetic datasets with different distribution, *Correlated(C)*, *Anti-correlated(A)* and *Equally(E)* respectively, were produced by the data generator in [15] and employed. Each dataset contains 100000 points (tuples). We used a cyclic buffer to represent the sliding window (SW), and set its size with 100, 200, 300 and 400 in every experiment. For comparison, we used BBS algorithm to get the full skyline result first and then calculate three convex skylines by adding a refinement phase, we denoted this method as BBCS and ours as ARR.

6.1 Effectiveness

The first experiment was conducted to evaluate the capability of CSQs to abstract dominant representatives from skyline points. We used a metric called **Dominant Representatives Ratio (DRR)** to quantify this estimation. For a dataset D,

$$DRR = \frac{|CSQ(D)|}{|FSQ(D)|} \times 100\% \qquad (1)$$

where |CSQ(D)| and |FSQ(D)| represent the counts of skyline points of convex skyline and full skyline respectively. As SW was totally updated, we issued CSQs and recorded the count of points in the result. Moreover, we used 100 reference points to query DCS and RCS. And the average value of partial DRRs was used to represent the final DDR. Figure 6 shows the effectiveness result of different CSQs. In Fig. 6(a), DRR of anti-correlated data is the lowest (26.2%), while that of correlated is highest (97.0%). The reason is correlated dataset contains few points (85% skyline queries return less than 3 points, 42% queries return only 1 point) in its skyline result, which gives few opportunities for BCSQ to prune such skyline that is combinatorial dominated by others. The result of anti-correlated dataset is promising, we can clear get that BCSQ greatly reduces the amount of skyline results as expected. Fig. 6 (b) and (c) demonstrate that the DRRs of DCSQ and RCSQ respectively.

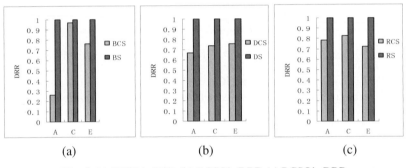

(a)   (b)   (c)

Fig. 6. (a) BCSQ's DRR (b) DCSQ's DRR (c) RCSQ's DRR

6.2 Efficiency

In this experiment, we evaluated ARR to answer CSQs and compared them with BBCS. To answer DCSQ, BBCS first maps points in the sliding window into reference point rp's 1st quadrant, and then calculates the basic skyline result in the transformed space; to answer RCSQ, each point in the sliding window firstly retrieves its DCS. If its DCS contains the reference point rp, it will be added into rp's RCSQ result. Notice the time of constructing indexes (arrangement and R-tree) is not taken into consideration for both methods. Since BBCS will spend huge time to retrieve RCSQ as the window size increases, we chose a relative small SW size 100 in this experiment. The results are presented in figure 7.

Figure 7 shows that our proposed algorithms outperform BBCS. Having lemma 1, the candidate sets of result can be easily obtained by sharing face checking over both the original and the SSTed arrangements in ARR. By contrary, BBCS has to retrieve the full skyline result first and then checks whether a skyline line is combinatorial dominated by other skylines in refinement phase, which is time-consuming.

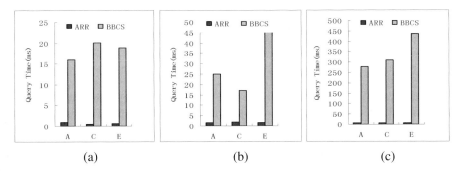

Fig. 7. (a) BCSQ Time Cost (b) DCSQ Time Cost (c) RCSQ Time Cost

7   Related Work

The skyline computation originates from the maximal vector problem in computational geometry, proposed by Kung et al. [24]. Borzsonyi et al. first introduce the skyline operator over large databases [15] and also propose a divide-and-conquer method. The method based on [24, 2] partitions the database into memory-fit partitions. The partial skyline objects in each partition is computed using a main-memory-based algorithm [25, 26], and the final skyline is obtained by merging the partial results. In [11], the authors proposed two progressive skyline computing methods. The current most efficient method is BBS (branch and bound skyline), proposed by Papadias et al., which is a progressive algorithm to find skyline with optimal times of node accesses [10, 23]. Balke et al. [18] in their paper show how to efficiently perform distributed skyline queries for querying Web information systems.

Arrangement, a concept from computation geometry, has many applications [3,4]. For instance, [6] uses it to answer ad top-k query. To our knowledge, [5] is the most relevant work to ours. It focuses on efficiently answer reverse skyline query. In that paper, the authors firstly expand the well-know BBS algorithm [11, 23] as BBRS to retrieve reverse skyline with respect to arbitrary reference points. To reduce the computational cost for determining whether a point belongs to the reverse skyline, an enhanced algorithm called RSSA are proposed.

8   Conclusion

In this paper, we investigate the problem of maintaining dominant representatives in data stream applications. Without specifying the number of representatives, we adaptively abstract sufficient points to represent the dominant situation of data in the sliding window. To prune less informative points in full skyline result, we introduce a novel dominate relationship called Combinatorial Dominate; based on it, we propose three novel convex skyline queries (CSQs). We design algorithms to efficiently answer CSQs by employing a geometry structure called arrangement for indexing data. By using this structure, we prune unqualified points to generate the candidate set for CSQs, which greatly improve the efficiency of our algorithms. We prove our algorithms can effectively get the dominant representatives of data stream by both theory and extensive experiments. The extension to high dimensional space of proposed methods and the optimization for sharing results is our future work.

References

1. Agarwal, P., Erickson, J.: Geometric Range Searching and Its Relatives. In: Advances in Discrete and Computational Geometry, Contemporary Mathematics, vol. 223, pp. 1–56 (1999)
2. Preparata, F.P., Shamos, M.I.: Computational geometry: An introduction. Springer, Heidelberg (1985)
3. Agarwal, P.K., Sharir, M.: Arrangements and Their Applications. In: Handbook of Computational Geometry, ch. 2, pp. 49–119. Elsevier, Amsterdam (2000)
4. De Berg, M., Cheong, O., Van Kreveld, M., Overmars, M.: Computational Geometry: Algorithms and Applications (March 2008)
5. Dellis, E., Seeger, B.: Efficient Computation of Reverse Skyline Queries. In: VLDB (2007)
6. Das, G., Gunopulos, D.: Ad hoc Top-k Query Answering for Data Streams. In: VLDB (2007)
7. Chan, C.Y., Jagadish, H.V., Tan, K.L., Tung, A.K.H., Zhang, Z.: Finding k-dominant skylines in high dimensional space. In: SIGMOD (2006)
8. Chan, C.Y., Jagadish, H.V., Tan, K.L., Tung, A.K.H., Zhang, Z.: On high dimensional skylines. In: Ioannidis, Y., Scholl, M.H., Schmidt, J.W., Matthes, F., Hatzopoulos, M., Böhm, K., Kemper, A., Grust, T., Böhm, C. (eds.) EDBT 2006. LNCS, vol. 3896, pp. 478–495. Springer, Heidelberg (2006)
9. Ramsak, F., Kossmann, D., Rost, S.: Shooting stars in the sky: An online algorithm for skyline queries. In: VLDB (2002)
10. Fu, G., Papadias, D., Tao, Y., Seeger, B.: An optimal and progressive algorithm for skyline queries. In: SIGMOD (2003)
11. Tan, K., et al.: Efficient progressive skyline computation. In: VLDB (2001)
12. Lin, X., Yuan, Y., Zhang, Q., Zhang, Y.: Selecting stars: The k most representative skyline operator. In: ICDE (2007)
13. Morse, M., Patel, J., Jagadish, H.V.: Efficient skyline computation over low-cardinality domains. In: VLDB (2007)
14. Pei, J., Jin, W., Ester, M., Tao, Y.: Catching the best views of skyline: a semantic approach based on decisive subspaces. In: VLDB (2005)
15. Kossmann, D., Borzsonyi, S., Stocker, K.: The skyline operator. In: ICDE (2001)
16. Sharifzadeh, M., Shahabi, C.: The spatial skyline queries. In: VLDB Conference (2006)
17. Anthony, K.H., Tung, L., Wang, X.S., Ooi, B.C.: Efficient skyline query processing on peer-to-peer networks. In: ICDE (2007)
18. Zheng, J.X., Balke, W.-T., Guntzer, U.: Efficient distributed skylining for web information systems. In: EBDT (2004)
19. Wang, W., Lin, X., Yuan, Y., Lu, H.: Stabbing the sky:efficient skyline computation over sliding windows. In: ICDE (2005)
20. Yuan, Y., Lin, X., Liu, Q., Wang, W., Yu, J.X., Zhang, Q.: Efficient computation of the skyline cube. In: VLDB (2005)
21. Kossmann, D., Ramsak, F., Rost, S.: Shooting stars in the sky: An online algorithm for skyline queries. In: Proc. of the Int'l Conf. in VLDB (2002)
22. Li, H.J., Tan, Q.Z., Lee, W.C.: Efficient progressive processing of skyline queries in peer-to-peer systems. In: Proc. of the 1st Int'l Conf., In INFOSCALE (2006)
23. Papadias, D., Tao, Y., Fu, G., Seeger, B.: Progressive skyline computation in database systems. ACM Trans. Database Syst. 30(1), 41–82 (2005)
24. Kung, H.T., Luccio, F., Preparata, F.P.: On finding the maxima of a set of vectors. JACM 22(4) (1975)
25. Stojmenovic, I., Miyakawa, M.: An optimal paralle lalgorithm for solving the maximal elements problem in the plane. In: Parallel Computing (1988)
26. Matousek, J.: Computing dominances in e^n. Inf. Process. Lett. (1991)

Modeling Complex Relationships

Mengchi Liu and Jie Hu

School of Computer, Wuhan University,
Hubei, China, 430072

Abstract. Real world objects have various natural and complex relationships with each other and via these relationships, objects play various roles that form their context and then have the corresponding context-dependent properties. Existing data models such as object-oriented models and role models cannot naturally and directly represent such complex relationships and context-dependent properties. In this paper, we present a method to provide such natural and direct support.

Keywords: Information modeling, complex relationships, context-dependent properties.

1 Introduction

Object-oriented models [1,2,3,4,5,6] have been proposed to model the real world objects and their relationships. They are mainly concerned about the static aspects of objects and normally require an object to be an instance of a most specific class. Thus they are not well suitable to modeling dynamic situations. To overcome these limitations, various role models [7,8,9,10,11,12,13,14] have been proposed to captures evolutionary aspects of real-world objects. The main problem with role models is that they just focus on roles of objects independently rather than the roles that objects play in the context of complex relationships.

In our view, real world objects have various natural and complex relationships with each other and via these relationships, objects play various roles that form their context, and then have the corresponding context-dependent properties. Existing data models oversimplify and ignore the complex relationships and context-dependent properties.

In this paper, we attempt to solve this problem by proposing a new method that can naturally and directly support complex relationships and context-dependent representation and access to object properties. It allows us to group not only static but also dynamic and context-dependent properties regarding objects into instances.

This paper is organized as follows. Section 2 discusses the related work. Section 3 proposes our method. Section 4 shows hierarchies and inheritance in our method. In Section 5, we conclude and comment on our future plans.

2 Related Work

Now let us consider company information modeling. A company involves several kinds of people such as customers, managers, department managers, and project

S.S. Bhowmick, J. Küng, and R. Wagner (Eds.): DEXA 2009, LNCS 5690, pp. 719–726, 2009.

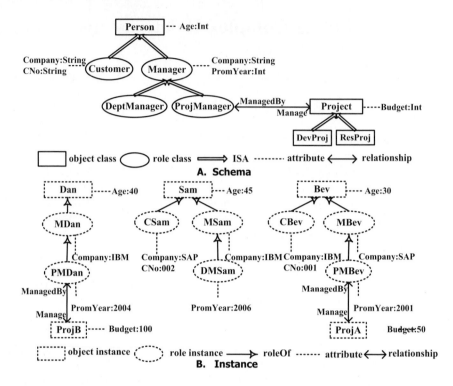

Fig. 1. Modeling Company Application in Role Model

managers. A manager has promotion year and is specialized into project manager and department manager; a project manager manages projects. Inversely, a project may be managed by a project manager. Also, a project may have a budget and is specialized into development project and research project. A customer has a customer number.

In object models, there are two kinds of classes: static class and dynamic class. They behave differently with respect to object migration. Instances of static subclass will never migrate but instances of dynamic subclasses can migrate. For example, if DevProj is a static subclass of Project, then a project that is not a development project will never migrate to DevProj subclass. If Customer is a dynamic subclass of Person, then a person that is not a customer may migrate to Customer subclass. In both cases, an instance of the subclass is also an instance of the superclass. That is, the instances of DevProj and Customer are also instances of their superclasses Project and Person respectively.

In role models, dynamic subclasses are modeled as role subclasses. If Customer is modeled as a role subclass of Person, then every customer differs from every person, but a Person instance can acquire one or more Customer instance as roles. Fig. 1 shows the schema and instance in role models for this application.

In the schema in Fig. 1-A, Person, Project, DevProj, and ResProj denoted graphically with rectangles are object classes. Customer, Manager, DeptManager, and ProjManager denoted graphically with ellipses are role classes. Object class Person has attribute Age of type Int and is specialized into two role subclass hierarchies Customer and Manager → {DeptManager, ProjManager}. Object class Project is specialized into subclasses DevProj and ResProj. Role class Customer has attributes Company and CNo of type String, Manager has attributes Company and PromYear of type String, ProjManager has relationship Manage with Project. Inversely, Project has relationship ManagedBy with ProjManager. Also, Project has attribute Budget of type Int.

The instance shown in Fig. 1-B contains five object instances identified by Dan, Sam, Bev, ProjA, and ProjB and eight role instances identified by MDan, PMDan, CSam, MSam, DMSam, CBev, MDev, and PMBev. Among them, Dan, Sam, Bev are instances of class Person, ProjA instance of class ResProj, ProjB instance of class DevProj, CSam and CBev instances of class Customer, MDan, MSam, and MDev instances of class Manager, DMSam instance of class DeptManager, PMDan and PMBev instances of class ProjManager.

Dan has value 40 for attribute Age and acquires two instances MDan and PMDan as roles. Role MDan has value IBM for attribute Company and PMDan has value 2004 for attribute PromYear and relationship Manage with ProjB. Inversely, ProjB has relationship ManagedBy with PMDan. Also, ProjB has value 100 for attribute Budget. Sam has value 45 for attribute Age and acquires three instances CSam, MSam, and DMSam as roles. Role CSam has values SAP for attribute Company and 002 for attribute CNo, MSam has value IBM for attribute Company, and DMSam has value 2006 for attribute PromYear. Similarly, Bev has value 30 for attribute Age and acquires three role instances CBev, MDev, and PMBev as roles. Role CBev has values IBM for attribute Company and 001 for attribute CNo, MDev has value SAP for attribute Company, and PMBev has value 2001 for attribute PromYear and relationship Manage with ProjA. Inversely, ProjA has relationship ManagedBy with PMBev. Also, ProjA has value 50 for attribute Budget.

Note that role models just treat Customer, Manager, DeptManager, and Proj-Manager as independent role subclasses of Person, the context such as Company just as attribute. They thus just support simple context representation. Also, the information about a person is scattered in a hierarchy of objects such as one Person instance Sam and three role instances CSam, MSam and DMSam, rather than a single object.

3 Our Method

In our method, we treat Company and Person as object classes, Customer, Manager, DeptManager, and ProjManager as role relationships from Company to Person. Project, DevProj, and ResProj are same as in Fig. 1, see Fig. 2-A.

The main novel feature of our method is the introduction of role relationships and novel mechanisms to represent complex relationships between objects and the context-dependent properties, and reflect the temporal, dynamic and many-faceted aspects of real-world objects in a natural and direct way.

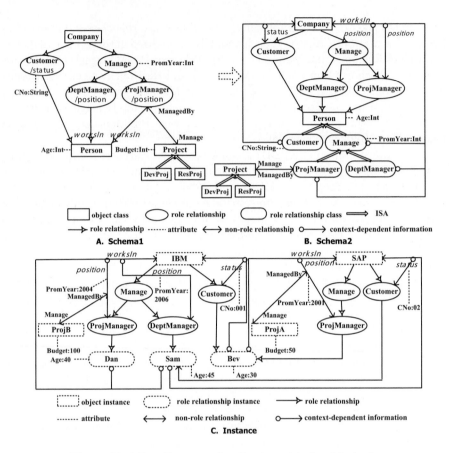

Fig. 2. Modeling Company Application with Our Method

A *role relationship* r represents the relationship from object class c_s to object class c_t, where c_s and c_t are called *source class* and *target class* of r respectively. It can have role sub-relationships and thus can form hierarchies in which every role relationship can have attributes and non-role relationships. For example, Manager→{DeptManager, ProjManager} forms a role relationship hierarchy from Company to Person. Also, every role relationship in the hierarchy has the same attributes and non-role relationships as in Fig. 1. The main difference is that we treat Company as the source class rather than attribute.

Like an object class that denotes a set of instances with common properties, a role relationship r also induces a set of instances of its target class that participate in the role relationship. We thus overload r to represent the class for this set of instances called role relationship class that is the subclass of target class c_t and automatically generate its properties based on the properties on the role relationship.

For example, Customer denoted graphically with ellipse in Fig. 2-A is a role relationship from Company to Person. Customer thus induces a set of instances of Person such as Sam and Bev participating in the role relationship Customer. So we overload Customer denoted graphically with round rectangle in Fig. 2-B to represent a role relationship class that is the subclass of Person and automatically generate property such as CNo. Thus, Sam and Bev are direct instances of role relationship class Customer. Similarly, Manager→{DeptManager, ProjManager} denoted graphically with ellipses in Fig. 2-A is a role relationship hierarchy from Company to Person. They thus induce a set of instances of Person such as Dan, Sam, and Bev participating in the role relationship hierarchy. We overload Manager→{DeptManager, ProjManager} denoted graphically with round rectangles in Fig. 2-B to represent the corresponding role relationship class hierarchy in which properties such as PromYear and Manage are generated automatically based on the properties specified on corresponding role relationship. Thus, Dan and Bev are direct instances of role relationship class ProjManager, Sam is a direct instance of role relationship class DeptManager, and they are all indirect instances of role relationship class Manager.

A role relationship r has two functions. On one hand, it is a directed relationship to connect objects in c_s to objects in c_t and may have inverse relationship from c_t to c_s as in ODMG [6]. On the other hand, it is a role that objects in c_t plays in objects in c_s and may have identification. Through specifying inverse relationship and identification on r, the context of the corresponding role relationship class can be directly and naturally represented.

For example, the inverse relationship worksIn and identification position specified on role relationship DeptManager in Fig. 2-A can generate the context of the corresponding role relationship class DeptManager as follows:

    class DeptManager [worksIn:Company[position:DeptManager]]

However, role relationship Customer just specifies identification status and inverse relationship is omitted. The context of the corresponding role relationship class Customer can be represented as follows:

    class Customer [status: Company.Customer]

Note that both inverse relationship and identification are optional and when they are both omitted, the corresponding role relationship class does not have context. Suppose role relationship Customer does not have any inverse relationship and identification, the corresponding role relationship class Customer does not have context. In this case, it is reduced to object-oriented models.

Moreover, attributes and non-role relationships of a role relationship class are allowed to be nested into the context to generate context-dependent properties and can be inherited, see Section 4. For example, considering the properties Manage and CNo on role relationship classes ProjManager and Customer respectively, context-dependent properties can be represented as follows:

    class ProjManager [worksIn:Company[position:ProjManager[Manage:Project]]]
    class Customer [status:Company.Customer[@CNo:String]]

Suppose Customer does not have context, context-dependent properties of role relationship class Customer can be represented as:

class Customer [@CNo:String]

Based on the schema, the instance shown in Fig. 2-C contains four object instances identified by IBM, SAP, ProjB, and ProjA and three role relationship instances identified by Dan, Sam, and Bev. Among them, IBM and SAP are instances of class Company, ProjA instance of class ResProj, ProjB instance of class DevProj, Bev and Sam instances of class Customer, Sam instance of class DeptManager, and Dan and Bev instances of class ProjManager. Also, three role relationship instances Dan, Sam, and Bev can be represented as follows:

```
ProjManager Dan [
  @age:40,
  worksIn:IBM[position:ProjManager[@PromYear:2004, Manage:ProjB]]]
DeptManager,Customer Sam [
  @age:45,
  worksIn:IBM[position:DeptManager[@PromYear:2006]],
  status:SAP.Customer[@CNo:02]]
ProjManager,Customer Bev [
  @age:30,
  worksIn:IBM[position:ProjManager[@PromYear:2001, Manage:ProjA]],
  status:IBM.Customer[@CNoR:01]]
```

Note that on one hand, when a role relationship class does not have any context, properties of corresponding role relationship instances are directly under the instances. On the other hand, even if a role relationship class has context and its properties are nested into the context. At instance level, the context of its role relationship instance can be omitted if we do not know its context. In this case, it is also reduced to object-oriented models. For example, if role relationship class Customer does not have context or we don't know the context of its instance Sam, then Sam can just be represented as:

```
DeptManager,Customer Sam [
  @age:45,
  worksIn:IBM[position:DeptManager[@PromYear:2006]],
  @CNo:02]
```

4   Hierarchies and Inheritance

In our method, both object classes and role relationship classes can form disjoint class hierarchies. The role relationship class hierarchies are induced by role relationship hierarchies.

We first discuss object class inheritance. Object classes correspond to static classes and can have class hierarchies and inherit attributes and relationships from their superclasses as in object models and role models. For example, object class Project in Fig 2 is specialized into subclasses DevProj and ResProj. Therefore, DevProj and ResProj inherit Budget, ManagedBy from their superclass Project.

Now we consider the role relationship class inheritance including the inheritance between the target class and its role relationship subclasses and the inheritance between role relationship classes.

As mentioned in Section 3, a role relationship class denotes a subset of instances of the target class participating in the corresponding role relationship in the context of source class. It is thus induced by the corresponding role relationship and is a subclass of the target class. Therefore, the role relationship class inherits the properties from its target class. For example, in Fig. 2, Customer is a role relationship class which is induced by role relationship Customer and the target class of role relationship Customer is Person. Therefore, Customer is the subclass of Person and inherits the attribute Age from Person.

As mentioned in Section 1, our method can naturally and directly support context-dependent representation and access to object properties. The context of a role relationship class is generated through the inverse relationship and identification specified on the corresponding role relationship. Role relationships can be specialized into role relationship hierarchies and can have a set of attributes and non-role relationships. Therefore, the role relationship subclass in a role relationship class hierarchy that is induced by a corresponding role relationship hierarchy inherits or overrides the context, attributes, and non-role relationships from its role relationship superclass. Moreover, the mechanism allows attributes and non-role relationships to be nested into the context to represent the context-dependent properties of a role relationship class.

For example, Manager→{DeptManager, ProjManager} denoted graphically with round rectangle in Fig. 2-B is a role relationship class hierarchy in which DeptManager and ProjManager are subclasses of Manager. DeptManager and ProjManager overrides the context but inherit the attribute PromYear from Manager. The context of Manager, DeptManager, ProjManager are worksIn:Company, worksIn:Company[position:DeptManager], worksIn:Company[position:ProjManager] respectively. For the above three role relationship classes, the context-dependent properties which is generated by nesting attributes and non-role relationships into the context can be represented as:

```
class Manager [worksIn:Company[@PromYear:Int]]
class DeptManager [worksIn:Company[position:DeptManager[@PromYear:Int]]]
class ProjManager [worksIn:Company[position:ProjManager[@PromYear:Int,
                Manage:Project]]]
```

5   Conclusion

In this paper, we have discussed limitations of existing data models such as object-oriented data models and role models in terms of complex relationship and context-dependent information representation and proposed our solutions to overcome these limitations.

Our method is suitable to deal with data having complex relationship and context-dependent information. With this method, the data modeling process can be greatly simplified. Every object in the real world is uniquely identified with its object identifier and is associated with exactly one instance that contains complete information about this object via all kinds of relationships. Context-dependent access to object properties is straightforward and the evolutionary, dynamic and many-faceted nature of real-world objects can be naturally reflected with this mechanism.

Due to space limitation, our presentation is concise and sketchy. We are working on a novel data model based on the method proposed here and would like to systematically implement a database management system based on this model and apply it in various areas.

References

1. Albano, A., Ghelli, G., Orsini, R.: A relationship mechanism for a strongly typed object-oriented database programming language. In: Proceedings of VLDB, Barcelona, Catalonia, Spain, September 1991, pp. 565–575 (1991)
2. Abiteboul, S., Bonner, A.: Objects and views. In: Proceedings of ACM SIGMOD, Denver, Colorado, May 1991, pp. 238–247 (1991)
3. Su, J.: Dynamic constraints and object migration. In: Proceedings of VLDB, Barcelona, Catalonia, Spain, September 1991, pp. 233–242 (1991)
4. Bancilhon, F., Delobel, C., Kanellakis, P.C. (eds.): Building an Object-Oriented Database System, The Story of O2. Morgan Kaufmann, San Francisco (1992)
5. Bertino, E., Guerrini, G.: Objects with multiple most specific classes. In: Olthoff, W. (ed.) ECOOP 1995. LNCS, vol. 952, pp. 102–126. Springer, Heidelberg (1995)
6. Cattell, R., Barry, D., Berler, M., Eastman, J., dan, D.J., Russel, C., Schadow, O., Stanienda, T., Velez, F.: The Object Data Standard: ODMG 3.0. Morgan Kaufmann Publishers, San Francisco (2000)
7. Richardson, J., Schwartz, I.: Aspects: Extending objects to support multiple, independent roles. In: Proceedings of ACM SIGMOD, Denver, Colorado, May 1991, pp. 298–307 (1991)
8. Wieringa, R.J., Jonge, W.D., Spruit, P.: Using dynamic classes and role classes to model object migration. Theory and Practice of Object Systems 1(1), 61–83 (1995)
9. Gottlob, G., Schrefl, M., Röck, B.: Extending object-oriented systems with roles. ACM Transaction on Office Information Systems 14(3), 268–296 (1996)
10. Wong, R.K., Chau, H.L., Lochovsky, F.H.: A data model and semantics of objects with dynamic roles. In: Proceedings of ICDE, Birmingham U.K, April 1997, pp. 402–411 (1997)
11. Chu, W.W., Zhang, G.: Associations and roles in object-oriented modeling. In: Proceedings of ER, Los Angeles, California, November 1997, pp. 257–270 (1997)
12. Steimann, F.: On the representation of roles in object-oriented and conceptual modelling. Data Knowledge Engineering 35(1), 83–106 (2000)
13. Dahchour, M., Pirotte, A., Zimányi, E.: A generic role model for dynamic objects. In: Pidduck, A.B., Mylopoulos, J., Woo, C.C., Ozsu, M.T. (eds.) CAiSE 2002. LNCS, vol. 2348, pp. 643–658. Springer, Heidelberg (2002)
14. Cabot, J., Raventós, R.: Roles as entity types: A conceptual modelling pattern. In: Proceedings of ER, Shanghai, China, November 2004, pp. 69–82 (2004)

Intuitive Visualization-Oriented Metamodeling

Dirk Draheim[2], Melanie Himsl[1], Daniel Jabornig[1], Werner Leithner[1],
Peter Regner[1], and Thomas Wiesinger[1]

[1] FAW-Institute, Johannes Kepler University, Linz, Austria
{mhimsl,djabornig,wleithner,pregner,twiesinger}@faw.at
[2] ZID, University of Innsbruck, Austria
draheim@acm.org

Abstract. In this article we present a metamodeling tool that is strictly oriented towards the needs of the working domain expert. The working domain expert longs for intuitive metamodeling features. In particular this concerns rich capabilities for specifying the visual appearance of models. In these efforts we have identified an important design rationale for metamodeling tools that we call visual reification – the notion that metamodels are visualized the same way as their instances. In our tool we support both, standard metamodeling features and new metamodeling features that are oriented towards the visual reification principle. We will start an unbiased discussion of the pragmatics of metamodeling tools against the background of this design rationale.

1 Introduction

In successful projects of today's enterprises we see modeling activities in business reengineering, logistics, supply chain management, industrial manufacturing and so on. Models foster the communication between stakeholders, because they enforce a certain standardization of the respective domain language. Therefore, they speed up requirement elicitation and then serve as a long-time documentation of system analysis efforts. Modeling is here to stay. Even if models are not intended as blueprints in software development projects they add value. For example, we currently see huge business process redocumentation projects in major enterprises. Research in model-driven engineering is important. In these efforts we have a different focus on modeling than model-driven engineering. We have a look at the working domain [12] expert. Often, it is necessary to adapt the modeling method and, in particular, to adapt the used modeling language to the current needs of the domain. It may become necessary to introduce new modeling elements, to deprecate an existing model element, to add attributes to an existing modeling element, to detail the semantics or to change the appearance of a model element.

2 Motivation and Requirements for a Visualization-Oriented Meta- and Instance Modeling Tool

Unlike most of the research done in domain specific modeling, metamodeling and model transformation we can place the origin of this work in the area of business or

S.S. Bhowmick, J. Küng, and R. Wagner (Eds.): DEXA 2009, LNCS 5690, pp. 727–734, 2009.

corporate modeling. In numerous projects from business process management to enterprise-wide IT architectures, modeling is an essential prerequisite for success. Moreover, it is hardly possible to achieve sustainable improvements without an appropriate abstraction of the real corporate structures and processes. This is where modeling has to take place. But on the other hand modeling must not become an end in itself. It has to be strongly focused on things that need to be analyzed. Otherwise it will be nearly impossible to maintain the results, considering that corporate structures and processes are frequently subject of changes.

As consequence of organizational changes the model repository and even modeling methods may have to be adapted to keep them suitable. This adaptation process is more than a tool function; it is moreover an organizational process that has to be implemented. We will discuss this organizational integration [1] in more depth later (see also [1]).

An important issue is that most popular corporate modeling tools do not allow the creation or adaptation of metamodels. This is quite interesting if we consider how much effort has been taken in "inventing" metamodeling methods and if we look at the list of tools supporting these metamodeling methods. Especially well-established (Meta)-CASE Tools [9][5][4] offer metamodeling features and the most recent developments like Eclipse GMF [21] and Microsoft DSL [22] offer outstanding possibilities to create domain specific languages.

To provide metamodeling features for the business domain and directly to end-users, we have seen that the first order principle is inituitivity. This may be the main reason why user-enabled metamodeling is de-facto currently not existent in business modeling tools. Methods like OMG's MOF [23] or proprietary methods implemented by e.g. (Meta)-CASE tools [24] are hardly accepted by users in this domain. Nevertheless, metamodeling features would add substantial value when applied in a user-friendly intuitive style.

In this work we are going to introduce a metamodeling methodology, which is strongly focused on the visual representation, in order to support what we coined the visual reification principle. With visual reification metamodeling is no longer an abstract visualization independent task, it is now intuitive WYSIWYG modeling.

3   Visual Reification

We discuss our tool against the background of a design rationale that we have coined 'visual reification'. Visual reification is the principle that the visual representation of the metamodel is at the same time also a visual representation of a model that adheres to the metamodel. Or to say it differently, in painting a metamodel the user also paints a correct and in particular a visually correct model. This design rationale is at the core of end-user oriented metamodeling targeted by our efforts. The basic argument is that metamodeling becomes more intuitive and less complex if the model specification mechanism, i.e., the metamodeling capability, is oriented towards the appearance of the model. The principle is so natural that meta case designers of leading meta case tools like MetaEdit+ [8][9], Atom3 [6][7], Kent [5], Moses[3], GME [20][2][10] have implicitly applied it with respect to a single most important concept, i.e., the meta association. Here is a choice between using the meta association as a specification of visual model elements or using it as a specification of connections between visual model elements. It is fair to say, that the latter one is the specification style found in the UML metamodel but all of the above tools have taken the first option.

In our tool we make the visual reification principle a first class citizen. We are not biased in favor of the visual reification principle. We rather want to understand under which circumstances and for which features it adds value. Therefore we make it available in our tool in order to make it available to sophisticated investigation and empirical evaluation in particular. We think that the visual reification principle is a contribution in its own right, because it helps to start a systematic discussion of the pragmatics of metamodeling features and their alternatives. We will delve into some example topics, i.e., meta associations, reference copies and abstract classes in the sequel. Furthermore, we will see that the principle is an ideal that we are sometimes tempted to violate in order to have the appropriate expressive metamodeling power and pragmatics at hand. All the available metamodeling tools somehow use the principle as a design rationale; however, they use it only implicitly. With our tool the design rationale becomes explicit.

4 Implementation

The concept has been implemented in a modeling platform by the use of open-source technologies from the Eclipse Project, Apache Software Foundation and Hibernate. The described intuitive and flexible metamodel definition language was integrated to allow either conceptual or visual-true graphical definition of metamodels and to support the iterative modeling process. The tool integrates a model adaptation engine for the adaptation of instances after metamodel changes and to enable model evolution.

A meta-layer is implemented for textual or graphical definition of metamodels and to enable the creation of metamodel instances an instance-layer has been developed on top of the meta-layer. For metamodel-based analysis on the repository of meta- and instance models an analysis and reporting module is available on a vertical analysis-layer.

All layers are integrated as modules within the platform and can be optionally removed to create either only a metamodeling- or instance modeling or analysis tool. Beside that the access to each module is role dependent and can be restricted by an administration module that manages roles, users and user groups.

The role specific access to modules and the central metamodel repository prevent from the decentralized definition or adaptation of metamodels by unauthorized users.

Right from the start it was always an issue to support a simple integration into a company's IT-infrastructure. This was the main reason to develop platform independent and to use JAVA technologies. Moreover relational databases are still de-facto standard in today's enterprises. To take this into account the persistence layer was designed generic to support different data stores. At the current state of development an implementation for relational databases using the object relational framework Hibernate is integrated. Nevertheless, other implementations like XMI flatfiles are possible.

Fig. 1 shows a screenshot of the tool's meta layer where metamodels can be created. As example, an organizational metamodel (organigram) is defined. The first (left) editor shows the metamodel in the conceptual style. The second editor visualizes the same metamodel but here the visual reification principle is applied. It is obvious

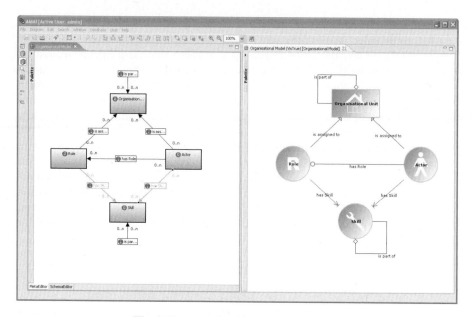

Fig. 1. Conceptual style vs. visual reification

that both metamodeling styles are structurally equal. In the first editor the MetaConnection "has Skill" is selected. You can see that there are two visual representations, created as reference copies. Instances of "has Skill" can now be drawn between instances of "Actor" and "Skill" as well as between instances of "Role" and "Skill". Multiplicities are defined for both reference copies and will be interpreted for each separately.

Fig. 2 demonstrates the use of MetaObjects to specify visual compartments (or container, compositions). The MetaObject acts as a container for child MetaObjects. For example we assume that an "Actor" can now be visually a child of an "Organizational Unit". Moreover we introduce the new MetaObject "Facility" and define that an "Actor" can be a child of a "Facility". The left editor once again shows the conceptual notation. You can see that it is possible to define that a MetaObject can be a child of several parent MetaObjects by the use of reference copies. The visual reification principle is applied in the second editor. Both visualizations are structurally equal. For the container layout the xy-layout is used, which allows to place child figures free inside its parent figure's bounds. Nevertheless, also stack layout, border layout and toolbar layout algorithms are available. The latter one can be used e.g. to define UML compartments like "Classes", "Attributes" and "Methods".

In Fig. 3. you can see a screenshot of the tool's instance layer where instance models based on metamodels are created. The editor visualizes a minimal process for an incoming order. For every selected element available attributes are shown in a property view. You can find the property view for the selected element "Calculate Capacity" in the lower part of the screen. Values and references to other model elements can be defined here. In this example references to incoming/outgoing information objects and documents have been created.

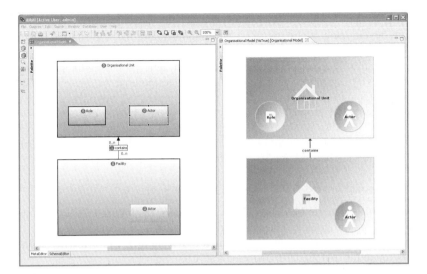

Fig. 2. Visual reification with compartments

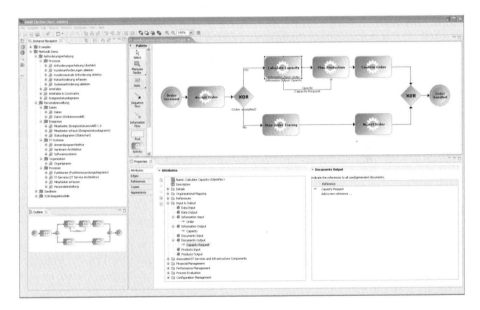

Fig. 3. Instance of a process model

5   Related Work

Adaptivity of modeling languages is a major driving issue in the community of Model-Driven Architecture [13][14] (MDA), which is the current automatic programming [15][16] metaphor. Modeling is pervasive in modern enterprises; however, it is so without automatic programming metaphor. Of course, modeling, and visual modeling in particular, is used in software development projects. With respect to software development, there are different opinions about the role and the importance of modeling. For example, the Rational Unified Process (RUP) [17] is based on modeling – it is model-driven. On the other hand, in agile processes like Extreme Programming (XP) [18] modeling is de-emphasized. Despite that we see severe modeling efforts in companies, in both vertical and horizontal projects, not only software development projects but projects [19] in general. That is why the current focus of our efforts is the working domain expert rather than model engineering. However, the techniques developed in the MDA community, i.e., model transformation techniques [11], are important, because model migration is an issue considered by us.

Related projects to the research area of this work are:

1. GME: GME (Generic Modeling Environment) is a toolkit used for domain specific modeling and program synthesis environments [20].
2. Atom3: Atom3 is a tool for multi-paradigm modeling with the two main tasks meta-modeling and model-transformation [6].
3. MetaEdit+ DSM environment: The MetaEdit+ DSM environment consists of two parts, the MetaEdit+ workbench and the tool MetaEdit+. It is used in the area of domain specific modeling [8][9]0.
4. META CASE [25][26].
5. KMF: The Kent Modeling Framework (KMF) is used in the area of model driven software development [5].
6. MOSES: Moses is a modeling and simulation environment [4].
7. Microsoft DSL: Microsoft's Domain Specific Modeling Tools are part of Visual Studio and allow developers to create their own graphical designers and code generation tools for domain specific languages [22].
8. Eclipse GMF: The Eclipse Graphical Modeling Framework is an Eclipse project that allows the creation of editors for domain specific languages based on the Eclipse EMF (Eclipse Modeling Framework) and GEF (Graphical Editing Framework) projects [21].

6 Conclusion

Our research institute has conducted many complex modeling projects with its company and research partners, for example in the domains of Business Process Management, IT-Service Management, IT-Architectures and IT-Landscapes. Based on our experience we can state:

- Modeling is here to stay in enterprises. A lot of modeling efforts exist even without model-driven approach. Models serve as intuitive system and process description at all levels of the enterprise. Metamodeling is always an issue, again also without a model-driven approach, and so is meta-model based model transformation, for example in order to deal with model migration.
- Without appropriate support for adaptivity, modeling projects lack agility and suffer the risk of a maintenance nightmare of their work products. Metamodeling enables adaptivity of modeling approach and therefore empowers modeling projects.

On basis of the above insights we have contributed a graphical modeling tool that is strictly oriented towards the pragmatics of dealing, i.e., defining and managing, visual artifacts. In this article we discussed the following:

- The requirements for a graphical metamodeling tool that meets the above objectives.
- The concrete features and modeling capabilities of our tool.
- The context of adaptive modeling and how it motivates our tool.

References

[1] Himsl, M., Jabornig, D., Leithner, W., Draheim, D., Regner, P., Wiesinger, T., Küng, J.: A Concept of an Adaptive and Iterative Meta- and Instance Modeling Process. In: Proceedings of DEXA 2007 - 18th International Conference on Database and Expert Systems Applications, September 2007. Springer, Heidelberg (2007)

[2] Agrawal, A., Karsai, G., Ledeczi, A.: An End-to-End Domain-Driven Development Framework, Domain-driven development track. In: 18th Annual ACM SIGPLAN Conference on Object-Oriented Programming, Systems, Languages, and Applications, Anaheim, California, October 26 (2003)

[3] Janneck, J.: Graph-type definition language (GTDL)—specification, Technical report, Computer, Engineering and Networks Laboratory, ETH Zurich (2000)

[4] Essar, R., Janneck, J., Naedele, M.: The Moses Tool Suite - A Tutorial. Version 1.2, Computer, Engineering and Networks Laboratory, ETH Zurich (2001)

[5] Kent, S., Patrascoiu, O.: Kent Modelling Framework Version – Tutorial, December 2002. Computing Laboratory, University of Kent, Canterbury (2002)

[6] Lara, J., Vangheluwe, H.: Using AToM as a Meta CASE Tool. In: 4th International Conference on Enterprise Information Systems, Universidad de Castilla-La Mancha, Ciudad Real (Spain), April 3-6 (2002)

[7] Lara, J., Vangheluwe, H.: Computer Aided Multi-Paradigm Modeling to Process Petri-Nets and Statecharts. In: 1st International Conference on Graph Transformation, Barcelona (Spain), October 7-12 (2002)

[8] MetaCase. ABC To Metacase Technology - White Paper. MetaCase Consulting, Finland (August 2000)

[9] MetaCase. Domain-Specific Modelling: 10 Times Faster Than UML. White Paper, MetaCase Consulting, Finland (January 200)

[10] Sprinkle, J., Karsai, G.: Model Migration through Visual Modeling, OOPSLA, Anaheim, CA, October 26 (2003)

[11] OMG, MOF 2.0 Query / Views / Transformations RFP (2002)

[12] Bjorner, D.: On Domains and Domain – Engineering Prerequisites for Trustworthy Software – A Necessity for Believable Project Management. Domain Engineering and Digital Rights Group (April 2006)

[13] Atkinson, C., Kühne, T.: The Role of Metamodeling in MDA. In: Proceedings of WISME@UML 2002 – International Workshop in Software Model Engineering (2002)

[14] Soley, R.: Model Driven Architecture, white paper formal/02-04-03, draft 3.2, Object Management Group (November 2003)

[15] Parnas, D.L.: Software Aspects of Strategic Defense Systems. In: Software Engineering Notes, ACM Sigsoft, October 1985, vol. 10(5). ACM Press, New York (1985)

[16] Czarnecki, K., Eisenecker, U.W.: Generative Programming – Methods, Tools, and Applications. Addison-Wesley, Reading (2000)
[17] Jacobson, I., Booch, G., Rumbaugh, J.: The Unified Software Development Process. Addison-Wesley, Reading (1999)
[18] Beck, K.: Extreme Programming Explained – Embrace Change. Addison-Wesley, Reading (2000)
[19] Duncan, W.R. (ed.): A Guide to the Project Management Body of Knowledge. Project Management Institute (1996)
[20] Ledeczi, A., Maroti, M., Bakay, A., Karsai, G., Garrett, J., Thomason IV, C., Nordstrom, G., Sprinkle, J., Volgyesi, P.: The Generic Modeling Environment, Workshop on Intelligent Signal Processing, accepted, Budapest, Hungary, May 17 (2001)
[21] Eclipse Graphical Modeling Framework (GMF), http://www.eclipse.org/gmf/
[22] Cook, S., Jones, G., Kent, S., Wills, A.C.: Domain Specific Development with Visual Studio DSL Tools. Addison-Wesley, Reading (2007)
[23] MOF, OMG's MetaObject Facility, http://www.omg.org/mof/
[24] Kelly, S.: GOPRR Description. PhD. dissertation, Appendix 1 (1997)
[25] Ebert, J., et al.: Meta-CASE in Practice: A Case for KOGGE. In: Proc. of CaiSE 1997 (1997)
[26] Costagliola, G., et al.: Constructing Meta-CASE Workbenches by Exploiting Visual Language Generators. IEEE TSE 32(3) (2006)

PISA: Federated Search in P2P Networks with Uncooperative Peers

Zujie Ren, Lidan Shou, Gang Chen, Chun Chen, and Yijun Bei

Zhejiang University, China
renzju@gmail.com, {should,cg,chenc}@zju.edu.cn,
alphabyj@yahoo.com.cn

Abstract. Recently, federated search in P2P networks has received much attention. Most of the previous work assumed a cooperative environment where each peer can actively participate in information publishing and distributed document indexing. However, little work has addressed the problem of incorporating uncooperative peers, which do not publish their own corpus statistics, into a network. This paper presents a P2P-based federated search framework called PISA which incorporates uncooperative peers as well as the normal ones. In order to address the indexing needs for uncooperative peers, we propose a novel heuristic query-based sampling approach which can obtain high-quality resource descriptions from uncooperative peers at relatively low communication cost. We also propose an effective method called RISE to merge the results returned by uncooperative peers. Our experimental results indicate that PISA can provide quality search results, while utilizing the uncooperative peers at a low cost.

Keywords: Federated search, P2P network, uncooperative peers.

1 Introduction

Federated search in peer-to-peer networks, as an alternative to centralized search engine, has attracted a lot of attention in the research community. Recent years have witnessed a number of systems for federated search in peer-to-peer(P2P) networks [1,2,3,4,5]. Typically, a P2P-based federated search engine consists of a number of autonomous and distributed peers, each of which contains a collection of documents, and responds to queries based on its local index.

Almost all P2P federated search systems assume a two-phase paradigm (or a variation) as described in the following. In the first phase, which is often referred to as the *directory construction* phase, each peer in the system publishes a number of *resource description* entries to some other peers in the P2P overlay. In the second phase, namely the *resource selection* phase, the distributed index structure is retrieved and the query is forwarded to a number of promising peers. In the latter case, the query is executed in the selected peers and the results are returned to the querying peer.

It is however important to note that the directory construction phase in the above may not be applicable to all peers. Specifically, the previous works assume that all peers are "cooperative" in the sense that they publish accurate *resource descriptions* of their own document repositories, most probably in the form of a list of terms (or some

S.S. Bhowmick, J. Küng, and R. Wagner (Eds.): DEXA 2009, LNCS 5690, pp. 735–744, 2009.
© Springer-Verlag Berlin Heidelberg 2009

more detailed structures). However, this assumption is not always valid in real applications. For example, despite its document-querying (and sharing) service, a hidden Web site such as an online digital library, may not be able to release accurate description information of its own archive. A more significant example is the majority of almost all commercial websites which provide local search services. If these websites are regarded as "peers", it simply seems to be impractical to request for resource descriptions from them via any peer-to-peer protocols. Such peers, which can answer queries but do not provide any resource descriptions to their own repositories, are often referred to as *uncooperative* peers in the literature. Regarding the massive quantity and coverage of such uncooperative peers in the Web, we would emphasize that the problem of incorporating uncooperative peers into a P2P search system is potentially significant but has been overlooked unfortunately in the past.

In this paper we shall look at novel techniques which integrate uncooperative peers into a P2P search system. Although the problem of handling uncooperative information sources has been addressed in the literature of distributed information retrieval(IR) [6,7,8], it is much more challenging in the context of P2P search and requires additional studies due to the P2P characteristics, such as the absence of centralized broker, limited bandwidth resource.

2   Related Work

A number of solutions to P2P search rely upon P2P networks. Lu et al. [1] propose a hybrid network model for federated search on digital libraries. Peers with similar contents connect to the same super-peer, called hub, to form a content-based cluster. Hubs are connected in an unstructured overlay, where each hub maintains connections to hubs covering similar content areas and to those serving dissimilar content areas. However, we argue that if P2P network adopts an unstructured overlay, it could be tolerant to churn, but its lack of direction in search could be a major barrier for the search effectiveness. Their research efforts mainly focus on how to improve the search accuracy and efficiency for federated digital libraries.

Si et al. [9] propose the SSL merging approach which relies on the overlap between the retrieved results and a centralized sampled database containing all samples for each peer. The SSL algorithm utilizes the documents in the overlap to train a linear regression model for each query. Then, the linear regression model is used to transform the local (peer-specific) scores into global (peer-independent) scores for each retrieved document. However, a centralized sample database is not available in PISA because the sampled documents are distributed on cooperative peers rather than being kept in a central broker. Moreover, we cannot assume all uncooperative peers in PISA system will return their results with the relevance scores. The only assumption is that they can provide ordered search results. In other words, the only information that we can utilize to merge results is the ranking of the results.

3   Structure of PISA

In PISA system, we assume each peer has its own document collection and a local search engine. In addition, each peer can accept a query issued from other peers and retrieve a set of document identifiers as results to be returned. All peers (both cooperative and uncooperative ones) are interconnected via a DHT network. Without loss of generality, we use Chord [10] as the overlay structure of PISA. For a cooperative data peer, it generates its own posts to be published, from its local inverted index, and publishes them into the DHT network, thereby populating the global index directory. A hash function is applied to every term to determine the responsible index peer respectively.

An uncooperative peer neither publishes its own resource description nor takes the responsibility of index directory maintenance. To utilize the search service provided by uncooperative peers, we propose a heuristic query-based sampling technique to estimate their resource description. We will discuss the details of this technique in Section 4.

4   Acquiring Resource Description of Uncooperative Peers

In this section, we will focus on how to acquire resource description of uncooperative peers. To acquire the resource description of an uncooperative peer P_u, the following steps should be performed: 1)first, we employ a query-based sampling mechanism to obtain a few samples of the documents stored in it. The term statistics of these samples can then be computed as *sampled* resource description; 2)second, the sampled resource description is scaled by a factor to ensure fairness between uncooperative and cooperative peers during query resolving.

Step 1: Heuristic query-based sampling: When an uncooperative peer P_u joins PISA, another peer P_s (a cooperative peer with the smallest ID no less than the one of P_u) will attempt to transfer a subset of index directory to P_u. If P_s do not receive *Accept* response from P_u, it can detect that the joining peer P_u is uncooperative. Then, P_s uses the Query-Based Sampling (QBS) technique [6] to acquire an approximate resource description of P_u, which works as follows:

(1) P_s initializes a query dictionary Q and randomly select a single-term q from Q. (2) P_s issues the query q to P_u, which returns the IDs of relevant documents as results. (3) P_s downloads the top m documents and adds them into the sample. (4) P_s updates Q with the terms in downloaded documents. (5) If a stopping criterion has not yet been reached, P_s selects a new query and go to step (2).

In the traditional QBS approach, the stopping criterion is that the number of sampled documents reaches a threshold (250 in our case). Callan et al. claimed in [6] that QBS can provide high-quality sampled resource descriptions. However, a sampling method using fixed stopping criterion will always suffer from either under-sampling or over-sampling problems when peer sizes have skewed distribution. A large threshold causes over-sampling for small scale peers, while a small one causes under-sampling for large peers.

To tackle the above problem, we propose a heuristic stopping criterion that automatically adjusts the sample size. Ideally, if we issue a sequence of queries $Q=\{q_1, q_2, \ldots, q_i, \ldots, q_n\}$ to an uncooperative peer P_u, we can observe a sequence of answer sets $\{D_1, D_2, \ldots, D_n\}$. If we compute a sequence $KLS=\{KL(D_1, D_{full}), KL(D_2, D_{full}), \ldots, KL(D_n, D_{full}), \ldots\}$, $KL(D_n, D_{full})$ should converge to zero, where $KL(D_i, D_j)$ represents the Kullback-leibler divergence of D_i and D_j. We call this sequence *direct K-L divergence*. Figure 1(a) shows the average KLS sequence values of 200 uncooperative peers, each of which owns a document set extracted from

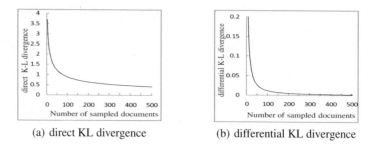

(a) direct KL divergence   (b) differential KL divergence

Fig. 1. Convergence of direct and differential KL divergence

TREC WT10g. Each simulated peer contains multiple collections, ranging from 1 to 2000 collections in the dataset. The results indicate that KLS values can be used as stopping criterion for the sampling process.

Unfortunately, the full description D_{full} is not available in real environment. Thus, we cannot use direct K-L divergence to determine when to stop sampling. Instead, we can compute sequence $KLS' = \{KL(D_1, D_2), KL(D_2, D_3), \ldots, KL(D_{n-1}, D_n)\}$, which also converges to zero, as shown in Figure 1(b). We call sequence KLS' the *differential K-L divergence*. Differential K-L divergence indicates the difference between two consecutive samples. Small differential K-L divergence implies that the corresponding samples are similar, which indicates that the sample becomes stable and has a good coverage of the actual vocabulary.

We propose a heuristic stopping criterion to keep query-based sampling until the *differential K-L divergence* becomes less than a predefined threshold. More specifically, after each query q_i probing step, the differential K-L divergence $KL(D_{i-1}, D_i)$ between D_{i-1} and D_i is measured. If the differential K-L divergences become less than a threshold τ for m consecutive sampling queries, the sampling process will be stopped. Using this stopping criterion, the number of samples becomes non-uniform for different peers, but generally in proportion to the peer size, which will minimize both over-sampling and under-sampling cases.

Step 2: Scaling the sampled resource description: The above heuristic query-based sampling method is designed to obtain a sampled resource description from an uncooperative peer. Remember that PISA needs to handle cooperative peers as well. If the sampled resource description generated from sampling is published to the DHT network indiscriminatively, a problem may arise as the sampled document frequencies (df_s) for each sampled term are usually much smaller than the actual values. As a result, an uncooperative peer is much less likely than a cooperative one to be selected during the process of peer selection.

For fairness in peer selection, we need to estimate the document frequencies for each term with respect to the whole collection in P_u. Si et al. [9] introduce a *scale factor*, which are defined as the ratio of estimated peer size and the sample size, to estimate the actual document frequencies. Influenced by their work, we decide to scale up the df_s for each sampled term by a factor of F, which is also defined as the ratio of estimated peer size and the number of sampled documents. For a given term t,

$$df\left(t, P_u\right) = df_s\left(t, P_u\right) \cdot \frac{|P_u|}{|Sample(P_u)|}$$

where: $df\left(t, P_u\right)$ is the estimated number of documents in P_u that contain t; $df_s\left(t, P_u\right)$ is number of documents in the sample of P_u that contain t; $|P_u|$ is the size of peer P_u, which can be estimated using the Sample-Resample method proposed in [9]; $|Sample(P_u)|$ is the number of documents in the sample of peer P_u.

5   Query Processing in PISA

Query processing in PISA involves three steps: First the query is resolved and a number of peers are selected to process the query; second, the query is executed in the selected peers; third, the query results from each peer are returned to the querying peer, where the final results are to be merged and ranked. We mainly focus on the query resolving and result merging/ranking techniques.

Query Resolving: In the process of query resolving, a certain number of peers need to be selected based on their relevance with the query. There are a lot of collection selection approaches proposed in distributed IR, including CORI [11], DTF[12]. CORI uses a Bayesian inference network model with an adapted Okapi term frequency normalization formula to rank available collections. DTF makes a collection selection decision that achieves a minimal overall cost, including retrieval accuracy and time consumption. PISA is not restricted to any particular peer selection method. We apply CORI algorithm in our PISA system due to its high performance and simplicity.

Merging Score-Absent Results: Merging the results returned from uncooperative peers is more complicated. Some of the uncooperative peers may return ranked results with no local relevance scores. Therefore, the result merging methods that rely on local relevance scores, such as CORI merging [11] and SSL [9] algorithm, are not applicable in PISA. In order to handle such results, we propose a novel scheme called *Result merg-Ing method in Score-absence Environments* (RISE), in which the peer-local relevance scores of retrieved results are not available.

The RISE method involves three steps:(1) Selectively download a limited number of documents as training data and calculate the relevance score of each downloaded document; (2) Train a regression model for each peer using the ranks and relevance scores of the downloaded documents; (3) Employ the trained regression model to approximate the relevance score of each retrieved document.

Downloading training data at limited cost: To obtain the training data for the regression model, we have to download a small fraction of the retrieved documents from an uncooperative peer in order to calculate the relevance score. In order to restrict network resource consumption, our algorithm must selectively download a limited number of documents only.

There are two issues to be addressed in the downloading process. One is that we should decide which documents to download. The other is that we should determine when to stop the downloading process. As most users are only interested in top-k results, our merging algorithm only concerns the relevance scores of the top-k result candidates. Firstly, the querying peer merges and reorders the results returned by the cooperative peers, if any, producing $\widehat{R}$. We assume the cooperative peers can provide necessary statistical information for each retrieved document. Merging the results of cooperative peers can be implemented with some existing algorithm, such as *TF*IDF* [13] or language modeling [9]. Due to space limit, we do not include the details here. We will only describe in detail the process of merging results returned by uncooperative peers in the rest of this section.

If $\widehat{R}$ contains more than k results, a relevance threshold T is set to the relevance score of the k-th document in $\widehat{R}$. Otherwise, the querying peer iteratively downloads documents ranked at $2^{\theta} (\theta = 0, 1, 2, \cdots)$ from each uncooperative peers respectively, until $2^{\theta} * |U| + |\widehat{R}| > k$, where $|U|$ is the number of uncooperative peers that have returned results, and $|\widehat{R}|$ indicates the number of results in $\widehat{R}$. The querying peer calculates the relevance score of downloaded document and uses them to estimate relevance score of the relevance score of the remaining documents for each peer. The documents ranged from 1 to 2^{θ} of each peer will be merged into $\widehat{R}$ and T is set to the relevance score of the k-th document in $\widehat{R}$.

The aim of the threshold score is to control the downloading process and therefore limit the number of downloaded documents. Using the T as a threshold, the querying peer continues to download the 2^{θ}-th document from each peer and calculates the corresponding relevance scores. As θ increments by one each time, this loop continues until the relevance score of the 2^{θ}-th document is less than the threshold T. For each uncooperative peer P_u, the querying peer uses the relevance scores of the documents downloaded from P_u as training data, to build a regression model for P_u. Using this regression model, the querying peer estimates the relevance score of all the documents returned by P_u and merge them into $\widehat{R}$. Finally, the top-k documents in $\widehat{R}$ are results.

Estimating the relevance scores: In this section, we will describe how to utilize the downloaded documents as training data to build a regression model for each peer, which are used for estimating the relevance score of all retrieved documents.

Regression is an efficient and effective mathematical tool for mapping ranks to relevance score. Inspired by the work in [14], we find that the correlation between the rank of a document and relevance score can be represented by a logistic function as following:

$$Y = \frac{e^{\alpha + \beta * \ln(X)}}{1 + e^{\alpha + \beta * \ln(X)}} \tag{1}$$

where $\ln(X)$ indicates the natural logarithm (noted ln) of the rank X for a retrieved document and Y indicates the corresponding relevance score. Equation 1 can be transformed as follows.

$$\ln(\frac{Y}{1 - Y}) = logit(Y) = \alpha + \beta * \ln(X) \tag{2}$$

In Equation 2, we can see that this equation is a linear one, as this equation can be represented in another form:

$$\widehat{Y} = \alpha + \beta * \widehat{X} \tag{3}$$

where $\widehat{Y}$ is $logit(Y)$ and $\widehat{X}$ is ln (X).

Therefore, this problem is transformed into a linear regression analysis. In order to fit the actual $\widehat{X}$ - $\widehat{Y}$ line, we need to estimate parameters α and β in the above equation using the training data. For a peer P_u and its results list RL, the training data includes the pairs of the rank x and the relevance score y of the downloaded documents in RL. The purpose of this model is to estimate parameter α and β that minimize the deviation, which represents the difference between the observed values of $\widehat{Y}$ and the ones estimated $\widehat{Y}$ (denoted as $\widehat{Y}^e$). More specifically, we aim at obtaining the values of parameter, which can make the sum of squared residuals S minimum:

$$S = \sum_{\theta=1}^{n} \left(\widehat{Y}_k - \widehat{Y}_k^e \right)^2$$

where n indicates the number of training documents and k represents 2^θ.

The regression over all training data from a selected peer can be shown in following equation.

$$\begin{bmatrix} \ln(x_1) & 1 \\ \ln(x_2) & 1 \\ \ln(x_4) & 1 \\ \cdots \\ \ln(x_{2^n}) & 1 \end{bmatrix} * [\alpha, \beta]^T = \begin{bmatrix} logit(y_1) \\ logit(y_2) \\ logit(y_4) \\ \cdots \\ logit(y_{2^n}) \end{bmatrix} \quad (4)$$

We denote these matrices as M (the first item on the left side of Equation 4, which is constructed from natural logarithm of ranks and constants), and N (the item on the right of Equation 4, which is the set of logit function on relevance score). The best way to calculate $[\alpha, \beta]$ is to employ the least (or minimum) square method (LSM). Using LSM, the optimal estimation for parameter α and β that minimize the S is given by:

$$[\alpha, \beta]^T = (M^T M)^{-1} (M^T N) \quad (5)$$

Using the Equation 5 with the relevance score and ranks of downloaded documents, we can obtain the values of α and β and construct a regression model for each peer, to map the ranks into relevance scores for each retrieved document. According to the estimated relevance scores, all the retrieved documents from each peer are merged into a single sorted list $\widehat{R}$. The top-k documents in $\widehat{R}$ are produced as the final results for the query.

6   Experimental Results

The experiments are conducted in three groups for studying 1) the effectiveness of HQBS. 2) the accuracy of RISE. 3)the performance of PISA system. Before presenting the results, we shall look at the experiment settings and the evaluation methodology.

Experiment Settings: We use the *TREC WT10g* collection for our experiments as it has also been used by a previous study in [1]. Each simulated peer contains n document collections, where n ranges from 1 to 2000. We randomly select 1000 queries from the query set[1], which is created by J. Callan and his colleagues based on WT10g data [1],

---

[1] http://boston.lti.cs.cmu.edu/callan/Data/P2P/trecwt10g-query-bydoc.v1.txt.gz

as our query set. In our experiments, the default number of requested results for each query is 50. In addition, we choose the following default parameter values: $\tau=0.01$ and $m=3$ for our heuristic query-based sampling process.

Evaluation Method: The search performance of PISA system is measured by search accuracy as well as efficiency. The *average precision at given document cut-off values* is standard rank-based measure commonly used to evaluate the accuracy of full-text ranked retrieval in distributed information retrieval, which computes the average precision over a set of queries when the 5, 10, 15, 20, 30 top-ranked documents have been seen for each query[11]. Besides precision, recall [15] is also chosen to evaluate the overall percentage of relevant documents that have been retrieved.

As described above, both precision and recall require a relevant answer set for each given query. We choose to use the retrieval results from a centralized search engine as the relevant documents. The centralized search engine is built on a *single large collection*, which aggregates all documents in the network. The top-k documents retrieved from this centralized search engine are treated as the set of "relevant" documents for the query. In our experiments, we use 50 as the default value for k.

Results of acquiring resources descriptions: In this paragraph, we will compare the effectiveness and efficiency of our proposed method for acquiring the resource descriptions to the traditional QBS on 200 uncooperative peers. The sampling process used by the traditional QBS technique will stop when the sample size reaches 250. However, in the proposed method, we employ HQBS to do the sampling, as well as the scaling techniques.

Figure 2(a) shows the results of search accuracy using different sampling techniques. The x-axis indicates the percentage of selected peers participating in each query. The y-axis indicates the average precision and recall of all queries. The results indicate that HQBS with the default parameters($\tau=0.01$ and $m=3$) outperforms the conventional QBS in both precision and recall is effective in improving search accuracy. Table 1 displays the sampling cost of QBS and HQBS. Figure 2(b) describes the results of search accuracy using HQBS with different stopping conditions. The parameters τ and m of HQBS for controlling the sample process are selected to (0.01,3), (0.01,5) and (0.1,3) in this evaluation. In one aspect, we can see that the more stricter is the stopping condition (lower τ and higher m), the higher search accuracy is performed by PISA, as the quality of samples is generally better. In the other aspect, the sampling cost (measured by number of sample documents) is also increased with the stopping condition becoming strict, as presented in Table 1. In a real system, tuning both parameters τ and m can be applied to obtain an optimal tradeoff between the search accuracy and the sampling cost.

Table 1. The number of sample documents using QBS and HQBS

	Sum	Max	Min	Avg
QBS (250 samples documents)	50000	250	250	250
HQBS(default:$\tau=0.01$,$m=3$)	42000	450	60	210
HQBS($\tau=0.01$,$m=5$)	53000	500	85	265
HQBS($\tau=0.1$,$m=3$)	10000	120	30	50

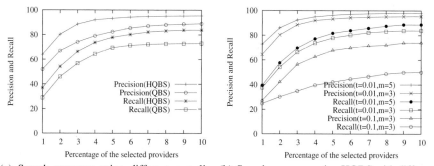

(a) Search accuracy using different sampling techniques.

(b) Search accuracy using HQBS with different stopping conditions.

Fig. 2. Search accuracy comparison

Results of merging: We evaluate the performance of our proposed result merging algorithm, namely RISE, on 200 uncooperative peers. We perform the same experiment on a well-known result merging algorithm called CORI [11], which has been shown to be effective in previous studies.

Figure 3 shows the results of accuracy for CORI and RISE. From the figure we see that RISE outperforms CORI in both precision and recall. The improvement achieved by RISE is slight compared to CORI. However, it is crucial to note that the most significant advantage of RISE is the applicability in score-absent environments.

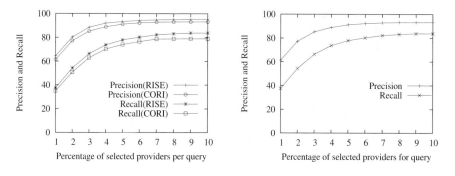

Fig. 3. Comparison of merging algorithms   **Fig. 4.** Search accuracy of PISA

Performance of PISA: Finally, we study the performance of a simulated PISA prototype which consists of 200 uncooperative peers and 800 cooperative ones. Figure 4 shows the results of search accuracy in PISA when different numbers of peers are selected to process queries. Note that the ratio between the number of uncooperative and cooperative peers is kept to be 2:8. We can see that the accuracy of PISA system increases as the number of selected peers grows.

7 Conclusion

In this paper, we presented a P2P-based federated search framework called PISA. PISA allowed the utilization of search service provided by uncooperative peers. The peers in PISA were interconnected via a Chord ring. We proposed an effective heuristic query-based sampling method (HQBS) to handle skewed data when acquiring approximate resource description from different uncooperative peers. We also presented a novel algorithm to merge results returned by uncooperative peers in absence of local relevance scores. Our extensive experiments showed that the proposed techniques were effective in supporting queries which involved both uncooperative and cooperative peers.

References

1. Lu, J.: Full-Text Federated Search in Peer-to-Peer Networks. PhD thesis, Carnegie Mellon University (2007)
2. Nejdl, W., Wolpers, M., Siberski, W., Schmitz, C.: Super-peer based routing and clustering strategies for rdf-based peer-to-peer networks. In: WWW, pp. 536–543 (2003)
3. Nottelmann, H., Fischer, G., Titarenko, A., Nurzenski, A.: An integrated approach for searching and browsing in heterogeneous peer-to-peer networks. In: ACM SIGIR WorkShop Hetergeneous and Distributed Information Retrieval (2006)
4. Suel, T., et al.: Odissea: A peer-to-peer architecture for scalable web search and information retrieval. In: WebDB, pp. 67–72 (2003)
5. Renda, M.E., Callan, J.: The robustness of content-based search in hierarchical peer to peer networks. In: CIKM, pp. 562–570 (2004)
6. Callan, J., Connell, M.: Query-based sampling of text databases. ACM Transaction of Information System, 97–130 (2001)
7. Craswell, N., Hawking, D., Thistlewaite, P.: Merging results from isolated search engines. In: Australasian Database Conference, pp. 189–200 (1999)
8. Thomas, P., Hawking, D.: Evaluating sampling methods for uncooperative collections. In: SIGIR, pp. 503–510 (2007)
9. Si, L.: Federated Search of Text Search Engines in Uncooperative Environments. PhD thesis, Carnegie Mellon University (2006)
10. Stoica, I., Morris, R., Karger, D., Kaashoek, F., Balakrishnan, H.: Chord: A scalable peer-to-peer lookup service for internet applications. In: ACM SIGCOMM, pp. 149–160 (2001)
11. Callan, J.: Distributed information retrieval. Advances in Information Retrieval (2000)
12. Nottelmann, H., Fuhr, N.: Decision-theoretic resource selection for different data types in mind. In: Distributed Multimedia Information Retrieval (2003)
13. Kirsch, S.T.: Distributed search patent. U.S. Patent 5,659,732 (1997)
14. Calv, A.L., Savoy, J.: Database merging strategy based on logistic regression. Information Process Manage (2000)
15. Baeza-Yates, R., Ribeiro-Neto, B.: Modern Information Retrieval. Addison-Wesley, Reading (1999)

Analysis of News Agencies' Descriptive Features of People and Organizations

Shin Ishida, Qiang Ma, and Masatoshi Yoshikawa

Department of Social Informatics, Graduate School of Informatics,
Kyoto University Yoshidahonmachi, Sakyo-ku,
Kyoto, 606-8501 Japan
ishida@db.soc.i.kyoto-u.ac.jp,
{qiang,yoshikawa}@i.kyoto-u.ac.jp

Abstract. News agencies report news from different viewpoints and with different writing styles. We propose a method to extract characteristic descriptions of a news agency written about people and organizations. To extract the characteristic descriptions of a given person or organization, we analyze words which appear in the same sentence on the basis of their SVO roles. We then extract a description that is often used by the news agency but not commonly used by the others. The experimental results show that our method can elucidate the different features of each agency's writing style.

1 Introduction

News articles are widely available on the Web from diverse sources of information, and news agencies are main sources. They report news from different viewpoints and styles of writing. To ensure that readers are not biased by what they read, they need to be aware of the views and styles of the agency. However, developing this awareness is difficult. There is therefore a need for a service that helps readers to understand the relationship between content and the position of the agency.

A news agency may reveal its lack of impartiality in its descriptions of people and organizations. For example, we find that a news agency, P, presents mostly positive descriptions of politician Z while almost all those of Q news agency are negative. We extracted typical characteristics of a news agency's descriptions of people and organizations to analyze the bias of news agencies. These characteristics can be used to ascertain the extent to which an article was written in the usual way or not. Intuitively, a news article having different description style is remarkable.

To extract the characteristic descriptions of a given person or organization by a news agency, we analyzed words which appear in the same sentence on the basis of whether they are a subject (S), object (O), or verb (V). When a given person or organization name appears in a sentence as S or O, the words of the other roles (O,V or S,V) are defined as descriptions of the given person or organization. We call such a description as a p-description (person-description).

S.S. Bhowmick, J. Küng, and R. Wagner (Eds.): DEXA 2009, LNCS 5690, pp. 745–752, 2009.

We then found a p-description that is prevalent in articles from a given news agency by comparing how frequently it appears in content from other agencies. To do this, we computed the appearance ratio and the inverse agency frequency of the p-description. We multiplied the appearance ratio by the inverse site frequency to obtain the feature score of the p-description. The p-description which has high feature score is the characteristic description of a given person or organization.

To estimate the effectiveness of our method, we carried out an experiment by using the articles of three major Japanese newspaper agencies. The experimental results show that our method can be used to elucidate the features of each agency.

2   Related Work

The major result from recent studies has been to show that there is bias and diversity in news contents.

NewsCube[1] presents multiple classified viewpoints (aspects) of a news event. It extracts the viewpoints of articles based on keyword selection and weight calculation considering the number of occurrences of keyword and the locations of occurrences. It then searches the elements of aspects by recognizing significant keywords and extracts the aspects of articles. Finally, NewsCube classifies these aspects regarding the similarity and present users multiple aspects of a news. Users can effectively compare diverse viewpoints and understand the event profoundly by using NewsCube. The Comparative Web Browser (CWB)[2], which is a system that allows users to concurrently browser different news articles for comparison, has been proposed. It discovers related contents from diverse information source based on a topic structure consisting of topic and content terms for comparison. TVBanc[3] is a tool to present bias and diversity of news contents by comparing topics and viewpoints presented in different articles. It extracts topics and viewpoints based on the structure of news articles: the topic is represented by the whole text while the viewpoint is often described in the conclusion part of a news article. Sentiment map[4] presents the emotional impact on the reader of news stories. It calculates the emotional impact of keywords in news content. Aoki et al[5] propose a method of extracting author intentions based on the peculiarity parts in related news contents. This work focuses on paragraphs having a peculiarity that is above a given threshold.

These studies deal with the bias and diversity of news contents; however, they are not concerned with ascertaining the features of different news in a long time span. To the best of our knowledge, writing style is not yet well studied for difference mining of news content. Our purpose is to mitigate bias caused by news agencies' distinct viewpoints based on the analysis of writing style.

3   Extraction of Descriptive Feature

We are concerned with the features of different news agencies, such as differing viewpoints or writing styles. Especially, in our current work, we focus on the descriptions of people and organizations as they often bring out the typical features of a news agency. In this section, we introduce our method to extract news agencies' characteristic descriptions of people and organizations.

3.1 SVO Structure

We define "SVO structure" as the structure in which subject, object, and verb make up a sentence. The description of a person or organization is constructed on the basis of this structure, in which the name of the person or organization is either a subject or object. We define such a description as p-description (person-description). The SVO structure has six patterns as follows.

1. Person is subject.
   - S-V structure (person-verb)
   - S-O structure (person-object)
   - S-O-V structure (person-object-verb)
2. Person is object.
   - O-S structure (person-subject)
   - O-V structure (person-verb)
   - O-S-V structure (person-subject-verb)

P-description is constructed by using these six patterns. Each of them represents a sentence pattern.

- The S-O and the O-S structure represents people or organizations or events which are often written with a given person or organization.
- The S-V and the O-V structure represents activities of a given person or organization.
- The S-O-V and the O-S-V structure represents people or organization and activities of a given person or organization.

3.2 Extracting the Characteristic Description

To extract p-descriptions, we collect articles in which a person or organization name appears. We extract the characteristic p-descriptions in period or in topic. A topic is an event or an affair related to a given person or organization. By extracting the characteristics in topic, we can elucidate how they change through various topics. We cluster articles by using keywords (person and organization names, etc.) and published time, related to a given person or organization in a predefined period will be grouped together.

   To extract the characteristic p-description, we analyze the structure of each sentence of articles and acquire the SVO structure for each one. We then get the SVO structure in which the name of a person or an organization appears. We divide SVO structure into 6 patterns (as described in 3.1). As the meaning of a description varies from pattern to pattern, we extract the characteristic p-description per pattern based on computing the local and global features.

S. Ishida, Q. Ma, and M. Yoshikawa

Local Feature of p-description. The description that often appears in the news agency represents the way that the agency typically reports the news. Such a feature is a local feature of the description and computed by the ratio r_x. The ratio r_x of p-description $d_i(p, x)$ in topic t written by news agency N_j is computed as

$$r_x(N_j, t, d_i(p, x)) = \frac{freq_x(N_j, t, d_i(p, x))}{\sum\limits_{i=0}^{n_x} freq_x(N_j, t, d_i(p, x))} \tag{1}$$

where p stands for a person or organization name and x denotes a pattern of SVO structure. n_x is the total number of p-descriptions in x. The $freq_x(N_j, t, d_i(p, x))$ is how frequently p-description $d_i(p, x)$ occurs in pattern x.

It is very likely that a description which has a high r_x value is characteristic.

Global Feature of p-description. The characteristics of descriptions of a news agency do not often appear in those of other agencies. This is a global feature of the news agency. We estimate how a news agency is distinct by comparing it's articles with other news agencies. The peculiarity of a description of a news agency by comparison with the others is computed by using the iaf (Inverse Agency Frequency). The iaf of p-description $d_i(p, x)$ in a topic t written about by news agency N_j is computed as

$$iaf(N_j, t, d_i(p, x)) = \frac{r_x(N_j, t, d_i(p, x))}{\sum\limits_{j=0}^{m} r_x(N_j, t, d_i(p, x))} \tag{2}$$

where m stands for the total number of news agencies.

Extracting the characteristic p-description. The feature score f of p-description $d_i(p, x)$ in topic t written by news agency N_j is computed by using r_x and iaf as follows.

$$f(N_j, t, d_i(p, x)) = r_x(N_j, t, d_i(p, x)) \times iaf(N_j, t, d_i(p, x)) \tag{3}$$

This formula is similar to tf-idf. We rank the p-description by using the feature score f per pattern. We define a p-description in the high rank in a certain pattern as a characteristic description of person p written by the news agency.

3.3 Presentation of Characteristic Descriptions

We propose a way of presenting the characteristic descriptions of different news agencies. The characteristic descriptions of a topic is represented by a graph such as Figure 1. A center node represents a given person or organization name, while a surrounding node represents 5 descriptions that have a high rank in a certain pattern. The example of Figure 1 shows the characteristic description of Abe[1] at the House of Councilors in Japan written by the Asahi Shimbun.

---

[1] Shinzo Abe: The ex-prime minister of Japan.

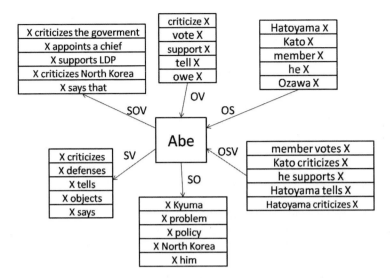

Fig. 1. The characteristic description of Shinzo Abe at the House of Councilors in Japan

4 Experimental Results

To evaluate the effectiveness of our method, we carried out experiments by using the articles of the Asahi, Yomiuri, and Mainichi Shimbun. We used articles for three months from January to March in 2007 and cluster them in a week period. We used Syncha [6] to analyze the Japanese morphology and extract the SVO structure. We then analyzed the description given by each agency of the following people and organizations: 1) Shizo Abe[2] , 2) the LDP[3], 3) Ichiro Ozawa[4] , and 4) Kitachosen[5].

To examine whether we could extract a unique description of each agency respectively, we estimate how accurately the extraction was by using the description coverage described below. The description coverage of person p in pattern x is computed as

$$cov(p, x) = \frac{1}{n} \times \sum_{k=1}^{n} \frac{|c_k(p, x)|}{|D_k(p, x)|} \tag{4}$$

where n is the number of articles in a topic or period (in this experiment, n is the number of articles in a week). $D_k(p, x)$ is the sets of characteristic descriptions of all agencies of person e in pattern x in cluster k, and $c_k(p.x)$ is the overlapped descriptions among them. If the description coverage is low, descriptions of each agency overlap less, and we can say that a unique description is extracted respectively.

---

[2] We input "Abe".
[3] The Liberal Democratic Party.
[4] Japanese politician. We input "Ozawa".
[5] North Korea.

Table 1. The description coverage

	SV	SO	OV	OS	SOV	OSV
Abe	0.08	0.05	0.06	0.04	0.04	0.03
the LDP	0.1	0.05	0	0.01	0.03	0
Ozawa	0.02	0.01	0.06	0.03	0.02	0
Kitachosen	0.08	0.11	0.03	0	0.03	0

Table 2. The extractive precision

	SV	SO	OV	OS	SOV	OSV
Abe	0.27	0.16	0.28	0.46	0.33	0.47
the LDP	0.26	0.28	0.3	0.58	0.29	0.72
Ozawa	0.21	0.23	0.11	0.39	0.22	0.42
Kitachosen	0.28	0.45	0.23	0.47	0.41	0.58
Average	0.255	0.28	0.23	0.475	0.313	0.548

The description coverage of each agency is shown in Table 1.

As shown in Table 1, the description coverage is less than 0.1 with the exception of that of Kitachosen in pattern S-O. The description of each agency overlaps by about one-tenth and we can say the unique descriptions are extracted.

To estimate how much the extracted descriptions express the style of news agencies, we calculated the extractive precision. The extractive precision of person p in topic t in pattern x is computed as

$$precision(N, t, p, x) = \frac{|D_s(N, t, p, x) \cap D_u(N, t, p, x)|}{|D_s(N, t, p, x)|} \qquad (5)$$

where $D_s(N, t, p, x)$ stands for the set of the characteristic descriptions which the system outputs. $D_u(N, t, p, x)$ stands for the set of the descriptions which we judged to express the features of news agencies.

As shown in Table 2, The average precision is various. The average precision is high in the OSV and the OS pattern, while low in the SV and OV ones. This is because proper nouns such as a person or organization name, which represents the features of a news agency tend to appear in the OSV and the OS pattern, and a general verb which does not represent the feature tends to appear in the SV and the OV pattern. We will discuss this tendency in near future.

We then analyzed whether the extracted description of each agency is characteristic.

We first analyzed the extracted description of Abe. The results of the analysis of descriptions of Abe are shown in Table 3. When analyzing the description about education like[Abe(S)-education(O)] or [Abe(S)-teacher(O)], P agency's had 9 descriptions in January 2007 , more than other agencies. However, there were gradually fewer, and, in March 2007, there were six descriptions from Q agency which is more than those of P agency. There were few descriptions from agency R . It turns out the that time and degree that a news agency picks up Abe and education differ from agency to agency.

When we analyzed negative descriptions, such as [Abe(O)-criticize(V)] or [Abe(O)-excuse(V)], we found that there were 15 such descriptions from P agency, a number that is significantly higher than those of the others'. The number of appearances of the word [criticize] in P descriptions is obviously larger than those of the others'. This is the feature of P agency to Abe.

Table 3. The descriptions to Abe

Descriptions about education		
	January	February	March
P agency	9	1	0
Q agency	1	1	6
R agency	1	2	1
Negative descriptions		
	January	February	March
P agency	2	7	6
Q agency	0	2	0
R agency	1	0	2
Number of person, organization, country names		
	January	February	March
P agency	9	15	25
Q agency	5	12	0
R agency	12	16	4

Finally, we analyzed the names of people, e.g., "Aso"[6], organizations, "the LDP" and countries, e.g., "Kitachosen" that appeared in descriptions with Abe. As shown in Table 3, P agency had the highest number with 41 while Q agency had the least with 17. Ozawa and Fukushima[7], head of a opposition party, appear for agency P, while relatively unknown people appear in reports for Q agency and Nakagawa secretary-general appears in R agency' reports.

It was possible to indicate the characteristic description of each news agency for descriptions of Abe in the way described above. We also analyzed the descriptions of other people and organizations. As a result, features of descriptions of each agency are seen in descriptions of Abe and the LDP and Kitachosen.

We can see both the consistent feature of descriptions and small difference between agencies. This is caused by the tendency for a news agency to write about the same topic for over a week. We mention several problems below and will discuss them in our future work.

- Meaningless descriptions are partly extracted [Abe appreciates Abe] and [Abe likes home]. This is because SVO structures are not precisely extracted. We should improve our method of analyzing a sentence.
- In our feature calculation technique, descriptions which have the same meanings are ranked more highly, such as [Abe responsibility] and [Abe has responsibility]. We should consider the inclusion of relations between the SVO structure patterns.

5 Conclusion

We propose a way of extracting characteristic description of a news agency. We used SVO structure to extract characteristic description of a person or orga-

---

[6] Taro Aso: Japanese minister.
[7] Mizuho Fukushima: Japanese politician.

nization. We used local and global features to measure how much description differs from others. The experimental results show that our method can be used to elucidate the different of each agency's style.

Our future work is to carry out further experiments that include the names of many entities and check how effective our method is. We should study on more effective and visual presentation method. We should let users find out how well our method helps them to ascertain the features of news agencies. In addition, opinion mining and sentiment analysis technologies will be studied in near future.

Acknowledgments

This research is partly supported by the research for the grant of Scientific Research (No.20700084, 20300042) made available by MEXT, Japan. This work is also supported in part by the National Institute of Information and Communications Technology, Japan. This work is in part supported by the Asahi Shimbun, the Mainichi Shimbun, and the Yomiuri Shimbun.

References

1. Souneil, P., Seungwoo, K., Sangyoung, C., Junehwa, S.: NewsCube: Delivering Multiple Aspects of News to Mitigate Media Bias. In: Proc. of the 27th international conference on Human factors in computing systems, pp. 443–452 (2009)
2. Nadamoto, A., Tanaka, K.: A comparative web browser (CWB) for browsing and comparing web pages. In: Proc. of the 12th international conference on World Wide Web, pp. 727–735 (2003)
3. Ma, Q., Yoshikawa, M.: Topic and Viewpoint Extraction for Diversity and Bias Analysis of News Contents. In: APWebWAIM 2009. LNCS, vol. 5446, pp. 152–160. Springer, Heidelberg (2009)
4. Hamasuna, Y., Kawai, Y., Kumamoto, T., Tanaka, K.: Using a Sentiment Map for Visualization of Web Site Distinction (in Japanese). In: Proceedings of Data Engineering Workshop, B6–B4 (2008)
5. Aoki, S., Yumoto, T., Sumiya, K., Nii, M., Takahashi, Y.: Extracting Author Intention based on Peculiarity Parts in Related News Articles (in Japanese), 2008-DBS-146, pp. 187–192 (2008)
6. syncha, http://cl.naist.jp/ryu-i/syncha/

Analyzing Document Retrievability in Patent Retrieval Settings

Shariq Bashir and Andreas Rauber

Institute of Software Technology and Interactive Systems,
Vienna University of Technology, Austria
{bashir,rauber}@ifs.tuwien.ac.at
http://www.ifs.tuwien.ac.at

Abstract. Most information retrieval settings, such as web search, are typically precision-oriented, i.e. they focus on retrieving a small number of highly relevant documents. However, in specific domains, such as patent retrieval or law, recall becomes more relevant than precision: in these cases the goal is to find all relevant documents, requiring algorithms to be tuned more towards recall at the cost of precision. This raises important questions with respect to retrievability and search engine bias: depending on how the similarity between a query and documents is measured, certain documents may be more or less retrievable in certain systems, up to some documents not being retrievable at all within common threshold settings. Biases may be oriented towards popularity of documents (increasing weight of references), towards length of documents, favour the use of rare or common words; rely on structural information such as metadata or headings, etc. Existing accessibility measurement techniques are limited as they measure retrievability with respect to all possible queries. In this paper, we improve accessibility measurement by considering sets of relevant and irrelevant queries for each document. This simulates how recall oriented users create their queries when searching for relevant information. We evaluate retrievability scores using a corpus of patents from US Patent and Trademark Office.

1 Introduction

In several information retrieval applications such as web search, e-commerce, scientific literature, patent applications etc., growing emphasis is put on the measurement of accessibility and retrievability of documents given an underlying information retrieval system [1,2]. In recent years measurement concepts like document *"retrievability"*, *"searchability"* and *"findability"* emerged [1]. These concepts measure, how retrievable each individual document is in the retrieval system, i.e. how likely it is that a document can be found at all given a specific set of queries. Any retrieval system is inherently biased towards certain document characteristics. This results in the risk that a certain number of documents cannot be found in the top-n ranked results via any query terms that they would actually be relevant for, which ultimately decreases the usability of the retrieval system [10]. This is specifically critical in recall oriented application scenarios, such as patent retrieval, or legal settings. In these cases, the focus of a system

S.S. Bhowmick, J. Küng, and R. Wagner (Eds.): DEXA 2009, LNCS 5690, pp. 753–760, 2009.

is not so much on providing the best document to answer a specific information need (as e.g. in Web search settings), but to retrieve all documents that are relevant [9]. Thus, all documents should at least potentially be retrievable via correct query terms.

In recent years, emphasis is put on designing retrieval systems for recall oriented tasks such as patent or legal documents search [4,9]. Before designing a new or using an existing retrieval system for recall oriented applications one needs to analyze the effects of the retrieval system bias as well as the overall retrievability of all documents in the collection using the retrieval function at hand.

In this paper, we take a closer look at document retrievability measurements particularly for patent retrieval applications. Section 2.1 and 2.2 introduce both the standard way of measuring retrievability as well as three novel, more fine-grained measures for assessing retrievability. Section 2.3 explains how queries are constructed, forming the basis for the experiments reported in this paper. Section 3 presents the experiments performed on the dentistry category of the US Patent and Trademark Office database, with conclusions as well as an outlook on future work being provided in Section 4.

2   Measuring Retrievability

2.1   Standard Retrievability Measurement

Given a retrieval system RS and a collection of documents D, the concept of retrievability [1,2] is to measure how much each and every document $d \in D$ is retrievable in top-n rank results of all queries, if RS is presented with a large set of queries $q \in Q$. Defined in this way, the retrievability of a document is essentially a cumulative score that is proportional to the number of times the document can be retrieved within that cut-off c over the set Q. A retrieval system is called best retrievable, if each document $d \in D$ has nearly the same retrievability score. More formally, retrievability $r(d)$ of $d \in D$ can be defined as follows.

$$r(d) = \sum_{q \in Q} f(k_{dq}, c) \tag{1}$$

Here, $f(k_{dq}, c)$ is a generalized utility/cost function, where k_{dq} is the rank of d in the result set of query q, c denotes the maximum rank that a user is willing to proceed down the ranked list. The function $f(k_{dq}, c)$ returns a value of 1 if $k_{dq} \leq c$, and 0 otherwise.

The work of Leif et al. [1] is pioneering in this regard. In their experiments using collections of *news* and *government web* documents, they analyze document retrievability, differentiating between highly retrievable and less retrievable documents.

2.2   Limitations of Standard Retrievability Measure

In this paper we argue that analyzing document retrievability using a single retrievability measure [1,2] has several limitations in terms of interpretability. For example, when using a single retrievability curve we cannot analyze accurately how large a gap exists between an optimal retrievable system and the current system; or what the effect of the query set is that is used for retrievability measurement. Other issues to be analyzed include whether highly retrievable documents are really highly retrievable, or whether they are simply more accessible from many irrelevant queries rather than from relevant queries.

Motivated by these limitations of existing retrievability measurement, the focus of our paper lies in understanding the following aspects: We identify four retrievability measurements rather than using just a single descriptor. These are

- How retrievable is each document using all queries, as done in [1]
- From how many relevant queries out of all queries each document is retrievable
- From how many irrelevant queries out of all queries each document $d \in D$ is retrievable, and
- What is the total number of relevant queries for each document.

The last measure provides an upper bound for by how much we can increase the retrievability score of all documents. The one but last indicates where we can decrease the relevance score of highly retrievable documents and thus potentially increase the relevance score of the other documents. In [1] queries used for retrievability measurements were selected using a sampling approach [3] without considering what type of queries are relevant and irrelevant to individual documents. From our experiments we learned that there is a significant difference in retrievability if queries are selected randomly or considering relevant and irrelevant queries seperately.

2.3   Query Generation Techniques

Clearly, it is impractical to calculate the absolute $r(d)$ scores because the set Q would be extremely large and require a significant amount of computation time as each query would have to be issued against the index for a given retrieval system. So, in order to perform the measurements in a practical way, a subset of all possible queries is commonly used that is sufficiently large and contains relatively probable queries. For generating reproducible and theoretically consistent queries, we try to reflect the way how patent examiners generate queries sets in *patent invalidity search* problems [5,8]. In invalidity search, the examiners have to find out all existing patent specifications that describe the same invention for collecting claims to make a particular patent invalid. In this search process, the examiners extract relevant query terms from a new patent application, particularly from the *Claim* sections for creating query sets [6,7]. We first extract all those frequent terms that are present in the Claim sections of each patent document and have a support greater than a certain threshold. Then, we combine the single frequent terms of each individual patent document into `two` and `three` terms combinations. After creating the query set, individual query terms of Q which appear in the claim section of a patent document $d \in D$ are separated for representing its relevant queries $\hat{Q}$, and all those query terms which do not exist in the claim section of d are used for representing their irrelevant queries.

2.4   Retrievability Measurement Using Relevant Queries

For analyzing the above factors, in this paper we conduct our retrievability measurements considering relevant queries for each document. In our approach, a set of relevant queries $q \in \hat{Q}$ for each document contains those terms or combinations of terms, that are considered most important for an individual document's accessibility. In our measurements, as a first step, we extract all possible relevant queries from the Claim sections of every document. The number of queries in $\hat{Q}$, can be considered as an upper bound for the retrievability score. If any document exhibits a much lower retrievability value than $\hat{Q}$, then it is called less retrievable.

In step two, relevant queries of all documents are used for constructing a single query set Q. In step three, using query sets Q and $\hat{Q}$ retrievability measurements are computed for every document according to Equation 1. Document retrievability in set Q minus document retrievability in set $\hat{Q}$, helps in determining the main cause behind low retrievability. It also identifies the list of queries, where we may be able to decrease the relevance of those documents which are wrongly listed in the top rank results set, for increasing the relevance of less retrievable documents. In short, rather than analyzing document retrievability from a single perspective we analyze retrievability using four factors.

(a) Document retrievability $r(d)$ in query set Q (Equation 1);
(b) Document retrievability $\hat{r}(d)$ in relevant queries $\hat{Q}$ (Equation 2);
(c) Document retrievability in irrelevant queries $\bar{r}(d)$ (Equation 3); and
(d) Total number of relevant queries for each document $\|\hat{Q}\|$.

$$\hat{r}(d) = \sum_{q \in \hat{Q}} f(k_{dq}, c) \tag{2}$$

$$\bar{r}(d) = \sum_{q \in Q} f(k_{dq}, c) - \sum_{q \in \hat{Q}} f(k_{dq}, c) \tag{3}$$

3   Experiments

3.1   Experiment Set-Up

For our experiments we use a collection of patents freely available from the US Patent and Trademark Office, downloaded from (http://www.uspto.gov/). We collected all patents that are listed under United States Patent Classification *(USPC)* class 433 *(Dentistry Domain)*. For query generation we consider only the *Claim* section of every document as this is the section that most professional patent searchers use as their basis for query formulation. However, for retrieval we index the full text of all documents *(Title, Abstract, Claim, Description)*. This reflects the default setting in a standard full-text retrieval engine. Some basic statistical properties of the data collection used are listed in Table1.

Four standard IR models are used for evaluating the retrievability bias. These are **tf-idf**, the OKAPI retrieval function **(BM25)** [11], the OKAPI field retrieval

Table 1. Properties of Patent Collection used for Retrievability Measurements

Total Docu-ments	Unique Terms	Average Doc-ument Length (words)	Average Title Sec-tion Length (words)	Average Abstract Sec-tion Length (words)	Average Claim Sec-tion Length (words)	Average De-scription Sec-tion Length (words)
7213	62343	2888	7.65	35.32	878.5	2234.5

Table 2. Query Collection Approaches

Approach	Total Queries	Average Retrievability Score	Average Relevant Queries/Doc-ument
Query Expansion (2-Terms)	67735	317.6	135
Query Expansion (3-Terms)	337200	248	150

function **(BM25F)** [12], and the *exact match model*. Before indexing, we remove stop words and apply stemming. For indexing and querying we use *Apache LUCENE*[1] IR toolkit. Each measurement graph depicts the four document retrievability indicators (cf. Section 2.4). (1) Document retrievability across all queries, (2) Document retrievability via relevant query set, (3) Document retrievability in irrelevant query set, and (4) Total number of relevant queries for each document in collection, correlating with the length of the respective Claim section.

In query generation approach we select all the single terms which are present in the Claim section of every document that have a term frequency greater than 2 *minimum support threshold*. There are a total of 9,751 single term queries extracted from all documents in the collection, with an average 25 terms per patent document. For creating longer length queries, we expand all the single term queries with *two* and *three* terms combinations, again extracted from the Claim sections. For documents which contain large number of single frequent terms, the different co-occurring term combinations of size `two` and `three` can become very large. Therefore, for generating similar number of queries for every document, we put an upper bound of 200 queries generated for every patent document. On average there are 135 queries per each document in *two* terms combinations, and 150 queries in *three* terms combinations. For generating the complete query set Q we remove all duplicate queries which are present in multiple documents. After generating these query sets for retrievability measurement, these were subdivided into relevant and irrelevant query sets for each document, depending on whether the query terms originated from the respective document. Table 2 shows the main properties of these query sets. For all experiments, the cut-off factor c is set to the top-35 documents in the ranked list, following the experiment set-up in [1].

3.2 Retrievability Results

Figures 1 and 2 show retrievability measurements on different types of query sets for the four different retrieval models. Following the presentation in [1], documents are sorted in ascending order in terms of overall retrievability. From all

---

[1] http://lucene.apache.org/java/docs/

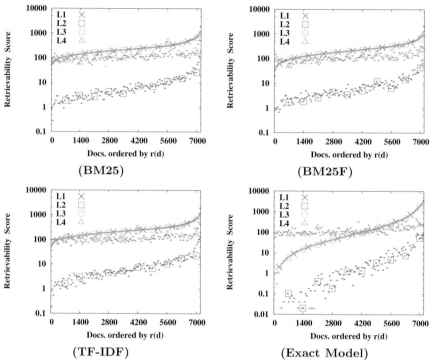

L1: *Retrievability in all Queries*, **L2:** *Retrievability in Relevant Queries*, **L3:** *Retrievability in Irrelevant Queries*, **L4:** *Total Relevant Queries*

Fig. 1. Retrievability Measurement Results using *Two Pairs* Query Expansion Approach

graphs it is clear, that there is a high difference in overall document retrievability scores between less and highly retrievable documents when using all queries (blue line / square symbols). This effect increases as the size of the query set $\|Q\|$ increases, specifically with the two and three terms query expansion approaches.

When using only the set of relevant queries per document (purple line / rhombus symbols), retrievability is almost constant, irrespective of document length or of the overall retrievability of documents across all queries. In most cases an overall high retrievability score is owed to high retrievability of documents via irrelevant queries (light blue line, triangular symbols). This means that most documents are frequently retrieved not because of high matches in the Claim section, as would be desired by patent retrieval experts, but via matches in other sections of the patent - at the cost of missing relevant matches in the Claim section for other documents. (Figures 1 and 2). Due to the bias of the given retrieval models, highly retrievable documents are not really highly accessible on their relevant queries, but on the other side decrease the accessibility of other documents.

The measurements depicted in Figures 1 and 2 show, that there is sufficient space for improving retrievability of less retrievable documents based on the number of potentially relevant queries (green line / circular symbols), which is

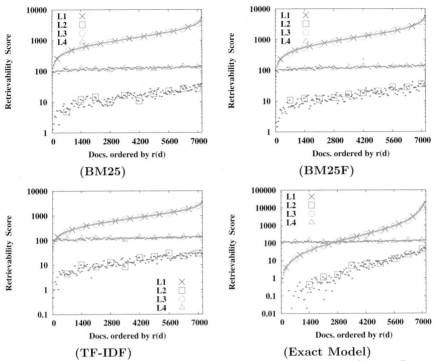

(BM25)

(BM25F)

(TF-IDF)

(Exact Model)

L1: *Retrievability in all Queries,* **L2:** *Retrievability in Relevant Queries,* **L3:** *Retrievability in Irrelevant Queries,* **L4:** *Total Relevant Queries*

Fig. 2. Retrievability Measurement Results using *Three Pairs* Query Expansion Approach

closer to the retrievability values on relevant queries for the highly retrievable documents on the right side of the graphs. When comparing different retrieval models, we see that the exact match model shows the worst performance on all query generation approaches. There are very few documents which are retrieved for almost all of their relevant queries (the optimal case). On the other hand, 22% of the documents cannot be retrieved by any of the queries they would be relevant for, result sets there being dominated by irrelevant documents in terms of query term presence in the Claim section. There is little difference between the *BM25* and *BM25F* (which considers individual sections) retrieval models. On almost all measurements they show comparable performance.

4   Conclusions

We use retrieval systems in order to access information. Therefore, it is important to measure how much different retrieval systems restrict us in accessing different information. Document Retrievability is a measurement, used for this purpose in order to analyze how much a given retrieval system makes individual documents in a collection easier to find ranked within top-n results. Existing document Retrievability *(Findability)* measurement techniques, which measure document accessibility with single factor analysis, are not suitable for understanding the complex aspects involved in documents retrievability. In this paper, we evaluate document retrievability by considering retrieval both for relevant and irrelevant query sets. Rather than taking random queries, we first model how expert users formulate their queries, identifying the Claim section as the relevant source of query terms. Extensive experiments reveal that 90% of documents which are highly retrievable considering all types of queries, are not highly retrievable on their relevant query sets. Furthermore, retrievability is rather constant across all documents when considering only relevant queries, as opposed to the rather large differences encountered when considering all potential queries. The number of relevant queries may also serve as a kind of upper bound of retrievability performance for every document. Further analysis is required to understand the effect of different query selection approaches and query expansion techniques as well as the characteristics hat make documents more or less retrievable under certain systems.

References

1. Azzopardi, L., Vinay, V.: Retrievability: An evaluation measure for higher order information access tasks. In: Proc. of CIKM 2008, Napa Valley, CA, USA, pp. 561–570 (2008)
2. Azzopardi, L., Vinay, V.: Accessibility in Information Retrieval. In: Macdonald, C., Ounis, I., Plachouras, V., Ruthven, I., White, R.W. (eds.) ECIR 2008. LNCS, vol. 4956, pp. 482–489. Springer, Heidelberg (2008)
3. Callen, J., Connell, M.: Query-based sampling of text databases. ACM Transactions on Information Systems 19(2), 97–130 (2001)
4. Fujii, A., Iwayama, M., Kando, N.: Introduction to the special issue on patent processing. Information Processing and Management: an International Journal 43(5), 1149–1153 (2007)
5. Fujita, S.: Technology survey and invalidity search: An comparative study of different tasks for Japanese patent document retrieval. Information Processing and Management: an International Journal 43(5), 1154–1172 (2007)
6. Itoh, H., Mano, H., Ogawa, Y.: Term distillation in patent retrieval. In: Proc. of ACL 2003, Sapporo, Japan, pp. 41–45 (2003)
7. Konishi, K., Kitauchi, A., Takaki, T.: Invalidity patent search system at NTT data. In: NTCIR 2004: Proceedings of NTCIR-4 Workshop Meeting, Tokyo, Japan (2004)
8. Konishi, K.: Query terms extraction from patent document for invalidity search. In: NTCIR 2005: Proceedings of NTCIR-5 Workshop Meeting, Tokyo, Japan (2005)
9. Kontostathis, A., Kulp, S.: The Effect of normalization when recall really matters. In: Proc. of IKE 2008, Las Vegas, Nevada, USA, pp. 96–101 (2008)
10. Baeza-Yates, R.: Applications of web query mining. In: Losada, D.E., Fernández-Luna, J.M. (eds.) ECIR 2005. LNCS, vol. 3408, pp. 7–22. Springer, Heidelberg (2005)
11. Robertson, S., Walker, S.: Some simple effective approximations to the 2-Poisson model for probabilistic weighted retrieval. In: Proc. of SIGIR 1994, Dublin, Ireland, pp. 345–354 (1994)
12. Robertson, S., Zaragoza, H., Taylor, M.: Simple BM25 extension to multiple weighted fields. In: Proc. of CIKM 2004, Washington, D. C., USA, pp. 42–49 (2004)

Classifying Web Pages by Using Knowledge Bases for Entity Retrieval

Yusuke Kiritani, Qiang Ma, and Masatoshi Yoshikawa

Department of Social Informatics, Graduate School of Informatices, Kyoto University
Yoshidahonmachi, Sakyo-ku, Kyoto 606–8501, Japan
y.kiritani@db.soc.i.kyoto-u.ac.jp,
{qiang,yoshikawa}@i.kyoto-u.ac.jp

Abstract. In this paper, we propose a novel method to classify Web pages by using knowledge bases for entity search, which is a kind of typical Web search for information related to a person, location or organization. First, we map a Web page to entities according to the similarities between the page and the entities. Various methods for computing such similarity are applied. For example, we can compute the similarity between a given page and a Wikipedia article describing a certain entity. The frequency of an entity appearing in the page is another factor used in computing the similarity. Second, we construct a directed acyclic graph, named PEC graph, based on the relations among Web pages, entities, and categories, by referring to YAGO, a knowledge base built on Wikipedia and WordNet. Finally, by analyzing the PEC graph, we classify Web pages into categories. The results of some preliminary experiments validate the methods proposed in this paper.

1 Introduction

Currently, there are many Web pages related to people, organizations and locations. Many users search for information related to these people, organizations, and locations by using search engines. Some statistics suggest that such searches account for as much as 5-10% of all search queries [1]. In other words, many Web pages describe certain entities, and many people search for these entities.

Conventional search engines return a list of Web pages ranked according to their estimated likelihood of relevance to the query, or according to an importance score computed by using the link structures in the search results. Users then have to judge whether a given page describes the desired entities by checking the page's URL, title, and snippet. It is still difficult, however, to find a suitable page among a large number of search results. Although users can try to modify their queries to improve the relevance of search results, it is not easy yet for them to specify their precise intentions by query modification to find relevant information. As one of many possible solutions, classification of Web pages has been widely studied.

Web directories(e.g., Open Directory Project[1]) are used for Web page categorization. A Web directory is a hand-crafted index that hierarchically categorizes

---

[1] Open Directory Project: http://www.dmoz.org/

S.S. Bhowmick, J. Küng, and R. Wagner (Eds.): DEXA 2009, LNCS 5690, pp. 761–768, 2009.

hierarchically Web pages in advance. The directory categorization is correct but has problems such as high maintenance cost and low coverage of Web pages. In addition, because these categorizations are not based on entities, they are still not effective in helping searches for people and organizations.

Recently, Wikipedia, a free encyclopedia on the Internet, is attracting attention because it has much information about entities and those categories. Knowledge bases (e.g., YAGO[2]) are built on this information and WordNet.

In this paper, we propose an approach to classify Web pages by using knowledge bases to support entity search. Concretely, we apply the following procedure (see also Figure 1).

Step 1. Web pages are mapped to entities according to a correspondence degree expressing how strongly a page relates to an entity. We can use methods based on the following four factors in order to compute the correspondence degree. (a) the similarity between the body text of the page and a Wikipedia article about the entity, (b) the similarity between the title and snippet of the page and the Wikipedia article about the entity, (c) the frequency of the entity in the body text of the page; and (d) the frequency of the entity in the title and snippet of the page.

Step 2. We construct a graph (called a PEC graph) based on the pages, entities, classes, and their relations, as obtained from the knowledge base YAGO.

Step 3. By analyzing the graph, the pages are classified into classes according to a correspondence degree expressing how strongly a page corresponds to a class.

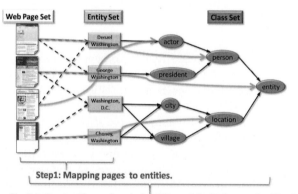

Fig. 1. Processing flow

The remainder of this paper is structured as follows. Related work is introduced in Section 2. The methods for mapping Web pages to entities and classifying pages into classes are introduced in Sections 3 and 4, respectively. Our preliminary experiment results are given in Section 5, and we conclude this paper in Section 6.

2   Related Work

Among studies of automated Web page classification, some have proposed, in addition to general document classification methods, methods using the distinct properties of Web documents, such as HTML structures and the anchor texts of hyperlinks[3]. In [4], Chakrabarti et al. proposed an approach using the link structures of Web pages, which increased the precision of classification by considering the contexts of the linked documents. Approaches using ontologies have also been proposed[5][6].

In constract, we propose an approach focusing on entities appearing in Web pages in order to support entity retrieval. Our approach applies classification focused on searching for entities by using the similarity for a Wikipedia page describing a certain entity or by using the frequency of the entity.

As for other approaches, the NAGA[7] knowledge base, which is a semantic search engine using YAGO, performs a ranked search of entities by using graph-based queries.

Shirakawa et al. proposed concept vectorization methods [8]. A vector value between two nodes on the a graph, consisting of the category network in Wikipedia, is defined according to the number of paths and the length of each path. In our study, on the other hand, the value between two nodes on a graph is constructed according to the relations among Web pages, entities, and categories.

3   Mapping of Web Pages and Entities

3.1   PE Correspondence Degree

Suppose that we have a Web page set P and an entity set E. The PE (page-entity) correspondence degree, $PE(p, e)$ indicates that how much a Web page $p(\in P)$ corresponds to an entity $e(\in E)$. Web pages are mapped to entities according to the degree. If p is similar to a text describing e or e frequently appears in p, p is a document describing e. Therefore, $PE(p, e)$ is computed as the similarity between the body text of p and a Wikipedia article describing e, or as the frequency with which e appears in the body text of p. In addition, to reduce time complexity, in each of two computation methods, we also approach methods using the title and snippet of p instead of the body text of p.

Then, the PE correspondence degree $PE(p, e)$ between a Web page $p(\in P)$ and an entity $e(\in E)$ is computed by one of the following four methods. Here, we let $sim(a, b)$ be the cosine similarity between the tf-idf vectors of a and b; let $wikip(e)$ be the text of a Wikipedia article describing e; let $text(p)$ be the body text, and $summary(p)$ the title and snippet of p; let $ef(p, e)$ be the number of occurrences e appears in p; and let $idf(e, P)$ be the idf(inverse document frequency) of e for document set P.

S-TW Method: $PE(p, e) = sim(text(p), wikip(e))$; $PE(p, e)$ is computed as the cosine similarity between the tf-idf vector of the body text of p and that of the text of a Wikipedia article describing e.

S-SW Method: $PE(p,e) = sim(summary(p), wikip(e))$; $PE(p,e)$ is computed as the cosine similarity between the tf-idf vector of the title and snippet of p and the text of a Wikipedia article describing e.

F-TE Method: $PE(p,e) = ef(text(p),e) \times idf(e,P)$; $PE(p,e)$ is computed by multiplying the frequency of e in the body text of p by the idf of e for document set P.

F-SE Method: $PE(p,e) = ef(summary(p),e) \times idf(e,P)$; $PE(p,e)$ is computed by multiplying the frequency of e in the title and snippet of p by the idf of e for document set P.

The mapping results vary greatly according to whether the similarity or the frequency is used. Using the similarity (S-TW, S-SW), the depth and accuracy of content in Wikipedia influence the precision of this mapping. Using the frequency (F-TE, F-SE), the entity term frequency influences the precision of the mapping. Also, using the title and snippet of a page, rather than the body text, requires less processing time; the precision of the mapping, however, will also decrease. In Section 5.1, we describe an experiment comparing these methods.

3.2   Mapping of Web Pages and Entities

Using the PE correspondence degree and two parameters, $\alpha \geq 1$ and $K \geq 1$, we map Web page to entities as follows: (1) The PE correspondence degrees of p and each entity in E are listed in descending order as $PE_1, PE_2, \cdots, PE_m$. (2) For each k-th degree and $(k+1)$-th degree, the minimum integer k satisfying $PE_k/PE_{k+1} \geq \alpha$, is chosen. (3) If $k > K$, then $k = K$ is set. (4) p is mapped to entities whose PE correspondence degree is ranked in the top k degrees.

We determine the number of mapped entities by adjusting α and K: α is the threshold value for reaching get a point where degrees vary greatly, while K is the maximum value limiting this number.

4   Classification of Web Pages

4.1   PEC Graph

To classify Web pages into classes, a PEC graph is constructed, based on the relations among pages, entities, and classes. The PEC graph G_{PEC} is constructed from three sub-graphs: G_{PE}, G_{EC}, and G_C. These are constructed as follows.

G_{PE}: Given that a Web page set P, an entity set E, and a correspondence relation between a page and an entity, $R_{PE} \subseteq P \times E$, a bipartite graph $G_{PE} = (P, E, R_{PE})$ is constructed. The left part in Figure 2 shows an example of this graph. Each of four Web pages is mapped to one or more entities.

G_{EC}: Given an entity set E, we obtain a set of classes C_0, and a type relation between an entity and a class, $R_{EC} \subseteq E \times C_0$, from knowledge base YAGO. Then, a bipartite graph $G_{EC} = (E, C, R_{EC})$ is constructed from the sets and the type relation. The middle part in Figure 2 shows an example of this graph.

The four entities are those in G_{PE}. The four classes, which are the types of each entity, are obtained from knowledge bases.

G_C: Given an entity set E, we obtain a class set $C(\supseteq C_0)$ and a subclass relation between two classes, $R_{CC} \subseteq C \times C$, from knowledge base information indicating that one class is a child class of another. Then, a tree $G_C = (C, R_{CC})$ is constructed from the sets and the subclass relation. The right part in Figure 2 shows an example of this tree. The four classes that are the leaves of the tree are those in G_{EC}. The other three classes are obtained from knowledge bases.

G_{PEC}: A graph $G = (P, E, C, R_{PE}, R_{EC}, R_{CC})$, called a PEC graph, is constructed by integrating the three graphs G_{PE}, G_{EC}, and G_C. Figure 2 shows an example of a PEC graph constructed from the graphs shown above.

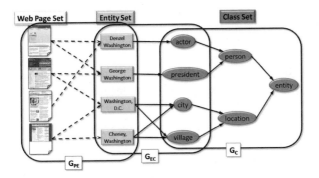

Fig. 2. Example of a PEC graph

Note that in knowledge base YAGO, some classes have relations to many other entities and classes, but these are invisible in the PEC graph. For example, as pictured in Figure 3, the class *person* has more than two relations that are not pictured in Figure 2. Let a sub-node of node n be a node with a relation to n, and let a super-node of n be a node with a relation from n.

4.2   Classification by PC Correspondence Degree

We compute the PC correspondence degree $PC(p, c)$ by analyzing a PEC graph constructed as described in the previous subsection. Here, $PC(p, c)$ indicates that how much a Web page p corresponds to a class c. We classify Web pages into classes according to the PC correspondence degree.

The PC correspondence degree is computed from three factors as follows: (1) The sum of all PC (PE) correspondence degrees between p and a class (entity) that is a sub-node of c. The higher this sum, the greater the PC correspondence degree; (2) The ratio r_c of a class (entity) related to p to the sub-nodes of c. We decided that the more p related to c, the greater the PC correspondence degree. This ratio expresses the coverage of p with respect to c. The more p covers c, the higher the ratio; (3) The ratio r_p of pages not related to c to P. We decided that the more p is identified by c, the greater the PC correspondence degree. This ratio expresses the distinguishability of p from other pages in terms of c. The fewer pages c relates to, the higher the ratio.

Concretely, the PC correspondence degree $PC(p,c)$ for $p \in P$ is computed according to the following equations. Let $subNode(c)$ be a node set whose elements are sub-nodes of c; let $subNode*(c)$ be a node set whose elements are either sub-nodes of c or sub-nodes of a node in $subNode*(c)$; let $superNode*(p)$ be a node set whose elements are either super-nodes of p or super-nodes of a node in $superNode*(p)$; and let $|S|$ be the size of set S. Then, we have

$$PC(p,c) = \left(\sum_{x \in subNode(c)} PX(p,x)\right) \times r_c \times r_p$$

where $PX(p,x)$ is $PE(p,x)$ if x is an entity and $PC(p,x)$ if x is a class, $r_c = |subNode(c) \cap superNode*(p)|/\log|subNode(c)|$, which is the ratio obtained by dividing the intersection of the sub-node set of c and the super-node* set of p (i.e., the number of classes related to p) by the logarithm of the size of the sub-node set of c, and $r_p = |P - subNode*(c)|/\log|P|$, which is the ratio obtained by dividing the size of the difference between P and the sub-node* set of c (i.e., the number of pages not related to c) by the logarithm of the size of P.

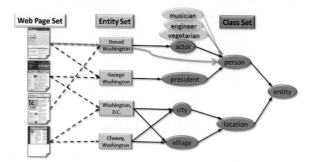

Fig. 3. Knowledge base with nodes and edges not visible in a PEC graph

For example, $PC(p_1, person)$ in Figure 3, which is the PC correspondence degree between page p_1 and class $person$, is computed as follows.

$PC(p_1, person) = \{PC(p_1, actor) + PC(p_1, president)\} \times (2/\log 5) \times (1/\log 4)$. Here, $r_c = 2/\log 5$ because $person$ has 5 sub-nodes and p_1 is related to 2 nodes ($actor$ and $president$) of these sub-nodes. $r_p = 1/\log 4$ because $person$ is not related to one page (p_1) among Web page set.

5 Experimental Results

5.1 Experiment of Mapping Web Pages and Entities

We acquired a Web page set consisting of the top 50 search results for the search word "washington" by using Yahoo! as the search engine. We also acquired

Table 1. Experimental results

	S-TW	S-SW	F-TE	F-SE
Mapping Precision	0.433	0.400	0.451	0.341
Classification Precision	0.80	0.55	0.74	0.48

an entity set consists of entities containing "washington" by using YAGO as the knowledge base. Relevant results E_{Ans} were created for each Web page by human judgment. Mapped entity sets E_{S-TW}, E_{S-SW}, E_{F-TE}, and E_{F-SE} were created for each Web page by applying each of the proposed methods. Based on some preliminary experimental results, we set the parameters α, k to 2.0 and 5, respectively. We then evaluated the average mapping precision for the 50 pages with each method. The mapping precision MP was computed as

$$MP = |E_{Ans} \cap (\text{the top } |E_{Ans}| \text{ entities of } E_{method})|/|E_{Ans}|,$$

where $|S|$ is the size of set S, and E_{method} is any one of E_{S-TW}, E_{S-SW}, E_{F-TE}, and E_{F-SE}. Table 1 lists the average mapping precisions with each of the four kinds of mapping methods.

It takes time to acquire a Wikipedia article if S-TW or S-SW is used. On the other hand, it takes time to acquire the body text of a Web page if S-TW or F-TE is used. Therefore, the order of processing time length for the mapping methods was S-TW>S-SW≃F-TE>F-SE. When this time was included, F-TE seemed better than S-SW.

S-TW and S-SW could correctly map Web pages to entities, because the similarity increased with the co-occurrence of specific terms (e.g., a person's name). On the other hand, these methods often mapped incorrectly to an entity having nothing to do with the mapped page or to an entity with little relation to the mapped page. One factor in this problem could be Wikipedia articles with a lack of description or an excess of description. These incorrect mappings occurred notably in more with S-TW as compared with S-SW.

5.2   Experiment on Classification

We also performed an experiment on classification, under the same conditions used for the above experiment on mapping. For each mapping result obtained by each of the four kinds of mapping methods, we created a set of answer classes by human judgment. We ranked the class sets by applying the proposed classification method and thus evaluated the average classification precision for 50 pages with each method. The classification precision CP was computed by the binary function: $CP = 1$ if classification is true and $CP = 0$ if classification is false. Table 1 lists the average classification precisions.

The precisions obtained by using S-TW and F-TE were better than those obtained with S-SW and F-SE. This implies that the precision was increased more by using body text of a Web page than by using the title and snippet, as expected. Also, for every mapping method, the classification precision was higher than the mapping precision. That is, many pages was mapped incorrectly to an entity but classified correctly.

It is necessary to use the different method to compute the PE correspondence degree according to the applications. For example, to classify search results online, F-TE method is better, while to classify Web pages in a Web directory based application, S-TW method is better.

6   Conclusion

In this paper, we have proposed a method to classify Web pages by using a knowledge base. To classify a page, we map it to an entity by using one of four methods to calculate the correspondence degree between the page and various entities. We then construct a graph based on the mapping and a knowledge base, and calculate the correspondence degree between the page and a class. We have also carried out some preliminary experiments to validate these methods. The results showed that the method computing entity frequency in the body text was the best of the methods, and it often happened that classification was correct even when mapping was incorrect. It is necessary to use the different method to compute the correspondence degree according to the applications.

In our future work, further experiments and evaluations are necessary, and we plan to improve the proposed methods. Also, we develop an interface to show the classification of search results.

Acknowledgments. This research is partly supported by the research for the grant of Scientific Research (No.20700084 and 20300036) made available by MEXT, Japan.

References

1. Guha, R., Garg, A.: Disambiguating People in Search. In: Proc. of WWW 2004 (2004)
2. Suchanek, F.M., Kasneci, G., Weikum, G.: Yago: A Core of Semantic Knowledge. In: Proc. of WWW 2007, pp. 697–706 (2007)
3. Sun, A., Lim, E.-P., Ng, W.-K.: Web Classification Using Support Vector Machine. In: WIDM 2002, pp. 96–99 (2002)
4. Chakrabarti, S., Dom, B.E., Indyk, P.: Enhanced hypertext categorization using hyperlinks. In: Proc. of the ACM SIGMOD 1998, pp. 307–318 (1998)
5. Prabowo, R., Jackson, M., Burden, P., Knoell, H.-D.: Ontology-Based Automatic Classification for the Web Pages: Design. In: Implementation and Evaluation Proc. of WISE 2002, pp. 182–191 (2002)
6. Song, M.-H., Lim, S.-Y., Kang, D.-J., Lee, S.-J.: Automatic Classification of Web Pages based on the Concept of Domain Ontology. In: Proc. of APSEC 2005, pp. 645–651 (2005)
7. Kasneci, G., Suchanek, F.M., Ifrim, G., Ramanath, M., Weikum, G.: NAGA: Searching and Ranking Knowledge. In: Proc. of ICDE 2008, pp. 1285–1288 (2008)
8. Shirakawa, M., Nakayama, K.: Concept Vector Extraction from Wikipedia Category Network. In: Proc. of ICUIMC 2009, pp. 71–79 (2009)

Terminology Extraction from Log Files

Hassan Saneifar[1,2], Stéphane Bonniol[2], Anne Laurent[1],
Pascal Poncelet[1], and Mathieu Roche[1]

[1] LIRMM - Université Montpellier 2 – CNRS
161 rue Ada, 34392 Montpellier Cedex 5, France
{saneifar,laurent,poncelet,mroche}@lirmm.fr
http://www.lirmm.fr/∼{saneifar,laurent,poncelet,mroche}
[2] Satin IP Technologies,
Cap Omega, RP Benjamin Franklin, 34960 Montpellier Cedex 2, France
stephane.bonniol@satin-ip.com
http://www.satin-ip.com/

Abstract. The log files generated by digital systems can be used in management information systems as the source of important information on the condition of systems. However, log files are not exhaustively exploited in order to extract information. The classical methods of information extraction such as terminology extraction methods are irrelevant to this context because of the specific characteristics of log files like their heterogeneous structure, the special vocabulary and the fact that they do not respect a natural language grammar. In this paper, we introduce our approach EXTERLOG to extract the terminology from log files. We detail how it deals with the particularity of such textual data.

1 Introduction

In many applications, automatic generated reports, known as system logs, represent the major source of information on the status of systems, products, or even causes of problems that can occur. In some areas, such as Integrated Circuit (IC) design systems, the log files generated by IC design tools, contain essential information on the conditions of production and the final products. In order to extract information from textual data, there exists the classic method of Natural Language Processing (NLP) and Information Extraction (IE) techniques. But the particularity of such textual data (*i.e.* log files) raise the new challenges. In this paper, we aim particularly at exploring the lexical structure of these log files in order to extract the terms of domain which will be used in creation of domain ontology. We thus study the relevance of two main methods of terminology extraction within our approach EXTERLOG (EXtraction of TERminology from LOGs), both of which extract co-occurrences with and without the use of syntactic patterns.

In Sect. 2 we present the characteristics and difficulties of this context. Our approach EXTERLOG is developed in Sect. 3. Section 4 describes and compares the various experiments that we performed to extract terms from the logs. Finally, we propose a comparison of EXTERLOG and TERMEXTRACTOR system.

S.S. Bhowmick, J. Küng, and R. Wagner (Eds.): DEXA 2009, LNCS 5690, pp. 769–776, 2009.

2 Context

In the domain of log file analysis, some logs like network monitoring logs or web usage logs are widely exploited [1][2]. These kinds of logs are based on the management of events. That is, the computing systems record the system events based on their occurring times. The contents of these logs comply with norms according to the nature of events and their global usage (*e.g.,* web usage area). However, in some areas such as IC design systems, rather than being some recorded events, the generated log files are digital reports on configuration, conditions and states of systems. The aim of the exploitation of these log files is not to analyze the events but to extract information about system configuration and especially about the final product's conditions. Hence, information extraction in log files generated by IC design tools has an attractive interest for automatic management and monitoring of production line. However, several aspects of these log files have been less emphasized in existing methods of information extraction and NLP. These specific characteristics pose several challenges that require more research.

The design of IC consists of several levels each corresponds to some design rules. At every level, several tools can be used. Despite the fact that the logs of the same design level report the same information, their structures can significantly differ depending on the design tool used. Specifically, each design tool often uses its own vocabulary to report the same information. For example, at the so-called verification level, two log files (*e.g.,* log "A" and log "B") are produced by two different tools. The information about, for example, the "Statement coverage" will be expressed as follows in the log "A":

```
TOTAL    COVERED PERCENT
statements    20      21    22
```

But the same information in the log "B", will be disclosed from this single line:

```
EC: 2.1%
```

As shown above, the same information in two log files produced by two different tools is represented by different structures and vocabulary. Moreover, the evolution of design tools changes the format of data in logs. The heterogeneity of data exists not only between the log files produced by different tools, but also within a given log file. For example, the symbols used to present an object, such as the header for tables, change in a given log. Similarly, there are several formats for punctuation, the separation lines and representation of missing data. To best generalize the extraction methods, we thus need to identify the terms used by each tool in order to create the domain ontology. This ontology allows us to better identify equivalent terms in the logs generated by different tools. The domain ontology can help to reduce the heterogeneity of terms existing in logs produced by different design tools. For instance, to check "Absence of Attributes" as a query on the logs, one must search for the following different sentences in the logs, depending on the version and type of design tool used:

```
"Do not use map_to_module attribute",
"Do not use one_cold or one_hot attributes",
"Do not use enum_encoding attribute",
```

Instead of using several patterns, each one adapted to a specific sentence, by associating the words "map_to_module attribute", "one_hot attributes" and "enum_ encoding attribute" to the concept "Absence of Attributes", we use a general pattern that expands automatically depending on the type of log. The ontology-driven expansion of query is studied in many work [3].

Moreover, the language used in these logs is a difficulty that affects the methods of information extraction. Although the language of log files are similar to English, the contents of these logs do not usually comply with *"classic"* grammar. Moreover, there exist words that are often constituted from alphanumeric and special characters.

Since the concepts used in domain ontology are the terms of log files, we aim at extracting the terminology of the log files. However, due to the particularity of log files described above, the methods of NLP, including the terminology extraction, developed for texts written in natural language, are not necessarily well suited to the log files. In this paper, we thus study these methods and their relevance in this specific context. Finally, we propose our approach EXTERLOG for extracting terminology from these log files.

3   Terminology Extraction from Log Files

The extraction of co-occurring words is an important step in identifying the terms. We explain at first some of approaches used to identify the co-occurrences and to extract the terminology of a corpus. Then, we introduce our approach of terminology extraction adapted to log files.

3.1   Related Work

Some approaches are based on syntactic techniques which rely initially on the grammatical tagging of words. The terminological candidates are then extracted using syntactic patterns (*e.g.*, adjective-noun, noun-noun). We develop the grammatical tagging of log files using our approach EXTERLOG in Sect. 3.2. Bigrams[1] are used in [4] as features to improve the performance of the text classification. Though, the series of three words (*i.e.* trigrams) or more is not always essential [5]. EXIT, introduced by [6] is an iterative approach that finds the terms in an incremental way. XTRACT is a terminology extraction system, which identifies lexical relations in the large corpus of English texts [7]. TERMEXTRACTOR, submitted by [8], extracts terminology consensually referred in a specific application domain. To select the relevant terms of domain, some measures based on entropy are used in TERMEXTRACTOR. The statistical methods are generally used for evaluating the adequacy of extracted terms [9]. In these methods, the

---

[1] N-grams are defined as the series of any "n" words.

occurrence frequency of candidates is a basic element. However, since the repetition of words is rare in log files, these statistical methods are not well suited. Indeed, statistical approaches can cope with high frequency terms but tend to miss low frequency ones [10].

Most of these studies are experimented on textual data which are classical texts written in natural language. Most of the experimented corpus are structured in a consistent way. In particular, they comply with the grammar of NL. However, the characteristics of logs such as their non compliance with NL grammar, their heterogeneous, and evolving structures (cf. Sect. 2) impose an adaptation of these methods to log files.

3.2   EXTERLOG

Our approach, EXTERLOG, is developed to extract the terminology in the log files. This process involves normalisation of log files, grammatical tagging of words and co-occurrences extraction.

Normalization. Given the specificity of our data, the normalization method, adapted to the logs, makes the vocabulary and structure of logs more consistent. We replace the punctuations, separation lines and the headers of the tables by special characters to limit ambiguity. Then, we tokenize the texts of logs, considering that certain words or structures do not have to be tokenized. For example, the technical word "Circuit4-LED3" is a single word which should not be tokenized into two words "Circuit4" and "LED3". Besides, we make the normalization method to distinguish the lines representing the header of tables from the lines which separate the parts. This normalization makes the structure of logs produced by different tools more homogeneous.

Grammatical Tagging. Grammatical tagging (also called *part-of-speech tagging*) is a method of NLP used to annotate words based on their grammatical roles. In our context, due to the particularity of log files described in Sect. 2, there are some difficulties and limitations for applying a grammatical tagging. Indeed, the classic techniques of POS tagging are developed and trained according to the standard grammar of a natural language. To identify the role of words in the log files, we use BRILL rule-based part-of-speech tagging method [11]. As existing taggers like BRILL are trained on general language corpora, they give inconsistent results on the specialized texts. [12] propose a semi-automatic approach for tagging corpora of specialty. They build a new tagger which corrects the base of rules obtained by BRILL tagger and adapt it to a corpus of specialty. In the context of log files, we also adapted BRILL tagger to our context by introducing the new *contextual* and *lexical* rules. Indeed, the classic rules of BRILL, which are defined according to the NL grammar, are not relevant to log files. For example, a word beginning with a number is considered a *"cardinal"* by BRILL. However, in the log files, there are many words like 12.1vSo10 that must not be labeled as *"cardinal"*. Therefore, we defined the special *lexical* and *contextual* rules in BRILL. Since the structures of log files can contribute important information for extracting the relevant patterns in future work, we preserve the structure of files during grammatical tagging. We introduce the new tags, called *"Document Structure Tags"*, which present the different structures in log files. For example, the tag "\TH" represents the header of tables or "\SPL" represents the lines separating the log parts. The special structures in log files are identified and normalized during preprocessing. Then, they are annotated during tagging according to the new specific contextual rules defined in BRILL. We use these tagged logs in next level to extract the co-occurrences.

Extraction of Co-occurrences. We are looking for co-occurrences in the log files with two different approaches: (1) using defined *part-of-speech* syntactic patterns, (2) without using the syntactic patterns.

We call the co-occurrences extracted by the first solution "POS-candidates"[2]. This approach consists of filtering words by the syntactic patterns. The syntactic patterns determine the adjacent words with the defined grammatical roles. The syntactic patterns are used in [9] to extract terminology. For complex terms identification, [9] defines syntactic structures which are potentially lexicalisable. As argued in [9], the base structures of syntactic patterns are not frozen structures and accept variations. According to the terms found in our context, the syntactic patterns "\JJ - \NN" (Adjective-Noun) and "\NN - \NN" (Noun-Noun) are used to extract the "POS-candidates" from log files.

The co-occurrences extracted by the second approach are called "bigrams". A bigram is extracted as a series of any two adjacent relevant words[3]. Bigrams are used in NLP approaches as representative features of a text [4]. However, the extraction of bigrams does not depend on the grammatical role of words. To extract significant bigrams, we normalize and tokenize the logs to reduce the rate of noise. We also eliminate the stop words existing in the logs. In this method, we thus do not filter the words according to their grammatical roles.

4   Experiments

We experimented two different approaches for the extraction of terminology from these logs: (1) extraction of *POS-candidates* and (2) extraction of *bigrams*. Here, we analyze the terminological candidates obtained by each one. The log corpus is composed of the logs of all IC design levels and its size is about 950 KB.

4.1   POS-Candidates vs. Bigrams

To analyze the performance of the two approaches chosen for the extraction of bigrams, we must evaluate the terms extracted. To automatically evaluate the relevance of the extracted terms, we compare the POS-candidates and bigrams with terms extracted from the reference documents. Indeed, for each level of design of integrated circuits, we use certain documents, which explain the principles

---

[2] POS: Part-Of-Speech.

[3] The relevant words, in our context, are all words of the vocabulary of this domain excluding the stop words like "have" or "the".

and the details of design tools. We use these documents as "reference experts" in the context of an automatic validation. Indeed, if a term extracted from logs is used in the reference documents, it is a valid term of domain. However, there are several terms in the logs especially the technical terms that are not used in the references. Therefore, a validation by an expert, carried out in our future work is needed to complete the automatic validation. We note that, to extract the domain terminology, we have to use log files and not the reference documents because, as described above, there are some terms that do not appear in reference documents according to their nature. Hence, we could use the references as a validation tool but not as the base of domain terminology.

Moreover, in order to select the most relevant and meaningful terms, we filter the extracted terminological candidates based on their frequency of occurrences in the logs. Therefore, we choose terminological candidates having a frequency of at least 2 (*i.e.* pruning task). We calculate the precision and recall of extracted candidates as shown below:

$$\text{Precision} = \frac{|Candidates \cap Terms\,of\,ref|}{|Candidates|} \qquad \text{Recall} = \frac{|Candidates \cap Terms\,of\,ref|}{|Terms\,of\,ref|}$$

Table 1 shows the precision and recall of POS-candidates and bigrams before and after pruning. To evaluate the terms extracted from logs, the precision is the most adapted measure to our context. Indeed, this measure gives a general tendency of the quality of terms extracted by our system. Note that to calculate a perfectly adapted precision, we should manually evaluate all the terms proposed by EXTERLOG. However, this task is difficult and costly to implement. The comparison of terminological candidates with the reference terms shows that the terminology extraction based on syntactic patterns is quite relevant to the context of log files. The precision of POS-candidates is indeed higher than the precision of bigrams. Our experiments show that an effort in normalization and tagging tasks is quite useful in order to extract quality terms. We note that the pruning of terms does not significantly improve results. As we have already explained, in our context, terms are not generally repeated in logs. Therefore, a *representative term* does *not* necessarily have a *high frequency*.

The low recall of terminological candidates is due to the large number of reference terms. The reference corpus is about five times larger than the logs corpus. In addition, we found that many extracted terminological candidates that

Table 1. Precision and recall of terminological candidates before and after pruning

		Level 1		Level 2		Level 3		Level 4		Level 5	
		POS	Bigrams	POS	Bigrams	POS	Bigrams	POS	Bigrams	POS	Bigrams
Before	Precision	67.7	11.3	20.7	6.5	37.8	9.9	40.1	6.5	19.6	5.1
	Recall	0.7	0.4	7.6	7.5	1.3	1.0	9.5	8.8	0.3	0.5
After	Precision	81.1	10.1	18.0	5.0	37.2	5.9	27.3	7.1	37.1	5.5
	Recall	0.1	0.1	3.0	2.0	0.1	0.4	1.6	2.2	0.2	0.1

have not been validated by reference terms are technical words or abbreviations, which are only found in the logs and not in the reference documents of domain. That is why the recall results are not entirely representative for evaluating the quality of EXTERLOG.

4.2 Validation by Experts

In order to validate the "automatic validation protocol" that we experimented using the reference documents, we asked two domain experts to evaluate the validated terms by our protocol. We calculate the percentage of terms extracted by EXTERLOG and validated using reference documents which are also annotated as relevant by experts. The results show that 84% to 98.1% of the terms validated by our protocol are really relevant terms according to experts[4].

4.3 Exterlog vs. TermExtractor

Here, we compare the results of our approach EXTERLOG with those obtained by TERMEXTRACTOR on the same corpus of logs. We chose TERMEXTRACTOR because it is well configurable and is evaluated by many users in many domains [8]. To adapt TERMEXTRACTOR to this context, we configured it according to characteristics of log files and especially the type of terms found in this context. Table 2 shows the results obtained by TERMEXTRACTOR compared with those obtained by EXTERLOG (*using syntactic patterns*). By analyzing the terms extracted by TERMEXTRACTOR, we find that the structure of logs has influenced the extraction of terms. That is, some terms extracted by TERMEXTRACTOR must not be considered as a term because of the position of words (*used in the term*) in text of logs. Furthermore, the technical terms of domain, normally constituted of special or alphanumeric characters, like "`ks_comp engine`" or "`rule b9`" are rarely found by TERMEXTRACTOR. According to Table 2, our approach EXTERLOG extracts more relevant terms than TERMEXTRACTOR. That is due to the special normalisation of logs and particularly due to the special contextual and lexical rules that we have defined using BRILL tagger.

Table 2. Precision and recall of terms extracted by EXTERLOG (EXT) and by TERMEXTRACTOR (TER)

	Level 1		Level 2		Level 3		Level 4		Level 5	
	EXT	TER	EXT	TER	EXT	TER	EXT	TER	EXT	TER
Precision	67.7	56.1	20.7	14.0	37.8	38.1	40.1	35.2	19.6	26.3
Recall	0.7	0.3	7.6	0.3	1.3	0.4	9.5	2.5	0.3	0.1

---

[4] This interval is due to some terms which are annotated as no idea by experts. If we consider the no idea terms as irrelevant, 84% of terms validated by our protocol are really relevant according to experts. If these terms are not taken into account in the calculation, we obtain 98.1% of terms really relevant.

5 Conclusion and Future Work

In this paper, we described a particular type of textual data: reporting log files. These textual data do not comply with the grammar of natural language, are highly heterogeneous and have evolving structures. To extract domain terminology from the log files, we extracted the co-occurrences with two different approaches: (1) using the syntactic patterns and (2) without syntactic patterns. The results show that terms obtained using the syntactic patterns are more relevant than those obtained without using syntactic patterns. Our experiments show that our approach extracts more relevant terms than other terminology extraction methods like TERMEXTRACTOR. Our future work will especially focus on the study of the more advanced protocols of automatic term evaluation.

References

1. Yamanishi, K., Maruyama, Y.: Dynamic syslog mining for network failure monitoring. In: KDD 2005, pp. 499–508. ACM, New York (2005)
2. Facca, F.M., Lanzi, P.L.: Mining interesting knowledge from weblogs: a survey. Data Knowl. Eng. 53(3), 225–241 (2005)
3. Dey, L., Singh, S., Rai, R., Gupta, S.: Ontology aided query expansion for retrieving relevant texts. In: Szczepaniak, P.S., Kacprzyk, J., Niewiadomski, A. (eds.) AWIC 2005. LNCS, vol. 3528, pp. 126–132. Springer, Heidelberg (2005)
4. Tan, C.M., Wang, Y.F., Lee, C.D.: The use of bigrams to enhance text categorization. Inf. Process. Manage. 38(4), 529–546 (2002)
5. Grobelnik, M.: Word sequences as features in text-learning. In: Proceedings of the 17th Electrotechnical and Computer Science Conference (ERK 1998), pp. 145–148 (1998)
6. Roche, M., Heitz, T., Matte-Tailliez, O., Kodratoff, Y.: Exit: Un système itératif pour l'extraction de la terminologie du domaine à partir de corpus spécialisés. In: Proceedings of JADT 2004, vol. 2, pp. 946–956 (2004)
7. Smadja, F.: Retrieving collocations from text: Xtract. Comput. Linguist. 19(1), 143–177 (1993)
8. Sclano, F., Velardi, P.: Termextractor: a web application to learn the shared terminology of emergent web communities. In: I-ESA 2007, Funchal, Portugal (2007)
9. Daille, B.: Conceptual structuring through term variations. In: Proceedings of the ACL 2003 workshop on Multiword expressions, Morristown, NJ, USA, pp. 9–16. Association for Computational Linguistics (2003)
10. Evans, D.A., Zhai, C.: Noun-phrase analysis in unrestricted text for information retrieval. In: Proceedings of the 34th annual meeting on Association for Computational Linguistics, Morristown, NJ, USA, pp. 17–24. Association for Computational Linguistics (1996)
11. Brill, E.: A simple rule-based part of speech tagger. In: Proceedings of the Third Conference on Applied Natural Language Processing, pp. 152–155 (1992)
12. Amrani, A., Kodratoff, Y., Matte-Tailliez, O.: A semi-automatic system for tagging specialized corpora. In: Dai, H., Srikant, R., Zhang, C. (eds.) PAKDD 2004. LNCS, vol. 3056, pp. 670–681. Springer, Heidelberg (2004)

Evaluating Non-In-Place Update Techniques for Flash-Based Transaction Processing Systems

Yongkun Wang, Kazuo Goda, and Masaru Kitsuregawa

Institute of Industrial Science, The University of Tokyo,
4–6–1 Komaba, Meguro–ku, Tokyo 153–8505 Japan
{yongkun,kgoda,kitsure}@tkl.iis.u-tokyo.ac.jp
http://www.tkl.iis.u-tokyo.ac.jp

Abstract. Recently, flash memory is emerging as the storage device. With price sliding fast, the cost per capacity is approaching to that of SATA disk drives. So far flash memory has been widely deployed in consumer electronics even partly in mobile computing environments. For enterprise systems, the deployment has been studied by many researchers and developers. In terms of the access performance characteristics, flash memory is quite different from disk drives. Without the mechanical components, flash memory has very high random read performance, whereas it has a limited random write performance because of the erase-before-write design. The random write performance of flash memory is comparable with or even worse than that of disk drives. Due to such a performance asymmetry, naive deployment to enterprise systems may not exploit the potential performance of flash memory at full blast. This paper studies the effectiveness of using non-in-place-update (NIPU) techniques through the IO path of flash-based transaction processing systems. Our deliberate experiments using both open-source DBMS and commercial DBMS validated the potential benefits; x3.0 to x6.6 performance improvement was confirmed by incorporating non-in-place-update techniques into file system without any modification of applications or storage devices.

Keywords: NAND Flash Memory, SSD, LFS, Transaction Processing.

1   Introduction

Flash memory is a recently emerging storage device. With price sliding fast, the cost per capacity of flash memory is approaching to that of low-end SATA disk drives. So far flash memory has been widely deployed in consumer electronics even partly in mobile computing environments. Extending the deployment of flash memory to enterprise systems looks a natural attempt. Actually many researchers and developers have been studying the idea of utilizing flash memory for enterprise systems. EMC is trying to incorporate flash-based SSDs into their enterprise-level storage products [3].

One big issue arising for deploying the flash memory to the enterprise systems is that flash memory is quite different from disk drives in terms of the

S.S. Bhowmick, J. Küng, and R. Wagner (Eds.): DEXA 2009, LNCS 5690, pp. 777–791, 2009.

access performance characteristics. Disk drives are mainly comprised of mechanical components, thus random access performance is poor due to the seek and rotational overheads. By contrast, flash memory is a solid-state device, without the mechanical components, yielding high random read performance. However, the flash memory cannot be written in place. When updating the data, the entire erase-block containing the data must be erased before the updated data is written there. Since such erase operations are often very time consuming compared with read/write operations, the random write performance of flash memory is relatively poor. Table 1 summarizes necessary time for each operation in Samsung 4GB flash memory chip [19]. In recent major products, the typical latency of random writes is several milliseconds, being comparable with or sometimes even worse than that of the latest high-end disk drives.

Table 1. Operational flash parameters of Samsung 4GB flash memory chip

Page Read to Register (4KB)	$25\mu s$
Page Write from Register (4KB)	$200\mu s$
Block Erase (256KB)	$1500\mu s$

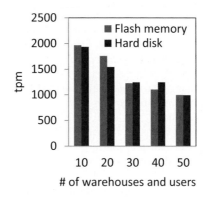

Fig. 1. Performance comparison between disk drive and flash memory (The details are described in Section 4.2)

Due to such a significant performance asymmetry, naive deployment of flash memory into enterprise systems may not exploit the potential performance of flash memory at full blast. Software components of existing enterprise systems are often designed and optimized for disk drives. Fig. 1 shows a typical example: we measured the obtainable throughput by the TPC-C benchmark on a commercial DBMS with the disk drive and flash memory. Contrary to our expectation, we could gain little or sometimes even lose by simply replacing the conventional disk drive with the flash-based SSD in this case study. That is, it may not be easy for existing enterprise system to directly enjoy the potential performance of recent flash memory.

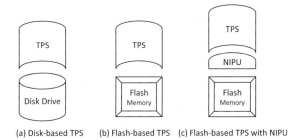

(a) Disk-based TPS   (b) Flash-based TPS   (c) Flash-based TPS with NIPU

Fig. 2. Comparison of transaction processing system designs

One solution is to redesign the system so that the system can be fully optimized for flash memory. For instance, if we were able to rewrite all the code of database engines and operating systems specially for flash memory, the system could derive the maximum performance. Such a solution may be possible for limited systems. But when it comes to enterprise systems that have a variety of customers, the huge cost of development may not be well accepted by many CIOs. In addition, a variety of succeeding solid-state technologies such as PCRAM [17] are about to emerge. Therefore, it may not be a good choice to invest huge cost on special development only based on flash memory.

Rather, if we could derive reasonable performance improvement of a flash-based enterprise system by simply incorporating optimization techniques into the IO path without modifying other components of the system, as shown in Fig. 2, it could be a good news for many CIOs even though it may not exploit the potential performance of flash memory at 100 percent. This paper studies the effectiveness of non-in-place update (NIPU) techniques through the IO path of transaction processing systems. NIPU techniques can convert a stream of in-place write operations into a stream of non-in-place write operations. Implementation of such techniques between DBMS and flash memory can significantly reduce the number of in-place writes, thus considerably improving the IO throughput of the flash memory. We built an experimental system using both open-source DBMS and commercial DBMS with a conventional hard disk drive and a flash-based SSD and then evaluated the effectiveness of deploying the NIPU techniques into transaction processing systems.

Our measurement-based analysis shows that x3.0 to x6.6 performance improvement can be expected by incorporating NIPU techniques into file systems without any modification of applications or storage devices. To the best of our knowledge, this finding has not yet been reported.

The rest of this paper will be organized as follows: Section 2 will briefly summarize the issue of flash memory for transaction processing system. In Section 3, we will discuss the deployment of the NIPU techniques on flash-based transaction processing system. Our deliberate experiments will be described in Section 4. Section 5 will summarize the related work. Finally, our conclusion and future work will be provided in Section 6.

2   Issue of Flash Memory for Transaction Processing

Unlike the traditional hard disk, which has an approximately symmetric read and write speed, flash memory, on the contrary, has substantial difference between the speeds of read and write, as shown in Table 2. The average response time of read, whatever in sequential or random mode, as well as that of the sequential write, is about two orders of magnitude faster than that of the hard disk. By contrast, the average response time of write in random mode, is comparable or even worse than that of the hard disk. This is primarily because the flash memory cannot be updated in place; a time-consuming block-erase operation has to be performed before the write operation, as disclosed in Table 1 [19]. For the sake of better performance, the size of erase block is usually large, about several hundred KB, leading to an expensive time cost of erase operation compared to that of flash read.

Table 2. Average response time of the flash memory and hard disk with the transfer request size of 4KB. Experiment setup is the same as that in Section 4.1 except here the hard disk and flash memory is bound as the raw device. Benchmark is Iometer 2006.07.27 [6].

	Hard Disk		Flash Memory	
	Read	Write	Read	Write
Sequential	$127\mu s$	$183\mu s$	$94\mu s$	$75\mu s$
Random	$13146\mu s$	$6738\mu s$	$106\mu s$	$8143\mu s$

The poor random write performance of flash memory could be painful for some transaction processing systems. In these systems, the intensive random write is often the main stream of disk IO. Though the operating system has an efficient buffer policy to cache the individual write operations into a bulk update, the performance characteristics of flash memory has been hardly considered here. Therefore, it would be problematic for the existing transaction processing systems to run on the flash memory directly, as reported in [2]. Our experiment also illustrates this points in Section 4.2 that the performance was not improved, even worse than that of the hard disk sometime by directly using the flash memory as the main storage media of data, though the flash memory has fine performance on read and sequential write. A better solution, such as NIPU techniques, is required to fully exploit the benefit of flash memory, as discussed in next section.

3   NIPU Techniques on Flash-Based Transaction Processing System

To utilize the flash memory efficiently, a tactful way is to introduce the NIPU techniques for enterprise system to improve the overall performance. Briefly, the

NIPU techniques convert the logical in-place updates into physical non-in-place updates, using special address table to manage the translation between logical address and physical address. An additional process called garbage collection is required to claw back the obsolete data blocks. A good example of the NIPU technique is the log-structured file system described in [18], with an implementation called *Sprite LFS*. Instead of seeking and updating in-place for each file and Inode, the LFS will collect all write operations and write them into a new address space continuously, as illustrated in Fig. 3. For such a NIPU-based file system, the principal feature is that a large number of data blocks are gathered in a cache before writing to disk in order to maximize the throughput of collocated write operations, thereby minimizing seek time and accelerating the performance of writes to small files. Though the write performance is optimized by some detriment of scan performance [4], this feature is greatly helpful on flash memory to make up for the inefficient random write performance since the random read performance is about two orders of magnitude higher than that of erase operations. The overall write performance is hereby improved.

- - - - - - - - → Apply changes and write to new address

Fig. 3. Non-In-Place Update techniques

Using such techniques for transaction processing systems on flash memory looks a good solution. In this case the flash memory is usually written sequentially through all the way, with a background process reclaiming the obsolete data blocks into the pool of available data block. On the basis of non-in-place update, all the update operations are performed by writing the data pages into the new flash pages, and the erase operations are not required right beforehand as long as the free flash pages are available. Thus, the overall throughput of transactions can be improved.

From a macro view of system, there are several possible places to implement the NIPU techniques through the IO path between DBMS and Flash memory. That is, the NIPU techniques can be incorporated into many places such as Flash Translation Layer (FTL), RAID controller, logical volume manager, file system, and database storage engine. Here arises a problem regarding which place we should implement the NIPU techniques. We are studying on this problem and would like to report it in another paper. In this paper, we focus on the potential benefits of file system. It would be a good choice to load a NIPU-based file system module to OS kernel without any changes to a variety of disk drivers, controllers and database applications.

It is to be noted here that a concern on the NIPU techniques is the design and settings of GC (Garbage Collection). Since the NIPU techniques consume free flash pages faster than other methods, the obsolete data pages (garbage) should be reclaimed by fine timing policy to the pool of available data blocks to ensure there are free pages available anytime when there are write requests. We will discuss the influence of the GC settings in Section 4.6.

4   Experimental Evaluation

We now describe a set of experiments that validate the effectiveness of the NIPU techniques and compare them against the traditional alternative. We use the popular TPC-C [22] as the benchmark, though it may not exactly emulate the real production workload [5], it discloses the general business process and workload, supported by the main hardware and software database system providers in the industry.

4.1   Experiment Setup

We build a database server on the Linux system. The flash memory is connected to the server with SATA 3.0Gbps hard drive controller as well as the hard disk driver. Fig. 4(a) gives the view of our experimental system.

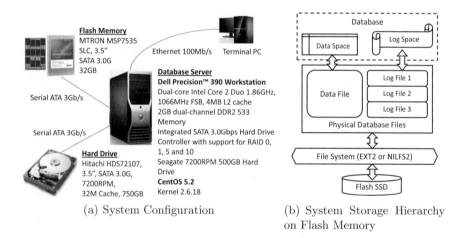

(a) System Configuration

(b) System Storage Hierarchy on Flash Memory

Fig. 4. Experiment Setup

We choose a commercial DBMS, as well as popular open source DBMS MySQL [13], as the database system for the TPC-C benchmark. In the commercial database system, the buffer cache is 8MB and log buffer is 5MB, with the block size of 4KB. This block size is set by our previous empirical experiment, in which we performed a low-level disk IO test, with a raw device test program written by us. We find that the optimal IO request size is 4KB for our flash memory. For MySQL, we use *InnoDB* storage engine, buffer cache is 4MB and log buffer is 2MB, with the block size of 16KB. The block size of MySQL is different from that of the commercial DBMS, because MySQL does not allow us to configure the block size, although 16KB might not be optimal.

As for the incorporation of the NIPU techniques into the IO path between the databases and devices, we choose a traditional log-structured file system, NILFS2 [16][11], a loadable kernel module without recompilation of the OS kernel, as an intermediate layer between the DBMS and flash memory. As a comparison, we choose the EXT2 file system as the representative of a conventional file system.

The storage hierarchy is simplified and shown in Fig. 4(b). We format the flash memory with EXT2 file system, on which we build the database instance, with all the related files together, such as the data files and log files, as well as the temporary files and system files. Thus, the main IO activities of this instance are confined within the flash memory. We refer this system as "Flash-EXT2". Similarly, we format the flash memory with NILFS2, on which we build the same instance as EXT2 system. We refer this system as "Flash-NILFS2" hereafter. As a comparison, we also build the same system on hard disk, denoted as "HDD-EXT2" and "HDD-NILFS2" respectively.

Unlike the EXT2 file system, NILFS2 file sytem has several settings of garbage collection. We set the interval of garbage collection to a very large value to disable this function firstly, so as to simplify the IO pattern. The influence of garbage collection will be discussed in Section 4.6.

4.2   Transaction Throughput

In this test, we create many threads to simulate the virtual users. Each virtual user will have a dedicated warehouse during the execution of transactions. Unlike the real users, virtual users in our test do not have the time for "Key and Think", for the purpose of getting intensive transaction workload. We gradually increase

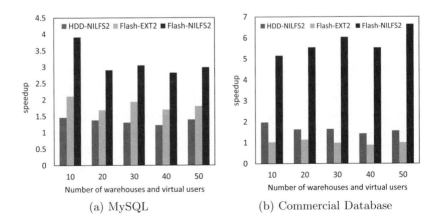

(a) MySQL   (b) Commercial Database

Fig. 5. Speedup of the transaction throughput on different systems based on "HDD-EXT2"

the number of warehouses as well as the number of virtual users to match. The speedup of transaction throughput based on "HDD-EXT2" is shown in Fig. 5.

In Fig. 5(a) we find that speedup of "Flash-EXT2" to "HDD-EXT2" is 1.8–2.1, which means that the naive replacement of flash memory to hard disk could have twofold transaction throughput on MySQL. The speedup of "Flash-NILFS2" shows that the NIPU-based flash memory system can have further improvement, 1.7–1.9 times to "Flash-EXT2", and about 3.0–3.9 times to the "HDD-EXT2". As for the commercial database system shown in Fig. 5(b), it is quite exciting.[1] We can find that the speedup of "Flash-EXT2" to "HDD-EXT2" is around 1.0, showing that the transaction throughput of "Flash-EXT2" is comparable with or sometimes even worse than that of "HDD-EXT2", which verifies our perspective that it is not beneficial for small-size transaction-intensive applications by directly utilizing the flash memory. Remarkably, a significant improvement can be found for "Flash-NILFS2"; the speedup is 5.2–6.6 times to "Flash-EXT2", which manifests that NIPU-based transaction processing system can undergo dramatic improvements on flash memory.

4.3   IO Performance

In our experiments regarding the IOPS, we examine the total number of transfers per second that were issued to the specific physical device. Here a transfer is an IO request to a physical device, and multiple logical requests can be combined into a single IO request to the device. So a transfer is of indeterminate size. Our trace result is shown in Fig. 6. As disclosed in Fig. 6(a) for MySQL, the IO request per second on "Flash-" side is much higher than that of "HDD-" side, which shows that flash memory can improve the IOPS. Meanwhile, the average response time of IO request, as shown in Fig. 6(b), is reduced significantly. Combined with the speedup of transaction throughput in Fig. 5(a), it implies that the NILF2 could coalesce more blocks into a single IO on MySQL, resulting in the higher performance. This can be confirmed in Fig. 6(c), which illustrates IO transfer rate. We can find that the total sector per second on "Flash-NILFS2" is about 6.2–8.4 times as many as that on "HDD-EXT2". On the commercial database system shown in Fig. 6(d), the total IO request per second of "HDD-EXT2" and "Flash-EXT2" is comparable. In sharp contrast, the total IO request per second of "Flash-NILFS2" is outstanding, and the average response time in Fig. 6(e) is also cut down greatly. It implies that the NIPU-based system can handle more requests at a time with shorter service time. Here the average response time includes the time spent by the requests in queue and the time spent servicing them. Since the response time is cut down greatly by NIPU techniques, the OLTP applications, which is required to respond immediately to user requests, could be benefited a lot. The number of sector per second of commercial DBMS shown in Fig. 6(f) follows the same trend as that of the IO request per second, except that on "Flash-NILFS2" it is about 18.6–22.2 times as

---

[1] We need further investigation regarding the difference between MySQL and the commercial DBMS.

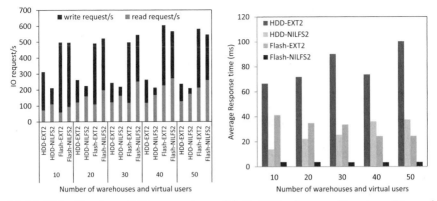

(a) MySQL: Number of IO request per second

(b) MySQL: Average Response Time of IO Request

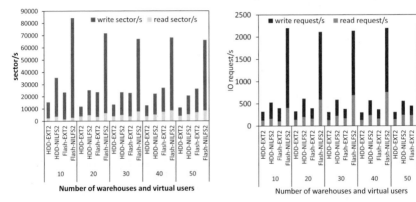

(c) MySQL: Number of sector per second

(d) Commercial DBMS: Number of IO request per second

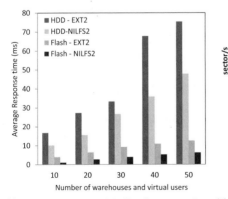

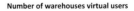

(e) Commercial DBMS: Average Response Time of IO Request

(f) Commercial DBMS: Number of sector per second

Fig. 6. IOPS and Average Response Time

many as that on "HDD-EXT2". Considered together with Fig. 6(d), the NIPU techniques tends to use relatively large IO request with the increasing of the number of sectors.

4.4   CPU Utilization

In this section we discuss the CPU Utilization in order to analysis the bottleneck of our experimental system. The CPU Utilization is traced when the transactions running in the steady state. The startup and terminate effect is eliminated. Trace result is shown in Fig. 7, in which the CPU Utilization is divided into four portions: *%user, %system, %iowait* and *%idle*. The main portion of CPU time on "HDD-EXT2", "HDD-NILFS2", and "Flash-NILFS2", is spent on waiting for the completion of IO, which implies that the system is possibly "IO-Bound". However, the CPU Utilization of "Flash-NILFS2" contrasts strongly in the ratio of four portions with the other cases: a uniform distribution of CPU time is observed, caused by the cutback of the CPU time spent on IO wait, and balanced by more CPU time moved to running the user applications, showing that "Flash-NILFS2" can utilize CPU time more efficiently.

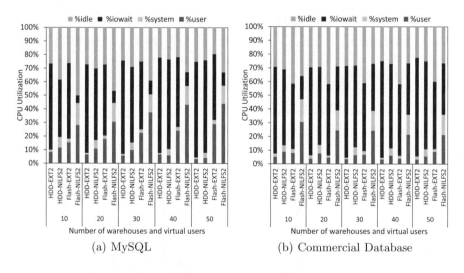

(a) MySQL   (b) Commercial Database

Fig. 7. CPU Utilization

4.5   Disk Buffer Cache

Although we have limited the buffer cache of the database system to a very small size, there is still some influence from the disk buffer cache, as long as we use the file system to manage the data blocks written to the storage device. At this moment, we cannot eliminate the influence of system buffer cache. A passive but efficient approach is to test the system with bound physical memory. Fig. 8

shows the result with 1GB and 512MB physical memory in the same experiment system described in Section 4.1. The *speedup* is the ratio of "Flash-NILFS2" to "Flash-EXT2", i.e. the improvement of NIPU techniques on *flash memory*. For MySQL shown in Fig 8(a), since it is memory efficient, the decreasing is not significant. As for the commercial database shown in the Fig. 8(b), the significant speedup is falling quickly with the very small memory size (512MB). However, with reasonable memory size (1GB), the "Flash-NILFS2" system can gain above fourfold.

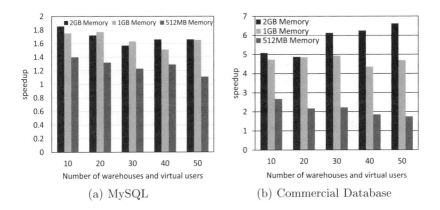

(a) MySQL   (b) Commercial Database

Fig. 8. Performance speedup with different amount of physical memory

4.6   Influence of Garbage Collection

We now discuss the influence introduced by the different settings of GC on NILFS2. In Section 4.2 to Section 4.5, no cleaning occurs during the execution, so the measurements represent the best-care performance. In fact, the GC function should be turned on to ensure the free data space. Therefore, the background cleaning processes of GC will consume the CPU time and IO bandwidth, producing some effect to the overall performance of system. A better GC strategy can emulate to the upper level system that the free data blocks are always available, with minimum cost of CPU time.

As indicated in [18], four issues must be addressed regarding the GC: (1) Cleaning Interval (CI), (2) Number of Segments Per Clean (NSPC), (3) Which segments to be clean, and (4) How to group the live blocks. Rosenblum and Ousterhout [18] analyzed and addressed the issue (3) and (4). In this paper, we will focus on analyzing the influence by (1) CI and (2) NSPC.

We can set the protection period (PP) to tell the daemon process how long the segments can be preserved for recovery. The NSPC can also be set when the device is mounted. With these settings, the experiment result in microscopic view is shown in Table 3. we use tuple $(PP, NSPC, CI)$ to denote the detailed settings. With the GC settings shown in Table 3, there is no appreciable change in the tpm (transactions-per-minute) and IOPS (either reads or writes) compared

Table 3. Performance Metrics of NILFS2-based transaction throughput of the commercial database on flash memory with GC

		10 warehouses, 10 virtual users	
	tpm	**IOPS**	**Average Response Time (ms) of I/O Request**
No GC	9983	reads: 406 writes: 1792 total: 2198	1.02
GC(1, 2, 5)	9933	reads: 384 writes: 1856 total: 2240	1.82

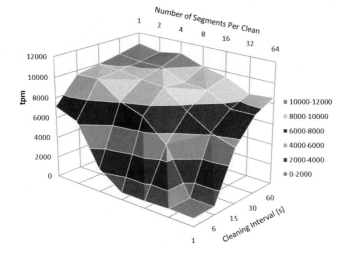

Fig. 9. Transaction throughput with $(0, NSPC, CI)$ on commercial DBMS with 10 warehouses and 10 virtual users

with that without GC. The IOPS includes the IOs issued for GC, so the average response time increases due to the additional IO added by the GC.

We set the protection period to 0, then the obsolete data blocks can be cleaned immediately when the cleaning process is invoked. The transaction throughput with $(0, NSPC, CI)$ is disclosed in Fig. 9. It shows that a greedy cleaning strategy (large NSPC and short CI) will have a detrimental effect to the transaction throughput, although the cleaning is very efficient. The maximum additional IO for GC can be roughly calculated by $\frac{NSPC \times SegmentSize}{CI}$. For example, in Fig. 9, when $NSPC = 4$, the CI should $\geq 30s$ to keep the transaction throughput from falling heavily. We use 8MB segment size,[2] thus the additional IO is about

---

[2] We use 4KB block size for the NILFS2 file system, and number of blocks per segment is 2048, so the segment size is 8MB.

$\frac{4 \times 8MB}{30s} \approx 1.07MB/s$. We should keep the additional IO less than $1.07MB/s$, then the performance will not be degraded. Therefore, carefully choosing the value of $(NSPC, CI)$ and Segment Size with heuristic method would ensure the high transaction throughput as well as the high utilization of the disk cleaned by GC.

5   Related Work

5.1   Non-In-Place Update Techniques

Continuous data protection (CDP) [21][24] is a backup technology automatically saving a copy of every change made to that data to a separate storage location in an enterprise storage system. Another successful example is the Sprite LFS [18], a log-structured file system. The LFS is designed to exploit fast sequential write performance of hard disk, by converting the random writes into sequential writes. However, the side effect is that the sequential reads may also be scattered into random reads. Overall, the performance can be improved to write-intensive applications. The LFS is also expected to improve the random write performance of flash memory, since the fast read performance of flash memory well mitigates the side effect. For the garbage collection of LFS, an adaptive method based on usage patterns is proposed in [15]. Shadow paging [20] is a copy-on-write technique for avoiding in-place updates of pages. It needs to modify indexes and block lists when the shadow pages are submitted. This procedure may recurse many times, becoming quite costly.

5.2   Flash-Based Technologies

By a systematical "Bottom-Up" view, the research on flash memory can be categorized as follow:

Hardware Interface. This is a layer to bridge the operating system and flash memory, usually called FTL (Flash Translation Layer). The main function of FTL is mapping the logical blocks to the physical flash data units, emulating flash memory to be a block device like hard disk. Early FTL using a simple but efficient page-to-page mapping [8] with a log-structured architecture [18]. However, it requires a lot of space to store the mapping table. In order to reduce the space for mapping table, the block mapping scheme is proposed, using the block mapping table with page offset to map the logical pages to flash pages [1]. However, the block-copy may happen frequently. To solve this problem, Kim improved the block mapping scheme to the hybrid scheme by using a log block mapping table [10].

File System. Most of the file system designs for flash memory are based on Log-structured file system [18], as a way to compensate for the write latency associated with erasures. JFFS, and its successor JFFS2 [7], are journaling file systems for flash. JFFS2 performs wear-leveling with the cleaner selecting a block with valid data at every 100th cleaning, and one with most invalid data at other times. YAFFS [23] is a flash file system for embedded devices.

Database System. Previous design for database system on flash memory mainly focused on the embedded systems or sensor networks in a log-structured behavior. FlashDB [14] is a self-tuning database system optimized for sensor networks, with two modes: disk mode for infrequent write, much like regular B^+-tree; log mode for frequent write, employed a log-structured approach. LGeDBMS [9], is a relational database system for mobile phone. For enterprise database design on flash memory, In-Page Logging [12] is proposed. The key idea is to co-locate a data page and its log records in the same physical location.

6   Conclusion and Future Work

For transaction processing system on flash memory, we describe non-in-place update techniques to improve the transaction throughput. In a system based on NIPU techniques, the write operations are performed sequentially; while the GC cleans the obsolete data in the background. This strategy greatly reduces time-consuming erase operations for applications with intensive write operations, thereby resulting in improved overall performance. We use a traditional log-structured file system to build a test model for examination. We then validated NIPU techniques with a set of experiments and showed that the NIPU-based systems can considerably speed up the transaction throughput by x3.0 to x6.6 on flash memory.

In the near future, we plan to apply the non-in-place update technique into different layers of the system and investigate appropriate algorithms for different context.

References

1. Ban, A.: Flash file system. US Patent No. 5404485 (April 1995)
2. Birrell, A., Isard, M., Thacker, C., Wobber, T.: A design for high-performance flash disks. Operating Systems Review 41(2), 88–93 (2007)
3. EMC: White Paper: Leveraging EMC CLARiiON CX4 with Enterprise Flash Drives for Oracle Database Deployments Applied Technology (December 2008)
4. Graefe, G.: Write-Optimized B-Trees. In: VLDB, pp. 672–683 (2004)
5. Hsu, W.W., Smith, A.J., Young, H.C.: Characteristics of production database workloads and the TPC benchmarks. IBM Systems Journal 40(3), 781–802 (2001)
6. Iometer, http://www.iometer.org
7. JFFS2: The Journalling Flash File System, Red Hat Corporation (2001), http://sources.redhat.com/jffs2/jffs2.pdf
8. Kawaguchi, A., Nishioka, S., Motoda, H.: A Flash-Memory Based File System. In: USENIX Winter, pp. 155–164 (1995)
9. Kim, G.J., Baek, S.C., Lee, H.S., Lee, H.D., Joe, M.J.: LGeDBMS: A Small DBMS for Embedded System with Flash Memory. In: VLDB, pp. 1255–1258 (2006)
10. Kim, J., Kim, J.M., Noh, S.H., Min, S.L., Cho, Y.: A space-efficient flash translation layer for CompactFlash systems. IEEE J CE 48(2), 366–375 (2002)

11. Konishi, R., Amagai, Y., Sato, K., Hifumi, H., Kihara, S., Moriai, S.: The Linux implementation of a log-structured file system. Operating Systems Review 40(3), 102–107 (2006)
12. Lee, S.W., Moon, B.: Design of flash-based DBMS: an in-page logging approach. In: SIGMOD Conference, pp. 55–66 (2007)
13. MySQL, http://www.mysql.com/
14. Nath, S., Kansal, A.: FlashDB: dynamic self-tuning database for NAND flash. In: IPSN, pp. 410–419 (2007)
15. Neefe, J.M., Roselli, D.S., Costello, A.M., Wang, R.Y., Anderson, T.E.: Improving the Performance of Log-Structured File Systems with AdaptiveMethods. In: SOSP, pp. 238–251 (1997)
16. NTT: New Implementation of a Log-structured File System, http://www.nilfs.org/en/about_nilfs.html
17. Pirovano, A., Redaelli, A., Pellizzer, F., Ottogalli, F., Tosi, M., Ielmini, D., Lacaita, A.L., Bez, R.: Reliability study of phase-change nonvolatile memories. IEEE_J_DMR 4(3), 422–427 (2004)
18. Rosenblum, M., Ousterhout, J.K.: The Design and Implementation of a Log-Structured File System. ACM Trans. Comput. Syst. 10(1), 26–52 (1992)
19. Samsung: K9XXG08XXM Flash Memory Specification (2007)
20. Shenai, K.: In: Introduction to database and knowledge-base systems, p. 223. World Scientific, Singapore (1992)
21. Strunk, J.D., Goodson, G.R., Scheinholtz, M.L., Soules, C.A.N., Ganger, G.R.: Self-Securing Storage: Protecting Data in Compromised Systems. In: OSDI, pp. 165–180 (2000)
22. TPC: Transaction Processing Performance Council: TPC BENCHMARK C, Standard Specification,Revision 5.10 (April 2008)
23. YAFFS: Yet Another Flash File System, http://www.yaffs.net
24. Zhu, N., Chiueh, T.: Portable and Efficient Continuous Data Protection for Network File Servers. In: DSN, pp. 687–697 (2007)

A Relational Encoding of a Conceptual Model with Multiple Temporal Dimensions

Donatella Gubiani and Angelo Montanari

Department of Mathematics and Computer Science,
University of Udine, Italy

Abstract. The theoretical interest and the practical relevance of a systematic treatment of multiple temporal dimensions is widely recognized in the database and information system communities. Nevertheless, most relational databases have no temporal support at all. A few of them provide a limited support, in terms of temporal data types and predicates, constructors, and functions for the management of time values (borrowed from the SQL standard). One (resp., two) temporal dimensions are supported by historical and transaction-time (resp., bitemporal) databases only. In this paper, we provide a relational encoding of a conceptual model featuring four temporal dimensions, namely, the classical valid and transaction times, plus the event and availability times. We focus our attention on the distinctive technical features of the proposed temporal extension of the relation model. In the last part of the paper, we briefly show how to implement it in a standard DBMS.

1 Introduction

Despite the pervasiveness of temporal information, most databases (and information systems) basically maintain information about the current state of the world only. Temporal databases can be viewed as an attempt to overcome this limitation, making it possible to keep track of the evolution of the domain of interest (valid time dimension) and/or of the database contents (transaction time dimension). The valid time of a fact can be defined as the time when the fact is true in the modeled domain, while its transaction time is the time when it is current in the database and may be retrieved. Historical (resp., transaction-time) relational databases support the valid (resp., transaction) time dimension. Relational databases that manage both dimensions are called bitemporal databases. In [3], two additional temporal dimensions, respectively called event and availability time, have been proposed to remedy to some weaknesses of valid and transaction times. The event time of a fact is defined as the pair of occurrence times of the real-world events that respectively initiate and terminate its validity interval, while its availability time is the time interval during which it is known and believed correct by the information system the database belongs to (in general, such an interval does not coincide with its transaction time interval). No effective support to these dimensions is provided by existing temporal relational databases. A comprehensive and up-to-date survey of temporal databases can be found in [6].

S.S. Bhowmick, J. Küng, and R. Wagner (Eds.): DEXA 2009, LNCS 5690, pp. 792–806, 2009.

The contribution of this paper is part of the work done within ChronoGeo-Graph (CGG) Project [2], which aims at developing a software framework for the conceptual and logical design of spatiotemporal databases. The core of the framework is the CGG model, a conceptual model that extends the classical Enhanced Entity-Relationship model (EER) with additional constructs for spatiotemporal information [5].

As for the spatial features, CGG supports a large set of representation primitives for spatial data. CGG distinguishes between spatial and non spatial entities. A spatial entity is characterized by a set of descriptive and spatial attributes plus a geometry of a given spatial data type (CGG supports 8 different spatial data types). Spatial attributes take their value over a spatial data type as well. A spatial dimension can be added to relations as well. CGG supports topological, metric and direction relations, and the relation of spatial aggregation (the part-of relation over spatial entities). Besides the usual relation of specialization, CGG introduces the relation of cartographic specialization, which supports different spatial representations of the same spatial entity. Finally, CGG supports the field-based view of spatial information by the notion of (spatial) field and the notion of schema territory, which defines the spatial domain over which all spatial elements of the schema are located.

As for the temporal features, CGG allows one to temporally qualify the various constructs by properly annotating them. One or more temporal dimensions can be associated with the schema territory, entities, attributes, relations, and fields. Different temporal dimensions are associated with different constructs. Entities can be provided with an existence time (which can be viewed as the valid time of the entity), possibly paired with a state diagram, a transaction time, an event time, and an availability time. The other constructs can be endowed with a valid time, a transaction time, an event time, and an availability time. Furthermore, CGG introduces a distinction between snapshot and lifespan cardinality constraints for attributes and relations. Snapshot cardinality constraints specify the minimum and maximum number of values that an attribute can take (resp., of instances of a given entity that may participate in a relation) at a given time, while lifespan cardinality constraints specify minimum and maximum bounds with respect to the whole existence of the entity instance (resp., the validity interval of the relation instance). As for attributes, CGG also allows one to collect sets of attributes of a given entity that change in a synchronous way (it defines a temporal collection as a set of entity attributes with a common temporal annotation). Finally, it explicitly keeps track of the events that affect a relevant element, e.g., events that change the state and/or the geometry of an entity, the validity of a relation, the value of an attribute.

The paper addresses the problem of providing a relational encoding of temporal information in CGG schemas. A special attention will be deserved to the management of temporal dimensions. The distinctive features of the proposed temporal extension of the relational model are the use of tuple timestampings, the partition of temporal schemas (resp., instances) into a current component and a historical one, and the development a number of constraints that guarantee the consistency of the values of the different temporal dimensions. In addition, we implemented the extended temporal model in a standard DBMS, taking advantage of SQL asssertions and triggers, and we developed a translation algorithm mapping CGG schemas into temporally-extended relational ones.

The rest of the paper is organized as follows. In Section 2 we give a short account of existing temporal relational models. In Section 3 we describe the basic features of the proposed temporal relational model supporting the temporal dimensions of valid, transaction, event, and availability times. We first consider the single temporal dimensions in isolation and then we analyze their interactions. In Section 4 we focus our attention on the specification of temporal keys. Section 5 provides some details about the implementation of the model in a specific DBMS. Finally, in Section 6 we briefly illustrate the translation of CGG schemas into the proposed model.

2   An Account of Existing Temporal Relational Models

The basic relational model only supports temporal data types, e.g., `Date` and `Timestamp`, and predicates, constructors, and functions for the management of time values. It provides no primitives to explicitly deal with temporal dimensions. Various extensions to the relational model have been proposed in the literature to support the valid and/or transaction time dimensions.

Temporal databases can be classified according to the granularity of timestamping, the nature of timestamps, and the temporal interpretation of the primary key. All temporal databases associate one or more timestamp attributes (timestamps for short) with facts, for every supported temporal dimension. The most common options are associating a single timestamp with the whole tuple (tuple timestamping) and a distinct timestamp with any temporal attribute (attribute timestamping). The former preserves First Normal Form (1NF) and its implementation is straightforward; in addition, it allows one to benefit from the standard relational database technology. However, the resulting tables suffer from two weaknesses: data redundancy and vertical anomaly (information about a domain object is not recorded in a single tuple, but it is spread over various tuples). The latter is not affected by the vertical anomaly, because it records the entire history of every domain object in a single tuple. However, in doing that it violates 1NF: for every tuple and every temporal dimension, the value of each temporal attribute is a set of pairs (value, timestamp). As for the nature of timestamps, three different choices of increasing complexity have been considered: time instants, time intervals, and temporal elements. In most cases (as an example, this is not the case with aggregations over time), time intervals are not interpreted as primitive temporal entities, but just as (convex) sets of time instants. In its turn, temporal elements are usually defined as a finite set of pairwise disjoint time intervals. Time intervals and temporal elements allow one to obtain a succinct representation of valid/transaction time periods, but their manipulation is more complicate: either it requires to transform them into time instants, to apply the necessary operations on such instants, and to provide an encoding of the result at the time interval/temporal element level or it imposes the introduction of additional non-trivial operations, such as coalescing.

Replacing a set of contiguous time instants with a single time interval makes it possible to overcome the problem of tuple-timestamping (vertical anomaly and redundancy) and attribute-timestamping (redundancy). However, these problems show up again as soon as a single time intervals must be replaced with two or more ones. The replacement of time intervals with temporal elements solves them, but it involves the violation of 1NF. Finally, there exist different ways of reinterpreting the notion of primary key in the temporal setting. Every temporal relation is obtained by extending an atemporal relation with one or more timestamps. Its temporal key can be defined as a set of (non-temporal) attributes which is a primary key for every temporal snapshot, as in [8,7]. As an alternative, one can introduce an explicit tuple identifier, which plays the same role of the object identifier in the object-oriented model. As a third possibility, one can define the temporal key as a combination of the primary key of the original atemporal relation and a suitable subset of timestamps, e.g., [1].

In the following, we will describe an original temporally-extended relational model supporting the four temporal dimensions described above. It opts for tuple-timestamping, to preserve 1NF, it assumes temporal homogeneity for all relations (a tuple holds over a given interval if and only if it holds at all time instants belonging to it), it makes use of time intervals (resp., time instants) to model valid, transaction, and availability times (resp., event time), and it defines temporal keys as suitable temporal extensions of the primary keys of the original atemporal relations. The closest relatives of such a model are the Time Relational Model (TiRM), the Temporal Relational Model (TRM), and the Historical DataBase Management System (HDBMS). Ben-Zvi's TiRM model [1] supports three temporal dimensions: (i) the *effective time* of a fact, which corresponds to valid time, (ii) the *registration time* of a fact, which is the pair of time instants at which the beginning and ending of its effective time interval are inserted into the database, and (iii) the *deletion time* of a fact, which is the time instant at which it is logically deleted (the combination of registration and deletion times can be viewed as a counterpart of transaction time). TiRM associates five timestamps with every tuple, namely, T_{es} and T_{ee} (for the beginning and ending of effective time), T_{rs} and T_{re} (for the beginning and ending of registration time), and T_d (for the deletion time). T_{es} and T_{ee} are specified by the user, while T_{rs}, T_{re}, and T_d are generated by system. The temporally-extended tuple is called *tuple version*. Tuples with the same value for the atemporal key are called *tuple version set*. A temporal relation is defined as a set of tuple version sets, rather than a set of tuples. Navathe and Ahmed's TRM model [10] supports one temporal dimension only, which corresponds to valid time. It distinguishes between the set R_s of static (atemporal) relations and the set R_t of time-varying (valid-time) relations. Every time-varying relation includes two timestamps t_s and t_e that record the left and right endpoints of valid-time intervals, respectively. The key of a time-varying relation consists of the primary key of its atemporal part (*time-invariant key*, TIK for short) and the timestamp t_s (since the value of t_e can be unknown, the pair (TIK,t_e) is not an alternative key). Sarda's HDBMS model [12] supports one temporal dimension only as well, called real valid time, which corresponds to valid time. The aim of Sarda was to develop a temporal DBMS that receives as input a set of atemporal relation schemas and provides a subset of them (specified by the designer) with a temporal extension. The model allows one to distinguish between properties (historical relations) and instantaneous events. The system automatically associates two timestamps, `from` and `to`, with historical relation schemas, to keep track of their historical evolution, and a single timestamp `at` with events, to record their occurrence time. It allows the timestamps of different temporal relations to refer to different time granularities. The tuples of each historical relation are partitioned in two classes: the *current segment*, which contains only tuples belonging to the current state (tuple whose timestamp `from` has value `null`), and the *history segment*, which contains tuples representing historical data (tuples such that `from` < `now`). New tuples are first inserted in the current segment and later, when their real valid interval ends, moved to the history one. The primary key of a relation in the current segment is defined as in the basic relational case. The key of a relation in the history fragment is defined as follows: a set of (atemporal) attributes $\mathbf{K}$ is a key for a relation $R(\mathbf{X})$, with $\mathbf{K} \subseteq \mathbf{X}$, if for any value $\mathbf{k}$ of $\mathbf{K}$ and any time instant t there is at most one tuple in $R(\mathbf{X})$ with value $\mathbf{k}$ for $\mathbf{K}$ whose real valid interval includes t (keys are time-unique, rather than tuple-unique as in the relational model).

3   A Relational Model with Four Temporal Dimensions

In the following, we describe a temporal extension to the relational model that supports the temporal dimensions of valid, transaction, availability, and event time. As a matter of fact, the resulting model can be viewed as the relation counterpart of the spatio-temporal conceptual model ChronoGeoGraph (CGG), a spatio-temporal model that pairs the classical features of the EER model with a large set of spatial and temporal constructs [5]. First, we take into consideration each temporal dimension in isolation; then, we will deal with their combination.

Valid time. The valid time of a tuple is the time when the fact it represents is true in the modeled domain. We encode valid time intervals by means of two distinct timestamps VT_start and VT_end. The extension of an atemporal relation $R(\mathbf{X})$ with valid time has the form:

$$R(\mathbf{X}, VT\_start, VT\_end) \tag{1}$$

Let $\mathbf{T}^g$ be the (discrete) temporal domain at granularity g over which timestamps VT_start, VT_end take their value. Any pair of values t_s for VT_start and t_e for VT_end identifies a time interval $[t_s, t_e) \subset \mathbf{T}^g$ (we assume valid time intervals, as well as transaction and availability time intervals, to be closed to the left and open to the right). Valid time intervals consisting of a single chronon are represented as degenerate intervals with coincident endpoints (notationally, $[t_s, t_s+1)$). While the left endpoint VT_start of a valid time interval must always exist, its right endpoint VT_end might be missing[1]. The intended semantics of valid time intervals is captured by the following constraints:

---

[1] To represent valid time intervals open to the right, most models assign to VT_end either the "value" `null` or the "value" `until change` (uc for short). Since `null` is used in a variety of contexts with different meanings, we opt for the second alternative.

$$
\begin{array}{ll}
(i) & \exists t_s \in \mathbf{T}^g \; t_s = VT\_start \\
(ii) & \exists t_e \in \mathbf{T}^g \; t_e = VT\_end \vee VT\_end = \mathtt{uc} \\
(iii) & VT\_start < VT\_end
\end{array}
\qquad (2)
$$

Transaction time. The *transaction time* of a tuple is the time when the tuple is current in the database. We represent transaction time intervals by means of two distinct timestamps TT_start and TT_end. A transaction time interval is generated whenever a database update is executed. For every interval associated with a tuple in the database, we have that the value of TT_start is less than the current instant and that of TT_end is either until change, if the tuple is current, or less than or equal to the current instant, if the tuple is not current. Since deletion of a tuple can never precede its insertion/modification, TT_start must obviously be less than TT_end. The intended semantics of transaction time intervals is captured by the following constraints:

$$
\begin{array}{ll}
(i) & \exists t \in \mathbf{T}^g \; t = TT\_start \\
(ii) & TT\_end \leq \mathtt{now} \vee TT\_end = \mathtt{uc} \\
(iv) & TT\_start < TT\_end
\end{array}
\qquad (3)
$$

As in the HDBMS model, the schema (resp., instance) of every temporal relation is partitioned into two distinct schemas (resp., instances). The first instance, called *current instance*, consists of all and only the tuples which are current in the database. It only features the timestamp TT_start, whose value records the time instant at which the tuple was added to the database (the value TT_end for all current tuples is equal to uc, and thus omitted). The second one, called *historical instance*, records the tuples which have been logically deleted from the database. It features the two timestamps TT_start and TT_end that respectively record the times at which insertion and deletion take place.

$$
R(\mathbf{X}, TT\_start) \text{ and } R\_history(\mathbf{X}, TT\_start, TT\_end) \qquad (4)
$$

Tuples are always inserted in the current instance. The time instant at which insertion is executed is automatically assigned to the timestamp TT_start. The logical deletion of a tuple simply moves the tuple from the current instance to the historical one, without changing the value of its attributes. The time instant at which deletion is executed is automatically assigned to the new timestamp TT_end. The update of a tuple in the current instance can be described as a logical deletion of the current tuple followed by the insertion of the updated one (deletion/insertion times are equal to the time instant at which the update is executed). Tuples in the historical instance cannot be deleted or updated.

There are several advantages in separating the historical schema/instance from the current one. First, TT_end can be omitted in the current schema. Second, transaction time management is fully automatized. Third, since tuples in the historical instance cannot be modified, constraint checking can be restricted to tuples in the current instance (we will come back to this in Section 4). Finally, an improvement in query performance is often achieved. Whenever a query refers to current information only (we expect it to be the most common case), its execution can ignore all tuples in the historical instance.

Availability time. The availability time of a tuple is the time interval during which the fact it represents is known and believed correct by the information system the database belongs to. Availability time intervals are encoded by a pair of timestamps AT_start, AT_end and must satisfy the same constraints that transaction time intervals must satisfy:

$$
\begin{array}{ll}
(i) & \exists t \in \mathbf{T}^g \, t = AT\_start \\
(ii) & AT\_end \leq \text{now} \vee AT\_end = \text{uc} \\
(iii) & AT\_start < AT\_end
\end{array}
\tag{5}
$$

Additional constraints are imposed on the relationships between availability and transaction times. First, a fact can be stored in the database only if it is or was known by the information system. Similarly, a fact can be (logically) deleted from the database only if it is not believed correct/up-to-date by the information system. Moreover, if a fact is known and believed correct by the information system and it has been added to the database ($AT\_end = uc$), then the corresponding tuple must belong to the current instance ($TT\_end = uc$); conversely, it can never happen that $AT\_end \neq uc$ and $TT\_end = uc$. Finally, we must consider the case in which the information systems acquires and discharges some fact before its insertion in the database. Such a situation can be modeled by letting $AT\_end \leq TT\_start$ (when inserted in the database, information was already out-of-date) if and only if $TT\_start = TT\_end$ (information never became current in the database).

$$
\begin{array}{ll}
(i) & AT\_start \leq TT\_start \\
(ii) & AT\_end \leq TT\_end \\
(iii) & AT\_end = uc \Rightarrow TT\_end = uc \\
(iv) & TT\_end = uc \Rightarrow AT\_end = uc \\
(v) & AT\_end \leq TT\_start \Leftrightarrow TT\_start = TT\_end
\end{array}
\tag{6}
$$

Since in any realistic scenario the choice of including availability time and excluding transaction time looks meaningless, we do not consider temporal relation schemas with availability time and without transaction time. In addition, we must find a way to deal with information that never becomes current in the database ($TT\_start = TT\_end$), preserving the condition that imposes to insert any new fact in the current instance and to move it to the historical one when it is logically deleted. To cope with this problem, we include both the AT_start and the AT_end timestamps in the current schema. As a result, we obtain the following schema:

$$
\begin{array}{l}
\text{R}(\mathbf{X}, \text{TT\_start}, \text{AT\_start}, \text{AT\_end}) \\
\text{R\_history}(\mathbf{X}, \text{TT\_start}, \text{TT\_end}, \text{AT\_start}, \text{AT\_end})
\end{array}
\tag{7}
$$

We must distinguish two different modalities of tuple insertion. The first one provides a value for AT_start, but no value for AT_end. In this case, the system assigns the specified value to AT_start, it sets TT_start to the current time, and it adds the tuple to the current instance. The second one deals with the case in which a value less than (or equal to) the current time is given to AT_end.

The system assigns to AT_start and AT_end the specified values, it sets both TT_start and TT_end to the current time (thus $TT\_start \geq AT\_end$), it inserts the tuple in the current instance, and it immediately moves it to the historical one. Two different modalities of (logical) tuple deletion must be considered as well, depending on the value of AT_end. The first case is that of synchronous deletion: both AT_end and TT_end are set to the current time and the tuple is automatically moved from the current instance to the historical one. The second case considers a possible delay in the registration of a deletion from the information system: the system replaces the value uc of AT_end with the deletion time (which is less than the current time), it sets TT_end to the current time, and it automatically moves the tuple from the current instance to the historical one.

Event time. The *event time* of a tuple consists of the occurrence times of the real-world events that respectively initiate and terminate the valid time interval of the fact it represents. To model it, we add two timestamps ET_start, ET_end to the relation schema. By definition, event time can be added only to relation schemas provided with valid time. No constraints are imposed on event time.

$$R(\mathbf{X}, VT\_start, VT\_end, ET\_start, ET\_end) \tag{8}$$

Relations with multiple temporal dimensions. We conclude the section with an analysis of temporal relations provided with two or more temporal dimensions. As a general rule, we start from an atemporal relational schema and we add the appropriate timestamps for every supported temporal dimension. However, we cannot add available (resp., event) time without adding transaction (resp., valid) time as well. The addition of transaction time forces the partition of the relation schema in a current schema and a historical one. As an example, a temporal schema with the four temporal dimensions can be obtained by an atemporal schema $R(\mathbf{X})$ as follows. First, we add valid time:

$$R(\mathbf{X}, VT\_start, VT\_end) \tag{9}$$

Then, we add event time:

$$R(\mathbf{X}, VT\_start, VT\_end, ET\_start, ET\_end) \tag{10}$$

The addition of transaction time forces the splitting of the table:

$$
\begin{aligned}
&\text{R } (\mathbf{X}, \text{VT\_start, VT\_end, ET\_start, ET\_end, TT\_start)} \\
&\text{R\_history } (\mathbf{X}, \text{VT\_start, VT\_end, ET\_start, ET\_end,TT\_start, TT\_end)}
\end{aligned} \tag{11}
$$

Finally, the addition of available time affects both schemas (in a different way):

$$
\begin{aligned}
&\text{R}(\mathbf{X}, \text{VT\_start, VT\_end, ET\_start, ET\_end, TT\_start, AT\_start, AT\_end)} \\
&\text{R\_history } (\mathbf{X}, \text{VT\_start, VT\_end, ET\_start, ET\_end, TT\_start, TT\_end,} \\
&\qquad\qquad \text{AT\_start, AT\_end)}
\end{aligned} \tag{12}
$$

A well-known problem in temporal databases is to assign a consistent value to missing temporal dimensions, thus providing every relation with a temporal interpretation with respect to all temporal dimensions. In such a way, no relations are ignored during (temporal) query evaluation. The assignment of a value to missing temporal dimensions is done according to the following rules.

- Transaction time is missing. Tuples are current in the database when they can be retrieved from it, that is, we assume the transaction time interval of tuples to be $[now, now]$).
- Valid time is missing. Valid time is assimilated to transaction time: if transaction time is present, then valid time intervals are equal to transaction time intervals; otherwise, tuples are valid at the time instant in which they are retrieved from the database, that is, we assume the valid time interval of tuples to be $[now, now]$).
- Event time is missing. We assume $ET\_start = VT\_start$ and $ET\_end = VT\_end$ (on-time events).
- Available time is missing. We assume $AT\_start = TT\_start$ and $AT\_end = TT\_end$ (no delay in registration).

Such rules can be turned into suitable projection functions (one for each temporal dimension) that, given a relation instance, return the temporal values it explicitly or implicitly takes on temporal dimensions. We consider transaction and valid times; the cases of event and availability times are similar. Given a (temporal) relation R, let r_c (resp., r_h) be the instance of its current (resp., historical) schema (if transaction time is missing, R has a current schema only).

Definition 1. *Let R be a (temporal) relation. If $TT\_start, TT\_end \in R$, then $\pi_{TT}(r_c) = [\pi_{TT\_start}(r_c), now]$ and $\pi_{TT}(r_h) = [\pi_{TT\_start}(r_h), \pi_{TT\_end}(r_h)]$. If $TT\_start, TT\_end \notin R$, then $\pi_{TT}(r_c) = [now, now]$.*

Definition 2. *Let R be a (temporal) relation. If both $VT\_start, VT\_end \in R$ and $TT\_start, TT\_end \in R$, then $\pi_{VT}(r_i) = [\pi_{VT\_start}(r_i), \pi_{VT\_end}(r_i)]$, for $i \in \{c, h\}$. If $VT\_start, VT\_end \in R$ and $TT\_start, TT\_end \notin R$, then $\pi_{VT}(r_c) = [\pi_{VT\_start}(r_c), \pi_{VT\_end}(r_c)]$. If $VT\_start, VT\_end \notin R$ and $TT\_start, TT\_end \in R$, then $\pi_{VT}(r_c) = \pi_{TT}(r_c)$ and $\pi_{VT}(r_h) = \pi_{TT}(r_h)$. If both $VT\_start, VT\_end \notin R$ and $TT\_start, TT\_end \notin R$, then $\pi_{VT}(r_c) = \pi_{TT}(r_c) = [now, now]$.*

4   Temporal Primary Keys and Functional Dependencies

In this section we deal with the problem of specifying primary key and functional dependencies of a temporal relation. As it happens in the relational setting, there is a close connection between them; however, the addition of multiple temporal dimensions introduces various technical intricacies.

Both problems have been already addressed in the temporal databases literature, but there are no consensus solutions to them. As for temporal keys, different alternatives have been proposed, which range from the addition of one or more temporal attributes to the primary key of the atemporal schema [8] to the introduction of explicit object identifiers that uniquely identify each tuple in the temporal relation [14]. As for temporal functional dependencies (TFDs), a short account of existing proposals can be found in [4]. The simplest ones define TFDs as classical functional dependencies on the temporal snapshots of the relation [8], the most complex ones allow TFDs to constrain the values of (atemporal) attributes at different time points [13,14].

Our goal is to guarantee an appropriate trade-off between expressiveness and effectiveness. In particular, we would like to maintain the notion of temporal key and temporal dependency as simple and easy to manage as possible. The solution we propose deals with multiple temporal dimensions retaining much of the simplicity of the relational model. In addition, the separation between current and historical schemas/instances makes it possible to simplify the process of constraint checking.

As a general rule, we define TFDs as temporal generalizations of (atemporal) functional dependencies (FDs), which are obtained by making the latter time dependent. From a notational point of view, we replace every FD $Z \rightarrow Y$ by the corresponding TFD $Z \rightarrow_T Y$. The role of the four temporal dimensions in TFDs is quite different. By means of TFDs, we constrain FDs to be satisfied by pairs of tuples at common valid time instants with respect to common transaction or availability time instants. As availability (resp., transaction) time intervals may start before (resp., end after) than the corresponding transaction (resp., availability) time intervals, this amounts to require functional dependency to be satisfied with respect to common availability/transaction time instants belonging to a time interval that starts when the availability interval starts and ends when the transaction time interval ends. Event time plays no role in the definition of TFDs.

Definition 3. *Given a temporal relation R with atemporal schema $R(X)$ and a TDF $Z \rightarrow_T Y$, with $Z, Y \subseteq X$, we say that an instance $r \in R$ satisfies the TFD if and only if, for each pair of tuples $a, b \in r$, if $a[Z] = b[Z]$, $\pi_{VT}(a) \cap \pi_{VT}(b) \neq \emptyset$ (their valid time intervals overlap), and $\pi_{TT}(a) \cap \pi_{TT}(b) \neq \emptyset \vee \pi_{AT}(a) \cap \pi_{AT}(b) \neq \emptyset$ (their transaction or availability time intervals overlap), then $a[Y] = b[Y]$.*

Missing temporal dimensions are implicitly added according to the assignment rules given in Section 3. The notion of violation of a TFD is defined in the obvious way. We say that two tuples are *temporally inconsistent* if they violate a TFD.

Let us consider now the problem of specifying the key of a temporal relation schema (temporal key for short). As anticipated in Section 2, we basically define the (primary) temporal key as a temporal extension of the primary key of the original atemporal relation. We distinguish between the current schema and the historical schema of a temporal relation: the temporal key of the current schema add valid time to the atemporal key, while the temporal key of the historical schema add both valid and transaction times to the atemporal key. If valid time is missing, the temporal key of the current schema coincides with the atemporal one, while that of the historical schema consists of the atemporal key extended with transaction time. The fact that we compactly represent both valid and transaction times by means of interval timestamps, instead of instant ones, introduces some complications. A simple example of these complications is given in Table 1.

Table 1. The current instance of a table *Employee* devoid of transaction time

SSN	Salary	VT_start	VT_end
XXXNNN88HH	1000	15/10/2000	31/07/2006
XXXNNN88HH	1200	01/10/2003	31/07/2007

The two tuples belonging to the relation in Table 1 are temporally inconsistent, because they assign both the value 1000 and the value 1200 to the salary of employee *XXXNNN88HH* over the valid time interval [01/10/2003,31/07/2006). Such an inconsistency can be obviously detected by replacing the interval timestamp (VT_start, VT_end) by the instant one VT, by choosing (SSN, VT) as the temporal key, and by replacing every tuple by a set of tuples, one for each time instant in the valid time interval. However, the resulting instance turns out to be extremely redundant: a single tuple is replaced by a number of tuples that only differ in their temporal value. To avoid to introduce such a redundancy, we decided to maintain the interval timestamp. Unfortunately, in such a case, all possible choices for the attributes of the temporal key, namely, (SSN, VT_start), (SSN, VT_end), and (SSN, VT_start, VT_end), do not detect the inconsistency in Table 1. As a consequence, the satisfaction of the key constraint (for any possible choice of the temporal key) does not suffice to conclude that there are not temporal inconsistencies and thus temporal consistency must be explicitly checked.

Table 2. Temporal keys for temporal relations: a summary

Cases	Temporal keys	Temporal dimensions
atemporal	$R(\underline{K}, ...)$	-
valid time	$R(\underline{K, VT\ start}, ...)$	V
		VE
transaction time	$R(\underline{K},...)$	T
	$R\_history(\underline{K, TT\ start},...)$	TA
		VT
valid and transaction	$R(\underline{K, VT\_start},...)$	VTE
times	$R\_history(\underline{K, VT\ start, TT\ start},...)$	VTA
		$VTAE$

Among the three possible choices for the temporal key of the current schema, we opt for the addition of VT_start to the atemporal key. It detects more temporal inconsistencies than the temporal key that includes both VT_start and VT_end and, unlike VT_end, VT_start does not assume the "value" uc. Analogously, for the historical schema we choose to add to the atemporal key VT_start and TT_start. A summary of the resulting cases is given in Table 2.

In principle, constraint checking can be executed whenever a tuple is inserted in the current database or moved from the current to the historical database (tuples in the historical database cannot change their values). However, when a tuple is transferred from the current to the historical database, it only changes the value of TT_end and, possibly, the value of AT_end. Such changes cannot cause any inconsistency in the historical database[2]. Hence, constraint checking can be confined to insertions in the current database. When a tuple is inserted in the current database, an inconsistency may arise with respect to both current and historical tuples. In the former case, according to the proposed model, the intersection of both transaction and availability time intervals associated with current tuples is always not empty and thus the only constraint one needs to check on the current database is:

$$\forall a, b \in R(\mathbf{X}) \forall \mathbf{Y} \subseteq \mathbf{X}(a[\mathbf{K}] = b[\mathbf{K}] \wedge \pi_{VT}(a) \cap \pi_{VT}(b) \neq \emptyset \rightarrow a[\mathbf{Y}] = b[\mathbf{Y}]) \quad (13)$$

where $\mathbf{X}$ is the set of atemporal attributes of R (and $R\_history$) and $\mathbf{K}$ is the atemporal key of R (and $R\_history$). In the latter case, the intersection of transaction time intervals associated with the inserted tuple and a historical one is always empty and thus the only constraint one needs to check is:

$$\forall a \in R(\mathbf{X}) \ \forall b \in R\_history(\mathbf{X}) \ \forall \mathbf{Y} \subseteq \mathbf{X}(a[\mathbf{K}] = b[\mathbf{K}] \wedge \\ \wedge \pi_{VT}(a) \cap \pi_{VT}(b) \neq \emptyset \wedge \pi_{AT}(a) \cap \pi_{AT}(b) \neq \emptyset \rightarrow a[\mathbf{Y}] = b[\mathbf{Y}]) \quad (14)$$

This constraint can be violated only if the relation R (and $R\_history$) includes the three dimensions VTA. An inconsistency may occur if and only if the value of AT_start for the inserted tuple a is less than the value of TT_start (if $\pi_{AT}(a) = \pi_{TT}(a)$, then the intersection of the availability time intervals for a and any historical tuple is empty).

5   Implementation

In this section, we briefly describe an implementation of the proposed model in the Oracle DBMS [11]. As for the definition of the relational schemas and of data types, we use the standard SQL facilities featured by Oracle SQL [9]. To deal with timestamps, we take advantage of the timestamp data type (the conventional value uc is represented by the null value). Temporal constraints are encoded either as generic SQL assertions, using the SQL construct check constraint (the simplest ones), or as triggers (the most complex ones). As a concrete example of constraint management, we describe the triggers that rule the transition of relation tuples from the current instance to the historical one (for the sake of simplicity, we assume the relations to be devoid of availability time).

When a tuple is inserted in the current instance, the trigger sets TT_start to the value systimestamp(0) (the current time of the system):

---

[2] As a matter of fact, this implies that temporal keys for historical schemas are not really necessary. We decided to keep them to comply with the relation model, but they could be removed without causing any problem.

```
CREATE OR REPLACE TRIGGER nameTable_insertTT
  BEFORE INSERT ON nameTable
  FOR EACH ROW
BEGIN
    SELECT systimestamp(0) INTO :new.TT_start
      FROM dual;
END;
```

The deletion of a tuple from the current instance consists of the assignment of the value systimestamp(0) to its *TT_end* and of its insertion in the historical instance:

```
CREATE OR REPLACE TRIGGER nameTable_delete
  BEFORE DELETE ON nameTable
  FOR EACH ROW
DECLARE
    now timestamp;
BEGIN
    SELECT systimestamp(0) INTO now
      FROM dual;
    INSERT INTO nameTable_history (A, TT_end)
      VALUES (:old.A, now);
END;
```

Tuple updates are implemented as a deletion followed by an insertion as usual:

```
CREATE OR REPLACE TRIGGER nameTable_update
  BEFORE UPDATE ON nameTable
  FOR EACH ROW
DECLARE
    now timestamp;
BEGIN
    SELECT systimestamp(0) INTO now
      FROM dual;
    INSERT INTO nameTable_history (A, TT_end)
      VALUES (:old.A, now);
    :new.TT_start := now;
END;
```

Finally, the following trigger disallows the execution of updates or deletions on the historical database (similar triggers have been added to prevent the user to execute other improper actions, e.g., to operate on transaction timestamps):

```
CREATE OR REPLACE TRIGGER nameTable_history_upde
  BEFORE UPDATE OR DELETE ON nameTable_history
  FOR EACH ROW
BEGIN
    raise_application_error(-20001,'Historical_tables_cannot
      _be_updated_or_deleted ');
END;
```

As an alternative, one can create one or more user views that specify the privileges of (different classes of) database users, e.g., information in the historical database can be queried, but not updated.

6   Mapping CGG Schemas into the Temporal Model

In [2] we define and implement a translation of CGG schemas into the above-described temporal model. On the one hand, the translation algorithm revises and extends the standard relational encoding of basic ER primitives (entities, relations, specializations,..); on the other hand, it introduces specific rules for the management of spatial and temporal information. Here, we briefly summarize the treatment of CGG temporal features.

The translation introduces a set of relation schemas for every temporal entity and relation in the CGG schema. Such a set consists of a root schema, called kernel, that plays the role of reference schema for all relation schemas generated by a given entity or relation. Each single relation schema is linked to the kernel by means of a suitable foreign key as shown in Figure 1.

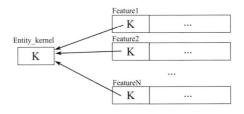

Fig. 1. The relational translation of a temporal entity

Let us consider the case of a temporal entity E with attributes $\mathbf{X} = \{k_1, \ldots, k_n, a_1, \ldots, a_m\}$, whose conceptual key is $\mathbf{K} = \{k_1, \ldots, k_n\}$, with $n \geq 1$ (the case of temporal relations is similar).

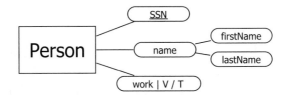

Fig. 2. A CGG entity with different types of attribute

The schema of the kernel consists of the key attributes $k_1, \ldots, k_{n,}$, that is, $E\_kernel(k_1, \ldots, k_n)$. It allows one to identify all entity instances. The temporal features of the entity (temporal qualification of the entity, sets of synchronized temporal attributes, ..) are distributed over different component relations. In addition, each component relation includes key attributes to allow one to merge information about an entity instance. All atemporal (single-valued) attributes $\{a_{i_1}, \ldots, a_{i_l}\}$ are collected in a single component relation $E_{a_{i_1}, \ldots, a_{i_l}}(\underline{k_1}, \ldots, k_n,$ $a_{i_1}, \ldots, a_{i_l})$. A distinct component relation is then added for each group of temporal attributes that change their values in a synchronous way. As an example, consider the entity in Figure 2. Its relational translation is as follows:

$Person\_kernel(\underline{SSN})$

$Person\_atemporal Nochange(\underline{SSN}, firstName, lastName)$

$Person\_work(\underline{SSN}, \underline{VT\_start}, VT\_end, TT\_start, work)$

$Person\_work\_history(\underline{SSN}, \underline{VT\_start}, VT\_end, \underline{TT\_start}, TT\_end, work)$.

References

1. Ben-Zvi, J.: The time relational model. PhD thesis, University of California, Los Angeles (1982)
2. ChronoGeoGraph (2009), http://dbms.dimi.uniud.it/cgg/
3. Combi, C., Montanari, A.: Data models with multiple temporal dimensions: Completing the picture. In: Dittrich, K.R., Geppert, A., Norrie, M.C. (eds.) CAiSE 2001. LNCS, vol. 2068, pp. 187–202. Springer, Heidelberg (2001)
4. Combi, C., Montanari, A., Rossato, R.: A uniform algebraic characterization of temporal functional dependencies. In: Proc. of the 12th International Symposium on Temporal Representation and Reasoning (TIME), pp. 91–99. IEEE Computer Society Press, Los Alamitos (2005)
5. Gubiani, D., Montanari, A.: ChronoGeoGraph: an expressive spatio-temporal conceptual model. In: Ceci, M., Malerba, D., Tanca, L. (eds.) Proc. of the 15th Italian Symposium on Advanced Database Systems (SEBD), pp. 160–171 (2007)
6. Jensen, C.S., Snodgrass, R.T. (eds.): Temporal Database Entries for the Springer Encyclopedia of Database Systems. Technical Report TIMECENTER TR-90 (2008)
7. Jensen, C.S., Snodgrass, R.T.: Semantics of time-varying attributes and their use for temporal database design. In: Papazoglou, M.P. (ed.) ER 1995 and OOER 1995. LNCS, vol. 1021, pp. 366–377. Springer, Heidelberg (1995)
8. Jensen, C.S., et al.: The consensus glossary of temporal database concepts - February 1998 version. In: Temporal Databases, Dagstuhl (TDB), pp. 367–405 (1997)
9. Lorentz, D., et al.: Oracle Database SQL Reference (2005)
10. Navathe, S.B., Ahmed, R.: Temporal extensions to the relational model and SQL. In: Uz Tansel, A., Clifford, J., Gadia, S., Jajodia, S., Segev, A., Snodgrass, R. (eds.) Temporal Databases: theory, design, and implementation, pp. 92–109. The Benjamin/Cummings Publishing Company (1993)
11. Oracle. Oracle 10g (2007), http://www.oracle.com
12. Sarda, N.L.: HSQL: A historical query language. In: Uz Tansel, A., Clifford, J., Gadia, S., Jajodia, S., Segev, A., Snodgrass, R. (eds.) Temporal Databases: theory, design, and implementation, pp. 110–140. The Benjamin/Cummings Publishing Company (1993)
13. Vianu, V.: Dynamic functional dependency and database aging. Journal of the ACM 34(1), 28–59 (1987)
14. Wijsen, J.: Temporal FDs on complex objects. ACM Transactions on Database Systems 24(1), 127–176 (1999)

Three Approximation Algorithms for Energy-Efficient Query Dissemination in Sensor Database System[*]

Zhao Zhang[1], Xiaofeng Gao[2], Xuefei Zhang[2], Weili Wu[2], and Hui Xiong[3]

[1] College of Mathematics and System Sciences, Xinjiang University, P.R. China
zhzhao@xju.edu.cn
[2] Department of Computer Science, University of Texas at Dallas, Richardson, USA
{xxg052000,xxz068000,weiliwu}@utdallas.edu
[3] Management Science & Information Systems Department,
The State University of New Jersey, Rutgers, USA
hxiong@rutgers.edu

Abstract. Sensor database is a type of database management system which offers sensor data and stored data in its data model and query languages. In this system, when a user poses a query to this sensor database, the query will be disseminated across the database. During this process, each sensor generates data that match the query from its covered area and then returns the data to the original sensor. In order to achieve an energy-efficient implementation, it will be useful to select a minimally sufficient subset of sensors to keep active at any given time. Thus, how to find a subset efficiently is an important problem for sensor database system. We define this problem as *sensor database coverage* (SDC) problem.

In this paper, we reduce the SDC problem to *connected set cover* problem, then present two approximation algorithms to select a minimum connected set cover for a given sensor database. Moreover, to guarantee robustness and accuracy, we require a fault-tolerant sensor database, which means that each target in a query region will be covered by at least m sensors, and the selected sensors will form a k-connected subgraph. We name this problem as (k,m)-SDC problem and design another approximation algorithm. These three algorithms are the first approximation algorithms with guaranteed approximation ratios to SDC problem. We also provide simulations to evaluate the performance of our algorithms. We compare the results with algorithms in [17]. The comparison proves the efficiency of our approximations. Thus, our algorithms will become a new efficient approach to solve coverage problem in sensor database systems.

Keywords: Sensor Database, Set Cover, Fault Tolerance.

1 Introduction

1.1 Background

Sensors are often deployed widely to monitor continuously changing entities such as temperature, sound, vibration, pressure, locations of moving objects and other interests.

---

[*] This work is supported by National Natural Science Foundations of China (10671152), NSFC (60603003), the National Science Foundation under grant CCF-0514796 and CNS-0524429. This work was completed when Dr. Zhao Zhang visiting Department of Computer Science, The University of Texas at Dallas.

S.S. Bhowmick, J. Küng, and R. Wagner (Eds.): DEXA 2009, LNCS 5690, pp. 807–821, 2009.
© Springer-Verlag Berlin Heidelberg 2009

The sensor readings are reported to a centralized database system, and are subsequently used to answer queries. Modern sensors not only respond to physical signals to produce data, but also embed computing and communication capabilities. They are able to store and process their productions locally, and transfer data through database system. Examples of monitoring applications include supervising items in a factory warehouse, gathering information in a disaster area, or organizing vehicle traffic in a large city [4]. These applications involve a combination of stored data, and we name them as *sensor databases* [2].

Sensor database system is a newly developed DBMS in recent years, which has been discussed in many literatures such as [2, 3, 21]. In a sensor database, users can issue database queries to one or more nodes in this database. Such process is called *sensor query*, which can also be defined as an acyclic graph of relational and sequence operators [2]. For instance, in a sensor database to measure temperature at regular interval, a typical sensor query can be shown like *"Return repeatedly the abnormal temperatures measured by all sensors"* or *"Every five minutes retrieve the maximum temperature measured over the last five minutes"* [2].

Sensor queries are long-running queries. During the span of a long-running query, relations and sensor sequences might be updated. The inputs of a relational operator are base sequences or the output of another sequence. We define R as a relation of a sensor database, and S as a sensor sequence. An update to R can be an insert, a delete, or modifications of record in R. An update to S is the insertion of a new record associated to a position greater than or equal to all undefined positions in S. There is a centralized realizations of a sensor database [6], where all data from each node in the sensor database is sent to a designated node within the database.

When a user (or an application) poses a query to the sensor database, the query is disseminated across the database. In response to this query, each node generates data that match this query, and transmits matching data to the original sensor. Each sensor can only generate data from its own covered area. As data routed through the database, intermediate sensors might apply one or more database operators. Then users can simply query this database. Such requirement means that users can get the result by querying at any sensor in the system. However, such process is impractical in the sensor database if every sensor can implement queries, since it requires significant communication and too many energy. Due to battery limitations, we need a minimally sufficient subset of sensors which can cover the whole query region at any given time. Since we need to transmit the query data outside the sensor network, such subset should also be connected. We define this problem as *sensor database coverage*(SDC) problem.

By Observation, SDC problem can be reduced to a *connected set cover* problem, which is proved to be NP-hard in general graph [16]. This problem can also be used in distributed Internet measurement systems for distributed agents to periodically measure the Internet by a tool called *traceroute* [5]. In this paper, we propose two approximation algorithms to select minimum connected set cover for a given sensor database. Moreover, to guarantee robustness and accuracy, we require a fault-tolerant sensor database, which means that each target in a query region will be covered by at least m sensors, and the selected sensors will form a k-connected subgraph. Under such constraints, we design another approximation algorithm for (k, m)-SDC problem. To make the algorithm practical, we set $k = 2$ specifically. Both of these algorithms are the first approximation algorithms with guaranteed approximation ratios in general sensor database systems. We also provide simulations to evaluate the performance of our algorithms. We compare the results with algorithms in [17]. The comparison proves the efficiency of our approximations.

1.2   Related Works

The COUGAR project at Cornell University [2] is one of the first attempts to model a sensor database system. It focused on the interaction between the sequence data produced in sensor networks and stored data in backend relational databases. It extended both the SEQ [8] sequence data model and the relational data model by introducing new operators between sequence data and relational data. In [3], R. Cheng and S. Prabhakar presented a framework that represents uncertainty of sensor data. They proposed a new kind of probabilistic queries called *Probabilistic Threshold Query*. Also, they studied techniques for evaluating queries under different details of uncertainty, and investigated the tradeoff between data uncertainty, answer accuracy and computation costs. Recently, A lot of techniques have been introduced to solve coverage problems in sensor networks (e.g., [10, 11, 17, 18, 19, 20, 22]). One of the commonly used approach is reduce sensor coverage problem into connected dominating set (CDS) problem [26]. For k-coverage problem, literatures [9, 27] etc. proposed several greedy algorithms, but did not regard connectivity properties. We can use these techniques to solve SDC problem.

In our paper, we use sensor database as our communication model, which is seldom discussed because of the complexity of problem requirements. Actually, it is well known that minimum set cover (SC) problem is NP-hard [16], and can not be approximated within a factor of $(1 - \varepsilon) \ln n$ for any $\varepsilon > 0$ unless $NP \subseteq DTIME(n^{\log \log n})$ [15], where $n = |V|$. Since SC is a special case of connect set cover (CSC) (taking G to be a complete graph), CSC is also NP-hard and is not $(1 - \varepsilon) \ln n$-approximable. Furthermore, Shuai et.al. [23] showed that even when at most one vertex of the graph G has degree greater than two, the CSC problem is still non-$(1 - \varepsilon) \ln n$-approximable. In the case that the graph is a path, Shuai et.al. gave two polynomial-time algorithms. In the case that the graph has exactly one vertex of degree greater than two, they proposed a $(1 + \ln n)$-approximation algorithm. For the general case, there is no known approximation algorithm with guaranteed performance ratio.

1.3   Our Contribution

In this paper, we provide three efficient approximation algorithms to solve the SDC problem and (k, m)-SDC problem for efficient query dissemination in sensor database systems. Those approximation algorithms are the first ones with approximation ratio analysis. We also provide simulations to evaluate the performance of our algorithms. We compare the results with algorithms in [17]. The comparison proves the efficiency of our approximations. The detailed technologies can be summarized as follows.

We first define a new generalization of the connected set cover (CSC) problem that is equivalent to the SDC problem, and give two approximation algorithms. Assume we have a set collection $\mathbf{S} = \{S_1, S_2, \cdots, S_k\}$. The goal of these two algorithms are finding a minimum size sub-collection $\mathbf{R} \subseteq \mathbf{S}$, such that all the target region is covered by $\mathbf{R}$, and $\mathbf{R}$ is connected. The approximation ratio is highly depends on a parameter $D_c(G)$, which can be defined as follows.

For any two sets $S_i, S_j \in \mathbf{S}$, $dist_G(S_i, S_j)$ is the length of a minimum (S_i, S_j)-path in G, where length refers to the number of edges on this path. Two sets $S_i, S_j \in \mathbf{S}$ are said to be *cover-adjacent* if $S_i \cap S_j \neq \emptyset$. Define $D_c(G) = \max\{dist_G(S_i, S_j) \mid S_i, S_j \in \mathbf{S}$ and S_i, S_j are cover-adjacent$\}$.

The first algorithm is a two-step algorithm. It finds an SC using an α-approximation algorithm, and then connects them with a Steiner Minimum Tree with Minimum Number of Steiner Points (SMT-MSP) using a β-approximation algorithm. The performance ratio of this algorithm is $\alpha + \beta + \alpha\beta(D_c(G) - 1)$. The second algorithm uses a greedy strategy, and the performance ratio is $1 + D_c(G) \cdot H(\gamma - 1)$, where H is the harmonic function, and $\gamma = \max\{|S| \mid S \in \mathbf{S}\}$. In many cases, $D_c = 1$. For example, if two reserves containing a same species are regarded to be adjacent, then $D_c = 1$. In such cases, the two algorithms given in this paper has performance ratio $\alpha + \beta$ and $1 + H(\gamma - 1)$ respectively.

Then, we consider the (k, m)-SDC problem. For a SDC $\mathbf{R}$, if the subgraph of G induced by $\mathbf{R}$ is k-connected, and every element of V is covered by at least m sets of $\mathbf{R}$, then $\mathbf{R}$ is a (k, m)-*connected set cover* ((k, m)-CSC for short). It is obvious that (k, m)-CSC problem is equivalent to (k, m)-SDC problem. If a reserve system takes the form of a (k, m)-SDC, then every species is represented at at least m reserves, and the connection among the reserves is more fault tolerant in face of disasters.

Specifically, in this paper, we present a greedy algorithm for the minimum $(2, m)$-SDC problem, using a parameter $PD(G)$. Given three vertices u, v, w in a graph G, define the *pair distance between u and* $\{v, w\}$, denoted by $dist(u; v, w)$, to be the shortest length of a pair of disjoint (u, v)-path and (u, w)-path. In another word, it is the length of a shortest (v, w)-path through vertex u. The *pair diameter* of a graph G is $PD(G) = \min\{dist(u; v, w)$, where u, v, w are three distinct vertices in $V(G)\}$. Our algorithm has performance ratio $(PD(G) - 1)(1 + H(\gamma - 1))$.

Then, we compared our algorithms to algorithms in [17] in several scenarios. We change the number of sensors in database and the radius of the sensors to exhibit the performance of our algorithms. The result showed that our algorithms are much better than these naive algorithms. The sizes of solutions we obtained are much closer to the corresponding optimum solutions.

The rest of this paper is organized as follows: Section 2 illuminates some basic concepts which may used in algorithm description and performance analysis. Section 3 presents the idea and detailed steps of our approximations for SDC problem. Section 4 provides a greedy algorithm to solve (k, m)-SDC problem. Proofs and performance analysis are also included in these two sections. Section 5 compares our performance with various previous works. Finally, Section 6 gives a brief conclusion of our work.

2 Preliminaries

We consider our communication model under general graphs, which can reflect any type of sensor database in practice, bringing benefits and efficiency to real-life applications.

Actually, we do not need to consider specific geometrical characteristics, since they are too strict to what dimension the models are built, and based on Euclidean formula (e.g., some of the rules are suitable in 2-dimensional space, but incorrect in 3-dimensional space). Therefore, our algorithm can be implemented in a wide range of environments. The following are basic definitions that we need to use in our algorithm descriptions.

Definition 1 (Query Region). *Query Region is the area of the entire sensor database that the end user (or an application) wants to issue a query.*

Definition 2 (Sensor Covering a Point). *A sensor in a sensor database system S is said to cover a point p, if the distance $d(p,S)$ between p and S is less than R_S, which is the sensing radius of the sensor (Here we assume that each sensor has the same R_S).*

Definition 3 (Sensor Database Coverage (SDC)). *Given a sensor database system with sensor set S, where $S = \{S_1, ..., S_k\}$. We need to find a minimum subset R of S to cover all the query region, such that the subgraph induced by R is connected.*

Definition 4 (Set Cover (SC)). *Let V be a set of elements, and S be a family of subsets of V such that $\bigcup_{S \in S} S = V$. A set cover (SC) with respect to (V, S) is a sub-family R of S such that every element $v \in V$ is in some set $S \in R$. We say that S covers v.*

Definition 5 (Connected Set Cover (CSC)). *Let G be a connected graph on vertex set S. A connected set cover with respect to (V, S, G) (abbreviated as CSC) is a set cover R with respect to (V, S) such that the subgraph of G induced by R is connected.*

Definition 6 ((k,m)-CSC). *A (k,m)-CSC is a set cover R with respect to (V, S) such that the subgraph of G induced by R is k-connected, and every element of V is covered by at least m sets of R, then R is a (k,m)-connected set cover ((k,m)-CSC for short).*

Note that we use terminology 'set' and 'vertex' interchangeably when talking about elements in S. Because the sensor database coverage problem is indeed a generalization of the connected set cover problem.

Now let us discuss how to reduce SDC problem to CSC problem (so that (k,m)-SDC is equivalent to (k,m)-CSC). SDC problem is considering cover a whole region, while CSC problem is considering cover a set of targets. Thus, we need to reduce region coverage into target coverage. Figure 1 is an example to illustrate this reduction.

The square in Fig. 1 is the potential region for sensing. For each point in this area, let A denote the set of sensors that can cover this point. Partition the area into different parts, each with different sensor coverage set. As a consequence, area coverage problem can be reduced to target coverage problem when we consider each part as a target. This problem can further reduced to a set cover problem. Consider each small division as a target and mark it with a number from 1 to 14, and then insert covered targets into each set of sensors. Say, $S_1 = \{7, 9, 11, 12, 13, 14\}$, $S_2 = \{6, 9, 13\}$, $S_3 = \{3, 6, 7, 8, 10, 13, 14\}$, $S_4 = \{3, 4, 5\}$, $S_5 = \{2, 1, 3, 4, 8, 12, 14\}$ and $S_6 = \{2\}$. The coverage problem can be reduced to the problem of finding a minimum set cover from S_i to cover $T = \{1, \cdots, 14\}$.

Next, we will introduce our approximation algorithm for SDC problem in the next section now.

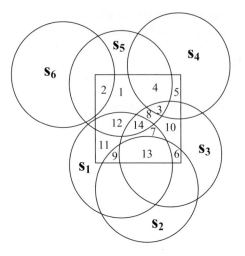

Fig. 1. An example to reduce region coverage to target coverage

3 Algorithm for SDC Problem

In this section, we will firstly exhibit a two-step SDC algorithm, and then give a modified greedy algorithm to deal with more general cases. In the next section, we will show algorithm to solve a special (k,m)-SDC problem where $k = 2$. Performance analysis and theory proofs are provided immediately after algorithm descriptions.

3.1 Two-Step SDC Algorithm

Firstly, let us depict the two-step algorithm as follows.

Algorithm 1. (Two-Step SDC)

Input: (V,S,G); an algorithm **A** computing a minimum set cover; an algorithm **B** computing a Steiner tree with minimum number of Steiner points.
Output: A connected set cover **R**.

 1: Use **A** to compute a set cover $\mathbf{R}_1$ with respect to (V,S).
 2: Use **B** to compute a Steiner tree T in G with terminal set $\mathbf{R}_1$. and Steiner points $\mathbf{R}_2$.
 3: Output $\mathbf{R} = \mathbf{R}_1 \cup \mathbf{R}_2$.

Theorem 1. *Suppose the approximation ratio of* **A** *and* **B** *are* α *and* β *respectively. Then the approximation ratio for Two-Step SDC is* $\alpha + \beta + \alpha\beta(D_c - 1)$.

Proof. Let $\mathbf{R}^*$ be an optimal solution to SDC, and $\mathbf{R}_2^*$ be a Steiner tree of G connecting terminal set $\mathbf{R}_1$ with minimum number of Steiner points. Since $\mathbf{R}^*$ is also a set cover with respect to (V,S), we have

$$|\mathbf{R}_1| \leq \alpha|\mathbf{R}^*|. \tag{1}$$

Let S be a set in $\mathbf{R}_1$. Suppose v is an element of V covered by S, and S^* is a set in $\mathbf{R}^*$ covering v. Then S, S^* are cover-adjacent, and thus $dist_G(S,S^*) \leq D_c$. By adding at most

$D_c - 1$ vertices of G connects S to S^*. It follows that by adding at most $(D_c - 1)|\mathbf{R}_1|$ vertices, all sets in $\mathbf{R}_1$ are connected to $\mathbf{R}^*$. Since $G[\mathbf{R}^*]$ is connected, we have

$$|\mathbf{R}_2^*| \le |\mathbf{R}^*| + (D_c - 1)|\mathbf{R}_1|. \tag{2}$$

Combining inequalities (1) and (2) with $|\mathbf{R}_2| \le \beta|\mathbf{R}_2^*|$, the approximation ratio follows.

In the Relay Node Problem, if $R \ge 2r$, then $D_c = 1$ and thus the approximation ratio is $\alpha + \beta$.

3.2 Best Candidate Path Algorithm (BCP) for SDC Problem

Now let us introduce the best candidate path algorithm (BCP) for SDC problem with better performance.

Definition 7 (Candidate Path). *Let $\mathbf{R}$ records the sets which have been chosen and U records the set of elements of V which have been covered. For $\mathbf{R} \ne \emptyset$ and a set $S \in \mathbf{S} \setminus \mathbf{R}$, an $\mathbf{R}$-S candidate path is a path in G such that its initial vertex is in $\mathbf{R}$ and its end vertex is S.*

For a shortest $\mathbf{R}$-S candidate path P_S, it has exactly $|P_S|$ vertices in $\mathbf{S} \setminus \mathbf{R}$, where $|P_S|$ is the number of edges in P_S. We use $C(P_S)$ to denote the set of elements of $V \setminus U$ which are covered by vertices on P_S. Define $e(P_S) = \frac{|P_S|}{|C(P_S)|}$. Then we have Algorithm 2.

Algorithm 2. (BCP Algorithm)

Input: $(V, \mathbf{S}, G)$.
Output: A connected set cover $\mathbf{R}$.
 1: Choose $S_1 \in \mathbf{S}$ such that $|S_1|$ is maximum. $\mathbf{R} = \{S_1\}, U = S_1$.
 2: **while** $V \setminus U \ne \emptyset$ **do**
 3:      For each $S \in \mathbf{S} \setminus \mathbf{R}$ which is cover-adjacent with a set in $\mathbf{R}$, compute a shortest $\mathbf{R}$-S path P_S.
         Choose S such that $e(P_S)$ is minimum. Add all sets on P_S except S into $\mathbf{R}$, $U = U \cup C(P_S)$.
 4: **end while**
 5: Output $\mathbf{R}$.

Clearly, the output $\mathbf{R}$ of Algorithm 2 is a connected set cover for $(V, \mathbf{S}, G)$. Next, we analyze the approximation ratio.

Theorem 2. *The BCP Algorithm has approximation ratio $1 + D_c(G) \cdot H(\gamma - 1)$, where $\gamma = \max\{|S| \mid S \in \mathbf{S}\}$, and H is the harmonic function.*

Proof. Suppose S_i is the set chosen in the ith iteration (S_1 is the initial set chosen in line 1). Let $\mathbf{S}_i$ be the set of sets added to $\mathbf{R}$ in the ith iteration (that is, the sets on P_{S_i} which is not already in $\mathbf{R}$). Then $\mathbf{R}_k = \bigcup_{i=1}^{k} \mathbf{S}_i$ is the set of sets chosen after the kth iteration. Suppose Algorithm 2 runs K rounds. Then $\mathbf{R}_K$ is the output of the algorithm. When S_i is chosen, we assign each element $v \in C(P_{S_i})$ a weight $w(v) = e(P_{S_i})$ for $i \ge 2$ and $w(v) = 1/|S_1|$ for $i = 1$. Then each element $v \in V$ is assigned a weight exactly once, and

$$\sum_{v \in V} w(v) = \sum_{k=1}^{K} \sum_{v \in C(P_{S_k})} w(v) = \sum_{k=1}^{K} \sum_{v \in C(P_{S_k})} \frac{|P_{S_k}|}{|C(P_{S_k})|} = \sum_{k=1}^{K} |P_{S_k}| = |\mathbf{R}_K|. \tag{3}$$

Suppose $\mathbf{R}^* = \{S_1^*, ..., S_{opt}^*\}$ is an optimal solution to the SDC problem. Set $N_1 = S_1^*$, and for $i = 2, ..., k$, set $N_i = S_i^* \setminus (\cup_{j=1}^{i-1} N_j)$. Since $\mathbf{R}^*$ covers all elements of V, we see that $N_1, ..., N_{opt}$ is a partition of V. It follows that

$$\sum_{v \in V} w(v) = \sum_{k=1}^{opt} \sum_{v \in N_k} w(v). \tag{4}$$

Next, we show that for each $k \in \{1, ..., opt\}$,

$$\sum_{v \in N_k} w(v) \leq 1 + D_c(G) \cdot H(\gamma - 1). \tag{5}$$

Let $n_0 = |N_k|$, and for $i = 1, ..., k$ let n_i be the number of elements in N_k which are not covered after the ith iteration. For $i = 1, ..., k$, after the ith iteration, $n_{i-1} - n_i$ elements of N_k are covered and each such an element is assigned a weight

$$e(P_{S_i}) \leq e(P_{S_k^*}) = \frac{|P_{S_k^*}|}{|C(P_{S_k^*})|} \leq \frac{D_c(G)}{n_{i-1}} \text{ for } i \geq 2, \tag{6}$$

and at most $1/(n_0 - n_1)$ for $i = 1$. There are something to be explained about (6).

(a) It is possible that $n_{i-1} - n_i > 0$. But only those i's with $n_{i-1} - n_i > 0$ works in the analysis.

(b) As a consequence of the above assumption, S_k^* is not chosen after the $(i-1)^{th}$ iteration since choosing S_k^* covers all the elements in N_k. Furthermore, $n_0 - n_1 > 0$ implies that

$$S_k^* \text{ is cover-adjacent with } S_1. \tag{7}$$

Hence S_k^* is a candidate to be chosen as S in the ith iteration for $i \geq 2$. By the choice of S_i, the first inequality of (6) holds.

(c) Also by observation (7), we have $|P_{S_k^*}| \leq D_c(G)$. Since choosing S_k^* could cover all the remaining elements in N_k, we have $|C(P_{S_k^*})| \geq n_{i-1}$. The second inequality in (6) holds.

Then by a standard analysis as in dealing with set cover problem (see for example [14] §35.3), we have

$$\sum_{v \in N_k} w(v) \leq (n_0 - n_1) \frac{1}{n_0 - n_1} + D_c(G) \sum_{i=2}^{opt} \frac{n_{i-1} - n_i}{n_{i-1}}$$
$$\leq 1 + D_c(G)(H(n_1) - H(n_{opt})).$$

Inequality (5) follows from the observation that $n_{opt} = 0$ and $n_1 < n_0 = |N_k| \leq |S_k^*| \leq \gamma$. Combining inequalities (3) (4) and (5), we have

$$|\mathbf{R}| = \sum_{k=1}^{opt} \sum_{v \in N_k} w(v) \leq (1 + D_c(G)H(\gamma - 1)) \cdot opt.$$

The theorem is proved.

4   Best Efficiency Ear Algorithm (BEE) for $(2, m)$-SDC Problem

In this section we provide another algorithm, best candidate ear algorithm (BEE) for (k, m)-SDC problem with fixed parameter $k = 2$. To compute a $(2, m)$-SDC, we make use of the *ear decomposition* of 2-connected graphs.

Definition 8 (Ear). *An ear of a graph G is a path P in G such that all internal vertices on P has degree two in G.*

An ear is *open* if its two ends are different, otherwise it is *closed*. A cycle is a closed ear. The ear decomposition theorem says that every 2-connected graph has an open ear P such that the graph obtained by deleting internal vertices of P from G is still 2-connected. In another word, a graph G is 2-connected if and only if G can be constructed in the following way: Starting from a cycle (that is a closed ear); Iteratively adding open ears to the graph.

The BEE Algorithm computes a $(2, m)$-SDC by greedy strategy. It starts from a 'most efficient' cycle, then iteratively adds 'most efficient' open ears to it until all the cover requirements are satisfied.

To compute the open ears, we use the concept of shortest (u, v)-cycle.

Definition 9 (Shortest (u, v)-cycle). *For two distinct vertices u, v in a graph G, a shortest (u, v)-cycle is a cycle in G through u and v such that the length of the cycle (that is, the number of edges in the cycle) is minimum.*

A shortest (u, v)-cycle can be computed by any algorithm finding *shortest pair of disjoint paths*. In fact, the union of a pair of disjoint (u, v)-paths is an (u, v)-cycle. There are many algorithms for shortest pair of disjoint paths problem, for example, [24].

For a subgraph H of G, a shortest open ear to H through a given vertex $v \in V(G) \setminus V(H)$ can be computed as follows: Add a new vertex s to G and connect s to every vertex in H; Compute a shortest (v, s)-cycle in the extended graph; Then the path obtained by deleting s from this cycle is as required.

4.1   Algorithm Description

Now let us give the detailed description of $(2, m)$-SDC algorithm in Algorithm 3.

In this algorithm, each element $v \in V$ is assigned a label $m(v)$ which records the remaining number of times element v is to be covered. Initially $m(v) = m$ for all v. When $m(v)$ decreases to zero, we say that the cover requirement on v is satisfied. The total number of remaining cover requirements is recorded by M. Initially $M = m|V|$. Set U is used to record the elements of V whose cover requirements has not been satisfied.

For an ear Q_S computed in the algorithm, we use $c(Q_S)$ to denote the number of cover requirements satisfied by adding Q_S to the currently constructed 2-connected subgraph. To speak it more concretely, for each element $v \in U$, let $m'(v)$ be the number of sets in $V(Q_S) \setminus \mathbf{R}$ which cover v, and set $\widetilde{m}(v) = \min\{m'(v), m(v)\}$. Then $\widetilde{m}(v)$ is the number of requirements newly satisfied at element v by adding Q_S, and $c(Q_S) = \sum_{v \in U} \widetilde{m}(v)$ is the total number of requirements newly satisfied by adding Q_S. Define the *efficiency* of Q_S to be

$$e(Q_S) = \frac{|V(Q_S) \setminus \mathbf{R}|}{c(Q_S)}.$$

Algorithm 3. (BEE Algorithm)

Input: (V, S, G), where G is 2-connected and every element in V is covered by at least m sets in **S**.

Output: A $(2, m)$-connected set cover **R**.

1: Set $M = m|V|$, $U = V$, and $m(v) = m$ for each $v \in V$.
2: Choose $S_1 \in S$ such that $|S_1|$ is maximum. $\mathbf{R} = \{S_1\}$. For each element $v \in S_1$, set $m(v) = m(v) - 1$. $M = M - |S_1|$. Remove all vertices v in U with $m(v) = 0$.
3: **if** $M = 0$ **then**
4:    Output **R**.
5: **else**
6:    For each $S \in \mathbf{S} \setminus \mathbf{R}$, compute a shortest (S_1, S)-cycle Q_S.
7:    Choose S such that $e(Q_S)$ is minimum.
8:    **for** each set $R \in V(Q_S) \setminus \mathbf{R}$ **do**
9:       $\mathbf{R} = \mathbf{R} \cup \{R\}$.
10:      For each element $v \in R \cap U$, $m(v) = m(v) - 1$, $M = M - 1$, and remove v from U if $m(v) = 0$.
11:   **end for**
12: **end if**
13: **while** $M > 0$ **do**
14:    Construct a graph $\widetilde{G}$ by adding a new vertex S_0 and connect S_0 to every vertex in **R**.
15:    For each $S \in \mathbf{S} \setminus \mathbf{R}$, compute a shortest (S_0, S)-cycle in $\widetilde{G}$. Let Q_S be the open ear to $G[\mathbf{R}]$ obtained by deleting S_0 from this cycle.
16:    Choose S such that $e(Q_S)$ is minimum.
17:    **for** each set $R \in V(Q_S) \setminus \mathbf{R}$ **do**
18:       $\mathbf{R} = \mathbf{R} \cup \{R\}$.
19:       For each element $v \in R \cap U$, $m(v) = m(v) - 1$, $M = M - 1$, and remove v from U if $m(v) = 0$.
20:    **end for**
21: **end while**
22: Output **R**.

Line 6 to 11 is constructing the initial cycle and line 14-20 is iteratively adding open ears. By the ear decomposition theorem, the output of Algorithm 3 is indeed a $(2, m)$-SDC.

4.2 Performance Analysis

To analyze the performance ratio of the BEE Algorithm, we define the concept of pair diameter. Given three vertices u, v, w in a graph G, define the *pair distance between u and $\{v, w\}$*, denoted by $dist(u; v, w)$, to be the shortest length of a pair of disjoint (u, v)-path and (u, w)-path. In another word, it is the length of a shortest (v, w)-path through vertex u. The *pair diameter* of a graph G is $PD(G) = \min\{dist(u; v, w) \mid u, v, w$ are three distinct vertices in $V(G)\}$.

Theorem 3. *The performance ratio of BEE Algorithm is $PD(G)(1 + H(\gamma - 1))$, where $\gamma = \max\{|S| \mid S \in \mathbf{S}\}$.*

Proof. The proof idea is similar to that of Theorem 2. The difference lies in dealing with the multiple covering of each element and estimating the length of added ear.

Suppose $V = \{v_1, ..., v_n\}$ where $n = |V|$. Duplicate each element v_i by m times. Denote by $V_i = \{v_i^{(1)}, ..., v_i^{(m)}\}$, where $v_i^{(1)}, ..., v_i^{(m)}$ are the duplicates of element v_i. Set $\mathbf{V} = \bigcup_{i=1}^{n} V_i$.

Use the notation S_i, $\mathbf{R}_k$ as in the proof of Theorem 2. Suppose Algorithm 3 runs K rounds. For $i \geq 2$, when S_i is chosen, sets in $V(Q_{S_i}) \setminus \mathbf{R}$ are added into $\mathbf{R}$ sequentially in line 8 to line 11. When it is the turn to deal with $R \in V(Q_{S_i}) \setminus \mathbf{R}$, a vertex $v \in R \cap U$ has its copy $v^{(m(v))}$ assigned a weight $e(Q_{S_i})$ (recall that $1 \leq m(v) \leq m$ is the remaining cover requirements on v just before R is added to $\mathbf{R}$). We may regard R to cover $v^{(m(v))}$. When $i = 1$, each element $v \in S_1$ has its copy $v^{(m)}$ assigned a weight $1/|S_1|$. Then each element $v^{(j)} \in \mathbf{V}$ is assigned a weight exactly once.

Suppose $\mathbf{R}^* = \{S_1^*, ..., S_{opt}^*\}$ is an optimal solution to the $(2, m)$-SDC problem. Define a partition $N_1, ..., N_{opt}$ of $\mathbf{V}$ as follows (write $\mathbf{N}_i = \bigcup_{k=1}^{i} N_k$ for simplicity): Set $N_1 = \{v^{(1)} \mid v \in S_1^*\}$, and for $i = 2, ..., opt$, set $N_i = \{v^{(j)} \mid v \in S_i^*, v^{(m)} \notin \mathbf{N}_{i-1}, j$ is the first index such that $v^{(1)}, ..., v^{(j-1)} \in \mathbf{N}_{i-1}$ and $v^{(j)} \notin \mathbf{N}_{i-1}\}$. Figure 2 illustrates the partition.

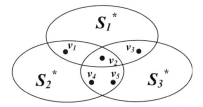

Fig. 2. An illustration of the partition. Here we have that $m = 2$, $N_1 = \{v_1^{(1)}, v_2^{(1)}, v_3^{(1)}\}$, $N_2 = \{v_1^{(2)}, v_2^{(2)}, v_4^{(1)}, v_5^{(1)}\}$, $N_3 = \{v_3^{(2)}, v_4^{(2)}, v_5^{(2)}\}$.

The following proof is similar to that in Theorem 2. The only difference is using $PD(G)$ to upper bound $|V(Q_{S_k^*}) \setminus \mathbf{R}_i|$. Note that for each $k \in \{1, ..., opt\}$, $|N_k| \leq \gamma$ since each element v has at most one copy in N_k.

5   Performance Evaluation

In this section, we present tow simulations to evaluate the performance of our approximation algorithms. Since the performance of BCP is better than two-step SDC, we will build one simulation for BCP algorithm, and another with (k, m)-SDC.

We ran our algorithms on a randomly generated sensor database system where a certain number of sensor nodes are placed randomly in an area of 40×40 unit square. We assume that the query region is the entire sensor database region. Each sensor has a uniform sensing radius of 4 units, which means that it can only detect targets in a cycle of 4 units with itself as the center. We vary the size n of this sensor database from 1000 to 4000 (which provides substantial redundancy), and deploy these sensors randomly. For each fixed topology, we calculate the required subset from our algorithm. Also, we determine the sensing radius R of sensor nodes from 2 units to 12 units. In both scenarios, we compare our outputs with the results calculated from K Times 1-Greedy algorithm, which is mentioned in [17].

5.1 Result for BCP Algorithm

We plot the size of the solution by different algorithms in Figure 3 for different values of sensor database sizes, and transmission radius. For each parameter, we run each scenario for 1000 times and take average value as our solution to avoid abnormal cases.

Figure 3 (a) shows the solution size delivered by two algorithms with different sensor database sizes. Note that the solution size is much smaller than the sensor database size (10^2 vs. 10^3), which means that selecting a subset of sensors to execute sensor queries for a sensor database do save many energy. Therefore, selecting a connected subset is an energy-efficient method for a sensor database system to improve its performance.

From this figure we can also observe that when the size of database increases, the solution size decreases. This is because when the target region fixed, if the density of sensors increases, it is easier to find a connected subset to cover the whole area. More- over, the solution size doesn't decrease so much when the sensor database size becomes larger, which means the solution obtained from 1000 sensors is quite close to the OPT. It also means that the sensors database provides substantial redundancy if the size in- creases. We can see that for a fixed sensing radius, the solution size returned by BCP algorithm is almost the half of the solution returned by K times 1-greedy algorithm, which proves that our algorithm can perform better results.

Figure 3 (b) shows the solution size delivered by various algorithms for different sensing radius. Note that when the sensing radius change from 2 to 4, the solution size has a sharp decrease, but if the sensing radius increase continuously, the solution size stays stable. That means when the sensing radius is bigger than 4, the query region is covered more than once. This property can guarantee robustness and accuracy. Also note that the solution size is quit small and good enough to reach the optimum solution when the sensing radius becomes larger.

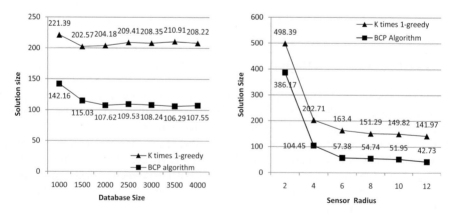

Fig. 3. Solution size of connected sensor cover delivered by various algorithms with different sensor database size and transmission radius

5.2   Result for BEE Algorithm

We plot the size of the solution from different algorithms in Figure 4 for different values of coverage degree. The coverage degree m denotes that each target should be covered by at least m sensors. Similarly as previous testing, under every parameter we run each algorithms for 1000 times, and choose the average value as our final result, so that we will avoid extreme cases (since the topology is randomly generated).

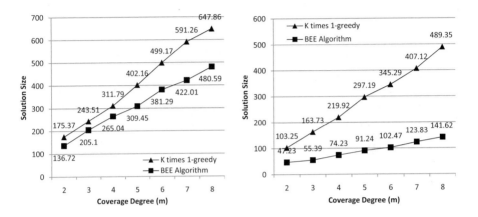

Fig. 4. Solution size of K-cover delivered by two algorithms with different K. Here the size of our sensor database 3000.

Figure 4 (a) shows the solution size delivered by two algorithms with different values of coverage degrees when the sensing radius is 4 units. Note that the results returned by each algorithm almost constitute a line in the 2-D plane. It means that to get larger coverage degree m, we need to select more sensors. Also we can see that the slope of BEE algorithm is much smaller than the slope of the K times 1-greedy algorithm, which proves the efficiency of BEE algorithm. This means the BEE algorithm uses less duplicated sensors than K times 1-greedy algorithm when the sensing radius increases.

Figure 4 (b) shows the solution size delivered by two algorithms with different values of coverage degree when the sensing radius is 8 units. We have similar conclusion as in Figure 4 (a). Note that the slope of BEE algorithm in (b)is smaller than in (a). From this, we can observe that the redundancy decreases when the sensing radius increases.

5.3   Summary

From the above two figures, we can see that the BCP algorithm and BEE algorithm did better that the K times 1-greedy algorithm for different size of the sensor database system, different sensing radius and different value of coverage degree. By apply these algorithms, we can have less redundant sensors in the system to guarantee an efficient, energy-saving and robust system. We can conclude that our algorithms are really efficient. Thus, our algorithms become a new approach to solve coverage problem in sensor database.

6 Conclusion

In this paper, to deal with coverage problem in sensor database system, we introduce minimum connected set cover (SDC) problem and k-connected m-set cover problem $((k,m)$-SDC$)$ for fault-tolerance. Moreover, we provide two approximation algorithms for SDC problem in general sensor database systems. Logarithm performance guarantee was obtained, incorporating a new parameter D_c which measures the maximum distance between two sets covering a common element. We also give a logarithm approximation algorithm for Minimum (k,m)-SDC problem with fixed $k=2$, using a new parameter $PD(G)$ which in fact measures the maximum length of an ear. These are the first algorithms for SDC problems in general graphs with guaranteed performance ratio. These two algorithms can become a new approach to deal with coverage problem in sensor database.

To improve the performance ratio is one of our future directions. To study the Minimum (k,m)-SDC problem for $k \geq 3$ is another direction. Weighted version of SDC problem is also an interesting topic. However, the methods used in this paper can not be generalized for that. A lot of deep insights and new ideas are needed.

References

1. Wilschut, A.N., Apers, P.M.G.: Dataflow Query Execution in a Parallel Main-Memory Environment. Distributed and Parallel Databases 1(1), 103–128 (1993)
2. Bonnet, P., Gehrke, J., Seshadri, P.: Towards Sensor Database Systems. Mobile Data Management, 3–14 (2001)
3. Cheng, R., Prabhakar, S.: Managing Uncertainty in Sensor Database. ACM SIGMOD 32(4), 41–46 (2003)
4. Estrin, D., Govindan, R., Heidemann, J.: Embedding the Internet: Introduction. Communications of the ACM Journal 43(5), 38–41 (2000)
5. Gonen, M., Shavitt, Y.: A Θ (log n)-Approximation for the Set Cover Problem with Set Ownership. Information Processing Letters 109, 183–186 (2009)
6. Govindan, R., Hellerstein, J.M., Hong, W., Madden, S., Franklin, M., Shenker, S.: The Sensor Network as a Database, USC Computer Science Department Technical Report (September 2002)
7. Hellerstein, J.M., Avnur, R., Ranman, V.: Informix under CONTROL: Online Query Processing. Data Mining and knowledge Discovery 4(4) (October 2000)
8. Seshadri, P., Livny, M., Ramaakrishman, R.: SEQ: A Model for Sequence Databases. In: Proceedings of the 11th International Conference on Data Engineering (ICDE), pp. 232–239 (1995)
9. Abrams, Z., Goel, A., Plotkin, S.: Set k-Cover Algorithms for Energy Efficient Monitoring in Wireless Sensor Networks. In: Proceedings of the 3rd Conference on Information Processing in Sensor Networks, IPSN 2004 (2004)
10. Cardei, M., Wu, J.: Energy-Efficient Coverage Problems in Wireless Ad-Hoc Sensor Networks. Computer Communications 29(4), 413–420 (2006)
11. Cardei, M., Thai, M., Li, Y., Wu, W.: Energy-Efficient Target Coverage in Wireless Sensor Networks. In: Proceedings of 24th Annual Joint Conference of the IEEE Computer and Communication Societies (INFOCOM 2005), Miami, Florida USA, March 13-17, pp. 1976–1984 (2005)

12. Cardei, M., Du, D.Z.: Improving Wireless Sensor Network Lifetime through Power Aware Organization. ACM Wireless Networks 11(3), 333–340 (2005)
13. Cerdeira, J.O., Pinto, L.S.: Requiring Connectivity in the Set Covering Problem. Journal of Combinatorial Optimization 9, 35–47 (2005)
14. Cormen, T.H., Leiserson, C.E., Rivest, R.L., Stein, C.: Introduction to Algorithms, 2nd edn (2002)
15. Feige, U.: A Threshold of $\ln n$ for Approximating Set Cover. In: Proceedings of the 28th ACM Symposium on Theory of Computing (ACM 1996), pp. 314–318 (1996)
16. Garey, M.R., Johnson, D.S.: Computers and Intractability. W.H. Freeman and Company, New York (1979)
17. Gupta, H., Das, S.R., Gu, Q.: Connected Sensor Cover: Self-Organization of Sensor Networks for Efficient Query Execution. In: Proceedings of the 4th ACM International Symposium on Mobile Ad Hoc Networking and Computing, MobiHoc 2003 (2003)
18. Huang, C.F., Tseng, Y.C.: The Coverage Problem in a Wireless Sensor Network. In: Proceedings of the 2nd ACM international conference on Wireless sensor networks and applications, pp. 115–121 (2003)
19. Jaggi, N., Abouzeid, A.A.: Energy-Efficient Connected Coverage in Wireless Sensor Networks. In: Proceedings of 4th Asian International Mobile Computing Conference, Kolkata, India, pp. 77–86 (2006)
20. Li, X.Y., Wan, P.J., Frieder, O.: Coverage in Wireless Ad-Hoc Sensor Networks. IEEE Transactions on Computers 52(6), 753–763 (2003)
21. Madden, S.R., Franklin, M.J., Hellerstein, J.M., Hong, W.: TAG: A Tiny Aggregation Service for Ad-Hoc Sensor Networks. In: OSDI (2002)
22. Meguerdichian, S., Koushanfar, F., Potkonjak, M., Srivastava, M.B.: Coverage Problems in Wireless Ad-Hoc Sensor Networks. In: Proceedings of Twentieth Annual Joint Conference of the IEEE Computer and Communications Societies (INFOCOM 2001), vol. 3, pp. 1380–1387 (2001)
23. Shuai, T.-P., Hu, X.: Connected Set Cover Problem and its Applications. In: Cheng, S.-W., Poon, C.K. (eds.) AAIM 2006. LNCS, vol. 4041, pp. 243–254. Springer, Heidelberg (2006)
24. Suurballe, J.W., Tarjan, R.E.: A Quick Method for Finding Shortest Pairs of Disjoint Paths. Networks 14, 325–336 (1984)
25. Tague, P., Lee, J., Poovendran, R.: A Set-Covering Approach for Modeling Attacks on Key Predistribution in Wireless Sensor Networks, Technical Report CACR, 41 (2005)
26. Thai, M.T., Wang, F., Du, H., Jia, X.: Coverage Problems in Wireless Sensor Networks: Designs and Analysis. International Journal of Sensor Networks, special issue on Coverage Problems in Sensor Networks 3(3), 191–200 (2008)
27. Zhou, Z.H., Das, S., Gupta, H.: Connected K-Coverage Problem in Sensor Networks. In: Proceedings of the 13th International Conference onComputer Communications and Networks (ICCCN 2004), pp. 373–378 (2004)

Top-k Answers to Fuzzy XPath Queries

Bettina Fazzinga, Sergio Flesca, and Andrea Pugliese

DEIS - Università della Calabria
Via Bucci - 87036 Rende (CS) Italy
{bfazzinga,flesca,apugliese}@deis.unical.it

Abstract. Data heterogeneity in XML repositories can be tackled by giving users the possibility to obtain approximate answers to their queries. In this setting, several approaches for XPath queries have been defined in the literature. In particular, fuzzy XPath queries have been recently introduced as a formalism to provide users with a clear understanding of the approximations that the query evaluation process introduces in the answers. However, in many cases, users are not a-priori aware of the maximum approximation degree they would allow in the answers; rather, they are interested in obtaining the first k answers ranked according to their approximation degrees. In this paper we investigate the problem of top-k fuzzy XPath querying, propose a query language and its associated semantics, and discuss query evaluation.

1 Introduction

An important issue is nowadays that of coping with heterogeneous representations of data about a certain domain of interest. In many cases, users need to retrieve information from a data source that adopts a schema which is not completely known. Moreover, the increasing adoption of XML [11] as a common data model "naturally" induces different data sources to employ their specific semi-structured schema – for instance, the first name of an author can be represented as a `firstname` element which is a direct child of an `author` element, or a child of a `name` element which, in turn, is a child of `author`.

Due to these intrinsic XML features, query languages for XML provide some degree of flexibility to resolve differences among the schema employed in a query and the one adopted for data. For instance, XPath [11] provides the **descendant** axis which allows users to select `firstname` elements which are direct or indirect children of `author`. However, XPath still requires some knowledge of the data schema; for instance, an XPath query must use the exact terms appearing in the data when specifying conditions on element names.

In this paper, we consider the scenario where the user is not aware of the schema used by the data source. The main problem to be dealt with is therefore the retrieval of information based on the specification of some properties of the objects to be retrieved. Several approaches have been proposed for XML querying that add flexibility to XPath by automatically adapting queries to the available data [1,3,4,5,9,10]. In particular, the relaxation of a tree pattern query and the use of "transformation costs" has been proposed in [1,9]. In [1] the goal is to identify all answers whose score exceeds a certain

S.S. Bhowmick, J. Küng, and R. Wagner (Eds.): DEXA 2009, LNCS 5690, pp. 822–829, 2009.

threshold. The evaluation algorithm works with complex join plans that embed all possible transformations. Early pruning is performed using branch-and-bound techniques and XML queries are adaptively processed through a lockstep strategy. The querying mechanism only supports node renamings w.r.t. fixed name hierarchies that must be provided. In [3,4] we additionally addressed the problem of combining partial information provided by different sources into a single representation of the objects the user is willing to retrieve.

The proposal presented in this paper is based on the following features:

- *Fuzzy querying.* Our query language gives users the possibility to assign *weights* to XPath steps, that express their relative importance, and an XML element can be part of a query answer "to a certain degree" if it just satisfies some of the conditions in the query. To model this, we employ *fuzzy sets*. A fuzzy set is a pair (P, m) where P is a set and $m : S \rightarrow [0..1]$ is a *membership grade* function that represents to which extent a given element belongs to the set. Thus, in our framework, the answer to a query is a fuzzy set of XML elements whose membership grades reflect the weights of the steps these elements satisfy.
- *Top-k query answering.* The semantics of our query language allows users to retrieve a subset of the query answers that only contains the best k answers, i.e., the k elements which better satisfy a given query.

Consider for instance the XML documents D_1 and D_2 in Fig. 1. A user interested in retrieving books authored by *Silberschatz* and not by *Galvin* could pose an XPath query of the form $exp = //book[//authors[/author[text() =' Silberschatz']]] [not(//authors[/author[text() =' Galvin']])]$ (depicted as a tree pattern in the figure). The exact evaluation of exp against D_1 and D_2 yields no answer. However, element e_2 is "almost" an answer to exp, since the only mismatch is in its

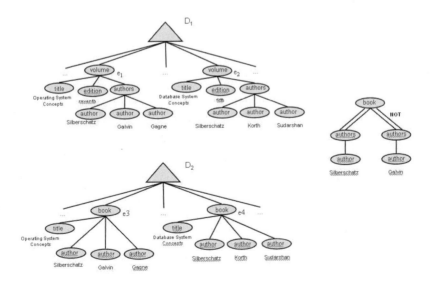

Fig. 1. Example XML documents and XPath query

label (volume instead of book); in element e_4, the only mismatch is the absence of an authors element between elements book and author. Thus, slightly relaxing some of the conditions in exp allows to retrieve potentially interesting elements. Our fuzzy evaluation of the answer to exp against D_1 and D_2 can indeed return all of the elements shown in the figure, along with information (in the form of membership grades) about how they satisfy exp. In addition, keeping only the first k results in the answer allows users to retrieve the highest-quality elements while enabling further optimization of the query evaluation process.

Fuzzy XPath queries have recently proved useful as a formalism to provide users with a clear understanding of the approximations that the XPath query evaluation process introduces in the answers when facing heterogeneous XML data [2,6,7]. For instance, [2] proposes a "deep-similar" function which aims at enhancing the typical XPath "deep-equal" function. Membership grades of XML elements in the results are computed by looking at *edit distances* (i.e., number of operations required to transform one element into the other). The results of the deep-similar function are ranked according to their membership grade, but all of them are always computed. The approach in [7] explicitly addresses the computation of the best k answers through a queue-based adaptive XML query processing technique; the main limitation of the approach is the adoption of a fixed *tf*idf* scoring function.

The main contributions of this paper are a fuzzy XML query language, its associated top-k semantics, and a general query evaluation algebra. Specifically, in Section 2 we give syntax and semantics of the proposed language; in Section 3 we outline our algebra for computing top-k answers and discuss incremental evaluation strategies; finally, Section 4 contains concluding remarks and discusses future work.

2   Top-k Fuzzy XPath Queries

We model an XML document D as a node-labeled tree. We denote the root of D as $root(D)$ and the label of an element $e \in D$ as $label(e)$. A leaf element e may also have an associated text content $text(e)$; we will (equivalently) model $text(e)$ as a child element of e with label $text(e)$. In our query language, a *step* is composed by an *axis* (child or descendant), a *node test* and zero or more *predicates* (we restrict the node tests to label and text content equality tests). That is, a step s is of the form $\tilde{s}[p_1] \ldots [p_n]$ where $\tilde{s}$ is of the form $axis : l$, l is a label, and each p_i is a predicate (whose complete form is defined below). A *simple step* is a step with $n = 0$. In our query language, a *weighted step* ws is a step of the form $\tilde{s}\{d_0\}[p_1]\{d_1\} \ldots [p_n]\{d_n\}$, where for each $i \in [0..n]$, $d_i \in [0..1]$ represents the relative *importance* of the associated term and $\sum_{i=0}^{n} d_i = 1$. In addition, the special symbol "*" can be used as a weight, to specify that the associated term (along with its sub-predicates, if any) must be *completely* satisfied by the elements to which it applies. A *weighted XPath expression* is an expression of the form $ws_1/ \cdots /ws_m$ or $not(ws_1/ \cdots /ws_m)$, where $ws_1, \cdots, ws_m$ are weighted steps. A predicate is a weighted XPath expression.

Example 1. The weighted step $ws = //book\{0.6\}[/author]\{0.1\}[/title]\{0.3\}$ (note that we adopt the abbreviated syntax of XPath axes – we will denote the descendant-or-self axis as ↓) expresses that an element satisfies the 60% of ws if it has label book, the 10% of ws if it has an author child, and the 30% of ws if it has a title child. Consequently, a book element with only an author child satisfies the 70% of ws. The same element does not satisfy a weighted step of the form $//book\{0.6\}[/author]\{0.4\}[/title]\{*\}$, which mandates a title child.   □

Fuzzy query evaluation permits to retrieve an element even if it just provides a certain degree of satisfaction w.r.t. some of the conditions specified in a given XPath expression. Thus, we define a measure of the degree of satisfaction by looking at how the conditions specified are satisfied by the element. We first introduce some notations. We denote as $semDist(l, l')$ the semantic dissimilarity between two labels, which has a value in $[0..1]$, if l is considered "similar enough" to l', or ∞, if l is considered "too different" from l'. Moreover, given a simple weighted step $\widetilde{ws} = axis : l\{d_0\}$ and two elements e and e', we write $axSat(axis, e', e) = \delta$, where $\delta \in [0..1]$ is equal to (i) 1, if e is reachable from e' through $axis$; (ii) d, if e is reachable from e' through a path corresponding to the replacement of $axis$ with $//$; (iii) 0, otherwise. The value of d is a tunable parameter of the query engine.

Let e, e' be two elements in an XML document D and exp be a weighted XPath expression. We define the *satisfaction degree* of exp both w.r.t a single element e ($sat_e(exp)$) and w.r.t. two elements e, e' ($sat_e(e', exp)$). In particular, $sat_e(exp) = \max_{e' \in D} sat_e(e', exp)$, if exp is positive, and $sat_e(not(exp)) = 1 - sat_e(exp)$. Moreover $sat_{e'}(e, exp)$ is defined as follows:

- If exp is a simple weighted step, i.e., $exp = axis : l\{d_0\}$, and $label(e) = l_e$, then

$$sat_{e'}(e, exp) = semDist(l, l_e) \cdot axSat(axis, e', e);$$

- If exp is a weighted step, i.e., $exp = \tilde{s}\{d_0\}[p_1]\{d_1\} \ldots [p_n]\{d_n\}$, then

$$sat_{e'}(e, exp) = d_0 \cdot sat_{e'}(e, \tilde{s}\{d_0\}) + \sum_{i \in [1..n]} d_i \cdot sat_e(p_i);$$

- If $exp = ws_1/\cdots/ws_m$, then $sat_{e'}(e, exp) =$

$$\max_{e_1, e_2, \ldots, e_{m-1} \in D} avg\big(sat_{e'}(e_1, \dot{w}s_1), sat_{e_1}(e_2, ws_2), \ldots sat_{e_{m-1}}(e, ws_m)\big).$$

Example 2. A weighted XPath expression corresponding to the XPath expression of Fig. 1 could have the following form: $exp = //book\{0.1\}[//authors\{0.2\}$ $[/author\{0.2\}[text() =' Silberschatz']\{0.8\}]\{0.8\}]\{0.4\}[not(//authors\{0.2\}$ $[/author\{0.2\} [text() =' Galvin']\{0.8\}]\{0.8\})]\{0.5\}$. Suppose we want to evaluate exp on the XML documents in Fig. 1. Consider elements e_1 and e_2. If the semantic distance between *book* and *volume* is 0.8, since there is no text node containing *Galvin* in the subtree rooted at e_2, elements e_1 and e_2 have the following satisfaction degrees: $sat_{root(D_1)}(e_1, exp) = 0.48$, $sat_{root(D_1)}(e_2, exp) = 0.8$. Now consider elements e_3 and e_4. Since there is no *authors* node in the subtrees rooted at e_3 and e_4, and there is no text node containing *Galvin* in the subtree rooted at e_4, elements e_3 and e_4 have the following satisfaction degrees: $sat_{root(D_2)}(e_3, exp) = 0.52$, $sat_{root(D_2)}(e_4, exp) = 0.84$.   □

We can now define the form of fuzzy queries we are interested in and their answers.

Definition 1 (Top-k fuzzy XPath query and query answer). *A top-k fuzzy XPath query is a pair $q = \langle exp, k \rangle$ where exp is a weighted XPath expression and k is a positive integer. Let D be an XML document and E the set of elements in D. An* answer *to q over D ($ans(q, D)$) is a set $\{e_1, \ldots, e_k\} \subseteq E$ such that for each element $e \in E \setminus \{e_1, \ldots, e_k\}$ it holds that $sat_{root(D)}(e, exp) \leq sat_{root(D)}(e_i, exp)$, $\forall i \in [1..k]$.*

Example 3. Consider the weighted XPath expression exp of Example 2 and the XML documents D_1 and D_2 of Fig. 1. The answers to the top-1 fuzzy XPath query corresponding to exp over D_1 and D_2 are the sets $\{e_2\}$ and $\{e_4\}$, respectively.   □

3   Query Evaluation

In this section we give a general algebra for computing answers to top-k fuzzy XPath queries, then we discuss incremental evaluation strategies that aim at increasing the efficiency of the query evaluation process. In the following, we only discuss the evaluation of positive top-k fuzzy Xpath queries, since the evaluation of negative predicates can be performed by first evaluating a top-1 query, taking the complement of the membership grade of the elements in their results and then exploiting these in the evaluation of the global query.

We start by presenting the data access model we employ. The access to XML data is represented by the Pt predicate which identifies pairs of elements connected by a path in an XML document (the latter is not explicitly represented). In particular, $Pt(ax, f)$, where ax is an XPath navigation axis and f is a node test function, is a set of element pairs $\langle e_1, e_2 \rangle$ such that e_2 is reachable from e_1 following a path specified by axis ax and e_2 satisfies f. Pt admits labels, boolean values, or the special symbols doc and sim_l as node test functions: doc is $true$ for the root of the XML document, whereas $Pt(ax, sim_l)$ is the set of pairs $\langle e_1, e_2 \rangle$ where $semDist(l, label(e_2)) < \infty$. Moreover, Pt predicates are annotated with a value in $[0..1]$ whose semantics will be explained in the following. When no subscript is specified a default value of 1 is implied.

In our query evaluation framework, partial results are maintained in fuzzy sets $C = (P_C, m_C)$ of pairs of XML elements whose membership grades represent the satisfaction degree with respect to a weighted XPath expression exp, i.e., $\forall \langle e', e \rangle \in P_C$, $m_C(\langle e', e \rangle) = sat_{e'}(e, exp)$. A pair $\langle e', e \rangle$ in $Pt_x(ax, f)$ has membership grade $x \cdot v$, where $v = semDist(l, label(e))$ if f is sim_l, and $v = 1$ otherwise.

The computation of an answer to a fuzzy XPath query requires to combine the basic information retrieved from the XML document in a proper way. To this end, we introduce the following operators:

- *Fuzzy left outer join:* $C_0 \bowtie_{w_0, w_1, \ldots, w_n} \{C_1, \ldots, C_n\}$. Given $n + 1$ fuzzy sets of partial results $C_0, \ldots, C_n$, $C_0 \bowtie_{w_0, w_1, \ldots, w_n} \{C_1, \ldots, C_n\}$ returns the fuzzy set C defined as follows:

- $P_C = \{\langle e, e_0 \rangle \mid \langle e, e_0 \rangle \in P_{C_0} \text{ s. t. } \forall i \in [1..n] \exists \langle e_0, e_i \rangle \in P_{C_i}\};$

- $\forall \langle e, e_0 \rangle \in P_C, m_C(\langle e, e_0 \rangle) =$

$$\max_{(\langle e_0, e_1 \rangle, \ldots, \langle e_0, e_n \rangle) \in P_{C_1} \times \ldots \times P_{C_n}} \left(\frac{w_0 \cdot m_{C_0}(\langle e, e_0 \rangle) + \sum_{i \in [1..n]} w_i \cdot m_{C_i}(\langle e_0, e_i \rangle)}{w_0 + \ldots + w_n} \right).$$

– *Fuzzy right outer join:* $C_1 \bowtie C_2$. Given two fuzzy sets of partial results C_1 and C_2, $C_1 \bowtie C_2$ returns the fuzzy set C defined as follows:
  - $P_C = \{\langle e, e_2 \rangle \mid \langle e, e_1 \rangle \in P_{C_1}, \langle e_1, e_2 \rangle \in P_{C_2}\};$

  - $\forall \langle e, e_2 \rangle \in P_C,$

$$m_C(\langle e, e_2 \rangle) = \max_{(\langle e, e_1 \rangle, \langle e_1, e_2 \rangle) \in P_{C_1} \times P_{C_2}} \left(\text{avg}(m_{C_1}(\langle e, e_1 \rangle), m_{C_2}(\langle e_1, e_2 \rangle)) \right).$$

– *Fuzzy union:* $\bigcup \{C_1, \ldots, C_n\}$. Given n fuzzy sets of partial results $C_1, \ldots, C_n$, $\bigcup \{C_1, \ldots, C_n\}$ returns the fuzzy set C defined as follows:
  - $P_C = \{\langle e, e' \rangle \mid \exists i \in [1..n] \text{ s. t. } \langle e, e' \rangle \in P_{C_i}\};$

  - $\forall \langle e, e' \rangle \in P_C, m_C(\langle e, e' \rangle) = \max_{i \in [1..n]} \left(m_{C_i}(\langle e, e' \rangle) \right).$

We introduce below an algebraic expression which corresponds to every possible way of partially evaluating the original XPath expression.

Definition 2 (Evaluation plan). *Let exp be a weighted XPath expression. The evaluation plan of exp, denoted as Plan(exp), is defined as follows:*

$$\begin{cases} \text{Plan}(ws_1) \bowtie \ldots \bowtie \text{Plan}(ws_m), & \text{if } exp = ws_1/\ldots/ws_m; \\[2ex] \text{Plan}(\tilde{s}) \bowtie_{d_0, d_1, \ldots, d_n} \{\text{Plan}(p_1), \ldots, \text{Plan}(p_n)\}, & \text{if } exp = \tilde{s}\{d_0\}[p_1]\{d_1\} \ldots [p_n]\{d_n\}; \\[2ex] \text{Pt}(ax, l) \cup \text{Pt}_d(//, l) \cup \text{Pt}(ax, \text{sim}_l) \cup \\ \quad \text{Pt}_d(//, \text{sim}_l) \cup \text{Pt}_0(\downarrow, *), & \text{if } exp = ax : l. \end{cases}$$

It is intuitive enough that the evaluation plan of an XPath expression permits to obtain every element which is in the answer of the corresponding fuzzy XPath query, as ensured by the following proposition.

Proposition 1. *Let D be an XML document, $q = \langle exp, k \rangle$ a top-k fuzzy XPath query, and $S = \{e \mid (root(D), e) \in Plan(exp)\}$. It holds that: (i) every answer to q over D is a subset of S; (ii) every subset of k elements of S such that there is no element in S with higher membership grade is an answer to q over D.*

3.1   Incremental Evaluation

In this section we briefly discuss an incremental evaluation technique for top-k fuzzy XPath queries and three different evaluation strategies.

Given a query $\langle exp, k \rangle$ and an evaluation plan $Plan(exp)$, we associate an *agent* α with the (incremental) evaluation of each operator in $Plan(exp)$. We denote the agents devoted to the evaluation of $\bowtie$, $\bowtie$, and $\cup$ operators with $\alpha_\bowtie, \alpha_\bowtie$ and $\alpha_\cup$, respectively.

The incremental evaluation of a query plan is activated by *evaluation requests* which are issued to $\alpha_{\cup}$ agents. Since an $\alpha_{\cup}$ agent is associated with an algebraic expression of the form $Pt(ax, l) \cup Pt_d(//, l) \cup Pt(ax, sim_l) \cup Pt_d(//, sim_l) \cup Pt_0(\downarrow, *)$ in the plan, an evaluation request is an expression of the form $r(\alpha, p)$, where α is the $\alpha_{\cup}$ agent and p is an integer identifying the position of the Pt predicate in the algebraic expression whose evaluation is being requested (positions are 0-based). When an $\alpha_{\cup}$ agents receives a request $r(\alpha_{\cup}, p)$, it performs the following steps: (*i*) evaluate the Pt predicate at position p; (*ii*) compute the union of the results obtained in the previous step with the previously computed ones; (*iii*) propagate upwards the results of the previous step which are new or have improved their membership grade.

$\alpha_{\bowtie}$ agents are dedicated to the evaluation of algebraic expressions of the form $Plan(\tilde{s}) \ltimes_{d_0, d_1, \ldots, d_n} \{Plan(p_1), \ldots, Plan(p_n)\}$. An agent $\alpha_{\bowtie}$ receives its inputs from agents $\alpha_0, \alpha_1, \ldots, \alpha_n$, which evaluate $Plan(\tilde{s}), Plan(p_1), \ldots, Plan(p_n)$, respectively. New (or improved) results coming from an agent α_i are denoted as $\Delta(\alpha_i)$ while existing ones are denoted as $R(\alpha_i)$. When agent $\alpha_{\bowtie}$ receives $\Delta(\alpha_i)$, it performs the following steps: (*i*) join the new results with the existing ones ($R(\alpha_0) \ltimes \Delta(\alpha_i)$ if $i \in [1..n]$; $\Delta(\alpha_0) \ltimes_i R(\alpha_i)$ if $i = 0$); (*ii*) if the previous step provides new (or improved) results, update the state and the statistics associated with the pairs in the result of α_0 and propagate upwards the new results.

Finally, $\alpha_{\bowtie}$ agents are dedicated to the evaluation of algebraic expressions of the form $Plan(ws_1) \bowtie \ldots \bowtie Plan(ws_m)$. An agent $\alpha_{\bowtie}$ receives its inputs from agents $\alpha_1, \ldots, \alpha_n$ which evaluate $Plan(ws_1), \ldots, Plan(ws_m)$, respectively. When an $\alpha_{\bowtie}$ agent receives new or improved results from agent α_i, it performs the following steps: (*i*) compute $R(\alpha_0) \bowtie \ldots \bowtie \Delta(\alpha_i) \bowtie \ldots \bowtie R(\alpha_n)$; (*ii*) if the previous step 1 provides new (or improved) results, propagate them upwards.

The answer to a top-k fuzzy XPath query $\langle exp, k \rangle$ consists of the first k element pairs (according to the membership grade) returned by agent α associated with exp. The result of α is computed by executing the evaluation requests relative to the $\alpha_{\cup}$ subagents of α. Clearly, the execution of evaluation requests can be terminated early if k element pairs have already been computed and there is no chance of computing a new pair having a higher membership grade.

We are currently studying three different incremental evaluation strategies, which differently order evaluation requests. According to the *naive* strategy, the evaluation of a top-k fuzzy XPath query is performed by ranking requests according to the improvement of the membership grade they could provide to the whole query result, then activating the requests in descending order. In this strategy, the improvement that could be provided by a request is evaluated by taking the membership grade of the elements which result from the evaluation of the query obtained by replacing those simple step in the exact query that are referred to by the request.

The *dynamic* strategy exploits the possibility of postponing the evaluation of requests that cannot immediately produce improved results. This strategy dynamically orders the requests according to the following ranking criterion: first it associates each request r with the set of element pairs which (*i*) could be added to the answer by evaluating r and (*ii*) already belonging to the answer whose membership grade may be improved by the evaluation of r; then, r is ranked according to the best membership grade of a pair in the above set. In this strategy, the requests regarding *Pt* predicates corresponding to the simple steps appearing in the exact query are executed first; then, the remaining requests are activated in descending rank order.

Finally, the *selectivity-based* strategy employs an extension of the *XSketch* XML data synopsis [8] to estimate the selectivity of an XPath expression, that is the number of XML elements that could be selected by the expression. The selectivity-based strategy dynamically orders the requests to $\alpha_\cup$ agents on the basis of a rank which takes into account both the membership grade improvement that they can provide to previously computed element pairs (i.e., those adopted as rank in the dynamic strategy) and the selectivity of the request w.r.t. the expression corresponding to the requests executed so far. Similarly to the dynamic strategy, the requests regarding *Pt* predicates corresponding to the simple steps appearing in the exact query are executed first; then, the remaining requests are activated in descending rank order.

4   Conclusions and Future Work

In this paper we proposed the syntax and semantics of an XML query language for fuzzy top-k querying. We provided a general query evaluation algebra and discussed possible evaluation strategies. As our future work we plan to investigate the expected benefits of our evaluation strategies (also combined with parallel execution) and to thoroughly assess the performance of the techniques through an extensive experimental validation.

References

1. Amer-Yahia, S., Cho, S., Srivastava, D.: Tree pattern relaxation. In: Jensen, C.S., Jeffery, K., Pokorný, J., Šaltenis, S., Bertino, E., Böhm, K., Jarke, M. (eds.) EDBT 2002. LNCS, vol. 2287, pp. 496–513. Springer, Heidelberg (2002)
2. Campi, A., Guinea, S., Spoletini, P.: A fuzzy extension for the xPath query language. In: Larsen, H.L., Pasi, G., Ortiz-Arroyo, D., Andreasen, T., Christiansen, H. (eds.) FQAS 2006. LNCS, vol. 4027, pp. 210–221. Springer, Heidelberg (2006)
3. Fazzinga, B., Flesca, S., Pugliese, A.: Vague queries on peer-to-peer XML databases. In: Wagner, R., Revell, N., Pernul, G. (eds.) DEXA 2007. LNCS, vol. 4653, pp. 287–297. Springer, Heidelberg (2007)
4. Fazzinga, B., Flesca, S., Pugliese, A.: Retrieving xml data from heterogeneous sources through vague querying. ACM Trans. Internet Techn. 9(2) (2009)
5. Fuhr, N., Großjohann, K.: Xirql: An xml query language based on information retrieval concepts. ACM Trans. Inf. Syst. 22(2), 313–356 (2004)
6. Kaushik, R., Krishnamurthy, R., Naughton, J.F., Ramakrishnan, R.: On the integration of structure indexes and inverted lists. In: SIGMOD Conference, pp. 779–790 (2004)
7. Marian, A., Amer-Yahia, S., Koudas, N., Srivastava, D.: Adaptive processing of top-k queries in xml. In: ICDE, pp. 162–173 (2005)
8. Polyzotis, N., Garofalakis, M.N.: Xsketch synopses for xml data graphs. ACM Trans. Database Syst. 31(3), 1014–1063 (2006)
9. Schlieder, T.: Schema-driven evaluation of approximate tree-pattern queries. In: Jensen, C.S., Jeffery, K., Pokorný, J., Šaltenis, S., Bertino, E., Böhm, K., Jarke, M. (eds.) EDBT 2002. LNCS, vol. 2287, pp. 514–532. Springer, Heidelberg (2002)
10. Theobald, A., Weikum, G.: Adding relevance to xml. In: WebDB (Selected Papers), pp. 105–124 (2000)
11. W3C. World wide web consortium (2007), http://www.w3.org

Deciding Query Entailment in Fuzzy Description Logic Knowledge Bases

Jingwei Cheng, Z.M. Ma, Fu Zhang, and Xing Wang

Northeastern University, Shenyang, 110004, China
cjingwei@gmail.com, mazongmin@ise.neu.edu.cn

Abstract. Existing fuzzy description logic (DL) reasoners either are not capable of answering conjunctive queries, or only apply to DLs with less expressivity. In this paper, we present an algorithm for answering expressive fuzzy conjunctive queries, which allows the occurrence of both lower bound and the upper bound of thresholds in a query atom, over the relative expressive DL, namely fuzzy $\mathcal{ALCN}$. Our algorithm is specially tailored for deciding conjunctive query entailment of negative role atoms in the form of $R(x, y) \leq n$ or $R(x, y) < n$ which, to the best of our knowledge, has not been touched on in other literatures.

1 Introduction

Description logics (DLs, for short) [1] are the logical foundation of the Semantic Web, which support knowledge representation and reasoning by means of the concepts and roles. However, the reasoning services that aim at accessing and querying the data underlying ontologies, such as *retrieval, realisation* and *instantiation*, are only in weak form and do not support complex queries (mainly conjunctive queries, CQs). Conjunctive queries originated from research in relational databases, and, more recently, have also been identified as a desirable form of querying DL knowledge bases (KBs). The first conjunctive query algorithm [2] over DLs was actually specified for the purpose of deciding conjunctive query containment for $\mathcal{DLR}_{reg}$. Recently, query entailment and answering have also been extensively studied both for tractable DLs [3][4] and for expressive DLs[5][6].

In order to capture and reason about vague and imprecise domain knowledge, there have been a substantial amount of work carried out in the context of fuzzy DLs. When querying over fuzzy DL KBs, as in the crisp case, same difficulties emerged in that existing fuzzy DL reasoners, such as fuzzyDL[7] and FiRE [8], are not capable of dealing with CQs either. In [9], A fuzzy extension of CARIN system[5] is provided, along with a decision procedure for answering union of conjunctive queries. Some other work mainly focuses on querying over lightweight ontologies, e.g. in [10][11]. However, these extensions allow only positive role atoms in a query, while the negative atoms are not touched on.

In this paper, we thus present a very first algorithm for answering expressive and fuzzy CQs, allowing in a query both positive atoms and negative atoms, over the relative expressive fuzzy DL, namely $f\text{-}\mathcal{ALCN}$.

S.S. Bhowmick, J. Küng, and R. Wagner (Eds.): DEXA 2009, LNCS 5690, pp. 830–837, 2009.

2 f-$\mathcal{ALCN}$

Definition 1. *(syntax) Let N_C, N_R, and N_I be countable infinite and pairwise disjoint sets of concept, role and individual names respectively. f-$\mathcal{ALCN}$ concepts (denoted by C and D) are formed out of concept names according to the following abstract syntax: $C, D \to \top |\bot| A |C \sqcap D| C \sqcup D| \neg C| \forall R.C| \exists R.C| \geq pR| \leq pR|$, where $A \in N_C$, $R \in N_R$, p is a nonnegative integer.*

An f-$\mathcal{ALCN}$ KB $\mathcal{K}$ can be partitioned into a terminological part called TBox and an assertional part called ABox, denoted by $\mathcal{K} = (\mathcal{T}, \mathcal{A})$. A TBox is a finite set of *general concept inclusion* axioms (GCIs) of the form $C \sqsubseteq D$, where C, D are concepts. An ABox consists of fuzzy assertions of the form $B(o) \bowtie n$, $R(o, o') \bowtie n$, or $o \neq o'$, where $o, o' \in N_I$, $\bowtie \in \{\geq, >, \leq, <\}$. We use $\rhd$ to denote $\geq$ or $>$, and $\lhd$ to denote $\leq$ or $<$. We call ABox assertions defined by $\rhd$ *positive assertions*, while those defined by $\lhd$ *negative assertions*. Moreover, for every operator $\bowtie$, we define (i) its symmetric operator $\bowtie^-$ as $\geq^-=\leq, >^-=<, \leq^-=\geq, <^-=>$, and (ii) its negation operator $\neg \bowtie$ as $\neg \geq=<, \neg >=\leq, \neg \leq=>, \neg <=\geq$.

Definition 2. *(semantics) The semantics of f-$\mathcal{ALCN}$ are provided by an interpretation, which is a pair $\mathcal{I} = (\Delta^{\mathcal{I}}, \cdot^{\mathcal{I}})$. Here $\Delta^{\mathcal{I}}$ is a non-empty set of objects, called the domain of interpretation, and $\cdot^{\mathcal{I}}$ is an interpretation function which maps different individual names into different elements in $\Delta^{\mathcal{I}}$, concept A into membership function $A^{\mathcal{I}}: \Delta^{\mathcal{I}} \to [0,1]$, role R into membership function $R^{\mathcal{I}}: \Delta^{\mathcal{I}} \times \Delta^{\mathcal{I}} \to [0,1]$. The semantics of f-$\mathcal{ALCN}$ concepts and roles are depicted as follows.*

- $\top^{\mathcal{I}}(o) = 1 \qquad \bot^{\mathcal{I}}(o) = 0 \qquad (\neg C)^{\mathcal{I}}(o) = 1 - C^{\mathcal{I}}(o)$
- $(C \sqcap D)^{\mathcal{I}}(o) = \min\{C^{\mathcal{I}}(o), D^{\mathcal{I}}(o)\} \qquad (C \sqcup D)^{\mathcal{I}}(o) = \max\{C^{\mathcal{I}}(o), D^{\mathcal{I}}(o)\}$
- $(\forall R.C)^{\mathcal{I}}(o) = \inf_{o' \in \Delta^{\mathcal{I}}}\{\max\{1 - R^{\mathcal{I}}(o, o'), C^{\mathcal{I}}(o')\}\}$
- $(\exists R.C)^{\mathcal{I}}(o) = \sup_{o' \in \Delta^{\mathcal{I}}}\{\min\{R^{\mathcal{I}}(o, o'), C^{\mathcal{I}}(o')\}\}$
- $(\geq pR)^{\mathcal{I}}(o) = \sup_{o_1,\ldots,o_p \in \Delta^{\mathcal{I}}} \min_{i=1}^{p}\{R^{\mathcal{I}}(o, o_i)\}$
- $(\leq pR)^{\mathcal{I}}(o) = \inf_{o_1,\ldots,o_{p+1} \in \Delta^{\mathcal{I}}} \max_{i=1}^{p+1}\{1 - R^{\mathcal{I}}(o, o_i)\}$

Given an interpretation $\mathcal{I}$ and an inclusion axiom $C \sqsubseteq D$, $\mathcal{I}$ is a model of $C \sqsubseteq D$, if $C^{\mathcal{I}}(o) \leq D^{\mathcal{I}}(o)$ for any $o \in \Delta^{\mathcal{I}}$, written as $\mathcal{I} \models A \sqsubseteq C$. Similarly, for ABox assertions, $\mathcal{I} \models B(o) \bowtie n$(resp. $\mathcal{I} \models R(o, o') \bowtie n$), iff $B^{\mathcal{I}}(o^{\mathcal{I}}) \bowtie n$(resp. $R^{\mathcal{I}}(o^{\mathcal{I}}, o'^{\mathcal{I}}) \bowtie n$). If an interpretation $\mathcal{I}$ is a model of all the axioms and assertions in a KB $\mathcal{K}$, we call it a model of $\mathcal{K}$. A KB is *satisfiable* iff it has at least one model. A KB $\mathcal{K}$ *entails* (logically implies) a fuzzy assertion φ, iff all the models of $\mathcal{K}$ are also models of φ, written as $\mathcal{K} \models \varphi$.

3 Querying Entailment Problems

3.1 Fuzzy Querying Language

Let N_V be a countable infinite set of variables and is disjoint with N_C, N_R, and N_I. A *term* t is either an individual name from N_I or a variable name from

N_V. Let C be a concept, R a role, and t, t' terms. An *fuzzy query atom* is an expression of the form $\langle C(t) \bowtie n\rangle$ or $\langle R(t,t') \bowtie m\rangle$, where n denotes the lower bound (which corresponds to $\rhd$) or upper bound (which corresponds to $\lhd$) of membership of being the term t a member of the fuzzy set C, m denotes the degree of membership of being the term pair (t,t') a member of the fuzzy role R. We refer to these two different types of atoms as *fuzzy concept atoms* and *fuzzy role atoms* respectively. We call fuzzy query atoms defined by $\rhd$ *fuzzy positive query atoms*, while those defined by $\lhd$ *fuzzy negative query atoms*. A *fuzzy conjunctive query* q is a conjunction of fuzzy query atoms which is of the form $q = \langle\rangle \leftarrow \wedge_{i=1}^{n}\langle at_i \bowtie n_i\rangle$, where $\langle at_i \bowtie n_i\rangle$ denotes the i-th fuzzy query atom of q. We use $\mathbf{Var}(q)$ to denote the set of variables occurring in q, $\mathbf{Ind}(q)$ to denote the set of individual names occurring in q, and $\mathbf{Term}(q)$ for the set of terms in q, where $\mathbf{Term}(q) = \mathbf{Var}(q) \cup \mathbf{Ind}(q)$.

The semantics of a fuzzy query is given in the same way as for the related fuzzy DL by means of interpretations consisting of an interpretation domain and a fuzzy interpretation function. Let $\mathcal{I}= (\Delta^{\mathcal{I}}, \cdot^{\mathcal{I}})$ be a fuzzy interpretation of f-$\mathcal{ALCN}$, q be a fuzzy conjunctive query and t, t' terms in q, and $\pi : \mathbf{Var}(q)\cup \mathbf{Ind}(q) \rightarrow \Delta^{\mathcal{I}}$ a total function (also called an *assignment*) such that $\pi(a) = a^{\mathcal{I}}$ for each $a \in \mathbf{Ind}(q)$. We say $\mathcal{I} \models^{\pi} \langle C(t) \bowtie n\rangle$ if $C^{\mathcal{I}}(\pi(t)) \bowtie n$, $\mathcal{I} \models^{\pi} \langle R(t,t') \bowtie n\rangle$ if $R^{\mathcal{I}}(\pi(t), \pi(t')) \bowtie n$. If $\mathcal{I} \models^{\pi} at$ for all atom $at \in q$, we write $\mathcal{I} \models^{\pi} q$. If there is a π, such that $\mathcal{I} \models^{\pi} q$, we say $\mathcal{I}$ satisfies q, written as $\mathcal{I} \models q$. We call such a π a *match* of q in $\mathcal{I}$. If $\mathcal{I} \models q$ for each model $\mathcal{I}$ of a KB $\mathcal{K}$, then we say $\mathcal{K}$ entails q, written as $\mathcal{K} \models q$. The *query entailment problem* is defined as follows: given a knowledge base $\mathcal{K}$ and a query q, decide whether $\mathcal{K} \models q$.

3.2   Deciding Query Entailment

The idea behind our algorithm is that it tries to decide fuzzy query entailment problem by (i) constructing (a representation of) all the models of an f-$\mathcal{ALCN}$ KB $\mathcal{K}$, then (ii) checks each of them for a match of a given conjunctive query q.

To construct models of an f-$\mathcal{ALCN}$ KB, we work with a data structure called completion forest. A *completion forest* consists of a labelled directed graph, each node of which is the root of a *completion tree*. Each node x in a completion tree is labelled with a set $\mathcal{L}(x) = \{\langle C, \bowtie, n\rangle\}$, and each edge (x,y) is labelled with a set $\mathcal{L}(x,y) = \{\langle R, \bowtie, n\rangle\}$. A node x is called an R-*predecessor* of a node y (and y is called an R-*successor* of x), if for some R, $\mathcal{L}(x,y) = \{\langle R, \bowtie, n\rangle\}$, *ancestor* is the transitive closure of predecessor.

Starting with an f-$\mathcal{ALCN}$ KB $\mathcal{K} = \langle \mathcal{T}, \mathcal{A}\rangle$, the completion forest $\mathcal{F}$ is initialized such that it contains a root node o, with $\mathcal{L}(o) = \{\langle C, \bowtie, n\rangle \mid \langle C(o) \bowtie n\rangle \in \mathcal{A}\}$, for each individual name o occurring in $\mathcal{A}$, and an edge $\langle o,o'\rangle$ with $\mathcal{L}(\langle o,o'\rangle) = \{\langle R, \bowtie, n\rangle \mid \langle R(o,o') \bowtie n\rangle \in \mathcal{A}\}$, for each pair $\langle o,o'\rangle$ of individual names for which the set $\{R \mid R(o,o') \bowtie n \in \mathcal{A}\}$ is non-empty.

An initial completion forest is expanded according to a set of *expansion rules*(see [12] and [13] for details) that reflect the constructors allowed in f-$\mathcal{ALCN}$. The expansion stops when there is a conjugated pair, called a *clash*, occurs within a node label, or when no more rules are applicable. In the latter case, the completion forest is called *complete*. Termination is guaranteed by a cycle-checking technique called *blocking*. The model that we can build from a complete and clash-free completion graph is called a *canonical model*. The expansion and blocking rules are such that we can build a model for the knowledge base from each complete and clash-free completion forest.

The query can be represented as a directed, labelled graph called a *query graph*. The nodes in the query graph correspond to the terms in q, and are labelled with a triple corresponds to a related fuzzy concept atom. The edges correspond to the role atoms in q and are labelled similarly.

In the following, we show how to check a completion forest for a match of a query graph. The following example may be helpful in illustrating our method. Given a query $q_1 = \langle\rangle \leftarrow \langle C(a) \geq n_1 \rangle \wedge \langle R(a, y_1) \geq n_2 \rangle \wedge \langle S(y_1, y_2) \geq n_3 \rangle \wedge \langle D(y_2) \geq n_4 \rangle$ and a KB $\mathcal{K}_1 = \{(\exists R.(\exists S.D) \sqcap C)(a) \geq n\}$ where $n \geq \max(\{n_i\})$ with $1 \leq i \leq 4$. We first build completion forest for $\mathcal{K}_1$. Initially, there is only one node a labelled with a triple $(\exists R.(\exists S.D) \sqcap C, \geq, n)$, then expansion rules are applied to label of nodes, and lead to the completion forest depicted in Fig. 1. The query graph of q_1 is shown in Fig. 2.

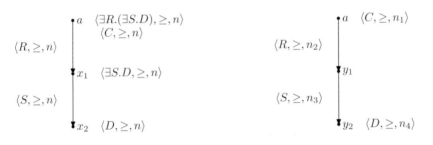

Fig. 1. The completion forest of $\mathcal{K}_1$ Fig. 2. The query graph of q_1

Now, we can compare the completion forest of $\mathcal{K}_1$ and query graph of q_1 for a match. We start from the root node in the query graph of q_1 and, at the same time, locate in the completion forest of $\mathcal{K}_1$ for a node whose label (a set of triples) *is more general than* the label (also a set of triples) of the node in the query graph. We say a label l_1 is subsumed by a label l_2 if for every triple tr in l_1, there exists a triple tr' in l_2, s.t. (i) tr' share the same concept or role name with tr, (ii) and the range identified by the sign of equality and the membership degrees in tr is *more general than* that of tr'. In Table 1, we list the conditions under which tr is more general than tr'. tr is listed in the leftmost column, while tr' is listed in the topmost row.

Then, we traverse the query graph along the outgoing edge to the next node, while in the completion forest we undeterministically select a branch to proceed (if any). The algorithm terminates when the leaf node in the query graph is encountered. In according to what they represent, we distinguish the nodes in the completion forest and query graph into *constant nodes* which represent individual names and *variable nodes* which represent variable names. During the traversal,

Table 1. The conditions under which tr is more general than tr'

tr	\multicolumn{4}{c}{tr'}			
	$(C,>,n)$	$(C,\geq,n)$	$(C,<,n)$	$(C,\leq,n)$
$(C,>,m)$	$m \leq n$	$m < n$	NULL	NULL
$(C,\geq,m)$	$m \leq n$	$m \leq n$	NULL	NULL
$(C,<,m)$	NULL	NULL	$m \geq n$	$m > n$
$(C,\leq,m)$	NULL	NULL	$m \geq n$	$m \geq n$

if (i) each constant node occurs in the query graph is mapped to a constant node in the completion forest that represents the same individual, and (ii) the label of every node and edge in query graph is more general than that of in the completion forest, we say that there is a match for a query in the knowledge base.

Given an $f\text{-}\mathcal{ALCN}$ knowledge base $\mathcal{K}$ and a query q, the algorithm answers "$\mathcal{K}$ entails q" if a match for q exists in each complete and clash-free completion forest and it answers "$\mathcal{K}$ does not entail q" otherwise.

The aforementioned idea is similar to the tableau algorithm for deciding $f\text{-}\mathcal{ALCN}$ KB satisfiability, but there are still three main problems need to be considered.

Firstly, for each concept C that occurs only in one of the fuzzy concept atoms of a query but not in the knowledge base, we must introduce into the knowledge base a concept inclusion axiom of the form $C \sqsubseteq C$. Clearly, this will exert no influence on the logical characteristics of the knowledge base, but it ensures that, for each node x in the completion forest, either $C(x) \rhd n$ or $C(x)\neg \rhd n$ holds.

Secondly, a $f\text{-}\mathcal{ALCN}$ KB may have infinitely many infinite models, whereas the tableau algorithm constructs only a subset of the finite models of the knowledge base. It can be shown that inspecting only the canonical models of the knowledge base is sufficient to decide query entailment.

Furthermore, we need to modify the standard blocking condition (for checking KB satisfiability) to make it applicable in the context of deciding query entailment. The standard blocking condition is to stop the cyclic expansion of a branch in the completion forest. For $f\text{-}\mathcal{ALCN}$ KBs, if there is a pair of nodes x and y such that x is an ancestor of y and the label of y is a subset of the label of x, then we say that x blocks y and no further expansion rules are applied to y. In the model obtained from the completion forest, a block corresponds to a cyclic path that links the predecessor of the blocked node y to the blocking node x. For query entailment, the blocking condition additionally has to take into account the length of the longest path in the query. We illustrate this by means of an example.

Let $\mathcal{K}_2 = (\mathcal{T}, \mathcal{A})$ be an $f\text{-}\mathcal{ALCN}$ KB with $\mathcal{T} = \{C \sqsubseteq \exists R.C\}$ and $\mathcal{A} = \{C(a) \geq n\}$ and let $q_2 = \langle\rangle \leftarrow \langle R(x,x) \geq n\rangle$ be a Boolean fuzzy conjunctive query. It is not hard to check that $\mathcal{K}_2 \nvDash q_2$. Figure 3 illustrates the only complete and clash-free completion forest for $\mathcal{K}_2$ that the algorithm would construct. Figure 4 shows a representation of a corresponding canonical model $\mathcal{I}$. The loop in the model $\mathcal{I}$ occurs since the node x_1 blocks the node x_2 in the completion forest.

It is not hard to check that $\mathcal{I} \models \mathcal{K}_2$ and that the mapping $\pi : x \longmapsto x_1$ is such that $\mathcal{I} \models^\pi q_2$. Since the completion forest shown in Fig. 3 is the only completion forest that the algorithm would construct, we would wrongly conclude that $\mathcal{K}_2$ entails q_2.

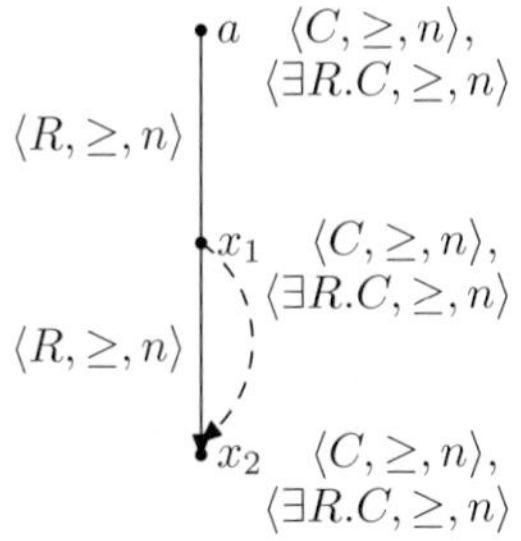

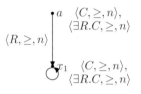

Fig. 3. A complete and clash-free completion forest for $\mathcal{K}_2$. The node x_2 is blocked by x_1, indicated by the dashed line.

Fig. 4. A canonical model $\mathcal{I}$ for $\mathcal{K}_2$

Instead of using a pair of nodes x and y in the blocking definition such that x is an ancestor of y and the label of y is a subset of the label of x, the new blocking condition, first introduced in [5], requires two isomorphic trees (instead of just two nodes) such that the depth of the trees is equal to the number of fuzzy role atoms in the query. The leaves of the descendant tree are then considered as blocked. In our example, the query contains only one role atom. Hence, the revised blocking condition requires just two trees of depth one.

Figure 5 shows an abstraction of a complete and clash-free completion forest using the modified blocking condition. In the canonical model that corresponds to the completion forest (see Fig. 6), we have a cycle from the element that corresponds to the predecessor of the blocked node to the element that corresponds to the blocking node.

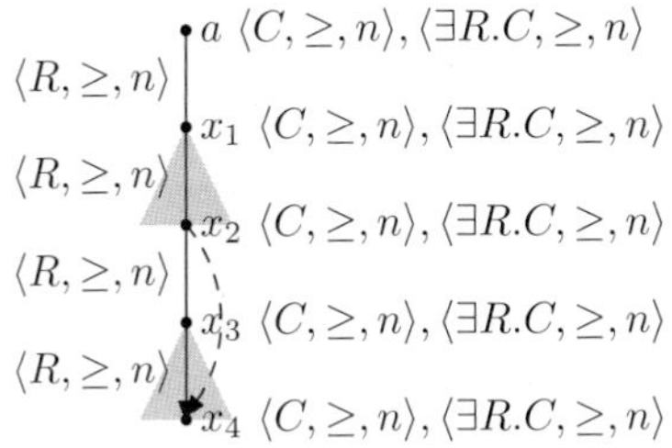

Fig. 5. A complete and clash-free completion forest for $\mathcal{K}_2$ under modified blocking condition

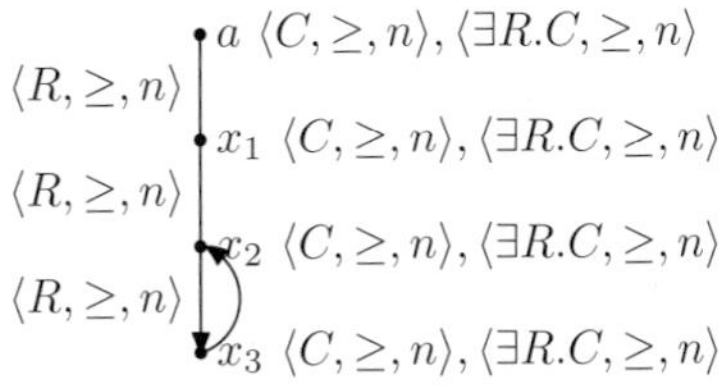

Fig. 6. A canonical model $\mathcal{I}$ for $\mathcal{K}_2$ corresponding to the completion forest in Fig 5

When building a model from a completion forest, one can, instead of building a cycle, also append infinitely many copies of the blocking tree and the path between the blocking and the blocked tree. In our example, this would result in a model that consists of an infinite R-chain.

Thirdly, the method cannot extend directly to the case where negative query atoms are allowed in a query. For example, given a KB $\mathcal{K}_3 = \{\forall R.D(a) \geq 0.6, D(b) < 0.3\}$ and a query $q_3 = \langle \rangle \leftarrow \langle R(a, y) \leq 0.5 \rangle \wedge \langle D(y) \leq 0.4 \rangle$, we can instinctively recognize that $\mathcal{K}_3$ entails q_3. However, existing algorithm cannot build a completion forest for $\mathcal{K}_3$, in which there exists a match of the query graph of q_3. We thus introduce two new rules for mending this. After applying these rules, the otherwise isolated two nodes are related by a fuzzy role R.

$\forall_{\rhd R}$-rule	$\exists_{\lhd R}$-rule
if 1.$\langle \forall R.C, \rhd, n \rangle \in \mathcal{L}(x)$, x is not blocked.	if 1. $\langle \exists R.C, \lhd, n \rangle \in \mathcal{L}(x)$, x is not blocked.
2.$\langle C, \neg\rhd, m \rangle \in \mathcal{L}(y)$,where $m \leq n$,	2. $\langle C, \neg\lhd, m \rangle \in \mathcal{L}(y)$, where $m \geq n$,
3.there is no $\langle R, \rhd^-, 1 - n \rangle \in \mathcal{L}(x, y)$	3. there is no $\langle R, \lhd, n \rangle \in \mathcal{L}(x, y)$
then $\mathcal{L}(x, y) \rightarrow \mathcal{L}(x, y) \cup \langle R, \rhd^-, 1 - n \rangle$.	then $\mathcal{L}(x, y) \rightarrow \mathcal{L}(x, y) \cup \langle R, \lhd, n \rangle$.

In our example, before applying $\forall_{\geq R}$ rule, the completion forest of $\mathcal{K}_3$ consists of two isolated nodes a and b labelled with $(\forall R.D, \geq, 0.6)$ and $(D, <, 0.3)$ respectively. The completion forest of $\mathcal{K}_3$ is shown in Fig. 7. The $\forall_{\geq R}$ rule then additionally adds an role assertion $\langle R(a, b) \leq 1 - 0.6 \rangle$ (simplified as $\langle R(a, b) \leq 0.4 \rangle$) into $\mathcal{K}_2$ and therefore connects the isolated a and b, resulting in the completion forest of $\mathcal{K}'_3$, which is also shown in Fig. 7. Note that the newly introduced role is depicted in Fig. 7 as a dashed line. We can easily find a match for the query graph of q_3(Fig. 8) in the completion forest of $\mathcal{K}'_3$.

Fig. 7. The completion forest of $\mathcal{K}_3$ and $\mathcal{K}'_3$ **Fig. 8.** The query graph of q_3

Theorem 1. *Let $\mathcal{K}$ be a f-$\mathcal{ALCN}$ KB, and $\mathcal{K}'$ a KB obtained by applying the $\forall_{\geq R}$ rule to $\mathcal{K}$. The $\forall_{\geq R}$ rule is sound if a model $\mathcal{I}$ of $\mathcal{K}$ is also a model of $\mathcal{K}'$.*

Proof. $\forall_{\geq R}$: For each $\mathcal{I} \models \mathcal{K}$, we show that $\mathcal{I} \models \mathcal{K}'$, where $\mathcal{K}'$ is obtained by applying $\forall_{\geq R}$ rule to $\mathcal{K}$, i.e. $\mathcal{K}' = \mathcal{K} \cup \{\langle R(x, y) < 1 - n \rangle\}$. For any assertion $\langle \forall R.C(x) \geq n \rangle \in \mathcal{A}$, we have $\mathcal{I} \models \langle \forall R.C(x) \geq n \rangle$, i.e., $\inf_{x' \in \Delta^{\mathcal{I}}} \{\max\{1 - R^{\mathcal{I}}(x, x'), C^{\mathcal{I}}(x')\}\} \geq n$. Since $y \in \Delta^{\mathcal{I}}$, we have $\max\{1 - R^{\mathcal{I}}(x, y), C^{\mathcal{I}}(y)\} \geq n$. Since there is an additional assertion of $\langle C(y) < m \rangle$ with $m \leq n$, the $1 - R^{\mathcal{I}}(x, y) \geq n$ (or its equivalence $R^{\mathcal{I}}(x, y) \leq n$) holds. The $\forall_{> R}$ and $\exists_{\lhd R}$ rules can be proved accordingly.

4   Conclusions

In this paper, we have presented a preliminary result of conjunctive query entailment over an expressive fuzzy DL knowledge base. Further and ongoing research will focus on the query answering or query entailment problems of more expressive DLs and the complexity analysis for them.

Acknowledgments. This work was supported by the National Natural Science Foundation of China (60873010).

References

1. Baader, F., Calvanese, D., McGuinness, D.L., Nardi, D., Patel-Schneider, P.F. (eds.): The description logic handbook: theory, implementation, and applications. Cambridge University Press, New York (2003)
2. Calvanese, D., De Giacomo, G., Lenzerini, M.: On the decidability of query containment under constraints. In: PODS 1998, pp. 149–158 (1998)
3. Calvanese, D., De Giacomo, G., Lembo, D., Lenzerini, M., Rosati, R.: Tractable reasoning and efficient query answering in description logics: The dl-lite family. J. of Automated Reasoning 39(3), 385–429 (2007)
4. Rosati, R.: On conjunctive query answering in EL. In: DL 2007, CEUR Electronic Workshop Proceedings (2007)
5. Levy, A.Y., Rousset, M.C.: Combining horn rules and description logics in carin. Artif. Intell. 104(1-2), 165–209 (1998)
6. Glimm, B., Horrocks, I., Lutz, C., Sattler, U.: Conjunctive query answering for the description logic shiq. In: IJCAI, pp. 399–404 (2007)
7. Bobillo, F., Straccia, U.: fuzzydl: An expressive fuzzy description logic reasoner. In: FUZZ-IEEE 2008, June 2008, pp. 923–930 (2008)
8. Stoilos, G., Simou, N., Stamou, G.B., Kollias, S.D.: Uncertainty and the semantic web. IEEE Intelligent Systems 21(5), 84–87 (2006)
9. Mailis, T., Stoilos, G., Stamou, G.: Expressive reasoning with horn rules and fuzzy description logics. In: Marchiori, M., Pan, J.Z., Marie, C.d.S. (eds.) RR 2007. LNCS, vol. 4524, pp. 43–57. Springer, Heidelberg (2007)
10. Straccia, U.: Answering vague queries in fuzzy dl-lite. In: IPMU 2006, pp. 2238–2245 (2006)
11. Pan, J.Z., Stamou, G.B., Stoilos, G., Taylor, S., Thomas, E.: Scalable querying services over fuzzy ontologies. In: WWW, pp. 575–584 (2008)
12. Stoilos, G., Stamou, G., Pan, J., Tzouvaras, V., Horrocks, I.: Reasoning with very expressive fuzzy description logics. JAIR 30(8), 273–320 (2007)
13. Stoilos, G., Straccia, U., Stamou, G.B., Pan, J.Z.: General concept inclusions in fuzzy description logics. In: ECAI, pp. 457–461 (2006)

An Optimization Technique for Multiple Continuous Multiple Joins over Data Streams

Changwoo Byun[1], Hunjoo Lee[2], YoungHa Ryu[2], and Seog Park[2]

[1] Department of Computer Systems and Engineering, Inha Technical College,
Incheon, 402-752 South Korea
cwbyun@inhatc.ac.kr
[2] Department of Computer Science, Sogang University,
Seoul, 121-742, South Korea
{hunz,ywenry,spark}@dblab.sogang.ac.kr

Abstract. Join queries having heavy cost are necessary to Data Stream Management System in the sensor network. In this paper, we propose an optimization algorithm for multiple continuous join operators over data streams using a heuristic strategy. First, we propose a solution of building the global shared query execution plan. Second, we solve the problems of updating a window size and routing for a join result. Our experimental results show that the proposed protocol can provide better throughputs than previous methods.

Keywords: Data Stream, Multiple Join, Multi-Join Queries Processing.

1 Introduction

In the sensor network [1], the system collects information from various sensors with a designated time interval, and sends the collected data to a central processing server. Data collected from a single sensor has a limited kind of information due to its processing capacity [2]. Thus, to obtain more general information, join operations are performed based on a specific time or location. Hash Table-based Join operators [3], [4], [5], Window-based Join operators [6], [7], and Hash Table-Window-based Join operators [8], [16] are the results of such efforts. MJoin operators allowing multiple inputs have been proposed [5]. In the processing of multiple queries, handling a single query with a single MJoin operator may be ineffective.

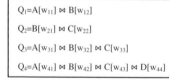
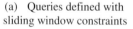

$Q_1 = A[w_{11}] \bowtie B[w_{12}]$

$Q_2 = B[w_{21}] \bowtie C[w_{22}]$

$Q_3 = A[w_{31}] \bowtie B[w_{32}] \bowtie C[w_{33}]$

$Q_4 = A[w_{41}] \bowtie B[w_{42}] \bowtie C[w_{43}] \bowtie D[w_{44}]$

(a) Queries defined with sliding window constraints

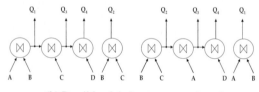
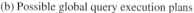

(b) Possible global query execution plans

Fig. 1. Example for the study motive

S.S. Bhowmick, J. Küng, and R. Wagner (Eds.): DEXA 2009, LNCS 5690, pp. 838–846, 2009.
© Springer-Verlag Berlin Heidelberg 2009

Figure 1 depicts the MJoin operators that can process the queries by sharing them. Query Q1 can be included in Q3 and Q4 and Q2 can also be included in Q3 and Q4.

The purpose of this paper is to analyze the containment relationship among operators when multiple MJoin operators are registered to a system, to establish a globally shared query execution plan, and to correctly perform the query execution plan. Our processing technique of Multiple MJoins is called MMJoin.

The paper is organized as follows. In Section 2, we evaluate related works. In Section 3, we introduce the MMJoin technique which achieves optimization in multiple MJoin operators. In Section 4, the MMJoin process performance and efficiency are analyzed through experiments. Section 5 contains the conclusion and provides future works.

2 Related Work

In relational and deductive databases, the most popular optimization technique of multiple queries is the A* greedy technique [10]. However, in the worst case, processing time which is proportional to the power of the number of all possible query execution plans is required. Also, this method is not suitable to the data stream environment, which requires constructing shared query execution plans in a prompt manner. Dynamic Regrouping [11] is one of the techniques used to construct shared query execution plans for continuous queries. This technique does not consider sliding window constraints.

In [12], to support the window update of the join operation result, the Negative Tuple technique is proposed. The Negative Tuple technique removes tuples that are identical to Negative Tuples through value-based searches. However, this technique has accuracy issues.

In [13], to resolve the routing problem for shared join operator, the Largest Window First (LWF) technique and the Smallest Window First (SWF) technique are proposed. In [14], regarding the sharing of join operators, the State-Slice strategy is proposed. All these techniques work in environments where the inputs are of original streams.

3 Optimization Technique

3.1 Approximation for Optimized Globally Shared Query Execution Plans

Sellis [10] discussed that selecting the most optimized globally shared execution plan in multiple queries is NP-Hard. So an approximation approach which necessitates the least processing cost shall be utilized to ensure suitability to the data stream environment. In this paper, the cost model of the join operators is defined as the following:

Definition 3.1 (Cost model). In an assumption where a certain MJoin operator uses n number of input $R = \{R_1, R_2, ..., R_n\}$, and r_i, and W_i represent the input rate and the window size of each R_i respectively, along with the Selectivity Factor for a join operation being f, the cost model is as follows:

$$\prod_{\substack{all \\ selectivities}} f \cdot \sum_{k=1}^{n} \left(r_k \cdot \prod_{\substack{i=1 \\ i \neq k}}^{n} W_i \right)$$

Also, the window size of each join operation result can be expected as follows. With this, we can evaluate the cost for upper join operators.

$$\prod_{all \atop selectivit\ ies} f \cdot \prod_{i=1}^{n} w_i$$

Lastly, sliding window constraints must be considered. Before operators are shared, evaluations to see whether the following conditions are met must be conducted.

Definition 3.2 (Sharing conditional expressions). In a supposition, there are m number of sharable queries Q_i ($1 \leq i \leq m$) with sharable MJoin operators, which has k number of input stream R_j ($1 \leq j \leq k$). In this supposition, each query Q_i has a window size of W_{ij} for each input R_j, and the largest window size from window sizes defined by all queries regarding R_j is W_j^*. With the sharing of operators, the following conditions must be met in order to decrease processing length.

$$\sum_{i=1}^{m} \prod_{j=1}^{k} W_{ij} \geq \prod_{j=1}^{k} W_j^*$$

For a simplified example, let us assume that the selectivity factor[1] for all joins is 0.1, and the window size for all input streams is 100 rows, all meeting the conditional expressions. When join operations are provided as shown in Figure 2.

Q_1 = R[Rows 100], S[Rows 100] = {R[100], S[100]}
Q_2 = R[Rows 100], T[Rows 100] = {R[100], T[100]}
Q_3 = S[Rows 100], T[Rows 100] = {S[100], T[100]}
Q_4 = R[Rows 100], S[Rows 100], T[Rows 100] = {R[100], S[100], T[100]}
Q_5 = R[Rows 100], S[Rows 100], T[Rows 100], U[Rows 100] = {R[100], S[100], T[100], U[100]}

Fig. 2. Example queries

In this example, after Q1 is selected(Figure 3(a)), the selection of a set that can be included in the other sets most frequently would be Q4, as Q4 is only included in Q5. The return of common elements in a set, including Q4 as a set of Q4, is shown in Figure 3(b). The remaining sets of Q2 and Q3 do not share containment relationships, and thus would construct independent query execution plans as shown in Figure 3(c).

Q_1 = {{R[100], S[100]}}	Q_1 = {{R[100], S[100]}}	Q_1 = {{R[100], S[100]}}
Q_2 = {R[100], T[100]}	Q_2 = {R[100], T[100]}	Q_2 = {{R[100], T[100]}}
Q_3 = {S[100], T[100]}	Q_3 = {S[100], T[100]}	Q_3 = {{S[100], T[100]}}
Q_4 = {{R[100], S[100]}, T[100]}	Q_4 = {{R[100], S[100]}, T[100]}}	Q_4 = {{R[100], S[100]}, T[100]}}
Q_5 = {{R[100], S[100]}, T[100], U[100]}	Q_5 = {{{R[100], S[100]}, T[100]}, U[100]}	Q_5 = {{{{R[100], S[100]}, T[100]}, U[100]}}
(a) The first phase	(b) The second phase	(c) The final phase

Fig. 3. Example after performing the algorithm

---

[1] In most cases, the window size and selection rate are unique. It is rare to have the same evaluation costs.

As shown in the algorithm of Figure 4, in each phase, when the most commonly shared sets selected are more than one, a set with the highest evaluation cost is selected. If the selected set does fulfill the conditional expression, the selected set is included, and parts of which the element of the selected set are located in all provided sets. These parts return as a single set. Finally, a single element set is excluded in the following phases as a single element set representing the query execution plan for such set is already completed. This series of tasks is iterated until all queries are excluded.

```
Input : a set of queries, QuerySet[]
Output : shared query plans, QuerySet[]
while QueryCount > 0
begin
    for i := 0 to QueryCount
    begin
        if Containing # of SelectedSet < Containg # of QuerySet[i] then  SelectedSet := QuerySet[i]
        else if Containing # of SelectedSet = Containg # of QuerySet[i] and
               Cost of QuerySet[i] > Cost of SelectedSet then SelectedSet := QuerySet[i]
    end
    if SelectedSet satisfies CONDITION OF SHARING then
    begin
        for i := 0 to QueryCount
        begin
            Replace the elements of QuerySet[i] to common elements of SelectedSet
        end
        for i := 0 to QueryCount
        begin
            if QuerySet[i] has only one element then
            begin
                exclude QuerySet[i]
                QueryCount := QueryCount - 1
            end
        end
    end
    else exclude SelectedSet
end
return QuerySet
```

Fig. 4. MMJoin optimization algorithm using heuristic greedy strategy

3.2 Window Update and Routing Technique for Join Operation Result

In Figure 5, should Q_1 and Q_3 define 2 tuples as the windows for input A and B, and should Q_4 have 4 tuples as windows for input A and B, all resultant tuples from the shared operators are not to be sent to Q_1, Q_3, and Q_4. If the number of recent tuples is higher than 2, the query results should not be directed to Q_1, and Q_3; instead, they should only be sent to Q_4. In operators performing join for original streams A and B, all tuples are sorted in sequential orders, and thus routings can be performed by using timestamps or by identifying the input sequential numbers. However, in operators performing the join of A⋈B and C, the tuples for A⋈B cannot be performed by employing such strategy of the original streams.

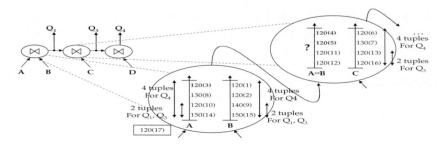

Fig. 5. Window update and routing in a globally shared query execution plan

In the original data stream, when a tuple is removed or routing information is added from the window update, all related resultant tuples from the join operations must be updated as well for correct windows updates and routing. In the paper, we propose using Purging Tuple and Dead Tuple for such purpose.

Definition 3.3 (Purging Tuple). When a tuple is removed during a window update, it is required to remove related tuples. The tuples generated for this particular purpose are called Purging Tuples.

Definition 3.4 (Dead Tuple). A Dead Tuple is proposed as a transport tuple for information.

Purging Tuple generation
In Figure 6(a), PostID has three index values which represent that this tuple was involved in the creation of three resultant tuples from a join operation. Also, a tuple in Figure 6(a) does not update the window by the Purging Tuple, but it is removed when the window is updated from an original stream. Consequently, the tuples of the upper join operators with identical PrevID among the indexes contained in this PostID must be removed, and the Purging Tuple has a role in routing the corresponding PostID values to the upper tuples. A Purging Tuple has a flag value(*) in the attribute of Attrs.

Key	Attrs	PrevID	PostID
120	...	NIL	ID_1, ID_2, ID_3

Key	Attrs	PrevID	PostID
120	*	NIL	ID_1, ID_2, ID_3

(a) Tuple removed from window update   (b) Purging Tuple

Fig. 6. An example of Purging Tuple generation

Creation of a Dead Tuple
As shown in Figure 7(a), when a certain original stream tuple is added as a Dead value of Q1, this value must be forwarded to all resultant tuples generated from this tuple. Figure 7(b) represents a Dead Tuple generated in this instance. A Dead Tuple has a flag value different from a Purging Tuple. Dead Tuples are indicated with double asterisks.

Key	Attrs	PrevID	PostID	Dead
120	...	NIL	ID_1, ID_2	Q_1

Key	Attrs	PrevID	PostID	Dead
120	**	NIL	ID_1, ID_2	Q_1

(a)A tuple with a new Dead Vector value added   (b) Dead Tuple

Fig. 7. An example of Dead Tuple creation

Searching of Purging and Dead Tuple

Figure 8(a) represents a Purging Tuple and a Dead Tuple generated either by tuple removal from a window update, or by a tuple with a new Dead value. When Purging Tuples and Dead Tuples are generated and sent to higher join operators, higher join operators need to read the PostID values of the Purging Tuple and the Dead Tuple, and remove a corresponding tuple from these values or add a new Dead value. These Purging/Dead Tuples, upon arriving at the upper join operators, need to have their PostID values extracted for the purpose of searching for corresponding PrevID values using the Index Hash Table, as shown in Figure 8(b). <I-Node> is the header information of the Index Hash Table, and maintains pointers for tuples with the first and last occurring PrevIDs.

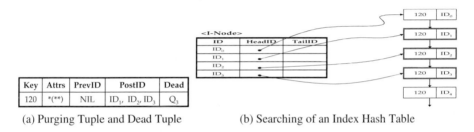

Key	Attrs	PrevID	PostID	Dead
120	*(**)	NIL	ID_1, ID_2, ID_3	Q_3

(a) Purging Tuple and Dead Tuple   (b) Searching of an Index Hash Table

Fig. 8. Index Hash Table via Purging Tuple and Dead Tuple

4 Experimental Evaluation

Both MJoin and MMJoin techniques described in the paper are implemented in Java, operating in Pentium IV 3.0Ghz processor with 1GB DDR2 memory. For the purpose of the experiment, 20 streams are assumed, each with the input rate of 300 tuples/sec. In all tuples, the key value can be of an integer value from 0 to 1000, randomly. Each query can have a maximum of 20 different input streams. The skewness for input streams for the entire queries is applied using Zipf distribution [15]. Also, for each input stream of each query, a window size is randomly selected in 500 tuples (rows), 1000 tuples (rows), and 1500 tuples (rows). All window sizes are generated in a uniform distribution. Table 1 lists the test parameters used in this test.

Table 1. Input stream and the test parameters for queries

Parameter	Range	Description
Q	10 to 100	The total number of distinct queries
SO	0 to 1	Skewness of input streams over queries
WS	500 to 1500	Window sizes for input stream

The first experiment is time comparison of globally shared query execution plan construction. The parameter SO is set to 0.5. Figure 9 represents the optimization time and the operator allocation time per independently processed Mjoin, and per MMJoin.

Measurement Unit. Let us define the operator allocation time for user-inputted queries to be t_{assign}, and the optimization time for the algorithm of Figure 4 to be $t_{optimizing}$. Here, Multiple-Query Setup Time is $t_{assign} + t_{optimizing}$.

In the MMJoin technique, as the optimization time for MMJoin is added, the number of queries increases, thus requiring more time. However, for the measurement aspect of this test, the query optimization and the operator allocation implementation is based on string, thus it is possible to achieve faster implementation via embedding parsing engine within the actual data stream management systems.

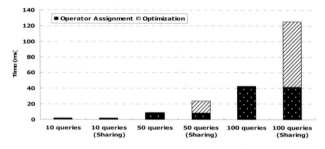

Fig. 9. MQST comparison to the number of queries (SO=0.5)

The second experiment is the throughput comparison between MJoin and MMJoin.

Measurement Unit. Let us define the total number of resultant tuples sent from all queries in a specific time period of i and to $i+1$ be the *throughput*$_i$, the total time of measurement to be t_{total}, and the total number of queries to be Q. The average throughput for each query AT is defined as $\sum_{i=0}^{t_{total}} throughput_i /(t_{total} \cdot Q)$.

In Figure 10(a), as the number of queries increases, the throughput of the MMJoin technique increases against that of independently processing MJoin, which represents that the probability of operators being shared is increasing as the number of queries increases. In Figure 10(b), as the input skewness increases, so does the common part of the queries, resulting in greater sharing and eventual increase in the relative average throughput of the MMJoin technique.

When either of the number of queries or input skewness reaches a certain point, the improvement rate drops, representing the points where a certain number of queries or input skewness creates an environment for sufficient sharing of operators.

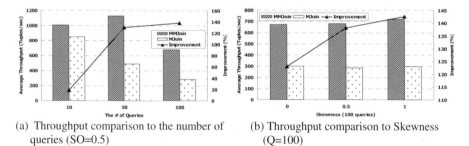

(a) Throughput comparison to the number of queries (SO=0.5)

(b) Throughput comparison to Skewness (Q=100)

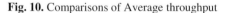

Fig. 10. Comparisons of Average throughput

5   Conclusion

In an environment where numerous users access a central system for information as in a Ubiquitous environment, a system should be able to handle numerous user queries. The MMJoin technique can prove to be useful in speeding up the processing time for such environments. The characteristics of the MMJoin technique are as follows.

First, MMJoin performs searches based on queries and sufficiently considers sliding window constraints to achieve query execution plan construction suitable for the data streaming environment. Second, index-based search is proposed so that it can be applied to both time-based windows and to a number of tuple-based windows. Also, the Purging Tuple technique is proposed to support less costly searches and to maintain accuracy during window updates when a frame is provided. Third, a routing technique is proposed for shared query execution plans. Along with this, additional Dead attributes are assigned for routing purposes to enhance the routing performance based on operations.

The technique proposed in this paper mostly assumes a static environment, and contains potential issues in re-establishing globally shared query execution plans when additional queries are added or deleted. Future studies in achieving globally shared query execution plans at a lesser cost in a dynamic environment are highly desired.

Acknowledgements

This work was supported by the second stage of the Brain Korea 21 Project in 2009.

References

1. Bonnet, P., Gehrke, J., Seshadri, P.: Towards Sensor Database Systems. In: Proc. 2th Int. Conf. on Mobile Data Management, pp. 3–14 (2001)
2. Schmidt, S., Fiedler, M., Lehner, W.: Source-aware Join Strategies of Sensor Data Streams. In: Proc. 17th Int. Conf. on Scientific and statistical database management, pp. 123–132 (2005)
3. Wilschut, N., Apers, P.M.G.: Pipelining in query execution. In: Conf. on Database. Parallel Architectures and their Applications, p. 562 (1991)
4. Urhan, T., Franklin, M.J.: XJoin: A reactively-scheduled pipelined join operator. IEEE Data Engineering Bulletin 23(2), 27–33 (2000)
5. Viglas, S.D., Naughton, J.F., Burger, J.: Maximizing the Output Rate of Multi-Way Join Queries over Streaming Information Sources. In: VLDB 2003, pp. 285–296 (2003)
6. Golab, L., Ozau, M.T.: Processing Sliding Window Multi-Joins in Continuous Queries over Data Streams. In: VLDB 2003, pp. 500–511 (2003)
7. Kang, J., Naughton, J.F., Viglas, S.D.: Evaluating Window Joins over unbounded Streams. In: ICDE 2003, pp. 341–352 (2003)
8. Ding, L., Rundensteiner, E.A.: Evaluating Window Joins over Punctuated Streams. In: Proc. 13th ACM Int. Conf. on Information and Knowledge Management, pp. 98–107 (2004)
9. Shim, K., Sellis, T.: Multiple-query optimization. ACM Transactions on Database Systems 13(1), 23–52 (1988)

10. Chen, J., DeWitt, D.J.: Dynamic Re-grouping of Continuous Queries. In: VLDB 2002, pp. 430–441 (2002)
11. Ghanem, T.M., Aref, W.G., Elmagarmid, A.K.: Exploiting Predicate-Window Semantics over Data Streams. ACM SIGMOD Record 35(1), 555–568 (2006)
12. Hammad, M., Franklin, M., Aref, W., Elmagarmid, A.: Scheduling for Shared Window Joins over Data Streams. In: VLDB 2003, pp. 297–308 (2003)
13. Wang, S., Rundensteiner, E., Ganguly, S., Bhatnagar, S.: State-Slice: New Paradigm of Multi-Query Optimization of Window-Based Stream Queries. In: VLDB 2006, pp. 619–630 (2006)
14. Krishnamurthy, S., Franklin, M.J., Hellerstein, J.M., Jacobson, G.: The Case for Precision Sharing. In: VLDB 2004, pp. 972–986 (2004)
15. Manning, C.D., Schütze, H.: Foundations of Statistical Natural Language Processing. The MIT Press, Cambridge (1999)
16. Li, H., Chen, S., Tatemura, J., Agrawal, D., Candan, K.S., Hsiung, W.: Safety Guarantee of Continuous Join Queries over Punctuated Data Streams. In: VLDB 2006, pp. 19–30 (2006)
17. Agarwal, P., Xie, J., Yang, J., Yu, H.: Scalable Continuous Query Processing by Tracking Hotspots. In: VLDB 2006 (2006)

Top-k Queries
with Contextual Fuzzy Preferences

Patrick Bosc, Olivier Pivert, and Amine Mokhtari

Irisa – Enssat, University of Rennes 1
Technopole Anticipa 22305 Lannion Cedex France
{bosc,pivert,mokhtari}@enssat.fr

Abstract. This paper deals with the interpretation of database queries with preference conditions of the form "attribute is low (resp. medium, high)" in the situation where the user is not aware of the actual content of the database but still wants to retrieve the best possible answers (relatively to that content). An approach to the definition of the terms "low", "medium" and "high" in a contextual and relative manner is introduced.

1 Introduction

The last two decades have witnessed an increasing interest in expressing preferences inside database queries. Motivations for such a concern are at least twofold. First, it has appeared to be desirable to offer more expressive query languages that can be more faithful to what a user intends to say. Second, the introduction of preferences in queries provides a basis for rank-ordering the retrieved items, which is especially valuable in case of large sets of items satisfying a query.

Fuzzy-set-based approaches are founded on the use of fuzzy set membership functions that describe the preference profiles of the user on each attribute domain involved in the query. Then satisfaction degrees associated with elementary conditions are combined using a panoply of fuzzy set connectives, which go much beyond conjunction and disjunction. It must be emphasized that fuzzy-set-based approaches rely on a *commensurability hypothesis* between the satisfaction degrees pertaining to the different attributes taking part in a query.

Another type of approach aims at the efficient computation of non Pareto-dominated answers, starting with the pioneering works of Börzsönyi et al. [3]. Clearly, this type of approach does not require any commensurability hypothesis between satisfaction degrees pertaining to elementary requirements. Notice that Pareto order yields a strict partial order only, while fuzzy set-based approaches lead to complete pre-orders. Kießling [9] has provided foundations for a Pareto-based preference model for database systems. A preference algebra including an operator called *winnow* has also been proposed by Chomicki [7].

In this paper, we deal with the situation where the user is not aware of the content of the database that he/she wants to query, and thus is unable to define selection predicates referring explicitly to some constants from the domains. Despite this absence of knowledge, the user may want to retrieve the best answers comparatively speaking. As an example, let us consider a user who wants

S.S. Bhowmick, J. Küng, and R. Wagner (Eds.): DEXA 2009, LNCS 5690, pp. 847–854, 2009.

to move to country X and would like to find the "best" cities to settle down according to the following preferences: population between 50,000 and 100,000 (hard constraint), low average price of the square meter, low crime rate, medium average annual temperature. We assume that the user does not have any idea of the values taken by the attributes involved in the cities of country X but still wants to find the best solutions relatively to the existing possibilities.

The idea that we advocate is to use the fuzzy-set-based framework and to define the fuzzy predicates "high", "medium" and "low" in a relative way, using the minimal, average and maximal values of the attribute values present in the associated query-defined context (in the example above: the cities whose population is between 50,000 and 100,000). There is a connection between this idea and the proposal by Tudorie *et al.* [10] where a fuzzy predicate can be defined relatively to the context created by another one. For example, in a query such as "find the inexpensive cars among the high speed ones", the authors suggest to adapt the definition of the fuzzy term "inexpensive" by taking into account the price values associated with high speed cars. However, the authors do not formalize the notion of context, while it is one of our primary aims. Even though the approach by Agrawal *et al.* [1] deals with preferences and contexts, it does not have much in common with the work presented here since: i) it deals with preferences over categorical attributes whereas we consider numerical ones, ii) it handles contexts and preferences which are given *explicitly* (and *statically*) by the user in the form of a set of conditions "attribute = value", whereas we deal with contexts which are specified *dynamically* by means of a specific clause in an extended SQL language, as well as with preferences which are expressed in an *implicit* and relative way, iii) the approach proposed in [1] does not combine contextual preferences with non-contextual ones, while the method presented here makes it possible to do.

The remainder of the paper is organized as follows. Section 2 is devoted to a reminder about fuzzy predicates and fuzzy queries. In Section 3, we present a method for constructing contextual fuzzy predicates automatically and describe different kinds of contextual fuzzy queries. Section 4 concludes the paper and outlines some perspectives for future work.

2   Fuzzy Predicates and Queries

2.1   Fuzzy Predicates

Regular sets allow for the definition of Boolean predicates. In an analogous way, gradual predicates (or conditions) can be associated with fuzzy sets [11] aimed at describing classes of objects with vague boundaries.

Often, elementary fuzzy predicates correspond to adjectives of the natural language, such as young, tall, cheap or well-paid. A fuzzy predicate P can be modeled as a function μ_P (usually of triangular or trapezoidal shape) from one (or several) domain(s) X to the unit interval. The degree $\mu_P(x)$ represents the extent to which element x satisfies the vague predicate P (or equivalently the extent to which x belongs to the fuzzy set of objects which match the fuzzy

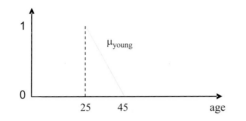

Fig. 1. A definition of the fuzzy predicate *young*

concept P). An exemple is given in Fig. 1. An elementary fuzzy predicate can also compare two attributes using a gradual comparison operator such as "more or less equal". It is assumed that a fuzzy querying system includes an interface (e.g., a GUI) that makes it possible to define his/her (non contextual) fuzzy predicates in a user-friendly way.

It is possible to alter the meaning of a given predicate using a modifier which is generally associated with an adverb (e.g., very, more or less, relatively). For instance, "very cheap" is more restrictive than "cheap" and "fairly high" is less demanding than "high". The meaning of the predicate $mod\ P$ (where mod is a modifier) may be defined in a compositional way and different approaches have been advocated, among which: $\mu_{modP}(x) = \mu_P(x)^n$ (see [6]).

Atomic and modified predicates can be involved in compound predicates which go far beyond those used in regular queries. Conjunction (resp. disjunction) is interpreted by means of a triangular norm $\top$ (resp. co-norm $\bot$), for instance the minimum or the product (resp. the maximum or the probabilistic sum). As to negation, it is interpreted as: $\forall x, \mu_{\neg P}(x) = 1 - \mu_P(x)$. Weighted conjunction and disjunction as well as weighted mean or OWA can be used to assign a different importance to each of the predicates (see [8] for more details).

2.2 Fuzzy Queries

The operations from the relational algebra can be straightforwardly extended to fuzzy relations by considering fuzzy relations as fuzzy sets on the one hand and by introducing gradual predicates in the appropriate operations on the other hand. The definitions of these extended relational operators can be found in [4]. As an illustration, we give the definition of the fuzzy selection hereafter, where r denotes a (fuzzy or crisp) relation and *cond* is a fuzzy predicate.

$$\mu_{sel(r,\ cond)}(t) = \top(\mu_r(t),\ \mu_{cond}(t)).$$

The language called SQLf described in [5] extends SQL so as to support fuzzy queries. Here, we just describe the base block in SQLf since this is all we need for our purpose. The principal differences w.r.t. SQL affect mainly two aspects:

- the calibration of the result since it is made with discriminated elements, which can be achieved through a number of desired answers (k), a minimal level of satisfaction (t), or both, and
- the nature of the authorized conditions as mentioned previously.

Therefore, the base block is expressed as:

select [**distinct**] [$k \mid t \mid k,\ t$] attributes **from** relations **where** fuzzy-cond

where "fuzzy-cond" may involve both Boolean and fuzzy conditions.

3   Top-k Queries with Contextual Fuzzy Predicates

In this section, we show how conditions of the form "A is low (resp. medium, high)" can be modeled in the framework of a fuzzy-set-based query language such as SQLf. The basic idea is to interpret these conditions relatively to a given *context* specified in the user query. First, let us clarify this notion.

Definition. A *query-defined context* is a referential of values returned by a (sub)query, on which a predicate can be defined in a relative manner.

For instance, considering a relation describing employees, one may define the predicate "young" (interpreted as "age is low") in the context of the *engineers'* ages, or the ages of those employees *whose monthly salary is less than $2500*, etc. Clearly, the *relative* meaning of "young" depends on the referential considered. Let us start with the simple case where a single context is given in the query.

3.1   Queries with One Level of Context

The syntax of the basic form of SQLf queries considered is given hereafter:

select k X **from** r **where** $cond_1$ **and** cfc
context {Boolean query | fuzzy query involving a threshold λ}

In this query, k denotes a desired number of answers (the best ones), X is a set of attributes from relation r, $cond_1$ is a selection condition which may involve Boolean predicates and explicit fuzzy predicates (i.e., predicates whose membership function is user-defined), and cfc is a contextual fuzzy condition. Notice that contextual fuzzy terms, which form the basis of contextual conditions, can be altered by means of fuzzy modifiers such as "very", "relatively", etc, so as to express a great variety of nuanced contextual predicates.

The new clause introduced by the keyword "context" aims at defining the referential of tuples that serves as a basis for constructing the fuzzy terms present in cfc. Two possibilities are offered: one may either

– use a Boolean subquery (not only an SPJ query, but any kind)
– a fuzzy query involving a qualitative threshold λ.

The context must include attributes of the same domains as those concerned by the contextual predicates. In the case of a fuzzy context query, the context is made of the tuples whose membership degree to the result of the fuzzy query is at least equal to λ. Hence, the context is always defined as a *crisp set*.

Example 1. Let us consider a relation FC(name, pop, crime, temp, univ) describing some French cities. Attribute "crime" corresponds to the crime rate and attribute "temp" to the annual average temperature in a given city, while "pop" gives the number of inhabitants. Attribute "univ" indicates whether there is a university in the city. Hereafter are two examples:

1. find the 10 best cities whose population is between 50,000 and 100,000, where a university is located, and where the crime rate is low relatively to the set of French cities in the same range of population.

    select 10 name **from** FC **where** pop **between** 50,000 **and** 100,000 **and**
            univ = "yes" **and** crime is low
    context (**select** * from FC **where** pop **between** 50,000 **and** 100,000)

2. find the 10 best cities whose population is around 50,000, and where the crime rate is low relatively to the set of French cities whose population fits the fuzzy predicate *around50k* at a degree at least equal to 0.7. Here, the fuzzy predicate *around50k* is supposed to be user-defined.

    select 10 name **from** FC
    where pop **is** around50k **and** crime **is** low
    context (**select** 0.7 * **from** FC **where** pop **is** around50k)   ◇

We propose to define the membership functions of "A is low", "A is medium" and "A is high" in terms of the minimum, average and maximum A-values in the result of the Boolean query which defines the context. For interpretating "low", "medium" and "high", we suggest the following default definitions (cf. Fig. 3):

- $lowA$ decreases linearly from 1 to 0 when A moves from $min(A)$ to $avg(A)$.
- $mediumA$ is represented by an isoceles triangle centered in $avg(A)$ and whose support is the interval $[avg(A) - \delta, avg(A) + \delta]$ where δ equals $min(avg(A) - min(A), max(A) - avg(A))$. This corresponds to interpreting "medium" as "close to the average."
- $highA$ increases linearly from 0 to 1 when A moves from $avg(A)$ to $max(A)$.

It is of course possible to argue about these definitions. One could choose for instance to express that full satisfaction is assumed *around* the minimum (for

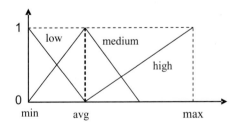

Fig. 2. Default definitions of implicit fuzzy terms

"low") or the average (for "medium") or the maximum (for "high"). In order to make the system more user-friendly, it is of course conceivable to let the user personalize these membership functions through an appropriate interface. Notice that the implicit assumption which is made here is that the context includes enough data to represent a statistically significant subset of the data (so as to compute significant values for min, avg and max).

Example 2. Let us consider the (partial) extension of a relation FC describing French cities represented in Table 3 and the query:

select 10 name **from** FC **where** pop **between** 50,000 **and** 100,000 **and**
           crime **is** low **and** temp **is** medium
context (**select** * **from** FC **where** pop **between** 50,000 **and** 10,000).

Let us suppose that the minimum (resp. average, resp. maximum) value for attribute "crime rate" over the set of cities in the specified population interval is 0.8 (resp. 4.1, resp. 11.3) and the minimum (resp. average, resp. maximum) value for attribute "average annual temperature" over the same set of cities is 8 (resp. 14, resp. 22). One can then define the fuzzy predicates low_{crime} and $medium_{temp}$ as described above, and the evaluation of the query on the first tuple (Montauban) provides the degrees 0.21 (for "low crime") and 0.75 (for "medium temperature"). If we assume that the conjunction is interpreted as a minimum, the first tuple gets the overall satisfaction degree 0.21.   ◇

Table 1. Extension of relation FC

name	pop	crime	temp
Montauban	53,300	4.8	20
Cannes	70,400	7.2	21
Rennes	210,200	4.3	14
Beauvais	54,100	2.4	12
Pau	82,500	5.3	18
...	...	...	...

Let us mention the possibility of introducing a notion of *default context* so as to make the user's task easier. Informally, the "context default" clause corresponds to a query involving the Boolean part of the "where" clause from the original user query (i.e., the "where" clause without its fuzzy conditions). For instance, the query from Example 2 could be expressed more simply as:

select 10 name **from** FC **where** pop **between** 50,000 **and** 100,000 **and**
           crime **is** low **and** temp **is** medium
context default.

3.2   Queries with Several Levels of Context

We now extend the approach presented above so as to consider several levels of context. Let us consider the relations: City (#id, name, pop, crime, temp) and Flat (#app, #city, category, price, surface), and the query: "find the 10 best apartments of category F5 which have a medium price and are located in a city whose population is between 50k and 100k and has a low crime rate." The term "low" is supposed to be defined relatively to the context of cities whose population is between 50k and 100k, while the term "medium" is defined in the context of the city where the apartment considered is located, for the category considered. In SQLf, this query can be expressed as:

select 10 #app **from** Flat f1
where price **is** medium **and** category = 'F5' **and** #city **in**
    (**select** #id **from** City
       where pop **between** 50,000 **and** 100,000 **and** crime **is** low
       context default)
context (**select** * **from** Flat **where** #city = f1.#city **and** category = 'F5').

It is also conceivable to jointly use fuzzy predicates defined on different contexts. For instance, let us consider a relation Emp (id, name, age, ed-level, salary) and the query: "find every employee who has a medium education level (relatively to the entire set of employees) and a high salary (relatively to the employees around his/her age at a degree ≥ 0.7)." This implies extending the "context" clause so as to deal with different referentials. A possible formulation is:

select id, name **from** Emp e
where ed-level **is** medium **and** salary **is** high
context (medium: **default**;
          high: (**select** 0.7 * **from** Emp **where** age around e.age)).

In this query, it is assumed that "around" is an explicit (i.e., user-defined) fuzzy comparator extending the equality.

Another extension consists in using the "top-k" mechanism in a nested manner. This would allow to restrict the contexts to the most relevant elements only. For instance, in the preceding query about flats, one might choose to assess only the flats which are located in a city which is sufficiently satisfactory (e.g., among the 20 best) w.r.t. the criterion on crime. Still another way of introducing nesting is to have a context clause which itself involves a context clause. Due to the lack of space, we do not detail these cases here.

4   Conclusion

In this paper, we have considered the interpretation of top-k fuzzy queries with preference conditions of the form "attribute is low (resp. medium, resp. high)" in the situation where the user is not aware of the actual content of the database but still wants to retrieve the best possible answers (relative to that content). We have pointed out an approach to the definition of the fuzzy terms "low", "medium" and "high" in a contextual and relative manner.

Among perspectives for future work, it would be interesting to study whether some optimization mechanisms proposed for top-k queries in a non-fuzzy framework, such as those described in [2], could be adapted to the processing of top-k queries with contextual fuzzy predicates.

References

1. Agarwal, R., Rantzau, R., Terzi, E.: Context-sensitive ranking. In: Proc. of SIG-MOD 2006, pp. 383–394 (2006)
2. Bast, H., Majumdar, D., Schenkel, R., Theobald, M., Weikum, G.: IO-top-k: Index-access optimized top-k query processing. In: Proc. of the VLDB 2006, pp. 475–486 (2006)
3. Bőrzsőnyi, S., Kossmann, D., Stocker, K.: The skyline operator. In: Proc. of the 17th IEEE Inter. Conf. on Data Engineering, April 2001, pp. 421–430 (2001)
4. Bosc, P., Buckles, B., Petry, F., Pivert, O.: Fuzzy Databases. In: Fuzzy Sets in Approximate Reasoning and Information Systems – The Handbook of Fuzzy Sets Series, pp. 403–468. Kluwer Academic Publishers, Dordrecht (1999)
5. Bosc, P., Pivert, O.: SQLf: a relational database language for fuzzy querying. IEEE Transactions on Fuzzy Systems 3(1), 1–17 (1995)
6. Bouchon-Meunier, B., Yao, J.: Linguistic modifiers and imprecise categories. International Journal of Intelligent Systems 7, 25–36 (1992)
7. Chomicki, J.: Preference formulas in relational queries. ACM Transactions on Database Systems 28, 1–40 (2003)
8. Fodor, J., Yager, R.: Fuzzy-set theoretic operators and quantifiers. In: Fundamentals of Fuzzy Sets – The Handbook of Fuzzy Sets Series, pp. 125–193. Kluwer Academic Publishers, Dordrecht (1999)
9. Kießling, W., Köstler, G.: Preference SQL — design, implementation, experiences. In: Proc. of the 2002 VLDB Conference, pp. 990–1001 (2002)
10. Tudorie, C., Bumbaru, S., Dumitriu, L.: Relative qualification in database flexible queries. In: Proc. 3rd Int. IEEE Conf. on Intelligent Syst. (IEEE IS 2006), pp. 83–88 (2006)
11. Zadeh, L.: Fuzzy sets. Information and Control 8, 338–353 (1965)

Reranking and Classifying Search Results Exhaustively Based on Edit-and-Propagate Operations

Takehiro Yamamoto, Satoshi Nakamura, and Katsumi Tanaka

Department of Social Informatics, Graduate School of Informatics, Kyoto University,
Yoshida-Honmachi, Sakyo, Kyoto 606-8501 Japan
{tyamamot,nakamura,tanaka}@dl.kuis.kyoto-u.ac.jp

Abstract. Search engines return a huge number of Web search results, and the user usually checks merely the top 5 or 10 results. However, the user sometimes must collect information exhaustively such as collecting all the publications which a certain person had written, or gathering a lot of useful information which assists the user to buy. In this case, the user must repeatedly check search results that are clearly irrelevant. We believe that people would use a search system which provides the reranking or classifying functions by the user's interaction. We have already proposed a reranking system based on the user's edit-and-propagate operations. In this paper, we introduce the drag-and-drop operation into our system to support the user's exhaustive search.

Keywords: Reranking, term-based-feedback, exhaustive search.

1 Introduction

Nowadays, many people seek information through search engines. In many information search tasks, it is enough for users to obtain one or two related search results, so browsing the top five or ten search results normally provides adequate information. However, in some information search tasks the user has to browse many search results in order to accomplish the task. For example, if the user must survey many publications related to his research topic, he must use publication search engines and check all the search results returned. If the user plans to buy something expensive, such as a car or TV set, the user has to gather a lot of information about the products to compare products in many aspects. In this work, we call this type of information search *"Exhaustive Search"*.

An *Exhaustive Search* can usually be split into two actions, *gathering* and *browsing*. Users perform an *Exhaustive Search* by combining these two actions:

Gathering: An action to create several queries and gather many search results.

Browsing: An action to browse obtained search results and judge if these search results are relevant or not.

When the user browses search results exhaustively, there are several problems:

Accuracy: The users' search intentions are diverse. This makes it difficult for search engines to return good search results that properly reflect the search intentions.

S.S. Bhowmick, J. Küng, and R. Wagner (Eds.): DEXA 2009, LNCS 5690, pp. 855–862, 2009.

Therefore, search engines often return search results that are not relevant to the user's intention. As a result, the user must often check search results that are not relevant.

Classification: When the user searches exhaustively, it is important to classify the search results. The simplest classification is to classify search results into "relevant" or "non-relevant". Other than this classification, when surveying papers, the user may want to classify these papers according to the genre of the paper or relevance to the user's research topic. However, in order to perform these classifications, the user needs to take actions like keeping windows open, taking down notes, or learning these classifications by heart. These actions impose a heavy burden to the user.

Because of these problems, the user needs to engage in several kinds of actions, such as judging whether search results are relevant or not, jumping to a page by clicking the URL, classifying search results, or creating new queries. For example, while the user is seeking information about a certain product by using search engines, the user creates a new query regarding another product in order to compare products, but then he forgets which search results have already been checked.

In our past work, we have proposed and implemented a reranking system that enables a user to rerank search results by using edit-and-propagate operations [1]. The system enables the user to edit any portion of a browsing page of Web search results at any time while searching. Our system detects the user's search intention from the editing operations. For example, if the user deletes a part of a search result, our system guesses that "*this user does not want this kind of result*". If the user emphasizes a part of the search result, our system guesses that "*this user wants more of this kind of result*". Our system propagates the user's search intentions based on his editing operations to all search results in order to rerank them. In this way, the user can easily obtain optimized search results.

In this paper, we introduce a drag-and-drop operation as an edit operation and some folders to support the *Exhaustive Search*. In our system, the user can increase the accuracy of search results and classify reranked search results easily by using three types of edit operations.

2 Exhaustive Search

We define *Exhaustive Search* as a series of actions performed to gather as efficiently as possible information related to the information the user wants to obtain. For example, collecting all the publications that a certain person has written or gathering a lot of information about cars can be classified as *Exhaustive Search* tasks. We can classify *Exhaustive Search* tasks into two categories: One is that tasks whose answers can be defined independently from the user's values. For example, in a task such as "*I want to collect all the books written by the author,*" or "*I have to collect all publications written by me,*" we can define the answers to these tasks. The other is that tasks whose answers *cannot* be defined independently of the users' values. Such a task might be "*I want to gather information about Kyoto's history.*", or "*I want to compare with these cars.*" The answers to these tasks differ depending the users' values. In this type of task, we cannot define the answers objectively.

Although there are two kinds of *Exhaustive Search* tasks, both the gathering and browsing of search results exhaustively are essential to both.

The user needs to create multiple queries that will return useful search results in an Exhaustive Search. One method of assisting users to gather many useful search results is query expansion [2]. Query expansion creates useful queries by using user feedback. Query expansion seems to be an effective way for users to create a useful query automatically. In this work, we do not focus only on how to gather search results effectively. The user inputs queries manually in our system, but it is effective to apply methods like query expansion.

As we mentioned in Section 1, the problems in browsing search results exhaustively are split into two groups: accuracy and classification. The most popular method for classifying a large set of search results is clustering [3,4]. It was shown in [5] that persons who share the same name can be separated with high accuracy. Therefore, some clustering methods may be effective for tasks such as gathering all publications written by a certain person. However, when the criterion for classifying clustered publication search results is their relevance to the user's research or to the field of the publication, classifying by the clustering method alone would be difficult and inadequate. Therefore, interaction between the user and the system is essential for classifying search results according to the user's own criteria. Moreover, we think such interaction may improve the clustering performance. In this paper, we therefore focus on how to support exhaustive browsing of search results.

3   Supporting an Exhaustive Search by Using Edit Operations

The interaction between our system and the user is as follows:

1. The user inputs a query to our system.
2. The system sends the query to a search engine.
3. The system receives results from the search engine and presents them to the user.
4. The user browses the search results.
5. The user uses the following three operations as necessary.
   a. Delete and emphasis operations proposed in [1] to improve the accuracy of the high ranked search results.
   b. Drag-and-drop operation to classify the search results.
   c. Adding a new query to the system to gather other search results.

After classifying an adequate number of search results, the user can check the classified or non-classified search results easily and obtains the desired information.

3.1   Improve Accuracy of Search Results

In our system, the user can rerank search results using two types of edit operations: *delete operation* and *emphasis operation* that were proposed in [1]. When the user emphasize/delete the term t at an attribute *attr* (e.g. title, summary, URL, publication year, and so on), then search results that contain term t at *attr* are ranked higher (or lower), with their previous rankings preserved.

3.2 Classify Search Results

Here we describe how our system supports users to classify search results.

Popular window systems use *folders* to classify and organize files. Storing multiple files in a single folder, we can handle these multiple files by manipulating the folder. In this work, therefore, we introduce *folder* as a function to organize multiple search results. In our system, the user classifies search results by using a *drag-and-drop* operation which is used in many applications. Drag-and-drop is an operation to move something to a certain proper position. For example, when we move a file to the trash can or move contents to another position while editing documents, we use drag-and-drop operation. We therefore introduce the drag-and-drop operation as an operation to classify search results.

- **Term:** If the user puts a term into a folder, the user might want to classify search results that contain the term in the folder.
- **Search results:** The user might want to classify some search results after reranking them. In this case, the user would select these search results and drag-and-drop them into a folder.
- **Folder:** A folder denotes a set of search results that the user has classified. If the user wants to merge two folders, the user might drag-and-drop a folder to another folder.

When the user drags-and-drops a term into a folder the classification algorithms work the same way as the reranking algorithms do [1]. When the user drags-and-drops a term t in an attribute *attr* to a folder f, the system moves search results containing term t in attribute *attr* to the target folder f.

When the user directly drags-and-drops some search results into a folder, the system moves only these search results to the destination folder. When the user drags-and-drops a certain folder into another folder the system moves all search results stored in the source folder to the destination folder.

4   Design and Implementation

In order to evaluate the effectiveness of our system, we implemented a prototype system. Our system enables users to search Web pages or publications. We used Google and GoogleScholar as Web and publication search engines.

A screenshot of our system is shown in Fig. 1. The system consists of the input area, the display area, and the classification area.

Input Area: The user gathers search results by inputting queries to the input area. When the user inputs a query, the system sends the query to a search engine and receives search results. The system then stores the obtained search results into the inbox folder, which is displayed in the top of the classification area and shows the search results to the user. When the user inputs another query, the system adds the obtained search results to the inbox folder, removing the search results that have already been stored in the folder. With this function, the user can handle multiple queries uniformly in one task.

Display Area: In the display area, the user can rerank search results by using delete and emphasis operations. When the user selects a term from a search result, two buttons for delete and emphasis operations will appear near the mouse cursor position. In the meantime, search results that contain the user-selected term will be highlighted. This function enables the user to understand visually how many search results contain the term. After either button is pressed, the system reranks search results accordingly. In order for the user to understand what topics search results contain, the system also shows *TagCloud* in the right side of the display area. In the *TagCloud*, more frequently appearing terms are shown in larger text. The user can edit terms displayed in *TagCloud* as well as terms in the search results.

Classification Area: The classification area consists of several folders. In the initial state, there are two folders. One is the inbox folder (displayed at the top of the classification area), which is for storing search results of the queries. The other is the trash folder (displayed at the bottom of the classification area), which is for storing search results that are not relevant to the user. When the user starts searching or creates a folder explicitly, the system creates a new folder in the classification area automatically.

The user is able to label the new folder explicitly. Meanwhile, when the user drags-and-drops a term or search results into a folder that he has not labeled, the system labels the folder automatically by extracting terms that appear with high frequency in the search results in the folder. In addition, when the user hovers the cursor over a folder, the system shows a pop up window near the cursor. This window displays a small *TagCloud* that consists of about ten terms extracted from the search results stored in the folder. These functions assist the user in appropriately understanding what search results the folder contains.

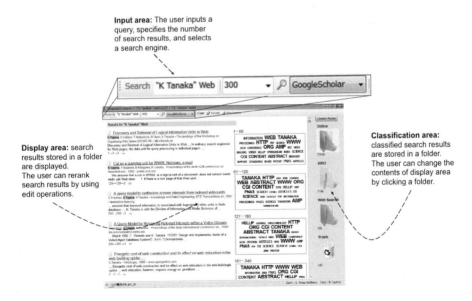

Fig. 1. Screenshot of our system

The user can drag-and-drop a term or some search results or a folder into a folder. When the user drags-and-drops these things into a folder, the system moves search results from the source folder to the destination folder. If the user performs a drag-and-drop operation with the control key pressed, the system copies search results to the destination folder instead of moving search results. This function is used when the user cannot classify the search results into one folder. When the user clicks a certain folder, the system displays search results stored in the folder. The user can switch among folders and view search results just by clicking a folder.

Fig. 2 shows a flowchart of our system in a task. The figure illustrates how a user reranks and classifies publication search results of query "KatsumiTanaka Web".

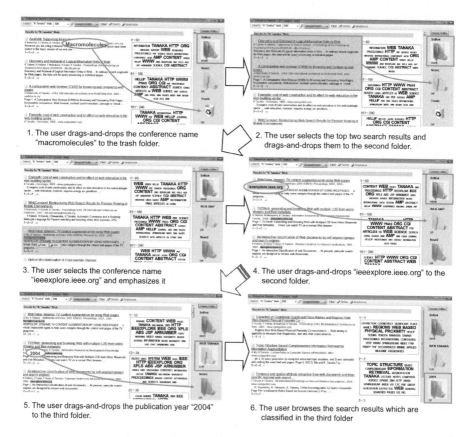

1. The user drags-and-drops the conference name "macromolecules" to the trash folder.

2. The user selects the top two search results and drags-and-drops them to the second folder.

3. The user selects the conference name "ieeexplore.ieee.org" and emphasizes it

4. The user drags-and-drops "ieeexplore.ieee.org" to the second folder.

5. The user drags-and-drops the publication year "2004" to the third folder.

6. The user browses the search results which are classified in the third folder

Fig. 2. How our system works

5   Experiments

In this section, we present the results of the experiment to evaluate the effectiveness of classification. In this experiment, the authors themselves performed the following task. First, the user inputed a prepared query and received its top 500 search results from the system. Then the user classified search results into two folders according to prepared criteria. In this experiment, the user classified search results of a query with multiple meanings into two folders. We prepared six queries and two classifications for each query. The user was allowed to drag-and-drop only keywords. For each query, the user performed drag-and-drop operation until the recall of classified search results in a folder exceeded 90%, or until the user had performed more than nine operations. After each drag-and-drop operation, we analyzed the precision and recall of search results in a folder. To calculate the recall rate, we assumed that all relevant search results were included in the top 500 search results returned by Google.

Fig. 3 shows the average of the precision and recall of the classified search results. After eight operations, the recall rate for three classifications did not become more than 90%. After three or four operations had been performed, many queries achieved a recall rate of more than 80%.

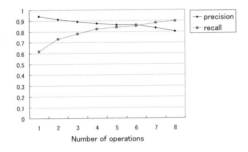

Fig. 3. Precision and recall obtained after each operation

6 Discussions

As mentioned earlier, our system mainly focuses on how to *browse* search results exhaustively. However, our system does not work sufficiently to support the *gathering* of search results exhaustively. The problem is that our system cannot obtain more than 1000 search results in one query because of the restrictions of Google and GoogleScholar. Therefore, if the total number of search results of a query is more than 1000, the system obtains only a subset of the search results. To solve this problem, we plan to implement a semi-automatic query generation system using query expansion or query suggestion techniques.

We found that our system may work effectively on some types of searches. The first example is a publication search. A publication search result has many *attributes,* such as author lists, publication year, and name of conference or journal. These *attributes* often have a strong relationship with the publications. Using these attributes, the user can rerank and classify publications with very high accuracy.

Another example is a recipe search. In a recipe search, the user usually uses cooking ingredients as a query like "pork AND carrot". By emphasizing some favorite ingredients or dragging-and-dropping some ingredients the user dislikes to the trash folder, the user can easily obtain good search results. In a recipe search, ingredients or seasoning are very important, and so the kinds of terms the user may edit are quite limited. Therefore, we think our system will work effectively on similar types of searches where the kinds of important keywords in the search are limited.

Finally, we found that our system may work well not only on Web or publication searches but also on other types of searches such as a product search in Amazon. In a publication search, we found that we can rerank and classify search results with high accuracy by editing attributes that a Web search result does not contain. This is because attributes, such as an author's list, conference name, and publication year, have a strong relationship with the publication. Similar to a publication search result, a product search result often has such *attributes*. For example, a search result of electronics in Amazon has a price, manufacturer name, release date, and so on.

7 Conclusion

We have proposed a system that supports exhaustive search tasks through edit operations. We have split the exhaustive search tasks into two categories: *exhaustive gathering* and *exhaustive browsing*. In this paper, we focused mainly on the exhaustive browsing and proposed a method to support it. In this system, the user can rerank search results by using delete or emphasis operations and classify search results by using drag-and-drop operations. We expect that we can apply this idea to other types of websites that are generated automatically, such as online shopping sites and bulletin boards. For example, the user will be able to rerank and classify products by their prices or brand. Fortunately, much research has focused on automatically extracting and detecting data from such websites. By applying these ideas to our system, we can handle many types of websites.

Acknowledgement

This work was supported in part by Grant-in-Aid for JSPS Fellows, "Informatics Education and Research Center for Knowledge-Circulating Society" (Project Leader: Katsumi Tanaka, MEXT Global COE Program, Kyoto University), MEXT Grant-in-Aid for Scientific Research on Priority Areas: "Cyber Infrastructure for the Informationexplosion Era", "Contents Fusion and Seamless Search for Information Explosion" (Project Leader: Katsumi Tanaka, A01-00-02, Grant#: 18049041), and by "Design and Development of Advanced IT Research Platform for Information" (Project Leader: Jun Adachi, Y00-01, Grant#: 18049073).

References

1. Yamamoto, T., Nakamura, S., Tanaka, K.: Rerank-by-example: Efficient browsing of web search results. In: Wagner, R., Revell, N., Pernul, G. (eds.) DEXA 2007. LNCS, vol. 4653, pp. 801–810. Springer, Heidelberg (2007)
2. Xu, J., Croft, W.: Query expansion using local global document analysis. In: Proc. of SIGIR 1996, pp. 4–11 (1996)
3. Hearst, M.A., Karger, D.R., Pedersen, J.O.: Scatter/Gather as a Tool for the Navigation of Retrieval Results. In: The proceedings of the 1995 AAAI Fall Symposium on Knowledge Navigation (1995)
4. Zeng, H., He, Q., Chen, Z., Ma, W., Ma, J.: Learning to cluster web search results. In: Proc. of SIGIR 2004, pp. 210–217 (2004)
5. Bilenko, M., Mooney, R., Cohen, W., Ravikumar, P., Fienberg, S.: Adaptive name matching in information integration. IEEE Intelligent Systems 18(5), 16–23 (2003)

Author Index

Printing: Mercedes-Druck, Berlin
Binding: Stein+Lehmann, Berlin